KENDALL'S

ADVANCED

Theory of

STATISTICS

Volume 2A

CLASSICAL INFERENCE
AND THE LINEAR MODEL

KENDALL'S ADVANCED THEORY OF STATISTICS

Volume 1: Distribution Theory
Alan Stuart and Keith Ord
ISBN 0 340 61430 7

1. Frequency Distributions
2. Measures of Location and Dispersion
3. Moments and Cumulants
4. Characteristic Functions
5. Standard Distribution
6. Systems of Distributions
7. The Theory of Probability
8. Probability and Statistical Inference
9. Random Sampling
10. Standard Errors
11. Exact Sampling Distributions
12. Cumulants of Sampling Distributions (Part 1)
13. Cumulants of Sampling Distributions (Part 2)
14. Order-Statistics
15. The Multinormal Distibution and Quadratic Forms
16. Distributions Associated with the Normal

Volume 2A: Classical Inference and the Linear Model
Alan Stuart, Keith Ord and Steven Arnold
ISBN 0 340 66230 1

17. Estimation and Sufficiency
18. Estimation: Maximum Likelihood and Other Methods
19. Interval Estimation
20. Tests of Hypotheses: Simple Null Hypotheses
21. Tests of Hypotheses: Composite Hypotheses
22. Likelihood Ratio Tests and Test Efficiency
23. Invariance and Equivariance
24. Sequential Methods
25. Tests of Fit
26. Comparative Statistical Inference
27. Statistical Relationship: Linear Regression and Correlation
28. Partial and Multiple Correlation
29. The General Linear Model
30. Fixed Effects Analysis of Variance
31. Other Analysis of Variance Models
32. Analysis and Diagnostics for the Linear Model

Volume 2B: Bayesian Inference
Anthony O'Hagan
ISBN 0 340 52922 9

1. The Bayesian Method
2. Inference and Decisions
3. General Principles and Theory
4. Subjective Probability
5. Non-subjective Theories
6. Subjective Prior Distributions
7. Robustness and Model Comparison
8. Computation
9. The Linear Model
10. Other Standard Models

KENDALL'S LIBRARY OF STATISTICS

1. Multivariate Analysis
   Part 1: Distributions, Ordination and Influence
   WJ Krzanowski (University of Exeter) and FHC Marriott (University of Oxford)
   1994, ISBN 0 340 59326 1

2. Multivariate Analysis
   Part 2: Classification, Covariance Structures and Repeated Measurements
   WJ Krzanowski (University of Exeter) and FHC Marriott (University of Oxford)
   1995, ISBN 0 340 59325 3

3. Multilevel Statistical Models
   Second edition
   H Goldstein (University of London)
   1995, ISBN 0 340 59529 9

4. The Analysis of Proximity Data
   BS Everitt (Institute of Psychiatry) and S Rabe-Hesketh (Institute of Psychiatry)
   1997, ISBN 0 340 67776 7

5. Robust Nonparametric Statistical Methods
   Thomas P Hettmansperger (Penn State University) and Joseph W McKean (Western Michigan University)
   1998, ISBN 0 340 54937 8

6. Statistical Regression with Measurement Error
   Chi-Lun Cheng (Academia Sinica, Republic of China) and John Van Ness (University of Texas at Dallas)
   1999 (in preparation), ISBN 0 340 61461 7

7. Latent Variable Models and Factor Analysis
   Second edition
   DJ Bartholomew (London School of Economics) and M Knott (London School of Economics)
   1999 (in preparation), ISBN 0 340 69243 X

8. Statistical Inference for Diffusion Type Processes
   BLS Prakasa Rao (Indian Statistical Institute)
   1999 (in preparation), ISBN 0 340 74149 X

KENDALL'S

ADVANCED
Theory of
STATISTICS

Volume 2A

CLASSICAL INFERENCE
AND THE LINEAR MODEL

Sixth edition

Alan Stuart
Formerly Professor of Statistics
London School of Economics

J. Keith Ord
Professor of Management Science and Statistics
The Pennsylvania State University, USA

Steven Arnold
Associate Professor of Statistics
The Pennsylvania State University, USA

ARNOLD

A member of the Hodder Headline Group
LONDON • SYDNEY • AUCKLAND
Co-published in the USA by Oxford University Press Inc., New York

First published in 1946 as *Classical Inference and Relationships*
Sixth edition published in 1999 by
Arnold, a member of the Hodder Headline Group,
338 Euston Road, London NW1 3BH
http://www.arnoldpublishers.com

Copublished in the United States of America by
Oxford University Press Inc.,
198 Madison Avenue,
New York, NY 10016

Publishing History

Volume 1	Distribution Theory		
		First edition	1943
		New editions and reprints	1945, 1947, 1948, 1952, 1958, 1963, 1969, 1977, 1987, 1994
Volume 2	Classical Inference and Relationships		
		First edition	1946
		New editions and reprints	1947, 1951, 1961, 1967, 1973, 1979, 1991
		As Volume 2A	1999
Volume 2B	Bayesian Inference		
		First edition	1994
Volume 3	Design and Analysis and Time Series		
		First edition	1966
		New editions	1968, 1976, 1983
	Now replaced by *Kendall's Library of Statistics*		

British Library Cataloguing in Publication Data
A catalogue record for this book is available from the British Library

Library of Congress Cataloging-in-Publication Data
A catalog record for this book is available from the Library of Congress

ISBN 0 340 66230 1

Publisher: Nicki Dennis
Production Editor: James Rabson
Production Controller: Sarah Kett
Cover designer: Terry Griffiths

Typeset in 10/12 pt Times by Focal Image, Torquay, and Academic and Technical, Bristol
Printed in Great Britain by St Edmundsbury Press, Bury St Edmunds, Suffolk, and bound by MPG Books, Bodmin, Cornwall

CONTENTS

PREFACE TO THE SIXTH EDITION

Some years ago, Alan Stuart decided that he wished to play a less active role in *The Advanced Theory,* so Steven Arnold joined Keith Ord as a principal author for the sixth edition of Volume 2. Regrettably, Alan did not live to see the completion of the new edition; his unexpected death in summer 1998, as the project neared completion, was a loss to all. We have tried to produce this volume in the best tradition of what will always be known to the statistical community as *Kendall and Stuart.*

The sixth edition differs from its predecessors in several key respects. At the beginning of the decade, we decided that *The Advanced Theory* should revert to its earlier two-volume format. Thus, this edition has been retitled Volume 2A: *Classical Inference and the Linear Model*, to emphasize the traditional emphasis on the frequentist approach. The complementary Volume 2B: *Bayesian Inference,* written by Tony O'Hagan, was published in 1994, thereby providing an overall balance to the coverage of modern inference. Volume 3 has been phased out and replaced by a series of monographs forming *Kendall's Library of Statistics;* five of these titles have already appeared and several more are in preparation.

With this change in direction, we decided to restructure the material in the present volume. Statistical inference appears in the first part of the book (Chapters 17–26) and the coverage of the linear model is in the second part (Chapters 27–32). These changes involved both the redesign of existing chapters and the creation of several new ones. Chapter 18 now covers both maximum likelihood and other approaches to inference, while Chapters 19–22 broadly correspond to Chapters 20–23 and 25 in the fifth edition. Steven Arnold wrote the new Chapter 23 on invariance and equivariance.

As before, the material on the linear model starts out with two chapters on correlation. The material on least squares estimation has been consolidated into Chapter 29 and leads into two new chapters (30 and 31) on the analysis of variance, developed by Steven Arnold. The volume closes with Chapter 32 on modeling and diagnostics.

The dramatic advances in word processing since the last edition led to a number of format changes. We now have separate author and subject indexes; the subject index retains its previous format of citations by section, but the author index uses page numbers. We felt that this distinction best served the different purposes of the two indexes.

The availability of computer programs for the evaluation of sampling distributions makes printed tables much less useful than before. After some discussion, we decided to retain the descriptive lists of tables, as such tables are often useful as benchmarks for new computational procedures. The appendix tables were also retained, although perhaps more for reasons of historical sentiment than utility.

We have made extensive changes to the list of references, dropping over a hundred older papers and adding about three hundred new ones. Doubtless, omissions remain, and we welcome suggestions for further improvements.

The production of this edition proved to be a major logistical undertaking. Our thanks go to Mr. Sandy Balkin, whose statistical and computing expertise brought the whole project to completion, and to Jane Auhl who mastered both the intricacies of LaTeX and the vagaries of the draft manuscripts. Our thanks go also to the team at Arnold, who provided just the right balance of encouragement, expertise, patience and good humor to enable us to complete the project.

Finally, we accept full responsibility for any and all remaining errors and trust that our colleagues will help us to identify such problems.

Keith Ord
Steven Arnold
December 1998

LIST OF EXAMPLES

GLOSSARY OF ABBREVIATIONS

a.e.	almost everywhere
ANCOVA	analysis of covariance
ANOVA	analysis of variance
ARE	asymptotic relative efficiency
ARL	average run length
ASN	average sample number
BAN	best asymptotically normal
BCR	best critical region
c.f.	characteristic function
d.f.	distribution function
d.fr.	degree(s) of freedom
d.p.	decimal place(s)
EE	estimating equation
EM	estimation–maximization (algorithm)
GMM	generalized method of moments
i.i.d.	independently and identically distributed
LAD	least absolute deviation
LF	likelihood function
LMS	least median of squares
LP	likelihood principle
LR	likelihood ratio
LS	least squares
MCB	multiple comparison with the best
MCML	Monte Carlo maximum likelihood
MCS	minimum chi-square
ML	maximum likelihood
m.s.e.	mean square error
MV	minimum variance
MVB	minimum variance bound
N (a,b)	(multi)normal distribution with mean (vector) a and variance (–covariance) matrix b
OC	operating characteristic
p.d.f.	probability density function
REML	restricted maximum likelihood
SLP	strong likelihood principle
SPR	sequential probability ratio
SS	sum of squares
UMA	uniformly most accurate

UMP	uniformly most powerful
UMPU	uniformly most powerful unbiased
WCP	weak conditionality principle
WLP	weak likelihood principle
WSP	weak sufficiency principle

CHAPTER 17

ESTIMATION AND SUFFICIENCY

The estimation problem

17.1 On several occasions in previous chapters we have encountered the problem of estimating from a sample the values of the population parameters. We have hitherto dealt on somewhat intuitive lines with such questions as arose – for example, in the theory of large samples we have taken the means and moments of the sample to be satisfactory estimates of the corresponding population quantities.

We now proceed to study this branch of the subject in more detail. In the present chapter, we shall examine the criteria that we require a 'good' estimate to satisfy, and discuss the question whether there exist 'best' estimates in an acceptable sense of the term. In the following chapters, we shall consider methods of obtaining estimates with the required properties.

17.2 It will be evident that if a sample is not random and nothing precise is known about the nature of the bias operating when it was chosen, very little can be inferred from it about the population. Certain conclusions of a trivial kind are sometimes possible – for instance, if we take ten turnips from a pile of 100 and find that the ten selected weigh 10 pounds altogether, the mean weight for all 100 turnips must be greater than 0.1 pounds; but such information is rarely of value, and estimation based on biased samples remains very much a matter of individual opinion and cannot be reduced to exact and objective terms. We shall therefore confine our attention to random samples only. Our general problem, in its simplest terms, is to estimate the value of a parameter of the population from the information given by the sample. Initially, we consider the case when only one parameter is to be estimated. The case of several parameters will be discussed later.

17.3 Let us first consider what we mean by 'estimation'. We know, or assume as a working hypothesis, that the population is distributed in a form that is completely known but for the value of some *parameter*[1] θ. We are given a sample of observations $x_1, \ldots, x_n$. We require to determine, from the observations, a number that can be taken to be the value of θ, or a range of numbers that can be taken to include that value.

Now the observations are random variables, and any function of the observations will also be a random variable. A function of the observations alone is called a *statistic*. If we use a statistic to estimate θ, it may on occasion differ considerably from the true value of θ. It appears, therefore, that we cannot expect to find any method of estimation that can be guaranteed to give us a close estimate of θ on every occasion and for every sample. We must content ourselves with formulating a rule that will give good results 'in the long run' or 'on average', or that has 'a high probability of success' – phrases that express the fundamental fact that we have to regard our method of estimation as generating a distribution of estimates and to assess its merits according

[1] The term 'parameter' was discussed in **2.2**. See also **20.3**.

to the properties of that distribution.

17.4 It will clarify our ideas if we draw a distinction between the method or rule of estimation, which we shall call an *estimator*, and the value to which it gives rise in a particular case, the *estimate*. The distinction is the same as that between a function $f(x)$, defined for a range of the variable x, and the particular value that the function assumes, say $f(a)$, for a specified value of x equal to a. Our problem is not to find estimates, but to find estimators. We do not reject an estimator because it may give a bad result in a particular case (in the sense that the estimate differs materially from the true value). We should only reject it if it gave bad results in the long run, that is to say, if the distribution of possible values of the estimator were seriously discrepant from the true value of θ. The merit of the estimator is judged by the distribution of estimates to which it gives rise, i.e. by the properties of its sampling distribution.

17.5 In the theory of large samples, we have often taken as an estimator of a parameter θ a statistic t calculated from the sample in exactly the same way as θ is calculated from the population: e.g. the sample mean is taken as an estimator of the population mean. Let us examine how this procedure can be justified. Suppose the population has p.d.f.

$$f(x) = (2\pi)^{-1/2}\exp\{-\tfrac{1}{2}(x-\theta)^2\}, \qquad -\infty < x < \infty. \tag{17.1}$$

Requiring an estimator for the parameter θ, which is the population mean, we take the sample mean

$$t = \sum_{j=1}^{n} x_j/n. \tag{17.2}$$

The sampling distribution of t has p.d.f. (Example 11.12)

$$f(t) = \{n/(2\pi)\}^{1/2}\exp\{-\tfrac{1}{2}n(t-\theta)^2\}, \tag{17.3}$$

that is to say, t is distributed normally about θ with variance $1/n$. We notice two things about this distribution: (a) it has a mean (and median and mode) at the true value θ, and (b) as n increases, the scatter of possible values of t about θ becomes smaller, so that the probability that a given t differs by more than a fixed amount from θ decreases. We may say that the *precision* of the estimator increases with n.

Generally, it will be clear that the phrase 'precision increasing with n' has a definite meaning whenever the sampling distribution of t has a variance that decreases with $1/n$ and a central value that is either identical to or differs from it by a quantity that also decreases with $1/n$. Many of the estimators with which we are commonly concerned are of this type, but there are exceptions. Consider, for example, the Cauchy distribution

$$f(x) = \frac{1}{\pi\{1+(x-\theta)^2\}}, \qquad -\infty < x < \infty. \tag{17.4}$$

If we estimate θ by the mean statistic t we have, for the distribution of t,

$$f(t) = \frac{1}{\pi\{1+(t-\theta)^2\}} \tag{17.5}$$

(cf. Example 11.1). In this case the distribution of t is the same as that of any single value of the sample, and does not increase in precision as n increases.

Identifiability of parameters

17.6 It is clear that as sample size increases indefinitely, we always obtain more information about the underlying distribution $f(x|\theta)$, but this will only be useful in estimating θ if θ is a single-valued function of f; if not, i.e. if more than one value of θ corresponds to a given value of f, not even an infinite-sized sample can distinguish between them. We then say that θ is *unidentifiable*. For example, consider a normal distribution with mean θ^2, (17.2) will estimate θ^2 with increasing precision as before, but there is no way to decide whether θ is positive or negative from even an exact knowledge of θ^2, for f is the same whatever the sign of θ. As this instance makes clear, even if θ is unidentifiable, some functions of θ may be identifiable – in this case, any single-valued function of θ^2 is.

Unless otherwise stated, we shall assume that parameters are *identifiable*, θ being uniquely determined by $f(x|\theta)$.

Consistency

17.7 The possession of the property of increasing precision is evidently very desirable; and indeed, if the variance of the sampling distribution of an estimator decreases with increasing n, it is necessary that its central value should tend to θ, for otherwise the estimator would have values differing systematically from the true value. We therefore formulate our first criterion for a suitable estimator as follows.

An estimator t_n, computed from a sample of n values, will be said to be a *consistent* estimator of θ if, for any positive ε and η, however small, there is some N such that the probability that

$$|t_n - \theta| < \varepsilon \tag{17.6}$$

is greater than $1 - \eta$ for all $n > N$. That is,

$$P\{|t_n - \theta| < \varepsilon\} > 1 - \eta, \qquad n > N. \tag{17.7}$$

The definition bears an obvious analogy to the definition of convergence in the mathematical sense. Given any fixed small quantity ε, we can find a large enough sample size such that, for all samples over that size, the probability that t differs from the true value by more than ε is as near zero as we please. t_n is said to *converge in probability*, or to *converge stochastically*, to θ. Thus t is a consistent estimator of θ if it converges to θ in probability.

The original definition of consistency by Fisher (1921a) was that when calculated from the whole population the estimators should be equal to the parameter, and it was no doubt this definition that inspired the choice of the term 'consistency'. This form is now known as the *Fisher consistency*.

Example 17.1 (Consistency of the sample mean for the mean of the normal distribution)

The sample mean is a consistent estimator of the parameter θ in the population (17.1). This we have already established in general argument, but formally the proof proceeds as follows.

Suppose we are given ε. From (17.3) we see that $(t-\theta)n^{1/2}$ is distributed normally about zero with unit variance. Thus the probability that $|(t-\theta)n^{1/2}| \leq \varepsilon n^{1/2}$ is the value of the normal integral between limits $\pm\varepsilon n^{1/2}$. Given any positive η, we can always take n large enough for this quantity to be greater than $1-\eta$ and it will continue to be so for any larger n. The number N may therefore be determined and the inequality (17.7) is satisfied.

Example 17.2 (Consistency of the sample mean for general distributions)

Suppose we have a statistic t_n whose mean value differs from θ by terms of order n^{-1}, whose variance v_n is of order n^{-1} and that tends to normality as n increases. Clearly, as in Example 17.1, $(t_n-\theta)/v_n^{1/2}$ will then tend to zero in probability and t_n will be consistent. This covers a great many statistics encountered in practice.

Even if the limiting distribution of t_n is unspecified, the result will still hold, as can be seen from a direct application of the Chebyshev inequality (3.94). In fact, if $E(t_n) = \theta + k_n$, and $\text{var}\, t_n = v_n$, where $\lim_{n\to\infty} k_n = \lim_{n\to\infty} v_n = 0$, we have at once

$$P\{|t_n-\theta| < \varepsilon\} \geq 1 - \frac{v_n + k_n^2}{\varepsilon^2} \underset{n\to\infty}{\longrightarrow} 1,$$

so that (17.7) will be satisfied.

17.8 The property of consistency is a limiting property, i.e. it concerns the behaviour of an estimator as the sample size tends to infinity. It requires nothing of the estimator's behaviour for finite n, and if there exists one consistent estimator t_n we may construct infinitely many others: e.g. for fixed a and b, $[(n-a)/(n-b)]t_n$ is also consistent. We have seen that in some circumstances a consistent estimator of the population mean is the sample mean $\bar{x} = (\sum x_j)/n$. But so is $\bar{x}' = (\sum x_j)/(n-1)$. Why do we prefer one to the other? Intuitively it seems absurd to divide the sum of n quantities by anything other than their number n. We shall see in a moment, however, that intuition is not a very reliable guide in such matters. There is reason for preferring

$$\frac{1}{(n-1)}\sum_{j=1}^{n}(x_j-\bar{x})^2$$

to

$$\frac{1}{n}\sum_{j=1}^{n}(x_j-\bar{x})^2$$

as an estimator of the population variance σ^2, notwithstanding the fact that the latter is the sample variance and is a consistent estimator of σ^2.

Unbiased estimators

17.9 Consider the sampling distribution of an estimator t. If the estimator is consistent, its distribution must, for large samples, have a central value in the neighbourhood of θ. We may choose among the class of consistent estimators by requiring that θ shall be equated to this central value not merely for large, but for all samples.

If we require that for all n and θ the mean value of t shall be θ, i.e. that

$$E(t) = \theta, \tag{17.8}$$

we call t an *unbiased* estimator of θ. This is an unfortunate word, like so many in statistics. The mean value is used, rather than the median or the mode, for its mathematical convenience. This is perfectly legitimate, but the term should not be allowed to convey non-technical overtones.

Example 17.3 (Unbiasedness of the sample mean)

Since

$$E\left\{\frac{1}{n}\sum x\right\} = \frac{1}{n}\sum E(x) = \mu_1',$$

the sample mean is an unbiased estimator of the population mean whenever the latter exists. But the sample variance is not an unbiased estimator of the population variance, for

$$E\{\sum(x_j - \bar{x})^2\} = E\Big\{\sum\Big[x_j - \sum_j x_j/n\Big]^2\Big\} = E\Big\{\frac{n-1}{n}\sum_j x_j^2 - \frac{1}{n}\sum\sum_{j\neq k} x_j x_k\Big\}$$
$$= (n-1)\mu_2' - (n-1)\mu_1'^2 = (n-1)\mu_2.$$

Thus $(1/n)\sum(x-\bar{x})^2$ has mean value $[(n-1)/n]\mu_2$. Generally, we prefer to use the unbiased estimator given by

$$s^2 = \frac{1}{n-1}\sum(x-\bar{x})^2.$$

In some later chapters, notably those on invariance (Chapter 23) and on linear models (Chapters 29–32) we refer to s^2 as the sample variance; further, the divisor is modified to $(n - p)$ when p parameters are estimated. The different usage reflects conventions in different areas of Statistics and we have tried to signal such changes clearly.

Our discussion in **17.8** shows that consistent estimators are not necessarily unbiased. We have already (Example 14.5) encountered an unbiased estimator that is not consistent. Thus neither property implies the other. But a consistent estimator whose asymptotic distribution has finite mean value must be asymptotically unbiased.

In certain circumstances, there may be no unbiased estimator (cf. Exercise 17.12). Even if there is one, it may be forced to give absurd estimates at times, or even always. For example, in estimating a parameter θ where $a \leq \theta \leq b$, no estimator t distributed on the same interval can be unbiased, for if $\theta = a$, we must have $E(t) > a$, and if $\theta = b$, $E(t) < b$. Thus 'impossible' estimates must sometimes arise if we insist on unbiasedness in this case. We shall meet an important example of this in **28.34**; another example is the estimation of the characteristic exponent α of a stable distribution (cf. **4.36**) – when $\alpha = 2$, no estimator in $(0, 2)$ can be unbiased (cf. Du Mouchel, 1983). See also Exercises 10.21–22.

Similarly, a bounded estimator cannot be unbiased for any unbounded function of a parameter, since its expectation is also bounded. This argument is enough to establish the non-existence of an unbiased estimator in Exercise 17.12 for negative a or b.

Further, in the sequential binomial sampling scheme discussed in Example 9.15, when $k = 1$ there is an unbiased estimator of a probability that only takes the values 0 or 1. An even more extreme instance, where the only unbiased estimate always gives impossible estimates, is contained in Exercise 17.26. As a matter of common sense, the final estimate should always lie in the feasible region, even if such a requirement leads us to abandon unbiasedness.

Corrections for bias: the jackknife and the bootstrap

17.10 If we have a biased estimator t, and wish to remove its bias, this may be possible by direct evaluation of its expected value and the application of a simple adjustment, as in Example 17.3. But sometimes the expected value is a rather complicated function of the parameter, θ, being estimated, and it is not obvious what the correction should be. Quenouille (1956) proposed an ingenious method of overcoming this difficulty in a fairly general class of situations.

We denote our biased estimator by t_n, its suffix being the number of observations from which t_n is calculated. Now suppose that

$$E(t_n) - \theta = \sum_{r=1}^{\infty} a_r/n^r, \tag{17.9}$$

where the a_r may be functions of θ but not of n. t_{n-1} may be calculated from each of the n possible subsets of $(n-1)$ observations – when the ith observation is omitted, we write it $t_{n-1,i}$. Let $\bar{t}_{n-1}$ denote the average of these n values, and consider the new statistic

$$t'_n = nt_n - (n-1)\bar{t}_{n-1} = t_n + (n-1)(t_n - \bar{t}_{n-1}). \tag{17.10}$$

It may happen that $t'_n \equiv t_n$ (as when t_n is the sample mean, or the sample median when n is even), but if not it follows at once from (17.9) that

$$\begin{aligned} E(t'_n) - \theta &= a_2\left(\frac{1}{n} - \frac{1}{n-1}\right) + a_3\left(\frac{1}{n^2} - \frac{1}{(n-1)^2}\right) + \cdots \\ &= -\frac{a_2}{n^2} - O(n^{-3}). \end{aligned} \tag{17.11}$$

Thus t'_n is only biased to order $1/n^2$. Similarly,

$$t''_n = \{n^2 t'_n - (n-1)^2 \bar{t}'_{n-1}\}/\{n^2 - (n-1)^2\}$$

will be biased only to order $1/n^3$, and so on. This method, called the *jackknife*, removes bias to any required degree.

Because t'_n differs from t_n by a quantity of order n^{-1}, we see that if t_n has variance of order n^{-1} the variance of t'_n is asymptotically the same as that of t_n so that reduction of bias carries no penalty in the variance – cf. Exercise 17.18; but this is not generally true for further-order corrections – cf. Robson and Whitlock (1964). Thorburn (1976) gives conditions for the asymptotic distributions of t'_n and t_n to be the same.

t'_n is the mean of n identically distributed, equally correlated variates $d_i = nt_n - (n-1)t_{n-1,i}$. If this correlation is negligible, the variance of t'_n is estimated by

$$\sum_{i=1}^{n} \{d_i - t'_n\}^2/\{n(n-1)\} = \frac{n-1}{n}\sum (t_{n-1,i} - \bar{t}_{n-1})^2,$$

using the final result in Exercise 17.17 – the corresponding estimator of the variance of t_n is shown by Efron and Stein (1981) to be biased upwards; but if not, this estimator may not be useful even as $n \to \infty$; see also Miller (1964), who also discusses asymptotic normality. Cressie (1981) discusses transformation of the observations before jackknifing estimators. Arvesen and Layard (1975) discuss the situation where the original observations have different distributions.

Booth and Hall (1993) provide a jackknife estimator for the distribution function. Earlier reviews of the topic are those by Miller (1974) and Parr and Schucany (1980); for detailed up-to-date treatments, see Peddada (1993), Shao and Tu (1995).

Instead of the jackknife's use of each subset of $n - 1$ distinct observations, Efron (1979) introduced the *bootstrap* resampling method that uses each possible sample of size m, drawn at random *with replacement* from the original sample, to estimate the sampling distribution of any random variable, based on $m \leq n$ observations. Thus the bootstrap is not restricted to estimating the mean and variance of such a variable as is the jackknife; indeed, the jackknife is shown to be a linear approximation to the bootstrap in those cases.

Since the number of possible samples, n^n, is prohibitively large for even moderate n, the method proceeds by using B replicates, or random samples with replacement, from the original sample. In general, B should be at least of order $n \log n$; then, as $n \to \infty$, the bootstrap sampling distribution converges to the true sampling distribution under fairly general conditions. For example, Efron (1981) compares the efficiency of these and other methods of estimating sampling variances in the case of the binormal correlation coefficient, and finds that the bootstrap performs best.

Young (1994) provides a good overview of the burgeoning literature on bootstrap methods. He concludes that the method usually performs well when the observations are independent, but less so for dependent data. In general, some (parametric) modelling of the dependence structure is needed for the bootstrap to be effective.

Much recent research has focused on developing bootstrap procedures for use in specific contexts; the simplicity of 'universal application' is thereby lost (if, indeed, it was ever there) but the resulting methods are more effective and grounded in a stronger theoretical base. Particular developments of note are bootstrap confidence intervals that use saddlepoint approximations to reduce computational effort (DiCiccio *et al.*, 1993); the bootstrap 'likelihood function' using a nested bootstrap with kernel smoothing (Davison *et al.*, 1992); use of the jackknife on bootstrapped data to assess the performance of accuracy measures (Efron, 1992); and work on missing data (Efron, 1994).

Recent monographs on bootstrapping include those of Hall (1992), Efron and Tibshirani (1993) and Shao and Tu (1995).

Example 17.4 (Jackknife estimator for the binomial distribution)

Suppose we wish to find an unbiased estimator of θ^2 in the binomial distribution

$$P\{x = r\} = \binom{n}{r}\theta^r(1-\theta)^{n-r}, \qquad r = 0, 1, 2, \ldots, n; \ n \geq 2.$$

The intuitive estimator is

$$t_n = \left(\frac{r}{n}\right)^2,$$

since r/n is an unbiased estimator of θ, but

$$E(t_n) = \text{var}(r/n) + \{E(r/n)\}^2 = \theta^2 + \theta(1-\theta)/n.$$

Now t_{n-1} can only take the values

$$\left(\frac{r-1}{n-1}\right)^2 \quad \text{or} \quad \left(\frac{r}{n-1}\right)^2$$

according to whether a 'success' or a 'failure' is omitted from the sample. Thus

$$\bar{t}_{n-1} = \frac{1}{n}\left\{r\left(\frac{r-1}{n-1}\right)^2 + (n-r)\left(\frac{r}{n-1}\right)^2\right\} = \frac{r^2(n-2)+r}{n(n-1)^2}.$$

Hence, from (17.10),

$$\begin{aligned} t'_n &= nt_n - (n-1)\bar{t}_{n-1} \\ &= \frac{r^2}{n} - \frac{r^2(n-2)+r}{n(n-1)} \\ &= \frac{r(r-1)}{n(n-1)}, \end{aligned} \tag{17.12}$$

which, it may be directly verified, is exactly unbiased for θ^2. Exercise 17.12 gives a general result of which this is a special case.

Exercises 17.17 and 17.13 give further applications of the method.

17.11 Often there will exist more than one consistent estimator of a parameter, even if we confine ourselves to unbiased estimators. Consider once again the estimation of the mean of a normal population with variance σ^2. The sample mean is consistent and unbiased. We will now prove that the same is true of the median. Consideration of symmetry is enough to show (cf. Exercise 14.7) that the median is an unbiased estimator of the population mean, which is, of course, the same as the population median. For large n the distribution of the median, t, tends to the normal form (cf. **14.12**)

$$f(t) \propto \exp\{-2nf_1^2(t-\theta)^2\}, \tag{17.13}$$

where f_1 is the value of the p.d.f. at $t = \theta$, here equal to $(2\pi\sigma^2)^{-1/2}$. The variance of the sample median is therefore, from (17.13), equal to $\pi\sigma^2/(2n)$ and tends to zero for large n. Hence the estimator is consistent.

17.12 We must therefore seek further criteria to choose between estimators with the common property of consistency. Such a criterion arises naturally if we consider the sampling variances of the estimators. We expect that the estimator with the smaller variance will be distributed more closely round the value θ; this will certainly be so for distributions of the normal type. An unbiased consistent estimator with a smaller variance will therefore deviate less, on average, from the true value than one with a larger variance. Hence we may reasonably regard it as better.

In the case of the mean and median of normal samples we have, for any n, from (17.3),

$$\text{var(mean)} = \sigma^2/n, \tag{17.14}$$

whereas (14.16) showed that for $n = 2r + 1$, the variance of the median is, putting $\sigma^2 = 1$,

$$\text{var}(x_{(r+1)}) = \frac{\pi}{2(n+2)} + \frac{\pi^2}{4(n+2)(n+4)} + o(n^{-2}).$$

Thus,

$$\text{var(mean)}/\,\text{var(median)} = \frac{2}{\pi} + \frac{1}{n}\left(\frac{4}{\pi} - 1\right) + o(n^{-1}). \tag{17.15}$$

Hodges (1967) gives the following exact values:

n:	1	3	5	7	9	11	13	15	17	19	∞
$\dfrac{\text{var(mean)}}{\text{var(median)}}$:	1	0.743	0.697	0.679	0.669	0.663	0.659	0.656	0.653	0.651	0.637

The approximation (17.15) is accurate to 2 d.p. for $n \geq 7$. The values for even n are always higher than the contiguous values for odd n – the definition of the median is different in this case – but tend monotonically to the same limit from the value 1 at $n = 2$. (The approximation (17.15) must now have the figure '4' replaced by '6'.) A few values of $\{\text{var(median)}/\,\text{var(mean)}\}^{1/2}$ for even n were given in **14.6**. Thus the mean always has smaller variance than the median in the normal case.

Example 17.5 (Sample median as an estimator for the Cauchy)

For the Cauchy distribution

$$f(x) = \frac{1}{\pi\{1 + (x-\theta)^2\}}, \qquad -\infty < x < \infty,$$

we have already seen (**17.6**) that the sample mean is not a consistent estimator of θ, the population median. However, for the sample median, t, we have, from (17.13), since f_1 is $1/\pi$, the large-sample variance

$$\text{var}\, t = \pi^2/4n.$$

It is seen that the median is consistent, and although direct comparison with the mean is not possible because the latter does not possess a sampling variance, the median is evidently a better estimator of θ than the mean. This provides an interesting contrast with the case of the normal distribution, particularly in view of the similarity of the shapes of the distributions.

Minimum variance bounds

17.13 It seems natural, then, to use the sampling variance of an estimator as a criterion of its acceptability; indeed, it has been so used since the days of Laplace and Gauss. But only in relatively recent times has it been established that, under fairly general conditions, there exists a bound below which the variance of an unbiased estimator cannot fall. In order to establish this bound, we first derive some preliminary results, which will also be useful in other connections later.

17.14 If the density function of the continuous or discrete population is $f(x|\theta)$, we define the *likelihood function*[2] of a sample of n independent observations by

$$L(x_1, x_2, \ldots, x_n|\theta) = f(x_1|\theta)\, f(x_2|\theta) \cdots f(x_n|\theta). \tag{17.16}$$

We shall often write this simply as L. Evidently, since L is the joint density function of the observations,

$$\int \cdots \int L\, \mathrm{d}x_1 \cdots \mathrm{d}x_n = 1. \tag{17.17}$$

Now suppose that the first two derivatives of L with respect to θ exist for all θ. If we differentiate both sides of (17.17) with respect to θ, and if we may interchange the operations of differentiation and integration on its left-hand side (see **17.16**), we obtain

$$\int \cdots \int \frac{\partial L}{\partial \theta} \mathrm{d}x_1 \cdots \mathrm{d}x_n = 0,$$

which we may rewrite

$$E\left(\frac{\partial \log L}{\partial \theta}\right) = \int \cdots \int \left(\frac{1}{L}\frac{\partial L}{\partial \theta}\right) L\, \mathrm{d}x_1 \cdots \mathrm{d}x_n = 0. \tag{17.18}$$

If we differentiate (17.18) again, we obtain, if we may again interchange operations,

$$\int \cdots \int \left\{\left(\frac{1}{L}\frac{\partial L}{\partial \theta}\right)\frac{\partial L}{\partial \theta} + L\frac{\partial}{\partial \theta}\left(\frac{1}{L}\frac{\partial L}{\partial \theta}\right)\right\} \mathrm{d}x_1 \cdots \mathrm{d}x_n = 0,$$

which becomes

$$\int \cdots \int \left\{\left(\frac{1}{L}\frac{\partial L}{\partial \theta}\right)^2 + \frac{\partial^2 \log L}{\partial \theta^2}\right\} L\, \mathrm{d}x_1 \cdots \mathrm{d}x_n = 0$$

or

$$E\left[\left(\frac{\partial \log L}{\partial \theta}\right)^2\right] = -E\left(\frac{\partial^2 \log L}{\partial \theta^2}\right). \tag{17.19}$$

17.15 Now consider an unbiased estimator, t, of some function of θ, say $\tau(\theta)$. This formulation allows us to consider unbiased and biased estimators of θ itself, and also permits us to consider, for example, the estimation of the standard deviation when the parameter is equal to the variance. We thus have

$$E(t) = \int \cdots \int tL\, \mathrm{d}x_1 \cdots \mathrm{d}x_n = \tau(\theta). \tag{17.20}$$

We now differentiate (17.20), the result being

$$\int \cdots \int t\frac{\partial \log L}{\partial \theta} L\, \mathrm{d}x_1 \cdots \mathrm{d}x_n = \tau'(\theta),$$

[2] Fisher (1921a) calls L the likelihood when regarded as a function of θ and the probability of the sample when it is regarded as a function of x for fixed θ. While appreciating this distinction, we use the term likelihood and the symbol L in both cases to preserve a single notation. The interpretation will be clear from the context.

which we may rewrite, using (17.18), as

$$\tau'(\theta) = \int \cdots \int \{t - \tau(\theta)\} \frac{\partial \log L}{\partial \theta} L \, dx_1 \cdots dx_n. \tag{17.21}$$

By the Cauchy–Schwarz inequality, we have from (17.21)[3]

$$\{\tau'(\theta)\}^2 \leq \int \cdots \int \{t - \tau(\theta)\}^2 L \, dx_1 \cdots dx_n \cdot \int \cdots \int \left(\frac{\partial \log L}{\partial \theta}\right)^2 L \, dx_1 \cdots dx_n,$$

which, on rearrangement, becomes

$$\operatorname{var} t = E\{t - \tau(\theta)\}^2 \geq \{\tau'(\theta)\}^2 / E\left[\left(\frac{\partial \log L}{\partial \theta}\right)^2\right]. \tag{17.22}$$

This is the fundamental inequality for the variance of an estimator, often known as the Cramér–Rao inequality, after two of its several discoverers (Rao, 1945; Cramér, 1946). It was also given implicitly by Aitken and Silverstone (1942), although particular cases of the result have been traced back much earlier. Using (17.19), it may be written in what is often, in practice, the more convenient form

$$\operatorname{var} t \geq -\{\tau'(\theta)\}^2 / E\left(\frac{\partial^2 \log L}{\partial \theta^2}\right). \tag{17.23}$$

When $\tau(\theta) \equiv \theta$, we have $\tau'(\theta) = 1$ in (17.22), so for an unbiased estimator of θ

$$\operatorname{var} t \geq 1 / E\left[\left(\frac{\partial \log L}{\partial \theta}\right)^2\right] = -1 / E\left(\frac{\partial^2 \log L}{\partial \theta^2}\right). \tag{17.24}$$

The quantity

$$I = E\left[\left(\frac{\partial \log L}{\partial \theta}\right)^2\right] \tag{17.25}$$

is sometimes called the *amount of information* or the *Fisher information* in the sample. We may similarly use (17.22) as a bound for I in terms of the moments of t,

$$I \geq \{\tau'(\theta)\}^2 / \operatorname{var} t. \tag{17.26}$$

We shall call (17.22) and (17.23) the minimum variance bound (MVB) for the estimation of $\tau(\theta)$. An estimator which attains this bound for all θ will be called an MVB estimator.

For the MVB (17.22) to follow from (17.20), it is only necessary that (17.18) holds. If (17.19) also holds, we may write the MVB in the form (17.23).

17.16 The interchange of the operations of differentiation and integration leading to (17.18) or (17.19) is permissible if, for example, the limits of integration (i.e. the limits of variation of x) are finite and independent of θ, and also if these limits are infinite, provided that the integral resulting from the interchange is uniformly convergent for all θ and its integrand is a continuous function of x and θ. These are sufficient sets of conditions – Exercises 17.21–22 show that they are not necessary.

---

[3]Expression (17.21) is the covariance between t and $\partial \log L / \partial \theta$, whose square cannot exceed the product of their variances, a result familiar in the theory of correlation (Chapter 27).

17.17 It is very easy to establish the condition under which the MVB is attained. The inequality in (17.22) arose purely from the use of the Cauchy–Schwarz inequality, and the necessary and sufficient condition that the Cauchy–Schwarz inequality becomes an equality is (cf. **2.29**) that $\{t - \tau(\theta)\}$ is proportional to $\partial \log L/\partial\theta$ for all sets of observations. We write this condition

$$\frac{\partial \log L}{\partial\theta} = A(\theta)\{t - \tau(\theta)\}, \tag{17.27}$$

where A is independent of the observations. Thus t is an MVB estimator if and only if it is a linear function of $\partial \log L/\partial\theta$.

Multiplying both sides of (17.27) by $\{t - \tau(\theta)\}$ and taking expectations, we find, using (17.21),

$$\tau'(\theta) = A(\theta) \operatorname{var} t. \tag{17.28}$$

Thus $A(\theta)$ has the same sign as $\tau'(\theta)$ and

$$\operatorname{var} t = \tau'(\theta)/A(\theta). \tag{17.29}$$

We thus conclude that if (17.27) is satisfied, t is an MVB estimator of $\tau(\theta)$, with variance (17.29), which is then equal to the right-hand side of (17.23). If $\tau(\theta) \equiv \theta$, $\operatorname{var} t$ is just $1/A(\theta)$, which is then equal to the right-hand side of (17.24).

Example 17.6 (The sample mean as MVB estimator for the normal)

We wish to estimate θ in the normal distribution

$$f(x) = \frac{1}{\sigma(2\pi)^{1/2}} \exp\left\{-\frac{1}{2}\left(\frac{x-\theta}{\sigma}\right)^2\right\}, \qquad -\infty < x < \infty,$$

where σ is known.

We have

$$\frac{\partial \log L}{\partial\theta} = \frac{n}{\sigma^2}(\bar{x} - \theta).$$

This is of the form (17.27) with

$$t = \bar{x}, \quad A(\theta) = n/\sigma^2 \quad \text{and} \quad \tau(\theta) = \theta.$$

Thus $\bar{x}$ is the MVB estimator of θ, with variance σ^2/n.

Example 17.7 (The lack of an MVB estimator for the Cauchy)

We wish to estimate θ in the Cauchy distribution

$$f(x) = \frac{1}{\pi\{1 + (x-\theta)^2\}}, \qquad -\infty < x < \infty.$$

We have

$$\frac{\partial \log L}{\partial\theta} = 2\sum \frac{x-\theta}{\{1 + (x-\theta)^2\}}.$$

This cannot be put in the form (17.27). Thus there is no MVB estimator in this case.

Example 17.8 (MVB estimator for the Poisson)
We wish to estimate θ in the Poisson distribution

$$f(x|\theta) = \mathrm{e}^{-\theta}\theta^x/x!, \qquad x = 0, 1, 2, \ldots,$$

we have

$$\frac{\partial \log L}{\partial \theta} = \frac{n}{\theta}(\bar{x} - \theta).$$

Thus $\bar{x}$ is the MVB estimator of θ, with variance θ/n.

Example 17.9 (MVB estimator for the binomial)
We wish to estimate θ in the binomial distribution, for which

$$L(r|\theta) = \binom{n}{r}\theta^r(1-\theta)^{n-r}, \qquad r = 0, 1, 2, \ldots, n.$$

We find

$$\frac{\partial \log L}{\partial \theta} = \frac{n}{\theta(1-\theta)}\left(\frac{r}{n} - \theta\right).$$

Hence r/n is the MVB estimator of θ, with variance $\theta(1-\theta)/n$.

17.18 It follows from our discussion that, where an MVB estimator exists, it will exist for one specific function $\tau(\theta)$ of the parameter θ, and for no other function of θ. The following example makes the point clear.

Example 17.10 (Existence of MVB estimator)
We wish to estimate θ in the normal distribution with mean zero

$$f(x) = \frac{1}{\theta(2\pi)^{1/2}}\exp\left(-\frac{x^2}{2\theta^2}\right), \qquad -\infty < x < \infty.$$

We find

$$\frac{\partial \log L}{\partial \theta} = -\frac{n}{\theta} + \frac{\sum x^2}{\theta^3} = \frac{n}{\theta^3}\left(\frac{1}{n}\sum x^2 - \theta^2\right).$$

We see at once that $(1/n)\sum x^2$ is an MVB estimator of θ^2 (the variance of the population) with sampling variance $(\theta^3/n)\,\mathrm{d}(\theta^2)/\mathrm{d}\theta = 2\theta^4/n$, by (17.29). But there is no MVB estimator of θ itself.

Equation (17.27) determines a condition on the density function under which an MVB estimator of some function of θ, $\tau(\theta)$, exists. If the density is not of this form, there may still be an estimator of $\tau(\theta)$ that has, uniformly in θ, smaller variance than any other estimator; we then call it a minimum variance (MV) estimator. In other words, the least *attainable* variance may be greater than the MVB. Further, if the regularity conditions leading to the MVB do not hold, the least attainable variance may be less than the (in this case inapplicable) MVB. In any case (17.27) demonstrates that there can only be one function of θ for which the MVB is attainable, namely, that function (if any) which is the expectation of a statistic t in terms of which $\partial \log L/\partial\theta$ may be linearly expressed.

17.19 From (17.27) we have on integration the necessary form for the likelihood function (continuing to write $A(\theta)$ for the integral of the arbitrary function $A(\theta)$ in (17.27))

$$\log L = tA(\theta) + P(\theta) + R(x_1, x_2, \ldots, x_n),$$

which we may rewrite in the density-function form

$$f(x|\theta) = \exp\{A(\theta)B(x) + C(x) + D(\theta)\}, \tag{17.30}$$

where $t = \sum_{i=1}^{n} B(x_i)$, $R(x_1, \ldots, x_n) = \sum_{i=1}^{n} C(x_i)$ and $P(\theta) = nD(\theta)$. (17.30) is the exponential family of distributions introduced in **5.47**. We shall return to it in **17.36**.

The regularity conditions required to derive (17.27) limit the generality of this result. In particular, Joshi (1976) showed that if $\log L$ is not absolutely continuous in θ the family (17.30) is no longer necessary – cf. Exercise 17.29.

Bhattacharyya bounds

17.20 We can find better (i.e. greater) lower bounds than the MVB (17.22) for the variance of an estimator in cases where the MVB is not attainable. The condition (17.27) for (17.22) to be an attainable bound is that there be an estimator t for which $t - \tau(\theta)$ is a linear function of

$$\frac{\partial \log L}{\partial \theta} = \frac{1}{L}\frac{\partial L}{\partial \theta}.$$

But even if no such estimator exists, there may still be one for which $t - \tau(\theta)$ is a linear function of

$$\frac{1}{L}\frac{\partial L}{\partial \theta} \quad \text{and} \quad \frac{1}{L}\frac{\partial^2 L}{\partial \theta^2}$$

or, in general, of $(1/L)\partial^r L/\partial\theta^r$. Following Bhattacharyya (1946), we therefore seek such a linear function that most closely approximates $t - \tau(\theta)$.

Estimating $\tau(\theta)$ by a statistic t as before, we write

$$L^{(r)} = \frac{\partial^r L}{\partial \theta^r}$$

and

$$\tau^{(r)} = \frac{\partial^r \tau(\theta)}{\partial \theta^r}.$$

We now construct the function

$$D_s = t - \tau(\theta) - \sum_{r=1}^{s} a_r L^{(r)}/L, \tag{17.31}$$

where the a_r are constants to be determined. Since, on differentiating (17.17) r times, we obtain

$$E(L^{(r)}/L) = 0 \tag{17.32}$$

under the same conditions as before, we have from (17.31) and (17.32)

$$E(D_s) = 0. \tag{17.33}$$

The variance of D_s is therefore

$$E(D_s^2) = \int \cdots \int \left\{t - \tau(\theta) - \sum_r a_r L^{(r)}/L\right\}^2 L\,\mathrm{d}x_1 \cdots \mathrm{d}x_n. \tag{17.34}$$

We minimize (17.34) for variation of the a_r by putting

$$\int \cdots \int \left\{t - \tau(\theta) - \sum_r a_r L^{(r)}/L\right\} \frac{L^{(p)}}{L} \cdot L\,\mathrm{d}x_1 \cdots \mathrm{d}x_n = 0 \tag{17.35}$$

for $p = 1, 2, \ldots, s$. This gives

$$\int \cdots \int (t - \tau(\theta))L^{(p)}\,\mathrm{d}x_1 \cdots \mathrm{d}x_n = \sum_r a_r \int \cdots \int \frac{L^{(r)}}{L} \cdot \frac{L^{(p)}}{L} L\,\mathrm{d}x_1 \cdots \mathrm{d}x_n. \tag{17.36}$$

The left-hand side of (17.36) is, from (17.32), equal to

$$\int \cdots \int tL^{(p)}\,\mathrm{d}x_1 \cdots \mathrm{d}x_n = \tau^{(p)}$$

on comparison with (17.20). The right-hand side of (17.36) is simply

$$\sum a_r E\left(\frac{L^{(r)}}{L} \cdot \frac{L^{(p)}}{L}\right).$$

On insertion of these values in (17.36) it becomes

$$\tau^{(p)} = \sum_{r=1}^{s} a_r E\left(\frac{L^{(r)}}{L} \cdot \frac{L^{(p)}}{L}\right), \qquad p = 1, 2, \ldots, s. \tag{17.37}$$

We may invert this set of linear equations, provided that the matrix of coefficients $J_{rp} = E(L^{(r)}/L \cdot L^{(p)}/L)$ is non-singular, to obtain

$$a_r = \sum_{p=1}^{s} \tau^{(p)} J_{rp}^{-1}, \qquad r = 1, 2, \ldots, s. \tag{17.38}$$

Thus, at the minimum of (17.34), (17.31) takes the value

$$D_s = t - \tau(\theta) - \sum_{r=1}^{s} \sum_{p=1}^{s} \tau^{(p)} J_{rp}^{-1} L^{(r)}/L, \tag{17.39}$$

and (17.34) itself has the value, from (17.39),

$$E(D_s^2) = \int \cdots \int \left\{(t - \tau(\theta)) - \sum_r \sum_p \tau^{(p)} J_{rp}^{-1} L^{(r)}/L\right\}^2 L\,\mathrm{d}x_1 \cdots \mathrm{d}x_n,$$

which, on using (17.35), becomes

$$= \int \cdots \int \{t - \tau(\theta)\}^2 L \, dx_1 \cdots dx_n$$
$$- \sum_r \sum_p \tau^{(p)} J_{rp}^{-1} \int \cdots \int t L^{(r)} \, dx_1 \cdots dx_n,$$

and finally we have

$$E(D_s^2) = \operatorname{var} t - \sum_r \sum_p \tau^{(p)} J_{rp}^{-1} \tau^{(r)}. \tag{17.40}$$

Since its left-hand side is non-negative, (17.40) gives the required inequality

$$\operatorname{var} t \geq \sum_{r=1}^{s} \sum_{p=1}^{s} \tau^{(p)} J_{rp}^{-1} \tau^{(r)}. \tag{17.41}$$

In the particular case $s = 1$, (17.41) reduces to the MVB (17.22).

17.21 The condition for the bound in (17.41) to be attained is simply that $E(D_s^2) = 0$, which with (17.33) leads to $D_s = 0$ or, from (17.39),

$$t - \tau(\theta) = \sum_{r=1}^{s} \sum_{p=1}^{s} \tau^{(p)} J_{rp}^{-1} L^{(r)}/L, \tag{17.42}$$

which is the generalization of (17.27). Equation (17.42) requires $t - \tau(\theta)$ to be a linear function of the quantities $L^{(r)}/L$. If it is a linear function of the first s such quantities, there is clearly nothing to be gained by adding further terms. On the other hand, the right-hand side of (17.41) is a non-decreasing function of s, as is easily seen by the consideration that the variance of D_s cannot be increased by allowing the optimum choice of a further coefficient a_r, in (17.34). Thus we may expect, in some cases where the MVB (17.22) is not attained for the estimation of a particular function $\tau(\theta)$, to find the greater bound in (17.41) attained for some value of $s > 1$.

Fend (1959) showed that if (17.27) holds, any polynomial of degree s in t attains the sth but not the $(s-1)$th variance bound (17.41), and is therefore the MVU estimator of its expectation.

If we seek similarly to improve the bound (17.26) for I, the natural analogue of the procedure of **17.20** is to approximate $\partial \log L/\partial\theta$ by a polynomial in $t - \tau(\theta)$, minimizing $\operatorname{var}\{\partial \log L/\partial\theta - \sum_{r=1}^{s} C_r(t - \tau(\theta))^r\}$.

This yields an increasing series of bounds in terms of the central moments μ_j and the skewness and kurtosis coefficients of t; Jarrett (1984) gives an explicit general result. $s = 1$ yields (17.26) of course; for $s = 2$, the result, previously given by Bhattacharyya (1946), is

$$I \geq \mu_2^{-1}\{(\tau')^2 + (\mu_2^{-1/2}\,\partial\mu_2/\partial\theta - \beta_1^{1/2}\tau')^2/(\beta_2 - \beta_1 - 1)\},$$

the second term being positive by Exercise 3.19.

17.22 We now investigate the improvement in the bound arising from taking $s = 2$ instead of $s = 1$ in (17.41). Remembering that we have defined

$$J_{rp} = E\left(\frac{L^{(r)}}{L} \cdot \frac{L^{(p)}}{L}\right), \tag{17.43}$$

we find that we may rewrite (17.41) in this case as

$$\begin{vmatrix} \text{var}\, t & \tau' & \tau'' \\ \tau' & J_{11} & J_{12} \\ \tau'' & J_{12} & J_{22} \end{vmatrix} \geq 0, \tag{17.44}$$

which becomes, on expansion,

$$\text{var}\, t \geq \frac{(\tau')^2}{J_{11}} + \frac{(\tau' J_{12} - \tau'' J_{11})^2}{J_{11}(J_{11}J_{22} - J_{12}^2)}. \tag{17.45}$$

The second term on the right of (17.45) is the improvement in the bound to the variance of t. It may easily be confirmed that, writing $J_{rp}(n)$ as a function of sample size,

$$\left.\begin{aligned} J_{11}(n) &= nJ_{11}(1), \\ J_{12}(n) &= nJ_{12}(1), \\ J_{22}(n) &= nJ_{22}(1) + 2n(n-1)\{J_{11}(1)\}^2, \end{aligned}\right\} \tag{17.46}$$

and using (17.46), (17.45) becomes

$$\text{var}\, t \geq \frac{(\tau')^2}{nJ_{11}(1)} + \frac{\{\tau'' - \tau' J_{12}(1)/J_{11}(1)\}^2}{2n^2\{J_{11}(1)\}^2} + o\left(\frac{1}{n^2}\right), \tag{17.47}$$

which makes it clear that the improvement in the bound is of order $1/n^2$ compared with the leading term of order $1/n$, and is only of importance when $\tau' = 0$ and the first term disappears. In the case where t is estimating θ itself, (17.45)–(17.47) give

$$\begin{aligned} \text{var}\, t &\geq \frac{1}{J_{11}} + \frac{J_{12}^2}{J_{11}(J_{11}J_{22} - J_{12}^2)} \\ &= \frac{1}{J_{11}} + \frac{J_{12}^2}{2J_{11}^4} + o\left(\frac{1}{n^2}\right). \end{aligned} \tag{17.48}$$

If the bound with $s = 2$ equals the MVB ($s = 1$), this does not generally imply that the latter is attainable. For the exponential family (17.30), however, this equality does imply attainability – cf. Patil and Shorrock (1965).

Example 17.11 (Bhattacharyya bound for the binomial)

To estimate $\theta(1-\theta)$ in the binomial distribution of Example 17.9, it is natural to take as our estimator an unbiased function of $r/n = p$, which is the MVB estimator of θ. We have seen in Example 17.4 that $r(r-1)/\{n(n-1)\}$ is an unbiased estimator of θ^2. Hence, or from Exercise 17.12,

$$t = \frac{r}{n} - \frac{r(r-1)}{n(n-1)} = \frac{r(n-r)}{n(n-1)} = \left(\frac{n}{n-1}\right)p(1-p)$$

is an unbiased estimator of $\theta(1-\theta)$. The exact variance of t is obtained by rewriting it as

$$\frac{n-1}{n}t \equiv \tfrac{1}{4} - \left(p - \tfrac{1}{2}\right)^2$$

so that

$$\left(\frac{n-1}{n}\right)^2 \operatorname{var} t = \operatorname{var}\left\{\left(p-\tfrac{1}{2}\right)^2\right\} = E\left\{\left(p-\tfrac{1}{2}\right)^4\right\} - \left[E\left\{\left(p-\tfrac{1}{2}\right)^2\right\}\right]^2$$
$$= \lambda_4' - (\lambda_2')^2$$

where λ_r', is the rth moment of p about the value $\frac{1}{2}$. Using (3.8), we express the λ_r' in terms of the central moments λ_r about $E(p) = \theta$:

$$\lambda_4' - (\lambda_2')^2 = \lambda_4 + 4\left(\theta - \tfrac{1}{2}\right)\lambda_3 + 4\left(\theta - \tfrac{1}{2}\right)^2\lambda_2 - \lambda_2^2.$$

Now λ_r is $n^{-r}\mu_r$, where μ_r is the moment of the binomial variable np, so

$$\left(\frac{n-1}{n}\right)^2 \operatorname{var} t = \frac{\mu_4 - \mu_2^2}{n^4} - \frac{2(1-2\theta)\mu_3}{n^3} + (1-2\theta)^2\frac{\mu_2}{n^2}$$

and we may substitute for the μ_r, from (5.4) – where p is written for our θ – obtaining finally

$$\operatorname{var} t = \frac{\theta(1-\theta)}{n}\left\{(1-2\theta)^2 + \frac{2\theta(1-\theta)}{n-1}\right\},$$

which is an order of magnitude in n smaller when $\theta = \frac{1}{2}$, when it equals $1/\{8n(n-1)\}$, than when $\theta \neq \frac{1}{2}$. We cannot attain the MVB for $\tau(\theta) = \theta(1-\theta)$, since it is attained for θ itself (cf. Example 17.9) and (cf. **17.18**) only one function can attain its MVB. However, when $\theta \neq \frac{1}{2}$ the term in n^{-1} in $\operatorname{var} t$ is $(\tau')^2/J_{11}$, the MVB, which is therefore *asymptotically* attained – cf. **17.23**. For any θ, $\operatorname{var} t$ exactly attains the bound (17.45) for $s = 2$, as may be seen from substituting the results $J_{12} = 0$, $J_{22} = 2n(n-1)/[\theta^2(1-\theta)^2]$ (which are left to the reader in Exercise 17.3) in (17.45). Alternatively, this result follows from the fact that t is a linear function of L'/L and L''/L – cf. Exercise 17.3.

Exercise 17.28 gives a general result for the bounds (17.41) in the binomial case.

17.23 Example 17.11 brings out a point of some importance. If we have an MVB estimator t for $\tau(\theta)$, i.e.

$$\operatorname{var} t = \{\tau'(\theta)\}^2/J_{11}, \qquad (17.49)$$

and we require to estimate some other function of θ, say $\psi\{\tau(\theta)\}$, we know from (10.14), that in large samples

$$\operatorname{var}\{\psi(t)\} \sim \left(\frac{\partial\psi}{\partial\tau}\right)^2 \operatorname{var} t$$
$$\sim \left(\frac{\partial\psi}{\partial\tau}\right)^2 \left(\frac{\partial\tau}{\partial\theta}\right)^2 \Big/ J_{11}$$

from (17.49), provided that ψ' is non-zero. But this may be rewritten

$$\text{var}\{\psi(t)\} \sim \left(\frac{\partial \psi}{\partial \tau}\right)^2 \bigg/ J_{11},$$

so that *any* such function of θ has an MVB estimator in large samples if some function of θ has such an estimator for all n. Further, the estimator is always the corresponding function of t. We shall, in **17.35**, be able to supplement this result by a more exact one concerning functions of the MVB estimator.

17.24 There is no difficulty in extending the results of this chapter on MVB estimation to the case when the distribution has more than one parameter, and we are interested in estimating a single function of them all, say $\tau(\theta_1, \theta_2, \ldots, \theta_k)$. In this case, the analogue of the simplest result, (17.22), is

$$\text{var}\, t \geq \sum_{i=1}^{k} \sum_{j=1}^{k} \frac{\partial \tau}{\partial \theta_i} \frac{\partial \tau}{\partial \theta_j} I_{ij}^{-1} \tag{17.50}$$

where the matrix which has to be inverted to obtain the terms in (17.50) is

$$\{I_{ij}\} = \left\{E\left(\frac{\partial \log L}{\partial \theta_i} \cdot \frac{\partial \log L}{\partial \theta_j}\right)\right\}. \tag{17.51}$$

As before, (17.50) only takes account of terms of order $1/n$, and a more complicated inequality is required if we are to take account of lower-order terms.

17.25 Even better lower bounds for the sampling variance of estimators can be obtained without imposing regularity conditions. See Kiefer (1952), Barankin (1949) and Blischke *et al.* (1969). Chapman and Robbins (1951) established that if $E(t) = \theta$,

$$\text{var}\, t \geq \frac{1}{\inf\limits_h \frac{1}{h^2}\{\int[(L(x|\theta+h))^2/L(x|\theta)]\,dx - 1\}}, \tag{17.52}$$

the infimum being for all $h \neq 0$ for which $L(x|\theta) = 0$ implies $L(x|\theta + h) = 0$. Expression (17.52) is in general at least as good a bound as (17.24), though more generally valid. For the denominator of the right-hand side of (17.24) is

$$E\left[\left(\frac{\partial \log L}{\partial \theta}\right)^2\right] = \int \lim_{h \to 0}\left[\frac{L(x|\theta+h) - L(x|\theta)}{hL(x|\theta)}\right]^2 L(x|\theta)\,dx,$$

and provided that we may interchange the integral and the limiting operation, this becomes

$$= \lim_{h \to 0} \frac{1}{h^2}\left[\int \frac{\{L(x|\theta+h)\}^2}{L(x|\theta)}\,dx - 1\right]. \tag{17.53}$$

The denominator on the right of (17.52) is the infimum of this quantity over all permissible values of h, and is consequently no greater than the limit as h tends to zero. Thus (17.52) is at least as good a bound as (17.24) – cf. Exercise 17.6.

Sen and Ghosh (1976) give a condition for the bound (17.52) to be attained, and investigate its relation to the bounds (17.42) with $s > 1$.

Minimum variance unbiased estimation

17.26 So far, we have been largely concerned with the uniform attainment of the MVB by a single estimator. In many cases, however, it will not be attained, even when the conditions under which it is derived are satisfied. When this is so, there may still be an MVU estimator for all values of θ. The question of the uniqueness of an MVU estimator now arises. We can easily show that if an MVU estimator exists, it is essentially unique, irrespective of whether any bound is attained.

Let t_1 and t_2 be MVU estimators of $\tau(\theta)$, each with variance V. Consider the new estimator

$$t_3 = \tfrac{1}{2}(t_1 + t_2)$$

which evidently also estimates $\tau(\theta)$ with variance

$$\operatorname{var} t_3 = \tfrac{1}{4}\{\operatorname{var} t_1 + \operatorname{var} t_2 + 2\operatorname{cov}(t_1, t_2)\}. \tag{17.54}$$

Now by the Cauchy–Schwarz inequality

$$\begin{aligned}\operatorname{cov}(t_1, t_2) &= \int (t_1 - \tau)(t_2 - \tau)\, L\, \mathrm{d}x_1 \cdots \mathrm{d}x_n \\ &\leq \left\{\int \cdots \int (t_1 - \tau)^2 L\, \mathrm{d}x_1 \cdots \mathrm{d}x_n \cdot \int \cdots \int (t_2 - \tau)^2 L\, \mathrm{d}x_1 \cdots \mathrm{d}x_n\right\}^{1/2} \\ &\leq (\operatorname{var} t_1 \operatorname{var} t_2)^{1/2} = V. \end{aligned} \tag{17.55}$$

Thus (17.54) and (17.55) give

$$\operatorname{var} t_3 \leq V,$$

which contradicts the assumption that t_1 and t_2 have minimum variance unless the equality holds. This implies that the equality sign holds in (17.55). This can only be so if

$$(t_1 - \tau) = k(\theta)(t_2 - \tau), \tag{17.56}$$

i.e. the variables are proportional. But (17.56) implies that

$$\operatorname{cov}(t_1, t_2) = k(\theta) \operatorname{var} t_2 = k(\theta) V$$

and this equals V since the equality sign holds in (17.55). Thus $k(\theta) = 1$ and hence, from (17.56), $t_1 = t_2$ identically. Thus an MVU estimator is unique.

We show in **21.16** that the bounded completeness of a distribution implies the existence of a minimal sufficient statistic. In turn, this property implies that such a sufficient statistic is unique and, therefore, that if an MVU estimator exists it must be unique.

17.27 The argument of the preceding section can be generalized to give an interesting relation for the correlation ρ between the MVU estimator and any other estimator, which, it will be remembered from **16.23**, is defined as the ratio of their covariance to the square root of the product of their variances.

Suppose that t_1 is the MVU estimator of $\tau(\theta)$, with variance V, and that t_2 is any other unbiased estimator of $\tau(\theta)$ with finite variance V_2, where $V/V_2 = E$. Consider a new estimator

$$t_3 = at_1 + (1-a)t_2. \qquad (17.57)$$

This will also estimate $\tau(\theta)$, with variance

$$\text{var}\, t_3 = a^2 \,\text{var}\, t_1 + (1-a)^2 \,\text{var}\, t_2 + 2a(1-a)\,\text{cov}(t_1, t_2). \qquad (17.58)$$

Writing

$$\rho = \frac{\text{cov}(t_1, t_2)}{(\text{var}\, t_1 \,\text{var}\, t_2)^{1/2}} = \frac{\text{cov}(t_1, t_2)E^{1/2}}{V},$$

we obtain from (17.58), since var $t_3 \geq V$,

$$a^2 + (1-a)^2 E^{-1} + 2a(1-a)\rho E^{-1/2} \geq 1. \qquad (17.59)$$

(17.59) may be rearranged as a quadratic inequality in a,

$$a^2\{1 + E^{-1} - 2\rho E^{-1/2}\} - 2a(E^{-1} - \rho E^{-1/2}) + (E^{-1} - 1) \geq 0,$$

and the discriminant of the left-hand side cannot be positive, since the roots of the equation are complex or equal. Thus

$$(E^{-1} - \rho E^{-1/2})^2 \leq (1 + E^{-1} - 2\rho E^{-1/2})(E^{-1} - 1)$$

which yields

$$(\rho E^{-1/2} - 1)^2 \leq 0. \qquad (17.60)$$

Hence

$$\rho = E^{1/2}, \qquad (17.61)$$

where E is the reciprocal of the relative variance of the other estimator, and in large samples is usually called its *efficiency*, for a reason to be discussed in **17.28–29**.

From the definition of ρ, it follows at once using (17.61) that the covariance of any unbiased finite-variance estimator u with an MVU estimator t is exactly var t. Since $u \equiv t + (u - t)$ this implies $\text{cov}(t, u-t) = 0$ – cf. Exercise 17.11. Hence t has zero covariance with *every* finite-variance function having zero expectation. Conversely, if $\text{cov}(t, u-t) = 0$ for all u with finite variance and $E(u) = E(t)$, then

$$\text{cov}(t, u) = \text{var}\, t$$

and their correlation, which must lie between 0 and 1, is $(\text{var}\, t/\,\text{var}\, u)^{1/2}$, so t must be the MVU estimator of its expectation.

Blom (1978) shows that n var t cannot increase with n – see Exercise 17.32. Bondesson (1975) shows that for a location parameter, when $f = f(x-\theta)$, no MVU estimator of θ exists if f tends to zero too rapidly, while no polynomial in x can be an MVU estimator unless f is normal.

Example 17.12 (Efficiency of the sample variance for the Poisson)

In Example 17.8, we saw that $\bar{x}$ is the MVB estimator of θ in the Poisson distribution. Since by Example 3.10 all the cumulants of the Poisson are equal to θ, $\bar{x}$ is the MVU estimator of each of them. If we were instead to estimate the variance unbiasedly by the general formula

$$\frac{1}{n-1}\sum_{j=1}^{n}(x_j - \bar{x})^2,$$

the efficiency of the latter, by (17.61), would be ρ^2, and the correlation ρ was found in Exercise 12.21 to be $(1+2\theta)^{-1/2}$. Thus $E = (1+2\theta)^{-1}$, tending to 1 as $\theta \to 0$, but tending to 0 as $\theta \to \infty$. At $\theta = 1$, the efficiency is only $\frac{1}{3}$.

Efficiency

17.28 So far, our discussion of MVU estimation has been exact, in the sense that it has not restricted sample size in any way. We now turn to consideration of large-sample properties. Even if there is no MVU estimator for finite n, an estimator may be MVU asymptotically. Since most of the estimators we deal with are asymptotically normally distributed by virtue of the central limit theorem, the distribution of such an estimator will depend for large samples on only two parameters – its mean and its variance. If it is a consistent estimator it will commonly be asymptotically unbiased – cf. **17.9**. This leaves the variance as the means of discriminating between consistent, asymptotically normal estimators of the same parametric function.

Among such estimators, one with MV in large samples[4] is called an *efficient estimator*, or simply *efficient*, the term being due to Fisher (1921a), although the term is sometimes also used for small samples.

Portnoy (1977) shows that for the exponential family (17.30), any MVU estimator is also efficient.

17.29 If we compare consistent asymptotically normal estimators in large samples, we may reasonably set up a *measure* of efficiency, something we have not attempted to do for small n and arbitrarily distributed estimators. We shall define the efficiency of any other estimator, relative to an efficient estimator, as the reciprocal of the ratio of sample sizes required to give the estimators equal sampling variances, i.e. to make them equally precise.

If t_1 is an efficient estimator, and t_2 is another estimator, and (as is commonly the case) the variances are, in large samples, simple inverse functions of sample size, we may translate our definition of efficiency into a simple form. Let us suppose that

$$\begin{aligned} V_1 &= \text{var}(t_1|n_1) \sim a_1/n_1^r \qquad (r > 0), \\ V_2 &= \text{var}(t_2|n_2) \sim a_2/n_2^s \qquad (s > 0), \end{aligned} \tag{17.62}$$

where a_1, a_2 are constants independent of n, and we have shown sample size as an argument in the variances. If we are to have $V_1 = V_2$, we must have

$$1 = \frac{V_1}{V_2} = \lim_{n_1,n_2\to\infty} \frac{a_1}{a_2}\frac{n_2^s}{n_1^r}.$$

[4]But see the discussion of 'superefficiency' in **18.16**.

Thus

$$\frac{a_2}{a_1} = \lim \frac{n_2^s}{n_1^r} = \lim \left(\frac{n_2}{n_1}\right)^s \cdot \frac{1}{n_1^{r-s}}. \tag{17.63}$$

Since t_1 is efficient, we must have $r \geq s$. If $r > s$, the last factor on the right of (17.63) will tend to zero, and hence, if the product is to remain equal to a_2/a_1, we must have

$$\frac{n_2}{n_1} \to \infty, \qquad r > s, \tag{17.64}$$

and we would thus say that t_2 has zero efficiency. If, in (17.63), $r = s$, we have at once

$$\lim \frac{n_2}{n_1} = \left(\frac{a_2}{a_1}\right)^{1/r},$$

which from (17.62) may be written

$$\lim \frac{n_2}{n_1} = \lim \left(\frac{V_2}{V_1}\right)^{1/r},$$

and the efficiency of t_2 is the reciprocal of this, namely

$$E = \lim \left(\frac{V_1}{V_2}\right)^{1/r}. \tag{17.65}$$

Note that if $r > s$, (17.65) gives the same result as (17.64). If $r = 1$, which is the most common case, (17.65) reduces to the inverse variance ratio encountered at the end of **17.27**. Thus, when we are comparing estimators with variances of order $1/n$, we measure efficiency relative to an efficient estimator by the inverse of the variance ratio.

If the variance of an efficient estimator is not of the simple form (17.62) – see, for example, Exercise 18.21 – the measurement of relative efficiency is not so simple.

Although it follows from the result of **17.26** that efficient estimators tend asymptotically to equivalence, there will in general be a multiplicity of efficient estimators, for if t_1 is efficient so is any $t_2 = t_1 + cn^{-p}$, where p is chosen large enough. If (17.62) holds for two efficient estimators t_1, t_2 we must have $r = s$ and $a_1 = a_2 = a$, say, so that $E = 1$ at (17.65). Now take (17.62) to a further term, so that

$$V_j = \frac{a}{n_j^r} + \frac{b_j}{n_j^{r+q}} + o(n_j^{-(r+q)}), \qquad j = 1, 2,$$

where $q > 0$ and we assume $b_1 \leq b_2$. If we now equate V_1 to V_2 we obtain, instead of (17.63),

$$\left(\frac{n_2}{n_1}\right)^r = \left(1 + \frac{b_2}{an_2^q}\right) \Big/ \left(1 + \frac{b_1}{an_1^q}\right)$$

and writing $d_n = n_2 - n_1$, this gives, on expanding the right-hand side,

$$1 + \frac{d_n}{n_1} = 1 + \frac{b_2}{ran_2^q} - \frac{b_1}{ran_1^q}.$$

Since $E = 1, n_1/n_2 \to 1$ and we obtain

$$d_n \to \begin{cases} \frac{b_2-b_1}{ra} & \text{for } q = 1, \\ \infty & \text{for } q < 1, \\ 0 & \text{for } q > 1. \end{cases}$$

The limiting value of d is called the *deficiency* of t_2 with respect to t_1 (Hodges and Lehmann, 1970). It is the number of additional observations that t_2 requires asymptotically to attain the second-order performance of t_1. The commonest case is $r = q = 1$, when $d = (b_2 - b_1)/a$. Exercise 17.8 applies the deficiency concept to the estimation of a variance.

Example 17.13 (Efficiency of estimators for the normal distribution)

We saw in Example 17.6 that the sample mean is an MVB estimator of the mean μ of a normal population, with variance σ^2/n. We saw in Example 11.12 that it is exactly normally distributed. *A fortiori*, it is an efficient estimator. In **17.11–12**, we saw that the sample median is asymptotically normal with mean μ and variance $\pi\sigma^2/(2n)$. Thus, from (17.65) with $r = 1$, the efficiency of the sample median is $2/\pi = 0.637$.

Consider now the estimation of the standard deviation. Two possible estimators are the standard deviation of the sample, say t_1, and the mean deviation of the sample multiplied by $(\pi/2)^{1/2}$ (cf. **5.42**), say t_2. The latter is easier to calculate, as a rule, and if we have plenty of observations (e.g. if we are using existing records and increasing sample size is merely a matter of turning up more records) it may be worth while using t_2 instead of t_1. Both estimators are asymptotically normally distributed.

In large samples the variance of the mean deviation is (cf. (10.39)) $\sigma^2(1 - 2/\pi)/n$. The variance of t_2 is then asymptotically $V_2 = \sigma^2(\pi - 2)/2n$. The asymptotic variance of the standard deviation (cf. **10.8**(d)) is $V_1 = \sigma^2/(2n)$, and we shall see later that it is an efficient estimator. Thus, using (17.65) with $r = 1$, the efficiency of t_2 is

$$E = \lim V_1/V_2 = 1/(\pi - 2) = 0.876.$$

The precision of the estimator from the mean deviation of a sample of 1000 is then about the same as that from the standard deviation of a sample of 876.

Minimum mean square error estimation

17.30 Our discussions of unbiasedness and the minimization of sampling variance have been conducted more or less independently. Sometimes, however, it is relevant to investigate both questions simultaneously. It is reasonable to argue that the presence of bias should not necessarily outweigh small sampling variance in an estimator. What we are really demanding of an estimator t is that it should be 'close' to the true value θ. Let us, therefore, consider its mean square error (m.s.e.) about that true value, instead of about its own expected value. We have at once

$$E(t - \theta)^2 = E\{(t - E(t)) + (E(t) - \theta)\}^2 = \text{var}\, t + \{E(t) - \theta\}^2,$$

the cross-product term on the right being equal to zero. The last term on the right is simply the square of the bias of t in estimating θ. If t is unbiased, this last term is zero, and the m.s.e. becomes the variance. In general, however, the minimization of the m.s.e. gives different results.

Exercise 17.15 shows that truncation of an estimator t to the known range of the parameter always improves m.s.e., whether or not t is unbiased – if it is, variance is also always improved.

Example 17.14 (Minimum m.s.e. estimation for the mean)

What multiple a of an estimator t estimates $E(t) = \theta$ with smallest m.s.e.? The m.s.e. of at is

$$M(a) = E(at - \theta)^2 = a^2 \operatorname{var} t + \theta^2 (a-1)^2,$$

and this can only be less than $\operatorname{var} t$ if $a < 1$: we must 'shrink' the unbiased estimator.

If V is the coefficient of variation of t, defined at (2.28), $M(a) < \operatorname{var} t$ if and only if $a > (1 - V^2)/(1 + V^2)$, and is minimized for variation in a when

$$a = \theta^2/(\theta^2 + \operatorname{var} t) = 1/(1 + V^2).$$

The reduction in m.s.e. is then

$$\operatorname{var} t - E(at - \theta)^2 = \operatorname{var} t - \theta^2 \operatorname{var} t/(\theta^2 + \operatorname{var} t) = \operatorname{var} t/(1 + 1/V^2),$$

which is a monotone function of V increasing from 0 to $\operatorname{var} t$ as V increases from 0 to ∞ and a decreases from 1 to 0. Large reductions occur when V is large. V is usually an inverse function of sample size, and tends to 0 as $n \to \infty$, while $a \to 1$.

In general, V (and therefore a) is a function of θ, so at is not a statistic usable for estimation. Thus, for example, if θ is the population mean and t is the sample mean, $\operatorname{var} t = \sigma^2/n$, where σ^2 is the population variance, and $V^2 = \sigma^2/(n\theta^2)$. For the Poisson distribution, $\sigma^2 = \theta$ and we can make no progress, but for the exponential distribution with $f(x) = \theta^{-1} \exp(-x/\theta)$, $x \geq 0$, $\theta > 0$, we have $\sigma^2 = \theta^2$ and $a = n/(n+1)$, independent of θ; the reader may show that if instead we are estimating $\lambda = \theta^{-1}$, $(n-1)/(nt)$ is unbiased with $V^2 = (n-2)^{-1}$, so that $a = (n-2)/(n-1)$ and $(n-2)/(nt)$ is the m.s.e.-minimizing multiple.

Thompson (1968) estimates a, replacing θ by t and σ^2 by its unbiased estimator s^2. Thus modified, at can have larger m.s.e. than t itself for moderate V as he shows for the normal, binomial, Poisson and gamma populations.

Exercise 17.16 deals with the estimation of powers of a normal standard deviation.

Minimum m.s.e. estimators have seen limited use, but it is as well to recognize that the objection to them is practical, rather than theoretical (cf. the comparative study by Johnson, 1950). MVU estimators are more tractable because they assume away the difficulty by insisting on unbiasedness.

Pitman (1938) showed that among estimators $t(\mathbf{x})$ of the location parameter θ in $f(x - \theta)$ that satisfy

$$t(\mathbf{x} + a) = t(\mathbf{x}) + a, \qquad (A)$$

minimum m.s.e. is attained by

$$t_L(\mathbf{x}) = \int_{-\infty}^{\infty} \theta L(\mathbf{x}|\theta)\, d\theta \Big/ \int_{-\infty}^{\infty} L(\mathbf{x}|\theta)\, d\theta,$$

which is always unbiased and is therefore MVU subject to (A). For the scale parameter θ in $\theta^{-1} f(x|\theta)$, $\theta > 0$, $t(\mathbf{x})$ satisfying

$$t(c\mathbf{x}) = ct(\mathbf{x}), \qquad c > 0 \qquad (B)$$

minimizes m.s.e. when

$$t_S(\mathbf{x}) = \int_0^{\infty} \theta^{-2} L(\mathbf{x}|\theta)\, d\theta \Big/ \int_0^{\infty} \theta^{-3} L(\mathbf{x}|\theta)\, d\theta,$$

and need not be unbiased – any multiple of t will also satisfy (B). Example 17.14 shows that we can sometimes improve on the unbiased estimator, the exponential distribution there being a case in point. For the location parameter, no multiple of t satisfies (A).

Hoaglin (1975) studies the small-sample variance of t_L in Cauchy and in logistic samples.

Pitman (1937) defined t to be 'closer' to θ than u is if $P\{[t-\theta| < |u-\theta|\} > \frac{1}{2}$, but this attractive concept is intransitive, as he pointed out. Exercise 17.34 shows that closeness is equivalent to comparing the variances of estimators that are binormally distributed. A recent detailed study of estimators based on Pitman's measure of closeness is provided by Keating *et al.* (1993). Pitman estimators are considered in detail in **23.34–7**.

Sufficient statistics

17.31 The criteria of estimation that we have so far discussed, namely consistency, unbiasedness and minimum variance or m.s.e., are reasonable guides in assessing the properties of an estimator. To permit a more fundamental discussion, we now introduce the concept of *sufficiency*, which is due to Fisher (1921a; 1925a).

Consider first the estimation of a single parameter θ by a single statistic t – we generalize in **17.38**. There is an unlimited number of possible estimators of θ, from among which we must choose. With a sample of $n \geq 2$ observations as before, consider the joint distribution of a set of r functionally independent statistics, $f_r(t, t_1, t_2, \ldots, t_{r-1}|\theta)$, $r = 2, 3, \ldots n$, where we have selected the statistic t for special consideration. We may write this as the product of the marginal distribution of t and the conditional distribution of the other statistics given t, i.e.

$$f_r(t, t_1, \ldots, t_{r-1}|\theta) = g(t|\theta)h_{r-1}(t_1, \ldots, t_{r-1}|t, \theta). \tag{17.66}$$

Now if the last factor on the right of (17.66) does not contain θ, we clearly have a situation in which, given t, the set $t_1, \ldots, t_{r-1}$ contributes nothing further to our knowledge of θ. If, further, this is true for every r and *any* set of $r-1$ statistics t_i, we may fairly say that t contains all the information in the sample about θ, and we therefore call it a (single) *sufficient* statistic for θ. We thus formally define t as sufficient for θ if and only if

$$f_r(t, t_1, \ldots, t_{r-1}|\theta) = g(t|\theta)h_{r-1}(t_1, \ldots, t_{r-1}|t), \tag{17.67}$$

where h_{r-1} is free of θ, for $r = 2, 3, \ldots, n$ and any choice of $t_1, \ldots, t_{r-1}$.[5]

17.32 As it stands, the definition of (17.67) does not enable us to see whether, in any given situation, a sufficient statistic exists. However, we may reduce it to a condition on the likelihood function. For if the latter may be written

$$L(x_1, \ldots, x_n|\theta) = g(t|\theta)k(x_1, \ldots, x_n), \tag{17.68}$$

where $g(t|\theta)$ is a function of t and θ alone[6] and k is free of θ, it is easy to see that (17.67) is deducible from (17.68). For any fixed r, and any set of $t_1, \ldots, t_{r-1}$, insert the differential

[5] This definition is usually given only for $r = 2$, but the definition for all r seems more natural. It adds no further restriction to the concept of sufficiency.

[6] We retain the notation $g(t|\theta)$, since the function of t and θ may always be expressed as the marginal distribution of t.

elements $dx_1, \ldots, dx_n$ on both sides of (17.68) and make the transformation

$$\begin{cases} t = t(x_1, \ldots, x_n), & \\ t_i = t_i(x_1, \ldots, x_n), & i = 1, 2, \ldots, r-1, \\ t_i = x_i, & i = r, \ldots, n-1. \end{cases}$$

The Jacobian of the transformation will not involve θ, and (17.68) will be transformed to

$$g(t|\theta)\, dt\, l(t, t_1, \ldots, t_{r-1}, t_r, \ldots, t_{n-1}) \prod_{i=1}^{n-1} dt_i, \tag{17.69}$$

and if we now integrate out the redundant variables $t_r \ldots, t_{n-1}$, we obtain, for the joint distribution of $t, t_1, \ldots, t_{r-1}$, precisely the form (17.67).

It should be noted that in performing the integration with respect to $t_r, \ldots, t_{n-1}$ for fixed t, no factor in θ is introduced. This is clearly so when the range of the distribution of the underlying variable is independent of θ; if the range depends on θ, this merely introduces a factor into L that forms part of $g(t|\theta)$ on the right of (17.68), while k remains free of θ – see Example 17.16 and **17.40–41**.

The converse result is also easily established. In (17.67) with $r = n$, put $t_i = x_i$ $(i = 1, 2, \ldots, n-1)$. We then have

$$f_n(t, x_1, x_2, \ldots, x_{n-1}|\theta) = g(t|\theta) h_{n-1}(x_1, \ldots, x_{n-1}|t). \tag{17.70}$$

On inserting the differential elements $dt, dx_1, \ldots, dx_{n-1}$ on either side of (17.70), the transformation

$$\begin{cases} x_n = x_n(t, x_1, \ldots, x_{n-1}), & \\ x_i = x_i, & i = 1, 2, \ldots, n-1 \end{cases}$$

applied to (17.70) yields (17.68) at once. Thus (17.67) is necessary and sufficient for (17.68). This proof deals only with the case when the variates are continuous. In the discrete case the argument simplifies, as the reader will find on retracing the steps of the proof. A very general proof of the equivalence of (17.67) and (17.68) is given by Halmos and Savage (1949).

We have discussed only the case $n \geq 2$. For $n = 1$ we take (17.68) as the definition of sufficiency.

Sufficiency and minimum variance

17.33 The necessary and sufficient condition for sufficiency at (17.68) has one immediate consequence of interest. On taking logarithms of both sides and differentiating, we have

$$\frac{\partial \log L}{\partial \theta} = \frac{\partial \log g(t|\theta)}{\partial \theta}. \tag{17.71}$$

On comparing (17.71) with (17.27), the condition that an MVB estimator of $\tau(\theta)$ exists, we see that such an estimator can only exist if there is a sufficient statistic. In fact, (17.27) is simply the special case of (17.71) when

$$\frac{\partial \log g(t|\theta)}{\partial \theta} = A(\theta)\{t - \tau(\theta)\}. \tag{17.72}$$

Thus sufficiency, which perhaps at first sight seems a more restrictive criterion than the attainment of the MVB, is in reality a less restrictive one. For whenever (17.27) holds, (17.71) also holds, while even if (17.27) does not hold we may still have a sufficient statistic.

Example 17.15 (MVB estimators and sufficiency)

The argument of this section implies that in all the cases (Examples 17.6, 17.8–17.10) where we have found MVB estimators to exist, they are also sufficient statistics. The reader should verify this in each case by direct factorization of L as at (17.68).

Example 17.16 (Sufficient statistics for the uniform distribution)

Consider the estimation of θ in

$$f(x) = \theta^{-1}, \qquad 0 \leq x \leq \theta.$$

The likelihood function (LF) is

$$L(x|\theta) = \begin{cases} \theta^{-n} & \text{if } x_{(n)} \leq \theta, \\ 0 & \text{otherwise,} \end{cases}$$

since we know that all the observations, including the largest of them, $x_{(n)}$, cannot exceed θ.

We may write this in the form

$$L(x|\theta) = \theta^{-n} u(\theta - x_{(n)}), \tag{17.73}$$

where

$$u(z) = \begin{cases} 1 & z \geq 0, \\ 0 & z < 0. \end{cases}$$

Expression (17.73) makes it clear at once that the LF can be factorized into a function of $x_{(n)}$ and θ alone,

$$g(x_{(n)}|\theta) = \theta^{-n} u(\theta - x_{(n)}),$$

and a second factor

$$k(x_1, \ldots, x_n) = 1.$$

Thus, from (17.68), $x_{(n)}$ is a sufficient statistic for θ.

17.34 If t is a sufficient statistic for θ, any one-to-one function of t, say u, will also be sufficient. For if $t(u)$ is one-to-one, we may write (17.68) as

$$\begin{aligned} L(x|\theta) &= g(t(u)|\theta)k(x) \\ &= g_1(u|\theta)k_1(x), \end{aligned} \tag{17.74}$$

where k_1 is independent of θ, so that u is also sufficient for θ. To resolve the estimation problem, we choose a function of t that is a consistent, and usually also an unbiased, estimator of θ.

Even if $t(u)$ is not one-to-one, the factorization (17.74) making u sufficient may still be possible in particular cases – Exercise 21.31 treats one of these.

Apart from such functional relationships, the sufficient statistic is unique. If there were two distinct sufficient statistics, t_1, and t_2, (17.67) with $r = 2$ would give

$$f_2(t_1, t_2|\theta) = g_1(t_1|\theta)h_1(t_2|t_1) = g_2(t_2|\theta)h_2(t_1|t_2)$$

so that, writing $h_3(t_1, t_2) = h_2(t_1|t_2)/h_1(t_2|t_1)$, we have

$$g_1(t_1|\theta) \equiv h_3(t_1, t_2)g_2(t_2|\theta) \tag{17.75}$$

identically in θ. This cannot hold unless t_1 and t_2 are functionally related.

17.35 We have seen in **17.33** that a sufficient statistic provides the MVB estimator, where there is one. We now prove a more general result, due to Rao (1945) and Blackwell (1947), that irrespective of the attainability of any variance bound, the MVU estimator of $\tau(\theta)$, if one exists, is always a function of the sufficient statistic.

We first prove a quite general result. Consider a statistic t and another statistic u with finite variance. To find $E(u)$, we carry out the expectation operation in two stages, first holding t constant and then allowing t to vary over its distribution. Symbolically

$$E(u) = E_t\{E(u|t)\}. \tag{17.76}$$

In general, $E(u|t)$ is not a statistic, since it depends on the parameter θ. For the variance of u, we have from (17.76)

$$\begin{aligned} \operatorname{var} u &= E\{u - E(u)\}^2 \\ &= E_t E(\{[u - E(u|t)] + [E(u|t) - E_t E(u|t)]\}^2|t). \end{aligned}$$

Now the contents of the second set of square brackets are a constant with respect to the conditional expectation given t. Thus, when the terms inside the braces are squared, the cross-product term is a constant times $u - E(u|t)$, whose conditional expectation is zero. We are left with

$$\operatorname{var} u = E_t E\{[u - E(u|t)]^2|t\} + E_t\{[E(u|t) - E_t E(u|t)]^2\},$$

which by the definition of variance may be written

$$\operatorname{var} u = E_t\{\operatorname{var}(u|t)\} + \operatorname{var}_t\{E(u|t)\}. \tag{17.77}$$

Thus, quite generally, the unconditional expectation is the expectation of conditional expectation, by (17.76), and the unconditional variance is the expectation of conditional variance plus the variance of conditional expectation. The variable $E(u|t)$ has the same expectation as u, and smaller variance than u by (17.77) unless $\operatorname{var}(u|t) \equiv 0$, i.e. u is a function of t.

Now let t be sufficient for θ. Expression (17.67) ensures that $E(u|t)$ cannot depend on θ, so that it is a function of t alone, say $p(t)$. (17.76) gives $E\{p(t)\} = E(u) = \tau(\theta)$, say, while (17.77) shows that

$$\operatorname{var}\{p(t)\} \leq \operatorname{var} u, \tag{17.78}$$

with equality if and only if $u = p(t)$.

Thus conditioning upon the value of a sufficient statistic t gives an estimator $p(t)$ with smaller variance. It follows that any MVU estimator must be a function of t.

We shall see in **21.9–13** that often there is only one function of a sufficient statistic with any given expectation. Then, whatever u we start from, $E(u|t)$ must be the same, and it is the *unique* MVU estimator of $\tau(\theta)$. It is not always easy to use this result constructively, since the conditional expectation $E(u|t)$ may be difficult to evaluate – Exercise 17.24 gives an important class of cases where explicit results can be obtained – but it does assure us that an unbiased estimator (with finite variance) that is a function of a sufficient statistic is the unique MVU estimator.

In the case of the exponential family (17.30), we can find the MV estimator based on the sufficient statistic directly in particular cases, as in Exercise 17.14. Since $p(t)$ has the same expectation as, and smaller variance than, u, it follows from **17.30** that its m.s.e. is also smaller than that of u no matter which function of θ is being estimated. We may be able to improve m.s.e. further by using a function of t with a different expectation. Example 17.14 implies that in estimating $\tau(\theta)$, even a constant multiple $ap(t)$, where $a < 1$, can have better m.s.e. than $p(t)$ itself – in the case given there,

$$f = \theta^{-1}\exp(-x/\theta), \qquad \theta > 0;\ x \geq 0,$$

it is easily seen that $t = \bar{x}$ is sufficient for θ and unbiased, while $nt/(n+1)$ has smaller m.s.e.

Distributions possessing a single sufficient statistic

17.36 We now seek to define the class of distributions in which a (single) sufficient statistic exists for a parameter. We first consider the case where the range of the variate does not depend on the parameter θ. From (17.71) we have, if t is sufficient for θ in a sample of n independent observations,

$$\frac{\partial \log L}{\partial \theta} = \sum_{j=1}^{n} \frac{\partial \log f(x_j|\theta)}{\partial \theta} = K(t,\theta), \tag{17.79}$$

where K is some function of t and θ. Regarding this as an equation in t, we see that since it remains true for any fixed value of θ and any u, t must be expressible in the form

$$t = M\left\{\sum_{j=1}^{n} k(x_j)\right\} = M(w), \tag{17.80}$$

where $w = \sum k(x_j)$ and M and k are arbitrary functions. Thus $K(M(w),\theta)$ is a function of w and θ only, say $N(w,\theta)$. We have then, from (17.79), if the derivatives exist,

$$\frac{\partial^2 \log L}{\partial\theta\,\partial x_j} = \frac{\partial N}{\partial w}\frac{\partial w}{\partial x_j}. \tag{17.81}$$

Now the left-hand side of (17.81) is a function of θ and x_j only and $\partial w/\partial x_j$ is a function of x_j only. Hence $\partial N/\partial w$ is a function of θ and x_j only. But it must be symmetrical in the xs, since w is, and hence is a function of θ only. Hence, integrating it with respect to w, we have

$$N(\theta, w) = wp(\theta) + q(\theta),$$

where p and q are arbitrary functions of θ. Thus (17.79) becomes

$$\frac{\partial}{\partial\theta}\log L = \frac{\partial}{\partial\theta}\sum_j \log f(x_j|\theta) = p(\theta)\sum k(x_j) + q(\theta), \tag{17.82}$$

whence

$$\frac{\partial}{\partial\theta}\log f(x|\theta) = p(\theta)k(x) + q(\theta)/n,$$

giving the necessary condition for a sufficient statistic to exist,

$$f(x|\theta) = \exp\{A(\theta)B(x) + C(x) + D(\theta)\}. \tag{17.83}$$

This result, which is due to Darmois (1935), Pitman (1936) and Koopman (1936), is precisely the form of the exponential family of distributions, seen at (17.30) to be a condition for the existence of an MVB estimator for some function of θ.

Brown (1964) gives a rigorous treatment of the regularity conditions sufficient for this result to hold, with references to related work. See also Denny (1967; 1972); Andersen (1970a) treats the discrete case.

If (17.83) holds, it is easily verified that if the range of $f(x|\theta)$ is independent of θ, the likelihood function yields a sufficient statistic for θ. Thus, under this condition, (17.83) is sufficient as well as necessary for the distribution to possess a sufficient statistic.

All the distributions of Example 17.15 are of the form (17.83).

17.37 Under regularity conditions, there is therefore a one-to-one correspondence between the existence of a sufficient statistic for θ and the existence of an MVB estimator of some function of θ. If (17.83) holds, a sufficient statistic exists for θ, and there will be just one function, t, of that statistic (itself sufficient) that will satisfy (17.27) and so estimate some function $\tau(\theta)$ with variance equal to the MVB. However, in large samples (cf. **17.23**), *any* function of the sufficient statistic will estimate its expected value with MVB precision. Finally, for any n (cf. **17.35**), any function of the sufficient statistic will have the minimum *attainable* variance in estimating its expected value.

Sufficient statistics for several parameters

17.38 All the ideas of the previous sections generalize immediately to the case where the distribution is dependent upon several parameters $\theta_1, \ldots, \theta_k$. It also makes no difference to the essentials if we have a multivariate distribution, rather than a univariate one. Thus if we define each x_i as a vector variate with p (≥ 1) components, $\mathbf{t}$ and $\mathbf{t}_i$ as vectors of statistics, and $\boldsymbol{\theta}$ as a vector of parameters with k components, **17.31–32** remain substantially unchanged. If we can write

$$L(\mathbf{x}|\boldsymbol{\theta}) = g(\mathbf{t}|\boldsymbol{\theta})h(\mathbf{x}) \tag{17.84}$$

we call the components of $\mathbf{t}$ a set of (jointly) sufficient statistics for $\boldsymbol{\theta}$. The property (17.67) follows as before.

If $\mathbf{t}$ has s components, we may have s greater than, equal to, or less than k. If $s = 1$, we call $\mathbf{t}$ a single sufficient statistic, as we have seen. If we put $\mathbf{t} = \mathbf{x}$ we see that the observations themselves always constitute a set of sufficient statistics for $\boldsymbol{\theta}$ with $s = n$. In order to reduce the problem of analysing the data as far as possible, we naturally desire s to be as small as possible. Even this is not quite restrictive enough – see Exercises 18.13 and 21.31 for cases in which we have alternative sufficient statistics with the same value of s. In Chapter 21 we shall define the concept of a *minimal* set of sufficient statistics for $\boldsymbol{\theta}$, which is a function of all other sets of sufficient statistics.

It evidently does not follow from the joint sufficiency of $\mathbf{t}$ for $\boldsymbol{\theta}$ that any particular component of $\mathbf{t}$, say $t^{(1)}$, is individually sufficient for θ_1. This will be so only if $g(\mathbf{t}|\boldsymbol{\theta})$ factorizes with $g_1(t^{(1)}|\theta_1)$ as one factor. Nor is the converse always true: individual sufficiency of all the $t^{(i)}$ when the others are known does not imply joint sufficiency.

If $k = 1$, the result of **17.35** holds unchanged if t is a vector with $s < n$ components (if $s = n$, the result is an empty one); the result is most easily applied with s as small as possible.

Example 17.17 (Joint sufficiency in the normal distribution)

Consider the estimation of the parameters μ and σ^2 in

$$f(x) = \frac{1}{\sigma(2\pi)^{1/2}} \exp\left\{-\frac{1}{2}\left(\frac{x-\mu}{\sigma}\right)^2\right\}, \qquad -\infty < x < \infty.$$

We have

$$L(x|\mu, \sigma^2) = \frac{1}{\sigma^n(2\pi)^{n/2}} \exp\left\{-\frac{1}{2}\sum_{i=1}^{n}\left(\frac{x_i-\mu}{\sigma}\right)^2\right\} \tag{17.85}$$

and we have seen (Example 11.7) that the joint distribution of $\bar{x}$ and s^2 in normal samples is

$$g(\bar{x}, s^2|\mu, \sigma^2) \propto \frac{1}{\sigma} \exp\left\{-\frac{n}{2\sigma^2}(\bar{x}-\mu)^2\right\} \cdot \frac{1}{\sigma^{n-1}} s^{n-3} \exp\left\{-\frac{ns^2}{2\sigma^2}\right\},$$

so that, remembering that $\sum(x-\mu)^2 = n\{s^2 + (\bar{x}-\mu)^2\}$, we have

$$L(x|\mu, \sigma^2) = g(\bar{x}, s^2|\mu, \sigma^2)\, k(x)$$

and therefore $\bar{x}$ and s^2 are jointly sufficient for μ and σ^2. We have already seen (Examples 17.6, 17.15) that $\bar{x}$ is sufficient for μ when σ^2 is known and (Examples 17.10, 17.15) that $(1/n)\sum(x-\mu)^2$ is sufficient for σ^2 when μ is known. It is easily seen directly from (17.85) that s^2 alone is not sufficient for σ^2 when μ is unknown.

17.39 The principal results for sufficient statistics generalize to the k-parameter case in a natural way. The condition (generalizing the exponential family (17.83)) for a distribution to possess a set of k jointly sufficient statistics for its k parameters becomes, under similar conditions of continuity and the existence of derivatives,

$$f(x|\boldsymbol{\theta}) = \exp\left\{\sum_{j=1}^{k} A_j(\boldsymbol{\theta})B_j(x) + C(x) + D(\boldsymbol{\theta})\right\}, \tag{17.86}$$

a direct extension of (17.81), again due to Darmois (1935), Koopman (1936) and Pitman (1936). More general results of this kind are given by Barankin and Maitra (1963). The result of **17.35** on the unique MV properties of functions of a sufficient statistic finds its generalization in a theorem due to Rao (1947): for the simultaneous estimation of $r(\leq k)$ functions τ_j of the k parameters θ_s, the unbiased functions of a minimal set of k sufficient statistics, say t_i, have the minimum attainable variances, and (if the range is independent of the θ_s) the (not necessarily attainable) lower bounds to their variances are given by

$$\text{var}\, t_i \geq \sum_{j=1}^{k}\sum_{l=1}^{k} \frac{\partial \tau_i}{\partial \theta_j}\frac{\partial \tau_i}{\partial \theta_l} I_{jl}^{-1}, \qquad i = 1, 2, \ldots, r, \tag{17.87}$$

where the *information matrix*, the generalization of (17.25),

$$(I_{jl}) = \left\{ E\left(\frac{\partial \log L}{\partial \theta_j} \cdot \frac{\partial \log L}{\partial \theta_l} \right) \right\} \tag{17.88}$$

is to be inverted. Inequality (17.87) is a further generalization of (17.22) and (17.50), its simplest cases. Like them, it takes account only of terms of order $1/n$ in the variance.

Exercise 17.20 shows that these lower bounds cannot be smaller than those obtained from (17.24) for a single estimator.

In random sampling from the exponential family (17.86), the joint distribution of the sufficient statistics is itself a member of the exponential family, as Exercise 17.14 shows. A good approximation to its density in the more general case where the observations need not be independent is outlined in Exercise 17.33.

Sufficiency when the range depends on the parameter

17.40 Now consider the situation when the range of the variable depends on θ. We omit the trivial case $n = 1$. First, we assume only one terminal or end-point of the range, say the lower terminal, depends on θ. We then have the density function $f(x|\theta)$, $a(\theta) \leq x \leq b$, where $a(\theta)$ is a monotone function of θ, and θ is in some non-degenerate interval.

Just as in Example 17.16, we have

$$L(x|\theta) = \prod_{i=1}^{n} f(x_i|\theta) u(x_{(1)} - a(\theta))$$

where

$$u(z) = \begin{cases} 1 & \text{if } z \geq 0, \\ 0 & \text{otherwise.} \end{cases}$$

It is clear from the likelihood function that the smallest observation $x_{(1)}$ cannot be factored away from θ in the u-function; thus, if there is a single sufficient statistic, it must be $x_{(1)}$. But $x_{(1)}$ can only be sufficient if $f(x_i|\theta)$ can be factored into a function of x_i alone and a function of θ alone, i.e. if

$$f(x|\theta) = g(x)/h(\theta). \tag{17.89}$$

Then and only then is

$$L(x|\theta) = \frac{u(x_{(1)} - a(\theta))}{\{h(\theta)\}^n} \cdot \prod_{i=1}^{n} g(x_i)$$

of the form required in (17.68) for $x_{(1)}$ to be sufficient.

Evidently the same result will hold, with $x_{(n)}$ instead of $x_{(1)}$, if the range is $a \leq x \leq b(\theta)$, where $b(\theta)$ is monotone in θ; if and only if (17.89) holds, $x_{(n)}$ is singly sufficient for θ.

In these situations, Exercise 17.24 uses the result of **17.35** to establish an explicit form for the unique MVU estimator of any function $\tau(\theta)$. The reader should note that there the problem is reparametrized so that the affected terminal is at θ itself. In these cases, the variance of the ML estimator is typically $O(n^{-2})$; see Exercise 17.36.

17.41 If both terminals of the range depend on θ, we have $a(\theta) \leq x \leq b(\theta)$ and

$$L(x|\theta) = \prod_{i=1}^{n} f(x_i|\theta)u(x_{(1)} - a(\theta))u(b(\theta) - x_{(n)}).$$

We see by exactly the argument in **17.40** that the extreme observations $x_{(1)}$ and $x_{(n)}$ are a pair of sufficient statistics for θ if and only if (17.89) holds. We now consider whether there can be a *single* sufficient statistic in this case; if there is it must clearly be a function of $x_{(1)}$ and $x_{(n)}$.

Essentially, we are asking whether we can find a single statistic that will tell us whether the product $u(x_{(1)} - a(\theta))u(b(\theta) - x_{(n)})$ in $L(x|\theta)$ is equal to 1 or 0. It is only equal to 1 if both

$$x_{(1)} \geq a(\theta), \qquad b(\theta) \geq x_{(n)}. \tag{17.90}$$

There are four possibilities:

(i) $a(\theta), b(\theta)$ are both increasing functions of θ and (17.90) becomes

$$a^{-1}(x_{(1)}) \geq \theta \geq b^{-1}(x_{(n)});$$

(ii) they are both decreasing functions of θ and (17.90) becomes

$$b^{-1}(x_{(n)}) \geq \theta \geq a^{-1}(x_{(1)});$$

(iii) $a(\theta)$ is increasing, $b(\theta)$ decreasing and (17.90) becomes

$$a^{-1}(x_{(1)}) \geq \theta, \qquad b^{-1}(x_{(n)}) \geq \theta; \tag{17.91}$$

(iv) $a(\theta)$ is decreasing, $b(\theta)$ increasing and (17.90) becomes

$$a^{-1}(x_{(1)}) \leq \theta, \qquad b^{-1}(x_{(n)}) \leq \theta. \tag{17.92}$$

In cases (i) and (ii), we need both $x_{(1)}$ and $x_{(n)}$, and no single sufficient statistic exists. But (17.91) shows that in case (iii), $u(t_1 - \theta)$ is equivalent to the product of the original u-functions, where

$$t_1 = \min\{a^{-1}(x_{(1)}), b^{-1}(x_{(n)})\}, \tag{17.93}$$

so t_1 is singly sufficient for θ. Similarly, in case (iv),

$$t_2 = \max\{a^{-1}(x_{(1)}), b^{-1}(x_{(n)})\} \tag{17.94}$$

is singly sufficient since $u(\theta - t_2)$ is equivalent to the product. We may summarize by saying that if the upper terminal is a monotone decreasing function of the lower terminal and (17.89) holds, there is a single sufficient statistic, given by t_1 or by t_2 according as the lower terminal is an increasing or a decreasing function of θ.

Exercise 19.6 gives the distribution of t_1 from which that of t_2 is immediately obtainable (by writing

$$a^*(\theta) = -a(\theta), \qquad b^*(\theta) = -b(\theta)$$

in case (iv), so that we return to case (iii), (17.94) becoming (17.93) in terms of a^* and b^*) and the unique MVU estimator of $\tau(\theta)$ in either case. Further, no generality is lost by making the lower terminal equal to θ itself, so we need only consider the range $(\theta, b(\theta))$ where $b(\theta)$ is a decreasing function and (17.93) is the single sufficient statistic.

These results were originally due to Pitman (1936) and Davis (1951). The condition that θ lies in a non-degenerate interval is important – cf. Example 17.23.

Example 17.18 (Single sufficient statistic for the uniform)

For the uniform distribution

$$f(x) = 1/(2\theta), \qquad -\theta \le x \le \theta,$$

we are in case (iv) above. The single sufficient statistic (17.94) is

$$t_2 = \max\{-x_{(1)}, x_{(n)}\}$$

and, since

$$x_{(1)} \le x_{(n)},$$

this is the same as

$$t_2 = \max\{|x_{(1)}|, |x_{(n)}|\}$$

which is intuitively acceptable.

Example 17.19 (Sufficient statistic for the lower end-point of the exponential)

The distribution $f(x) = \exp\{-(x-\alpha)\}$, $\alpha \le x < \infty$, is of the form (17.89), since it may be written

$$f(x) = \exp(-x)/\exp(-\alpha).$$

Here the smallest observation, $x_{(1)}$, is sufficient for the lower terminal α.

Example 17.20 (Sufficient statistic for the gamma distribution)

The distribution

$$f(x) = \frac{1}{\Gamma(p)}(x-\alpha)^{p-1}\exp\{-(x-\alpha)\}, \qquad \alpha \le x < \infty;\ p > 1,$$

evidently cannot be put in the form (17.89). Thus there is no single sufficient statistic for α when $n > 2$. Example 17.19 is the case $p = 1$.

Example 17.21 (Jointly sufficient statistics for end-points)

In the two-parameter distribution

$$f(x) = g(x)/h(\alpha, \beta), \qquad \alpha \leq x \leq \beta,$$

it is clear that, given β, $x_{(1)}$ is sufficient for α; and given α, $x_{(n)}$ is sufficient for β. $x_{(1)}$ and $x_{(n)}$ are a set of jointly sufficient statistics for α and β, as is confirmed by observing that the joint distribution of $x_{(1)}$ and $x_{(n)}$ is, by (14.2) with $r = 1, s = n$,

$$g(x_{(1)}, x_{(n)}) = p(x_{(1)}, x_{(n)})/\{h(\alpha, \beta)\}^n,$$

so that when g and L are non-zero

$$L(x|\alpha, \beta) = g(x_{(1)}, x_{(n)})k(x_1, x_2, \ldots, x_n).$$

Example 17.22 (Jointly sufficient statistics for the uniform)

The uniform distribution $f(x) = 1/\theta, k\theta \leq x \leq (k+1)\theta; k > 0$ comes under case (i), as started after (17.90). No single sufficient statistic exists, but $(x_{(1)}, x_{(n)})$ are a sufficient pair since (17.89) holds.

Example 17.23 (Sufficiency for special uniform distributions)

The uniform distribution $f(x) = 1, \theta \leq x < \theta + 1; \theta = 0, 1, 2, \ldots$, in which θ is confined to integer values, does not satisfy the condition that the upper terminal be a monotone decreasing function of the lower. But, evidently, *any* single observation, x_i, in a sample of n is a single sufficient statistic for θ. In fact, $[x_i]$ estimates θ with zero variance. If the integer restriction on θ is removed, no single sufficient statistic exists, in accordance with case (i).

Statistical curvature

17.42 Efron (1975) – see also Efron (1978) and Amari (1982a; 1982b) – obtains results for the MVB using a concept of statistical curvature that measures departure from the exponential family by

$$\gamma^2(\theta) = \frac{\text{var } f''}{\{\text{var}(f')\}^2}\left[1 - \frac{\{\text{cov}(f', f'')\}^2}{\text{var}(f')\,\text{var}(f'')}\right], \tag{17.95}$$

where $f^{(r)} = \partial^r \log f(x|\theta)/\partial\theta^r$.

By the Cauchy–Schwarz inequality, the term in square brackets lies between 0 and 1, attaining 0 if and only if f' is a linear function of f''. Exercise 17.44 asks the reader to show that this is true for any member of the exponential family (17.30). Interpreting (17.95) as in Exercise 26.23 or (27.27), we see that $\gamma^2\{\text{var}(f')\}^2$ is the variance of f'' after regression upon f'.

If we reparametrize from θ to a monotone function $\tau(\theta)$, γ^2 is invariant, while if we replace $f(x|\theta)$ by the LF $L(x|\theta)$ based on n observations, γ^2 becomes γ^2/n – cf. Exercise 17.35.

The Bhattacharyya bound to order n^{-2} for the estimation of θ at (17.47) may be written

$$\text{var } t \geq \text{MVB} + \frac{B(\theta)}{n^2} + o\left(\frac{1}{n^2}\right). \tag{17.96}$$

As we saw in **17.18** and Example 17.11, attainability of the MVB and of the Bhattacharyya bound depends on the parametrization used – for the binomial discussed in Example 17.11, the MVB is attained for θ itself, while the Bhattacharyya bound is attained for $\theta(1-\theta)$, and neither is attained for an arbitrary function of θ. Efron (1975) obtains an essentially different variance bound for efficient estimators t of θ biased to at most $O(n^{-2})$,

$$\operatorname{var} t \geq \text{MVB} + \frac{1}{n^2}\{B(\theta) + \gamma^2(\theta) + \Delta(t,\theta)\} + o\left(\frac{1}{n^2}\right), \tag{17.97}$$

where γ^2 is as in (17.95) and Δ is also non-negative. Since γ^2 is invariant under reparametrization, its effect cannot be removed from (17.97), but Δ may be reduced to zero if we put $t = \hat{\theta}$, the ML estimator. The $o(n^{-2})$ term in (17.97) is then minimized, giving the ML estimator optimum second-order efficiency.

Although the bound for efficient estimators (17.97) is at least as great as the MVB and the Bhattacharyya bounds, it does not prevent attainment of these, any more than the second of these prevents the attainment of the first. In the exponential family, (17.97) reduces to the Bhattacharyya bound if $\hat{\theta}$ is used.

17.43 We have now concluded our discussion of the basic ideas of the theory of estimation. In Chapter 21 we shall be developing the theory of sufficient statistics further. Meanwhile, in the next chapter we study estimation from another point of view, by studying the properties of the estimators given by the method of maximum likelihood.

EXERCISES

17.1 Show that in samples from

$$f(x) = \frac{1}{\Gamma(p)\theta^p} x^{p-1} e^{-x/\theta}, \qquad p > 0;\ 0 \le x < \infty,$$

the MVB estimator of θ for known p is $\bar{x}/p$, with variance θ^2/np, while if $\theta = 1$ that of $\partial \log \Gamma(p)/\partial p$ is $(1/n)\sum_{i=1}^{n} \log x_i$ with variance $\{\partial^2 \log \Gamma(p)/\partial p^2\}/n$.

17.2 A random variable x has p.d.f.

$$f(x|\theta) = \theta f_1(x) + (1-\theta) f_2(x),$$

where $0 < \theta < 1$ and f_1, f_2 are completely specified p.d.f.s whose ranges of variation do not involve θ. Show that the MVB in estimating θ from a sample of n observations is not generally attainable, but is equal to $\theta(1-\theta)/\{n(1-I)\}$, where

$$I = \int_{-\infty}^{\infty} \frac{f_1(x) f_2(x)}{f(x|\theta)} \,dx.$$

$I = 0$ if the ranges of f_1 and f_2 do not overlap and the MVB then is that for the binomial distribution (Example 17.9), which is the special case $f_1(x) = x$, $f_2(x) = 1 - x$, $x = 0, 1$.

(Hill, 1963b)

17.3 In Example 17.11, show that

$$L'/L = (r - n\theta)/\{\theta(1-\theta)\}$$

and

$$L''/L = \frac{\partial^2 \log L}{\partial \theta^2} + \left(\frac{\partial \log L}{\partial \theta}\right)^2$$

$$= \frac{1}{\theta^2(1-\theta)^2}\left[\left\{(r - n\theta) - \tfrac{1}{2}(1-2\theta)\right\}^2 - \left\{\tfrac{1}{4} + (n-1)\theta(1-\theta)\right\}\right].$$

Hence show that $J_{12} = 0$. Using (17.19), show that $E(L''/L) = 0$ and that

$$J_{22} = \text{var}(L''/L) = 2n(n-1)/\{\theta^2(1-\theta)^2\}.$$

Use these results to verify that var t coincides with (17.45). Show that t is a linear function of the uncorrelated variables L'/L and L''/L.

17.4 Writing (17.27) as

$$(t - \tau) = \frac{\tau'(\theta)}{J_{11}} \cdot \frac{L'}{L},$$

and the c.f. of t about its mean as $\phi(z)$ ($z = iu$), show that for an MVB estimator

$$\frac{\partial \phi(z)}{\partial z} = \frac{\tau'(\theta)}{J_{11}}\left\{\frac{\partial \phi(z)}{\partial \theta} + z\tau'(\theta)\phi(z)\right\}$$

and that its cumulants are given by

$$\kappa_{r+1} = \frac{\partial \tau}{\partial \theta} \cdot \frac{\partial \kappa_r}{\partial \theta} \Big/ J_{11}, \qquad r = 2, 3, \ldots. \qquad (A)$$

Hence show that the covariance between t and an unbiased estimator of its rth cumulant is equal to its $(r+1)$th cumulant.
Establish the inequality (17.50) for an estimated function of several parameters, and show that (A) holds in this case also when the bound is attained.

(Bhattacharyya, 1946–47)

17.5 Show that for the estimation of θ in the logistic distribution

$$f(x) = e^{-(x-\theta)}\{1 + e^{-(x-\theta)}\}^{-2},$$

the MVB is exactly $3/n$, whereas the sample mean has exact variance $\pi^2/3n$ and the sample median has asymptotic variance $4/n$.

17.6 Show that in estimating σ in the distribution

$$f(x) = \frac{1}{\sigma(2\pi)^{1/2}} \exp\left(-\frac{1}{2}\frac{x^2}{\sigma^2}\right), \qquad -\infty < x < \infty,$$

$$s_1 = \left(\frac{1}{2}\sum_{i=1}^{n} x_i^2\right)^{1/2} \Gamma\left(\frac{1}{2}n\right) \Big/ \Gamma\left\{\frac{1}{2}(n+1)\right\}$$

and

$$s_2 = \left\{\frac{1}{2}\sum_{i=1}^{n}(x_i - \bar{x})^2\right\}^{1/2} \Gamma\left\{\frac{1}{2}(n-1)\right\} \Big/ \Gamma\left(\frac{1}{2}n\right)$$

are both unbiased. Hence show that s in Example 17.3 has expectation $\sigma(1 - (1/4n))$ approximately, agreeing with the result of Exercise 10.20.
Show that (17.52) generally gives a greater bound than (17.24), which gives $\sigma^2/2n$, but, by considering the case $n = 2$, that even this greater bound is not attained for small n by s_1.

(Cf. Chapman and Robbins, 1951)

17.7 In estimating μ^2 in

$$f(x) = \frac{1}{(2\pi)^{1/2}} \exp\left\{-\frac{1}{2}(x-\mu)^2\right\},$$

show that $(\bar{x}^2 - 1/n - \mu^2)$ is a linear function of

$$\frac{1}{L}\frac{\partial L}{\partial \mu} \quad \text{and} \quad \frac{1}{L}\frac{\partial^2 L}{\partial \mu^2},$$

and hence, from (17.42), that $\bar{x}^2 - 1/n$ is an unbiased estimator of μ^2 with minimum attainable variance.

17.8 In estimating the variance σ^2 of a distribution with known mean μ from a sample of n observations x_i, consider the two unbiased estimators

$$t_1 = \frac{1}{n}\sum_{i=1}^{n}(x_i - \mu)^2, \qquad t_2 = \frac{1}{n-1}\sum_{i=1}^{n}(x_i - \bar{x})^2.$$

Using (10.5) and (12.35), show that in the terminology of **17.29** they have the same efficiency, and that the deficiency of t_2, with respect to t_1, is $d = 2/(2 + \gamma_2)$, where $\gamma_2 = \kappa_4/\kappa_2^2$ is the kurtosis coefficient of the distribution. From Exercise 3.19, $\gamma_2 \geq -2$, so that $0 \leq d < \infty$, with $d = 1$ in the normal case, reflecting the single degree of freedom lost in passing from t_1 to t_2 (cf. Example 11.7).

(Hodges and Lehmann, 1970)

17.9 For the three-parameter distribution

$$f(x) = \frac{1}{\sigma\Gamma(p)}\exp\left\{-\left(\frac{x-\alpha}{\sigma}\right)\right\}\left(\frac{x-\alpha}{\sigma}\right)^{p-1}, \qquad p,\sigma > 0;\ \alpha \le x < \infty,$$

show that there are single sufficient statistics for p and σ individually when the other two parameters are known; that there is a pair of sufficient statistics for p and σ jointly if α is known; and that if σ is known and $p = 1$, there is a single sufficient statistic for α.

17.10 In samples of size n from a normal distribution with cumulants κ_1, κ_2, show that the sample mean, k_1, has moments

$$E(k_1^r) = \sum_{j=0}^{r}\binom{r}{j}\kappa_1^{r-j}E(k_1-\kappa_1)^j = \sum_{i=0}^{[\frac{1}{2}r]}\frac{r!}{i!(r-2i)!}\kappa_1^{r-2i}\left(\frac{\kappa_2}{2n}\right)^i, \quad r = 1, 2, \ldots$$

Hence show that

$$\kappa_1^r = \sum_{i=0}^{[\frac{1}{2}r]}(-1)^i\frac{r!}{i!(r-2i)!}E(k_1^{r-2i})\left(\frac{\kappa_2}{2n}\right)^i,$$

and finally, using (16.7), that for $n \ge 2$ and $m > -\frac{1}{2}(n-1)$ the MV unbiased estimator of $\kappa_1^r\kappa_2^m$ is

$$\sum_{i=0}^{[\frac{1}{2}r]}(-1)^i\frac{r!}{i!(r-2i)!}\frac{\Gamma\{\frac{1}{2}(n-1)\}}{\Gamma\{\frac{1}{2}(n-1)+m+i\}}\frac{k_1^{r-2i}}{(2n)^i}\left\{\frac{(n-1)k_2}{2}\right\}^{m+i}$$

where k_2 is the second k statistic.

(Cf. Hoyle, 1968)

17.11 Show that if t is the MV unbiased estimator, and u another unbiased estimator, of θ, the covariance of t and $(u-t)$ is zero. Hence show that we may regard the variation of $u-\theta$ as composed of two orthogonal parts, one being the variation of $t-\theta$ and the other a component due to inefficiency of estimation.

(Cf. Fisher, 1925)

17.12 In the binomial distribution of Example 17.4, let a and b be integers or zero and define $\tau_{ab} = \theta^a(1-\theta)^b$. Show that if $a, b \ge 0$ and $a+b \le n$, τ_{ab} has the unbiased estimator $r^{(a)}(n-r)^{(b)}/n^{(a+b)}$, but that otherwise no unbiased estimator of τ_{ab} exists.

17.13 For a sample of n observations from a distribution of form (17.89) with $a \le x \le \theta$, show that the sufficient statistic $x_{(n)}$ is not an unbiased estimator of θ. Using the jackknife of **17.10**, show that

$$t' = x_{(n)} + (x_{(n)} - x_{(n-1)})(n-1)/n \sim 2x_{(n)} - x_{(n-1)}$$

is unbiased to order n^{-1} and has the same m.s.e. as $x_{(n)}$ to order n^{-2}, namely

$$\frac{2}{n^2}\left\{\frac{h(\theta)}{h'(\theta)}\right\}^2.$$

(Cf. Robson and Whitlock, 1964)

17.14 For a sample of n independent observations from a distribution with density function (17.86), and the range of the variates independent of the parameters, show that the statistics

$$t_j = \sum_{i=1}^{n} B_j(x_i), \qquad j = 1, 2, \ldots, k,$$

are jointly sufficient for the k parameters $\theta_1, \ldots, \theta_k$, and that their joint p.d.f. is

$$g(t_1, t_2, \ldots, t_k|\theta) = \exp\{nD(\theta)\}h(t_1, t_2, \ldots, t_k) \exp\left\{\sum_{j=1}^{k} A_j(\theta)t_j\right\}$$

which is itself of the form (17.86). When $k = 1$, write this as

$$g_n(t|\theta) = \{d(\theta)\}^n \cdot h_n(t) \exp\{tA(\theta)\}$$

and show that if m is an integer, $0 \leq m < n$, and c is a constant, then

$$E\{h_{n-m}(t-c)/h_n(t)\} = \{d(\theta)\}^m \exp\{cA(\theta)\}.$$

Show that if

$$f(x|\theta) = \theta e^{-\theta x}, \qquad 0 \leq x < \infty,$$

the MV unbiased estimator of θ^m is $(n-1)^{(m)}/\sum_{j=1}^{n} x_j$; Example 17.14 gave the result for $m = 1$.

17.15 t is an estimator of a parameter θ known to lie in the interval (a, b). u is defined equal to t if $a \leq t \leq b$, but equal to a if $t < a$ and to b if $t > b$. Show that u has better m.s.e. than t in estimating θ. If $E(t) = \theta$, show that u also has better variance than t.
Show that if $\nu > 0$ is known in the distribution

$$f(x) = \frac{\exp\{-\frac{1}{2}(x+\theta)\}x^{(\nu-2)/2}}{2^{\nu/2}} \sum_{r=0}^{\infty} \frac{(\theta x)^r}{r!2^{2r}\Gamma(\frac{1}{2}\nu + r)}, \qquad x \geq 0;\ \theta \geq 0,$$

then $t = x - \nu$ is an unbiased estimator of θ, and that

$$u = \begin{cases} t, & x \geq \nu \\ 0, & x < \nu \end{cases}$$

has smaller m.s.e. than t. ($f(x)$ is the non-central χ^2 distribution to be treated in Chapter 22.)

17.16 In a sample of size $n = \nu + 1$ from a normal population with variance σ^2, show that σ^p is unbiasedly estimated if $\nu + p > 0$ by

$$t_p = S^p 2^{-p/2}\Gamma(\tfrac{1}{2}\nu)/\Gamma\{\tfrac{1}{2}(\nu + p)\},$$

where $S^2 = \sum(x_i - \bar{x})^2$. Verify that $p = 1$ gives the second estimator of σ in Exercise 17.6 and $p = 2$ gives the estimator of σ^2 in Example 17.3.
Using Example 17.14, show that the multiple of t_p with smallest m.s.e. in estimating σ^p is exactly t_p with ν replaced by $\nu + p$. Verify that when $p = 2$, this gives $n + 1$ as the divisor of S^2.

17.17 $\bar{x}$ is the mean and s^2 the variance of a random sample of n observations $x_1, \ldots, x_n$ from a population with mean μ and variance σ^2. Show that use of the jackknife of **17.10** to correct the bias of (a) $t_{na} = s^2$ as an estimator of σ^2 yields the exactly unbiased $t'_{na} = ns^2/(n-1)$ and of (b) $t_{nb} = \bar{x}^2$ as an estimator of μ^2 yields the exactly unbiased

$$t'_{nb} = \sum\sum_{i \neq j = 1} x_j x_j/\{n(n-1)\}.$$

Show that

$$\hat{v}(\bar{x}) = \bar{x}^2 - t'_{nb} = \sum_{i=1}^{n}(x_i - \bar{x})^2/\{n(n-1)\}$$

is an unbiased estimator of the variance of $\bar{x}$ even if the x_j have different distributions, provided that they all have the same mean and are all uncorrelated in pairs.

17.18 Show that if the jackknife of **17.10** is used to correct for bias, and $\text{var}\, t_n \sim c/n$, $\text{var}\, t'_n \sim \text{var}\, t_n$. Show that the m.s.e. of t'_n is consequently no greater than that of t_n.

(Cf. Quenouille, 1956)

17.19 In a sample of n observations x_i from a Poisson distribution with parameter λ, we write $X = \sum_{i=1}^{n} x_i$, $\bar{x} = X/n$ and k_p for the pth k statistic of the sample, with expectation κ_p, the pth cumulant. Show using **17.35** that $E(k_p|X) = \bar{x}$ and that $\text{var}(k_p - \bar{x}) = \sum_{r=1}^{p} c_{rp}\lambda^r$ is unbiasedly estimated by

$$\hat{V}_p = \text{var}(k_p|X) = \sum_{r=1}^{p} c_{rp} X^{(r)} n^{-r}$$

(cf. Example 11.17 and Exercise 5.11). Hence, using the results in Chapter 12, show that

$$\text{var}(k_2|X) = \frac{2}{n}\frac{X^{(2)}}{n^{(2)}}, \qquad \text{var}(k_3|X) = \frac{6}{n}\left\{\frac{3X^{(2)}}{n^{(2)}} + \frac{X^{(3)}}{n^{(3)}}\right\}$$

and similarly that

$$\mu_3(k_2|X) = \frac{4}{n^3(n-1)^2}\{(n-2)X^{(2)} + 2X^{(3)}\}.$$

(Gart and Pettigrew, 1970)

17.20 For a (positive definite) covariance matrix, the product of any diagonal element with the corresponding diagonal element of the inverse matrix cannot be less than unity. Hence show that if, in (17.87), we have $r = k$ and $\tau_i = \theta_i$ (all i), the resulting bound for an estimator of θ_i is not less than the bound given by (17.24). Give a reason for this result.

(Rao, 1952)

17.21 The MVB (17.22) holds good for a distribution $f(x|\theta)$ whose range (a, b) depends on θ, provided that (17.18) remains true. Show that this is so if

$$f(a|\theta) = f(b|\theta) = 0,$$

and that if in addition

$$\left[\frac{\partial f(x|\theta)}{\partial\theta}\right]_{x=a} = \left[\frac{\partial f(x|\theta)}{\partial\theta}\right]_{x=b} = 0,$$

then (17.19) also remains true and we may write the MVB in the form (17.23).

17.22 Apply the result of Exercise 17.21 to show that the MVB holds for the estimation of θ in

$$f(x) = \frac{1}{\Gamma(p)}(x-\theta)^{p-1}\exp\{-(x-\theta)\}, \qquad \theta \le x < \infty; \quad p > 2,$$

and equals $(p-2)/n$, but is not attainable since there is no single sufficient statistic for θ.

17.23 x is a random variable in the range (a, b) (which may depend on θ) whose distribution is $f(x|\theta)$. If $E\{t(x, \theta)\} = \tau(\theta)$ and f and $\partial f/\partial x$ vanish at a and at b, show that

$$\operatorname{var} t \geq \left[E\left(\frac{\partial t}{\partial x}\right)\right]^2 \Big/ E\left\{\left(\frac{\partial \log f}{\partial x}\right)^2\right\} = \left[E\left(\frac{\partial t}{\partial x}\right)\right]^2 \Big/ - E\left(\frac{\partial^2 \log f}{\partial x^2}\right).$$

(Cf. Exercise 17.21 to establish (17.22)–(17.23).)

(B.R. Rao (1958). The analogue of (17.45) also holds, $\tau^{(r)}$ being replaced by $E(\partial^r t/\partial x^r)$ and $L^{(r)}/L$ by $(1/f)\partial^r f/\partial x^r$ in J_{rp} – see Sankaran (1964))

17.24 For a distribution of the form

$$f(x) = g(x)/h(\theta), \qquad \theta \leq x \leq b,$$

show that if a function $t(x)$ of a single observation is an unbiased estimator of $\tau(\theta)$, then

$$-t(x) = \{\tau(x)h'(x) + \tau'(x)h(x)\}/g(x).$$

Hence show, using **17.35**, that the unique MV unbiased estimator of $\tau(\theta)$ in samples of size n is

$$p(x_{(1)}) = \tau(x_{(1)}) - \frac{\tau'(x_{(1)})}{n} \frac{h(x_{(1)})}{g(x_{(1)})},$$

and similarly, if $a \leq x \leq \theta$, that it is

$$p(x_{(n)}) = \tau(x_{(n)}) + \frac{\tau'(x_{(n)})}{n} \frac{h(x_{(n)})}{g(x_{(n)})}.$$

Show that for $f(x) = 1/\theta, 0 \leq x \leq \theta$, $([n+s]/n)x_{(n)}^s$ is the estimator of θ^s, while for $f(x) = \exp -\{(x-\theta)\}, \theta \leq x < \infty$, $x_{(1)}^s - (s/n)x_{(1)}^{s-1}$ estimates θ^s.
(Tate (1959). Karakostas (1985) generalizes to the two-parameter case of Example 17.21; Lwin (1975) generalizes to the case where a scale parameter is also unknown in the truncated exponential family (17.30).)

17.25 If the pair of statistics (t_1, t_2) is jointly sufficient for two parameters (τ_1, τ_2), and t_1 is sufficient for τ_1 when τ_2 is known, show that the conditional distribution of t_2, given t_1, is independent of τ_1. As an illustration, consider c independent binomial distributions with sample sizes n_i $(i = 1, 2, \ldots, c)$ and parameters θ_i connected by the relation

$$\lambda_i = \log\left(\frac{\theta_i}{1-\theta_i}\right) = \alpha + \beta x_i,$$

where the x_i are known constants, and show that if y_i is the number of 'successes' in the ith sample, the conditional distribution of $\sum_i x_i y_i$, given $\sum_i y_i$, is independent of α.

(D.R. Cox, 1958a)

17.26 Show that from a single observation x $(x = 1, 2, \ldots)$, on a zero-truncated Poisson distribution with parameter θ (cf. Exercise 3.29), the only unbiased estimator of $1 - e^{-\theta}$ takes the values 0 when x is odd, 2 when x is even, whereas $1 - e^{-\theta}$ always lies between these values.

(Cf. Lehmann (1983a) for a discussion and further examples.)

17.27 As in Example 17.13, show that the efficiency of the estimator of σ based on the mean difference discussed in **10.14** is 0.978 in normal samples.

17.28 In Exercise 17.2, show that (17.43) is given by

$$J_{rp} = \begin{cases} 0, & r \neq p, \\ (r!)^2 \binom{n}{r} \left\{ \dfrac{1-I}{\theta(1-\theta)} \right\}^r, & r = p, \end{cases}$$

reducing to the binomial case if $I = 0$, when (17.41) becomes

$$\text{var}\, t \geq \sum_{r=1}^{s} (\tau^{(r)})^2 \{\theta(1-\theta)\}^r \Big/ \left\{ (r!)^2 \binom{n}{r} \right\}.$$

Hence verify the result of Exercise 17.3 that t of Example 17.11 attains this bound with $s = 2$.

(Whittaker, 1973)

17.29 From an $N(\theta, 1)$ variate x, form a new variate y by doubling the density for $a \leq |x - \theta| \leq b$, where $0 < a < 1 < b$ and $\int_a^b (u^2 - 1) \exp(-\frac{1}{2}u^2) du = 0$, and adjusting the constant to ensure the p.d.f. integrates to unity. Show that y attains the MVB (which is 1) for θ.

(Joshi, 1976)

17.30 Using 17.27, show that if t is an MV unbiased estimator, and u is any other unbiased estimator, of θ, the statistic v defined by $au + (1-a)v = t$, where a is a constant, $0 < a < 1$, has variance

$$\text{var}\, v \geq (\leq)\, \text{var}\, u \qquad \text{when } a \geq (\leq) \frac{1}{2}.$$

Hence show that in estimating the variance θ of a normal distribution with zero mean, the statistic

$$v = \left\{ (n-2) \sum_{j=1}^{n} x_j^2 + (n\bar{x})^2 \right\} / \{n(n-1)\}$$

has the same mean and variance as

$$u = \sum_{j=1}^{n} (x_j - \bar{x})^2 / (n-1).$$

17.31 For the natural exponential family given by (17.30) with $A(\theta) = \theta$, $B(x) = x$, show that the c.f. of x is $\phi(t) = \exp\{D(\theta) - D(\theta + \mathrm{i}t)\}$ and that $E(x) = -D'(\theta)$. Show further that $-C'(x)$ is the MV unbiased estimator of θ.

17.32 $x_1, x_2, x_3, \ldots$ are independent variates with p.d.f. $f(x|\theta)$. If t_n is an MV unbiased estimator of θ based on the first n x_j and $\bar{t}_{n-1}$ is defined, as in **17.10**, as the average of n $t_{n-1,i}$, show that

$$\text{cov}(t_{n-1,i}, t_{n-1,j}) / \text{var}\, t_{n-1} \leq (n-2)/(n-1), \qquad i \neq j,$$

and that $\text{var}\, \bar{t}_{n-1} \leq (n-1)\, \text{var}\, t_{n-1}/n$.
Hence show that $n\, \text{var}\, t_n \leq (n-1)\, \text{var}\, t_{n-1}$, with equality only if $t_n = \sum_{j=1}^{n} g(x_j)$.

(Cf. Blom, 1978)

17.33 Given that **t** is sufficient for $\boldsymbol{\theta}$, show that its density when $\boldsymbol{\theta} = \boldsymbol{\theta}_0$ is

$$g(\mathbf{t}|\boldsymbol{\theta}_0) = g(\mathbf{t}|\boldsymbol{\theta})L(\mathbf{x}|\boldsymbol{\theta}_0)/L(\mathbf{x}|\boldsymbol{\theta}),$$

so that if **t** is close to $\boldsymbol{\theta}$, and we approximate $g(\mathbf{t}|\mathbf{t})$ by its asymptotic multinormal density, we obtain

$$g(\mathbf{t}|\boldsymbol{\theta}_0) \doteq (2\pi)^{-k/2}|\mathbf{V}(\mathbf{t})|^{-1/2}L(\mathbf{x}|\boldsymbol{\theta}_0)/L(\mathbf{x}|\mathbf{t}),$$

where $\mathbf{V}(\mathbf{t})$ is the inverse of (17.88).
(Durbin (1980) gives details of analytical conditions that make this approximation correct to $O(n^{-1})$ or to $O(n^{-3/2})$ if its constant term is properly adjusted, just as in (11.97), which is the special case of a sum of independent univariate observations from an exponential family; independence is not required here.)

17.34 t_1 and t_2 are unbiased estimators of θ with variances σ_1^2, σ_2^2 and correlation coefficient ρ, and $u_j = t_j - \theta$. Show that $|u_1| < |u_2|$ if and only if $V = u_2 + u_1$ and $W = u_2 - u_1$ have the same sign. Show that the correlation between V and W is

$$R = (\sigma_2^2 - \sigma_1^2)/\{(\sigma_2^2 - \sigma_1^2)^2 + 4\sigma_1^2\sigma_2^2(1-\rho^2)\}^{1/2}.$$

Hence, using (15.28), show that if t_1 and t_2 are binormally distributed,

$$P = P\{|u_1| < |u_2|\} = \frac{1}{2} + \frac{1}{\pi}\arcsin R,$$

and that if $\sigma_1^2 < \sigma_2^2$, $P > \frac{1}{2}$, so that the more efficient estimator is closer to θ in Pitman's sense of **17.30**.
If t_1 is an MV estimator, use (17.61) to show that

$$R = \left(\frac{1-\rho^2}{1+3\rho^2}\right)^{1/2},$$

where $\rho^2 = \sigma_1^2/\sigma_2^2$.

(Cf. Geary, 1944)

17.35 Show that the statistical curvature defined at (17.95) is zero for the exponential family (17.30), that quite generally it is invariant under monotone reparametrization, and that for an LF based on n independent observations from $f(x|\theta)$, it becomes equal to γ^2/n.

17.36 When the p.d.f. of x is given by (17.89), use (14.21) to show that the ML estimator of θ has large-sample variance $[h(\theta)/ng'(x)]^2$.

CHAPTER 18

ESTIMATION: MAXIMUM LIKELIHOOD AND OTHER METHODS

18.1 We have already (**8.22–27**) encountered the maximum likelihood (ML) principle in its general form. In this chapter we shall be concerned with its properties when used as a method of estimation. We shall confine our discussion for the most part to the case of samples of n independent observations from the same distribution. The joint probability of the observations, regarded as a function of a single unknown parameter θ, is called (cf. **17.14**) the likelihood function (LF) of the sample, and is written

$$L(x|\theta) = f(x_1|\theta)f(x_2|\theta)\cdots f(x_n|\theta), \tag{18.1}$$

where we write $f(x|\theta)$ indifferently for a univariate or multivariate, continuous or discrete distribution.

The ML principle, whose extensive use in statistical theory dates from the work of Fisher[1] (1921a), directs us to take as our estimator of θ that value (say, $\hat{\theta}$) within the admissible range of θ which makes the LF as large as possible. That is, we choose $\hat{\theta}$ so that for any admissible value θ

$$L(x|\hat{\theta}) \geq L(x|\theta). \tag{18.2}$$

Unless otherwise specified, we assume that θ may take any real value in an interval (which may be infinite in either or both directions). In determining $\hat{\theta}$, we may clearly ignore factors in L not involving θ.

18.2 The determination of the form of the ML estimator becomes relatively simple in one general situation. If the LF is a twice-differentiable function of θ throughout its range, stationary values of the LF within the admissible range of θ will, if they exist, be given by roots of

$$L'(x|\theta) = \frac{\partial L(x|\theta)}{\partial\theta} = 0. \tag{18.3}$$

A sufficient (though not a necessary) condition for any of these stationary values (say, $\tilde{\theta}$) to be a local maximum is that

$$L''(x|\tilde{\theta}) < 0. \tag{18.4}$$

If we find all the local maxima of the LF in this way (and, given more than one, choose the largest of them) we shall have found the solution(s) of (18.2), provided that there is no terminal maximum of the LF at the boundary of the parameter space.

In practice, it is often simpler to work with the logarithm of the LF than with the function itself. Under the conditions above, they will have maxima together, since

$$\frac{\partial}{\partial\theta}\log L = L'/L$$

[1] F.Y. Edgeworth was also a major contributor to these ideas; see Pratt (1976).

and $L > 0$. We therefore seek solutions of

$$(\log L)' = 0 \tag{18.5}$$

for which

$$(\log L)'' < 0, \tag{18.6}$$

if these are simpler to solve than (18.3) and (18.4). Expression (18.5) is often called the likelihood equation.

If $L \to 0$ as θ tends to its permissible extremes, and *all* roots of (18.5) satisfy (18.6), there can be only one such root, since there must be a minimum (contradicting (18.6)) between any two maxima; and there must be one such root because $L > 0$. Thus a unique maximum of the LF must exist.

18.3 Suppose now that we are interested in the estimation of some function of θ, $\tau(\theta)$ say, where τ is a single-valued and twice-differentiable function of θ such that $[\mathrm{d}\tau/\mathrm{d}\theta] \neq 0$ for any θ in the parameter space. Reworking (18.5)–(18.6) with respect to τ, we have

$$\begin{aligned}\frac{\partial \log L}{\partial \tau} &= \frac{\partial \log L}{\partial \theta} \cdot \frac{\mathrm{d}\theta}{\mathrm{d}\tau} \\ &= (\log L)'/[\mathrm{d}\tau/\mathrm{d}\theta],\end{aligned} \tag{18.5a}$$

$$\frac{\partial^2 \log L}{\partial \tau^2} = (\log L)''\left[\frac{\mathrm{d}\tau}{\mathrm{d}\theta}\right]^{-2} + (\log L)'\left[\frac{\mathrm{d}^2\theta}{\mathrm{d}\tau^2}\right]. \tag{18.6a}$$

It follows immediately from (18.5a) that the solution(s) for τ will be the same as those for θ, i.e. they occur at $(\log L)' = 0$. Substituting into (18.6a), we see immediately that the second-order conditions are also the same.

Thus, if $\hat{\theta}$ is the ML estimator for θ, $\tau(\hat{\theta})$ is the ML estimator for $\tau(\theta)$, i.e. the ML estimator is *transformation-invariant*. The conditions we have given on $\tau(\theta)$ are sufficient but can be relaxed. For example, the parameter space for θ may be discrete (e.g. integers) so that τ is not differentiable, but the property still holds.

Maximum likelihood and sufficiency

18.4 If a single sufficient statistic exists for θ, we see at once that the ML estimator of θ, if unique, must be a function of it. For sufficiency of t for θ implies the factorization of the LF (17.84). That is,

$$L(x|\theta) = g(t|\theta)h(x), \tag{18.7}$$

the second factor on the right of (18.7) being independent of θ. Thus choice of $\hat{\theta}$ to maximize $L(x|\theta)$ is equivalent to choosing $\hat{\theta}$ to maximize $g(t|\theta)$, and hence $\hat{\theta}$ will be a function of t alone.

18.5 If an MVB estimator t exists for $\tau(\theta)$, and the likelihood equation (18.5) has a solution $\hat{\theta}$, then $t = \tau(\hat{\theta})$ and the solution $\hat{\theta}$ is unique, occurring at a maximum of the LF. For we have seen (**17.33**) that, when there is a single sufficient statistic, the LF is of the form in which MVB estimation of some function $\tau(\theta)$ is possible. Thus, as at (17.27), the LF is of the form

$$(\log L)' = A(\theta)\{t - \tau(\theta)\}, \tag{18.8}$$

so that the solutions of (18.5) are of the form

$$t = \tau(\hat{\theta}). \tag{18.9}$$

Differentiating (18.8) again, we have

$$(\log L)'' = A'(\theta)\{t - \tau(\theta)\} - A(\theta)\tau'(\theta). \tag{18.10}$$

But since, from (17.29), $\tau'(\theta)/A(\theta) = \text{var}\, t$, the last term in (18.10) may be written

$$-A(\theta)\tau'(\theta) = -\{A(\theta)\}^2 \,\text{var}\, t. \tag{18.11}$$

Moreover, at $\hat{\theta}$ the first term on the right of (18.10) is zero from (18.9). Hence (18.10) becomes, on using (18.11),

$$(\log L)''_{\hat{\theta}} = -A(\hat{\theta})\tau'(\hat{\theta}) = \{A(\hat{\theta})\}^2 \,\text{var}\, t < 0. \tag{18.12}$$

By (18.12), every solution of (18.5) is a maximum of the LF. But under regularity conditions there must be a minimum between successive maxima. Since there is no minimum, it follows that there cannot be more than one maximum. This is otherwise obvious from the uniqueness of the MVB estimator t.

Expression (18.9) shows that where an MVB (unbiased) estimator exists, it is given by the ML method. This result does not extend to the more general variance bound (17.41) – cf. Exercise 18.37.

18.6 The uniqueness of the ML estimator where a single sufficient statistic exists extends to the case where the range of $f(x|\theta)$ depends upon θ, but the argument is somewhat different in this case. We have seen (**17.40–41**) that a single sufficient statistic can exist only if

$$f(x|\theta) = g(x)/h(\theta). \tag{18.13}$$

The LF is thus also of the form

$$L(x|\theta) = \prod_{i=1}^{n} g(x_i)/\{h(\theta)\}^n \tag{18.14}$$

when it is non-zero, and (18.14) is as large as possible if $h(\theta)$ is as small as possible. Now from (18.13)

$$1 = \int f(x|\theta)\,\mathrm{d}x = \int g(x)\,\mathrm{d}x/h(\theta),$$

where integration is over the whole range of x. Hence

$$h(\theta) = \int g(x)\,\mathrm{d}x. \tag{18.15}$$

From (18.15) it follows that to make $h(\theta)$ as small as possible, we must choose $\hat{\theta}$ so that the value of the integral on the right (one or both of whose limits of integration will depend on θ) is minimized.

Now a single sufficient statistic t for θ exists (**17.40–41**) only if one terminal of the range is independent of θ or if the upper terminal is a monotone decreasing function of the lower terminal, and is then given by $x_{(1)}, x_{(n)}$ or a function of them. In either of these situations, the value of (18.15) is a monotone function of the range of integration on its right-hand side, reaching a unique terminal minimum if that range is as small as is possible, consistent with the observations, when it is a function of t. The ML estimator $\hat{\theta}$ obtained by minimizing this range is thus unique and is a function of t, and the LF (18.14) has a terminal maximum at $L(x|\hat{\theta})$.

The results of this and the previous section were originally obtained by Huzurbazar (1948), who used a different method in the 'regular' case of **18.5**.

18.7 Thus we have seen that where a single sufficient statistic t exists for a parameter θ, the ML estimator $\hat{\theta}$ of θ is a function of t alone. Further, $\hat{\theta}$ is unique, the LF having a single maximum in this case. The maximum is a stationary value (under regularity conditions) or a terminal maximum according to whether the range is independent of or dependent upon θ.

18.8 It follows from our results that all the optimum properties of single sufficient statistics are conferred upon ML estimators that are one-to-one functions of them. For example, we may obtain the solution of the likelihood equation and find a function of it which is unbiased for the parameter. It then follows from the results of **17.35** that this will be the unique MV estimator of the parameter, attaining the MVB (17.22) if this is possible.

The sufficient statistics derived in Examples 17.8–10, 17.16, 17.18 and 17.19 are all easily obtained by the ML method.

Example 18.1 (ML estimator for the uniform distribution)

To estimate θ in

$$f(x) = \theta^{-1}, \qquad 0 \leq x \leq \theta,$$

we see at once from the LF in Example 17.16 that $\hat{\theta} = x_{(n)}$, the sufficient statistic, the LF having a sharp (non-differentiable) maximum there.

Obviously, $\hat{\theta}$ is not an unbiased estimator of θ, since $\hat{\theta} \leq \theta$. An unbiased estimator is

$$t = (n+1)x_{(n)}/n.$$

Exercise 18.17 considers a one-parameter uniform distribution for which no single sufficient statistic exists.

Example 18.2 (ML estimator for the normal mean)

To estimate the mean θ of a normal distribution with known variance σ^2, we see that the LF consists of a factor free of θ multiplied by $\exp\{-n(\bar{x}-\theta)^2/(2\sigma^2)\}$, which tends to 0 as $\theta \to \pm\infty$ and has a single maximum when $\theta = \bar{x}$. We may alternatively see (Example 17.6) that

$$(\log L)' = \frac{n}{\sigma^2}(\bar{x}-\theta)$$

and obtain the ML estimator by equating this to zero, and find

$$\hat{\theta} = \bar{x}.$$

In this case, $\hat{\theta}$ is unbiased for θ.

Suppose now that θ is known to lie in the interval (a, b). If $\bar{x}$ is in this interval, $\hat{\theta} = \bar{x}$ as before; but if $\bar{x} < a$, the LF has a sharp maximum at a, and if $b < \bar{x}$ it has one at b, since L declines monotonically on either side of $\bar{x}$.

Exercise 18.2 gives a general form of distribution f for which $\hat{\theta} = \bar{x}$ – if θ is a location parameter, f must be normal, and if θ is a scale parameter, f must be exponential.

The general one-parameter case

18.9 If no single sufficient statistic for θ exists, the LF no longer necessarily has a unique maximum value if the conditions of **18.3** do not hold (cf. Examples 18.7, 18.9 and Exercises 18.17, 18.33), and we choose the ML estimator to satisfy (18.2).

We now have to consider the properties of the estimators obtained by this method. We shall see that, under very broad conditions, the ML estimator is consistent; and that under regularity conditions, the most important of which is that the range of $f(x|\theta)$ does not depend on θ, the ML estimator is consistent, is asymptotically normally distributed and is an efficient estimator. These, however, are large-sample properties and, important as they are, it should be borne in mind that they are not such powerful recommendations of the ML method as the properties, inherited from sufficient statistics, that we have discussed in **18.4** onwards. Perhaps it would be unreasonable to expect any method of estimation to produce 'best' results under all circumstances and for all sample sizes. However that may be, the fact remains that, outside the field of sufficient statistics, the optimum properties of ML estimators are asymptotic ones.

Shenton and Bowman (1977) deal with small-sample properties.

Example 18.3 (ML estimator for the correlation parameter of the binormal)

As an example of the general situation, consider the estimation of the correlation parameter ρ in samples of n from the standardized binormal distribution

$$f(x, y) = \frac{1}{2\pi(1-\rho^2)^{1/2}} \exp\left\{-\frac{1}{2(1-\rho^2)}(x^2 - 2\rho xy + y^2)\right\}, \qquad -\infty < x,\ y < \infty;\ |\rho| < 1.$$

We find

$$\log L = -n\log(2\pi) - \frac{1}{2}n\log(1-\rho^2) - \frac{1}{2(1-\rho^2)}\left(\sum x^2 - 2\rho\sum xy + \sum y^2\right),$$

and thus the three statistics $(\sum x^2, \sum xy, \sum y^2)$ are sufficient for ρ – indeed, the pair of statistics $(\sum x^2 + \sum y^2, \sum xy)$ is sufficient. There is no single sufficient statistic for ρ. (See Exercise 18.13.) At the turning-points of L, $\partial \log L/\partial\rho = 0$ and we have the cubic equation

$$g(\rho) = \frac{(1-\rho^2)^2}{n}\frac{\partial \log L}{\partial \rho} = \rho(1-\rho^2) + (1+\rho^2)\frac{1}{n}\sum xy - \rho\left(\frac{1}{n}\sum x^2 + \frac{1}{n}\sum y^2\right) = 0.$$

This has three roots, two of which may be complex. No more than two real roots can be at maxima of $\log L$, since two maxima must be separated by a minimum. There is always a maximum of the LF in the admissible interval $-1 < \rho < 1$, for $L > 0$ in the interval and $L \to 0$ as $\rho \to \pm 1$; in fact

$$g(-1) = -\frac{1}{n}\sum(x+y)^2 > 0 \qquad \text{and} \qquad g(+1) = -\frac{1}{n}\sum(x-y)^2 < 0,$$

so $g(\rho)$ crosses the axis in the interval and is declining from a maximum at $+1$. Since $g(0) = (1/n)\sum xy$, this maximum's root will have the sign of $\sum xy$.

Moreover,

$$g'(\rho) = \frac{(1-\rho^2)^2}{n}\frac{\partial^2 \log L}{\partial \rho^2} - 4\rho\frac{(1-\rho^2)}{n}\frac{\partial \log L}{\partial \rho}$$

so when

$$\frac{\partial \log L}{\partial \rho} = 0$$

we have

$$g'(\rho) = \frac{(1-\rho^2)^2}{n}\frac{\partial^2 \log L}{\partial \rho^2} = 1 - 3\rho^2 - \frac{1}{n}\left(\sum x^2 - 2\rho\sum xy + \sum y^2\right)$$

and substitution from $g(\rho) = 0$ makes this

$$g'(\rho) = -\left\{2\rho^2 + \frac{(1-\rho^2)}{\rho}\sum\frac{xy}{n}\right\}.$$

If a root ρ_1 has the same sign as $\sum xy$, $g'(\rho_1)$ is negative and the root is at a maximum; we have shown that there must be one such root in $[-1, +1]$. We now see that if there is a second maximum there, it must be at a root ρ_3 of opposite sign to $\sum xy$, for the intervening minimum at ρ_2 must be so to make $g'(\rho_2) > 0$, and ρ_3 must be further from ρ_1, satisfying

$$2|\rho_3|^3/(1-\rho_3^2) > \frac{1}{n}\left|\sum xy\right|,$$

while for ρ_2 this inequality is reversed. If ρ_1 and ρ_3 are both in $[-1, +1]$, that with the larger value of the LF is the ML estimator $\hat{\rho}$ in accordance with (18.2). At any root of the likelihood equation, we find

$$L(x|g(\rho) = 0) = \{2\pi(1-\rho^2)^{1/2}\}^{-n} \cdot \exp\left\{-\frac{1}{2}\left(n + \frac{\sum xy}{\rho}\right)\right\}.$$

The first factor on the right increases with $|\rho|$, and the second factor increases with $\sum xy/\rho$, i.e. with $\rho \operatorname{sgn}(\sum xy)$. Since

$$\rho_1 \operatorname{sgn}\left(\sum xy\right) = \rho_1 \operatorname{sgn}(\rho_1) = |\rho_1|,$$

and $\rho_3 \operatorname{sgn}(\sum xy) = -|\rho_3|$, we see that if $|\rho_1| > |\rho_3|$, both factors in the LF will be no less at ρ_1 than at ρ_3, so that $\hat{\rho} = \rho_1$; if the real root with the sign of $\sum xy$ is the largest root in absolute value, it is the ML estimator.

If there are three real roots of $g(\rho) = 0$, $g(\rho)$ has two real turning-points satisfying $g'(\rho) = 0$, a quadratic equation that has distinct real roots if and only if

$$\left(\frac{1}{n}\sum xy\right)^2 - 3\left(\frac{1}{n}\sum x^2 + \frac{1}{n}\sum y^2 - 1\right) > 0, \tag{18.16}$$

one root of $g'(\rho) = 0$ then lying on either side of $\rho_0 = \frac{1}{3}(1/n)\sum xy$. It follows that at least one root of $g(\rho) = 0$ lies on each side of ρ_0, and, since we may have $|\rho_0| > 1$, that there may be real roots of $g(\rho) = 0$ outside the interval $(-1, 1)$. Such real roots, and complex roots, are inadmissible.

Since, by the results of **10.3** and **10.9**, the sample moments on the left-hand side of (18.16) are consistent estimators of the corresponding population moments, that left-hand side will converge in probability to $\rho^2 - 3(1 + 1 - 1) < 0$. Thus, as $n \to \infty$, there will tend to be only one real root of the likelihood equation, and, by our argument above, it must be at a maximum.

Consistency of ML estimators

18.10 We now show that, under very general conditions, ML estimators are consistent. We begin with a heuristic version of Wald's (1949) argument.

As at (18.2), we consider the case of n independent observations from a distribution $f(x|\theta)$, and for each n we choose the ML estimator $\hat{\theta}$ so that, if θ is any admissible value of the parameter, we have[2]

$$\log L(x|\hat{\theta}) \geq \log L(x|\theta). \tag{18.17}$$

We denote the true value of θ by θ_0, and let E_0 represent the operation of taking expectations when the true value θ_0 holds. Consider the random variable $L(x|\theta)/L(x|\theta_0)$. Since the geometric mean of a non-degenerate distribution cannot exceed its arithmetic mean, we have, for all $\theta^* \neq \theta_0$,

$$E_0\left\{\log \frac{L(x|\theta^*)}{L(x|\theta_0)}\right\} < \log E_0\left\{\frac{L(x|\theta^*)}{L(x|\theta_0)}\right\}. \tag{18.18}$$

Now the expectation on the right of (18.18) is

$$\int \cdots \int \frac{L(x|\theta^*)}{L(x|\theta_0)} L(x|\theta_0)\, dx_1 \cdots dx_n = 1.$$

Thus (18.18) becomes

$$E_0\left\{\log \frac{L(x|\theta^*)}{L(x|\theta_0)}\right\} < 0$$

or, inserting a factor $1/n$,

$$E_0\left\{\frac{1}{n} \log L(x|\theta^*)\right\} < E_0\left\{\frac{1}{n} \log L(x|\theta_0)\right\} \tag{18.19}$$

provided that the expectation on the right exists.

[2]Because of the equality in (18.17), the sequence of values of $\hat{\theta}$ may be determinable in more than one way. See **18.12–13**.

If the right-hand side of (18.19) does not exist, examples can be given in which the ML estimator is not consistent, converging to a fixed value whatever the true value of θ may be – cf. Ferguson (1982).

Now for any value of θ,

$$\frac{1}{n}\log L(x|\theta) = \frac{1}{n}\sum_{i=1}^{n}\log f(x_i|\theta)$$

is the mean of a set of n independent identical variates with expectation

$$E_0\{\log f(x|\theta)\} = E_0\left\{\frac{1}{n}\log L(x|\theta)\right\}.$$

By the strong law of large numbers of **8.46**, therefore, $(1/n)\log L(x|\theta)$ converges with probability one to its expectation, as n increases. Thus as $n \to \infty$ we have, from (18.19), with probability one,

$$\frac{1}{n}\log L(x|\theta^*) < \frac{1}{n}\log L(x|\theta_0)$$

or

$$\lim_{n\to\infty} P\{\log L(x|\theta^*) < \log L(x|\theta_0)\} = 1, \qquad \theta^* \neq \theta_0. \tag{18.20}$$

On the other hand, (18.17) with $\theta = \theta_0$ gives

$$\log L(x|\hat{\theta}) \geq \log L(x|\theta_0). \tag{18.21}$$

Inequalities (18.20) and (18.21) imply that, as $n \to \infty$, $L(x|\hat{\theta})$ cannot take any other value than $L(x|\theta_0)$. If $L(x|\theta)$ is identifiable (cf. **17.6**), this implies that

$$P\{\lim_{n\to\infty} \hat{\theta} = \theta_0\} = 1. \tag{18.22}$$

Wald's (1949) rigorous proof of the consistency of ML estimators requires further conditions, which are often difficult to verify – see also extensions by Huber (1967) and Perlman (1972).

18.11 Other proofs of consistency, e.g. by Cramér (1946), concentrate on solutions to the likelihood equation, rather than on direct maximization of the LF – cf. **18.16**.

18.12 There may be multiple roots of the likelihood equation, but Huzurbazar (1948) has shown under regularity conditions that if there is a consistent root, it is unique as $n \to \infty$.

Suppose that the LF possesses two derivatives. It follows from the convergence in probability of $\hat{\theta}$ to θ_0 that

$$\frac{1}{n}\left[\frac{\partial^2}{\partial\theta^2}\log L(x|\theta)\right]_{\theta=\hat{\theta}} \underset{n\to\infty}{\longrightarrow} \frac{1}{n}\left[\frac{\partial^2}{\partial\theta^2}\log L(x|\theta)\right]_{\theta=\theta_0}. \tag{18.23}$$

Now by the strong law of large numbers, once more,

$$\frac{1}{n}\frac{\partial^2}{\partial\theta^2}\log L(x|\theta) = \frac{1}{n}\sum_{i=1}^{n}\frac{\partial^2}{\partial\theta^2}\log f(x_i|\theta)$$

is the mean of n independent identical variates and converges with probability one to its expectation. Thus we may write (18.23) as

$$\lim_{n\to\infty} P\left\{\left[\frac{\partial^2}{\partial\theta^2}\log L(x|\theta)\right]_{\theta=\hat{\theta}} = E_0\left[\frac{\partial^2}{\partial\theta^2}\log L(x|\theta)\right]_{\theta=\theta_0}\right\} = 1. \tag{18.24}$$

But we have seen in (17.19) that, under regularity conditions

$$E\left[\frac{\partial^2}{\partial\theta^2}\log L(x|\theta)\right] = -E\left\{\left(\frac{\partial\log L(x|\theta)}{\partial\theta}\right)^2\right\} < 0. \tag{18.25}$$

Thus (18.24) becomes

$$\lim_{n\to\infty} P\left\{\left[\frac{\partial^2}{\partial\theta^2}\log L(x|\theta)\right]_{\theta=\hat{\theta}} < 0\right\} = 1. \tag{18.26}$$

Now suppose that the conditions of **18.2** hold, and that two local maxima of the LF, at $\hat{\theta}_1$ and $\hat{\theta}_2$, are roots of (18.5) satisfying (18.6). If $\log L(x|\theta)$ has a second derivative everywhere, as we have assumed, there must be a minimum between the maxima at $\hat{\theta}_1$ and $\hat{\theta}_2$. If this is at $\hat{\theta}_3$, we must have

$$\left[\frac{\partial^2\log L(x|\theta)}{\partial\theta^2}\right]_{\theta=\hat{\theta}_3} \geq 0. \tag{18.27}$$

If both $\hat{\theta}_1$ and $\hat{\theta}_2$ are consistent estimators, $\hat{\theta}_3$, which lies between them in value, must also be consistent and must satisfy (18.26). Since (18.26) and (18.27) directly contradict each other, it follows that we can only have one consistent estimator $\hat{\theta}$ obtained as a root of the likelihood equation (18.5).

Non-uniqueness of ML estimators

18.13 A point that should be discussed in connection with the consistency of ML estimators is that, for particular samples, there is the possibility that the LF has two (or more) equal suprema, i.e. that the equality holds in (18.2). How can we choose between the values $\hat{\theta}_1$, $\hat{\theta}_2$, etc., at which they occur? There is an essential indeterminacy here. The difficulty usually arises only when particular configurations of sample values are realized which have small probability of occurrence, but it may arise in *all* samples – see Examples 18.7, 18.9 and Exercises 18.17, 18.33. If the parameter itself is unidentifiable, the difficulty *must* arise in *all* samples, as the following example makes clear.

In Example 18.3 put

$$\cos\theta = \rho.$$

To each real solution of the cubic likelihood equation, say $\hat{\rho}$, there will now correspond an infinity of estimators of θ, of the form

$$\hat{\theta}_r = \arccos\hat{\rho} + 2r\pi$$

where r is any integer. The parameter θ is essentially incapable of estimation, even when $n \to \infty$. Considered as a function of θ, the LF is periodic, with an infinite number of equal maxima at $\hat{\theta}_r$, and the $\hat{\theta}_r$ differ by multiples of 2π. There can be only one consistent estimator of θ_0, the true value of θ, but we have no means of deciding which $\hat{\theta}_r$ is consistent. In such a case, we must recognize that only $\cos\theta$ is directly estimable. Since θ is *unidentifiable*, neither ML nor any other method can be effective.

Consistency and bias of ML estimators

18.14 Although, under the conditions of **18.10**, the ML estimator is consistent, it is not unbiased generally. We have already seen in Example 18.1 that there may be bias even when the ML estimator is a function of a single sufficient statistic. In general, we must expect bias, for if the ML estimator is $\hat{\theta}$ and we seek to estimate a function $\tau(\theta)$, we have seen in **18.3** that the ML estimator of $\tau(\theta)$ is $\tau(\hat{\theta})$. But in general

$$E\{\tau(\hat{\theta})\} \neq \tau\{E(\hat{\theta})\}, \tag{18.28}$$

so that if $\hat{\theta}$ is unbiased for θ, $\tau(\hat{\theta})$ cannot be unbiased for $\tau(\theta)$. If the ML estimator is consistent, the remark concerning asymptotic unbiasedness below Example 17.3 may apply.

If the bias of the ML estimator is

$$b(\theta) = \frac{b_1(\theta)}{n} + \frac{b_2(\theta)}{n^2} + \cdots,$$

Firth (1993) shows that the bias of $O(n^{-1})$ may be eliminated by using the modified version of (18.5):

$$U'(\theta) = (\log L)' - I(\theta)b_1(\theta)/n = 0,$$

where

$$I(\theta) = I_1 = -(\log L)'', \quad \text{or} \quad I(\theta) = I_2 = -E[(\log L)''].$$

The choice between these two forms is discussed in **18.21**. Both produce consistent estimators that are unbiased to $O(n^{-2})$, and other asymptotic properties are not affected.

Example 18.4 (Bias correction for ML estimator (Firth, 1993))

Suppose we wish to estimate $\theta = \mu^{-1}$, where μ is the parameter of the Poisson distribution with probability function

$$f_j = \mathrm{e}^{-\mu}\mu^y/y!.$$

The ML estimator is $\hat{\theta} = 1/\bar{y}$ with bias $\theta^2/n + O(n^{-2})$, conditional on $\bar{y} > 0$, since

$$(\log L)' = n\theta^{-2}(1 - \theta\bar{y}).$$

Also

$$-(\log L)'' = I_1 = n\theta^{-3}$$

so that

$$I_2 = n\theta^{-3}.$$

The solutions to $U'(\theta) = 0$ yield,

$$\text{using } I_1: \quad \theta_1^* = n/(\bar{y}n + 1)$$

$$\text{using } I_2: \quad \theta_2^* = \begin{cases} (n/2\bar{y})\{\bar{y} + 2n^{-1} - (\bar{y}^2 + 4n^{-2})^{1/2}\}, & \bar{y} > 0 \\ n/2, & y = 0. \end{cases}$$

Both estimators are finite for all values for $\bar{y}$ and have bias $O(n^{-2})$.

Consistency, asymptotic normality and efficiency of ML estimators

18.15 When we turn to the discussion of the normality and efficiency of ML estimators, the following example is enough to show that we must make restrictions before we can obtain optimal results.

Example 18.5 (ML estimator and sufficiency)

We saw in Example 17.22 that in the distribution

$$f(x) = \theta^{-1}, \qquad k\theta \leq x \leq (k+1)\theta; \; k > 0,$$

there is no single sufficient statistic for θ, but that the extreme observations $x_{(1)}$ and $x_{(n)}$ are a pair of jointly sufficient statistics for θ. Let us now find the ML estimator of θ. We maximize the LF as in Example 18.1. Here

$$L(x|\theta) = \theta^{-n} u(x_{(1)} - k\theta) u((k+1)\theta - x_{(n)})$$

is non-zero only if $x_{(n)}/(k+1) \leq \theta \leq x_{(1)}/k$, and then is monotone decreasing in θ. Thus

$$\hat{\theta} = x_{(n)}/(k+1)$$

is the ML estimator. We see at once that $\hat{\theta}$ is a function of $x_{(n)}$ only, although $x_{(1)}$ and $x_{(n)}$ are both required for sufficiency.

Since $x_{(n)}$ will have an extreme-value distribution as in **14.17**, $\hat{\theta}$ is not asymptotically normal here. Nor does it have the smallest possible variance asymptotically, for by symmetry, $x_{(1)}$ and $x_{(n)}$ have the same variance, say V. The ML estimator has variance

$$\operatorname{var} \hat{\theta} = V/(k+1)^2,$$

and the estimator

$$\theta^* = x_{(1)}/k$$

has variance

$$\operatorname{var} \theta^* = V/k^2.$$

Since $x_{(1)}$ and $x_{(n)}$ are asymptotically independently distributed (**14.23**), the function

$$\bar{\theta} = a\hat{\theta} + (1-a)\theta^*$$

will, like $\hat{\theta}$ and θ^*, be a consistent estimator of θ, and its variance is

$$\operatorname{var} \bar{\theta} = V\left\{\frac{a^2}{(k+1)^2} + \frac{(1-a)^2}{k^2}\right\},$$

which is minimized for variation in a when

$$a = \frac{(k+1)^2}{k^2 + (k+1)^2}.$$

Then

$$\operatorname{var}\bar{\theta} = V/\{k^2 + (k+1)^2\}.$$

Thus, for all $k > 0$,

$$\frac{\operatorname{var}\bar{\theta}}{\operatorname{var}\hat{\theta}} = \frac{(k+1)^2}{k^2 + (k+1)^2} < 1$$

and the ML estimator has the larger variance. If k is large, the variance of $\hat{\theta}$ is nearly twice that of the other estimator.

18.16 To prove the consistency, asymptotic normality and efficiency of $\hat{\theta}$, we shall assume that the first two derivatives of $\log L(x|\theta)$ exist, and that (17.18)–(17.19) hold, i.e. that

$$E\left(\frac{\partial \log L}{\partial \theta}\right) = 0 \tag{18.29}$$

and

$$R^2(\theta) = -E\left(\frac{\partial^2 \log L}{\partial \theta^2}\right) = E\left\{\left(\frac{\partial \log L}{\partial \theta}\right)^2\right\}, \tag{18.30}$$

where $R^2(\theta) > 0$. As we pointed out in **18.2**, our differentiability assumptions imply that $\hat{\theta}$ is a root of the likelihood equation $\partial \log L/\partial \theta = 0$, and in this section we use the symbol $\hat{\theta}$ to denote such a root.

Using Taylor's theorem, we have

$$\left(\frac{\partial \log L}{\partial \theta}\right)_{\hat{\theta}} = \left(\frac{\partial \log L}{\partial \theta}\right)_{\theta_0} + (\hat{\theta} - \theta_0)\left(\frac{\partial^2 \log L}{\partial \theta^2}\right)_{\theta^*}, \tag{18.31}$$

where θ^* is some value between $\hat{\theta}$ and θ_0. Thus the left-hand side of (18.31) is zero. On its right-hand side, both $\partial \log L/\partial \theta$ and $\partial^2 \log L/\partial \theta^2$ are sums of independent identical variates, and as $n \to \infty$ each therefore converges to its expectation by the strong law of large numbers, as in the argument of **18.10**. The first of these expectations is zero by (18.29) and the second non-zero by (18.30). Since the right-hand side of (18.31) as a whole must converge to zero, to remain equal to the left, we see that we must have $(\hat{\theta} - \theta_0)$ converging to zero as $n \to \infty$, so that $\hat{\theta}$ is a consistent estimator under our assumptions.

We now rewrite (18.31) in the form

$$(\hat{\theta} - \theta_0)R(\theta_0) = \frac{\left(\frac{\partial \log L}{\partial \theta}\right)_{\theta_0} / R(\theta_0)}{\left(\frac{\partial^2 \log L}{\partial \theta^2}\right)_{\theta^*} / \{-R^2(\theta_0)\}}. \tag{18.32}$$

In the denominator on the right of (18.32) we have, since $\hat{\theta}$ is consistent for θ_0 and θ^* lies between them, from (18.23), (18.24) and (18.30),

$$\lim_{n\to\infty} P\left\{\left[\frac{\partial^2 \log L}{\partial \theta^2}\right]_{\theta^*} = -R^2(\theta_0)\right\} = 1 \tag{18.33}$$

so that the denominator converges to unity. The numerator on the right of (18.32) is the ratio to $R(\theta_0)$ of the sum of the n independent identical variates $\partial \log f(x_i|\theta_0)/\partial\theta$. This sum has zero mean by (18.29) and variance defined at (18.30) to be $R^2(\theta_0)$. The central limit theorem (**8.47–50**) therefore applies, and the numerator is asymptotically a standardized normal variate; the same is therefore true of the right-hand side as a whole. That is, we have the important result that the left-hand side of (18.32) is asymptotically standard normal: the ML estimator $\hat{\theta}$ is asymptotically normally distributed with mean θ_0 and variance $1/R^2(\theta_0)$.

This result, which gives the ML estimator an asymptotic variance equal to the MVB (17.24), implies that under these regularity conditions the ML estimator is efficient. Since the MVB can only be attained if there is a single sufficient statistic (cf. **17.33**) we are also justified in saying that the ML estimator is 'asymptotically sufficient'.

A more rigorous proof on these lines is given by Cramér (1946). Daniels (1961) – see the note by Williamson (1984) – relaxes the conditions for the asymptotic normality and efficiency of ML estimators. See also Huber (1967) and LeCam (1970). Weiss and Wolfowitz (1973) prove efficiency in a class of non-regular estimators of a location parameter – cf. Exercise 18.5 for an instance.

Although a root of the likelihood equation will be consistent, the ML estimator may be a different root – Kraft and LeCam (1956) give examples in which the ML estimator exists, is unique and is not consistent while a consistent root exists, the conditions above being satisfied. In such a case, we could proceed iteratively starting with another consistent estimator, as in **18.21**.

LeCam (1953) has objected to the use of the term 'efficient' because it implies absolute minimization of variance in large samples, and in the strict sense this is not achieved by the ML (or any other) estimator. For example, consider a consistent estimator t of θ, asymptotically normally distributed with variance of order n^{-1}. Define a new statistic

$$t' = \begin{cases} t & \text{if } |t| \geq n^{-1/4}, \\ kt & \text{if } |t| < n^{-1/4}. \end{cases} \tag{18.34}$$

We have

$$\lim_{n\to\infty} \operatorname{var} t' / \operatorname{var} t = \begin{cases} 1 & \text{if } \theta \neq 0, \\ k^2 & \text{if } \theta = 0, \end{cases}$$

and k may be taken very small, so that at one point t' is more efficient than t, and nowhere is it worse. LeCam has shown (cf. also Bahadur, 1964) that such 'superefficiency' can arise only for a set of θ-values of measure zero. In view of this, we shall retain the term 'efficiency' in its ordinary use. However, Rao (1962b) shows that even this limited paradox can be avoided by redefining the efficiency of an estimator in terms of its correlation with $\partial \log L/\partial\theta$ – cf. **17.15–17** and (17.61). Walker (1963) gives sufficient regularity conditions for the asymptotic variances of all asymptotically normal estimators to be bounded by the MVB.

Brillinger (1964) shows that if $\hat{\theta}$ is jackknifed by dividing the sample into a fixed number of groups and omitting one group at a time, then as $n \to \infty$, t'_n in (17.10) leads to normality under regularity conditions; he gives expansions for its reduced bias and its m.s.e. Reeds (1978) shows that under regularity conditions, the ordinary jackknifed $\hat{\theta}$ has the same asymptotic distribution as $\hat{\theta}$ itself, and that the jackknifed estimator of $\operatorname{var}\hat{\theta}$ is consistent.

The bootstrap estimators are consistent provided the functions in question are well-behaved (Efron, 1979); see **10.19–20**. The bootstrap does not work for certain problems, such as the estimation of end-points of the range.

Example 18.6 (Variance of the ML estimator for the correlation parameter)

In Example 18.3 we found that the ML estimator $\hat{\rho}$ of the correlation parameter in a standardized bivariate normal distribution is a root of the cubic equation

$$g(\rho) = \frac{(1-\rho^2)^2}{n}\frac{\partial \log L}{\partial \rho} = 0,$$

and that

$$g'(\rho) = \frac{(1-\rho^2)^2}{n}\frac{\partial^2 \log L}{\partial \rho^2} - \frac{4\rho(1-\rho^2)}{n}\frac{\partial \log L}{\partial \rho}$$

with

$$\frac{(1-\rho^2)^2}{n}\frac{\partial^2 \log L}{\partial \rho^2} = 1 - 3\rho^3 - \frac{1}{n}\left(\sum x^2 - 2\rho\sum xy + \sum y^2\right).$$

Taking expectations in $g'(\rho)$, (17.18) removes the second term on the right, and we find, since $E(x^2) = E(y^2) = 1$, $E(xy) = \rho$,

$$E\{g'(\rho)\} = \frac{(1-\rho^2)^2}{n}E\left\{\frac{\partial^2 \log L}{\partial \rho^2}\right\} = 1 - 3\rho^2 - 2(1-\rho^2)$$
$$= -(1+\rho^2)$$

so that

$$E\left(\frac{\partial^2 \log L}{\partial \rho^2}\right) = -\frac{n(1+\rho^2)}{(1-\rho^2)^2}.$$

Hence, from (18.30), we have asymptotically

$$\text{var}\,\hat{\rho} = \frac{(1-\rho^2)^2}{n(1+\rho^2)}.$$

Exercise 18.12 uses this result to establish the efficiency of the sample correlation coefficient.

Example 18.7 (ML estimator for the double exponential distribution)

The distribution

$$f(x) = \frac{1}{2}\exp\{-|x-\theta|\}, \qquad -\infty < x < \infty,$$

yields the log likelihood

$$\log L(x|\theta) = -n\log 2 - \sum_{i=1}^{n}|x_i - \theta|.$$

This is maximized when $\sum_i |x_i - \theta|$ is minimized, and by the result of Exercise 2.1 this occurs when θ is the median of the n values of x. (If n is odd, the value of the middle observation is the median; if n is even, any value in the interval between the two middle observations is a median.) Thus the ML estimator is $\hat{\theta} = \tilde{x}$, the sample median, and is not unique (because the median is not) when n is even. It is easily seen from Example 10.7 that its asymptotic variance in this case is

$$\text{var}\,\hat{\theta} = 1/n.$$

We cannot use the argument behind (18.30)–(18.34) to check the efficiency of $\hat{\theta}$, since the differentiability conditions there imposed do not hold for this distribution. But since

$$\frac{\partial \log f(x|\theta)}{\partial\theta} = \begin{cases} +1 & \text{if } x > \theta \\ -1 & \text{if } x < \theta \end{cases}$$

fails to exist only at $x = \theta$, we have

$$\left(\frac{\partial \log f(x|\theta)}{\partial\theta}\right)^2 = 1, \qquad x \neq \theta.$$

For $\varepsilon > 0$, we now interpret $E[(\partial \log f(x|\theta)/\partial\theta)^2]$ as

$$\lim_{\varepsilon\to 0}\left\{\int_{-\infty}^{\theta-\varepsilon} + \int_{\theta+\varepsilon}^{\infty}\right\}\left(\frac{\partial \log f(x|\theta)}{\partial\theta}\right)^2 \mathrm{d}F(x) = 1.$$

Thus we have

$$E\left\{\left(\frac{\partial \log L}{\partial\theta}\right)^2\right\} = nE\left\{\left(\frac{\partial \log f(x|\theta)}{\partial\theta}\right)^2\right\} = n,$$

so that the MVB for an estimator of θ is

$$\operatorname{var} t \geq 1/n,$$

which is attained asymptotically by $\hat{\theta}$.

Here $\log L$ consists of a continuous set of straight-line segments with joins at the observed values x_i. Exercise 18.38 gives a $\log L$ with cusps at these points.

18.17 The result of **18.16** simplifies for a distribution admitting a single sufficient statistic for the parameter. For in that case, from (18.10) and (18.12),

$$E\left(\frac{\partial^2 \log L}{\partial\theta^2}\right) = -A(\theta)\tau'(\theta) = \left(\frac{\partial^2 \log L}{\partial\theta^2}\right)_{\hat{\theta}=\theta}, \qquad (18.35)$$

so that there is no need to evaluate the expectation in this case: the MVB becomes simply $-1/(\partial^2 \log L/\partial\theta^2)_{\hat{\theta}=\theta}$; it is attained exactly when $\hat{\theta}$ is unbiased for θ, and asymptotically in any case under the conditions of **18.16**.

Estimation of the asymptotic variance

18.18 To estimate the variance of the asymptotic distribution of $\hat{\theta}$, there are two obvious estimators,

$$\widehat{\operatorname{var}}_1(\hat{\theta}) = 1\Big/\left\{-E\left(\frac{\partial^2 \log L}{\partial\theta^2}\right)\right\}_{\theta=\hat{\theta}} \qquad (18.36)$$

and

$$\widehat{\operatorname{var}}_2(\hat{\theta}) = 1\Big/\left\{-\frac{\partial^2 \log L}{\partial\theta^2}\right\}_{\theta=\hat{\theta}}, \qquad (18.37)$$

each of which is consistent since its denominator converges to the true $R^2(\theta)$ at (18.30). Expressions (18.36) and (18.37) will coincide when (18.35) holds, that is when there is a single sufficient statistic for θ, but in general they will differ.

Efron and Hinkley (1978) show that (18.37) is a first approximation to $\text{var}(\hat{\theta}|u)$ where u is an ancillary statistic as in **26.12** (i.e. a component of the minimal sufficient statistics, with marginal distribution free of θ), so that the conditionality principle of **26.15** is satisfied asymptotically; further, a first approximation to u is (18.37) itself. See also Hinkley (1980) and Skovgaard (1985).

Other estimators of $\text{var}\,\hat{\theta}$ may also be used. For example, if $\hat{\theta}$ may be expressed as a function of the sample moments, then whether or not $L(x|\theta)$ is correctly specified, $\hat{\theta}$ will be a consistent estimator of the corresponding function of the population moments as in Chapter 10, and $\text{var}\,\hat{\theta}$, another such function, will be consistently estimated by the corresponding function of the sample moments. We thus obtain a more robust estimator of $\text{var}\,\hat{\theta}$, although of course it is not in general as efficient as (18.36) and (18.37) if $L(x|\theta)$ is the true LF.

Example 18.8 (Robust estimation of the sampling variance)

We wish to estimate the standard deviation σ of a normal distribution with known mean, taken as the origin,

$$f(x) = \frac{1}{\sigma(2\pi)^{1/2}} \exp\left(-\frac{x^2}{2\sigma^2}\right), \qquad -\infty < x < \infty$$

We have, from Example 17.10,

$$(\log L)' = -n/\sigma + \sum x^2/\sigma^3,$$

so that the ML estimator, which is sufficient, is $\hat{\sigma} = \sqrt{\sum x^2/n}$.

Exercise 17.6 demonstrated the bias in $\hat{\sigma}$ and other estimators. (However, the ML estimator $\hat{\sigma}^2$ of the variance is unbiased, since $\sum x^2/\sigma^2$ has a χ^2_n distribution with expectation n – cf. **18.14**.) We have

$$(\log L)'' = n/\sigma^2 - 3\sum x^2/\sigma^4 = \frac{n}{\sigma^2}\left(1 - \frac{3\hat{\sigma}^2}{\sigma^2}\right).$$

Thus, using (18.35), we have as n increases

$$\text{var}\,\hat{\sigma} \to -1/(\log L)''_{\hat{\sigma}=\sigma} = \sigma^2/(2n),$$

which is estimated from either (18.36) or (18.37) as

$$\widehat{\text{var}}\hat{\sigma} = \hat{\sigma}^2/(2n)$$

In terms of the sample moments, $\hat{\sigma} = (m_2')^{1/2}$ is estimating $\sigma = (\mu_2')^{1/2}$, and from (10.5) and (10.14) we have

$$\text{var}\,\hat{\sigma} = \left(\frac{1}{2\mu_2'^{1/2}}\right)^2 \frac{\mu_4' - \mu_2'}{n} = \frac{\mu_4' - \mu_2'^2}{4n\mu_2'},$$

reducing to $\sigma^2/(2n)$ as above if the distribution is normal with $\mu'_1 = 0$, when $\mu'_4 = 3\mu'^2_2$. The more robust estimator of var $\hat{\sigma}$ is

$$\frac{m'_4 - m'^2_2}{4nm'_2} = \frac{m'_2}{4n}(\hat{\gamma}_2 + 2),$$

where $\hat{\gamma}_2 = m'_4/(m'_2)^2 - 3$, and is less efficient than $\hat{\sigma}^2/(2n) = m'_2/(2n)$ in the normal case because the latter is based on the sufficient statistic and is not affected by sampling variation in $\hat{\gamma}_2$.

18.19 Even when the regularity conditions for its efficiency do not hold, the ML estimator may have MV properties conferred upon it by the sufficient statistic(s) of which it is always a function. Thus, for example, in Example 18.1 $\hat{\theta} = x_{(n)}$ is a multiple $n/(n+1)$ of the unbiased estimator t with minimum variance (cf. Exercise 17.24), and consequently $\hat{\theta}$ will be asymptotically unbiased with the same asymptotic variance as t. However, we saw in Example 11.4 that here

$$\text{var}\, x_{(n)} = \frac{n\theta^2}{(n+1)^2(n+2)}.$$

Thus when we consider the m.s.e. of $x_{(n)}$ as in **17.30**, the square of its bias, $\theta^2/(n+1)^2$, is of the same order of magnitude as its variance. If we apply the result of Example 17.14, we find that

$$\frac{n(n+2)}{(n+1)^2}t = \left(\frac{n+2}{n+1}\right)x_{(n)}$$

has m.s.e. $\theta^2/(n+1)^2$, against $\theta^2/\{n(n+2)\}$ for the unbiased t and $2\theta^2/\{(n+1)(n+2)\}$ for $\hat{\theta}$. Thus $\hat{\theta}$ in this case has asymptotically, and indeed almost exactly, twice as large an m.s.e. as its best multiple has.

The cumulants of an ML estimator

18.20 Haldane and Smith (1956) obtained expressions for the first four cumulants of an ML estimator. Suppose that the distribution sampled, $f(x|\theta)$, is grouped into k classes, and that the probability of an observation falling into the rth class is π_r $(r = 1, 2, \ldots, k)$. We thus reduce any distribution to a multinomial distribution (**7.7**), and if the range of the original distribution is independent of the unknown parameter θ, we seek solutions of the likelihood equation (18.5). Since the probabilities π_r are functions of θ, we write, from **7.15**,

$$L(x|\theta) \propto \prod_r \pi_r^{n_r},$$

where n_r is the number of observations in the rth class and $\sum_r n_r = n$, the sample size. The likelihood equation is

$$\frac{\partial \log L}{\partial \theta} = \sum_r n_r \frac{\pi'_r}{\pi_r} = 0,$$

where a prime denotes differentiation with respect to θ. Of course, the ML estimator $\hat{\theta}$ will now differ from what it would be for the ungrouped observations, when more information is available.

Exercises 18.24–25 discuss the difference, which is important in some contexts – cf. **25.15–19**. However, as $k \to \infty$ the difference disappears.

Now, using Taylor's theorem, we expand π_r and π_r' about the true value θ_0, and obtain

$$\begin{aligned}\pi_r(\hat{\theta}) &= \pi_r(\theta_0) + (\hat{\theta}-\theta_0)\pi_r'(\theta_0) + \tfrac{1}{2}(\hat{\theta}-\theta_0)^2\pi_r''(\theta_0) + \cdots \\ \pi_r'(\hat{\theta}) &= \pi_r'(\theta_0) + (\hat{\theta}-\theta_0)\pi_r''(\theta_0) + \tfrac{1}{2}(\hat{\theta}-\theta_0)^2\pi_r'''(\theta_0) + \cdots.\end{aligned} \tag{18.38}$$

If we insert (18.38) into the likelihood equation, expand binomially, and sum the series, we have, writing

$$\begin{aligned}
A_i &= \sum_r \{\pi_r'(\theta_0)\}^{i+1}/\{\pi_r(\theta_0)\}^i, \\
B_i &= \sum_r \{\pi_r'(\theta_0)\}^i \pi_r''(\theta_0)/\{\pi_r(\theta_0)\}^i, \\
C_i &= \sum_r \{\pi_r'(\theta_0)\}^{i-1}\{\pi_r''(\theta_0)\}^2/\{\pi_r(\theta_0)\}^i, \\
D_i &= \sum_r \{\pi_r'(\theta_0)\}^i \pi_r'''(\theta_0)/\{\pi_r(\theta_0)\}^i, \\
\alpha_i &= \sum_r \{\pi_r'(\theta_0)\}^i \left\{\frac{n_r}{n} - \pi_r(\theta_0)\right\}/\{\pi_r(\theta_0)\}^i, \\
\beta_i &= \sum_r \{\pi_r'(\theta_0)\}^{i-1}\pi_r''(\theta_0)\left\{\frac{n_r}{n} - \pi_r(\theta_0)\right\}\bigg/\{\pi_r(\theta_0)\}^i, \\
\delta_i &= \sum_r \{\pi_r'(\theta_0)\}^{i-1}\pi_r'''(\theta_0)\left\{\frac{n_r}{n} - \pi_r(\theta_0)\right\}\bigg/\{\pi_r(\theta_0)\}^i,
\end{aligned}$$

the expansion

$$\begin{aligned}&\alpha_1 - (A_1 + \alpha_2 - \beta_1)(\hat{\theta}-\theta_0) + \tfrac{1}{2}(2A_2 - 3B_1 + 2\alpha_3 - 3\beta_2 + \delta_1)(\hat{\theta}-\theta_0)^2 \\ &\quad - \frac{1}{6}(6A_3 - 12B_2 + 3C_1 + 4D_1)(\hat{\theta}-\theta_0)^3 + \cdots = 0.\end{aligned} \tag{18.39}$$

For large n, (18.39) may be inverted by Lagrange's theorem to give

$$\begin{aligned}\hat{\theta}-\theta_0 = {}& A_1^{-1}\alpha_1 + A_1^{-3}\alpha_1[(A_2 - \tfrac{3}{2}B_1)\alpha_1 - A_1(\alpha_2-\beta_1)] \\ & + A_1^{-5}\alpha_1[\{2(A_2 - \tfrac{3}{2}B_1)^2 - A_1(A_3 - 2B_2 + \tfrac{1}{2}C_1 + \frac{2}{3}D_1)\}\alpha_1^2 \\ & - 3A_1(A_2 - \tfrac{3}{2}B_1)\alpha_1(\alpha_2-\beta_1) + \tfrac{1}{2}A_1^2\alpha_1(2\alpha_3 - 3\beta_2 + \delta_1) \\ & + A_1^2(\alpha_2-\beta_1)^2] + O(n^{-3}).\end{aligned} \tag{18.40}$$

Expression (18.40) enables us to obtain the moments of $\hat{\theta}$ as series in powers of n^{-1}.

Consider the sampling distribution of the sum

$$W = \sum_r h_r\left\{\frac{n_r}{n} - \pi_r(\theta_0)\right\},$$

where the h_r are any constant weights. From the moments of the multinomial distribution (cf. **7.7**), we obtain for the moments of W, writing $S_i = \sum_r h_r^i \pi_r(\theta_0)$,

$$\begin{aligned}
\mu_1'(W) &= 0,\\
\mu_2(W) &= n^{-1}(S_2 - S_1^2),\\
\mu_3(W) &= n^{-2}(S_3 - S_1 S_2 + 2S_1^3),\\
\mu_4(W) &= 3n^{-2}(S_2 - S_1^2)^2 + n^{-3}(S_4 - 4S_1S_3 - 3S_2^2 + 12S_1^2S_2 - 6S_1^4),\\
\mu_5(W) &= 10n^{-3}(S_2 - S_1^2)(S_3 - 3S_1S_2 + 2S_1^3) + O(n^{-4}),\\
\mu_6(W) &= 15n^{-3}(S_2 - S_1^2)^3 + O(n^{-4}).
\end{aligned} \tag{18.41}$$

From (18.41) we can derive the moments and product moments of the random variables α_i, β_i and δ_i appearing in (18.40), for all of these are functions of the form W. Finally, we substitute these moments into the powers of (18.40) to obtain the moments of $\hat{\theta}$. Expressed as cumulants, these are

$$\begin{aligned}
\kappa_1 &= \theta_0 - \tfrac{1}{2}n^{-1}A_1^{-2}B_1 + O(n^{-2}),\\
\kappa_2 &= n^{-1}A_1^{-1} + n^{-2}A_1^{-4}[-A_2^2 + \tfrac{7}{2}B_1^2 + A_1(A_3 - B_2 - D_1) - A_1^3] + O(n^{-3}),\\
\kappa_3 &= n^{-2}A_1^{-3}(A_2 - 3B_1) + O(n^{-3}),\\
\kappa_4 &= n^{-3}A_1^{-5}[-12B_1(A_2 - 2B_1) + A_1(A_3 - 4D_1) - 3A_1^3] + O(n^{-4}),
\end{aligned} \tag{18.42}$$

whence

$$\begin{aligned}
\gamma_1 &= \kappa_3/\kappa_2^{3/2} = n^{-1/2}A_1^{-3/2}(A_2 - 3B_1) + o(n^{-1/2}),\\
\gamma_2 &= \kappa_4/\kappa_2^2 = n^{-1}A_1^{-3}[-12B_1(A_2 - 2B_1) + A_1(A_3 - 4D_1) - 3A_1^3] + o(n^{-1}).
\end{aligned} \tag{18.43}$$

The first cumulant in (18.42) shows that the bias in $\hat{\theta}$ is of order of magnitude n^{-1} unless $B_1 = 0$, when it is of order n^{-2}, as may be confirmed by calculating a further term in the first cumulant. If $k \to \infty$, $A_i \to E\{(\partial \log f/\partial\theta)^{i+1}\}$ and so on, and thus the leading term in the second cumulant corresponds to the asymptotic variance previously established in **18.16**. Expressions (18.43) illustrate the tendency to normality, established in **18.16** for ungrouped observations.

If the terms in (18.42) were all evaluated, and unbiased estimates made of each of the first four moments of $\hat{\theta}$, a Pearson distribution or Edgeworth expansion could be fitted and an estimate of the small-sample distribution of $\hat{\theta}$ obtained which would provide a better approximation than the ultimate normal approximation of **18.16**. Pfanzagl (1973) gives an Edgeworth expansion, to improve the normal approximation in the continuous case, which is equivalent to using the cumulants (18.42).

Winterbottom (1979) shows how to use (18.42) in a Cornish–Fisher expansion to find approximate confidence intervals for θ.

The next higher-order terms in (18.42)–(18.43) are derived by Shenton and Bowman (1963).

Peers (1978) obtained the asymptotic joint cumulants of

$$n^{1/2}\hat{\theta} \quad \text{and} \quad t_r = n^{-r/2}(\partial^r \log L/\partial\theta^r)_{\theta=\hat{\theta}}, \quad r = 2, 3, 4,$$

in terms of $n^{-s/2}\partial^s \log L/\partial\theta^s$, $s = 1, 2, 3, 4$. The set of statistics $(\hat{\theta}, t_2, t_3, t_4)$ is 'second-order sufficient' for θ, since the Taylor expansion of $\log L$ about $\hat{\theta}$ contains only these statistics to the second order of approximation in n.

Successive approximation to ML estimators

18.21 In most of the examples we have considered, the ML estimator has been obtained in explicit form. An exception was in Example 18.3, where we were left with a cubic equation to solve for the estimator, and this can be done without much trouble when the sample is given. Sometimes, however, the likelihood equation can only be solved numerically, starting from some trial value t.

As at (18.31), we expand $\partial \log L/\partial\theta$ in a Taylor series, but this time about its value at t, obtaining

$$0 = \left(\frac{\partial \log L}{\partial\theta}\right)_{\hat{\theta}} = \left(\frac{\partial \log L}{\partial\theta}\right)_t + (\hat{\theta} - t)\left(\frac{\partial^2 \log L}{\partial\theta^2}\right)_{\theta^*},$$

where θ^* lies between $\hat{\theta}$ and t. Thus

$$\hat{\theta} = t - \left(\frac{\partial \log L}{\partial\theta}\right)_t \Big/ \left(\frac{\partial^2 \log L}{\partial\theta^2}\right)_{\theta^*}. \tag{18.44}$$

If we can choose t so that it is likely to be in the neighbourhood of $\hat{\theta}$, we can replace θ^* in (18.44) by t and obtain

$$\hat{\theta} = t - \left(\frac{\partial \log L}{\partial\theta}\right)_t \Big/ \left(\frac{\partial^2 \log L}{\partial\theta^2}\right)_t, \tag{18.45}$$

which will give a closer approximation to $\hat{\theta}$. The process can be repeated until no further correction is achieved to the desired degree of accuracy.

The most common method for choice of t is to take it as the value of some (preferably simply calculated) consistent estimator of θ, ideally one with high efficiency, so that, by (17.61), it will be highly correlated with the efficient $\hat{\theta}$. Then, as $n \to \infty$, we shall have the two consistent estimators t and $\hat{\theta}$ converging to θ_0, and θ^* consequently also doing so. The three random variables $(\partial^2 \log L/\partial\theta^2)_{\theta^*}$, $(\partial^2 \log L/\partial\theta^2)_t$ and $[E(\partial^2 \log L/\partial\theta^2)]_t$ will all converge to $[E(\partial^2 \log L/\partial\theta^2)]_{\theta_0}$. Use of the second of these variables, instead of the first, in (18.44) gives (18.45) above: use of the third instead of the first gives the alternative iterative procedure

$$\hat{\theta} = t - \left(\frac{\partial \log L}{\partial\theta}\right)_t \Big/ \left[E\left(\frac{\partial^2 \log L}{\partial\theta^2}\right)\right]_t = t + \left(\frac{\partial \log L}{\partial\theta}\right)_t (\operatorname{var}\hat{\theta})_t, \tag{18.46}$$

$\operatorname{var}\hat{\theta}$ being the asymptotic variance obtained in **18.16**. Expression (18.45) is known as the Newton–Raphson iterative process; (18.46) is sometimes known as 'the *method of scoring* for parameters', due to Fisher (1925), because $(\partial \log L/\partial\theta)_{\hat{\theta}}$ is called the *score* function. Kale (1961) shows that although (18.45) ultimately converges faster, (18.46) will often give better results for the first few iterations when n is large. It is usually less laborious.

Both (18.45) and (18.46) may fail to converge in particular cases. Even when they do converge, if the likelihood equation has multiple roots there is no guarantee that they will converge to the root corresponding to the absolute maximum of the LF; this should be verified by examining the changes in sign of $\partial \log L/\partial\theta$ from positive to negative and searching the intervals in which these changes occur to locate, evaluate and compare the maxima. Barnett (1966) discusses a systematic method of doing this. As with all such methods, the location of the maximum should be verified by repeating the analysis with several different starting-points.

It should be observed that the first iterate obtained from either (18.46) or (18.45) is itself an estimator of θ, with the same asymptotic properties as $\hat{\theta}$ itself, for the correction term is of order $n^{-1/2}$, as may be seen from (17.18)–(17.19).

Example 18.9 (ML estimation for the Cauchy distribution)

We wish to estimate the location parameter θ in the Cauchy distribution

$$f(x) = \frac{1}{\pi\{1+(x-\theta)^2\}}, \qquad -\infty < x < \infty.$$

The likelihood equation is

$$\frac{\partial \log L}{\partial \theta} = 2\sum_{i=1}^{n} \frac{(x_i - \theta)}{\{1+(x_i-\theta)^2\}} = 0,$$

an equation of degree $2n-1$ in θ.

Since $L \to 0$ as $\theta \to \pm\infty$, r of the real roots of the likelihood equation must be at maxima of the LF, alternating with $r-1$ at minima, where $1 \leq r \leq n$.

For $n = 1$, the equation is linear, and the unique maximum of the LF is at $\hat{\theta} = x_1$. But even for $n = 2$, a problem arises. We assume $x_1 < x_2$ and write $M = \frac{1}{2}(x_1 + x_2)$, $R = x_2 - x_1$. The solutions of the likelihood equation are the real roots of the cubic

$$\begin{aligned} P &= (x_1-\theta)\{1+(x_2-\theta)^2\} + (x_2-\theta)\{1+(x_1-\theta)^2\} \\ &= 2(M-\theta)\{1+(x_1-\theta)(x_2-\theta)\} \equiv 2LQ, \end{aligned}$$

say. The linear factor L changes from positive to negative at its unique root $\theta = M$, while the quadratic Q has real roots only if $R \geq 2$, when they are $\theta = M \pm \{(R/2)^2 - 1\}^{1/2}$, coinciding at M when $R = 2$. If $R < 2$, $Q > 0$, so P has the sign of $(M - \theta)$, and thus there is a unique real root at a maximum

$$\hat{\theta} = M, \qquad R \leq 2.$$

If $R > 2$, P has three real roots,

$$\theta_1 = M - \left\{\left(\frac{R}{2}\right)^2 - 1\right\}^{1/2}, \quad \theta_2 = M, \quad \theta_3 = M + \left\{\left(\frac{R}{2}\right)^2 - 1\right\}^{1/2},$$

and Q changes sign from positive to negative at θ_1, and back to positive at θ_3. Thus

$$P \left\{ \begin{array}{l} > 0, \ \theta < \theta_1, \\ < 0, \ \theta_1 < \theta < \theta_2, \\ > 0, \ \theta_2 < \theta < \theta_3, \\ < 0, \ \theta_3 < \theta, \end{array} \right\} R > 2$$

and $\theta_2 = M$ is at a local minimum since it is the middle root. Further, it is easily verified that $L(x|\theta_1) \equiv L(x|\theta_3)$; the two maxima of the LF are *always* equal, and we have no reason for choosing between them. Unlike the cases in Example 18.7 and Exercises 18.17, 18.33, the equal maxima of the LF are separated by an interval that may be arbitrarily large.

For $n \geq 3$, we must iterate. From **18.16** the asymptotic variance of $\hat{\theta}$ is given by

$$-\frac{1}{\operatorname{var}\hat{\theta}} \sim E\left(\frac{\partial^2 \log L}{\partial\theta^2}\right) = nE\left(\frac{\partial^2 \log f}{\partial\theta^2}\right)$$
$$= \frac{n}{\pi}\int_{-\infty}^{\infty} \frac{2(x-\theta)^2 - 2}{\{1+(x-\theta)^2\}^3}\,dx$$
$$= \frac{4n}{\pi}\int_0^{\infty} \frac{(x^2-1)}{(1+x^2)^3}\,dx$$
$$= -n/2.$$

Hence

$$\operatorname{var}\hat{\theta} = 2/n.$$

For small n, (18.45) or (18.46) may not converge – cf. Barnett (1966). For $n \geq 9$, however, $\hat{\theta}$ is almost always – cf. Haas *et al.* (1970) – the nearest maximum to the sample median t, which has large-sample variance (Example 17.5) $\operatorname{var} t = \pi^2/(4n)$ and thus has efficiency $8/\pi^2 = 0.81$ approximately. For $n \geq 15$, we may confidently use the median as our starting-point in seeking the value of $\hat{\theta}$, and solve (18.46), which here becomes

$$\hat{\theta} = t + \frac{4}{n}\sum_i \frac{(x_i - t)}{\{1+(x_i-t)^2\}}.$$

This is our first approximation to $\hat{\theta}$, which we may improve by further iterations of the process.

Reeds (1985) shows that $r - 1$ is asymptotically a Poisson variate with parameter π^{-1}, so that there is a unique maximum ($r = 1$) in about 70 per cent of samples, as Barnett (1966) had verified empirically – even for $n = 3$, it was 65 per cent.

More efficient simple starting-points for ML iteration are available. Bloch (1966) gives an estimator with efficiency greater than 0.95, namely the linear combination of five order statistics $x_{(r)}$ ($r = 0.13n, 0.4n, 0.5n, 0.6n, 0.87n$) with weights $-0.052, 0.3485, 0.407, 0.3485, -0.052$. Rothenberg *et al.* (1964) show that the mean of the central 24 per cent of a Cauchy sample has asymptotic variance $2.28/n$, its efficiency therefore being 0.88.

If a scale parameter σ alone is, or both θ and σ are, to be estimated, the LF is generally unimodal, as Exercise 18.39 shows. Exercise 18.40 gives explicit estimators for $n = 3$ and 4. Haas *et al.* (1970) provide numerical procedures for finding the ML estimator in the two parameter case.

McCullagh (1992; 1993) derives some general properties for the ML estimators and obtains explicit expressions for the marginal densities of the estimators in the two-parameter case when $n = 3$ and $n = 4$.

Example 18.10 (Iterative solution to ML equations)

We now examine the iterative method of solution in more detail, using data due to Fisher (1925b, Chapter 9).

Consider a multinomial distribution (cf. **7.7**) with four classes, their probabilities being

$$p_1 = (2+\theta)/4,$$
$$p_2 = p_3 = (1-\theta)/4,$$
$$p_4 = \theta/4.$$

The parameter θ, which lies in the range $(0, 1)$, is to be estimated from the observed frequencies (a, b, c, d) falling into the classes, the sample size n being equal to $a + b + c + d$. We have

$$L((a, b, c, d)|\theta) \propto (2+\theta)^a(1-\theta)^{b+c}\theta^d,$$

so that

$$\frac{\partial \log L}{\partial \theta} = \frac{a}{2+\theta} - \frac{(b+c)}{1-\theta} + \frac{d}{\theta},$$

and if this is set equal to zero, we obtain the quadratic equation in θ

$$n\theta^2 + \{2(b+c) + d - a\}\theta - 2d = 0.$$

Since the product of the coefficient of θ^2 and the constant term is negative, the product of the roots of the quadratic must also be negative, and only one root can be positive. Only this positive root falls into the permissible range for θ. Its value $\hat{\theta}$ is given by

$$2n\hat{\theta} = \{a - d - 2(b+c)\} + [\{a + 2(b+c) + 3d\}^2 - 8a(b+c)]^{1/2}.$$

For Fisher's (genetics) example, the observed frequencies are

$$a = 1997,\ b = 906,\ c = 904,\ d = 32,\ n = 3839$$

and the value of $\hat{\theta}$ is 0.0357.

Further, we have

$$\operatorname{var}\hat{\theta} \sim -\frac{1}{E\left(\frac{\partial^2 \log L}{\partial \theta^2}\right)} = \frac{2\theta(1-\theta)(2+\theta)}{n(1+2\theta)},$$

the value being 0.000 033 6 in this case, when $\hat{\theta}$ is substituted for θ in $\operatorname{var}\hat{\theta}$.

For illustrative purposes, we now suppose that we wish to find $\hat{\theta}$ iteratively in this case, starting from the value of an inefficient estimator. A simple inefficient estimator that was proposed by Fisher is

$$t = \{a + d - (b+c)\}/n,$$

which is easily seen to be consistent and has variance

$$\operatorname{var} t = (1-\theta^2)/n.$$

The value of t for the genetics data is

$$t = \{1997 + 32 - (906 + 904)\}/3839 = 0.0570.$$

This is a long way from the value of $\hat{\theta}$, 0.0357, which we seek. Using (18.46) we have, for our first approximation to $\hat{\theta}$,

$$\hat{\theta}_1 = 0.0570 + \left(\frac{\partial \log L}{\partial \theta}\right)_{\theta=t} (\operatorname{var}\hat{\theta})_{\theta=t}.$$

Now

$$\left(\frac{\partial \log L}{\partial \theta}\right)_{\theta=0.0570} = \frac{1997}{2.0570} - \frac{1810}{0.9430} + \frac{32}{0.0570} = -387.1713,$$

$$(\text{var}\,\hat{\theta})_{\theta=0.0570} = \frac{2 \times 0.057 \times 0.943 \times 2.057}{3839 \times 1.114} = 0.000\,051\,706\,78,$$

so that our improved estimator is

$$\begin{aligned}\hat{\theta}_1 &= 0.0570 - 387.1713 \times 0.000\,051\,706\,78 = 0.0570 - 0.0200 \\ &= 0.0370\end{aligned}$$

which is in fairly close agreement with the value sought, 0.0357. A second iteration gives

$$\hat{\theta}_2 = 0.0358,$$

which is very close to the value sought. At least one further iteration would be required to bring the value to 0.0357 correct to 4 d.p., and a further iteration to confirm that the iterative procedure had converged. The reader should verify that use of (18.45) instead of (18.46) gives a much worse value of $\hat{\theta}$, and that it converges much more slowly in this case.

This example makes it clear that care must be taken to carry the iterative process far enough for practical purposes. It is a somewhat unfavourable example, in that t has an efficiency of $2\theta(2+\theta)/[(1+\theta)(1+2\theta)]$, which takes the value 0.13 when $\hat{\theta} = 0.0357$ is substituted for θ. One would usually seek to start from the value of an estimator with greater efficiency.

Osborne (1992) shows the importance of working with consistent estimators in the method of scoring (18.46) to ensure rapid convergence and also demonstrates that convergence may be very slow if the data do not conform to the assumed model.

Exercise 18.35 shows that a single iteration is always enough using (18.46), but not using (18.45), if there is an MVB estimator of θ. See also Exercise 18.36.

ML estimators for several parameters

18.22 We now turn to discussion of the general case, in which a numbers of parameters are to be estimated simultaneously, whether in a univariate or multivariate distribution. If we interpret θ, and possibly also x, as a vector, the formulation of the ML principle at (18.2) holds good: we have to choose the set of admissible values of the parameters $\theta_1, \ldots, \theta_k$ that makes the LF an absolute maximum. We assume here that the admissible range for any θ_j is independent of that of the other parameters – see **18.38** for the contrary case. Under the regularity conditions of **18.2–3**, the necessary condition for a local turning-point in the LF is that

$$\frac{\partial}{\partial \theta_r} \log L(x|\theta_1, \ldots, \theta_k) = 0, \qquad r = 1, 2, \ldots, k, \tag{18.47}$$

and a sufficient condition for this to be a maximum is that when (18.47) holds the matrix

$$\left(\frac{\partial^2 \log L}{\partial \theta_r \, \partial \theta_s}\right) \tag{18.48}$$

is negative definite. The k equations (18.47) are to be solved for the k ML estimators $\hat{\theta}_1, \ldots, \hat{\theta}_k$.

Just as in **18.3**, if $L \to 0$ as $\boldsymbol{\theta}$ tends to its boundary and each set of roots of (18.47) has negative definite (18.48), only one such set can exist, giving a unique maximum – cf. Mäkeläinen *et al.* (1981).

The case of joint sufficiency

18.23 Just as in **18.4**, we see that if there exists a set of s statistics $t_1, \ldots, t_s$ that are jointly sufficient for the parameters $\theta_1, \ldots, \theta_k$, the ML estimators $\hat{\theta}_1, \ldots, \hat{\theta}_k$, if unique, must be functions of the sufficient statistics. As before, this follows immediately from the factorization at (17.84),

$$L(x|\theta_1, \ldots, \theta_k) = g(t_1, \ldots, t_s|\theta_1, \ldots, \theta_k)h(x), \tag{18.49}$$

since $h(x)$ does not contain $\theta_1, \ldots, \theta_k$.

However, the ML estimators need not be one-to-one functions of the sufficient set of statistics, and are therefore not necessarily themselves a sufficient set. In Example 18.5, we have already met a case where the ML estimator of a single parameter is a function of only one of the jointly sufficient pair of statistics, there being no single sufficient statistic.

18.24 The uniqueness of the solution of the likelihood equation in the presence of sufficiency (**18.5**) extends to the multiparameter case, as Huzurbazar (1949) pointed out. Under regularity conditions, the most general form of distribution admitting a set of k jointly sufficient statistics (17.86) yields an LF whose logarithm is

$$\log L = \sum_{j=1}^{k} A_j(\boldsymbol{\theta}) \sum_{i=1}^{n} B_j(x_i) + \sum_{i=1}^{n} C(x_i) + nD(\boldsymbol{\theta}), \tag{18.50}$$

where $\boldsymbol{\theta}$ is written for $(\theta_1, \ldots, \theta_k)$ and x is possibly multivariate. The likelihood equations are therefore

$$\frac{\partial \log L}{\partial \theta_r} = \sum_j \frac{\partial A_j}{\partial \theta_r} \sum_i B_j(x_i) + n\frac{\partial D}{\partial \theta_r} = 0, \qquad r = 1, 2, \ldots, k, \tag{18.51}$$

and a solution of $\hat{\boldsymbol{\theta}} = (\hat{\theta}_1, \hat{\theta}_2, \ldots, \hat{\theta}_k)$ of (18.51) is a maximum if

$$\left(\frac{\partial^2 \log L}{\partial \theta_r \, \partial \theta_s}\right)_{\hat{\theta}} = \sum_j \left(\frac{\partial^2 A_j}{\partial \theta_r \, \partial \theta_s}\right)_{\hat{\theta}} \sum_i B_j(x_i) + n\left(\frac{\partial^2 D}{\partial \theta_r \, \partial \theta_s}\right)_{\hat{\theta}} \tag{18.52}$$

forms a negative definite matrix (18.48).

From (17.18) we have

$$E\left(\frac{\partial \log L}{\partial \theta_r}\right) = \sum_j \frac{\partial A_j}{\partial \theta_r} E\left(\sum_i B_j(x_i)\right) + n\frac{\partial D}{\partial \theta_r} = 0, \tag{18.53}$$

and further

$$E\left(\frac{\partial^2 \log L}{\partial \theta_r \, \partial \theta_s}\right) = \sum_j \frac{\partial^2 A_j}{\partial \theta_r \, \partial \theta_s} E\left(\sum_i B_j(x_i)\right) + n\frac{\partial^2 D}{\partial \theta_r \, \partial \theta_s}. \tag{18.54}$$

On their right-hand sides, (18.53) and (18.54) have exactly the same form as (18.51) and (18.52), the difference being only that $T = \sum_i B_j(x_i)$ is replaced by its expectation and $\hat{\theta}$ by the true value θ. If we eliminate T from (18.52), using (18.51), and replace $\hat{\theta}$ by θ, we obtain exactly the same result as if we eliminate $E(T)$ from (18.54), using (18.53). We thus have

$$\left(\frac{\partial^2 \log L}{\partial\theta_r\,\partial\theta_s}\right)_{\hat{\theta}=\theta} = E\left(\frac{\partial^2 \log L}{\partial\theta_r\,\partial\theta_s}\right), \tag{18.55}$$

which is the generalization of (18.35). Moreover, from (17.19),

$$E\left(\frac{\partial^2 \log L}{\partial\theta_r^2}\right) = -E\left\{\left(\frac{\partial \log L}{\partial\theta_r}\right)^2\right\}, \tag{18.56}$$

and from (17.18) we find analogously

$$E\left(\frac{\partial^2 \log L}{\partial\theta_r\,\partial\theta_s}\right) = -E\left\{\frac{\partial \log L}{\partial\theta_r}\cdot\frac{\partial \log L}{\partial\theta_s}\right\}, \tag{18.57}$$

so that the matrix

$$\begin{aligned}-\left\{E\left(\frac{\partial^2 \log L}{\partial\theta_r\,\partial\theta_s}\right)\right\} &= \left\{E\left(\frac{\partial \log L}{\partial\theta_r}\cdot\frac{\partial \log L}{\partial\theta_s}\right)\right\}\\ &= \left\{\operatorname{cov}\left(\frac{\partial \log L}{\partial\theta_r}, \frac{\partial \log L}{\partial\theta_s}\right)\right\}\end{aligned} \tag{18.58}$$

is the covariance matrix $\mathbf{D}$ of the variates $\partial \log L/\partial\theta_r$, and this is non-negative definite by **15.3**. If we rule out linear dependencies among the variates, $\mathbf{D}$ is positive definite and the matrix on the left of (18.58) is negative definite. Thus, from (18.55), the matrix

$$\left\{\left(\frac{\partial^2 \log L}{\partial\theta_r\,\partial\theta_s}\right)_{\hat{\theta}=\theta}\right\}$$

is also negative definite, and hence any solution of (18.51) is a maximum. But under regularity conditions, there must be a minimum between any two maxima. Since there is no minimum, there can be only one maximum. Thus, under regularity conditions, joint sufficiency ensures that the likelihood equations have a unique solution, and that this is at a maximum of the LF.

Foutz (1977) gives a more general proof of uniqueness, together with consistency.

Example 18.11 (ML estimators for the normal mean and variance)

We have seen in Example 17.17 that in samples from a univariate normal distribution the sample mean and variance, $\bar{x}$ and s^2, are jointly sufficient for the population mean and variance, μ and σ^2. It follows from **18.23** that the ML estimators must be functions of $\bar{x}$ and s^2. We may confirm directly that $\bar{x}$ and s^2 are themselves the ML estimators. The LF is given by

$$\log L = -\frac{1}{2}n\log(2\pi) - \frac{1}{2}n\log(\sigma^2) - \sum_i (x_i - \mu)^2/(2\sigma^2),$$

whence the likelihood equations are

$$\frac{\partial \log L}{\partial \mu} = \frac{\sum(x-\mu)}{\sigma^2} = \frac{n(\bar{x}-\mu)}{\sigma^2} = 0,$$
$$\frac{\partial \log L}{\partial (\sigma^2)} = -\frac{n}{2\sigma^2} + \frac{\sum(x-\mu)^2}{2\sigma^4} = 0.$$

The solution of these equations is

$$\hat{\mu} = \bar{x}$$
$$\hat{\sigma}^2 = \frac{1}{n}\sum(x-\bar{x})^2 = s^2.$$

$\hat{\mu}$ is unchanged from Example 18.2, where σ^2 was known, but $\hat{\sigma}^2$ differs from Example 18.8 with known mean μ (there taken as origin), where $\hat{\sigma}^2 = (1/n)\sum(x-\mu)^2$. While $\hat{\mu}$ is unbiased, $\hat{\sigma}^2$ is now biased, having expected value $(n-1)\sigma^2/n$. As in the one-parameter case (**18.14**), ML estimators need not be unbiased.

18.25 In the case where the terminals of the range of a distribution depend on more than one parameter, there has not, so far as we know, been any general investigation of the uniqueness of the ML estimator in the presence of sufficient statistics, corresponding to that for the one-parameter case in **18.6**. But if the statistics are individually, as well as jointly, sufficient for the parameters on which the terminals of the range depend, the result of **18.6** obviously holds good, as in the following example.

Example 18.12 (ML estimation with range dependent on parameters)
From Example 17.21, we see that for the distribution

$$f(x) = 1/(\beta-\alpha), \qquad \alpha \leq x \leq \beta, \tag{18.59}$$

the extreme observations $x_{(1)}$ and $x_{(n)}$ are a pair of jointly sufficient statistics for α and β. In this case, it is clear that the ML estimators

$$\hat{\alpha} = x_{(1)}, \qquad \hat{\beta} = x_{(n)},$$

maximize the LF uniquely, and the same will be true whenever each terminal of the range of a distribution depends on a different parameter.

The general multiparameter case

18.26 In the general case, where there is not necessarily a set of k sufficient statistics for the k parameters, the joint ML estimators may not exist (cf. Exercises 18.23, 18.34), and we saw in **18.9** that even in the one-parameter case, $\hat{\theta}$ may not be unique when finite. However, $\hat{\boldsymbol{\theta}}$ has similar optimum properties, in large samples, to those in the single-parameter case.

In the first place, we note that the consistency results given in **18.10** and **18.16** hold good for the multiparameter case if we there interpret θ as a vector of parameters $(\theta_1, \ldots, \theta_k)$ and $\hat{\theta}$,

θ^* as vectors of estimators of θ. We therefore have the result that under regularity conditions the joint ML estimators converge in probability, as a set, to the true set of parameter values θ_0.

Further, by an immediate generalization of the method of **18.16**, we may show – see, for example, Wald (1943a) – that the joint ML estimators tend, under regularity conditions, to a multivariate normal distribution, with covariance matrix whose inverse is given by

$$(V_{rs}^{-1}) = -E\left(\frac{\partial^2 \log L}{\partial\theta_r\,\partial\theta_s}\right) = E\left(\frac{\partial \log L}{\partial\theta_r}\cdot\frac{\partial \log L}{\partial\theta_s}\right) \tag{18.60}$$

using (18.57). We shall only sketch the essentials of the proof. The analogue of the Taylor expansion of (18.31) becomes, on putting the left-hand side equal to zero,

$$\left(\frac{\partial \log L}{\partial\theta_r}\right)_{\theta_0} = \sum_{s=1}^{k}(\hat{\theta}_s - \theta_{s0})\left(\frac{\partial^2 \log L}{\partial\theta_r\,\partial\theta_s}\right)_{\theta^*}, \qquad r = 1, 2, \ldots, k. \tag{18.61}$$

Since θ^* is a value converging in probability to θ_0, and the second derivatives on the right-hand side of (18.61) converge in probability to their expectations, we may regard (18.61) as a set of linear equations in the quantities $(\hat{\theta}_r - \theta_{r0})$, which we may rewrite as

$$\mathbf{y} = \mathbf{V}^{-1}\mathbf{z} \tag{18.62}$$

where $\mathbf{y} = \partial \log L/\partial\boldsymbol{\theta}$, $\mathbf{z} = \hat{\boldsymbol{\theta}} - \boldsymbol{\theta}_0$ and $\mathbf{V}^{-1}$ is defined in (18.60).

By the multivariate central limit theorem, the vector $\mathbf{y}$ will tend to be multinormally distributed, with zero mean if (18.29) holds for each θ_r, and hence so will the vector $\mathbf{z}$ be. The covariance matrix of $\mathbf{y}$ is $\mathbf{V}^{-1}$ of (18.60), by definition, so that the exponent of its multinormal distribution will be the quadratic form (cf. **15.3**)

$$-\tfrac{1}{2}\mathbf{y}^T\mathbf{V}\mathbf{y}. \tag{18.63}$$

The transformation (18.62) gives the quadratic form for $\mathbf{z}$,

$$-\tfrac{1}{2}\mathbf{z}^T\mathbf{V}^{-1}\mathbf{z},$$

so that the covariance matrix of $\mathbf{z}$ is $(\mathbf{V}^{-1})^{-1} = \mathbf{V}$, as stated at (18.60). From (17.87)–(17.88), we see that each $\hat{\theta}_r$ asymptotically attains the minimum attainable variance, which by Exercise 17.20 is no less than if a single parameter was being estimated.

Sweeting (1992) extends these asymptotic results to cases where a particular parameter value approaches the boundary of the parameter space (e.g. $V^{-1} \to 0$, where V denotes the d.fr. in Student's t distribution).

Cheng and Amin (1983) show that if, instead of the LF, we maximize the product of the spacings of the order statistics, transformed as in Exercise 14.14, the asymptotic properties of the estimators are no worse than those of the ML estimators.

18.27 If there is a set of k jointly sufficient statistics for the k parameters, we may use (18.55) in (18.60) to obtain, for the inverse of the covariance matrix of the ML estimators in large samples,

$$(V_{rs}^{-1}) = -\left(\frac{\partial^2 \log L}{\partial\theta_r\, \partial\theta_s}\right)_{\hat{\theta}=\theta}. \tag{18.64}$$

This generalization of the result of **18.17** removes the need to find expectations.

If there is no set of k sufficient statistics, the elements of the covariance matrix may be estimated from the sample by standard methods.

Kale (1962) considers alternative iterative methods for solving the likelihood equations for several parameters. Expression (18.62) is used with $\boldsymbol{\theta}_0$ replaced by a trial vector $\mathbf{t}$ (cf. **18.21**) so that $\hat{\boldsymbol{\theta}} = \mathbf{t}+\mathbf{V}\mathbf{y}$ may be iterated as often as necessary.

Exercise 18.34, where no ML estimator exists, satisfies the two-parameter generalization of the conditions of **18.16**, so that there are consistent roots of the likelihood equations nonetheless, and we may use iterative methods as in **18.16**.

18.28 We have seen in **18.16** that under regularity conditions the ML estimator of a single parameter is efficient. Now, in the multiparameter case, consider any linear function

$$l = \mathbf{c}^T\hat{\boldsymbol{\theta}} \tag{18.65}$$

of the joint ML estimators $\hat{\boldsymbol{\theta}}$, which have covariance matrix $\mathbf{V}$ as before. We may reformulate the problem in terms of a new set of parameters $\boldsymbol{\phi}$ chosen so that $\phi_1 = \mathbf{c}^T\boldsymbol{\theta}$. From the invariance of ML estimators under transformations (cf. **18.3**) we see that $\hat{\phi}_1 = l$. It follows from **18.26** that $\text{var}(\hat{\phi}_1)$ is asymptotically minimized, and in particular that

$$\text{var}(\mathbf{c}^T\hat{\boldsymbol{\theta}}) \leq \text{var}(\mathbf{c}^T\mathbf{u}), \tag{18.66}$$

where $\mathbf{u}$ is any other set of consistent estimators of $\boldsymbol{\theta}$, with covariance matrix $\mathbf{D}$. Inequality (18.66) holds for any choice of $\mathbf{c}$.

We may rewrite (18.66) as

$$\mathbf{c}^T\mathbf{V}\mathbf{c} \leq \mathbf{c}^T\mathbf{D}\mathbf{c}.$$

We may simultaneously diagonalize $\mathbf{V}$ and $\mathbf{D}$ by a real non-singular transformation $\mathbf{c} = \mathbf{A}\mathbf{b}$ to give

$$\mathbf{b}^T(\mathbf{A}^T\mathbf{V}\mathbf{A})\mathbf{b} \leq \mathbf{b}^T(\mathbf{A}^T\mathbf{D}\mathbf{A})\mathbf{b},$$

the bracketed matrices being diagonal. Since $\mathbf{b}$ is arbitrary, suitable choice of its elements ensures that any element of $\mathbf{A}^T\mathbf{V}\mathbf{A}$ cannot exceed the corresponding element of $\mathbf{A}^T\mathbf{D}\mathbf{A}$. Thus their determinants satisfy

$$|A^TVA| \leq |A^TDA|, \qquad \text{i.e. } |A^T||V||A| \leq |A^T||D||A|,$$

or

$$|V| \leq |D|. \tag{18.67}$$

The determinant of a covariance matrix is called the *generalized variance* of the estimators. Thus the ML estimators minimize the generalized variance in large samples, a result due originally to Geary (1942). Our proof was given by Daniels (1951–52).

Example 18.13 (Covariance matrix for ML estimators in the normal case)
Consider again the ML estimators $\bar{x}$ and s^2 in Example 18.11. We have

$$\frac{\partial^2 \log L}{\partial \mu^2} = -\frac{n}{\sigma^2},$$

$$\frac{\partial^2 \log L}{\partial(\sigma^2)^2} = \frac{n}{2\sigma^4} - \frac{\sum(x-\mu)^2}{\sigma^6}, \qquad \frac{\partial^2 \log L}{\partial \mu \, \partial(\sigma^2)} = -\frac{n(\bar{x}-\mu)}{\sigma^4}.$$

Remembering that the ML estimators $\bar{x}$ and s^2 are sufficient for μ and σ^2, we use (18.64) and obtain the inverse of their covariance matrix in large samples by putting $\bar{x} = \mu$ and $\sum(x-\mu)^2 = n\sigma^2$ in these second derivatives. We find

$$\mathbf{V}^{-1} = \begin{pmatrix} n/\sigma^2 & 0 \\ 0 & n/2\sigma^4 \end{pmatrix},$$

so that

$$\mathbf{V} = \begin{pmatrix} \sigma^2/n & 0 \\ 0 & 2\sigma^4/n \end{pmatrix}.$$

We see from this that $\bar{x}$ and s^2 are asymptotically normally and *independently* distributed with the variances given. However, we know that the independence property and the normality and variance of $\bar{x}$ are exact for any n (Examples 11.3, 11.12); but the normality property and the variance of s^2 are strictly limiting ones, for we have seen (Example 11.7) that ns^2/σ^2 is distributed exactly as χ^2 with $(n-1)$ d.fr., the variance of s^2 therefore, from (16.5), being exactly $(\sigma^2/n)^2 \cdot 2(n-1) = 2\sigma^4(n-1)/n^2$.

Although μ as well as σ^2 is being estimated here, this variance is smaller than the variance of the ML estimator of σ^2 with μ known, which Example 18.8 (with origin at μ) gives as $\hat{\sigma}^2 = (1/n)\sum(x-\mu)^2$, distributed as a multiple σ^2/n of a χ^2_n variate, with exact variance $2\sigma^4/n$. The difference disappears asymptotically, in accordance with the statement at the end of **18.26**.

18.29 Where a distribution depends on k parameters, we may be interested in estimating any number of them from 1 to k, the others being known. Under regularity conditions, the ML estimators of the parameters concerned will be obtained by selecting the appropriate subset of the k likelihood equations (18.47) and solving them. By the nature of this process, it is not to be expected that the ML estimator of a particular parameter will be unaffected by knowledge of the other parameters of the distribution. The form of the ML estimator depends on the company it keeps, as is made clear by the following example.

Example 18.14 (ML estimator for the binormal distribution)

For the binormal distribution

$$f(x, y) = \frac{1}{2\pi\sigma_1\sigma_2(1-\rho^2)^{1/2}} \exp\left[-\frac{1}{2(1-\rho^2)}\left\{\left(\frac{x-\mu_1}{\sigma_1}\right)^2 - 2\rho\left(\frac{x-\mu_1}{\sigma_1}\right)\right.\right.$$
$$\left.\left.\times\left(\frac{y-\mu_2}{\sigma_2}\right) + \left(\frac{y-\mu_2}{\sigma_2}\right)^2\right\}\right], \qquad -\infty < x, y < \infty;\ \sigma_1, \sigma_2 > 0;\ |\rho| < 1$$

we obtain the logarithm of the LF

$$\log L(x, y|\mu_1, \mu_2, \sigma_1^2, \sigma_2^2, \rho) = -n\log(2\pi) - \tfrac{1}{2}n\{\log\sigma_1^2 + \log\sigma_2^2 + \log(1-\rho^2)\}$$
$$-\frac{1}{2(1-\rho^2)}\sum\left\{\left(\frac{x-\mu_1}{\sigma_1}\right)^2 - 2\rho\left(\frac{x-\mu_1}{\sigma_1}\right)\left(\frac{y-\mu_2}{\sigma_2}\right) + \left(\frac{y-\mu_2}{\sigma_2}\right)^2\right\},$$

from which the five likelihood equations are

$$\left.\begin{aligned}\frac{\partial\log L}{\partial\mu_1} &= \frac{n}{\sigma_1(1-\rho^2)}\left\{\frac{(\bar{x}-\mu_1)}{\sigma_1} - \rho\frac{(\bar{y}-\mu_2)}{\sigma_2}\right\} = 0,\\ \frac{\partial\log L}{\partial\mu_2} &= \frac{n}{\sigma_2(1-\rho^2)}\left\{\frac{(\bar{y}-\mu_2)}{\sigma_2} - \rho\frac{(\bar{x}-\mu_1)}{\sigma_1}\right\} = 0,\end{aligned}\right\} \quad (18.68)$$

$$\left.\begin{aligned}\frac{\partial\log L}{\partial(\sigma_1^2)} &= \frac{1}{2\sigma_1^2(1-\rho^2)}\left\{n(1-\rho^2) - \frac{\sum(x-\mu_1)^2}{\sigma_1^2} + \rho\frac{\sum(x-\mu_1)(y-\mu_2)}{\sigma_1\sigma_2}\right\} = 0,\\ \frac{\partial\log L}{\partial(\sigma_2^2)} &= \frac{1}{2\sigma_2^2(1-\rho^2)}\left\{n(1-\rho^2) - \frac{\sum(y-\mu_2)^2}{\sigma_2^2} + \rho\frac{\sum(x-\mu_1)(y-\mu_2)}{\sigma_1\sigma_2}\right\} = 0,\end{aligned}\right\} \quad (18.69)$$

$$\frac{\partial\log L}{\partial\rho} = \frac{1}{(1-\rho^2)}\left\{n\rho - \frac{1}{(1-\rho^2)}\left[\rho\left(\frac{\sum(x-\mu_1)^2}{\sigma_1^2} + \frac{\sum(y-\mu_2)^2}{\sigma_2^2}\right)\right.\right.$$
$$\left.\left. - (1+\rho^2)\frac{\sum(x-\mu_1)(y-\mu_2)}{\sigma_1\sigma_2}\right]\right\} = 0. \quad (18.70)$$

(a) Suppose first that we wish to estimate ρ alone, the other four parameters being known. We then solve (18.70) alone. We have already dealt with this case, in standardized form, in Example 18.3: (18.70) yields a cubic equation for the ML estimator $\hat{\rho}$.

(b) Suppose now that we wish to estimate σ_1^2, σ_2^2 and ρ, μ_1, and μ_2 being known. We notice that the same three statistics that were jointly sufficient for ρ alone in Example 18.3 remain sufficient for these three parameters, but that we cannot, as there, find a pair of sufficient statistics, let alone a single sufficient statistic, for them. We have to solve the three likelihood equations (18.69) and (18.70). Dropping the non-zero factors outside the braces, these equations become, after a slight rearrangement,

$$\left.\begin{aligned}n(1-\rho^2) &= \frac{\sum(x-\mu_1)^2}{\sigma_1^2} - \rho\frac{\sum(x-\mu_1)(y-\mu_2)}{\sigma_1\sigma_2},\\ n(1-\rho^2) &= \frac{\sum(y-\mu_2)^2}{\sigma_2^2} - \rho\frac{\sum(x-\mu_1)(y-\mu_2)}{\sigma_1\sigma_2},\end{aligned}\right\} \quad (18.71)$$

and

$$n(1-\rho^2) = \frac{\sum(x-\mu_1)^2}{\sigma_1^2} + \frac{\sum(y-\mu_2)^2}{\sigma_2^2} - \frac{1+\rho^2}{\rho}\frac{\sum(x-\mu_1)(y-\mu_2)}{\sigma_1\sigma_2}. \quad (18.72)$$

Equations (18.71)–(18.72) yield the solutions

$$\rho = \frac{\frac{1}{n}\sum(x-\mu_1)(y-\mu_2)}{\sigma_1\sigma_2}, \quad (18.73)$$

$$\hat{\sigma}_1^2 = \frac{1}{n}\sum(x-\mu_1)^2 \quad (18.74a)$$

and

$$\hat{\sigma}_2^2 = \frac{1}{n}\sum(x-\mu_2)^2. \quad (18.74b)$$

Hence, from (18.73)

$$\hat{\rho} = \frac{\frac{1}{n}\sum(x-\mu_1)(y-\mu_2)}{\hat{\sigma}_1\hat{\sigma}_2}. \quad (18.75)$$

In this case, therefore, all three ML estimators use the known population means, and $\hat{\rho}$ is the sample correlation coefficient calculated about them.

(c) Finally, suppose that we wish to estimate all five parameters of the distribution. We solve (18.68), (18.69) and (18.70) together. (18.68) reduces to

$$\begin{aligned} \frac{(\bar{x}-\mu_1)}{\sigma_1} &= \rho\frac{(\bar{y}-\mu_2)}{\sigma_2}, \\ \frac{(\bar{y}-\mu_2)}{\sigma_2} &= \rho\frac{(\bar{x}-\mu_1)}{\sigma_1}, \end{aligned} \quad (18.76)$$

a pair of equations whose only solution is

$$\bar{x}-\mu_1 = \bar{y}-\mu_2 = 0. \quad (18.77)$$

Taken with (18.74) and (18.75), which are the solutions of (18.69) and (18.70), (18.77) gives for the set of five ML estimators

$$\begin{aligned} \hat{\mu}_1 &= \bar{x}, \quad \hat{\sigma}_1^2 = \frac{1}{n}\sum(x-\bar{x})^2, \\ \hat{\mu}_2 &= \bar{y}, \quad \hat{\sigma}_2^2 = \frac{1}{n}\sum(y-\bar{y})^2, \\ \hat{\rho} &= \frac{(1/n)\sum(x-\bar{x})(y-\bar{y})}{\hat{\sigma}_1\hat{\sigma}_2}. \end{aligned} \quad (18.78)$$

Thus the ML estimators of all five parameters are the corresponding sample moments and its correlation coefficient. They are jointly sufficient statistics, as is easily verified from the LF.

Exercise 18.14 treats other combinations of parameters of the binormal distribution.

18.30 Since the ML estimator of a parameter is a different function of the observations, according to which of the other parameters of the distribution is known, its large-sample variance will also differ. To facilitate the evaluation of the covariance matrices of ML estimators, we recall that if a distribution admits a set of k sufficient statistics for its k parameters, we may avail ourselves of the form (18.64) for the inverse of the covariance matrix of the ML estimators in large samples.

Example 18.15 (Variances of ML estimators with some parameters known)

We now evaluate the large-sample covariance matrices of the ML estimators in each of the three cases considered in Example 18.14.

(a) When we are estimating ρ alone, $\hat{\rho}$ is not sufficient. But we have already evaluated its large-sample variance in Example 18.6, finding

$$\operatorname{var}\hat{\rho} = \frac{(1-\rho^2)^2}{n(1+\rho^2)}. \qquad (18.79)$$

The fact that we were there dealing with a standardized distribution is irrelevant, since ρ is invariant under changes of location and scale.

(b) In estimating the three parameters σ_1^2, σ_2^2 and ρ, the three ML estimators given by (18.74) and (18.75) are jointly sufficient, and we therefore make use of (18.64). Writing the parameters in the order above, we find for the 3×3 inverse covariance matrix

$$\mathbf{V}_3^{-1} = (V_{rs}^{-1}) = -\left(\frac{\partial^2 \log L}{\partial\theta_r\,\partial\theta_s}\right)_{\hat{\theta}=\theta}$$
$$= \frac{n}{(1-\rho^2)}\begin{pmatrix} \dfrac{2-\rho^2}{4\sigma_1^4} & \dfrac{-\rho^2}{4\sigma_1^2\sigma_2^2} & \dfrac{-\rho}{2\sigma_1^2} \\ \dfrac{-\rho^2}{4\sigma_1^2\sigma_2^2} & \dfrac{2-\rho^2}{4\sigma_2^4} & \dfrac{-\rho}{2\sigma_2^2} \\ \dfrac{-\rho}{2\sigma_1^2} & \dfrac{-\rho}{2\sigma_2^2} & \dfrac{1+\rho^2}{1-\rho^2} \end{pmatrix}. \qquad (18.80)$$

Inversion of (18.80) gives for the large-sample covariance matrix

$$\mathbf{V}_3 = \frac{1}{n}\begin{pmatrix} 2\sigma_1^4 & 2\rho^2\sigma_1^2\sigma_2^2 & \rho(1-\rho^2)\sigma_1^2 \\ 2\rho^2\sigma_1^2\sigma_2^2 & 2\sigma_2^4 & \rho(1-\rho^2)\sigma_2^2 \\ \rho(1-\rho^2)\sigma_1^2 & \rho(1-\rho^2)\sigma_2^2 & (1-\rho^2)^2 \end{pmatrix}. \qquad (18.81)$$

Only if $\rho = 0$ does (18.81) become diagonal, when the three ML estimators are uncorrelated and asymptotically independent because of their approach to normality.

(c) In estimating all five parameters, the ML estimators (18.78) are a sufficient set. Moreover, the 3×3 matrix $\mathbf{V}_3^{-1}$ in (18.80) will form part of the 5×5 inverse variance matrix $\mathbf{V}_5^{-1}$ which we now seek. Writing the parameters in the order $\mu_1, \mu_2, \sigma_1^2, \sigma_2^2, \rho$, (18.80) will be the lower 3×3 principal minor of $\mathbf{V}_5^{-1}$. For the elements involving derivatives with respect to μ_1 and μ_2, we find from (18.68)

$$\frac{\partial^2 \log L}{\partial \mu_1^2} = \frac{-n}{\sigma_1^2(1-\rho^2)},$$
$$\frac{\partial^2 \log L}{\partial \mu_1\, \partial \mu_2} = \frac{-n\rho}{\sigma_1\sigma_2(1-\rho^2)}, \qquad \frac{\partial^2 \log L}{\partial \mu_2^2} = \frac{-n}{\sigma_2^2(1-\rho^2)}, \tag{18.82}$$

while

$$\frac{\partial^2 \log L}{\partial \mu_i\, \partial \sigma_1^2} = \frac{\partial^2 \log L}{\partial \mu_i\, \partial \sigma_2^2} = \frac{\partial^2 \log L}{\partial \mu_i\, \partial \rho} = 0, \qquad i = 1, 2, \tag{18.83}$$

at $\bar{x} = \mu_1$, $\bar{y} = \mu_2$. Thus if we write, for the inverse of the covariance matrix of the ML estimators of μ_1 and μ_2,

$$\mathbf{V}_2^{-1} = \frac{-n}{(1-\rho^2)} \begin{pmatrix} \dfrac{1}{\sigma_1^2} & \dfrac{-\rho}{\sigma_1\sigma_2} \\ \dfrac{-\rho}{\sigma_1\sigma_2} & \dfrac{1}{\sigma_2^2} \end{pmatrix} \tag{18.84}$$

we have

$$\mathbf{V}_5^{-1} = \begin{pmatrix} \mathbf{V}_2^{-1} & \mathbf{0} \\ \mathbf{0} & \mathbf{V}_3^{-1} \end{pmatrix} \tag{18.85}$$

and we may invert $\mathbf{V}_2^{-1}$ and $\mathbf{V}_3^{-1}$ separately to obtain the non-zero elements of the inverse of $\mathbf{V}_5^{-1}$. We have already inverted $\mathbf{V}_3^{-1}$ in (18.81). The inverse of $\mathbf{V}_2^{-1}$ is, from (18.84),

$$\mathbf{V}_2 = \frac{1}{n} \begin{pmatrix} \sigma_1^2 & \rho\sigma_1\sigma_2 \\ \rho\sigma_1\sigma_2 & \sigma_2^2 \end{pmatrix}, \tag{18.86}$$

so

$$\mathbf{V}_5 = \begin{pmatrix} \mathbf{V}_2 & \mathbf{0} \\ \mathbf{0} & \mathbf{V}_3 \end{pmatrix}, \tag{18.87}$$

with $\mathbf{V}_2^{-1}$ and $\mathbf{V}_3^{-1}$ defined in (18.86) and (18.81).

We see from this result what we have already observed (cf. **16.25**) to be true for any sample size: that the sample means are distributed independently of the variances and covariance in bivariate normal samples; that the correlation between the sample means is ρ; and that the correlation between the sample variances is ρ^2 (Example 13.1).

It follows from the changes in covariance matrices with the set of parameters being estimated, that if an inappropriate ML estimator is used, it may be inefficient – Exercise 18.12 treats the binormal correlation coefficient.

Shenton and Bowman (1977) give methods of evaluating the performance of ML estimators in small samples, using series expansions.

Non-identical distributions

18.31 Throughout our discussions so far, we have been considering problems of estimation when all the observations come from the same underlying distribution. We now examine the behaviour of ML estimators when this condition no longer holds. In fact, we replace (18.1) by the more general LF

$$L(x|\theta_1, \ldots, \theta_k) = f_1(x_1|\theta) f_2(x_1|\theta) \cdots f_n(x_n|\theta), \tag{18.88}$$

where the different factors f_i on the right of (18.88) depend on possibly different functions of the set of parameters $\theta_1, \ldots, \theta_k$.

It is not now necessarily true even that ML estimators are consistent – Basu and Ghosh (1980) give sufficient conditions for asymptotic normality and consistency. In one particular class of cases, in which the number of parameters increases with the number of observations (k being a function of n), the ML method may become ineffective, as in Example 18.16 below.

Example 18.16 (ML estimation when the number of parameters increases with sample size)

(a) Suppose that x_i is a normal variate with mean θ_i and variance $\sigma^2 > 0$ ($i = 1, 2, \ldots, n$). The LF (18.88) gives

$$\log L = -\frac{1}{2} n \log(2\pi) - \frac{1}{2} n \log(\sigma^2) - \frac{1}{2\sigma^2} \sum (x_i - \theta_i)^2. \tag{18.89}$$

The final term in (18.89) has its numerator reduced to zero if we put $\theta_i = x_i$, for all i, so that as $\sigma^2 \to 0$ the preceding term makes $\log L \to \infty$. Thus no ML estimators of (θ_i, σ^2) exist. Nonetheless, we can estimate the θ_i since $E(x) = \theta_i$, and since $E(x_i^2) = \sigma^2 + \theta_i^2$, $(1/n) \sum_i x_i^2$ is an upward-biased estimator of σ^2, so that the sample does yield some information concerning σ^2, even if $n = 1$.

(b) The ML estimator exists, but is seriously deficient if we have two observations from each of the n normal distributions. We then have

$$\log L = -n \log(2\pi) - n \log(\sigma^2) - \frac{1}{2\sigma^2} \sum_{i=1}^{n} \sum_{j=1}^{2} (x_{ij} - \theta_i)^2$$

and

$$\hat{\theta}_i = \tfrac{1}{2}(x_{i1} + x_{i2}) = \bar{x}_i,$$
$$\hat{\sigma}^2 = \tfrac{1}{2n} \sum_{i=1}^{n} \sum_{j=1}^{2} (x_{ij} - \bar{x}_i)^2. \tag{18.90}$$

But since

$$E\left\{\frac{1}{2} \sum_{j=1}^{2} (x_{ij} - \bar{x}_i)^2\right\} = \frac{1}{2}\sigma^2$$

(cf. Example 17.3) we have from (18.90), for all n,

$$E(\hat{\sigma}^2) = \frac{1}{2}\sigma^2,$$

so that $\hat{\sigma}^2$ not consistent, as Neyman and Scott (1948) pointed out. What has happened is that the small-sample bias of ML estimators (**18.14**) persists in this example as n increases, for the number of parameters also increases with n.

(c) Finally, suppose that we interchange the roles of mean and variance in (a), and consider $n = 2$ independent normal observations x_i with the same mean θ and variances $\sigma_i^2, i = 1, 2$. As in (a), we see that the ML estimator does not exist by putting $\theta = x_i$ and letting $\sigma_i^2 \to \infty$ for either value of i. Here, however, we can estimate not only θ (say, by $t = \frac{1}{2}(x_1 + x_2)$), but also the σ_i^2, for

$$E\{x_1(x_1 - x_2)\} = \sigma_1^2, \qquad E\{x_2(x_2 - x_1)\} = \sigma_2^2.$$

The information about the common mean contained in the other observation permits us to estimate each variance from only a single observation. These estimators are not free from problems, for one of them must be negative if x_1 and x_2 have the same sign, but they illustrate that all the parameters may be estimated even when the ML method will estimate none of them. It is worth adding that we may unbiasedly estimate $\sigma_1^2 + \sigma_2^2$, which is 4 var t, by $(x_1 - x_2)^2$, and there is clearly no difficulty here.

The situation is not essentially changed if $n > 2$, as the reader may verify.

Careful investigation of the properties of ML estimators is necessary in non-standard situations – it cannot be assumed that the large-sample optimum properties will persist.

Marginal and conditional likelihood

18.32 We may correct the problem that arises in Example 18.16 by considering the variates $y_i = x_{i1} - x_{i2}$, which are $N(0, 2\sigma^2)$; the ML estimator is $\hat{\sigma}^2 = \sum y_i^2/2n$, using the likelihood given by $(y_1, \ldots, y_n)$.

We now assume that x may be partitioned into (y, z) by some transformation that is independent of the unknown parameters. The joint p.d.f. of (y, z) may be written as

$$f(y, z|\theta, \phi) = f_{z|y}(z|y, \theta, \phi) f_y(y|\theta, \phi). \tag{18.91}$$

When f_y does not depend on ϕ, we may define the *marginal likelihood* as $L(y|\theta) = f_y(y|\theta)$. That is, we operate as though only y had been observed. The use of the marginal likelihood may entail some loss of information, but it may provide sensible estimators in situations where the ML estimator is unsatisfactory.

Marginal likelihood and the conditional form (in **18.33**) were introduced by Bartlett (1936b; 1937); the theory is developed by Andersen (1970b) and Kalbfleisch and Sprott (1970).

18.33 If we make the same transformation $x \to (y, z)$, we may define the conditional likelihood for z given y, when $f_{z|y}$ does not depend on ϕ, as

$$L(z|y, \theta) = f_{z|y}(z|y, \theta).$$

Again, there may be a loss of information, but the benefit comes in terms of a simpler analysis.

Example 18.17 (Conditional likelihood for a binary data model (Cox and Hinkley, 1974, Chapter 9))

Let (x_{i1}, x_{i2}) be pairs of binary random variables, $i = 1, \dots, n$ such that $\theta = (\gamma, \beta)$ and

$$P(x_{i1} = 1) = \frac{\gamma\beta_j}{1 + \gamma\beta_j}, \quad P(x_{i2} = 1) = \frac{\beta_j}{1 + \beta_j}.$$

If we condition upon $y_i = x_{i1} + x_{i2}$ and set $z_i = x_{i1}$, we have

$$P(z_i = 1|y_i = 0) = 0, \quad P(z_i = 1|y_i = 2) = 1,$$
$$P(z_i = 1|y_i = 1) = \beta/(1 + \beta),$$

so that the conditional likelihood becomes

$$L(z|y, \theta) = L(z|y, \beta) = \beta^Y/(1 + \beta)^n,$$

where $Y = y_1 + \cdots + y_n$. Clearly, the conditional estimator is $\hat{\beta} = Y/(n - Y)$, provided $0 < Y < n$.

18.34 A number of other estimation procedures exist with the label 'likelihood' attached. In particular, we briefly examine the *profile likelihood* in **18.47** and *pseudo-likelihood* in **18.58**. We defer discussion of *restricted maximum* likelihood (REML) to Chapter 32 as its primary application (see **32.19**) is to the general linear model.

Dependent observations

18.35 Hitherto, we have assumed the observations to be independent. However, in both stochastic processes and time series analysis, dependence between successive observations is assumed. That is, the log likelihood becomes

$$\log L = \sum_{j=1}^{n} \log f_j(x_j|\theta, \mathbf{X}_{(j-1)})$$
$$= \sum l_j, \tag{18.92}$$

say, where $\mathbf{X}_{(j-1)} = (x_1, \dots, x_{j-1})$ denotes the previous history. $\mathbf{X}_{(j-1)}$ might also include starting values $(x_0, x_{-1}, \dots)$ but we define $\log L$ conditional upon these. Provided the number of initial values is finite, the asymptotic arguments are not affected.

The ML estimator for θ remains consistent under the conditions given in **18.16** plus the requirement that

$$\lim_{|t|\to\infty} \operatorname{cov}(l_j, l_{j+t}) = 0; \tag{18.93}$$

that is, we require the dependence between observations to diminish sufficiently quickly. Furthermore, asymptotic normality can be demonstrated under the additional conditions that

$$\lim_{|t|\to\infty} \operatorname{cov}\left[\frac{\partial l_j}{\partial\theta}, \frac{\partial l_{j+t}}{\partial\theta}\right] = 0 \tag{18.94}$$

and

$$\lim_{n\to\infty} \frac{1}{2} \sum_{j=1}^{n} \frac{\partial^2 l_j}{\partial\theta^2} > 0. \tag{18.95}$$

When θ is a vector, the matrix in (18.95) must be non-singular. These conditions can be stated more formally in terms of strong mixing conditions; see Lehmann (1983b).

Example 18.18 (ML estimation for a Markov chain)

Consider the two-state Markov chain with transition probabilities

$$P(X_j = r | X_{j-1} = s) = q_{sr}, \qquad r = 0, 1; \; s = 0, 1,$$

usually written as the matrix of transition probabilities

$$Q = \begin{bmatrix} q_{00} & q_{01} \\ q_{10} & q_{11} \end{bmatrix}. \tag{18.96}$$

Let

$$Z_j(s, r) = \begin{cases} 1, & \text{if } X_j = r, X_{j-1} = s \\ 0, & \text{otherwise,} \end{cases}$$

so that the log likelihood (18.92) is

$$\log L = \sum_{j=1}^{n} \left(\sum \sum Z_j(s, r) \log q_{sr} \right) \tag{18.97}$$

which reduces to

$$Z(0, 0) \log q_{00} + Z(0, 1) \log q_{01} + Z(1, 0) \log q_{10} + Z(1, 1) \log q_{11}, \tag{18.98}$$

where

$$Z(s, r) = \sum_{j=1}^{n} Z_j(s, r).$$

It is clear from (18.98) that the $Z(s, r)$ are jointly sufficient for the q_{sr}. Further, it follows directly that the ML estimators are

$$\hat{q}_{s1} = \frac{Z(s, 1)}{Z(s, 0) + Z(s, 1)}, \qquad s = 0, 1. \tag{18.99}$$

The covariances in (18.93) and (18.94) both depend upon the covariances between $Z_j(s, r)$ and $Z_{j+t}(s', r')$. By way of example we consider $s = s' = 1$, $r = r' = 1$. It suffices to consider $j = 1$ and to condition on the initial event $X_0 = 1$. Then

$$\begin{aligned} P[Z_1(1, 1) = 1] &= q_{11} \\ P[Z_{t+1}(1, 1) = 1] &= P(X_{t+1} = 1 | X_t = 1) P(X_t = 1 | X_0 = 1) \\ &= q_{11} q_{11}^{(t)} \end{aligned}$$

where $q_{11}^{(t)}$ is the (1, 1)th element of Q^t. Finally,

$$\begin{aligned} P[Z(1,1) = 1, Z_{t+1}(1,1) = 1] \\ &= P(X_{t+1} = 1|X_t = 1)P(X_t = 1|X_1 = 1)P(X_t = 1|X_0 = 1) \\ &= q_{11}^2 q_{11}^{(t-1)}, \end{aligned}$$

so that the covariance is

$$q_{11}^2[q_{11}^{(t-1)} - q_{11}^{(t)}].$$

It is readily established that this tends to zero as $t \to \infty$ provided that $0 < q_{sr} < 1$, for all s, r; that is, the Markov chain must be *ergodic*. Finally, we have from (18.98) that

$$\frac{\partial^2 \log L}{\partial q_{s1}^2} = \frac{-Z(s,0)}{(1-q_{s1})^2} - \frac{Z(s,1)}{q_{s1}^2}, \qquad s = 0, 1,$$

so that, conditionally upon $Z(s,0) + Z(s,1) = n_s$,

$$-E\left(\frac{\partial^2 \log L}{\partial q_{s1}^2}\right) = \frac{n_s}{q_{s1}(1-q_{s1})}.$$

Since $\partial^2 \log L/\partial q_{01}\,\partial q_{11} = 0$, conditions (18.95) are satisfied and $\hat{q}_{s1}$ is asymptotically $N(q_{s1}, q_{s1}(1-q_{s1})/n_s)$. (For a general discussion of estimation in stochastic processes, see Billingsley (1961) and for time series, Fuller (1995).)

ML estimation of location and scale parameters

18.36 We may use ML methods to solve, following Fisher (1921a), the problem of finding efficient estimators of location and scale parameters for any given form of distribution.

Consider a density function

$$f(x) = \frac{1}{\beta} f\left(\frac{x-\alpha}{\beta}\right), \qquad \beta > 0. \qquad (18.100)$$

The parameter α locates the distribution and β is a scale parameter. We may rewrite (18.100) as

$$f(y) = \exp\{g(y)\}, \qquad (18.101)$$

where

$$y = (x-\alpha)/\beta, \quad g(y) = \log f(y).$$

In samples of size n, the LF is

$$\log L(x|\alpha,\beta) = \sum_{i=1}^{n} g(y_i) - n\log\beta, \qquad (18.102)$$

yielding the likelihood equations

$$\begin{aligned}\frac{\partial \log L}{\partial \alpha} &= -\tfrac{1}{\beta}\textstyle\sum_i g'(y_i) = 0,\\ \frac{\partial \log L}{\partial \beta} &= -\tfrac{1}{\beta}\left\{\textstyle\sum_i y_i g'(y_i) + n\right\} = 0,\end{aligned} \tag{18.103}$$

where $g'(y) = \partial g(y)/\partial y$. Under regularity conditions, solution of (18.103) gives the ML estimators $\hat{\alpha}$ and $\hat{\beta}$.

If, in $\hat{y} = (x - \hat{a})/\hat{\beta}$, we replace x by $(x - \alpha)/\beta$, $\hat{\alpha}$ by $(\hat{\alpha} - \alpha)/\beta$ and $\hat{\beta}$ by $\hat{\beta}/\beta$, $\hat{y}$ is unaffected. Hence, whatever the true values (α, β), the solutions of (18.103) will be the same. Thus the joint distribution of the variables $((\hat{\alpha} - \alpha)/\beta, \hat{\beta}/\beta)$, and that of their ratio $(\hat{\alpha} - \alpha)/\hat{\beta}$, cannot depend upon the parameters (α, β) – cf. Haas *et al.* (1970) – but can be functions of n alone. This gives a direct method of removing the biases in $\hat{\alpha}$ and $\hat{\beta}$, for if

$$E\left(\frac{\hat{\alpha} - \alpha}{\beta}\right) = a(n), \quad E\left(\frac{\hat{\beta}}{\beta}\right) = b(n),$$

we find at once

$$E\left(\frac{\hat{\beta}}{b(n)}\right) = \beta, \quad E\left(\hat{\alpha} - \frac{a(n)\hat{\beta}}{b(n)}\right) = \alpha.$$

We now assume that for all permissible values of α and β, (17.18) holds, so that

$$\begin{aligned}E\left(\frac{\partial \log L}{\partial \alpha}\right) &= -\tfrac{n}{\beta}E\{g'(y)\} = 0,\\ E\left(\frac{\partial \log L}{\partial \beta}\right) &= -\tfrac{n}{\beta}[E\{yg'(y)\} + 1] = 0,\end{aligned} \tag{18.104}$$

and that (17.19) and its generalization (18.57) also hold. We may rewrite (18.104) as

$$E\{g'(y)\} = 0, \tag{18.105}$$

$$E\{yg'(y)\} = -1. \tag{18.106}$$

We now evaluate the elements of the inverse covariance matrix (18.60). From (18.103), dropping the argument of $g(y)$, we have

$$\begin{aligned}E\left(\frac{\partial^2 \log L}{\partial \alpha^2}\right) &= \frac{n}{\beta^2}E(g''),\\ E\left(\frac{\partial^2 \log L}{\partial \beta^2}\right) &= \frac{n}{\beta^2}E(g''y^2 + 2g'y + 1),\end{aligned} \tag{18.107}$$

which on using (18.106) becomes

$$E\left(\frac{\partial^2 \log L}{\partial \beta^2}\right) = \frac{n}{\beta^2}E(g''y^2 - 1). \tag{18.108}$$

Also

$$E\left(\frac{\partial^2 \log L}{\partial\alpha\,\partial\beta}\right) = \frac{n}{\beta^2}E(g' + g''y),$$

which from (18.105) becomes

$$E\left(\frac{\partial^2 \log L}{\partial\alpha\,\partial\beta}\right) = \frac{n}{\beta^2}E(g''y). \tag{18.109}$$

Equations (18.107)–(18.109) give, for the matrix (18.60),

$$\mathbf{V}^{-1} = -\frac{n}{\beta^2}E\begin{pmatrix} g'' & g''y \\ g''y & g''y^2 - 1 \end{pmatrix}, \tag{18.110}$$

from which the variances and covariance may be determined by inversion. Of course, if α or β alone is being estimated, the variance of the ML estimator will be the reciprocal of the appropriate term in the leading diagonal of $\mathbf{V}^{-1}$ and, by the first sentence of Exercise 17.20, cannot be larger than when both parameters are being estimated.

18.37 If $g(y)$ is an even function of its argument, i.e. the distribution is symmetric about α, (18.110) simplifies. For then

$$\begin{aligned} g(y) &= g(-y), \\ g'(y) &= -g'(-y), \\ g''(y) &= g''(-y). \end{aligned} \tag{18.111}$$

Using (18.111), we see that the off-diagonal term in (18.110)

$$E\{g''(y)y\} = 0, \tag{18.112}$$

so that (18.110) is a diagonal matrix for symmetric distributions. Hence the ML estimators of the location and scale parameters of any symmetric distribution obeying our regularity conditions will be asymptotically uncorrelated, and (since they are asymptotically bivariate normally distributed by **18.26**) asymptotically *independent*. In particular, this applies to the normal distribution, for which we have already derived this result directly in Example 18.13.

18.38 Even for asymmetrical distributions, we can make the off-diagonal term in (18.110) zero by a simple change of origin. Put

$$z = y - \frac{E(g''y)}{E(g'')}. \tag{18.113}$$

Then

$$\begin{aligned} E(g''y) &= E\left\{g''\left[z + \frac{E(g''y)}{E(g'')}\right]\right\} \\ &= E(g''z) + E(g''y), \end{aligned}$$

so that

$$E(g''z) = 0.$$

Thus if instead of y we use z as in (18.113), we reduce (18.110) to a diagonal matrix, and we obtain the variances of the estimators easily by taking the reciprocals of the terms in the diagonal.

The variance of $\hat{\beta}$ is unaffected by the change of origin since β is a scale parameter. The origin that makes the estimators uncorrelated is called the *centre of location* of the distribution. A consequence of choosing the centre of location as origin is that, where iterative procedures are necessary, the estimation process may be simpler. Parameters whose estimators are uncorrelated (orthogonal) are sometimes called *orthogonal* parameters.

Example 18.19 (Orthogonal parameters for the gamma distribution)

The gamma distribution

$$f(x) = \frac{1}{\beta\Gamma(p)}\left(\frac{x-\alpha}{\beta}\right)^{p-1}\exp\left\{-\frac{(x-\alpha)}{\beta}\right\}, \qquad \alpha \leq x < \infty;\ \beta > 0;\ p > 2,$$

has its range dependent upon α, but is zero and has a zero first derivative with respect to α at its lower terminal for $p > 2$ (cf. Exercise 17.22), so our regularity conditions hold. Here

$$g(y) = -\log\Gamma(p) + (p-1)\log y - y,$$

and

$$E(g'') = E\left\{-\frac{(p-1)}{y^2}\right\} = -\frac{1}{(p-2)},$$
$$E(g''y) = E\left\{-\frac{(p-1)}{y}\right\} = -1,$$
$$E(g''y^2) = E\{-(p-1)\} = -(p-1),$$

Thus the centre of location is, from (18.113),

$$\frac{E(g''y)}{E(g'')} = p-2.$$

The inverse covariance matrix (18.110) is

$$\mathbf{V}^{-1} = \frac{n}{\beta^2}\begin{pmatrix} \frac{1}{p-2} & 1 \\ 1 & p \end{pmatrix}, \qquad (18.114)$$

and its inverse, the covariance matrix of $\hat{\alpha}$ and $\hat{\beta}$, is easily obtained directly as

$$\mathbf{V} = \frac{(p-2)\beta^2}{2n}\begin{pmatrix} p & -1 \\ -1 & \frac{1}{p-2} \end{pmatrix}. \qquad (18.115)$$

If we measure from the centre of location, we have for the uncorrelated estimators $\hat{\alpha}_u$, $\hat{\beta}_u$,

$$\operatorname{var}\hat{\alpha}_u = (p-2)\beta^2/n, \qquad \operatorname{var}\hat{\beta}_u = \beta^2/(2n).$$

Comparing these results, we see that var $\hat{\beta}$ is unaffected by the change of origin, while var $\hat{\alpha}_u$ is reduced from its value in (18.115) to what is seen from (18.114) to be its value when α alone is being estimated. If β alone is estimated, (18.114) shows that var $\hat{\beta}$ is smaller than when both parameters are unknown.

Cox and Reid (1987) consider the general (i.e. not only the location-scale) problem of choosing a parametrization so that the ML estimators of the parameters of interest, θ, are uncorrelated with those of unwanted ('nuisance') parameters ψ. In general, this cannot be achieved for all values of the parameters, but it can be if θ is a scalar, the case they discuss in detail.

Estimation with misspecified models

18.39 Thus far, we have always assumed the form of the density function to be known except, of course, for its parameters. However, it is often the case that we do not have such information (a question we examined briefly in **18.18**). We now consider this issue in greater detail for the multiparameter case.

We assume that the true but unknown density, $g(x)$, where x may be scalar- or vector-valued, is well behaved and that $E[\log g]$ exists. We shall use the LF based upon the assumed density function $f(x, \boldsymbol{\theta})$, which is also taken to be well behaved, with bounded first and second derivatives with respect to the elements of $\boldsymbol{\theta}$. Finally, as usual, we assume that $\boldsymbol{\theta}_0$, the true value of $\boldsymbol{\theta}$, lies in the interior of the parameter space.

Under these conditions, which are stated formally in the original development by White (1981, 1982), it may be shown that the resulting estimator, $\boldsymbol{\theta}^*$ say, based on the LF for f, is asymptotically normal, or

$$n^{1/2}(\boldsymbol{\theta}^* - \boldsymbol{\theta}_0) \sim N[\mathbf{0}, \mathbf{C}],$$

where $\mathbf{C} = \mathbf{A}^{-1}\mathbf{B}\mathbf{A}^{-1}$ is a function of $\boldsymbol{\theta}_0$ defined by

$$\mathbf{A} \equiv \mathbf{A}(\boldsymbol{\theta}) = -E[\partial^2 \log f(x, \boldsymbol{\theta})/\partial\boldsymbol{\theta}\, \partial\boldsymbol{\theta}^T] \tag{18.116}$$

$$\mathbf{B} \equiv \mathbf{B}(\boldsymbol{\theta}) = E\{[\partial \log f(x, \boldsymbol{\theta})/\partial\boldsymbol{\theta}]\cdot[\partial \log f(x, \boldsymbol{\theta})/\partial\boldsymbol{\theta}^T]\}. \tag{18.117}$$

When $g \equiv f$, $\mathbf{B} = \mathbf{A}$ and the usual result given by (18.60) holds.

Example 18.20 (Estimation with a misspecified model (White, 1982))

We take $f(x, \boldsymbol{\theta})$ to be the normal p.d.f., with $\boldsymbol{\theta} = (\mu, \sigma^2)$ as usual; g is unknown. The ML estimators are as in Example 18.11 and from (18.80)–(18.83) we have, from (18.116),

$$\mathbf{A}(\boldsymbol{\theta}) = \begin{bmatrix} \sigma^{-2} & 0 \\ 0 & 2\sigma^{-4} \end{bmatrix}.$$

After some simplification, we obtain from (18.117)

$$\mathbf{B}(\boldsymbol{\theta}) = \begin{bmatrix} \sigma^{-2} & \gamma_1\sigma^{-3}/2 \\ \gamma_1\sigma^{-3}/2 & (\gamma_2+2)\sigma^{-4}/4 \end{bmatrix}$$

where $\gamma_1 = \mu_3/\sigma^3$ and $\gamma_2 = (\mu_4/\sigma^4) - 3$ are the usual skewness and kurtosis measures. Hence

$$\mathbf{C}(\boldsymbol{\theta}) = \mathbf{A}^{-1}\mathbf{B}\mathbf{A}^{-1} = \begin{bmatrix} \sigma^2 & \sigma^3\gamma_1 \\ \sigma^3\gamma_1 & (\gamma_2+2)\sigma^4 \end{bmatrix}.$$

The variance of $\hat{\mu}$ is unchanged and that of $\hat{\sigma}^2$ is consistent with the form for var $\hat{\sigma}$ given in Example 18.8 earlier. If g is normal, $\gamma_1 = \gamma_2 = 0$ and the usual form of **C** results. In general, we would estimate **C** using the higher-order sample moments.

Efficiency of the method of moments

18.40 In Chapter 6 we discussed distributions of the Pearson family. We were there mainly concerned with the properties of populations only and no question of the reliability of estimates arose. If, however, the observations are a *sample* from a population, the question arises whether fitting by moments provides the most efficient estimators of the unknown parameters. As we shall see presently, in general it does not.

Consider a distribution dependent on four parameters. If the ML estimators of these parameters are to be obtained in terms of linear functions of the moments (as in the fitting of Pearson curves), we must have

$$\frac{\partial \log L}{\partial \theta_r} = a_0 + a_1 \sum x + a_2 \sum x^2 + a_3 \sum x^3 + a_4 \sum x^4, \qquad r = 1, \ldots, 4, \qquad (18.118)$$

and consequently

$$f(x|\theta_1, \ldots, \theta_4) = \exp(b_0 + b_1 x + b_2 x^2 + b_3 x^3 + b_4 x^4), \qquad (18.119)$$

where the bs depend on the θs. This is the most general form for which the method of moments gives ML estimators. The bs are, of course, conditioned by the requirement that $f(x)$ is a valid p.d.f.

Without loss of generality we may take $b_1 = 0$. If, then, b_3 and b_4 are zero, the distribution is normal and the method of moments is efficient. In other cases, (18.119) does not yield a Pearson distribution except as an approximation. For example, consider

$$\frac{\partial \log f}{\partial x} = 2b_2 x + 3b_3 x^2 + 4b_4 x^3. \qquad (18.120)$$

If b_3 and b_4 are small, this is approximately

$$\frac{\partial \log f}{\partial x} = \frac{2b_2 x}{1 - \frac{3b_3}{2b_2}x - \frac{2b_4}{b_2}x^2},$$

which is one form of the equation defining Pearson distributions (cf. (6.1)). Only when b_3 and b_4 are small compared with b_2 can we expect the method of moments to give estimators of high efficiency.

18.41 A detailed discussion of the efficiency of moments in determining the parameters of a Pearson distribution has been given by Fisher (1921a). Exercise 18.16 deals with the Type IV distribution and Exercises 18.26–28 with some discrete distributions. As an illustration, we discuss the three-parameter gamma (Type III) distribution.

Example 18.21 (Method of moments for the gamma distribution)

Consider the gamma distribution with three parameters, α, σ, p,

$$f(x) = \frac{1}{\sigma\Gamma(p)}\left(\frac{x-\alpha}{\sigma}\right)^{p-1} \exp\left\{-\left(\frac{x-\alpha}{\sigma}\right)\right\}, \qquad \alpha \leq x < \infty;\ \sigma > 0;\ p > 2.$$

For the LF, we have

$$\log L = -np\log\sigma - n\log\Gamma(p) + (p-1)\sum\log(x-\alpha) - \sum(x-\alpha)/\sigma.$$

The three likelihood equations are

$$\begin{aligned}
\frac{\partial \log L}{\partial \alpha} &= -(p-1)\sum\frac{1}{(x-\alpha)} + n/\sigma = 0,\\
\frac{\partial \log L}{\partial \sigma} &= -np/\sigma + \sum(x-\alpha)/\sigma^2 = 0,\\
\frac{\partial \log L}{\partial p} &= -n\log\sigma - n\frac{\mathrm{d}}{\mathrm{d}p}\log\Gamma(p) + \sum\log(x-\alpha) = 0.
\end{aligned}$$

Taking the parameters in the above order, the inverse covariance matrix (18.60) is

$$\mathbf{V}^{-1} = n\begin{pmatrix} \dfrac{1}{\sigma^2(p-2)} & \dfrac{1}{\sigma^2} & \dfrac{1}{\sigma(p-1)} \\ \dfrac{1}{\sigma^2} & \dfrac{p}{\sigma^2} & \dfrac{1}{\sigma} \\ \dfrac{1}{\sigma(p-1)} & \dfrac{1}{\sigma} & \dfrac{\mathrm{d}^2\log\Gamma(p)}{\mathrm{d}p^2} \end{pmatrix}.$$

The leading (2×2) sub-matrix of $\mathbf{V}^{-1}$ is, of course, (18.114) with σ written for β. If we multiply the determinant by n^{-3}, we may write

$$|\mathbf{V}^{-1}|/n^3 = \Delta = \frac{1}{(p-2)\sigma^4}\left\{2\frac{\mathrm{d}^2\log\Gamma(p)}{\mathrm{d}p^2} - \frac{2}{p-1} + \frac{1}{(p-1)^2}\right\}.$$

From this the sampling variances are found to be

$$\begin{aligned}
\operatorname{var}\hat{a} &= \frac{1}{n\Delta\sigma^2}\left\{p\frac{\mathrm{d}^2\log\Gamma(p)}{\mathrm{d}p^2} - 1\right\},\\
\operatorname{var}\hat{\sigma} &= \frac{1}{n\Delta\sigma^2}\left\{\frac{1}{p-2}\frac{\mathrm{d}^2\log\Gamma(p)}{\mathrm{d}p^2} - \frac{1}{(p-1)^2}\right\},\\
\operatorname{var}\hat{p} &= \frac{2}{n\Delta(p-2)\sigma^4} = \frac{2}{n}\Big/\left\{2\frac{\mathrm{d}^2\log\Gamma(p)}{\mathrm{d}p^2} - \frac{2}{p-1} + \frac{1}{(p-1)^2}\right\}.
\end{aligned} \qquad (18.121)$$

Now for large p, using Stirling's series (3.63), we find when p is large

$$2\frac{\mathrm{d}^2}{\mathrm{d}p^2}\log\Gamma(1+p) - \frac{2}{p} + \frac{1}{p^2} = \frac{1}{3}\left\{\frac{1}{p^3} - \frac{1}{5p^5} + \frac{1}{7p^7} - \cdots\right\}$$

and hence approximately, from (18.121),

$$\operatorname{var} \hat{p} \doteq \frac{6}{n}\left\{(p-1)^3 + \frac{1}{5}(p-1)\right\}. \tag{18.122}$$

If instead we estimate the parameters by equating sample moments to the population moments in terms of parameters, we find

$$\begin{aligned} \alpha + \sigma p &= m_1, \\ \sigma^2 p &= m_2, \\ 2\sigma^3 p &= m_3, \end{aligned}$$

so that we shall have in the sample

$$b_1 = m_3^2/m_2^3 = 4/p, \tag{18.123}$$

which equates the sample skewness coefficient to its population value $\beta_1 = 4/p$. Now, from Exercise 10.26 and (10.14), we have

$$\operatorname{var} b_1 = \frac{4\beta_1}{n}\left\{\frac{\mu_6}{\mu_2^3} - 6\beta_2 + 9 + \frac{\beta_1}{4}(9\beta_2 + 35) - \frac{3\mu_5\mu_3}{\mu_2^4}\right\},$$

which, using (16.5) with $\nu = 2p$, reduces to

$$\operatorname{var} b_1 = \frac{96(p+1)(p+5)}{np^3}. \tag{18.124}$$

Hence, from (18.123) and (10.14), we have for $\tilde{p}$ the estimator by the method of moments,

$$\operatorname{var} \tilde{p} \sim \left(\frac{4}{\beta_1^2}\right)^2 \operatorname{var} b_1 = \frac{6}{n} p(p+1)(p+5).$$

For large p the efficiency of this estimator is then, from (18.122),

$$\frac{\operatorname{var} \hat{p}}{\operatorname{var} \tilde{p}} = \frac{\{(p-1)^3 + \frac{1}{5}(p-1)\}}{p(p+1)(p+5)},$$

which is evidently less than 1. When p exceeds 39.1 ($\beta_1 = 0.102$), the efficiency is over 80 per cent. For $p = 20$ ($\beta_1 = 0.20$), it is 65 per cent. For $p = 5$, a more exact calculation based on the tables of the trigamma function $\mathrm{d}^2 \log \Gamma(1+p)/\mathrm{d}p^2$ shows that the efficiency is only 22 per cent.

Bowman and Shenton (1988) study the ML estimators of the gamma distribution in detail. Example 18.17 treated the case where p is known, and Exercise 18.15 treats the case $\alpha = 0$, while Exercise 17.1 in effect considered two single-parameter cases.

Order restrictions on parameters: isotonic estimation

18.42 We have assumed since **18.22** that the admissible values of any parameter do not depend on those of another. However, problems do arise in which we know that the parameters occur in a certain order, e.g. in the simplest two-parameter case that $\theta_1 \leq \theta_2$, and we must require their estimators to satisfy the same order restrictions.

Example 18.22 (Estimation with order restrictions on the mean)

Independent random samples of equal size n are taken from two normal distributions with the same known variance (say, unity) and means θ_1, θ_2 where $\theta_1 \leq \theta_2$. The LF to be maximized is given by

$$\log L = \text{constant} - \frac{1}{2}\left\{\sum_{j=1}^{n}(x_{1j} - \theta_1)^2 + \sum_{j=1}^{n}(x_{2j} - \theta_2)^2\right\},$$

which on factorizing the sums of squares is

$$\log L = \text{constant} - \tfrac{1}{2}n\{(\bar{x}_1 - \theta_1)^2 + (\bar{x}_2 - \theta_2)^2\}.$$

We thus require to minimize the function in braces for choice of θ_1, θ_2, subject to the order restriction $\theta_1 \leq \theta_2$. Clearly the unconditional minimum ignoring this order restriction is

$$\theta_1 = \bar{x}_1, \quad \theta_2 = \bar{x}_2,$$

so that if $\bar{x}_1 \leq \bar{x}_2$ we have the ML estimators

$$\hat{\theta}_1 = \bar{x}_1, \quad \hat{\theta}_2 = \bar{x}_2, \quad \bar{x}_1 \leq \bar{x}_2. \tag{18.125}$$

However, if $\bar{x}_1 > \bar{x}_2$, this solution breaches the order restriction. Since $\log L$ is constant on any circle with centre $(\bar{x}_1, \bar{x}_2)$ and decreases as the circle's radius increases, the circle that touches the line $\theta_1 = \theta_2$ will have a single point in the region $\theta_1 \leq \theta_2$ and the maximum value of $\log L$ in that region. It is easy to see that it touches at the point $(\frac{1}{2}(\bar{x}_1 + \bar{x}_2), \frac{1}{2}(\bar{x}_1 + \bar{x}_2))$. Thus the ML estimators are

$$\hat{\theta}_1 = \hat{\theta}_2 = \tfrac{1}{2}(\bar{x}_1 + \bar{x}_2), \qquad \bar{x}_1 > \bar{x}_2 \tag{18.126}$$

and (18.125)–(18.126) are the complete ML solution.

18.43 In Example 18.22, the ML solution is an intuitively acceptable one: if $\bar{x}_1$ and $\bar{x}_2$ are in the wrong order, we pool and average them. If we now generalize to three normal distributions, with the order restriction

$$\theta_1 \leq \theta_2 \leq \theta_3,$$

the analogous geometrical representation is of $\log L$ constant on spheres centred at $(\bar{x}_1, \bar{x}_2, \bar{x}_3)$ and decreasing as the radius increases. When $\bar{x}_1 \leq \bar{x}_2 \leq \bar{x}_3$, the ML estimator is $\theta_j = \bar{x}_j$, at the unconditional maximum of the LF as before. If $\bar{x}_1 > \bar{x}_2$ or $\bar{x}_2 > \bar{x}_3$ holds, the result (18.126) applies to the pair of θs involved, while the other θ is unaffected by the single inversion of order and may be estimated unconditionally as in (18.125). But if $\bar{x}_1 > \bar{x}_3$ we must also have $\bar{x}_1 > \bar{x}_2$ or $\bar{x}_2 > \bar{x}_3$ or both, and there is a new problem. We have to determine the sphere of smallest radius that has a point in the region $\theta_1 \leq \theta_2 \leq \theta_3$.

General methods of solving this type of problem are given by Barlow *et al.* (1972), who show that the ML solution can be found for the class of distributions

$$f(x) = \exp\{A(\theta) + A'(\theta)(x - \theta) + B(x)\}, \tag{18.127}$$

of which the normal distribution is a special case, by finding the *isotonic regression*[3] of the sample means (i.e. the unconditional ML estimators – cf. Exercise 18.2), which is the upper envelope of all convex functions lying below the cumulative graph of the sample means. Barlow *et al.* (1972) give algorithms for carrying out the computations – the pooling of adjacent inverted-order means in Example 18.22 and its generalization above correspond to the simplest examples of such a convex envelope. For recent developments in order restricted inference, see Robertson *et al.* (1988).

ML estimation for incomplete data: the EM algorithm

18.44 Thus far, we have assumed the data set to be complete, that is, none of the observations is missing. Yet there are many reasons why particular values may not be available: responses to a questionnaire may be omitted, experiments may fail or the records get mixed up, values may be reported incorrectly or not at all or the data may be abridged in some way (censored, truncated or grouped). In addition, it may sometimes be desirable to consider observations that, were they to be available, would simplify the analysis, either by providing a more regular structure (in an unbalanced experiment design) or by simplifying algebraic or numerical procedures. Whether such observations could be made or whether the investigator planned to make them is of no consequence. Example 18.23 and Exercise 18.42 illustrate these points. All we need to know is that the complete data set x is an augmented form of the observed, or incomplete, set y such that the density function for y may be determined from that for x, as a marginal p.d.f.; that is,

$$g(\mathbf{y}|\boldsymbol{\theta}) = \int f(\mathbf{x}|\boldsymbol{\theta})\, d\mathbf{x}, \tag{18.128}$$

where the integral is taken over $\{\mathbf{x} : \mathbf{y} = y(\mathbf{x})\}$ and is replaced by a sum when the variates are discrete. In general, there are many possible sets $\mathbf{x}$ in which we may choose to embed $\mathbf{y}$, but there may be a natural choice in some situations.

18.45 These problems may be treated in a unified way through the EM algorithm, first presented in full generality[4] by Dempster *et al.* (1977a). The algorithm, an iterative procedure, consists of two conceptually distinct steps at each stage of the iteration, those of expectation (E) and maximization (M). We assume that

$$Q(\boldsymbol{\theta}^*|\boldsymbol{\theta}) = E\{\log L(\mathbf{x}|\boldsymbol{\theta}^*)|\mathbf{y}, \boldsymbol{\theta}\} \tag{18.129}$$

exists for all $\mathbf{x}$ and $\boldsymbol{\theta}$, where L is the likelihood to be maximized. Let $\boldsymbol{\theta}_{(k)}$ denote the estimate of $\boldsymbol{\theta}$ obtained at the kth iteration. Then we update it as follows:

[3] 'Isotonic' is used here to mean 'order-preserving', since 'monotonic' has an ambiguity of direction; its earlier and etymological sense 'having the same tension' has particular reference to osmotic pressure.

[4] Dempster *et al.* trace the original notion back as far as McKendrick (1926).

E-step: Evaluate $Q(\boldsymbol{\theta}|\boldsymbol{\theta}_{(k)})$, which is the conditional expectation taken over the unknown (missing) elements of $\mathbf{x}$. This step may be performed algebraically, when feasible, or numerically.

M-step: Determine $\boldsymbol{\theta}_{(k+1)}$ as that value of $\boldsymbol{\theta}$ that maximizes $Q(\boldsymbol{\theta}|\boldsymbol{\theta}_{(k)})$.

The rationale is that we wish to maximize $\log L$ to find the ML estimator of $\boldsymbol{\theta}$. Since we cannot do this because of the incompleteness of the data we maximize instead the conditional expectation of $\log L$, given the observed $\mathbf{y}$ and the current estimate of $\boldsymbol{\theta}$. The process is illustrated in Example 18.23.

Example 18.23 (Use of the EM algorithm)

In Example 18.10, suppose that we now have the additional information that the first class, with observed frequency $a = 1997$, is actually the merger of two classes with unknown frequencies a_1 and a_2 and probabilities $p_{11} = \theta/4$, $p_{12} = \frac{1}{2}$ adding to the previous p_1. We now use the EM algorithm to evaluate the ML estimator $\hat{\boldsymbol{\theta}}$, which we know to be equal to 0.0357. We start from $\boldsymbol{\theta}_{(1)} = 0.0570$ as before.

Expression (18.129) gives, at the E-step,

$$\begin{aligned} &E\{\log L(x|\theta)|\mathbf{y},\theta_{(1)}\} \\ &= E\left\{\text{constant} + a_1 \log\frac{\theta_{(1)}}{4} + (1997 - a_1)\log\frac{1}{2}\middle| a = 1997\right\} \\ &+ (b+c)\log\frac{1-\theta_{(1)}}{4} + d\log\frac{\theta_{(1)}}{4} \end{aligned} \tag{18.130}$$

since b, c and d are components of y. Now clearly

$$E\{a_1|a = 1997\} = 1997 \times \frac{\theta_{(1)}/4}{\theta_{(1)}/4 + \frac{1}{2}} = 1997 \times \frac{0.0570}{2.025\,70} = 55.3373.$$

For the M-step, we have to maximize (18.130) for this value of a_1. Dropping terms free of θ, (18.130) becomes

$$(a_1 + d)\log\theta + (b + c)\log(1 - \theta).$$

Differentiating and equating to zero, we find, with $b = 906$, $c = 904$, $d = 32$

$$\theta_{(2)} = \frac{a_1 + d}{a_1 + d + b + c} = \frac{55.3373 + 32}{87.3373 + 1810} = 0.046\,03.$$

Successive further iterations give

$$\begin{aligned} \theta_{(3)} &= 0.040\,77 \\ \theta_{(4)} &= 0.038\,20 \\ \theta_{(5)} &= 0.036\,94 \\ \theta_{(6)} &= 0.036\,31 \\ \theta_{(7)} &= 0.036\,01 \\ \theta_{(8)} &= 0.035\,86 \\ \theta_{(9)} &= 0.035\,79 \\ \theta_{(10)} &= 0.035\,71. \end{aligned}$$

It has taken 10 iterations to reach the correct value, illustrating that convergence may be slow. The scoring method in Example 18.10 was much faster. Nevertheless, the generality of the method and advances in computing capability make the EM algorithm a very attractive option for many problems.

18.46 In addition to its generality, an attraction of the method is its simplicity, since it is a gradient method that requires only first derivatives. However, as we have seen in Example 18.23, this property implies that the rate of convergence is linear, rather than the quadratic rate achieved by the Newton–Raphson procedure in **18.21**. Modifications to improve the rate of convergence in particular cases were noted in the discussion in Dempster *et al.* (1977). A general procedure to speed up the algorithm is given in Louis (1982); this approach has been simplified and extended by Laird *et al.* (1987). Wu (1983) and Boyles (1983) discuss the convergence properties of the algorithm; when the function $Q(\phi, \theta)$ is continuous in both arguments, the procedure converges to a stationary point of the LF. However, a global maximum is guaranteed only when the LF is unimodal. As with other methods, the use of several sets of starting values is recommended to avoid problems with multiple local maxima.

Titterington (1984) discusses alternatives to the EM algorithm for incomplete data. Meng and Rubin (1993) break the maximization step down into several (computationally simpler) conditional maximizations, which yields a simpler analysis, may provide greater computational speed and yet achieve the same rate of convergence. Laird (1993) provides a survey of recent developments.

Monte Carlo maximum likelihood

18.47 For many problems involving dependent data, the likelihood function may be intractable. For example, when the n variates are binary, the normalizing constant has 2^n terms, rapidly becoming unmanageable as n increases. Geyer and Thompson (1992) develop a method known as *Monte Carlo maximum likelihood* (MCML) which provides a highly effective computational procedure to resolve such difficulties.

Suppose we have n observations, denoted by $\mathbf{x}$, drawn from a distribution in the exponential family as in (17.83); the parameter vector $\boldsymbol{\theta}$ contains $p \geq 1$ elements. Slightly modifying the argument of **5.48**, we see that the density function

$$f(\mathbf{x}|\boldsymbol{\theta}) = \exp\left[\sum \theta_j t_j(x) + C(\mathbf{x}) + D(\boldsymbol{\theta})\right], \tag{18.131}$$

taken in canonical form, yields the relationship:

$$K(\boldsymbol{\theta}) = \exp[D(\boldsymbol{\psi}) - D(\boldsymbol{\theta})] = \int \exp\left[\sum (\theta_j - \psi_j) t_j(\mathbf{x})\right] \mathrm{d}F(\mathbf{x}|\boldsymbol{\psi}), \tag{18.132}$$

where $\mathrm{d}F$ is used to denote the n-dimensional Lebesgue integral so that both continuous and discrete distributions are covered, and $\boldsymbol{\psi}$ is some convenient choice of parameter values. We may now generate N realizations from the distribution with density $f(\mathbf{z}|\boldsymbol{\psi})$ in (18.131). From (18.132), for a given value of $\boldsymbol{\theta}$, we may estimate $K(\boldsymbol{\theta})$ by

$$K_N(\boldsymbol{\theta}) = N^{-1} \sum_{i=1}^{N} \exp\left[\sum (\theta_j - \psi_j) t_j(z_j)\right]; \tag{18.133}$$

clearly K_N will be a consistent estimator for K provided successive realizations of $\mathbf{z}$ are not too strongly dependent. In fact, the key to the operation lies in setting up an ergodic Markov chain to generate successive realizations (see Geyer and Thompson, 1992). Thus, the approach has a strong similarity to Gibbs sampling which has become a very powerful tool in Bayesian analysis; see Geman and Geman (1984) and O'Hagan (1994, sections **8.48–58**) for further details.

18.48 From (18.131), the log LF may be written as

$$l(\mathbf{x}|\boldsymbol{\theta}) = \sum \theta_j t_j(\mathbf{x}) + D(\boldsymbol{\theta}) + C(\mathbf{x})$$

which has the same maximum and second-order properties as the modified function:

$$\begin{aligned} l(\mathbf{x}|\boldsymbol{\theta}) &= \sum \theta_j t_j(\mathbf{x}) + D(\boldsymbol{\theta}) - D(\boldsymbol{\psi}) \\ &= \sum \theta_j t_j(\mathbf{x}) - \log K(\boldsymbol{\theta}). \end{aligned} \tag{18.134}$$

MCML now proceeds iteratively as follows:

(i) Select a starting value, $\boldsymbol{\psi}$.
(ii) Generate N realizations from $f(\mathbf{z}|\boldsymbol{\psi})$.
(iii) Working with (18.133) and (18.134), estimate $K(\boldsymbol{\theta})$ by $K_N(\boldsymbol{\theta})$ for specified $\boldsymbol{\theta}$ values and determine the maximum of the log LF, given by $\boldsymbol{\theta}_{(r)}$ say.
(iv) Use $\boldsymbol{\theta}_{(r)}$ as the new value of $\boldsymbol{\psi}$ and repeat steps (ii) and (iii) until convergence is obtained.

Geyer and Thompson (1992) show that $\boldsymbol{\theta}_{(r)}$ will converge to the ML estimator $\hat{\boldsymbol{\theta}}$ under reasonable conditions. The following example serves to illustrate the process.

Example 18.24 (MCML estimation for the gamma distribution)

Suppose we have n observations from the gamma distribution with

$$f(x|\alpha) = x^{\alpha-1}e^{-x}/\Gamma(\alpha)$$

and log LF

$$l(\mathbf{x}|\alpha) = \alpha \sum \log x_i - n \log \Gamma(\alpha) - C(\mathbf{x}).$$

We select a starting value $\psi = \alpha_0$ and generate N observations from the corresponding gamma distribution (see Example 9.10 for details). The corresponding form of (18.133) is

$$K_N(\alpha) = N^{-1} \sum_{i=1}^{N} \exp[(\alpha - \alpha_0) \log z_i]$$

which may be used in (18.134) to determine α_1; the process is then repeated as needed.

18.49 Example 18.24 was chosen to illustrate the method in a straightforward context. The strength of MCML lies in dealing with dependent data when $D(\boldsymbol{\theta})$ is intractable. Geyer and Thompson (1992) recommend starting with a smaller value of N (e.g. 500) and then increasing N as the steps in $\boldsymbol{\theta}_{(r)}$ become smaller and greater numerical accuracy is required.

Finally, we note that the method is not restricted to the exponential family, but can work quite generally provided a suitable function K can be found and the ML estimator is well defined.

The use of the likelihood function

18.50 It is natural, as Fisher (1956) recommends, to examine the course of the LF throughout the permissible range of variation of θ, and to draw a graph of the LF. This may be generally informative, but is not of any immediate value in the estimation problem. The LF contains all the information in the sample, and is thus a comprehensive summary of the data, precisely in the sense that, as we remarked in **17.38**, the observations themselves constitute a set of jointly sufficient statistics for the parameters of any problem. This way of putting it has the merit of drawing attention to the fact that the functional form of the distribution(s) generating the observations must be specified before the LF can be used at all, whether for ML estimation or otherwise. In other words, some information (in a general sense) must be supplied by the statistician: if he or she is unable or unwilling to supply it, resort must be had to non-parametric methods; see, for example, Hettmansperger and McKean (1998).

Efron (1982) discusses the virtues of the *ML summary* $L(x|\hat{\theta})$ of the data, and distinguishes them from those of ML estimation that we have discussed.

Generally, if we partition the parameters into $(\boldsymbol{\beta}, \boldsymbol{\theta})$ and primary interest focuses upon estimation of $\boldsymbol{\theta}$, we may define the *profile likelihood* as

$$L(\mathbf{x}|\hat{\boldsymbol{\beta}}, \boldsymbol{\theta}) = \max_{\boldsymbol{\beta}} L(\mathbf{x}|\boldsymbol{\beta}, \boldsymbol{\theta}).$$

Typically $\boldsymbol{\theta}$ has small dimensionality and plots of the profile likelihood enable the investigator to get a feel for the form of the likelihood surface.

18.51 As we have seen earlier in this chapter, ML estimators are consistent, efficient, asymptotically unbiased and functions of the sufficient statistics. Thus, in most circumstances, we would prefer ML estimators to other options. However, circumstances arise when the ML estimators are not available:

(1) we are not prepared (or able) to specify the functional form of the likelihood function (LF);
(2) the LF is too complex (usually when observations are dependent);
(3) the combined effects of the size of the data set and the complexity of the LF make simpler methods adequate;
(4) ML estimators are inadequate in some way.

Advances in statistical computing constantly roll back the boundaries on (2) and (3), but interest in alternatives to ML is unlikely to wane. Finally, there may be alternatives to ML with similar properties, a topic we explore in **18.53**.

The method of least squares

18.52 The principle of least squares (LS) states that if we wish to estimate the parameter vector $\boldsymbol{\theta}$ from a set of n observations y_j, $j = 1, 2, \ldots, n$, which are related to $\boldsymbol{\theta}$ through their expectations

$$E(y_j) = g_j(\boldsymbol{\theta}), \tag{18.135}$$

then we should choose our estimator $\hat{\boldsymbol{\theta}}$ to minimize the sum of squared deviations between the y_j and their estimated expectations, i.e. to minimize

$$S = \sum_{j=1}^{n} \{y_j - g_j(\hat{\boldsymbol{\theta}})\}^2. \tag{18.136}$$

If the distributions of the y_j about their expectations are independently normal with the same variance σ^2, the LS estimator will be identical with the ML estimator of $\boldsymbol{\theta}$. Exercise 18.43 shows that the identity persists if the y_j are distributed in a sub-class of the exponential family (17.30), which includes the normal case. In general, however, the estimators will differ.

As with any other systematic principle of estimation, the acceptability of the LS method depends on the properties of the estimators to which it leads. Like the ML method, it has some asymptotic optimum properties but does not generally yield unbiased estimators. However, in an extremely important class of situations, it has the property, even in small samples, that it provides unbiased estimators, linear in the observations, that have minimum variance. This situation is usually described as the *linear model*, in which observations are distributed with constant variance about (possibly differing) mean values which are linear functions of the unknown parameters, and in which the observations are all uncorrelated in pairs. This subject will be studied in detail in Chapters 29–32.

Other methods of estimation

18.53 We have seen that, apart from the fact that they are functions of sufficient statistics for the parameters being estimated, the desirable properties of the ML estimators are all asymptotic, namely:

1. consistency;
2. asymptotic normality; and
3. efficiency.

As we saw in **17.29**, the ML estimator, $\hat{\theta}$, cannot be unique in the possession of these properties. For example, the addition to $\hat{\theta}$ of an arbitrary constant C/n^r will make no difference to its first-order properties if r is large enough. It is thus natural to enquire, as Neyman (1949) did, concerning the class of estimators which share the asymptotic properties of $\hat{\theta}$.

18.54 Suppose that we have s (≥ 1) samples, with n_i observations in the ith sample. As in **18.19**, we simplify the problem by supposing that each observation in the ith sample is classified into one of k_i mutually exclusive and exhaustive classes. If π_{ij} is the probability of an observation in the ith sample falling into the jth class, we therefore have

$$\sum_{j=1}^{k_i} \pi_{ij} = 1,$$

and we have reduced the problem to one concerning a set of s multinomial distributions. Let n_{ij} be the number of ith sample observations actually falling into the jth class, and $p_{ij} = n_{ij}/n_i$ the corresponding proportion. The probabilities π_{ij} are functions of a set of unknown parameters $(\theta_1, \ldots, \theta_r)$.

A continuous function T of the random variables p_{ij} is called a best asymptotically normal (BAN) estimator of θ_1, one of the unknown parameters, if:

1. $T(\{p_{ij}\})$ is consistent for θ_1;
2. T is asymptotically normal as $N = \sum_{i=1}^{s} n_i \to \infty$;
3. T is efficient; and
4. $\partial T/\partial p_{ij}$ exists and is continuous in p_{ij} for all i, j.

The first three of these conditions are precisely those we have already proved for the ML estimator. It is easily verified that the ML estimator also possesses the fourth property in this multinomial situation. Thus the class of BAN estimators contains the ML estimator as a special case.

18.55 Neyman showed that a set of necessary and sufficient conditions for an estimator to be BAN is that:

1. $T(\{\pi_{ij}\}) \equiv \theta_1$;
2. condition 4 of **18.54** holds; and
3. $\sum_{i=1}^{s}(1/n_i)\sum_{j=1}^{k_i}[(\partial T/\partial p_{ij})_{p_{ij}=\pi_{ij}}]^2\pi_{ij}$ is minimized for variation in $\partial T/\partial p_{ij}$.

Condition 1 is enough to ensure consistency: it is, in general, a stronger condition than consistency.[5] Since the statistic T is a continuous function of the p_{ij}, and the p_{ij} converge in probability to the π_{ij}, T converges in probability to $T(\{\pi_{ij}\})$, that is, to θ_1.

Condition 3 is simply the efficiency condition, for the function there to be minimized is simply the variance of T subject to the necessary condition for a minimum $\sum_j(\partial T/\partial p_{ij})_{p_{ij}=\pi_{ij}}\pi_{ij} = 0$.

As they stand, these three conditions are not of much practical value. However, Neyman also showed that a sufficient set of conditions is obtainable by replacing (3) by a direct condition on $\partial T/\partial p_{ij}$, which we shall not give here. From this he deduced that:

(a) the ML estimator is a BAN estimator, as we have already seen;
(b) the class of estimators known as minimum chi-square estimators are also BAN estimators.

We now proceed to examine this second class of estimators.

For later work on BAN estimators, see Bemis and Bhapkar (1983).

Minimum chi-square estimators

18.56 Referring to the situation described in **18.54**, a statistic T is called a minimum chi-square (MCS) estimator of θ_1, if it is obtained by minimizing, with respect to θ_1, the expression

$$\chi^2 = \sum_{i=1}^{s} n_i \sum_{j=1}^{k_i} \frac{(p_{ij}-\pi_{ij})^2}{\pi_{ij}} = \sum_{i=1}^{s} n_i \left(\sum_{j=1}^{k_i} \frac{p_{ij}^2}{\pi_{ij}} - 1 \right), \qquad (18.137)$$

[5] In fact condition 1 is the form in which consistency was originally defined by Fisher (1921a), as we mentioned in **17.7**.

where the π_{ij} are functions of $\theta_1, \ldots, \theta_r$. To minimize (18.137), we put

$$\frac{\partial \chi^2}{\partial \theta_1} = -\sum_i n_i \sum_j \left(\frac{p_{ij}}{\pi_{ij}}\right)^2 \frac{\partial \pi_{ij}}{\partial \theta_1} = 0, \tag{18.138}$$

and a root of (18.138), regarded as an equation in θ_1, is the MCS estimator of θ_1. Evidently, we may generalize (18.138) to a set of r equations to be solved jointly to find the MCS estimators of $\theta_1, \ldots, \theta_r$.

The procedure for finding MCS estimators is quite analogous to that for finding ML estimators. Moreover, the (asymptotic) properties of MCS estimators are similar to those of ML estimators. In fact, there is, with probability one, a unique consistent root of the MCS equations, and this corresponds to the absolute minimum value (infimum) of (18.137). The proofs are given, for the commonest case $s = 1$, by Rao (1957).

18.57 A modified form of MCS estimator is obtained by minimizing

$$(\chi')^2 = \sum_{i=1}^{s} n_i \sum_{j=1}^{k_i} \frac{(p_{ij} - \pi_{ij})^2}{p_{ij}} = \sum_i n_i \left(\sum_j \frac{\pi_{ij}^2}{p_{ij}} - 1\right) \tag{18.139}$$

instead of (18.137). In (18.139), we assume that no $p_{ij} = 0$. To obtain an estimator for θ_1, we solve

$$\frac{\partial (\chi')^2}{\partial \theta_1} = 2\sum_i n_i \sum_j \left(\frac{\pi_{ij}}{p_{ij}}\right) \frac{\partial \pi_{ij}}{\partial \theta_1} = 0. \tag{18.140}$$

These modified MCS estimators have also been shown to be BAN estimators by Neyman (1949).

Choice between methods

18.58 Since the ML, the MCS and the modified MCS methods all have the same asymptotic properties, the choice between them must rest, in any particular case, either on the grounds of computational convenience, or on those of superior sampling properties in small samples, or on both. As to the first grounds, there is little that can be said in general. Sometimes the ML, and sometimes the MCS, equation is the more difficult to solve. But when dealing with a continuous distribution, the observations *must* be grouped in order to make use of the MCS method, which will cause some loss of efficiency. Our own view is therefore that the traditional leaning towards ML estimation is fairly generally justifiable on computational grounds. The following example illustrates the MCS computational procedure in a relatively simple case.

Example 18.25 (MCS estimation for the Poisson)

Consider the estimation, from a single sample of n observations, of the parameter θ of a Poisson distribution. We have seen (Examples 17.8, 17.15) that the sample mean $\bar{x}$ is an MVB sufficient estimator of θ, and it follows from **18.5** that $\bar{x}$ is also the ML estimator.

The MCS estimator of θ, however, is not equal to $\bar{x}$, illustrating the point that MCS methods do not necessarily yield a single sufficient statistic if one exists.

The theoretical probabilities here are

$$\pi_j = \mathrm{e}^{-\theta}\theta^j/j!, \qquad j = 0, 1, 2, \ldots,$$

so that

$$\frac{\partial \pi_j}{\partial \theta} = \pi_j \left(\frac{j}{\theta} - 1 \right).$$

The minimizing equation (18.138) is therefore, dropping the factor n,

$$\sum_j \frac{p_j^2}{\pi_j} \left(1 - \frac{j}{\theta} \right) = 0. \tag{18.141}$$

This is the equation to be solved for θ, and we use an iterative method of solution similar to that used for the ML estimator in **18.21**. We expand the left-hand side of (18.141) in a Taylor series as a function of θ about the sample mean $\bar{x}$, regarded as a trial value. We obtain, to the first order of approximation,

$$\sum_j \frac{p_j^2}{\pi_j} \left(1 - \frac{j}{\theta} \right) = \sum_j \frac{p_j^2}{m_j} \left(1 - \frac{j}{\bar{x}} \right) + (\theta - \bar{x}) \sum_j \frac{p_j^2}{\pi_j} \left\{ \frac{j}{\bar{x}^2} + \left(1 - \frac{j}{\bar{x}} \right)^2 \right\}, \tag{18.142}$$

where we have written $m_j = \mathrm{e}^{-\bar{x}} \bar{x}^j / j!$. From (18.141), we find

$$(\theta - \bar{x}) = \bar{x} \cdot \frac{\sum_j \frac{p_j^2}{m_j} (j - \bar{x})}{\sum_j \frac{p_j^2}{m_j} \{ j + (j - \bar{x})^2 \}}. \tag{18.143}$$

We may use (18.143) to find an improved estimate of θ from $\bar{x}$, and repeat the process as necessary – cf. (18.45) for the ML estimator.

As a numerical example, we use Whitaker's (1914) data on the number of deaths of women over 85 years old reported in *The Times* newspaper for each day of 1910–12, 1096 days in all:

No. of deaths (j)	0	1	2	3	4	5	6	7
Reported frequency	364	376	218	89	33	13	2	1

The mean number of deaths reported is found to be $\bar{x} = 1295/1096 = 1.181$. This is therefore the ML estimator, and we use it as our first trial value for the MCS estimator.

From (18.143) we obtain $\theta^* = 1.198$ as our improved value. Smith (1916) reported a value of 1.1969 when working to greater accuracy, with more than one iteration of this procedure.

Smith also gives details of the computational procedure when we are estimating the parameters of a continuous distribution.

18.59 Small-sample properties, the second grounds for choice between the ML and MCS methods, are more amenable to general inquiry. Rao (1961; 1962a) defines a concept of second-order efficiency – cf. the concept of deficiency in **17.29** – and shows that in the multinomial model of **18.20**, the ML is the only BAN estimator with optimum second-order efficiency, under regularity conditions; essentially, the $O(n^{-2})$ term in the variance in (18.42) is minimized.

We have already seen an instance of an ML estimator with poor m.s.e. properties in **18.19**. Berkson (1955; 1956) has given another, in which the ML estimator is sometimes infinite, while another BAN estimator has smaller m.s.e. (These papers should be read with Silverstone (1957), which points out some errors in them.) Ghosh and Subramanyam (1974) extend Rao's results to the exponential family, and show that if the ML estimator in Berkson's problem is adjusted to have the same first-order bias as the other BAN estimator, it has the smaller variance. Amemiya (1980) confirmed both Berkson's and Ghosh and Subramanyam's results in detail. Ghosh and Sinha (1981) give a necessary and sufficient condition for it to be possible to improve the m.s.e. of the ML estimator, and Davis (1984) gives examples in which the ML estimator has the smaller m.s.e. Fountain and Rao (1993) show that improvements in the m.s.e. of the estimators are achieved using Amemiya's corrections, relative to both the Berkson and ML estimators.

Estimating equations

18.60 The method of moments discussed in **18.36–37** may be written in terms of a set of estimating equations (EEs),

$$n^{-1}\sum x_i^j - \mu'_j(\tilde{\boldsymbol{\theta}}) = 0, \qquad j = 1, \ldots, k, \tag{18.144}$$

when $E(x^j) = \mu'_j(\boldsymbol{\theta})$, and $\boldsymbol{\theta}$ denotes a vector of k parameters. However, we need not restrict attention only to moments. Godambe (1960; 1976) defines an EE in general terms as

$$\sum_{i=1}^{n} g_j(x_i, \boldsymbol{\theta}) = 0, \tag{18.145}$$

where

$$E[g_j(x, \boldsymbol{\theta})] = 0. \tag{18.146}$$

Any set of functions may be selected provided they satisfy the expectation condition (18.146).

The optimum EE is defined to be that with minimum variance for θ, subject to (18.146), which results when the EE is the score function given in **18.21**; see Exercise 18.32. However, the EE approach is applicable in a broad range of situations where ML may be infeasible. The extension to the multiparameter case is straightforward.

Durbin (1960) considers the class of unbiased EEs defined by $E[g_1(x) - \theta g(x)] = 0$.

18.61 A set of EEs of particular interest arises when the conditional distribution $f(x_j|X_{(j)}, \boldsymbol{\theta})$ is specified, where $X_{(j)} = \{x_i : i \neq j\}$. The product

$$L^*(\mathbf{x}|\boldsymbol{\theta}) = \prod_{j=1}^{n} f(x_j|X_{(j)}, \boldsymbol{\theta}) \tag{18.147}$$

is known as the *pseudo-likelihood function*[6] for $\boldsymbol{\theta}$; the concept was first introduced by Besag (1975). The estimators from (18.147) may be evaluated by standard methods, but the second derivatives cannot be used to define asymptotic variances.

Lele (1991) provides a jackknife procedure for estimating the variances of EE estimators, by leaving out one g_j in (18.145), or one conditional density in (18.147), at each step.

[6]The term pseudo-likelihood is also used, rather confusingly, to describe the class of likelihood procedures that focus on a subset of the parameters, including marginal, conditional and profile likelihoods.

The volume edited by Godambe (1991) surveys recent developments in EE methods.

18.62 We may extend the EE approach to consider p equations in k unknowns $(p > k)$. We may then define $\tilde{\boldsymbol{\theta}}$ as the vector minimizing the quadratic form $\mathbf{g}^T\mathbf{W}\mathbf{g}$, where $\mathbf{g}^T = (g_1, \ldots, g_p)$, $g_j = g_j(\mathbf{x}, \boldsymbol{\theta})$, and $\mathbf{W}$ is a symmetric positive definite weighting matrix. This approach, known as the *generalized method of moments*[7] (GMM) was introduced by Hansen (1982). The framework is very general and encompasses a number of estimation procedures used for simultaneous equation models in econometrics (cf. Hamilton, 1994, Chapter 14).

Assuming that $E(g_j) = 0$ as in (18.146), that the $\{g_j\}$ are differentiable functions of $\boldsymbol{\theta}$ and that regularity conditions are met to ensure the validity of the weak law of large numbers as needed, it follows that $\tilde{\boldsymbol{\theta}}$ is a consistent estimator, asymptotically normally distributed with mean $\boldsymbol{\theta}$ and covariance matrix $n^{-1}\mathbf{V}$, where

$$\mathbf{V} = (\mathbf{G}^T\mathbf{W}\mathbf{G})^{-1}\mathbf{G}^T\mathbf{W}\mathbf{H}\mathbf{W}\mathbf{G}(\mathbf{G}^T\mathbf{W}\mathbf{G})^{-1}, \tag{18.148}$$

in which $\mathbf{G} = E[\partial\mathbf{g}/\partial\boldsymbol{\theta}]$ is a $(p \times k)$ matrix assumed to be of full rank and

$$\mathbf{H} = \lim_{n\to\infty} E\{n^{-1}[\textstyle\sum \mathbf{g}(x_i, \boldsymbol{\theta})][\textstyle\sum \mathbf{g}(x_i, \boldsymbol{\theta})]^T\}. \tag{18.149}$$

Equation (18.148) follows by application of the generalized least squares approach (see **29.15**) to the Taylor series expansion for $\mathbf{g}$.

18.63 It follows directly from **29.15** that the best choice for $\mathbf{W}$ is the large-sample covariance matrix for $\mathbf{g}$, namely $\mathbf{H}^{-1}$, when (18.147) reduces to $\mathbf{V} = (\mathbf{G}^T\mathbf{H}^{-1}\mathbf{G})^{-1}$.

Example 18.26 (GMM estimation for the variance)

Let $(x_1, \ldots, x_n)$ denote a random sample from an $N(c\sigma, \sigma^2)$ distribution, where c is known. By standard arguments, the log LF yields

$$\frac{\partial \log L}{\partial \sigma} = \frac{-n}{\sigma} + \frac{\sum x_i^2}{\sigma^3} - \frac{c\sum x_i}{\sigma^2}$$

and

$$\frac{\partial^2 \log L}{\partial \sigma^2} = \frac{n}{\sigma^2} - \frac{3\sum x_i^2}{\sigma^4} + \frac{2c\sum x_i}{\sigma^3}$$

so that $\hat{\sigma}$ is the positive root of the quadratic (cf. Exercise 18.31):

$$\sigma^2 - s^2 + \bar{x}(\bar{x} - c\sigma) = 0,$$

where $ns^2 = \sum(x_i - \bar{x})^2$ with large-sample variance

$$V(\hat{\sigma}) = \sigma^2/[n(2 + c^2)].$$

[7]The method is also known as *minimum chi-square*, but we do not use this term to avoid confusion with the MCS estimators considered in **18.56–59**.

We know that $V(\bar{x}) = \sigma^2/n$, that $\bar{x}$ and s are independent and that, in large samples, $V(s) = \sigma^2/2n$ (see **10.15**), so we may consider the GMM estimator given by minimizing the quadratic form:

$$(\bar{x} - c\sigma)^2 \cdot \frac{n}{\sigma^2} + (s - \sigma)^2 \cdot \frac{2n}{\sigma^2};$$

that is, $g_1 = \bar{x} - c\sigma$, $g_2 = s - \sigma$ and $\mathbf{W} = \mathbf{H}^{-1} = \frac{\sigma^2}{n}\begin{bmatrix} 1 & 0 \\ 0 & 0.5 \end{bmatrix}$. Thus, we obtain the solution:

$$\tilde{\sigma} = (c\bar{x} + s)/(c^2 + 2).$$

The large-sample variance is readily shown to be

$$V(\sigma) = \sigma^2/[n(2 + c^2)],$$

the same as that for the ML estimator. In this case, the GMM is based upon the sufficient statistics and the optimal weighting matrix, so an asymptotically efficient estimator results. The GMM estimator may be improved by using exact expressions for the moments of s for finite n; cf. Exercise 17.6.

18.64 In Example 18.26, the structure of **H** was such that no estimation was required. In general, **H** will depend upon $\boldsymbol{\theta}$ and a two-step GMM estimator must be used. First, a consistent estimator for $\boldsymbol{\theta}$ is found using a 'sensible' **W** and then **H** is estimated before generating the final GMM estimates of $\boldsymbol{\theta}$. For discussion of such methods in the context of econometric models, see Newey and West (1987; 1994).

EXERCISES

18.1 In Example 18.7 show by considering the case $n = 1$ that the ML estimator does not attain the MVB for small samples, and deduce that for this distribution the efficiency of the sample 1 mean compared with the ML estimator is $\frac{1}{2}$.

18.2 If the ML estimator $\hat{\theta}$ is a root of $\partial \log L/\partial\theta = 0$, show that the most general form of distribution differentiable in θ, for which $\hat{\theta} = \bar{x}$, the sample arithmetic mean, is

$$f(x|\theta) = \exp\{A(\theta) + A'(\theta)(x-\theta) + B(x)\},$$

a member of the natural exponential family (5.117), and hence that $\bar{x}$ is sufficient for θ, with MVB variance $\{nA''(\theta)\}^{-1}$. Show that if θ is a location parameter, then f is a normal distribution with mean θ (a result going back to Gauss), while if θ is a scale parameter, then $f = \theta^{-1}\exp(-x/\theta)$.

(Cf. Keynes, 1911; Teicher, 1961)

18.3 Show that the most general continuous distribution for which the ML estimator of a parameter θ is the geometric mean of the sample is

$$f(x|\theta) = \left(\frac{x}{\theta}\right)^{\theta A'(\theta)} \exp\{A(\theta) + B(x)\}.$$

Show further that the corresponding distribution having the harmonic mean as ML estimator of θ is

$$f(x|\theta) = \exp\left[\frac{1}{x}\{\theta A'(\theta) - A(\theta)\} - A'(\theta) + B(x)\right].$$

(Keynes, 1911)

18.4 In Exercise 18.3, show in each case that the ML estimator is sufficient for θ, but that it is not an MVB estimator of θ, in contrast to the case of the arithmetic mean in Exercise 18.2. Find in each case the function of θ that is estimable with variance equal to its MVB, and evaluate the MVB.

18.5 In samples of n observations from

$$f(x|\theta) = \frac{1}{\Gamma(p)}(x-\theta)^{p-1}\exp\{\theta - x\}, \qquad \theta \le x < \infty;\ p \ge 2,$$

show that the smallest observation $x_{(1)}$ is neither a sufficient statistic for θ nor the ML estimator of $\hat{\theta}$ of θ, but that $\hat{\theta}$ is unique and satisfies

$$x_{(1)} - (p-1) \le \hat{\theta} < x_{(1)},$$

the equality holding when $n = 1$.
(For general f of asymptotic form $\alpha c(x-\theta)^{\alpha-1}$ as $x \to \theta$ (where α and c may be unknown parameters), Smith (1985) shows that for $\alpha > 2$, $\hat{\theta}$ has the standard ML properties, that for $\alpha = 2$, $\hat{\theta}$ is asymptotically normal and efficient with variance of lower order, but that for $\alpha < 2$, the normality is lost, while for $\alpha \le 1$, ML estimators may not exist.)

18.6 Show that for samples of n from the extreme-value distribution (cf. (14.66))

$$f(x) = \alpha\exp\{-\alpha(x-\mu) - \exp[-\alpha(x-\mu)]\}, \qquad -\infty < x < \infty,$$

the ML estimators $\hat{\alpha}$ and $\hat{\mu}$ are given by

$$\frac{1}{\hat{\alpha}} = \bar{x} - \frac{\sum x e^{-\hat{\alpha}x}}{\sum e^{-\hat{\alpha}x}},$$
$$e^{-\hat{\alpha}\hat{\mu}} = \frac{1}{n}\sum e^{-\hat{\alpha}x},$$

and that in large samples

$$n \operatorname{var} \hat{\alpha} = \alpha^2/(\pi^2/6),$$
$$n \operatorname{var} \hat{\mu} = \frac{1}{\alpha^2}\left\{1 + \frac{(1-\gamma)^2}{\pi^2/6}\right\},$$
$$n \operatorname{cov}(\hat{\alpha}, \hat{\mu}) = -(1-\gamma)/(\pi^2/6),$$

where γ is Euler's constant 0.5772....

(Kimball, 1946)

18.7 If x is distributed in the normal form

$$f(x) = \frac{1}{\sigma\sqrt{2\pi}} \exp\left\{-\frac{1}{2}\left(\frac{x-\mu}{\sigma}\right)^2\right\}, \qquad -\infty < x < \infty,$$

the lognormal distribution of $y = e^x$ has mean $\theta_1 = \exp(\mu + \frac{1}{2}\sigma^2)$, and its variance is $\theta_2 = \exp(2\mu + \sigma^2)\{\exp(\sigma^2) - 1\}$ (cf. (6.63)). Show that the ML estimator of θ_1 is

$$\hat{\theta}_1 = \exp(\bar{x} + \tfrac{1}{2}s^2),$$

where $\bar{x}$ and s^2 are the sample mean and variance of x, and that

$$E(\hat{\theta}_1) = E\{\exp(\bar{x})\}E\{\exp(\tfrac{1}{2}s^2)\} = \theta_1 \exp\left\{-\frac{(n-1)}{n}\frac{\sigma^2}{2}\right\}\left(1 - \frac{\sigma^2}{n}\right)^{-(n-1)/2} > \theta_1,$$

so that $\hat{\theta}_1$ is biased upwards. Show that $E(\hat{\theta}_1) \to_{n\to\infty} \theta_1$, so θ_1 is asymptotically unbiased.

18.8 In Exercise 18.7, define the series

$$f(t) = 1 + t + \frac{n-1}{n+1}\frac{t^2}{2!} + \frac{(n-1)^2}{(n+1)(n+3)}\frac{t^3}{3!} + \cdots.$$

Show that the adjusted ML estimator

$$\bar{\theta}_1 = \exp(\bar{x}) f(\tfrac{1}{2}s^2)$$

is strictly unbiased and that it is an MV unbiased estimator. Show further that $\hat{\theta}_1 > \bar{\theta}_1$ for all samples, so that the bias of $\hat{\theta}_1$ over $\bar{\theta}_1$ is uniform.

(Rukhin (1986) gives estimates with smaller m.s.e. than both $\bar{\theta}_1$ and $\hat{\theta}_1$ and similarly improved estimates for moments about zero, the median and the mode, which are all of the form $\exp(a\mu + b\sigma^2)$; cf. **6.29–30**.)

18.9 In Exercise 18.7, show that

$$\begin{aligned}\operatorname{var}\hat{\theta}_1 &= E\{\exp(2\bar{x})\}E\{\exp(s^2)\} - \{E(\hat{\theta}_1)\}^2 \\ &= \exp(2\mu+\sigma^2/n)\left[\exp\{\sigma^2/n\}\left(1-\frac{2\sigma^2}{n}\right)^{-(n-1)/2} - \left(1-\frac{\sigma^2}{n}\right)^{-(n-1)}\right]\end{aligned}$$

exactly, with asymptotic variance

$$\operatorname{var}\hat{\theta}_1 \sim \exp\left(2\mu+\sigma^2\right)\cdot\frac{1}{n}\left(\sigma^2+\frac{1}{2}\sigma^4\right),$$

and that this is also the asymptotic variance of $\bar{\theta}_1$ in Exercise 18.8. Hence show that the unbiased moment estimator of θ_1,

$$\bar{y} = \frac{1}{n}\sum y,$$

has efficiency

$$(\sigma^2+\tfrac{1}{2}\sigma^4)/\{\exp(\sigma^2)-1\}.$$

(Exercises 18.7–9 are due to Finney (1941) and Sichel (1951–52))

18.10 A multinomial distribution has n classes, each of which has equal probability $1/n$ of occurring. In a sample of N observations, k classes occur. Show that the LF for the estimation of n is

$$L(k|n) = \frac{N!}{\prod_{i=1}^{n}(r_i!)}\left(\frac{1}{n}\right)^N\cdot\binom{n}{k}\cdot\frac{k!}{\prod_{j=1}^{N}(m_j!)},$$

where r_i (≥ 0) is the number of observations in the ith class and m_j is the number of classes with j (≥ 1) observations in the sample. Show that the ML estimator of n is $\hat{n}$, where

$$\frac{N}{\hat{n}} = \sum_{j=\hat{n}-k+1}^{\hat{n}}\frac{1}{j},$$

and hence that approximately

$$\frac{N}{\hat{n}} = \log\left(\frac{\hat{n}}{\hat{n}-k+1}\right),$$

and that k is sufficient for n. Show that for large N,

$$\operatorname{var}\hat{n} \sim \frac{n}{\exp\left(\frac{N}{n}\right)-\left(1+\frac{N}{n}\right)}.$$

(Lewontin and Prout, 1956)

18.11 In Example 18.14, verify that the ML estimators (18.78) are jointly sufficient for the five parameters of the distribution, and that the ML estimators (18.74)–(18.75) are jointly sufficient for σ_1^2, σ_2^2 and ρ and also sufficient for ρ alone.

18.12 In estimating the correlation parameter ρ of the bivariate normal population, the other four parameters being known (Examples 18.14 and 18.15, case (a)), show that the sample correlation coefficient (which is the ML estimator of ρ if all five parameters are being estimated – cf. (18.78)) has estimating efficiency $1/(1+\rho^2)$. Show further that if the estimator

$$r' = \frac{\frac{1}{n}\sum(x-\mu_1)(y-\mu_2)}{\sigma_1\sigma_2}$$

based on the true means and variances is used, the efficiency drops even further, to

$$\left(\frac{1-\rho^2}{1+\rho^2}\right)^2.$$

(Stuart, 1955)

18.13 In Example 18.3, show that $(\sum x^2 + \sum y^2, \sum xy)$ is a pair of sufficient statistics for the single parameter ρ, and that the ML estimator $\hat{\rho}$ is a function of this pair. Show that when $n = 1$, this sufficient statistic is a single-valued function of $\theta(x, y)$, itself sufficient, but not conversely.

18.14 In Examples 18.14 and 18.15, find the ML estimators of μ_1 when all other parameters are known, and of σ_1^2 similarly, and show that their large-sample variances are respectively $\sigma_1^2(1-\rho^2)/n$ and $4\sigma_1^4(1-\rho^2)/[n(2-\rho^2)]$. Find the joint ML estimators of μ_1 and σ_1^2 when the other three parameters are known, and evaluate their large-sample covariance matrix.

18.15 In Example 18.19, show that if $\alpha = 0$, $\bar{x}$ and g are jointly sufficient statistics for σ and p, and that the ML estimators of σ and p are roots of the equations

$$\bar{x} = \sigma p, \quad \log(\bar{x}/g) = \log p - \frac{\mathrm{d}}{\mathrm{d}p}\log\Gamma(p),$$

where $\bar{x}$, g are respectively the arithmetic and geometric means of the observations. Find the asymptotic covariance matrix of $\hat{\sigma}$ and $\hat{p}$.

(Choi and Wette (1969) and Box (1971) discuss the biases of $\hat{p}$ and $1/\hat{\sigma}$. The *Biometrika Tables*, Vol. II (Pearson and Hartley, 1972), give tables for the computation of $\hat{p}$ its bias and its variance. See also Grice and Bain (1980) for inferences concerning the mean σp. For the c.f. of $\log(g/\bar{x})$, see Exercise 11.3.)

18.16 Show that the centre of location of the Pearson Type IV distribution

$$f(x) \propto \exp\left\{-\nu\arctan\left(\frac{x-\alpha}{\beta}\right)\right\}\left\{1+\left(\frac{x-\alpha}{\beta}\right)^2\right\}^{-(\rho+2)/2}, \qquad -\infty < x < \infty,\ \rho > 1,$$

where ν and ρ are assumed known, is distant $\nu\beta/(\rho+4)$ below the mode of the distribution; that the variance of the ML estimator $\hat{\alpha}$ in large samples is

$$\frac{\beta^2}{n}\cdot\frac{(\rho+4)^2+\nu^2}{(\rho+1)(\rho+2)(\rho+4)};$$

and that the efficiency of the method of moments in locating the curve is therefore

$$\frac{\rho^2(\rho-1)\{(\rho+4)^2+\nu^2\}}{(\rho+1)(\rho+2)(\rho+4)(\rho^2+\nu^2)}.$$

(Fisher, 1921a)

18.17 For n independent observations from the uniform distribution

$$f(x) = 1, \qquad \theta - \frac{1}{2} \leq x \leq \theta + \frac{1}{2},$$

show that $(x_{(1)}, x_{(n)})$ is sufficient for θ, that no single sufficient statistic exists, and that the LF is maximum at any value in the interval $(x_{(n)} - \frac{1}{2}, x_{(1)} + \frac{1}{2})$. Show that the mid-point of this interval is an unbiased estimator of θ.

18.18 Members are drawn from an infinite population in which the proportion bearing a given attribute is π, the drawing proceeding until a members bearing that attribute have appeared. The sample number then attained is n. Show that the distribution of n is given by

$$\binom{n-1}{a-1}\pi^a(1-\pi)^{n-a}, \qquad n = a, a+1, a+2, \ldots,$$

and that the ML estimator of π is a/n. Show also that this is biased and that its asymptotic variance is $\pi^2(1-\pi)/a$.

18.19 In the lognormal distribution of Exercises 18.7–8, consider the estimation of the variance θ_2. Show that the ML estimator

$$\hat{\theta}_2 = \exp(2\bar{x} + s^2)\{\exp(s^2) - 1\}$$

is biased and that the adjusted ML estimator

$$\bar{\theta}_2 = \exp(2\bar{x})\left\{f(2s^2) - f\left(\frac{n-2}{n-1}s^2\right)\right\}$$

is unbiased with minimum variance.

18.20 In Exercise 18.19, show that asymptotically

$$\operatorname{var}\hat{\theta}_2 \sim \frac{2\sigma^2}{n}\exp(4\mu + 2\sigma^2)[2\{\exp(\sigma^2) - 1\}^2 + \sigma^2\{2\exp(\sigma^2) - 1\}^2],$$

and hence that the efficiency of the unbiased moment estimator $s_y^2 = \sum(y - \bar{y})^2/(n-1)$ is

$$\frac{2\sigma^2[2\{\exp(\sigma^2) - 1\}^2 + \sigma^2\{2\exp(\sigma^2) - 1\}^2]}{\{\exp(\sigma^2) - 1\}^2\{\exp(4\sigma^2) - 2\exp(3\sigma^2) + 3\exp(2\sigma^2) - 4\}}.$$

(Finney, 1941)

18.21 Show that if θ is an integer-valued parameter and the LF is unimodal, the ML estimator $\hat{\theta}$ is the integer part of θ^*, where $L(x|\theta^*) = L(x|\theta^* - 1)$. Given that the LF has m modes and tends to zero at extreme values of θ, show that there are $2m - 1$ solutions of θ^*, the ML estimates being the integer part of one of the θ^*_{2i-1}, $i = 1, 2, \ldots, m$. For a normal distribution with integer mean μ, show that $\mu = [\bar{x} + \frac{1}{2}]$, the integer nearest to $\bar{x}$.

(Cf. Dahiya (1981). Hammersley (1950) showed that if the variance is known then $\hat{\mu}$ is unbiased and consistent, with asymptotic variance

$$\operatorname{var}\hat{\mu} \sim \left(\frac{8\sigma^2}{n\pi}\right)^{1/2}\exp\left(-\frac{n}{8\sigma^2}\right),$$

decreasing exponentially as n increases. Lindsay and Roeder (1987) treat a general class of integer-parameter distributions.)

18.22 In the previous exercise, show that for an integer Poisson parameter

$$\hat{\lambda} = [\{1 - \exp(-\bar{x}^{-1})\}^{-1}].$$

(Cf. Hammersley, 1950; Stark, 1975)

18.23 Suppose $\log(t-\gamma)$ is normally distributed with mean μ and variance σ^2, where $\gamma < t < \infty$, and the three parameters (γ, μ, σ^2) are to be estimated. Defining

$$\hat{\mu}(\gamma) = \frac{1}{n}\sum_{i=1}^{n} \log(t_i - \gamma), \quad \hat{\sigma}^2(\gamma) = \frac{1}{n}\sum_{i=1}^{n} \{\log(t_i - \gamma) - \hat{\mu}(\gamma)\}^2,$$

show that

$$L^{**}(\gamma) = \sup_{\mu,\sigma^2} L(x|\gamma, \mu, \sigma^2) \propto \{\hat{\sigma}(\gamma)\}^{-n} \prod_{i=1}^{n} (t_i - \gamma)^{-1}$$

and that if $t_{(1)}$ is the smallest observed value of t

$$\lim_{\gamma \to t_{(1)}} L^{**}(\gamma) = +\infty,$$

so that the ML estimator of (γ, μ, σ^2) for this three-parameter lognormal distribution is always $(t_{(1)}, -\infty, +\infty)$.

(Hill (1963a). Voorn (1981) explains the source of the unbounded LF.)

18.24 For a sample of n observations from $f(x|\theta)$ grouped into intervals of width h, write

$$f(x|\theta) = \int_{x-h/2}^{x+h/2} f(y|\theta)\, dy.$$

Show by a Taylor expansion that

$$f(x|\theta, h) = hf(x|\theta)\left\{1 + \frac{h^2}{24}\frac{\partial^2 f(x|\theta)/\partial x^2}{f(x|\theta)} + \cdots\right\}$$

and hence, to the first approximation, that the correction to be made to the ML estimator $\hat{\theta}$ to allow for grouping is

$$\Delta = -\frac{1}{24}h^2 \frac{\sum_{i=1}^{n} \partial((\partial^2 f/\partial x^2)/f)/\partial\theta}{\sum_{i=1}^{n} \partial^2(\log f)/\partial\theta^2},$$

the value of the right-hand side being taken at $\hat{\theta}$.

(Lindley, 1950)

18.25 Using the previous exercise, show that for estimating the mean of a normal population with known variance, $\Delta = 0$, while in estimating the variance with known mean, $\Delta = -h^2/12$.
Each of these corrections is exactly the Sheppard grouping correction (cf. (3.54)–(3.55)) to the corresponding population moment. To show that the ML grouping correction does not generally coincide with the Sheppard correction, consider the distribution

$$f(x) = \theta^{-1}e^{-x/\theta}, \qquad \theta > 0;\ 0 \le x < \infty,$$

where $\hat{\theta} = \bar{x}$, the sample mean, and the correction to it is

$$\Delta = -\frac{1}{12}\frac{h^2}{\bar{x}},$$

whereas the Sheppard correction to the population mean is zero.

(Lindley (1950). The normal case is studied in detail by Gjeddebaek (1949–61) and Kulldorf (1958) and the general theory by Kulldorf (1961) and Stadje (1985).)

18.26 Rewriting the negative binomial distribution (5.43) with m for the mean $k(1-p)/p$,

$$f_r = \left(1+\frac{m}{k}\right)^{-k}\binom{k+r-1}{r}\left(\frac{m}{m+k}\right)^r, \qquad r = 0, 1, 2, \ldots,$$

show that for a sample of n independent observations, with n_r observations at the value r and $n_0 < n$, the ML estimator of m is the sample mean

$$\hat{m} = \bar{r},$$

while that of k is a root of

$$n\log\left(1+\frac{\bar{r}}{k}\right) = \sum_{r=1}^{\infty} n_r \sum_{i=0}^{r-1}\frac{1}{k+i}.$$

Show that as k decreases towards zero, the right-hand side of this equation exceeds the left, and that if the sample variance, s_r^2, exceeds $\bar{r}$ the left-hand side exceeds the right as $k \to \infty$, and hence that the equation has at least one finite positive root. On the other hand, given that $s_r^2 < \bar{r}$; show that the two sides of the equation tend to equality as $k \to \infty$, so that $\hat{k} = \infty$, and f_r reduces to a Poisson distribution with parameter m.

(Anscombe, 1950)

18.27 In the previous exercise, show that

$$\operatorname{var}\hat{m} = (m + m^2/k)/n,$$

$$\operatorname{var}\hat{k} \sim \left\{\frac{2k(k+1)}{n(\frac{m}{m+k})^2}\right\}\Big/\left\{1 + 2\sum_{j=2}^{\infty}\frac{(\frac{j}{j+1})(\frac{m}{m+k})^{j-1}}{\binom{k+j}{j-1}}\right\},$$

$$\operatorname{cov}(\hat{m}, \hat{k}) \sim 0.$$

(Anscombe (1950). Fisher (1941) investigated the efficiency of the method of moments in this case.)

18.28 For the Neyman Type A contagious distribution of Exercise 5.7, with probability function f_r, show that the ML estimators of λ_1, λ_2 are given by the roots of the equations

$$\hat{\lambda}_1\hat{\lambda}_2 = \bar{r} = \sum_r n_r(r+1)f_{r+1}/(nf_r),$$

where $n_r, \bar{r}$ have the same meanings as in Exercise 18.26.

(Cf. Shenton (1949), who investigated the efficiency of the method of moments in this case and found it to be relatively low (< 70 per cent) for small λ_1 (< 3) and large λ_2 ($\geq$ 1), and examined (Shenton, 1950; 1951) the efficiency of using moments in the general case and in the particular case of the Gram–Charlier Type A series. See also Katti and Gurland (1962).)

18.29 For n observations from the logistic distribution, with distribution function

$$F(x) = \frac{1}{1+\exp\{-(x-\alpha)/\beta\}}; \qquad -\infty < x < \infty,\ \sigma > 0,$$

show that the ML estimators of α and β are roots of the equations

$$\sum_{i=1}^{n} \frac{1}{1+e^{-y_i}} = \frac{1}{2}n, \tag{18.150}$$

$$\sum_{i=1}^{n} \frac{y_i}{1+e^{-y_i}} - \frac{1}{2}\sum_{i=1}^{n} y_i = \frac{1}{2}n, \tag{18.151}$$

where $y_i = (x_i - \alpha)/\beta$, and that the elements of their asymptotic inverse covariance matrix (18.110) are $E(g'') = -\frac{1}{3}$, $E(g'y) = 0$, $E(g''y^2 - 1) = -\frac{1}{3}(\pi^2/3 + 1)$.
Show that when β is known, there is a unique solution of (A) for α. When $n = 2$, show that the unique solutions of (A) and (B) are $\hat{\alpha} = \frac{1}{2}(x_1 + x_2)$, $\hat{\beta} = |x_1 - x_2|/(4L)$, where $L \doteq 0.7717$ is the unique positive root of $\tanh L = 1/(2L)$.

(Cf. Antle *et al.* (1970), who show that the solutions of (A) and (B) are unique for any n if the x_i are not all equal.)

18.30 Independent samples of sizes n_1, n_2 are taken from two normal populations with equal means μ and variances respectively equal to $\lambda\sigma^2, \sigma^2$. Find the ML estimator of μ, and show that its large-sample variance is

$$\text{var}(\hat{\mu}) = \sigma^2 \Big/ \left(\frac{n_1}{\lambda} + n_2\right).$$

Hence show that the unbiased estimator

$$t = (n_1\bar{x}_1 + n_2\bar{x}_2)/(n_1 + n_2)$$

has efficiency

$$\frac{\lambda(n_1 + n_2)^2}{(n_1\lambda + n_2)(n_1 + n_2\lambda)},$$

which attains the value 1 if and only if $\lambda = 1$.

18.31 For a sample of n observations from a normal distribution with mean θ and variance $V(\theta)$, show that the ML estimator $\hat{\theta}$ is a root of

$$V' = 2(\bar{x} - \theta) + \frac{V'}{V}\frac{1}{n}\sum(x - \theta)^2,$$

and hence that if $V(\theta) = \sigma^2\theta^k$, where σ^2 is known, $\hat{\theta}$ is a function of both $\bar{x}$ and $\sum x^2$ unless $k = 0$ (when $\hat{\theta} = X$ is the single sufficient statistic) or $k = 1$ (when $\hat{\theta}$ is a function of $\sum x^2$ only). When $k = 2$, show that the distribution of $\sum x^2/(\sum x)^2$ does not depend on θ.

18.32 A parameter θ is estimated by θ^*, a root of the equation $g(x, \theta) = 0$. Given that regularity conditions for the MVB in **17.14–15** hold, and $E\{g(x, \theta)\} = 0$, show that

$$\frac{E(g^2)}{\left\{E\left(\frac{\partial g}{\partial \theta}\right)\right\}^2} \geq \frac{1}{E\left\{\left(\frac{\partial \log L}{\partial \theta}\right)^2\right\}},$$

the equality holding only when $g(x, \theta)$ is a constant multiple of $\partial \log L/\partial\theta$, when $\theta^* = \hat{\theta}$, the ML estimator. This generalization reduces to (17.22) when $g(x, \theta) = t - \tau(\theta)$.

(Godambe, 1960; Durbin, 1960)

18.33 For a random sample from the distribution

$$f(x|\theta) = \begin{cases} \frac{1}{3}\exp\{-|x-\theta|\}, & x < \theta, \\ \frac{1}{3}, & \theta \le x \le \theta + 1, \\ \frac{1}{3}\exp\{-|x-(\theta+1)|\}, & \theta + 1 < x, \end{cases}$$

show that the ML estimator of θ is never unique. (Consider the cases $n = 1, n = 2$ in particular.)

(Daniels, 1961)

18.34 A sample of n observations x_i is drawn from a normal population with mean μ and a positive variance that has equal probabilities of being 1 or σ^2. By considering $L(x|\mu, \sigma^2)$ when $\mu = x_j$ and $\sigma^2 \to 0$, show that as $n \to \infty$ no ML estimator of (μ, σ^2) exists.

(Kiefer and Wolfowitz, 1956)

18.35 Show that if an MVB estimator exists for a parameter θ, the method of scoring for parameters given in (18.46) will reach the ML estimator $\hat{\theta}$ in a single iteration, no matter what trial value is used, but that this is not true of (18.45).

18.36 In Example 18.8, show that if (18.46) were used to evaluate $\hat{\sigma}$ from a trial value $t > 0$, the first iteration would give the value $\theta_1 = \frac{1}{2}(t + \hat{\theta}^2/t)$, that $\theta_1 > \hat{\theta}$ always, and that θ_1 is closer to $\hat{\theta}$ than t is if $t > \frac{1}{3}\hat{\theta}$.

18.37 Given that a single observation x is distributed in the form

$$f(x|\theta) = \theta^{-1/m}\exp(-x\theta^{-1/m}), \qquad x > 0;\ \theta > 0;\ m = 1, 2, 3, \ldots,$$

show that the ML estimator of θ is $\hat{\theta} = x^m$, that it is biased unless $m = 1$, and that $u = \hat{\theta}/m!$ is unbiased. Show from **17.21** that u is the MV unbiased estimator of θ, attaining the bound (17.41) with $s = m$.

(Fend, 1959)

18.38 A sample of n observations from the triangular distribution

$$f(x) = \begin{cases} 2x/\theta, & 0 < x \le \theta, \\ 2(1-x)/(1-\theta), & \theta \le x \le 1, \end{cases}$$

is ordered so that $0 \equiv x_{(0)} < x_{(1)} < \cdots < x_{(n)} < x_{(n+1)} \equiv 1$. Show that $\log L(x|\theta)$ is continuous in θ, and differentiable except where θ is equal to some $x_{(r)}$. Show that in the open interval $(x_{(r)}, x_{(r+1)})$, $r = 0, 1, \ldots, n$,

$$\frac{\partial \log L}{\partial \theta} = \frac{n\theta - r}{\theta(1-\theta)}, \qquad \frac{\partial^2 \log L}{\partial \theta^2} = \frac{r}{\theta^2} + \frac{n-r}{(1-\theta)^2},$$

and hence that $\log L$ as no regular maximum, but has a cusp at each $x_{(r)}$, one of which is the ML estimator $\hat{\theta}$.

Show that as $n \to \infty$, $\hat{\theta} \sim x_{(n\theta)}$.

(Cf. Oliver, 1972)

18.39 For the two-parameter Cauchy distribution

$$f(x) = \frac{1}{\pi\sigma\{1 + (\frac{x-\theta}{\sigma})^2\}}, \qquad -\infty < x < \infty;\ \sigma > 0,$$

show that the ML estimators of θ and σ from a sample of n observations are given by the roots of the equations

$$\frac{1}{2}\frac{\partial \log L}{\partial \theta} = \sum_{i=1}^{n} \frac{x_i - \theta}{\sigma^2 + (x_i - \theta)^2} = 0, \tag{A}$$

$$\frac{\sigma}{n}\frac{\partial \log L}{\partial \theta} = 1 - \frac{2\sigma^2}{n}\sum_{i=1}^{n} \frac{1}{\sigma^2 + (x_i - \theta)^2} = 0, \tag{B}$$

provided that (B) has a solution (as it always has if fewer than $\frac{1}{2}n$ observations coincide in value, and thus almost certainly when $n \geq 3$) and that otherwise $\hat{\sigma} = 0$ is to be used in (A). Show that $\partial^2 \log L/\partial\sigma^2 < 0$ when (B) holds, and hence that if θ is known, the solution of (B) for $\hat{\sigma}$ is at a unique maximum of the LF.

Writing $y_i = x_i - \theta$; $D_i = y_i^2 - \sigma^2$; $S_i = y_i^2 + \sigma^2$, show by considering the positive and the negative values of D_i separately that

$$\frac{\partial^2 \log L}{\partial \theta^2} \leq \frac{1}{\sigma^2}\sum_{i=1}^{n} \frac{D_i}{S_i},$$

this upper bound being zero when (B) holds. Further, show that when (A) and (B) both hold,

$$-2\sigma^2 \sum_i y_i S_i^{-2} = \sum y_i D_i S_i^{-2}$$

and

$$-2\sigma^2 \sum_i y_i D_i S_i^{-2} = \sum_i y_i D_i S_i^{-2}$$

so that

$$\left(\frac{\partial^2 \log L}{\partial\theta\,\partial\sigma}\right)^2 - \frac{\partial^2 \log L}{\partial\theta^2}\cdot\frac{\partial^2 \log L}{\partial\sigma^2} = \frac{4}{\sigma^2}\left\{\left(\sum_i y_i D_i S_i^{-2}\right)^2 - \left(\sum_i D_i^2 S_i^{-2}\right)\left(\sum y_i^2 S_i^{-2}\right)\right\} < 0,$$

and hence that the solution of (A) and (B) for $(\hat{\theta}, \hat{\sigma})$ is at a unique maximum of the LF.

(Copas, 1975)

18.40 In Exercise 18.39, when $n = 3$ and $x_{(1)} < x_{(2)} < x_{(3)}$, show (writing (i) for $x_{(i)}$ and $(i - j)$ for $(x_{(i)} - x_{(j)})$) that the ML estimators are

$$\hat{\theta} = \frac{(1)(3-2)^2 + (2)(3-1)^2 + (3)(2-1)^2}{(3-2)^2 + (3-1)^2 + (2-1)^2},$$

$$3^{-1/2}\hat{\sigma} = (3-2)(3-1)(2-1)/\{(3-2)^2 + (3-1)^2 + (2-1)^2\}.$$

When $n = 4$, show similarly that

$$\hat{\theta} = \{(2)(4) - (1)(3)\}/\{(4-3) + (2-1)\}$$

$$\hat{\sigma}^2 = (4-3)(3-2)(2-1)(4-1)/\{(4-3) + (2-1)\}^2.$$

Show that as $x_{(n)} \to \infty$ with the other observations held fixed,

$$\hat{\theta} \to \begin{cases} \frac{1}{2}\{(1) + (2)\} & \text{for } n = 3 \\ (2) & \text{for } n = 4. \end{cases}$$

(Cf. Ferguson, 1978)

18.41 Show that

$$\begin{aligned} \textstyle\sum_{j=1}^{n}(x_i - \mu)^2/(x_j\mu^2) &\equiv n(\bar{x} - \mu)^2/(x\mu^2) + \textstyle\sum_{j=1}^{n}(x_j^{-1} - \bar{x}^{-1}) \\ &\equiv u \qquad\qquad\qquad + v, \end{aligned}$$

so that the LF of a sample from the inverse Gaussian distribution of Exercise 11.28 may be written

$$L(x|\mu, \lambda) = \left(\frac{\lambda}{2\pi}\right)^{-n/2}\left(\prod_{j=1}^{n} x_j\right)^{-3/2} \exp\left\{\frac{-\lambda}{2}(u + v)\right\}.$$

Hence show that if λ is known, $\bar{x}$ is a complete sufficient statistic for μ, and also its ML estimator, λu having an exact χ_1^2 distribution (cf. Exercise 11.29). Show that λv is distributed independently of $\bar{x}$, having an exact χ_{n-1}^2 distribution, and that $\bar{x}$ and v are jointly complete sufficient statistics for μ and λ, the ML estimators then being

$$\hat{\mu} = \bar{x}, \quad \hat{\lambda} = nv^{-1}.$$

Finally, show that $\hat{\mu}$ is unbiased, while $(n - 3)\hat{\lambda}/n$ is unbiased for λ.

(Cf. Folks and Chhikara (1978). Iwase and Seto (1983) give MV unbiased estimates of the cumulants – cf. Exercise 11.28 – and of other functions of the parameters, and find their estimator of κ_2 has smaller m.s.e. than the ML estimator of κ_2 to which it is asymptotically equivalent.)

18.42 A sample of n observations is drawn from a Poisson population with probabilities

$$\pi_j = e^{-\lambda}\lambda^j/j!, \qquad j = 0, 1, \ldots.$$

However, the counts are recorded only for m groups $\{y_1, \ldots, y_m\}$; let x_j correspond to the number of times j occurred, which is now merged into the rth group $(j \in r)$. Applying the EM algorithm of **18.44–45**, show that at the kth iteration

$$x_{j(k)} = y_r\pi_{j(k)}\Big/\sum_{i\in r}\pi_{i(k)}$$

and

$$\lambda_{(k+1)} = \sum_j x_{j(k)}/n.$$

(Note that the mth group must be closed off at some appropriate upper value.)

18.43 Show that if $\mathbf{y}$ is distributed in the form $f(y|\theta) = \exp(\theta y)B(y)C(\theta)$, its c.f. is $C(\theta)/C(\theta + it)$ with cumulants $\kappa_1 = -C'(\theta)/C(\theta) = p(\theta)$, $\kappa_2 = \partial\kappa_1/\partial\theta$. Hence show that if for n such independent y_i with differing means $p_j(\theta)$ the distributions are re-scaled to have the same variance, the LS estimator of θ is identical with the ML estimator.

(Charnes *et al.* (1976) generalize this)

18.44 y_i is a Poisson variable with parameter θx_i, where x_i is a constant observed with y_i, $(i = 1, 2, \ldots, n)$. Show that the ML estimator of θ is $\sum y_i / \sum x_i$, with asymptotic variance $\theta / \sum x_i$, and that the LS estimator $\sum y_i x_i / \sum x_i^2$ has exact variance $\theta \sum x_i^3 / (\sum x_i^2)^2$. Hence show that the LS estimator is inefficient unless the x_i are all equal. Explain the result.

18.45 Show that when all the n_i are large the minimization of the chi-square expression (18.137) or (18.139) gives the same estimator as the ML method.

18.46 In the case $s = 1$, show that the first two moments of the statistic (18.137) are given by

$$n_1^2 E(\chi^2) = k_1 - 1,$$

$$n_1^4 \operatorname{var}(\chi^2) = 2(k_1 - 1)\left(1 - \frac{1}{n_1}\right) + \frac{1}{n_1} \sum_{j=1}^{k_1} \frac{1}{\pi_{1j}} - \frac{k_1^2}{n_1},$$

and that for any $c > 0$, the generalization of (18.139) has expectation

$$E\left\{\sum_{j=1}^{k_1} \frac{(n_{1j} - n_1)^2 + b}{n_{1j} + c}\right\} =$$

$$k_1 - 1 + \frac{1}{n_1}\left[(b - c + 2) \sum_{j=1}^{k_1} \frac{1}{\pi_{1j}} - (3 - c)k_1 + 1\right] + O\left(\frac{1}{n_1^2}\right).$$

Thus, to the second order at least, the π_{1j} disappear from the expectation if $b = c - 2$. If $b = 0$, $c = 2$, it is $(k_1 - 1)(1 - (1/n_1))$ and if $b = 1, c = 3$, it is $(k_1 - 1) + (1/n_1)$, which for $k_1 > 2$ is even closer to the expectation of (18.137).

(David, 1950; Haldane, 1955)

18.47 For a binomial distribution with probability of success equal to π, show that the MCS estimator of π obtained from (18.137) is identical to the ML estimator for any n, and that if the number of successes is not 0 or n, the modified MCS estimator obtained from (18.139) is also identical to the ML estimator.

CHAPTER 19

INTERVAL ESTIMATION

19.1 In the two previous chapters we have been concerned with point estimation. It was recognized that the estimate might differ from the parameter in any particular case, and hence that there was a margin of uncertainty, which was expressed in terms of the sampling variance of the estimator. For example, if the asymptotic distribution of t is normal, we could say that with probability (near) 0.95 the random interval $t \pm 1.96\sqrt{\text{var}\, t}$ includes the true but unknown value of θ. That is, we locate θ in a range and not at a particular point, although one point in the range, the value of t itself, is the estimate of θ.

19.2 In the present chapter we shall examine this procedure more closely and look at the problem of estimation from a different point of view. We now consider the specification of a range in which θ lies. The method we shall first discuss, which uses *confidence intervals*, relies only on the theory of probability without importing any new principle of inference. Other methods, discussed at the end of the chapter, explicitly require additional assumptions.

Confidence intervals

19.3 Consider first a distribution dependent on a single unknown parameter θ and suppose that we are given a random sample of n values $x_1, \ldots, x_n$ from it. Let z be a random variable dependent on the xs and on θ, with distribution function (d.f.) $F(z)$ which we shall initially assume does not depend on θ. Then (at least in the case where z is continuously distributed) we can find a fixed value z_1 such that

$$F(z_1) = P\{z(x_1, \ldots, x_n, \theta) \leq z_1\} = 1 - \alpha. \tag{19.1}$$

The inequality $z(x_1, \ldots, x_n, \theta) \leq z_1$ implies a restriction on the value of θ. Now it may happen that this restriction takes the form $t_0 \leq \theta \leq t_1$ where t_0 and t_1 are values (possibly infinite) of a statistic $t(x_1, \ldots, x_n)$ not dependent on θ. Then (19.1) is equivalent to

$$P(t_0 \leq \theta \leq t_1) = 1 - \alpha. \tag{19.2}$$

For example, if $z = \bar{x} - \theta$, the inequality $\bar{x} - \theta \leq z_1$ is equivalent to $\bar{x} - z_1 \leq \theta$, so (19.1) implies

$$P(\bar{x} - z_1 \leq \theta < \infty) = 1 - \alpha. \tag{19.3}$$

Here t_1 is infinite, just as t_0 would be negatively infinite if we were to reverse the inequality in (19.1).

More generally, even if $F(z)$ depends on θ, we may still construct probabilistic statements like (19.2), as we shall see in Example 19.3.

19.4 In (19.2), we have the probability that the random interval (t_0, t_1) covers the parameter, θ. Thus, if we state that the inequality $t_0 \leq \theta \leq t_1$ in (19.2) is true for any sample, we shall be correct in a proportion $(1-\alpha)$ of samples *in the long run*. We stress that this remains true however θ may vary – not merely for repeated samples from a population with fixed θ, but for repeated samples from populations with varying θ.

19.5 The interval (t_0, t_1) is called a *confidence interval* for θ, while t_0 and t_1 are called the lower and upper *confidence limits*. Intervals like (19.2) are called two-sided, and those like (19.3) are called one-sided. For fixed *confidence coefficient* (or *confidence level*) $1-\alpha$, the totality of the confidence intervals for θ over all possible samples determines a zone called a *confidence belt*, which will be graphically illustrated in Example 19.3.

The ideas and methods of confidence interval estimation were first developed by J. Neyman – see especially Neyman (1937).

Example 19.1 (Confidence interval for the normal mean with variance known)

Suppose we have a sample of size n from the normal distribution with known variance (taken without loss of generality to be unity)

$$f(x) = \frac{1}{\sqrt{2\pi}} \exp\{-\tfrac{1}{2}(x-\mu)^2\}, \qquad -\infty < x < \infty.$$

The distribution of the sample mean $\bar{x}$ is exactly normal,

$$f(\bar{x}) = \left(\frac{n}{2\pi}\right)^{1/2} \exp\left\{-\frac{n}{2}(\bar{x}-\mu)^2\right\}, \qquad -\infty < \bar{x} < \infty.$$

Here $\bar{x}-\mu$ has a distribution that does not depend on μ.

From the normal d.f. we obtain

$$P(\bar{x}-\mu \le 2/\sqrt{n}) = 0.977\,25,$$

which is equivalent to

$$P(\bar{x}-2/\sqrt{n} \le \mu) = 0.977\,25.$$

Thus, if we assert that μ is greater than or equal to $\bar{x}-2/\sqrt{n}$ we shall be right in about 97.7 per cent of cases in the long run. Similarly, we have

$$P(\bar{x}-\mu \ge -2/\sqrt{n}) = P(\mu \le \bar{x}+2/\sqrt{n}) = 0.977\,25.$$

Thus, combining the two results,

$$\begin{aligned} P(\bar{x}-2/\sqrt{n} \le \mu \le \bar{x}+2/\sqrt{n}) &= 1-2(1-0.977\,25) \\ &= 0.9545. \end{aligned} \qquad (19.4)$$

Hence, if we assert that μ lies in the range $\bar{x} \pm 2/\sqrt{n}$, we shall be right in about 95.45 per cent of cases in the long run, the confidence level being 0.9545.

Conversely, given the confidence level, we can easily find the value d such that

$$P(\bar{x}-d/\sqrt{n} \le \mu \le \bar{x}+d/\sqrt{n}) = 1-\alpha.$$

The reader to whom this approach is new will probably ask: but is this not a roundabout method of using the standard error to set limits to an estimate of the mean? In a way, it is. Effectively, what we have done in this example is to show how the use of the standard error of the mean in normal samples may be justified. We must remember, however, that we are dealing here with a very special case where $\bar{x}$ is exactly normally distributed, something that is certainly not required by the theory of confidence intervals.

Another point of interest in this example is that the upper and lower confidence limits derived above are equidistant from the mean $\bar{x}$. This is not by any means necessary, and it is easy to see that we can derive any number of alternative limits for the same confidence level $1-\alpha$. Suppose, for instance, we take $1-\alpha = 0.9545$, and select two numbers α_0 and α_1 that satisfy the condition

$$\alpha_0 + \alpha_1 = \alpha = 0.0455,$$

say $\alpha_0 = 0.01$ and $\alpha_1 = 0.0355$. From the normal d.f. we arrive at the confidence interval:

$$P\left(\bar{x} - \frac{2.326}{\sqrt{n}} \leq \mu \leq \bar{x} + \frac{1.806}{\sqrt{n}}\right) = 1 - \alpha_0 - \alpha_1 = 0.9545. \tag{19.5}$$

Thus, with the same level of confidence we may assert that μ lies in the interval $\bar{x} - 2/\sqrt{n}$ to $\bar{x} + 2/\sqrt{n}$, or in the interval $\bar{x} - 2.326/\sqrt{n}$ to $\bar{x} + 1.806/\sqrt{n}$. In either case we shall be right in about 95.45 per cent of cases in the long run.

We note that in the first case the interval has length $4/\sqrt{n}$, while in the second case its length is $4.132/\sqrt{n}$. Intuitively, we should choose the first set of limits since they locate the parameter in a narrower range with the same level of confidence. We shall consider this point in more detail later in this chapter.

Graphical representation

19.6 In a number of simple cases, including that of Example 19.1, the confidence limits can be represented in graphical form. We take two orthogonal axes, OX relating to the observed $\bar{x}$ and OY to μ (see Fig. 19.1), and for simplicity consider $n = 1$ initially.

The two straight lines shown have as their equations

$$\mu = \bar{x} + 2, \quad \mu = \bar{x} - 2.$$

Consequently for any point between the lines,

$$\bar{x} - 2 \leq \mu \leq \bar{x} + 2.$$

Hence, for any observed $\bar{x}$ we may read off the confidence limits for μ on the vertical axis. The vertical interval between the limits is the confidence interval (shown in the diagram for $\bar{x} = 1$), and the total zone between the lines is the confidence belt. We may refer to the two lines as upper and lower confidence lines, respectively.

For different values of n and a fixed confidence coefficient $1-\alpha$, there will be different confidence lines, all parallel to $\mu = \bar{x}$, and getting closer to each other as n increases.

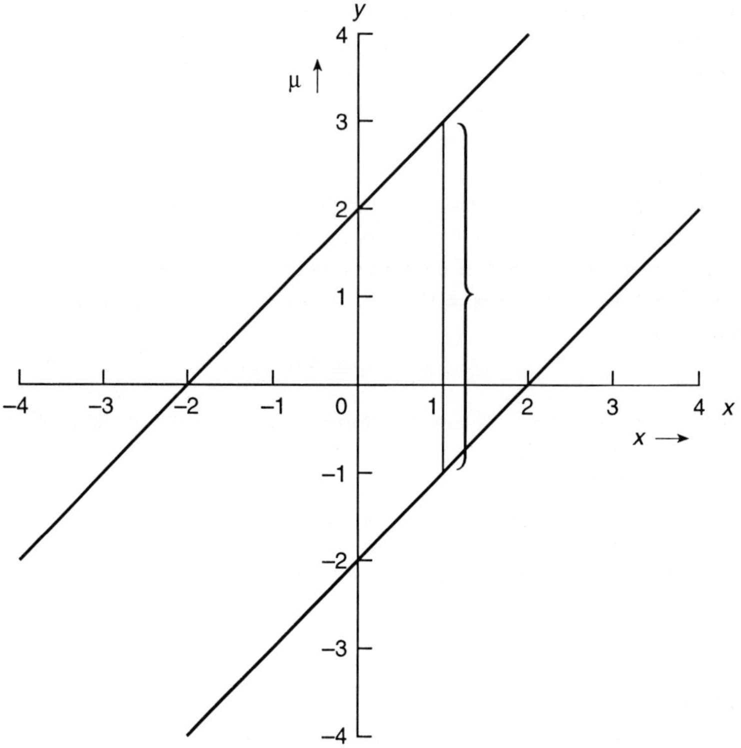

Fig. 19.1 Confidence limits in Example 19.1 for $n = 1$

Alternatively, we may wish to vary the confidence coefficient, and we may show a series of pairs of confidence lines, each pair corresponding to a selected value of $1 - \alpha$, on a single diagram relating to some fixed value of n. In this case, of course, the lines grow further apart with increasing $1 - \alpha$. Indeed, we may validly make assertions of the form (19.2) simultaneously for a number of values of α: each will be true in the corresponding proportion of cases in the long run. Indeed, this procedure may be taken to its extreme form, when we consider *all* values of $1 - \alpha$ in (0, 1) simultaneously, and thus generate a *confidence distribution* of the parameter – the term is due to Cox (e.g. 1958a): we then have an infinite sequence of simultaneous confidence statements, each containing the preceding one, with increasing values of $1 - \alpha$.

Example 19.2 (Confidence interval for the normal mean with variance unknown)

Surprisingly little changes in Example 19.1 if we now assume that the variance σ^2 is unknown. The statistic

$$t = (n - 1)^{1/2}(\bar{x} - \mu)/s,$$

where s^2 is the sample variance, follows Student's t distribution ('Student', 1908); see (11.44) in Example 11.8. Given α, we can now find t_0 and t_1, where $\frac{1}{2}\alpha$ of the distribution lies below $-t_1$ and $\frac{1}{2}\alpha$ above t_0, so that

$$P(-t_1 \leq t \leq t_0) = 1 - \alpha,$$

which is equivalent to

$$P(\bar{x} - st_0(n-1)^{-1/2} \leq \mu \leq \bar{x} + st_1(n-1)^{-1/2}) = 1 - \alpha.$$

Thus we may assert that μ lies in this interval with confidence level $1 - \alpha$, and the interval is independent of μ and σ^2. Because of the symmetry of the t distribution, the confidence limits are equidistant from $\bar{x}$.

As in Example 19.1, the confidence limits are linear in the statistic $\bar{x}$, and the confidence lines are parallel straight lines as in Fig. 19.1. The difference is that whereas, with known variance, the vertical distance between the lines is the same for all samples, here that distance is a random variable, being a multiple of s. Thus we cannot here fix the length of the confidence interval in advance of the observations.

Goutis and Casella (1992) show that by conditioning on certain events, such as whether $|\bar{x}/s|$ is less than some constant in Example 19.2, it is possible to achieve a confidence level strictly greater than $1 - \alpha$ when the condition is satisfied; when the condition is not satisfied, the level remains at $1 - \alpha$. Thus, it is possible to improve the average coverage of such intervals by conditioning on selected reference subsets.

Central and non-central intervals

19.7 In Examples 19.1 and 19.2, the sampling distribution on which the confidence intervals were based was symmetrical, and by taking equal deviations from the mean we obtained equal values of

$$1 - \alpha_0 = P(t_0 \leq \theta) \quad \text{and} \quad 1 - \alpha_1 = P(\theta \leq t_1).$$

In general, we cannot achieve this result with equal deviations, but subject to the condition $\alpha_0 + \alpha_1 = \alpha$, α_0 and α_1 may still be chosen arbitrarily.

If α_0 and α_1 are taken to be equal, we shall say that the intervals are *central*. In such a case we have

$$P(t_0 > \theta) = P(\theta > t_1) = \alpha/2. \tag{19.6}$$

Otherwise, the intervals are *non-central*. In general, the confidence limits are equidistant from the sample statistic only if its sampling distribution is symmetrical.

19.8 In the absence of other considerations it is usually convenient to employ central intervals, but circumstances sometimes arise in which non-central intervals are more serviceable. Suppose, for instance, we are estimating the proportion of some drug in a medicinal preparation and the drug is toxic in large doses. We must then clearly err on the safe side, an excess of the true value over our estimate being more serious than a deficiency. In such a case we might use the *one-sided* interval given by

$$P(\theta \leq t_1) = 1 - \alpha.$$

Conservative confidence intervals and discontinuities

19.9 On a somewhat similar point, it may be remarked that in certain circumstances it is enough to know that $P(t_0 \leq \theta \leq t_1) \geq 1 - \alpha$. Such an interval is often called *conservative*. We then know that in asserting θ to lie in the range t_0 to t_1 we shall be right in *at least* a proportion $1 - \alpha$ of the cases. Mathematical difficulties in ascertaining confidence limits exactly for given $1 - \alpha$, or theoretical difficulties when the distribution is discrete, may, for example, lead us to be content with this inequality rather than the equality of (19.2).

Example 19.3 (Confidence intervals for the binomial distribution)

We now construct confidence intervals for the probability π of success in a binomial sample.

We shall determine the limits for the case $n = 20$ and confidence coefficient 0.95. The obvious statistic to use is the observed proportion of successes, p, but we see that here we cannot find a function of p whose distribution does not depend on π. Nonetheless, confidence intervals can be constructed.

Although the variate p is discrete, π can take any value in (0, 1). For given π we cannot in general find limits to p for which $1 - \alpha$ is exactly 0.95; but we will take p to be the sample proportion that gives a confidence level at least equal to 0.95, so the intervals will be conservative. We will consider only central intervals, so that for given p we have to find π_0 and π_1 such that

$$P(p \geq \pi_0) \geq 0.975$$
$$P(p \leq \pi_1) \geq 0.975,$$

the inequalities for P being as near to equality as we can make them.

We now illustrate the construction of the confidence intervals using the values of the binomial distribution function given in Table 19.1.

For example, when $\pi = 0.3$ the greatest value of π_0 such that $P(p \geq \pi_0) > 0.975$ is 0.10. By the time π has increased to 0.4 the value of π_0 has increased to 0.20. Somewhere between is the value of π such that $P(p \geq 0.1)$ is exactly 0.975. The coarse grid search based on Table 19.1 gives the step functions shown in Fig. 19.2. An accurate search yields the two boundaries shown as dashed lines in that diagram.

The zone between the stepped lines is the confidence belt. It will be observed that it was constructed horizontally. In applying it, however, we read it *vertically*, that is to say, with observed p we read off two values p_0 and p_1 where the vertical through p intersects the boundaries of the confidence belt, and assert that $p_0 \leq \pi \leq p_1$. We may see at once that this gives the required confidence interval. Considering the diagram horizontally, we see that, for any given π, an observation falls in the confidence belt with probability at least $1 - \alpha$. If and only if the observation is in the belt, the pair of values (p_0, p_1) will contain between them the true value of π. Thus the latter event has probability at least $1 - \alpha$, whatever the true value of π.

19.10 In Example 19.3, we remarked that, as the number of successes (say, c) is necessarily integral, and the proportion of successes p $(= c/n)$ therefore discrete, the confidence belt yields conservative confidence statements. By a randomization device, we can always make exact statements of form $P = 1 - \alpha$ even in the presence of discontinuity. The method was given by Stevens (1950).

Table 19.1 Cumulative probabilities for the binomial distribution ($n = 20$)

Proportion of successes p	$\pi = 0.1$	$\pi = 0.2$	$\pi = 0.3$	$\pi = 0.4$	$\pi = 0.5$
0.00	0.1216	0.0115	0.0008	—	—
0.05	0.3917	0.0692	0.0076	0.0005	—
0.10	0.6769	0.2061	0.0355	0.0036	0.0002
0.15	0.8670	0.4114	0.1071	0.0160	0.0013
0.20	0.9568	0.6296	0.2375	0.0510	0.0059
0.25	0.9887	0.8042	0.4164	0.1256	0.0207
0.30	0.9976	0.9133	0.6080	0.2500	0.0577
0.35	0.9996	0.9679	0.7723	0.4159	0.1316
0.40	0.9999	0.9900	0.8867	0.5956	0.2517
0.45		0.9974	0.9520	0.7553	0.4119
0.50		0.9994	0.9829	0.8725	0.5881
0.55		0.9999	0.9949	0.9435	0.7483
0.60		1.0000	0.9987	0.9790	0.8684
0.65			0.9997	0.9935	0.9423
0.70			1.0000	0.9984	0.9793
0.75				0.9997	0.9941
0.80				1.0000	0.9987
0.85					0.9998
0.90					1.0000

In fact, after we have drawn our sample and observed c successes, let us independently draw a random number x from the uniform distribution, where $0 \leq x \leq 1$. Then the variate

$$y = c + x \tag{19.7}$$

can take all values in the range 0 to $n + 1$. If y_0 is some given value $c_0 + x_0$, we have, writing π for the probability to be estimated,

$$\begin{aligned} P(y \geq y_0) &= P(c > c_0) + P(c = c_0)P(x \geq x_0) \\ &= \sum_{j=c_0+1}^{n} \binom{n}{j} \pi^j (1-\pi)^{n-j} + \binom{n}{c_0} \pi^{c_0} (1-\pi)^{n-c_0} (1 - x_0) \\ &= x_0 \sum_{j=c_0+1}^{n} \binom{n}{j} \pi^j (1-\pi)^{n-j} + (1 - x_0) \sum_{j=c_0}^{n} \binom{n}{j} \pi^j (1-\pi)^{n-j}. \end{aligned} \tag{19.8}$$

This expression defines a continuous probability distribution for y. We can therefore use this distribution to set confidence limits for π and our confidence statements based upon them will be exact.

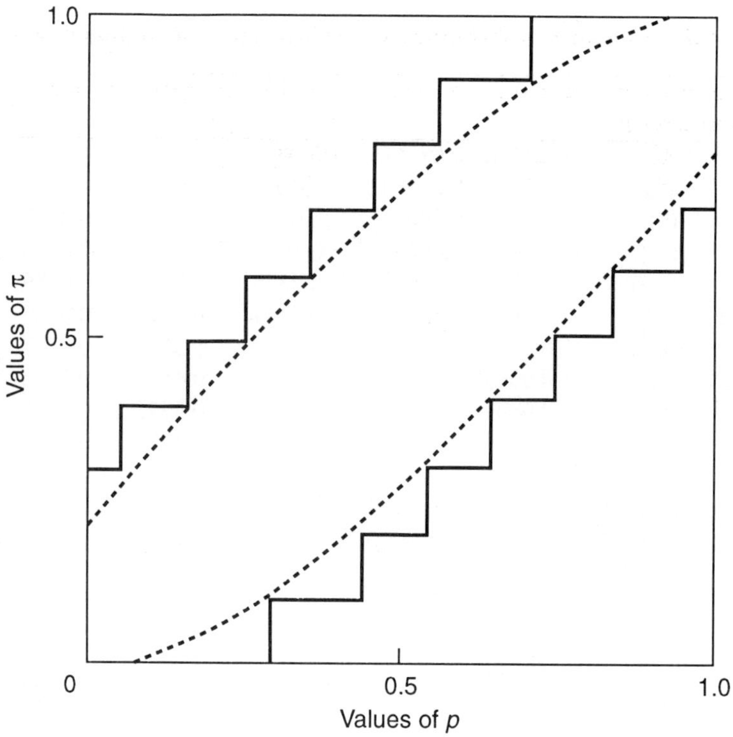

Fig. 19.2 Confidence limits for a binomial parameter ($n = 20$)

The confidence intervals are of the type exhibited in Fig. 19.3. The upper limit is now shifted to the right by amounts which, in effect, join up the discontinuities by a series of arcs. The lower limit also has a series of arcs, but there is no displacement to the right, and we have therefore shown on the diagram only the (dashed) approximate upper limit of Fig. 19.2. On our scale the lower approximate limit would almost coincide with the lower series of arcs. The general effect is to shorten the confidence interval.

19.11 It is at first sight surprising that the intervals set up in this way lie inside the conservative step intervals of Fig. 19.2, and are therefore no less accurate; for by taking an additional random number x we have imported additional uncertainty into the situation. A little reflection will show, however, that we have not got something for nothing. We have removed one uncertainty, associated with the inequality in $P \geq 1 - \alpha$, by bringing in another so as to make statements of the kind $P = 1 - \alpha$; what we lose on the second we more than offset by removing the first.

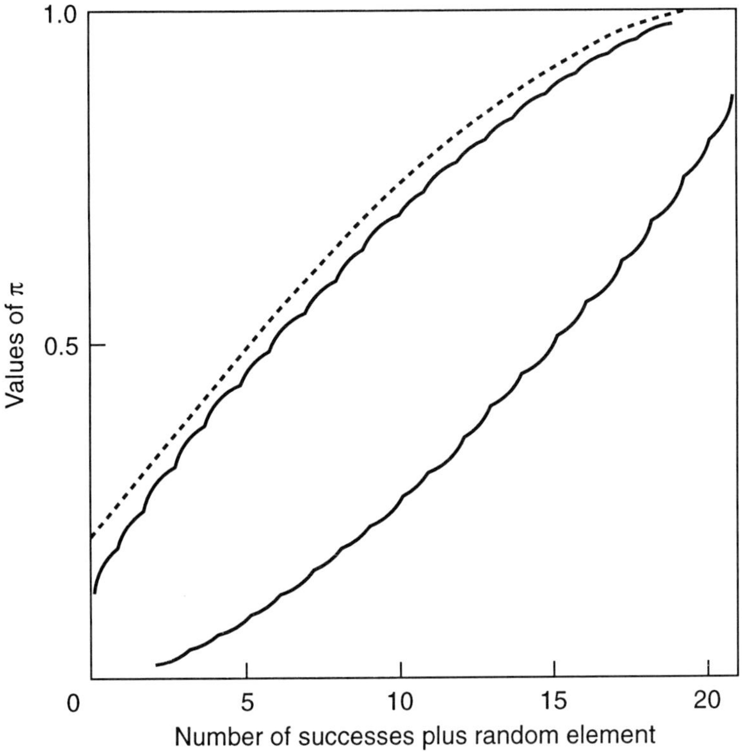

Fig. 19.3 Randomized confidence intervals for a binomial parameter

19.12 In **19.3** we saw that a central element in the construction of a confidence interval is that the probability statement

$$P\{z(x_1, \ldots, x_n, \theta) \leq z_1\} = 1 - \alpha \tag{19.9}$$

should be equivalent to

$$P\{t_0(x_1, \ldots, x_n) \leq \theta \leq t_1(x_1, \ldots, x_n)\} = 1 - \alpha. \tag{19.10}$$

In general, if we can find a function $z = z(x, \theta)$ such that the distribution of z is functionally independent of θ, z is called a *pivotal quantity*. In such circumstances we may write (19.9) as

$$P\{z \leq z_1\} = 1 - \alpha$$

without any dependence on θ and (19.10) follows, at least in principle. A full development of pivotal quantities and equivariant confidence intervals is given in **23.16–28**.

In the cases of Examples 19.1–3, the inversion caused no difficulty, the function z being essentially linear in θ. However, it may be far from straightforward if z is a more complicated function. Even in the relatively simple quadratic case, the results are somewhat unexpected. Suppose that

$$z(x_1, \ldots, x_n, \theta) - z_1$$

is quadratic in θ, and call it $Q(\theta, x)$. To pass from (19.9) to (19.10) we need to solve for θ in $Q(\theta, x) = 0$. The nature of the solution depends upon whether the roots of Q are real and upon the sign S of θ^2 in Q. If S is positive, the values of θ satisfying (19.9) lie between the real roots of $Q = 0$ (which are therefore t_0 and t_1 in (19.10)), but if S is negative, (19.9) requires θ to lie outside the interval between the real roots; while if roots of $Q = 0$ are complex, θ may take any value on the real line if S is negative, and if S is positive there is no real θ that satisfies (19.9).

In Example 19.4, this last event alone cannot occur.

Example 19.4 (Estimation of the ratio of means of two normal variables)

Let x, y be binormally distributed as in Example 18.14, all five parameters of the distribution being unknown, as in case (c) of that example. We make a convenient reparametrization and write $\rho\sigma_1\sigma_2 = \sigma_{12}$.

From a sample of n observations on (x, y), we now wish to estimate the ratio of means $\theta = \mu_2/\mu_1$, where we assume $\mu_1 \neq 0$. From Example 18.14(c) it follows that the ML estimator of σ_{12} is

$$\hat{\sigma}_{12} = \hat{\sigma}_1\hat{\sigma}_2\hat{\rho} = \frac{1}{n}\sum(x - \bar{x})(y - \bar{y}) = s_{12},$$

say. Similarly, we now write s_1^2, s_2^2 for the ML estimators of σ_1^2, σ_2^2 found in (18.78). The ML estimator of θ is $\hat{\theta} = \hat{\mu}_2/\hat{\mu}_1 = \bar{y}/\bar{x}$. It is intuitively clear that if $|\bar{x}|$ is small, we must expect poor precision in estimating θ. We now turn to the problem of finding confidence intervals for θ, where we shall see that if $\bar{x}^2/s_1^2$ is small, difficulties of the kind discussed above arise.

Consider the new variable $z_i = y_i - \theta x_i$ $(i = 1, 2, \ldots, n)$, which is normally distributed. The sample mean of the z_i is

$$\bar{z} = \bar{y} - \theta\bar{x},$$

and also has zero mean. For fixed θ, the variance of $\bar{z}$ is

$$\mathbf{V}_\theta(\bar{z}) = \sigma_z^2/n = \frac{1}{n}(\sigma_2^2 - 2\theta\sigma_{12} + \theta^2\sigma_1^2)$$

with unbiased estimator

$$\hat{\mathbf{V}}_\theta = \frac{1}{n(n-1)}\sum_{i=1}^{n}(z_i - \bar{z})^2 = \frac{1}{n-1}(s_2^2 - 2\theta s_{12} + \theta^2 s_1^2)$$

and $\hat{\mathbf{V}}_\theta/\mathbf{V}_\theta(\bar{z})$ is distributed, independently of $\bar{z}$, as a $\chi^2_{n-1}/(n-1)$ variable. Thus

$$t = \bar{z}/\hat{\mathbf{V}}_\theta^{1/2} = (\bar{y} - \theta\bar{x})\Big/\left\{\frac{1}{n-1}(s_2^2 - 2\theta s_{12} + \theta^2 s_1^2)\right\}^{1/2} \tag{19.11}$$

Table 19.2 The four principle cases for Example 19.4

a	$b^2 - ac$	Turning point	Form of interval
> 0	< 0	minimum	This case is impossible since (19.13) is always satisfied by $\theta = \hat{\theta} = \bar{y}/\bar{x}$.
> 0	≥ 0	minimum	Finite interval lying between roots given by (19.14); reduces to simple point θ at $b^2 = ac$.
< 0	> 0	maximum	Confidence region is all values lying outside the finite interval defined by the two roots.
< 0	< 0	maximum	Parabola (19.13) lies entirely below θ-axis, so the confidence region is the whole real line.

has a Student's t distribution with $\nu = n - 1$ degrees of freedom, a result due to Fieller (1940). Equivalently, $\bar{z}^2/\hat{\mathbf{V}}_\theta$ has an F-distribution with $(1, n-1)$ degrees of freedom (cf. **16.15**). From the d.f. of Student's t distribution, we may now find critical values t_α such that, for any θ,

$$P\left\{\frac{\bar{z}^2}{\hat{\mathbf{V}}_\theta} \leq t_\alpha^2\right\} = 1 - \alpha. \tag{19.12}$$

We now examine the nature of the confidence intervals for θ that emerge from (19.12).

The event $\bar{z}^2/\hat{\mathbf{V}}_\theta \leq t_\alpha^2$ in (19.12) can be rewritten in the form of a quadratic inequality in θ:

$$\left(\bar{x}^2 - \frac{t_\alpha^2}{n-1}s_1^2\right)\theta^2 - 2\left(\bar{x}\bar{y} - \frac{t_\alpha^2}{n-1}s_{12}\right)\theta + \left(\bar{y}^2 - \frac{t_\alpha^2}{n-1}s_2^2\right) \leq 0 \tag{19.13}$$

and the problem is to determine the values of θ that satisfy this inequality – these will constitute the $100(1-\alpha)$ per cent confidence interval for θ.

The left-hand side of (19.13) is a parabola in θ, of standard type $a\theta^2 - 2b\theta + c$.

The possible solutions depend upon the relative values of a and of $b^2 - ac$. We ignore degenerate cases such as $a = 0$. When they exist, the roots are given, as usual, by

$$\theta = \{b \pm (b^2 - ac)^{1/2}\}/a \tag{19.14}$$

The various cases are summarized in Table 19.2 (Fieller, 1992). The conditions $a > 0$ and $b^2 - ac > 0$ may always be achieved by choosing a sufficiently small value of $g = t_\alpha^2/(n-1)$. Thus, violation of one or both conditions is an indication that the information content of the data set is too weak to support construction of an interval at the desired confidence level, and a lower level must be chosen. Conversely, Koschat (1987) has shown that for this problem there is no procedure yielding confidence intervals that are always bounded when the confidence level is prespecified.

Finally, we observe that the condition $a < 0$ which led to our difficulties is, from (10.13), simply $\bar{x}^2/s_1^2 < t_\alpha^2/(n-1)$, so that, as we anticipated at the beginning of this example, it is the relative smallness of $|\bar{x}|$ that produces the trouble.

Scheffé (1970a) avoids the whole-real-line confidence interval by reformulating the problem slightly. In (19.13), he replaces $t_\alpha^2 = F_\alpha(1, n-1)$ by a positive monotone function of an F statistic which decreases from $2F$ at $F = F_\alpha(2, n-1)$ to $F_\alpha(1, n-1)$ as $F \to \infty$. As a result, one obtains *either* a finite interval *or* its complement for θ *or* an elliptical region for (μ_1, μ_2), with pre-assigned overall probability.

Shortest sets of confidence intervals

19.13 We have seen in Example 19.1 that for some problems there may be many sets of confidence intervals, and we must now consider what criteria to use in choosing among the possibilities. The problem is analogous to that of estimation, where we found that in general there are many different estimators for a parameter, but that we could sometimes find one (such as that with minimum variance) which was superior to the rest.

In Example 19.1 the problem presented itself in rather a limited form. We found that for intervals based on a normal sample mean $\bar{x}$ there were infinitely many sets of intervals with confidence coefficient $(1-\alpha)$ according to the way in which we selected α_0 and α_1, (subject to the condition that $\alpha_0 + \alpha_1 = \alpha$). Among these the central intervals are obviously the shortest, for since the distribution of $\bar{x}$ is unimodal and symmetric, a given range will include the greatest area of the normal distribution if it is centred at the mean. We might reasonably say that the central intervals are the best among those determined by $\bar{x}$.

But it does not follow that they are the shortest of all possible intervals, or even that a set exists that is uniformly the shortest for all values of θ. In general, for two sets of intervals c_1 and c_2, those of c_1 may be shorter than those of c_2 for some samples and longer in others.

Exercise 19.5 shows that the simplicity of Example 19.1 does not persist even in the case of the normal variance, discussed in Example 19.6; essentially this is because s^2, unlike $\bar{x}$, is asymmetrically distributed. A method for constructing physically shortest intervals is given in Exercise 19.25.

19.14 Although the idea of calculating the expected length of a confidence interval is intuitively appealing, and often useful for comparative purposes (e.g. in Exercise 19.5 and in **19.33**), expected length is not the principal criterion used in the theory of confidence intervals, for it is not merely the physical length of the interval that matters. The probability that an interval covers the true value of the parameter θ is fixed by the confidence coefficient, but we must also consider the probability that it covers other values of θ than the true one. If we can minimize this, we shall be taking account of the distribution of the interval over all values of θ. If there is a confidence interval that has smallest probability of covering every value of θ but the true one, we call it the *uniformly most accurate (UMA) interval.* Neyman (1937), who introduced this criterion, rather confusingly called it the 'shortest' interval. UMA intervals are considered in detail in **23.16–28**. In particular, we note that UMA intervals are invariant to monotone transformations of the parameter, whereas physically shortest intervals are not.

Further, it seems highly desirable that a good confidence interval should cover a value of θ with higher probability when it is the true value than when it is not, so that the confidence coefficient will exceed the probability of covering any false value. Such a confidence interval is called *unbiased* – this use of the term is unconnected with estimation bias, discussed in **17.9**.

Formally, if θ_T is the true value of the parameter, and $I(x)$ is the confidence interval, we say that $I(x)$ is *unbiased* if

$$P[\theta \in I(x)|\theta_T] \leq 1 - \alpha$$

for all $\theta \neq \theta_T$. $I(x)$ is UMA if

$$P[\theta \in I(x)|\theta_T] \leq P[\theta \in I^*(x)|\theta_T]$$

for all intervals $I^* \neq I$.

These ideas of accuracy and bias for intervals are, in effect, translations into confidence interval terms of fundamental notions in the theory of testing hypotheses, which will be treated in Chapters 20 and 21 – cf. **20.9**. The results to be obtained in those chapters on various optimum properties of tests will therefore translate into equivalent results for confidence intervals. Here we only add that their mathematical tractability ensured that the accuracy and bias criteria became central to the theory of confidence intervals.

Exercises 19.10–11 deal with a relationship between interval length and accuracy, and Madansky (1962) illuminates their difference with an example. Harter (1964) discusses other criteria for intervals.

19.15 Plante (1991) shows that even for UMA unbiased confidence intervals, anomalous outcomes are always a possibility (and may always occur(!); see Exercise 19.26) unless the confidence belt is *inclusion-consistent*. Suppose we have two confidence belts, $C(\mathbf{x})$ and $C'(\mathbf{x})$ with confidence levels $1 - \alpha$ and $1 - \alpha'$, for some $\mathbf{x}$; the belts are inclusion-consistent if

$$C(\mathbf{x}) \subset C'(\mathbf{x}) \quad \text{implies} \quad 1 - \alpha \leq 1 - \alpha'. \tag{IC}$$

Further, Plante (1991) shows that condition (IC) is equivalent to the condition that

> a vector of proper pivotal quantities exists for θ, with the range of x independent of θ and the range of θ values in the parameter space independent of x.

Further, each of these conditions is equivalent to the requirement that a confidence distribution should exist; see **19.6**. Thus the UMA unbiased interval may be inappropriate; in such circumstances, a better approach will often be to use a large-sample argument for pivotal quantities that exist only in an asymptotic sense.

Example 19.4 (Continued.)

Plante (1991) shows that the pivotal quantity defined by (19.11) does not satisfy the (IC) requirement and thus may lead to unsatisfactory results.

It should be noted that the condition (IC) is sufficient for intervals to be well behaved; as Example 19.4 shows, the interval may be well behaved even when it is not inclusion-consistent.

Choice of confidence intervals

19.16 The confidence intervals that we discussed in Examples 19.1 and 19.3 were in each case based on an unbiased sufficient statistic for the parameter, while in Examples 19.2 and 19.4 they were based on a set of jointly sufficient statistics. We shall see in **21.3** that (translated into confidence interval terms) there is no loss of accuracy if we restrict ourselves to functions of a sufficient statistic.

However, the exact distribution of the sufficient statistics may not be tractable, or problems may arise as in **19.15**. Therefore, we are led to consider procedures for large samples that will be generally applicable.

Confidence intervals for large samples

19.17 We have seen (**18.16**) that the first derivative of the logarithm of the likelihood function is, under regularity conditions, asymptotically normally distributed with zero mean and

$$\operatorname{var}\left(\frac{\partial \log L}{\partial \theta}\right) = E\left\{\left(\frac{\partial \log L}{\partial \theta}\right)^2\right\} = -E\left\{\frac{\partial^2 \log L}{\partial \theta^2}\right\}. \tag{19.15}$$

We may use this fact to set confidence intervals for θ in large samples. Writing

$$\psi(x, \theta) = \frac{\partial \log L}{\partial \theta} \bigg/ \left[E\left\{\left(\frac{\partial \log L}{\partial \theta}\right)^2\right\}\right]^{1/2}, \tag{19.16}$$

so that ψ is a standardized normal variate in large samples, we may determine confidence limits for θ in large samples if ψ is a monotonic function of θ, so that inequalities in one may be transformed to inequalities in the other. The following example illustrates the procedure.

Example 19.5 (Large-sample confidence intervals for the Poisson)

Consider the Poisson distribution whose general term is

$$f(x, \lambda) = \frac{e^{-\lambda}\lambda^x}{x!}, \qquad x = 0, 1, \ldots.$$

Here, as in Example 19.3 for the binomial, there is no function of x whose distribution does not depend on λ, but we could, as there, construct a conservative confidence belt – the analogue of Fig. 19.2 would extend over the whole positive quadrant for x and λ. We shall instead discuss the large-sample case as an illustration.

We have seen in Example 17.8 that

$$\frac{\partial \log L}{\partial \lambda} = \frac{n}{\lambda}(\bar{x} - \lambda). \tag{19.17}$$

Hence

$$-\frac{\partial^2 \log L}{\partial \lambda^2} = \frac{n\bar{x}}{\lambda^2}$$

and

$$E\left(-\frac{\partial^2 \log L}{\partial \lambda^2}\right) = \frac{n}{\lambda}. \tag{19.18}$$

Thus, from (19.15)–(19.16),

$$\psi = (\bar{x} - \lambda)\sqrt{(n/\lambda)}. \tag{19.19}$$

For example, with $1-\alpha = 0.95$, corresponding to a normal deviate ± 1.96, we have, for the central confidence limits,

$$(\bar{x}-\lambda)\surd(n/\lambda) = \pm 1.96,$$

giving, on squaring,

$$\lambda^2 - \left(2\bar{x} + \frac{3.84}{n}\right)\lambda + \bar{x}^2 = 0, \qquad (19.20)$$

a quadratic with roots always real. Thus

$$\lambda = \bar{x} + \frac{1.92}{n} \pm \left(\frac{3.84\bar{x}}{n} + \frac{3.69}{n^2}\right)^{1/2}, \qquad (19.21)$$

the confidence interval having the two roots as its upper and lower limits.

To order $n^{-1/2}$ this is equivalent to

$$\lambda = \bar{x} \pm 1.96\surd(\bar{x}/n), \qquad (19.22)$$

when the upper and lower limits are seen to be equidistant from the mean $\bar{x}$, as we should expect. Essentially, (19.22) is the confidence band obtained by estimating the Poisson mean and variance λ by $\bar{x}$. However, we note that the CI given by (19.21) contains only non-negative values, whereas (19.22) may include negative values. A similar large-sample result follows in the binomial case of Example 19.3 – see Exercise 19,3.

Hall (1982) gives improved one-sided intervals based on use of a continuity correction.

19.18 We may obtain a closer approximation, following Bartlett (1953), by finding higher cumulants of ψ, and using a Cornish–Fisher expansion as in **6.25–26**. Writing

$$\frac{\partial^r \log L}{\partial\theta^r} = d_r \quad\text{and}\quad E(d_r^s) = (r^s),$$
$$E(d_r^s d_t^u) = (r^s t^u),$$

(17.19) and (17.25) give

$$I = (1^2), \qquad (19.23)$$
$$= (-2) \qquad (19.24)$$

and from (17.18) and (17.19) the first two cumulants of $d_1 = \partial \log L/\partial\theta$ are

$$\kappa_1 = 0, \qquad (19.25)$$
$$\kappa_2 = I. \qquad (19.26)$$

We now seek

$$\kappa_3 = \mu_3 = (1^3) \quad\text{and}\quad \kappa_4 = \mu_4 - 3\mu_2^2 = (1^4) - 3I^2.$$

Differentiating (19.23) with respect to θ, we find

$$I' = 2(21) + (1^3) \qquad (19.27)$$

while differentiating (19.24) gives

$$I' = -(3) - (21). \tag{19.28}$$

Eliminating (21) from (19.27)–(19.28) gives

$$\kappa_3 = (1^3) = 3I' + 2(3) = -(3) - 3(21). \tag{19.29}$$

Differentiating each of (19.27) and (19.28) once more, we have respectively

$$I'' = 2(31) + 2(2^2) + 5(21^2) + (1^4) \tag{19.30}$$

and

$$I'' = -(4) - 2(31) - (2^2) - (21^2). \tag{19.31}$$

Eliminating (21^2) from (19.30) and (19.31), we find

$$(1^4) = 6I'' + 5(4) + 8(31) + 3(2^2) \tag{19.32}$$

while eliminating I'' instead gives the alternative form

$$(1^4) = -(4) - 4(31) - 3(2^2) - 6(21^2) \tag{19.33}$$

and

$$\kappa_4 = (1^4) - 3I^2.$$

Using the first four terms in (6.54) on $\psi = d_1/I^{1/2}$ with $l_1 = l_2 = 0$, we then have the statistic

$$\begin{aligned} T(\theta) = \frac{1}{\sqrt{I}}\Bigg[\frac{\partial \log L}{\partial\theta} &- \frac{1}{6}\frac{\kappa_3(d_1)}{I^2}\left\{\left(\frac{\partial \log L}{\partial\theta}\right)^2 - I\right\} \\ &- \frac{1}{24}\frac{\kappa_4(d_1)}{I^3}\left\{\left(\frac{\partial \log L}{\partial\theta}\right)^3 - 3I\frac{\partial \log L}{\partial\theta}\right\} \\ &+ \frac{1}{36}\frac{\kappa_3^2(d_1)}{I^4}\left\{4\left(\frac{\partial \log L}{\partial\theta}\right)^3 - 7I\frac{\partial \log L}{\partial\theta}\right\}\Bigg], \end{aligned} \tag{19.34}$$

which is, to the next order of approximation, normal with zero mean and unit variance. The first term is the quantity we have called ψ. The correction terms involve the standardized cumulants of d_1 which are equivalent to the cumulants of ψ.

Example 19.6 (Approximate confidence interval for the normal variance)

Let us consider the problem of setting confidence intervals for the variance of a normal distribution with mean μ and variance θ. The distribution of the sample variance is known to be skew, and we can compare the exact results with those given by the foregoing approximation. We first derive exact results. We know that

$$s^2 = \frac{1}{n}\sum(x - \bar{x})^2 \qquad \text{and} \qquad u^2 = \frac{1}{n}\sum(x - \mu)^2$$

are distributed as multiples θ/n of a chi-square variable with $n-1$ and n degrees of freedom respectively. If we take $n = 10$, Appendix Table 3 gives the lower and upper 5 per cent points of a χ^2_9 variate as 3.325 and 16.919, so

$$P\left(3.325 \leq \frac{10s^2}{\theta} \leq 16.919\right) = 0.90,$$

which we may invert to give

$$P(0.591s^2 \leq \theta \leq 3.008s^2) = 0.90.$$

Turning now to the large-sample approximation, we recall from Example 17.10 (where our present θ was called θ^2) that

$$\frac{\partial \log L}{\partial \theta} = d_1 = \frac{n}{2\theta^2}(u^2 - \theta). \tag{19.35}$$

Differentiating twice further we find

$$d_2 = \frac{-n}{\theta^3}\left(u^2 - \tfrac{1}{2}\theta\right), \tag{19.36}$$

$$d_3 = \frac{n}{\theta^4}(3u^2 - \theta), \tag{19.37}$$

so that by (19.24) and (19.36)

$$I = -(2) = \frac{n}{2\theta^2}$$

whence

$$I' = -\frac{n}{\theta^3}.$$

From (19.37) on taking expectations,

$$(3) = \frac{2n}{\theta^3}.$$

Using these results in (19.29), we find

$$\kappa_3 = 3I' + 2(3) = \frac{n}{\theta^3}. \tag{19.38}$$

Taking the expansion in (19.34) as far as κ_3 only, we find

$$\begin{aligned} T(\theta) &= \left(\frac{2\theta^2}{n}\right)^{1/2}\left[\frac{n}{2\theta^2}(u^2 - \theta) - \frac{n}{6\theta^3}\left(\frac{2\theta^2}{n}\right)^2\left\{\left(\frac{n}{2\theta^2}\right)^2(u^2-\theta)^2 - \frac{n}{2\theta^2}\right\}\right] \\ &= \left(\frac{2}{n}\right)^{1/2}\left[\frac{n}{2\theta}(u^2-\theta) - \frac{n}{6\theta^2}(u^2-\theta)^2 + \tfrac{1}{3}\right]. \end{aligned} \tag{19.39}$$

Replacing u^2 by $ns^2/(n-1)$, which has the same expected value, we obtain

$$T = \left(\frac{2}{n}\right)^{1/2}\left[\frac{n}{2}\left(\frac{ns^2}{(n-1)\theta} - 1\right) - \frac{n}{6}\left(\frac{ns^2}{(n-1)\theta} - 1\right)^2 + \frac{1}{3}\right]. \tag{19.40}$$

The first term gives us the confidence limits for θ based on ψ alone. The other terms will be corrections of lower order in n. We then have approximately, from (19.40),

$$T = \left(\frac{2}{n}\right)^{1/2}\left[\frac{n}{2}\left(\frac{ns^2}{(n-1)\theta}-1\right)-\frac{n}{6}\frac{2}{n}T^2+\frac{1}{3}\right],$$

giving

$$\frac{ns^2}{(n-1)\theta} = 1 + T\left(\frac{2}{n}\right)^{1/2} + \frac{2}{3n}T^2 - \frac{2}{3n}. \tag{19.41}$$

For example, with $n = 10$, $1-\alpha = 0.90$, $s^2 = 1$ and $T = \pm 1.6449$ (the 5 per cent points of the standardized normal distribution), we find for the limits of s^2/θ the values 0.3403 and 1.6644, and hence limits for θ of 0.601 and 2.939. The true values, as we saw above, are 0.591 and 3.008. For so low a value as $n = 10$ the approximation seems very fair.

19.19 We now prove a result due to Wilks (1938b), which is analogous to that of **18.16** showing that ML estimators have minimum variance asymptotically. Let x have density $f(x|\theta)$ and let the monotone function of θ

$$\zeta(x,\theta) = \frac{\sum_{j=1}^{n} h(x_j,\theta)}{\{n\,\mathrm{var}[h(x,\theta)]\}^{1/2}} \tag{19.42}$$

be a standardized sum of independent variables $h(x_j,\theta)$ with $E(h) = 0$, asymptotically normally distributed by the central limit theorem. We shall show that

$$h(x,\theta) = \frac{\partial \log f(x|\theta)}{\partial\theta},$$

which gives $\psi(x,\theta)$ of (19.16), yields confidence limits for θ that are asymptotically the shortest obtainable from any member of the class (19.42).

First, we differentiate (19.42) and find

$$\frac{\partial\zeta}{\partial\theta} = \sum_j \frac{\partial h(x_j,\theta)}{\partial\theta}\Big/\{n\,\mathrm{var}[h(x,\theta)]\}^{1/2} - \frac{\partial\,\mathrm{var}[h(x,\theta)]/\partial\theta}{2\,\mathrm{var}[h(x,\theta)]}\zeta(x,\theta). \tag{19.43}$$

Taking expectations, the second term on the right disappears since $E(\zeta) = 0$ and

$$E\left(\frac{\partial\zeta}{\partial\theta}\right) = \sum_j E\left\{\frac{\partial h(x,\theta)}{\partial\theta}\right\}\Big/\{n\,\mathrm{var}[h(x,\theta)]\}^{1/2}. \tag{19.44}$$

Now, just as in **17.14**, we differentiate

$$0 = E\{h(x,\theta)\} = \int_{-\infty}^{\infty} h(x,\theta) f(x|\theta)\,\mathrm{d}x$$

under the integral sign to obtain

$$0 = E\left\{\frac{\partial h(x,\theta)}{\partial\theta}\right\} + E\left\{h(x,\theta)\frac{\partial \log f(x|\theta)}{\partial\theta}\right\},$$

so that (19.44) becomes

$$E\left(\frac{\partial\zeta}{\partial\theta}\right) = -\left(\frac{n}{\operatorname{var} h}\right)^{1/2} E\left[h\frac{\partial \log f}{\partial\theta}\right]. \tag{19.45}$$

By the Cauchy–Schwarz inequality (2.49), (19.45) becomes

$$\left|E\left(\frac{\partial\zeta}{\partial\theta}\right)\right| \leq \left(\frac{n}{\operatorname{var} h}\right)^{1/2} \left\{E(h^2)E\left[\left(\frac{\partial \log f}{\partial\theta}\right)^2\right]\right\}^{1/2},$$

which since $E(h) = E(\partial \log f/\partial\theta) = 0$ gives

$$\left|E\left(\frac{\partial\zeta}{\partial\theta}\right)\right| \leq \left\{nE\left[\left(\frac{\partial \log f}{\partial\theta}\right)^2\right]\right\}^{1/2}, \tag{19.46}$$

the equality holding only when $h \propto \partial \log f/\partial\theta$. Apart from this case, we have

$$\left|E\left(\frac{\partial\zeta}{\partial\theta}\right)\right| < \left|E\left(\frac{\partial\psi}{\partial\theta}\right)\right|, \tag{19.47}$$

where ψ is defined at (19.16). Thus the absolute average rate of change of ψ with respect to θ is greater than that of ζ. We note that (19.42) is an EE in the sense used in **18.57** and that the condition for equality in (19.46) is precisely that for optimality of the EE in Exercise 18.32.

19.20 Because ψ is the most sensitive member of the class ζ to change in θ, we expect that the interval for θ based on it will be shorter. We now show that this is true asymptotically in n.

The central range of the distribution of ζ with probability $1 - \alpha$, is

$$-d_{\alpha/2} \leq \zeta(x, \theta) \leq +d_{\alpha/2}, \tag{19.48}$$

where G is the standardized normal d.f. and $G(-d_\alpha) = \alpha$. From this, we obtain the confidence interval

$$u_0 \leq \theta \leq u_1.$$

In the special case where $\zeta \equiv \psi$, we write the confidence interval as

$$t_0 \leq \theta \leq t_1.$$

At the end-points of (19.48), which correspond to u_0 and u_1, we expand ζ as a Taylor series about the true value θ_0:

$$\zeta(x, u_j) = \zeta(x, \theta_0) + (u_j - \theta_0)\left(\frac{\partial\zeta}{\partial\theta}\right)_{\theta'} = \pm d_{\alpha/2}, \tag{19.49}$$

where θ' lies between u and θ_0; similarly, for ψ we have

$$\psi(x, t_j) = \psi(x, \theta_0) + (t_j - \theta_0)\left(\frac{\partial\psi}{\partial\theta}\right)_{\theta''} = \pm d_{\alpha/2}, \tag{19.50}$$

where θ'' lies between t and θ_0. We now find, subtracting the lower end-point result from the upper in each of (19.49) and (19.50),

$$\begin{aligned}
&\zeta(x, u_1) - \zeta(x, u_0) \\
&= (u_1 - \theta_0)\left(\frac{\partial \zeta}{\partial \theta}\right)_{\theta'(1)} - (u_0 - \theta_0)\left(\frac{\partial \zeta}{\partial \theta}\right)_{\theta'_{(0)}} \\
&= \psi(x, t_1) - \psi(x, t_0) = (t_1 - \theta_0)\left(\frac{\partial \psi}{\partial \theta}\right)_{\theta''_{(1)}} - (t_0 - \theta_0)\left(\frac{\partial \psi}{\partial \theta}\right)_{\theta''_{(0)}} \\
&= 2d_{\alpha/2}.
\end{aligned} \tag{19.51}$$

Here we have used suffixes (1) and (0) to identify θ' and θ'' at upper and lower end-points.

As $n \to \infty$, u_1 and u_0, t_1 and t_0 all converge to the true value θ_0 and carry the θ's and θ''s with them since these are bounded by u and t. Moreover, the derivatives at θ_0, being sums of identical independent variates, tend to their expectations by the weak law of large numbers (**8.44–46**). Thus we find, from (19.51),

$$(u_1 - u_0)E\left(\frac{\partial \zeta}{\partial \theta}\right) = (t_1 - t_0)E\left(\frac{\partial \psi}{\partial \theta}\right). \tag{19.52}$$

Together with (19.47), (19.52) implies

$$u_1 - u_0 > t_1 - t_0, \tag{19.53}$$

so that asymptotically ψ gives shorter intervals.

Simultaneous confidence intervals for several parameters

19.21 We have already, in Examples 19.2 and 19.4, discussed cases involving several parameters when we require a confidence interval for one, or for a single function, of them, and we shall shortly be treating another such problem, that of the difference between two normal means. First, we briefly discuss the construction of intervals for several parameters simultaneously.

Given two parameters θ_1 and θ_2 and two statistics t and u, one approach would be to try to make simultaneous interval assertions of the type

$$P\{t_0 \leq \theta_1 \leq t_1 \text{ and } u_0 \leq \theta_2 \leq u_1\} = 1 - \alpha. \tag{19.54}$$

This, however, is rarely possible and may produce larger confidence regions (rectangles in this case) than could be achieved by recognizing the dependence between t and u. If, of course, the statistics are genuinely independent, we could satisfy (19.54) by constructing separate intervals each with confidence level $(1 - \alpha)^{1/2}$.

More commonly, we may establish simultaneous confidence regions of the form

$$P\{Q(\mathbf{t}, \boldsymbol{\theta}) \leq q\} = 1 - \alpha,$$

where $\mathbf{t}$ and $\boldsymbol{\theta}$ are the $p \times 1$ vectors of estimators and parameters, respectively. For example, if $\mathbf{x}$ is multivariate normal with mean vector $\boldsymbol{\theta}$ and covariance matrix $\Sigma = \mathbf{A}^{-1}$ (cf. **15.1–3**) it follows (**15.15**) that the quadratic form

$$Q(t,\theta) \equiv Q = n(\mathbf{t}-\boldsymbol{\theta})^{\mathrm{T}}\mathbf{A}(\mathbf{t}-\boldsymbol{\theta}) \tag{19.55}$$

is chi-square with p d.fr., where $\mathbf{t} \equiv \bar{\mathbf{x}}$. Thus Q is a pivotal quantity and the region

$$\{\boldsymbol{\theta} : Q(t,\theta) \le q_\alpha\}$$

forms a p-dimensional confidence ellipsoid with q_α determined by the χ^2 distribution.

Draper and Guttman (1995) discuss the relative sizes of confidence regions given by the ellipsoid (E) and by separate intervals for each parameter (the rectangular region R) and show how to equate the volumes of E and R by adjusting the confidence coefficient as p increases.

19.22 When the estimators are not exactly multinormal, we may develop large-sample intervals akin to (19.55) in which $\mathbf{A}$ is replaced by

$$\mathbf{I} = -E\left[\frac{\partial^2 \log L}{\partial\boldsymbol{\theta}\,\partial\boldsymbol{\theta}^{\mathrm{T}}}\right], \tag{19.56}$$

where $\mathbf{I}$ is the information matrix given in (17.88).

The theorem of **19.19–20** concerning shortest intervals was generalized by Wilks and Daly (1939). Under fairly general conditions the large-sample regions for k parameters that are smallest on average are given by

$$\sum_{i=1}^{k}\sum_{j=1}^{k}\left\{I_{ij}^{-1}\frac{\partial \log L}{\partial\theta_i}\frac{\partial \log L}{\partial\theta_j}\right\} \le d_\alpha \tag{19.57}$$

where $\mathbf{I}^{-1} = \{I_{ij}^{-1}\}$ is the inverse of the information matrix given in (19.56) and $P\{\chi_k^2 \le q_\alpha\} = 1-\alpha$. This is clearly related to the result of **17.39** giving the minimum attainable variances (and, by a simple extension, covariances) of a set of unbiased estimators of several parametric functions.

The theory of simultaneous confidence intervals was first developed in Roy and Bose (1953) and Roy (1954). Bartlett (1953; 1955) discussed the generalization of the method of **19.18** to the case of two or more unknown parameters. A number of authors consider large-sample intervals for non-linear functions of the parameters; see, for example, the discussion on curvature measures in **17.43** and Clarke (1987a; 1987b), Hamilton *et al.* (1982) and Cook and Witmer (1985).

Bootstrap confidence intervals

19.23 An alternative to large-sample approximation is the use of the bootstrap; the basic computational details are given in **10.17–22**. Following Efron (1985; 1987) we may draw B samples of size n with replacement form the original sample of n observations and generate the parameter estimates $\theta_j^*\{j = 1, \ldots, B\}$ from each such sample. The bootstrap confidence interval is then determined as the interval between the $\alpha/2$ and $(1-\alpha/2)$ quantities of the empirical distribution. A number of improvements have been developed by Efron and others; see, for example, Hall (1988). In general, values of B of 2000 or higher appear to be desirable for interval estimation, as opposed to the 200 or so adequate for point estimation.

19.24 Building on these ideas, DiCiccio and Efron (1992) develop what they term approximate bootstrap confidence intervals that do not require Monte Carlo simulations. The large-sample intervals based upon (19.56) approximate the confidence level to $O(n^{-1/2})$ whereas the approximate bootstrap confidence intervals improve the approximation to $O(n^{-1})$ by taking account of the curvature in non-linear problems.

The problem of two means

19.25 We now turn to the problem of finding a confidence interval for the difference between the means of two normal distributions.

Suppose that we have two normal distributions, the first with mean and variance parameters μ_1, σ_1^2 and the second with parameters μ_2, σ_2^2. Samples of size n_1, n_2 respectively are taken, and the sample means and variances observed are $\bar{x}_1$, s_1^2 and $\bar{x}_2$, s_2^2. Without loss of generality, we assume $n_1 \leq n_2$.

When $\sigma_1^2 = \sigma_2^2 = \sigma^2$, the problem of finding an interval for $\mu_1 - \mu_2 = \delta$ is simple. For in this case $d = \bar{x}_1 - \bar{x}_2$ is normally distributed with

$$E(d) = \delta$$
$$\operatorname{var} d = \sigma^2\left(\frac{1}{n_1} + \frac{1}{n_2}\right),$$

and $n_1 s_1^2/\sigma_1^2$, $n_2 s_2^2/\sigma_2^2$ are each distributed as χ^2 with $n_1 - 1$, $n_2 - 1$ d.fr., respectively. Since the two samples are independent, $(n_1 s_1^2 + n_2 s_2^2)/\sigma^2$ will be distributed as χ^2 with $n_1 + n_2 - 2$ d.fr., and hence, writing

$$s^2 = (n_1 s_1^2 + n_2 s_2^2)/(n_1 + n_2 - 2),$$

we have

$$E(s^2) = \sigma^2.$$

Now

$$y = \frac{(d-\delta)/(s^2/\sigma^2)^{1/2}}{\left\{\sigma^2\left(\frac{1}{n_1} + \frac{1}{n_2}\right)\right\}^{1/2}} = \frac{d-\delta}{\left\{s^2\left(\frac{1}{n_1} + \frac{1}{n_2}\right)\right\}^{1/2}} \qquad (19.58)$$

is the ratio of a standardized normal variate to the square root of an unbiased estimator of its sampling variance, which is distributed independently of it (since s_1^2 and s_2^2 are independent of $\bar{x}_1$ and $\bar{x}_2$). Moreover, $(n_1 + n_2 - 2)s^2/\sigma^2$ is a χ^2 variable with $n_1 + n_2 - 2$ d.fr. Thus y is of exactly the same form as the one-sample ratio

$$\frac{\bar{x}_1 - \mu_1}{\{s_1^2/(n_1 - 1)\}^{1/2}} = \frac{\bar{x}_1 - \mu_1}{(\sigma^2/n_1)^{1/2}} \Big/ \left\{\frac{n_1 s_1^2/(n_1 - 1)}{\sigma^2}\right\}^{1/2}$$

which we have on several occasions (e.g. Example 11.8) seen to follow Student's t distribution with $n_1 - 1$ d.fr. Hence (19.58) is also a Student's t variable, but with $n_1 + n_2 - 2$ d.fr., a result which may easily be proved directly.

There is therefore no difficulty in setting confidence intervals for δ in this case – it is a straightforward analogue of the one-sample case treated in Example 19.2.

19.26 When we leave the case $\sigma_1^2 = \sigma_2^2$, complications arise. The variate distributed in Student's t form, with $n_1 + n_2 - 2$ d.fr., by analogy with (19.58), is now

$$t = \frac{d-\delta}{\left\{\frac{\sigma_1^2}{n_1} + \frac{\sigma_2^2}{n_2}\right\}^{1/2}} \Bigg/ \left\{\frac{\frac{n_1 s_1^2}{\sigma_1^2} + \frac{n_1 s_2^2}{\sigma_2^2}}{n_1 + n_2 - 2}\right\}^{1/2}. \tag{19.59}$$

The numerator of (19.59) is a standardized normal variate, and its denominator is the square root of an independently distributed χ^2 variate divided by its degrees of freedom, as for (19.58). The difficulty is that (19.59) involves the unknown ratio of variances $\theta = \sigma_1^2/\sigma_2^2$. If we also define $u = s_1^2/s_2^2$, $N = n_1/n_2$, we may rewrite (19.59) as

$$t = \frac{(d-\delta)(n_1 + n_2 - 2)^{1/2}}{s_2\left\{\left(1 + \frac{\theta}{N}\right)\left(1 + \frac{Nu}{\theta}\right)\right\}^{1/2}}, \tag{19.60}$$

which clearly displays its dependence upon the unknown θ. If θ is known (19.60) essentially reduces to (19.58).

We now have to consider methods by which the nuisance parameter θ can be eliminated from interval statements concerning δ. We must clearly seek some statistic other than t of (19.60). One possibility suggests itself immediately from inspection of (19.58). The statistic

$$z = \frac{d-\delta}{\left(\frac{s_1^2}{n_1-1} + \frac{s_2^2}{n_2-1}\right)^{1/2}} \tag{19.61}$$

is, like (19.58), the ratio of a normal variate with zero mean to the square root of an independently distributed unbiased estimator of its sampling variance. However, as we shall see in **19.27**, that estimator is not a multiple of a χ^2 variate, and hence z is not distributed in Student's t form. The statistic z is the basis of an approximate confidence interval approach to this problem, as we shall see below, as well as of another (fiducial) method, to be discussed later.

An alternative possibility is to investigate the distribution of (19.58) itself, i.e. to see how far the statistic appropriate to the case $\theta = 1$ retains its properties when $\theta \neq 1$. This, too, has been investigated from the confidence interval standpoint.

However, before discussing these approaches, we examine an exact confidence interval solution to this problem, based on Student's t distribution, and its properties. The results are due to Scheffé (1943; 1944).

Exact confidence intervals based on Student's t distribution

19.27 If we seek an exact confidence interval for δ based on Student's t distribution, it will be sufficient if we can find a linear function of the observations, L, and a quadratic function of them, Q, such that, for all values of σ_1^2, σ_2^2:

(1) L and Q are independently distributed;

(2) $E(L) = \delta$ and $\operatorname{var} L = V$; and
(3) Q/V has a χ^2 distribution with k d.fr.

Then

$$t = \frac{L - \delta}{(Q/k)^{1/2}} \tag{19.62}$$

has Student's t distribution with k d.fr.

Scheffé (1944) showed that no statistic of the form (19.62) can be a symmetric function of the observations in each sample; that is to say, t cannot be invariant under permutation of the first sample members x_{1i} $(i = 1, 2, \ldots, n_1)$ among themselves and of the second sample members x_{2i} $(i = 1, 2, \ldots, n_2)$ among themselves.

For suppose that t is symmetric in the sense indicated. Then we must have

$$\begin{aligned} L &= c_1 \sum_i x_{1i} + c_2 \sum_i x_{2i}, \\ Q &= c_3 \sum x_{1i}^2 + c_4 \sum_{i \neq j} x_{1i} x_{1j} + c_5 \sum x_{2i}^2 + c_6 \sum_{i \neq j} x_{2i} x_{2j} + c_7 \sum_{i,j} x_{1i} x_{2j}, \end{aligned} \tag{19.63}$$

where the cs are constants independent of the parameters.

Now, from (2) above,

$$E(L) = \delta = \mu_1 - \mu_2, \tag{19.64}$$

while, from (19.63),

$$E(L) = c_1 n_1 \mu_1 + c_2 n_2 \mu_2. \tag{19.65}$$

Expressions (19.64) and (19.65) are identities in μ_1 and μ_2; hence

$$c_1 n_1 \mu_1 = \mu_1, \quad c_2 n_2 \mu_2 = -\mu_2,$$

so that

$$c_1 = 1/n_1, \quad c_2 = -1/n_2. \tag{19.66}$$

From (19.66) and (19.63),

$$L = \bar{x}_1 - \bar{x}_2 = d, \tag{19.67}$$

and hence

$$\operatorname{var} L = V = \sigma_1^2/n_1 + \sigma_2^2/n_2. \tag{19.68}$$

Since Q/V has a χ^2 distribution with k d.fr.,

$$E(Q/V) = k,$$

so that, using (19.68),

$$E(Q) = k(\sigma_1^2/n_1 + \sigma_2^2/n_2), \tag{19.69}$$

while, from (19.63),

$$\begin{aligned} E(Q) = {} & c_3 n_1 (\sigma_1^2 + \mu_1^2) + c_4 n_1 (n_1 - 1) \mu_1^2 + c_5 n_2 (\sigma_2^2 + \mu_2^2) \\ & + c_6 n_2 (n_2 - 1) \mu_2^2 + c_7 n_1 n_2 \mu_1 \mu_2. \end{aligned} \tag{19.70}$$

Equating (19.69) and (19.70), we obtain expressions for the cs, and thence find

$$Q = k\left\{\frac{s_1^2}{n_1 - 1} + \frac{s_2^2}{n_2 - 1}\right\}. \tag{19.71}$$

Expressions (19.67) and (19.71) reduce (19.62) to (19.61). Now a linear function of two independent χ^2 variates can only itself have a χ^2 distribution if it is a simple sum of them, and $n_1 s_1^2/\sigma_1^2$ and $n_2 s_2^2/\sigma_2^2$ are independent χ^2 variates. Thus, from (19.71), Q will only be a χ^2 variate if

$$\frac{k\sigma_1^2}{n_1(n_1 - 1)} = \frac{k\sigma_2^2}{n_2(n_2 - 1)} = 1$$

or

$$\theta = \frac{\sigma_1^2}{\sigma_2^2} = \frac{n_1(n_1 - 1)}{n_2(n_2 - 1)}. \tag{19.72}$$

Given n_1, n_2, this is only true for special values of σ_1^2, σ_2^2; but we require it to be true for *all* values of σ_1^2, σ_2^2 and thus t cannot be a symmetric function in the sense stated.

19.28 Since we cannot find a symmetric function of the desired type having Student's t distribution, we now consider others. We specialize (19.62) to the situation where

$$\left.\begin{aligned} L &= \sum_{i=1}^{n_1} d_i/n_1, \\ Q &= \frac{1}{n_1}\sum_{i=1}^{n_1}(d_i - L)^2, \end{aligned}\right. \tag{19.73}$$

and the d_i are independent identical normal variates with

$$E(d_i) = \delta, \quad \operatorname{var} d_i = \sigma^2, \tag{19.74}$$

for all i. It will be remembered that we have taken $n_1 \leq n_2$. Expression (19.62) now becomes

$$t = \frac{L - \delta}{\{Q/(n_1 - 1)\}^{1/2}} = (L - \delta)\left\{\frac{n_1(n_1 - 1)}{\sum(d_i - L)^2}\right\}^{1/2}, \tag{19.75}$$

which is a Student's t variate with $n_1 - 1$ d.fr.

Suppose now that, in terms of the original observations,

$$d_i = x_{1i} - \sum_{j=1}^{n_2} c_{ij}x_{2j}. \tag{19.76}$$

The d_i are multinormally distributed, since they are linear functions of normal variates (cf. **15.4**). Necessary and sufficient conditions for (19.74) to hold are

$$\left.\begin{aligned} &\textstyle\sum_j c_{ij} = 1, \\ &\textstyle\sum_j c_{ij}^2 = c^2, \\ \text{and, for } i \neq k, \quad &\textstyle\sum_j c_{ij}c_{kj} = 0. \end{aligned}\right\} \tag{19.77}$$

Thus, from (19.76) and (19.77),

$$\operatorname{var} d_i = \sigma^2 = \sigma_1^2 + c^2\sigma_2^2. \tag{19.78}$$

19.29 The central confidence interval, with confidence coefficient $1-\alpha$, derived from (19.75) is

$$|L-\delta| \leq t_{n_1-1,\alpha}\{Q/(n_1-1)\}^{1/2}, \tag{19.79}$$

where $t_{n_1-1,\alpha}$ is the appropriate deviate for n_1-1 d.fr. The interval length l has expected value, from (19.79),

$$E(l) = 2t_{n_1-1,\alpha}\frac{\sigma}{\{n_1(n_1-1)\}^{1/2}}E\left\{\left(\frac{n_1 Q}{\sigma^2}\right)^{1/2}\right\}, \tag{19.80}$$

the last factor on the right being found, from (16.7), since n_1Q/σ^2 has a χ^2 distribution with n_1-1 d.fr., to be

$$E\left\{\left(\frac{n_1 Q}{\sigma^2}\right)^{1/2}\right\} = \frac{2^{1/2}\Gamma(\frac{1}{2}n_1)}{\Gamma\{\frac{1}{2}(n_1-1)\}}. \tag{19.81}$$

To minimize the expected length (19.80), we must minimize σ, or equivalently, minimize c^2 in (19.78), subject to (19.77). The problem may be visualized geometrically as follows: consider a space of n_2 dimensions, with one axis for each *second* suffix of the c_{ij}. Then $\sum_j c_{ij} = 1$ is a hyperplane, and $\sum_j c_{ij}^2 = c^2$ is an n_2-dimensional hypersphere which is intersected by the plane in an (n_2-1)-dimensional hypersphere. We must locate $n_1 \leq n_2$ vectors through the origin which touch this latter hypersphere and (to satisfy the last condition of (19.77)) are mutually orthogonal, in such a way that the radius of the n_2-dimensional hypersphere is minimized. This can be done by making our vectors coincide with n_1 axes, and then $c^2 = 1$. But if $n_1 < n_2$, we can improve upon this procedure, for we can, while keeping the vectors orthogonal, space them symmetrically about the equiangular vector, and reduce c^2 from 1 to its minimum value n_1/n_2, as we shall now show.

19.30 Written in vector form, the conditions (19.77) are

$$\left.\begin{aligned} \mathbf{c}_i^T\mathbf{u} &= 1 & & \\ \mathbf{c}_i^T\mathbf{c}_k &= c^2 & & i=k, \\ &= 0 & & i\neq k, \end{aligned}\right\} \tag{19.82}$$

where $\mathbf{c}_i$ is the ith column vector of the matrix $\{c_{ij}\}$ and $\mathbf{u}$ is the unit column vector.

If the n_1 vectors $\mathbf{c}_i$ satisfy (19.82), we can add another n_2-n_1 vectors, satisfying the normalizing and orthogonalizing condition of (19.82), so that the augmented set forms a basis for an n_2-space. We may therefore express $\mathbf{u}$ as a linear function of the n_2 $\mathbf{c}$-vectors,

$$\mathbf{u} = \sum_{k=1}^{n_2} g_k\mathbf{c}_k, \tag{19.83}$$

where the g_k are scalars. Now, using (19.82) and (19.83),

$$\begin{aligned}1 = \mathbf{c}_i^T \mathbf{u} = \mathbf{c}_i^T \sum_{k=1}^{n_2} g_k \mathbf{c}_k &= \sum g_k \mathbf{c}_i^T \mathbf{c}_k \\ &= g_i c^2.\end{aligned}$$

Thus

$$g_i = 1/c^2, \qquad i = 1, 2, \ldots, n_1. \tag{19.84}$$

Also, since $\mathbf{u}$ is the unit column vector,

$$n_2 = \mathbf{u}^T \mathbf{u} = \Big(\sum_k g_k \mathbf{c}_k^T\Big)\Big(\sum_k g_k \mathbf{c}_k\Big),$$

which, on using (19.82), becomes

$$\begin{aligned}n_2 &= \sum_{k=1}^{n_2} g_k^2 \mathbf{c}_k^T \mathbf{c}_k \\ &= c^2 \Big(\sum_{k=1}^{n_1} + \sum_{n_1+1}^{n_2}\Big) g_k^2.\end{aligned} \tag{19.85}$$

Use of (19.84) gives, from (19.85),

$$n_2 = c^2 \left\{ n_1/c^4 + \sum_{n_1+1}^{n_2} g_k^2 \right\},$$

so

$$n_2 \geq \frac{n_1}{c^2}.$$

Hence

$$c^2 \geq n_1/n_2, \tag{19.86}$$

the required result.

19.31 The equality sign holds in (19.86) whenever $g_k = 0$ for $k = n_1 + 1, \ldots, n_2$. Then the equiangular vector $\mathbf{u}$ lies entirely in the space spanned by the original n_1 $\mathbf{c}$-vectors. From (19.84), these will be symmetrically disposed around it. Evidently, there is an infinite number of ways of determining c_{ij}, merely by rotating the set of n_1 vectors. Scheffé (1943) obtained the particularly appealing solution

$$\begin{aligned} c_{jj} &= (n_1/n_2)^{1/2} - (n_1 n_2)^{-1/2} + 1/n_2, && j = 1, 2, \ldots, n_1, \\ c_{ij} &= -(n_1 n_2)^{-1/2} + 1/n_2, && j\ (\neq i) = 1, 2, \ldots, n_1, \\ c_{ij} &= 1/n_2, && j = n_1 + 1, \ldots, n_2. \end{aligned} \tag{19.87}$$

It may easily be confirmed that (19.87) satisfies the conditions (19.77) with $c^2 = n_1/n_2$. Substituted into (19.76), (19.87) gives

$$d_i = x_{1i} - (n_1/n_2)^{1/2} x_{2i} + (n_1 n_2)^{-1/2} \sum_{j=1}^{n_1} x_{2j} + (1/n_2) \sum_{j=1}^{n_2} x_{2j}, \tag{19.88}$$

which yields in (19.73)

$$\begin{aligned} L &= \bar{x}_1 - \bar{x}_2, \\ Q &= \frac{1}{n_1} \sum_{i=1}^{n_1} (u_i - \bar{u})^2, \end{aligned} \tag{19.89}$$

where

$$\begin{aligned} u_i &= x_{1i} - (n_1/n_2)^{1/2} x_{2i}, \\ \bar{u} &= \sum_{i=1}^{n_1} u_i / n_1. \end{aligned} \tag{19.90}$$

Hence, from (19.75) and (19.88)–(19.90),

$$\{\bar{x}_1 - \bar{x}_2 - \delta\} \left\{ \frac{n_1(n_1 - 1)}{\sum (u_i - \bar{u})^2} \right\}^{1/2} \sim t_{n_1} - 1, \tag{19.91}$$

a Student's t variate with $n_1 - 1$ d.fr., and we may proceed to set confidence limits for $\delta = \mu_1 - \mu_2$.

Bain (1967) derives this procedure by a purely algebraic method, and also applies the method to other problems.

19.32 It is rather remarkable that we have been able to find an exact solution of the confidence interval problem in this case only by abandoning the natural requirement of symmetry. Expressions (19.91) holds for *any* randomly selected subset of n_1 of the n_2 variates in the second sample. Just as, in **19.10**, we resorted to randomization to remove the difficulty in making exact confidence interval statements about a discrete variable, so we find here that randomization allows us to bypass the nuisance parameter θ. But the extent of the randomisation should not be exaggerated. The numerator of (19.91) uses the sample means of both samples, complete; only the denominator varies with different random selections of the subset in the second sample. It is impossible to assess intuitively how much efficiency is lost by this procedure. We now examine the length of the confidence intervals it provides.

19.33 From (19.78) and (19.86), we have, for the solution (19.88),

$$\operatorname{var} d_i = \sigma^2 = \sigma_1^2 + (n_1/n_2)\sigma_2^2. \tag{19.92}$$

Putting (19.92) into (19.80), and using (19.81), we have for the expected length of the confidence interval

$$E(l) = 2t_{n_1-1,\alpha} \left\{ \frac{\sigma_1^2 + (n_1/n_2)\sigma_2^2}{n_1(n_1 - 1)} \right\}^{1/2} \frac{2^{1/2}\Gamma(\frac{1}{2}n_1)}{\Gamma\{\frac{1}{2}(n_1 - 1)\}}. \tag{19.93}$$

Table 19.3 E(l)/E(L) (from Scheffé, 1943)

		$1-\alpha=0.95$					$1-\alpha=0.99$				
	n_2-1	5	10	20	40	∞	5	10	20	40	∞
n_1-1	5	1.15	1.20	1.23	1.25	1.28	1.27	1.36	1.42	1.47	1.52
	10		1.05	1.07	1.09	1.11		1.10	1.13	1.16	1.20
	20			1.03	1.03	1.05			1.05	1.06	1.09
	40				1.01	1.02				1.02	1.04
	∞					1					1

We now compare this interval l with the interval L obtained from (19.59) if $\theta=\sigma_1^2/\sigma_2^2$ is known. The latter has expected length

$$E(L)=2t_{n_1+n_2-2,\alpha}\left\{\frac{\sigma_1^2/n_1+\sigma_2^2/n_2}{n_1+n_2-2}\right\}^{1/2}E\left\{\frac{n_1s_1^2}{\sigma_1^2}+\frac{n_2s_2^2}{\sigma_2^2}\right\}^{1/2}, \tag{19.94}$$

the last factor being evaluated by (16.7) from the χ^2 distribution with (n_1+n_2-2) d.fr. as

$$\frac{2^{1/2}\Gamma\{\frac{1}{2}(n_1+n_2-1)\}}{\Gamma\{\frac{1}{2}(n_1+n_2-2)\}}. \tag{19.95}$$

Expressions (19.93)–(19.95) give for the ratio of expected lengths

$$E(l)/E(L)=\frac{t_{n_1-1,\alpha}}{t_{n_1+n_2-2,\alpha}}\left(\frac{n_1+n_2-2}{n_1-1}\right)^{1/2}\frac{\Gamma(\frac{1}{2}n_1)\Gamma\{\frac{1}{2}(n_1+n_2-2)\}}{\Gamma\{\frac{1}{2}(n_1-1)\}\Gamma\{\frac{1}{2}(n_1+n_2-1)\}}. \tag{19.96}$$

As $n_1\to\infty$ with n_2/n_1 fixed, each of the three factors of (19.96) tends to 1, and therefore the ratio of expected interval length does so, as is intuitively reasonable. For small n_1, the first two factors exceed 1, but the last is less than 1. Table 19.3 gives the exact values of (19.96) for $1-\alpha=0.95, 0.99$ and a few sample sizes.

Evidently, l is a very efficient interval even for moderate sample sizes, having an expected length no more than 11 per cent greater than that of L for $n_1-1\geq 10$ at $1-\alpha=0.95$, and no more than 9 per cent greater than for $n_1-1\geq 20$ at $1-\alpha=0.99$. Furthermore, we are comparing it with an interval *based on knowledge of* θ. Taking this into account, we may fairly say that the element of randomization cannot have resulted in much loss of efficiency.

In addition to this solution to the two-means problem there are also approximate confidence interval solutions, which we now summarize.

Approximate confidence interval solutions

19.34 Welch (1938) has investigated the approximate distribution of the statistic (19.58), which is a Student's t variate when $\sigma_1^2=\sigma_2^2$, in the case $\sigma_1^2\neq\sigma_2^2$. In this case, the sampling variance of its numerator is

$$\text{var}(d-\delta)=\sigma_1^2/n_1+\sigma_2^2/n_2,$$

so that, writing

$$u = (d - \delta)/(\sigma_1^2/n_1 + \sigma_2^2/n_2)^{1/2}$$
$$w^2 = s^2\left(\frac{1}{n_1} + \frac{1}{n_2}\right)\bigg/\left(\frac{\sigma_1^2}{n_1} + \frac{\sigma_2^2}{n_2}\right), \tag{19.97}$$

(19.58) may be written

$$y = u/w. \tag{19.98}$$

The difficulty now is that w^2, although distributed independently of u, is not a multiple of a χ^2 variate when $\theta = \sigma_1^2/\sigma_2^2 \neq 1$. However, by equating its first two moments to those of a χ^2 variate, we can determine a number of degrees of freedom, ν, for which it is *approximately* a χ^2 variate. Its mean and variance are, from (19.97),

$$E(w^2) = b(\nu_1\theta + \nu_2),$$
$$\text{var}(w^2) = 2b^2(\nu_1\theta^2 + \nu_2), \tag{19.99}$$

where we have written

$$\nu_1 = n_1 - 1, \qquad \nu_2 = n_2 - 1,$$
$$b = (n_1 + n_2)\sigma_2^2/\{(n_1 + n_2 - 2)(n_2\sigma_1^2 + n_1\sigma_2^2)\}. \tag{19.100}$$

If we identify (19.99) with the moments of a multiple g of a χ^2 variate with ν d.fr.,

$$\mu_1' = g\nu, \qquad \mu_2 = 2g^2\nu, \tag{19.101}$$

we find

$$g = b(\theta^2\nu_1 + \nu_2)/(\theta\nu_1 + \nu_2),$$
$$\nu = (\theta\nu_1 + \nu_2)^2/(\theta^2\nu_1 + \nu_2). \tag{19.102}$$

With these values of g and ν, w^2/g is approximately a χ^2 variate with ν degrees of freedom, and hence, from (19.97),

$$t = u\bigg/\left\{\frac{w^2}{g\nu}\right\}^{1/2} \tag{19.103}$$

is a Student's t variate with ν d.fr. If $\theta = 1$, $\nu = \nu_1 + \nu_2 = n_1 + n_2 - 2$, $g = b = 1/\nu$, and (19.103) reduces to (19.58), as it should. But in general, g and ν depend upon θ.

ν is never greater than at $\theta = 1$ – cf. Exercise 19.22.

19.35 Welch (1938) investigated the extent to which the assumption that $\theta = 1$ in (19.103), when in reality it takes some other value, leads to erroneous conclusions. (His discussion was couched in terms of testing hypotheses rather than of interval estimation, which is our present concern.) He found that, so long as $n_1 = n_2$, no great harm was done by ignorance of the true value of θ, but that if $n_1 \neq n_2$, serious errors could arise. To overcome this difficulty, he used the technique of **19.34** to approximate the distribution of the statistic z of (19.61). In this case he found that, whatever the values of n_1 and n_2, z approximately followed Student's t distribution with

$$\nu = \left(\frac{\theta}{n_1} + \frac{1}{n_2}\right)^2\bigg/\left(\frac{\theta^2}{n_1^2(n_1 - 1)} + \frac{1}{n_2^2(n_2 - 1)}\right) \tag{19.104}$$

degrees of freedom, and that the influence of a wrongly assumed value of θ was now very much smaller. This is what we should expect, since the denominator of z at (19.61) estimates the variances σ_1^2, σ_2^2 separately, while that of t at (19.103) uses a 'pooled' estimate s^2, that is clearly not appropriate when $\sigma_1^2 \neq \sigma_2^2$.

Mickey and Brown (1966) show that the exact distribution of z is bounded by Student's t distributions with $n_1 + n_2 - 2$ and $\min(n_1 - 1, n_2 - 1)$ d.fr. – cf. Exercise 19.22.

Lawton (1965) obtains close bounds upon the confidence coefficient and the coverage probability of central intervals based on z.

A simple but reasonably accurate approximate procedure, due to Satterthwaite (1946), is to use (19.61) with modified degrees of freedom:

$$\nu^* = (\omega_1 + \omega_2)^2 / [\omega_1^2/\nu_1 + \omega_2^2/\nu_2]$$

where

$$\omega_i = s_i^2/(n_i - 1), \qquad \nu_i = n_i - 1, \; i = 1, 2.$$

Recalling that $\theta = \sigma_1^2/\sigma_2^2$, we note that the approximation involves replacing (σ_i^2/n_i) in (19.104) by the estimator $\sum_j (x_{ij} - \bar{x}_i)^2/[n_i(n_i - 1)], i = 1, 2$.

19.36 Welch (1947) refined the approximate approach of **19.35**. His argument is a general one, but for the present problem may be summarized as follows. Defining s_1^2, s_2^2 with $n_1 - 1$, $n_2 - 1$ as divisors respectively, so that they are unbiased estimators of variances, we seek a statistic $h(s_1^2, s_2^2, P)$ such that

$$P\{(d - \delta) < h(s_1^2, s_2^2, P)\} = P \tag{19.105}$$

whatever the value of θ. Now since $d - \delta$ is normally distributed independently of s_1^2, s_2^2 with zero mean and variance $\sigma_1^2/n_1 + \sigma_2^2/n_2 = D^2$, we have

$$P\{(d - \delta) \leq h(s_1^2, s_2^2, P) \mid s_1^2, s_2^2\} = I\left(\frac{h}{D}\right), \tag{19.106}$$

where $I(x) = \int_{-\infty}^{x} (2\pi)^{-1/2} \exp(-\frac{1}{2}t^2)\,dt$. Thus, from (19.105)–(19.106),

$$P = \int\int I(h/D) f(s_1^2) f(s_2^2)\,ds_1^2\,ds_2^2. \tag{19.107}$$

Now we may expand $I(h/D)$, which is a function of s_1^2, s_2^2, in a Taylor series about the true values σ_1^2, σ_2^2. We write this symbolically

$$I\left\{\frac{h(s_1^2, s_2^2, P)}{D}\right\} = \exp\left\{\sum_{i=1}^{2} (s_i^2 - \sigma_i^2)\partial_i\right\} I\left\{\frac{h(s_1^2, s_2^2, P)}{s}\right\}, \tag{19.108}$$

where the operator ∂_i represents differentiation with respect to s_i^2 and then putting $s_i^2 = \sigma_i^2$ and $s^2 = s_1^2/n_1 + s_2^2/n_2$. We put (19.108) into (19.107) to obtain

$$P = \prod_{i=1}^{2} \left[\int \exp\{(s_i^2 - \sigma_i^2)\partial_i\} f(s_i^2)\,ds_i^2\right] I\left\{\frac{h(s_1^2, s_2^2, P)}{s}\right\}. \tag{19.109}$$

Now since we have

$$f(s_i^2)\,\mathrm{d}s_i^2 = \frac{1}{\Gamma(\frac{1}{2}\nu_i)}\left(\frac{\nu_i s_i^2}{2\sigma_i^2}\right)^{(\nu_i/2)-1}\exp\left(-\frac{\nu_i s_i^2}{2\sigma_i^2}\right)\mathrm{d}\left(\frac{\nu_i s_i^2}{\sigma_i^2}\right),$$

on carrying out each integration in the symbolic expression (19.109) we find

$$\int \exp\{(s_i^2-\sigma_i^2)\partial_i\} f(s_i^2)\,\mathrm{d}s_i^2 = \left(1-\frac{2\sigma_i^2\partial_i}{\nu_i}\right)^{-\nu_i/2}\exp(-\sigma_i^2\partial_i),$$

so that (19.109) becomes

$$P = \prod_{i=1}^{2}\left[\left(1-\frac{2\sigma_i^2\partial_i}{\nu_i}\right)^{-\nu_i/2}\exp(-\sigma_i^2\partial_i)\right] I\left\{\frac{h(s_1^2, s_2^2, P)}{s}\right\}. \tag{19.110}$$

We can solve (19.110) to obtain the form of the function h, and hence find $h(s_1^2, s_2^2, P)$, for any known P.

Welch gave a series expansion for h, which in our special case becomes

$$\frac{h(s_1^2, s_2^2, P)}{s} = \xi\left[1+\frac{(1+\xi)^2}{4}\sum_{i=1}^{2} c_i^2/\nu_i - \frac{(1+\xi^2)}{2}\sum_{i=1}^{2} c_i^2/\nu_i^2 + \cdots\right], \tag{19.111}$$

where $c_i = (s_i^2/n_i)((s_1^2/n_1)+(s_2^2/n_2))^{-1}$, $\nu_i = n_i - 1$ and ξ is defined by $I(\xi) = P$.

Asymptotic expressions of this type have been justified by Chernoff (1949) and Wallace (1958) – each succeeding term on the right is an order lower in ν_i.

Since $(d-\delta)/s = z$ of (19.61), (19.111) gives the distribution function of z.

Following further work by Welch, Aspin (1948; 1949) and Trickett *et al.* (1956) tabled (19.111) as a function of ν_1, ν_2 and c_1, for $P = 0.95, 0.975, 0.99$ and 0.995.

Press (1966) shows that for $1-\alpha = 0.90$, $n_1 \le n_2 \le 30$, Welch's intervals have smaller expected length than (19.93) if θ is small (when (19.91) discards information about the more variable population) but not if θ is large. The two sets of intervals are shown to be asymptotically equivalent, and never differ by more than 10 per cent in expected length when $n_1 > 10$ if $0.01 \le \theta \le 100$. See also some comparisons with other methods by Mehta and Srinivasan (1970) and Scheffé (1970b). Lee and Gurland (1975) show that Welch's method has almost exactly the prescribed value of α, irrespective of θ, and has very high accuracy; they propose a simpler method with similar properties. Pfanzagl (1974) proves an asymptotic optimum property of (19.111).

In Chapters 30–31, where we discuss the analysis of variance, we shall consider setting confidence limits for an arbitrary number of normal means simultaneously. A problem of related interest is setting a confidence interval for an entire regression line; see **32.14**.

Tables and charts of confidence intervals

19.37 (1) *Binomial distribution.* Clopper and Pearson (1934) give two central confidence interval charts for the parameter, for $\alpha = 0.01$ and 0.05; each gives contours for $n = 8\ (2)\ 12\ (4)\ 24, 30, 40, 60, 100, 200, 400$ and 1000. The charts are reproduced in the *Biometrika Tables*. Upper or lower one-sided intervals can be obtained for half these values of α. Incomplete B-function tables may also be used – see **5.7** and the *Biometrika Tables*.

Pachares (1960) gives central limits for $\alpha = 0.01, 0.02, 0.05, 0.10$ and $n = 55$ (5) 100, and references to other tables, including those of Clark (1953) for the same values of α and $n = 10$ (1) 50.

Blyth (1986) compares several normal approximations in detail and gives exact tables of upper one-sided limits for $\alpha = 0.005, 0.01, 0.025, 0.05$ and $n = 3$ (1) 20.

Sterne (1954) has proposed an alternative method of setting confidence limits for a proportion. Instead of being central, the belt contains the values of p with the largest probabilities of occurrence. Since the distribution of p is skew in general, we clearly shorten the interval in this way. Walton (1970) extended Sterne's tables. Crow (1956) has shown that these intervals constitute a confidence belt with minimum total area, and has tabulated a slightly modified set of intervals for $n \leq 30$ and $1 - \alpha = 0.90, 0.95$ and 0.99. Blyth and Still (1983) give intervals of Crow's type that are approximately central and approximately unbiased and have other desirable regularity properties, for $\alpha = 0.01, 0.05$.

Central intervals for the binomial parameter may easily be derived by use of (19.8), but they are not the most accurate unbiased randomized intervals, which are tabulated by Blyth and Hutchinson (1960) for $\alpha = 0.01, 0.05$ and $n = 2$ (1) 24 (2) 50.

(2) *Poisson distribution.* (a) The *Biometrika Tables*, using the work of Garwood (1936), give central confidence intervals for the parameter, for observed values $x = 0$ (1) 30 (5) 50 and $\alpha = 0.002, 0.01, 0.02, 0.05, 0.10$. As in (1), one-sided intervals are available for $\alpha/2$. (b) Woodcock and Eames (1970) give similar tables for $x = 0$ (1) 100 (2) 200 (5) 500 (10) 1200 and α, 10α, 100α or 1000α equal to 0.1, 0.2 and 0.5. (c) Przyborowski and Wilénski (1935) give upper confidence limits only for $x = 0$ (1) 50, $\alpha = 0.001, 0.005, 0.01, 0.02, 0.05, 0.10$. (d) Walton (1970) tabulates Sterne-type intervals and Crow and Gardner (1959) tabulate modified intervals of the Sterne–Crow binomial type for $x = 0$ (1) 300 and $\alpha = 0.001, 0.01, 0.05, 0.10, 0.20$. (e) Blyth and Hutchinson (1961) give most accurate unbiased randomized intervals for observed x up to 250 and $\alpha = 0.01, 0.05$.

(3) *Variance of a normal distribution.* Tate and Klett (1959) give the most accurate unbiased confidence intervals, and the physically shortest intervals (cf. Exercise 19.5), based on the sample variance s^2 for $\alpha = 0.001, 0.005, 0.01, 0.05, 0.10$ and $n = 3$ (1) 30. The former are also given by Pachares (1961) for $\alpha = 0.01, 0.05, 0.10$ and $n - 1 = 1$ (1) 20, 24, 30, 40, 60, 120; by Lindley *et al.* (1960) for $\alpha = 0.001, 0.01, 0.05$ and $n - 1 = 1$ (1) 100; and by G.R. Murdak and W.O. Williford in Owen and Odeh (1977).

(4) *Ratio of normal variances.* John (1975) gives the most accurate unbiased intervals to 3 d.p. for $\alpha = 0.001, 0.01, 0.05, 0.1$ and $n_1 - 1, n_2 - 1 = 1$ (1) 20 (2) 30, 36, 45, 60, 90, 180, ∞. He tabulates in terms of the ratio of one estimator of variance to the pooled estimator.

(5) *Correlation parameter.* David (1938) gives four central confidence interval charts for the correlation parameter ρ of a bivariate normal population, for $\alpha = 0.01, 0.02, 0.05, 0.10$; each gives contours for $n = 3$ (1) 8, 10, 12, 15, 20, 25, 50, 100, 200 and 400. The *Biometrika Tables* reproduce the $\alpha = 0.01$ and $\alpha = 0.05$ charts. One-sided intervals may be obtained as in (1).

Odeh (1982, reproduced in Shah and Odeh, 1986) provides comprehensive tables of intervals for $\alpha = 0.005, 0.01, 0.025, 0.05, 0.10, 0.25$, $n = 3$ (1) 60 (2) 80 (5) 100 (10) 200 (25) 300 (50) 600 (100) 1000 and observed $r = -0.98$ (0.02) 0.98.

(6) *Difference of two normal means.* The *Biometrika Tables* give tables for setting central confidence intervals based on the expressions (19.110)–(19.111), for $1 - \alpha = 0.90, 0.95, 0.98, 0.99$.

Tolerance intervals

19.38 Throughout this chapter we have been discussing the setting of confidence intervals for parameters, but such intervals can be set for other quantities than parameters. We shall see in **19.40–41** that intervals can be found for any quantile of a distribution without any assumption other than continuity, and we shall see in **25.38** that we can find distribution-free intervals for the entire d.f. of a continuous distribution.

Also, in **32.8–11** we consider prediction intervals for future observations. Following convention, that treatment focuses on prediction intervals in regression analysis, although such intervals are more broadly applicable.

There is another type of problem, met in practical sampling, which may be solved by these methods. Suppose that, on the basis of a sample of n independent observations from a distribution, we wish to find two limits, L_1 and L_2, between which at least a given proportion γ of the distribution may be asserted to lie. Clearly, we can only make such an assertion in probabilistic form, i.e. we assert that, with given probability β, at least a proportion γ of the distribution lies between L_1 and L_2. L_1 and L_2 are called *tolerance limits* for the distribution; we shall call them the (β, γ) tolerance limits. The interval (L_1, L_2) is called a *tolerance interval.* In **19.42–43** we shall see that tolerance limits, also, may be set without assumptions (except continuity) on the form of the underlying distribution. First, however, we shall discuss the derivation of tolerance limits for a normal distribution, due to Wald and Wolfowitz (1946).

Tolerance intervals for a normal distribution

19.39 Since the sample mean and variance are a pair of sufficient statistics for the parameters of a normal distribution (Example 17.17), it is natural to base tolerance limits for the distribution upon them. In a sample of size n, we work with the unbiased statistics

$$\bar{x} = \sum x/n, \qquad s'^2 = \sum (x - \bar{x})^2/(n-1),$$

and define

$$A(\bar{x}, s', \lambda) = \int_{\bar{x}-\lambda s'}^{\bar{x}+\lambda s'} f(t)\,dt, \tag{19.112}$$

where $f(t)$ is the normal density. We now seek to determine the value λ so that

$$P\{A(\bar{x}, s', \lambda) > \gamma\} = \beta. \tag{19.113}$$

$L_1 = \bar{x} - \lambda s'$ and $L_2 = \bar{x} + \lambda s'$ will then be a pair of central (β, γ) tolerance limits for the normal distribution. Since we are concerned only with the proportion of that distribution covered by the interval (L_1, L_2), we may without any loss of generality standardize the population mean at 0 and its variance at 1. Thus

$$f(t) = (2\pi)^{-1/2} \exp(-\tfrac{1}{2}t^2).$$

Consider first the conditional probability, given $\bar{x}$, that $A(\bar{x}, s', \lambda)$ exceeds γ. We denote this by $P\{A > \gamma|\bar{x}\}$. Now A is a monotone increasing function of s', and the equation in s'

$$A(\bar{x}, s', \lambda) = \gamma \tag{19.114}$$

has just one root, which we denote by $s'(\bar{x}, \gamma, \lambda)$. Let

$$\lambda s'(\bar{x}, \gamma, \lambda) = r. \tag{19.115}$$

The value r satisfying

$$\int_{\bar{x}-r}^{\bar{x}+r} f(t)\,\mathrm{d}t = \gamma \tag{19.116}$$

may be obtained for any $\bar{x}$ and γ, and we write it $r(\bar{x}, y)$. Moreover, since A is monotone increasing in s', the inequality $A > \gamma$ is equivalent to

$$s' > s'(\bar{x}, \gamma, \lambda) = r(\bar{x}, \gamma)/\lambda.$$

Thus we may write

$$P\{A > \gamma|\bar{x}\} = P\left\{s' > \frac{r}{\lambda}\middle|\bar{x}\right\}, \tag{19.117}$$

and since $\bar{x}$ and s' are independently distributed, the right-hand side of (19.117) may have the conditioning on $\bar{x}$ suppressed so that (19.117) becomes

$$P\{A > \gamma|\bar{x}\} = P\{(n-1)s'^2 > (n-1)r^2/\lambda^2\}. \tag{19.118}$$

Since $(n-1)s'^2 = \sum(x-\bar{x})^2$ is a χ^2_{n-1} variate, we have

$$P\{A > \gamma|\bar{x}\} = P\{\chi^2_{n-1} > (n-1)r^2/\lambda^2\}. \tag{19.119}$$

To obtain the unconditional probability $P(A > \gamma) = \beta$ from (19.119), we must integrate it over the distribution of $\bar{x}$, which is normal with zero mean and variance $1/n$. This is a tedious numerical operation, but fortunately an excellent approximation is available. We form the Taylor expansion of $P(A > \gamma|\bar{x})$ about $\bar{x} = \mu = 0$, and since it is an even function of $\bar{x}$, the odd powers in the expansion vanish,[1] leaving

$$P(A > \gamma|\bar{x}) = P(A > \gamma|0) + \frac{\bar{x}^2}{2!}P''(A > \gamma|0) + \cdots. \tag{19.120}$$

Integrating over the distribution of $\bar{x}$, we have, using the moments of $\bar{x}$,

$$P(A > \gamma) = P(A > \gamma|0) + \frac{1}{2n}P''(A > \gamma|0) + O(n^{-2})\cdots. \tag{19.121}$$

But from (19.120) with $\bar{x} = 1/\sqrt{n}$, we also have

$$P\left(A > \gamma\middle|\frac{1}{\sqrt{n}}\right) = P(A > \gamma|0) + \frac{1}{2n}P''(A > \gamma|0) + O(n^{-2}). \tag{19.122}$$

[1]This is because the interval is symmetric about $\bar{x}$; it would not happen otherwise.

From (19.121) and (19.122) we obtain the approximation

$$P(A > \gamma) \doteq P\left(A > \gamma \,\Big|\, \frac{1}{\sqrt{n}}\right) \tag{19.123}$$

and (19.113), (19.123) and (19.119) yield, finally,

$$\beta = P\{A > \gamma\} \doteq P\{\chi^2_{n-1} > (n-1)r^2/\lambda^2\}, \tag{19.124}$$

where r is defined from (19.116) by

$$\int_{n^{-(1/2)}-r}^{n^{-(1/2)}+r} f(t)\,\mathrm{d}t = \gamma. \tag{19.125}$$

Given γ, β, n and $\bar{x}$, we can determine λ approximately from (19.124) and (19.125), and hence the tolerance limits $\bar{x} \pm \lambda s'$. Wald and Wolfowitz (1946) showed that the approximation is extremely good even for values of n as low as 2 if β and γ are at least 0.95, as they usually are in practice. Bowker (1947) gave tables of λ (his k) for β (his γ) $= 0.75, 0.90, 0.95, 0.99$ and γ (his P) $= 0.75, 0.90, 0.99$ and 0.999, for sample sizes $n = 2\,(1)\,102\,(2)\,180\,(5)\,300\,(10)\,400\,(25)\,750\,(50)\,1000$.

On examination of the argument above, it will be seen to hold if $\bar{x}$ is replaced by any estimator $\hat{\mu}$ of the mean, and s'^2 by any independent estimator $\hat{\sigma}^2$ of the variance, of a normal population, as pointed out by Wallis (1951). Ellison (1964) has shown that if the mean is estimated from n observations and the variance estimate has ν degrees of freedom, the approximation corresponding to (19.124) is valid only to order ν/n^2 (reducing to $1/n$ for (19.124) itself, where $\nu = n-1$), and Howe (1969) gives better tolerance limits for the case where ν/n^2 is large. In the contrary case, there are some useful tables. Taguti (1958) gives tables of λ (his k) for β (his $1-\alpha$) and γ (his P) $= 0.90, 0.95$ and 0.99 as well as $n = 0.5\,(0.5)\,2\,(1)\,10\,(2)\,20\,(5)\,30\,(10)\,60\,(20)\,100, 200, 500, 1000, \infty$ and $\nu = 1\,(1)\,20\,(2)\,30\,(5)\,100\,(100)\,1000, \infty$. The small fractional values of n are useful in some applications discussed by Taguti. Weissberg and Beatty (1960) give tables of λ/r (their u) for ν (their f) $= 1\,(1)\,150\,(2)\,250\,(5)\,500\,(10)\,1000\,(1000)\,10\,000, \infty$ and β (their γ) $= 0.90, 0.95, 0.99$; and of r for $n = 1\,(1)\,100\,(5)\,200\,(10)\,300\,(20)\,500\,(100)\,1000\,(1000)\,10\,000, \infty$ and γ (their P) $= 0.5, 0.75, 0.9, 0.95, 0.99, 0.999$.

Fraser and Guttman (1956) and Guttman (1957) consider tolerance intervals that cover a given proportion of a normal distribution *on average*. Sharpe (1970) studies the robustness of both kinds of tolerance intervals.

Zacks (1970) considers tolerance limits, and their relationship with confidence limits, for a large class of discrete distributions including the binomial, the negative binomial and the Poisson.

Distribution-free confidence intervals for quantiles

19.40 The joint distribution of the order statistics depends directly upon the population d.f. (cf., for example, (14.1) and (14.2)) and therefore point estimation of the population quantiles by order statistics also does so. Remarkably enough, however, pairs of order statistics may be used to set confidence intervals for any population quantile that are *distribution-free*, i.e. that do not depend on the form of the underlying distribution, provided that it is continuous.

Consider the pair of order statistics $x_{(r)}$ and $x_{(s)}$, $r < s$, in a sample of n observations from the continuous d.f. $F(x)$. Expression (14.2) gives the joint distribution of $F_r = F(x_{(r)})$ and

$F_s = F(x_{(s)})$ as

$$dG_{r,s} = \frac{F_r^{r-1}(F_s - F_r)^{s-r-1}(1-F_s)^{n-s}\,dF_r\,dF_s}{B(r, s-r)B(s, n-s+1)}. \quad (19.126)$$

X_p, the p-quantile of $F(x)$, is defined by $F(X_p) = p$. For the probability that the interval $(x_{(r)}, x_{(s)})$ covers X_p we have

$$P\{x_{(r)} \leq X_p \leq x_{(s)}\} = P\{x_{(r)} \leq X_p\} - P\{x_{(s)} < X_p | x_{(r)} \leq X_p\}. \quad (19.127)$$

Since $x_{(r)} \leq x_{(s)}$, we can *only* have $x_{(s)} < X_p$ if $x_{(r)} < X_p$, so that

$$P\{x_{(s)} < X_p | x_{(r)} \leq X_p\} = P\{x_{(s)} \leq X_p\}, \quad (19.128)$$

the equality on the right having zero probability because of the continuity of F. From (19.127) and (19.128) we obtain

$$P\{x_{(r)} \leq X_p \leq x_{(s)}\} = P\{x_{(r)} \leq X_p\} - P\{x_{(s)} \leq X_p\}. \quad (19.129)$$

The d.f. of $x_{(r)}$ given at (14.4) and (14.5) here yields, since $F(X_p) = p$,

$$P\{x_{(r)} \leq X_p\} = G_r(X_p) = I_p(r, n-r+1), \quad (19.130)$$

and similarly for $x_{(s)}$. Expression (19.129) therefore becomes

$$P\{x_{(r)} \leq X_p \leq x_{(s)}\} = I_p(r, n-r+1) - I_p(s, n-s+1) = 1-\alpha, \quad (19.131)$$

say, independent of the form of the continuous distribution F.

19.41 We see from (19.131) that the interval $(x_{(r)}, x_{(s)})$ covers the quantile X_p with a confidence coefficient that does not depend on $F(x)$ at all, and we thus have a distribution-free confidence interval for X_p. Since $I_p(a, b) = 1 - I_{1-p}(b, a)$, we may also write the confidence coefficient as

$$1-\alpha = I_{1-p}(n-s+1, s) - I_{1-p}(n-r+1, r). \quad (19.132)$$

By the incomplete beta relation with the binomial expansion given at (5.16), (19.132) may be expressed as

$$1-\alpha = \left\{\sum_{i=0}^{s-1} - \sum_{i=0}^{r-1}\right\}\binom{n}{i}p^i q^{n-i} = \sum_{i=r}^{s-1}\binom{n}{i}p^i q^{n-i}, \quad (19.133)$$

where $q = 1-p$. The confidence coefficient is therefore the sum of the terms in the binomial $(q+p)^n$ from the $(r+1)$th to the sth inclusive.

If we choose a pair of symmetrically placed order statistics we have $s = n-r+1$, and find in (19.131)–(19.133)

$$\begin{aligned} 1-\alpha &= I_p(r, n-r+1) - I_p(n-r+1, r) \\ &= 1 - \{I_{1-p}(n-r+1, r) + I_p(n-r+1, r)\}, \quad (19.134) \\ &= \sum_{i=r}^{n-r}\binom{n}{i}p^i q^{n-i}, \quad (19.135) \end{aligned}$$

so that the confidence coefficient is the sum of the central $(n - 2r + 1)$ terms of the binomial, r terms at each end being omitted.

For any values of r and n, the confidence coefficient attaching to the interval $(x_{(r)}, x_{(n-r+1)})$ may be calculated from (19.134)–(19.135), if necessary using the *Tables of the Incomplete Beta Function*. The tables of the binomial distribution listed in **5.6** may also be used. Exercise 19.23 provides a numerical example.

Scheffé and Tukey (1945) show that if the underlying distribution is discrete, the confidence intervals above cover X_p with probability at least $1 - \alpha$.

In the special case of the population median $X_{0.5}$, (19.134) and (19.135) reduce to

$$1 - \alpha = 1 - 2I_{0.5}(n - r + 1, r) = 2^{-n} \sum_{i=r}^{n-r} \binom{n}{i}, \qquad (19.136)$$

a particularly simple form. This confidence interval procedure for the median was first proposed by W. R. Thompson (1936).

MacKinnon (1964) gives tables of r for $n = 1\ (1)\ 1000$ and α as little as possible below 0.50, 0.10, 0.05, 0.02, 0.01 and 0.001. Van der Parren (1970) gives similar tables for $n = 3\ (1)\ 150$ and $\alpha = 0.3, 0.2, 0.1, 0.05, 0.02$ and 0.01, together with the exact value of $\alpha/2$ in each case.

One may also obtain confidence intervals for the quantile interval (X_p, X_q), $p < q$; cf. Reiss and Rüschendorf (1976) and Sathe and Lingras (1981).

Distribution-free tolerance intervals

19.42 In **19.39** we discussed the problem of finding tolerance intervals for a normal d.f. Suppose now that we require such intervals without making assumptions beyond continuity on the underlying distributional form. We seek a random interval (l, u) such that

$$P\left\{\int_l^u f(x)\,dx \geq \gamma\right\} = \beta, \qquad (19.137)$$

where $f(x)$ is the unknown continuous density function. It is not obvious that such a distribution-free procedure is possible, but Wilks (1941; 1942) showed that the order statistics $x_{(r)}, x_{(s)}$ provide distribution-free tolerance intervals, and Robbins (1944) showed that *only* the order statistics do so.

If we write $l = x_{(r)}$, $u = x_{(s)}$ in (19.137), we may rewrite it as

$$P[\{F(x_{(s)}) - F(x_{(r)})\} \geq \gamma] = \beta. \qquad (19.138)$$

We may obtain the exact distribution of the random variable $F(x_{(s)}) - F(x_{(r)})$ from (19.126) by the transformation $y = F(x_{(s)}) - F(x_{(r)})$, $z = F(x_{(r)})$, with Jacobian equal to 1. Expression (19.126) becomes

$$dH_{y,z} = \frac{z^{r-1} y^{s-r-1} (1 - y - z)^{n-s}\, dy\, dz}{B(r, s - r) B(s, n - s + 1)}, \qquad 0 \leq y + z \leq 1. \qquad (19.139)$$

In (19.139) we integrate out z over its range $(0, 1-y)$, obtaining for the marginal distribution of y,

$$\mathrm{d}J_{r,s} = \frac{y^{s-r-1}\,\mathrm{d}y}{B(r, s-r)B(s, n-s+1)} \int_0^{1-y} z^{r-1}(1-y-z)^{n-s}\,\mathrm{d}z. \tag{19.140}$$

We put $z = (1-y)t$, reducing (19.140) to

$$\begin{aligned} \mathrm{d}J_{r,s} &= \frac{y^{s-r-1}(1-y)^{n-s+r}\,\mathrm{d}y}{B(r, s-r)B(s, n-s+1)} \int_0^1 t^{r-1}(1-t)^{n-s}\,\mathrm{d}t \\ &= y^{s-r-1}(1-y)^{n-s+r}\,\mathrm{d}y \frac{B(r, n-s+1)}{B(r, s-r)B(s, n-s+1)} \\ &= \frac{y^{s-r-1}(1-y)^{n-s+r}\,\mathrm{d}y}{B(s-r, n-s+r+1)}, \qquad 0 \le y \le 1. \end{aligned} \tag{19.141}$$

Thus $y = F(x_{(s)}) - F(x_{(r)})$ is distributed as a beta variate of the first kind with parameters depending only on the difference $s - r$. If we put $r = 0$ in (19.141) and interpret $F(x_{(0)})$ as zero (so that $x_{(0)} = -\infty$), (19.141) reduces to (14.1), with s written for r.

19.43 From (19.141), we see that (19.138) becomes

$$P\{y \ge \gamma\} = \int_\gamma^1 \frac{y^{s-r-1}(1-y)^{n-s+r}\,\mathrm{d}y}{B(s-r, n-s+r+1)} = \beta, \tag{19.142}$$

which we may rewrite in terms of the incomplete beta function as

$$P\{F(x_{(s)}) - F(x_{(r)}) \ge \gamma\} = 1 - I_\gamma(s-r, n-s+r+1) = \beta. \tag{19.143}$$

The relationship (19.143) for the distribution-free tolerance interval $(x_{(r)}, x_{(s)})$ contains five quantities: γ (the minimum proportion of $F(x)$ it is desired to cover), β (the probability with which we desire to do this), the sample size n, and the order statistics' positions in the sample, r and s. Given any four of these, we can solve (19.143) for the fifth. In practice, β and γ are usually fixed at levels required by the problem, and r and s symmetrically chosen, so that $s = n - r + 1$. Expression (19.143) then reduces to

$$I_\gamma(n - 2r + 1, 2r) = 1 - \beta. \tag{19.144}$$

The left-hand side of (19.144) is a monotone increasing function of n, and for any fixed β, γ, r we can choose n large enough so that (19.144) is satisfied. In practice, we must choose n as the nearest integer above the solution of (19.144). If $r = 1$, so that the extreme values in the sample are being used, (19.144) reduces to

$$I_\gamma(n-1, 2) = 1 - \beta, \tag{19.145}$$

which gives the probability β that the range of the sample of n observations covers at least a proportion γ of the population d.f.

Murphy (1948) gives graphs of γ as a function of n for $\beta = 0.90, 0.95$ and 0.99 and $r+(n-s+1) =$ 1 (1) 6 (2) 10 (5) 30 (10) 60 (20) 100; these are exact for $n \le 100$, and approximate up to $n = 500$.

Example 19.7 (Determination of a distribution-free tolerance interval)

We consider the numerical solution of (19.145) for n with $\beta = 0.99$. It may be rewritten

$$1-\beta = \frac{1}{B(n-1,2)}\int_0^{\gamma} y^{n-2}(1-y)\,dy = n(n-1)\left\{\frac{\gamma^{n-1}}{n-1}-\frac{\gamma^n}{n}\right\}$$
$$= n\gamma^{n-1}-(n-1)\gamma^n. \qquad (19.146)$$

A numerical search yields $n = 645$ as the solution. That is, we should take $n = 645$ in order to get a 99 per cent tolerance interval for 99 per cent of the population d.f. Exercise 19.24 gives further numerical results.

We have discussed only the simplest case of setting distribution-free tolerance intervals for a univariate continuous distribution. Extensions to multivariate tolerance regions, including the discontinuous case, have been made by numerous authors. Classic papers include those of Wald (1943b), Scheffé and Tukey (1945), Tukey (1947, 1948), and Fraser (1953). Guttman (1970) provides a comprehensive review of the subject.

Scheffé and Tukey (1945) and Tukey (1948) show that if the underlying distribution is discrete, the above tolerance intervals and regions have probability at least $1-\alpha$.

Fiducial intervals

19.44 Let us reconsider Example 19.1 from a different standpoint. The sampling distribution of the statistic $\bar{x}$ given there,

$$f(x) = \left(\frac{n}{2\pi}\right)^{1/2} \exp\{-\tfrac{1}{2}n(\bar{x}-\mu)^2\}, \qquad (19.147)$$

leads directly to the likelihood function since $\bar{x}$ is sufficient for μ. If we are prepared intuitively to use (19.147) to express our credence in a particular value of μ given a fixed observed value $\bar{x}$, we may consider the density function:

$$f(\mu) = \left(\frac{n}{2\pi}\right)^{1/2} \exp\{-\tfrac{1}{2}n(\bar{x}-\mu)^2\}, \qquad (19.148)$$

which, following Fisher (1935b), we shall call the *fiducial distribution* of the parameter μ. We note that the integral of (19.148) with respect to μ over the real line is 1, so that the constant needs no adjustment in this particular case.

This fiducial distribution is not a probability distribution in the usual sense. It is a new concept, expressing the intensity of our belief in the various possible values of a parameter. It may be regarded as a distribution of probability in the sense of degree of belief; the consequent link with interval estimation based on the use of Bayes' theorem will be discussed below. Or it may be regarded as a new concept, giving formal expression to somewhat intuitive ideas about the extent to which we place credence in various values of μ. It so happens, in this case, that the algebraic form of the density function for μ in (19.148) is the same as that in (19.147). This is not essential, though it is not infrequent.

19.45 The fiducial distribution may now be used to determine intervals within which μ is located. For example, we may select the probability levels 0.022 75 and 0.977 25, which correspond to deviations of $\pm 2\sigma$ from the mean of a normal distribution yielding the interval:

$$-2 \leq (\bar{x} - \mu)\sqrt{n} \leq 2,$$

which is equivalent to

$$\bar{x} - 2/\sqrt{n} \leq \mu \leq \bar{x} + 2/\sqrt{n}. \qquad (19.149)$$

This, as it happens, is the same inequality as in the central confidence intervals (19.4) in Example 19.1. But it is essential to note that it is not reached by the same line of thought. The confidence approach says that if we assert (19.149) we shall be right in about 95.45 per cent of the cases *in the long run*. Under the fiducial approach the assertion of (19.149) is that (in some sense not defined) we are 95.45 per cent sure of being right *in this particular case*. The shift of emphasis is similar to that encountered in considering the likelihood function itself, where the function $L(x|\theta)$ can be considered as a probability in which θ is fixed and x varies, or as a likelihood in which x is fixed and θ varies. However, for the likelihood function, all we needed was to compare values for different θ, so as to find the maximum, and hence the ML estimator. By contrast, the fiducial argument requires that we define a probability measure on the parameter space. The fiducial argument is of interest, perhaps mostly historical, as an attempt to solve the problem of scientific induction by developing a probability (or corroboration) measure for a hypothesis, given the observations. At this stage, we focus on a brief development of fiducial theory, reserving evaluative comments for the discussion on comparative inference; see **26.26–29**.

There is one further fundamental distinction between the two methods. We have seen earlier in this chapter that in confidence theory it is possible to have different sets of intervals for the same parameter based on different statistics (although we may discriminate between the different sets, and choose the shortest or most selective set). This is explicitly ruled out in fiducial theory (even in the sense that we may choose central or non-central intervals for the same distribution when using both its tails). We must, in fact, use all the information about the parameter that the likelihood function contains. This implies that if we are to set limits to θ by a single statistic t, the latter must be sufficient for θ. For confidence intervals, on the other hand, sufficiency is desirable to maximize accuracy – cf. **19.16**.

As we pointed out in **17.38**, there is always a *set* of jointly sufficient statistics for an unknown parameter, namely the n observations themselves. But this tautology offers little consolation: even a sufficient set of two statistics would be difficult enough to handle; a larger set is almost certainly practically useless. As to what should be done to construct an interval for a single parameter θ where a single sufficient statistic does not exist, writers on fiducial theory are for the most part silent.

19.46 We now examine the fiducial argument in greater detail. Let $f(t, \theta)$ be a continuous density and $F(t, \theta)$ the d.f. of a statistic t that is sufficient for θ. Consider the behaviour of f for some fixed t, as θ varies. Suppose also that we know beforehand that θ must lie in a certain range, which may in particular be $(-\infty, \infty)$. Take some critical probability $1 - \alpha$ (a fiducial coefficient, by analogy with a confidence coefficient) and let θ_α be the value of θ for which $F(t, \theta) = 1 - \alpha$.

Now suppose also that over the permissible range of θ, $f(t, \theta)$ is a monotonic non-increasing function of θ for any t. Then for all $\theta \leq \theta_\alpha$ the observed t has at least as large a density as $f(t, \theta_\alpha)$, and for $\theta > \theta_\alpha$ it has a lower density. We then choose $\theta \leq \theta_\alpha$ as our fiducial interval. It includes all those values of the parameter that give to the density a value greater than or equal to $f(t, \theta_\alpha)$.

If we require a fiducial interval of type

$$\theta_{\alpha_1} \leq \theta \leq \theta_{\alpha_2},$$

we look for two values of θ such that $f(t, \theta_{\alpha_1}) = f(t, \theta_{\alpha_2})$ and $F(t, \theta_{\alpha_2}) - F(t, \theta_{\alpha_1}) = 1 - \alpha$. If, between these values, $f(t, \theta)$ is greater than the extreme values $f(t, \theta_{\alpha_1})$ and $f(t, \theta_{\alpha_2})$, and is less than those values outside it, the interval again comprises values for which the density is at least as great as the density at the critical points.

If the distribution of t is symmetrical this involves taking a range that cuts off equal tail areas on it. For a non-symmetrical distribution the tails are to be such that their total probability content is α, but the contents of the two tails are not equal. Similar considerations have already been discussed in connection with central confidence intervals in **19.7**.

On this understanding, if our fiducial interval is increased by an element $\mathrm{d}\theta$ at each end, the probability density at the end decreases by $(\partial F(t, \theta)/\partial\theta)\,\mathrm{d}\theta$. For the fiducial distribution we then have

$$f^*(\theta|t)\frac{\mathrm{d}F^*}{\alpha\theta} = -\frac{\partial F(t, \theta)}{\partial\theta}. \tag{19.150}$$

This formula, however, requires that $f(t, \theta)$ shall be a non-decreasing function of θ at the lower end and a non-increasing function of θ at the upper end of the interval.

Example 19.8 (Fiducial interval for the gamma distribution)

As an example of a non-symmetrical sampling distribution, consider the gamma distribution

$$f(x) = \frac{x^{p-1}\mathrm{e}^{-x/\theta}}{\theta^p\Gamma(p)}, \qquad p > 0;\ 0 \leq x < \infty. \tag{19.151}$$

If p is known, $t \equiv \bar{x}/p$ is sufficient for θ (cf. Exercise 17.1) and its sampling distribution is easily seen to be

$$f(x) = \left(\frac{\beta}{\theta}\right)^\beta \frac{t^{\beta-1}\mathrm{e}^{-\beta t/\theta}}{\Gamma(\beta)}, \tag{19.152}$$

where $\beta = np$. Now in this case θ may vary only from 0 to ∞. As it does so the value of $f(x)$ in (19.152) for fixed t rises monotonically from zero to a maximum and then falls again to zero. Thus, if we determine θ_{α_1} and θ_{α_2} so that the densities at these two values are equal and the integral of (19.152) between them has the assigned value $1 - \alpha$, the fiducial range is given by $\theta_{\alpha_1} \leq \theta \leq \theta_{\alpha_2}$.

Integrating (19.152) and putting $u = \beta t/\theta$, we have

$$F(t, \theta) = \int_0^{\beta t/\theta} \frac{u^{\beta-1}\mathrm{e}^{-u}}{\Gamma(\beta)}\,\mathrm{d}u. \tag{19.153}$$

Thus the fiducial distribution of θ is

$$f^*(\theta|t) = -\frac{\partial F}{\partial \theta} = -\left[\frac{u^{\beta-1}\mathrm{e}^{-u}}{\Gamma(\beta)}\right]_{u=\beta t/\theta} \frac{\partial}{\partial \theta}\left(\frac{\beta t}{\theta}\right)$$
$$= \left(\frac{\beta t}{\theta}\right)^{\beta-1} \frac{\mathrm{e}^{-\beta t/\theta}}{\Gamma(\beta)} \frac{\beta t}{\theta^2} = \left(\frac{\beta t}{\theta}\right)^{\beta} \frac{\mathrm{e}^{-\beta t/\theta}}{\theta \Gamma(\beta)}. \tag{19.154}$$

This is a Pearson Type V distribution (Exercise 6.6), so its integral from $\theta = 0$ to $\theta = \infty$ is unity. In Example 19.6, $ns^2/2\theta$ was a gamma variate with parameter $\frac{1}{2}(n-1)$, whereas in (19.152) and (19.154), $\beta t/\theta$ is a gamma variate with parameter β.

It is evident from (19.153), and the earlier discussion on the normal mean, that the key to developing a fiducial element is finding a *pivotal quantity*, $g = g(t, \theta)$ that enables us to write

$$F(t, \theta) = \int_L^{g(t,\theta)} f(u)\,\mathrm{d}u \tag{19.155}$$

where L is some lower limit (often 0 or $-\infty$) that is independent of θ; $f(u)$ is some density function independent of θ. Then (19.150) yields

$$\frac{\mathrm{d}F^*}{\mathrm{d}\theta} = -[f(u)]_{u=g(t,\theta)} \frac{\partial g(t,\theta)}{\partial \theta}. \tag{19.156}$$

For the normal mean, the pivot is

$$g(\bar{x}, \mu) = \bar{x} - \mu \quad \text{with} \quad \partial g/\partial \mu = -1$$

whereas for the gamma in Example 19.8 the pivot is

$$g(t, \theta) = \beta t/\theta \quad \text{with} \quad \partial g/\partial \theta = -\beta t/\theta^2.$$

One of the major difficulties for the fiducial argument is that exact pivotal quantities are rarely available, although the argument may be applicable to large-sample distributions.

19.47 When two or more parameters are involved, the fiducial argument begins to meet difficulties. We shall concentrate on two important standard cases, the estimation of the mean in normal samples where the variance is unknown, and the estimation of the difference of two means in samples from two normal populations with unequal variances.

Example 19.9 (Fiducial interval for a normal mean with population variance unknown)

From Example 17.17, the sample mean $\bar{x}$ and the sample variance s^2 $(= \sum(x-\bar{x})^2/n)$ are jointly sufficient for the mean μ and variance σ^2 of the normal population, with joint density

$$f(\bar{x}, s) \propto \frac{1}{\sigma} \exp\left\{-\frac{n}{2\sigma^2}(\bar{x}-\mu)^2\right\} \frac{s^{n-2}}{\sigma^{n-1}} \exp\left\{-\frac{ns^2}{2\sigma^2}\right\}, \tag{19.157}$$

which we have expressed in terms of $\bar{x}$ and s. If we were considering fiducial limits for μ with known σ we should use the first factor on the right of (19.157); but if we were considering limits for σ with known μ we should *not* use the second factor, since σ enters into the first factor. In fact (cf. Example 17.17), the sufficient statistic in this case is not s^2 but $\sum(x-\mu)^2/n = s'^2$ whose distribution is obtained by merging the two factors in (19.157).

For known σ, we should, as in **19.44**, use the pivot $g_1 = \bar{x} - \mu$ to obtain the fiducial distribution of μ. For known μ, we should use the fact that s'^2 is distributed like t in (19.152), with $\beta = n$ and $\theta = \sigma^2$, and hence, as in Example 19.8, use the pivot $g_2 = \beta t/\theta$. In (19.157), s is distributed as s', but with $\beta = n-1$. The question is, can we assume that the two pivots (g_1, g_2) may be applied independently to obtain the joint fiducial distribution of μ and σ?

Fiducialists assume that this is so. The question appears to us to be very debatable.[2] However, let us make the assumption and see where it leads us. For the fiducial distribution we shall then have

$$f^* = f^*(\mu, \sigma|\bar{x}, s) \propto \frac{1}{\sigma}\exp\left\{-\frac{n}{2\sigma^2}(\bar{x}-\mu)^2\right\}\left(\frac{s^{n-1}}{\sigma^n}\right)\exp\left\{-\frac{ns^2}{2\sigma^2}\right\}.$$

We now integrate for σ to obtain the fiducial distribution of μ and arrive at

$$f^*(\mu|\bar{x}, s) \propto \frac{s^{-1}}{\{1+\frac{(\bar{x}-\mu)^2}{s^2}\}^{n/2}}.$$

This is Student's t distribution, with $t = (\mu - \bar{x})(n-1)^{1/2}/s$ and $n-1$ d.fr. Thus, given α, we can find two values of t, t_0 and t_1, so that

$$P\{-t_0 \le t \le t_1\} = 1 - \alpha,$$

where

$$P\{\bar{x} - st_0/(n-1)^{1/2} \le \mu \le \bar{x} + st_1/(n-1)^{1/2}\} = 1 - \alpha.$$

Thus we obtain the same interval as in Example 19.2, but this equivalence is by no means essential to the fiducial argument, as Example 19.10 will show.

Example 19.10 (Fiducial intervals for the problem of two means)

The fiducial solution of the problem of two means starts from the joint distribution of sample means and variances, which may be written, from Example 19.9,

$$f \propto \frac{1}{\sigma_1\sigma_2}\exp\left\{-\frac{n_1}{2\sigma_1^2}(\bar{x}_1-\mu_1)^2 - \frac{n_2}{2\sigma_2^2}(\bar{x}_2-\mu_2)^2\right\}\frac{s_1^{n_1-2}}{\sigma_1^{n_1-1}}\frac{s_2^{n_2-2}}{\sigma_2^{n_2-1}}\exp\left\{-\frac{n_1}{2}\frac{s_1^2}{\sigma_1^2} - \frac{n_2}{2}\frac{s_2^2}{\sigma_2^2}\right\}.$$

In accordance with the fiducial argument, we use the pivots $\bar{x}_i - \mu_i$, s_i/σ_i $(i = 1, 2)$ as in Example 19.9. Then for the fiducial distribution (omitting powers of s_1 and s_2, which are now constants) we have, as in Example 19.9,

$$f^* \propto \frac{1}{\sigma_1^{n_1+1}\sigma_2^{n_2+1}}\exp\left\{-\frac{n_1}{2\sigma_1^2}(\bar{x}_1-\mu_1)^2 - \frac{n_2}{2\sigma_2^2}(\bar{x}_2-\mu_2)^2\right\}\exp\left\{-\frac{n_1s_1^2}{2\sigma_1^2} - \frac{n_2s_2^2}{2\sigma_2^2}\right\}.$$

[2]Although $\bar{x}$ and s are statistically independent, μ and σ are not independent in any fiducial sense. Some support for the replacement can be derived *a posteriori* from the reflection that it leads to Student's t distribution, but if fiducial theory is to be accepted on its own merits, something more is required.

Writing

$$t_1 = \frac{(\mu_1 - \bar{x}_1)\surd(n_1 - 1)}{s_1}, \qquad t_2 = \frac{(\mu_2 - \bar{x}_2)\surd(n_2 - 1)}{s_2}, \tag{19.158}$$

we find, as in Example 19.9, the joint fiducial density for μ_1 and μ_2 is

$$f^* \propto \{1 + t_1^2/(n_1 - 1)\}^{-n_1/2}\{1 + t_2^2/(n_2 - 1)\}^{-n_2/2}. \tag{19.159}$$

We cannot proceed at once to find an interval for $\delta = \mu_1 - \mu_2$ In fact, from (19.158) we have

$$(\mu_1 - \bar{x}_1) - (\mu_2 - \bar{x}_2) = \delta - d = t_1 s_1/\surd(n_1 - 1) - t_2 s_2/\surd(n_2 - 1), \tag{19.160}$$

and to set limits to δ we require the fiducial distribution of the right-hand side of (19.160) or some convenient function of it. This is a linear function of t_1 and t_2, whose fiducial distribution is given by (19.159). In fact Fisher (1935b; 1939), following Behrens (1929), chose the statistic (19.61)

$$z = \frac{d - \delta}{\left(\frac{s_1^2}{n_1 - 1} + \frac{s_2^2}{n_2 - 1}\right)^{1/2}}$$

as the most convenient function. We have

$$z = t_1 \cos\psi - t_2 \sin\psi,$$

where

$$\tan^2\psi = \frac{s_2^2}{n_2 - 1}\Big/\frac{s_1^2}{n_1 - 1}.$$

For given ψ, the distribution of z (known as the Fisher–Behrens distribution) can be found from (19.160). It has no simple form, but Rahman and Saleh (1974) express it in series forms and provide 4 d.p. tables of its 97.5 and 95 percentage points for $\nu_1 = 6\,(1)\,15$ and $\nu_2 = 6\,(1)\,9$. Earlier tables had been given by Fisher and Yates' (1963) *Statistical Tables*. In using these tables (and in consulting Fisher's papers generally) the reader should note that our $s^2/(n-1)$ is written by him as s^2. Linssen (1991) provides tables for each combination of $n_1, n_2 = 2, 3, 5, 9, 17, 33, \infty$ and $\psi = 0°\,(15°)\,90°$.

In this case, the most important yet noticed, the fiducial argument does not give the same result as the approach from confidence intervals. That is to say, if we determine from a probability $1-\alpha$ the corresponding points of z, say z_0 and z_1, and then assert

$$\bar{x}_1 - \bar{x}_2 - z_0\left(\frac{s_1^2}{n_1 - 1} + \frac{s_2^2}{n_2 - 1}\right)^{1/2} \leq \mu_1 - \mu_2 \leq \bar{x}_1 - \bar{x}_2 + z_1\left(\frac{s_1^2}{n_1 - 1} + \frac{s_2^2}{n_2 - 1}\right)^{1/2}, \tag{19.161}$$

we shall not be correct in a proportion $1 - \alpha$ of cases in the long run, whatever the values of σ_1^2, σ_2^2, as is obvious from the fact that z may be expressed from (19.60) and (19.61) as

$$z = t\left\{\frac{s_2^2(1 + \theta/N)(1 + Nu/\theta)}{(n_2 - 1)s_1^2 + (n_1 - 1)s_2^2}\right\}^{1/2}\left\{\frac{(n_1 - 1)(n_2 - 1)}{n_1 + n_2 - 2}\right\}^{1/2}$$

where t, defined by (19.60), has an exact Student's t distribution. t is distributed independently of θ, but z is not.

The fiducialist view is that there is no particular reason why such statements should be correct in a known proportion of cases in the long run; and that to impose such a desideratum is to miss the point of the fiducial approach. We return to this point in Chapter 26.

Davis and Scott (1971) use the symbolic method of **19.33** to obtain an asymptotic expansion of the confidence coefficient corresponding to (19.161), which always exceeds $1-\alpha$; see also Robinson (1976).

Bayesian intervals

19.48 We now consider the relation between fiducial theory and Bayesian influence. Our discussion is, by design, very brief as the Bayesian approach is fully developed in the companion volume by O'Hagan (1994). Bayes' theorem, given in (8.10), may be rewritten in our present context as

$$g(\theta|x, H) \propto f(\theta|H)L(x|\theta, H), \tag{19.162}$$

where g is the posterior distribution, f is the prior distribution and L is the likelihood.

The major problem, as we saw in Chapter 8, is to choose a prior distribution $f(\theta|H)$. Jeffreys (1961) extended Bayes' postulate (which stated that if nothing is known about θ and its range is finite, the prior distribution should be proportional to $d\theta$) to take account of various situations. In particular, (1) if the range of θ is infinite in both directions the prior probability is still taken as proportional to $d\theta$; (2) if θ ranges from 0 to ∞ the prior distribution is taken as proportional to $d\theta/\theta$.

Paradoxes that can arise from the use of such improper prior distributions for more than one parameter are discussed, with many examples, by Dawid *et al.* (1973) and by Stone (1976). Diaconis and Freedman (1986a; 1986b) show that problems also arise with respect to estimation consistency, especially when there are many parameters.

Example 19.11 (Bayesian interval for the normal mean)

In the case of the normal distribution considered in **19.44** we have, with $\bar{x}$ sufficient for μ,

$$L(\bar{x}|\mu, H) = \frac{n^{1/2}}{(2\pi)^{1/2}} \exp\left\{-\frac{n}{2}(\bar{x}-\mu)^2\right\},$$

and if μ can lie anywhere in $(-\infty, +\infty)$, the improper prior distribution is taken as

$$f(\mu|H) = \text{constant}.$$

Hence, for the posterior distribution of μ,

$$g(\mu|\bar{x}, H) \propto \frac{n^{1/2}}{(2\pi)^{1/2}} \exp\left\{-\frac{n}{2}(\bar{x}-\mu)^2\right\}.$$

Integration shows that the constant of proportionality is unity. Thus we may, for any given level of probability, determine the range of μ. The result is the same as that given by confidence interval theory or fiducial theory.

If we assume that the prior distribution for μ is $N(\theta, \omega^2)$ and then let $\omega^2 \to \infty$, the same final result is achieved.

On the other hand, for the distribution of Example 19.8 we take the prior distribution of θ, which is in $(0, \infty)$, to be

$$f(\theta|H) \propto \theta^{-1}.$$

The essential similarity to the fiducial procedure in Example 19.8 will be evident. Using (19.152) we have the posterior distribution

$$g(\theta|t, H) \propto \left(\frac{\beta}{\theta}\right)^{\beta} \frac{t^{\beta-1} e^{-\beta t/\theta}}{\theta \Gamma(\beta)}.$$

The constant of proportionality is t, so that

$$g(\theta|t, H) = \left(\frac{\beta}{\theta}\right)^{\beta} \frac{t^{\beta} e^{-\beta t/\theta}}{\theta \Gamma(\beta)}, \tag{19.163}$$

which is (19.154) again. The results with a proper prior distribution are now somewhat different; see Exercise 19.27.

Example 19.12 (Bayesian interval for the normal mean, with variance unknown)

Let us now consider the case of setting limits to the mean in normal samples when the variance is unknown. For Student's t distribution we have

$$L(t|\mu, \sigma, H) = \frac{k}{(1 + t^2/\nu)^{(\nu+1)/2}},$$

where k is some constant and $\nu = n - 1$. The parameters μ and σ do not appear on the right and hence are irrelevant and may be suppressed. Thus we have

$$L(t|H) = \frac{k}{(1 + t^2/\nu)^{(\nu+1)/2}}.$$

Suppose now that we *assume* that

$$L(t|\bar{x}, s, H) = f(t). \tag{19.164}$$

Then, as before, $\bar{x}$ and s may be suppressed, and we have

$$L(t|H) = f(t),$$

and hence, by comparison with (19.162),

$$L(t|\bar{x}, s, H) = \frac{k}{(1 + t^2/\nu)^{(\nu+1)/2}}.$$

We can then find limits to t, given $\bar{x}$ and s, in the usual way. Jeffreys (1961) emphasized, however, that this depends on a new postulate expressed by (19.164) which, though natural, is not trivial. It amounts to an assumption that if we are comparing different distributions, samples from which give different $\bar{x}$s and ss, the scale of the distribution of μ must be taken proportional to s and its mean displaced by the difference of sample means.

Example 19.13 (Bayesian interval for the problem of two means)

In a similar way to Example 19.12, it will be found that to arrive at the Fisher–Behrens distribution it is necessary to postulate that

$$L(t_1, t_2|\bar{x}_1, \bar{x}_2, s_1, s_2, H) = f_1(t_1)f_2(t_2).$$

Jeffreys' (1961) derivation of the Fisher–Behrens form from Bayes' theorem is as follows.

The prior probability of the four parameters is

$$f(\mu_1, \mu_2, \sigma_1, \sigma_2|H) \propto (\sigma_1\sigma_2)^{-1}.$$

The likelihood (denoting the data by D) is

$$L\{D|\mu_1, \mu_2, \sigma_1, \sigma_2, H\} \propto \frac{1}{\sigma_1^{n_1}\sigma_2^{n_2}} \exp\left[-\frac{n_1}{2\sigma_1^2}\{(\mu_1-\bar{x}_1)^2+s_1^2\} - \frac{n_2}{2\sigma_2^2}\{(\mu_2-\bar{x}_2)^2+s_2^2\}\right].$$

Hence, by Bayes' theorem, the posterior distribution is

$$g(\mu_1, \mu_2, \sigma_1, \sigma_2|D, H) = \frac{1}{\sigma_1^{n_1+1}\sigma_2^{n_2+1}} \times \exp\left[-\frac{n_1}{2\sigma_1^2}\{(\mu_1-\bar{x}_1)^2+s_1^2\} - \frac{n_2}{2\sigma_2^2}\{(\mu_2-\bar{x}_2)^2+s_2^2\}\right].$$

Integrating out the values of σ_1 and σ_2, we find for the posterior distribution of μ_1 and μ_2 a form that reduces to (19.159).

19.49 We saw in Examples 19.10–13 that in special circumstances there is a close correspondence between fiducial intervals and those derived by the Bayes–Jeffreys argument, and that the latter is more explicit about the postulates on which it depends.

Fiducial theory plays a minor role in modern statistical inference, but continues to attract some attention. Seidenfeld (1992) and Zabell (1992) review recent research and point to the need for a step-by-step development of the fiducial argument for two or more parameters, since a t-like pivot can lead to inconsistencies. Barnard (1995) also stresses the need for correct pivotal arguments.

By contrast, interest in Bayesian methods has flourished, as can be seen in the companion volume by O'Hagan (1994). We return to further discussion of these methods in Chapter 26.

EXERCISES

19.1 In setting confidence limits for the variance of a normal population by the use of the distribution of the sample variance (Example 19.6), sketch the confidence belt for some value of the confidence coefficient, and show graphically that it always provides a connected range within which σ^2 is located.

19.2 Show how to set confidence limits to the ratio of variances σ_1^2/σ_2^2 in two normal populations, based on independent samples of n_1 observations from the first and n_2 observations from the second. (Use the distribution of the ratio of sample variances at (16.24).)

19.3 Use the method of **19.17** to show that large-sample 95 per cent confidence limits for π in the binomial distribution of Example 19.2 are given by

$$\frac{1}{1+(1.96)^2/n}\left(p+\frac{(1.96)^2}{2n}\pm 1.96\left(\frac{p(1-p)}{n}+\frac{(1.96)^2}{4n^2}\right)^{1/2}\right\}.$$

(Ghosh (1979) shows that this interval has high probability of being shorter than the simpler $p \pm 1.96\{[p(1-p)/n]\}^{1/2}$ obtained by using only the leading terms above, as well as controlling α better. Fujino (1980) confirms this and gives an even better approximation. Hall (1982) gives improved one-sided corrections using a continuity correction.)

19.4 Using Geary's theorem (Exercise 11.11), show that large-sample 95 per cent confidence limits for the ratio π_2/π_1 of the parameters of two binomial distributions based on independent samples of size n_2, n_1 respectively are given by

$$\frac{p_2/p_1}{1+(1.96)^2/n_2}\left\{1+\frac{(1.96)^2}{2n_2p_2}+1.96\left[\frac{1-p_1}{n_1p_1}+\frac{1-p_2}{n_2p_2}+\frac{(1.96)^2}{4}\left(\frac{1}{n_2^2p_2^2}+\frac{4(1-p_1)}{n_1n_2p_1}\right)\right]^{1/2}\right\}.$$

(Noether, 1957)

19.5 In Example 19.6 show that a confidence interval of the form

$$P\left\{\frac{ns^2}{b_n}\leq\sigma^2\leq\frac{ns^2}{a_n}\right\}=1-\alpha,$$

where $t = ns^2/\sigma^2$ has the χ^2 distribution $f_{n-1}(t)$ with $(n-1)$ d.fr., has minimum length if a_n, b_n, satisfy $f_{n+3}(a_n) = f_{n+3}(b_n)$, and hence does not coincide with the central interval used in Example 19.6.

(Tate and Klett (1959); Cohen (1972) showed that no other interval based on s^2 is shorter than this, but that if the pair of sufficient statistics $(\bar{x}, s^2)$ is used, small improvements can be made.)

19.6 Given that $f(x|\theta) = g(x)/h(\theta)$, $(a(\theta) \leq x \leq b(\theta))$, $a(\theta)$ is a monotone increasing function and $b(\theta)$ is a monotone decreasing function of θ, show (cf. **17.40–41**) that the extreme observations $x_{(1)}$ and $x_{(n)}$ are a pair of jointly sufficient statistics for θ. From their joint distribution, show that the single sufficient statistic for θ,

$$t=\min\{a^{-1}(x_{(1)}), b^{-1}(x_{(n)})\},$$

has density function

$$f(t)=\frac{n\{h(t)\}^{n-1}}{\{h(\theta)\}^n}|h'(t)|,\qquad \theta\leq t\leq c,$$

where the constant c is defined by $a(c) = b(c)$. Show that this is of the form $f(t|\theta) = g^*(t)/h^*(\theta)$, $\theta \leq t \leq c$, with

$$g^*(t) = -n\{h(t)\}^{n-1}h'(t), \qquad h^*(\theta) = \{h(\theta)\}^n,$$

so that from Exercise 17.14 the unique MV unbiased estimator of $\tau(\theta)$ is

$$\tau(t) - \tau'(t)h^*(t)/g^*(t) = \tau(t) + \tau'(t)h(t)/\{nh'(t)\}.$$

Show that these results are unchanged (except that $c \leq t \leq \theta$) when $a(\theta)$ is decreasing, $b(\theta)$ is increasing and $t = \max\{a^{-1}(x_{(1)}), b^{-1}(x_{(n)})\}$.

19.7 In Exercise 19.6, show that $\psi = h(t)/h(\theta)$ has density function

$$f(\psi) = n\psi^{n-1}, \qquad 0 \leq \psi \leq 1.$$

Show that

$$P\{\alpha^{1/n} \leq \psi \leq 1\} = 1 - \alpha,$$

and hence set a confidence interval for θ. Show that this is shorter than any other interval based on the distribution of ψ.

(Huzurbazar, 1955)

19.8 Apply the result of Exercise 19.7 to show that a confidence interval for θ in

$$f(x) = \theta^{-1}, \qquad 0 \leq x \leq \theta,$$

is obtainable from

$$P\{x_{(n)} \leq \theta \leq x_{(n)}\alpha^{-1/n}\} = 1 - \alpha.$$

19.9 Use the result of Exercise 19.7 to show that a confidence interval for θ in

$$f(x) = e^{-(x-\theta)}, \qquad \theta \leq x \leq \infty$$

is obtainable from

$$P\left\{x_{(1)} + \frac{1}{n}\log\alpha \leq \theta \leq x_{(1)}\right\} = 1 - \alpha.$$

(Huzurbazar, 1955)

19.10 Given that $I(x)$ is a confidence interval for θ calculated from the distribution of a sample, $f(x|\theta)$, and that θ_0 is the true value of θ, show that the expected length of $I(x)$ may be written as

$$E(L) = \int\left\{\int_{\theta\in I(x)} d\theta\right\} dF(x|\theta_0)$$

and that

$$E(L) = \int_{\theta\neq\theta_0} P\{\theta \in I(x)|\theta_0\}\, d\theta,$$

the integral over all false values of θ of the probability of inclusion in the confidence interval.

(Pratt, 1961; 1963)

19.11 Given that $x \in A(\theta)$ if and only if $\theta \in I(x)$ in Exercise 19.10, show that $E(L)$ is minimized by choosing $A(\theta)$ for each θ so that $P\{x \in A(\theta)|\theta_0)$ is minimized. (This is equivalent to choosing the most powerful test for each θ against the alternative value θ_0 – cf. **21.9** below.)

(Pratt, 1961; 1963)

19.12 x and y have a bivariate normal distribution with variances σ_1^2, σ_2^2, and correlation parameter ρ. Show that the variables

$$u = \frac{x}{\sigma_1} + \frac{y}{\sigma_2}, \quad v = \frac{x}{\sigma_1} - \frac{y}{\sigma_2},$$

are independently normally distributed. In a sample of n observations with sample variances s_x^2 and s_y^2 and coefficient correlation r_{xy}, show that the sample correlation coefficient of u and v may be written

$$r_{uv}^2 = \frac{(l-\lambda)^2}{(l+\lambda)^2 - 4r_{xy}^2 l\lambda},$$

where $l = s_1^2/s_2^2$ and $\lambda = \sigma_1^2/\sigma_2^2$. Hence show that, whatever the value of ρ, confidence limits for λ are given by

$$l\{K - (K^2-1)^{1/2}\}, \quad l\{K + (K^2-1)^{1/2}\},$$

where

$$K = 1 + \frac{2(1-r_{xy}^2)}{n-2} F_\alpha$$

and F_α is the 100α per cent point of the F distribution with $(1, n-1)$ d.fr.

(Pitman, 1939a)

19.13 Using the asymptotic multivariate normal distribution of maximum likelihood estimators (**18.26**) and the χ^2 distribution of the exponent of a multivariate normal distribution (**15.14**), show that (19.56) gives a large-sample confidence region for a set of parameters. From it, derive a confidence region for the mean and variance of a univariate normal distribution.

19.14 In **19.39**, show that $r(\bar{x}, y)$ defined at (19.116) is, asymptotically in n,

$$r(\bar{x}, \gamma) \sim r(0, \gamma)\left(1 + \frac{1}{2n}\right).$$

(Bowker, 1946)

19.15 Using the method of Example 6.4, show that for a χ^2 distribution with ν degrees of freedom, the value above which 100β per cent of the distribution lies is χ_β^2, given by

$$\frac{\chi_\beta^2}{\nu} \sim 1 + \left(\frac{2}{\nu}\right)^{1/2} d_{1-\beta} + \frac{2}{3\nu}(d_{1-\beta}^2 - 1) + o\left(\frac{1}{\nu}\right),$$

where

$$\int_{-\infty}^{d_{1-\beta}} (2\pi)^{-1/2} \exp\left(-\tfrac{1}{2}t^2\right) dt = 1 - \beta.$$

19.16 Combine the results of Exercises 19.14–15 to show that, from (19.119),

$$\lambda \sim r(0, \gamma)\left\{1 + \frac{d\beta}{(2n)^{1/2}} + \frac{5(d_\beta^2 + 2)}{12n}\right\}.$$

(Bowker, 1946)

19.17 Let $l_{11}, l_{12}, \ldots, l_{1,n-1}$ be $n-1$ linear functions of the observations in a sample of size n that are orthogonal to one another and to $\bar{x}_1$, and let them have zero mean and variance σ_1^2. Similarly define $l_{21}, l_{22}, \ldots, l_{2,n-1}$.

Then, in two samples of size n from normal populations with equal means and variances σ_1^2, and σ_2^2 the function

$$\frac{(\bar{x}_1 - \bar{x}_2)n^{1/2}}{\left\{\sum(l_{1j} + l_{2j})^2/(n-1)\right\}^{1/2}}$$

is distributed as Student's t with $n-1$ degrees of freedom. Show how to set confidence intervals for the difference of two means by this result, and show that the solution (19.91) is a member of this class of statistics when $n_1 = n_2$.

19.18 Given two samples of n_1, n_2 members from normal populations with unequal variances, show that by picking n_1 members at random from the n_2 (where $n_1 \leq n_2$) and pairing them at random with the members of the first sample, confidence intervals for the difference of means can be based on Student's t distribution independently of the variance ratio in the populations. Show that this step is equivalent to putting $c_{ij} = 0, i \neq j$, in (19.76) and hence that this is an inefficient solution of the two-means problem.

19.19 Use the method of **19.34** to show that the statistic z of (19.61) is distributed approximately in Student's form with degrees of freedom given by (19.104), and show that if we take the first two terms in the expansion on the right of (19.111), (19.105) gives the same approximation to order n^{-1}.

19.20 From Fisher's F distribution (16.24), find the fiducial distribution of $\theta = \sigma_1^2/\sigma_2^2$, and show that if we regard the Student's distribution of the statistic (19.60) as the joint fiducial distribution of δ and θ, and integrate out θ over its fiducial distribution, we arrive at the result of Example 19.10 for the distribution of z.

(Fisher, 1939)

19.21 Prove the statement in **19.24** to the effect that if $ax + by = z$, where x and y are independent random variables and x, y, z are all χ^2 variates, the constants $a = b = 1$.

19.22 Show that ν, the approximate degrees of freedom of z given at (19.104), increases steadily from the value $n_2 - 1$ at $\theta = 0$ to its unique maximum value $n_1 + n_2 - 2$ at $\theta = n_1(n_1-1)/\{n_2(n_2-1)\}$ and then decreases steadily to $n_1 - 1$ as $\theta \to \infty$. Show that the approximate d.fr. for t at (19.103) has the same extreme values and maximum, but that the latter is now at $\theta = 1$.

19.23 In setting confidence intervals for the median of a continuous distribution using the symmetrically spaced order statistics $x_{(r)}$ and $x_{(n-r+1)}$, show from (19.136) that for $n = 30$, the values of r listed below give the confidence coefficients shown:

r	$1-\alpha$	r	$1-\alpha$
8	0.995	12	0.80
9	0.98	13	0.64
10	0.96	14	0.42
11	0.90	15	0.14

19.24 Show from Example 19.7 that the range of a sample of size $n = 100$ from a continuous distribution has probability exceeding 0.95 of covering at least 95 per cent of the distribution, but that if we wish to cover at least 99 per cent with probability 0.95, n must be about 475 or more.

19.25 Let $q = q(\mathbf{x})$ be a pivotal quantity with p.d.f. $f(q)$. Denote the corresponding confidence interval for θ by $[G(a), G(b)]$ where $P[a \leq q \leq b] = 1 - \alpha$. The physically shortest confidence interval is given by minimizing

$$\int_C g(q)\, dq \quad \text{such that} \quad \int_C f(q)\, dq = 1 - \alpha,$$

where $g(q) = \mathrm{d}G/\mathrm{d}q$. Show that the optimal choice of the set $C = \{q | f(q)/g(q) > \lambda\}$ where λ is chosen to ensure the constraint is satisfied.
Show that for the normal mean, $g(q) = 1$, whereas for the variance $g(q) = q^{-2}$ (where the interval becomes $[G(b), G(a)]$ since G is a decreasing function of q). Hence find the shortest intervals in each case.

(Juola, 1993)

19.26 Consider the mixture distribution

$$f(x, \theta) = \theta f(x|0) + (1 - \theta) f(x|1)$$

where $f(x|\mu)$ is the p.d.f. of a normal distribution with mean μ and variance 1, and $0 < \theta < 1$. When $n = 1$, show that the 95 per cent confidence interval for θ is

$$\begin{cases} 0 < \theta < 1, & \text{if } -1.6815 \leq x \leq 2.6815 \\ \emptyset & \text{otherwise.} \end{cases}$$

Show that this interval is nevertheless uniformly most accurate and unbiased for θ. Also show that the ML estimator for θ is

$$\hat{\theta} = \begin{cases} 0 & \text{if } x < 0.5 \\ 1 & \text{if } x > 0.5. \end{cases}$$

Show that the confidence interval is not inclusion-consistent (in the sense of **19.15**).

(Plante, 1991)

19.27 Suppose that t has a gamma distribution with density

$$f(t|\theta) = \theta^{-\beta} t^{\beta-1} \mathrm{e}^{-t/\theta} / \Gamma(\beta)$$

and that the prior distribution for θ is a Pearson Type V distribution with density

$$f(\theta) = \theta^{-\alpha-1} \omega^{\alpha} \mathrm{e}^{-\omega/\theta} \Gamma(\alpha).$$

Show that the posterior distribution for θ is Pearson Type V with density

$$g(\theta|t) = (t + \omega)^{\beta+\alpha} \frac{\theta^{-\beta-\alpha-1} \exp[-(t + \omega)/\theta]}{\Gamma(\beta + \alpha)}.$$

Show that $g(\theta|t)$ reduces to the result in (19.163) only if $\omega \to 0$ as $\alpha \to 0$.

CHAPTER 20

TESTS OF HYPOTHESES: SIMPLE NULL HYPOTHESES

20.1 We now pass from the problems of estimating parameters to those of testing hypotheses concerning parameters. Instead of seeking the best (point or interval) estimator of an unknown parameter, we shall now be concerned with deciding whether some predesignated value is acceptable in the light of the observations.

In a sense, the testing problem is logically prior to that of estimation. If, for example, we are examining the difference between the means of two normal populations, our first question is whether the observations indicate that there is *any* true difference between the means. In other words, we have to compare the observed differences between the two samples with what might be expected under the hypothesis that there is no true difference at all, but only random sampling variation. If this hypothesis is not sustained, we might proceed to the second step of estimating the *magnitude* of the difference between the population means.

Quite obviously, the problems of testing hypotheses and of estimation are closely related (cf. **20.9**) but it is nevertheless useful to preserve a distinction between them, if only for expository purposes. The monograph by Lehmann (1986) treats the testing problem in full detail. Many of the ideas are due to Neyman and Pearson, whose remarkable series of papers (1928; 1933a; 1933b; 1936a; 1936b; 1938) is fundamental.

20.2 The kind of hypothesis that we test in statistics is more restricted than the general scientific hypothesis. It is a scientific hypothesis that every particle of matter in the universe attracts every other particle, or that life exists on Mars; but these are not hypotheses that we can test statistically. Statistical hypotheses concern the behaviour of observable random variables; that is, we consider only hypotheses that are *testable* on the basis of observing a set of random variables, $\mathbf{x} = \{x_1, \ldots, x_n\}$ say. Such hypotheses specify to a greater or lesser degree the nature of the distribution function corresponding to the random variable x. Further, no hypothesis can be tested in isolation; there must be at least two competing hypotheses (and we usually consider exactly two) even if one asserts proposition A and the other asserts 'not-A'.

Given that $\mathbf{x}$ describes an n-dimensional sample space, W say, we may consider any subset, w say, and for hypothesis H consider the probability $P(x \in w|H)$. Such probabilities form the basis of statistical hypothesis testing. In general, any hypothesis concerning the generating mechanisms for observable random variables is a statistical hypothesis.

For example, the hypothesis (a) that a normal distribution has a specified mean and variance is statistical; so is the hypothesis (b) that it has a given mean but unspecified variance; so is the hypothesis (c) that a distribution is of normal form with both mean and variance unspecified; finally, so is the hypothesis (d) that two unspecified continuous distributions are identical. Each of these four examples implies certain behaviour of random variables in the sample space, which may be tested by comparison with observation.

Parametric and non-parametric hypotheses

20.3 It will have been noticed that in the examples (a) and (b) in **20.2** the distribution underlying the observations was taken to be of a certain form (the normal) and the hypothesis was concerned entirely with the value of one or both of its parameters. Such a hypothesis, for obvious reasons, is called *parametric*.

Hypothesis (c) was of a different nature, as no parameter values are specified in the statement of the hypothesis; we might reasonably call such a hypothesis *non-parametric*. Hypothesis (d) is also non-parametric but, in addition, it does not even specify the underlying form of the distribution and may reasonably be termed *distribution-free*. Notwithstanding these distinctions, the statistical literature now commonly applies the label 'non-parametric' to test procedures that we have just termed 'distribution-free', thereby losing a useful classification. Having noted the problem, we will henceforth abide by the standard definitions.

Non-parametric procedures (as commonly labelled) are considered in detail in the companion volume by Hettmansperger and McKean (1998). Our discussion in this chapter and the next embraces hypotheses such as (c) along with (a) and (b), although most of our particular discussions will focus upon parametric hypotheses.

Simple and composite hypotheses

20.4 There is a distinction between the hypotheses (a) and (b) in **20.2**. In (a), the values of *all* the parameters of the distribution were specified by the hypothesis; in (b) only a subset of the parameters was specified by the hypothesis. This distinction is important for the theory. In general, if a distribution depends upon l parameters, and a hypothesis specifies unique values for k of these parameters, we call the hypothesis *simple* if $k = l$ and *composite* if $k < l$. In geometrical terms, we can represent the possible values of the parameters as a region in a space of l dimensions, one for each parameter. If the hypothesis considered selects a unique point in this parameter space, it is simple; if the hypothesis selects a subregion of the parameter space that contains more than one point, it is composite.

k is called the number of *constraints* imposed by the hypothesis, and $l - k$ is known as the number of *degrees of freedom* of the hypothesis, a terminology obviously related to the geometrical picture in the previous paragraph.

In this chapter we examine tests of simple hypotheses, and turn to tests of composite hypotheses in Chapter 21. Chapter 22 then deals with likelihood ratio tests, an important element of the Neyman–Pearson theory, and with measures of test efficiency. Finally, in Chapter 23, we examine inferential procedures that are invariant under certain classes of transformation. Such procedures have a lot of intuitive appeal; for example, it is clearly desirable that two investigators studying changes in body temperature as a result of exercise should reach the same conclusions from a given data set even when one uses the Fahrenheit scale and the other the Centigrade (Celsius) scale; that is, we would wish to use procedures that were location- and scale-invariant.

Critical regions and alternative hypotheses

20.5 To test any hypothesis, say H_0, on the basis of a random sample of observations, we must divide the sample space (i.e. all possible sets of observations) into two regions. If the observed sample point $\mathbf{x}$ falls into one of these regions, say w, we shall reject H_0; if $\mathbf{x}$ falls into the complementary region, $W - w$, we shall accept H_0. w is known as the *critical region* of the test, and $W - w$ is called the *acceptance region.*

The rather peremptory terms 'reject' and 'accept' that we have used of a hypothesis under test are now conventional usage, to which we shall adhere, and are not intended to imply that any hypothesis is ever finally accepted or rejected. Indeed, the classical theory of hypothesis testing owes much to Karl Popper's ideas on 'falsification' (Popper, 1963). Popper argues that a scientific hypothesis can never be fully corroborated, but is always open to rejection. Thus, 'do not reject' is preferable to 'accept'. We are concerned to investigate procedures that make such decisions, not with certainty but with calculable probabilities of error, in a sense to be explained.

20.6 Now if we know the probability distribution of the observations under H_0, we can determine the critical region w so that, given H_0, the probability of rejecting H_0 (i.e. the probability that $\mathbf{x}$ falls in w) is equal to a pre-assigned value α, i.e.

$$P\{x \in w | H_0\} = \alpha. \tag{20.1}$$

If we are dealing with a discrete distribution, it may not be possible to satisfy (20.1) for every α in the interval (0, 1). The value α is called the *size* of the test.[1] For the moment, we shall regard α as determined in some way. We shall discuss the choice of α later.

Evidently, we can in general find many, and often even an infinity of, subregions w of the sample space, all obeying (20.1). Which of them should we prefer to the others? This is the fundamental problem of the theory of testing hypotheses.

We cannot answer the question based only upon the specification of H_0, but we must consider some alternative explanation, known as the alternative hypothesis, H_1 say. If we are forced to choose between H_0 and H_1, we will favour the hypothesis that is in some sense better supported by the data. We will now develop these ideas formally, but before doing so, we should observe that the notion of contrasting H_0 and H_1 has not always been accepted. The introduction of the alternative hypothesis is due to Neyman and Pearson. Prior to their work, Fisher (1925b) had developed the idea of *pure significance tests*, wherein we record the probability

$$p = P(\mathbf{x} \in w | H_0),$$

commonly known as the p-value. Fisher's view (cf. Lehmann, 1993) was that the researcher could be relied upon to identify a sensible critical region w for the hypothesis H_0 under investigation. Formally, we could arrive in the same place under both the Fisher and Neyman–Pearson schemes if we agree to reject H_0 when $p < \alpha$, but the key difference is, as we shall see, that only the Neyman–Pearson theory provides an objective way to select the most appropriate w.

In practice, as we shall see in this and subsequent chapters, heavy use is made of tests where w falls in the tails of the distribution (under H_0) of the test statistic. When the test is one-tailed, the definition of the p-value is unambiguous, but for two-sided tests the picture is more murky. Often, the one-tailed value is simply doubled to provide the two-tailed value for p; however, this procedure can be misleading if the distribution under H_0 is not symmetric, or at least approximately so.

We consider this topic from other perspectives in **26.34** and **26.46**.

---

[1] The hypothesis under test is often called the 'null hypothesis', and the size of the test the 'level of significance'.

The power of a test

20.7 The discussion of **20.6** leads us to the recognition that a test must be judged by its properties both when H_0 is true and when it is false. Thus we may say that the errors made in testing a statistical hypothesis are of two types:

(I) We may wrongly reject the null hypothesis, H_0, when it is true.
(II) We may wrongly accept H_0, when it is false.

These are known as *Type I* and *Type II errors*, respectively. The probability of a Type I error is equal to the size of the critical region used, α. The probability of a Type II error is, of course, a function of the alternative hypothesis (say, H_1) considered, and is usually denoted by β. Thus

$$P\{\mathbf{x} \in W - w | H_1\} = \beta$$

or

$$P\{\mathbf{x} \in w | H_1\} = 1 - \beta. \tag{20.2}$$

This complementary probability, $1 - \beta$ is called the *power* of the test of the hypothesis H_0 against the alternative hypothesis H_1. The specification of H_1 is critical, since power is a function of H_1.

Example 20.1 (Test for the mean of the normal distribution with variance known)
Consider a normal distribution with known (unit) variance. Given

$$f(x) = (2\pi)^{-1/2} \exp\{-\tfrac{1}{2}(x - \mu)^2\}, \qquad -\infty < x < \infty,$$

we wish to test the hypothesis

$$H_0 : \mu = \mu_0.$$

This is a simple hypothesis, since it specifies $f(x)$ completely. The alternative hypothesis will also be taken as the simple

$$H_1 : \mu = \mu_1 > \mu_0.$$

Thus we must choose between a smaller given value (μ_0) and a larger one (μ_1) for the mean of the distribution.

We may represent the situation diagrammatically for a sample of $n = 2$ observations. In Fig. 20.1 we show the scatters of sample points that might arise, the lower cluster arising when H_0 is true, and the higher when H_1 is true.

In practice, we would only observe one such cluster, depending upon which hypothesis was correct. However, when considering the design of a test we may conceive of both samples (or, more formally, the probability density contours of both distributions) being available for consideration.

To choose a critical region, we need, in accordance with (20.1), to choose a region in the plane containing a proportion α of the distribution given H_0. One such region is represented by the area above the line PQ, which is perpendicular to the line AB connecting the hypothetical means. (A is the point (μ_0, μ_0), and B the point (μ_1, μ_1).) Another possible critical region of size α is the region CAD.

We see at once from the circular symmetry of the clusters that the first of these critical regions contains a very much larger proportion of the H_1 cluster than does the CAD region. The first

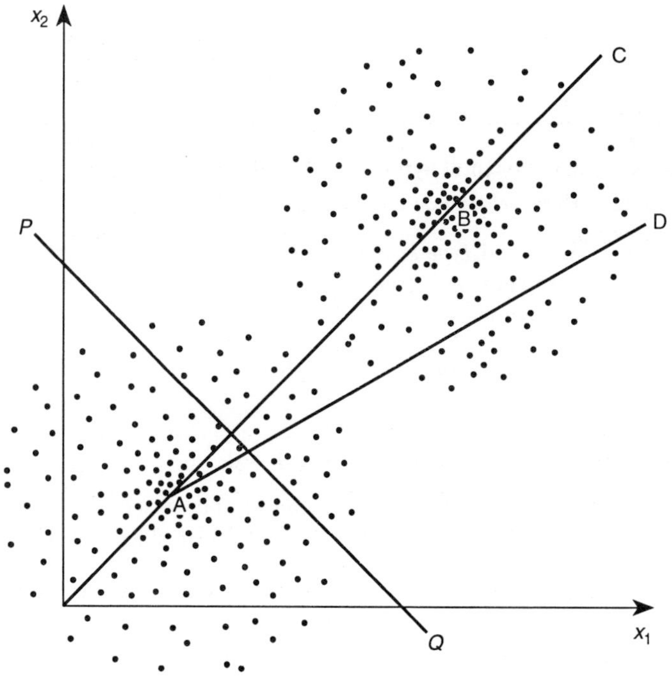

Fig. 20.1 Critical regions for $n = 2$ (see text)

region will reject H_0 rightly, when H_1 is true, in a higher proportion of cases than will the second region. Consequently, its value of $1 - \beta$ in (20.2), or in other words its power, will be greater.

20.8 Example 20.1 directs us to an obvious criterion for choosing among critical regions, all satisfying (20.1). We seek a critical region w such that its power, defined in (20.2), is as large as possible. Then, in addition to having controlled the probability of Type I errors at α, we shall have minimized the probability of a Type II error, β. This is the fundamental idea, first expressed explicitly by Neyman and Pearson, that underlies the theory of this and following chapters.

A critical region whose power is no smaller than that of any other region of the same size for testing a hypothesis H_0 against the alternative H_1, is called a *best critical region* (BCR), and a test based on a BCR is called a *most powerful* test.

At this stage, the reader may well ask why we should treat H_0 and H_1 in a non-symmetrical fashion and why α should be controlled. These are very legitimate concerns and we return to consider them in detail in **20.29**.

Tests and confidence intervals

20.9 We now take the opportunity of translating the ideas of the theory of hypothesis testing into those of the theory of confidence intervals, as promised in **19.15**.

If a sample is observed, we may ask the question: for which values of θ does the sample point $\mathbf{x}$

Table 20.1 Relationships between hypothesis testing and interval estimation

Property of test		Property of corresponding confidence interval
Size = α		Confidence coefficient = $1 - \alpha$
Power = probability of rejecting a false value of $\theta = 1 - \beta$		Probability of not covering a false value of $\theta = 1 - \beta$
Most powerful		Uniformly most accurate
$\longleftarrow$	*Unbiased* $1 - \beta \geq \alpha$	$\longrightarrow$
Equal-tails test $\alpha_1 = \alpha_2 = \frac{1}{2}\alpha$		Central interval

form part of the acceptance region A complementary to the size-α critical region for a certain test on the parameter θ? If we aggregate these 'acceptable' values of θ, we obtain the level-$(1 - \alpha)$ confidence interval C for θ corresponding to that test, for θ is in C if and only if $\mathbf{x}$ is in A, i.e. with probability $1 - \alpha$. We used this method of constructing confidence intervals in **19.3**, and indeed throughout Chapter 19.

There is thus no need to derive optimum properties separately for tests and for intervals: there is a one-to-one correspondence between the problems as in the dictionary in Table 20.1.

In particular, we say that a test of $H_0 : \theta = \theta_0$ against $H_1 : \theta = \theta_1 \neq \theta_0$ is *unbiased* if

$$p = P(\mathbf{x} \in w | H_1) \geq \alpha$$

for all choices of θ_1. Test unbiasedness is considered in detail in **21.23–24**. Relating back to point estimation, we say that a test is consistent if

$$\lim_{n \to \infty} P(\mathbf{x} \in w | H_1, \alpha) = 1;$$

that is, as the sample size increases and $(\theta_1 \neq \theta_0, \alpha)$ are held fixed, the power should approach one. This definition is in accord with that in **17.7** for point estimators, indicating that the sampling distribution of the statistic should become more and more concentrated about the parameter value as n increases.

Testing a simple H_0 against a simple H_1

20.10 If we are testing a simple null hypothesis against a simple alternative hypothesis, i.e. choosing between two completely specified distributions, the problem of finding a BCR of size α is straightforward. Its solution is given by the Neyman–Pearson lemma (Neyman and Pearson, 1933b), which we now prove.

As in earlier chapters, we write $L(x|H_i)$ for the likelihood function given the hypothesis H_i $(i = 0, 1)$, and write a single integral to represent n-fold integration in the sample space. Our problem is to select w, to maximize the power, given in (20.2):

$$1 - \beta = \int_w L(x|H_1)\,\mathrm{d}x, \tag{20.3}$$

subject to the condition (20.1), which we write as

$$\alpha = \int_w L(x|H_0)\,\mathrm{d}x. \tag{20.4}$$

The critical region w should obviously include all points $\mathbf{x}$ for which $L(x|H_0) = 0$, $L(x|H_1) > 0$; these points contribute nothing to the integral in (20.4). For the other points in w, we may rewrite (20.3) as

$$1 - \beta = \int_w \frac{L(x|H_1)}{L(x|H_0)} L(x|H_0)\,\mathrm{d}x. \tag{20.5}$$

From (20.4)–(20.5), we see that $(1-\beta)/\alpha$ is the mean within w, when H_0 holds, of $L(x|H_1)/L(x|H_0)$. Clearly this will be maximized if and only if w consists of that fraction α of the sample space containing the largest values of $L(x|H_1)/L(x|H_0)$. Thus we have the result of the Neyman–Pearson lemma, that the BCR consists of the points in W satisfying

$$\frac{L(x|H_0)}{L(x|H_1)} \leq k_\alpha \tag{20.6}$$

when H_0 holds. To any constant k_α in (20.6) there corresponds a value α for the size (20.4). If the xs are continuously distributed, we can also find a k_α for any α. The left-hand side of (20.6) defines a likelihood ratio (LR); generalizations of this concept lead to the class of LR tests considered in Chapter 22.

20.11 If the distribution of the xs is not continuous, we may effectively render it so by a randomization device (cf. **19.10**). In this case,

$$\frac{L(x|H_0)}{L(x|H_1)} = k_\alpha \tag{20.7}$$

with some non-zero probability q, while in general, owing to discreteness, we can only choose k_α in (20.6) to make the size of the test equal to $\alpha - q$ $(0 < q < p)$. To convert the test into one of exact size α, we reject H_0 with probability q/p while we accept H_0 with probability $1 - (q/p)$. The overall probability of rejection will then be $(\alpha - q) + q \cdot q/q = \alpha$, as required, whatever the value of α desired. In this case, the BCR is clearly not unique, being subject to random sampling fluctuation.

Example 20.2 (Test of the normal mean using the Neyman–Pearson lemma)
Consider again the normal distribution of Example 20.1,

$$f(x) = (2\pi)^{-1/2}\exp\{-\tfrac{1}{2}(x-\mu)^2\} \qquad -\infty < x < \infty, \tag{20.8}$$

where we now test $H_0 : \mu = \mu_0$ against the alternative $H_1 : \mu = \mu_1 (\neq \mu_0)$. We have

$$\begin{aligned} L(x|H_i) &= (2\pi)^{-1/2}\exp\left\{-\tfrac{1}{2}\sum_{j=1}^{n}(x_j-\mu_i)^2\right\}, \qquad i = 0, 1, \\ &= (2\pi)^{-n/2}\exp\left[-\frac{n}{2}\{s^2 + (\bar{x}-\mu_i)^2\}\right], \end{aligned} \tag{20.9}$$

where $\bar{x}$, s^2 are the sample mean and variance, respectively. Thus, for the BCR, we have from (20.6)

$$\frac{L(x|H_0)}{L(x|H_1)} = \exp\left[\frac{n}{2}\{(\bar{x}-\mu_1)^2 - (\bar{x}-\mu_0)^2\}\right]$$
$$= \exp\left[\frac{n}{2}\{(\mu_0-\mu_1)2\bar{x} - (\mu_1^2 - \mu_0^2)\}\right] \leq k_\alpha, \quad (20.10)$$

or

$$(\mu_0-\mu_1)\bar{x} \leq \tfrac{1}{2}(\mu_0^2-\mu_1^2) + \frac{1}{n}\log k_\alpha. \quad (20.11)$$

Thus, given μ_0, μ_1 and α, the BCR is determined by the value of the sample mean $\bar{x}$ alone. This is what we might have expected since (cf. Examples 17.6, 17.15) $\bar{x}$ is an MVB sufficient statistic for μ. Further, from (20.11), we see that if $\mu_0 > \mu_1$ the BCR is

$$\bar{x} \leq \tfrac{1}{2}(\mu_0+\mu_1) + \log k_\alpha/\{n(\mu_0-\mu_1)\}, \quad (20.12)$$

whereas if $\mu_0 < \mu_1$ it is

$$\bar{x} \geq \tfrac{1}{2}(\mu_0+\mu_1) - \log k_\alpha/\{n(\mu_1-\mu_0)\}, \quad (20.13)$$

which is again intuitively reasonable: in testing μ_0 against a smaller value μ_1, we reject μ_0 if the sample mean falls below a certain value, which depends on α, the size of the test; in testing μ_0 against a larger value μ_1, we reject μ_0 if the sample mean exceeds a certain value.

20.12 A feature of Example 20.2 that is worth noting, since it occurs in a number of problems, is that the BCR turns out to be determined by a single statistic, rather than by the whole configuration of sample values. This simplification permits us to carry on our discussion entirely in terms of the sampling distribution of that statistic, called the test statistic, and to avoid the complexities of n-dimensional distributions.

Example 20.3 (BCR for testing the normal mean)

In Example 20.2, we know (cf. Example 11.12) that whatever the value of μ, $\bar{x}$ is itself exactly normally distributed with mean μ and variance $1/n$. Thus, to obtain the BCR (20.13) of size α for testing μ_0 against $\mu_1 > \mu_0$, we determine $\bar{x}_\alpha$ so that

$$\int_{\bar{x}_\alpha}^{\infty}\left(\frac{n}{2\pi}\right)^{1/2}\exp\left\{-\frac{n}{2}(\bar{x}-\mu_0)^2\right\}d\bar{x} = \alpha.$$

Writing

$$G(x) = \int_{-\infty}^{x}(2\pi)^{-1/2}\exp(-\tfrac{1}{2}y^2)\,dy \quad (20.14)$$

for the standardized normal d.f., we see by substituting $y = n^{1/2}(\bar{x}-\mu_0)$ that

$$\alpha = 1 - G\{n^{1/2}(\bar{x}_\alpha-\mu_0)\} = G\{-n^{1/2}(\bar{x}_\alpha-\mu_0)\}$$

since $G(x) = 1 - G(-x)$ by symmetry. If we now write

$$d_\alpha = n^{1/2}(\bar{x} - \mu_0) \tag{20.15}$$

we have

$$G(-d_\alpha) = \alpha. \tag{20.16}$$

For example, when $\alpha = 0.05$, $d_{0.05} = 1.6449$, so that when $\mu_0 = 2$ and $n = 25$, we have from (20.15)

$$\bar{x}_\alpha = 2 + 1.6449/5 = 2.3290.$$

In this normal case, the power of the test, $\beta \equiv \beta(\mu_1)$, may be written down explicitly. It is

$$\int_{\bar{x}_\alpha}^{\infty} \left(\frac{n}{2\pi}\right)^{1/2} \exp\left\{-\frac{n}{2}(\bar{x} - \mu_1)^2\right\} \mathrm{d}\bar{x} = 1 - \beta. \tag{20.17}$$

Substituting $y = n^{1/2}(\bar{x} - \mu_1)$, this becomes

$$1 - \beta = 1 - G\{n^{1/2}(\mu_0 - \mu_1) + d_\alpha\} = G\{n^{1/2}(\mu_1 - \mu_0) - d_\alpha\}, \tag{20.18}$$

again using symmetry. From (20.18) it is clear that the power is a monotone increasing function both of n, the sample size, and of $\mu_1 - \mu_0$. It should be observed that although (20.13) is only the BCR for $\mu_1 > \mu_0$ the expression for its power in (20.18) remains valid for any μ_1 whatsoever. As $\mu_1 \to -\infty$, the power tends to 0, and it remains less than α for all $\mu_1 < \mu_0$.

Example 20.4 (BCR for testing the location parameter of the Cauchy distribution)

As a contrast, consider the Cauchy distribution

$$f(x) = \frac{1}{\pi\{1 + (x - \theta)^2\}}, \qquad -\infty < x < \infty,$$

and suppose that we wish to test

$$H_0 : \theta = 0$$

against

$$H_1 : \theta = \theta_1.$$

For simplicity, we shall confine ourselves to the case $n = 1$. According to (20.6), the BCR is given by

$$\frac{L(x|H_0)}{L(x|H_1)} = \frac{1 + (x - \theta_1)^2}{1 + x^2} \leq k_\alpha.$$

This is equivalent to

$$x^2(k_\alpha - 1) + 2\theta_1 x + (k_\alpha - 1 - \theta_1^2) \geq 0. \tag{20.19}$$

The form of the BCR thus defined depends upon the value of α chosen. If $k_\alpha = 1$, (20.19) reduces to $\theta_1 x \geq \frac{1}{2}\theta_1^2$, i.e. to $x \geq \frac{1}{2}\theta_1$ if $\theta_1 > 0$ and to $x \leq \frac{1}{2}\theta_1$ if $\theta_1 < 0$ (cf. Example 20.2). But if $k_\alpha < 1$, the quadratic on the left of (20.19) can be non-negative only within an interval for x, while for $k_\alpha > 1$ it can be so only outside an interval for x. Thus the BCR changes its form with α. Indeed, if $(k_\alpha - 1)^2/k_\alpha > \theta_1^2$, the quadratic has no real root and (20.19) is satisfied either for all or for no values of x.

Since the Cauchy distribution is a Student's t distribution with one degree of freedom, and $F(x) = \frac{1}{2} + (1/\pi)\arctan(x - \theta)$ by Example 16.1, we may calculate the size of the test for any k_α and θ_1 by putting $\theta = 0$ in $F(x)$. Thus, for $\theta_1 = 1$ and $k_\alpha = 1$, the size is

$$P(x \geq \tfrac{1}{2}) = 0.352,$$

while for $\theta_1 = 1$, $k_\alpha = 0.5$, (20.19) holds when $1 \leq x \leq 3$ and

$$P(1 \leq x \leq 3) = 0.148.$$

This method may also be used to determine the power of these tests. We leave the details to the reader as Exercise 20.4.

20.13 The examples we have given so far of the use of the Neyman–Pearson lemma have related to the testing of a parametric hypothesis for some given form of distribution. But, as will be seen on inspection of the proof in **20.10**, (20.6) gives the BCR for *any* test of a simple hypothesis against a simple alternative. For instance, we might be concerned to test the *form* of a distribution with known location parameter, as in the following example.

Example 20.5 (BCR for a test of distributional form)

Suppose that we know that a distribution is standardized, but wish to investigate its form. We wish to choose between the alternative forms

$$\left.\begin{aligned} H_0 &: f_0 = (2\pi)^{-1/2}\exp\left(-\tfrac{1}{2}x^2\right), \\ H_1 &: f_1 = 2^{-1/2}\exp(-2^{1/2}|x|), \end{aligned}\right\} \qquad -\infty < x < \infty.$$

For simplicity, we again take sample size $n = 1$.

Using (20.6), the BCR is given by

$$\frac{L(x|H_0)}{L(x|H_1)} = \pi^{-1/2}\exp(2^{1/2}|x| - \tfrac{1}{2}x^2) \leq k_\alpha.$$

Thus we reject H_0 when

$$2^{1/2}|x| - \tfrac{1}{2}x^2 \leq \log(k_\alpha \pi^{1/2}) = c_\alpha.$$

The BCR therefore consists of extreme positive and negative values of the observation, supplemented, if $k_\alpha > \pi^{-1/2}$ (i.e. $c_\alpha > 0$), by values in the neighbourhood of $x = 0$. Just as in Example 20.4, the form of the BCR depends upon α. The reader should verify this by drawing a diagram. Again, the test procedure is intuitively reasonable since the double exponential distribution (H_1) has heavier tails.

BCR and sufficient statistics

20.14 If both hypotheses being compared refer to the value of a parameter θ, and there is a single sufficient statistic t for θ, it follows from the factorization of the likelihood function at (17.68) that (20.6) becomes

$$\frac{L(x|\theta_0)}{L(x|\theta_1)} = \frac{g(t|\theta_0)}{g(t|\theta_1)} \leq k_\alpha, \qquad (20.20)$$

so that the BCR is a function of the value of the sufficient statistic t, as might be expected. We have already encountered an instance of this property in Example 20.2. (The same result evidently holds if θ is a set of parameters for which t is a jointly sufficient set of statistics.) Exercise 20.13 shows that the ratio of likelihoods on the left of (20.20) that determines the BCR is itself a sufficient statistic.

However, it will not always be the case that the BCR will, as in Example 20.2, be of the form $t \geq a_\alpha$ or $t \leq b_\alpha$: Example 20.4, in which the single observation x is a sufficient statistic for θ, is a counter-example. Inspection of (20.20) makes it clear that the BCR will be of this particularly simple form if $g(t|\theta_0)/g(t|\theta_1)$ is a non-decreasing function of t for $\theta_0 > \theta_1$. This will certainly be true if

$$\frac{\partial^2}{\partial\theta\,\partial t}\log g(t|\theta) \geq 0, \tag{20.21}$$

a condition which is satisfied by nearly all the distributions met in statistics.

Example 20.6 (BCR for a test of the exponential location parameter)

For the exponential distribution

$$f(x) = \begin{cases} \exp\{-(x-\theta)\}, & \theta \leq x < \infty, \\ 0 & \text{elsewhere,} \end{cases}$$

the smallest sample observation $x_{(1)}$ is sufficient for θ (cf. Example 17.19). For a sample of n observations, we have, for testing θ_0 against $\theta_1 > \theta_0$,

$$\frac{L(x|\theta_0)}{L(x|\theta_1)} = \begin{cases} \infty & \text{if } x_{(1)} < \theta_1 \\ \exp\{n(\theta_0-\theta_1)\} & \text{otherwise.} \end{cases}$$

Thus we require for a BCR

$$\exp\{n(\theta_0-\theta_1)\} \leq k_\alpha. \tag{20.22}$$

Now the left-hand side of (20.22) does not depend on the observations at all, being a constant, and (20.22) will therefore be satisfied by *every* critical region of size α with $x_{(1)} \geq \theta_1$. Thus every such critical region is of equal power, and is therefore a BCR.

If we allow θ_1 to be greater or less than θ_0 we find

$$\frac{L(x|\theta_0)}{L(x|\theta_1)} = \begin{cases} \infty & \text{if } \theta_0 \leq x_{(1)} < \theta_1, \\ \exp\{n(\theta_0-\theta_1)\} > 1 & \text{if } x_{(1)} \geq \theta_0 > \theta_1, \\ \exp\{n(\theta_0-\theta_1)\} < 1 & \text{if } x_{(1)} \geq \theta_1 > \theta_0, \\ 0 & \text{if } \theta_1 \leq x_{(1)} < \theta_0. \end{cases}$$

Thus the BCR is given by

$$(x_{(1)}-\theta_0) < 0, \quad (x_{(1)}-\theta_0) > c_\alpha.$$

The first of these events has probability zero on H_0. The value of c_α is determined to give probability α that the second event occurs when H_0 is true.

Estimating efficiency and power

20.15 The use of a statistic which is efficient in estimation (cf. **17.28–29**) does not imply that a more powerful test will be obtained than if a less efficient estimator had been used for testing purposes. This result is due to Sundrum (1954).

Let t_1 and t_2 be two asymptotically normally distributed estimators of a parameter θ, and suppose that, at least asymptotically,

$$\left.\begin{aligned} E(t_1) = E(t_2) &= \theta, \\ \text{var}(t_i|\theta = \theta_0) &= \sigma_{i0}^2, \\ \text{var}(t_i|\theta = \theta_1) &= \sigma_{i1}^2 \end{aligned}\right\} \qquad i = 1, 2.$$

We now test $H_0 : \theta = \theta_0$ against $H_1 : \theta = \theta_1 > \theta_0$. Exactly as in (20.15) in Example 20.3, we have the critical regions, one for each test,

$$t_i \geq \theta_0 + d_\alpha \sigma_{i0}, \qquad i = 1, 2, \tag{20.23}$$

where d_α is the normal deviate defined by (20.14) and (20.16). The powers of the tests are (generalizing (20.18) which dealt with a case where $\sigma_{i0} = \sigma_{i1} = n^{-1}$)

$$1 - \beta(t_i) = G\left\{\frac{(\theta_1 - \theta_0) - d_\alpha \sigma_{i0}}{\sigma_{i1}}\right\}. \tag{20.24}$$

Since $G(x)$ is a monotone increasing function of its argument, t_1 will provide a more powerful test than t_2 if and only if, from (20.24),

$$\frac{(\theta_1 - \theta_0) - d_\alpha \sigma_{10}}{\sigma_{11}} > \frac{(\theta_1 - \theta_0) - d_\alpha \sigma_{20}}{\sigma_{21}},$$

i.e. if

$$\theta_1 - \theta_0 > d_\alpha \left(\frac{\sigma_{10}\sigma_{21} - \sigma_{20}\sigma_{11}}{\sigma_{21} - \sigma_{11}}\right). \tag{20.25}$$

If we put $E_j = \sigma_{2j}/\sigma_{1j} (j = 0, 1)$, (20.25) becomes

$$\theta_1 - \theta_0 > d_\alpha \left(\frac{E_1 - E_0}{E_1 - 1}\right)\sigma_{10}. \tag{20.26}$$

E_0, E_1 are simply powers (usually square roots) of the estimating efficiency of t_1 relative to t_2 when H_0 and H_1 respectively hold. Now if

$$E_0 = E_1 > 1, \tag{20.27}$$

the right-hand side of (20.25) is zero, and (20.26) always holds. Thus if the estimating efficiency of t_1 exceeds that of t_2 *by the same amount* on both hypotheses, the more efficient statistic t_1 always provides a more powerful test, whatever value α or $\theta_1 - \theta_0$ takes. But if

$$E_1 > E_0 \geq 1 \tag{20.28}$$

we can always find a test size α small enough for (20.26) to be falsified. Hence, the less efficient estimator t_2 will provide a more powerful test if (20.28) holds, i.e. if its relative efficiency is greater on H_0 than on H_1. Alternatively, if $E_0 > E_1 > 1$, we can find σ *large* enough to falsify (20.26). If E_1 is continuous in θ, $E_1 \to E_0$ as $\theta_1 \to \theta_0$, so that (20.26) is not falsified in the immediate neighbourhood of θ_0.

This result, though a restrictive one, is enough to show that the relation between estimating efficiency and test power is rather loose. In Chapter 22 we shall again consider this relationship when we discuss the measurement of test efficiency.

Example 20.7 (Use of inefficient estimators to test hypotheses)

In Examples 18.3 and 18.6 we saw that in estimating the parameter ρ of a standardized bivariate normal distribution, the ML estimator $\hat{\rho}$ is a root of a cubic equation, with large-sample variance equal to $(1-\rho^2)^2/\{n(1+\rho^2)\}$, while the sample correlation coefficient r has large-sample variance $(1-\rho^2)^2/n$. Both estimators are consistent and asymptotically normal, and the ML estimator is efficient. In the notation of **20.15**,

$$E = (1+\rho^2)^{1/2}.$$

If we test $H_0 : \rho = 0$ against $H_1 : \rho = 0.1$, we have $E_0 = 1$, and (20.26) simplifies to

$$0.1 > d_\alpha \sigma_{10} = d_\alpha n^{-1/2}. \qquad (20.29)$$

If we choose n to be, say, 400, so that the normal approximations are adequate, we require

$$d_\alpha > 2$$

to falsify (20.29). This corresponds to $\alpha < 0.023$, so that for tests of size < 0.023, the inefficient estimator r has greater power asymptotically in this case than the efficient $\hat{\rho}$. Since tests of size 0.01, 0.05 are quite commonly used, this is not merely a theoretical example: it cannot be assumed in practice that 'good' estimators are 'good' test statistics.

Testing a simple H_0 against a class of alternatives

20.16 So far we have been discussing the most elementary problem, where in effect we have only to choose between two completely specified competing hypotheses. For such a problem, there is a certain symmetry about the situation – it is only a matter of convention or convenience which of the two hypotheses we regard as being under test and which as the alternative, provide that we do not arbitrarily fix α without regard for β. As soon as we generalize the testing situation, this symmetry disappears.

Consider now the case where H_0 is simple, but H_1 is composite and consists of a class of simple alternatives. The most frequently occurring case is that where Ω is a class of simple parametric hypotheses of which H_0 is one and H_1 comprises the remainder; for example, the hypothesis H_0 may be that the mean of a certain distribution has some value μ_0 and the hypothesis H_1 that $\mu \neq \mu_0$.

For each of these other values we may apply the foregoing results and find, for given α, corresponding to any particular member of H_1 (say H_t) a BCR w_t. But this region in general will vary from one H_t to another. It is clearly impossible to use a different region for all the unspecified possibilities and we are therefore led to enquire whether there exists one BCR which is the best for all H_t in H_1. Such a region is called *uniformly most powerful* (UMP) and the test based on it a UMP test.

20.17 Unfortunately, as we shall find below, a UMP test does not usually exist unless we restrict Ω in certain ways. Consider, for instance, the case dealt with in Example 20.2. We found there that for $\mu_1 < \mu_0$ the BCR for a simple alternative was defined by

$$\bar{x} \leq a_\alpha. \tag{20.30}$$

Provided $\mu_1 < \mu_0$, the regions determined by (20.30) do not depend on μ_1 and can be found directly from the sampling distribution of $\bar{x}$ when the test size, α, is given. Consequently the test based on (20.30) is UMP for the class of hypotheses $\mu_1 < \mu_0$.

However, from Example 20.2, if $\mu_1 > \mu_0$, the BCR is defined by $\bar{x} \geq b_\alpha$. Here again, if we restrict attention to values of μ_1 greater than μ_0 the test is UMP. But if μ_1 can be either greater or less than μ_0, no UMP test is possible, for one or other of the two BCRs that we have just discussed will be better than any other region against this class of alternatives.

20.18 We now prove that for a simple $H_0 : \theta = \theta_0$ concerning a parameter θ defining a class of hypotheses, no UMP test exists in general against an interval including positive and negative values of $\theta - \theta_0$, under regularity conditions, in particular that the derivative of the likelihood with respect to θ is continuous in θ.

We form the Taylor expansion of the likelihood function about θ_0, obtaining

$$L(x|\theta_1) = L(x|\theta_0) + (\theta_1 - \theta_0)L'(x|\theta^*) \tag{20.31}$$

where θ^* is some value between θ_0 and θ_1. For the BCR, if any, we must have, from (20.6) and (20.31),

$$\frac{L(x|\theta_1)}{L(x|\theta_0)} = 1 + \frac{(\theta_1 - \theta_0)L'(x|\theta^*)}{L(x|\theta_0)} \geq k_\alpha(\theta_1). \tag{20.32}$$

Thus the BCR is defined by

$$\frac{L'(x|\theta^*)}{L(x|\theta_0)} \begin{cases} \geq a_\alpha, & \theta_1 > \theta_0, \quad (20.33) \\ \leq b_\alpha, & \theta_1 < \theta_0. \quad (20.34) \end{cases}$$

Now let θ_1 approach θ_0. θ^* necessarily does the same, and in the immediate neighbourhood of θ_0 (20.33)–(20.34) become, since L' is continuous in θ,

$$\frac{L'(x|\theta_0)}{L(x|\theta_0)} = \left[\frac{\partial \log L}{\partial \theta}\right]_{\theta=\theta_0} \begin{cases} \geq a_\alpha, & \theta > \theta_0, \quad (20.35) \\ \leq b_\alpha, & \theta < \theta_0. \quad (20.36) \end{cases}$$

We thus establish, incidentally, that in the immediate neighbourhood of θ_0, one-sided tests based on $\left[\partial \log L/\partial\theta\right]_{\theta=\theta_0}$ are UMP. This is a testing analogue of the confidence interval result obtained in **19.19–20**.

Our main result now follows at once. If we are considering an interval of alternatives including positive and negative values of $(\theta_1 - \theta_0)$, (20.35) and (20.36) cannot both hold (and there can therefore be no BCR) unless

$$\left[\frac{\partial \log L}{\partial \theta}\right]_{\theta=\theta_0} = \text{constant}, \tag{20.37}$$

which is the essential condition for the existence of a two-sided BCR. It cannot be satisfied if (17.18) holds (e.g. for distributions with range independent of θ) unless the constant is zero, for the condition $E(\partial \log L/\partial\theta) = 0$ with (20.37) implies $[\partial \log L/\partial\theta]_{\theta=\theta_0} = 0$.

In Example 20.6, we have already encountered an instance where a two-sided BCR exists. The reader should verify that for that distribution $[\partial \log L/\partial\theta]_{\theta=\theta_0} = n$ exactly, so that (20.37) is satisfied.

UMP tests of more than one parameter

20.19 If the distribution considered has more than one parameter, and we are testing a simple hypothesis, it remains possible that a common BCR exists for a class of alternatives varying with these parameters. The following two examples discuss the case of the two-parameter normal distribution, where we might expect to find such a BCR, but where none exists, and the two-parameter exponential distribution, where a BCR does exist.

Example 20.8 (Lack of a common BCR for the two-parameter normal distribution)

Consider the normal distribution with mean μ and variance σ^2. The hypothesis to be tested is

$$H_0 : \mu = \mu_0, \quad \sigma = \sigma_0,$$

and the alternative, H_1, is restricted only in that it must differ from H_0. For any such

$$H_1 : \mu = \mu_1, \quad \sigma = \sigma_1,$$

the BCR is, from (20.6), given by

$$\frac{L(x|H_0)}{L(x|H_1)} = \left(\frac{\sigma_1}{\sigma_0}\right)^n \exp\left[-\frac{1}{2}\left\{\sum\left(\frac{x-\mu_0}{\sigma_0}\right)^2 - \sum\left(\frac{x-\mu_1}{\sigma_1}\right)^2\right\}\right] \leq k_\alpha.$$

This may be written in the form

$$s^2\left(\frac{1}{\sigma_1^2} - \frac{1}{\sigma_0^2}\right) + \frac{(\bar{x}-\mu_1)^2}{\sigma_1^2} - \frac{(\bar{x}-\mu_0)^2}{\sigma_0^2} \leq \frac{2}{n}\log\left\{\left(\frac{\sigma_0}{\sigma_1}\right)^n k_\alpha\right\}, \tag{20.38}$$

where $\bar{x}, s^2$ are the sample mean and variance, respectively. If $\sigma_0 \neq \sigma_1$, (20.38) becomes

$$\left(\frac{\sigma_1^2 - \sigma_0^2}{\sigma_0^2\sigma_1^2}\right)\sum(x-\rho)^2 \geq c_\alpha, \tag{20.39}$$

where c_α is independent of the observations, and

$$\rho = \frac{\mu_0\sigma_1^2 - \mu_1\sigma_0^2}{\sigma_1^2 - \sigma_0^2}.$$

When a strict equality, (20.39) is the equation of a hypersphere, centred at $x_1 = x_2 = \cdots = x_n = \rho$. Thus the BCR is always bounded by a hypersphere. When $\sigma_1 > \sigma_0$, (20.39) yields

$$\sum(x-\rho)^2 \geq a_\alpha,$$

so that the BCR lies outside the sphere; when $\sigma_1 < \sigma_0$, we find from (20.39)

$$\sum(x-\rho)^2 \leq b_\alpha,$$

and the BCR is inside the sphere.

Since ρ is a function of μ_1 and σ_1, it is clear that there will not generally be a common BCR for different members of H_1, even if we limit ourselves by $\sigma_1 < \sigma_0$ and $\mu_1 < \mu_0$ or similar restrictions. We may illustrate the situation by a diagram of the $(\bar{x}, s)$ plane, for

$$\begin{aligned}\sum(x-\rho)^2 &= \sum(x-\bar{x})^2 + n(\bar{x}-\rho)^2 \\ &= n\{s^2 + (\bar{x}-\rho)^2\},\end{aligned}$$

and if this is held constant, we obtain a circle with centre $(\rho, 0)$ and fixed radius a function of α.

Figure 20.2 (adapted from Neyman and Pearson, 1933b) illustrates some of the contours for particular cases. A single curve, corresponding to a fixed value of k_α in (20.38), is shown in each case.

Cases (1) and (2): $\sigma_1 = \sigma_0$ and $\rho \to \pm\infty$. The BCR lies on the right of the line (1) if $\mu_1 > \mu_0$ and on the left of (2) if $\mu_1 < \mu_0$ This is the case discussed in Example 20.2 where we put $\sigma_1 = \sigma_0 = 1$.

Case (3): $\sigma_1 < \sigma_0$, say $\sigma_1 = \frac{1}{2}\sigma_0$. Then $\rho = \mu_0 + \frac{4}{3}(\mu_1 - \mu_0)$ and the BCR lies inside the semicircle marked (3).

Case (4): $\sigma_1 < \sigma_0$ and $\mu_1 = \mu_0$. The BCR is inside the semicircle (4).

Case (5): $\sigma_1 > \sigma_0$ and $\mu_1 = \mu_0$. The BCR is outside the semicircle (5).

There is evidently no common BCR for these cases. The regions of acceptance, however, may have a common part, centred on the value (μ_0, σ_0), and we should expect them to do so. Let us find the envelope of the BCR, which is, of course, the same as that of the regions of acceptance. The ratio of likelihoods (20.38) is differentiated with respect to μ_1 and to σ_1, and these derivatives equated to zero. This gives precisely the ML solutions of Example 18.11,

$$\begin{aligned}\hat{\mu}_1 &= \bar{x}, \\ \hat{\sigma}_1 &= s.\end{aligned}$$

Substituting in (20.38), we find for the envelope

$$1 - \frac{s^2}{\sigma_0^2} - \left(\frac{\bar{x}-\mu_0}{\sigma_0}\right)^2 = \frac{2}{n}\log\left\{\left(\frac{\sigma_0}{s}\right)^n k_\alpha\right\}$$

or

$$\left(\frac{\bar{x}-\mu_0}{\sigma_0}\right)^2 - \log\left(\frac{s^2}{\sigma_0^2}\right) + \frac{s^2}{\sigma_0^2} = 1 - \frac{2}{n}\log k_\alpha. \tag{20.40}$$

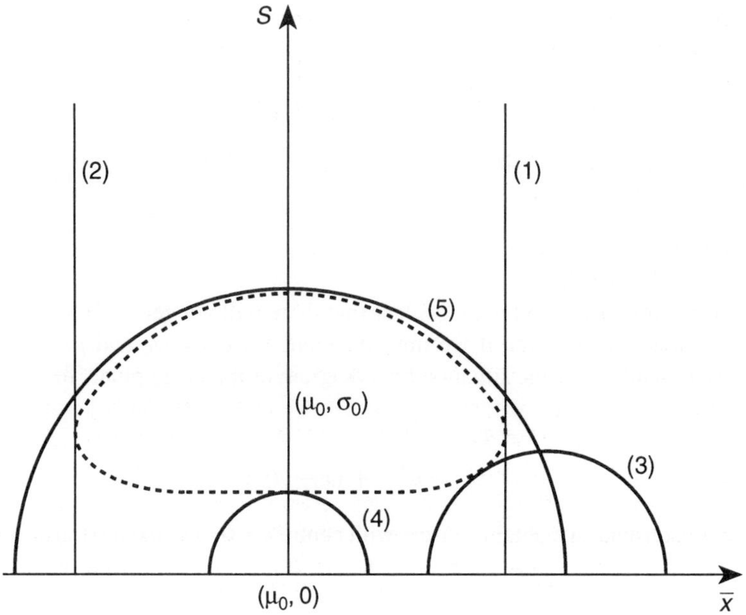

Fig. 20.2 Contours of constant likelihood ratio k (see text)

The dotted curve in Fig. 20.2 shows one such envelope. It touches the boundaries of all the BCRs that have the same k (and hence are not of the same size α). The space inside may be regarded as a 'good' region of acceptance and the space outside accordingly as a good critical region. There is no BCR for all alternatives, but the regions determined by (20.40) effect a compromise by picking out and amalgamating parts of critical regions that are best for individual alternatives.

Example 20.9 (UMP test for the two-parameter exponential)

We wish to test the simple hypothesis

$$H_0 : \theta = \theta_0, \quad \sigma = \sigma_0$$

against the alternative

$$H_1 : \theta = \theta_1 < \theta_0, \quad \sigma = \sigma_1 < \sigma_0,$$

for the exponential distribution

$$f(x) = \sigma^{-1} \exp\left\{-\left(\frac{x-\theta}{\sigma}\right)\right\}, \qquad \theta \leq x < \infty; \sigma > 0. \tag{20.41}$$

From (20.6), the BCR is given by

$$\frac{L_0}{L_1} = \left(\frac{\sigma_1}{\sigma_0}\right)^n \exp\left\{-\frac{n(\bar{x}-\theta_0)}{\sigma_0} + \frac{n(\bar{x}-\theta_1)}{\sigma_1}\right\} \leq k_\alpha,$$

so that whatever the values of θ_1, σ_1 in H_1, the BCR is of the form

$$x_{(1)} \leq \theta_0, \quad \bar{x} \leq \frac{\frac{1}{n}\log\left\{k_\alpha\left(\frac{\sigma_0}{\sigma_1}\right)^n\right\} + \left(\frac{\theta_1}{\sigma_1} - \frac{\theta_0}{\sigma_0}\right)}{\left(\frac{1}{\sigma_1} - \frac{1}{\sigma_0}\right)}. \tag{20.42}$$

The first of these events has probability zero when H_0 holds. There is therefore a common BCR for the whole class of alternatives H_1, on which a UMP test may be based.

We have already dealt with the case $\sigma_1 = \sigma_0 = 1$ in Example 20.6.

UMP tests and sufficient statistics

20.20 In **20.14** we saw that in testing a simple parametric hypothesis against a simple alternative, the BCR is necessarily a function of any (jointly) sufficient statistics for the parameter(s). In testing a simple H_0 against a composite H_1 consisting of a class of simple parametric alternatives, it evidently follows from the argument of **20.14** that if a common BCR exists, providing a UMP test against H_1, and if t is a sufficient statistic for the parameter(s), then the BCR will be a function of t. But, since a UMP test does not always exist, new questions now arise. Does the existence of a UMP test imply the existence of a corresponding sufficient statistic? And, conversely, does the existence of a sufficient statistic guarantee the existence of a corresponding UMP test?

20.21 The first of these questions may be affirmatively answered if an additional condition is imposed. In fact, as Neyman and Pearson (1936a) showed, if

(1) there is a common BCR for, and therefore a UMP test of, H_0 against H_1 for every size α in an interval $0 < \alpha \leq \alpha_0$ (where α_0 is not necessarily equal to 1), and
(2) every point in the sample space W (save possibly a set of measure zero) forms part of the boundary of the BCR for at least one value of α, and then corresponds to a value of $L(x|H_0) > 0$,

then a single sufficient statistic exists for the parameter(s) whose variation provides the class of admissible alternatives H_1.

To establish this result, we first note that, if a common BCR exists for H_0 against H_1 for two test sizes α_1, and $\alpha_2 < \alpha_1$, a common BCR of size α_2 can always be formed as a subregion of that of size α_1. This follows from the fact that any common BCR satisfies (20.6). We may therefore, without loss of generality, take it that as α decreases, the BCR is adjusted simply by exclusion of some of its points.[2]

Now, suppose that conditions (1) and (2) are satisfied. If a point (say, x) of w forms part of the boundary of the BCR for only one value of α, we define the statistic

$$t(x) = \alpha. \tag{20.43}$$

If a point x forms part of the BCR boundary for more than one value of α, we define

$$t(x) = \tfrac{1}{2}(\alpha_1 + \alpha_2), \tag{20.44}$$

[2] This is not true of critical regions in general – see, for example, Chernoff (1951).

where α_1 and α_2 are the smallest and largest values of α for which it does so: it follows from the remark of the previous paragraph that x will also be part of the BCR boundary for all α in the interval (α_1, α_2). The statistic t is thus defined by (20.43) and (20.44) for all points in W (except possibly a zero-measure set). Further, if t has the same value at two points, they must lie on the same boundary. Thus, from (20.6), we have

$$\frac{L(x|\theta_0)}{L(x|\theta_1)} = k(t, \theta),$$

where k does not contain the observations except in the statistic t. Thus we must have

$$L(x|\theta) = g(t|\theta)h(x) \tag{20.45}$$

so that the single statistic t is sufficient for θ, the set of parameters concerned.

20.22 We have already considered in Example 20.2 a situation where single sufficiency and a UMP test exist together. Exercises 20.1–3 give further instances. But condition (2) of **20.21** is not always fulfilled, and then the existence of a single sufficient statistic may not follow from that of a UMP test. The following example illustrates the point.

Example 20.10 (A UMP test without a single sufficient statistic)

In Example 20.9, we showed that the distribution (20.41) admits a UMP test of the H_0 against the H_1 there described. The UMP test is based on the BCR (20.42), depending on $x_{(1)}$ and $\bar{x}$.

We have already seen (cf. Example 17.19 and Exercise 17.9) that the smallest observation $x_{(1)}$ is sufficient for θ if σ is known, and that $\bar{x}$ is sufficient for σ if θ is known. The pair of statistics $x_{(1)}$ and $\bar{x}$ are jointly sufficient, but there is no *single* sufficient statistic for θ and σ.

20.23 On the other hand, the possibility that a single sufficient statistic exists without a one-sided UMP test, even where only a single parameter is involved, is made clear by Example 20.11.

Example 20.11 (A single sufficient statistic without a UMP test)

Consider the multinormal distribution of n variates $x_1, \ldots, x_n$ with

$$\begin{aligned} E(x_1) &= n\theta, & \theta &> 0, \\ E(x_r) &= 0, & r &> 1, \end{aligned}$$

and covariance matrix

$$\mathbf{V} = \begin{pmatrix} n-1+\theta^2 & -1 \cdots -1 \\ -1 & \\ \vdots & \mathbf{I}_{n-1} \\ -1 & \end{pmatrix} \tag{20.46}$$

where $\mathbf{I}_{n-1}$ is the identity matrix of order $n-1$. The determinant of (20.46) is easily seen to be

$$|\mathbf{V}| = \theta^2$$

and its inverse is

$$\mathbf{V}^{-1} = \frac{1}{\theta^2}\begin{pmatrix} 1 & & & \mathbf{1} \\ & 1+\theta^2 & & \\ & \mathbf{1} & \ddots & \\ & & & 1+\theta^2 \end{pmatrix} \tag{20.47}$$

with every off-diagonal element equal to unity. Thus, from **15.3**, the joint distribution is

$$f(\mathbf{x}) = \frac{1}{\theta(2\pi)^{n/2}}\exp\left\{-\frac{1}{2}\left[\frac{n^2}{\theta^2}(\bar{x}-\theta)^2 + \sum_{i=2}^{n} x_i^2\right]\right\}. \tag{20.48}$$

Consider now testing the hypothesis $H_0 : \theta = \theta_0 > 0$ against $H_1 : \theta = \theta_1 > 0$ on the basis of a single observation. From (20.6), the BCR is given by

$$\frac{L(x|\theta_0)}{L(x|\theta_1)} = \left(\frac{\theta_1}{\theta_0}\right)\exp\left\{-\frac{n^2}{2}\left[\frac{(\bar{x}-\theta_0)^2}{\theta_0^2} - \frac{(\bar{x}-\theta_1)^2}{\theta_1^2}\right]\right\} \leq k_\alpha,$$

which reduces to

$$\frac{(\bar{x}-\theta_1)^2}{\theta_1^2} - \frac{(\bar{x}-\theta_0)^2}{\theta_0^2} \leq \frac{2}{n^2}\log(k_\alpha\theta_0/\theta_1)$$

or

$$\bar{x}^2(\theta_0^2-\theta_1^2) - 2\bar{x}\theta_0\theta_1(\theta_0-\theta_1) \leq \frac{2\theta_0^2\theta_1^2}{n^2}\log(k_\alpha\theta_0/\theta_1).$$

If $\theta_0 > \theta_1$, this is of form

$$\bar{x}^2(\theta_0+\theta_1) - 2\bar{x}\theta_0\theta_1 \leq a_\alpha, \tag{20.49}$$

which implies

$$b_\alpha \leq \bar{x} \leq c_\alpha. \tag{20.50}$$

If $\theta_0 < \theta_1$ the BCR is of form

$$\bar{x}^2(\theta_0+\theta_1) - 2\bar{x}\theta_0\theta_1 \geq d_\alpha, \tag{20.51}$$

implying

$$\bar{x} \leq e_\alpha \quad \text{or} \quad \bar{x} \geq f_\alpha. \tag{20.52}$$

In both (20.50) and (20.52), the limits between which (or outside which) $\bar{x}$ has to lie are functions of the exact value of θ_1. This difficulty, which arises from the fact that θ_1 appears in the coefficient of $\bar{x}^2$ in the quadratics (20.49) and (20.51), means that there is no BCR even for a one-sided set of alternatives, and therefore no UMP test.

It is easily verified from (20.48) that $\bar{x}$ is a single sufficient statistic for θ, and this completes the demonstration that single sufficiency does not imply the existence of a UMP test.

The power function

20.24 Now that we are considering the testing of a simple H_0 against a composite H_1, we generalize the idea of the power of a test defined at (20.2). As stated there, the power is an explicit function of H_1. If H_1 specifies a set of alternative values for θ, we may consider the power of a test of $H_0 : \theta = \theta_0$ against the simple alternative $H_1' : \theta = \theta_1 > \theta_0$ for each θ_1. The power for each H_1' depends on the value of θ_1, so that the set of all such values defines the *power function* of the test of H_0 against the class of alternatives $H_1 : \theta > \theta_0$. We indicate the composite nature of H_1 by writing it in this way, instead of the form used for a simple $H_1 : \theta = \theta_1 > \theta_0$. For instance, we saw in Example 20.3 that the power of the most powerful test of $H_0 : \mu = \mu_0$ against $H_1 : \mu > \mu_0$ for a normal population is given by the power function (20.18), a monotone increasing function of μ_1.

The evaluation of a power function is rarely as easy as in Example 20.3, since even if the sampling distribution of the test statistic is known exactly for both H_0 and the class of alternatives H_1 (and more commonly only approximations are available, especially for H_1), there is still the problem of evaluating (20.2) for each value of θ in H_1, which usually is a matter of numerical integration. Asymptotically, however, the central limit theorem comes to our aid: the distributions of many test statistics tend to normality, given either H_0 or H_1, as the sample size increases, so that the asymptotic power function will be of the form (20.18), as we shall see when we come to the comparison of tests in Chapter 22.

Example 20.12 (Form of the power function for a test of the normal mean)

The power function (20.18) in Example 20.3 follows a normal distribution function. It increases from the valuc

$$G\{-d_\alpha\} = \alpha$$

at $\mu = \mu_0$ (in accordance with the size requirement) to the value

$$G\{0\} = 0.5$$

at $\mu = \mu_0 + d_\alpha/n^{1/2}$, the first derivative G' increasing up to this point; as μ increases beyond it, G' declines to its limiting value of zero as G approaches 1.0 and $\mu \to \infty$.

20.25 Once the power function of a test has been determined, it may be used to determine how large the sample should be in order to test H_0 with given size and power. The procedure is illustrated in the next example.

Example 20.13 (Choice of sample size for a test)

How many observations should be taken in Example 20.3 so that we may test $H_0 : \mu = 3$ with $\alpha = 0.05$ (i.e. $d_\alpha = 1.6449$) and power of at least 0.75 against alternatives such that $\mu \geq 3.5$? Put another way, how large should n be to ensure that the probability of a Type I error is 0.05, and that of a Type II error at most 0.25 for $\mu \geq 3.5$?

From (20.18), we require n large enough to make

$$G\{n^{1/2}(3.5 - 3) - 1.6449\} = 0.75, \tag{20.53}$$

since the power will be greater than this for $\mu > 3.5$. Now,

$$G\{0.6745\} = 0.75, \tag{20.54}$$

and hence, from (20.53) and (20.54),

$$0.5n^{1/2} - 1.6449 = 0.6745,$$

whence

$$n = (4.6388)^2 = 21.5,$$

so that $n = 22$ will suffice to give the test the required power.

One- and two-sided tests

20.26 We have seen in **20.18** that in general no UMP test exists when we test a parametric hypothesis $H_0 : \theta = \theta_0$ against a two-sided alternative hypothesis, i.e. one in which $\theta - \theta_0$ changes sign. Nevertheless, situations often occur when departures from H_0 in either direction are considered important. In such circumstances, it is tempting to continue to use as our test statistic one which is known to give a UMP test against one-sided alternatives ($\theta > \theta_0$ or $\theta < \theta_0$) but to modify the critical region in the distribution of the statistic by compromising between the BCR for $\theta > \theta_0$ and that for $\theta < \theta_0$.

20.27 For instance, in Example 20.2 and in **20.17** we saw that the mean $\bar{x}$, used to test $H_0 : \mu = \mu_0$ for the mean μ of a normal population, gives a UMP test against $\mu_1 < \mu_0$ with common BCR $\bar{x} \leq a_\alpha$, and a UMP test for $\mu_1 > \mu_0$ with common BCR $\bar{x} \geq b_\alpha$. Suppose, then, that for the two-sided alternative $H_1 : \mu \neq \mu_0$ we construct a compromise *equal-tails* critical region defined by

$$\bar{x} \leq a_{\alpha/2}, \bar{x} \geq b_{\alpha/2}. \tag{20.55}$$

In other words, we combine the one-sided critical regions and make each of them of size $\frac{1}{2}\alpha$, so that the overall Type I error remains fixed at α.

We know that the critical region defined by (20.55) will always be less powerful than one or other of the one-sided BCRs, but we also know that it will always be more powerful than the other. For its power function will be, exactly as in Example 20.3,

$$G\{n^{1/2}(\mu - \mu_0) - d_{\alpha/2}\} + G\{n^{1/2}(\mu_0 - \mu) - d_{\alpha/2}\}. \tag{20.56}$$

Expression (20.56) is an even function of $\mu - \mu_0$ with a minimum at $\mu = \mu_0$ and thus is greater than or equal to α; in the terminology of **20.9**, the equal-tails test is *unbiased*. Hence it is always intermediate in value between $G\{n^{1/2}(\mu - \mu_0) - d_\alpha\}$ and $G\{n^{1/2}(\mu_0 - \mu) - d_\alpha\}$, which are the power functions of the one-sided BCR, except when $\mu = \mu_0$, when all three expressions are equal. The comparison is worth making diagrammatically, in Fig. 20.3, where a single fixed value of n and of α is illustrated.

20.28 We shall see in **21.23** onwards that other, less intuitive, justifications can be given for splitting the critical region in this way between the tails of the distribution of the test statistic.

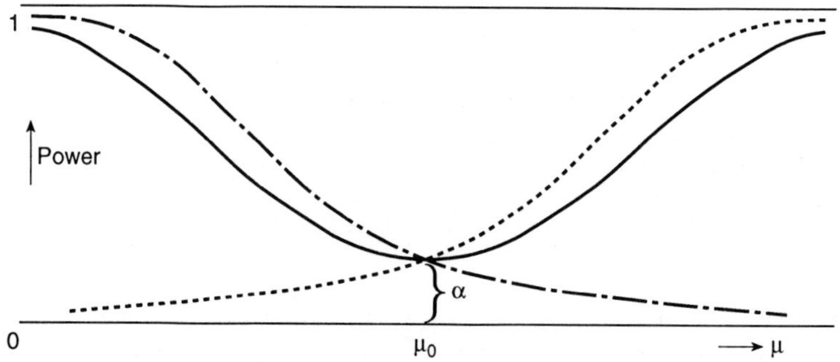

Fig. 20.3 Power functions of three tests based on $\bar{x}$. (—— Critical region in both tails equally. - - - - Critical region in upper tail. — · — Critical region in lower tail.)

For the moment, the procedure is to be regarded as simply a common-sense way of insuring against the risk of extreme loss of power which, as Fig. 20.3 makes clear, would be the result if we located the critical region in the wrong tail of the distribution. The choice between one- and two-sided tests lies with the investigator. In some contexts, such as quality control or testing a new drug against as existing one, a one-sided test may be desirable or even mandatory. However, without a compelling reason to favour one direction over the other, a two-sided test should be used, notwithstanding the reduction in power.

Choice of test size

20.29 Thus far we have assumed that the test size α has been fixed in some way. All our results are valid however that choice was made. We now turn to the question of how α is to be determined.

In the first place, it is natural to suggest that α should be made 'small' according to some acceptable criterion, and indeed it is customary to use certain conventional values of α, such as 0.05, 0.01 or 0.001. But we must take care not to go too far in this direction. We can only fix two of the quantities n, α and β, even in testing a simple H_0 against a simple H_1. If n is fixed, we can only in general decrease the value of α, the probability of Type I error, by increasing the value of β, the probability of Type II error. In other words, reduction in the size of a test decreases its power.

This point is well illustrated in Example 20.3 by the expression (20.18) for the power of the BCR in a one-sided test for a normal population mean. We see there that as $\alpha \to 0$, by (20.16) $d_\alpha \to \infty$, and consequently the power (20.18) tends to 0.

Thus, for a fixed sample size, we have must reconcile the size and power of the test. If the practical risks attaching to a Type I error are great, while those attaching to a Type II error are small, there is a case for reducing α, at the expense of increasing β, if n is fixed. If, however, sample size is at our disposal, we may, as in Example 20.13, ensure that n is large enough to reduce both α and β to any pre-assigned levels. These levels have still to be fixed, but unless we have supplementary information in the form of the *costs* (in money or other common terms) of the two types of error, and the costs of making observations, we cannot obtain an optimum combination of α, β and n for any given problem. It is sufficient for us to note that, however α is determined, we shall obtain a *valid* test.

These comments appear straightforward and, indeed, they follow the original Neyman–Pearson recommendations. Likewise, in his pure tests of significance, Fisher (cf. **20.6**) recommended recording the actual value of p. However, as pointed out by Lehmann (1993), the availability of tables of fixed percentage points led practitioners to use 'conventional' levels such as $\alpha = 0.05$ or $\alpha = 0.01$, rather than go to the (then considerable) effort of computing the p-value and power function explicitly.

This numerical convenience has persisted long beyond its hour of need. Fortunately, some relief is at hand, since most computer programs now report p-values, making their use increasingly popular. However, the computation of power against selected alternatives of scientific interest remains all too rare; many scientists (and journal editors!) persist in treating the $\alpha = 0.05$ or 0.01 levels as sacrosanct and potentially ignoring the real conclusions to be drawn from the study. As computer programs continue to evolve, we hope that this dismal practice will become a relic of the past.

For further discussion of p-values, see **25.33** and Barnard (1990).

20.30 The point discussed in **20.29** is reflected in another, which has sometimes been made the basis of criticism of the theory of testing hypotheses.

Suppose that we carry out a test with α fixed, no matter how, and n extremely large. The power of a reasonable test will be very near 1 in detecting departure of any sort from the hypothesis tested. Indeed, this statement essentially corresponds to the definition of consistency in **20.9**. The argument (formulated by Berkson, 1938) runs as follows. Nobody really supposes that any hypothesis holds precisely; we are simply setting up an abstract model of real events that is bound to be some way, if only a little, from the truth. Nevertheless, as we have seen, an enormous sample would almost certainly (i.e. with probability approaching 1 as n increases beyond any bound) reject the hypothesis tested at any pre-assigned size α. Why, then, do we bother to test the hypothesis at all with a smaller sample, whose verdict is less reliable than that of the larger one?

This paradox is really concerned with two points. In the first place, if n is fixed, and we are not concerned with the exactness of the hypothesis tested, but only with its approximate validity, our alternative hypothesis should embody this fact by being sufficiently distant from the hypothesis tested to make the difference of practical interest. This in itself would tend to increase the power of the test. But if we had no wish to reject the hypothesis tested on the evidence of small deviations from it, we should want the power of the test to be very low against these small deviations, and this would imply a small α and a correspondingly high β and low power.

But the crux of the paradox is the argument about increasing sample size. The hypothesis tested will only be rejected with probability near 1 if we keep α fixed as n increases. There is no reason why we should do this: we can determine α in any way we please, and it is rational, in the light of the discussion of **20.29**, to apply the gain in sensitivity arising from increased sample size to the reduction of α as well as of β. It is only the habit of fixing α at certain conventional levels that leads to the paradox. If we allow α to decline as n increases, it is no longer certain that a very small departure from H_0 will cause H_0 to be rejected: this now depends on the rate at which α declines. A reasonable, though arbitrary, solution is to make α equal to the β value at the smallest departure from H_0 that is of practical importance.

Table 20.2 Power function calculated from (20.56). The entries in the first row of the table give the sizes of the tests

Value of $\|\mu - \mu_0\|$	Sample size (n)						
	10	10	10	20	100	100	100
0	0.050	0.072	0.111	0.050	0.050	0.019	0.0056
0.1				0.073	0.170	0.088	0.038
0.2	0.097	0.129	0.181	0.145	0.516	0.362	0.221
0.3				0.269	0.851	0.741	0.592
0.4	0.244	0.298	0.373	0.432	0.979	0.950	0.891
0.5				0.609	0.999	0.996	0.987
0.6	0.475	0.539	0.619	0.765			

Another view of the problem is to observe that, in many contexts, an interval estimate is more appropriate than a hypothesis test. Thus, reporting a confidence interval (presumably narrow since n is large) allows the investigator to decide whether an effect is of practical importance.

20.31 There is a converse to the paradox discussed in **20.30**. Just as, for large n, inflexible use of conventional values of α will lead to very high power, which may possibly be too high for the problem in hand, so for very small fixed n their use will lead to very low power, perhaps too low. Again, the situation can be remedied by allowing α to rise and consequently reducing β. It is always incumbent upon the statistician to satisfy herself that, for the conditions of her problem, she is not sacrificing sensitivity in one direction to sensitivity in another.

Example 20.14 (Power function for tests of the normal mean)

In the discussion on Lindley (1953), E.S. Pearson calculated a few values of the power function (20.56) of the two-sided test for a normal mean, which we reproduce in Table 20.2 to illustrate our discussion. It will be seen that when the sample size is increased from 20 to 100, the reductions of α from 0.050 to 0.019 and 0.0056 successively reduce the power of the test for each value of $|\mu - \mu_0|$. In fact, for $\alpha = 0.0056$ and $|\mu - \mu_0| = 0.1$, the power actually falls below the value attained at $n = 20$ with the $\alpha = 0.05$. Conversely, on reduction of sample size from 20 to 10, the increase in α to 0.072 and 0.111 increases the power correspondingly, though only in the case $\alpha = 0.111$, $|\mu - \mu_0| = 0.2$, does it exceed the power at $n = 20$, $\alpha = 0.05$.

20.32 Bartholomew (1967) discusses the choice of α and β when n is a random variable, distributed free of θ. Inequality (20.6) then remains valid, but the distribution of n enters into the determination of k_α, which remains the same whatever the value of n observed. See also the discussion following Bartholomew's paper.

EXERCISES

20.1 Show directly by use of (20.6) that the BCR for testing a simple hypothesis $H_0 : \mu = \mu_0$ concerning the mean μ of a Poisson distribution against a simple alternative $H_1 : \mu = \mu_1$ is of the form

$$\bar{x} \begin{cases} \leq a_\alpha & \text{if } \mu_0 > \mu_1, \\ \geq b_\alpha & \text{if } \mu_0 < \mu_1, \end{cases}$$

where $\bar{x}$ is the sample mean and a_α, b_α are constants.

20.2 Show similarly that for the parameter π of a binomial distribution, the BCR is of the form

$$p \begin{cases} \leq a_\alpha & \text{if } \pi_0 > \pi_1, \\ \geq b_\alpha & \text{if } \pi_0 < \pi_1, \end{cases}$$

where p is the observed proportion of successes in the sample.

20.3 Show that for the normal distribution with zero mean and variance σ^2, the BCR for $H_0 : \sigma = \sigma_0$ against the alternative $H_1 : \sigma = \sigma_1$ is of the form

$$\sum_{i=1}^{n} x_i^2 \begin{cases} \leq a_\alpha & \text{if } \sigma_0 > \sigma_1, \\ \geq b_\alpha & \text{if } \sigma_0 < \sigma_1. \end{cases}$$

Show that the power of the BCR when $\sigma_0 > \sigma_1$ is $F\{(\sigma_0^2/\sigma_1^2)\chi^2_{\alpha,n}\}$, where $\chi^2_{\alpha,n}$ is the lower 100α per cent point and F is the d.f. of the χ^2 distribution with n degrees of freedom.

20.4 In Example 20.4, show that the power of the test when $\theta_1 = 1$ is 0.648 when $k_\alpha = 1$ and 0.352 when $k_\alpha = 0.5$. Draw a diagram of the two Cauchy distributions to illustrate the power and size of each test. Show that when $\theta_1 = 1$, $k_\alpha = 1.5$, the BCR consists of values of x outside the interval $(-2 - \surd 5, -2 + \surd 5)$.

20.5 In Exercise 20.3, verify that the power is a monotone increasing function of σ_0^2/σ_1^2, and also verify numerically from a table of the χ^2 distribution that the power is a monotone increasing function of n.

20.6 Confirm that (20.21) holds for the sufficient statistics on which the BCRs of Example 20.2 and Exercises 20.1–20.3 are based.

20.7 In **20.15** show that the more efficient estimator always gives the more powerful test if its test power exceeds 0.5.

(Sundrum, 1954)

20.8 Show that for testing $H_0 : \mu = \mu_0$ in samples from the distribution

$$f(x) = 1, \qquad \mu \leq x \leq \mu + 1,$$

there is a pair of UMP one-sided tests, and hence no UMP test for all alternatives.

20.9 In Example 20.11, show that $\bar{x}$ is normally distributed with mean θ and variance θ^2/n^2, and that it is a sufficient statistic for θ.

20.10 Verify that the distribution of Example 20.9 does not satisfy condition (2) of **20.21**.

20.11 In Example 20.9, let σ be any positive increasing function of θ. Show that there is still a common BCR in testing $H_0 : \theta = \theta_0$ against $H_1 : \theta = \theta_1 < \theta_0$.

(Neyman and Pearson, 1936a)

20.12 Generalizing the discussion of **20.27**, write down the power function of any test based on the distribution of $\bar{x}$ with its critical region of the form

$$\bar{x} \leq a_{\alpha_1}, \quad \bar{x} \leq b_{\alpha_2},$$

where $\alpha_1 + \alpha_2 = \alpha$ (α_1 and α_2 not necessarily being equal). Show that the power function of any such test lies completely between those for the cases $\alpha_1 = 0$, $\alpha_2 = 0$, illustrated in Fig. 20.3.

20.13 Referring to the discussion of **20.14**, show that the likelihood ratio (for testing a simple $H_0 : \theta = \theta_0$ against a simple $H_1 : \theta = \theta_1$) is a sufficient statistic for θ on either hypothesis by writing the likelihood function as

$$L(x|\theta) = L(x|\theta_1)\left[\frac{L(x|\theta_0)}{L(x|\theta_1)}\right]^{(\theta-\theta_1)/(\theta_0-\theta_1)}$$

(Pitman, 1957)

20.14 For the most powerful test based on the BCR (20.6), show from (20.5) that $(1-\beta)/\alpha \geq 1/k_\alpha$ and hence by interchanging H_0 and H_1 that $\beta/(1-\alpha) \leq 1/k_\alpha \leq (1-\beta)\alpha$.

20.15 From Exercise 20.14, show that as $n \to \infty$ with α fixed, the inequality $\beta \leq (1-\alpha)/k_\alpha$ becomes an equality and $\log \beta \sim E\{\log(L_1/L_0)\}$.

(Cf. Rao (1962a). Efron (1967) obtains a fuller result.)

20.16 Following on from the discussion of **20.18**, show that the test of $H_0 : \theta = \theta_0$ against $H_1 : \theta \neq \theta_0$ for the normal distribution, $N(\theta, 1)$, is UMP within the class of tests with critical regions that are symmetric about H_0.

(Edelman, 1990)

CHAPTER 21

TESTS OF HYPOTHESES: COMPOSITE HYPOTHESES

21.1 We have seen in Chapter 20 that, when the null hypothesis tested is simple (specifying the distribution completely), there is *always* a BCR, providing a most powerful test, against a simple alternative hypothesis; that there *may* be a UMP test against a class of simple hypotheses constituting a composite parametric alternative hypothesis; and that there will not, in general, be a UMP test if the alternative hypothesis is two-sided.

If the hypothesis tested is composite, leaving at least one parameter unspecified, it is to be expected that UMP tests will be even rarer than for simple hypotheses, but we shall find that progress can be made if we are prepared to restrict the class of tests considered in certain ways.

Composite hypotheses

21.2 First, we formally define the problem. We suppose that the n observations have a distribution dependent upon the values of l ($\leq n$) parameters, which we shall write

$$L(x|\theta_1, \ldots, \theta_l)$$

as before. The hypothesis to be tested is

$$H_0 : \theta_1 = \theta_{10}; \ \theta_2 = \theta_{20}; \ldots; \ \theta_k = \theta_{k0}, \tag{21.1}$$

where $k \leq l$, and the second subscript 0 denotes the hypothesis. We lose no generality by labelling the k parameters whose values are specified by H_0 as the first k of the set of l parameters. H_0 as defined in (21.1) is said to impose k constraints on the parameter space.

Hypotheses of the form

$$H_0 : \theta_1 = \theta_2; \ \theta_3 = \theta_4; \ \ldots,$$

which do not specify the values of parameters whose equality we are testing, are transformable into the form (21.1) by reparametrizing the problem in terms of $\theta_1 - \theta_2$, $\theta_3 - \theta_4$, etc., and testing the hypothesis that these new parameters have zero values. Thus (21.1) is a more general composite hypothesis than at first appears.

To keep the our notation simple, we shall write $L(x|\theta_r, \theta_s)$ and

$$H_0 : \theta_r = \theta_{r0}, \tag{21.2}$$

where it is to be understood that θ_r, θ_s may each consist of more than one parameter, the *nuisance parameters* θ_s being left unspecified by the hypothesis tested. In general, a nuisance parameter is defined as a parameter that is required for full specification of the model, but one that is not of intrinsic interest to the analyst.

An optimum property of sufficient statistics

21.3 This is a convenient place to prove an optimum test property of sufficient statistics analogous to the estimation result proved in **17.35**. There we saw that if u is an unbiased estimator of θ and t is a sufficient statistic for θ, then the statistic $E(u|t)$ is unbiased for θ with variance no greater than that of u. We now prove a result due to Lehmann (1950): if w is a critical region for testing H_0, a hypothesis concerning θ in $L(x|\theta)$, against some alternative H_1, and t is a sufficient statistic, both on H_0 and on H_1, for θ, then there is a test of the same size, based on a function of t, which has the same power as w.

We first define the indicator function[1]

$$c(w) = \begin{cases} 1 & \text{if the sample point is in } w, \\ 0 & \text{otherwise.} \end{cases} \tag{21.3}$$

Then the integral

$$\int c(w)L(x|\theta)\,\mathrm{d}x = E\{c(w)\} \tag{21.4}$$

gives the probability that the sample point falls in w, and is therefore equal to the size (α) of the test when H_0 is true and to the power of the test when H_1 is true. Using the factorization property (17.68) of the likelihood function in the presence of a sufficient statistic, (21.4) becomes

$$\begin{aligned} E\{c(w)\} &= \int c(w)h(x|t)g(t|\theta)\,\mathrm{d}x \\ &= E\{E(c(w)|t)\}, \end{aligned} \tag{21.5}$$

the expectation operation outside the braces being with respect to the distribution of t. Thus the particular function of t, $E(c(w)|t)$, not dependent upon θ since t is sufficient, has the same expectation as $c(w)$. There is therefore a test based on the sufficient statistic t that has the same size and power as the original region w. We may therefore, without loss of power, confine the discussion of any test problem to functions of a sufficient statistic.

This result is quite general, and therefore covers the case of a simple H_0 discussed in Chapter 20. The argument also extends to cover randomized tests, defined in **20.11**.

Test size for composite hypotheses: similar regions

21.4 Since a composite hypothesis leaves some parameter values unspecified, a new problem immediately arises, for the size of the test of H_0 will obviously be a function, in general, of these unspecified parameter values, θ_s.

If we wish to keep Type I errors down to some pre-assigned level, we must seek critical regions whose size can be kept down to that level for all possible values of θ_s. Thus we require

$$\alpha(\theta_s) \leq \alpha, \qquad \text{for all } \theta_s. \tag{21.6}$$

If a critical region has

$$\alpha(\theta_s) = \alpha \tag{21.7}$$

[1] $c(w)$ is known in measure theory as the characteristic function of the set of points w. We shall avoid this terminology, since there is some possibility of confusion with the use of 'characteristic function' for the Fourier transform of a distribution function, with which we have been familiar since Chapter 4.

as a strict equality for all θ_s, it is called a (critical) region *similar to the sample space*[2] with respect to θ_s, or, more briefly, a similar (critical) region. The test based on a similar critical region is called a similar size-α test.

21.5 It is not obvious that similar regions exist at all generally, but in one sense they exist whenever we are dealing with a set of n independent identically distributed observations on a continuous variate x. For no matter what the distribution or its parameters, we have

$$P\{x_1 < x_2 < x_3 < \cdots < x_n\} = 1/n! \tag{21.8}$$

(cf. **11.4**), since any of the $n!$ permutations of the x_i is equally likely. Thus, for α an integral multiple of $1/n!$, there are similar regions based on the $n!$ hypersimplices in the sample space obtained by permuting the n subscripts in (21.8).

21.6 If we confine attention to regions defined by symmetric functions of the observations (so that similar regions based on (21.8) are excluded) it is easy to see that, where similar regions do exist, they need not exist for all sample sizes. For example, for a sample of n observations from the normal distribution

$$f(x) = (2\pi)^{-1/2}\exp\{-\tfrac{1}{2}(x-\theta)^2\},$$

there is no similar region with respect to θ for $n = 1$, but for $n \geq 2$ the fact that $ns^2 = \sum_{i=1}^{n}(x_i - \bar{x})^2$ has a chi-square distribution with $n-1$ degrees of freedom, whatever the value of θ, ensures that similar regions of any size can be found from the distribution of s^2. This is because $\bar{x}$ is a single sufficient statistic for θ, and to find a similar region we must, by Exercise 21.3, find a statistic uncorrelated with

$$\frac{\partial \log L}{\partial \theta} = n(\bar{x} - \theta).$$

This is impossible when $n = 1$, since $\bar{x} = x$ is then the whole sample; but for $n \geq 2$, $\sum(x - \bar{x})^2$ is distributed independently of $\bar{x}$ and thus gives similar regions. The same argument holds in Exercise 21.1, where there is a pair of sufficient statistics for two parameters, and at least three observations are required so that we may have a statistic independent of both.

21.7 Even if n is large, symmetric similar regions will not exist if each observation brings a new parameter with it, as in the following example, due to Feller (1938).

Example 21.1 (Possible non-existence of similar regions)

Consider a sample of n observations, where the ith observation has distribution

$$f(x_i) = (2\pi)^{-1/2}\exp\{-\tfrac{1}{2}(x_i - \theta_i)^2\},$$

so that

$$L(x|\theta) = (2\pi)^{-n/2}\exp\{-\tfrac{1}{2}\sum(x_i - \theta_i)^2\}.$$

[2]The term arose because, trivially, the entire sample space is a similar region with $\alpha = 1$.

For a similar region w of size α, we require, identically in θ,

$$\int_w L(x|\theta)\,\mathrm{d}x = \alpha.$$

Using (21.3), we may rewrite this size condition as

$$\int_W L(x|\theta)\frac{c(w)}{\alpha}\,\mathrm{d}x = 1, \tag{21.9}$$

where W is the whole sample space. Differentiating (21.9) with respect to θ_i, we find

$$\int_W L(x|\theta)\frac{c(w)}{\alpha}(x_i - \theta_i)\,\mathrm{d}x = 0. \tag{21.10}$$

A second differentiation with respect to θ_i gives

$$\int_W L(x|\theta)\frac{c(w)}{\alpha}\{(x_i - \theta_i)^2 - 1\}\,\mathrm{d}x = 0. \tag{21.11}$$

Now from the definition of $c(w)$,

$$g(x|\theta) = L(x|\theta)\frac{c(w)}{\alpha} \tag{21.12}$$

is a (joint) density function. Expressions (21.10) and (21.11) indicate that the marginal distribution of x_i in $g(x|\theta)$ has

$$E(x_i) = \theta_i, \qquad \operatorname{var} x_i = 1,$$

just as it has in the initial distribution of x_i.

If we examine the form of $g(x|\theta)$, we see that if we were to proceed with further differentiations, we should find that *all* the moments and product moments of $g(x|\theta)$ are identical to those of $L(x|\theta)$, which is uniquely determined by its moments. Thus, from (21.12), $c(w)/\alpha = 1$ identically. But since $c(w)$ is either 0 or 1, we see finally that the trivial values $\alpha = 0$ or 1 are the only values for which similar regions can exist. The difficulty here is that all n observations are required to form a sufficient set for the n parameters, and we can find no statistic independent of them all.

21.8 It is nevertheless true that for many problems of testing composite hypotheses, similar regions exist for any size α and any sample size n. We now consider how they are to be found.

Let t be a sufficient statistic for the parameter θ_s unspecified by the hypothesis H_0, and suppose that we have a critical region w such that for all values of t, when H_0 is true, w is composed of a fraction α of the probability content of each contour of constant t, i.e.

$$E\{c(w)|t\} = \alpha. \tag{21.13}$$

Then, on taking expectations with respect to t, we have, as in (21.5),

$$E\{c(w)\} = E\{E(c(w)|t)\} = \alpha, \tag{21.14}$$

so that the original critical region w is similar of size α, as Neyman (1937) and Bartlett (1937) pointed out.

It should be noticed that here t need be sufficient only for the unspecified parameter θ_s, and only when H_0 is true. This should be contrasted with the more demanding requirements of **21.3**.

Our argument has shown that (21.13) is a sufficient condition for w to be similar. We shall show in **21.19** that it is necessary and sufficient, provided that a further condition is fulfilled, and in order to state that condition we must now introduce, following Lehmann and Scheffé (1950), the concept of the *completeness* of a parametric family of distributions, a concept that also permits us to supplement the discussion of sufficient statistics in Chapter 17.

We shall see in **21.22** that restriction to similar tests may involve a loss of power compared with other tests satisfying (21.6).

Complete parametric families and complete statistics

21.9 Consider a parametric family of (univariate or multivariate) distributions, $f(x|\theta)$, depending on the value of a vector of parameters θ. Let $h(x)$ be any statistic, independent of θ. If

$$E\{h(x)\} = \int h(x)f(x|\theta)\,\mathrm{d}x = 0 \tag{21.15}$$

for all θ implies that

$$h(x) = 0 \tag{21.16}$$

identically (save possibly on a zero-measure set), then the family $f(x|\theta)$ is called *complete*. The term is apt, since no non-zero function can be found that is orthogonal to all members of the family. If (21.15) implies (21.16) only for all bounded $h(x)$, $f(x|\theta)$ is called *boundedly complete*.

In statistical applications of the concept of completeness, the family of distributions we are interested in is often the sampling distribution of a (possibly vector) statistic t, say $g(t|\theta)$. We then call t a complete (or boundedly complete) statistic if, for all θ, $E\{h(t)\} = 0$ implies $h(t) = 0$ identically, for all functions (or bounded functions) $h(t)$. In other words, we label the statistic t with the completeness property of its distribution.

An immediate consequence of the completeness of a statistic t is that only one function of that statistic can have a given expected value. Thus if one function of t is an unbiased estimator of a certain function of θ, no other function of t will be. Completeness confers a uniqueness property upon an estimator.

Bounded completeness of f implies that its c.f. $\phi(u)$ can have no zeros, for otherwise $0 = \phi(u) = E(\mathrm{e}^{iux})$. If θ is a location parameter, i.e. $f \equiv f(x-\theta)$, the converse also holds – cf. Ghosh and Singh (1966). Thus, for example, the Cauchy distribution of Example 17.7 is boundedly complete – the c.f. is given in Example 4.2.

The completeness of sufficient statistics

21.10 The special case of the exponential family (17.83) with $A(\theta) = \theta$, $B(x) = x$ has

$$f(x|\theta) = \exp\{\theta x + C(x) + D(\theta)\}, \qquad -\infty < x < \infty. \tag{21.17}$$

If, for all θ,

$$\int h(x)f(x|\theta)\,\mathrm{d}x = 0,$$

we must have

$$\int [h(x)\exp\{C(x)\}]\exp(\theta x)\,\mathrm{d}x = 0. \tag{21.18}$$

The integral in (21.18) is the two-sided Laplace transform[3] of the function in square brackets in the integrand. By the uniqueness property of the transform, the only function having a transform equal to zero is zero itself; i.e.

$$h(x)\exp\{C(x)\} = 0$$

identically, whence

$$h(x) = 0$$

identically. Thus $f(x|\theta)$ is complete.

This result generalizes to the multiparameter case, as shown by Lehmann and Scheffé (1955): the k-parameter, k-variate exponential family

$$f(\mathbf{x}|\boldsymbol{\theta}) = \exp\left\{\sum_{j=1}^{k}\theta_j x_j + C(\mathbf{x}) + D(\boldsymbol{\theta})\right\} \tag{21.19}$$

is a complete family. We have seen (Exercise 17.14) that the joint distribution of the set of k sufficient statistics for the k parameters of the general univariate exponential form (17.86) takes a form of which (21.19) is the special case, with $A_j(\theta) = \theta_j$. (We have replaced nD and h of Exercise 17.14 by D and $\exp(C)$, respectively.) By **21.3**, we may confine ourselves, in testing hypotheses about the parameters, to the sufficient statistics.

Example 21.2 (Completeness properties for the normal distribution)

Consider the family of normal distributions

$$f(x|\theta_1,\theta_2) = (2\pi\theta_2)^{-1/2}\exp\left\{-\frac{1}{2\theta_2}(x-\theta_1)^2\right\}, \qquad -\infty < x < \infty;\ \theta_2 > 0.$$

(a) If θ_2 is known (say $= 1$), the family is complete with respect to θ_1, for we are then considering a special case of (21.17) with

$$\theta = \theta_1, \quad \exp\{C(x)\} = (2\pi)^{-1/2}\exp(-\tfrac{1}{2}x^2)$$

and

$$D(\theta) = -\tfrac{1}{2}\theta_1^2.$$

[3] The two-sided Laplace transform of a function $g(x)$ is defined by

$$\lambda(\theta) = \int_{-\infty}^{\infty}\exp(\theta x)g(x)\,\mathrm{d}x.$$

The integral converges in a strip of the complex plane $\alpha < R(\theta) < \beta$, where one or both of α, β may be infinite. (The strip may degenerate to a line.) Except possibly for a zero-measure set, there is a one-to-one correspondence between $g(x)$ and $\lambda(\theta)$. Compare the inversion theorem for c.f.s in **4.3**.

(b) If, on the other hand, θ_1 is known (say $= 0$), the family is not even boundedly complete with respect to θ_2, for $f(x|0, \theta_2)$ is an even function of x, so that any odd function $h(x)$ will have zero expectation without being identically zero. However, if we transform to $y = x^2$, we see that the distribution of y is complete, since $g(y|\theta_2) = (2\pi\theta_2 y)^{-1/2}\exp\{-y/(2\theta_2)\}$ is again a special case of (21.17). Cf. the remark in **21.16** below.

21.11 In **21.10** we discussed the completeness of the characteristic form of the joint distribution of sufficient statistics in samples from a distribution with range independent of the parameters. Hogg and Craig (1956) have established the completeness of the sufficient statistic for distributions whose range is a function of a single parameter θ and that possess a single sufficient statistic for θ. We recall from **17.40–41** that the distribution must then be of the form

$$f(x|\theta) = g(x)/h(\theta) \tag{21.20}$$

and that

(i) if a single terminal of $f(x|\theta)$ (which may be taken to be θ itself without loss of generality) is a function of θ, the corresponding extreme order statistic is sufficient;
(ii) if both terminals are functions of θ, the upper terminal ($b(\theta)$) must be a monotone decreasing function of the lower terminal (θ) for a single sufficient statistic

$$\min\{x_{(1)}, b^{-1}(x_{(n)})\}. \tag{21.21}$$

to exist.

We consider cases (i) and (ii) in turn.

21.12 In case (i), take the upper terminal equal to θ, the lower equal to a constant a. $x_{(n)}$ is then sufficient for θ. Its distribution is, from (11.34) and (21.20),

$$\begin{aligned} \mathrm{d}G(x_{(n)}) &= n\{F(x_{(n)})\}^{n-1} f(x_{(n)})\,\mathrm{d}x_{(n)} \\ &= \frac{n\{\int_a^{x_{(n)}} g(x)\,\mathrm{d}x\}^{n-1} g(x_{(n)})}{\{h(\theta)\}^n}\,\mathrm{d}x_{(n)}, \qquad a \le x_{(n)} \le \theta. \end{aligned} \tag{21.22}$$

Now suppose that for a statistic $u(x_{(n)})$ we have

$$\int_a^\theta u(x_{(n)})\,\mathrm{d}G(x_{(n)}) = 0,$$

or, substituting from (21.22), and dropping the factor in $h(\theta)$,

$$\int_a^\theta u(x_{(n)}) \left\{ \int_a^{x_{(n)}} g(x)\,\mathrm{d}x \right\}^{n-1} g(x_{(n)})\,\mathrm{d}x_{(n)} = 0. \tag{21.23}$$

If we differentiate (21.23) with respect to θ, we find

$$u(\theta) \left\{ \int_a^\theta g(x)\,\mathrm{d}x \right\}^{n-1} g(\theta) = 0, \tag{21.24}$$

and since the integral in braces equals $h(\theta)$, while $g(\theta) \neq 0 \neq h(\theta)$, (21.24) implies

$$u(\theta) = 0$$

for any θ. Hence the function $u(x_{(n)})$ is identically zero, and the distribution of $x_{(n)}$, given at (21.22), is complete. Exactly the same argument holds for the lower terminal and $x_{(1)}$.

21.13 In case (ii), the distribution function of the sufficient statistic (21.21) is

$$\begin{aligned} G(t) &= 1 - P\{x_{(1)}, b^{-1}(x_{(n)}) \geq t\} \\ &= 1 - P\{x_{(1)} \geq t, x_{(n)} \leq b(t)\} \\ &= 1 - \left\{\int_t^{b(t)} \frac{g(x)}{h(\theta)}\,\mathrm{d}x\right\}^n. \end{aligned} \qquad (21.25)$$

Differentiating (21.25) with respect to t, we obtain the density of the sufficient statistic,

$$g(t) = \frac{-n}{\{h(\theta)\}^n}\left\{\int_t^{b(t)} g(x)\,\mathrm{d}x\right\}^{n-1}[g\{b(t)\}b'(t) - g(t)], \qquad \theta \leq t \leq c, \qquad (21.26)$$

where c is determined by $c = b(c)$ – this is the result of Exercise 19.6. If there is a statistic $u(t)$ for which

$$\int_\theta^c u(t)g(t)\,\mathrm{d}t = 0, \qquad (21.27)$$

we find, on differentiating (21.27) with respect to θ and by following through the argument of **21.12**, that $u(\theta) = 0$ for any θ, as before. Thus $u(t) = 0$ identically and $g(t)$ in (21.26) is complete.

21.14 The following example is of a non-complete single sufficient statistic.

Example 21.3 (Non-completeness in a uniform distribution)

Consider a sample of a single observation x from the uniform distribution

$$f(x) = 1, \qquad \theta \leq x \leq \theta + 1.$$

x is evidently a sufficient statistic. (There would be no single sufficient statistic for $n \geq 2$, since condition (ii) of **21.11** is not satisfied.)

Any bounded periodic function $h(x)$ of period 1 which satisfies

$$\int_0^1 h(x)\,\mathrm{d}x = 0$$

will give

$$\int_\theta^{\theta+1} h(x)\,\mathrm{d}F = \int_\theta^{\theta+1} h(x)\,\mathrm{d}x = \int_0^1 h(x)\,\mathrm{d}x = 0,$$

so that the distribution is not even boundedly complete, since $h(x)$ is not identically zero.

Minimal sufficiency

21.15 We recall from **17.38** that, when we consider the problem of sufficient statistics in general (i.e. without restricting ourselves, as we did initially in Chapter 17, to the case of a single sufficient statistic), we have to consider the choice between alternative sets of sufficient statistics. In a sample of n observations we *always* have a set of n sufficient statistics (namely, the observations themselves) for the k (≥ 1) parameters of the distribution. Sometimes, though not always, there will be a set of s ($< n$) statistics sufficient for the parameters. Often, $s = k$; e.g. all the cases of sufficiency discussed in Examples 17.15–16 have $s = k = 1$, while in Example 17.17 we have $s = k = 2$. By contrast, the following is an example in which $s < k$.[4]

Example 21.4 (A case where the number of parameters exceeds the number of sufficient statistics)

Consider again the problem of Example 20.11, with the alteration that

$$E(x_1) = n\mu$$

instead of $n\theta$ as previously. As in (20.48), we find for the joint distribution

$$f(\mathbf{x}) = \frac{1}{\theta(2\pi)^{n/2}} \exp\left[-\frac{1}{2}\left\{\frac{n^2}{\theta^2}(\bar{x}-\mu)^2 + \sum_{i=2}^{n} x_i^2\right\}\right].$$

Here it is clear that the single statistic $\bar{x}$ is sufficient for the parameters μ, θ. Note that $\bar{x}$ does not provide any information on θ since the ML estimators are $\hat{\mu} = \bar{x}, \theta \to 0$; furthermore $L \to \infty$ as $\theta \to 0$.

21.16 We thus have to ask: what is the *smallest* number s of statistics that constitute a sufficient set in any problem? With this in mind, Lehmann and Scheffé (1950) define a vector of statistics as *minimal sufficient* if it is a single-valued function of all other vectors of statistics that are sufficient for the parameters of the distribution.[5] The problems that now arise are: how can we be sure, in any particular situation, that a sufficient vector is the minimal sufficient vector? And can we find a construction that yields the minimal sufficient vector?

A partial answer to the first of these questions is supplied by the following result: if the vector $\mathbf{t}$ is a boundedly complete sufficient statistic for $\boldsymbol{\theta}$, and the vector $\mathbf{u}$ is a minimal sufficient statistic for $\boldsymbol{\theta}$, then $\mathbf{t}$ is equivalent to $\mathbf{u}$, i.e. they are identical, except possibly for a set of zero measure.

The proof is simple. Let w be a region in the sample space for which

$$D = E(c(w)|\mathbf{t}) - E(c(w)|\mathbf{u}) \neq 0, \tag{21.28}$$

where the function $c(w)$ is defined at (21.3). From (21.28), we find, on taking expectations over the entire sample space,

$$E(D) = 0. \tag{21.29}$$

[4] Fisher (e.g. 1956) called a set of statistics 'sufficient' only if $s = k$ and 'exhaustive' if $s > k$.

[5] That this is for practical purposes equivalent to a sufficient statistic with minimum number of components is shown by Barankin and Katz (1959). See also Fraser (1963).

Now since **u** is minimal sufficient, it is a function of **t**, another sufficient statistic, by definition. Hence we may write (21.28) as

$$D = h(\mathbf{t}) \neq 0. \tag{21.30}$$

Since D is a bounded function, (21.29) and (21.30) contradict the assumed bounded completeness of **t**, and thus there can be no region w for which (21.28) holds. Hence **t** and **u** are equivalent statistics, i.e. **t** is minimal sufficient.

The converse does not hold: while bounded completeness implies minimal sufficiency, we can have minimal sufficiency without bounded completeness. An important instance is discussed in Example 21.10 below.

A consequence of the result of this section is that there cannot be more than one boundedly complete sufficient statistic for a parameter. Thus in Example 21.2(b) x^2 is minimal sufficient and complete, while x is sufficient and not complete – cf. also Exercises 18.13 and 21.31 for other instances of minimal and non-minimal single sufficient statistics.

An alternative formulation of the problem of minimal sufficiency is given by Dynkin (1951), who uses the term 'necessary' rather than 'minimal'.

21.17 In view of the results of **21.10–13** concerning the completeness of sufficient statistics, a consequence of **21.16** is that all the examples of sufficient statistics we have discussed in earlier chapters are minimal sufficient, as one would expect on intuitive grounds.

Since, by **18.23**, a unique ML estimator $\hat{\boldsymbol{\theta}}$ of $\boldsymbol{\theta}$ is a function of any sufficient statistic for $\boldsymbol{\theta}$, it will necessarily be a function of the minimal sufficient statistic. Hence, if $\hat{\boldsymbol{\theta}}$ is itself sufficient, it must be minimal sufficient.

21.18 The result of **21.16**, though useful, is less direct than the following procedure for finding a minimal sufficient statistic, given by Lehmann and Scheffé (1950).

We have seen in **20.14** and **20.20** that in testing a simple hypothesis, the ratio of likelihoods is a function of the sufficient (set of) statistic(s). We may now, so to speak, put this result into reverse, and use it to find the minimal sufficient set. Writing $L(x|\theta)$ for the LF as before, where x and θ may be vectors, consider a particular set of values x_0 and select all those values of x within the permissible range for which $L(x|\theta)$ is non-zero and

$$\frac{L(x|\theta)}{L(x_0|\theta)} = k(x, x_0) \tag{21.31}$$

is independent of θ. Now any sufficient statistic t (possibly a vector) will satisfy (17.68), whence

$$\frac{L(x|\theta)}{L(x_0|\theta)} = \frac{g(t|\theta)}{g(t_0|\theta)} \cdot \frac{h(x)}{h(x_0)}, \tag{21.32}$$

so that if $t = t_0$, (21.32) reduces to the form (21.31). Conversely, if (21.31) holds for all θ, this implies the constancy of the sufficient statistic t at the value t_0. This may be used to identify sufficient statistics, and to select the minimal set, as illustrated in the following examples.

Example 21.5 (Minimal sufficiency and the normal distribution)

We saw in Example 17.17 that the set of statistics $(\bar{x}, s^2)$ is jointly sufficient for the parameters (μ, σ^2) of a normal distribution. For this distribution, $L(x|\theta)$ is non-zero for all $\sigma^2 > 0$, and condition (21.31) is, on taking logarithms, that

$$-\frac{1}{2\sigma^2}\left\{\left(\sum_i x_i^2 - \sum_i x_{0i}^2\right) - 2\mu n(\bar{x} - \bar{x}_0)\right\} \tag{21.33}$$

is independent of (μ, σ^2), i.e. that the term in braces is equal to zero. This will be so, for example, if every x_i is equal to the corresponding x_{0i}, confirming that the set of n observations is a jointly sufficient set, as we have remarked that they always are.

It will also be so if the x_i are any rearrangement (permutation) of the x_{0i}: thus the set of *order statistics* is sufficient, as it is again obvious that they always are. But in this example, we can go further, for (21.33) will be zero if we divide the observations into any l subsets and within each subset have

$$\sum x_i = \sum x_{0i} \quad \text{and} \quad \sum x_i^2 = \sum x_{0i}^2.$$

Thus the $2l$-dimensional statistic formed of the l subset pairs $(\sum x_i, \sum x_i^2)$ will be jointly sufficient. In particular, with $l = 1$, the condition on (21.33) will be satisfied if

$$\bar{x} = \bar{x}_0, \qquad \sum_i x_i^2 = \sum_i x_{0i}^2, \tag{21.34}$$

and clearly, from inspection, nothing less than this will do. Thus the pair $(\bar{x}, \sum x^2)$ is minimal sufficient: equivalently, since $ns^2 = \sum x^2 - n\bar{x}^2$, $(\bar{x}, s^2)$ is minimal sufficient.

Example 21.6 (Minimal sufficiency and the Cauchy distribution)

As a contrast, consider the Cauchy distribution of Example 17.7. $L(x|\theta)$ is everywhere non-zero and (21.31) requires

$$\prod_{i=1}^{n}\{1 + (x_{0i} - \theta)^2\} \Big/ \prod_{i=1}^{n}\{1 + (x_i - \theta)^2\} \tag{21.35}$$

to be independent of θ. As in the previous example, the set of order statistics is sufficient, but nothing less will do here, for (21.35) is the ratio of two polynomials, each of degree $2n$ in θ. If the ratio is to be independent of θ, each polynomial must have the same set of roots, possibly permuted. Thus we are thrown back on the order statistics as the minimal sufficient set.

Completeness and similar regions

21.19 After our lengthy excursus on completeness, we return to the discussion of similar regions in **21.8**. We may now show that if, given H_0, the sufficient statistic t is boundedly complete, *all* size-α similar regions must satisfy (21.13). For any such region, (21.14) holds and may be rewritten as

$$E\{E(c(w)|t) - \alpha\} = 0. \tag{21.36}$$

The expression in braces in (21.36) is bounded. Thus if t is boundedly complete, (21.36) implies that $E(c(w)|t) - \alpha = 0$ identically, i.e. that (21.13) holds.

The converse result also holds: if all similar regions satisfy (21.13), then Lehmann and Scheffé (1950) proved that t must be boundedly complete. Thus the bounded completeness of a sufficient statistic is equivalent to the condition that all similar regions w satisfy (21.13).

The choice of most powerful similar regions

21.20 The importance of the result of **21.19** is that it permits us to reduce the problem of finding most powerful similar regions for a composite hypothesis to the familiar problem of finding a BCR for a simple hypothesis.

By **21.19**, the bounded completeness of the statistic t, sufficient for θ_s on H_0, implies that all similar regions w satisfy (21.13), i.e. every similar region is composed of a fraction α of the probability content of each contour of constant t. We therefore may conduct our discussion with t held constant. Constancy of the sufficient statistic, t, for θ_s implies from (17.68) that the conditional distribution of the observations in the sample space will be independent of θ_s. Thus the composite H_0 with θ_s unspecified is reduced to a simple H_0 with t held constant. If t is also sufficient for θ_s when H_1 holds, the composite H_1 is also reduced to a simple H_1 with t constant (and, incidentally, the power of any critical region with t constant, as well as its size, will be independent of θ_s). If, however, t is not sufficient for θ_s when H_1 holds, we consider H_1 as a class of simple alternatives to the simple H_0, as in the previous chapter.

Thus, by keeping t constant, we reduce the problem to that of testing a simple H_0 concerning θ_r against a simple H_1 (or a class of simple alternatives constituting H_1). We use the methods of the previous chapter, based on the Neyman–Pearson lemma (21.6), to seek a BCR (or common BCR) for H_0 against H_1. If there is such a BCR for each fixed value of t, it will evidently be an unconditional BCR, and gives the most powerful similar test of H_0 against H_1. As before, if this test remains most powerful against a class of alternative values of θ_r, it is a UMP similar test.

Example 21.7 (UMP test for a composite null hypothesis)

We wish to test $H_0 : \mu = \mu_0$ against $H_1 : \mu = \mu_1$ for the normal distribution

$$f(x) = \frac{1}{\sigma(2\pi)^{1/2}} \exp\left\{-\frac{1}{2}\left(\frac{x-\mu}{\sigma}\right)^2\right\}, \qquad -\infty < x < \infty.$$

H_0 and H_1 are composite, σ^2 being unspecified.

From Examples 17.10 and 17.15, the statistic (calculated from a sample of n independent observations) $u = \sum_{i=1}^{n}(x_i - \mu_0)^2$ is sufficient for σ^2 when H_0 holds, but not otherwise. From **21.10**, u is a complete statistic. All similar regions for H_0 therefore consist of fractions α of each contour of constant u.

Holding u fixed, we now test

$$H_0 : \mu = \mu_0 \quad \text{against} \quad H_1' : \mu = \mu_1, \quad \sigma = \sigma_1,$$

both hypotheses being simple. The BCR obtained from (21.6) is that for which

$$\frac{L(x|H_0)}{L(x|H_1')} \leq k_\alpha.$$

This reduces, on simplification, to the condition

$$\bar{x}(\mu_1 - \mu_0) \geq C(\mu_0, \mu_1, \sigma^2, \sigma_1^2, k_\alpha, u) \tag{21.37}$$

where C is a constant containing no function of x except u. Thus the BCR consists of large values of $\bar{x}$ if $\mu_1 - \mu_0 > 0$ and of small values of $\bar{x}$ if $\mu_1 - \mu_0 < 0$, and this is true whatever the values of σ^2 and σ_1^2, and whatever the magnitude of $|\mu_1 - \mu_0|$. Thus we have a common BCR for the class of alternatives $H_1 : \mu = \mu_1$ for each one-sided situation $\mu_1 > \mu_0$ and $\mu_1 < \mu_0$.

We have been holding u fixed. Now

$$u = \sum(x - \mu_0)^2 = \sum(x - \bar{x})^2 + n(\bar{x} - \mu_0)^2 \tag{21.38}$$

$$= \sum(x - \bar{x})^2 \left\{1 + \frac{n(\bar{x} - \mu_0)^2}{\sum(x - \bar{x})^2}\right\}. \tag{21.39}$$

Since the BCR for fixed u consists of extreme values of $\bar{x}$, (21.38) implies that the BCR consists of small values of $\sum(x - \bar{x})^2$, which by (21.39) implies large values of

$$\frac{t^2}{n-1} = \frac{n(\bar{x} - \mu_0)^2}{\sum(x - \bar{x})^2}. \tag{21.40}$$

t^2 as defined by (21.40) is the square of the t statistic ('Student', 1908) whose distribution was derived in Example 11.8. By Exercise 21.7, t, which is distributed free of σ^2, is distributed independently of the complete sufficient statistic, u, for σ^2. Taking into account (21.37), we have finally that the unconditional UMP similar test of H_0 against H_1 is to reject the largest or smallest 100α per cent of the distribution of t according to whether $\mu_1 > \mu_0$ or $\mu_1 < \mu_0$.

As we have seen, the distribution of t does not depend on σ^2. The power of the UMP similar test, however, does depend on σ^2, for u is not sufficient for σ^2 when H_0 does not hold. Since every similar region for H_0 consists of fractions α of each contour of constant u, and the distribution on any such contour is a function of σ^2 when H_1 holds, there can be no similar region for H_0 with power independent of σ^2, a result first established by Dantzig (1940). It is for this reason that we found in Example 20.2 that the length of the corresponding confidence interval for μ could not be fixed in advance.

Example 21.8 (UMP test of two means, with variances unknown but equal)

For two normal distributions with means μ, $\mu + \theta$ and common variance σ^2, we now consider a test of

$$H_0 : \theta = \theta_0 \ (= 0, \text{ without loss of generality})$$

against

$$H_1 : \theta = \theta_1$$

on the basis of independent samples of size n_1, n_2 with means $\bar{x}_1, \bar{x}_2$. Write

$$\begin{aligned} n &= n_1 + n_2, \\ n\bar{x} &= n_1\bar{x}_1 + n_2\bar{x}_2, \\ s^2 &= \textstyle\sum_{i=1}^{2} \sum_{j=1}^{n_i} (x_{ij} - \bar{x})^2 = \sum\sum x_{ij}^2 - n\bar{x}^2. \end{aligned} \tag{21.41}$$

The hypotheses are composite, two parameters being unspecified. When H_0 holds, but not otherwise, the pair of statistics $(\bar{x}, s^2)$ is sufficient for the unspecified parameters (μ, σ^2), and it follows from **21.10** that $(\bar{x}, s^2)$ is complete. Thus all similar regions for H_0 satisfy (21.13), and we hold $(\bar{x}, s^2)$ fixed, and test the simple

$$H_0 : \theta = 0$$

against

$$H_1' : \theta = \theta_1, \quad \mu = \mu_1, \quad \sigma = \sigma_1.$$

Our original H_1 consists of the class of H_1' for all μ_1, σ_1. The BCR obtained from (21.6) reduces, on simplification, to

$$\bar{x}_2\theta_1 \geq g_\alpha,$$

where g_α is a constant function of all the parameters, and of $\bar{x}$ and s^2, but not otherwise of the observations. For fixed $\bar{x}$, s^2, the BCR is therefore characterized by extreme values of $\bar{x}_2$ of the same sign as θ_1, and this is true whatever the values of the other parameters. Expressions (21.41) then imply that for each fixed $(\bar{x}, s^2)$, the BCR will consist of large values of $(\bar{x}_1 - \bar{x}_2)^2/s^2$, and hence of the equivalent monotone increasing function

$$\frac{(\bar{x}_1 - \bar{x}_2)^2}{\sum(x_1 - \bar{x}_1)^2 + \sum(x_2 - \bar{x}_2)^2} = \frac{t^2}{(n-2)} \cdot \frac{n}{n_1 n_2}. \tag{21.42}$$

This is the definition of the usual Student's t^2 statistic for this problem, which we have encountered in the corresponding interval estimation problem in **19.25**. By Exercise 21.7, t^2, which is distributed free of μ and σ^2, is distributed independently of the complete sufficient statistic $(\bar{x}, s^2)$ for (μ, σ^2) Thus, unconditionally, the UMP similar test of H_0 against H_1 is given by rejecting the 100α per cent largest or smallest values in the distribution of t, according to whether θ_1 (or, more generally, $\theta_1 - \theta_0$) is positive or negative.

Here, as in the previous example, the power of the BCR depends on (μ, σ^2), since $(\bar{x}, s^2)$ is not sufficient when H_0 does not hold.

Example 21.9 (UMP test for the scale parameter of the exponential)

We wish to test the composite $H_0 : \sigma = \sigma_0$ against $H_1 : \sigma = \sigma_1$ for the distribution

$$f(x) = \sigma^{-1} \exp\left\{-\left(\frac{x-\theta}{\sigma}\right)\right\}, \qquad \theta \leq x < \infty; \ \sigma > 0.$$

We have seen (Example 17.19) that $x_{(1)}$, the smallest order statistic, is sufficient for the unspecified parameter θ, whether H_0 or H_1 holds. By **21.12** it is also complete. Thus all similar regions consist of fractions α of each contour of constant $x_{(1)}$.

The comprehensive sufficiency of $x_{(1)}$ renders both H_0 and H_1 simple when $x_{(1)}$ is fixed. The BCR obtained from (21.6) consists of points satisfying

$$\sum_{i=1}^{n} x_i \left(\frac{1}{\sigma_1} - \frac{1}{\sigma_0}\right) \leq g_\alpha,$$

where g_α is a constant, a function of σ_0, σ_1. For each fixed $x_{(1)}$, we therefore have the BCR defined by

$$\sum_{i=1}^{n} x_i \begin{cases} \leq a_\alpha & \text{if } \sigma_1 < \sigma_0, \\ \geq b_\alpha & \text{if } \sigma_1 > \sigma_0. \end{cases} \tag{21.43}$$

The statistic in (21.43), $\sum x_i$, is not distributed independently of $x_{(1)}$. To put (21.43) in a form of more practical value, we observe that the statistic

$$z = \sum_{i=1}^{n} (x_{(i)} - x_{(1)})$$

is distributed independently of $x_{(1)}$. (This is a consequence of the completeness and sufficiency of $x_{(1)}$ – see Exercise 21.7.) Thus if we rewrite (21.43) for fixed $x_{(1)}$ as

$$z \begin{cases} \leq c_\alpha, & \text{if } \sigma_1 < \sigma_0, \\ \geq d_\alpha, & \text{if } \sigma_1 > \sigma_0, \end{cases} \tag{21.44}$$

where $c_\alpha = a_\alpha - nx_{(1)}$, $d_\alpha = b_\alpha - nx_{(1)}$, we have on the left of (21.44) a statistic which for every fixed $x_{(1)}$ determines the BCR by its extreme values and whose distribution does not depend on $x_{(1)}$. Thus (21.44) gives an unconditional BCR for each of the one-sided situations $\sigma_1 < \sigma_0$, $\sigma_1 > \sigma_0$, and we have the usual pair of UMP tests.

Note that in this example, the comprehensive sufficiency of $x_{(1)}$ makes the power of the UMP tests independent of the location parameter θ.

21.21 Examples 21.7 and 21.8 afford a sophisticated justification for two of the standard normal distribution test procedures for means. Exercises 21.13 and 21.14, by following through the same argument, similarly justify two other standard procedures for variances, arriving in each case at a pair of UMP similar one-sided tests. Unfortunately, not all the problems of normal test theory are so tractable: the thorniest of them, the problem of two means that we discussed at length in Chapter 19, does not yield to the present approach, as the next example shows.

Example 21.10 (Lack of similar regions for the Behrens–Fisher problem)

For two normal distributions with means and variances (θ, σ_1^2), $(\theta + \mu, \sigma_2^2)$, we wish to test $H_0 : \mu = 0$ on the basis of independent samples of n_1 and n_2 observations.

Given H_0, the sample means and variances $(\bar{x}_1, \bar{x}_2, s_1^2, s_2^2) = \mathbf{t}$ form a set of four jointly sufficient statistics for the three parameters $\theta, \sigma_1^2, \sigma_2^2$ left unspecified by H_0. They may be seen to be minimal sufficient by use of (21.31) – cf. Lehmann and Scheffé (1950). But $\boldsymbol{t}$ is not boundedly complete, since $\bar{x}_1$, $\bar{x}_2$ are normally distributed independently of s_1^2, s_2^2 and of each other, so that any bounded odd function of $\bar{x}_1 - \bar{x}_2$ alone will have zero expectation. We therefore cannot rely on (21.13) to find all similar regions, though regions satisfying (21.13) would certainly be similar, by **21.8**. But it is easy to see, from the fact that the likelihood function contains the four components of $\mathbf{t}$ and no other functions of the observations, that any region consisting entirely of a fraction α of each surface of constant $\mathbf{t}$ will have the same probability content in the sample space *whatever the value of* μ, and will therefore be an ineffective critical region with power exactly equal to its size. This disconcerting aspect of a familiar and useful property of normal distributions was pointed out by Watson (1957a).

No useful exact unrandomized similar regions exist for this problem – see Linnik (1964; 1967) and Pfanzagl (1974) – but for all practical purposes α is constant for Welch's method, developed in **19.36** as an interval estimation technique. If we are prepared to introduce an element of randomization, Scheffé's method of **19.27–33** is also available.

21.22 The discussion of **21.20** and Examples 21.8–10 make it clear that, if there is a complete sufficient statistic for the unspecified parameter, the problem of selecting a most powerful test for a composite hypothesis is considerably reduced if we restrict our choice to similar regions. But something may be lost by this – for specific alternatives there may be a non-similar test, satisfying (21.6), with power greater than the most powerful similar test.

Lehmann and Stein (1948) considered this problem for the composite hypotheses considered in Example 21.7 and Exercise 21.13. In the former, where we are testing the mean of a normal distribution, they found that if $\alpha \geq \frac{1}{2}$ there is no non-similar test more powerful than Student's t, whatever the true values μ_1, σ_1, but that for $\alpha < \frac{1}{2}$ (as in practice it always is) there is a more powerful critical region, which is of the form

$$\sum_i \{x_i - c_\alpha(\mu_1, \sigma_1)\}^2 \leq k_\alpha(\mu_1, \sigma_1). \tag{21.45}$$

Similarly, for the variance of a normal distribution (Exercise 21.13), they found that if $\sigma_1 > \sigma_0$ no more powerful non-similar test exists, but if $\sigma_1 < \sigma_0$ the region

$$\sum_i (x_i - \mu_1)^2 \leq k_\alpha \tag{21.46}$$

is more powerful than the best similar critical region.

Thus if we restrict the alternative class H_1 sufficiently, we can sometimes improve the power of the test, while reducing the average value of the Type I error below the size α, by abandoning the requirement of similarity. In practice, this is not a very strong argument against using similar regions, precisely because we are not usually in a position to be very restrictive about the alternatives to a composite hypothesis. At the same time we must admit that one of the reasons why the criterion of similarity was so widely adopted was that it simplified the mathematical problems, just as did use of the arithmetic mean and variance rather than other measures of location and dispersion.

Bias in tests

21.23 In **20.26–28**, we briefly discussed the problem of testing a simple H_0 against a two-sided class of alternatives, where no UMP test generally exists. We now return to this subject from another viewpoint, although the two-sided nature of the alternative hypothesis is not essential to our discussion, as we shall see.

Example 21.11 (Bias in tests of normal means)

We now consider the problem of Examples 20.2–3 and of **20.27**, that of testing the mean μ of a normal population with known variance, taken as unity for convenience. Suppose that we restrict ourselves to tests based on the distribution of the sample mean $\bar{x}$, as we may do by **21.3** since $\bar{x}$ is sufficient. Generalizing (20.55), consider the size-α region defined by

$$\bar{x} \leq a_{\alpha_1}, \qquad \bar{x} \geq b_{\alpha_2}, \tag{21.47}$$

where $\alpha_1 + \alpha_2 = \alpha$, and α_1 is not now necessarily equal to α_2. a and b are defined as in (20.15), by

$$a_\alpha = \mu_0 - d_\alpha/n^{1/2}, \quad b_\alpha = \mu_0 + d_\alpha/n^{1/2},$$

and

$$G(-d_\alpha) = \int_{-\infty}^{-d_\alpha} (2\pi)^{-1/2} \exp\left(-\tfrac{1}{2}y^2\right) dy = \alpha.$$

We take $d_\alpha > 0$ without loss of generality.

Exactly as in (20.56), the power of the critical region (21.47) is seen to be

$$P = G\{n^{1/2}\Delta - d_{\alpha_2}\} + G\{-n^{1/2}\Delta - d_{\alpha_1}\}, \tag{21.48}$$

where $\Delta = \mu_1 - \mu_0$.

We consider the power (21.48) as a function of Δ. Its first two derivatives are

$$P' = \left(\frac{n}{2\pi}\right)^{1/2} \left[\exp\left\{-\tfrac{1}{2}(n^{1/2}\Delta - d_{\alpha_2})^2\right\} - \exp\{-\tfrac{1}{2}(n^{1/2}\Delta + d_{\alpha_1})^2\}\right] \tag{21.49}$$

and

$$P'' = \frac{n}{(2\pi)^{1/2}}[(d_{\alpha_2} - n^{1/2}\Delta)\exp\{-\tfrac{1}{2}(n^{1/2}\Delta - d_{\alpha_2})^2\} \\ +(n^{1/2}\Delta + d_{\alpha_1})\exp\{-\tfrac{1}{2}(n^{1/2}\Delta + d_{\alpha_1})^2\}]. \tag{21.50}$$

From (21.49), we can only have $P' = 0$ if

$$\Delta = (d_{\alpha_2} - d_{\alpha_1})/(2n^{1/2}). \tag{21.51}$$

When (21.51) holds, we have from (21.50)

$$P'' = \frac{n}{(2\pi)^{1/2}}(d_{\alpha_1} + d_{\alpha_2})\exp\{-\tfrac{1}{8}(d_{\alpha_2} + d_{\alpha_1})^2\}. \tag{21.52}$$

Since we have taken d_α always positive, we therefore have $P'' > 0$ at the stationary value, which is therefore a minimum. From (21.51), it occurs at $\Delta = 0$ only when $\alpha_1 = \alpha_2$, the case discussed in **20.27**. Otherwise, the unique minimum occurs at some value μ_m where

$$\mu_m > \mu_0 \quad \text{if } \alpha_1 > \alpha_2, \qquad \mu_m < \mu_0 \quad \text{if } \alpha_1 < \alpha_2.$$

21.24 The implication of Example 21.11 is that, except when $\alpha_1 = \alpha_2$, there exist values of μ in the alternative class H_1 for which the probability of rejecting H_0 is actually smaller when H_0 is false than when it is true. (Note that if we were considering a one-sided class of alternatives (say, $\mu_1 > \mu_0$), the same situation would arise if we used the critical region located in the wrong tail of the distribution of $\bar{x}$ (say, $\bar{x} \leq a_\alpha$).) It is clearly undesirable to use a test which is more likely to reject the hypothesis when it is true than when it is false. Indeed, we may improve on such a test by using a table of random numbers to reject the hypothesis with probability α – the power of this procedure will always be α.

We may now generalize our discussion. If a size-α critical region w for $H_0 : \theta = \theta_0$ against the simple $H_1 : \theta = \theta_1$ is such that it has power

$$P\{\mathbf{x} \in w | \theta_1\} \geq \alpha, \tag{21.53}$$

it is said to give an *unbiased* test of H_0 against H_1; in the contrary case, the region w, and the test it provides, are said to be *biased*.[6] If H_1 is composite, and (21.53) holds for every member of H_1, w is said to be an unbiased critical region against H_1. It should be noted that unbiasedness does not require that the power function should actually have a regular minimum at θ_0, as we found to be the case in Example 21.11 when $\alpha_1 = \alpha_2$, although this is often true in practice. Figure 20.3 in **20.27** illustrates the appearance of the power function for an unbiased test (the full line) and two biased tests.

If no unbiased test exists, there may be a 'locally unbiased Type M' test (Krishnan, 1966) which has average power at least α in a neighbourhood of H_0.

The criterion of unbiasedness for tests has such strong intuitive appeal that it is natural to restrict attention to the class of unbiased tests when investigating a problem, and to seek UMP unbiased (UMPU) tests, especially for tests against two-sided alternative hypotheses, where UMP tests rarely exist without some restriction on the class of tests considered. Thus, in Example 21.11, the equal-tails test based on $\bar{x}$ is at once seen to be UMPU in the class of tests there considered. That it is actually UMPU among all tests of H_0 will be seen in **21.33**.

Example 21.12 (Unbiased similar region test for the normal variance)

We leave as Exercise 21.13 the result that, for a normal distribution with mean μ and variance σ^2, the statistic $z = \sum_{i=1}^{n}(x_i - \bar{x})^2$ gives a pair of one-sided UMP similar tests of the hypothesis $H_0 : \sigma^2 = \sigma_0^2$, the BCR being

$$z \geq a_{1-\alpha} \quad \text{if } \sigma_1 > \sigma_0, \qquad z \leq a_\alpha \quad \text{if } \sigma_1 < \sigma_0.$$

Now consider the two-sided alternative hypothesis

$$H_1 : \sigma^2 \neq \sigma_0^2.$$

---

[6] This use of 'bias' is unconnected with that of the theory of estimation, and is only prevented from being confusing by the fortunate fact that the context rarely makes confusion possible.

By **21.18** there is no UMP test of H_0 against H_1, but we are intuitively tempted to use the statistic z, splitting the critical region equally between its tails in the hope of achieving unbiasedness, as in Example 21.11. Thus we reject H_0 if

$$z \geq a_{1-\alpha/2} \quad \text{or} \quad z \leq a_{\alpha/2}.$$

This critical region is certainly similar, for the distribution of z is not dependent on μ, the nuisance parameter. Since z/σ^2 has a chi-square distribution with $n-1$ d.fr., whether H_0 or H_1 holds, we have

$$a_{1-\alpha/2} = \sigma_0^2 \chi^2_{1-\alpha/2}, \quad a_{\alpha/2} = \sigma_0^2 \chi^2_{\alpha/2},$$

where χ^2_α is the 100α per cent point of that chi-square distribution. When H_1 holds, it is z/σ_1^2 that has this distribution, and H_0 will then be rejected when

$$\frac{z}{\sigma_1^2} \geq \frac{\sigma_0^2}{\sigma_1^2}\chi^2_{1-\alpha/2} \quad \text{or} \quad \frac{z}{\sigma_1^2} \leq \frac{\sigma_0^2}{\sigma_1^2}\chi^2_{\alpha/2}.$$

The power of the test against any alternative value σ_1^2 is the sum of the probabilities of these two events. We thus require the probability that a chi-square variable will fall outside its $100(\frac{1}{2}\alpha)$ per cent and $100(1-\frac{1}{2}\alpha)$ per cent points each multiplied by a constant σ_0^2/σ_1^2. For each value of α and $n-1$, the degrees of freedom, this probability can be calculated for each value of σ_0^2/σ_1^2. Figure 21.1 shows the power function resulting from such calculations by Neyman and Pearson (1936b) for the case $n = 3$, $\alpha = 0.02$. The power is less than α in this case when $0.5 < \sigma_1^2/\sigma_0^2 < 1$, and the test is therefore biased.

We now enquire whether, by modifying the apportionment of the critical region between the tails of the distribution of z, we can remove the bias. Suppose that the critical region is

$$z \geq a_{1-\alpha_1} \quad \text{or} \quad z \leq a_{\alpha_2},$$

where $\alpha_1 + \alpha_2 = \alpha$. As before, the power of the test is the probability that a chi-square variable with $n-1$ degrees of freedom, say y_{n-1}, falls outside the range of its $100\alpha_2$ per cent and $100(1-\alpha_1)$ per cent points, each multiplied by the constant $\theta = \sigma_0^2/\sigma_1^2$. Writing F for the distribution function of y_{n-1}, we have

$$P = F(\theta\chi^2_{\alpha_2}) + 1 - F(\theta\chi^2_{1-\alpha_1}). \tag{21.54}$$

Regarded as a function of θ, this is the power function. We now choose α_1 and α_2 so that this power function has a regular minimum at $\theta = 1$, where it equals the size of the test. Differentiating (21.54), we have

$$P' = \chi^2_{\alpha_2} f(\theta\chi^2_{\alpha_2}) - \chi^2_{1-\alpha_1} f(\theta\chi^2_{1-\alpha_1}), \tag{21.55}$$

where f is the density of y_{n-1}. If this is to be zero when $\theta = 1$, we require

$$\chi^2_{\alpha_2} f(\chi^2_{\alpha_2}) = \chi^2_{1-\alpha_1} f(\chi^2_{1-\alpha_1}). \tag{21.56}$$

Substituting

$$f(y) \propto e^{-y/2} y^{(n-3)/2}, \tag{21.57}$$

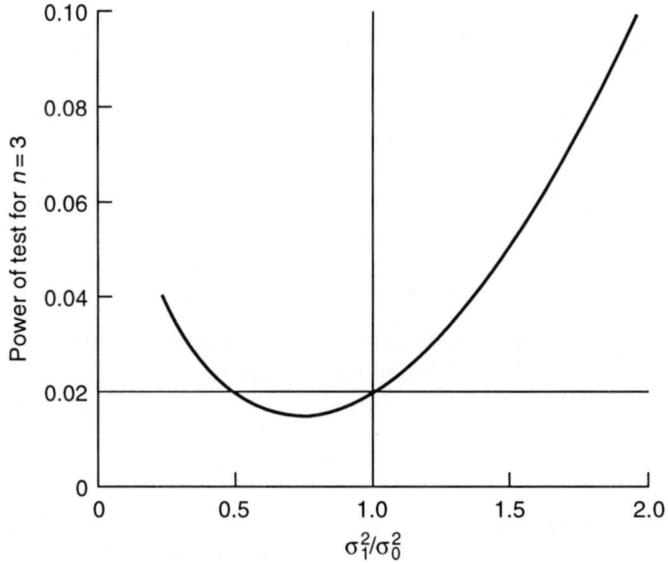

Fig. 21.1 Power function of a test for a normal distribution variance (see text)

we have finally from (21.56) the condition for unbiasedness,

$$\left\{\frac{\chi^2_{1-\alpha_1}}{\chi^2_{\alpha_2}}\right\}^{(n-1)/2} = \exp\{\tfrac{1}{2}(\chi^2_{1-\alpha_1} - \chi^2_{\alpha_2})\}. \tag{21.58}$$

Values of α_1 and α_2 satisfying (21.58) will give a test whose power function has zero derivative at the origin. To investigate whether it is strictly unbiased, we write (21.55), using (21.57) and (21.58), as

$$P' = c\theta^{(n-3)/2}\chi^{n-1}_{\alpha_2}\exp(-\tfrac{1}{2}\chi^2_{\alpha_2})\big[\exp\{\tfrac{1}{2}\chi^2_{\alpha_2}(1-\theta)\} - \exp\{\tfrac{1}{2}\chi^2_{1-\alpha_1}(1-\theta)\}\big], \tag{21.59}$$

where c is a positive constant. Since $\chi^2_{1-\alpha_1} > \chi^2_{\alpha_2}$, we have from (21.59)

$$P' \begin{cases} < 0, & \theta < 1, \\ = 0, & \theta = 1 \\ > 0, & \theta > 1. \end{cases} \tag{21.60}$$

This shows that the test with α_1, α_2 determined by (21.58) is unbiased in the strict sense, for the power function is monotonic decreasing as θ increases from 0 to 1 and monotonic increasing as θ increases from 1 to ∞.

Table 21.1 Limits outside which the chi-square variable $\sum(\mathrm{x}-\bar{\mathrm{x}})^2/\sigma_0^2$ must fall for H_0 : $\sigma^2=\sigma_0^2$ to be rejected ($\alpha = 0.05$)

Degrees of freedom $(n-1)$	Unbiased test limits	Equal-tails test limits	Differences
2	(0.08, 9.53)	(0.05, 7.38)	(0.03, 2.15)
5	(0.99, 14.37)	(0.83, 12.83)	(0.16, 1.54)
10	(3.52, 21.73)	(3.25, 20.48)	(0.27, 1.25)
20	(9.96, 35.23)	(9.59, 34.17)	(0.37, 1.06)
30	(17.21, 47.96)	(16.79, 46.98)	(0.42, 0.98)
40	(24.86, 60.32)	(24.43, 59.34)	(0.43, 0.98)
60	(40.93, 84.23)	(40.48, 83.30)	(0.45, 0.93)

Tables of the values of $\chi^2_{\alpha_2}$ and $\chi^2_{1-\alpha_1}$ satisfying (21.58) are given by Ramachandran (1958) for $\alpha = 0.05$ and $n-1 = 2\,(1)\,8\,(2)\,24, 30, 40$ and 60; other tables are described in **19.34**(3), where the terminology of confidence intervals is used. Table 21.1 compares some of Ramachandran's values with the corresponding limits for the biased equal-tails test that we have considered, obtained from the *Biometrika Tables*.

It will be seen that the differences in both limits are proportionately large for small n, that the lower limit difference increases steadily with n, and the larger limit difference decreases steadily with n. At $n-1=60$, both differences are just over 1 per cent of the values of the limits.

We defer the question whether the unbiased test is UMPU to Example 21.14.

Unbiased tests and similar tests

21.25 There is a close connection between unbiasedness and similarity, which often leads to the best unbiased test emerging directly from an analysis of the similar regions for a problem.

We consider a more general form of hypothesis than (21.2), namely

$$H_0' : \theta_r \leq \theta_{r0}, \tag{21.61}$$

which is to be tested against

$$H_1 : \theta_r > \theta_{r0}. \tag{21.62}$$

If we can find a critical region w satisfying (21.6) for all θ_r in H_0' as well as for all values of the unspecified parameters θ_s, i.e.

$$P(H_0', \theta_s) \leq \alpha \tag{21.63}$$

(where P is the power function whose value is the probability of rejecting H_0), the test based on w will be of size α as before. If it is also unbiased, we have from (21.53)

$$P(H_1, \theta_s) \geq \alpha. \tag{21.64}$$

Now if the power function P is a continuous function of θ_r, (21.63) and (21.64) imply, in view of the form of H_0' and H_1,

$$P(\theta_{r0}, \theta_s) = \alpha, \tag{21.65}$$

i.e. that w is a similar critical region for the 'boundary' hypothesis

$$H_0 : \theta_r = \theta_{r0}.$$

All unbiased tests of H_0' are similar tests of H_0. If we confine our discussions to similar tests of H_0, using the methods we have encountered, and find a test with optimum properties – e.g. a UMP similar test – then *provided that this test is unbiased* it will retain the optimum properties in the class of unbiased tests of H_0' – e.g. it will be a UMPU test.

Exactly the same argument holds if H_0' specifies that the parameter point θ_r lies within a certain region R (which may consist of a number of subregions) in the parameter space, and H_1 that the θ_r lies in the remainder of that space: if the power function is continuous in θ_r, then if a critical region w is unbiased for testing H_0' it is a similar region for testing the hypothesis H_0 that θ_r lies on the boundary of R. If w gives an unbiased test of H_0', it will carry over into the class of unbiased tests of H_0' any optimum properties it may have as a similar test of H_0. There will not always be a UMP similar test of H_0 if the alternatives are two-sided: a UMPU test may exist against such alternatives, but it must be found by other methods.

The early work on unbiased tests was largely carried out by Neyman and Pearson in the series of papers mentioned in **20.1**, and by Neyman (1935; 1938b), Scheffé (1942a) and Lehmann (1947). Much of the detail of their work has now been superseded, as pioneering work usually is, but their terminology is still commonly used.

Example 21.13 (UMPU test for the normal variance)

We return to the discussion in Example 21.12. One-sided critical regions based on the statistic $z \geq a_{1-\alpha}$, $z \leq a_\alpha$, give UMP similar tests against one-sided alternatives. Each of them is easily seen to be unbiased in testing one of

$$H_0' : \sigma^2 \leq \sigma_0^2, \quad H_0'' : \sigma^2 \geq \sigma_0^2$$

respectively against

$$H_1' : \sigma^2 > \sigma_0^2, \quad H_1'' : \sigma^2 < \sigma_0^2.$$

Thus they are, by the argument of **21.25**, UMPU tests for these one-sided situations.

For the two-sided alternative $H_1 : \sigma^2 \neq \sigma_0^2$ the unbiased test based on (21.58) cannot be shown to be UMPU by this method, since we have not shown it to be UMP similar – see Example 21.14.

UMPU tests for the exponential family

21.26 We now give an account of some remarkably comprehensive results, due to Lehmann and Scheffé (1955), which establish the existence of, and give a construction for, UMPU tests for a variety of parametric hypotheses in distributions belonging to the exponential family (17.86).

We write the joint distribution of n independent observations from such a distribution as

$$f(\mathbf{x}) = D(\boldsymbol{\tau})h(\mathbf{x}) \exp\left\{\sum_{j=1}^{r+1} b_j(\boldsymbol{\tau})u_j(\mathbf{x})\right\}, \tag{21.66}$$

where $\mathbf{x}$ is the column vector $(x_1, \ldots, x_n)$ and $\boldsymbol{\tau}$ is a vector of $r+1$ parameters $(\tau_1, \ldots, \tau_{r+1})$. In matrix notation, the exponent in (21.66) may be concisely written $\mathbf{u}^T\mathbf{b}$, where $\mathbf{u}$ and $\mathbf{b}$ are column vectors.

Suppose now that we are interested in the particular linear function of the parameter

$$\theta = \sum_{j=1}^{r+1} a_{j1} b_j(\boldsymbol{\tau}), \tag{21.67}$$

where $\sum_{j=1}^{r+1} a_{j1}^2 = 1$. Write $\mathbf{A}$ for an orthogonal matrix (a_{uv}) whose first column contains the coefficients in (21.67), and transform to a new vector of $r+1$ parameters $(\theta, \boldsymbol{\psi})$, where $\boldsymbol{\psi}$ is the column vector $(\psi_1, \ldots, \psi_r)$, by the equation

$$\begin{pmatrix} \theta \\ \boldsymbol{\psi} \end{pmatrix} = \mathbf{A}^T \mathbf{b}. \tag{21.68}$$

The first row of (21.68) is (21.67). We now suppose that there is a column vector of statistics $\mathbf{T} = (s, t_1, \ldots, t_r)$ defined by the relation

$$\mathbf{T}^T \begin{pmatrix} \theta \\ \boldsymbol{\psi} \end{pmatrix} = \mathbf{u}^T \mathbf{b}, \tag{21.69}$$

i.e. we suppose that the exponent in (21.66) may be expressed as $\theta s(\mathbf{x}) + \sum_{j=1}^{r} \psi_j t_j(\mathbf{x})$. Using (21.68), (21.69) becomes

$$\mathbf{T}^T \begin{pmatrix} \theta \\ \boldsymbol{\psi} \end{pmatrix} = \mathbf{u}^T \mathbf{A} \begin{pmatrix} \theta \\ \boldsymbol{\psi} \end{pmatrix}. \tag{21.70}$$

Expression (21.70) is an identity in $(\theta, \boldsymbol{\psi})$, so we have $\mathbf{T}^T = \mathbf{u}^T \mathbf{A}$ or

$$\mathbf{T} = \mathbf{A}^T \mathbf{u}. \tag{21.71}$$

Comparing (21.71) with (21.68), we see that each component of $\mathbf{T}$ is the same function of the $u_j(\mathbf{x})$ as the corresponding component of $(\theta, \boldsymbol{\psi})$ is of the $b_j(\boldsymbol{\tau})$. In particular, the first component is, from (21.67),

$$s(x) = \sum_{j=1}^{r+1} a_{j1} u_j(\mathbf{x}), \tag{21.72}$$

while the $t_j(\mathbf{x})$, $j = 1, 2, \ldots, r$, are orthogonal to $s(\mathbf{x})$.

Note that the orthogonality condition $\sum_{j=1}^{r+1} a_{j1}^2 = 1$ does not hamper us in testing hypotheses about θ defined by (21.67), since only a constant factor need be changed and the hypothesis adjusted accordingly.

21.27 If, therefore, we can reduce a hypothesis-testing problem (usually through its sufficient statistics) to the standard form of one concerning θ in

$$f(\mathbf{x}|\theta, \boldsymbol{\psi}) = C(\theta, \boldsymbol{\psi}) h(\mathbf{x}) \exp\left\{\theta s(\mathbf{x}) + \sum_{i=1}^{r} \psi_i t_i(\mathbf{x})\right\}, \tag{21.73}$$

by the device of the previous section, we can avail ourselves of the results summarized in **21.10**: given a hypothesis value for θ, the r-component vector $\mathbf{t} = (t_1, \ldots, t_r)$ will be a complete sufficient statistic for the r-component parameter $\boldsymbol{\psi} = (\psi_1, \ldots, \psi_r)$, and we now consider the problem of using s and $\mathbf{t}$ to test various composite hypotheses concerning θ, $\boldsymbol{\psi}$ being an unspecified nuisance parameter. Simple hypotheses are the special case when $r = 0$, with no nuisance parameters.

21.28 For this purpose we shall need an extended form of the Neyman–Pearson lemma of **20.10**. Let $f(\mathbf{x} \mid \boldsymbol{\theta})$ be a density function, and $\boldsymbol{\theta}_i$ a subset of admissible values of the vector of parameters $\boldsymbol{\theta}$ $(i = 1, 2, \ldots, k)$. A specific element of $\boldsymbol{\theta}_i$ is written $\boldsymbol{\theta}_i^0$. $\boldsymbol{\theta}^*$ is a particular value of $\boldsymbol{\theta}$. The vector $\mathbf{u}_i(\mathbf{x})$ is sufficient for $\boldsymbol{\theta}$ when $\boldsymbol{\theta}$ is in $\boldsymbol{\theta}_i$ and its distribution is $g_i(\mathbf{u}_i|\boldsymbol{\theta}_i)$. Since the likelihood function factorizes in the presence of sufficiency, the conditional value of $f(\mathbf{x}|\boldsymbol{\theta}_i)$, given $\mathbf{u}_i$, will be independent of $\boldsymbol{\theta}_i$, and we write it as $f(\mathbf{x}|\mathbf{u}_i)$. Finally, we define $l_i(\mathbf{x})$, $m_i(\mathbf{u}_i)$ to be non-negative functions of the observations and of $\mathbf{u}_i$, respectively.

21.29 Now suppose we have a critical region w for which

$$\int_w \{l_i(\mathbf{x}) f(\mathbf{x}|\mathbf{u}_i)\} \, d\mathbf{x} = \alpha_i. \tag{21.74}$$

Since the product in braces in (21.74) is non-negative, it may be regarded as a density function, and we may say that the conditional size of w, given $\mathbf{u}_i$, is α_i with respect to this distribution. We now write

$$\begin{aligned} \beta_i &= \alpha_i \int m_i(\mathbf{u}_i) g_i(\mathbf{u}_i|\boldsymbol{\theta}_i^0) \, d\mathbf{u}_i \\ &= \int_w l_i(\mathbf{x}) m_i(\mathbf{u}_i) \left\{ \int f(\mathbf{x}|\mathbf{u}_i) g_i(\mathbf{u}_i|\boldsymbol{\theta}_i^0) \, d\mathbf{u}_i \right\} d\mathbf{x} \\ &= \int_w \{l_i(\mathbf{x}) m_i(\mathbf{u}_i) f(\mathbf{x}|\boldsymbol{\theta}_i^0)\} \, d\mathbf{x}. \end{aligned} \tag{21.75}$$

The product in braces is again essentially a density function, say $p(\mathbf{x}|\boldsymbol{\theta}_i^0)$. To test the simple hypothesis that $p(\mathbf{x}|\boldsymbol{\theta}_i^0)$ holds against the simple alternative that $f(\mathbf{x}|\boldsymbol{\theta}^*)$ holds, we use (20.6) and find that the BCR w of size β_i consists of points satisfying

$$[f(\mathbf{x}|\boldsymbol{\theta}^*)]/[p(\mathbf{x}|\boldsymbol{\theta}_i^0)] \geq c_i(\beta_i), \tag{21.76}$$

where c_i is a non-negative function, of β_i only. Expression (21.76) will hold for every value of i. Thus for testing the composite hypothesis that *any* of $p(\mathbf{x}|\boldsymbol{\theta}_i^0)$ holds $(i = 1, 2, \ldots, k)$, we require all k of the inequalities (21.76) to be satisfied by w. If we now write $km_i(\mathbf{u}_i)/c_i(\beta_i)$ for $m_i(\mathbf{u}_i)$ in $p(\mathbf{x}|\boldsymbol{\theta}_i^0)$, as we may since $m_i(\mathbf{u}_i)$ is arbitrary, we have from (21.76), adding the inequalities for $i = 1, 2, \ldots, k$, the necessary and sufficient condition for a BCR

$$f(\mathbf{x}|\boldsymbol{\theta}^*) \geq \sum_{i=1}^{k} l_i(\mathbf{x}) m_i(\mathbf{u}_i) f(\mathbf{x}|\boldsymbol{\theta}_i^0). \tag{21.77}$$

This is the required generalization. Expression (20.6) is the special case with $k = 1$, $l_1(\mathbf{x}) = k_\alpha$ (constant), $m_1(u_1) \equiv 1$. Expression (21.77) will play a role for composite hypotheses similar to that of (20.6) for simple hypotheses.

One-sided alternatives

21.30 Reverting to (21.73), we now investigate the problem of testing

$$H_0^{(1)} : \theta \le \theta_0$$

against

$$H_1^{(1)} : \theta > \theta_0,$$

which we discussed in general terms in **21.25**. Now that we are dealing with the exponential form (21.73), we can show that there is always a UMPU test of $H_0^{(1)}$ against $H_1^{(1)}$. By our discussion in **21.25**, if a size-α critical region is unbiased for $H_0^{(1)}$ against $H_1^{(1)}$, it is a similar region for testing $\theta = \theta_0$.

Consider testing the simple

$$H_0 : \theta = \theta_0, \qquad \boldsymbol{\psi} = \boldsymbol{\psi}^0$$

against the simple

$$H_1 : \theta = \theta^* > \theta_0, \qquad \boldsymbol{\psi} = \boldsymbol{\psi}^*.$$

We now apply the result of **21.29**. Putting $k = 1$, $l_1(\mathbf{x}) \equiv 1$, $\alpha_1 = \alpha$, $\boldsymbol{\theta} = (\theta, \boldsymbol{\psi})$, $\boldsymbol{\theta}_1 = (\theta_0, \boldsymbol{\psi})$, $\boldsymbol{\theta}^* = (\theta^*, \boldsymbol{\psi}^*)$, $\boldsymbol{\theta}_1^0 = (\theta_0, \boldsymbol{\psi}^0)$, $u_1 = \mathbf{t}$, we have the result that the BCR for testing H_0 against H_1 is defined from (21.77) and (21.73) as

$$\frac{C(\theta^*, \boldsymbol{\psi}^*)\exp\{\theta^* s(\mathbf{x}) + \sum_{i=1}^r \psi_i^* t_i(\mathbf{x})\}}{C(\theta_0, \boldsymbol{\psi}^0)\exp\{\theta_0 s(\mathbf{x}) + \sum_{i=1}^r \psi_i^0 t_i(\mathbf{x})\}} \ge m_1(\mathbf{t}). \tag{21.78}$$

This may be rewritten

$$s(\mathbf{x})(\theta^* - \theta_0) \ge c_\alpha(\mathbf{t}, \theta^*, \theta_0, \boldsymbol{\psi}^*, \boldsymbol{\psi}^0). \tag{21.79}$$

We now see that c_α is not a function of $\boldsymbol{\psi}$ since, by **21.27**, $\mathbf{t}$ is a sufficient statistic for $\boldsymbol{\psi}$ when H_0 holds. Thus the value of c_α for given $\mathbf{t}$ will be independent of $\boldsymbol{\psi}^0$, $\boldsymbol{\psi}^*$. Further, from (21.79) we see that so long as the sign of $\theta^* - \theta_0$ does not change, the BCR will consist of the largest 100α per cent of the distribution of $s(\mathbf{x})$ given θ_0. We thus have a BCR for testing $\theta = \theta_0$ against $\theta > \theta_0$ giving a UMP test. This UMP test cannot have smaller power than a randomized test against $\theta > \theta_0$ which ignores the observations. The latter test has power equal to its size α, so the UMP test is unbiased against $\theta > \theta_0$, i.e. by **21.25** it is UMPU. Its size for $\theta < \theta_0$ will not exceed its size at θ_0 as is evident from the consideration that the critical region (21.79) has *minimum* power against $\theta < \theta_0$ and therefore its power (size) there is less than α. Thus finally we have shown that the largest 100α per cent of the conditional distribution of $s(\mathbf{x})$, given $\mathbf{t}$, gives a UMPU size-α test of $H_0^{(1)}$ against $H_1^{(1)}$.

Two-sided alternatives

21.31 We now consider the problem of testing

$$H_0^{(2)} : \theta = \theta_0$$

against

$$H_1^{(2)} : \theta \neq \theta_0.$$

Our earlier examples stopped short of establishing UMPU tests for two-sided hypotheses of this kind (cf. Examples 21.12 and 21.13). Nevertheless a UMPU test does exist for the exponential form (21.73).

From **21.25** we have that if the power function of a critical region is continuous in θ, and unbiased, it is similar for $H_0^{(2)}$. Now for any region w, the power function is

$$P(w|\theta) = \int_w f(\mathbf{x}|\theta, \boldsymbol{\psi})\,d\mathbf{x}, \tag{21.80}$$

where f is defined by (21.73). Expression (21.80) is continuous and differentiable under the integral sign with respect to θ. For the test based on the critical region w to be unbiased we must therefore have, for each value of $\boldsymbol{\psi}$, the necessary condition

$$P'(w|\theta_0) = 0. \tag{21.81}$$

Differentiating (21.80) under the integral sign and using (21.73) and (21.81), we find the condition for unbiasedness,

$$0 = \int_w \left[s(\mathbf{x}) + \frac{C'(\theta_0, \boldsymbol{\psi})}{C(\theta_0, \boldsymbol{\psi})} \right] f(\mathbf{x}|\theta_0, \boldsymbol{\psi})\,d\mathbf{x}$$

or

$$E\{s(\mathbf{x})c(w)\} = -\alpha C'(\theta_0, \boldsymbol{\psi})/C(\theta_0, \boldsymbol{\psi}). \tag{21.82}$$

Since, from (21.73),

$$1/C(\theta, \boldsymbol{\psi}) = \int h(\mathbf{x}) \exp\left\{\theta s(\mathbf{x}) + \sum_i \psi_i t_i(\mathbf{x})\right\} d\mathbf{x},$$

we have

$$\frac{C'(\theta, \boldsymbol{\psi})}{C(\theta, \boldsymbol{\psi})} = -E\{s(\mathbf{x})\}, \tag{21.83}$$

and putting (21.83) into (21.82) gives

$$E\{s(\mathbf{x})c(w)\} = \alpha E\{s(\mathbf{x})\}. \tag{21.84}$$

Taking the expectation first conditionally upon the value of $\mathbf{t}$, and then unconditionally, (21.84) gives

$$E_t[E\{s(\mathbf{x})c(w) - \alpha s(\mathbf{x})|\mathbf{t}\}] = 0. \tag{21.85}$$

Since $\mathbf{t}$ is complete, (21.85) implies

$$E\{s(\mathbf{x})c(w) - \alpha s(\mathbf{x})|\mathbf{t}\} = 0, \tag{21.86}$$

and since all similar regions for H_0 satisfy

$$E\{c(w)|\mathbf{t}\} = \alpha, \tag{21.87}$$

(21.86) and (21.87) combine into

$$E\{s^{i-1}(\mathbf{x})c(w)|\mathbf{t}\} = \alpha E\{s^{i-1}(\mathbf{x})|\mathbf{t}\} = \alpha_i, \qquad i = 1, 2. \tag{21.88}$$

All our expectations are taken when H_0 holds.

Now consider a test of the simple

$$H_0 : \theta = \theta_0, \qquad \boldsymbol{\psi} = \boldsymbol{\psi}^0$$

against the simple

$$H_1 : \theta = \theta^* \neq \theta_0, \qquad \boldsymbol{\psi} = \boldsymbol{\psi}^*$$

and apply the result of **21.29** with $k = 2$, α_i as in (21.88), $\boldsymbol{\theta} = (\theta, \boldsymbol{\psi})$, $\boldsymbol{\theta}_1 = \boldsymbol{\theta}_2 = (\theta_0, \boldsymbol{\psi})$, $\boldsymbol{\theta}^* = (\theta^*, \boldsymbol{\psi}^*)$, $\boldsymbol{\theta}_1^0 = \boldsymbol{\theta}_2^0 = (\theta_0, \boldsymbol{\psi}^0)$, $l_i(\mathbf{x}) = s^{i-1}(\mathbf{x})$, $u_1 = u_2 = \mathbf{t}$. We find that the BCR w for testing H_0 against H_1 is given by (21.77) and (21.73) as

$$\frac{C(\theta^*, \boldsymbol{\psi}^*)\exp\{\theta^* s(\mathbf{x}) + \sum_{i=1}^{r} \psi_i^* t_i(\mathbf{x})\}}{C(\theta_0, \boldsymbol{\psi}^0)\exp\{\theta_0 s(\mathbf{x}) + \sum_{i=1}^{r} \psi_i^0 t_i(\mathbf{x})\}} \geq m_1(\mathbf{t}) + s(\mathbf{x})m_2(\mathbf{t}). \tag{21.89}$$

This reduces to

$$\exp\{s(\mathbf{x})(\theta^* - \theta_0)\} \geq c_1(\mathbf{t}, \theta^*, \theta_0, \boldsymbol{\psi}^*, \boldsymbol{\psi}^0) + s(\mathbf{x})c_2(\mathbf{t}, \theta^*, \theta_0, \boldsymbol{\psi}^*, \boldsymbol{\psi}^0)$$

or

$$\exp\{s(\mathbf{x})(\theta^* - \theta_0)\} - s(\mathbf{x})c_2 \geq c_1. \tag{21.90}$$

Inequality (21.90) is equivalent to $s(\mathbf{x})$ lying outside an interval, i.e.

$$s(\mathbf{x}) \leq v(\mathbf{t}), \quad s(\mathbf{x}) \geq w(\mathbf{t}), \tag{21.91}$$

where $v(\mathbf{t}) < w(\mathbf{t})$ are possibly functions also of the parameters. We now show that they are not dependent on the parameters other than θ_0. As before, the sufficiency of $\mathbf{t}$ for $\boldsymbol{\psi}$ rules out the dependence of v and w on $\boldsymbol{\psi}$ when $\mathbf{t}$ is given. That they do not depend on θ^* follows at once from (21.86), which states that when H_0 holds

$$\int_w \{s(\mathbf{x})|\mathbf{t}\} f \, d\mathbf{x} = \alpha \int_W \{s(\mathbf{x})|\mathbf{t}\} f \, d\mathbf{x}. \tag{21.92}$$

The right-hand side of (21.92), which is integrated over the whole sample space, clearly does not depend on θ^* at all. Hence the left-hand side is also independent of θ^*, so that the BCR w defined by (21.91) depends only on θ_0, as it must. The BCR therefore gives a UMP test of $H_0^{(2)}$ against $H_1^{(2)}$. Its unbiasedness follows by precisely the argument at the end of **21.30**. Thus, finally, we have established that the BCR defined by (21.91) gives a UMPU test of $H_0^{(2)}$ against $H_1^{(2)}$. If we determine from the conditional distribution of $s(\mathbf{x})$, given $\mathbf{t}$, an interval that excludes 100α per cent of the distribution when $H_0^{(2)}$ holds, and take the excluded values as our critical region, then if the region is unbiased it gives the UMPU size-α test.

Finite-interval hypotheses

21.32 We may also consider the hypothesis

$$H_0^{(3)} : \theta_0 \leq \theta \leq \theta_1$$

against

$$H_1^{(3)} : \theta < \theta_0 \quad \text{or} \quad \theta > \theta_1,$$

or the complementary

$$H_0^{(4)} : \theta \leq \theta_0 \quad \text{or} \quad \theta \geq \theta_1,$$

against

$$H_1^{(4)} : \theta_0 < \theta < \theta_1.$$

We now set up two hypotheses

$$H_0' : \theta = \theta_0, \quad \boldsymbol{\psi} = \boldsymbol{\psi}^0, \quad H_0'' : \theta = \theta_1, \qquad \boldsymbol{\psi} = \boldsymbol{\psi}^1,$$

to be tested against

$$H_1 : \theta = \theta^*, \quad \boldsymbol{\psi} = \boldsymbol{\psi}^*,$$

where $\theta_0 \neq \theta^* \neq \theta_1$. We use the result of **21.29** again, this time with $k = 2$, $\alpha_1 = \alpha_2 = \alpha$, $\boldsymbol{\theta} = (\theta, \boldsymbol{\psi}), \boldsymbol{\theta}_1 = (\theta_0, \boldsymbol{\psi})$, $\boldsymbol{\theta}_2 = (\theta_1, \boldsymbol{\psi})$, $\boldsymbol{\theta}^* = (\theta^*, \boldsymbol{\psi}^*)$, $\boldsymbol{\theta}_1^0 = (\theta_0, \boldsymbol{\psi}^0)$, $\boldsymbol{\theta}_2^0 = (\theta_1, \boldsymbol{\psi}^1)$, $l_i(\mathbf{x}) \equiv 1$, $u_1 = u_2 = \mathbf{t}$. We find that the BCR w for testing H_0' or H_0'' against H_1 is defined by

$$f(\mathbf{x}|\theta^*, \boldsymbol{\psi}^*) \geq m_1(\mathbf{t}) f(\mathbf{x}|\theta_0, \boldsymbol{\psi}^0) + m_2(\mathbf{t}) f(\mathbf{x}|\theta_1, \boldsymbol{\psi}^1). \tag{21.93}$$

On substituting $f(\mathbf{x})$ from (21.73), (21.93) is equivalent to

$$H(s) = c_1 \exp\{(\theta_0 - \theta^*)s(\mathbf{x})\} + c_2 \exp\{(\theta_1 - \theta^*)s(\mathbf{x})\} < 1, \tag{21.94}$$

where c_1, c_2 may be functions of all the parameters and of $\mathbf{t}$. If $\theta_0 < \theta^* < \theta_1$, (21.94) requires that $s(\mathbf{x})$ lie inside an interval, i.e.

$$v(\mathbf{t}) \leq s(\mathbf{x}) \leq w(\mathbf{t}). \tag{21.95}$$

On the other hand, if $\theta^* < \theta_0$ or $\theta^* > \theta_1$, (21.94) requires that $s(\mathbf{x})$ lie *outside* the interval $(v(\mathbf{t}), w(\mathbf{t}))$. The proof that the end-points of the interval are not dependent on the values of the parameters, other than θ_0 and θ_1, follows the same lines as before, as does the proof of unbiasedness. Thus we have a UMPU test for $H_0^{(3)}$ and another for $H_0^{(4)}$. The test is similar at values θ_0 and θ_1 as follows from **21.25**. To obtain a UMPU test for $H_0^{(3)}$ ($H_0^{(4)}$), we determine an interval in the distribution of $s(\mathbf{x})$ for given $\mathbf{t}$ which excludes (includes) 100α per cent of the distribution both when $\theta = \theta_0$ and $\theta = \theta_1$. The excluded (included) region, if unbiased, will give a UMPU test of $H_0^{(3)}$ ($H_0^{(4)}$).

21.33 We now turn to some applications of the fundamental results of **21.30–32** concerning UMPU tests for the exponential family of distributions. We first mention briefly that in Example 21.11 and Exercises 20.1–3, UMPU tests for all four types of hypothesis are obtained directly from the distribution of the single sufficient statistic, no conditional distribution being involved since there is no nuisance parameter.

Example 21.14 (UMPU tests for the two-parameter normal distribution)

For n independent observations from a normal distribution, the statistics $(\bar{x}, s^2)$ are jointly sufficient for (μ, σ^2), with joint density (cf. Example 17.17)

$$g(\bar{x}, s^2|\mu, \sigma^2) \propto \frac{s^{n-3}}{\sigma^n} \exp\left\{-\frac{\sum(x-\mu)^2}{2\sigma^2}\right\}. \tag{21.96}$$

Expression (21.96) may be written

$$g \propto C(\mu, \sigma^2) \exp\left\{\left(-\tfrac{1}{2}\sum x^2\right)\left(\frac{1}{\sigma^2}\right) + \left(\sum x\right)\left(\frac{\mu}{\sigma^2}\right)\right\}, \tag{21.97}$$

which is of the form (21.73). Remembering the discussion of **21.26**, we now consider a linear form in the parameters of (21.97). We put

$$\theta = A\left(\frac{1}{\sigma^2}\right) + B\left(\frac{\mu}{\sigma^2}\right), \tag{21.98}$$

where A and B are arbitrary known constants. We specialize A and B to obtain from the results of **21.30–32** UMPU tests for the following hypotheses:

(1) Put $A = 1$, $B = 0$ and test hypotheses concerning $\theta = 1/\sigma^2$, with $\psi = \mu/\sigma^2$ as nuisance parameter. Here $s(\mathbf{x}) = -\frac{1}{2}\sum x^2$ and $t(\mathbf{x}) = \sum x$. From (21.97) there is a UMPU test of $H_0^{(1)}$, $H_0^{(2)}$, $H_0^{(3)}$ and $H_0^{(4)}$ concerning $1/\sigma^2$, and hence concerning σ^2, based on the conditional distribution of $\sum x^2$ given $\sum x$, i.e. of $\sum(x-\bar{x})^2$ given $\sum x$. Since these two statistics are independently distributed, we may use the unconditional distribution of $\sum(x-\bar{x})^2$, or of $\sum(x-\bar{x})^2/\sigma^2$, which has a χ^2 distribution with $n-1$ degrees of freedom. $H_0^{(2)}$ was discussed in Example 21.12, where the UMP similar test was given for $\theta = \theta_0$ against one-sided alternatives and an unbiased test based on $\sum(x-\bar{x})^2$ given for $H_0^{(2)}$; it now follows that this is a UMPU test for $H_0^{(2)}$, while the one-sided test is UMPU for $H_0^{(1)}$, as we saw in Example 21.13.

Graphs of the critical values of the UMPU tests of $H_0^{(3)}$ and $H_0^{(4)}$ for $\alpha = 0.05, 0.10$ are given by Guenther and Whitcomb (1966).

(2) To test hypotheses concerning μ, invert (21.98) into $\mu = (\theta\sigma^2 - A)/B$. If we specify a value μ_0 for μ, we cannot choose A and B to make this correspond uniquely to a value θ_0 for θ (without knowledge of σ^2) if $\theta_0 \neq 0$. But if $\theta_0 = 0$ we have $\mu_0 = -A/B$. Thus from our UMPU tests for $H_0^{(1)} : \theta \leq 0$, $H_0^{(2)} : \theta = 0$, we obtain UMPU tests of $\mu \leq \mu_0$ and of $\mu = \mu_0$, We use (21.71) to see that the test statistic $s(\mathbf{x})|\mathbf{t}$ is here $(-\frac{1}{2}\sum x^2)A + (\sum x)B$ given an orthogonal function, say $(-\frac{1}{2}\sum x^2)B - (\sum x)A$. This reduces to the conditional distribution of $\sum x$ given $\sum x^2$. Clearly we cannot obtain tests of $H_0^{(3)}$ or $H_0^{(4)}$ for μ in this case.

The test of $\mu = \mu_0$ against one-sided alternatives has been discussed in Example 21.7, where we saw that the Student's t test to which it reduces is the UMP similar test of $\mu = \mu_0$ against one-sided alternatives. This test is now seen to be UMPU for $H_0^{(1)}$. It also follows that the two-sided equal-tails Student's t test, which is unbiased for $H_0^{(2)}$ against $H_1^{(2)}$ is the UMPU test of $H_0^{(2)}$.

For the power of these tests, see **21.35**.

Example 21.15 (UMPU tests for k normal means)

Consider k independent samples of n_i $(i = 1, 2, \ldots, k)$ observations from normal distributions with means μ_i and common variance σ^2. Write $n = \sum_{i=1}^{k} n_i$. It is easily confirmed that the k sample means $\bar{x}_i$ and the pooled sum of squares $S^2 = \sum_{i=1}^{k}\sum_{j=1}^{n_i}(x_{ij} - \bar{x}_i)^2$ are jointly sufficient for the $k+1$ parameters. The distribution of the sufficient statistics has joint density function

$$g(\bar{x}_1, \ldots, \bar{x}_k, S^2) \propto \frac{S^{n-k-2}}{\sigma^n} \exp\left\{-\frac{1}{2\sigma^2}\sum_i\sum_j(x_{ij} - \mu_i)^2\right\}. \tag{21.99}$$

This is a simple generalization of (21.96), obtained by using the independence of the $\bar{x}_i$ of each other and of S^2, and the fact that S^2/σ^2 has a χ^2 distribution with $n-k$ degrees of freedom. Expression (21.99) may be rewritten

$$g \propto C(\mu_i, \sigma^2) \exp\left\{\left(-\frac{1}{2}\sum_i\sum_j x_{ij}^2\right)\left(\frac{1}{\sigma^2}\right) + \sum_i\left(\sum_j x_{ij}\right)\left(\frac{\mu_i}{\sigma^2}\right)\right\}, \tag{21.100}$$

in the form (21.73). We now consider the linear function

$$\theta = A\left(\frac{1}{\sigma^2}\right) + \sum_{i=1}^{k} B_i\left(\frac{\mu_i}{\sigma^2}\right). \tag{21.101}$$

(1) Put $A = 1$, $B_i = 0$ (all i). Then $\theta = 1/\sigma^2$, and $\psi_i = \mu_i/\sigma^2$ $(i = 1, \ldots, k)$ is the set of nuisance parameters. There is a UMPU test of each of the four $H_0^{(r)}$ discussed in **21.30–32** for $1/\sigma^2$ and therefore for σ^2. The tests are based on the conditional distribution of $\sum\sum x_{ij}^2$ given the vector $(\sum_j x_{1j}, \sum_j x_{2j}, \ldots, \sum_j x_{kj})$, i.e. of $S^2 = \sum_i\sum_j(x_{ij} - \bar{x}_i)^2$ given that vector. Just as in Example 21.14, this leads to the use of the unconditional distribution of S^2 to obtain the UMPU tests.

(2) Exactly analogous considerations to those of Example 21.14(2) show that by putting $\theta_0 = 0$, we obtain UMPU tests of $\sum_{i=1}^{k} c_i\mu_i \leq c_0$, $\sum c_i\mu_i = c_0$, where c_0 is any constant (cf. Exercise 21.19). Just as before, no 'interval' hypothesis can be tested, using this method, concerning the linear form $\sum c_i\mu_i$.

(3) The substitution $k = 2, c_1 = 1, c_2 = -1, c_0 = 0$, reduces (2) to testing $H_0^{(r)} : \mu_1 - \mu_2 \leq 0$, $H_0^{(2)} : \mu_1 - \mu_2 = 0$. The test of $\mu_1 - \mu_2 = 0$ has been discussed in Example 21.8, where it was shown to reduce to a Student's t test and to be UMP similar. It is now seen to be UMPU for $H_0^{(1)}$. The equal-tails two-sided Student's t test, which is unbiased, is also seen to be UMPU for $H_0^{(2)}$.

Example 21.16 (UMPU tests of variances, and the lack of a UMPU test for the Behrens–Fisher problem)

We generalize the situation in Example 21.15 by allowing the variances of the k normal distributions to differ. We now have a set of $2k$ sufficient statistics for the $2k$ parameters, which are the sample sums and sums of squares $\sum_{j=1}^{n_i} x_{ij}, \sum_{j=1}^{n_i} x_{ij}^2, i = 1, 2, \ldots, k$. We write

$$\theta = \sum_{i=1}^{k} A_i\left(\frac{1}{\sigma_i^2}\right) + \sum_{i=1}^{k} B_i\left(\frac{\mu_i}{\sigma_i^2}\right). \tag{21.102}$$

(1) Put $B_i = 0$ (all i). We get UMPU tests for all four hypotheses concerning

$$\theta = \sum_i A_i\left(\frac{1}{\sigma_i^2}\right),$$

a weighted sum of the reciprocals of the population variances. The case $k = 2$ reduces this to

$$\theta = \frac{A_1}{\sigma_1^2} + \frac{A_2}{\sigma_2^2}.$$

If we want to test hypotheses concerning the variance ratio σ_2^2/σ_1^2, then just as in (2) of Examples 21.14–15, we have to put $\theta = 0$ to make any progress. If we do this, the UMPU tests of $\theta = (\leq)\ 0$ reduce to those of

$$\frac{\sigma_2^2}{\sigma_1^2} = (\leq) -\frac{A_2}{A_1},$$

and we therefore have UMPU tests of $H_0^{(1)}$ and $H_0^{(2)}$ concerning the variance ratio. The joint distribution of the four sufficient statistics may be written

$$g\left(\sum x_{1j}, \sum x_{2j}, \sum x_{1j}^2, \sum x_{2j}^2\right) \propto C(\mu_i, \sigma_i^2)$$
$$\times \exp\left\{-\frac{1}{2}\left(\frac{1}{\sigma_1^2}\sum x_{1j}^2 + \frac{1}{\sigma_2^2}\sum x_{2j}^2\right) + \frac{\mu_1}{\sigma_1^2}\sum x_{1j} + \frac{\mu_2}{\sigma_2^2}\sum x_{2j}\right\}.$$

By **21.26**, the coefficients $s(\mathbf{x})$ of θ when this is transformed to make θ one of its parameters will be the same function of $-\frac{1}{2}\sum x_{1j}^2, -\frac{1}{2}\sum x_{2j}^2$ as θ is of $1/\sigma_1^2, 1/\sigma_2^2$, i.e.

$$-2s(\mathbf{x}) = A_1 \sum x_{1j}^2 + A_2 \sum x_{2j}^2,$$

and the UMPU tests will be based on the conditional distribution of $s(\mathbf{x})$ given any three functions of the sufficient statistics, orthogonal to $s(\mathbf{x})$ and to each other, say

$$\sum x_{1j}, \sum x_{2j} \quad \text{and} \quad A_2 \sum x_{1j}^2 - A_1 \sum x_{2j}^2.$$

This is equivalent to holding $\bar{x}_1, \bar{x}_2$ and

$$t = \sum (x_{1j} - \bar{x}_1)^2 - \frac{A_1}{A_2}\sum (x_{2j} - \bar{x}_2)^2$$

fixed, so that $s(\mathbf{x})$ is equivalent to

$$\sum(x_{1j}-\bar{x}_1)^2+\frac{A_2}{A_1}\sum(x_{2j}-\bar{x}_2)^2$$

for fixed t. In turn, this is equivalent to considering the distribution of the ratio $\sum(x_{1j}-\bar{x}_1)^2/\sum(x_{2j}-\bar{x}_2)^2$, so that the UMPU tests of $H_0^{(1)}$, $H_0^{(2)}$ are based on the distribution of the sample variance ratio – cf. Exercises 21.14 and 21.17.

(2) We cannot get UMPU tests concerning functions of the μ_i free of the σ_i^2, as is obvious from (21.102). In the case $k=2$, this precludes us from finding a solution to the problem of two means by this method.

Geometrical interpretation

21.34 The results of **21.26–33** may be better appreciated with the help of a partly geometrical explanation. From **5.48**, the characteristic function of $s(\mathbf{x})$ in (21.73) is

$$\phi(u)=E\{\exp(\mathrm{i}us)\}=\frac{C(\theta)}{C(\theta+\mathrm{i}u)},\tag{21.103}$$

where we suppress the second argument of C for convenience. The cumulant generating function is

$$\psi(u)=\log\phi(u)=\log C(\theta)-\log C(\theta+\mathrm{i}u),\tag{21.104}$$

with rth cumulant

$$\kappa_r=\left[\frac{\partial^r}{\partial(\mathrm{i}u)^r}\psi(u)\right]_{u=0}=-\frac{\partial^r}{\partial\theta^r}\log C(\theta),\tag{21.105}$$

and

$$E(s)=\kappa_1=-\frac{\partial}{\partial\theta}\log C(\theta),\tag{21.106}$$

$$\kappa_r=\frac{\partial^{r-1}}{\partial\theta^{r-1}}E(s),\qquad r\geq 2.\tag{21.107}$$

Consider the derivative

$$\mathbf{D}^q f\equiv\frac{\partial^q}{\partial\theta^q}f(x|\theta,\psi).$$

From (21.73) and (21.106),

$$\mathbf{D}f=\left\{s+\frac{C'(\theta)}{C(\theta)}\right\}f=\{s-E(s)\}f.\tag{21.108}$$

By Leibniz's rule, we have from (21.108)

$$\begin{aligned}\mathbf{D}^q f&=\mathbf{D}^{q-1}[\{s-E(s)\}f]\\&=\{s-E(s)\}\mathbf{D}^{q-1}f+\sum_{i=1}^{q-1}\binom{q-1}{i}[\mathbf{D}^i\{s-E(s)\}][\mathbf{D}^{q-1-i}f],\end{aligned}\tag{21.109}$$

which, using (21.107), may be written

$$\mathbf{D}^q f = \{s - E(s)\}\mathbf{D}^{q-1} f - \sum_{i=1}^{q-1} \binom{q-1}{i} \kappa_{i+1} \mathbf{D}^{q-1-i} f. \tag{21.110}$$

21.35 Now consider any critical region w of size α. Its power function is defined in (21.80), and we may alternatively express this as an integral in the sample space of the sufficient statistics $(s, \mathbf{t})$ by

$$P(w|\theta) = \int_w f \, ds \, d\mathbf{t}, \tag{21.111}$$

where f now stands for the joint density function of $(s, \mathbf{t})$, which is of the form (21.73) as we have seen. The derivatives of the power function (21.111) are

$$P^{(q)}(w|\theta) = \int_w \mathbf{D}^q f \, ds \, d\mathbf{t}, \tag{21.112}$$

since we may differentiate under the integral sign in (21.111). Using (21.108) and (21.110), (21.111) gives

$$P'(w|\theta) = \int_w \{s - E(s)\} f \, ds \, d\mathbf{t} = \text{cov}\{s, c(w)\}, \tag{21.113}$$

and

$$P^{(q)}(w|\theta) = \int_w \{s - E(s)\}\mathbf{D}^{q-1} f \, ds \, d\mathbf{t} - \sum_{i=1}^{q-1} \binom{q-1}{i} \kappa_{i+1} P^{(q-1-i)}(w|\theta), \qquad q \geq 2, \tag{21.114}$$

a recurrence relation that enables us to build the value of any derivative from lower derivatives. In particular, (21.114) gives

$$P''(w|\theta) = \text{cov}\{[s - E(s)]^2, c(w)\}, \tag{21.115}$$

$$\left.\begin{aligned} P'''(w|\theta) &= \text{cov}\{[s - E(s)]^3, c(w)\} - 3\kappa_2 P'(w|\theta), \\ P^{(\text{iv})}(w|\theta) &= \text{cov}\{[s - E(s)]^4, c(w)\} - 6\kappa_2 P''(w|\theta) - 4\kappa_3 P'(w|\theta). \end{aligned}\right\} \tag{21.116}$$

Expressions (21.113) and (21.115) show that the first two derivatives are simply the covariances of $c(w)$ with s, and with the squared deviation of s from its mean, respectively. The third and fourth derivatives given by (21.116) are more complicated functions of covariances and of the cumulants of s, as are the higher derivatives.

21.36 We are now in a position to interpret geometrically some of the results of **21.26–33**. To maximize the power we must choose w to maximize, for all admissible alternatives, the covariance of $c(w)$ with s, or some function of $s - E(s)$, in accordance with (21.113) and (21.114). In the $(r + 1)$-dimensional space of the sufficient statistics, $(s, \mathbf{t})$, maximum power is obtained by restricting attention to the subspace orthogonal to the r-dimensional sample space of $\mathbf{t}$, i.e. by considering the conditional distribution of s given $\mathbf{t}$.

If we are testing $\theta = \theta_0$ against $\theta > \theta_0$ we maximize $P(w|\theta)$ for all $\theta > \theta_0$ by maximizing $P'(w|\theta)$, i.e. by maximizing $\text{cov}(s, c(w))$ for all $\theta > \theta_0$. This is easily seen to be done if w

consists of the 100α per cent *largest* values of the distribution of s given $\mathbf{t}$. Similarly, for testing $\theta = \theta_0$ against $\theta < \theta_0$, we maximize P by *minimizing* P', and this is done if w consists of the 100α per cent *smallest* values of the distribution of s given $\mathbf{t}$. Since $P'(w|\theta)$ is always of the same sign, the one-sided tests are unbiased.

For the two-sided $H_0^{(2)}$ of **21.31**, (21.81) and (21.115) require us to maximize $P''(w|\theta)$, i.e. $\text{cov}\{[s - E(s)]^2, c(w)\}$. By exactly the same argument as in the one-sided case, we choose w to include the 100α per cent largest values of $\{s - E(s)\}^2$ given $\mathbf{t}$, so that we obtain a two-sided test, which is an equal-tails test only if the distribution of s given $\mathbf{t}$ is symmetrical. It will then follow that the boundaries of the UMPU critical region are equidistant from $E(s|\mathbf{t})$.

Spjøtvoll (1968) gives results on most powerful tests for some non-exponential families of distributions.

Testing with the bootstrap

21.37 Our discussions in **17.10** and **19.23** showed how the bootstrap might be used to generate an empirical estimator for the sampling distribution, from which confidence intervals and other useful measures may be derived. When we turn to testing a hypothesis, the approach must be somewhat modified, as noted by Fisher and Hall (1990). First, the bootstrap operations must be performed on quantities that conform to the hypothesis under test, H_0. Further, Fisher and Hall recommend that the bootstrap is performed on statistics that are pivotal, or at least asymptotically so. The reader is referred to their paper for further details, but the following example illustrates their approach. Earlier discussions of the Behrens–Fisher problem in **19.25–36** and **21.16** have documented the difficulty of making inferences in this context.

Example 21.17 (A bootstrapping approach to the Behrens–Fisher problem)

Given random samples of size n_i from two normal populations, $N(\mu_i, \sigma_i^2)$, $i = 1, 2$, we wish to test

$$H_0 : \mu_1 = \mu_2 \quad \text{vs.} \quad H_1 : \mu_1 \neq \mu_2,$$

by means of the bootstrap approach.

We first define the statistics:

$$\bar{x}_i = \sum x_{ij}/n_i \quad \bar{x} = (n_1\bar{x}_1 + n_2\bar{x}_2)/(n_1 + n_2) \text{ and } S_i = \sum (x_{ij} - \bar{x}_i)^2, \quad i = 1, 2. \quad (21.117)$$

We now draw samples of size n_i with replacement from each of the original samples:

$$x_i^* = \{x_{i1}^*, \ldots, x_{i,n_i}^*\}$$

and compute

$$y_{ij}^* = x_{ij}^* - \bar{x}_i, \qquad i = 1, 2.$$

Finally, we define the test statistic as

$$T = \sum_{i=1}^{2} [n_i(n_i - 1)(\bar{y}_i^* - \bar{y}^*)^2 / S_i^*]^2$$

where the starred quantities are defined for the bootstrap samples by analogy with (21.117). Clearly (21.118) extends immediately to the k-sample case. When $k = 2$, (21.118) reduces to $T = t^2$, where

$$t = \frac{(n_1 n_2)^{1/2}}{(n_1 + n_2)}(\bar{y}_1^* - \bar{y}_2^*)\left[\frac{n_2(n_1 - 1)}{S_1^*} + \frac{n_1(n_2 - 1)}{S_2^*}\right]^{1/2}$$

may be used for one-sided tests.

Following standard methods, $B - 1$ bootstrap samples may be generated and a p-value computed by ranking the observed T, computed using (21.117) in (21.118), along with bootstrap values.

EXERCISES

21.1 Show that for samples of n observations from a normal distribution with mean θ and variance σ^2, no symmetric similar region with respect to θ and σ^2 exists for $n \leq 2$, but that such regions do exist for $n \geq 3$.

(Feller, 1938)

21.2 Show, as in Example 21.1, that for a sample of n observations. The ith of which has density function

$$f_i = \frac{1}{\Gamma(\theta_i)} e^{-x_i} x_i^{\theta_i - 1}, \qquad 0 \leq x_i < \infty; \; \theta_i > 0,$$

no similar size-α region exists for $0 < \alpha < 1$.

(Feller, 1938)

21.3 Given that $L(x|\theta)$ is a likelihood function and $E(\partial \log L/\partial\theta) = 0$, show that if the distribution of a statistic z does not depend on θ, then $\text{cov}(z, \partial \log L/\partial\theta) = 0$. As a corollary, show that no similar region with respect to θ exists if no statistic exists that is uncorrelated with $\partial \log L/\partial\theta$.

(Neyman, 1938a)

21.4 Show, using the c.f. of z, that the converses of the result and the corollary of Exercise 21.3 are true. Together, this and the previous exercise state that $\text{cov}(z, \partial \log L/\partial\theta) = 0$ is a necessary and sufficient condition for $\text{cov}(e^{iuz}, \partial \log L/\partial\theta) = 0$, where u is a dummy variable.

(Neyman, 1938a)

21.5 Show that the Cauchy family of distributions

$$f(x) = \frac{1}{\pi\theta^{1/2}(1 + \frac{x^2}{\theta})}, \qquad -\infty < x < \infty,$$

is not complete.

(Lehmann and Scheffé, 1950)

21.6 Show that if θ is fixed and a statistic z is distributed independently of t, a sufficient statistic for θ, then the distribution of z does not depend on θ.
In Example 17.23, with $n = 1$, show that $t = [x]$ is sufficient for θ and that, for each fixed θ, $z = x$ is distributed independently of t, but that the distribution of z depends on θ.

(Here the range of z is different for every θ; the difficulty does not arise if the ranges are identical for all θ – cf. Basu (1955; 1958).)

21.7 In Exercise 21.6, write $H_1(z)$ for the d.f. of z, $H_2(z|t)$ for its conditional d.f. given t, and $g(t|\theta)$ for the density function of t. Show that

$$\int \{H_1(z) - H_2(z|t)\} g(t|\theta)\, dt = 0$$

for all θ. Hence show that if t is a boundedly complete sufficient statistic for θ, the converse of the result of Exercise 21.6 holds, namely, if the distribution of z does not depend upon θ, z is distributed independently of t.

(Basu, 1955)

21.8 Use the result of Exercise 21.7 to show directly that, in univariate normal samples:

(a) any moment about the sample mean $\bar{x}$ is distributed independently of $\bar{x}$;
(b) the quadratic form $\mathbf{x}^T\mathbf{A}\mathbf{x}$ is distributed independently of $\bar{x}$ if and only if the elements of each row of the matrix $\mathbf{A}$ add to zero (cf. **15.15**);
(c) the sample range is distributed independently of $\bar{x}$;
(d) $(x_{(n)} - \bar{x})/(x_{(n)} - x_{(1)})$ is distributed independently both of $\bar{x}$ and of s^2, the sample variance.

(Hogg and Craig, 1956)

21.9 Use Exercise 21.7 to show that:

(a) in samples from a bivariate normal distribution with $\rho = 0$, the sample correlation coefficient is distributed independently of the sample means and variances (cf. **16.28**);
(b) in independent samples from two univariate normal populations with the same variance σ^2, the statistic

$$F = \frac{\sum_j (x_{1j} - \bar{x}_1)^2/(n_1 - 1)}{\sum_j (x_{2j} - \bar{x}_2)^2/(n_2 - 1)}$$

is distributed independently of the set of three jointly sufficient statistics

$$\bar{x}_1, \bar{x}_2, \sum_j (x_{1j} - \bar{x}_1)^2 + \sum_j (x_{2j} - \bar{x}_2)^2$$

and therefore of the statistic

$$t^2 = \frac{(\bar{x}_1 - \bar{x}_2)^2}{\sum(x_{1j} - \bar{x}_1)^2 + \sum(x_{2j} - \bar{x}_2)^2}\left\{\frac{n_1 n_2(n_1 + n_2 - 2)}{n_1 + n_2}\right\}$$

which is a function of the sufficient statistics. This holds whether or not the population means are equal.

(Hogg and Craig, 1956)

21.10 Use Exercise 21.7 to show that:

(a) in samples of size n from the distribution

$$f(x) = \exp\{-(x - \theta)\}, \qquad \theta \leq x < \infty,$$

$x_{(1)}$ is distributed independently of

$$z = \sum_{i=1}^{r} (x_{(i)} - x_{(1)}) + (n - r)(x_{(r)} - x_{(1)}), \qquad r \leq n;$$

(b) if $f(x_j) = \sigma_j \exp\{-\sigma_j(x_j - \theta)\}, x_j, \sigma_j > 0, j = 1, 2, x_{(2)} - x_{(1)}$ is distributed independently of $x_{(1)}$, and $(\sigma_1 + \sigma_2)(x_{(1)} - \theta)$ is exponentially distributed.

(cf. Epstein and Sobel, 1954)

21.11 In Exercise 18.15, use Exercise 21.7 to show that $\bar{x}$ and $\bar{x}/g$ are independently distributed.

21.12 Show that for the binomial distribution with parameter π, the sample proportion p is minimal sufficient for π, while for the uniform distribution

$$dF = dx, \qquad \theta - \tfrac{1}{2} \leq x \leq \theta + \tfrac{1}{2},$$

the pair of statistics $(x_{(1)}, x_{(n)})$ is minimal sufficient for θ.

(Lehmann and Scheffé, 1950)

21.13 For a normal distribution with variance σ^2 and unspecified mean μ, show by the method of **21.20** that the UMP similar test of $H_0 : \sigma^2 = \sigma_0^2$ against $H_1 : \sigma^2 = \sigma_1^2$ takes the form

$$\sum (x - \bar{x})^2 \begin{cases} \geq a_{1-\alpha} & \text{if } \sigma_1^2 > \sigma_0^2 \\ \leq a_\alpha & \text{if } \sigma_1^2 < \sigma_0^2. \end{cases}$$

21.14 Two normal distributions have unspecified means and variances $\sigma^2, \theta\sigma^2$. From independent samples of sizes n_1, n_2, show by the method of **21.20** that the UMP similar test of $H_0 : \theta = 1$ against $H_1 : \theta = \theta_1$ takes the form

$$s_1^2/s_2^2 \begin{cases} \geq a_{1-\alpha} & \text{if } \theta_1 < 1, \\ \leq a_\alpha & \text{if } \theta_1 > 1, \end{cases}$$

where s_1^2, s_2^2 are the sample variances.

(Harter (1963) shows that a test based on the ratio of sample ranges is almost as powerful as the UMP similar test.)

21.15 Independent samples, each of size n, are taken from the distributions

$$\left.\begin{aligned} dF &= \exp(-x/\theta_1)\, dx/\theta_1, \\ dG &= \exp(-y\theta_2)\theta_2\, dy, \end{aligned}\right\} \qquad \begin{aligned} &\theta_1, \theta_2 > 0, \\ &0 \leq x, y < \infty. \end{aligned}$$

Show that $t = (\sum x, \sum y) = (X, Y)$ is minimal sufficient for (θ_1, θ_2) and remains so if $H_0 : \theta_1 = \theta_2 = \theta$ holds. By considering the function $XY - E(XY)$ show that the distribution of t is not boundedly complete given H_0, so that not all similar regions satisfy (21.13). Finally, show that the statistic XY is then distributed independently of θ, so that H_0 may be tested by similar regions from it.

(Watson, 1957a)

21.16 In the problem of the ratio of two normal means (Example 19.4), assume that $\sigma_{xy} = 0, \sigma_x^2 = \sigma_y^2 = 1$. To test the composite $H_0 : \theta = \theta_0$ against $H_1 : \theta = \theta_1$ show that when H_0 holds, the nuisance parameter μ_1 has a complete sufficient statistic $u = \theta_0\bar{y} + \bar{x}$. Hence show, using **21.20**, that the UMP similar test of H_0 against H_1 is based on large (small) values of $\bar{y} - \theta_0\bar{x}$ if $(\theta_1 - \theta_0)\mu_1 > 0\ (< 0)$.

(Cf. Cox, 1967)

21.17 In Exercise 21.14, show that the critical region

$$s_1^2/s_2^2 \geq a_{1-\alpha/2}, \quad \leq a_{\alpha/2},$$

is biased against the two-sided alternative $H_1 : \theta \neq 1$ unless $n_1 = n_2$. By exactly the same argument as in Example 21.12, show that an unbiased critical region

$$t = s_1^2/s_2^2 \geq a_{1-\alpha_1}, \quad \leq a_{\alpha_2}, \quad \alpha_1 + \alpha_2 = \alpha,$$

is determined by the condition (cf. 21.56))

$$V_{\alpha_2} f(V_{\alpha_2}) = V_{1-\alpha_1} f(V_{1-\alpha_1}),$$

where f is the density of the variance-ratio statistic t and V_α its 100α per cent point. Show that the power function of the unbiased test is monotone increasing for $\theta > 1$, and monotone decreasing for $\theta < 1$.

(John's (1975) tables described in **19.37** (4) give critical values for a simple function of s_1^2/s_2^2.)

21.18 In Exercise 21.17, show that the unbiased confidence interval for θ given by $(t/V_{\alpha_2}, t/V_{1-\alpha_1})$ minimizes the expectation of $\log U - \log L$ for confidence intervals (L, U) based on the tails of the distribution of t.

(Scheffé, 1942b)

21.19 In Example 21.15, use **21.26** to show that the UMPU tests for $\sum_i c_i\mu_i$ are based on the distribution of

$$t = \sum c_i(\bar{x} - \mu_i) \Big/ \left\{ \frac{s^2}{n-k} \sum \frac{c_i^2}{n_i} \right\}^{1/2},$$

which is a Student's t with $n - k$ degrees of freedom.

21.20 In Example 21.16, show that there is a UMPU test of the hypothesis

$$\frac{\mu_i}{\mu_j} = a\frac{\sigma_i^2}{\sigma_j^2}, \qquad i \neq j, \ \mu_j \neq 0.$$

21.21 For independent samples from two Poisson distributions with parameters μ_1, μ_2, show that there are UMPU tests for all four hypotheses considered in **21.30–32** concerning μ_1/μ_2, and that the test of $\mu_1/\mu_2 = 1$ consists of testing whether the sum of the observations is binomially distributed between the samples with equal probabilities.

(Lehmann and Scheffé, 1955)

21.22 For independent binomial distributions with parameters θ_1, θ_2, find the UMPU tests for all four hypotheses in **21.30–32** concerning the 'odds ratio' $(\theta_1/(1-\theta_1))/(\theta_2/(1-\theta_2))$, and the UMPU tests for $\theta_1 = \theta_2, \theta_1 \leq \theta_2$.

(Lehmann and Scheffé, 1955)

21.23 For the uniform distribution

$$f(x) = 1, \qquad \theta - \tfrac{1}{2} \leq x \leq \theta + \tfrac{1}{2},$$

the conditional distribution of the mid-range M given the range R, and the marginal distribution of M, are given by the results of Exercise 14.12. For testing $H_0 : \theta = \theta_0$ against the two-sided alternative $H_1 : \theta \neq \theta_0$ show that the equal-tails test based on M given R, when integrated over all values of R, gives uniformly less power than the equal-tails test based on the marginal distribution of M; use the value $\alpha = 0.08$ for convenience.

(Welch, 1939)

21.24 In Example 21.9, show that the UMPU test of $H_0 : \sigma = \sigma_0$ against $H_1 : \sigma \neq \sigma_0$ is of the form

$$\sum_{i=1}^{n} x_i \geq a_{\alpha_1}, \qquad \leq b_{\alpha_2}.$$

(Lehmann, 1947)

21.25 For the distribution of Example 21.9, show that the UMP similar test of $H_0 : \theta = \theta_0$ against $H_1 : \theta \neq \theta_0$ is of the form

$$\frac{x_{(1)} - \theta_0}{\bar{x} - x_{(1)}} < 0, \quad \geq c_\alpha.$$

(Lehmann (1947); see also Takeuchi (1969) for $H_1 : \theta < \theta_0$)

21.26 For the uniform distribution

$$f(x) = \theta^{-1}, \qquad \mu \leq x \leq \mu + \theta$$

show that the UMP similar test of $H_0 : \mu = \mu_0$ against $H_1 : \mu \neq \mu_0$ is of the form

$$\frac{x_{(1)} - \mu_0}{x_{(n)} - x_{(1)}} < 0, \quad \geq c_\alpha.$$

Cf. the simple hypothesis with $\theta = 1$, where it was seen in Exercise 20.8 that no UMP test exists.

(Lehmann, 1947)

21.27 Given that $x_1, \ldots, x_n$ are independent observations from the distribution

$$f(x) = \frac{1}{\theta^p \Gamma(p)} \exp(-x/\theta) x^{p-1}, \qquad p > 0;\ 0 \leq x < \infty,$$

use Exercises 21.6 and 21.7 to show that a necessary and sufficient condition for a statistic $h(x_1, \ldots, x_n)$ to be independent of $S = \sum_{i=1}^{n} x_i$ is that $h(x_1, \ldots, x_n)$ is homogeneous of degree zero in x. (Cf. references to Exercise 15.22.)

21.28 From (21.113) and (21.114), show that if the first non-zero derivative of the power function is the mth, then

$$P^{(m)}(w|\theta) = \text{cov}\{[s - E(s)]^m, c(w)\}$$

and

$$\frac{\{P^{(m)}(w|\theta)\}^2}{\mu_{2m}} \leq \frac{1}{4},$$

where μ_r is the rth central moment of s. In particular,

$$|P'(w|\theta)| \leq \tfrac{1}{2}\mu_2^{1/2}.$$

21.29 From **21.35**, show that w is a similar region for a hypothesis for which θ is a nuisance parameter if and only if

$$\text{cov}\{s, c(w)\} = 0$$

identically in θ. (Cf. Exercises 21.3–4.)

21.30 Generalize the argument of the last paragraph of Example 21.7 to show that for any distribution of the form

$$f(x) = \sigma^{-1} f\left(\frac{x-\mu}{\sigma}\right),$$

admitting a complete sufficient statistic for σ when μ is known, there can be no similar critical region for $H_0 : \mu = \mu_0$ against $H_1 : \mu = \mu_1$ with power independent of σ.

21.31 For a normal distribution with mean and variance both equal to θ, show that for a single observation, x and x^2 are each singly sufficient for θ, x^2 being minimal. Hence it follows that single sufficiency does not imply minimal sufficiency. (Exercise 18.13 gives a bivariate instance of the same phenomenon.)

21.32 Let (T_r, T_s) be the set of $r+s$ minimal sufficient statistics for $k+l$ parameters (θ_k, θ_l) where $k \geq 1$, $l \geq 0$. Use Exercise 21.7 to show that the subset T_s can only be distributed free of θ if (T_r, T_s) is not boundedly complete. In Exercise 21.23, where $l = 0$, show that the pair of statistics (M, R) is minimal sufficient for θ, but not boundedly complete, and that R is distributed free of θ.

CHAPTER 22

LIKELIHOOD RATIO TESTS AND TEST EFFICIENCY

The LR statistic

22.1 The ML method discussed in Chapter 18 is a constructive method of obtaining estimators which, under certain conditions, have desirable properties. A method of test construction closely allied to it is the likelihood ratio (LR) method, proposed by Neyman and Pearson (1928). It has played a role in the theory of tests analogous to that of the ML method in the theory of estimation.

As before, we have the LF

$$L(x|\boldsymbol{\theta}) = \prod_{i=1}^{n} f(x_i|\boldsymbol{\theta}),$$

where $\boldsymbol{\theta} = (\boldsymbol{\theta}_r, \boldsymbol{\theta}_s)$ is a vector of $r + s = k$ parameters ($r \geq 1, s \geq 0$) and x may also be a vector. We wish to test the hypothesis

$$H_0 : \boldsymbol{\theta}_r = \boldsymbol{\theta}_{r0}, \tag{22.1}$$

which is composite unless $s = 0$, against

$$H_1 : \boldsymbol{\theta}_r \neq \boldsymbol{\theta}_{r0}.$$

We know that there is generally no UMP test in this situation, but that there may be a UMPU test – cf. **21.31**.

The LR method first requires us to find the ML estimators of $(\boldsymbol{\theta}_r, \boldsymbol{\theta}_s)$, giving the unconditional maximum of the LF

$$L(x|\hat{\boldsymbol{\theta}}_{r0}, \hat{\boldsymbol{\theta}}_s), \tag{22.2}$$

and also to find the ML estimators of $\boldsymbol{\theta}_s$, when H_0 holds,[1] giving the conditional maximum of the LF

$$L(x|\boldsymbol{\theta}_{r0}, \hat{\hat{\boldsymbol{\theta}}}_s). \tag{22.3}$$

$\hat{\hat{\boldsymbol{\theta}}}_s$ in (22.3) has been given a double circumflex to emphasize that it does not in general coincide with $\hat{\boldsymbol{\theta}}_s$ in (22.2). Now consider the likelihood ratio[2]

$$l = \frac{L(x|\boldsymbol{\theta}_{r0}, \hat{\hat{\boldsymbol{\theta}}}_s)}{L(x|\hat{\boldsymbol{\theta}}_r, \hat{\boldsymbol{\theta}}_s)}. \tag{22.4}$$

Since (22.4) is the ratio of a conditional maximum of the LF to its unconditional maximum, we clearly have

$$0 \leq l \leq 1. \tag{22.5}$$

---

[1] When $s = 0$, H_0 being simple, no maximization process is needed, for L is uniquely determined.

[2] The ratio is often denoted by λ, and the LR statistic is sometimes called 'the lambda criterion', but we use the Roman letter in accordance with the convention that Greek symbols are reserved for parameters.

Intuitively, l is a reasonable test statistic for H_0: it is the maximum likelihood under H_0 as a fraction of its largest possible value, and large values of l signify that H_0 is reasonably acceptable. The critical region for the test statistic is therefore

$$l \leq c_\alpha, \tag{22.6}$$

where c_α is determined from the distribution $g(l)$ of l to give a size-α test, that is,

$$\int_0^{c_\alpha} g(l)\,\mathrm{d}l = \alpha. \tag{22.7}$$

Neither maximum value of the LF is affected by a change of parameter from $\boldsymbol{\theta}$ to $\tau(\boldsymbol{\theta})$, the ML estimator of $\tau(\boldsymbol{\theta})$ being $\tau(\hat{\boldsymbol{\theta}})$ – cf. **18.3**. Thus the LR statistic is invariant under reparametrization.

22.2 For the LR method to be useful in the construction of similar tests, i.e. tests based on similar critical regions, the distribution of l should be free of nuisance parameters, and for many common statistical problems this is so. The next two examples illustrate the method in cases where it does and does not lead to a similar test.

Example 22.1 (LR test for the normal mean)

For the normal distribution

$$f(x) = (2\pi\sigma^2)^{-1/2} \exp\left\{-\frac{1}{2}\left(\frac{x-\mu}{\sigma}\right)^2\right\},$$

we wish to test

$$H_0 : \mu = \mu_0.$$

Here

$$L(x|\mu, \sigma^2) = (2\pi\sigma^2)^{-n/2} \exp\left\{-\frac{1}{2}\sum\left(\frac{x-\mu}{\sigma}\right)^2\right\}.$$

Using Example 18.11, we have for the unconditional ML estimators

$$\hat{\mu} = \bar{x},$$
$$\hat{\sigma}^2 = \frac{1}{n}\sum(x-\bar{x})^2 = s^2,$$

so that

$$L(x|\hat{\mu}, \hat{\sigma}^2) = (2\pi s^2)^{-n/2} \exp(-\tfrac{1}{2}n). \tag{22.8}$$

When H_0 holds, the ML estimator is (cf. Example 18.8)

$$\hat{\hat{\sigma}}^2 = \frac{1}{n}\sum(x-\mu_0)^2 = s^2 + (\bar{x}-\mu_0)^2,$$

so that

$$L(x|\mu_0, \hat{\hat{\sigma}}^2) = [2\pi\{s^2 + (\bar{x}-\mu_0)^2\}]^{-n/2} \exp(-\tfrac{1}{2}n). \tag{22.9}$$

From (22.4), (22.8) and (22.9), we find

$$l = \left\{\frac{s^2}{s^2 + (\bar{x} - \mu_0)^2}\right\}^{n/2}$$

or

$$l^{-2/n} = 1 + \frac{t^2}{n-1},$$

where t is Student's t statistic with $n - 1$ degrees of freedom. Thus l is a monotone decreasing function of t^2. Hence we may use the known exact distribution of t^2 as equivalent to that of l, rejecting the 100α per cent largest values of t^2, which correspond to the 100α per cent smallest values of l. We thus obtain an equal-tails test based on the distribution of Student's t. We have seen that this test is UMPU for H_0 in Example 21.14.

Example 22.2 (LR test for equality of two means)

Consider again the problem of two means, extensively discussed in Chapters 19 and 21. We have samples of sizes n_1, n_2 from normal distributions with means and variances (μ_1, σ_1^2), (μ_2, σ_2^2), and wish to test $H_0 : \mu_1 = \mu_2$, which we may reparametrize (cf. **21.2**) as $H_0 : \theta \equiv \mu_1 - \mu_2 = 0$. We call the common unknown value of the means μ. We have

$$L(x|\mu_1, \mu_2, \sigma_1^2, \sigma_2^2) = (2\pi)^{-(n_1+n_2)/2}\sigma_1^{-n_1}\sigma_2^{-n_2} \times \exp\left\{-\frac{1}{2}\left(\sum_{j=1}^{n_1}\frac{(x_{1j} - \mu_1)^2}{\sigma_1^2} + \sum_{j=1}^{n_2}\frac{(x_{2j} - \mu_2)^2}{\sigma_2^2}\right)\right\}.$$

The unconditional ML estimators are

$$\hat{\mu}_1 = \bar{x}_1, \quad \hat{\mu}_2 = \bar{x}_2, \quad \hat{\sigma}_1^2 = s_1^2, \quad \hat{\sigma}_2^2 = s_2^2,$$

so that

$$L(x|\hat{\mu}_1, \hat{\mu}_2, \hat{\sigma}_1^2, \hat{\sigma}_2^2) = (2\pi)^{-(n_1+n_2)/2}s_1^{-n_1}s_2^{-n_2}\exp\{-\tfrac{1}{2}(n_1 + n_2)\}.$$

When H_0 holds, the ML estimators are roots of the set of three equations

$$\begin{aligned} \frac{n_1(\bar{x}_1 - \mu)}{\sigma_1^2} + \frac{n_2(\bar{x} - \mu)}{\sigma_2^2} &= 0, \\ \sigma_1^2 = \frac{1}{n_1}\sum_{j=1}^{n_1}(x_{1j} - \mu)^2 &= s_1^2 + (\bar{x}_1 - \mu)^2, \\ \sigma_2^2 = \frac{1}{n_2}\sum_{j=1}^{n_2}(x_{2j} - \mu)^2 &= s_2^2 + (\bar{x}_2 - \mu)^2. \end{aligned} \qquad (22.10)$$

When the solutions of (22.10) are substituted into the LF, we obtain

$$L(x|\hat{\hat{\mu}}, \hat{\hat{\sigma}}_1^2, \hat{\hat{\sigma}}_2^2) = (2\pi)^{-(n_1+n_2)/2}\hat{\hat{\sigma}}_1^{-n_1}\hat{\hat{\sigma}}_2^{-n_2}\exp\{-\tfrac{1}{2}(n_1 + n_2)\},$$

and the likelihood ratio is

$$l = \left(\frac{s_1}{\hat{\hat{\sigma}}_1}\right)^{n_1}\left(\frac{s_2}{\hat{\hat{\sigma}}_2}\right)^{n_2} = \left\{\frac{s_1^2}{s_1^2 + (\bar{x}_1 - \hat{\hat{\mu}})^2}\right\}^{n_1/2}\left\{\frac{s_2^2}{s_2^2 + (\bar{x}_2 - \hat{\hat{\mu}})^2}\right\}^{n_2/2}. \qquad (22.11)$$

We need then only to determine $\hat{\hat{\mu}}$ to be able to use (22.11). From (22.10), we see that $\hat{\hat{\mu}}$ is a solution of a cubic equation in μ whose coefficients are functions of the n_i and of the sums and sums of squares of the two sets of observations. We cannot therefore write down $\hat{\hat{\mu}}$ as an explicit function, though we can solve for it numerically in any given case. Its distribution is not independent of the ratio σ_1^2/σ_2^2, for $\hat{\hat{\mu}}$ is a function of both s_1^2 and s_2^2, and l is therefore of the form

$$l = g(s_1^2, s_2^2)\, h(s_1^2, s_2^2).$$

Thus the LR method fails in this case to give us a similar test; cf. Example 21.10.

22.3 If, as in Example 22.1, we find that the LR test statistic is a one-to-one function of some statistic whose distribution is either known exactly (as in that example) or can be found, there is no difficulty in constructing a valid test of H_0, though we must ask what desirable properties LR tests as a class possess. Frequently, the LR test statistic is a complicated function of the observations whose exact distribution cannot be derived, as in Example 22.2. In such cases, we have to resort to approximations.

Since l is distributed on the interval (0, 1), we see that for any fixed constant $c > 0$, $w = -2c\log l$ will be distributed on the interval $(0, \infty)$. In **22.7**, we prove that, as n increases, the distribution of $-2\log l$ when H_0 holds tends to a chi-square distribution with r degrees of freedom. We also determine the asymptotic distribution of $-2\log l$ when H_1 holds, but in order to do this we must introduce a generalization of the chi-square distribution.

The non-central chi-square distribution

22.4 We have seen in **16.2–3** that the sum of squares of n independent standardized normal variates is distributed as χ^2 with n degrees of freedom (16.1) and c.f. given by (16.3). We now consider the distribution of the statistic

$$z = \sum_{i=1}^{n} x_i^2$$

where the x_i are still independent normal variates with unit variance, but where their means can differ from zero and

$$E(x_i) = \mu_i, \qquad \sum \mu_i^2 = \lambda. \tag{22.12}$$

We write the joint distribution of the x_i as

$$f(\mathbf{x}) \propto \exp\{-\tfrac{1}{2}(\mathbf{x}-\boldsymbol{\mu})^T(\mathbf{x}-\boldsymbol{\mu})\},$$

and make an orthogonal transformation to a new set of independent normal variates with unit variances,

$$\mathbf{y} = \mathbf{B}^T\mathbf{x}.$$

Since

$$E(\mathbf{x}) = \boldsymbol{\mu},$$

we have

$$\boldsymbol{\theta} = E(\mathbf{y}) = \mathbf{B}^T\boldsymbol{\mu},$$

so that

$$\boldsymbol{\theta}^T\boldsymbol{\theta} = \boldsymbol{\mu}^T\boldsymbol{\mu} = \lambda, \tag{22.13}$$

since $\mathbf{BB}^T = \mathbf{I}$. We now set the first $n-1$ components of $\boldsymbol{\theta}$ equal to zero. Then by (22.13),

$$\theta_n^2 = \lambda.$$

Thus

$$z = \mathbf{x}^T\mathbf{x} = \mathbf{y}^T\mathbf{y}$$

is a sum of squares of n independent normal variates, the first $n-1$ of which are standardized, and the last of which has mean $\lambda^{1/2}$ and variance 1. We write

$$u = \sum_{i=1}^{n-1} y_i^2, \qquad v = y_n^2,$$

and we know that u is distributed as χ^2 with $n-1$ degrees of freedom. The distribution of y_n is

$$f(y_n) \propto \exp\{-\tfrac{1}{2}(y_n - \lambda^{1/2})^2\},$$

so that the distribution of v is

$$\begin{aligned} f_1(v) &\propto \frac{1}{2v^{1/2}}[\exp\{-\tfrac{1}{2}(v^{1/2} - \lambda^{1/2})^2\} + \exp\{-\tfrac{1}{2}(-v^{1/2} - \lambda^{1/2})^2\}] \\ &\propto v^{-1/2}\exp\{-\tfrac{1}{2}(v+\lambda)\}\sum_{r=0}^{\infty}\frac{(v\lambda)^r}{(2r)!} \end{aligned} \tag{22.14}$$

The joint distribution of v and u is

$$g(u, v) \propto f_1(v)f_2(u), \tag{22.15}$$

where f_2 is the χ^2 distribution with $n-1$ degrees of freedom

$$f_2(u) \propto \mathrm{e}^{-u/2}u^{(n-3)/2}. \tag{22.16}$$

We put (22.14) and (22.16) into (22.15) and make the transformation

$$z = u + v, \qquad w = \frac{u}{u+v},$$

with Jacobian equal to z. We find for the joint distribution of z and w

$$g(z, w) \propto \mathrm{e}^{-(z+\lambda)/2}z^{(n-2)/2}w^{(n-3)/2}(1-w)^{-1/2}\sum_{r=0}^{\infty}\frac{\lambda^r z^r}{(2r)!}(1-w)^r.$$

We now integrate out w over its range from 0 to 1, yielding the marginal distribution of z:

$$h(z) \propto \mathrm{e}^{-(z+\lambda)/2}z^{(n-2)/2}\sum_{r=0}^{\infty}\frac{\lambda^r z^r}{(2r)!}B\{\tfrac{1}{2}(n-1), \tfrac{1}{2}+r\}. \tag{22.17}$$

To obtain the constant in (22.17), we recall that it does not depend on λ, and put $\lambda = 0$. Expression (22.17) should then reduce to (16.1), which is the ordinary χ^2 distribution with n degrees of freedom. The non-constant factors agree, but whereas (16.1) has a constant term $1/[2^{n/2}\Gamma(\frac{1}{2}n)]$, the constant term in (22.17) may be shown to be

$$B\{\tfrac{1}{2}(n-1), \tfrac{1}{2}\} = \frac{\Gamma\{\frac{1}{2}(n-1)\}\Gamma(\frac{1}{2})}{\Gamma(\frac{1}{2}n)}.$$

We must therefore divide (22.17) by the factor $2^{n/2}\Gamma\{\frac{1}{2}(n-1)\}\Gamma(\frac{1}{2})$ so that $B\{\frac{1}{2}(n-1), \frac{1}{2}+r\}$ becomes $\Gamma(\frac{1}{2}+r)/\{2^{n/2}\Gamma(\frac{1}{2})\Gamma(\frac{1}{2}n+r)\}$. Since by (3.66)

$$\Gamma(\tfrac{1}{2}+r)/\{\Gamma(\tfrac{1}{2})(2r)!\} = 1/\{2^{2r}r!\},$$

we have for any λ, writing ν for n,

$$h(z) = \frac{e^{-(z+\lambda)/2}z^{(\nu-2)/2}}{2^{\nu/2}} \sum_{r=0}^{\infty} \frac{\lambda^r z^r}{2^{2r}r!\Gamma(\frac{1}{2}\nu+r)}. \tag{22.18}$$

Guenther (1964) gives a simple geometric derivation of (22.18) with references to other geometric proofs. McNolty (1962) obtains it by inverting its c.f., which the reader is asked to find (and to invert more simply) in Exercise 22.1.

As for central χ^2 in **16.2**, we see that (22.18) is valid for any $\nu > 0$ even if not an integer, and $\lambda \geq 0$; indeed, we can here (as we could not in **16.2**) allow $\nu = 0$ also if $\lambda > 0$ – see Exercise 22.19, which expresses (22.18) as a Poisson mixture of central χ^2 variables: for $\nu = 0$, (22.18) holds for $z > 0$, with a discrete probability $e^{-\lambda/2}$ at $z = 0$.

22.5 The distribution (22.18) is known as the *non-central χ^2 distribution* with ν degrees of freedom and non-central parameter λ, and written $\chi'^2(\nu, \lambda)$. It was first given by Fisher (1928a), and has been studied by Wishart (1932), Patnaik (1949), Tiku (1965) and Han (1975).

The d.f. of z is discussed in Exercise 22.21. Johnson and Pearson (1969) give the 0.5, 1, 2.5, 5, 95, 97.5, 99 and 99.5 percentage points of the distribution of $\sqrt{z}$ for $\sqrt{\lambda} = 0.2\ (0.2)\ 6.0$ and $\nu = 1\ (1)\ 12, 15, 20$ to four significant figures, and also approximations for $\sqrt{\lambda} = 8, 10$. Their tables are reproduced in the *Biometrika Tables*, Vol. II, Tables 24 and 29. Ding (1992) gives an algorithm for computation of the non-central chi-square distribution. Chattamvelli and Shanmugam (1995) compare the performance of different algorithms.

Since the first two cumulants of χ'^2 are (cf. Exercise 22.1)

$$\begin{aligned} \kappa_1 &= \nu + \lambda, \\ \kappa_2 &= 2(\nu + 2\lambda), \end{aligned} \tag{22.19}$$

it can be approximated by a (central) χ^2 distribution as follows. The first two cumulants of a χ^2 with ν^* degrees of freedom are (putting $\lambda = 0$, $\nu = \nu^*$ in (22.19))

$$\kappa_1 = \nu^*, \qquad \kappa_2 = 2\nu^*. \tag{22.20}$$

If we equate the first two cumulants of χ'^2 with those of $\rho\chi^2$, where ρ is a constant to be determined, we have, from (22.19) and (22.20),

$$\nu + \lambda = \rho\nu^*,$$
$$2(\nu + 2\lambda) = 2\rho^2\nu^*,$$

so that χ'^2/ρ is approximately a central χ^2 variate with

$$\left.\begin{aligned}\rho &= \frac{\nu+2\lambda}{\nu+\lambda} = 1 + \frac{\lambda}{\nu+\lambda},\\ \nu^* &= \frac{(\nu+\lambda)^2}{\nu+2\lambda} = \nu + \frac{\lambda^2}{\nu+2\lambda},\end{aligned}\right\} \tag{22.21}$$

ν^* in general being fractional. If $\nu \to \infty$, then $\rho \to 1$ and $\nu^* \sim \nu$, and it is easily seen in Exercise 22.1 that the c.f. of the approximation tends to the exact result; but if $\lambda \to \infty$, $\rho \to 2$ and $\nu^* \sim \frac{1}{2}\lambda$.

Patnaik (1949) shows that this approximation to the d.f. of χ'^2 is adequate for many purposes, but he also gives better approximations obtained from Edgeworth series expansions.

If ν^* is large, we may make the approximation simpler by approximating the χ^2 approximating distribution itself, for (cf. **16.6**) $(2\chi'^2/\rho)^{1/2}$ tends to normality with mean $(2\nu^* - 1)^{1/2}$ and variance 1, while, more slowly, χ'^2/ρ becomes normal with mean ν^* and variance $2\nu^*$.

Pearson (1959) gives a more accurate central χ^2 approximation by equating three moments. Johnson and Pearson (1969) fit a Pearson Type I distribution using four moments. Jensen and Solomon (1972) use the Wilson–Hilferty–Haldane method of **16.7** to find the fractional power that best normalizes the distribution of χ'^2 – it is

$$h = \frac{1}{3}\left\{1 + 2\Big/\left(\frac{\nu}{\lambda}+2\right)^2\right\},$$

reducing to $h = \frac{1}{3}$ as in **16.7** when $\lambda = 0$.

Just as we saw in **15.15** that a weighted sum of central χ^2 variates may be approximated by a single central χ^2, Muller and Barton (1989) show that a positively weighted sum of χ'^2 variates may be approximated by a single χ'^2.

22.6 We may now generalize our derivation of **22.4**. Suppose that $\mathbf{x}$ is a vector of n multinormal variates with mean $\boldsymbol{\mu}$ and non-singular covariance matrix $\mathbf{V}$. We can find an orthogonal transformation $\mathbf{x} = \mathbf{B}\mathbf{y}$ that reduces the quadratic form $\mathbf{x}^T\mathbf{V}^{-1}\mathbf{x}$ to the diagonal form $\mathbf{y}^T\mathbf{B}^T\mathbf{V}^{-1}\mathbf{B}\mathbf{y} = \mathbf{y}^T\mathbf{C}\mathbf{y}$, the elements of the diagonal of $\mathbf{C}$ being the eigenvalues of $\mathbf{V}^{-1}$. To $\mathbf{y}^T\mathbf{C}\mathbf{y}$ we apply a further scaling transformation $\mathbf{y} = \mathbf{D}\mathbf{z}$, where the leading diagonal elements of the diagonal matrix $\mathbf{D}$ are the reciprocals of the square roots of the corresponding elements of $\mathbf{C}$, so that $\mathbf{D}^2 = \mathbf{C}^{-1}$. Thus $\mathbf{x}^T\mathbf{V}^{-1}\mathbf{x} = \mathbf{y}^T\mathbf{C}\mathbf{y} = \mathbf{z}^T\mathbf{z}$, and $\mathbf{z}$ is a vector of n independent normal variates with unit variances and mean vector $\boldsymbol{\theta}$ satisfying $\boldsymbol{\mu} = \mathbf{B}\mathbf{D}\boldsymbol{\theta}$, where $\lambda = \boldsymbol{\theta}^T\boldsymbol{\theta} = \boldsymbol{\mu}^T\mathbf{V}^{-1}\boldsymbol{\mu}$. We have now reduced our problem to that considered in **22.4**. We see that the distribution of $\mathbf{x}^T\mathbf{V}^{-1}\mathbf{x}$, where $\mathbf{x}$ is a multinormal vector with covariance matrix $\mathbf{V}$ and mean vector $\boldsymbol{\mu}$, is a non-central χ^2 distribution with n degrees of freedom and non-central parameter $\boldsymbol{\mu}^T\mathbf{V}^{-1}\boldsymbol{\mu}$. This generalizes the result of **15.10** for multinormal variates with zero means.

Graybill and Marsaglia (1957) have generalized the theorems on the distribution of quadratic forms in normal variates, discussed in **15.14–27** and Exercises 15.13–18, to the case where $\mathbf{x}$ has mean $\boldsymbol{\mu} \neq \mathbf{0}$. Idempotency of a matrix is then a necessary and sufficient condition that its quadratic form has a non-central χ^2 distribution, and all the theorems of Chapter 15 hold with this modification.

The asymptotic distribution of the LR statistic

22.7 We saw in **18.16** that under regularity conditions the ML estimator (temporarily written as t) of a single parameter θ attains the MVB asymptotically. It follows from **17.17** that the LF is asymptotically of the form

$$\frac{\partial \log L}{\partial \theta} = -E\left(\frac{\partial^2 \log L}{\partial \theta^2}\right)(t-\theta), \tag{22.22}$$

which is the leading term (of order $n^{1/2}$) obtained by differentiating the logarithm of

$$L \propto \exp\left\{\tfrac{1}{2}E\left(\frac{\partial^2 \log L}{\partial \theta^2}\right)(t-\theta)^2\right\}, \tag{22.23}$$

showing that the LF reduces to the normal distribution of the 'asymptotically sufficient' statistic t.

For a k-component vector of parameters $\boldsymbol{\theta}$, the matrix analogue of (22.22) is

$$\frac{\partial \log L}{\partial \theta} = (\mathbf{t}-\boldsymbol{\theta})^T \mathbf{V}^{-1}, \tag{22.24}$$

where $\mathbf{V}^{-1}$ is defined by (cf. **18.26**)

$$V_{ij}^{-1} = -E\left(\frac{\partial^2 \log L}{\partial \theta_i \, \partial \theta_j}\right).$$

When integrated, (22.24) gives the analogue of (22.23),

$$L \propto \exp\{\tfrac{1}{2}(\mathbf{t}-\boldsymbol{\theta})^T \mathbf{V}^{-1}(\mathbf{t}-\boldsymbol{\theta})\}. \tag{22.25}$$

We saw in **18.26** that under regularity conditions the vector of ML estimators $\mathbf{t}$ is asymptotically multinormally distributed with covariance matrix $\mathbf{V}$. Thus the LF reduces to the multinormal distribution of $\mathbf{t}$. This result was rigorously proved by Wald (1943a).

We may now easily establish the asymptotic distribution of the LR statistic l defined at (22.4). From (22.25), we may reduce the problem to considering the ratio of the maximum of the right-hand side of (22.25) given H_0 to its maximum given H_1. When H_1 holds, the maximum of (22.25) occurs when $\boldsymbol{\theta} = \hat{\boldsymbol{\theta}} = \mathbf{t}$, so that every component of $\mathbf{t}-\boldsymbol{\theta}$ is equal to zero and we have

$$L(x|\hat{\boldsymbol{\theta}}_r, \hat{\boldsymbol{\theta}}_s) \propto \text{constant}. \tag{22.26}$$

When H_0 holds, the s components of $\mathbf{t}-\boldsymbol{\theta}$ corresponding to $\boldsymbol{\theta}_s$ will still be zero, for the maximum of (22.25) occurs at $\boldsymbol{\theta}_s = \hat{\hat{\boldsymbol{\theta}}}_s = \mathbf{t}_s - \mathbf{B}(\mathbf{t}_r - \boldsymbol{\theta}_{r0})$, where $\mathbf{t}^T = (\mathbf{t}_r^T, \mathbf{t}_s^T)$ and $\mathbf{B}$ is determined by the conditional distribution of $\mathbf{t}_s$ given $\mathbf{t}_r$. Expression (22.25) reduces to

$$L(x|\boldsymbol{\theta}_{r0}, \hat{\hat{\boldsymbol{\theta}}}_s) \propto \exp\{-\tfrac{1}{2}(\mathbf{t}_r - \boldsymbol{\theta}_{r0})^T \mathbf{V}_r^{-1}(\mathbf{t}_r - \boldsymbol{\theta}_{r0})\}, \tag{22.27}$$

the subscript r signifying that we are now confined to an r-dimensional distribution. Thus, from (22.26) and (22.27),

$$l = \frac{L(x|\boldsymbol{\theta}_{r0}, \hat{\boldsymbol{\theta}}_s)}{L(x|\hat{\boldsymbol{\theta}}_r, \hat{\boldsymbol{\theta}}_s)} = \exp\{-\tfrac{1}{2}(\mathbf{t}_r - \boldsymbol{\theta}_{r0})^T \mathbf{V}_r^{-1}(\mathbf{t}_r - \boldsymbol{\theta}_{r0})\},$$

and

$$-2\log l = (\mathbf{t}_r - \boldsymbol{\theta}_{r0})^T \mathbf{V}_r^{-1}(\mathbf{t}_r - \boldsymbol{\theta}_{r0}).$$

Now we have seen that $\mathbf{t}_r$ is multinormal with covariance matrix $\mathbf{V}_r$ and mean vector $\boldsymbol{\theta}_{r0}$. Thus, by the result of **22.6**, $-2\log l$ for a hypothesis imposing r constraints is asymptotically distributed as non-central χ^2 with r degrees of freedom and non-central parameter

$$\lambda = (\boldsymbol{\theta}_r - \boldsymbol{\theta}_{r0})^T \mathbf{V}_r^{-1}(\boldsymbol{\theta}_r - \boldsymbol{\theta}_{r0}), \qquad (22.28)$$

a result due to Wald (1943a). If λ is to be bounded away from infinity, we must, since $\mathbf{V}_r^{-1}$ is of order n, restrict $\boldsymbol{\theta}_r - \boldsymbol{\theta}_{r0}$ to be of order $n^{-1/2}$. For the case of fixed alternatives, see Stroud (1972). When H_0 holds, $\lambda = 0$ and $-2\log l$ has a central χ^2 distribution with r degrees of freedom, a result originally due to Wilks (1938a).

A simple rigorous proof of the H_0 result is given by Roy (1957). Stroud (1973) proves the non-central result when sampling is from a member of the exponential family. The singular case is discussed by Moore (1977).

Woodroofe (1978) studies the tail of the H_0 central approximation, and shows that it is accurate for deviations of $o(n)$.

Using the LR test is therefore asymptotically equivalent to basing a test on the ML estimators of the parameters tested. It should be emphasized that these results only hold if the conditions for the asymptotic normality and efficiency of the ML estimators are satisfied.

See also the generalizations by Chernoff (1954) and Feder (1968; 1975). Chant (1974) considers the asymptotic equivalence of other tests. Self and Liang (1987) obtain the asymptotic distribution of ML and LR statistics when the parameter has a boundary value. Foutz and Srivastava (1977) and Kent (1982) study the asymptotic distribution of l when the form of the p.d.f. is misspecified.

The asymptotic power of LR tests

22.8 The result of **22.7** makes it possible to calculate the asymptotic power function of the LR test in any case satisfying the conditions for that result to be valid. We have first to evaluate the matrix $\mathbf{V}_r^{-1}$, and then to evaluate the integral

$$P = \int_{\chi_\alpha'^2(\nu,0)}^{\infty} d\chi'^2(\nu, \lambda) \qquad (22.29)$$

where $\chi_\alpha'^2(\nu, 0)$ is the $100(1-\alpha)$ per cent point of the central χ^2 distribution. P is the power of the test, and its size when $\lambda = 0$.

Harter and Owen (1970) give tables, due to G.E. Rayman, Z. Govindarajulu and F.C. Leone, of P to 4 d.p. for $\alpha = 0.1, 0.05, 0.025, 0.01, 0.005, 0.001$; $\lambda = 0$ (0.1) 1 (0.2) 3 (0.5) 5 (1) 40 (2) 50 (5) 100; and $\nu = 1$ (1) 30 (2) 50 (5) 100; they also give to 3 d.p., for the same values of α and ν, inverse tables of the values of λ required to attain power $P = 0.1$ (0.02) 0.7 (0.01) 0.99. The *Biometrika*

Tables, Vol. II, Table 25, give 3 d.p. inverse tables, compiled from the same source, of the values of λ for $\alpha = 0.05, 0.01$, P (their β) $= 0.25, 0.50, 0.60, 0.70\ (0.05)\ 0.95, 0.97, 0.99$ and ν as above. Fix (1949) gives inverse tables of λ for $\nu = 1\ (1)\ 20\ (2)\ 40\ (5)\ 60\ (10)\ 100$, $\alpha = 0.05, 0.01$ and P (her β) $= 0.1\ (0.1)\ 0.9$.

Peers (1971) obtains an asymptotic series expansion for the power function of l against local alternatives when H_0 is simple, and compares it with some asymptotically equivalent tests.

If we use the approximation of **22.5** for the non-central distribution, (22.29) becomes, using (22.21) with $\nu = r$,

$$P = \int_{[(r+\lambda)/(r+2\lambda)]\chi^2_\alpha(r)}^{\infty} \mathrm{d}\chi^2\left(r + \frac{\lambda^2}{r+2\lambda}\right), \tag{22.30}$$

where $\chi^2(r)$ is the central χ^2 distribution with r degrees of freedom and $\chi^2_\alpha(r)$ its $100(1-\alpha)$ per cent point. Putting $\lambda = 0$ in (22.30) gives the size of the test.

Since the non-central parameter λ defined by (22.28) is, under the regularity conditions assumed, a quadratic form with the elements of $\mathbf{V}_r^{-1}$ as coefficients and the variances and covariances are of the order n^{-1}, it follows that λ will have a factor n and hence that the power (22.29) tends to 1 as n increases – this is evident from the approximation (22.30). On the other hand, Dasgupta and Perlman (1974) show that for fixed λ, (22.29) is a decreasing function of ν – cf. Exercise 22.3.

Example 22.3 (Power for the test of the normal variance)

We wish to test $H_0 : \sigma^2 = \sigma_0^2$ for the normal distribution of Example 22.1. The unconditional ML estimators are as given there, so that (22.8) remains the unconditional maximum of the LF. Given our present H_0, the ML estimator of μ is $\hat{\mu} = \bar{x}$ (Example 18.2). Thus

$$L(x|\hat{\mu}, \sigma_0^2) = (2\pi\sigma_0^2)^{-n/2}\exp\left\{-\frac{1}{2}\frac{\sum(x-\bar{x})^2}{\sigma_0^2}\right\}. \tag{22.31}$$

The ratio of (22.31) to (22.8) gives

$$l = \left(\frac{s^2}{\sigma_0^2}\right)^{n/2}\exp\left[-\tfrac{1}{2}n\left\{\frac{s^2}{\sigma_0^2} - 1\right\}\right],$$

so that

$$z = \mathrm{e}^{-1}l^{2/n} = \frac{t}{n}\mathrm{e}^{-t/n}, \tag{22.32}$$

where $t = ns^2/\sigma_0^2$. z is a monotone function of l, but is not a monotone function of t/n, its derivative being

$$\frac{\mathrm{d}z}{\mathrm{d}t} = \frac{1}{n}\left(1 - \frac{t}{n}\right)\mathrm{e}^{-t/n},$$

so that z increases steadily for $t < n$ to a maximum at $t = n$ and then decreases steadily. Putting $l \leq c_\alpha$ is therefore equivalent to putting

$$t \leq a_\alpha, \qquad t \geq b_\alpha,$$

where a_α, b_α are determined, using (22.32), by

$$\begin{aligned} P\{t \leq a_\alpha\} + P\{t \geq b_\alpha\} &= \alpha, \\ a_\alpha e^{-a_\alpha/n} &= b_\alpha e^{-b_\alpha/n}. \end{aligned} \tag{22.33}$$

Since the statistic t has a χ^2 distribution with $n-1$ d.fr. when H_0 holds, we may solve (22.33) for a_α and b_α.

Now consider the approximate distribution of

$$-2\log l = (t-n) - n\log(t/n).$$

Since $E(t) = n-1$, $\operatorname{var} t = 2(n-1)$, we may write

$$\begin{aligned} -2\log l &= (t-n) - n\log\left\{1 + \frac{t-n}{n}\right\} \\ &= (t-n) - n\sum_{r=1}^{\infty}(-1)^{r-1}\left(\frac{t-n}{n}\right)^r \Big/ r \\ &= (t-n) - n\left\{\frac{t-n}{n} - \frac{(t-n)^2}{2n^2} + o_p(n^{-2})\right\} \\ &= \frac{(t-n)^2}{2n} + o_p(n^{-1}). \end{aligned} \tag{22.34}$$

We have seen (**16.3**) that a χ^2 distribution with $n-1$ degrees of freedom is asymptotically normally distributed with mean $n-1$ and variance $2(n-1)$; or equivalently, that $(t-n)/(2n)^{1/2}$ tends to a standardized normal variate. Its square, the first term on the right of (22.34), is therefore a χ^2 variate with 1 d.fr. This is precisely the distribution of $-2\log l$ given by the general result of **22.7** when H_0 holds. This result also tells us that when H_0 is false, $-2\log l$ has a non-central χ^2 distribution with 1 degree of freedom and non-central parameter, by (22.28),

$$\lambda = -E\left\{\frac{\partial^2 \log L}{\partial(\sigma^2)^2}\right\}(\sigma^2 - \sigma_0^2) = \frac{n}{2\sigma^4}(\sigma^2 - \sigma_0^2) = \frac{n}{2}\left(1 - \frac{\sigma_0^2}{\sigma^2}\right)^2.$$

Thus the expression (22.30) for the approximate power of the LR test in this case, where $r = 1$, is

$$P = \int_{[(1+\lambda)/(1+2\lambda)]\chi_\alpha^2(1)}^{\infty} d\chi^2\left(1 + \frac{\lambda^2}{1+2\lambda}\right). \tag{22.35}$$

For example, suppose that $\alpha = 0.05$, $n = 50$ and consider the alternative $\sigma^2 = \sigma_1^2 = 1.25\sigma_0^2$. We then have $\chi_\alpha^2(1) = 3.84$ and $\lambda = 0.02n = 1$, so that (22.35) yields

$$P = \int_{2.56}^{\infty} d\chi^2(\tfrac{4}{3}),$$

or $P = 0.166$ approximately. The exact power may be obtained from the normal d.f.: it is the power of an equal-tails size-α test against an alternative with mean $\lambda^{1/2} = 1$ standard deviations distant from the mean on H_0, i.e. the proportion of the alternative distribution lying outside the interval $(-2.96, +0.96)$ standard deviations from its mean. The normal tables give the value $P = 0.170$. The approximation to the power function is thus quite accurate enough.

Closer approximations to the distribution of the LR statistic

22.9 We now confine attention to the distribution of l when H_0 holds, and seek closer approximations than the asymptotic result of **22.7**. As indicated in **22.3**, if we wish to find χ^2 approximations to the distribution of a function of l, we can gain some flexibility by considering the distribution of $w = -2c \log l$ and adjusting c to improve the approximation.

The simplest way of doing this would be to find the expected value of w and adjust c so that

$$E(w) = r,$$

the expectation of a χ^2 variate with r degrees of freedom. An approximation of this kind was first given by Bartlett (1937), and a general method for deriving the value of c has been given by Lawley (1956), who uses essentially the methods of **19.18** to investigate the moments of $-2 \log l$. If

$$E(-2\log l) = r\left\{1 + \frac{a}{n} + O\left(\frac{1}{n^2}\right)\right\}, \tag{22.36}$$

Lawley shows that by putting either

$$\begin{aligned} w_1 &= -2\left(\tfrac{1}{1+\frac{a}{n}}\right) \log l \\ \text{or} \qquad w_2 &= -2\left(1 + \tfrac{a}{n}\right) \log l \end{aligned} \tag{22.37}$$

we not only have

$$E(w) = r + O\left(\frac{1}{n^2}\right),$$

which follows immediately from (22.36) and (22.37), but also that *all* the cumulants of w conform, to order n^{-1}, to those of a χ^2 distribution with r degrees of freedom. In the continuous case, this simple scaling correction which adjusts the mean of w to the correct value is an unequivocal improvement. Barndorff-Nielsen and Cox (1984) demonstrate this by a quite different argument, using the saddlepoint method of **11.13–16**. Barndorff-Nielsen and Hall (1988) and McCullagh and Cox (1986) show that the correction can be expressed in terms of the cumulants of the first- and second-order derivatives of the log LF and that the χ^2 approximation is indeed accurate to $O(n^{-2})$ as conjectured by Lawley. Ross (1987) obtains related results. Frydenberg and Jensen (1989) show that in the discrete case, the improvement is less consistent.

If even closer approximations are required, they can be obtained in a large class of situations by methods due to Box (1949), who gives improved χ^2 approximations and shows how to derive a function of $-2\log l$ that follows an F distribution. Box also derives an asymptotic expansion for the distribution function in terms of incomplete gamma functions. Asymptotic expansions of the distribution of l under local alternatives to H_0 are given for simple H_0 by Peers (1971) and for composite H_0 by Hayakawa (1975; 1977); see also Harris (1985; 1986).

Cordeiro (1983, 1987) obtains the expected value of $-2\log l$ for members of the natural exponential family.

Example 22.4 (Bartlett's correction for tests of variances)

Suppose k independent samples of sizes n_i $(i = 1, 2, \ldots, k; n_i \geq 2)$ are taken from different normal populations with means μ_i and variances σ_i^2, and we wish to test

$$H_0 : \sigma_1^2 = \sigma_2^2 = \cdots = \sigma_k^2,$$

a composite hypothesis imposing the $r = k - 1$ constraints

$$\frac{\sigma_2^2}{\sigma_1^2} = \frac{\sigma_3^2}{\sigma_1^2} = \cdots = \frac{\sigma_k^2}{\sigma_1^2} = 1.$$

Call the common unknown value of the variances σ^2.

The unconditional maximum of the LF is obtained, as in Example 22.1, by putting

$$\hat{\mu}_i = \bar{x}_i,$$
$$\hat{\sigma}_i^2 = \frac{1}{n_i}\sum_{j=1}^{n_i}(x_{ij} - \bar{x}_i)^2 = s_i^2,$$

giving

$$L(x|\hat{\mu}_1, \ldots, \hat{\mu}_k, \hat{\sigma}_1^2, \ldots, \hat{\sigma}_k^2) = (2\pi)^{-n/2}\prod_{i=1}^{k}(s_i^2)^{-n_i/2}e^{-n/2}, \tag{22.38}$$

where

$$n = \sum_{i=1}^{k} n_i.$$

Given H_0, the ML estimators of the means and the common variance σ^2 are

$$\hat{\hat{\mu}}_i = \bar{x}_i,$$
$$\hat{\hat{\sigma}}^2 = \frac{1}{n}\sum_{i=1}^{k} n_i s_i^2 = s^2,$$

so that

$$L(x|\hat{\hat{\mu}}_1, \ldots, \hat{\hat{\mu}}_k, \hat{\hat{\sigma}}^2) = (2\pi)^{-n/2}(s^2)^{-n/2}e^{-n/2}. \tag{22.39}$$

From (22.4), (22.38) and (22.39),

$$l = \prod_{i=1}^{k}\left(\frac{s_i^2}{s^2}\right)^{-n_i/2}, \tag{22.40}$$

so that

$$-2\log l = n\log(s^2) - \sum_{i=1}^{k} n_i \log(s_i^2). \tag{22.41}$$

Now when H_0 holds, each of the statistics $n_i s_i^2/2\sigma^2$ has a gamma distribution with parameter $\frac{1}{2}(n_i - 1)$, and their sum $ns^2/2\sigma^2$ has the same distribution with parameter $\sum_{i=1}^{k} \frac{1}{2}(n_i - 1) = \frac{1}{2}(n-k)$. For a gamma variate x with parameter p, we have

$$\begin{aligned} E\{\log(ax)\} &= \frac{1}{\Gamma(p)} \int_0^\infty \log(ax) \mathrm{e}^{-x} x^{p-1}\, \mathrm{d}x \\ &= \log a + \frac{\mathrm{d}}{\mathrm{d}p} \log \Gamma(p), \end{aligned}$$

which, using Stirling's series (3.63), becomes

$$E\{\log(ax)\} = \log a + \log p - \frac{1}{2p} - \frac{1}{12p^2} + O\left(\frac{1}{p^3}\right). \tag{22.42}$$

Using (22.42) in (22.41), we have

$$\begin{aligned} E\{-2\log l\} &= n\left\{\log\left(\frac{2\sigma^2}{n}\right) + \log\{\tfrac{1}{2}(n-k)\} - \frac{1}{n-k} - \frac{1}{3(n-k)^2} + O\left(\frac{1}{n^3}\right)\right\} \\ &\quad - \sum_{i=1}^{k} n_i \left\{\log\left(\frac{2\sigma^2}{n_i}\right) + \log\{\tfrac{1}{2}(n_i - 1)\} - \frac{1}{n_i - 1} - \frac{1}{3(n_i-1)^2} + O\left(\frac{1}{n_i^3}\right)\right\} \\ &= n\left\{\log\left(1 - \frac{k}{n}\right) - \frac{1}{n-k} - \frac{1}{3(n-k)^2} + O\left(\frac{1}{n^3}\right)\right\} \\ &\quad - \sum_{i=1}^{k} n_i \left\{\log\left(1 - \frac{1}{n_i}\right) - \frac{1}{n_i - 1} - \frac{1}{3(n_i-1)^2} + O\left(\frac{1}{n_i^3}\right)\right\} \\ &= (k-1) + \left[\left(\sum_{i=1}^{k} \frac{1}{n_i - 1} - \frac{k}{n-k}\right) + \frac{1}{2}\left(\sum_{i=1}^{k} \frac{1}{n_i} - \frac{k^2}{n}\right)\right. \\ &\quad \left. + \frac{1}{3}\left\{\sum_{i=1}^{k} \frac{n_i}{(n_i-1)^2} - \frac{n}{(n-k)^2}\right\}\right] + O\left(\frac{1}{N^3}\right), \end{aligned} \tag{22.43}$$

where we now write N indifferently for n_i and n. We could now improve the χ^2 approximation, in accordance with (22.37), with the expression in square brackets in (22.43) as $(k-1)a/n$.

Now consider Bartlett's (1937) modification of the LR statistic (22.40) in which n_i is replaced throughout by the degrees of freedom $n_i - 1 = \nu_i$, so that n is replaced by $\nu = \sum_{i=1}^{k}(n_i - 1) = n - k$. We write this

$$l^* = \prod_{i=1}^{k} \left(\frac{s_i^2}{s^2}\right)^{\nu_i/2},$$

where now

$$s_i^2 = \frac{1}{\nu_i} \sum_{j=1}^{n_i} (x_{ij} - \bar{x}_i)^2,$$

$$s^2 = \frac{1}{\nu}\sum_{i=1}^{k} \nu_i s_i^2.$$

Thus

$$-2\log l^* = \nu \log s^2 - \sum_{i=1}^{k} \nu_i \log s_i^2. \tag{22.44}$$

We shall see in Example 22.6 that l^* has the advantage over l that it gives an unbiased test for any values of the n_i. If we retrace the passage from (22.42) to (22.43), we find that

$$\begin{aligned} E(-2\log l^*) &= -\nu\left\{\frac{1}{\nu} + \frac{1}{3\nu^2} + O\left(\frac{1}{\nu^3}\right)\right\} + \sum_{i=1}^{k} \nu_i \left\{\frac{1}{\nu_i} + \frac{1}{3\nu_i^2} + O\left(\frac{1}{\nu_i^3}\right)\right\} \\ &= (k-1) + \frac{1}{3}\left(\sum_{i=1}^{k} \frac{1}{\nu_i} - \frac{1}{\nu}\right) + O\left(\frac{1}{\nu_i^3}\right). \end{aligned} \tag{22.45}$$

From (22.37) and (22.45) it follows that $-2\log l^*$ defined at (22.44) should be divided by the scaling constant

$$1 + \frac{1}{3(k-1)}\left(\sum_{i=1}^{k} \frac{1}{\nu_i} - \frac{1}{\nu}\right)$$

to give a closer approximation to a χ^2 distribution with $k-1$ degrees of freedom.

For $k = 3\,(1)\,10$ and all $\nu_i = 4\,(1)\,11, 14\,(5)\,29, 49, 99$, Glaser (1976) gives exact critical values of l^* for $\alpha = 0.01, 0.05, 0.10$.

For $k = 3\,(1)\,12$ and all $\nu_i = 1\,(1)\,10$, Harsaae (1969) gives exact 3 d.p. critical values of $-2\log l^*$ for α and $1-\alpha = 0.001, 0.01, 0.05, 0.10$; Glaser (1976) gives 4 d.p. values of $(l^*)^{2/\nu}$ for $k = 3\,(1)\,10$, all $\nu_i = 4\,(1)\,11, 14\,(5)\,29, 49, 99$ and $\alpha = 0.01, 0.05, 0.10$, as do Dyer and Keating (1980) for $k = 2\,(1)\,10$, all $\nu_i = 2\,(1)\,29\,(10)\,59\,(20)\,99$ and $\alpha = 0.01, 0.05, 0.10, 0.25$. For unequal ν_i Chao and Glaser (1975) obtained the exact probability of l^* in usable form, and Glaser (1980) gave a computable algorithm; Nagarsenker (1984) obtained a simpler representation of the distribution in terms of incomplete beta functions, yielding an asymptotic expansion that gives a close approximation; Dyer and Keating (1980) give an excellent approximation based on a weighted average of the equal – ν_i critical values. Approximate critical values are given in the *Biometrika Tables*, Vol. I, Tables 32–33, for $k = 3\,(1)\,15$, $(\sum_{i=1}^{k} \nu_i^{-1} - \nu^{-1}) = 0\,(0.5)\,5\,(1)\,10\,(2)\,14$ and $\alpha = 0.01, 0.05$.

Rivest (1986) shows that the test size assuming normality is an understatement for various classes of distribution when all ν_i are equal.

LR tests when the range depends upon the parameter

22.10 The asymptotic distribution of the LR statistic, given in **22.7**, depends essentially on the regularity conditions necessary to establish the asymptotic normality of ML estimators. We have seen in Example 18.5 that these conditions break down where the range of the underlying distribution is a function of the parameter θ. What can be said about the distribution of the LR statistic in such cases? It is remarkable that, as Hogg (1956) showed, for certain hypotheses concerning such distributions the statistic $-2\log l$ is distributed *exactly* as χ^2, but with $2r$ degrees of freedom, i.e. twice as many as there are constraints imposed by the hypothesis.

22.11 We first derive some preliminary results concerning uniform distributions. If k variables z_i are independently distributed as

$$f(z_i) = 1, \qquad 0 \leq z_i \leq 1, \tag{22.46}$$

the distribution of

$$t_i = -2 \log z_i$$

has density function

$$f(t_i) = \tfrac{1}{2} \exp(-\tfrac{1}{2} t_i), \qquad 0 \leq t_i < \infty,$$

a χ^2 distribution with 2 degrees of freedom, so that the sum of k such independent variates

$$t = \sum_{i=1}^{k} t_i = -2 \sum_{i=1}^{k} \log z_i = -2 \log \prod_{i=1}^{k} z_i$$

has a χ^2 distribution with $2k$ degrees of freedom.

It follows from (14.1) that the distribution function of $y_{(n_i)}$ the largest among n_i independent observations from a uniform distribution on the interval (0, 1), is

$$H = y_{(n_i)}^{n_i}, \qquad 0 \leq y_{(n_i)} \leq 1, \tag{22.47}$$

and hence that $y_{(n_i)}^{n_i} = z_i$ is uniformly distributed as in (22.46). Hence for k independent samples of size n_i, $t = -2 \log \prod_{i=1}^{k} y_{(n_i)}^{n_i}$ has a χ^2 distribution with $2k$ degrees of freedom.

Now consider the distribution of the largest of the k largest values $y_{(n_i)}$. Since all the observations are independent, this is simply the largest of $n = \sum_{i=1}^{k} n_i$ observations from the original uniform distribution. If we denote this largest value by $y_{(n)}$ the distribution of $-2 \log y_{(n)}^{n}$ will, by the argument above, be a χ^2 distribution with 2 degrees of freedom. We now show that the statistics $y_{(n)}$ and $u = \prod_{i=1}^{k} y_{(n_i)}^{n_i} / y_{(n)}^{n}$ are independently distributed. Introduce the parameter θ, so that the original uniform distribution is on the interval $(0, \theta)$. The joint density of the $y_{(n_i)}$ then becomes, from (22.47),

$$f = \prod_{i=1}^{k} \{n_i y_{(n_i)}^{n_i - 1} / \theta^{n_i}\} = \frac{1}{\theta^n} \prod_{i=1}^{k} n_i y_{(n_i)}^{n_i}.$$

By **17.40**, $y_{(n)}$ is sufficient for θ, and by **21.12** its distribution is complete. Thus by Exercise 21.7 we need only observe that the distribution of u is free of the parameter θ to establish the result that u is distributed independently of the complete sufficient statistic $y_{(n)}$. We see that

$$(-2 \log u) + (-2 \log y_{(n)}^{n}) = -2 \log \prod_{i=1}^{k} y_{(n_i)}^{n_i},$$

and since $y_{(n)}$ and u are independent, so are $y_{(n)}^{n}$ and u. The c.f. of a sum of independent variates is the product of their c.f.s (cf. **8.40–42**), so that if $\phi(t)$ is the c.f. of $-2 \log u$, we have

$$\phi(t) \cdot (1 - 2it)^{-1} = (1 - 2it)^{-k},$$

using the χ_2^2 and χ_{2k}^2 results just derived, whence

$$\phi(t) = (1 - 2\mathrm{i}t)^{-(k-1)},$$

so that $-2\log u$ has a χ^2 distribution with $2(k-1)$ degrees of freedom.

Collecting our results, finally, we have established that if we have k variates $y_{(n_i)}$ independently distributed as in (22.47), then $-2\log\prod_{i=1}^{k} y_{(n_i)}^{n}$ has a χ^2 distribution with $2k$ degrees of freedom, while, if $y_{(n)}$ is the largest of the $y_{(n_i)}$,

$$-2\log\left\{\prod_{i=1}^{k} y_{(n_i)}^{n_i} \Big/ y_{(n)}^{n}\right\}$$

has a χ^2 distribution with $2(k-1)$ degrees of freedom.

22.12 We now consider in turn the two situations where a single sufficient statistic exists for θ when the range is a function of θ, taking first the case when only one terminal (say, the upper) depends on θ. We then have, from **17.40**, the necessary form for the density function

$$f(x|\theta) = g(x)/h(\theta), \qquad a \le x \le \theta. \tag{22.48}$$

Now suppose we have k (≥ 1) separate populations of this form, $f(x_i|\theta_i)$, and wish to test

$$H_0 : \theta_1 = \theta_2 = \cdots = \theta_k = \theta_0,$$

a simple hypothesis imposing k constraints, on the basis of samples of sizes n_i ($i = 1, 2, \ldots, k$). We now find the LR criterion for H_0. The unconditional ML estimator of θ_i is the largest observation $x_{(n_i)}$ (cf. Example 18.1). Thus

$$L(x|\hat{\theta}_1, \ldots, \hat{\theta}_k) = \prod_{i=1}^{k}\prod_{j=1}^{n_i}\{g(x_{ij})/[h(x_{(n_i)})]^{n_i}\}. \tag{22.49}$$

Since H_0 is simple, the LF is fully determined and no ML estimation is necessary. We have

$$L(x|\theta_0, \theta_0, \ldots, \theta_0) = \prod_{i=1}^{k}\prod_{j=1}^{n_i}\{g(x_{ij})\}/[h(\theta_0)]^{n}.$$

Hence the LR statistic is

$$l = \frac{L(x|\theta_0, \ldots, \theta_0)}{L(x|\hat{\theta}_1, \ldots, \hat{\theta}_k)} = \prod_{i=1}^{k}\left[\frac{h(x_{(n_i)})}{h(\theta_0)}\right]^{n_i}. \tag{22.50}$$

When H_0 holds, $y_i = h(x_{(n_i)})/h(\theta_0)$ is the probability that an observation falls below or at $x_{(n_i)}$ and is itself a random variable with distribution function obtained from that of $x_{(n_i)}$ as

$$F = y_i^{n_i}, \qquad 0 \le y_i \le 1,$$

of the form (22.47). Thus, from the result of the previous section,

$$-2\log\prod_{i=1}^{k} y_i^{n_i} = -2\log l$$

has a χ^2 distribution with $2k$ degrees of freedom.

22.13 We now investigate the composite hypothesis, for $k \geq 2$ populations,

$$H_0 : \theta_1 = \theta_2 = \cdots = \theta_k$$

which imposes $k-1$ constraints, leaving the common value of θ unspecified. The unconditional maximum of the LF is given by (22.49) as before. The maximum under our present H_0 is $L(x|\hat{\hat{\theta}}, \hat{\hat{\theta}}, \ldots, \hat{\hat{\theta}})$, where $\hat{\hat{\theta}} = x_{(n)}$, is the ML estimator for the pooled samples. Thus we have the LR statistic

$$l = \frac{L(x|\hat{\hat{\theta}}, \ldots, \hat{\hat{\theta}})}{L(x|\hat{\theta}_1, \ldots, \hat{\theta}_k)} = \prod_{i=1}^{k} \frac{[h(x_{(n_i)})]^{n_i}}{[h(x_{(n)})]^{n}}. \tag{22.51}$$

By writing this as

$$l = \prod_{i=1}^{k}\left[\frac{h(x_{(n_i)})}{h(\theta)}\right]^{n_i} \Big/ \left[\frac{h(x_{(n)})}{h(\theta)}\right]^{n},$$

where θ is the common unspecified value of the θ_i, we see that in the notation of the previous section,

$$l = \left[\prod_{i=1}^{k} y_{(n_i)}^{n_i}\right] \Big/ y_{(n)}^{n},$$

so that by **22.11** we have that in this case $-2\log l$ is distributed as χ^2 with $2(k-1)$ degrees of freedom.

22.14 When both terminals of the range are functions of θ, we have from **17.41** that if there is a single sufficient statistic for θ, then

$$f(x|\theta) = g(x)/h(\theta), \qquad \theta \leq x \leq b(\theta), \tag{22.52}$$

where $b(\theta)$ must be a monotone decreasing function of θ. For $k \geq 1$ such populations $f(x_i|\theta_i)$, we again test the simple

$$H_0 : \theta_1 = \theta_2 = \cdots = \theta_k = \theta_0$$

on the basis of samples of sizes n_i. The unconditional ML estimator of θ_i is the sufficient statistic

$$t_i = \min\{x_{(1i)}, b^{-1}(x_{(ni)})\},$$

where $x_{(1i)}, x_{(ni)}$ are respectively the smallest and largest observations in the ith sample. When H_0 holds, the LF is specified by $L(x|\theta_0, \ldots, \theta_0)$. Thus the LR statistic

$$l = \frac{L(x|\theta_0, \ldots, \theta_0)}{L(x|\hat{\theta}_1, \ldots, \hat{\theta}_k)} = \prod_{i=1}^{k} \left[\frac{h(t_i)}{h(\theta_0)}\right]^{n_i}. \tag{22.53}$$

Just as for (22.50), we see that

$$l = \prod_{i=1}^{k} y_i^{n_i},$$

where the y_i are distributed in the form (22.47), and hence $-2\log l$ again has a χ^2 distribution with $2k$ degrees of freedom.

Similarly, for the composite hypothesis with $k-1$ constraints ($k \geq 2$)

$$H_0 : \theta_1 = \theta_2 = \cdots = \theta_k,$$

we find, as in **22.13**, that the LR statistic is

$$l = \frac{L(x|\hat{\hat{\theta}}, \ldots, \hat{\hat{\theta}})}{L(x|\hat{\theta}_1, \ldots, \hat{\theta}_k)} = \prod_{i=1}^{k} [h(t_i)^{n_i} / h(t)]^n$$

where $t = \min\{t_i\}$ is the combined ML estimator $\hat{\hat{\theta}}$, so that by writing

$$l = \prod_{i=1}^{k} \left[\frac{h(t_i)}{h(\theta)}\right]^{n_i} \Big/ \left[\frac{h(t)}{h(\theta)}\right]^n$$

we again reduce l to the form required in **22.11** for $-2\log l$ to be distributed as χ^2 with $2(k-1)$ degrees of freedom.

22.15 We have thus obtained exact χ^2 distributions for two classes of hypotheses concerning distributions whose terminals depend upon the parameter being tested. Exercises 22.8 and 22.9 give further examples, one exact and one asymptotic, of LR tests for which $-2\log l$ has a χ^2 distribution with twice as many degrees of freedom as there are constraints imposed by the hypothesis tested. It should be noted that these χ^2 forms spring not from any tendency to multinormality on the part of the ML estimators, as did the limiting results of **22.7** for 'regular' situations, but from the intimate connection between the uniform and χ^2 distributions explored in **22.11**. One effect of this difference is that the power functions of the tests take a quite different form from the non-central χ^2 distribution in **22.8**.

Barr (1966) finds the power function of the test of the simple hypothesis based on the LR statistic (22.50), and shows that the test is unbiased. When $k = 1$, the test is shown to be UMP (cf. Example 20.6 and the condition (20.37)), but there is no UMP test for $k > 1$, and the LR test is not even UMPU. For the composite hypothesis of **22.13** with $k = 2$, the power function of the LR test statistic (22.51) is used to show that the LR test is UMPU.

The properties of LR tests

22.16 So far, we have been concerned entirely with the problems of determining the distribution of the LR statistic, or a function of it. We now have to enquire into the properties of LR tests, in particular the question of their unbiasedness and whether they are optimum tests in any sense. First, however, we consider a weaker property, that of consistency, introduced briefly in **20.9**.

Test consistency

22.17 A test of a hypothesis H_0 against a class of alternatives H_1 is said to be consistent if, when any member of H_1 holds, the probability of rejecting H_0 tends to 1 as sample size(s) tend to infinity. If w is the critical region, and $\mathbf{x}$ the sample point, we write this property as

$$\lim_{n\to\infty} P\{\mathbf{x} \in w|H_1\} = 1. \tag{22.54}$$

The idea of test consistency, which is a simple and natural one, was first introduced by Wald and Wolfowitz (1940). It seems reasonable to require that, as the number of observations increases, any test worth considering should reject a false hypothesis with increasing certainty, and in the limit with complete certainty. Test consistency is as intrinsically acceptable as is consistency in estimation (**17.7**), of which it is in one sense a generalization. For if a test concerning the value of θ is based on a statistic that is a consistent estimator of θ, it is immediately obvious that the test will be consistent too. But an inconsistent estimator may still provide a consistent test. For example, if t tends in probability to $a\theta$, t will give a consistent test of hypotheses about θ. In general, it is clear that it is sufficient for test consistency that the test statistic, when regarded as an estimator, should tend in probability to some one-to-one function of θ.

Since the condition for a size-α test to be unbiased is (cf. (21.53)) that

$$P\{\mathbf{x} \in w|H_1\} \geq \alpha, \tag{22.55}$$

it is clear from (22.54) and (22.55) that a consistent test will lose its bias, if any, as $n \to \infty$. However, an unbiased test need not be consistent.[3]

The consistency and unbiasedness of LR tests

22.18 We saw in **18.10, 18.16** and **18.26** that under certain conditions, the ML estimator $\hat{\boldsymbol{\theta}}$ of a parameter vector $\boldsymbol{\theta}$ is consistent, though in other circumstances it need not be. When all the ML estimators are consistent, we see from the definition of the LR statistic at (22.4) that, as sample sizes increase,

$$l \to \frac{L(x|\boldsymbol{\theta}_{r0}, \boldsymbol{\theta}_s)}{L(x|\boldsymbol{\theta}_r, \boldsymbol{\theta}_s)}, \tag{22.56}$$

where $\boldsymbol{\theta}_r, \boldsymbol{\theta}_s$ are the true values of those parameters, and $\boldsymbol{\theta}_{r0}$ is the hypothetical value of $\boldsymbol{\theta}_r$ being tested. Thus, when H_0 holds, $l \to 1$ in probability, and the critical region (22.6) will therefore have its boundary c_α approaching 1. When H_0 does not hold, the limiting value of l in (22.56) will be some constant k satisfying (cf. (18.20))

$$0 \leq k < 1$$

[3]Cf. the remark in **17.9** on consistent and asymptotically unbiased estimators.

and thus we have

$$P\{l \leq c_\alpha\} \to 1 \tag{22.57}$$

and the LR test is consistent.

In **22.8** we confirmed from the approximate power function that LR tests are consistent under regularity conditions, in conformity with our present discussion.

22.19 When we turn to the question of unbiasesdness, we recall the penultimate sentence of **22.17** which, coupled with the result of **22.18**, ensures that most LR tests are asymptotically unbiased. Of itself, this is not very comforting (though it would be so if it could be shown under reasonable restrictions that the maximum extent of the bias is always small), for the criterion of unbiasedness in tests is intuitively attractive enough to impose itself as a necessity for all sample sizes. Example 22.5 shows that an important LR test is biased.

Example 22.5 (Bias in the LR test for the variance)

Consider again the hypothesis H_0 of Example 22.3. The LR test uses as its critical region the tails of the χ^2_{n-1} distribution of $t = ns^2/\sigma_0^2$ determined by (22.33). Now in Examples 21.12 and 21.14 we saw that the unbiased (actually UMPU) test of H_0 was determined from the distribution of t by the relations

$$\begin{aligned} P\{t \leq a_\alpha\} + P\{t \geq b_\alpha\} &= \alpha, \\ a_\alpha^{(n-1)/2} \exp(-a_\alpha/2) &= b_\alpha^{(n-1)/2} \exp(-b_\alpha/2), \end{aligned} \tag{22.58}$$

the second of which is (21.58) in the notation of Example 22.3. It is clear on comparison of (22.58) with (22.33) that they would only give the same result if $a_\alpha = b_\alpha$, which cannot hold except in the trivial case $a_\alpha - b_\alpha = 0$, $\alpha = 1$. In all other cases, the tests have different critical regions, the LR test having higher values of a_α and b_α than the unbiased test, i.e. a larger fraction of α concentrated in the lower tail. It is easy to see that for alternative values of σ^2 just larger than σ_0^2, for which the distribution of t is slightly displaced towards higher values of t compared to its H_0 distribution, the probability content lost to the critical region of the LR test through its larger value of b_α will exceed the gain due to the larger value of a_α; and thus the LR test will be biased.

It will be seen by reference to Example 21.12 that whereas the LR test has values of a_α, b_α too large for unbiasedness, the equal-tails test there discussed has values too small for unbiasedness. Thus the two more or less intuitively acceptable critical regions 'bracket' the unbiased critical region.

If in (22.33) we replace n by $n - 1$, it becomes precisely equivalent to the unbiased (22.58), illustrating the general result that the LR test loses its bias asymptotically. It is suggestive to trace this bias to its source. If, in constructing the LR statistic in Example 22.3, we had adjusted the unconditional ML estimator of σ^2 to be unbiased, s^2 would have been replaced by $[n/(n-1)]s^2$, and the adjusted LR test would have been unbiased: the estimation bias of the ML estimator lies behind the test bias of the LR test. Alternatively, we may observe that the bias disappears if we use the marginal likelihood for σ^2, as discussed in **18.32**.

The test corresponding to the physically shortest confidence intervals based on s^2, given in Exercise 19.15, replaces n in (22.58) by $n + 2$, so its critical values will be above even those of the LR test.

Spjøtvoll (1972b) showed that LR tests – he discussed the corresponding integrals – are unbiased under certain regularity conditions if there is no nuisance parameter. Peers (1971) showed in this simple H_0 case that l is unbiased against local alternatives, and compared it with asymptotically equivalent tests, but Hayakawa (1975; 1977) (see also a correction by Harris, 1986) and Harris and Peers (1980) showed that the result does not hold generally for composite H_0.

Unbiased invariant tests for location and scale parameters

22.20 Example 22.5 suggests that a good principle in constructing an LR test is to adjust all the ML estimators used in the process so that they are unbiased. A further confirmation of this principle is contained in Example 22.4, where we stated that the adjusted LR statistic l^* gives an unbiased test. We now verify this principle, developing a method due to Pitman (1939b) for this purpose.

If the hypothesis being tested concerns a set of k location parameters θ_i $(i = 1, 2, \ldots, k)$, we write the joint density function as

$$f = f(x_1 - \theta_1, x_2 - \theta_2, \ldots, x_k - \theta_k). \tag{22.59}$$

We wish to test

$$H_0 : \theta_1 = \theta_2 = \cdots = \theta_k. \tag{22.60}$$

Any test statistic t, to be satisfactory intuitively, must satisfy the location-invariance condition (cf. **23.4**)

$$t(x_1, x_2, \ldots, x_k) = t(x_1 - \lambda, x_2 - \lambda, \ldots, x_k - \lambda). \tag{22.61}$$

Therefore, without loss of generality, we may take the common value of the θ_i in (22.60) to be zero. We suppose that $t > 0$, and that w_0, the size-α critical region based on the distribution of t, is defined by

$$t \leq c_\alpha; \tag{22.62}$$

if either or both of these statements were not true, we could transform to a function of t for which they were.

Because of its invariance property (22.61), t must be constant in the k-dimensional sample space W on any line L parallel to the equiangular vector V defined by $x_1 = x_2 = \cdots = x_k$. When H_0 holds, the content of w_0 is its size

$$\alpha = \int_{w_0} \mathrm{d}F(x_1, x_2, \ldots, x_k), \tag{22.63}$$

and when H_0 is not true the content of w_0 is its power

$$1 - \beta = \int_{w_0} \mathrm{d}F(x_1 - \theta_1, x_2 - \theta_2, \ldots, x_k - \theta_k) = \int_{w_1} \mathrm{d}F(x_1, x_2, \ldots, x_k), \tag{22.64}$$

where w_1 is derived from w_0 by translation in W without rotation. We define the integral, on any line L parallel to V,

$$P(L) = \int_L f(x_1, x_2, \ldots, x_k)\, \mathrm{d}\bar{x}; \tag{22.65}$$

the variation along any L being the same for each coordinate x_j, it can be summarized in the variation of the mean coordinate $\bar{x}$. Since the aggregate of lines L is the whole of W, we have

$$\int P(L)\,\mathrm{d}L = \int \left\{ \int_L f\,\mathrm{d}\bar{x} \right\} \mathrm{d}L = \int \cdots \int f\,\mathrm{d}x_1 \cdots \mathrm{d}x_k = 1. \quad (22.66)$$

We now determine w_0 as the aggregate of all lines L for which the statistic $P(L)$ is at most some constant h. Then $P(L)$ will exceed h on any L which is in w_1 but not in w_0. Hence, from (22.63), (22.64) and (22.66),

$$\alpha = \int_{w_0} \mathrm{d}F \leq \int_{w_1} \mathrm{d}F = 1 - \beta,$$

so that the test is unbiased. We therefore define the test statistic t so that at any point on a line L, parallel to V, it is equal to $P(L)$. Now using the invariance property (22.61) with $\lambda = \bar{x}$, we have from (22.65)

$$t(x) = P(L) = \int_L f(x_1 - \bar{x}, x_2 - \bar{x}, \ldots, x_k - \bar{x})\,\mathrm{d}\bar{x},$$

and replacing $\bar{x}$ by u, this is

$$t(x) = \int_{-\infty}^{\infty} f(x_1 - u, x_2 - u, \ldots, x_k - u)\,\mathrm{d}u, \quad (22.67)$$

the unbiased size-α region being defined by (22.62). It will be seen that the unbiased test thus obtained is unique. An example of the use of (22.67) is given in Exercise 22.15.

22.21 Turning now to tests concerning scale parameters, suppose that the joint density function is

$$g = g(y_1/\phi_1, y_2/\phi_2, \cdots, y_k/\phi_k)/\phi_1\phi_2\cdots\phi_k, \quad (22.68)$$

where all the scale parameters ϕ_i are positive. We make the transformation

$$x_i = \log|y_i|, \quad \theta_i = \log\phi_i,$$

and find for the p.d.f. of the x_i

$$f = g\{\exp(x_1 - \theta_1), \exp(x_2 - \theta_2), \ldots, \exp(x_k - \theta_k)\} \exp\left\{\sum_{i=1}^{k}(x_i - \theta_i)\right\}. \quad (22.69)$$

Expression (22.69) is of the form (22.59), already discussed. To test

$$H_0' : \phi_1 = \phi_2 = \cdots = \phi_k \quad (22.70)$$

is the same as to test H_0 of (22.60). The statistic (22.67) becomes

$$t(x) = \int_{-\infty}^{\infty} g\{\exp(x_1 - u), \exp(x_2 - u), \ldots, \exp(x_k - u)\} \exp\left\{\sum_{i=1}^{k} x_i - ku\right\} \mathrm{d}u,$$

which when expressed in terms of the y_i becomes

$$t(y) = \prod_{i=1}^{k} |y_i| \int_0^\infty g\left(\frac{y_1}{v}, \frac{y_2}{v}, \ldots, \frac{y_k}{v}\right) \frac{dv}{v^{k+1}}. \tag{22.71}$$

22.22 Now consider the special case of k independently distributed gamma variates y_i/ϕ_i with parameters m_i. Their joint density is

$$g = \prod_{i=1}^{k} \left\{ \frac{1}{\Gamma(m_i)} \left(\frac{y_i}{\phi_i}\right)^{m_i - 1} \right\} \exp\left(-\sum_{i=1}^{k} \frac{y_i}{\phi_i}\right) \prod_{i=1}^{k} \phi_i^{-1}. \tag{22.72}$$

To test H_0' of (22.70), we use (22.71) and obtain

$$t(y) = \prod_{i=1}^{k} \left\{ \frac{y_i^{m_i}}{\Gamma(m_i)} \right\} \int_0^\infty \exp\left(-\sum_{i=1}^{k} y_i/v\right) \frac{dv}{v^{m+1}}, \tag{22.73}$$

where $m = \sum_{i=1}^{k} m_i$. On substituting $u = \sum_{i=1}^{k} y_i/v$ in (22.73), we find

$$t(y) = \left[\frac{\Gamma(m)}{\Pi_i \Gamma(m_i)}\right] \frac{\Pi_i y_i^{m_i}}{\left(\sum_i y_i\right)^m}. \tag{22.74}$$

We now ignore the constant factor in square brackets in (22.74). T, the maximum attainable value of t, occurs when $y_i / \sum_i y_i = m_i/m$; that is

$$T = \prod m_i^{m_i} / m^m. \tag{22.75}$$

We now write

$$t^* = -\log\left(\frac{t}{T}\right) = m \log\left(\frac{\sum_i y_i}{m}\right) - \sum_i m_i \log\left(\frac{y_i}{m_i}\right). \tag{22.76}$$

t^* will be unbiased for H_0', and will range from 0 to ∞, large values being rejected.

Example 22.6 (Unbiased invariant test for variances)

We may now apply (22.76) to the problem of testing the equality of k normal variances, discussed in Example 22.4. Each of the quantities $\sum_j (x_{ij} - \bar{x}_i)^2/(2\sigma^2)$ is, when H_0 holds, a gamma variate with parameter $\frac{1}{2}(n_i - 1)$. We thus have to substitute in (22.76)

$$\begin{aligned} y_i &= \sum (x_{ij} - \bar{x}_i)^2 = \nu_i s_i^2, \\ m_i &= \frac{1}{2}(n_i - 1) = \frac{1}{2}\nu_i, \\ m &= \sum_i m_i = \frac{1}{2}(n - k) = \frac{1}{2}\nu, \end{aligned} \tag{22.77}$$

and we find for the unbiased test statistic

$$2t^* = \nu \log\left(\frac{\sum_i \nu_i s_i^2}{\nu}\right) - \sum \nu_i \log s_i^2. \tag{22.78}$$

Expression (22.78) is identical with (22.44), so that $2t^*$ is simply $-2\log l^*$ which we discussed there. Thus the l^* test is unbiased, as stated in Example 22.4. From this, it is fairly evident that the unadjusted LR test statistic l of (22.40), which employs another weighting system, cannot also be unbiased in general. When all sample sizes are equal, the two tests are equivalent, as Exercise 22.7 shows. Even in the case $k = 2$, the unadjusted LR test is biased when $n_1 \neq n_2$: this is left to the reader to prove in Exercise 22.14.

A quite different proof is given by A. Cohen and Strawderman (1971).

22.23 Paulson (1941) investigated the bias of a number of LR tests for exponential distributions – some of his results are given in Exercises 22.16 and 22.18.

Other properties of LR tests

22.24 Apart from questions of consistency and unbiasedness, what can be said in general concerning the properties of LR tests? In the first place, we know that unique ML estimators are functions of the sufficient statistics (cf. **18.4**) so that the LR statistic (22.4) may be rewritten

$$l = \frac{L(x|\boldsymbol{\theta}_{r0}, \boldsymbol{t}_s)}{L(x|\boldsymbol{T}_{r+s})}, \tag{22.79}$$

where $\boldsymbol{t}_s$ is the statistic minimal sufficient for $\boldsymbol{\theta}_s$ when H_0 holds and $\boldsymbol{T}_{r+s}$ is the statistic sufficient for all the parameters when H_0 does not hold. As we have seen in **17.38**, it is not true in general that the components of $\boldsymbol{T}_{r+s}$ include the components of $\boldsymbol{t}_s$ – the sufficient statistic for $\boldsymbol{\theta}_s$ when H_0 holds may no longer form part of the sufficient set when H_0 does not hold, and even when it does may not then be separately sufficient for $\boldsymbol{\theta}_s$, merely forming part of $\boldsymbol{T}_{r+s}$ which is sufficient for $(\boldsymbol{\theta}_r, \boldsymbol{\theta}_s)$. Thus all that we can say of l is that it is *some* function of the two sets of sufficient statistics involved. There is, in general, no reason to suppose that it will be the right function of them.

Exercise 22.20 shows that the LR method does not necessarily produce a UMP test when one exists in the case of testing a simple H_0 against a simple H_1. However, Birkes (1990) gives conditions for LR tests of exponential family distributions to be UMP.

If we are seeking a UMPU test, the LR method is handicapped by its own general biasedness, but we have seen that a simple bias adjustment will sometimes remove this difficulty. The adjustment takes the form of a reweighting of the test statistic by substituting unbiased estimators for the ordinary ML estimators (Examples 22.4–6), or sometimes equivalently of an adjustment of the critical region of the statistic to which the LR method leads (Exercise 22.14). Exercise 22.16 shows that two UMP similar tests derived in Exercises 22.25–26 for an exponential and a uniform distribution are produced by the LR method, while the UMPU test for an exponential distribution given in Exercise 22.24 is not equivalent to the LR test, which is biased.

Wald (1943a) shows that the LR test *asymptotically* has a number of optimum power properties under regularity conditions – but see **22.27** and Example 22.8.

The LR principle is intuitively appealing when there is no 'optimum' test. It is of particular value in tests of linear hypotheses (which we shall discuss in Chapter 29) for which, in general, no UMPU test exists. But it is as well to be reminded of the possible fallibility of the LR method in exceptional circumstances, and the following example, adapted from one due to C. Stein and given by Lehmann (1950), is a salutary warning against using the method without investigation of its properties in the particular situation concerned.

Example 22.7 (Potential lack of unbiased tests using the LR procedure)

A discrete random variable x is defined at the values $0, \pm 1, \pm 2$, and the probabilities at these points given a hypothesis H_1 are:

$$\begin{array}{llllll} x: & 0 & \pm 1 & +2 & -2 \\ P|H_1 & \alpha\left(\dfrac{1-\theta_1}{1-\alpha}\right) & (\frac{1}{2}-\alpha)\left(\dfrac{1-\theta_1}{1-\alpha}\right) & \theta_1\theta_2 & \theta_1(1-\theta_2) \end{array} \tag{22.80}$$

The parameters θ_1, θ_2, are restricted by the inequalities

$$0 \leq \theta_1 \leq \alpha < \tfrac{1}{2}, \qquad 0 \leq \theta_2 \leq 1,$$

where α is a known constant. We wish to test the simple

$$H_0 : \theta_1 = \alpha, \qquad \theta_2 = \tfrac{1}{2},$$

H_1 being the general alternative (22.80), on the evidence of a single observation. The probabilities on H_0 are:

$$\begin{array}{llll} x: & 0 & \pm 1 & \pm 2 \\ P|H_0: & \alpha & \frac{1}{2}-\alpha & \frac{1}{2}\alpha \end{array} \tag{22.81}$$

The LF (22.80) is independent of θ_2 when $x = 0, \pm 1$, and is maximized unconditionally by making θ_1 as small as possible, i.e. putting $\hat{\theta}_1 = 0$. The LR statistic is therefore

$$l = \frac{L(x|H_0)}{L(x|\hat{\theta}_1, \hat{\theta}_2)} = 1 - \alpha, \qquad x = 0, \pm 1. \tag{22.82}$$

When $x = +2$ or -2, the LF is maximized unconditionally by choosing θ_2 respectively as large or as small as possible, i.e. $\hat{\theta}_2 = 1, 0$, respectively; and by choosing θ_1 as large as possible, i.e. $\hat{\theta}_1 = \alpha$. The maximum value of the LF is therefore α and the LR statistic is

$$l = \tfrac{1}{2}, \qquad x = \pm 2. \tag{22.83}$$

Since $\alpha < \frac{1}{2}$, it follows from (22.82) and (22.83) that the LR test consists of rejecting H_0 when $x = \pm 2$. From (22.81) this test is seen to be of size α. But from (22.80) its power is seen to be θ_1 exactly, so for any value of θ_1 in

$$0 \leq \theta_1 < \alpha \tag{22.84}$$

the LR test will be biased for all θ_2, while for $\theta_1 = \alpha$ the test will have power equal to its size α for all θ_2. In this latter extreme case the test is useless, but in the former case it is worse than useless, for we can get a test of size and power α by using a table of random numbers as the basis for our decision concerning H_0. Furthermore, a useful test exists, for if we reject H_0 when $x = 0$, we still have a size-α test by (22.81), the power of which, from (22.80), is $\alpha(1-\theta_1)/(1-\alpha)$ which exceeds α when (22.84) holds and equals α when $\theta_1 = \alpha$.

Apart from the fact that the random variable is discrete, the noteworthy feature of this cautionary example is that the range of one of the parameters is determined by α, the size of the test.

Cox (1961; 1962) considers the distribution of LR test statistics when H_0 and H_1 are entirely separate families of composite hypotheses (so that (22.5) no longer holds) and obtains some large-sample results. Pereira (1977) gives an annotated bibliography. These tests are not of controlled size: Loh (1985) proposes a method based on the bootstrap of **17.10** to remedy this deficiency. See also Kent (1986).

The relative efficiency of tests

22.25 Thus far we have been concerned with the selection of best tests (e.g. UMP, UMPU) rather than comparing the performance of two or more tests. We now seek measures of the relative efficiency of tests.

In **17.29** we examined the relative efficiency of estimators, a relatively straightforward task. By contrast, measures of test efficiency are more complex and, historically, were not considered until 1935, when the Neyman–Pearson theory was largely established. Likewise, the concept of test consistency was not introduced until the work of Wald and Wolfowitz (1940), nearly 20 years after Fisher's (1921a) development of consistent estimation; such properties only became relevant as 'inefficient' tests were explored more fully.

22.26 Comparisons between tests must take into account three factors: α, β and n. For example, if an *efficient* test requires n_1 observations to achieve power $1-\beta$ for given α, and a second test requires n_2 $(> n_1)$ observations to do the same, the ratio n_1/n_2 is a measure of the efficiency of the second test relative to the first.

Although comprehensive, in that the power function summarizes all the information about test performance for a given size α and sample n, such a measure is cumbersome, and often unavailable analytically. Furthermore, the results are often rather similar over a range of (α, β, n) values, indicating that simpler summary (asymptotic) measures may suffice.

22.27 One approach is to let sample sizes tend to infinity, as in **17.29**, and take the ratio of the powers of the tests as our measure of test efficiency. However, if the tests we are considering are both size-α consistent tests against the class of alternative hypotheses in the problem (and henceforth we shall always assume this to be so), it follows by definition that the power function of each tends to 1 as sample size increases. If we compare the tests against some fixed alternative value of θ, it follows that the efficiency thus defined will always tend to 1 as sample size increases. The suggested measure of test efficiency is therefore quite useless.

More generally, it is easy to see that consideration of the power functions of consistent tests asymptotically in n is of limited value. For instance, Wald (1941) defined an asymptotically most powerful test as one whose power function cannot be bettered as sample size tends to infinity, i.e. which is UMP asymptotically. The following example, due to Lehmann (1949), shows that one asymptotically UMP test may in fact be decidedly inferior to another such test, even asymptotically.

Example 22.8 (Relative performance of asymptotically UMP tests)

Consider again the problem, discussed in Examples 20.1 and 20.2, of testing the mean θ of a normal distribution with known variance, taken to be equal to 1. We wish to test $H_0 : \theta = \theta_0$

against the one-sided alternative $H_1 : \theta = \theta_1 > \theta_0$. In **20.17**, we saw that a UMP test of H_0 against H_1 is given by the critical region $\bar{x} \geq \theta_0 + d_\alpha/n^{1/2}$, and in Example 20.3 that its power function is

$$P_1 = G\{\Delta n^{1/2} - d_\alpha\} = 1 - G\{d_\alpha - \Delta n^{1/2}\}, \tag{22.85}$$

where $\Delta = \theta_1 - \theta_0$ and the fixed value d_α defines the size α of the test as in (20.16).

We now construct a two-tailed size-α test, rejecting H_0 when

$$\bar{x} \geq \theta_0 + d_{\alpha_2}/n^{1/2} \quad \text{or} \quad \bar{x} \leq \theta_0 - d_{\alpha_1}/n^{1/2},$$

where d_{α_1} and d_{α_2}, functions of n, may be chosen arbitrarily subject to the condition $\alpha_1 + \alpha_2 = \alpha$, which implies that d_{α_1} and d_{α_2} both exceed d_α. Equation (21.48) shows that the power function of this second test is

$$P_2 = G\{\Delta n^{1/2} - d_{\alpha_2}\} + G\{-\Delta n^{1/2} - d_{\alpha_1}\}, \tag{22.86}$$

and since G is always positive, it follows that

$$P_2 > G\{\Delta n^{1/2} - d_{\alpha_2}\} = 1 - G\{d_{\alpha_2} - \Delta n^{1/2}\}. \tag{22.87}$$

Since the first test is UMP, we have, from (22.85) and (22.87),

$$G\{d_{\alpha_2} - \Delta n^{1/2}\} - G\{d_\alpha - \Delta n^{1/2}\} > P_1 - P_2 \geq 0. \tag{22.88}$$

It is easily seen that the difference between $G\{x\}$ and $G\{y\}$ for fixed $x - y$ is maximized when x and y are symmetrically placed about zero; i.e. when $x = -y$, so that

$$G\{\tfrac{1}{2}(x-y)\} - G\{-\tfrac{1}{2}(x-y)\} \geq G\{x\} - G\{y\}. \tag{22.89}$$

Applying (22.89) to (22.88), we have

$$G\{\tfrac{1}{2}(d_{\alpha_2} - d_\alpha)\} - G\{-\tfrac{1}{2}(d_{\alpha_2} - d_\alpha)\} > P_1 - P_2 \geq 0. \tag{22.90}$$

Thus if we choose d_{α_2} for each n so that

$$\lim_{n\to\infty} d_{\alpha_2} = d_\alpha, \tag{22.91}$$

the left-hand side of (22.89) will tend to zero, whence $P_1 - P_2$ will tend to zero uniformly in Δ. The two-tailed test will therefore be asymptotically UMP.

Now consider the ratio of Type II errors of the tests. From (22.85) and (22.86), we have

$$\frac{1 - P_2}{1 - P_1} = \frac{G\{d_{\alpha_2} - \Delta n^{1/2}\} - G\{-d_{\alpha_1} - \Delta n^{1/2}\}}{G\{d_\alpha - \Delta n^{1/2}\}}. \tag{22.92}$$

As $n^{1/2} \to \infty$, both the numerator and denominator of (22.92) tend to zero. Using l'Hôpital's rule, we find, using a prime to denote differentiation with respect to $n^{1/2}$ and writing g for the normal p.d.f.,

$$\lim_{n^{1/2}\to\infty}\frac{1-P_2}{1-P_1}=\lim_{n^{1/2}\to\infty}\left[\frac{(d'_{\alpha_2}-\Delta)g\{d_{\alpha_2}-\Delta n^{1/2}\}}{-\Delta g\{d_\alpha-\Delta n^{1/2}\}}+\frac{(d'_{\alpha_1}+\Delta)g\{-d_{\alpha_1}-\Delta n^{1/2}\}}{-\Delta g\{d_\alpha-\Delta n^{1/2}\}}\right]. \tag{22.93}$$

Now (22.91) implies that $d_{\alpha_1}\to\infty$ with n, and therefore that the second term on the right of (22.93) tends to zero: (22.91) also implies that the first term on the right of (22.93) will tend to infinity if

$$\begin{aligned}\lim_{n\to\infty}\frac{-d'_{\alpha_2}g\{d_{\alpha_2}-\Delta n^{1/2}\}}{g\{d_\alpha-\Delta n^{1/2}\}}&=\lim_{n\to\infty}-d'_{\alpha_2}\frac{\exp\{-\frac{1}{2}(d_{\alpha_2}-\Delta n^{1/2})^2\}}{\exp\{-\frac{1}{2}(d_\alpha-\Delta n^{1/2})^2\}}\\&=\lim_{n\to\infty}-d'_{\alpha_2}\exp\{-\tfrac{1}{2}(d^2_{\alpha_2}-d^2_\alpha)+\Delta n^{1/2}(d_{\alpha_2}-d_\alpha)\}\end{aligned} \tag{22.94}$$

does so. By (22.91), the first term in the exponent on the right of (22.94) tends to zero. If we put

$$d_{\alpha_2}=d_\alpha+n^{-\delta},\qquad 0<\delta<\tfrac{1}{2}, \tag{22.95}$$

(22.91) is satisfied and (22.94) tends to infinity with n. Thus, although both tests are asymptotically UMP, the ratio of Type II errors (22.92) tends to infinity with n. It is clear, therefore, that the criterion of being asymptotically UMP is not very selective.

Asymptotic relative efficiency

22.28 In order to obtain a useful asymptotic measure of test efficiency from the relative efficiency, we consider the limiting relative efficiency of tests against a sequence of alternative hypotheses in which θ approaches the value tested, θ_0, as n increases. We do this in order to avoid forcing the powers of tests to be nearly 1, as in **22.27**. This type of alternative was first investigated by Pitman (1948), whose work was generalized by Noether (1955). Other types of limiting process on relative efficiency are considered by Dixon (1953) and Hodges and Lehmann (1956).

Let t_1 and t_2 be consistent test statistics for the hypothesis $H_0:\theta=\theta_0$ against the one-sided alternative $H_1:\theta>\theta_0$. We assume for the moment that t_1 and t_2 are asymptotically normally distributed whatever the value of θ – we shall relax this restriction in **22.37–38**. For brevity, we shall write

$$\left.\begin{aligned}E(t_i|H_j)&=E_{ij},\\ \operatorname{var}(t_i|H_j)&=D^2_{ij},\\ E_i^{(r)}(\theta)&=\frac{\partial^r}{\partial\theta^r}E_{i1},\\ D^{(r)}_{i1}&=\frac{\partial^r}{\partial\theta^r}D_{i1},\qquad D^{(r)}_{i0}=D^{(r)}_{i1}(\theta_0),\end{aligned}\right\}\qquad i=1,2;\ j=0,1.$$

Large-sample size-α tests are defined by the critical regions

$$t_i > E_{i0} + \lambda_\alpha D_{i0} \tag{22.96}$$

(the sign of t_i being changed if necessary to make the region of this form), where λ_α is the normal deviate defined by $G\{-\lambda_\alpha\} = \alpha$, G being the standardized normal d.f. as before. As in Example 20.3, the asymptotic power function of t_i is

$$P_i(\theta) = G\{[E_{i1} - (E_{i0} + \lambda_\alpha D_{i0})]/D_{i1}\}. \tag{22.97}$$

Writing $u_i(\theta, \lambda_\alpha)$ for the argument of G in (22.97), we form the Taylor expansion of $E_{i1} - E_{i0}$, obtaining

$$u_i(\theta, \lambda_\alpha) = \left[E_i^{(m_i)}(\theta_i^*) \frac{(\theta - \theta_0)^{m_i}}{m_i!} - \lambda_\alpha D_{i0} \right] \Big/ D_{i1}, \tag{22.98}$$

where $\theta_0 < \theta_i^* < \theta$ and m_i is the first non-zero derivative at θ_0, i.e. m_i, is defined by

$$\begin{aligned} E_i^{(r)}(\theta_0) &= 0, \qquad r = 1, 2, \ldots, m_i - 1, \\ E_i^{(m_i)}(\theta_0) &\neq 0. \end{aligned} \tag{22.99}$$

In order to define the alternative hypothesis, we assume that, as $n \to \infty$,

$$R_i = [E_i^{(m_i)}(\theta_0)/D_{i0}] \sim c_i n^{m_i \delta_i}. \tag{22.100}$$

Expression (22.100) defines the constants $\delta_i > 0$ and c_i. Now consider the sequences of alternatives, approaching θ_0 as $n \to \infty$,

$$\theta = \theta_0 + \frac{k_i}{n^{\delta_i}}, \tag{22.101}$$

where k_i is an arbitrary positive constant. If the regularity conditions

$$\lim_{n \to \infty} \frac{E_i^{(m_i)}(\theta)}{E_i^{(m_i)}(\theta_0)} = 1, \qquad \lim_{n \to \infty} \frac{D_{i1}}{D_{i0}} = 1 \tag{22.102}$$

are satisfied, (22.100)–(22.102) and (22.98) reduce to

$$u_i(\theta, \lambda_\alpha) = \frac{c_i k_i^{m_i}}{m_i!} - \lambda_\alpha, \tag{22.103}$$

and the asymptotic powers of the tests are $G\{u_i\}$ from (22.97).

22.29 If the two tests are to have equal power against the same sequence of alternatives for any fixed α, we must have, from (22.101) and (22.103),

$$\frac{k_1}{n_1^{\delta_1}} = \frac{k_2}{n_2^{\delta_2}} \tag{22.104}$$

and

$$\frac{c_1 k_1^{m_1}}{m_1!} = \frac{c_2 k_2^{m_2}}{m_2!}, \tag{22.105}$$

where n_1 and n_2 are the sample sizes upon which t_1 and t_2 are based. We combine (22.104) and (22.105) into

$$\frac{n_1^{\delta_1}}{n_2^{\delta_2}} = \left(\frac{c_2}{c_1}\frac{m_1!}{m_2!}k_2^{m_2-m_1}\right)^{1/m_1}. \tag{22.106}$$

The right-hand side of (22.106) is a positive constant. Thus if we let $n_1, n_2 \to \infty$, the ratio n_1/n_2 will tend to a constant if and only if $\delta_1 = \delta_2$. If $\delta_1 > \delta_2$, we must have $n_1/n_2 \to 0$, while if $\delta_1 < \delta_2$ we have $n_1/n_2 \to \infty$. If we define the *asymptotic relative efficiency* (ARE) of t_2 compared to t_1 as

$$A_{21} = \lim \frac{n_1}{n_2}, \tag{22.107}$$

we therefore have the result

$$A_{21} = 0, \qquad \delta_1 > \delta_2. \tag{22.108}$$

Thus to compare two tests by the criterion of ARE, we first compare their values of δ: if one has a smaller δ than the other, it has ARE of zero compared to the other. The value of δ plays the same role here as the order of magnitude of the variance plays in measuring efficiency of estimation (cf. **17.29**).

We may now confine ourselves to the case $\delta_1 = \delta_2 = \delta$. Equations (22.106) and (22.107) then give

$$A_{21} = \lim \frac{n_1}{n_2} = \left(\frac{c_2}{c_1}\frac{m_1!}{m_2!}k_2^{m_2-m_1}\right)^{1/m_1\delta} \tag{22.109}$$

If, in addition,

$$m_1 = m_2 = m, \tag{22.110}$$

(22.109) reduces to

$$A_{21} = \left(\frac{c_2}{c_1}\right)^{1/m\delta},$$

which on using (22.100) becomes

$$A_{21} = \lim_{n\to\infty}\left\{\frac{E_2^{(m)}(\theta_0)/D_{20}}{E_1^{(m)}(\theta_0)/D_{10}}\right\}^{1/(m\delta)}. \tag{22.111}$$

Expression (22.111) is simple to evaluate in most cases. Most commonly, $\delta = \frac{1}{2}$ (corresponding to an estimation variance of order n^{-1}) and $m = 1$. For an interpretation of the value of m, see **22.33**.

In passing, we may note that if $m_2 \neq m_1$, (22.109) is indeterminate, depending as it does on the arbitrary constant k_2. We therefore see that tests with equal values of δ do not have the same ARE against all sequences of alternatives (22.101) unless they also have equal values of m. We comment on the reasons for this in **22.33**.

22.30 If we wish to test H_0 against the two-sided $H_1 : \theta \neq \theta_0$, our results for the ARE are unaffected if we use 'equal-tails' critical regions of the form

$$t_i > E_{i0} + \lambda_{\alpha/2} D_{i0} \quad \text{or} \quad t_i > E_{i0} - \lambda_{\alpha/2} D_{i0},$$

for the asymptotic power functions (22.97) are replaced by

$$Q_i(\theta) = G\{u_i(\theta, \lambda_{\alpha/2})\} + 1 - G\{u_i(\theta, -\lambda_{\alpha/2})\}, \tag{22.112}$$

and $Q_1 = Q_2$ against the alternative (22.101) (where k_i need no longer be positive) if (22.104) and (22.105) hold, as before. Konijn (1956, 1958) gives a more general treatment of two-sided tests, which need not necessarily be 'equal-tails' tests.

Example 22.9 (Relative efficiency of median to mean for the normal distribution)

Let us compare the sample median $\tilde{x}$ with the UMP sample mean $\bar{x}$ in testing the mean θ of a normal distribution with known variance σ^2. Both statistics are asymptotically normally distributed. We know that

$$E(\bar{x}) = \theta, \qquad D^2(\bar{x}|\theta) = \sigma^2/n$$

and $\tilde{x}$ is a consistent estimator of θ, symmetrically distributed about θ, with

$$E(\tilde{x}) = \theta, \qquad D^2(\tilde{x}|\theta) \sim \pi\sigma^2/(2n)$$

(cf. Example 10.7). Thus we have

$$E'(\theta_0) = 1$$

for both tests, so that $m_1 = m_2 = 1$, while from (22.100), $\delta_1 = \delta_2 = \frac{1}{2}$. Thus, from (22.111),

$$A_{\tilde{x},\bar{x}} = \lim_{n\to\infty} \left\{ \frac{1/(\pi\sigma^2/2n)^{1/2}}{1/(\sigma^2/n)^{1/2}} \right\}^2 = \frac{2}{\pi}.$$

This is precisely the result we obtained in Example 17.13 for the efficiency of $\tilde{x}$ in estimating θ. We shall see in **22.36** that this is a special case of a general relationship between estimating efficiency and ARE for tests.

ARE and derivatives of power functions

22.31 The nature of the sequence of alternative hypotheses (22.101), which approaches θ_0 as $n \to \infty$, makes it clear that the ARE is in some way related to the behaviour, near θ_0, of the power functions of the tests being compared. We shall make this relationship more precise by showing that, under certain conditions, the ARE is a simple function of the ratio of derivatives of the power functions.

We first treat the case of the one-sided H_1 discussed in **22.28–29**, where the power functions of the tests are asymptotically given by (22.97), which we write, as before,

$$P_i(\theta) = G\{u_i(\theta, \lambda_\alpha)\}. \tag{22.113}$$

Differentiating with respect to θ, we have

$$P_i'(\theta) = g\{u_i\}u_i'(\theta, \lambda_\alpha), \tag{22.114}$$

where g is the normal density function. From (22.97) we find

$$u_i'(\theta, \lambda_\alpha) = \frac{E_{i1}'}{D_{i1}} - \frac{D_{i1}'}{D_{i1}^2}(E_{i1} - E_{i0} - \lambda_\alpha D_{i0}). \tag{22.115}$$

As $n \to \infty$, we find, using (22.102), and the further regularity conditions

$$\lim_{n\to\infty} \frac{D_{i1}'}{D_{i0}'} = 1, \qquad \lim_{n\to\infty} \frac{E_{i1}}{E_{i0}} = 1,$$

that (22.115) becomes

$$u_i'(\theta, \lambda_\alpha) = \frac{E_i'(\theta_0)}{D_{i0}} + \frac{D_{i0}'}{D_{i0}}\lambda_\alpha, \tag{22.116}$$

so that if $m_i = 1$ in (22.99) and if

$$\lim_{n\to\infty} \frac{D_{i0}'}{E_i'(\theta_0)} = 0, \tag{22.117}$$

then (22.116) reduces at θ_0 to

$$u_i'(\theta, \lambda_\alpha) \sim E_i'(\theta_0)/D_{i0}. \tag{22.118}$$

Since, from (22.97),

$$g\{u_i(\theta_0, \lambda_\alpha)\} = g\{-\lambda_\alpha\}, \tag{22.119}$$

(22.114) becomes, on substituting (22.118) and (22.119),

$$P_i'(\theta_0) = P_i'(\theta_0, \lambda_\alpha) \sim g\{-\lambda_\alpha\}E_i'(\theta_0)/D_{i0}. \tag{22.120}$$

Remembering that $m_i = 1$, we therefore have, from (22.120) and (22.110),

$$\lim_{n\to\infty} \frac{P_2'(\theta_0)}{P_1'(\theta_0)} = A_{21}^{\delta}, \qquad m_1 = m_2 = 1, \tag{22.121}$$

so that the limiting ratio of the first derivatives of the power functions of the tests at θ_0 is simply the ARE raised to the power δ (commonly $\frac{1}{2}$). Thus if we were to use this ratio as a criterion of asymptotic efficiency of tests, we should get precisely the same results as by using the ARE. This criterion was, in fact, proposed (under the name 'asymptotic local efficiency') by Blomqvist (1950).

22.32 If $m_i > 1$, i.e. $E_i(\theta_0) = 0$, (22.120) is zero to our order of approximation and the result of **22.31** is of no use. The differentiation process has to be taken further to yield useful results.

From (22.114), we obtain

$$P_i''(\theta) = \frac{\partial g\{u_i\}}{\partial u_i}[u_i'(\theta, \lambda_\alpha)]^2 + g\{u_i\}u_i''(\theta, \lambda_\alpha). \tag{22.122}$$

From (22.115),

$$u_i''(\theta,\lambda_\alpha)=\frac{E_{i1}''}{D_{i1}}-\frac{2E_{i1}'D_{i1}'}{D_{i1}^2}-(E_{i1}-E_{i0}-\lambda_\alpha D_{i0})\left[\frac{D_{i1}''}{D_{i1}^2}-\frac{2(D_{i1}')^2}{D_{i1}^3}\right]. \tag{22.123}$$

If (22.102) holds with $m_i=2$ along with the regularity conditions below (22.115) and

$$\lim_{n\to\infty}E_{i1}'=E_i'(\theta_0)=0,\qquad \lim_{n\to\infty}\frac{D_{i1}'}{D_{i0}'}=1,\qquad \lim_{n\to\infty}\frac{D_{i1}''}{D_{i0}''}=1,\qquad \lim_{n\to\infty}\frac{E_{i1}}{E_{i0}}=1, \tag{22.124}$$

then (22.123) gives

$$u_i''(\theta_0,\lambda_\alpha)=\frac{E_i''(\theta_0)}{D_{i0}}+\lambda_\alpha\left[\frac{D_{i0}''}{D_{i0}}-2\left(\frac{D_{i0}'}{D_{i0}}\right)^2\right]. \tag{22.125}$$

Instead of (22.117), we now assume the conditions

$$\lim_{n\to\infty}\frac{D_{i0}''}{E_i''(\theta_0)}=0,\qquad \lim_{n\to\infty}\frac{(D_{i0}')^2}{D_{i0}E_i''(\theta_0)}=0. \tag{22.126}$$

Conditions (22.126) reduce (22.125) to

$$u_i''(\theta_0,\lambda_\alpha)\sim E_i''(\theta_0)/D_{i0}. \tag{22.127}$$

Returning now to (22.122), we see that since

$$\frac{\partial g\{u_i\}}{\partial u_i}=-u_i g\{u_i\},$$

we have, using (22.116), (22.119) and (22.127) in (22.122),

$$P_i''(\theta_0)\sim g\{-\lambda_\alpha\}\left\{\lambda_\alpha\left[\frac{E_i'(\theta_0)}{D_{i0}}+\frac{D_{i0}'}{D_{i0}}\lambda_\alpha\right]^2+\frac{E_i''(\theta_0)}{D_{i0}}\right\}. \tag{22.128}$$

Since we are considering the case $m_i=2$ here, the term in $E_i'(\theta_0)$ is zero, and from the second condition of (22.120), (22.128) may finally be written

$$P_i''(\theta_0)\sim g\{-\lambda_\alpha\}E_i''(\theta_0)/D_{i0}, \tag{22.129}$$

whence, with $m=2$, (22.111) gives

$$\lim_{n\to\infty}\frac{P_2''(\theta_0)}{P_1''(\theta_0)}=A_{21}^{2\delta} \tag{22.130}$$

for the limiting ratio of the second derivatives.

We may express (22.121) and (22.130) concisely by the statement that for $m=1,2$, the ratio of the mth derivatives of the power functions of one-sided tests is asymptotically equal to the ARE raised to the power $m\delta$.

If, instead of (22.117) and (22.126), we impose the stronger conditions

$$\lim_{n\to\infty} D'_{i0}/D_{i0} = 0, \qquad \lim_{n\to\infty} D''_{i0}/D_{i0} = 0, \tag{22.131}$$

we arrive at the same results; (22.131) may be easier to verify in particular cases.

The interpretation of the value of *m*

22.33 We now discuss the general conditions under which m will take the value 1 or 2. Consider again the asymptotic power function (22.97) for a one-sided alternative $H_1 : \theta > \theta_0$ and a one-tailed test (22.96). For brevity, we drop the subscript i in this section. If $\theta \to \theta_0$ and $D_1 \to D_0$ by (22.102), (22.97) becomes

$$P(\theta) = G\left\{\frac{E_1 - E_0}{D_0} - \lambda_\alpha\right\},$$

a monotone increasing function of $E_1 - E_0$.

If $E_1 - E_0$ is an increasing function of $\theta - \theta_0$, $P(\theta) \to 0$ as $\theta \to -\infty$ (which implies that the other 'tail' of the distribution of the test statistic would be used as a critical region if $\theta < \theta_0$). If $E'(\theta_0)$ exists, it is non-zero and $m = 1$, and $P'(\theta_0) \neq 0$ also, by (22.120).

If, on the other hand, $E_1 - E_0$ is an even function of $\theta - \theta_0$ (which implies that the same 'tail' would be used as critical region whatever the sign of $\theta - \theta_0$), and an increasing function of $|\theta - \theta_0|$, and $E'(\theta_0)$ exists, it must under regularity conditions equal zero, and $m > 1$ – in practice, we find $m = 2$. By (22.120), $P'(\theta_0) = 0$ also to this order of approximation.

We are now in a position to see why, as remarked at the end of **22.29**, the ARE is not useful in comparing tests with differing values of m, which in practice are 1 and 2. For we are then comparing tests whose power functions behave essentially differently at θ_0, one having a regular minimum there and the other not. The indeterminacy of (22.109) in such circumstances is not really surprising.

Example 22.10 (Relative efficiency of tests for normal mean)

Consider the problem of testing $H_0 : \theta = \theta_0$ for a normal distribution with mean θ and variance 1. The pair of one-tailed tests based on the sample mean $\bar{x}$ are UMP (cf. **20.17**), the upper or lower tail being selected according to whether H_1 is $\theta > \theta_0$ or $\theta < \theta_0$. From Example 22.9, $\delta = \frac{1}{2}$ and $m = 1$ for $\bar{x}$.

We could also use as a test statistic

$$S = \sum_{i=1}^{n} (x_i - \theta_0)^2.$$

S has a non-central chi-square distribution with n degrees of freedom and non-central parameter $n(\theta - \theta_0)^2$, so that (cf. Exercise 22.1)

$$E(S|\theta) = n\{1 + (\theta - \theta_0)^2\},$$
$$D^2(S|\theta_0) = 2n,$$

and as $n \to \infty$, S is asymptotically normally distributed. We have $E'(\theta) = 2n(\theta - \theta_0)$, $E'(\theta_0) = 0$, $E''(\theta) = 2n = E''(\theta_0)$, so that $m = 2$ and

$$\frac{E''(\theta_0)}{D_0} = \frac{2n}{(2n)^{1/2}} = (2n)^{1/2}.$$

From (22.100), since $m = 2$, $\delta = \frac{1}{4}$. Since $\delta = \frac{1}{2}$ for $\bar{x}$, the ARE of S compared to $\bar{x}$ is zero by (22.108). The critical region for S consists of the upper tail, whatever the value of θ.

22.34 We now turn to the case of the two-sided alternative $H_1 : \theta \neq \theta_0$. The power function of the 'equal-tails' test is given asymptotically by (22.112). Its derivative at θ_0 is

$$Q_i'(\theta_0) = P_i'(\theta_0, \lambda_{\alpha/2}) - P_i'(\theta_0, -\lambda_{\alpha/2}), \tag{22.132}$$

where P_i' is given by (22.120) if $m_i = 1$ and (22.117) or (22.131) holds. Since $g\{-\lambda_\alpha\}$ in (22.120) is an even function of λ_α, (22.132) immediately gives the asymptotic result

$$Q_i'(\theta_0) \sim 0,$$

so that the slope of the power function at θ_0 is asymptotically zero. This result is also implied (under regularity conditions) by the remark in **22.17** concerning the asymptotic unbiasedness of consistent tests.

The second derivative of the power function (22.112) is

$$Q_i''(\theta_0) = P_i''(\theta_0, \lambda_{\alpha/2}) - P_i''(\theta_0, -\lambda_{\alpha/2}). \tag{22.133}$$

We have evaluated P_i'' at (22.128) where we had $m_i = 2$. Expression (22.128) still holds for $m_i = 1$ if we strengthen the first condition in (22.131) to

$$D_{i0}'/D_{i0} = o(n^{-\delta}), \tag{22.134}$$

for then by (22.100) the second term on the right of (22.123) may be neglected and we obtain (22.128) as before. Substituted into (22.133), it gives

$$Q_i''(\theta_0) \sim 2\lambda_{\alpha/2}\, g\{-\lambda_{\alpha/2}\}\left\{\left(\frac{E_i'(\theta_0)}{D_{i0}}\right)^2 + \left(\frac{D_{i0}'}{D_{i0}}\lambda_\alpha\right)^2\right\},$$

and (22.134) reduces this to

$$Q_i''(\theta_0) \sim 2\lambda_{\alpha/2}\, g\{-\lambda_{\alpha/2}\}\left(\frac{E_i'(\theta_0)}{D_{i0}}\right)^2. \tag{22.135}$$

In this case, therefore, (22.111) and (22.135) give

$$\frac{Q_2''(\theta_0)}{Q_1''(\theta_0)} = A_{21}^{2\delta}. \tag{22.136}$$

Thus for $m = 1$, the asymptotic ratio of second derivatives of the power functions of two-sided tests is exactly that given by (22.130) for one-sided tests when $m = 2$, and exactly the square of the one-sided test result for $m = 1$ at (22.121).

The case $m = 2$ does not seem of much importance for two-tailed tests: the remarks in **22.33** suggest that where $m = 2$ a one-tailed test would often be used even against a two-sided H_1.

Example 22.11 (ARE of two-sided tests using the median)

Reverting to Example 22.9, we saw that both tests have $\delta = \frac{1}{2}$, $m = 1$ and $E'(\theta_0) = 1$. Since the variance of each statistic is independent of θ, at least asymptotically, we see that (22.117) and (22.134) are satisfied and, the regularity conditions being satisfied, it follows from (22.121) that for one-sided tests

$$\lim_{n\to\infty} \frac{P'_{\tilde{x}}(\theta_0)}{P'_{\bar{x}}(\theta_0)} = A^{1/2}_{\tilde{x},\bar{x}} = \left(\frac{2}{\pi}\right)^{1/2},$$

while for two-sided tests, from (22.136),

$$\lim_{n\to\infty} \frac{Q''_{\tilde{x}}(\theta_0)}{Q''_{\bar{x}}(\theta_0)} = A_{\tilde{x},\bar{x}} = \frac{2}{\pi}.$$

The maximum power loss and the ARE

22.35 Although the ARE of tests essentially reflects their power properties in the neighbourhood of θ_0, it does have some implications for the asymptotic power function as a whole, at least for the case $m = 1$, to which we now confine attention.

The power function $P_i(\theta)$ of a one-sided test is $G\{u_i(\theta)\}$, where $u_i(\theta)$, given in (22.98), is asymptotically equal, under regularity conditions (22.102), to

$$u_i(\theta) = \frac{E'_i(\theta_0)}{D_{i0}}(\theta - \theta_0) - \lambda_\alpha, \tag{22.137}$$

when $m_i = 1$. Thus $u_i(0)$ is asymptotically linear in θ. If we write $R_i = E'_i(\theta_0)/D_{i0}$ as in (22.100), we may write the difference between two such power functions as

$$d(\theta) = P_2(\theta) - P_1(\theta) = G\{(\theta - \theta_0)R_2 - \lambda_\alpha\} - G\left\{(\theta - \theta_0)R_2\frac{R_1}{R_2} - \lambda_\alpha\right\}, \tag{22.138}$$

where we assume $R_2 > R_1$ without loss of generality. Consider the behaviour of $d(\theta)$ as a function of θ. When $\theta = \theta_0$, $d = 0$, and again as θ tends to infinity P_1 and P_2 both tend to 1 and d to zero. The maximum value of $d(\theta)$ depends only on the ratio R_1/R_2, for although R_2 appears in the right-hand side of (22.138) it is always the coefficient of $\theta - \theta_0$, which is being varied from 0 to ∞, so that $R_2(\theta - \theta_0)$ also goes from 0 to ∞ whatever the value of R_2. We therefore write $\Delta = R_2(\theta - \theta_0)$ in (22.138), obtaining

$$d(\Delta) = G\{\Delta - \lambda_\alpha\} - G\left\{\Delta\frac{R_1}{R_2} - \lambda_\alpha\right\}. \tag{22.139}$$

The first derivative of (22.139) with respect to Δ is

$$d'(\Delta) = g\{\Delta - \lambda_\alpha\} - \frac{R_1}{R_2}g\left\{\Delta\frac{R_1}{R_2} - \lambda_\alpha\right\},$$

and if this is equated to zero, we have

$$\frac{R_1}{R_2} = \frac{g\{\Delta - \lambda_\alpha\}}{g\{\Delta\frac{R_1}{R_2} - \lambda_\alpha\}} = \exp\left\{-\tfrac{1}{2}(\Delta - \lambda_\alpha)^2 + \tfrac{1}{2}\left(\Delta\frac{R_1}{R_2} - \lambda_\alpha\right)^2\right\}$$

$$= \exp\left\{-\tfrac{1}{2}\Delta^2\left(1 - \frac{R_1^2}{R_2^2}\right) + \lambda_\alpha\Delta\left(1 - \frac{R_1}{R_2}\right)\right\}. \qquad (22.140)$$

Equation (22.140) is a quadratic in Δ, whose only positive root is

$$\Delta = \frac{\lambda_\alpha + \left\{\lambda_\alpha^2 + 2\left[\left(1 + \frac{R_1}{R_2}\right)/\left(1 - \frac{R_1}{R_2}\right)\right]\log\frac{R_2}{R_1}\right\}^{1/2}}{\left(1 + \frac{R_1}{R_2}\right)}. \qquad (22.141)$$

This is the value at which (22.139) is maximized. Consider, for example, the case $\alpha = 0.05$ ($\lambda_\alpha = 1.645$) and $R_1/R_2 = 0.5$: (22.141) gives

$$\Delta = \frac{1.645 + \{1.645^2 + 6\log_e 2\}^{1/2}}{1.5} = 2.85.$$

Equation (22.139) then gives, using the normal d.f.,

$$P_2 = G\{2.85 - 1.64\} = G\{1.21\} = 0.89,$$
$$P_1 = G\{1.42 - 1.64\} = G\{-0.22\} = 0.41.$$

Cox and Stuart (1955) gave values of P_2 and P_1 at the point of maximum difference, obtained by the graphical equivalent of the above method, for a range of values of α and R_1/R_2. Their table is reproduced as Table 22.1, our worked example above corresponding to one of the entries.

Table 22.1 Asymptotic percentage powers of tests at the point of greatest difference (Cox and Stuart, 1955)

α	0.10		0.05		0.01		0.001	
R_1/R_2	P_1	P_2	P_1	P_2	P_1	P_2	P_1	P_2
0.9	67	73	63	71	49	60	54	67
0.8	61	74	56	72	49	71	43	72
0.7	59	80	51	77	42	77	39	83
0.6	54	84	47	84	39	86	29	87
0.5	48	88	41	89	30	90	20	93
0.3	35	96	27	96	14	97	7	99

It will be seen from the table that as α decreases for fixed R_1/R_2, the maximum difference between the asymptotic power functions increases steadily – it can, in fact, be made as near to 1 as desired by taking α small enough. Similarly, for fixed α, the maximum difference increases steadily as R_1/R_2 falls.

The practical consequence of the table is that if R_1/R_2 is 0.9 or more, the loss of power *along the whole course* of the asymptotic power function will not exceed 0.08 for $\alpha = 0.05$,

0.11 for $\alpha = 0.01$, and 0.13 for $\alpha = 0.001$, the most commonly used test sizes. Since R_1/R_2 is, from (22.120), the ratio of first derivatives of the power functions, we have from (22.121) that $(R_1/R_2)^{1/\delta} = A_{12}$, where δ is commonly $\frac{1}{2}$, and thus the ARE needs to be $(0.9)^{1/\delta}$ for the statements above to be true.

ARE and estimating efficiency

22.36 There is a simple connection between the ARE and estimating efficiency. If we have two consistent test statistics t_i as before, we define functions f_i, independent of n, such that the statistics

$$T_i = f_i(t_i) \tag{22.142}$$

are consistent *estimators* of θ. If we write

$$\theta = f_i(\tau_i), \tag{22.143}$$

it follows from (22.142) that since $T_i \to \theta$ in probability, $t_i \to \tau_i$, and $E(t_i)$ if it exists also tends to τ_i. Expanding (22.142) about τ_i by Taylor's theorem, we have, using (22.143),

$$T_i = \theta + (t_i - \tau_i)\left[\frac{\partial f(t_i)}{\partial \tau_i}\right]_{t_i = t_i^*} \tag{22.144}$$

where t_i^*, intermediate in value between t_i and τ_i, tends to τ_i as n increases. Thus (22.144) may be written

$$T_i - \theta \sim (t_i - \tau_i)\left[\frac{\partial \theta}{\partial E(t_i)}\right],$$

whence

$$\operatorname{var} T_i \sim \operatorname{var} t_i \bigg/ \left(\frac{\partial E(t_i)}{\partial \theta}\right)^2. \tag{22.145}$$

If 2δ is the order of magnitude in n of the variances of the T_i, the estimating efficiency of T_2 compared to T_1 is, by (17.65) and (22.145),

$$\lim_{n\to\infty}\left(\frac{\operatorname{var} T_1}{\operatorname{var} T_2}\right)^{1/(2\delta)} = \left[\frac{\{\partial E(t_2)/\partial\theta\}^2/\operatorname{var} t_2}{\{\partial E(t_1)/\partial\theta\}^2/\operatorname{var} t_1}\right]^{1/(2\delta)}. \tag{22.146}$$

At θ_0, (22.146) is precisely equal to the ARE (22.111) when $m_i = 1$. Thus the ARE essentially gives the relative estimating efficiencies of transformations of the test statistics that are consistent estimators of the parameter concerned. But this correspondence is a local one: in **20.15** we saw that the connection between estimating efficiency and power is not strong in general. It follows at once that tests based upon efficient estimators have maximum ARE and (from **22.31–34**) that the derivatives of their power functions at θ_0 are maximized. Equations (22.146) and (17.61) also imply that if T_1 is efficient, $A_{21} = \{\rho(T_1, T_2)\}^{1/\delta}$. A more general result in terms of $\rho(t_1, t_2)$ is given in Exercise 22.29. From this argument we also see that the ARE is directly related to the relative (squared) lengths of the asymptotic confidence intervals.

Our result explains the finding given in Example 22.9, that the ARE of the sample median, compared to the sample mean, in testing the mean of a normal distribution has exactly the same value as its estimating efficiency for that parameter.

Non-normal cases

22.37 From **22.28** onwards, we have confined ourselves to the case of asymptotically normally distributed test statistics. However, examination of **22.28–30** will show that in deriving the ARE we made no specific use of the normality assumption. We were concerned to establish the conditions under which the arguments u_i, of the power functions $G\{u_i\}$ in (22.103), would be equal against the sequence of alternatives (22.101). G played no role in the discussion other than of ensuring that the asymptotic power functions were of the same form, and we need only require that G is a regularly behaved d.f.

It follows that if two tests have asymptotic power functions of any two-parameter form G, only one of whose parameters is a function of θ, the results of **22.28–30** will hold, for (22.101) will fix this parameter and u_i in (22.103) then determines the other. Given the form G, the critical region for one-tailed tests can always be put in the form (22.96), where λ_α is more generally interpreted as the multiple of D_{i0} required to make (22.96) a size-α critical region.

22.38 The only important limiting distributions other than the normal are the non-central χ^2 distributions whose properties were discussed in **22.4–5**. Suppose that for the hypothesis H_0 : $\theta = \theta_0$ we have two test statistics t_i with such distributions, the degrees of freedom being ν_i (independent of θ), and the non-central parameters $\lambda_i(\theta)$, where $\lambda_i(\theta_0) = 0$, so that the χ^2 distributions are central when H_0 holds. We have (cf. Exercise 22.1)

$$E_{i1} = \nu_i + \lambda_i(\theta), \qquad D_{i0}^2 = 2\nu_i. \tag{22.147}$$

All the results of **22.28–29** for one-sided tests therefore hold for the comparison of test statistics distributed as non-central χ^2 (central when H_0 holds) with degrees of freedom independent of θ. In particular, when $\delta_1 = \delta_2 = \delta$ and $m_1 = m_2 = m$, (22.147) substituted into (22.111) gives

$$A_{21} = \lim_{n\to\infty} \left\{ \frac{\lambda_2^{(m)}(\theta_0)/\nu_2^{1/2}}{\lambda_1^{(m)}(\theta_0)/\nu_1^{1/2}} \right\}^{1/(m\delta)}. \tag{22.148}$$

A different derivation of this result is given by Hannan (1956).

Rothe (1981) gives a generalization.

Bahadur efficiency

22.39 Bahadur (1967) reviews results (largely his own) on the comparison of tests by means of the rate of convergence to zero of the minimum size of critical region that includes the observed value of the test statistic. Under conditions given by Lambert and Hall (1982), the distribution of this 'tail area' is asymptotically lognormal. We sketch only the simple case when the standardized test statistic $z = n^{1/2}(t_n - b)$ has a limiting standardized normal d.f. $F(z)$. By (5.105),

$$1 - F(z) \doteq F'(z)/z$$

with

$$F'(z) = (2\pi)^{-1/2} \exp(-\tfrac{1}{2}z^2).$$

Consider

$$\begin{aligned} u_n &= \log[n^{1/2}\{1 - F(n^{1/2}t_n)\}] \\ &\doteq \log\{F'(n^{1/2}t_n)/t_n\} \\ &\doteq -\tfrac{1}{2}nt_n^2 - \log t_n - \tfrac{1}{2}\log(2\pi) \doteq -\tfrac{1}{2}nt_n^2. \end{aligned}$$

Now nt_n^2 tends to $\chi'^2(1, nb^2)$, so from Exercise 22.1 it has mean $1+nb^2$ and variance $2(1+2nb^2)$, and it tends to normality by **22.5** as n increases. Thus

$$n^{-1/2}u_n \doteq -\tfrac{1}{2}n^{1/2}t_n^2 \sim N(-\tfrac{1}{2}n^{1/2}b^2, b^2),$$

and finally we see that the tail area $\{1 - F(n^{1/2}t_n)\}$ asymptotically behaves as a multiple of a lognormal variable.

Dempster and Schatzoff (1965) use the expected value of this tail area as a criterion for choosing between tests. See also Joiner (1969), who compares these with other methods including that of Exercise 22.30, and Sievers (1969). Groeneboom and Oosterhoff (1981) show that Bahadur's efficiency measure is essentially the limit of the relative efficiency n_1/n_2 in **22.26** as size $\alpha \to 0$ with H_1 fixed (instead of keeping α fixed and letting $H_1 \to H_0$ as in the derivation of Pitman's ARE in **22.28**) and find it a worse approximation to relative efficiency in moderate-sized samples than is the ARE. Wieand (1976) gives conditions under which the limiting approximate Bahadur efficiency, as $H_1 \to H_0$, equals the limiting ARE as size $\alpha \to 0$, as do Kallenberg and Ledwina (1987) for the exact Bahadur efficiency. Brown (1971) shows that appropriate LR tests are asymptotically optimum in Bahadur's sense. See also Fu (1973). Koziol (1978) discusses using the optimality of the LR test to prove the optimality of other tests without evaluating small-tail probabilities.

Kallenberg (1983) discusses a measure of efficiency intermediate between Pitman's and Bahadur's. Walsh (1946) and Chernoff (1952) discuss other measures of test efficiency.

Efficient score tests

22.40 The local nature of the ARE criterion suggests that it may be possible to construct local approximations to LR tests that are asymptotically efficient yet may remain analytically tractable when the original statistic is not. We begin by considering a single parameter, testing $H_0 : \theta = \theta_0$ against the composite alternative $\theta \neq \theta_0$. Using a Taylor series expansion we may write, approximately,

$$\begin{aligned} \log l &= \log\{L(x|\theta_0)/L(x|\hat{\theta})\} \\ &= (\hat{\theta} - \theta_0)u(\hat{\theta}) + \tfrac{1}{2}(\hat{\theta} - \theta_0)^2 u^{(1)}(\theta^*) \end{aligned} \qquad (22.149)$$

(cf. 18.21), where the score

$$\begin{aligned} u(\hat{\theta}) &= [\mathrm{d}\log L(x|\theta)/\mathrm{d}\theta]_{\theta=\hat{\theta}}, \\ u^{(1)}(\theta^*) &= [\mathrm{d}u(\theta)/\mathrm{d}\theta]_{\theta=\theta^*} \end{aligned}$$

and $|\theta^* - \theta_0| \leq |\hat{\theta} - \theta_0|$.

The expression $u(\hat{\theta})$ was termed the *efficient score* by Rao (1948), and the test statistics discussed in the following sections are sometimes known as Rao's score test statistics.

Since the score function $u(\theta)$ is zero when $\theta = \hat{\theta}$, (22.149) reduces to

$$-2\log l = -n(\hat{\theta} - \theta_0)^2\{n^{-1}u^{(1)}(\theta^*)\}.$$

Since $\hat{\theta}$ and thus θ^* are both consistent estimators of θ under H_0 it follows from **18.12** and **18.16** that

$$n^{-1}u^{(1)}(\theta^*) \to n^{-1}u^{(1)}(\theta_0) \to n^{-1}I(\theta_0) = -n^{-1}E(\mathrm{d}^2\log L/\mathrm{d}\theta^2)_{\theta=\theta_0}.$$

Then, from **18.16** and **22.7**, the test statistic

$$W_1 = (\hat{\theta} - \theta_0)I(\theta_0)(\hat{\theta} - \theta_0) \tag{22.150}$$

is asymptotically distributed as χ^2 with 1 d.fr. Further, from (18.31), we know that, asymptotically,

$$\hat{\theta} - \theta_0 = [I(\theta_0)]^{-1}u(\theta_0)$$

so that the statistic

$$W_2 = u(\theta_0)[I(\theta_0)]^{-1}u(\theta_0) \tag{22.151}$$

is also asymptotically equivalent to W_1 and the LR statistic.

22.41 The statistics (22.150) and (22.151) extend immediately to cover the case where H_0 is simple and $\boldsymbol{\theta}$ is a vector, upon replacing the scalar quantities by the appropriate vector and matrix forms. When H_0 is composite with $\boldsymbol{\theta} = (\boldsymbol{\theta}_1, \boldsymbol{\theta}_2)$ and $H_0 : \boldsymbol{\theta}_1 = \boldsymbol{\theta}_{10}$, it follows that

$$W_1 = (\hat{\boldsymbol{\theta}}_1 - \boldsymbol{\theta}_{10})^T[I^{11}(\hat{\boldsymbol{\theta}})]^{-1}(\hat{\boldsymbol{\theta}}_1 - \boldsymbol{\theta}_{10}),$$

where $I^{11}(\hat{\boldsymbol{\theta}})$ is the top left element of the inverse of

$$I(\hat{\boldsymbol{\theta}}) = -E\left(\frac{\partial^2\log L}{\partial\boldsymbol{\theta}\,\partial\boldsymbol{\theta}^T}\right)_{\boldsymbol{\theta}=\hat{\boldsymbol{\theta}}} = \begin{bmatrix} I_{11}(\hat{\boldsymbol{\theta}}) & I_{12}(\hat{\boldsymbol{\theta}}) \\ I_{21}(\hat{\boldsymbol{\theta}}) & I_{22}(\hat{\boldsymbol{\theta}}) \end{bmatrix},$$

or

$$[I^{11}(\hat{\boldsymbol{\theta}})]^{-1} = I_{11} - I_{12}I_{22}^{-1}I_{21}. \tag{22.152}$$

Further, if the score vector is partitioned as (u_1, u_2), the corresponding version of W_2 becomes, using (22.152),

$$W_2 = \tilde{\mathbf{u}}_1^T I^{11}(\boldsymbol{\theta}_0)\tilde{\mathbf{u}}_1 \tag{22.153}$$

where

$$\tilde{\mathbf{u}}_1 = u_1(\hat{\boldsymbol{\theta}}_0) - I_{12}(\hat{\boldsymbol{\theta}}_0)\{I_{22}(\hat{\boldsymbol{\theta}}_0)\}^{-1}u_2(\hat{\boldsymbol{\theta}}_0) \tag{22.154}$$

and $\hat{\boldsymbol{\theta}}_0 = (\hat{\boldsymbol{\theta}}_{10}, \hat{\boldsymbol{\theta}}_{20})$, where $\hat{\boldsymbol{\theta}}_{20}$ is the ML estimator for $\boldsymbol{\theta}_2$ under H_0. In each case, the statistic will be distributed as χ^2 when H_0 is true, with s degrees of freedom, where s is the dimensionality of $\boldsymbol{\theta}_2$. The form of (22.154) is clearly simplified if $\boldsymbol{\theta}_2$ can be chosen so that $I_{12}(\hat{\boldsymbol{\theta}}_0) = \mathbf{0}$.

22.42 Wald (1943a) first established that tests based upon the score function have an ARE of one; thus such tests may be termed *locally most powerful*. The tests are also known as $C(\alpha)$ tests following the work of Neyman (1959); see also Moran (1970) and Cox and Hinkley (1974, Chapter 9). The value of such tests is that they will often provide simple criteria in otherwise complex situations.

Example 22.12 (Score test for the Poisson distribution)

Suppose $x_1, \ldots, x_n$ are observations from Poisson distributions with parameters $\lambda_j = \exp(\theta_2 + \theta_1 z_j)$, where $\sum z_j = 0$ and θ_2 is known. We have that

$$u(\theta_1) = \frac{\mathrm{d}\log L}{\mathrm{d}\theta_1} = \sum x_j z_j - \sum z_j \exp(\theta_2 + \theta_1 z_j),$$

and

$$\frac{\mathrm{d}^2 \log L}{\mathrm{d}\theta_1^2} = -\sum z_j^2 \exp(\theta_2 + \theta_1 z_j).$$

Thus

$$u(\theta_{10}) = \sum x_j z_j, \qquad I(\theta_{10}) = \left(\sum z_j^2\right) \exp(\theta_2)$$

and, from (22.154),

$$W_2 \propto \sum x_j z_j \Big/ \sum z_j^2. \tag{22.155}$$

When θ_2 is unknown,

$$u(\theta_2) = \sum x_j - \sum \exp(\theta_2 + \theta_1 z_j)$$

so that $\exp(\hat{\theta}_{20}) = \bar{x}$. Also, since $\sum z_j = 0$,

$$I(\hat{\boldsymbol{\theta}}_0) = \exp(\theta_2) \begin{bmatrix} \sum z_j^2 & 0 \\ 0 & n \end{bmatrix}$$

and W_2 is of the same form as before.

The statistic (22.155) is of a simple form because the ML estimators under H_0 are easy to find. Conversely, W_1 is useful when the full set of ML estimators is easy to find but the restricted set under H_0 is not.

22.43 A similar class of tests, also asymptotically equivalent to the LR procedures, are the Lagrange multiplier tests developed by Aitchison and Silvey (1958) and Silvey (1959). Assume that H_0 can be represented by the set of restrictions

$$\boldsymbol{h}(\boldsymbol{\theta}) = \boldsymbol{0}$$

and write, to the first order of approximation,

$$\boldsymbol{h}(\hat{\boldsymbol{\theta}}) = \boldsymbol{h}(\boldsymbol{\theta}) + \boldsymbol{H}'(\boldsymbol{\theta})(\hat{\boldsymbol{\theta}} - \boldsymbol{\theta}),$$

where

$$\boldsymbol{H}(\boldsymbol{\theta}) = \{\partial h_j(\boldsymbol{\theta})/\partial\theta_i\}.$$

Then, under H_0, $n^{1/2}h(\hat{\boldsymbol{\theta}})$ is asymptotically normally distributed with mean zero and variance

$$V(\boldsymbol{\theta}) = H^T(\boldsymbol{\theta})\{I(\boldsymbol{\theta})\}^{-1}H(\boldsymbol{\theta}).$$

We may test H_0 using the statistic

$$W_1^* = h^T(\hat{\boldsymbol{\theta}})\{V(\hat{\boldsymbol{\theta}})\}^{-1}h(\hat{\boldsymbol{\theta}}),$$

which is approximately distributed as χ^2 with s degrees of freedom. When the ML estimators are evaluated under the restrictions in H_0, the Lagrange multiplier test reduces to W_2.

22.44 As progressively more complex hypotheses are tested, score tests have found application in a variety of fields. The *Journal of Statistical Planning and Inference* is producing a special issue devoted to score tests in 1998.

EXERCISES

22.1 Show that the c.f. of the non-central χ^2 distribution (22.18) is

$$\phi(t) = (1-2\mathrm{i}t)^{-\nu/2}\exp\left\{\frac{\lambda \mathrm{i}t}{1-2\mathrm{i}t}\right\},$$

giving cumulants $\kappa_r = (\nu + r\lambda)2^{r-1}(r-1)!$. In particular,

$$\kappa_1 = \nu+\lambda, \qquad \kappa_2 = 2(\nu+2\lambda),$$
$$\kappa_3 = 8(\nu+3\lambda), \qquad \kappa_4 = 48(\nu+4\lambda).$$

Hence show that the sum of two independent non-central χ^2 variates is another such, with both degrees of freedom and non-central parameter equal to the sum of those of the component distributions. Show that the c.f. of the central χ^2 approximation at (22.21) tends to the exact result as $\nu \to \infty$. By writing $\frac{1}{2}\lambda\{(1-2\mathrm{i}t)^{-1}-1\}$ for $\lambda \mathrm{i}t(1-2\mathrm{i}t)^{-1}$, use the inversion theorem for c.f.s to deduce (22.18) from the central χ^2 density.

22.2 Show that if the non-central normal variates x_i of **22.4** are subjected to k orthogonal linear constraints

$$\sum_{i=1}^{n} a_{ij}x_i = b_j, \qquad j = 1, 2, \ldots, k,$$

where

$$\sum_{i=1}^{n} a_{ij}^2 = 1, \qquad \sum_{i=1}^{n} a_{ij}a_{il} = 0, \quad j \neq l,$$

then

$$y^2 = \sum_{i=1}^{n} x_i^2 - \sum_{j=1}^{k} b_j^2$$

has the non-central χ^2 distribution with $n-k$ degrees of freedom and non-central parameter

$$\lambda = \sum_{i=1}^{n} \mu_i^2 - \sum_{j=1}^{k}\left(\sum_{i=1}^{n} a_{ij}\mu_i\right)^2.$$

(Patnaik, 1949. Cf. also Bateman, 1949)

22.3 Show that for any fixed r, the first r moments of a non-central χ^2 distribution with fixed λ tend, as degrees of freedom increase, to the corresponding moments of the central χ^2 distribution with the same degrees of freedom. Hence show that, in testing a hypothesis H_0, if the test statistic has a non-central χ^2 distribution with degrees of freedom an increasing function of sample size n, and non-central parameter λ a non-increasing function of n such that $\lambda = 0$ when H_0 holds, then the test will become ineffective as $n \to \infty$, i.e. its power will tend to its size α.

22.4 Show that the LR statistic l defined by (22.40) for testing the equality of k normal variances has moments about zero given by

$$\mu_r' = \frac{n^{rn/2}\Gamma\{\frac{1}{2}(n-k)\}}{\Gamma\{\frac{1}{2}[(r+1)n-k]\}}\prod_{i=1}^{k}\frac{\Gamma\{\frac{1}{2}[(r+1)n_i-1]\}}{n_i^{rn/2}\Gamma\{\frac{1}{2}(n_i-1)\}}.$$

(Neyman and Pearson, 1931)

22.5 For testing the hypothesis H_0 that k normal distributions are identical in mean and variance, show that the LR statistic is, for sample sizes $n_i \geq 2$,

$$l_0 = \prod_{i=1}^{k} \left(\frac{s_i^2}{s_0^2} \right)^{n_i/2},$$

where

$$s_i^2 = \frac{1}{n_i} \sum_{j=1}^{n_i} (x_{ij} - \bar{x}_i)^2, \qquad \bar{x} = \frac{1}{n} \sum_{i=1}^{k} n_i \bar{x}_i$$

and

$$s_0^2 = \frac{1}{n} \sum_{i=1}^{k} n_i \{ s_i^2 + (\bar{x}_i - \bar{x})^2 \},$$

and that its moments about zero are

$$\mu_r' = \frac{n^{rn/2} \Gamma\{\frac{1}{2}(n-1)\}}{\Gamma\{\frac{1}{2}[(r+1)n - 1]\}} \prod_{i=1}^{k} \frac{\Gamma\{\frac{1}{2}[(r+1)n_i - 1]\}}{n_i^{rn/2} \Gamma\{\frac{1}{2}(n_i - 1)\}}.$$

(Neyman and Pearson, 1931)

22.6 For testing the hypothesis H_2 that k normal distributions with the same variance have equal means, show that the LR statistic (with sample sizes $n_i \geq 2$) is

$$l_2 = l_0 / l,$$

where l and l_0 are as defined for Exercises 22.4 and 22.5, and that the exact distribution of $z = 1 - l_2^{2/n}$ when H_2 holds has p.d.f.

$$f(z) \propto z^{(k-3)/2} (1-z)^{(n-k-2)/2}, \qquad 0 \leq z \leq 1.$$

Find the moments of l_2 and hence show that when the hypothesis H_0 of Exercise 22.5 holds, l and l_2 are independently distributed.

(Neyman and Pearson (1931); Hogg (1961). See also Hogg (1962) for a test of H_2)

22.7 Show that when all the sample sizes n_i are equal, the LR statistic l of (22.40) and its modified form l^* of (22.44) are connected by the relation

$$n \log l^* = (n-k) \log l,$$

so that in this case the tests based on l and l^* are equivalent.

22.8 For samples from k distributions of form (22.48) or (22.52), show that if l is the LR statistic for testing the hypothesis

$$H_0 : \theta_1 = \theta_2 = \cdots = \theta_{p_1}; \theta_{p_1+1} = \theta_{p_1+2} = \cdots = \theta_{p_2}; \theta_{p_2+1} = \cdots = \theta_{p_3}$$
$$\cdots ; \theta_{p_{r-1}+1} = \cdots = \theta_{p_r},$$

so that the θ_i fall into r distinct groups (not necessarily of equal size) within which they are equal, then $-2 \log l$ is distributed exactly as χ^2 with $2(n-r)$ degrees of freedom.

(Hogg, 1956)

22.9 In a sample of n observations from the uniform distribution with p.d.f.

$$f(x) = (2\theta)^{-1}, \qquad \mu - \theta \le x \le \mu + \theta,$$

show that the LR statistic for testing $H_0 : \mu = 0$ is

$$l = \left(\frac{x_{(n)} - x_{(1)}}{2z}\right)^n = \left(\frac{R}{2z}\right)^n$$

where $z = \max\{-x_{(1)}, x_{(n)}\}$. Using Exercise 21.7, show that l and z are independently distributed, so that we have the factorization of c.f.s

$$E[\exp\{(-2\log R^n)it\}] = E[\exp\{(-2\log l)it\}]E[\exp\{[-2\log(2z)^n]it\}].$$

Hence show that the c.f. of $-2\log l$ is

$$\phi(t) = \frac{n-1}{n(1-2it)-1}$$

so that, as $n \to \infty$, $-2\log l$ is distributed as χ^2 with 2 degrees of freedom.

(Hogg and Craig, 1956)

22.10 In **22.6**, show that a quadratic form $\mathbf{x}^T\mathbf{A}\mathbf{x}$ has a non-central χ^2 distribution if and only if $\mathbf{AV}$ is idempotent, and that if the distribution has n degrees of freedom this implies $\mathbf{A} = \mathbf{V}^{-1}$.

22.11 k independent samples, of sizes $n_i \ge 2$, $\sum_{i=1}^{k} n_i = n$, are taken from exponential populations

$$f_i(x) = \exp\left\{-\left(\frac{x-\theta_i}{\sigma_i}\right)\right\}\bigg/\sigma_i, \qquad \theta_i \le x < \infty.$$

Show that the LR statistic for testing

$$H_0 : \theta_1 = \theta_2 = \cdots = \theta_k; \quad \sigma_1 = \sigma_2 = \cdots = \sigma_k$$

is

$$l_0 = \prod_{i=1}^{k} d_i^{n_i}/d^n,$$

where $d_i = \bar{x}_i - x_{(1)i}$ is the difference between the mean and smallest observation in the ith sample, and d is the same function of the combined samples, i.e.

$$d = \bar{x} - x_{(1)}.$$

Show that the moments of $l_0^{1/n}$ are

$$\mu'_p = \frac{n^p\Gamma(n-1)}{\Gamma(n+p-1)} \prod_{i=1}^{k} \frac{\Gamma\{(n_i-1)+pn_i/n\}}{n_i^{pn_i/n}\Gamma(n_i-1)}.$$

(Sukhatme, 1936)

22.12 In Exercise 22.11, show that for testing

$$H_1 : \sigma_1 = \sigma_2 = \cdots = \sigma_k,$$

the θ_i being unspecified, the LR statistic is

$$l_1 = \frac{\prod_{i=1}^k d_i^{n_i}}{\left(\frac{1}{n}\sum_{i=1}^k n_i d_i\right)^n},$$

and that the moments of $l_1^{1/n}$ are

$$\mu'_p = \frac{n^p \Gamma(n-k)}{\Gamma(n+p-k)} \prod_{i=1}^k \frac{\Gamma\{(n_i - 1) + pn_i/n\}}{n_i^{pn_i/n}\Gamma(n_i - 1)}$$

Given that the θ_i are all known to be zero, show that the LR statistic for H_1 is l_1 with d_i replaced by $\bar{x}_i$ and that the moments μ'_p have $n_i - 1$ replaced by n_i and $n - k$ replaced by n. Show that when $k = 2$, the LR test of H_1 is equivalent to a variance-ratio (F) test with $2n_1$ and $2n_2$ d.fr.

(Nagarsenker (1980) gives 5 per cent and 1 per cent points when each $n_i = 4\ (1)\ 10, 20, 40, 80, 100$ and $k = 3\ (1)\ 6$.)

22.13 In Exercise 22.11, show that if it is known that the σ_i are all equal, the LR statistic for testing

$$H_2 : \theta_1 = \theta_2 = \cdots = \theta_k$$

is

$$l_2 = l_0 / l_1,$$

where l_0 and l_1 are defined in Exercises 22.11–12. Show that the exact distribution of $l_2^{1/n} = u$ is

$$f(u) = \frac{1}{B(n-k, k-1)} u^{n-k-1}(1-u)^{k-2}, \qquad 0 \leq u \leq 1,$$

and find the moments of u. Show that when H_0 of Exercise 22.11 holds, l_1 and l_2 are independently distributed.

(Sukhatme, 1936; Hogg, 1961. Cf. also Hogg and Tanis (1963) for other tests of these hypotheses.)

22.14 Show by comparison with the unbiased test of Exercise 21.17 that the LR test for the hypothesis that two normal populations have equal variances is biased for unequal sample sizes n_1, n_2.

22.15 Show by using (22.67) that an unbiased similar size-α test of the hypothesis H_0 that k independent observations x_i $(i = 1, 2, \ldots, k)$ from normal populations with unit variance have equal means is given by the critical region

$$\sum_{i=1}^k (x_i - \bar{x})^2 \geq c_\alpha,$$

where c_α is the $100(1-\alpha)$ per cent point of the distribution of χ^2 with $n - 1$ degrees of freedom. Show that this is also the LR test.

22.16 Show that the three test statistics of Exercises 21.24–26 are equivalent to the LR statistics in the situations given; that the critical region of the LR test in Exercise 21.24 is not the UMPU region and is in fact biased; but that in the other two exercises the LR test coincides with the UMP similar test.

22.17 Extending the results of **21.10–13**, show that if a distribution is of the form

$$f(x|\theta_1, \theta_2, \ldots, \theta_k) = Q(\boldsymbol{\theta})M(x)\exp\left\{\sum_{j=3}^{k} B_j(x)A_j(\theta_3, \theta_4, \ldots, \theta_k)\right\}, \quad a(\theta_1, \theta_2) \leq x \leq b(\theta_1, \theta_2)$$

(the terminals of the distribution depending only on the two parameters not entering into the exponential term), the statistics $t_1 = x_{(1)}$, $t_2 = x_{(n)}$, $t_j = \sum_{i=1}^{n} B_j(x_i)$ are jointly sufficient for θ in a sample of n observations, and that their distribution is complete.

(Hogg and Craig, 1956)

22.18 Using the result of Exercise 22.17, show that in independent samples of sizes n_1, n_2 from two distributions

$$f(x_i) = \exp\{-(x_i - \theta_i)/\sigma\}/\sigma, \qquad \sigma > 0;\ x_i \geq \theta_i;\ i = 1, 2,$$

the statistics

$$z_1 = \min\{x_{i(1)}\},$$
$$z_2 = \sum_{j=1}^{n_1} x_{1j} + \sum_{j=1}^{n_2} x_{2j}$$

are sufficient for θ_1 and θ_2 and complete.
Show that the LR statistic for $H_0 : \theta_1 = \theta_2$ is

$$l = \left\{\frac{z_2 - (n_1 x_{1(1)} + n_2 x_{2(1)})}{z_2 - (n_1 + n_2)z_1}\right\}^{n_1+n_2}$$

and that l is distributed independently of z_1, z_2 and hence of its denominator. Show that l gives an unbiased test of H_0.

(Paulson, 1941)

22.19 Show that (22.18) may be written as a mixture (cf. **5.22**) of central χ^2 distributions with $\nu + 2r$ d.fr., with Poisson probabilities as the mixing distribution. By representing these χ^2 distributions as the sum of a $\chi^2(\nu)$ and $r\chi^2(2)$ distributions, all independent, use Exercise 5.22 to establish the c.f. of (22.18) given in Exercise 22.1; show that if for $\lambda > 0$ we put $\nu = 0$, the c.f., cumulants and additive property there remain valid while the distribution has discrete probability $e^{-\lambda/2}$ at $z = 0$ and p.d.f. given by (22.18) for $z > 0$.
Show further that (22.18) satisfies the relations

$$\chi'^2(\nu, \lambda) = \left(1 + 2\frac{\mathrm{d}}{\mathrm{d}\lambda}\right)\chi'^2(\nu - 2, \lambda)$$
$$= \left(1 + 2\frac{\mathrm{d}}{\mathrm{d}z}\right)\chi'^2(\nu + 2, \lambda),$$

this last result generalizing the first one of Exercise 16.7 in the central case.

(Cf. Siegel, 1979; Cohen, 1988)

22.20 In testing a simple $H_0 : L = L_0$ against a simple $H_1 : L = L_1$, show that the LR statistic (22.4) is $l = L_0/\max(L_0, L_1) = \min(1, L_0/L_1)$ with critical region (22.6), as against the BCR (21.6). Show that the two tests coincide if their size

$$\alpha \leq P(L_1 > L_0)$$

but that the LR test cannot achieve the BCR if $P(L_0 \geq L_1) > 1 - \alpha > 0$.

22.21 Using Exercise 22.19 and Exercise 16.7, show that the d.f. of (22.18) may be written, for even ν,

$$H(z) = P\{u - v \geq \tfrac{1}{2}\nu\},$$

where u and v are independent Poisson variates with parameters $\frac{1}{2}z$ and $\frac{1}{2}\lambda$, respectively; and that for any ν we have

$$\begin{aligned} H(z) &= \sum_{r=0}^{\infty} e^{\lambda/2} \frac{(\frac{1}{2}\lambda)^r}{r!} \left\{ F_\nu(z) - 2 \sum_{p=1}^{r} f_{\nu+2p}(z) \right\} \\ &= F_\nu(z) - 2 \sum_{p=1}^{\infty} f_{\nu+2p}(z) F_{2p}(\lambda), \end{aligned}$$

where $F_r(z)$, $f_\nu(z)$ are the d.f. and p.d.f. of a central χ^2_ν variable. Show that, regarded as a function of λ for fixed ν and z, $1 - H(z) = G(\lambda)$, say, is a non-decreasing function of λ. Regarding this as a d.f., show that its moment generating function is

$$M_\lambda(t) = (1 - 2t)^{(\nu/2)-1} \exp\left(\frac{zt}{1-2t}\right) F_\nu\left(\frac{z}{1-2t}\right),$$

and that near $t = 0$, if $1 - F_\nu(z)$ is negligible, we may approximate it by the power moment generating function

$$M^*_\lambda(t) = (1 - 2t)^{(\nu/2)-1} \exp\left(\frac{zt}{1-2t}\right)$$

with cumulants (cf. Exercise 22.1) $\kappa^*_r = \{rz - (\nu - 2)\}2^{r-1}(r-1)!$.

(Cf. Johnson (1959a) and Venables (1975), who uses the approximate cumulants in a Cornish–Fisher expansion (**6.25–26**) to find percentiles of $G(\lambda)$ for setting confidence intervals, and extends the analysis to λ in the non-central F' distribution discussed in **29.26**. See also Winterbottom (1979))

22.22 For the inverse Gaussian distribution of Exercises 18.41 and 11.28–29, show that if λ is known the LR test of $H_0 : \mu = \mu_0$ against $H_1 : \mu \neq \mu_0$ rejects H_0 when

$$l_1 = n\lambda(\bar{x} - \mu_0)/(\bar{x}\mu_0^2)$$

exceeds the $100(1 - \alpha)$ per cent point of its χ^2_1 distribution. Show that if λ is unknown, the LR test rejects H_0 when

$$l_2 = l_1/\{\lambda v/(n-1)\}$$

exceeds the $100(1 - \alpha)$ per cent point of its distribution, which is Student's t^2 with $n - 1$ d.fr. as a consequence of the results of Exercise 18.41.

(Chhikara and Folks (1976) show that these tests are UMPU.)

22.23 The sign test for the hypothesis H_0 that a population median takes a specified value θ_0 consists of counting the number of sample observations exceeding θ_0 and rejecting H_0 when this number is too large. Show that for a normal population this test has ARE $2/\pi$ compared to the Student's t test for H_0, and connect this with the result of Example 22.9.

(Cochran, 1937)

22.24 Generalizing the result of Exercise 22.23, show that for any continuous density function f with variance σ^2, the ARE of the sign test compared to the t test is $4\sigma^2\{f(\theta_0)\}^2$. Show that the sign test has ARE $= 2$ for the double exponential distribution given in Exercise 4.3.

(Pitman, 1948)

22.25 The difference between the means of two normal populations with equal variances is tested from two independent samples by comparing every observation y_j in the second sample with every observation x_i in the first sample, and counting the number of times a y_j exceeds an x_i. Show that the ARE of this test, known as the Wilcoxon test, compared to the two-sample Student's t test is $3/\pi$.

(Pitman, 1948)

22.26 Generalizing Exercise 22.25, show that if any two continuous densities $f(x)$, $f(x-\theta)$, differ only by a location parameter θ, and have variance σ^2, the ARE of the Wilcoxon test compared to the t test is

$$12\sigma^2\left\{\int_{-\infty}^{\infty}\{f(x)\}^2\,\mathrm{d}x\right\}^2.$$

Show that this reduces to 1.50 for the double exponential distribution of Exercise 4.3.

(Pitman, 1948)

22.27 t_1 and t_2 are unbiased estimators of θ, jointly normally distributed in large samples with variances σ^2, σ^2/e, respectively ($0 < e \leq 1$). Using the results of **16.23** and **17.29**, show that

$$E(t_1|t_2) = \theta(1-e) + t_2 e,$$

and hence that if t_2 is observed to differ from θ by a multiple d of its standard deviation we expect t_1 to differ from θ by a multiple $de^{1/2}$ of its standard deviation.

(Cox, 1956)

22.28 Using Exercise 22.27, show that if t_2 is used to test $H_0 : \theta = \theta_0$, we may calculate the 'expected result' of a test based on the more efficient statistic t_1. In particular, show that if a one-tailed test of size 0.01, using t_2, rejects H_0, we should expect a one-tailed test of size 0.05, using t_1, to do so if $e > 0.50$; while if an equal-tails size-0.0l test on t_2 rejects H_0, we should expect an equal-tails size-0.05 test on t_1 to do so if $e > 0.58$.

22.29 Let t_1 be a statistic with maximum ARE and t_2 any other test statistic for the same problem with δ and m as for t_1. By considering

$$t_3 = a\frac{t_1}{D_{10}} + (1-a)\frac{t_2}{D_{20}},$$

show that, in (22.100),

$$R_3 = \{aR_1 + (1-a)R_2\}/\{a^2 + (1-a)^2 + 2a(1-a)\rho\}^{1/2},$$

where ρ is the asymptotic correlation coefficient of t_1 and t_2. Hence show that $\rho = R_2/R_1$, the $(m\delta)$th power of the ARE (22.111). Cf. (17.61) for estimators.

(Cf. van Eeden, 1963)

22.30 In **22.28**, let $\delta_1 = \delta_2 = \delta$ and $m_1 = m_2 = m$. It is proposed to measure the efficiency of the tests t_1, t_2 by the reciprocal of the ratio of the distances $\theta - \theta_0$ which they require for $E(t|\theta)$ to fall on the boundary of the critical region (22.96). Show that this is approximately the same as using the δth power of the ARE.

(This *average critical value* method is due to Geary – cf. Stuart (1967).)

CHAPTER 23

INVARIANCE AND EQUIVARIANCE

23.1 Many statistical models can be transformed without changing the model. For example, the family of normal distributions is unchanged by location and scale changes. In such models, it is often sensible to require procedures to behave 'appropriately' under such changes. We say that a procedure is *invariant* under a transformation if it is unchanged by that transformation and *equivariant* under the transformation if it changes in an appropriate fashion. In this chapter we discuss procedures which are optimal in the class of invariant or equivariant procedures. Note that many early authors used the word 'invariance' for both invariance and equivariance as defined above (Ferguson, 1967; Berger, 1985; and Arnold, 1981). More recent authors have felt it important to distinguish between those procedures which change appropriately under the transformations and those which do not change at all (Lehmann, 1983b; and Eaton, 1983).

One nice aspect of invariance and equivariance is that they can be applied to many different aspects of a statistical model. For this reason, in **23.3–6** we illustrate how invariance and equivariance apply to the (univariate) one-sample normal model. In **23.7–18** we present the basic theory of invariance and equivariance, continuing to use the one-sample normal model as the primary example. In later sections we apply invariance and equivariance to many other univariate and multivariate parametric models, ending with a discussion of the Pitman estimators of location and scale parameters and a very brief discussion of the relationship between equivariant procedures and generalized Bayes procedures for certain 'non-informative' priors.

Equivariance in estimation was first suggested for location and scale parameters in Pitman (1939b) and extended to general problems in Piesakoff (1950) and Kiefer (1957). Invariant tests were suggested for certain multivariate problems in Hotelling (1936) and extended to general problems in Hunt and Stein (1946). The relationship between sufficiency and invariance was established for testing problems in Hall *et al.* (1965) and extended to equivariant procedures in Arnold (1985). The invariant pivotal quantity and its relationship to equivariant confidence regions were developed in Wijsman (1980) and Arnold (1984).

For simpicity of presentation we essentially limit discussion in this chapter to classical (non-Bayesian) fixed sample parametric models. (For further such parametric examples, see Arnold, 1981; Eaton, 1983.) However, invariance has been applied in many other settings, such as non-parametric models (Lehmann, 1986, pp. 314–326, 334–337; Ferguson, 1967, pp. 191–196), finite populations (Lehmann, 1983, pp. 207–218), sequential models (Ferguson, 1967, pp. 340–348) and Bayesian models (Berger, 1985, pp. 82–90, 406–422). In this chapter, we have also attempted to keep the presentation as elementary as possible. For a more rigorous presentation of the basic theory of invariance and equivariance, see Eaton (1983b, pp. 184-296).

23.2 Before discussing an example we recall two distributions which are useful in this chapter. We write

$$T \sim t_k(\delta) \quad \text{and} \quad F \sim F_{k,n}(\gamma)$$

to mean that T has a non-central t distribution with k degrees of freedom and non-centrality parameter δ and that F has a non-central F distribution with k and n degrees of freedom and non-centrality parameter γ. Let $f(t;\delta)$ and $g(f;\gamma)$ be the density functions of T and F. The only property of these distributions we need in this chapter is that

$$\frac{f(t;\delta)}{f(t;0)} \quad \text{and} \quad \frac{g(f;\gamma)}{g(f;0)}$$

are increasing functions of t and F respectively for all $\delta > 0$ and $\gamma > 0$ (see Exercises 23.1–23.2). These facts imply that if we have the model in which we observe $T \sim t_k(\delta)$ and want to test the null hypothesis that $\delta = 0$ against the alternative hypothesis that $\delta > 0$, the test which rejects if $t > t_k^\alpha$ is the UMP size-α test. Similarly, the test which rejects if $F > F_{k,n}^\alpha$ is the UMP size-α test for testing the null hypothesis that $\gamma = 0$ against the alternative hypothesis that $\gamma > 0$ for the model in which we observe $F \sim F_{k,n}(\gamma)$. For more details on these non-central distributions see **29.29–30**.

An example

23.3 Let $X_1, \ldots, X_n$ be independent,

$$X_i \sim N(\mu, \sigma^2).$$

As usual, let

$$\overline{X} = \frac{\sum X_i}{n}, \qquad S^2 = \frac{\sum(X_i - \overline{X})^2}{n-1}$$

be the sample mean and sample variance of the X_i, a sufficient statistic for this model.

There are three families of transformations which take normal random variables to normal random variables:

1. (location changes)

$$X_i \to X_i + a, (\mu, \sigma^2) \to (\mu + a, \sigma^2), \quad (\overline{X}, S^2) \to (\overline{X} + a, S^2), \quad a \in \mathbf{R};$$

2. (scale changes)

$$X_i \to bX_i, (\mu, \sigma^2) \to (b\mu, b^2\sigma^2), \quad (\overline{X}, S^2) \to (b\overline{X}, b^2S^2), \quad b > 0;$$

3. (sign changes)

$$X_i \to \pm X_i, \quad (\mu, \sigma^2) \to (\pm\mu, \sigma^2), \quad (\overline{X}, S^2) \to (\pm\overline{X}, S^2).$$

23.4 Consider the problem of estimating μ with loss function

$$L(d, (\mu, \sigma^2)) = \frac{(d-\mu)^2}{\sigma^2}.$$

We say that an estimator $T_1(X_1, \ldots, X_n)$ of μ is *(location and scale) equivariant* if

$$T_1(bX_1 + a, \ldots, bX_n + a) = bT_1(X_1, \ldots, X_n) + a,$$

for all $(X_1, \ldots, X_n) \in \mathbf{R}^n$, $b > 0$, $a \in \mathbf{R}$. The *best (location and scale) equivariant estimator* of μ is the equivariant estimator with smallest expected loss. We shall see that if $T_1(X_1, \ldots, X_n)$ is an equivariant estimator then so is the Rao–Blackwellized estimator (see **17.35**)

$$T_1^*(\overline{X}, S^2) = E(T_1(X_1, \ldots, X_n)|(\overline{X}, S^2));$$

Therefore, we consider only estimators which are functions of the sufficient statistic $(\overline{X}, S^2)$. Such a function is equivariant if and only if

$$T_1^*(b\overline{X} + a, b^2S^2) = bT_1^*(\overline{X}, S^2) + a.$$

If we choose $b = 1/S$, $a = -\overline{X}/S$, we see that an equivariant estimator must satisfy

$$T_1^*(0, 1) = \frac{T_1^*(\overline{X}, S^2) - \overline{X}}{S} \Leftrightarrow T_1^*(\overline{X}, S^2) = \overline{X} + ST_1^*(0, 1).$$

Conversely, if T_1^* satisfies this condition and $b > 0$, then

$$T_1^*(b\overline{X} + a, b^2S^2) = b\overline{X} + a + bST_1^*(0, 1) = bT_1^*(\overline{X}, S^2) + a$$

so that an estimator is equivariant if and only if it has this form. Therefore, the best equivariant estimator is the estimator of this form with the smallest expected loss. However,

$$R(T_1^*, (\mu, \sigma^2)) = \frac{E(\overline{X} + ST_1^*(0, 1) - \mu)^2}{\sigma^2} =$$
$$\frac{E(\overline{X} - \mu)^2 + (T_1^*(0, 1))^2 ES^2}{\sigma^2} = n^{-1} + (T_1^*(0, 1))^2$$

(using the independence of $\overline{X}$ and S^2) which is clearly minimized by taking $T_1^*(0, 1) = 0$. Therefore $\overline{X}$ is the best (location and scale) equivariant estimator of μ. Note that $\overline{X}$ is also the ML estimator and MV unbiased estimator of μ for this model.

Now consider estimating σ^2 with the loss function

$$L_2(d_2; (\mu, \sigma^2)) = \frac{(d_2 - \sigma^2)^2}{\sigma^4}.$$

An estimator $T_2(X_1, \ldots, X_n)$ of σ^2 is (*location and scale*) *equivariant* if

$$T_2(bX_1 + a, \ldots, bX_n + a) = b^2T_2(X_1, \ldots, X_n).$$

An equivariant estimator of σ^2 is the *best (location and scale) equivariant estimator* if it has the lowest expected loss of any equivariant estimator. We may again reduce attention to estimators which are functions of the sufficient statistic $(\overline{X}, S^2)$. Such an estimator $T_2^*(\overline{X}, S^2)$ is equivariant if and only if

$$T_2^*(b\overline{X} + a, b^2S^2) = b^2T_2^*(\overline{X}, S^2)$$

for all $a \in \mathbf{R}$, $b > 0$. By arguments similar to those in the last paragraph we see that this condition is satisfied if and only if

$$T_2^*(\overline{X}, S^2) = S^2 T_2^*(0, 1).$$

Direct calculation shows that the equivariant estimator with the smallest expected loss has $T_2^*(0, 1) = (n-1)/(n+1)$ and hence the best (location and scale) equivariant estimator of σ^2 is

$$\frac{n-1}{n+1} S^2$$

(see Exercise 23.3). Note that the ML estimator is an equivariant estimator with $T_2^*(0, 1) = (n-1)/n$ and the MV unbiased estimator is an equivariant estimator with $T_2^*(0, 1) = 1$. Therefore, the best equivariant estimator is better than the ML estimator and the MV unbiased estimator (at least for this quadratic loss function).

We shall see that under fairly general conditions, a unique ML estimator or MV unbiased estimator must be equivariant, so that the best equivariant estimator is best in a class which includes the ML estimator and MV unbiased estimator.

23.5 Now, consider testing the null hypothesis that $\mu = 0$ against the one-sided alternative hypothesis $\mu > 0$. Note that these hypotheses are unaffected by scale changes. That is, if $b > 0$, then

$$\mu = 0 \Leftrightarrow b\mu = 0 \quad \text{and} \quad \mu > 0 \Leftrightarrow b\mu > 0.$$

(The hypotheses are affected by location and sign changes, however.) This fact suggests that it is reasonable to require a test $\Phi_1(X_1, \ldots, X_n)$ to be invariant under scale changes, i.e. to require that

$$\Phi_1(bX_1, \ldots, bX_n) = \Phi_1(X_1, \ldots, X_n)$$

for all $b > 0$. As above, we may reduce to the sufficient statistic $(\overline{X}, S^2)$. A critical function $\Phi_1^*(\overline{X}, S^2)$ is scale invariant if and only if

$$\Phi_1^*(b\overline{X}, b^2 S^2) = \Phi_1^*(\overline{X}, S^2)$$

for all $b > 0$. Take $b = S^{-1}$ to verify that a scale invariant function satisfies

$$\Phi_1^*(\overline{X}, S^2) = \Phi_1^*(\overline{X}/S, 1) = \Phi_1^{**}(t), \quad t = \sqrt{n}\frac{\overline{X}}{S}.$$

It is also easily verified that if Φ_1^* satisfies this equation, it is scale invariant. Hence a test based on $(\overline{X}, S^2)$ is scale invariant if and only if it based on the Student's statistic t. Let $\delta = \sqrt{n}\mu/\sigma$. Then

$$t \sim t_{n-1}(\delta), \quad \mu = 0 \Leftrightarrow \delta = 0, \quad \mu > 0 \Leftrightarrow \delta > 0.$$

By the result stated in **23.2**, the test which is most powerful among all size-α tests based on t for testing the null hypothesis that $\delta = 0$ against the alternative hypothesis that $\delta > 0$ rejects if

$$t > t_{n-1}^{\alpha},$$

i.e. the usual one-sided one-sample t test. This test is UMP among all (scale) invariant size-α tests and is called the *UMP (scale) invariant size-α test* for this problem.

Now, consider testing the null hypothesis that $\mu = 0$ against the two-sided alternative hypothesis that $\mu \neq 0$. These hypotheses are unaffected by changes in either scale or sign. Therefore, it may be reasonable to require a critical function Φ_2 be scale and sign invariant. By arguments in the above paragraph, we see that the critical function (based on the sufficient statistic $(\overline{X}, S^2)$) is scale invariant if and only if it depends only on the Student's statistic t. A scale invariant critical function $\Phi_2^{**}(t)$ is sign invariant if and only if

$$\Phi_2^{**}(-t) = \Phi_2^{**}(t).$$

That is, a critical function is scale and sign invariant if and only if it is a function of

$$F = t^2 = n\frac{\overline{X}^2}{S^2}.$$

Let $\gamma = \delta^2 = n\mu^2/\sigma^2$. Then

$$F \sim F_{1,n-1}(\gamma), \qquad \mu = 0 \Leftrightarrow \gamma = 0, \qquad \mu \neq 0 \Leftrightarrow \gamma > 0.$$

(Note that invariance has reduced the two-sided alternative $\mu \neq 0$ to the one-sided alternative $\gamma > 0$.) By the comments in **23.2**, the UMP size-α test based on F for testing $\gamma = 0$ against $\gamma > 0$ rejects if

$$F > F_{1,n-1}^{\alpha} \Leftrightarrow t > t_{n-1}^{\frac{\alpha}{2}} \quad \text{or} \quad t < -t_{n-1}^{\frac{\alpha}{2}},$$

i.e. the usual two-sided one-sample t test. This test is UMP among size-α tests which are scale and sign invariant. We call it the *UMP (scale and sign) invariant size-α test* for this problem.

23.6 Now consider a confidence region $R(\overline{X}, S^2)$ for μ. Again, it seems reasonable to require that

$$\mu \in R(\overline{X}, S^2) \Leftrightarrow b\mu + a \in R(b\overline{X} + a, b^2S^2)$$

for all $a \in \mathbf{R}$, $b > 0$. We call a confidence region with this property a *(location and scale) equivariant confidence region for* μ. Let $b = 1/S$, $a = -\overline{X}/S$. Then

$$\mu \in R(\overline{X}, S^2) \Leftrightarrow -t(\mu) \in R(0, 1), \qquad t(\mu) = \sqrt{n}\frac{\overline{X} - \mu}{S} \sim t_{n-1}.$$

Conversely, it is easily seen that a confidence region of this form is location and scale equivariant (see Exercise 23.5). That is, a confidence region for μ is location and scale equivariant if and only if it is based on the pivotal quantity $t(\mu)$. Unfortunately, there is no confidence region which is optimal among all $1 - \alpha$ confidence regions based on $t(\mu)$. Therefore we add the additional requirement that the region is sign invariant, i.e. that

$$\mu \in R(\overline{X}, S^2) \Leftrightarrow -\mu \in R(-\overline{X}, S^2).$$

By similar arguments to those given above, we can show that a confidence region is location, scale and sign equivariant if and only if it is based on the pivotal quantity

$$F(\mu) = (t(\mu))^2 = n\frac{(\overline{X}-\mu)^2}{S^2} \sim F_{1,n-1}.$$

Let

$$R_1(\overline{X}, S^2) = \{\mu : F(\mu) \le F^{\alpha}_{1,n-1}\}, \qquad R_2(\overline{X}, S^2) = \{\mu : F(\mu) \in C\}$$

be equivariant $1-\alpha$ confidence regions for μ. Let $a \ne \mu$. Then

$$\begin{aligned} P_{\mu,\sigma}(a \in R_2(\overline{X}, S^2)) &= P_{\mu,\sigma}(0 \in R_2(\overline{X}-a, S^2)) = P_{\mu-a,\sigma}(0 \in R_2(\overline{X}, S^2)) \\ &= P_{\mu-a,\sigma}(F(0) \in C) \le P_{\mu-a,\sigma}(F(0) \le F^{\alpha}_{1,n-1}) = P_{\mu,\sigma}(a \in R_1(\overline{X}, S^2)) \end{aligned}$$

(where the inequality follows because the test which accepts the null hypothesis $\mu = 0$ when $F(0) \in C$ is a size-α (scale and sign) invariant test and the test which accepts $\mu = 0$ when $F(0) \le F^{\alpha}_{1,n-1}$ is UMP among all size-α (scale and sign) invariant tests for this hypothesis.) Therefore, $R_1(\overline{X}, S^2)$ is more accurate than any other location, scale and sign equivariant $1-\alpha$ confidence region for μ. (Recall that R_1 is more accurate than R_2 if it has smaller probability of covering false values.) Hence R_1 is uniformly most accurate (UMA) among all confidence regions which are location, scale and sign equivariant. We call this confidence region the *UMA (location, scale and sign) equivariant confidence region* for μ. Note that

$$\mu \in R_1(\overline{X}, S^2) \Leftrightarrow \mu \in \overline{X} \pm t^{\alpha/2}_{n-1}\frac{S}{\sqrt{n}},$$

the usual equal-tailed confidence interval for μ.

Basic definitions: maximal invariants

23.7 In this chapter, we define a *transformation* on a set W as an invertible function from W to W. (Technically we also require the function to be measurable. All the examples in this chapter are continuous functions so this assumption is trivially satisfied and will not be mentioned again.) If $g(w)$ is a transformation on W, we let $g^{-1}(w)$ be its inverse function. If $g_1(w)$ and $g_2(w)$ are transformations on W, we let $g_1 \circ g_2$ be the composition of g_1 and g_2. That is, $(g_1 \circ g_2)(w) = g_1(g_2(w))$. A collection G of transformations on W is a *group* if it is closed under compositions and inverses. A group of transformations on W is *transitive* if for any w_1 and w_2 in W, there exists $g \in G$ such that

$$w_2 = g(w_1).$$

A function $h(w)$ from W to another set D, is *invariant* under G if

$$h(g(w)) = h(w) \qquad \text{for all } g \in G, w \in W$$

and $h(w)$ is *equivariant* under G if for all $g \in G$, there is a transformation $\tilde{g}(d)$ on D such that

$$h(g(w)) = \tilde{g}(h(w)) \qquad \text{for all } w \in W.$$

Note that $\widetilde{g_1 \circ g_2} = \widetilde{g}_1 \circ \widetilde{g}_2$ and that $\widetilde{g^{-1}} = \widetilde{g}^{-1}$ so that the set $\widetilde{G}$ of $\widetilde{g}$ is also a group. Note also that invariance is a particular case of equivariance in which the group $\widetilde{G}$ consists only of the identity transformation.

Example 23.1 (Some transitive and intransitive groups)

Let $W = \mathbf{R}$, $g_b(w) = bw, b > 0$. The set G of such g_b is a group. It is not transitive, because there is no b such that $g_b(-1) = 1$. Let $f_1(w)$ be the sign of w. Since $b > 0$, $f_1(w)$ is an invariant function under G_1. The function $f_2(w) = \exp(w)$ is an equivariant function under G_1 because

$$f_2(bw) = (f_2(w))^b.$$

Let $h_a(w) = w + a$, $a \in \mathbf{R}$. The set H of such h_a is also a group. It is transitive, because

$$w_2 = h_a(w_1), a = w_2 - w_1.$$

The function $f_2(w) = \exp(w)$ is also equivariant under H because

$$f_2(w + a) = \exp(a) f_2(w).$$

The function $f_3(w) = w^2$ is not equivariant under H because there is no way to write $f_3(w+a) = (w + a)^2$ as a function of $f_3(w) = w^2$.

23.8 A *maximal invariant* under G is a function $T(w)$ from W to another set D such that:

1. $T(g(w)) = T(w)$ for all $w \in W$, $g \in G$ (invariance)
2. if $T(w_2) = T(w_1)$ then there exists $g \in G$ such that $w_2 = g(w_1)$ (maximality).

Some elementary facts about maximal invariants are the following (see Exercises 23.7–8):

1. If G is not transitive, then a maximal invariant always exists. An invertible function of a maximal invariant is also a maximal invariant. Two maximal invariants are invertible functions of each other.
2. If G is transitive, then the only invariant functions are constant functions, and we make the convention that the maximal invariant is empty. In this case, we write $T(w) = \emptyset$.

The basic property of maximal invariants is the following:

Theorem 23.1.

Let $T(w)$ be a maximal invariant under G. A function $f(w)$ is invariant under G if and only if there is a function $q(t)$ such that

$$f(w) = q(T(w)),$$

i.e. if and only if f is a function of $T(w)$.

Proof. If $T(w_1) = T(w_2)$, then there exists $g \in G$ such that $w_2 = g(w_1)$ and hence $f(w_2) = f(w_1)$. This implies that $f(w)$ depends on w only through $T(w)$. □

Note that this theorem only applies to invariant functions. Equivariant functions are functions of the maximal invariant only if they are invariant functions. (Recall that invariant functions are a particular subset of equivariant functions.) For this reason, maximal invariants are useful in finding optimal invariant tests and the invariant pivotal quantity. They are not useful in finding equivariant sufficient statistics or optimal equivariant estimators. (Optimal equivariant confidence regions are functions of the invariant pivotal quantity so involve maximal invariants indirectly.) This property of invariant functions is the most important reason to have separate words for invariance and equivariance.

We now give a theorem which is often helpful in finding maximal invariants. If G_1 and G_2 are two groups of transformations on a set W, the *union* of G_1 and G_2 is the smallest group containing G_1 and G_2. Such a union always exists.

Theorem 23.2.

Let G_1 and G_2 be two groups of transformations on W. Let G be the union of G_1 and G_2. Let $T_1(w)$ be a maximal invariant under G_1. Suppose that $T_1(w)$ is equivariant under G_2, $T_1(g_2(w)) = \widehat{g}_2(T_1(w))$. Let $T_2(t)$ be a maximal invariant under $\widehat{G}_2$ (the group of transformations $\widehat{g}_2$). Then $T(w) = T_2(T_1(w))$ is a maximal invariant under G.

Proof. (Invariance.) Let $g_1 \in G_1$. Then

$$T(g_1(w)) = T_2(T_1(g_1(w))) = T_2(T_1(w)) = T(w).$$

Let $g_2 \in G_2$. Then

$$T(g_2(w)) = T_2(T_1(g_2(w))) = T_2(\widehat{g}_2(T_1(w))) = T_2(T_1(w)) = T(w).$$

Therefore, $T(w)$ is invariant under both G_1 and G_2. Since the set of transformations under which a function is invariant is a group, $T(w)$ must be invariant under G.

(Maximality.) Suppose that $T(w_1) = T(w_2)$. Then by the maximality of T_2 under $\widehat{G}_2$

$$T_2(T_1(w_2)) = T_2(T_1(w_1)) \Rightarrow T_1(w_2) = \widehat{g}_2(T_1(w_1)) = T_1(g_2(w_1))$$

for some $g_2 \in G_2$. By the maximality of T_1 under G_1

$$w_2 = g_1(g_2(w_1)) = (g_1 \circ g_2)(w_1).$$

Since G is a group containing g_1 and g_2, $g_1 \circ g_2 \in G$, and hence $T(w)$ is maximal. □

If we have a group which consists of several subgroups, we can find a maximal invariant under the first group. If this maximal invariant is compatible with the second group, we then find the maximal invariant under the second group applied to the maximal invariant under the first group. If this new maximal invariant is compatible with the third group, we find the maximal invariant under the third group applied to the maximal invariant under the first two, etc.

Example 23.2 (Some maximal invariants)

Let $W = \{(x, y, s); x \in \mathbf{R}, y \in \mathbf{R}, s > 0\}$. Let G_1, G_2 and G be groups of transformations

$$g_1^a(x, y, s) = (x + a, y + a, s), \qquad a \in \mathbf{R},$$

$$g_2^b(x, y, s) = (bx, by, bs), \qquad b > 0,$$
$$g^{a,b}(x, y, s) = (bx + a, by + a, bs).$$

Then G is the union of G_1 and G_2. A maximal invariant under G_1 is

$$T_1(x, y, s) = (x - y, s)$$

(see Exercise 23.9(a)). Now,

$$T_1(g_2^b(x, y, s)) = (b(x - y), bs) = bT_1(x, y, s),$$

so that $T_1(x, y, s)$ is equivariant under G_2 and

$$\hat{g}_2^b(u, s) = (bu, bs).$$

A maximal invariant under $\hat{G}_2$ is

$$T_2(u, s) = \frac{u}{s},$$

(see Exercise 23.9(b)). Therefore, a maximal invariant under G is

$$T(x, y, s) = T_2(T_1(x, y, s)) = T_2(x - y, s) = \frac{x - y}{s}.$$

(In Exercise 23.9(c) it is shown directly that T is a maximal invariant under G.) Now, suppose that we reduce first by G_2. A maximal invariant under G_2 is

$$T_2^*(x, y, s) = \left(\frac{x}{s}, \frac{y}{s}\right), \qquad T_2^*(g_1^a(x, y, s)) = \left(\frac{x + a}{s}, \frac{y + a}{s}\right),$$

which cannot be written as a function of $T_2^*(x, y, s)$. Therefore, Theorem 23.2 is not applicable when we reduce first by G_2, but does work when we reduce first by G_1.

Sometimes Theorem 23.2 works for one order of application, but not another, as this example shows. If it works in two different orders, then it does not matter which order is used. The maximal invariants for the two orders will be invertible functions of each other.

Comment. The maximal invariant under G_1 is equivariant under G_2 (so that Theorem 23.2 may be used) if and only if G_1 is a normal subgroup in G. If G_1 is a normal subgroup, then G_2 is isomorphic to the quotient group, G/G_1. For a discussion of normal subgroups and quotient groups, see Herstein (1964, pp. 41–46).

Most groups used in parametric statistics are unions of several of the 13 groups listed (with their maximal invariants) in Table 23.1. (In the table X, $S^2 > 0$ and $T^2 > 0$ are real variables, $\mathbf{X}$ and $\mathbf{Y}$ are arbitrary matrices of the same dimension, $\mathbf{S}$ and $\mathbf{T}$ are arbitrary positive definite matrices of the same dimension, and $\mathbf{H}$ is an arbitrary matrix of appropriate dimension.) If $\mathbf{C}$ is replaced by $\mathbf{C}^T$ in the last two groups, we can just interchange the subscripts to find the maximal invariant. These maximal invariants are all derived in Arnold (1981).

Table 23.1 Some maximial invariants.

	group	MI
	Location	
	group	*MI*
L1	$\mathbf{X} \to \mathbf{X}+\mathbf{a}$	ϕ
L2	$(\mathbf{X},\mathbf{Y}) \to (\mathbf{X}+\mathbf{a},\mathbf{Y}+\mathbf{a})$	$\mathbf{X}-\mathbf{Y}$
	Univariate Scale ($b > 0$)	
	group	*MI*
S1	$S^2 \to b^2S^2$	ϕ
S2	$(T^2, S^2) \to (b^2T^2, b^2S^2)$	T^2/S^2
S3	$(X, S^2) \to (bX, b^2S^2)$	X/S
	Multivariate Scale (A invertible)	
	group	*MI*
M1	$\mathbf{S} \to \mathbf{ASA}^T$	ϕ
M2	$(\mathbf{T},\mathbf{S}) \to (\mathbf{ATA}^T, \mathbf{ASA}^T)$	eigenvalues of $\mathbf{S}^{-1}\mathbf{T}$
M3	$(\mathbf{X},\mathbf{S}) \to (\mathbf{AX}, \mathbf{ASA}^T)$	$\mathbf{X}^T\mathbf{S}^{-1}\mathbf{X}$
	Orthogonal ($\boldsymbol{\Gamma}$ orthogonal)	
	group	*MI*
O1	$X \to \pm X$	$\mathbf{X}^2$
O2	$\mathbf{X} \to \boldsymbol{\Gamma}\mathbf{X}$	$\mathbf{X}^T\mathbf{X}$
O3	$\mathbf{S} \to \boldsymbol{\Gamma}\mathbf{S}\boldsymbol{\Gamma}^T$	eigenvalues of $\mathbf{S}$
	Block ($\mathbf{C} = \begin{pmatrix} \mathbf{I} & \mathbf{0} \\ \mathbf{H} & \mathbf{I} \end{pmatrix}$)	
	group	*MI*
B1	$\mathbf{S} \to \mathbf{CSC}^T$	$\mathbf{S}_{11}, \mathbf{S}_{22} - \mathbf{S}_{21}\mathbf{S}_{11}^{-1}\mathbf{S}_{12}$
B2	$(\mathbf{X},\mathbf{S}) \to (\mathbf{CX}, \mathbf{CSC}^T)$	$\mathbf{X}_1, \mathbf{X}_2 - \mathbf{S}_{21}\mathbf{S}_{11}^{-1}\mathbf{X}_1$ $\mathbf{S}_{11}, \mathbf{S}_{22} - \mathbf{S}_{21}\mathbf{S}_{11}^{-1}\mathbf{S}_{12}$

In future sections, we shall use Table 23.1 together with Theorem 23.2 when we need to find maximal invariants. Suppose $\mathbf{X} = (\mathbf{Y}, \mathbf{Z})$, $g(\mathbf{X}) = (g_1(\mathbf{Y}), \mathbf{Z})$ and $\mathbf{T}_1(\mathbf{Y})$ is a maximal invariant under G_1. Then $\mathbf{T}(\mathbf{X}) = (\mathbf{T}_1(\mathbf{Y}), \mathbf{Z})$ is a maximal invariant under G (see Exercise 23.10). If $\mathbf{T}_1(\mathbf{Y}) = \emptyset$, then $\mathbf{Z}$ is a maximal invariant under G. Often when we find a maximal invariant for a problem, we multiply by a constant to obtain one which is easier to analyse. Note that such a multiplication leads to an invertible function of the maximal invariant which is also a maximal invariant.

23.9 We now discuss a technical detail which may be skipped on first reading. Suppose we have a measure λ on the set W. We say that an event A happens *almost everywhere* (a.e.) with respect to λ if $\lambda(A^c) = 0$. We say that function $f(w)$ is *almost invariant under G* if

$$f(g(w)) = f(w) \qquad \text{a.e. for all } g \in G,$$

and $f(w)$ is *almost equivariant* under G if

$$\text{for all } g \in G, \text{ there exists } \tilde{g} \text{ such that } f(g(w)) = \tilde{g}(f(w)) \text{ a.e.}$$

Note that for both almost invariance and almost equivariance, the exceptional sets may depend on the particular g chosen.

Theorem 23.3.

Under fairly general conditions, if $f(w)$ is almost invariant (equivariant), then there exists an invariant (equivariant) function $f^*(w)$ such that

$$f(w) = f^*(w) \text{ a.e.}$$

Proof. See Lehmann (1986, pp. 297–298) for the statement of conditions and a proof for almost invariance. The proof for almost equivariance is similar. □

The conditions for this theorem are quite complicated, but they are general enough to cover all the examples given here. Note that if $f(w) = f^*(w)$ a.e. then the exceptional set cannot depend on g.

Invariant distributions and equivariant sufficient statistics

23.10 Consider a statistical model in which we observe a random vector $\mathbf{X} \in \chi$, having joint density function $f(\mathbf{X}; \boldsymbol{\theta})$ depending on the unknown parameter vector $\boldsymbol{\theta} \in \Omega$. We call χ the *sample space* and Ω the *parameter space.* Let G be a group of transformations on χ. We say that this model is *invariant* under G if, for all $g(\mathbf{x}) \in G$, there exists a transformation $\overline{g}(\boldsymbol{\theta})$ on Ω such that

$$\mathbf{Y} = g(\mathbf{X}) \text{ has joint density } f(\mathbf{y} : \overline{g}(\boldsymbol{\theta})).$$

Note that

$$\overline{g_1 \circ g_2} = \overline{g_1} \circ \overline{g_2}, \quad \overline{g^{-1}} = \overline{g}^{-1}$$

so that the set $\overline{G}$ of transformations $\overline{g}$ is also a group. We call $\overline{G}$ the *induced* group on the parameter space. Two elementary properties which follow directly from this definition are

$$P_{\boldsymbol{\theta}}(g(\mathbf{X}) \in A) = P_{\overline{g}(\boldsymbol{\theta})}(\mathbf{X} \in A), \qquad E_{\boldsymbol{\theta}}(h(g(\mathbf{X})) = E_{\overline{g}(\boldsymbol{\theta})}(h(\mathbf{X})). \tag{23.1}$$

Example 23.3 (Invariant groups for the one-sample model)

Let $X_1, \ldots, X_n$ be independent, $X_i \sim N(\mu, \sigma^2)$. For this model,

$$\mathbf{X} = (X_1, \ldots, X_n), \qquad \boldsymbol{\theta} = (\mu, \sigma^2).$$

As discussed in **23.3**, this model is invariant under three groups:

$$\begin{aligned}
g_1(X_1, \ldots, X_n) &= (X_1 + a, \ldots, X_n + a), & \overline{g}_1(\mu, \sigma^2) &= (\mu + a, \sigma^2), & a &\in \mathbf{R}, \\
g_2(X_1, \ldots, X_n) &= (bX_1, \ldots, bX_n), & \overline{g}_2(\mu, \sigma^2) &= (b\mu, b^2\sigma^2), & b &> 0, \\
g_3(X_1, \ldots, X_n) &= \pm(X_1, \ldots, X_n), & \overline{g}_3(\mu, \sigma^2) &= (\pm\mu, \sigma^2).
\end{aligned}$$

We now give a theorem which we shall use to establish the invariance or equivariance of likelihood procedures.

Theorem 23.4.

Suppose that a model is invariant under the group G with induced group $\overline{G}$, and that $g(\mathbf{X})$ is a differentiable function with Jacobian J_g. Then

$$f(\mathbf{x}; \overline{g}^{-1}(\boldsymbol{\theta})) = f(g(\mathbf{x}); \boldsymbol{\theta})|J_g(\mathbf{x})|.$$

Proof. Let $\mathbf{Y}$ have joint density $f(\mathbf{y}; \boldsymbol{\theta})$ and let $\mathbf{X} = g^{-1}(\mathbf{Y})$. By the definition of invariance, $\mathbf{X}$ has joint density $f(\mathbf{x}; \overline{g}^{-1}(\boldsymbol{\theta}))$. By the usual result on transformations of random vectors, $\mathbf{X}$ has joint density $f(g(\mathbf{X}); \boldsymbol{\theta})|J_g|$ and the result is proved. □

In the remaining sections, we shall have groups defined on many different spaces. In developing the theory, it is important to keep track of which transformations are acting on which space. In the theoretical sections, we use $g \in G$ for transformations on the sample space, $\overline{g} \in \overline{G}$ for transformations on the parameter space, $g^* \in G^*$ for transformations on the sufficient statistic, and $\widetilde{g} \in \widetilde{G}$ for transformations on a parameter $\boldsymbol{\tau}(\boldsymbol{\theta})$. (We also use $\widehat{g} \in \widehat{G}$ and $g^{\#} \in G^{\#}$ for groups on other spaces.) After developing the general theory we look at many examples. When we look at these later examples we shall use $\rightarrow$ notation and not worry about which space the transformation is operating on.

23.11 Let $\mathbf{S}(\mathbf{X})$ be a sufficient statistic for $\boldsymbol{\theta}$. We say that $\mathbf{S}(\mathbf{X})$ is *equivariant* under G if for all $g \in G$ there exists a transformation $g^*(\mathbf{s})$ such that

$$\mathbf{S}(g(\mathbf{X})) = g^*(\mathbf{S}(\mathbf{X})).$$

Note that the set G^* of all g^* is a group.

Not all sufficient statistics are equivariant, as the following example shows.

Example 23.4 (Equivariant sufficient statistic for the one-sample model)

In the one-sample normal model, $(\overline{X}, S^2)$ is a sufficient statistic which is equivariant under location, scale and sign changes (as we indicated in **23.3**), but $(\overline{X}, S^2, X_1/X_2)$ is a sufficient statistic which is not equivariant under location changes.

As the example above indicates, a minimal sufficient statistic is typically equivariant (see Arnold, 1985 for conditions). Non-equivariance of a sufficient statistic usually occurs when we add non-equivariant pieces to the minimal sufficient statistic. In all the examples in this chapter the sufficient statistics given are equivariant (as we shall see by computing the induced group G^*).

For any function $d(\mathbf{X})$ from χ to $\mathbf{R}^k$ let

$$d^*(\mathbf{S}) = E(d(\mathbf{X})|\mathbf{S})$$

be its Rao–Blackwellized version. For many statistical problems, the Rao–Blackwellized procedure $d^*(\mathbf{S})$ is at least as good as the original procedure $d(\mathbf{X})$. For such problems, the next theorem implies (under the assumed affineness), that for any equivariant procedure based on the original data, there is an equivariant procedure based on the sufficient statistic which is at least as good. Note that an invariant procedure is an equivariant procedure with $\widetilde{g}(d) = d$, an affine function, so the theorem also implies that for any invariant procedure based on the original data there is an invariant procedure based on the sufficient statistic which is just as good.

Theorem 23.5.

Let $\mathbf{S}(\mathbf{X})$ *be an equivariant sufficient statistic. Let* $d(\mathbf{X})$ *be an equivariant function under* G *from* χ *to* $\mathbf{R}^k$ *so that* $d(g(\mathbf{X})) = \widetilde{g}(d(\mathbf{X}))$. *If* $\widetilde{g}(\mathbf{d})$ *is an affine function, then* $d^*(\mathbf{S}) = E(d(\mathbf{X})|\mathbf{S})$ *is an equivariant function under* G^*.

Proof. We need to show that $d^*(g^*(\mathbf{S})) = \widetilde{g}(d^*(\mathbf{S}))$, or equivalently that

$$\widetilde{g}^{-1}(d^*(g^*(S))) = E_{\boldsymbol{\theta}}(d(\mathbf{X})|\mathbf{S}).$$

By the measure-theoretic definition of conditional expectation (see Billingsley, 1979, pp. 395–405), it is enough to show that

$$E_{\boldsymbol{\theta}}(\widetilde{g}^{-1}(d^*(g^*(\mathbf{S})))h(\mathbf{S})) = E_{\boldsymbol{\theta}}(d(\mathbf{X})h(\mathbf{S}(\mathbf{X})))$$

for all bounded functions $h(\mathbf{S})$. To verify this fact, note first that (23.1) implies

$$E_{\boldsymbol{\theta}}k(g^*(\mathbf{S})) = E_{\boldsymbol{\theta}}k(\mathbf{S}(g(\mathbf{X}))) = E_{\overline{g}(\boldsymbol{\theta})}k(\mathbf{S}(\mathbf{X})) = E_{\overline{g}(\boldsymbol{\theta})}k(\mathbf{S}).$$

Therefore, we see that

$$\begin{aligned} E_{\boldsymbol{\theta}}(\widetilde{g}^{-1}(d^*(g^*(\mathbf{S})))h(\mathbf{S})) &= \widetilde{g}^{-1}(E_{\boldsymbol{\theta}}(d^*(g^*(\mathbf{S}))h(g^{*-1}(g^*(\mathbf{S}))))) \\ &= \widetilde{g}^{-1}(E_{\overline{g}(\boldsymbol{\theta})}(d^*(\mathbf{S})h(g^{*-1}(\mathbf{S})))) \\ &= \widetilde{g}^{-1}(E_{\overline{g}(\boldsymbol{\theta})}(d(\mathbf{X})h(g^{*-1}(\mathbf{S}(\mathbf{X}))))) \\ &= E_{\boldsymbol{\theta}}(\widetilde{g}^{-1}(d(g(\mathbf{X})))h(g^{*-1}(\mathbf{S}(g(\mathbf{X}))))) \\ &= E_{\boldsymbol{\theta}}(d(\mathbf{X})h(\mathbf{S}(\mathbf{X}))), \end{aligned}$$

where the first equality follows from the affineness of $\widetilde{g}$, the second equality from the definition of invariance of distribution, the third equality from the definition of conditional expectation (and the fact that $d^*(\mathbf{S})$ does not depend on $\boldsymbol{\theta}$), the fourth equality from the affineness of $\widetilde{g}$ and the last equality from the equivariance of d and $\mathbf{S}$. □

Comment. Technically, conditional expectation is only defined up to sets of measure 0. Therefore, a more careful statement of Theorem 23.5 would say that $d^*(\mathbf{S})$ is almost equivariant. However, Theorem 23.3 implies that there is a version of $d^*(\mathbf{S})$ which is equivariant.

We finish this section with a theorem similar to Basu's theorem.

Theorem 23.6.

Let $\mathbf{S}(\mathbf{X})$ *be a sufficient statistic which is equivariant under a group* G *such that the induced group* G^* *on the sufficient statistic is transitive. Let* $\mathbf{T}(\mathbf{X})$ *be an invariant function. Then* $\mathbf{S}(\mathbf{X})$ *and* $\mathbf{T}(\mathbf{X})$ *are independent.*

Proof. For any bounded function, $h(\mathbf{T})$ let $h^*(\mathbf{S}) = E(h(\mathbf{T})|\mathbf{S})$. By Theorem 23.5 (with $\widetilde{g}(d) = d$), $h^*(g^*(\mathbf{S})) = h^*(\mathbf{S})$. Since G^* is transitive, $h^*(\mathbf{S})$ is constant. This implies that $\mathbf{S}$ and $\mathbf{T}$ are independent. □

Example 23.5 (Using invariance to establish independence of sample mean and variance)

Let $X_1, \ldots, X_n$ be independent, $X_i \sim N(\mu, c^2)$, where c is a known constant. This model is invariant under the group $g(X_1, \ldots, X_n) = (X_1 + a, \ldots, X_n + a)$. $\overline{X}$ is an equivariant sufficient statistic for this model and S^2 is an invariant function. The set of transformations on $\overline{X}$, $g^*(\overline{X}) = \overline{X} + a$, is transitive. Therefore $\overline{X}$ and S^2 are independent.

Equivariant estimators

23.12 Let $\boldsymbol{\tau} = \tau(\theta)$ be a (possibly vector-valued) parameter. Consider the problem of estimating $\boldsymbol{\tau}$ with loss function $L(\mathbf{d}; \boldsymbol{\theta})$. If $\mathbf{T} = \mathbf{T}(\mathbf{X})$ is an estimator of $\boldsymbol{\tau}$, we let $R(\mathbf{T}; \boldsymbol{\theta}) = E_{\boldsymbol{\theta}} L(\mathbf{T}(\mathbf{X}); \boldsymbol{\theta})$ be the risk function of $\mathbf{T}$. We say that $\boldsymbol{\tau}$ is *equivariant* and $L(\mathbf{d}; \boldsymbol{\theta})$ is *invariant* if, for all $g \in G$, there exists a transformation $\widetilde{g}$ on τ such that

$$\boldsymbol{\tau}(\overline{g}(\boldsymbol{\theta})) = \widetilde{g}(\boldsymbol{\tau}(\boldsymbol{\theta})), \qquad L(\widetilde{g}(\mathbf{d}); \overline{g}(\boldsymbol{\theta})) = L(\mathbf{d}; \boldsymbol{\theta}).$$

(Note that the set $\widetilde{G}$ of all $\widetilde{g}$ is a group.) We say that an estimator $\mathbf{T}(\mathbf{X})$ of $\boldsymbol{\tau}$ is *equivariant* if for all $g \in G$

$$\mathbf{T}(g(\mathbf{X})) = \widetilde{g}(\mathbf{T}(\mathbf{X})).$$

An equivariant estimator is the *best equivariant estimator* of $\boldsymbol{\tau}$ if no other equivariant estimator has smaller risk.

One argument for using best equivariant estimators is that both the ML and MV unbiased estimators are equivariant and hence the best equivariant estimator is the best in a class that includes these two estimators.

We now give some basic properties of equivariant estimators.

Theorem 23.7.

Let $\mathbf{T} = \mathbf{T}(\mathbf{X})$ *be an equivariant estimator of the equivariant parameter* $\boldsymbol{\tau} = \boldsymbol{\tau}(\boldsymbol{\theta})$ *for the invariant loss function* $L(\mathbf{d}; \boldsymbol{\theta})$.

(a) If G *is transitive, then* $\mathbf{T}(\mathbf{X})$ *is completely determined by its value at any single point* $\mathbf{x}$.

(b) $R(\mathbf{T}; \boldsymbol{\theta})$ *is an invariant function of* $\boldsymbol{\theta}$ *under* $\overline{G}$. *If* $\overline{G}$ *is transitive, then* $R(\mathbf{T}; \boldsymbol{\theta})$ *is a constant function of* $\boldsymbol{\theta}$.

(c) If $L(\mathbf{d}; \boldsymbol{\theta})$ *is a convex function of* $\mathbf{d}$, $\widetilde{g}$ *is affine for all* $\widetilde{g} \in \widetilde{G}$, *and* $\mathbf{S} = \mathbf{S}(\mathbf{X})$ *is an equivariant sufficient statistic, then for any equivariant (under* G*) estimator* $\mathbf{T}(\mathbf{X})$ *of* $\boldsymbol{\tau}$ *based on the original data* $\mathbf{X}$, *there is an equivariant (under* G^**) estimator* $\mathbf{T}^*(\mathbf{S})$ *of* $\boldsymbol{\tau}$ *based on the sufficient statistic* $\mathbf{S}$ *which is just as good.*

Proof.

(a) Fix $\mathbf{x}_0 \in \chi$. For any $\mathbf{x} \in \chi$, let $g_{\mathbf{x}}$ be a transformation such that $g_{\mathbf{x}}(\mathbf{x}_0) = \mathbf{x}$. (Such a transformation exists because G is transitive.) Then

$$\mathbf{T}(\mathbf{x}) = \mathbf{T}(g_{\mathbf{x}}(\mathbf{x}_0)) = \widetilde{g}_{\mathbf{x}}(\mathbf{T}(\mathbf{x}_0)),$$

and hence $\mathbf{T}(\mathbf{X})$ is completely determined by its value at $\mathbf{x}_0$.

(b) Note that

$$\begin{aligned} R(\mathbf{T}; \overline{g}(\boldsymbol{\theta})) &= E_{\overline{g}(\boldsymbol{\theta})}(L(\mathbf{T}(\mathbf{X}); \overline{g}(\boldsymbol{\theta}))) = E_{\boldsymbol{\theta}}(L(\mathbf{T}(g(\mathbf{X})); \overline{g}(\boldsymbol{\theta}))) \\ &= E_{\boldsymbol{\theta}}(L(\widetilde{g}(\mathbf{T}(\mathbf{X})); \overline{g}(\boldsymbol{\theta}))) = E_{\boldsymbol{\theta}}(L(\mathbf{T}(\mathbf{X}); \boldsymbol{\theta})) = R(\mathbf{T}; \boldsymbol{\theta}) \end{aligned}$$

so that $R(\mathbf{T}; \boldsymbol{\theta})$ is invariant under $\overline{G}$. (The second equality follows from the invariance of the distribution (see (23.1)), the third equality from the equivariance of $\mathbf{T}$ and the fourth equality from the invariance of L.) If $\overline{G}$ is transitive, then any function which is invariant under $\overline{G}$ must be constant.

(c) This follows from Theorem 23.5 and the Rao–Blackwell theorem. □

Part (c) of this theorem implies that for most estimation problems, we can reduce to a sufficient statistic before considering equivariant estimators. Part (b) shows why it is often possible to find best equivariant estimators. Since all equivariant estimators have constant risk functions, either the risk functions for two equivariant estimators are the same or one is lower than the other for all $\boldsymbol{\theta}$. (That is, the risk functions cannot cross.) Part (a) of the theorem is what often causes it to be easy to find a best equivariant estimator, as the following example indicates.

Example 23.6 (Best equivariant estimators for the one-sample model)

Let $X_1, \ldots, X_n$ be independent, $X_i \sim N(\mu, \sigma^2)$. This model is invariant under location and scale changes

$$g(X_1, \ldots, X_n) = (bX_1 + a, \ldots, bX_n + a), \qquad \overline{g}(\mu, \sigma^2) = (b\mu + a, b^2\sigma^2).$$

Let $\overline{X}$ and S^2 be the sample mean and the sample variance of the X_i. Then $(\overline{X}, S^2)$ is an equivariant sufficient statistic,

$$g^*(\overline{X}, S^2) = (b\overline{X} + a, b^2 S^2).$$

(a) Consider first estimating $\tau_1(\mu, \sigma^2) = \mu$ with loss function

$$L_1(d_1; (\mu, \sigma^2)) = \frac{(d_1 - \mu)^2}{\sigma^2}.$$

Then

$$\tau_1(\overline{g}(\mu, \sigma^2)) = \tau_1(b\mu + a, b^2\sigma^2) = b\mu + a = b\tau_1(\mu, \sigma^2) + a.$$

Therefore, μ is equivariant and

$$\widetilde{g}_1(\mu) = b\mu + a.$$

In addition,

$$L_1(\tilde{g}_1(d); \overline{g}(\mu, \sigma^2)) = \frac{(bd + a - (b\mu + a))^2}{b^2\sigma^2} = \frac{(d - \mu)^2}{\sigma^2} = L_1(d_1; (\mu, \sigma^2)),$$

so that the loss function is invariant. In **23.4**, we showed that an estimator $T_1(\overline{X}, S^2)$ is equivariant if and only if

$$T_1(\overline{X}, S^2) = \overline{X} + T_1(0, 1)S.$$

Note that this estimator is completely determined by its value at (0, 1), because G^* is a transitive group. We also showed that the risk function of this estimator is

$$R(T_1; (\mu, \sigma^2)) = n^{-1} + (T_1(0, 1))^2,$$

which does not depend on (μ, σ^2) since $\overline{G}$ is transitive. Finally, we noted that this risk function is minimized when $T_1(0, 1) = 0$, so that $\overline{X}$ is the best equivariant estimator of μ. As mentioned in **23.3**, this model is also invariant under sign changes in the X_i. The induced groups on the sufficient statistic and μ are

$$g(\overline{X}, S^2) = (\pm\overline{X}, S^2), \qquad \tilde{g}(\mu) = \pm\mu.$$

Note that the loss function is also invariant under these transformations. If we require $T_1(\overline{X}, S^2)$ also to be equivariant under this group, then we must have $T_1(0, 1) = 0$, and hence $\overline{X}$ is the only estimator which is equivariant under location, scale and sign changes. Since the ML estimator and MV unbiased estimator are equivariant, they must also be $\overline{X}$.

(b) Now consider estimating $\tau_2(\mu, \sigma^2) = \sigma^2$ with the loss function

$$L_2(d_2; (\mu, \sigma^2)) = \frac{(d_2 - \sigma^2)^2}{\sigma^4}.$$

(Note that the usual squared error loss $(d_2 - \sigma^2)^2$ is not invariant for this problem.) This parameter is also (location and scale) equivariant with

$$\tilde{g}_2(\sigma^2) = b^2\sigma^2.$$

It is easily verified that this loss function is invariant and that an estimator $T_2(\overline{X}, S^2)$ is equivariant if $T_2(b\overline{X} + a, b^2S^2) = b^2T_2(\overline{X}, S^2)$. In **23.4**, we showed that T_2 is equivariant if and only if

$$T_2(\overline{X}, S^2) = T_2(0, 1)S^2.$$

Note that an equivariant estimator of σ^2 is also completely determined by its value at (0, 1). In **23.4**, the risk function of all equivariant estimators was also derived and the optimal choice was found to be $T(0, 1) = (n - 1)/(n + 1)$. Note that the ML estimator for σ^2 is an equivariant estimator with $T(0, 1) = (n - 1)/n$ and the MV unbiased estimator of σ^2 is an equivariant estimator $T(0, 1) = 1$.

(c) Now consider estimating σ^2 with the invariant loss function

$$(\log(d_2) - \log(\sigma^2))^2 = (\log(d_2/\sigma^2))^2.$$

Note that the class of equivariant estimators is still the class of estimators of the form $T_2^*(\overline{X}, S^2) = T_2^*(0, 1)S^2$. In Ferguson (1967, pp. 178–180) it is shown that the best equivariant estimator for this loss function has

$$T_2^*(0, 1) = \frac{n-1}{2}\exp\left(-\Psi\left(\frac{n-1}{2}\right)\right),$$

where $\Psi(t) = \Gamma^T(t)/\Gamma(t)$ is the digamma function. For $n \geq 3$,

$$T_2^*(0, 1) \approx \frac{n-1}{n-2}$$

so that the best equivariant estimator for this loss function is different from the best equivariant loss function for the quadratic loss function in part (b).

As illustrated by the example above, for any estimation problem, the class of equivariant estimators is unaffected by the loss function. The loss function merely determines which equivariant estimator is best. (Note that when there is a complete sufficient statistic for a model, the best (MV) unbiased estimator is the same for any convex loss function, essentially because after reducing to a sufficient statistic, it is best in a class of one. When there is a complete sufficient statistic, we should probably call the best unbiased estimator just the unbiased estimator.)

23.13 We now discuss the relationship between equivariance and other concepts in estimation.

Theorem 23.8.

(a) A unique ML estimator is an equivariant estimator.

(b) Let $\mathbf{S}$ *be a complete sufficient statistic and let* $\mathbf{T} = \mathbf{T}(\mathbf{S})$ *be the MV unbiased estimator of* $\boldsymbol{\tau}$. *If* $\tilde{g}$ *is an affine function, then* $\mathbf{T}$ *is equivariant.*

Proof.

(a) Let $\widehat{\boldsymbol{\theta}} = \widehat{\boldsymbol{\theta}}(\mathbf{X})$ be the ML estimator of $\boldsymbol{\theta}$. We first show that $\widehat{\boldsymbol{\theta}}(g(\mathbf{X})) = \overline{g}(\widehat{\boldsymbol{\theta}}(\mathbf{X}))$. Let

$$\widetilde{\boldsymbol{\theta}} = \widetilde{\boldsymbol{\theta}}(\mathbf{X}) = \overline{g}^{-1}(\widehat{\boldsymbol{\theta}}(g(\mathbf{X}))).$$

The definition of ML estimator implies that

$$L_{\mathbf{X}}(\widehat{\boldsymbol{\theta}}(\mathbf{X})) \geq L_{\mathbf{X}}(\boldsymbol{\theta}), \qquad \text{for all } \boldsymbol{\theta} \in \Omega,$$

and hence

$$L_{g(\mathbf{X})}(\widehat{\boldsymbol{\theta}}(g(\mathbf{X}))) \geq L_{g(\mathbf{X})}(\overline{g}^{-1}(\boldsymbol{\theta})), \qquad \text{for all } \boldsymbol{\theta} \in \Omega.$$

Using Theorem 23.4, we see that

$$\begin{aligned} L_{\mathbf{X}}(\widetilde{\boldsymbol{\theta}}(\mathbf{X})) &= L_{\mathbf{X}}(\overline{g}^{-1}(\widehat{\boldsymbol{\theta}}(g(\mathbf{X})))) = L_{g(\mathbf{X})}(\widehat{\boldsymbol{\theta}}(g(\mathbf{X})))|J_g| \\ &\geq L_{g(\mathbf{X})}(\overline{g}^{-1}(\boldsymbol{\theta}))|J_g| = L_{\mathbf{X}}(\boldsymbol{\theta}), \end{aligned}$$

for all $\boldsymbol{\theta} \in \Omega$. Therefore, $\widetilde{\boldsymbol{\theta}}(\mathbf{X})$ is also an ML estimator of $\boldsymbol{\theta}$. Since the ML estimator is unique

$$\widehat{\boldsymbol{\theta}}(\mathbf{X}) = \widetilde{\boldsymbol{\theta}}(\mathbf{X}) = \overline{g}^{-1}(\widehat{\boldsymbol{\theta}}(g(\mathbf{X}))),$$

which implies that $\widehat{\boldsymbol{\theta}}(\mathbf{X})$ is equivariant. The ML estimator of $\boldsymbol{\tau}(\boldsymbol{\theta})$ is $\widehat{\boldsymbol{\tau}}(\mathbf{X}) = \boldsymbol{\tau}(\widehat{\boldsymbol{\theta}}(\mathbf{X}))$. Now

$$\begin{aligned}\widehat{\boldsymbol{\tau}}(g(\mathbf{X})) &= \boldsymbol{\tau}(\widehat{\boldsymbol{\theta}}(g(\mathbf{X}))) = \boldsymbol{\tau}(\overline{g}(\widehat{\boldsymbol{\theta}}(\mathbf{X}))) \\ &= \widetilde{g}(\boldsymbol{\tau}(\widehat{\boldsymbol{\theta}}(\mathbf{X}))) = \widetilde{g}(\widehat{\boldsymbol{\tau}}(\mathbf{X})),\end{aligned}$$

and therefore $\widehat{\boldsymbol{\tau}}$ is an equivariant estimator of $\boldsymbol{\tau}$.

(b) Let

$$\mathbf{T}^* = \mathbf{T}^*(\mathbf{S}) = \widetilde{g}^{-1}(\mathbf{T}(g^*(\mathbf{S}))).$$

Then

$$\begin{aligned}E_{\boldsymbol{\theta}}\mathbf{T}^*(\mathbf{S}) &= \widetilde{g}^{-1}(E_{\boldsymbol{\theta}}\mathbf{T}(g(\mathbf{S}))) = \widetilde{g}^{-1}(E_{\overline{g}(\boldsymbol{\theta})}\mathbf{T}(\mathbf{S})) \\ &= \widetilde{g}^{-1}(\boldsymbol{\tau}(\overline{g}(\boldsymbol{\theta}))) = \boldsymbol{\tau}(\boldsymbol{\theta}).\end{aligned}$$

Therefore, $\mathbf{T}^*(\mathbf{S})$ is another unbiased estimator of $\boldsymbol{\tau}$. By the completeness of $\mathbf{S}$, $\mathbf{T}^*(\mathbf{S}) = \mathbf{T}(\mathbf{S})$ and hence $\mathbf{T}(\mathbf{S})$ is equivariant. □

Comment. The likelihood function is only defined up to sets of measure 0. Therefore, the derivation in part (a) only implies that the ML estimator is almost equivariant. Theorem 23.3 can then be used to show that there is a version of the ML estimator which is equivariant. Similarly, in part (b), completeness only implies that $\mathbf{T}^* = \mathbf{T}$ a.e., and hence that the MV unbiased estimator is almost equivariant. Theorem 23.3 can be again used to show that there is a version of the MV unbiased estimator which is equivariant.

For estimation problems involving a single parameter, best equivariant estimators are typically admissible. For three or more parameters, a best equivariant estimator is typically inadmissible even for the loss function for which it is 'best' (and hence so are the ML and MV unbiased estimators). The case of two parameters is a borderline case. See Brown (1966) and Zidek (1976) for some general theorems on admissibility of best equivariant rules. (Best equivariant estimators of single parameters may also be inadmissible. For example, the best equivariant estimator of σ^2 for quadratic loss in the one-sample model (with unknown mean) is inadmissible. See Stein (1964).)

Kiefer (1957) shows under fairly general conditions that the best equivariant estimator is minimax. The conditions for minimaxity are quite complicated and this result is rarely used to prove minimaxity. Rather equivariance is used to find a good constant risk estimator which is then proved minimax by other means. See Bondar and Milnes (1981) for a survey paper on the relation between equivariance and minimaxity.

Invariant tests

23.14 We continue assuming we have a family of distributions which is invariant under a group G of transformations on χ with induced group of transformation $\overline{G}$ on Ω. Consider now the testing problem in which we test

$$H_0 : \boldsymbol{\theta} \in N \text{ vs. } H_1 : \boldsymbol{\theta} \in A.$$

We say that this problem is *invariant* under the group G if, for all $\overline{g} \in \overline{G}$,

$$\overline{g}(N) = N, \qquad \overline{g}(A) = A$$

(where $\overline{g}(S) = \{\overline{g}(s) : s \in S\}$). Note that a group which leaves a testing problem invariant can never be transitive on the parameter space because there is no $\overline{g} \in \overline{G}$ such that $\overline{g}(n) = a$ where $n \in N$ and $a \in A$. We say that a test with critical function $\Phi(\mathbf{X})$ is *invariant* under G if, for all $g \in G$,

$$\Phi(g(\mathbf{X})) = \Phi(\mathbf{X}).$$

We say that a size-α invariant test $\Phi(\mathbf{X})$ is the *UMP invariant size-α test* if $\Phi(\mathbf{X})$ is more powerful than any other invariant size-α test.

One argument for UMP invariant tests is that the UMP invariant test is always unbiased. Also, both likelihood ratio tests and UMP unbiased tests are invariant tests so that a UMP invariant test is best in a class which includes the LR test and the UMP unbiased test.

Let $\mathbf{T} = \mathbf{T}(\mathbf{X})$ be a maximal invariant under G and let $\boldsymbol{\delta} = \boldsymbol{\delta}(\boldsymbol{\theta})$ be a maximal invariant under $\overline{G}$. We call $\mathbf{T}$ the *(sample) maximal invariant* and $\boldsymbol{\delta}$ the *parameter maximal invariant* for this problem.

Theorem 23.9.

(a) A test $\Phi(\mathbf{X})$ is invariant if and only if there exists $\Phi^*(\mathbf{T})$ such that

$$\Phi(\mathbf{X}) = \Phi^*(\mathbf{T}(\mathbf{X})).$$

(b) The distribution of $\mathbf{T}(\mathbf{X})$ depends only on $\boldsymbol{\delta}$.

(c) Let $\mathbf{S}(\mathbf{X})$ be an equivariant sufficient statistic. For any invariant (under G) test $\Phi(\mathbf{X})$ based on the original data, there is an invariant (under G^*) test $\Phi^*(\mathbf{S})$ based on $\mathbf{S}$ which has the same power function.

Proof.

(a) This follows directly from Theorem 23.1.

(b) Let $\mathbf{X}^* = g(\mathbf{X}), \theta^* = \overline{g}(\boldsymbol{\theta})$. Then, $\mathbf{X}^*$ has density $f(\mathbf{x}^*; \boldsymbol{\theta}^*)$. Also

$$\mathbf{T}(\mathbf{X}^*) = \mathbf{T}(g(\mathbf{X})) = \mathbf{T}(\mathbf{X}),$$

since $\mathbf{T}$ is invariant. Therefore, the distribution of $\mathbf{T}$ is the same whether computed from $\boldsymbol{\theta}^*$ or $\boldsymbol{\theta}$ and is hence invariant under $\overline{G}$. Therefore, this distribution depends on $\boldsymbol{\theta}$ only through the maximal invariant under $\overline{G}$.

(c) Let $\Phi^*(\mathbf{S}) = E(\Phi(\mathbf{X})|\mathbf{S})$. Then $\Phi^*(\mathbf{S})$ has the same power function as $\Phi(\mathbf{X})$. If $\Phi(\mathbf{X})$ is invariant then $\Phi^*(\mathbf{S})$ is also invariant by Theorem 23.5 with $\tilde{g}(d) = d$. □

Let the maximal invariant $\mathbf{T}$ have density function $f^*(\mathbf{t}; \boldsymbol{\delta})$. Let

$$N^* = \boldsymbol{\delta}(N) = \{\boldsymbol{\delta}(\boldsymbol{\theta}) : \boldsymbol{\theta} \in N\}, \qquad A^* = \boldsymbol{\delta}(A) = \{\boldsymbol{\delta}(\boldsymbol{\theta}) : \boldsymbol{\theta} \in A\}.$$

Consider the testing problem in which we observe $\mathbf{T}$ having density $f^*(\mathbf{t}; \boldsymbol{\delta})$ and we are testing the null hypothesis that $\boldsymbol{\delta} \in N^*$ against the alternative hypothesis that $\boldsymbol{\delta} \in A^*$. We call this testing problem the *reduced* problem.

Corollary 23.10

A test which is UMP size-α for the reduced problem is the UMP invariant size-α test for the original problem.

Example 23.7 (UMP invariant tests for the one-sample model)

We return to the example in which we observe $X_1, \ldots, X_n$ independent, $X_i \sim N(\mu, \sigma^2)$. By part (c) of Theorem 23.9, we can reduce to a sufficient statistic $(\overline{X}, S^2)$ before considering invariance. This model is invariant under three groups, as we have seen previously:

$$\begin{aligned}
&g_1^*(\overline{X}, S^2) = (\overline{X} + a, S^2), && \overline{g}_1(\mu, \sigma^2) = (\mu + a, \sigma^2), && a \in \mathbf{R}, \\
&g_2^*(\overline{X}, S^2) = (b\overline{X}, b^2 S^2), && \overline{g}_2(\mu, \sigma^2) = (b\mu, b^2\sigma^2), && b > 0, \\
&g_3^*(\overline{X}, S^2) = (\pm\overline{X}, S^2), && \overline{g}_2(\mu, \sigma^2) = (\pm\mu, \sigma^2).
\end{aligned}$$

(a) Consider first testing the null hypothesis that $\mu = 0$ against the alternative hypothesis that $\mu > 0$. This testing problem is only invariant under the group G_2^* of scale changes. Maximal invariants under G_2^* and $\overline{G}_2$ are

$$T_1 = \sqrt{n}\frac{\overline{X}}{S}, \qquad \delta_1 = \sqrt{n}\frac{\mu}{\sigma}$$

(see group S3 of Table 23.1). Therefore T_1 and δ_1 are the sample maximal invariant and the parameter maximal invariant. Furthermore,

$$T_1 \sim t_{n-1}(\delta_1),$$

and we are testing the null hypothesis that $\delta_1 = 0$ against the alternative hypothesis that $\delta_1 > 0$. In **23.2**, we argued that the UMP size-α test for this reduced problem rejects if $T_1 > t_{n-1}^{\alpha}$, the usual one-sided one-sample t test. Since this test is the UMP size-α test for the reduced problem, it is UMP invariant size-α test for the original model.

(b) Now, consider testing the null hypothesis that $\mu = 0$ against the alternative hypothesis that $\mu \neq 0$. In addition to G_2^*, this problem is invariant under the group G_3^* of sign changes. In terms of T_1 this becomes

$$\widehat{g}_3(T_1) = \pm T_1.$$

Maximal invariants under $\widehat{G}_3$ and its induced group on the parameter space are

$$T_2 = T_1^2 = n\frac{\overline{X}^2}{S^2}, \qquad \delta_2 = \delta_1^2 = n\frac{\mu^2}{\sigma^2},$$

(see group O1) and

$$T_2 \sim F_{1,n-1}(\delta_2).$$

In the reduced problem, we are testing the null hypothesis that $\delta_2 = 0$ against the alternative hypothesis that $\delta_2 > 0$. In **23.2**, we argued that the UMP size-α test for this reduced problem rejects if

$$T_2 > F_{1,n-p}^{\alpha} \Longleftrightarrow T_1 > t_{n-1}^{\alpha/2} \quad \text{or} \quad T_1 < -t_{n-1}^{\alpha/2},$$

the usual two-sided one-sample t test. This test is therefore UMP invariant size-α for the original model. (Note that Theorem 23.2 implies that T_2 is the maximal invariant under the union of G_2^* and G_3^*.)

(c) Now consider testing that $\sigma^2 = 1$ against $\sigma^2 > 1$. This problem is invariant under the group G_1^* of location changes. A maximal invariant and parameter maximal invariant for this problem are

$$T_3 = S^2, \quad \delta_3 = \sigma^2$$

(see group L1). Furthermore,

$$T_3/\delta_3 \sim \chi^2_{n-1}.$$

In the reduced problem we are testing the null hypothesis that $\delta_3 = 1$ against the alternative hypothesis that $\delta_3 > 1$. The distribution of T_3 has monotone likelihood ratio (see Exercise 23.12) and hence the UMP size-α test for this reduced problem rejects if $T_3 > \chi^{2,\alpha}_{n-1}$. Since this test is the UMP size-α test for the reduced problem, it is UMP invariant size-α for the original problem. (Note that this problem is also invariant under the group G_3^* of sign changes. However, T_3 and δ_3 are already invariant under this group, so that no further reduction occurs.)

(d) Finally, consider testing the null hypothesis that $\sigma^2 = 1$ against the alternative hypothesis that $\sigma^2 \neq 1$. Unfortunately, this problem is still only invariant under G_1^* and G_3^*. The maximal invariant and parameter maximal invariant are therefore still T_3 and δ_3. In the reduced problem, we are testing the null hypothesis that $\delta_3 = 1$ against the alternative hypothesis that $\delta_3 \neq 1$, which is a two-sided problem for which no UMP tests exist. Therefore, there is no UMP invariant size-α test for this testing problem.

(Note that to test the null hypothesis that $\mu = c$ against one-sided or two-sided alternatives, we merely transform to $Y_i = X_i - c$ and proceed as above. Similarly, to test $\sigma^2 = d^2$, we let $Y_i = X_i/d$.)

23.15 Sometimes it is possible to reduce a problem by either sufficiency or invariance, as the following example illustrates.

Example 23.8 (Equal reductions by sufficiency and invariance)

Suppose we observe $\mathbf{X} = (X_1, \ldots, X_n)$, where the X_i are independent, $X_i \sim N(0, \sigma^2)$. We want to test the null hypothesis that $\sigma^2 = 1$ against the alternative hypothesis that $\sigma^2 > 1$. This problem is invariant under the group of transformations $g(\mathbf{X}) = \mathbf{\Gamma X}$, where $\mathbf{\Gamma}$ is an arbitrary orthogonal $n \times n$ matrix. A maximal invariant and parameter maximal invariant under this group are $U = \sum X_i^2$ (see group O2) and σ^2, and $U \sim \sigma^2 \chi_n^2$. This model has monotone likelihood ratio and hence the UMP size-α test for the reduced model rejects if $U > \chi_n^{2\alpha}$; this test is UMP invariant size-α for the original model. Note that U is also a sufficient statistic for this model, and hence the test given above is actually UMP size-α for the original model.

Since it is a stronger statement to be UMP among all size-α tests than to be UMP among all invariant size-α tests, the reduction by sufficiency is preferred to the reduction by invariance. For this reason, in later models we reduce first to a sufficient statistic before reducing by invariance.

We now give the relationship between invariant procedures and other properties of tests.

Theorem 23.11.

(a) The likelihood ratio test is an invariant test.

(b) A unique UMP unbiased size-α test is invariant.

(c) A UMP invariant size-α test is unbiased.

Proof.

(a) Note that invariance of the problem implies that $\overline{g}(N) = N$ and $\overline{g}(A \cup N) = A \cup N$. Let $\Lambda(\mathbf{X})$ be the likelihood ratio test statistic. Then by Theorem 23.4

$$\Lambda(g(\mathbf{X})) = \frac{\sup_{\boldsymbol{\theta} \in N} L_{g(\mathbf{X})}(\boldsymbol{\theta})}{\sup_{\boldsymbol{\theta} \in N \cup A} L_{g(\mathbf{X})}(\boldsymbol{\theta})} = \frac{\sup_{\boldsymbol{\theta} \in N} L_{\mathbf{X}}(\overline{g}(\boldsymbol{\theta}))}{\sup_{\boldsymbol{\theta} \in N \cup A} L_{\mathbf{X}}(\overline{g}(\boldsymbol{\theta}))}$$
$$= \frac{\sup_{\boldsymbol{\theta} \in N} L_{\mathbf{X}}(\boldsymbol{\theta})}{\sup_{\boldsymbol{\theta} \in N \cup A} L_{\mathbf{X}}(\boldsymbol{\theta})} = \Lambda(\mathbf{X}).$$

Therefore, $\Lambda(\mathbf{X})$ is an invariant test statistic and the LR test is an invariant test.

(b) Let $\Phi(\mathbf{X})$ be a unique UMP unbiased size-α test with power function $K(\boldsymbol{\theta})$, and let

$$\Phi^*(X) = \Phi(g(\mathbf{X})).$$

Then Φ^* has power function

$$K^*(\boldsymbol{\theta}) = E_{\boldsymbol{\theta}}\Phi^*(\mathbf{X}) = E_{\boldsymbol{\theta}}\Phi(g(\mathbf{X})) = E_{\overline{g}(\boldsymbol{\theta})}\Phi(\mathbf{X}) = K(\overline{g}(\boldsymbol{\theta})).$$

Now, $\overline{g}(N) = N$, $\overline{g}(A) = A$. Therefore,

$$\sup_{\boldsymbol{\theta} \in N} K^*(\boldsymbol{\theta}) = \sup_{\boldsymbol{\theta} \in N} K(\boldsymbol{\theta}),\ \inf_{\boldsymbol{\theta} \in A} K^*(\boldsymbol{\theta}) = \inf_{\boldsymbol{\theta} \in A} K(\boldsymbol{\theta}).$$

Hence, Φ^* is an unbiased size-α test. Since Φ is the UMP unbiased size-α test,

$$K(\boldsymbol{\theta}) \le K^*(\boldsymbol{\theta}) = K(\overline{g}(\boldsymbol{\theta}))$$

for all $\overline{g} \in \overline{G}$, $\boldsymbol{\theta} \in A$. In particular, this equality is true for $\boldsymbol{\theta} = \overline{g}^*(\boldsymbol{\theta}^*)$, $\overline{g} = \overline{g}^{*-1}$ and hence

$$K(\overline{g}^*(\boldsymbol{\theta}^*)) \le K(\boldsymbol{\theta}^*)$$

for all $\overline{g}^* \in \overline{G}$, $\boldsymbol{\theta}^* \in A$. Therefore,

$$K(\boldsymbol{\theta}) = K(\overline{g}(\boldsymbol{\theta})) = K^*(\boldsymbol{\theta}),$$

and hence Φ^* is also a UMP unbiased size-α test. By the uniqueness of the UMP unbiased size-α test, $\Phi^* = \Phi$ and hence Φ is invariant.

(c) Let $\Phi(\mathbf{X})$ be the UMP invariant size-α test and let $\Phi_0(\mathbf{X}) \equiv \alpha$. Then Φ_0 is an invariant size-α test, so that Φ is more powerful that Φ_0. Hence

$$E_{\boldsymbol{\theta}}\Phi(\mathbf{X}) \ge E_{\boldsymbol{\theta}}\Phi_0(\mathbf{X}) = \alpha,$$

for all $\boldsymbol{\theta} \in A$, and therefore Φ is unbiased. □

Comment. As discussed previously, the likelihood is only defined up to sets of measure 0, so that the argument in part (a) actually implies that the LR test is almost invariant. Theorem 23.3 can then be used to show that there is a version of the LR test which is invariant. Similarly, if Φ is a UMP unbiased size-α test and $\Phi^* = \Phi$ a.e., then Φ^* is also a UMP unbiased size-α test, so that the UMP unbiased size-α test is only unique up to sets of measure 0, and hence the argument above only implies that the UMP unbiased test is almost invariant. However, Theorem 23.3 can be used to argue that there is a version of the UMP unbiased test which is invariant.

Parts (b) and (c) of this theorem imply that if there are both a UMP unbiased and a UMP invariant test, then they must be the same. Since the testing problems considered in Example 23.7 involve only a single parameter in an exponential family, we know there must be UMP unbiased tests for those models. We have derived UMP invariant tests. By the theorem above, these tests must also be UMP unbiased. It is often easier to show a test is UMP invariant than to show it is UMP unbiased, so this method is often a useful way to show that a test is UMP unbiased.

Hunt and Stein (1946) show that under fairly general conditions the UMP invariant test has certain minimaxity properties. See Lehmann (1986, pp. 516–527) for a discussion of these results.

UMP invariant tests are typically admissible, although there does not appear to be a general theorem in this direction, as the following example (due to Stein) indicates.

Example 23.9 (Inadmissible UMP invariance test)

Let $\mathbf{X}$ and $\mathbf{Y}$ be independent random vectors,

$$\mathbf{X} \sim N_2(\mathbf{0}, \mathbf{\Sigma}), \qquad \mathbf{Y} \sim N_2(\mathbf{0}, \Delta\mathbf{\Sigma}),$$

where $\Delta > 0$ is an unknown scalar and $\mathbf{\Sigma} > 0$ is an unknown 2×2 matrix. Consider testing the null hypothesis that $\Delta = 1$ against the alternative that $\Delta > 1$. This problem is invariant under the group of transformations

$$g(\mathbf{X}, \mathbf{Y}) = (\mathbf{AX}, \mathbf{AY}),$$

where $\mathbf{A}$ is an arbitrary 2×2 invertible matrix. This group is transitive and hence the only invariant functions are constant. Therefore the UMP invariant size-α test is the only invariant test size-α test which is given by $\Phi(\mathbf{X}, \mathbf{Y}) \equiv \alpha$. Let X_1 and Y_1 be the first components of $\mathbf{X}$ and $\mathbf{Y}$. Then the test which rejects if $Y_1^2/X_1^2 > F_{1,1}^{\alpha}$ is an unbiased size-α test and is hence more powerful than the UMP invariant size-α test Φ. Hence the UMP invariant size-α test for this model is inadmissible.

The invariant pivotal quantity and equivariant confidence regions

23.16 Let $\boldsymbol{\tau}(\boldsymbol{\theta})$ be an equivariant parameter and let $\mathbf{S}(\mathbf{X})$ be an equivariant sufficient statistic. A *pivotal quantity* for $\boldsymbol{\tau}$ is a function $q(\mathbf{X}, \boldsymbol{\tau})$ whose distribution does not depend on any unknown parameters. Pivotal quantities are very useful for finding confidence intervals and tests. In this section, we present a very general way to construct pivotal quantities.

Define

$$g^{\#}(\mathbf{S}, \boldsymbol{\tau}) = (g^*(\mathbf{S}), \tilde{g}(\boldsymbol{\tau})).$$

Let $W = W(\mathbf{S}, \boldsymbol{\tau})$ be a maximal invariant under $g^{\#}$.

Theorem 23.12.

If the group $\overline{G}$ of transformations on the parameter space is transitive, then $W(\mathbf{S}, \boldsymbol{\tau})$ is a pivotal quantity for $\boldsymbol{\tau}$

Proof. Let $J(\boldsymbol{\theta}) = P_{\boldsymbol{\theta}}(W(\mathbf{S}, \boldsymbol{\tau}(\boldsymbol{\theta})) \in A)$ for an arbitrary set A. Then

$$\begin{aligned} J(\overline{g}(\boldsymbol{\theta})) &= P_{\overline{g}(\boldsymbol{\theta})}(W(\mathbf{S}, \boldsymbol{\tau}(\overline{g}(\boldsymbol{\theta}))) \in A) = P_{\boldsymbol{\theta}}(W(g^*(\mathbf{S}), \widetilde{g}(\boldsymbol{\tau}(\boldsymbol{\theta}))) \in A) \\ &= P_{\boldsymbol{\theta}}(W(\mathbf{S}, \boldsymbol{\tau}(\boldsymbol{\theta})) \in A) = J(\boldsymbol{\theta}). \end{aligned}$$

Therefore $J(\boldsymbol{\theta})$ is an invariant function under the transitive group $\overline{G}$ and must be constant. Therefore the distribution of W does not depend on $\boldsymbol{\theta}$. □

We call the function $W = W(\mathbf{S}, \boldsymbol{\tau})$ the *invariant pivotal quantity* (under G).

Example 23.10 (Invariant pivotal quantities for the one-sample normal model)

Let $X_1, \ldots, X_n$ be independent, $X_i \sim N(\mu, \sigma^2)$. As we have seen, this model is invariant under location and scale changes, and the group of location and scale changes, is transitive on the parameter space.

(a) Consider first a pivotal quantity for μ. The group of location changes leads to

$$g_1^{\#}(\overline{X}, S^2, \mu) = (\overline{X} + a, S^2, \mu + a).$$

A maximal invariant under this group is $T_1(\overline{X}, S^2, \mu) = (\overline{X} - \mu, S^2)$ (see group L2). The group of scale changes acts on T_1 by

$$\widehat{g}_2(\overline{X} - \mu, S^2) = (b(\overline{X} - \mu), b^2 S^2).$$

A maximal invariant under this group is

$$t = t(\overline{X}, S^2, \mu) = \sqrt{n}\frac{\overline{X} - \mu}{S} \sim t_{n-1},$$

(see group S3) which is therefore a maximal invariant under the group of location and scale changes by Theorem 23.2. Note that this is the usual pivotal quantity for μ. This model is also invariant under the group of sign changes which operates on t by

$$\widehat{g}_3(t) = \pm t,$$

and a maximal invariant is

$$F(\overline{X}, S^2, \mu) = t^2 = n\frac{(\overline{X} - \mu)^2}{S^2} \sim F_{1,n-1}.$$

(see group O1).

(b) Now consider σ^2. The group of location changes gives

$$g_1^{\#}(\overline{X}, S^2, \sigma^2) = (\overline{X} + a, S^2, \sigma^2),$$

and a maximal invariant is $T_1(\overline{X}, S^2, \sigma^2) = (S^2, \sigma^2)$ (see group L1). The group of scale changes operates on this maximal invariant by

$$\widehat{g}_2(S^2, \sigma^2) = (b^2 S^2, b^2 \sigma^2),$$

and a maximal invariant is

$$(n-1)\frac{S^2}{\sigma^2} \sim \chi^2_{n-1},$$

(see group S2) the usual pivotal quantity for σ^2.

Many parameters have no pivotal quantities even in continuous models. (For example, there is no pivotal quantity for the correlation coefficient in a bivariate normal distribution.) The theorem above may explain why some parameters have pivotal quantities and others do not. It seems that in order to have a pivotal quantity, a parameter must be equivariant under a transitive group.

Choose a possible value $\mathbf{a}_0$ for $\boldsymbol{\tau}$ and consider the problem of testing the null hypothesis that $\boldsymbol{\tau} = \mathbf{a}_0$ against the alternative $\boldsymbol{\tau} \neq \mathbf{a}_0$. A natural test statistic for this problem is $W(\mathbf{S}, \mathbf{a}_0)$. We now show that this test statistic is the appropriate test statistic for this testing problem. Let $G_0 \subset G$ be the subgroup of transformations such that

$$\widetilde{g}(\mathbf{a}_0) = \mathbf{a}_0$$

(i.e. the subgroup which leaves this testing problem invariant).

Theorem 23.13.

$W(\mathbf{S}, \mathbf{a}_0)$ *is a maximal invariant quantity under* G_0.

Proof. $W(g_0^*(\mathbf{S}), \mathbf{a}_0) = W(g_0^*(\mathbf{S}), \widetilde{g}_0(\mathbf{a}_0)) = W(\mathbf{S}, \mathbf{a}_0)$, so that $W(\mathbf{S}, \mathbf{a}_0)$ is invariant. Furthermore, if $W(\mathbf{S}, \mathbf{a}_0) = W(\mathbf{S}^*, \mathbf{a}_0)$, then there exists $g \in G$ such that $(\mathbf{S}^*, \mathbf{a}_0) = (g^*(\mathbf{S}), \widetilde{g}(\mathbf{a}_0))$. Since $\widetilde{g}(\mathbf{a}_0) = \mathbf{a}_0$, $g \in G_0$. □

23.17 We continue to assume that $\boldsymbol{\tau} = \boldsymbol{\tau}(\boldsymbol{\theta})$ is an equivariant parameter and $\mathbf{S} = \mathbf{S}(\mathbf{X})$ is an equivariant sufficient statistic. We also assume that the induced group $\overline{G}$ is transitive. We say that a confidence region $R(\mathbf{S})$ for $\boldsymbol{\tau}$ is *equivariant* if, for all $g \in G$,

$$\widetilde{g}(\mathbf{a}) \in R(g^*(\mathbf{S})) \Leftrightarrow \mathbf{a} \in R(\mathbf{S}).$$

An equivariant $1-\alpha$ confidence region for $\boldsymbol{\tau}$ is the *UMA equivariant* $1-\alpha$ *confidence region* if it is more accurate than any other equivariant $1-\alpha$ confidence region. We say that a confidence region $R(\mathbf{S})$ is *based on the pivotal quantity* $W(\mathbf{S}, \boldsymbol{\tau})$ if there exists a set C such that

$$R(\mathbf{S}) = \{\boldsymbol{\tau} : W(\mathbf{S}, \boldsymbol{\tau}) \in C\}.$$

Theorem 23.14.

A confidence region is equivariant if and only if it is based on the invariant pivotal quantity $W = W(\mathbf{S}, \boldsymbol{\tau})$.

Proof. Let

$$I(\mathbf{S}, \mathbf{a}) = \begin{cases} 1 & \text{if } \mathbf{a} \in R(\mathbf{S}) \\ 0 & \text{if } \mathbf{a} \notin R(\mathbf{S}). \end{cases}$$

Then $R(\mathbf{S})$ is equivariant if and only if $I(\mathbf{S}, \mathbf{a})$ is invariant under $g^{\#}(\mathbf{S}, \mathbf{a}) = (g^*(\mathbf{S}), \tilde{g}(\mathbf{a}))$ if and only if $I(\mathbf{S}, \mathbf{a})$ is a function of the maximal invariant W under $G^{\#}$. But this happens if and only if $R(\mathbf{S})$ is based on W. □

By this theorem an equivariant confidence region $R(\mathbf{S})$ must have the form

$$R(\mathbf{S}) = \{\boldsymbol{\tau} : W(\mathbf{S}, \boldsymbol{\tau}) \in C\}$$

for some set C. We now find the optimal C. Choose a possible value $\mathbf{a}_0$ for $\boldsymbol{\tau}$ and consider testing $\boldsymbol{\tau} = \mathbf{a}_0$ against $\boldsymbol{\tau} \neq \mathbf{a}_0$. Let G_0 be the subgroup of G which leaves this testing problem invariant. By Theorem 23.13 $W(\mathbf{S}, \mathbf{a}_0)$ is a maximal invariant under G_0. Therefore, a test is invariant under G_0 if an only if it is based on the test statistic $W(\mathbf{S}, \mathbf{a}_0)$. The invariant test *associated* with the equivariant confidence region $R(\mathbf{S})$ is the test which accepts $\boldsymbol{\tau} = \mathbf{a}_0$ if and only if $W(\mathbf{S}, \mathbf{a}_0) \in C$ (i.e. if and only if $\mathbf{a}_0$ is in the confidence region $R(\mathbf{S})$).

Theorem 23.15.

The equivariant confidence region $R(\mathbf{S})$ is the UMA equivariant (under G) $1 - \alpha$ confidence region for $\boldsymbol{\tau}$ if and only if its associated test is the UMP invariant (under G_0) size-α test of the null hypothesis $\boldsymbol{\tau} = \mathbf{a}_0$ against the alternative hypothesis $\boldsymbol{\tau} \neq \mathbf{a}_0$.

Proof. For any equivariant $1 - \alpha$ confidence region

$$Q(\mathbf{S}) = \{\boldsymbol{\tau} : W(\mathbf{S}, \boldsymbol{\tau}) \in D\},$$

let $K_Q(\boldsymbol{\tau})$ be the power function of the test associated with Q. For all $\mathbf{a}$, let $\tilde{g}_{\mathbf{a}}$ be the transformation such that $\tilde{g}_{\mathbf{a}}(\mathbf{a}) = \mathbf{a}_0$. (Such a transformation is always possible because $\overline{G}$ is transitive and hence $\tilde{G}$ is transitive.) Then

$$\begin{aligned} P_{\tau}(\mathbf{a} \in Q(\mathbf{S})) &= P_{\tau}(\tilde{g}_{\mathbf{a}}(\mathbf{a}) \in Q(g_a^*(\mathbf{S}))) = P_{\overline{g}_a(\tau)}(\mathbf{a}_0 \in Q(\mathbf{S})) \\ &= P_{\overline{g}_a(\tau)}(W(\mathbf{S}, \mathbf{a}_0) \in D) = 1 - K_Q(\overline{g}_a(\boldsymbol{\theta})). \end{aligned} \tag{23.2}$$

Let $N = \{\boldsymbol{\theta} : \boldsymbol{\tau}(\boldsymbol{\theta}) = \mathbf{a}_0\}$ be the null hypothesis for the testing problem. Note that

$$\boldsymbol{\tau}(\boldsymbol{\theta}) = \mathbf{a} \Leftrightarrow \mathbf{a}_0 = \tilde{g}_{\mathbf{a}}(\mathbf{a}) = \boldsymbol{\tau}(\overline{g}_{\mathbf{a}}(\boldsymbol{\theta})) \Leftrightarrow \overline{g}_{\mathbf{a}}(\boldsymbol{\theta}) \in N.$$

Therefore, $Q(\mathbf{S})$ is a $1 - \alpha$ confidence region if and only if its associated test is a size-α test. Now, suppose that $R(\mathbf{S})$ is a UMA equivariant $1 - \alpha$ confidence region and that $\boldsymbol{\tau} \notin N$ so that $\boldsymbol{\tau}(\boldsymbol{\theta}) \neq \mathbf{a}_0$. Then for any other $1 - \alpha$ equivariant confidence region $Q(\mathbf{S})$,

$$K_Q(\boldsymbol{\theta}) = 1 - P_{\boldsymbol{\theta}}(\mathbf{a}_0 \in Q(\mathbf{S})) \leq 1 - P_{\boldsymbol{\theta}}(\mathbf{a}_0 \in R(\mathbf{S})) = K_R(\boldsymbol{\theta})$$

and hence the test associated with $R(\mathbf{S})$ is more powerful that the test associated with any other $1 - \alpha$ equivariant confidence region and is therefore UMP among all non-randomized invariant size-α tests. Now, suppose that the test associated with $R(\mathbf{S})$ is UMP among all non-randomized invariant size-α tests. Then if $\boldsymbol{\tau}(\boldsymbol{\theta}) \neq \mathbf{a}$ (and hence $\overline{g}_{\mathbf{a}}(\boldsymbol{\theta}) \notin N$),

$$P_{\boldsymbol{\theta}}(\mathbf{a} \in Q(\mathbf{S})) = 1 - K_Q(\overline{g}_{\mathbf{a}}(\boldsymbol{\theta})) \leq 1 - K_R(\overline{g}_{\mathbf{a}}(\boldsymbol{\theta})) = P_{\boldsymbol{\theta}}(\mathbf{a} \in R(\mathbf{S}))$$

(since the test associated with R is more powerful than the test associated with Q). Therefore $R(\mathbf{S})$ is more accurate than $Q(\mathbf{S})$. □

Comment. Technically, what we have shown is that the confidence region is UMA equivariant if and only if the test is UMP among all invariant non-randomized size-α tests. In all practical examples of invariance, a UMP invariant size-α test is non-randomized, so this point is of little consequence here. Note also that to use this theorem, it is only necessary to find the UMP invariant test for testing $\boldsymbol{\tau} = \mathbf{a}_0$ for a single $\mathbf{a}_0$.

Example 23.11 (UMA equivariant confidence region for the one-sample model)

Let $X_1, \ldots, X_n$ be independent $X_i \sim N(\mu, \sigma^2)$.

(a) Consider first the optimal confidence region for μ. The location, scale and sign invariant pivotal quantity for μ is

$$F(\overline{X}, S^2, \mu) = n\frac{(\overline{X} - \mu)^2}{S^2} \sim F_{1,n-1}.$$

Consider testing the null hypothesis that $\mu = 0$ against the alternative hypothesis that $\mu \neq 0$. The subgroup G_0 which leaves this testing problem invariant is the group of scale and sign changes. In a previous example, it was shown that the UMP invariant size-α test for this group accepts if

$$F(\overline{X}, S^2, 0) = n\frac{\overline{X}^2}{S^2} \leq F^{\alpha}_{1,n-1}.$$

Therefore, the UMA (location, scale and sign) equivariant $1-\alpha$ confidence region for μ is given by

$$R(\overline{X}, S^2) = \left\{\mu : F(\overline{X}, S^2, \mu) \leq F^{\alpha}_{1,n-1}\right\} = \left\{\mu : \mu \in \overline{X} \pm t^{\alpha/2}_{n-1}\frac{S}{\sqrt{n}}\right\}.$$

(b) Now consider confidence regions for σ^2. The location, sign and scale invariant pivotal quantity for σ^2 is

$$U(\overline{X}, S^2, \sigma^2) = (n-1)\frac{S^2}{\sigma^2} \sim \chi^2_{n-1}.$$

Now, consider testing the null hypothesis that $\sigma^2 = 1$ against the alternative that $\sigma^2 \neq 1$. This testing problem is invariant under the subgroup of location and sign changes. Unfortunately, there is no UMP (location and sign) invariant test for this problem, so that there is no UMA (location, scale and sign) equivariant confidence region for σ^2. (Note, however, that any confidence region based on $U(\overline{X}, S^2, \sigma^2)$ is location, scale and sign equivariant.)

23.18 We now give some properties of equivariant confidence regions. We recall that a $1-\alpha$ confidence region $R(\mathbf{S})$ is unbiased if, for all $\mathbf{a}$ such that $\mathbf{a} \neq \boldsymbol{\tau}(\boldsymbol{\theta})$, $P_{\boldsymbol{\theta}}(\mathbf{a} \in R(\mathbf{S})) \leq 1 - \alpha$ (i.e. if the probability of covering false values is smaller than the probability of covering true values).

Theorem 23.16.

(a) An equivariant confidence region is unbiased if and only if its associated test is unbiased.

(b) A UMA equivariant confidence region is unbiased.

(c) A unique UMA unbiased $1-\alpha$ confidence region is equivariant.

Proof.

(a) Note that $\mathbf{a} \neq \boldsymbol{\tau}(\boldsymbol{\theta}) \Leftrightarrow \overline{g}(\boldsymbol{\theta}) \notin N$. Therefore this result follows from (23.2).

(b) The test associated with a UMA equivariant region is UMP invariant and hence unbiased. Therefore this result follows from part (a).

(c) Let $R(\mathbf{S})$ be the unique UMA unbiased confidence region for $\boldsymbol{\tau}$, let $g \in G$ and let

$$R^*(\mathbf{S}) = \widetilde{g}^{-1}(R(g^*(\mathbf{S}))) = \{\widetilde{g}^{-1}(\mathbf{a}) : \mathbf{a} \in R(g^*(\mathbf{S}))\}.$$

Then $R(\mathbf{S})$ is equivariant if and only if $R^*(\mathbf{S}) = R(\mathbf{S})$ for all $g \in G$. Define

$$P(\mathbf{a}, \boldsymbol{\theta}) = P_{\boldsymbol{\theta}}(\mathbf{a} \in R(\mathbf{S})),\ P^*(\mathbf{a}, \boldsymbol{\theta}) = P_{\boldsymbol{\theta}}(\mathbf{a} \in R^*(\mathbf{S})).$$

The crucial identity of this proof is that

$$\begin{aligned} P^*(\mathbf{a}, \boldsymbol{\theta}) &= P_{\boldsymbol{\theta}}(\mathbf{a} \in \widetilde{g}^{-1}(R(g^*(\mathbf{S})))) \\ &= P_{\overline{g}(\boldsymbol{\theta})}(\widetilde{g}(\mathbf{a}) \in R(\mathbf{S})) \\ &= P(\widetilde{g}(\mathbf{a}), \overline{g}(\boldsymbol{\theta})). \end{aligned}$$

If $\boldsymbol{\tau}(\boldsymbol{\theta}) = \mathbf{a}$, then $\boldsymbol{\tau}(\overline{g}(\boldsymbol{\theta})) = \widetilde{g}(\boldsymbol{\tau}(\boldsymbol{\theta})) = \widetilde{g}(\mathbf{a})$. Therefore in this case

$$P^*(\mathbf{a}, \boldsymbol{\theta}) = P(\widetilde{g}(\mathbf{a}), \overline{g}(\boldsymbol{\theta})) \geq 1-\alpha$$

(because R is a $1-\alpha$ confidence region). Hence R^* is a $1-\alpha$ confidence region. If $\boldsymbol{\tau}(\boldsymbol{\theta}) \neq \mathbf{a}$, then $\boldsymbol{\tau}(\overline{g}(\boldsymbol{\theta})) \neq \widetilde{g}(\mathbf{a})$ and hence in this case

$$P^*(\mathbf{a}, \boldsymbol{\theta}) = P(\widetilde{g}(\mathbf{a}), \overline{g}(\boldsymbol{\theta})) \leq 1-\alpha$$

(because R is unbiased) and hence R^* is unbiased. Therefore

$$\boldsymbol{\tau}(\boldsymbol{\theta}) \neq \mathbf{a} \Rightarrow P(\mathbf{a}, \boldsymbol{\theta}) \geq P^*(\mathbf{a}, \boldsymbol{\theta}) = P(\widetilde{g}(\mathbf{a}), \overline{g}(\boldsymbol{\theta}))$$

(because R is UMA unbiased). Since this is true for all such $\mathbf{a}$ and $\boldsymbol{\theta}$, and all $g \in G$, it is true for $\widetilde{g}(\mathbf{a}), \overline{g}(\boldsymbol{\theta})$ and g^{-1}, so that

$$P(\widetilde{g}(\mathbf{a}), \overline{g}(\boldsymbol{\theta})) \leq P(\widetilde{g}^{-1}(\widetilde{g}(\mathbf{a})), \overline{g}^{-1}(\overline{g}(\boldsymbol{\theta}))) = P(\mathbf{a}, \boldsymbol{\theta}).$$

Therefore

$$\boldsymbol{\tau}(\boldsymbol{\theta}) \neq \mathbf{a} \Rightarrow P(\mathbf{a}, \boldsymbol{\theta}) = P^*(\mathbf{a}, \boldsymbol{\theta})$$

and R^* is also a UMP unbiased $1-\alpha$ confidence region for $\boldsymbol{\tau}$. By the assumed uniqueness, $R^* = R$ and hence R is equivariant. □

Parts (b) and (c) imply that if there are both UMA equivariant and UMA unbiased confidence regions, then they must be the same.

A $1-\alpha$ confidence region is *admissible* if there is no other $1-\alpha$ confidence region which is more accurate. A size-α test is admissible if there is no other size-α which is more powerful. Arnold (1984) has shown that an equivariant confidence region is admissible if and only if its associated test is admissible. Most UMP invariant tests are admissible, so that most UMA equivariant confidence regions are also. As we have noted earlier, however, in high dimensions, most best equivariant estimators are inadmissible. Therefore, although the confidence regions are typically admissible, the estimators associated with these confidence intervals are often not admissible.

Other normal examples

23.19 We have now finished presenting the basic theory of invariant and equivariant procedures. The remaining sections of this chapter present further examples of these concepts.

In the next example, we observe $X_1, \ldots, X_m, Y_1, \ldots, Y_n$ independent,

$$X_i \sim N(\mu, \sigma^2), \quad Y_j \sim N(\nu, \sigma^2), \qquad \mu \in \mathbf{R},\ \nu \in \mathbf{R},\ \sigma^2 > 0.$$

(Note that we are assuming that the variances are equal for the two populations.) We consider inference about $\delta = \mu - \nu$. A sufficient statistic for this model is $(\overline{X}, \overline{Y}, S_p^2)$ where S_p^2 is the pooled sample variance estimator. This model is invariant under the groups $G_1 = G_{11} \cup G_{12}$, G_2 and G_3 given by

$$\begin{aligned} G_{11} &: (X_i, Y_j) \to (X_i + a, Y_j + a), \qquad G_{12} : (X_i, Y_j) \to (X_i + b, Y_j), \\ G_2 &: (X_i, Y_j) \to (cX_i, cY_j), c > 0; \\ G_3 &: (X_i, Y_j) \to \pm(X_i, Y_j). \end{aligned}$$

Note that under G_1, $\delta \to \delta + b$, under G_2, $\delta \to c\delta$ and under G_3, $\delta \to \pm\delta$. Note that the group G_1 could also be written as $(X_i, Y_j) \to (X_i + d_1, Y_j + d_2)$ (with $d_1 = a + b$ and $d_2 = b$). The form given above is more convenient, however. (Notice that this model is not invariant under different scale changes on the X_i and Y_j because of the assumed equal variances. Note also that δ is not equivariant under different sign changes in the X_i and Y_j.)

Consider estimating δ with the invariant loss function $L(d; (\mu, \nu, \sigma^2)) = (d - \delta)^2/\sigma^2$. Let

$$\widehat{\delta} = \overline{X} - \overline{Y}.$$

An estimator $T(\overline{X}, \overline{Y}, S_p^2)$ is equivariant under G_1 if

$$T(\overline{X} + a + b, \overline{Y} + a, S_p^2) = T(\overline{X}, \overline{Y}, S_p^2) + b$$

for all a and b. This condition is satisfied if and only if

$$T(\overline{X}, \overline{Y}, S_p^2) = \widehat{\delta} + T(0, 0, S_p^2)$$

(see Exercise 23.13). Since $\widehat{\delta}$ and S_p^2 are independent, the risk function for such an equivariant estimator is

$$R(T;(\mu,\nu,\sigma^2)) = \frac{E(\widehat{\delta}-\delta)^2}{\sigma^2} + \frac{E(T(0,0,S_p^2))^2}{\sigma^2},$$

which is minimized when $T(0,0,S_p^2) = 0$. Therefore $\widehat{\delta}$ is the best equivariant estimator of δ under the group G_1. ($\widehat{\delta}$ is the only estimator of δ which is invariant under all of the groups given above. Therefore, $\widehat{\delta}$ is also the ML and MV unbiased estimator of δ.)

Now, consider testing the null hypothesis that $\delta = 0$ against the one-sided alternative hypothesis that $\delta > 0$. This problem is invariant under the groups G_{11} and G_2 given above (but not under G_{12} because this group would change the null hypothesis of equal means). Let

$$t = \frac{\widehat{\delta}}{S_p\sqrt{\frac{1}{m}+\frac{1}{n}}}, \qquad \gamma = \frac{\delta}{\sigma\sqrt{\frac{1}{m}+\frac{1}{n}}}.$$

Under G_{11}, $(\overline{X},\overline{Y},S_p^2) \to (\overline{X}+a,\overline{Y}+a,S_p^2)$ and a maximal invariant is $(\widehat{\delta},S_p^2)$ (see group L2). G_2 operates on this maximal invariant by $(\widehat{\delta},S_p^2) \to (c\widehat{\delta},c^2S_p^2)$ and a maximal invariant is t (group S3). Similarly, γ is a parameter maximal invariant. Finally

$$t \sim t_{m+n-2}(\gamma), \delta = 0 \Leftrightarrow \gamma = 0, \delta > 0 \Leftrightarrow \gamma > 0.$$

The UMP size-α test for this reduced problem rejects if $t > t^\alpha$. This test (the usual one-sided, two-sample t test) is therefore the UMP invariant size-α test for the original model.

Consider next the problem of testing the null hypothesis that $\delta = 0$ against the two-sided alternative that $\delta \neq 0$. This problem is invariant under the groups G_{11}, G_2 and G_3. Let

$$F = t^2 = \frac{\widehat{\delta}^2}{S_p^2\left(\frac{1}{m}+\frac{1}{n}\right)}, \qquad \zeta = \gamma^2 = \frac{\delta^2}{\sigma^2\left(\frac{1}{m}+\frac{1}{n}\right)}.$$

G_3 operates on t, the maximal invariant under G_{11} and G_2, by $t \to \pm t$ and a maximal invariant is F (group O1). Similarly, ζ is a parameter maximal invariant. Finally,

$$F \sim F_{1,m+n-2}(\zeta), \delta = 0 \Leftrightarrow \zeta = 0, \delta \neq 0 \Leftrightarrow \zeta > 0.$$

The UMP size-α test for this reduced problem rejects if

$$F > F^\alpha \Leftrightarrow t > t^{\alpha/2} \text{ or } t < -t^{\alpha/2}.$$

This test (the usual two-sided, two-sample t test) is therefore the UMP invariant size-α test for the original model.

Finally, consider a confidence region for δ. We use all four groups to find the invariant pivotal quantity. G_{11} takes $(\overline{X},\overline{Y},S_p^2,\delta) \to (\overline{X}+a,\overline{Y}+a,S_p^2,\delta)$ and a maximal invariant (group L2) is $(\widehat{\delta},S_p^2,\delta) \to (\widehat{\delta}+b,S_p^2,\delta+b)$ under G_{12}. A maximal invariant under this group (L2 again) is $(\widehat{\delta}-\delta,S_p^2) \to (c(\widehat{\delta}-\delta),c^2S_p^2)$ under G_2 with maximal invariant (group S3)

$$t(\delta) = \frac{\widehat{\delta}-\delta}{S_p\sqrt{\frac{1}{m}+\frac{1}{n}}} \sim t_{n-1}.$$

(Note that the union of G_1 and G_2 is transitive on the parameter space and hence $t(\delta)$ is a pivotal quantity. There is, however, no UMA equivariant confidence region based on $t(\delta)$.) G_3 takes $t(\delta) \to \pm t(\delta)$ and a maximal invariant (group O1) is

$$F(\delta) = (t(\delta))^2 = \frac{(\widehat{\delta} - \delta)^2}{S_p^2\left(\frac{1}{m} + \frac{1}{n}\right)} \sim F_{1,m+n-2},$$

which is the invariant pivotal quantity under G_1, G_2 and G_3. Now consider testing that $\delta = 0$ against $\delta \neq 0$. The subgroup leaving this problem invariant is the union of G_{11}, G_2 and G_3. We showed above that the UMP invariant size-α test for this testing problem accepts if $F(0) \leq F^\alpha$. Therefore, the UMA equivariant $1 - \alpha$ confidence region for δ is given by

$$\{\delta : F(\delta) \leq F^\alpha\} = \left\{\delta \in \widehat{\delta} \pm t^{\alpha/2} S_p \sqrt{\frac{1}{m} + \frac{1}{n}}\right\},$$

(the usual two-sample confidence interval for δ).

23.20 We now consider the two-sample model with unequal variances. In this model, we observe $X_1, \ldots, X_m, Y_1, \ldots, Y_n$ independent such that

$$X_i \sim N(\mu, \sigma^2), \quad Y_i \sim N(\nu, \tau^2), \; \mu \in \mathbf{R}, \qquad \nu \in \mathbf{R}, \; \sigma^2 > 0, \; \tau^2 > 0.$$

A sufficient statistic for this model is $(\overline{X}, \overline{Y}, S^2, T^2)$, where S^2 and T^2 are the sample variances of the Xs and Ys.

We first consider inference about λ where $\lambda = \sigma^2/\tau^2, \widehat{\lambda} = S^2/T^2$. We use the groups G_1 and $G_2 = G_{21} \cup G_{22}$, given by

$$G_1 : (X_i, Y_j) \to (X_i + a + b, Y_j + a),$$
$$G_{21} : (X_i, Y_j) \to (cX_i, cY_j), c > 0, \qquad G_{22} : (X_i, Y_j) \to (dX_i, Y_j), d > 0.$$

Under G_1, $\lambda \to \lambda$ and under G_2, $\lambda \to d\lambda$. (Note that this model is invariant under different scale changes for the X_i and Y_j because we are allowing the variances to be different.) An estimator $Q(\overline{X}, \overline{Y}, S^2, T^2)$ is equivariant under these four groups if and only if it has the form

$$Q(\overline{X}, \overline{Y}, S^2, T^2) = Q(0, 0, 1, 1)\widehat{\lambda}$$

(see Exercise 23.14(a)). The optimal equivariant estimator depends on the loss function.

Consider testing that $\lambda = 1$ against $\lambda > 1$. This problem is invariant under the groups G_1 and G_{21}. The maximal invariant and parameter maximal invariant under these groups are $\widehat{\lambda}$ and λ. Furthermore,

$$\widehat{\lambda}/\lambda \sim F_{m-1,n-1}.$$

The distribution of $\widehat{\lambda}$ has monotone LR (see Exercise 23.14(b)), and hence the UMP size-α test for the reduced problem of testing the null hypothesis $\lambda = 1$ against the alternative hypothesis $\lambda > 1$ rejects if $\widehat{\lambda} > F^\alpha$. This test is therefore the UMP invariant size-α test for the original problem. Now, consider testing the null hypothesis $\lambda = 1$ against the two-sided alternative that

$\lambda \neq 1$. In general, there is no further reduction in the problem, and hence no UMP invariant size-α test for this problem. However, if $m = n$, then the problem is further invariant under

$$G_3 : (X_i, Y_j) \to (Y_j, X_i)$$

(i.e. this problem is invariant under interchanging the two samples). A maximal invariant and parameter maximal invariant under this third group (applied to the maximal invariant under the union of G_1 and G_{21}) are

$$U = (\log(\widehat{\lambda}))^2, \quad \varpi = (\log(\lambda))^2 \qquad \lambda = 1 \Leftrightarrow \varpi = 0, \quad \lambda \neq 1 \Leftrightarrow \varpi > 0$$

(see Exercise 23.14(c)). Note that the reduced problem is the one-sided problem of testing that $\varpi = 1$ against $\varpi > 1$. The distribution of U has monotone LR (see Exercise 23.14(d)), so that when $m = n$, the UMP invariant size-α test for this problem rejects if U is too large, i.e. if $\widehat{\lambda} > F^{\alpha/2}$ or $\widehat{\lambda} < 1/F^{\alpha/2}$.

Finally, consider a confidence region for λ. The invariant pivotal quantity (under G_1 and G_2) for λ is $F(\lambda) = \widehat{\lambda}/\lambda \sim F_{m-1,n-1}$. Unless $m = n$, there is no UMP invariant test that $\lambda = 1$ against $\lambda \neq 1$, and hence there is no UMA equivariant confidence region for λ. When $m = n$, we can use G_3 to get a UMP invariant test for this problem, which accepts if $1/F^{\alpha/2} \leq F(1) \leq F^{\alpha/2}$. Therefore, when $m = n$, the UMA equivariant $1 - \alpha$ confidence region for λ is

$$\{\lambda : 1/F^{\alpha/2} \leq F(\lambda) \leq F^{\alpha/2}\} = \{\lambda : \widehat{\lambda}/F^{\alpha/2} \leq \lambda \leq \widehat{\lambda}F^{\alpha/2}\}.$$

Now consider $\delta = \mu - \nu$. This parameter is not equivariant under the group G_{22} given above. In fact the group of transformations for which δ is equivariant is not transitive on the parameter space, which may account for the lack of a suitable pivotal quantity for δ. See **19.25–26** and **21.21–23** for some exact and approximate procedures for this problem. (The problem of making exact inference about δ in the presence of unequal variances is called the Behrens–Fisher problem and has vexed statisticians for a long time.)

23.21 Suppose we observe $(X_1, Y_1), \ldots, (X_n, Y_n)$ independent bivariate random vectors such that

$$\begin{pmatrix} X_i \\ Y_i \end{pmatrix} \sim N_2\left(\begin{pmatrix} \mu \\ \nu \end{pmatrix}, \begin{pmatrix} \sigma_{11} & \sigma_{12} \\ \sigma_{21} & \sigma_{22} \end{pmatrix}\right), \qquad \mu \in \mathbf{R},\ \nu \in \mathbf{R},\ \sigma_{ii} > 0,\ \sigma_{12}^2 < \sigma_{11}\sigma_{22}.$$

We are again concerned with inference about $\delta = \mu - \nu$. Before considering invariance for this model, we transform it by letting $U_i = X_i - Y_i$, $V_i = X_i + Y_i$. Then

$$\begin{pmatrix} U_i \\ V_i \end{pmatrix} \sim N_2\left(\begin{pmatrix} \delta \\ \gamma \end{pmatrix}, \begin{pmatrix} \tau_{11} & \tau_{12} \\ \tau_{21} & \tau_{22} \end{pmatrix}\right),$$

for some $\gamma \in \mathbf{R}$, $\tau_{ii} > 0$, $\tau_{12}^2 < \tau_{11}\tau_{22}$. This model is invariant under the groups $G_1 = G_{11} \cup G_{12}$, G_2, $G_3 = G_{31} \cup G_{32}$ and G_4, where

$$G_{11} : (U_i, V_i) \to (U_i + a_1, V_i); \qquad G_{12} : (U_i, V_i) \to (U_i, V_i + a_2);$$

$$G_2 : (U_i, V_i) \to (U_i, V_i + cU_i); \qquad G_{31} : (U_i, V_i) \to (b_1 U_i, V_i), b_1 > 0;$$
$$G_{32} : (U_i, V_i) \to (U_i, b_2 V_i), b_2 > 0; \qquad G_4 : (U_i, V_i) \to (\pm U_i, V_i).$$

Note that under G_{11}, $\delta \to \delta + a_1$, under G_{31}, $\delta \to b_1\delta$ and under G_4, $\delta \to \pm\delta$, but that δ is unchanged by the other groups. (This model is also invariant under $(U_i, V_i) \to (U_i + dV_i, V_i)$. Under this group, however, $\delta \to \delta + d\gamma$. Hence δ is not an equivariant parameter for this group.) Let T_{11} and T_{22} be the sample variances of the U_i and V_i and let T_{12} be the sample covariance between the U_i and V_i. A complete sufficient statistic for this model is $(\overline{U}, \overline{V}, T_{11}, T_{22}, T_{12})$. Under G_1, the best equivariant estimator of δ is $\overline{U} = \overline{X} - \overline{Y}$ for quadratic loss functions (see Exercise 23.15).

We now find the invariant pivotal quantity for δ. It is easily seen that the maximal invariant under G_1 is

$$(\overline{U} - \delta, T_{11}, T_{22}, T_{12}) \to (\overline{U} - \delta, T_{11}, cT_{11} + T_{12}, c^2 T_{11} + cT_{12} + T_{22})$$

under G_2. A maximal invariant (group B1) is

$$(\overline{U} - \delta, T_{11}, T_{22} - T_{12}^2/T_{11}) \to (\overline{U} - \delta, T_{11}, b_2^2(T_{22} - T_{21}^2/T_{11}))$$

under G_{32}. A maximal invariant (group S1) is $(\overline{U} - \delta, T_{11}) \to (b_1(\overline{U} - \delta), b_1^2 T_{11})$ under G_{31}. A maximal invariant (group S3) is

$$t(\delta) = \frac{\sqrt{n}(\overline{U} - \delta)}{\sqrt{T_{11}}} \sim t_{n-1}.$$

(Note that these groups together are transitive on the parameter space, so that $t(\delta)$ is a pivotal quantity. However, there is no UMA equivariant confidence region under these groups.) Under G_4, $t(\delta) \to \pm t(\delta)$. A maximal invariant under this group (O1) is

$$F(\delta) = (t(\delta))^2 = n\frac{(\overline{U} - \delta)^2}{T_{11}} \sim F_{1,n-1}.$$

Now, consider the problem of testing the null hypothesis that $\delta = 0$ against the alternative hypothesis that $\delta \neq 0$. Let $\theta = n\delta^2/\tau_{11}$. This testing problem is invariant under all the groups listed above except G_{11}. By Theorem 23.13, the maximal invariant under these groups is

$$F(0) \sim F_{1,n-1}(\theta), \qquad \delta = 0 \Leftrightarrow \theta = 0, \quad \delta \neq 0 \Leftrightarrow \theta > 0.$$

For the reduced problem, the UMP size-α test rejects if

$$F(0) > F_{1,n-1}^{\alpha} \Leftrightarrow t(0) > t^{\alpha/2} \text{ or } t(0) < -t^{\alpha/2},$$

and hence this test (the usual two-sided paired comparison test) is the UMP invariant size-α test for the original problem. Since this test accepts the null hypothesis when $F(0) \leq F^{\alpha}$, the UMA equivariant confidence region for δ is

$$\{\delta : F(\delta) \leq F^{\alpha}\} = \left\{\delta \in \overline{U} \pm t^{\alpha/2}\sqrt{\frac{T_{11}}{n}}\right\},$$

the usual paired confidence interval for δ. The problem of testing the null hypothesis that $\delta = 0$ against the one-sided alternative that $\delta > 0$ is invariant under all the groups given above except G_{11} and G_4. By arguments similar to those in previous examples, it is straightforward to see that the UMP invariant size-α test for this problem rejects if $t(0) > t^{\alpha}$, the usual one-sided paired comparison test.

Let ρ and r be the correlation coefficient and sample correlation coefficient for the (U_i, V_i),

$$\rho = \frac{\tau_{12}}{\sqrt{\tau_{11}\tau_{22}}}, \qquad r = \frac{T_{12}}{\sqrt{T_{11}T_{22}}}.$$

This parameter ρ is not equivariant under G_2. In fact, the group of transformations for which ρ is equivariant is not transitive on the parameter space, which may explain the lack of a pivotal quantity for ρ. Consider testing the null hypothesis that $\rho = 0$ against the alternative hypothesis that $\rho > 0$. This problem is invariant under the groups G_1 and G_3. The maximal invariant and parameter maximal invariant for these groups are r and ρ (see Exercise 23.16). Therefore the distribution of r depends only on ρ. Let $h(r; \rho)$ be the density function of r. Then $h(r; \rho)/h(r; 0)$ is an increasing function of r. Therefore, the UMP test of these hypotheses based on r rejects if r is too large. Recalling that

$$t(r) = \frac{r\sqrt{n-2}}{\sqrt{1-r^2}} \sim t_{n-2}$$

under the null hypothesis that $\rho = 0$, we see that the UMP invariant size-α test for this problem rejects if $t(r) > t^{\alpha}$. Similarly for the two-sided problem, the equal-tailed two-sided test based on $t(r)$ is the UMP invariant size-α test under G_1, G_3 and G_4. (This example illustrates that UMP invariant one-sided and two-sided tests are possible even for parameters which have no pivotal quantity.)

Note that r is unchanged by G_1 and G_3. Therefore an estimator of ρ is equivariant under G_1 and G_3 if it is invariant under those groups, that is, if and only if it is a function of the maximal invariant r. An estimator $T(r)$ is equivariant under G_4 if and only if $T(-r) = -T(r)$. Since the induced group $\overline{G}$ on the parameter space induced by $G = G_1 \cup G_3 \cup G_4$ is not transitive, risk functions of equivariant estimators are not constant and hence there need not be a best equivariant estimator of ρ for many invariant loss functions.

23.22 We now give an example in which invariance leads to some less obvious procedures. This example is a special case of a growth curves model. Suppose we observe $(U_1, V_1), \ldots, (U_n, V_n)$ independent random vectors such that

$$\begin{pmatrix} U_i \\ V_i \end{pmatrix} \sim N_2\left(\begin{pmatrix} \delta \\ 0 \end{pmatrix}, \begin{pmatrix} \tau_{11} & \tau_{12} \\ \tau_{21} & \tau_{22} \end{pmatrix}\right), \qquad \delta \in \mathbf{R},\ \tau_{ii} > 0,\ \tau_{12}^2 < \tau_{11}\tau_{22}.$$

We are interested in inference about δ. A minimal sufficient statistic for this model is $(\overline{U}, \overline{V}, T_{11}, T_{22}, T_{12})$, where the T_{ij} are defined in the above example. Note that this statistic is not complete because $E\overline{V} = 0$. Note also that the sufficient statistic is the same as for the example above even though we have one parameter less (because $\gamma = 0$ in this model). An obvious estimator of δ

is $\overline{U}$ and an obvious pivotal quantity is

$$t(\delta) = \sqrt{n}(\overline{U} - \delta)/\sqrt{T_{11}} \sim t_{n-1}.$$

(Note that these are the estimator and pivotal quantity derived in the example above when we made no assumption about γ.) We shall show, however, that invariance considerations lead to different procedures.

To motivate these procedures, note that

$$U_i | V_i \sim N(\delta + \theta V_i, \xi^2), \qquad \theta = \frac{\tau_{12}}{\tau_{22}}, \quad \xi^2 = \tau_{11} - \tau_{12}^2/\tau_{22}$$

(see Exercise 23.17(a)). Since the (U_i, V_i) are independent, this conditional model is a simple regression model. Another possible estimator for δ is the ordinary least squares estimator for the intercept in the conditional simple linear regression model,

$$\widehat{\delta} = \overline{U} - \frac{T_{12}\overline{V}}{T_{22}}, \qquad \widehat{\delta}|\mathbf{V} \sim N\left(\delta, \frac{\xi^2}{n}(1+Q)\right), \qquad (n-1)Q = \frac{n\overline{V}^2}{T_{22}} \sim F_{1,n-1}$$

(see Exercise 23.17(b)). Let $\widehat{\xi^2}$ be the usual unbiased estimator of ξ^2 for the conditional linear model,

$$\widehat{\xi^2} = \frac{(n-1)(T_{11} - T_{12}^2/T_{22})}{n-2}.$$

Another pivotal quantity for δ can be constructed from the conditional linear model,

$$t^*(\delta) = \frac{\sqrt{n}(\widehat{\delta} - \delta)}{\widehat{\xi}\sqrt{1+Q}} \sim t_{n-2}.$$

(see Exercise 23.17(c)).

This model is invariant under the groups G_1, G_2, $G_3 = G_{31} \cup G_{32}$ and $G_4 = G_{41} \cup G_{42}$:

$$\begin{aligned} G_1 &: (U_i, V_i) \to (U_i + cV_i, V_i); & G_2 &: (U_i, V_i) \to (U_i + a_1, V_i); \\ G_{31} &: (U_i, V_i) \to (b_1 U_i, V_i), b_1 > 0; & G_{32} &: (U_i, V_i) \to (U_i, b_2 V_i), b_2 > 0; \\ G_{41} &: (U_i, V_i) \to (\pm U_i, V_i); & G_{42} &: (U_i, V_i) \to (U_i, \pm V_i). \end{aligned}$$

Note that G_1 takes $\delta \to \delta$, G_2 takes $\delta \to \delta + a$, G_3 takes $\delta \to b_1\delta$ and G_4 takes $\delta \to \pm\delta$. Note also that this model is not invariant under location changes in the V_i because their mean is assumed to be 0.

We first find the invariant pivotal quantity for δ. G_1 takes

$$(\overline{U}, \overline{V}, T_{11}, T_{22}, T_{12}, \delta) \to (\overline{U} + c\overline{V}, \overline{V}, T_{11} + 2cT_{12} + c^2 T_{22}, T_{22}, T_{12} + cT_{22}, \delta)$$

and a maximal invariant (group B2) is $(\widehat{\delta}, \widehat{\xi}^2, \overline{V}, T_{22}, \delta)$. The reduction by G_2, G_3 and G_4 is straightforward, leading to the pivotal quantity

$$(F^*(\delta), Q), \qquad F^*(\delta) = (t^*(\delta))^2 = \frac{n(\widehat{\delta} - \delta)^2}{\widehat{\xi}^2(1+Q)}, \qquad Q = \frac{n\overline{V}^2}{(n-1)T_{22}}$$

(see Exercise 23.18(a)). Note that

$$F^*(\delta)|Q \sim F_{1,n-2}, \qquad (n-1)Q \sim F_{1,n-1}$$

(see Exercise 23.18(b)) so that F^* and Q are independent. Consider testing the null hypothesis that $\delta = 0$ against the two-sided alternative that $\delta \neq 0$. This problem is invariant under G_1, G_3 and G_4. By Theorem 23.13, a maximal invariant for this problem is $(F^*(0), Q)$. Let $\eta = n\delta^2/\xi^2$. Then

$$F^*(0)|Q \sim F_{1,n-2}\left(\frac{\eta}{1+Q}\right), \qquad (n-1)Q \sim F_{1,n-1},$$

$$\delta = 0 \Leftrightarrow \eta = 0, \qquad \delta \neq 0 \Leftrightarrow \eta > 0$$

(see Exercise 23.18(c)). Unfortunately, there is no UMP test for this reduced model. Note, however, that the distribution of Q does not depend on any unknown parameters and hence Q is an ancillary statistic for this problem. The ancillarity principle implies that we should make our inference conditionally on Q. Conditionally, this model is one in which we are testing the null hypothesis that a non-central F distribution has non-centrality parameter 0 against the alternative that it is positive. The UMP test for this reduced conditional problem rejects if $F^*(0) > F^\alpha$. This test is therefore UMP among invariant conditional size-α tests. Since this test accepts when $F^*(0) \leq F^\alpha$, the confidence region which is UMA among equivariant conditional $1-\alpha$ confidence regions is given by

$$\{\delta : F^*(\delta) \leq F^\alpha\} = \left\{\delta \in \widehat{\delta} \pm t^{\frac{\alpha}{2}}\widehat{\xi}\sqrt{\frac{1+Q}{n}}\right\},$$

the confidence interval based on the conditional pivotal quantity $t^*(\delta)$. (Note that the test based on the naive pivotal quantity $t(\delta)$ is not invariant under the groups which leave this model invariant nor is the associated confidence interval equivariant under those groups.)

Since δ is unaffected by G_1, an estimator of δ is equivariant under G_1 if it is invariant under G_1, that is, if it is a function of $(\widehat{\delta}, \widehat{\xi}^2, \overline{V}, T_{22})$. Such an estimator is equivariant under G_2 if it has the form

$$K(\widehat{\delta}, \widehat{\xi}^2, \overline{V}, T_{22}) = \widehat{\delta} + K(0, \widehat{\xi}^2, \overline{V}, T_{22}).$$

Using the fact that

$$E((\widehat{\delta} - \delta)K(0, \widehat{\xi}^2, \overline{V}, T_{22})) = E(K(0, \widehat{\xi}^2, \overline{V}, T_{22})E((\widehat{\delta} - \delta)|(\widehat{\xi}^2, \overline{V}, T_{22}))) = 0,$$

we see that $\widehat{\delta}$ is the best equivariant (under G_1 and G_2) estimator of δ for quadratic loss functions. If we ask that the estimator also be equivariant under G_4, then $\widehat{\delta}$ is the only equivariant estimator of δ and hence must also be the ML estimator of δ. (Note that there is no MV unbiased estimator for this model, essentially because the minimal sufficient statistic is not complete.) It is interesting to compare the performance of $\widehat{\delta}$ with that of the naive estimator $\overline{U}$ (which is not equivariant for this model). Note that both estimators are unbiased and that

$$\text{var}(\overline{U}) = \frac{\tau_{11}}{n},$$

$$\mathrm{var}(\widehat{\delta}) = E(\mathrm{var}(\widehat{\delta}|\mathbf{V})) + \mathrm{var}(E(\widehat{\delta}|\mathbf{V})) = \frac{\xi^2(1+EQ)}{n} + 0 = \frac{\xi^2(n-2)}{n(n-3)},$$

$$\mathit{Eff} = \frac{\mathrm{var}(\widehat{\delta})}{\mathrm{var}(\overline{U})} = \frac{\xi^2(n-2)}{\tau_{11}(n-3)} = (1-\rho^2)\frac{n-2}{n-3},$$

where ρ is the correlation coefficient between U_i and V_i. Note that

$$0 < \mathit{Eff} \leq \frac{n-2}{n-3} = 1 + \frac{1}{n-3},$$

so that Eff cannot be much more that 1, but can be arbitrarily close to 0 (as ρ approaches 1). Similarly, the expected squared lengths of the confidence intervals, EL^2 and EL^{*2}, based on the marginal pivotal quantity $t(\delta)$ and the conditional pivotal quantity $t^*(\delta)$, are

$$EL^2 = F^{\alpha}_{1,n-1}\frac{\tau_{11}}{n},\ EL^{*2} = F^{\alpha}_{1,n-2}\frac{\xi^2(1+EQ)}{n} = F^{\alpha}_{1,n-2}\frac{\xi^2(n-2)}{n(n-3)},$$

$$\frac{EL^{*2}}{EL^2} = (1-\rho^2)\frac{(n-2)F^{\alpha}_{1,n-2}}{(n-3)F^{\alpha}_{1,n-1}}.$$

Note again that the efficiency can never be much more than 1, but can be arbitrarily near 0. Both these calculations indicate that in the absence of information about ρ, the conditional procedures based on $\widehat{\delta}$ and $t^*(\delta)$ are preferable to the marginal procedures based on $\overline{U}$ and $t(\delta)$. Note that neither $\overline{U}$ nor $t(\delta)$ behaves correctly under G_1. It is interesting that invariance, which involves no conditioning, leads to the conditional procedures based on $\widehat{\delta}$ and $t^*(\delta)$ rather than the marginal procedures based on $\overline{U}$ and $t(\delta)$.

23.23 We now give one of the classical counter-examples in statistics due to Neymann and Scott (1948). Suppose we observe $X_1, \ldots, X_n, Y_1, \ldots, Y_n$ independent,

$$X_i \sim N(\varepsilon_i, \sigma^2), \qquad Y_i \sim N(-\varepsilon_i, \sigma^2).$$

We want to estimate σ^2 with the loss function $(d-\sigma^2)^2/\sigma^4$. This problem is invariant under the groups

$$G_1 : (X_i, Y_i) \to (X_i + \tfrac{1}{2}a_i, Y_i - \tfrac{1}{2}a_i), \qquad G_2 : (X_i, Y_i) \to (bX_i, bY_i), b > 0.$$

A sufficient statistic for this model is

$$(U_1, \ldots, U_n, W), \qquad U_i = X_i - Y_i, \quad W = \sum(X_i + Y_i)^2.$$

An estimator $T(U_1, \ldots, U_n, W)$ of σ^2 is equivariant if

$$\forall b > 0, a_i, T(bU_1 + a_1, \ldots, bU_n + a_n, b^2W) = b^2T(U_1, \ldots, U_n, W)$$
$$\Leftrightarrow T(U_1, \ldots, U_n, W) = dW$$

for some $d > 0$ (see Exercise 23.19). Straightforward calculations show that the risk function for such equivariant estimators is minimized when $d = (n+2)^{-1}$, so the best equivariant estimator is $W/(n+2)$. Note that the ML estimator for this model is the equivariant estimator with $d = (2n)^{-1}$. The best equivariant estimator of σ^2 is consistent, but the ML estimator is badly inconsistent, converging to $\sigma^2/2$ as $n \to \infty$. This is one example in which the best equivariant estimator is much better than the ML estimator.

Non-normal examples

23.24 All the examples so far have involved normal random variables. We now give examples involving non-normal models.

Let $X_1, \ldots, X_n$ be independent random variables which are exponentially distributed with mean μ. This model is invariant under the group G of scale changes

$$G : X_i \to cX_i, \quad \mu \to c\mu, \quad c > 0$$

(note that the model is not invariant under location changes). A sufficient statistic for this model is $\overline{X}$. Consider first estimating μ with the invariant loss function

$$L(d, \mu) = (d - \mu)^2/\mu^2.$$

An estimator $T(\overline{X})$ of μ is equivariant if

$$\forall c > 0, \qquad T(c\overline{X}) = cT(\overline{X}) \Leftrightarrow T(\overline{X}) = \overline{X}T(1) = d\overline{X}$$

for some d. The risk for such an equivariant estimator is

$$\frac{E(d\overline{X} - \mu)^2}{\mu^2} = d^2\frac{n+1}{n} - 2d + 1,$$

which is minimized for $d = n/(n+1)$, so that the best equivariant estimator of μ is

$$\frac{n}{n+1}\overline{X}$$

(see Exercise 23.20). Note that the ML estimator and MV unbiased estimator is $\overline{X}$. Note also that when $n = 1$, the best equivariant estimator is half of the ML estimator.

The invariant pivotal quantity for this model is

$$\frac{2n\overline{X}}{\mu} \sim \chi^2_{2n}$$

(see Exercise 23.20).There are no UMA confidence regions or UMP two-sided tests. The one-sided test based on this pivotal quantity is actually UMP, not just UMP invariant.

Now, suppose we observe $X_1, \ldots, X_m, Y_1, \ldots, Y_n$ independently exponentially distributed with means $\mu = EX_i$ and $\nu = EY_j$. This model is invariant under the following groups

$$G_1 : (X_i, Y_j) \to (cX_i, cY_j), c > 0; \qquad G_2 : (X_i, Y_j) \to (dX_i, Y_j), d > 0.$$

The parameter $\delta = \mu - \nu$ is not equivariant under these groups and has no pivotal quantity. Instead, we study the parameter $\gamma = \mu/\nu$. Let $\widehat{\gamma} = \overline{X}/\overline{Y}$. Under G_1, $(\overline{X}, \overline{Y}, \gamma) \to (c\overline{X}, c\overline{Y}, \gamma)$.

A maximal invariant (group S2) is $(\widehat{\gamma}, \gamma) \to (d\widehat{\gamma}, d\gamma)$ under G_2. A maximal invariant (group S2 again) is

$$F(\gamma) = \widehat{\gamma}/\gamma \sim F_{2m,2n},$$

which is the invariant pivotal quantity for γ. When $m = n$, the model is also invariant under the group which interchanges the two samples. In this case, the UMP invariant size-α test that $\gamma = 1$ against $\gamma \neq 1$ is the equal-tailed test based on the test statistic $F(1)$. Therefore, when $m = n$, the UMA equivariant confidence region for γ is the equal-tailed region based on $F(\gamma)$. When $m \neq n$, there is no UMP invariant test for the two-sided problem and hence no UMA equivariant confidence region for γ. For the one-sided problem of testing that $\gamma = 1$ against $\gamma > 1$, the one-sided test based on the test statistic $F(1)$ is UMP invariant even when $m \neq n$ (see Exercise 23.21).

23.25 Let $X_1, \ldots, X_n$ be independently distributed uniformly on the interval $(0, \theta)$. This model is invariant under the group G of scale changes $X_i \to cX_i$. A sufficient statistic for this model is $S = \max(X_i)$.

Consider estimating θ with the invariant quadratic loss function $(d - \theta)^2/\theta^2$. An estimator $T(S)$ is equivariant if and only if $T(S) = dS$ for some d. Direct calculation shows that

$$\frac{E(dS - \theta)^2}{\theta^2} = d^2\frac{n}{n+2} + 2d\frac{n}{n+1} + 1,$$

which is minimized by $d = (n + 2)/(n + 1)$, so that the best equivariant estimator of θ is

$$\frac{n+2}{n+1}S$$

(see Exercise 23.22). Note that the ML estimator is an equivariant estimator with $d = 1$ and the MV unbiased estimator is equivariant with $d = (n + 1)/n$.

The invariant pivotal quantity for θ is S/θ, which can be used to construct tests and confidence intervals in the usual way. The one-sided test is UMP. Interestingly enough for this model, there is a non-randomized UMP size-α test for the problem of testing the null hypothesis that $\theta = c$ against the two-sided alternative that $\theta \neq c$ (see Ferguson, 1967, p. 213) which can be used to construct a confidence interval which is UMA among all $1 - \alpha$ confidence intervals for θ which is given by

$$S \leq \theta \leq S/\alpha^{1/n}.$$

This confidence interval is based on the invariant pivotal quantity. This model appears to be the only example in which a UMP size-α test exists for a two-sided alternative and hence appears to be the only one for which a UMA confidence interval exists. (For other models, the best we can hope for is a UMA equivariant or UMA unbiased confidence region.)

23.26 Let $X_1, \ldots, X_n$ be independently uniformly distributed on the interval $(\mu - \delta, \mu + \delta)$. This model is invariant under location, scale and sign changes:

$$G_1 : X_i \to X_i + a, \qquad G_2 : X_i \to bX_i, \quad b > 0, \qquad G_3 : X_i \to \pm X_i.$$

Let

$$Q = \min(X_i), \qquad S = \max(X_i), \qquad M = \frac{Q+S}{2}, \qquad H = \frac{S-Q}{2}$$

(that is, M is the mid-range and H is the half-range of the observations). Then (M, H) is a sufficient statistic for this model.

We first consider estimating μ with the quadratic loss function $(d - \mu)^2/\delta^2$. An estimator $T_1(M, H)$ is equivariant under G_1 if

$$\forall a \in \mathbf{R},\ T_1(M + a, H) = a + T_1(M, H) \Leftrightarrow T_1(M, H) = M + T_1(0, H).$$

It can be shown that

$$E((M - \mu)T_1(0, H)) = 0$$

(see Exercise 23.23(b)) and hence M is the best equivariant (under G_1) estimator of μ for this quadratic loss function. If we also require T_1 to be equivariant under G_3, then $T_1(0, H)$ must be 0, and M is the only equivariant estimator of μ and hence M is the ML estimator and MV unbiased estimator of μ.

Now, consider estimating δ with quadratic loss function $(d_2 - \delta)^2/\delta^2$. An estimator $T_2(M, H)$ is equivariant under G_1 and G_2 if

$$\forall a \in R, b > 0,\ T_2(bM + a, bH) = bT_2(M, H) \Leftrightarrow T_2(M, H) = dH,$$

for some d. The risk function for an estimator of this form is

$$\frac{E(dH - \delta)^2}{\delta^2} = \frac{n(n-1)}{(n+1)(n+2)}d^2 + 2\frac{n-1}{n+1}d + 1$$

which is minimized when $d = (n + 2)/n$, so that the best equivariant estimator of δ for this quadratic loss function is $T_2(M, H) = (n + 2)H/n$ (see Exercise 23.23(d)). Note that the ML estimator and the MV unbiased estimator of δ are equivariant estimators with $d = 1$ and $d = (n + 1)/(n - 1)$, respectively. When $n = 2$, the best equivariant estimator is twice the ML estimator.

We now find the invariant pivotal quantity for μ. G_1 takes $(M, H, \mu) \to (M + a, H, \mu + a)$. A maximal invariant is $(M - \mu, H) \to (b(M - \mu), bH)$ under G_2. A maximal invariant is

$$q(\mu) = \frac{M - \mu}{H}$$

which is a pivotal quantity for μ. In order to find optimal procedures, it is necessary, however, to reduce further by G_3 which takes $q(\mu) \to \pm q(\mu)$ and a maximal invariant is $p(\mu) = (q(\mu))^2$. Now, consider testing the null hypothesis that $\mu = 0$ against the alternative hypothesis that $\mu \neq 0$. By arguments similar to those in previous examples, the UMP invariant size-α test rejects if

$$p(0) > c^2 \Leftrightarrow q(0) > c \text{ or } q(0) < -c,$$

where c is chosen so the test has size α (see Exercise 23.24(c)). Therefore, the UMA equivariant $1 - \alpha$ confidence region for μ is

$$\{\mu : p(\mu) \leq c^2\} = \{\mu \in M \pm cH\}.$$

By similar arguments to those given above, we can show that the one-sided test based on $q(\mu)$ is UMP invariant for the one-sided testing problem.

Similarly, we can show that the invariant pivotal quantity for δ is $r(\delta) = H/\delta$. One-sided tests based on $r(\delta)$ are UMP invariant for one-sided problems involving δ. There is no UMP invariant test for the two-sided problem about δ and hence no UMA equivariant confidence region for δ (see Exercise 23.24 for details).

23.27 Let $X_1, \ldots, X_n$ be independently uniformly distributed on the interval $(\mu - \frac{1}{2}, \mu + \frac{1}{2})$. This model is invariant under location and sign changes (G_1 and G_3 above). A minimal sufficient statistic for this model is (M, H) defined above. Note that this statistic is not complete because

$$E(H - (n-1)/2(n+1)) = 0.$$

An estimator $T(M, H)$ of μ is equivariant if

$$\forall a \in \mathbf{R}, T(M + a, H) = T(M, H) + a \Leftrightarrow T(M, H) = M + T(0, H).$$

As mentioned above, $E((M - \mu)T(0, H)) = 0$. Therefore M is the best location equivariant estimator of μ. (It is the only estimator which is location and sign equivariant). Note also that there is no MV unbiased estimator for μ and the ML estimator is not unique. Any point in

$$(M - (\tfrac{1}{2} - H), M + \tfrac{1}{2} - H)$$

is an ML estimator. This model gives one situation in which a best equivariant estimator is unique, but no unique MV unbiased or ML estimator exists.

The invariant pivotal quantity for this model is $(M - \mu, H)$. Note that the distribution of H does not depend on μ so that H is ancillary. The ancillarity principle implies that we should make our inference conditionally on H. Now conditionally on H, $M - \mu$ is uniformly distributed on $(-(\frac{1}{2} - H), \frac{1}{2} - H)$. Therefore, an equivariant conditionally $1 - \alpha$ confidence interval for μ is

$$\mu \in M \pm (1 - \alpha)(\tfrac{1}{2} - H).$$

Note that, using the obvious marginal pivotal quantity, $M - \mu$ leads to confidence intervals with serious interpretation problems (see, for example, Arnold, 1990, pp. 402–404).

23.28 For all the testing examples considered so far the relevant parameters are univariate, and for most such parameters it can be shown that there must be a UMP unbiased test. Since a UMP unbiased test and a UMP invariant test must be the same (if they both exist), most of the UMP invariant tests given above must also be UMP unbiased. Similarly, most of the UMA equivariant confidence regions are also UMA unbiased. It is often easier to show procedures are UMP invariant or UMA equivariant than to show directly that they are UMP unbiased or UMA unbiased so invariance considerations can often lead to simpler derivations of optimal unbiased procedures.

In the following examples, the parameters of interest are vectors and hence optimal unbiased tests or confidence regions are not available. Optimal invariant tests or equivariant confidence regions often are available, however. Note that all UMP invariant tests and UMA equivariant confidence regions are unbiased, so that even in vector cases, invariance considerations often lead to a fairly simple proof of unbiasedness.

Linear model examples

23.29 Suppose we observe

$$\mathbf{Y} \sim N_n(\mathbf{X}\boldsymbol{\beta}, \sigma^2\mathbf{I}),$$

where $\mathbf{X}$ is a known $n \times p$ matrix of rank p, and $\boldsymbol{\beta} \in \mathbf{R}^p$ and $\sigma^2 > 0$ are unknown parameters. Suppose we want to estimate $\boldsymbol{\beta}$ with a quadratic loss function

$$L(\mathbf{d}, (\boldsymbol{\beta}, \sigma^2)) = \frac{(\mathbf{d} - \boldsymbol{\beta})^T \mathbf{A}(\mathbf{d} - \boldsymbol{\beta})}{\sigma^2}.$$

(Two common choices for $\mathbf{A}$ are $\mathbf{A} = \mathbf{I}$ and $\mathbf{A} = \mathbf{X}^T\mathbf{X}$.) This model is invariant under the group of transformations

$$G : \mathbf{Y} \to \mathbf{Y} + \mathbf{X}\mathbf{b}, \qquad \mathbf{b} \in \mathbf{R}^p.$$

A sufficient statistic for this model is $(\widehat{\boldsymbol{\beta}}, \widehat{\sigma}^2)$, where

$$\widehat{\boldsymbol{\beta}} = (\mathbf{X}^T\mathbf{X})^{-1}\mathbf{X}^T\mathbf{Y}, \qquad \widehat{\sigma}^2 = \frac{\|\mathbf{Y} - \mathbf{X}\widehat{\boldsymbol{\beta}}\|^2}{n - p}.$$

An estimator $\mathbf{T}(\widehat{\boldsymbol{\beta}}, \widehat{\sigma}^2)$ is equivariant under G if

$$\forall \mathbf{b} \in \mathbf{R}^p, \mathbf{T}(\widehat{\boldsymbol{\beta}} + \mathbf{b}, \widehat{\sigma}^2) = \mathbf{T}(\widehat{\boldsymbol{\beta}}, \widehat{\sigma}^2) + \mathbf{b}, \Leftrightarrow \mathbf{T}(\widehat{\boldsymbol{\beta}}, \widehat{\sigma}^2) = \widehat{\boldsymbol{\beta}} + \mathbf{T}(\mathbf{0}, \widehat{\sigma}^2).$$

Using the independence of $\widehat{\boldsymbol{\beta}}$ and $\widehat{\sigma}^2$, we see that $\widehat{\boldsymbol{\beta}}$ is the best equivariant estimator of $\boldsymbol{\beta}$ for quadratic loss functions. $\widehat{\boldsymbol{\beta}}$ is also the MV unbiased and ML estimator of $\boldsymbol{\beta}$. Note, however, that when $p > 2$, this estimator is inadmissible for this loss function. (For a derivation of this fact when $\mathbf{A} = \mathbf{X}^T\mathbf{X}$, see Arnold, 1981, pp. 159–180). Typically best equivariant estimators of vectors with more than two components are inadmissible, even for the loss function for which they were derived.) Testing and confidence region problems about $\boldsymbol{\beta}$ can be transformed to the canonical form given in **23.31**, in order to show that the usual F tests are UMP invariant and that the confidence band for the response surface is equivalent to the UMA equivariant confidence region for $\boldsymbol{\beta}$ (see **30.35**).

23.30 Consider the one-way analysis of variance (ANOVA) model in which we observe Y_{ij} independent,

$$Y_{ij} \sim N(\theta + \alpha_i, \sigma^2), \qquad i = 1, \ldots, k; \; j = 1, \ldots, n_i, \sum n_i\alpha_i = 0, N = \sum n_i.$$

We are interested in inference about $\boldsymbol{\alpha} = (\alpha_1, \ldots, \alpha_k)$. This model is invariant under the following groups:

$$G_1 : Y_{ij} \to Y_{ij} + a; \qquad G_2 : Y_{ij} \to Y_{ij} + c_i, \quad \sum n_i c_i = 0;$$
$$G_3 : Y_{ij} \to bY_{ij}.$$

A sufficient statistic for this model is $(\overline{Y}, U_1, \ldots, U_k, V)$, where

$$\overline{Y} = \frac{\sum\sum Y_{ij}}{N}, \qquad \overline{Y}_{i.} = \frac{\sum_j Y_{ij}}{n_i}, \qquad U_i = \overline{Y}_{i.} - \overline{Y}_{..}, \qquad V^2 = \frac{\sum\sum(Y_{ij} - \overline{Y}_{i.})}{N-k}.$$

Under G_1, $(\overline{Y}, U_1, \ldots, U_k, V^2, \alpha_1, \ldots, \alpha_k) \to (\overline{Y}+a, U_1, \ldots, U_k, V^2, \alpha_1, \ldots, \alpha_k)$. A maximal invariant is

$$(U_1, \ldots, U_k, V^2, \alpha_1, \ldots, \alpha_k) \to (U_1 + c_1, \ldots, U_k + c_k, V^2, \alpha_1 + c_1, \ldots, \alpha_k + c_k).$$

A maximal invariant is

$$(U_1 - \alpha_1, \ldots, U_k - \alpha_k, V^2) \to (b(U_1 - \alpha_1), \ldots, b(U_k - \alpha_k), b^2V^2)$$

under G_3. A maximal invariant here is

$$\mathbf{W}(\boldsymbol{\alpha}) = (W_1(\alpha_1), \ldots, W_k(\alpha_k)), \qquad W_i(\alpha_i) = \frac{U_i - \alpha_i}{V}.$$

The union of G_1, G_2 and G_3 is transitive on the parameter space and hence this is a pivotal quantity. The problem of testing the null hypothesis that $\boldsymbol{\alpha} = \mathbf{0}$ against the alternative that $\boldsymbol{\alpha} \neq \mathbf{0}$ is invariant under the groups G_1 and G_3. By Theorem 23.13, the maximal invariant under that group is $\mathbf{W}(\mathbf{0})$. We now discuss some possible invariant tests and equivariant confidence regions. Recall that any function of a pivotal quantity is a pivotal quantity. Let

$$F(\boldsymbol{\alpha}) = \frac{\sum n_i (W_i(\alpha_i))^2}{k-1} \sim F_{k-1,N-k}.$$

Then the region

$$\{\boldsymbol{\alpha} : F(\boldsymbol{\alpha}) \leq F^{\alpha}\} = \left\{\boldsymbol{\alpha} : \sum d_i \alpha_i \in \sum d_i U_i \pm V\left((k-1)F^{\alpha}\frac{\sum d_i^2}{n_i}\right)^{1/2}, \ \forall d_i \ni \sum d_i = 0\right\}$$

is the region associated with the Scheffé simultaneous confidence intervals for contrasts. The F statistic for testing that $\boldsymbol{\alpha} = \mathbf{0}$ is $F(\mathbf{0})$. When the $n_i = n$ are all equal, let

$$q(\boldsymbol{\alpha}) = \sqrt{n} \max_{i,j} |W_i(\alpha_i) - W_j(\alpha_j)| \sim q_{k,k(n-1)}$$

(where $q_{k,k(n-1)}$ is the usual studentized range distribution). Then the region

$$\{\boldsymbol{\alpha} : q(\boldsymbol{\alpha}) < q^{\alpha}\} = \left\{\boldsymbol{\alpha} : \alpha_i - \alpha_j \in U_i - U_j \pm \frac{Vq^{\alpha}}{\sqrt{n}} \ \forall i \neq j\right\}$$

is the region associated with the Tukey simultaneous confidence intervals. We can also use $q(\mathbf{0})$ as a test statistic for testing that $\boldsymbol{\alpha} = \mathbf{0}$. Using only the groups in this example, there is no UMP invariant test for this problem and hence no UMA equivariant confidence region. In **30.35**, this model is put into the canonical form of **23.31**, in which a larger group is listed for which the F test and the Scheffé confidence region are shown to be UMP invariant and UMA equivariant.

However, the studentized range test and Tukey confidence region are not invariant or equivariant under this larger group, so that the F test and Scheffé confidence region are not more powerful or more accurate that the studentized range test and Tukey confidence region. Note that both the Scheffé and Tukey intervals are based on the invariant pivotal quantity $\mathbf{W}(\boldsymbol{\alpha})$ and hence their associated confidence regions are equivariant under the three groups given above.

23.31 A canonical form for the general linear model (including tests in multiple regression, fixed effects analysis of variance and analysis of covariance) is one in which we observe

$$\mathbf{Y} = \begin{pmatrix} \mathbf{Y}_1 \\ \mathbf{Y}_2 \\ \mathbf{Y}_3 \end{pmatrix} \sim N_K\left(\begin{pmatrix} \boldsymbol{\nu}_1 \\ \boldsymbol{\nu}_2 \\ \mathbf{0} \end{pmatrix}, \sigma^2\mathbf{I} \right),$$

where $\mathbf{Y}_i$ and $\boldsymbol{\nu}_i$ are k_i-dimensional vectors, $K = \sum k_i$ (see **30.35**). We want to make inferences about $\boldsymbol{\nu}_2$. We use the groups

$$G_1 : (\mathbf{Y}_1, \mathbf{Y}_2, \mathbf{Y}_3) \rightarrow (\mathbf{Y}_1 + \mathbf{a}_1, \mathbf{Y}_2, \mathbf{Y}_3); \qquad G_2 : (\mathbf{Y}_1, \mathbf{Y}_2, \mathbf{Y}_3) \rightarrow (\mathbf{Y}_1, \mathbf{Y}_2 + \mathbf{a}_2, \mathbf{Y}_3);$$
$$G_3 : (\mathbf{Y}_1, \mathbf{Y}_2, \mathbf{Y}_3) \rightarrow (\mathbf{Y}_1, \boldsymbol{\Gamma}\mathbf{Y}_2, \mathbf{Y}_3); \qquad G_4 : (\mathbf{Y}_1, \mathbf{Y}_2, \mathbf{Y}_3) \rightarrow (c\mathbf{Y}_1, c\mathbf{Y}_2, c\mathbf{Y}_3), c > 0,$$

where $\boldsymbol{\Gamma}$ is an arbitrary orthogonal $k_2 \times k_2$ matrix. (Note that this model is also invariant under multiplication of $\mathbf{Y}_1$ and $\mathbf{Y}_3$ by orthogonal matrices, but these groups provide no additional reduction and are not listed here. The model is not, however, invariant under location changes in $\mathbf{Y}_3$.) We first find the invariant pivotal quantity for $\boldsymbol{\nu}_2$. Let $W_3 = \|\mathbf{Y}_3\|^2$. A sufficient statistic for this model is $(\mathbf{Y}_1, \mathbf{Y}_2, W_3)$. G_1 takes $(\mathbf{Y}_1, \mathbf{Y}_2, W_3, \boldsymbol{\nu}_2) \rightarrow (\mathbf{Y}_1 + \mathbf{a}_1, \mathbf{Y}_2, W_3, \boldsymbol{\nu}_2)$. A maximal invariant under this group (L1) is $(\mathbf{Y}_2, W_3, \boldsymbol{\nu}_2) \rightarrow (\mathbf{Y}_2 + \mathbf{a}_2, W_3, \boldsymbol{\nu}_2 + \mathbf{a}_2)$ under G_2. A maximal invariant (group L2) is $(\mathbf{Y}_2 - \boldsymbol{\nu}_2, W_3) \rightarrow (\boldsymbol{\Gamma}(\mathbf{Y}_2 - \boldsymbol{\nu}_2), W_3)$ under G_3. A maximal invariant (group O2) is $(W_2(\boldsymbol{\nu}_2), W_3)$ where $W_2(\boldsymbol{\nu}_2) = \|\mathbf{Y}_2 - \boldsymbol{\nu}_2\|^2$. Under G_4, $(W_2(\boldsymbol{\nu}_2), W_3) \rightarrow (c^2 W_2(\boldsymbol{\nu}_2), c^2 W_3)$ and a maximal invariant (group S2) is

$$F(\boldsymbol{\nu}_2) = \frac{W_2(\boldsymbol{\nu}_2)/k_2}{W_3/k_3} \sim F_{k_2,k_3},$$

which is a pivotal quantity for $\boldsymbol{\nu}_2$. Now, consider testing the null hypothesis that $\boldsymbol{\nu}_2 = \mathbf{0}$ against the alternative hypothesis that $\boldsymbol{\nu}_2 \neq \mathbf{0}$. By Theorem 23.13, a maximal invariant for this testing problem is $F(\mathbf{0})$. A parameter maximal invariant is $\delta = \|\boldsymbol{\nu}_2\|^2/\sigma^2$. Furthermore,

$$F(\mathbf{0}) \sim F_{k_2,k_3}(\delta),\ \boldsymbol{\nu}_2 = 0 \Leftrightarrow \delta = 0,\ \boldsymbol{\nu}_2 \neq 0 \Leftrightarrow \delta > 0.$$

Therefore the UMP size-α test for the reduced problem rejects the null hypothesis when $F(\mathbf{0}) > F^\alpha$, which is the UMP invariant size-α test for the original problem. (Note that in **30.35**, this test is shown to be the usual F test for this problem.) This test accepts the null hypothesis when $F(\mathbf{0}) \leq F^\alpha$, and hence the UMA equivariant confidence region for $\boldsymbol{\nu}_2$ is given by

$$\{\boldsymbol{\nu}_2 : F(\boldsymbol{\nu}_2) \leq F^\alpha\} = \left\{\boldsymbol{\nu}_2 : \mathbf{t}^T\boldsymbol{\nu}_2 \in \mathbf{t}^T\mathbf{Y}_2 \pm \|\mathbf{t}\|\left(k_2 F^\alpha \frac{W_3}{k_3}\right)^{1/2}, \forall \mathbf{t} \in \mathbf{R}^{k_2}\right\}.$$

(Note that in **30.35**, these intervals are shown to be the Scheffé simultaneous confidence intervals for the set of all contrasts in $\boldsymbol{\nu}_2$.)

In the special case of the one-way ANOVA model given in **23.20**, the groups G_1 and G_2 of this example reduce to G_1 and G_2 given there. Group G_4 here reduces to G_3 in the previous example. Group G_3 in this example is somewhat hard to interpret for the ANOVA model given above (and other ANOVA models). However, without that group it is not possible to find optimal invariant or equivariant procedures.

Multivariate models

23.32 Suppose we observe $\mathbf{X}_1, \ldots, \mathbf{X}_n$ independent,

$$\mathbf{X}_i \sim N_r(\boldsymbol{\mu}, \boldsymbol{\Sigma}), \qquad \boldsymbol{\nu} \in \mathbf{R}^r, \quad \boldsymbol{\Sigma} > 0.$$

A sufficient statistic for this model is $(\overline{\mathbf{X}}, \mathbf{S})$, where $\overline{\mathbf{X}}$ and $\mathbf{S}$ are the sample mean vector and sample covariance matrix of the $\mathbf{X}_i$. We assume that $n > r$ so that $\mathbf{S} > 0$. This model is invariant under the groups

$$G_1 : \mathbf{X}_i \to \mathbf{X}_i + \mathbf{a}, \qquad G_2 : \mathbf{X}_i \to \mathbf{B}\mathbf{X}_i,$$

where $\mathbf{a} \in \mathbf{R}^p$ and $\mathbf{B}$ is an arbitrary invertible $r \times r$ matrix. (Note that when $r = 1$, G_2 is the union of the scale and sign groups.) In this example, we consider inference about $\boldsymbol{\mu}$.

An estimator $T(\overline{\mathbf{X}}, \mathbf{S})$ for $\boldsymbol{\mu}$ is equivariant under G_1 if and only if

$$\forall \mathbf{a} \in \mathbf{R}^p,\ T_1(\overline{\mathbf{X}} + \mathbf{a}, \mathbf{S}) = T_1(\overline{\mathbf{X}}, \mathbf{S}) + \mathbf{a} \Leftrightarrow T_1(\overline{\mathbf{X}}, \mathbf{S}) = \overline{\mathbf{X}} + T_1(\mathbf{0}, \mathbf{S}).$$

Using the independence of $\overline{\mathbf{X}}$ and $\mathbf{S}$, we see that the best equivariant estimator of $\boldsymbol{\mu}$ under G_1 for the invariant loss function $(\mathbf{d} - \boldsymbol{\mu})^T \boldsymbol{\Sigma}^{-1} (\mathbf{d} - \boldsymbol{\mu})$ is $\overline{\mathbf{X}}$, which is also the MV unbiased and ML estimator, but which is unfortunately inadmissible for this loss function when $r > 2$ (see, for example, Arnold, 1981, pp. 332–335).

We find the invariant pivotal quantity for $\boldsymbol{\mu}$. Under G_1, $(\overline{\mathbf{X}}, \mathbf{S}, \boldsymbol{\mu}) \to (\overline{\mathbf{X}} + \mathbf{a}, \mathbf{S}, \boldsymbol{\mu} + \mathbf{a})$. A maximal invariant (group L2) is $(\overline{\mathbf{X}} - \boldsymbol{\mu}, \mathbf{S}) \to (\mathbf{B}(\overline{\mathbf{X}} - \boldsymbol{\mu}), \mathbf{B}\mathbf{S}\mathbf{B}^T)$ under G_2. A maximal invariant (M3) under G_2 is

$$F(\boldsymbol{\mu}) = c(\overline{\mathbf{X}} - \boldsymbol{\mu})^T \mathbf{S}^{-1} (\overline{\mathbf{X}} - \boldsymbol{\mu}) \sim F_{r,n-r}, \qquad c = \frac{n(n-r)}{r(n-1)},$$

which is the invariant pivotal quantity for $\boldsymbol{\mu}$. Now consider testing the null hypothesis that $\boldsymbol{\nu} = \mathbf{0}$ against the alternative hypothesis that $\boldsymbol{\nu} \neq \mathbf{0}$. The group which leaves this testing problem invariant is G_2. By Theorem 23.13 $F(\mathbf{0}) = c\overline{\mathbf{X}}^T \mathbf{S}^{-1} \overline{\mathbf{X}}$ is a maximal invariant under this group. A parameter maximal invariant is $\delta = n\boldsymbol{\mu}^T \boldsymbol{\Sigma}^{-1} \boldsymbol{\mu}$. Furthermore,

$$F(\mathbf{0}) \sim F_{r,n-r}(\delta),\ \boldsymbol{\mu} = \mathbf{0} \Leftrightarrow \delta = 0,\ \boldsymbol{\mu} \neq \mathbf{0} \Leftrightarrow \delta > 0.$$

Hence, the UMP size-α test for the reduced problem rejects the null hypothesis if $F(\mathbf{0}) > F^{\alpha}$. This test (Hotelling's one-sample T^2 test) is therefore the UMP invariant size-α test for the original problem. It is also the LR test for this problem, although its derivation as the LR test is much less straightforward than its derivation as UMP invariant. Note that being a UMP invariant test is a stronger result than being an LR test (since the LR test is a particular invariant test and a UMP invariant test must be unbiased).

This test accepts the null hypothesis if $F(\mathbf{0}) \le F^\alpha$. Therefore the UMA equivariant confidence region for $\boldsymbol{\mu}$ is given by

$$\{\boldsymbol{\mu} : F(\boldsymbol{\mu}) \le F^\alpha\} = \left\{\boldsymbol{\mu} : \mathbf{t}^T\boldsymbol{\mu} \in \mathbf{t}^T\overline{\mathbf{X}} \pm \left(\frac{F^\alpha}{c}\mathbf{t}^T\mathbf{S}\mathbf{t}\right)^{1/2}, \ \forall \mathbf{t} \in \mathbf{R}^r\right\}.$$

That is, the UMA equivariant confidence region for $\boldsymbol{\mu}$ is the region associated with the simultaneous confidence intervals of Roy and Bose (1953) for linear functions of $\boldsymbol{\mu}$.

The results of this example can be extended in an elementary way to the multivariate two-sample model with equal covariance matrices (see Exercise 23.25).

23.33 Let $\mathbf{Y}$ and $\mathbf{A}$ be $n \times r$ matrices with rows $\mathbf{Y}_i$ and $\mathbf{A}_i$ and let $\mathbf{B} > 0$ be $r \times r$. We say that

$$\mathbf{Y} \sim N_{n\times r}(\mathbf{A}, \mathbf{B}) \Leftrightarrow \mathbf{Y}_i \text{ are independent, } \mathbf{Y}_i^T \sim N_r(\mathbf{A}_i^T, \mathbf{B}).$$

A canonical form for the multivariate linear model is one in which we observe

$$\mathbf{Y} = \begin{pmatrix} \mathbf{Y}_1 \\ \mathbf{Y}_2 \\ \mathbf{Y}_3 \end{pmatrix} \sim N_{K\times r}\left(\begin{pmatrix} \boldsymbol{\nu}_1 \\ \boldsymbol{\nu}_2 \\ \mathbf{0} \end{pmatrix}, \boldsymbol{\Sigma}\right),$$

where $\mathbf{Y}_i$ and $\boldsymbol{\nu}_i$ are $k_i \times r$ matrices, $K = \sum k_i$ and $\boldsymbol{\Sigma} > 0$ is $r \times r$. As in the univariate linear model (**23.31**), we are interested in inference about $\boldsymbol{\nu}_2$. This model is invariant under the groups

$$G_1 : (\mathbf{Y}_1, \mathbf{Y}_2, \mathbf{Y}_3) \to (\mathbf{Y}_1 + \mathbf{a}_1, \mathbf{Y}_2, \mathbf{Y}_3), \qquad G_2 : (\mathbf{Y}_1, \mathbf{Y}_2, \mathbf{Y}_3) \to (\mathbf{Y}_1, \mathbf{Y}_2 + \mathbf{a}_2, \mathbf{Y}_3),$$
$$G_3 : (\mathbf{Y}_1, \mathbf{Y}_2, \mathbf{Y}_3) \to (\mathbf{Y}_1, \boldsymbol{\Gamma}\mathbf{Y}_2, \mathbf{Y}_3), \qquad G_4 : (\mathbf{Y}_1, \mathbf{Y}_2, \mathbf{Y}_3) \to (\mathbf{Y}_1\mathbf{A}, \mathbf{Y}_2\mathbf{A}, \mathbf{Y}_3\mathbf{A}),$$

where $\boldsymbol{\Gamma}$ is an arbitrary $k_2 \times k_2$ orthogonal matrix and $\mathbf{A}$ is an arbitrary invertible $r \times r$ matrix. (Note that G_1, G_2 and G_3 are formally the same as for the univariate linear model, and that G_4 is a natural extension of the group in that example.)

Let

$$\mathbf{W}_2(\boldsymbol{\nu}_2) = (\mathbf{Y}_2 - \boldsymbol{\nu}_2)^T(\mathbf{Y}_2 - \boldsymbol{\nu}_2), \qquad \mathbf{W}_3 = \mathbf{Y}_3^T\mathbf{Y}_3,$$

as in the univariate case. (Note, however, that $\mathbf{W}_2(\boldsymbol{\nu}_2)$ and $\mathbf{W}_3$ are $r \times r$ matrices in this example.) We assume that $k_3 \ge r$ so that $\mathbf{W}_3 > 0$. A sufficient statistic for this model is $(\mathbf{Y}_1, \mathbf{Y}_2, \mathbf{W}_3)$. We now find the invariant pivotal quantity. As in the univariate case, the maximal invariant after G_1, G_2 and G_3 is

$$(\mathbf{W}_2(\boldsymbol{\nu}_2), \mathbf{W}_3) \to (\mathbf{A}^T\mathbf{W}_2(\boldsymbol{\nu}_2)\mathbf{A}, \mathbf{A}^T\mathbf{W}_3\mathbf{A})$$

under G_4. A maximal invariant (group M2) is the set $t_1(\boldsymbol{\nu}_2) \ge \cdots \ge t_b(\boldsymbol{\nu}_2)$ of b non-zero eigenvalues of $\mathbf{W}_3^{-1}\mathbf{W}_2(\boldsymbol{\nu}_2)$ ($b = \text{rank}(\mathbf{W}_2(\boldsymbol{\nu}_2)) = \min(k_2, r)$).

We first look at the cases in which $b = 1$ (i.e. when $r = 1$ or $k_2 = 1$).

1. Suppose $r = 1$. Then

$$t_1(\boldsymbol{\nu}_2) = \frac{W_2(\boldsymbol{\nu}_2)}{W_3}, \qquad F(\boldsymbol{\nu}_2) = \frac{k_3}{k_2}t_1(\boldsymbol{\nu}_2) \sim F_{k_2,k_3}.$$

This model is just the univariate linear model and we have already shown that the UMP invariant test for testing $\boldsymbol{\nu}_2 = \mathbf{0}$ against $\boldsymbol{\nu}_2 \neq \mathbf{0}$ rejects if $F(\mathbf{0}) > F^\alpha$ and the UMA equivariant confidence region for $\boldsymbol{\nu}_2$ is given by

$$\{\boldsymbol{\nu}_2 : F(\boldsymbol{\nu}_2) \leq F^\alpha\} = \left\{\boldsymbol{\nu}_2 : \mathbf{d}^T \boldsymbol{\nu}_2 \in \mathbf{d}^T \mathbf{Y}_2 \pm \left(\frac{k_2 F^\alpha}{k_3} W_3 \mathbf{d}^T \mathbf{d}\right)^{1/2}, \forall \mathbf{d} \in \mathbf{R}^{k_2}\right\},$$

the region associated with the Scheffé simultaneous confidence intervals.

2. Suppose $k_2 = 1$. Then it can be shown that

$$t_1(\boldsymbol{\nu}_2) = (\mathbf{Y}_2 - \boldsymbol{\nu}_2)\mathbf{W}_3^{-1}(\mathbf{Y}_2 - \boldsymbol{\nu}_2)^T, \qquad F(\boldsymbol{\nu}_2) = \frac{k_3 - r + 1}{r} t_1(\boldsymbol{\nu}_2) \sim F_{r, k_3 - r + 1}.$$

By calculations similar to those in the multivariate one-sample model above, it can be show that the UMP invariant test for testing $\boldsymbol{\nu}_2 = \mathbf{0}$ against $\boldsymbol{\nu}_2 \neq \mathbf{0}$ rejects if $F(0) > F^\alpha$ and the UMA equivariant confidence region for $\boldsymbol{\nu}_2$ is given by

$$\{\boldsymbol{\nu}_2 : F(\boldsymbol{\nu}_2) \leq F^\alpha\} = \left\{\boldsymbol{\nu}_2 : \boldsymbol{\nu}_2 \mathbf{q} \in \mathbf{Y}_2 \mathbf{q} \pm \left(\frac{r F^\alpha}{k_3 - r + 1} \mathbf{q}^T \mathbf{W}_3 \mathbf{q}\right)^{1/2}, \forall \mathbf{q} \in \mathbf{R}^r\right\},$$

the region associated with the Roy–Bose simultaneous confidence intervals. (Note that Hotelling's one-sample T^2 given above is a particular case of this test.)

Now, suppose that both $r > 1$ and $k_2 > 1$. Consider testing the null hypothesis that $\boldsymbol{\nu}_2 = \mathbf{0}$ against the alternative hypothesis that $\boldsymbol{\nu}_2 \neq \mathbf{0}$. There is no UMP invariant test for this problem. By Theorem 23.13, a maximal invariant is the set $t_1, \ldots, t_b$ of eigenvalues of $\mathbf{W}_3^{-1}\mathbf{W}_2(\mathbf{0})$. Several invariant test statistics have been suggested for this problem:

$$\lambda_1 = \frac{|\mathbf{W}_2(\mathbf{0}) + \mathbf{W}_3|}{|\mathbf{W}_3|} = \prod (1 + t_i), \qquad \lambda_2 = \text{tr}(\mathbf{W}_2(\mathbf{0})\mathbf{W}_3^{-1}) = \sum t_i,$$

$$\lambda_3 = \text{tr}(\mathbf{W}_2(\mathbf{0})(\mathbf{W}_2(\mathbf{0}) + \mathbf{W}_3)^{-1}) = \sum \frac{t_i}{1 + t_i}, \qquad \lambda_4 = t_1.$$

For each of these test statistics, we reject if λ_i is too large.

(a) $\lambda_1^{-\frac{n}{2}}$ is the LR test statistic and hence the first test is the LR test. The test statistic λ_1 was suggested by Wilks (1932) and is called Wilks' λ.

(b) When $\boldsymbol{\Sigma}$ is known, the optimal (UMP invariant and LR test) rejects if $U = \text{tr}(\mathbf{W}_2(\mathbf{0})\boldsymbol{\Sigma}^{-1})$ is too large.

1. The test statistic λ_2 is motivated by substituting the estimator

$$\widehat{\boldsymbol{\Sigma}} = k_3^{-1}\mathbf{W}_3$$

for $\boldsymbol{\Sigma}$ in U. It was suggested by Lawley (1938) and Hotelling (1951) and is called the Lawley–Hotelling trace.

2. The test statistic λ_3 is motivated by substituting the null hypothesis estimator

$$\widehat{\widehat{\boldsymbol{\Sigma}}} = (k_2 + k_3)^{-1}(\mathbf{W}_2(\mathbf{0}) + \mathbf{W}_3)$$

for $\boldsymbol{\Sigma}$ in U. It was suggested by Pillai (1955) and is called Pillai's trace.

(c) λ_4 is the union-intersection test for this problem. It was suggested by Roy (1953) and is called Roy's largest root test.

All four of the tests have been shown to be unbiased (Dasgupta *et al.*, 1964) and admissible (Schwartz, 1967). The test based on λ_4 has the advantage of a relatively simple associated confidence region

$$\{\boldsymbol{\nu}_2 : t_1(\boldsymbol{\nu}_2) \le c\} = \{\boldsymbol{\nu}_2 : \mathbf{d}^T \boldsymbol{\nu}_2 \mathbf{q} \in \mathbf{d}^T \mathbf{Y}_2 \mathbf{q} \pm (c\mathbf{d}^T \mathbf{d} \mathbf{q}^T \mathbf{W}_3 \mathbf{q})^{\frac{1}{2}}, \forall \mathbf{d} \in \mathbf{R}^{k_2}, \mathbf{q} \in \mathbf{R}^r\}$$

which gives the Roy–Bose simultaneous confidence intervals for linear functions of $\boldsymbol{\nu}_2$.

Pitman estimators

23.34 In the discussion of Pitman estimators, we use often the following facts. Let Q and R be random variables and let $\mathbf{U}$ be a random vector. Then

$$E(Q - Rh(\mathbf{U}))^2$$

is minimized by

$$h(\mathbf{U}) = \frac{E(QR|\mathbf{U})}{E(R^2|\mathbf{U})}$$

(see Exercise 23.26(b)). Similarly the Bayes estimator of a univariate parameter $\tau(\boldsymbol{\theta})$ for the loss function

$$L(a, \boldsymbol{\theta}) = (M(\boldsymbol{\theta})(\tau(\boldsymbol{\theta}) - a))^2$$

is given by

$$\frac{E(M^2(\boldsymbol{\theta})\tau(\boldsymbol{\theta})|\mathbf{X})}{E(M^2(\boldsymbol{\theta})|\mathbf{X})}$$

(see Exercise 23.26(c)). Both these results are of course dependent on the finiteness of the expectations involved, an assumption we make without further mention throughout this discussion.

23.35 Let $\mathbf{X} = (X_1, \ldots, X_n)$ be a continuous random vector whose distribution depends on the (univariate) parameter δ. We say that δ is a *location parameter* for $\mathbf{X}$ if the distribution of $(X_1 - \delta, \ldots, X_n - \delta)$ does not depend on any unknown parameters, that is if $\mathbf{X}$ has joint density function

$$f_\delta(x_1, \ldots, x_n) = f_0(x_1 - \delta, \ldots, x_n - \delta)$$

where f_0 is known. This family is invariant under the group of location changes

$$(X_1, \ldots, X_n) \to (X_1 + c, \ldots, X_n + c), \qquad \delta \to \delta + c,$$

where c is an arbitrary real number. Since we are allowing f to be an arbitrary joint density function, there is no possible reduction by sufficiency. Note, however, that the induced group on the parameter space is transitive, so that all equivariant estimators have constant risk for any invariant loss function.

Consider estimating δ with the invariant loss function

$$L(a, \delta) = (a - \delta)^2.$$

An estimator $T(\mathbf{X})$ is equivariant if for all a,

$$T(X_1 + c, \ldots, X_n + c) = T(X_1, \ldots, X_n) + c.$$

Let $U_i = X_i - X_n, i = 1, \ldots, n-1, \mathbf{U} = (U_1, \ldots, U_{n-1})$. Choosing $c = -X_n$, we see that an equivariant estimator must satisfy

$$T(\mathbf{X}) = X_n - h(\mathbf{U})$$

for some function h. Conversely, if $T(\mathbf{X}) = X_n - h(\mathbf{U})$, then

$$T(X_1 + c, \ldots, X_n + c) = X_n + c + h(\mathbf{U}) = T(X_1, \ldots, X_n) + c$$

so that an estimator $T(\mathbf{X})$ is equivariant if and only if $T(\mathbf{X}) = X_n - h(\mathbf{U})$ for some function $h(\mathbf{U})$. To find the best equivariant estimator of δ, we must minimize

$$R(T, \delta) = R(T, 0) = E_0((X_n - h(\mathbf{U}))^2)$$

which is minimized by

$$h(\mathbf{U}) = E_0(X_n|U).$$

Therefore, the best equivariant estimator of δ is given by

$$T(\mathbf{X}) = X_n - E_0(X_n|U) = \frac{\int_{-\infty}^{\infty} \delta f_\delta(\mathbf{X})\,\mathrm{d}\delta}{\int_{-\infty}^{\infty} f_\delta(\mathbf{X})\,\mathrm{d}\delta}$$

(see Exercise 23.27(b)). This estimator is called the Pitman estimator for the location parameter δ.

Note that the Pitman estimator of δ is the generalized Bayes estimator for δ for the (improper) prior distribution which is uniformly distributed on the whole line, the Jeffreys (1961) prior for this problem (see Exercise 23.27(c)).

23.36 We say that $\tau > 0$ is a *scale parameter* for the continuous random vector $\mathbf{X}$ if the joint distribution of $(X_1/\tau, \ldots, X_n/\tau)$ is known, that is, if

$$f_\tau(X_1, \ldots, X_n) = \tau^{-n} f_1(x_1/\tau, \ldots, x_n/\tau),$$

where f_1 is a known function. This family is invariant under scale changes in the X_i. This group is again transitive on the parameter space so that risk functions of equivariant estimators are constant for invariant loss functions. In this section we outline the derivation of the best equivariant (Pitman) estimator for τ. See Exercise 23.28 for details.

An estimator $T(\mathbf{X})$ of τ is equivariant if, for all $b > 0$,

$$T(bX_1, \ldots, bX_n) = bT(X_1, \ldots, X_n).$$

Then T is equivariant if and only

$$T(\mathbf{X}) = |X_n|h(\mathbf{V}),$$

where $\mathbf{V} = (V_1, \ldots, V_n)$, $V_i = X_i/|X_n|$. (Note that $V_n = \text{sign}(X_n)$ but that the other V_i are continuous random variables.) Consider estimating τ with the invariant loss function

$$L(a, \tau) = \frac{(a-\tau)^2}{\tau^2}.$$

(The squared error loss function $L(a, \tau) = (a - \tau)^2$ is not invariant for this problem.) The risk function for equivariant estimators

$$R(T, \tau) = R(T, 1) = E_1(|X_n|h(\mathbf{V}) - 1)^2$$

is minimized for

$$h(\mathbf{V}) = \frac{E_1(|X_n||\mathbf{V})}{E_1(|X_n|^2|\mathbf{V})}$$

and the best equivariant estimator is

$$T(\mathbf{X}) = |X_n|\frac{E_1(|X_n||\mathbf{V})}{E_1(|X_n|^2|\mathbf{V})} = \frac{\int_0^\infty \tau^{-2} f_\tau(\mathbf{X})\, d\tau}{\int_0^\infty \tau^{-3} f_\tau(\mathbf{X})\, d\tau}$$

which is called the Pitman estimator for the scale parameter τ.

The Pitman estimator of scale is the generalized Bayes estimator for the improper prior distribution $\pi(\tau) = \tau^{-1}$ which is the Jeffreys prior for this situation. This prior is equivalent to a uniform distribution for $\log(\tau)$. (Note that $\log(\tau)$ is a location parameter for the $\log(|X_i|)$.)

23.37 Suppose now that the distribution of the continuous random vector $\mathbf{X}$ depends on the bivariate parameter (δ, τ). We say this parameter is a *location–scale parameter* if the joint distribution of $((X_1 - \delta)/\tau, \ldots, (X_n - \delta)/\tau)$ does not depend on any unknown parameters, that is, if

$$f_{\delta,\tau}(X_1, \ldots, X_n) = \tau^{-n} f_{0,1}\left(\frac{X_1 - \delta}{\tau}, \ldots, \frac{X_n - \delta}{\tau}\right),$$

where $f_{0,1}$ is a known function. (Note that δ is a location parameter for $\mathbf{X}$ for each τ, but that τ is not a scale parameter unless $\delta = 0$.) This family of distributions is invariant under location and scale changes. Note that this group is transitive on the parameter space so that equivariant estimators have constant risk functions for invariant loss functions.

In finding optimal equivariant estimators, it is helpful to define the following random variables. Let

$$U_{n-1} = X_{n-1} - X_n, \qquad \mathbf{W} = (W_1, \ldots, W_{n-1}),\ W_i = (X_i - X_n)/|U_{n-1}|.$$

(Note that $W_{n-1} = \text{sign}(U_{n-1})$ but that the other W_i are continuous random variables.)

Consider first equivariant estimation of δ for the invariant loss function

$$L(a, (\delta, \tau)) = \frac{(a-\delta)^2}{\tau^2}.$$

We outline the derivation of the best equivariant estimator for δ. See Exercise 23.29 for details. An estimator $T(\mathbf{X})$ is a location and scale equivariant estimator of δ if, for all c and all $b > 0$,

$$T(bX_1 + c, \ldots, bX_n + c) = bT(X_1, \ldots, X_n) + c,$$

which occurs if and only if

$$T(X_1, \ldots, X_n) = X_n - |U_{n-1}|h(\mathbf{W})$$

for some function $h(\mathbf{W})$. The best equivariant estimator minimizes

$$R(T, (\delta, \tau)) = R(T, (0, 1)) = E_{0,1}(X_n - |U_{n-1}|h(\mathbf{W}))^2$$

and hence has

$$h(\mathbf{W}) = \frac{E_{0,1}(X_n|U_{n-1}||\mathbf{W})}{E_{0,1}(|U_{n-1}|^2|\mathbf{W})},$$

so that the best equivariant estimator of δ is

$$T(\mathbf{X}) = X_n - |U_{n-1}|\frac{E_{0,1}(X_n|U_{n-1}||\mathbf{W})}{E_{0,1}(|U_{n-1}|^2|\mathbf{W})} = \frac{\int_{-\infty}^{\infty}\int_0^{\infty} \delta\tau^{-3} f_{\delta,\tau}(\mathbf{X})\, \mathrm{d}\tau\, \mathrm{d}\delta}{\int_{-\infty}^{\infty}\int_0^{\infty} \tau^{-3} f_{\delta,\tau}(\mathbf{X})\, \mathrm{d}\tau\, \mathrm{d}\delta},$$

which we call the Pitman estimator of δ. Note that for this loss function, the Pitman estimator is the generalized Bayes estimator for the improper prior $\pi(\delta, \tau) = \tau^{-1}$.

Now consider estimating τ with the loss function

$$L^*(a^*, (\delta, \tau)) = \frac{(a^* - \tau)^2}{\tau^2},$$

(see Exercise 23.30 for details). An estimator $T^*(\mathbf{X})$ is a location and scale equivariant estimator of τ if, for all c and all $b > 0$,

$$T^*(bX_1 + c, \ldots, bX_n + c) = bT^*(X_1, \ldots, X_n),$$

which occurs if and only if

$$T^*(X_1, \ldots, X_n) = |U_{n-1}|h^*(\mathbf{W})$$

for some function $h^*(\mathbf{W})$. The best equivariant estimator for this loss function minimizes

$$R^*(T^*, (\delta, \tau)) = R^*(T^*, (0, 1)) = E_{0,1}((|U_{n-1}|h^*(\mathbf{W}) - 1)^2)$$

and therefore chooses

$$h^*(\mathbf{W}) = \frac{E_{0,1}(|U_{n-1}||\mathbf{W})}{E_{0,1}(|U_{n-1}|^2|\mathbf{W})}.$$

The best equivariant estimator of τ is therefore

$$T^*(\mathbf{X}) = |U_{n-1}|\frac{E_{0,1}(|U_{n-1}||\mathbf{W})}{E_{0,1}(|U_{n-1}|^2|\mathbf{W})} = \frac{\int_{-\infty}^{\infty}\int_0^{\infty}\tau^{-2}f_{\delta,\tau}(\mathbf{X})\,\mathrm{d}\tau\,\mathrm{d}\delta}{\int_{-\infty}^{\infty}\int_0^{\infty}\tau^{-3}f_{\delta,\tau}(\mathbf{X})\,\mathrm{d}\tau\,\mathrm{d}\delta},$$

which is called the Pitman estimator of τ. It is also the generalized Bayes estimator for the prior $\pi(\delta, \tau) = \tau^{-1}$.

For both these estimation problems we have seen that the best equivariant estimator is the generalized Bayes estimator with respect to the improper prior

$$\pi(\delta, \tau) = \frac{1}{\tau}.$$

Note that the Jeffreys prior for this model is

$$\pi^*(\delta, \tau) = \frac{1}{\tau^2}$$

(see Exercise 23.31) so that in this situation, the best equivariant procedure is generalized Bayes with respect to an improper prior which is different from the Jeffreys prior.

Invariant prior distributions

23.38 As illustrated by the Pitman estimators given above, there is a strong relationship between the best equivariant procedures and certain generalized Bayes procedures. Suppose we have a model in which we observe the random vector $\mathbf{X}$ whose distribution depends on the (possibly vector-valued) parameter $\boldsymbol{\theta}$. We assume that the model is invariant under the group G with induced group $\overline{G}$ on the parameter space. We further assume that the group $\overline{G}$ is uniquely transitive in the sense that, for any two points $\boldsymbol{\theta}_1$ and $\boldsymbol{\theta}_2$ in the parameter space, there is a unique element in $\overline{G}$ which maps $\boldsymbol{\theta}_1$ to $\boldsymbol{\theta}_2$. In this case the parameter space and the group are really the same space (i.e. $\overline{G}$ is isomorphic to the parameter space Ω).

We say that a prior on Ω is invariant under $\overline{G}$ if $\overline{g}(\boldsymbol{\theta})$ has the same distribution as $\boldsymbol{\theta}$. Such a measure always exists and is often called the left Haar measure and is unique (up to scalar multiplication), but cannot be a probability measure (unless the parameter space is compact). A measure satisfying this invariance property seems like a natural candidate for a 'non-informative' prior. In Exercises 23.32–33, it is shown that the Jeffreys prior given by

$$\pi(\boldsymbol{\theta}) \approx |\mathbf{I}(\boldsymbol{\theta})|^{1/2}$$

(where $\mathbf{I}(\boldsymbol{\theta})$ is the Fisher information matrix) is an invariant prior distribution. (Note that the Jeffreys prior is typically an improper prior so that the scale factor is not important.)

Under fairly general conditions, the best equivariant estimator is the generalized Bayes estimator for a less natural measure, called the right Haar measure. In addition, many classical confidence intervals are generalized Bayes credible intervals with respect to the right Haar measure. (See Berger, 1985b, pp. 406–418), for derivations of these facts.)

We say that the group $\overline{G}$ is abelian if, for all $\overline{g}_1, \overline{g}_2 \in \overline{G}$,

$$\overline{g}_1 \circ \overline{g}_2 = \overline{g}_2 \circ \overline{g}_1.$$

As long as $\overline{G}$ is abelian then the left and right Haar measures are the same and the best equivariant estimator and the classical confidence regions are generalized Bayes with respect to the invariant (Jeffreys) prior. For non-abelian groups, the best equivariant estimator and classical confidence regions are generalized Bayes with respect to a different improper prior. Note that the group of location changes is abelian, as is the group of scale changes, but the larger group of both location and scale changes is not abelian (see Exercise 23.34). For the location–scale parameter the left Haar measure (invariant or Jeffreys prior) is $\pi(\delta, \tau) = \tau^{-2}$ and the right Haar measure is $\pi^*(\delta, \tau) = \tau^{-1}$.

For non-abelian groups we have two possible 'non-informative' priors, the left Haar measure (Jeffreys prior), which seems natural from invariance considerations, and the right Haar measure, which leads to the classical confidence regions and best equivariant procedures. Berger (1985b, p. 413) argues that the right Haar measure is the correct choice. (See also his comments on pp. 88-89 for the particular case of location–scale models.) The problem of choosing a non-informative prior for even so simple a situation as a location–scale parameter illustrates the difficulty in defining non-informative priors for general problems. In practice, however, it often makes little difference whether we use the right or left Haar measures. As long as there is a reasonable amount of data, the actual procedures will not depend too heavily on the prior chosen provided it is fairly flat.

EXERCISES

23.1 Let $T \sim t_k(\delta)$ so that T has density function

$$f(t;\delta) \approx (t^2+k)^{-\frac{k+1}{2}} \exp\left(-\frac{k\delta^2}{2(t^2+k)}\right) \int_0^\infty x^k \exp\left(-\frac{1}{2}\left(x - \frac{\delta t}{\sqrt{t^2+k}}\right)^2\right) \mathrm{d}x,$$

and hence $f(t;0) \approx (t^2+k)^{-\frac{k+1}{2}}$. Show that if $\delta > 0$, then $f(t;\delta)/f(t;0)$ is an increasing function of t. (*Hint*: if $h(x;t)$ is increasing in t for all x, then $\int_0^\infty h(x,t)\,\mathrm{d}x$ is increasing in t. Why?)

23.2 Let $F \sim F_{m,n}(\gamma)$ so that F has density function

$$g(f;\gamma) \approx \sum_{k=0}^{\infty} \frac{\left(\frac{m\gamma}{2n}\right)^k}{k!} \frac{\Gamma\left(\frac{m+n}{2}+k\right)}{\Gamma\left(\frac{m}{2}+k\right)} \frac{f^{\frac{m}{2}+k-1}}{\left(1+\frac{m}{n}f\right)^{\frac{m+n}{2}+k}}$$

so that $g(F;0) \approx f^{\frac{m}{2}-1}/(1+\frac{m}{n}f)^{\frac{m+n}{2}}$. Show that if $\gamma > 0$, then $g(f;\gamma)/g(f;0)$ is an increasing function of f. (*Hint*: if $h(k,f)$ is increasing in f for each k, then $\sum_{k=0}^\infty h(k,f)$ is increasing in f. Why?)

23.3 For the univariate one-sample normal model of **23.3–23.7**, consider estimating σ^2 with the quadratic loss function $(d_2-\sigma^2)^2/\sigma^4$. An estimator $T_2^*(\overline{X}, S^2)$ is location, scale and equivariant if $T_2^*(b\overline{X}+a, b^2S^2) = b^2T_2^*(\overline{X}, S^2)$.
(a) Show that $T_2^*(\overline{X}, S^2)$ is location, scale and equivariant if and only if $T_2^*(\overline{X}, S^2) = S^2T_2^*(0,1)$.
(b) Find the risk function of a general equivariant estimator of σ^2.
(c) Show that this risk function is minimized when $T_2^*(0,1) = (n-1)/(n+1)$.

23.4 For the univariate one-sample normal model, a test $\Phi_1^*(\overline{X}, S^2)$ is scale invariant if and only if $\Phi_1^*(b\overline{X}, b^2S^2) = \Phi_1^*(\overline{X}, S^2)$. Verify that Φ_1^* is scale invariant if and only if Φ_1^* is actually only a function of $t = \sqrt{n}\overline{X}/S$.

23.5 (a) For the univariate one-sample normal model, a confidence region $R(\overline{X}, S^2)$ for μ is location and scale equivariant if $\mu \in R(\overline{X}, S^2) \Leftrightarrow b\mu + a \in R(\overline{X}, S^2)$. Show that R is location and scale equivariant if it is a function of $t(\mu) = \sqrt{n}(\overline{X}-\mu)/S$.
(b) A location and scale equivariant confidence region for μ is location, scale and sign equivariant if $\mu \in R(\overline{X}, S^2) \Leftrightarrow -\mu \in R(-\overline{X}, S^2)$. Show that $R(\overline{X}, S^2)$ is location, scale and sign equivariant if and only if R depends only on $F(\mu) = (t(\mu))^2 = n(\overline{X}-\mu)^2/S^2$.

23.6 (a) Is the function $\text{sign}(X)$ equivariant under location changes?
(b) Is the function $\log(X)$ equivariant under location changes?
(c) Is the function $\log(X)$ equivariant under scale changes?

23.7 Let G be a transitive group on W. Show that the only functions which are invariant under G are constant functions.

23.8 Let G be a non-transitive group on a set W. Define a relationship on W by $w_1 \approx w_2$ if there exists $g \in G$ such that $w_2 = g(w_1)$.
(a) Show that this relationship is an equivalence relationship (i.e. that for all $w \in W$, $w \approx w$ (reflexive); (b) $w_1 \approx w_2$ implies $w_2 \approx w_1$ (symmetric); and (c) $w_1 \approx w_2$ and $w_2 \approx w_3 \Rightarrow w_1 \approx w_3$ (transitive).
(b) Show that there is a maximal invariant under G. (*Hint*: the equivalence relationship divides W into equivalence classes. Let $T(w)$ be constant on equivalence classes and different on different equivalence classes.)

23.9 Consider Example 23.2.
(a) Show that a maximal invariant under G_1 is $T_1(x, y, s) = (x - y, s)$. (That is, show that $T_1(x + a, y + a, s) = T_1(x, y, s)$ and that if $T_1(x, y, s) = T_1(x^*, y^*, s^*)$ then there exist a such that $x^* = x + a$, $y^* = y + a$ and $s^* = s$. (Note that $T_1(x, y, s) = T_1(x^*, y^*, s)$ implies that $x^* - x = y^* - y$ and $s^* = s$.))
(b) Show that a maximal invariant under $\widetilde{G}_2$ is $T_2(u, s) = u/s$. (That is, show that for $b > 0$, $T_2(bu, bs) = T_2(u, s)$ and that if $T_2(u, s) = T_2(u^*, s^*)$ then there exists $b > 0$ such that $u^* = bu$ and $s^* = bs$. (Note that $T_2(u, s) = T_2(u^*, s^*)$ implies that $u^*/u = s^*/s$. Take $b = s^*/s$.))
(c) Show directly that a maximal invariant under G is $T(x, y, s) = (x - y)/s$. (That is show that if $b > 0$, $T(bx + a, by + a, bs) = T(x, y, s)$ and that if $T(x, y, s) = T(x^*, y^*, s^*)$ then there exist a and $b > 0$ such that $x^* = bx + a$, $y^* = by + a$ and $s^* = bs$. (Note that $T(x, y, s) = T(x^*, y^*, s^*)$ implies that $(x^* - y^*)/(x - y) = s^*/s$. Take $b = s^*/s$.))

23.10 Suppose that $\mathbf{X} = (\mathbf{Y}, \mathbf{Z})$ and that $g(\mathbf{Y}, \mathbf{Z}) = (g_1(\mathbf{Y}), \mathbf{Z})$. Let $\mathbf{T}_1(\mathbf{Y})$ be a maximal invariant under G_1 (the set of g_1). Show that $\mathbf{T}(\mathbf{Y}, \mathbf{Z}) = (\mathbf{T}_1(\mathbf{Y}), \mathbf{Z})$ is a maximal invariant under G (the set of g). (That is, show that $\mathbf{T}(g(\mathbf{Y}, \mathbf{Z})) = \mathbf{T}(g_1(\mathbf{Y}), \mathbf{Z}) = \mathbf{T}(\mathbf{Y}, \mathbf{Z})$, and if $\mathbf{T}(\mathbf{Y}, \mathbf{Z}) = \mathbf{T}(\mathbf{Y}^*, \mathbf{Z}^*)$, then there exists g such that $(\mathbf{Y}^*, \mathbf{Z}^*) = g(\mathbf{Y}, \mathbf{Z})$ or equivalently there exists g_1 such that $\mathbf{Y}^* = g_1(\mathbf{Y})$ and $\mathbf{Z}^* = \mathbf{Z}$.)

23.11 Let $X_1, \ldots, X_n$ be independently and identically distributed, with density function $f(x; (\delta, \tau)) = \delta^{-1} \exp((x - \delta)/\tau)$, $x > \delta$.

(a) Show that a sufficient statistic for this model is (U, Q), where $U = \min(X_i)$, $Q = \sum(X_i - U)$.
(b) Show that U and Q are independent. (Use Theorem 23.6, with τ assumed known.)

23.12 Suppose that $T_3/\sigma^2 \sim \chi^2_{n-1}$.

(a) Find the density function of T_3.
(b) Verify that this density function has monotone LR.

23.13 Consider the problem of estimating $\delta = \mu - \nu$ in the two-sample equal variance example of **23.19**.

(a) Show that an estimator $T(\overline{X}, \overline{Y}, S^2)$ of δ is equivariant under G_1 if and only if $T(\overline{X}, \overline{Y}, S_p^2) = \widehat{\delta} + T(0, 0, S_p^2)$.
(b) Show that for quadratic loss, the risk of such an equivariant estimator is $m^{-1} + n^{-1} + E(T(0, 0, S_p^2))/\sigma^2$ and hence that $\widehat{\delta}$ is the best equivariant estimator (under G_1).
(c) Show that the only equivariant estimator under G_1 and G_3 is $\widehat{\delta}$.

23.14 Consider the problem of drawing inference about $\lambda = \sigma^2/\tau^2$ in the two-sample model of **23.20**.

(a) Show that an estimator $Q(\overline{X}, \overline{Y}, S^2, T^2)$ of λ is equivariant under G_1 and G_2 if and only if $Q(\overline{X}, \overline{Y}, S^2, T^2) = k\widehat{\lambda}$ for some k.
(b) Find the density of $\widehat{\lambda}$ and show that it has monotone LR.
(c) Suppose that $m = n$. Show that a maximal invariant and parameter maximal invariant under G_1, G_{21} and G_3 are $U = (\log(\widehat{\lambda}))^2$ and $\varpi = (\log(\lambda))^2$.
(d) In this case show that the distribution of U has monotone LR.

23.15 For the paired model of **23.21**, show that the best equivariant estimator of $\delta = \mu - \nu$ under G_1 for quadratic loss is $\widehat{\delta} = \overline{X} - \overline{Y}$.

23.16 Consider inference for the parameter ρ in the paired model of **23.21**.

(a) Show that r and ρ are the maximal invariant and parameter maximal invariant under G_1 and G_3.
(b) Show that r^2 and ρ^2 are the maximal invariant and parameter maximal invariant under G_1, G_3 and G_4.

23.17 For the model of **23.22**:

(a) Verify that $U_i|V_i \sim N(\delta + \theta V_i, \xi^2)$.
(b) Verify that $\widehat{\delta}|\mathbf{V} \sim N(\delta, \xi^2(1+Q)/n)$.
(c) Verify that $t^*(\delta) \sim t_{n-2}$.
(d) Verify that $t(\delta) \sim t_{n-1}$.

23.18 For the model of **23.22**:

(a) Verify that a maximal invariant under G_1, G_2, G_3 and G_4 is $(F^*(\delta), Q)$.
(b) Verify that $F^*(\delta)|Q \sim F_{1,n-2}$ and that $(n-1)Q \sim F_{1,n-1}$.
(c) Verify that $F^*(0)|Q \sim F_{1,n-2}(\eta/(1+Q))$.

23.19 For the Neyman–Scott example of **23.23** show that an estimator $T(\mathbf{U}, W)$ of σ^2 is equivariant if and only if it has the form dW.

23.20 For the one-sample exponential model of **23.24**:

(a) Verify that an estimator $T(\overline{X})$ of μ is equivariant if and only if it has the form $d\overline{X}$.
(b) Verify the formula given there for the risk of such an equivariant estimator.
(c) Verify that the best equivariant estimator of μ is $n\overline{X}/(n+1)$.
(d) Show that the invariant pivotal quantity for μ is $2n\overline{X}/\mu \sim \chi^2_{2n}$.

23.21 For the two-sample exponential model of **23.24**:

(a) Verify that the invariant pivotal quantity under G_1 and G_2 for γ is $\widehat{\gamma}/\gamma \sim F_{2m,2n}$.
(b) Verify that the one-sided test based on $\widehat{\gamma}$ is the UMP invariant test (under G_1) for testing the null hypothesis that $\gamma = 1$ against the alternative hypothesis that $\gamma > 1$.

23.22 For the scale parameter uniform model of **23.25**:

(a) Verify the form for the equivariant estimators of θ.
(b) Verify the risk function for equivariant estimators and the form of the best equivariant estimator.
(c) Show that the UMP size-α test for testing that $\theta = c$ against $\theta > c$ rejects if S/c is too large.

23.23 For the location–scale parameter uniform model of **23.26**:

(a) Verify that an estimator $T_1(M, H)$ of μ is equivariant under G_1 if and only if $T_1(M, H) = M + T_1(0, H)$.
(b) Verify that $E((M - \mu)|H) = 0$ and hence that M is the best equivariant estimator of μ for quadratic loss.
(c) Verify the form for equivariant estimators of δ.
(d) Verify the formula for the risk function for such estimators and the formula for the best equivariant estimator of δ.

23.24 For the location–scale parameter uniform model of **23.26**:

(a) Verify that the maximal invariant and parameter maximal invariant (under G_2) are $Q = M/H$ and $\kappa = \mu/\delta$.
(b) Find the density function for Q and show that it has monotone LR. Show that the UMP invariant (under G_2) test for testing $\mu = 0$ against $\mu > 0$ rejects if Q is too large. What is the critical value for this test?
(c) Verify that the UMP invariant test (under G_2 and G_3) for testing $\mu = 0$ against $\mu \neq 0$ rejects if $P = Q^2$ is too large. What is the critical value for this test?
(d) Verify that the invariant pivotal quantity for δ is $r(\delta) = H/\delta$.
(e) Show that the UMP invariant test (under G_1) for testing that $\delta = d$ against $\delta > d$ rejects if $r(d)$ is too large. What is the critical value for this test?

23.25 Consider the multivariate two-sample model (with equal covariance matrices) in which we observe $\mathbf{X}_1, \ldots, \mathbf{X}_m, \mathbf{Y}_1, \ldots, \mathbf{Y}_n$ independent, $\mathbf{X}_i \sim N_r(\boldsymbol{\mu}, \boldsymbol{\Sigma})$, $\mathbf{Y}_j \sim N_r(\boldsymbol{\nu}, \boldsymbol{\Sigma})$. Consider inference for $\boldsymbol{\delta} = \boldsymbol{\mu} - \boldsymbol{\nu}$. A sufficient statistic for this model is $(\overline{\mathbf{X}}, \overline{\mathbf{Y}}, \mathbf{S}_p)$, where $\mathbf{S}_p$ is the pooled sample covariance matrix. Let $\widehat{\boldsymbol{\delta}} = \overline{\mathbf{X}} - \overline{\mathbf{Y}}$.

(a) Show that this model is invariant under the three groups G_{11}, G_{12} and G_2, where $G_{11}: (\mathbf{X}_i, \mathbf{Y}_i) \to (\mathbf{X}_i + \mathbf{a}, \mathbf{Y}_i + \mathbf{a})$, $G_{12}: (\mathbf{X}_i, \mathbf{Y}_i) \to (\mathbf{X}_i + \mathbf{b}, \mathbf{Y}_i)$, $G_2: (\mathbf{X}_i, \mathbf{Y}_i) \to (\mathbf{C}\mathbf{X}_i, \mathbf{C}\mathbf{Y}_j)$, in which $\mathbf{a}$ and $\mathbf{b}$ are arbitrary vectors in $\mathbf{R}^r$ and $\mathbf{C}$ is an arbitrary $r \times r$ invertible matrix. (Why isn't this model invariant under $(\mathbf{X}_i, \mathbf{Y}_i) \to (\mathbf{D}\mathbf{X}_i, \mathbf{Y}_j)$?)

(b) Show that an estimator $\mathbf{T}(\overline{\mathbf{X}}, \overline{\mathbf{Y}}, \mathbf{S}_p)$ of $\boldsymbol{\delta}$ is equivariant under G_{11} and G_{12} if and only if $\mathbf{T}(\overline{\mathbf{X}}, \overline{\mathbf{Y}}, \mathbf{S}_p) = \widehat{\boldsymbol{\delta}} + \mathbf{T}(\mathbf{0}, \mathbf{0}, \mathbf{S}_p)$. Show that $\widehat{\boldsymbol{\delta}}$ is the best equivariant estimator of $\boldsymbol{\delta}$ for the invariant loss function $L(\mathbf{d}, (\boldsymbol{\mu}, \boldsymbol{\nu}, \boldsymbol{\Sigma})) = (\mathbf{d} - \boldsymbol{\delta})^T \boldsymbol{\Sigma}^{-1} (\mathbf{d} - \boldsymbol{\delta})$. (*Hint*: Use the fact that $\widehat{\boldsymbol{\delta}}$ and $\mathbf{S}_p$ are independent.)

(c) Show that the invariant pivotal quantity (under all three groups) is $F(\boldsymbol{\delta}) = c(\widehat{\boldsymbol{\delta}} - \boldsymbol{\delta})^T \mathbf{S}_p^{-1} (\widehat{\boldsymbol{\delta}} - \boldsymbol{\delta})$ where c is chosen so that $F(\boldsymbol{\delta}) \sim F_{r,m+n-1-r}$.

(d) Show that the problem of testing that $\boldsymbol{\delta} = \mathbf{0}$ against $\boldsymbol{\delta} \neq \mathbf{0}$ is invariant under the groups G_{11} and G_2 and that the maximal invariant and parameter maximal invariant are $F = F(\mathbf{0}) = c\widehat{\boldsymbol{\delta}}^T \mathbf{S}_p^{-1} \widehat{\boldsymbol{\delta}}$ and $\gamma = d\boldsymbol{\delta}^T \boldsymbol{\Sigma}^{-1} \boldsymbol{\delta}$, $d = mn/(m+n)$.

(e) Use the fact that $F \sim F_{r,m+n-1-r}(\gamma)$ to show that the UMP invariant size-α test for this hypothesis rejects if $F > F^\alpha$.

(f) Find the UMA equivariant $1 - \alpha$ confidence region for $\boldsymbol{\delta}$ and show that it is equivalent to the (Roy–Bose) confidence region given by

$$\{\boldsymbol{\delta} : \mathbf{d}^T \boldsymbol{\delta} \in \mathbf{d}^T \widehat{\boldsymbol{\delta}} \pm (\mathbf{d}^T \mathbf{S}_p \mathbf{d} F^\alpha / c)^{1/2}, \forall \mathbf{d} \in \mathbf{R}^r\}.$$

(*Hint*: Use the fact that $F(\boldsymbol{\delta}) = c \sup_{\mathbf{d} \neq \mathbf{0}}((\mathbf{d}^T(\widehat{\boldsymbol{\delta}} - \boldsymbol{\delta}))^2 / \mathbf{d}^T \mathbf{S}_p \mathbf{d})$.)

23.26 Let Q and R be arbitrary square integrable (possibly constant) random variables and let $\mathbf{U}$ be an arbitrary random vector.

(a) Show that the h which minimizes $E(Q - hR)^2$ is $h = EQR/ER^2$.

(b) Show that the function $h(\mathbf{U})$ which minimizes $E(Q - h(\mathbf{U})R)^2$ is $E(QR|\mathbf{U})/E(R^2|\mathbf{U})$. (*Hint*: $E(Q - h(\mathbf{U})R)^2 = E(E((Q - h(\mathbf{U})R)^2|\mathbf{U}))$. Use part (a).)

(c) Suppose we want to estimate a univariate parameter $\tau(\boldsymbol{\theta})$ with the loss function $L(a, \boldsymbol{\theta}) = (M(\boldsymbol{\theta})(a - \tau(\boldsymbol{\theta}))^2)$. Show that the Bayes estimator for this loss function (i.e. the function $T(\mathbf{X})$ which minimizes $E(L(T(\mathbf{X}), \boldsymbol{\theta})|\mathbf{X})$) is given by

$$T(\mathbf{X}) = E(M^2(\boldsymbol{\theta})\tau(\boldsymbol{\theta})|\mathbf{X}) / E(M^2(\boldsymbol{\theta})|\mathbf{X}).$$

23.27 Consider the general location parameter model of **23.35**.

(a) Show that for each $\mathbf{U}$, $E_0(X_n - h(\mathbf{U}))^2$ is minimized for $h(\mathbf{U}) = E_0(X_n|\mathbf{U})$ (use Exercise 26.26(b)).

(b) Show that the conditional expectation of X_n given $\mathbf{U}$ when $\delta = 0$ is given by

$$\frac{\int_{-\infty}^{\infty} x f_0(U_1 + x, \ldots, U_{n-1} + x, x)\,\mathrm{d}x}{\int_{-\infty}^{\infty} f_0(U_1 + x, \ldots, U_{n-1} + x, x)\,\mathrm{d}x} = X_n - \frac{\int_{-\infty}^{\infty} q f_q(X_1, \ldots, X_n)\,\mathrm{d}q}{\int_{-\infty}^{\infty} f_q(X_1, \ldots, X_n)\,\mathrm{d}q}.$$

(The joint density of X_n and U is $f_0(u_1 + x_n, \ldots, u_{n-1} + x_n, x_n$.) To compute the second formula, make the change of variable $x = X_n - q$ $\mathrm{d}x = -\mathrm{d}q$ in the integrals and recall that $U_i + X_n = X_i$.)

(c) Verify that the Fisher information for this model is constant and hence that the Jeffreys prior for this model is a uniform distribution.

23.28 Consider the general scale parameter model of **23.36**. Note that the group is transitive on the parameter space so that the risk is constant (and hence may be computed under the assumption that $\tau = 1$). Note also that the continuity of the distribution implies that $P(X_n = 0) = 0$ so that we can assume that $X_n \neq 0$.

(a) Show that an estimator $T(\mathbf{X})$ is equivariant if and only if $T(\mathbf{X}) = |X_n|h(\mathbf{V})$ for some function $h(\mathbf{V})$. (Take $b = |X_n|^{-1}$ to show necessity.)
(b) Show that for each fixed $\mathbf{V}$, $E_1(|X_n|h(\mathbf{V}) - 1)^2$ is minimized for $h(\mathbf{V}) = E_1(|X_n||\mathbf{V})/E_1(|X_n|^2|\mathbf{V})$ (see Exercise 23.26(b)).
(c) Show that $E_1(|X_n|^k|\mathbf{V}) =$

$$\frac{\int_0^\infty x^{k+n-1} f_1(V_1 x, \ldots, V_{n-1}x, V_n x)\,dx}{\int_0^\infty x^{n-1} f_1(V_1 x, \ldots, V_n x, V_n x)\,dx} = |X_n|^k \frac{\int_0^\infty t^{-k-1} f_t(X_1, \ldots, X_n)\,dt}{\int_0^\infty t^{-1} f_t(X_1, \ldots, X_n)\,dt}.$$

(*Hint*: the density function of $(X_n, v_1, \ldots, v_{n-1})$ is $|x_n|^{n-1} f_1(v_1|x_n|, \ldots, v_{n-1}|x_n|, x_n)$. Why? To compute the first formula, construct two arguments depending on whether $V_n = 1$ or -1 (i.e. depending on whether $X_n > 0$ or not). To compute the second formula, make the substitution $x = |X_n|/t$, $dx = -(|X_n|/t^2)\,dt$ in the integral and recall that $V_i|X_n| = X_i$.) (Note that this argument is more elementary when we assume that the $X_i > 0$ so that $V_n \equiv 1$.)
(d) Verify the formula given for the best equivariant estimator.
(e) Verify that the best equivariant estimator is the generalized Bayes estimator for the improper prior $\pi(\tau) = \tau^{-1}$. (Note that for this loss function, the Bayes estimator is given by $E(\tau^{-1}|\mathbf{X})/E(\tau^{-2}|\mathbf{X})$. See Exercise 23.26(b).)
(f) Show that the Fisher information for this model is proportional to τ^{-2} so that the Jeffreys prior for this model is proportional to τ^{-1}. (The score function for this model is $S(\mathbf{X}, \tau) = \frac{\partial}{\partial \tau} \log(f_\tau(\mathbf{X})) = \tau^{-1} h(X_1/\tau, \ldots, X_n/\tau)$, $h(\mathbf{U}) = \sum U_i \frac{\partial}{\partial U_i} f_1(\mathbf{U})$. Note that the distribution of $(X_1/\tau_1, \ldots, X_n/\tau)$ does not depend on τ.)
(g) Verify that if θ has a uniform distribution on the whole line and $\theta = \log(\tau)$, then τ has density $\pi(\tau) = \tau^{-1}$.

23.29 Consider the problem of estimating δ in the general location–scale model of **23.37** with the loss function given there.

(a) Show that an estimator $T(\mathbf{X})$ of δ is location and scale equivariant if and only if $T(\mathbf{X}) = X_n - |U_{n-1}|h(\mathbf{W})$ for some function $h(\mathbf{W})$. (To show necessity for this condition, consider $X_i \to bX_i + a$ with $b = |U_{n-1}|^{-1}$, $a = -X_n/|U_{n-1}|$.)
(b) Show that $E_{0,1}(X_n - |U_{n-1}|h(\mathbf{W}))^2$ is minimized by

$$h(\mathbf{W}) = E_{0,1}(X_n|U_{n-1}||\mathbf{W})/E_{0,1}(|U_{n-1}|^2|\mathbf{W}).$$

(c) Show that $E_{0,1}(X_n^k|U_{n-1}|^j|\mathbf{W}) =$

$$\frac{\int_{-\infty}^\infty \int_0^\infty x^k u^{n-2+j} f_{0,1}(W_1 u + x, \ldots, W_{n-2}u + x, W_{n-1}u + x, x)\,du\,dx}{\int_{-\infty}^\infty \int_0^\infty u^{n-2} f_{0,1}(W_1 u + x, \ldots, W_{n-2}u + x, W_{n-1}u + x, x)\,du\,dx}$$

$$= |U_{n-1}|^j \frac{\int_{-\infty}^\infty \int_0^\infty t^{-k-j-1}(X_n - q)^k f_{q,t}(\mathbf{X})\,dt\,dq}{\int_{-\infty}^\infty \int_0^\infty t^{-1} f_{q,t}(\mathbf{X})\,dt\,dq}.$$

The joint density of X_n, U_{n-1} and $W_1, \ldots, W_{n-2}$ is

$$|u_{n-1}|^{n-2} f_{0,1}(w_1|u_{n-1}| + x_n, \ldots, w_{n-2}|u_{n-1}| + x_n, u_{n-1} + x_n, x_n).$$

To obtain the first formula, construct two arguments depending on whether $W_{n-1} = 1$ or -1 (i.e. whether $U_{n-1} > 0$ or not). To obtain the second formula, make the change of variable $x = (X_n - q)/t$, $u = |U_{n-1}|/t$, $du\,dx = (|U_{n-1}|/t^3)\,dt\,dq$. Use the fact that $W_i|U_{n-1}| + X_n = X_i$.)

(d) Verify the formula for the best equivariant estimator of δ.

(e) Verify that this estimator is the generalized Bayes estimator for the improper prior $\pi(\delta, \tau) = \tau^{-1}$. (Note that the Bayes estimator for this loss function is $E(\delta\tau^{-2}|\mathbf{X})/E(\tau^{-2}|\mathbf{X})$. See Exercise 23.26(b).)

23.30 Consider the problem of estimating τ for the general location–scale model given in **23.37** with the loss function given there.

(a) Show that an estimator $T(\mathbf{X})$ of τ is location and scale equivariant if and only if $T(\mathbf{X}) = |U_{n-1}|h^*(\mathbf{W})$ for some $h^*(\mathbf{W})$.

(b) Show that $E_{0,1}(|U_{n-1}|h^*(\mathbf{W}) - 1)^2$ is minimized when $h^*(\mathbf{W}) = E_{0,1}(|U_{n-1}|\;|\mathbf{W})/E_{0,1}(|U_{n-1}|^2|\mathbf{W})$. (See Exercise 23.26(b).)

(c) Show that

$$E(|U_{n-1}|^j|\mathbf{W}) = |U_{n-1}|^j \frac{\int_{-\infty}^{\infty}\int_0^{\infty} t^{-j-1} f_{q,t}(\mathbf{X})\,dq\,dt}{\int_{-\infty}^{\infty}\int_0^{\infty} t^{-1} f_{q,t}(\mathbf{X})\,dq\,dt}$$

(Use Exercise 23.28(c) with $k = 0$.)

(d) Verify the formula for the best equivariant estimator.

(e) Verify that this estimator is generalized Bayes for the improper prior distribution $\pi(\delta, \tau) = \tau^{-1}$. (Note that the Bayes estimator for this loss function is $E(\tau^{-1}|\mathbf{X})/E(\tau^{-2}|\mathbf{X})$. See Exercise 23.26(c).)

23.31 Consider the general location–scale model of **23.37**.

(a) Show that the score function for this model is

$$\mathbf{S}(\mathbf{X}, \delta, \tau) = \begin{pmatrix} \frac{\partial}{\partial\delta}\log(f_{\delta,\tau}(\mathbf{X})) \\ \frac{\partial}{\partial\tau}\log(f_{\delta,\tau}(\mathbf{X})) \end{pmatrix} = \begin{pmatrix} \tau^{-1}h((X_1-\delta)/\tau, \ldots, (X_n-\delta)/\tau) \\ \tau^{-1}k((X_1-\delta)/\tau, \ldots, (X_n-\delta)/\tau) \end{pmatrix},$$

where $h(\mathbf{U}) = \sum \frac{\partial}{\partial U_i}\log(f_{0,1}(\mathbf{U}))$, $k(\mathbf{U}) = \sum U_i \frac{\partial}{\partial U_i}\log(f_{0,1}(\mathbf{U}))$.

(b) Show that Fisher information matrix for this model (the covariance matrix of the score function) has the form $\tau^{-2}\mathbf{A}$, where $\mathbf{A}$ is a fixed 2×2 matrix independent of τ, and hence the Jeffreys prior for this model is proportional to $|\mathbf{I}(\delta, \tau)|^{1/2} = \tau^{-2}|\mathbf{A}|^{1/2}$.

23.32 Consider a general model with a univariate parameter which is invariant under a group G of transformations with induced group $\overline{G}$ on the parameter space. Let $S(\mathbf{X}, \theta) = \frac{\partial}{\partial\theta}\log(f(\mathbf{X}, \theta))$ (i.e. $S(\mathbf{X}, \theta)$ is the score function so that the Fisher information $I(\theta)$ is $\mathrm{var}_\theta(S(\mathbf{X}, \theta))$.

(a) Show that $S(\mathbf{X}, \overline{g}^{-1}(\theta))\overline{g}^{-1\prime}(\theta) = S(g(\mathbf{X}), \theta)$. (Note that $S(\mathbf{X}, \overline{g}^{-1}(\theta))\overline{g}^{-1\prime}(\theta) = \frac{\partial}{\partial\theta}\log(f(\mathbf{X}, \overline{g}^{-1}(\theta))) = \frac{\partial}{\partial\theta}\log(f(g(\mathbf{X}), \theta)|J_g(\mathbf{X})|)$.

(b) Verify that $I(\overline{g}^{-1}(\theta))(\overline{g}^{-1\prime}(\theta))^2 = I(\theta)$. (Recall that $\mathrm{var}_{\overline{g}^{-1}(\theta)}(h(g(\mathbf{X}))) = \mathrm{var}_\theta(h(\mathbf{X}))$.)

(c) Verify that if θ has density function $(I(\theta))^{1/2}$, then $\delta = \overline{g}(\theta)$ has density function $(I(\delta))^{1/2}$ so that the Jeffreys prior is invariant.

23.33 Consider a general model with a multivariate parameter which is invariant under a group G of transformations with induced group $\overline{G}$ on the parameter space. Let $\mathbf{S}(\mathbf{X}, \theta)$ be the score function (i.e. the vector of partial derivatives (with respect to θ_i) of $\log(f(\mathbf{X}, \boldsymbol{\theta}))$) so that the Fisher information matrix $\mathbf{I}(\boldsymbol{\theta})$ is the covariance matrix of $\mathbf{S}(\mathbf{X}, \boldsymbol{\theta})$.

(a) Show that $\mathbf{J}_{\overline{g}^{-1}}(\boldsymbol{\theta})\mathbf{S}(\mathbf{X}, \overline{g}^{-1}(\boldsymbol{\theta})) = \mathbf{S}(g(\mathbf{X}), \boldsymbol{\theta})$ (where $\mathbf{J}_{\overline{g}^{-1}}$ is the Jacobian matrix of $\overline{g}^{-1}$).

(b) Show that $|\mathbf{I}(\overline{g}^{-1}(\boldsymbol{\theta}))|^{1/2}|\mathbf{J}_{\overline{g}^{-1}}(\boldsymbol{\theta})| = |\mathbf{I}(\boldsymbol{\theta})|^{1/2}$.

(c) Show that if $\boldsymbol{\theta}$ has the Jeffreys prior distribution, then so does $\boldsymbol{\delta} = \overline{g}(\boldsymbol{\theta})$ and hence that the Jeffreys prior is invariant.

23.34 (a) Let $\overline{g}_i(\delta) = \delta + c_i$. Show that $\overline{g}_1(\overline{g}_2(\delta)) = \delta + (c_1 + c_2) = \overline{g}_2(\overline{g}_1(\delta))$.

(b) Let $\overline{g}_i(\tau) = b_i\tau$. Show that $\overline{g}_1(\overline{g}_2(\tau)) = (b_1 b_2)\tau = \overline{g}_2(\overline{g}_1(\tau))$.

(c) Let $\overline{g}_i(\delta, \tau) = (b_i\delta + c_i, b_i\tau)$. Show that $\overline{g}_1(\overline{g}_2(\delta, \tau)) = (b_1 b_2\delta + b_1 c_2 + c_1, b_1 b_2\tau) \neq \overline{g}_2(\overline{g}_1(\delta, \tau))$.

CHAPTER 24

SEQUENTIAL METHODS

24.1 When considering sampling problems in the foregoing chapters we have usually assumed that the sample size n was fixed. This assumption may hold because we chose n beforehand; or it may not be by choice, as for example when we are presented with the results of a finished experiment; or it may be that the sample size was determined by some other criterion, as when we decide to observe for a given period of time. For example, in setting a standard error to an estimate, we were making probability statements within a field of samples all of size n. If n is determined in some way that is independent of the values of the observations, such a conditional argument seems reasonable, but it is as well to realize that there is nothing automatic about accepting it. We hope we will not be thought quite cynical if we add that, in our view, the reason why so many statistical arguments are made conditionally upon an observed value of n is that this procedure is very convenient mathematically – cf. **20.32**.

Even when n is fixed, we are sometimes able to improve upon even the most powerful test procedures discussed in earlier chapters. For example, if π is the parameter of a binomial distribution, Exercise 20.2 showed that the UMP test of $H_0 : \pi = \frac{1}{2}$ against $H_1 : \pi > \frac{1}{2}$ rejects H_0 when the number of successes in the sample, x, is large enough. Using the normal approximation to the distribution of x, and assuming a test size of 0.05 based on $n = 100$ observations, H_0 is rejected when

$$x > n\pi_0 + 1.645\{n\pi_0(1 - \pi_0)\}^{1/2} = \tfrac{1}{2}n + 0.822n^{1/2},$$

i.e. $x > 58$. Now suppose that we had paused after collecting three-quarters of the sample, and noticed that we already had $x = 65$. We are clearly bound to reject H_0 for $n = 100$, and indeed for any n in $75 \leq n \leq 100$, since $an + bn^{1/2}$ is an increasing function of n. Thus it is clearly wasteful to collect the last 25 observations at all. It evidently pays, in this example at least, to make the sample size a function of the values so far observed.

Sequential procedures

24.2 Occasionally, the sample size is a random variable explicitly dependent upon the values of the observations. One of the simplest cases is one we have already touched upon in Example 9.15. Suppose we are sampling human beings one by one to discover the proportion belonging to a rare blood group. Instead of sampling, say, 1000 individuals and counting the number of members of that blood group we may go on sampling until 20 such members have occurred. We shall see later why this may be a preferable procedure; for the moment we take for granted that it is worth considering. In successive trials of such an investigation we should find that for a fixed number of successes, say 20, the sample size n required to achieve them varied considerably. It must be at least 20, but it might be much larger.

24.3 Such procedures are called *sequential*. Their typical feature is a *sampling scheme*, which lays down a *stopping rule* under which we decide after each observation whether to stop or to continue sampling. In our present example the rule is very simple: if we draw a failure, continue; if we draw a success, continue unless 19 successes have previously occurred, in which event, stop. The decision at any point is, in general, dependent on the observations made up to that point. Thus, for a sequence of values $x_1, x_2, \ldots, x_n$, the sample size at which we stop is not independent of the xs. This dependence gives sequential analysis its characteristic features.

Sequential methods were first developed during the Second World War, principally by Wald (1947) in the USA, and simultaneously in England by Barnard (1946); for a historical perspective, see Ghosh (1991).

A useful general reference is the set of papers edited by Ghosh and Sen (1991).

24.4 If the probability is one that the procedure will terminate, the scheme is said to be *closed*. If there is a non-zero probability that sampling can continue indefinitely, the scheme is called *open*. Open schemes are obviously of little practical use compared to closed schemes, and we usually have to reduce them to closed form by putting an upper limit on the size of the sample. Such truncation often makes their properties difficult to determine exactly.

Usage in this matter is not entirely uniform. 'Closed' sometimes means 'truncated', that is to say, applies to the case where the stopping rule puts an upper limit to the sample size, and 'open' then means 'non-truncated'.

Fixing the sample size beforehand may be regarded as a very special case of a sequential scheme with stopping rule: go on until you have obtained n members, irrespective of what values arise.

Example 24.1 (Sequential sampling for Bernoulli trials)

As an example of a sequential scheme let us consider sequential sampling for attributes (Example 9.15), where we proceed until k successes are observed and then stop. It scarcely needs proof that such a scheme is closed. The probability that n exceeds some large value n_0 approaches zero as $n_0 \to \infty$; this is always true when $E(n) < \infty$.

From (9.15),

$$\binom{n-1}{k-1}\pi^k(1-\pi)^{n-k}, \qquad n = k, k+1, \ldots, \tag{24.1}$$

is the distribution of n. The probability generating function of n (with the origin at zero) is given by (5.45) as, in our present notation,

$$[\pi t/\{1-(1-\pi)t\}]^k, \tag{24.2}$$

while for the cumulant generating function we have from (5.47)

$$\psi(t) = k\log[\pi e^t/\{1-(1-\pi)e^t\}].$$

From (5.48), we find

$$\kappa_1 = k/\pi, \tag{24.3}$$

$$\kappa_2 = k(1-\pi)/\pi^2. \tag{24.4}$$

From (9.16), an unbiased estimator of π is

$$p = (k-1)/(n-1). \tag{24.5}$$

For $k > 1$, the variance of p is expressible as a series. We have

$$\begin{aligned}
E(p^2) &= (k-1)\pi^{k-1}(1-\pi)^{1-k}\sum_{n=k}^{\infty}\binom{n-2}{k-2}\frac{(1-\pi)^{n-1}}{n-1} \\
&= (k-1)\pi^{k}(1-\pi)^{1-k}\int_0^{(1-\pi)}\sum_{n=k}^{\infty}\binom{n-2}{k-2}t^{n-2}\,\mathrm{d}t \\
&= (k-1)\pi^{k}(1-\pi)^{1-k}\int_0^{(1-\pi)} t^{k-2}(1-t)^{1-k}\,\mathrm{d}t.
\end{aligned} \tag{24.6}$$

Putting $u = \pi t/\{(1-\pi)(1-t)\}$, we find

$$E(p^2) = (k-1)\pi^2\int_0^1 \frac{u^{k-2}\,\mathrm{d}u}{\pi+(1-\pi)u} \tag{24.7}$$

$$\begin{aligned}
&= (k-1)\pi^2\int_0^1 u^{k-2}\Big\{\sum_{j=0}^{\infty}(1-\pi)^j(1-u)^j\Big\}\,\mathrm{d}u \\
&= (k-1)\pi^2\sum_{j=0}^{\infty}(1-\pi)^j B(k-1, j+1) \\
&= \pi^2\sum_{j=0}^{\infty}\binom{k+j-1}{j}^{-1}(1-\pi)^j.
\end{aligned} \tag{24.8}$$

Hence, subtracting π^2, we have

$$\operatorname{var} p = \pi^2\sum_{j=1}^{\infty}\binom{k+j-1}{j}^{-1}(1-\pi)^j. \tag{24.9}$$

As $\pi \to 0$ in (24.7), if $k > 2$, $E(p^2)/\pi^2 \to (k-1)/(k-2)$, and the coefficient of variation of p, $(\operatorname{var} p)^{1/2}/E(p) \to (k-2)^{-1/2}$.

Prasad and Sahai (1982) give close bounds for var p – see Exercise 24.23. Best (1974) gives a finite series for var p, and finds its m.s.e. numerically.

We may obtain an unbiased estimator of var p in a simple closed form since

$$E\left\{\frac{(k-1)(k-2)}{(n-1)(n-2)}\right\} = \pi^2.$$

Hence

$$E\left\{\left(\frac{k-1}{n-1}\right)^2 - \frac{(k-1)(k-2)}{(n-1)(n-2)}\right\} = E\left(\frac{k-1}{n-1}\right)^2 - \pi^2 = \operatorname{var} p.$$

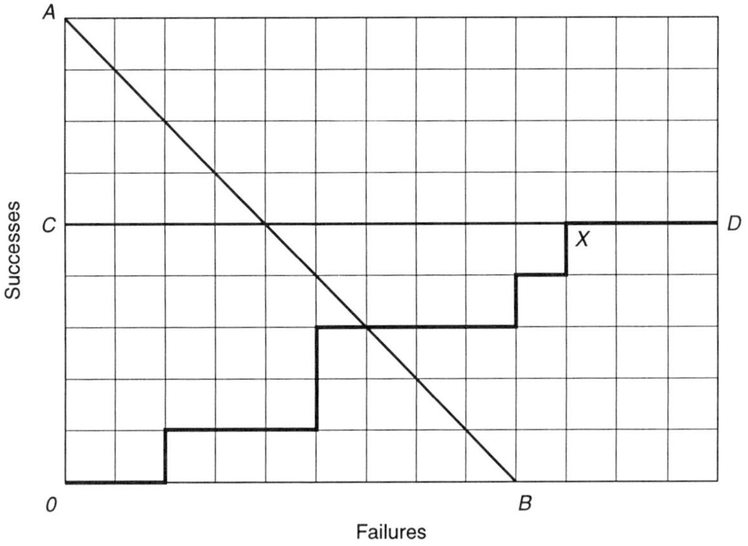

Fig. 24.1 Diagrammatic representation of sample path

Thus the expression in braces on the left is an unbiased estimator for var p; it simplifies to

$$\frac{(k-1)(n-k)}{(n-1)^2(n-2)} = \frac{p(1-p)}{n-2}. \tag{24.10}$$

Result (24.10) may be contrasted with the corresponding result for fixed sample size n, given in Exercise 17.11, which has $n-1$ rather than $n-2$ in the denominator.

Since $n-2 = (k-1-p)/p$, (24.10) is equal to $p^2(1-p)/(k-1-p)$, and for small p is approximately $p^2/(k-1)$. Thus the coefficient of variation of p is (biasedly) estimated by $\{(1-p)/(k-1-p)\}^{1/2}$ and for small p this is approximately $(k-1)^{-1/2}$. For this sequential procedure, the relative sampling error of p is thus approximately constant for small p.

This sequential scheme is often called inverse binomial sampling. Knight (1965) provides a unified theory for the binomial, Poisson, hypergeometric and exponential distributions. For further details of the distribution and its sampling properties, see Johnson *et al.* (1993, Chapter 5).

24.5 The sampling of attributes plays such a large part in sequential analysis that we may usefully discuss a diagrammatic representation of the procedure.

Consider a grid such as that of Fig 24.1 and measure the number of failures and successes along the horizontal and vertical axes, respectively. The sample may be represented on this grid by a path from the origin, moving one step to the right for a failure F and one step upwards for a success S. The path OX corresponds, for example, to the sequence $FFSFFFSSFFFFSFS$. A stopping rule is equivalent to some sort of barrier on the diagram. For example, the line AB is such that $S+F=9$ and thus corresponds to the case of fixed sample size $n=9$. The line CD corresponds to $S=5$ and is a rule of the type considered in Example 24.1 with $k=5$. The path

OX, involving a sample of 15, is then one sample that would terminate at X. If X is the point whose coordinates are (x, y) the number of different paths from O to X with the last observation a success is the number of ways in which x can be selected from $(x + y - 1)$. The probability of terminating at X is

$$\binom{x+y-1}{x}\pi^x(1-\pi)^y.$$

Example 24.2 (The gambler's ruin problem)

One of the oldest problems in the theory of probability concerns a sequential procedure. Consider two players, A and B, playing a series of independent games at each of which A's chance of success is π and B's is $1 - \pi$. The loser of each game pays the winner one unit. If A starts with a units and B with b units, what are their chances of ruin (a player being ruined when he has lost his last unit)?

The series of games is representable on a diagram like Fig. 24.1. We may take A's winning as a success. The game continues so long as both A and B have any money left but stops when A or B has $a + b$ (when the other player has lost all his initial stake). The boundaries of the scheme are therefore the lines $y - x = -a$ and $y - x = b$.

Figure 24.2 shows the situation for the case $a = 5, b = 3$. The lines AB, CD are at 45° to the axes and go through $F = 0$, $S = 3$ and $F = 5$, $S = 0$, respectively. For any point between these lines $S - F$ is less than 3 and $F - S$ is less than 5. On AB, $S - F$ is 3, and if a path arrives at that line B has lost three more games than A and is ruined; similarly, if the path arrives at CD, A is ruined. The stopping rule is, then: if the point lies between the lines, continue sampling; if it reaches AB, stop with the ruin of B; if it reaches CD, stop with the ruin of A. The actual probabilities are easily obtained. Let u_x be the probability that A will be ruined when he possesses x units. By considering a single game we see that

$$u_x = \pi u_{x+1} + (1-\pi)u_{x-1}, \tag{24.11}$$

with boundary conditions

$$u_0 = 1, \qquad u_{a+b} = 0. \tag{24.12}$$

The general solution of (24.11) is

$$u_x = At_1^x + Bt_2^x$$

where t_1 and t_2 are the roots of

$$\pi t^2 - t + (1-\pi) = 0,$$

namely

$$t = 1 \text{ and } t = (1-\pi)/\pi.$$

Provided that $1 - \pi \neq \pi$, the solution is, from (24.12),

$$u_x = \frac{\left(\frac{1-\pi}{\pi}\right)^{a+b} - \left(\frac{1-\pi}{\pi}\right)^x}{\left(\frac{1-\pi}{\pi}\right)^{a+b} - 1}, \qquad \pi \neq \frac{1}{2}. \tag{24.13}$$

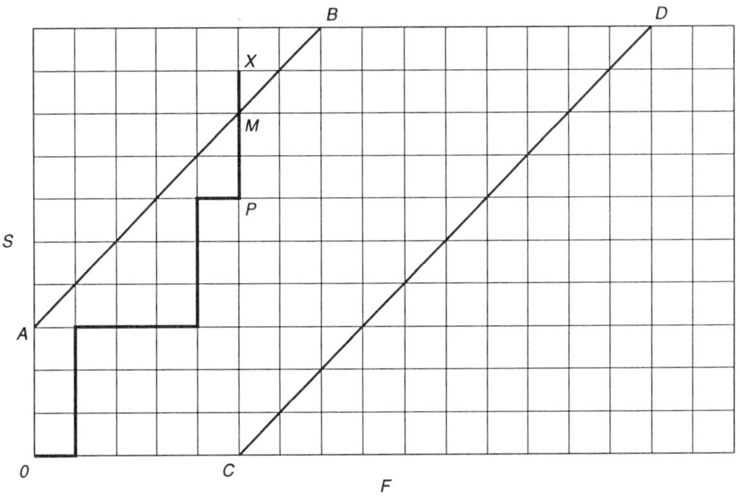

Fig. 24.2 Boundaries for the gambler's ruin problem

If, however, $\pi = \frac{1}{2}$, the solution is

$$u_x = \frac{a+b-x}{a+b}. \tag{24.14}$$

In particular, at the start of the game, for $\pi = \frac{1}{2}$, $x = a$,

$$u_a = \frac{b}{a+b}. \tag{24.15}$$

24.6 We can obviously generalize this kind of situation in many ways and, in particular, can set up various types of boundary. A closed scheme is one for which the boundary will be reached with probability one.

Suppose, in particular, that the scheme specifies that if A loses he pays one unit but if B loses he pays k units. The path on Fig. 24.2 representing a series then consists of horizontal and vertical steps of sizes 1 and k respectively. This scheme enables us to emphasize a point that constantly bedevils the mathematics of sequential sampling: a path may not end exactly on a boundary, but may cross it. For example, with $k = 3$ such a path might be OX in Fig. 24.2. After two successes and five failures we arrive at P. Another success would take us to X, crossing the boundary at M. We stop, of course, at this stage, whether the boundary is reached or crossed. The point of the example is that the exact probability of reaching the boundary at M is zero – in fact, this point is inaccessible. As we shall see, such discontinuities sometimes make it difficult to put forward exact and concise statements about the probabilities of what we are doing. We refer to such situations as 'end-effects'; perhaps surprisingly, we find that in most practical circumstances they may be neglected.

Woodroofe (1982) and Siegmund (1985) provide improved approximations that further reduce the impact of end-effects.

Sequential tests of hypotheses

24.7 We now apply the ideas of sequential analysis to testing hypotheses and, in the first instance, to choosing between H_0 and H_1. Suppose that these hypotheses concern a parameter θ that may take values θ_0 and θ_1, respectively; i.e. H_0 and H_1 are simple. We seek a sampling scheme that divides the sample space into three mutually exclusive domains: (a) domain ω_a, such that if the sample point falls within it we accept H_0 (and reject H_1); (b) domain ω_r, such that if the sample point falls within it we accept H_1 (and reject H_0); (c) the remainder of the sample space, ω_c, such that if the sample point falls within it we continue sampling. In Fig. 24.2, taking A's ruin as H_0, B's ruin as H_1, the region ω_a is the region to the right of CD, including the line itself; ω_r is the region above AB, including the line itself; ω_c is the region between the lines.

The operating characteristic

24.8 The probability of accepting H_0 when H_1 is true is a function of θ_1, which we shall denote by $K(\theta_1)$. If the scheme is closed, the probability of rejecting H_0 when H_1 is true is then $1 - K(\theta_1)$. Considered as a function of θ_1, this is simply the power function. As in our previous work we could, of course, work in terms of power; but in sequential analysis it has become customary to work with $K(\theta_1)$ itself.

$K(\theta)$ considered as a function of θ is called the operating characteristic (OC) of the scheme. The plot of $K(\theta)$ against θ gives the OC curve, the complement of the power function.

The average sample number

24.9 A second function used to describe the performance of a sequential test is the average sample number (ASN). The ASN is the expected sample size required to reach a decision to accept H_0 or H_1 and therefore to discontinue sampling. The OC for H_0 and H_1 does not depend on the sample number, but only on constants determined initially by the sampling scheme. The ASN is the expected amount of sampling that we have to do to implement the scheme, given the value of θ.

Example 24.3 (A simple sequential scheme for attributes)

Consider sampling from a population of attributes of which a proportion π are successes, and let π be small. We are interested in the possibility that π is less than some given value π_0. For example, such a situation arises when a manufacturer wishes to guarantee that the proportion of rejects in a batch of articles is below some declared figure. Consider first the alternative $\pi_1 > \pi_0$.

We will take a very simple scheme. If no success appears we proceed to sample until a pre-assigned sample number n_0 has appeared and accept π_0. If, however, a success appears we accept π_1 and stop sampling.

If the true probability of success is π, the probability that we accept the hypothesis is $(1-\pi)^{n_0}$; this is the OC. It is a J-shaped curve decreasing monotonically from $\pi = 0$ to $\pi = 1$.

Common sense dictates that we should accept the smaller of π_0 and π_1 if no success appears, and the larger if a success does appear. Let π_0 be the smaller; then the probability α of a Type I error equals $1-(1-\pi_0)^{n_0}$ and that of an error of Type II, β, equals $(1-\pi_1)^{n_0}$. If we were to interchange π_0 and π_1, the α-error would be $1-(1-\pi_1)^{n_0}$ and the β-error $(1-\pi_0)^{n_0}$, both of which are greater than before.

We can use the OC in this particular case to provide a test of the composite hypothesis H_0 : $\pi \leq \pi_0$ against $H_1 : \pi > \pi_0$. If $\pi < \pi_0$ the probability of a Type I error is less than $1-(1-\pi_0)^{n_0}$ and the probability of a Type II error is less than $(1-\pi_0)^{n_0}$.

The ASN is the expectation of m, the sample number at which we terminate:

$$\sum_{m=1}^{n_0-1} m\pi(1-\pi)^{m-1} + n_0(1-\pi)^{n_0-1} = -\pi\frac{\partial}{\partial\pi}\sum_{0}^{n_0-1}(1-\pi)^m + n_0(1-\pi)^{n_0-1}$$
$$= \frac{1-(1-\pi)^{n_0}}{\pi}. \tag{24.16}$$

The ASN in this case is a decreasing function of π since it is equal to

$$\frac{1-(1-\pi)^{n_0}}{1-(1-\pi)} = 1+(1-\pi)+(1-\pi)^2+\cdots+(1-\pi)^{n_0-1}.$$

We observe that the ASN will differ according to whether π_0 or π_1 is the true value.

A comparison of the results of the sequential procedure with those of an ordinary fixed sample size is not easy to make for discrete distributions, especially as we have to compare two kinds of error. Consider, however, $\pi_0 = 0.1$ and $n = 30$. From tables of the binomial (e.g. *Biometrika Tables*, Vol. I, Table 37) we see that the probability of five successes or more is about 0.18. Thus on a fixed sample-size basis we may reject $\pi = 0.1$ in a sample of 30 with a Type I error of 0.18. For the alternative $\pi = 0.2$ the probability of four or fewer successes is 0.26, which is the Type II error.

With the sequential test, the Type I error is $1-(1-\pi_0)^{n_0}$ and the Type II error is $(1-\pi_1)^{n_0}$. When $n_0 = 2$ the Type I error is 0.19 and the Type II error 0.64. For a sample of size 6 the errors are 0.47 and 0.26, respectively. We clearly cannot make both types of errors correspond in this simple case, but it is evident that samples of smaller size are needed in the sequential case to fix either type of error at a given level. With more flexible sequential schemes, both types of error can be fixed at given levels with smaller ASN than the fixed-size sample number. In fact, their economy in sample number is one of their principal recommendations – cf. Example 24.10.

Wald's sequential probability ratio test

24.10 Suppose we take a sample of n observations $x_1, x_2, \ldots, x_n$ from a population with p.d.f. $f(x, \theta)$. As usual, $f(x, \theta)$ represents probabilities when x is discrete and the derivative of the distribution function when x is continuous. At any stage the ratio of the probabilities of the sample on hypotheses $H_0(\theta = \theta_0)$ and $H_1(\theta = \theta_1)$ is

$$L_n = \prod_{i=1}^{n} f(x_i, \theta_1) \Big/ \prod_{i=1}^{n} f(x_i, \theta_0). \tag{24.17}$$

We select two numbers A and B, related to the desired α- and β-errors in a manner to be described later, and set up a sequential test as follows: so long as $B < L_n < A$ we continue sampling; at the first occasion when $L_n \geq A$ we accept H_1; at the first occasion when $L_n \leq B$ we accept H_0.

An equivalent but more convenient form for computation is the logarithm of L_n, the critical inequality then being

$$\log B < \sum_{i=1}^{n} \log f(x_i, \theta_1) - \sum_{i=1}^{n} \log f(x_i, \theta_0) < \log A. \tag{24.18}$$

We shall refer to this family of tests as sequential probability ratio (SPR) tests.

24.11 If we write

$$z_i = \log\{f(x_i, \theta_1)/f(x_i, \theta_0)\}, \tag{24.19}$$

the critical inequality (24.18) becomes a statement concerning the cumulative sums of z_is. We first prove that an SPR test terminates with probability one, that is, is closed.

Sampling terminates if either

$$\sum z_i \geq \log A$$

or

$$\sum z_i \leq \log B.$$

The z_is are independent random variables with variance, say, $\sigma^2 > 0$. $\sum_{i=1}^{n} z_i$ then has variance $n\sigma^2$. The variance increases with n and the probability that a value of $\sum z_i$ remains within the finite limits $\log B$ and $\log A$ tends to zero. More precisely, the mean $\bar{z}$ tends under the central limit effect to a (normal) distribution with variance σ^2/n, and hence the probability that it falls between $(\log B)/n$ and $(\log A)/n$ tends to zero.

It was shown by Stein (1946) that $E(e^{nt})$ exists for any complex number t whose real part is less than some $t_0 > 0$. It follows that the random variable n has moments of all orders; n is said to be exponentially bounded. For a more general, rigorous proof and further discussion, see Gut (1988).

Example 24.4 (The SPR test for the binomial)

Consider again the binomial distribution, the probability of success being π. If there are k successes in the first n trials the SPR criterion is given by

$$\log L_n = k \log \frac{\pi_1}{\pi_0} + (n-k) \log \frac{1-\pi_1}{1-\pi_0}. \tag{24.20}$$

Sampling continues until we reach one of the boundary values $\log B$ or $\log A$. We now consider the determination of A and B.

24.12 It is remarkable that the numbers A and B can be derived very simply (at least to an acceptable degree of approximation) from the probabilities of errors of the first and second kinds, α and β, without knowledge of the population. This does not mean that the sequential process is unaffected by the underlying distribution. All that is happening is that our knowledge of the density function is put into the criterion L_n of (24.17) and we work with this ratio directly. It will not, then, come as a surprise to find that SPR tests have certain optimum properties; for they use all the available information, including the order in which the sample values occur.

Consider a sample for which L_n lies between A and B for the first $n-1$ trials and then becomes equal to or greater than A at the nth trial, so that we accept H_1 (and reject H_0). From (24.18), the joint density for $x = (x_1, \ldots, x_n)$ under H_1 must be at least A times that for H_0. This property holds for all samples that result in the acceptance of H_1. Since the probability of accepting H_1 when H_0 is true is α, and that of accepting H_1 when H_1 is true is $1-\beta$, we have

$$A \leq \frac{1-\beta}{\alpha}. \tag{24.21}$$

Likewise, if we accept H_0 we find that

$$B \geq \frac{\beta}{1-\alpha}. \tag{24.22}$$

24.13 If our boundaries were such that A and B were exactly attained when attained at all, i.e. if there were no end-effects, we could write

$$A = \frac{1-\beta}{\alpha}, \qquad B = \frac{\beta}{1-\alpha}. \tag{24.23}$$

Wald (1947) showed that for all practical purposes these equalities may be assumed to hold. Suppose that we have exactly

$$a = \frac{1-\beta}{\alpha}, \qquad b = \frac{\beta}{1-\alpha} \tag{24.24}$$

and that the true errors of first and second kind for the limits a and b are α', β'. We then have, from (24.21),

$$\frac{\alpha'}{1-\beta'} \leq \frac{1}{a} = \frac{\alpha}{1-\beta}, \tag{24.25}$$

and, from (24.22),

$$\frac{\beta'}{1-\alpha'} \leq b = \frac{\beta}{1-\alpha}. \tag{24.26}$$

Hence

$$\alpha' \leq \frac{\alpha(1-\beta')}{1-\beta} \leq \frac{\alpha}{1-\beta}, \tag{24.27}$$

$$\beta' \leq \frac{\beta(1-\alpha')}{1-\alpha} \leq \frac{\beta}{1-\alpha}. \tag{24.28}$$

Furthermore, from (24.25) and (24.26),

$$\alpha'(1-\beta) + \beta'(1-\alpha) \leq \alpha(1-\beta') + \beta(1-\alpha')$$

or

$$\alpha' + \beta' \leq \alpha + \beta. \tag{24.29}$$

Now in practice α and β are small, often conventionally 0.01 or 0.05. It follows from (24.27) and (24.28) that the amount by which α' can exceed α, or β' exceed β, is negligible. Moreover, from (24.29) we see that either $\alpha' \leq \alpha$ or $\beta' \leq \beta$. Hence by using a and b in place of A and B, the worst we can do is to increase one of the errors, and then only by a very small amount. Such a procedure, then, will always be conservative in that it will not increase the error probabilities. To avoid tedious repetition we shall henceforward use the equalities (24.23) except where the contrary is specified.

The inequalities for A and B were also derived for the critical value $1/k_\alpha$ of the fixed sample size probability ratio test in Exercise 20.14. The fact that A and B practically attain their limits implies that the two sequential critical values enclose the single fixed-n critical value, as is intuitively acceptable.

Example 24.5 (More on the binomial SPR test)

Consider again the binomial distribution of Example 24.4 with $\alpha = 0.01, \beta = 0.10, \pi_0 = 0.01$ and $\pi_1 = 0.03$. We have, for k successes and $n-k$ failures ,

$$\log\frac{\beta}{1-\alpha} \leq (n-k)\log\frac{1-\pi_1}{1-\pi_0} + k\log\frac{\pi_1}{\pi_0} \leq \log\frac{1-\beta}{\alpha}$$

or

$$\log\frac{10}{99} \leq (n-k)\log\frac{97}{99} + k\log 3 \leq \log 90$$

or, to the nearest integer,

$$-112 \leq 54k - n \leq 220. \tag{24.30}$$

For a test of this kind, for example, if no success occurred in the first 112 drawings we should accept π_0. If one occurred at the 100th drawing and another at the 200th, we could not accept before the 220th (i.e. $112 + (2 \times 54)$) drawing. And if, by the 200th drawing, six successes had occurred, say at the 50th, 100th, 125th, 150th, 175th, 200th, we could not reject, $54k - n$ being 124 at the 200th drawing; but if that experience was then repeated, the quantity $54k - n$ would exceed 220 and we should accept π_1.

The OC of the SPR test

24.14 Consider the function

$$L^h = \left\{\frac{f(x,\theta_1)}{f(x,\theta_0)}\right\}^h, \tag{24.31}$$

where h is a function of θ. $L^h f(x,\theta)$, say $g(x,\theta)$, is a density function for any value of θ provided that

$$E(L^h) = \int\left\{\frac{f(x,\theta_1)}{f(x,\theta_0)}\right\}^h f(x,\theta)\,\mathrm{d}x = 1. \tag{24.32}$$

It may be shown (cf. Exercise 24.4) that there is at most one non-zero value of h satisfying this equation. Consider the rule: accept H_0, continue sampling, or accept H_1 according to the inequality

$$B^h \leq \frac{\prod\{L^h f(x,\theta)\}}{\prod\{f(x,\theta)\}} \leq A^h. \tag{24.33}$$

Expression (24.33) is evidently equal to the ordinary rule of (24.18) provided that $h > 0$. Consider testing H_0: that the true distribution is $f(x, \theta)$, against H_1: that the true distribution is $g(x, \theta)$. If α', β' are the two errors, the probability ratio is given by (24.33), and we then have, using (24.23),

$$A^h = \frac{1-\beta'}{\alpha'}, \qquad B^h = \frac{\beta'}{1-\alpha'} \tag{24.34}$$

and hence

$$\alpha' = \frac{1-B^h}{A^h - B^h}$$

Since α' is the power function when H_1 holds, its complement, the OC, is given by

$$1-\alpha' = K(\theta) = \frac{A^h - 1}{A^h - B^h}. \tag{24.35}$$

The same formula holds if $h < 0$.

We can now find the OC of the test. When $h(\theta) = 1$ we have the performance at $\theta = \theta_0$. When $h(\theta) = -1$ we have the performance at $\theta = \theta_1$. For other values we must solve (24.32) for θ and then substitute in (24.35). But this step is not needed to plot the OC curve of $K(\theta)$ against θ. We can take $h(\theta)$ itself as a parameter and plot (24.35) against it.

Example 24.6 (The OC curve for the binomial SPR test)

Consider once again the binomial distribution of previous examples. We may write for the discrete values 1 (success) and 0 (failure)

$$f(1, \pi) = \pi,$$
$$f(0, \pi) = 1 - \pi.$$

Then (24.32) becomes

$$\pi\left(\frac{\pi_1}{\pi_0}\right)^h + (1-\pi)\left(\frac{1-\pi_1}{1-\pi_0}\right)^h = 1,$$

or

$$\pi = \frac{1-\left(\frac{1-\pi_1}{1-\pi_0}\right)^h}{\left(\frac{\pi_1}{\pi_0}\right)^h - \left(\frac{1-\pi_1}{1-\pi_0}\right)^h}. \tag{24.36}$$

For $A = (1-\beta)/\alpha$, $B = \beta/(1-\alpha)$ we then have from (24.35)

$$K(\pi) = \frac{\left(\frac{1-\beta}{\alpha}\right)^h - 1}{\left(\frac{1-\beta}{\alpha}\right)^h - \left(\frac{\beta}{1-\alpha}\right)^h}. \tag{24.37}$$

We can now plot $K(\pi)$ against π by using (24.36) and (24.37) as parametric equations in h.

The ASN of the SPR test

24.15 Consider a sequence of n random variables z_i. If n was a fixed number we would have

$$E\left(\sum_{i=1}^{n} z_i\right) = nE(z).$$

This is not true for sequential sampling, but we have instead the result

$$E\left(\sum_{i=1}^{n} z_i\right) = E(n)E(z), \tag{24.38}$$

which is not quite as obvious as it looks. The result is due to Wald (1947) and to Blackwell (1946), and may be demonstrated as follows.

Let each z_i have mean value μ, $E|z_i| \leq C < \infty$, and let

$$Z_n = \sum_{i=1}^{n} z_i. \tag{24.39}$$

Then

$$\begin{aligned} E(Z_n) &= E_n\{E_{z|n}(\textstyle\sum z_i|n)\} \qquad (24.40)\\ &= E_n\{n\mu\} = \mu E(n)\\ &= E(n)E(z), \qquad (24.41)\end{aligned}$$

which is (24.38).

We then have

$$E(n) = \frac{E(Z_n)}{E(z)}. \tag{24.42}$$

But, to our usual approximation, Z_n can take only two values for the sampling to terminate, $\log A$ with probability $1 - K(\theta)$ and $\log B$ with probability $K(\theta)$. Thus

$$E(n) = \frac{K \log B + (1 - K) \log A}{E(z)}, \tag{24.43}$$

which is the approximate formula for the average sample number.

Siegmund (1975) improves on the approximations (24.35) and (24.43). For a general discussion on the moments of n, see Wijsman (1991).

Example 24.7 (ASN for the binomial SPR test)

For the binomial we find

$$\begin{aligned} E(z) &= E \log\left(\frac{f(x, \pi_1)}{f(x, \pi_0)}\right)\\ &= \pi \log \frac{\pi_1}{\pi_0} + (1 - \pi) \log \frac{1 - \pi_1}{1 - \pi_0}. \qquad (24.44)\end{aligned}$$

The ASN can then be calculated from (24.43) when π_0, π_1, A and B (or α and β) are given. It is, of course, a function of π.

24.16 For practical application, sequential testing for attributes is often expressed in such a way that the calculations are in terms of integers. Equation (24.30) is a case in point. We may rewrite it as

$$332 \geq 220 + (n-k) - 53k \geq 0.$$

We may imagine a game in which we start with a score of 220. If a failure occurs we add one to the score; if a success occurs we lose 53 units. The game stops as soon as the score falls to zero or rises to 332, corresponding to acceptance of the values π_0 and π_1 respectively.

24.17 In such a scheme, suppose that we start with a score S_2. For every failure we gain one unit, but for every success we lose b units. If the score rises by S_1 so as to reach $S_1 + S_2$ $(= 2S$, say) we accept one hypothesis; if it falls to zero we accept the other. Let the score at any point be x and the probability be u_x that it will ultimately reach $2S$ without in the meantime falling to zero. Consider the outcome of the next trial. A failure increases the score by unity to $x+1$, a success diminishes it by b to $x-b$. This general form of the gambler's ruin problem yields the recurrence relations:

$$u_x = (1-\pi)u_{x+1} + \pi u_{x-b}, \tag{24.45}$$

with initial conditions

$$u_0 = u_{-1} = u_{-2} = \cdots = u_{-b+1} = 0, \tag{24.46}$$

$$u_{2S} = 1. \tag{24.47}$$

For $b = 1$ this equation is easy to solve, as in Example 24.2. For integer $b > 1$ the solution is more cumbersome. We quote without proof the solution obtained by Burman (1946),

$$u_x = \frac{F(x)}{F(2S)}, \tag{24.48}$$

where

$$F(x) = \begin{cases} (1-\pi)^{-x}\left\{1 - \binom{x-b-1}{1}\pi(1-\pi)^b + \binom{x-2b-1}{2} \right. \\ \left. \quad \times[\pi(1-\pi)^b]^2 + \binom{x-3b-1}{3}[\pi(1-\pi)^b]^3 + \cdots\right\}, & x > 0, \\ 0, & x \leq 0. \end{cases} \tag{24.49}$$

Here the series continues as long as $x - kb - 1$ is positive. Burman also gave expressions for the ASN and the variance of the sample number.

Anscombe (1949a) tabulated functions of this kind for various values of the errors α, β $1-\alpha$ and β) and the ratio S_1/S_2.

SPR test for continuous distributions

24.18 We now consider SPR tests for various continuous distribution, starting with tests for the mean and variance of the normal, which are given in the following examples.

Example 24.8 (SPR test for the mean of the normal distribution)

Consider testing H_0 ($\mu = \mu_0$) against H_1 ($\mu = \mu_1$) for the mean of a normal distribution with unit variance. With z defined as at (24.19) we have

$$z_i = -\tfrac{1}{2}(x_i - \mu_1)^2 + \tfrac{1}{2}(x_i - \mu_0)^2$$
$$= (\mu_1 - \mu_0)x_i - \tfrac{1}{2}(\mu_1^2 - \mu_0^2), \qquad (24.50)$$
$$Z_n = \sum_{i=1}^{n} z_i = n(\mu_1 - \mu_0)\bar{x} - \tfrac{1}{2}n(\mu_1^2 - \mu_0^2). \qquad (24.51)$$

We accept H_0 or H_1 according as $Z_n \leq \log B$ or $Z_n \geq \log A$. For the appropriate OC curve we have, from (24.35),

$$K(\mu) = \frac{A^h - 1}{A^h - B^h}, \qquad (24.52)$$

where h is given by

$$\frac{1}{\sqrt{2\pi}} \int_{-\infty}^{\infty} \exp[h\{(\mu_1 - \mu_0)x - \tfrac{1}{2}(\mu_1^2 - \mu_0^2)\}] \exp\{-\tfrac{1}{2}(x - \mu)^2\}\,dx = 1, \qquad (24.53)$$

which is equivalent to

$$\exp\{\mu^2 - h\mu_1^2 + h\mu_0^2 - (\mu - h\mu_1 + h\mu_0)^2\} = 1$$

or to

$$h = \frac{\mu_1 + \mu_0 - 2\mu}{\mu_1 - \mu_0}, \quad \mu_1 \neq \mu_0, \quad h \neq 0. \qquad (24.54)$$

We can then construct the OC curve by calculating h from (24.54) and substituting in (24.52).

For the ASN we first find

$$E(z) = \frac{1}{\sqrt{2\pi}} \int_{-\infty}^{\infty} \exp\{-\tfrac{1}{2}(x - \mu)^2\{(\mu_1 - \mu_0)x - \tfrac{1}{2}(\mu_1^2 - \mu_0^2)\}\,dx$$
$$= (\mu_1 - \mu_0)\mu - \tfrac{1}{2}(\mu_1^2 - \mu_0^2). \qquad (24.55)$$

For a range of μ the ASN can be determined from this equation in conjunction with (24.52) and

$$E(n) = \frac{K \log B + (1 - K)\log A}{E(z)}. \qquad (24.56)$$

Manly (1970a) gives charts that permit choice of A and B for specified α, β, and also give the value of $E(n)$. Harter and Owen (1974) give tables, due to Blyth and Hutchinson, of a truncated SPR test for the normal mean.

Example 24.9 (SPR test for the variance of the normal distribution)

Suppose that the mean of a normal distribution is known to be μ. To test a hypothesis concerning its variance, $H_0 : \sigma^2 = \sigma_0^2$ against $H_1 : \sigma^2 = \sigma_1^2$, we have

$$Z_n = \sum z_i = -n \log \sigma_1 - \frac{1}{2\sigma_1^2} \sum (x - \mu)^2 + n \log \sigma_0 + \frac{1}{2\sigma_0^2} \sum (x - \mu)^2. \qquad (24.57)$$

This quantity lies between the limits $\log\{\beta/(1-\alpha)\}$ and $\log\{(1-\beta)/\alpha\}$ if

$$\log\frac{\beta}{1-\alpha} < -n\log\frac{\sigma_1}{\sigma_0} - \frac{1}{2}\left(\frac{1}{\sigma_1^2} - \frac{1}{\sigma_0^2}\right)\sum(x-\mu)^2 < \log\frac{1-\beta}{\alpha} \tag{24.58}$$

or

$$\frac{2\log\frac{\beta}{1-\alpha} + n\log\frac{\sigma_1^2}{\sigma_0^2}}{\frac{1}{\sigma_0^2} - \frac{1}{\sigma_1^2}} \leq \sum(x-\mu)^2 \leq \frac{2\log\frac{1-\beta}{\alpha} + n\log\frac{\sigma_1^2}{\sigma_0^2}}{\frac{1}{\sigma_0^2} - \frac{1}{\sigma_1^2}}. \tag{24.59}$$

The OC and ASN are given in Exercises 24.18 and 24.19. See also Exercise 24.20.

SPR tests in the exponential family

24.19 The natural exponential family, introduced in **5.47**, has p.d.f.

$$f(x) = \exp\{xA(\theta) + C(x) + D(\theta)\} \tag{24.60}$$

so that

$$z_i = [A(\theta_1) - A(\theta_0)]x_i + [D(\theta_1) - D(\theta_0)].$$

Working with $S_n = \sum x_i$ rather than $\sum z_i$, the rules for the SPR test become

$$\text{accept } H_1 \text{ if } S_n \geq \{\log(A) - n[D(\theta_1) - D(\theta_0)]\}/[A(\theta_1) - A(\theta_0)]$$
$$\text{accept } H_0 \text{ if } S_n \leq \{\log(B) - n[D(\theta_1) - D(\theta_0)]\}/[A(\theta_1) - A(\theta_0)].$$

In particular, we have the following $A(\theta)$ and $D(\theta)$ functions for standard distributions from Chapter 5:

Distribution	θ	$A(\theta)$	$D(\theta)$
normal, $\sigma = 1$	μ	θ	$-\frac{1}{2}\theta^2$
normal, $\mu = 0$ $(y = x^2)$	σ^2	$-1/2\theta$	$-\frac{1}{2}\ln\theta$
Bernoulli	π	$\log[\theta/(1-\theta)]$	$\log(1-\theta)$
Poisson (5.18)	λ	$\log\theta$	$-\theta$
exponential (5.82)	λ	$-\theta$	$\log\theta$

The efficiency of a sequential test

24.20 In general, many different tests may be derived for given α and β, θ_0 and θ_1. There is no point in comparing their power for given sample sizes because they are constructed to have the same β. We may, however, define efficiency in terms of sample size or ASN. The test with the smaller ASN may reasonably be said to be the more efficient. Following Wald (1947), we shall prove that when end-effects are negligible the SPR test is a most efficient test. More precisely, if S' is an SPR test and S is some other test based on the sum of logarithms of identically distributed variables, then

$$E_i(n|S) \geq E_i(n|S'), \qquad i = 0, 1, \tag{24.61}$$

where E_i denotes the expected value of n on hypothesis H_i.

Note first of all that if u is any random variable, we have by Jensen's inequality **(2.31)** that

$$E(u) \leq \log E\{\exp(u)\}. \tag{24.62}$$

We also have, from (24.42), for *any* closed sequential test based on the sums of type Z_n,

$$E_i(n|S) = \frac{E_i(\log L_n|S)}{E_i(z)} \tag{24.63}$$

If E^* denotes the conditional expectation when H_0 is true, and E^{**} the conditional expectation when H_1 is true, we have, as at (24.22), neglecting end-effects,

$$E^*(L_n|S) = \frac{\beta}{1-\alpha}, \tag{24.64}$$

and similarly, as at (24.21),

$$E^{**}(L_n|S) = \frac{1-\beta}{\alpha}. \tag{24.65}$$

Hence

$$E_0(n|S) = \frac{1}{E_0(z)}\{(1-\alpha)E^*(\log L_n|S) + \alpha E^{**}(\log L_n|S)\}. \tag{24.66}$$

From (24.62), (24.64) and (24.65) we have $E_0(z) < 0$ and

$$E_0(n|S) \geq \frac{1}{E_0(z)}\left\{(1-\alpha)\log\frac{\beta}{1-\alpha} + \alpha\log\frac{1-\beta}{\alpha}\right\}. \tag{24.67}$$

Interchanging H_0 and H_1, α and β in (24.67) gives, with $E_1(z) > 0$,

$$E_1(n|S) \geq \frac{1}{E_0(z)}\left\{\beta\log\frac{\beta}{1-\alpha} + (1-\beta)\log\frac{1-\beta}{\alpha}\right\}. \tag{24.68}$$

When $S = S'$ these inequalities, as at (24.43), are replaced (neglecting end-effects) by equalities. Result (24.61) follows.

Example 24.10 (Fixed versus sequential tests for the normal mean)

One of the hoped-for properties of the sequential method (see **24.1**) is that for a given (α, β), it requires a smaller sample on average than the method employing a fixed sample size. General formulae comparing the two would be difficult to derive, but we may illustrate the point by the test of a normal mean (Example 24.8).

For fixed n and α the test consists of finding a value d such that

$$P\{\mu_0 - d \leq \bar{x} \leq \mu_0 + d|H_0\} = 1-\alpha,$$
$$P\{\mu_0 - d \leq \bar{x} \leq \mu_0 + d|H_1\} = \beta,$$

and, putting

$$\lambda_0 = \sqrt{n}(d-\mu_0),$$
$$\lambda_1 = \sqrt{n}(d-\mu_1),$$

we have

$$n = \frac{(\lambda_1 - \lambda_0)^2}{(\mu_0 - \mu_1)^2}. \tag{24.69}$$

Using (24.69), let us compare it with the ASN of an SPR test. Taking the approximate formula (24.43), which is

$$E_i(n) = \frac{1}{E_i(z)}[K(\mu) \log B + \{1 - K(\mu)\} \log A], \tag{24.70}$$

we find, since from (24.55)

$$E_1(z) = \tfrac{1}{2}(\mu_0 - \mu_1)^2$$
$$E_0(z) = -\tfrac{1}{2}(\mu_0 - \mu_1)^2,$$

that

$$\frac{E_1(n)}{n} = \frac{2}{(\lambda_1 - \lambda_0)^2}\{\beta \log B + (1 - \beta) \log A\}. \tag{24.71}$$

Likewise we find

$$\frac{E_0(n)}{n} = \frac{2}{(\lambda_1 - \lambda_0)^2}\{(1 - \alpha) \log B + \alpha \log A\}. \tag{24.72}$$

Thus, for $\alpha = 0.01$, $\beta = 0.03$, $A = 97$, $B = 3/99$ and we find $\lambda_0 = 2.5758$, $\lambda_1 = -1.8808$. The ratio $E_0(n)/n$ is then 0.43 and $E_1(n)/n = 0.55$. We thus require in the sequential case, on average, either 43 or 55 per cent of the fixed sample size needed to attain the same performance.

A reduced ASN is only obtained using an SPR test when either H_0 or H_1 is true, and not necessarily for intermediate values. Kiefer and Weiss (1957) provided a general basis for improving performance by allowing A and B to be functions of n. The original solution for the normal distribution was obtained by Anderson (1960) and extended by Lorden (1976). Huffman (1983) generalizes these results to cover the one-parameter exponential family. For a review of recent developments, see Lai (1991).

Even when comparing the SPR test with the fixed sample size test, the former is not always more efficient, though usually so. Bechhofer (1960) studies the case of the normal mean (Example 24.10); when $\beta = c\alpha \to 0$, $c > 0$, the ratio of the ASN to the fixed n is $|(\mu_1 - \mu_0)/\{4(\mu_0 + \mu_1 - 2\mu)\}|$ which $= \frac{1}{4}$ if $\mu = \mu_0$ or μ_1 and $\to 0$ as $\mu \to \pm\infty$ but is very large near $\mu = \frac{1}{2}(\mu_0 - \mu_1)$, as is intuitively reasonable. Berk (1975; 1976; 1978) devises general methods of computing efficiency in such comparisons (one of which is an analogue of ARE in **22.29**). Lai (1978) obtains a direct sequential equivalent of asymptotic efficiency, and shows that under general conditions it takes the same value as in the fixed sample size case. Berk and Brown (1978) extend another efficiency criterion to the sequential case.

Composite sequential hypotheses

24.21 Although we have considered the test of a simple H_0 against a simple H_1, the OC and ASN functions are, in effect, calculated against a range of alternatives and therefore give us the performance of the test for a simple H_0 against a composite H_1. We now consider the case of a composite H_0. Suppose that θ may vary in some domain Ω. We wish to test that it lies in some subdomain ω_a against the alternatives that it lies either in a rejection subdomain ω_r, or in a region of indifference $\Omega - \omega_a - \omega_r$ (which may be empty). We could require that the probability of a Type I error, $\alpha(\theta)$, which in general varies with θ, should not exceed some fixed number α for all θ in ω_a; and that the probability of a Type II error, $\beta(\theta)$, should not exceed β for all θ in ω_r.

There are three principal approaches to this problem using asymptotic, Bayesian and invariance arguments, respectively. The asymptotic approach, developed by Bartlett (1946), is based on ML theory and requires no more than the usual replacement of the nuisance parameters by their ML estimates; details are given in Exercise 24.21.

Cox (1963) developed an alternative procedure for testing the normal mean with unknown variance (a one-sided sequential t test) based on the joint estimation of both parameters. However, Joanes (1972) shows the Bartlett approach to be superior.

24.22 We now return to exact arguments. The requirements that $\alpha(\theta) \leq \alpha$ and $\beta(\theta) \leq \beta$ for all θ in the specified regions may lead to an unduly conservative test. This observation led Wald (1947) to suggest that it might be better to consider some average of $\alpha(\theta)$ over ω_a and of $\beta(\theta)$ over ω_r. Wald introduces two weighting functions, $w_a(\theta)$ and $w_r(\theta)$, such that

$$\int_{\omega_a} w_a(\theta)\,\mathrm{d}\theta = 1, \qquad \int_{\omega_r} w_r(\theta)\,\mathrm{d}\theta = 1, \tag{24.73}$$

and then defines

$$\int_{\omega_a} w_a(\theta)\alpha(\theta)\,\mathrm{d}\theta = \alpha, \tag{24.74}$$

$$\int_{\omega_r} w_r(\theta)\beta(\theta)\,\mathrm{d}\theta = \beta. \tag{24.75}$$

If we let

$$L_{0n} = \int_{\omega_a} f(x_1,\theta)f(x_2,\theta)\cdots f(x_n,\theta)\alpha(\theta)\,\mathrm{d}\theta, \tag{24.76}$$

$$L_{1n} = \int_{\omega_r} f(x_1,\theta)f(x_2,\theta)\cdots f(x_n,\theta)\beta(\theta)\,\mathrm{d}\theta, \tag{24.77}$$

the probability ratio L_{0n}/L_{1n} can be used in the ordinary way with errors α and β. Thus we reduce the problem to one of testing simple hypotheses. Although we have followed Wald's terminology of weighting functions, we may regard (24.76) and (24.77) as the posterior probabilities of the sample when θ has prior probabilities $w_a(\theta)$ and $w_r(\theta)$. Indeed, integration over the parameter space to eliminate a nuisance parameter is standard practice in Bayesian analysis; see O'Hagan (1994, sections 3.13–14).

We must now specify $w_a(\theta)$ and $w_r(\theta)$. We may apply some form of Bayes' postulate, e.g. by assuming that $w_a(\theta)$ = constant everywhere in ω_a. Another possibility is to choose $w_a(\theta)$ and $w_r(\theta)$ to optimize some properties of the test.

We now give sufficient conditions for a test to be optimized in that we minimize the maximum values of $\alpha(\theta)$ and $\beta(\theta)$.

Theorem 24.1.

If a test based on weight functions $v_a(\theta)$ and $v_r(\theta)$ is such that

(i) $\alpha(\theta)$ is constant in ω_a,
(ii) $\beta(\theta)$ is constant over the boundary of ω_r, and
(iii) $\beta(\theta)$ does not exceed its boundary value for any θ inside ω_r,

then the test is optimal in the sense that for any other weight functions $\omega_a(\theta)$, $\omega_r(\theta)$ with error functions $\alpha^*(\theta)$, $\beta^*(\theta)$,

(a) $\max \alpha^*(\theta) \geq \max \alpha(\theta)$ *in* ω_a *and*
(b) $\max \beta^*(\theta) \geq \max \beta(\theta)$ *in* ω_r.

Proof. For the ν-weights, we have from (24.84)–(24.85) $\max[\alpha(\theta)] = \alpha$ and $\max[\beta(\theta)] = \beta$, given assumption (i). For the ω-weights

$$\alpha = \int_{\omega_a} \alpha^*(\theta)\omega_a(\theta)\,\mathrm{d}\theta \leq \int_{\omega_a} \max[\alpha^*(\theta)]\omega_a(\theta)\,\mathrm{d}\theta = \max[\alpha^*(\theta)],$$

proving (a). Part (b) follows in similar fashion using assumptions (ii) and (iii).

These conditions are sufficient but by no means necessary. For a recent discussion of the asymptotic optimality of generalized SPR tests, see Lai (1991).

An early application of this result appears in Wald (1947), who proposed a two-sided SPR test for the normal mean μ with σ^2 known (cf. Examples 24.8, 24.10), giving weight $\frac{1}{2}$ to each of the two alternative hypothesis values of μ. Billard (1972) defined a more general procedure whose α and ASN can be closely approximated, and evaluated them for Wald's test. See also Borgan (1979).

A sequential *t* test

24.23 A test proposed by Wald (1947) and, in a modified form, by other writers sets out to test the mean μ of a normal distribution when the variance is unknown. It is known as the sequential t test because it deals with the same problem as Student's t in the fixed sample size case; but since the scale parameter σ is removed by integration rather than estimation the name is, perhaps, somewhat misleading.

Specifically, we wish to test that, compared to some value μ_0, the deviation $(\mu - \mu_0)/\sigma$ exceeds some level, say δ. The three subdomains of **24.21** are then as follows:

(a) ω_a consists of (μ_0, σ) for all σ;
(b) ω_r consists of values for which $|\mu - \mu_0| \geq \sigma\delta$, for all σ;
(c) $\Omega - \omega_a - \omega_r$, consists of values for which $0 \leq |\mu - \mu_0| < \sigma\delta$, for all σ.

We define weight functions for σ:

$$v_{ac} = \begin{cases} \frac{1}{c}, & 0 \leq \sigma \leq c, \\ 0 & \text{elsewhere.} \end{cases} \tag{24.78}$$

$$v_{rc} = \begin{cases} \frac{1}{2c}, & 0 \leq \sigma \leq c, \quad |\mu - \mu_0| = \delta\sigma \\ 0 & \text{elsewhere.} \end{cases} \tag{24.79}$$

Then

$$\begin{aligned} L_{1n} &= \int_0^c v_{rc} \frac{1}{(2\pi)^{n/2}\sigma^n} \exp\left\{-\frac{1}{2\sigma^2}\sum(x_i - \mu)^2\right\} \mathrm{d}\sigma \\ &= \frac{1}{(2\pi)^{n/2}} \frac{1}{2c} \int_0^c \left[\frac{1}{\sigma^n} \exp\left\{-\frac{1}{2\sigma^2}\sum(x_i - \mu_0 - \delta\sigma)^2\right\}\right. \end{aligned}$$

$$+\frac{1}{\sigma^n}\exp\left\{-\frac{1}{2\sigma^2}\sum(x_i-\mu_0+\delta\sigma)^2\right\}\Bigg]\,d\sigma. \tag{24.80}$$

$$L_{0n}=\frac{1}{(2\pi)^{n/2}}\frac{1}{c}\int_0^c\frac{1}{\sigma^n}\exp\left\{-\frac{1}{2\sigma^2}\sum(x_i-\mu_0)^2\right\}d\sigma. \tag{24.81}$$

As c tends to infinity, the limit of the ratio L_{1n}/L_{0n} is

$$\lim L_{1n}/L_{0n}=\frac{\frac{1}{2}\int_0^\infty\frac{d\sigma}{\sigma^n}\left[\exp\{-\frac{1}{2\sigma^2}\sum(x_i-\mu_0-\delta\sigma)^2+\exp\{-\frac{1}{2\sigma^2}\sum(x_i-\mu_0+\delta\sigma)^2\}\right]}{\int_0^\infty\frac{d\sigma}{\sigma^n}\exp\{-\frac{1}{2\sigma^2}\sum(x_i-\mu_0)^2\}} \tag{24.82}$$

This ratio depends on the xs, which are observed, and on μ_0 and δ, which are given, but not on σ, which we have integrated out of the problem by the weight functions (24.78) and (24.79). If we can evaluate the integrals in (24.82) we can apply this ratio to give a sequential test.

24.24 To establish that (24.78)–(24.79) yield an optimal test, we must show: (a) that $\alpha(\mu,\sigma)$ is constant in ω_a; (b) that $\beta(\mu,\sigma)$ is a function of $|(\mu-\mu_0)/\sigma|$ alone; and (c) that $\beta(\mu,\sigma)$ is monotonically decreasing in $|(\mu-\mu_0)/\sigma|$.

If $\bar{x}$ is the sample mean and S_2 is the sum $\sum(x_i-\bar{x})^2$, the distribution of the ratio $(\bar{x}-\mu_0)/S$ depends only on $(\mu-\mu_0)/\sigma$. If we can show that (24.82) is a single-valued function of $(\bar{x}-\mu_0)/S$, then properties (a) and (b) will follow, for $(\mu-\mu_0)/\sigma$ is zero in ω_a and $\beta(\mu,\sigma)$ depends only on the distribution of (24.82). The ratio (24.82) is invariant under scale transformations, as may be verified by putting $x_i=\lambda_i$, $\mu_0=\lambda\mu_0$, $\sigma=\lambda\sigma$. Further, it is a function of $\sum(x-\mu_0)^2$ and $\sum(x-\mu_0)$ only, and hence we do not change it by putting $(x_i-\mu_0)/\sqrt{\sum(x_i-\mu_0)^2}$ for $x_i-\mu_0$. That is, the ratio is a function only of $\sum(x_i-\mu_0)/[\sum(x_i-\mu_0)^2]^{1/2}$, or rather the square of that quantity, namely of

$$\frac{(\bar{x}-\mu_0)^2}{\sum(x_i-\mu_0)^2}=\frac{(\bar{x}-\mu_0)^2}{n(\bar{x}-\mu_0)^2+S^2}.$$

It is therefore a single-valued function of $(\bar{x}-\mu_0)/S$.

To show that $\beta(\mu,\sigma)$ is monotonically decreasing in $|(\mu-\mu_0)/\sigma|$ it is sufficient to show that the ratio (24.82) is a strictly increasing function of $|(\bar{x}-\mu_0)/S|$, or equivalently of $(\bar{x}-\mu_0)^2/\sum(x_i-\mu_0)^2$. Now for fixed $\sum(x_i-\mu_0)^2$ the denominator of (24.82) is fixed and the numerator is an increasing function of $(\bar{x}-\mu_0)^2$. Thus the whole ratio is increasing in $(\bar{x}-\mu_0)^2$ for fixed $\sum(x_i-\mu_0)^2$ and the required result follows. Finally, we see from Theorem 24.1 that the sequential t test is optimal.

Some tables for the use of this test were provided by Armitage (1947). The integrals occurring in (24.82) are, in fact, expressible in terms of the confluent hypergeometric function and, in turn, in terms of the distribution of non-central t. Suitable tables are provided by the US Bureau of Standards (1970). See also work by Armitage and others, in particular Myers *et al.* (1966).

24.25 An alternative method of attack is given by Cox (1952a; 1952b) for the case where the distribution of a set of sufficient statistics factorizes as in (31.14). An SPR test of θ_1 can then be based on T_s, free of the nuisance (often scale) parameters θ_k, in the ordinary way. Invariant SPR tests may be developed by the following argument due to Hall *et al.* (1965). Suppose that the observations $(x_1, \ldots, x_n)$ have the joint density $f(x_1, \ldots, x_n; \theta)$. Let $\mathcal{G}$ denote the group of transformations which leaves both f and the hypotheses unchanged; $\mathcal{G}$ may involve location, scale, sign or permutation changes. We determine a maximal invariant $T_n = T_n(x_1, \ldots, x_n)$ under $\mathcal{G}$ and from there the invariant sufficient statistic, v_n say.

The SPR test is based upon the ratio

$$L_n = f(v_n; \phi_1)/f(v_n; \phi_0) \tag{24.83}$$

where $\phi = \phi(\theta)$ is the invariant function of θ under $\mathcal{G}$; see Chapter 23 for a fuller discussion of invariance arguments. The advantage of the invariance argument is that the v_n can sometimes be determined by other approaches. In particular, if S_n denotes the set of sufficient statistics for θ based upon $(x_1, \ldots, x_n)$, $\mathcal{G}$ induces a set of transformations on S_n. A result of Stein, quoted by Hall *et al.* (1965), states that, under very broad conditions, v_n is any maximal invariant for S_n under $\mathcal{G}$. We now derive the invariant t test using this argument.

Example 24.11 (Invariant SPR test for the normal with unknown variance)

The sufficient statistic for the normal is

$$S_n = \left[\sum x_i, \sum (x_i - \bar{x})^2\right].$$

The scale transformation $x \to cx$ leaves the problem invariant and induces the transformation

$$S_n(c) = \left[c \sum x_i, c^2 \sum (x_i - \bar{x})^2\right],$$

so that a maximal invariant for S_0 under $\mathcal{G}$ is

$$V_n = \sum x_i / \left[\sum (x_i - \bar{x})^2\right]^{\frac{1}{2}}. \tag{24.84}$$

To complete the development we must find $f(v_n)$. However, it may be shown that $f(v_n)$ yields a monotone likelihood so that the continuation region of the test may be expressed as $B_n^* < v_n < A_n^*$.

For further discussion of invariant SPR tests, see Eisenberg and Ghosh (1991), on which the development in this section is based.

Hajnal (1961) develops a two-sample sequential t test using similar arguments. Weiss and Wolfowitz (1972) give a simple class of sequential tests for the normal mean with variance unknown, all of which are asymptotically efficient.

Köllerstrom and Wetherill (1979) compare SPR tests for the bivariate normal correlation parameter.

Sequential estimation: the moments and distribution of n

24.26 In testing hypotheses we usually fix the error probabilities in advance and continue the sampling until acceptance or rejection is reached. We may also use the sequential process for estimation, but our problems are then more difficult to solve and may even be difficult to formulate. We draw a sequence of observations with the object of estimating some parameter of the distribution; but in general it is not easy to decide what is the appropriate estimator, what biases are present, what are the sampling errors or what should be the rules determining the ending of the sampling process. The basic difficulty is that the sample size has a complicated distribution. A secondary nuisance is the end-effect to which we have already referred.[1]

24.27 We gave the mean of n in (24.42) and now rewrite it in the form

$$E\{Z_n - nE(z)\} = 0. \tag{24.85}$$

Assume that absolute second moments exist, that the variance of each z_i is equal to σ^2, and that $E(n^2)$ exists. The variance of $\{Z_n - nE(z)\}$ may then be derived (the proof is left to the reader as Exercise 24.5) as

$$E\{Z_n - nE(z)\}^2 = \sigma^2 E(n) \tag{24.86}$$

and it follows that, with $E(z) = \mu$ as before,

$$\mu^2 E(n^2) = \sigma^2 E(n) + 2\mu E(nZ_n) - E(Z_n^2).$$

If Z_n and n are uncorrelated, this simplifies to

$$\mu^2 \operatorname{var} n = \sigma^2 E(n) - \operatorname{var} Z_n. \tag{24.87}$$

Similar results for higher moments have been obtained by Wolfowitz (1947) – cf. Exercise 24.6.

Approximate expressions for the first four moments of n are obtained by Ghosh (1969) by differentiating its c.f., given in Exercise 24.11. He applies his results to the cases of the normal mean, normal variance, the binomial, the Poisson and the exponential scale parameter. The distribution of n is also discussed – Exercise 24.13 below treats the case of the normal mean. See Wijsman (1991) for recent developments.

24.28 Now let

$$Y_n = \sum_{i=1}^{n} \frac{\partial}{\partial\theta} \log f(x_i, \theta). \tag{24.88}$$

Then, under regularity conditions, $E(\partial \log f/\partial\theta) = 0$ as at (17.18) and we have

$$E(Y_n) = 0 \tag{24.89}$$

and

$$\operatorname{var} Y_n = E\left(\sum \frac{\partial \log f}{\partial\theta}\right)^2 = E(n)E\left\{\left(\frac{\partial \log f}{\partial\theta}\right)^2\right\} \tag{24.90}$$

as in **24.15**. If t is an estimator of θ with bias $b(\theta)$, i.e. is such that

$$E(t) = \theta + b(\theta)$$

[1]Lehmann and Stein (1950) considered the concept of completeness (cf. **21.9**) in the sequential case, but general criteria are not easy to apply even in attribute sampling – cf. de Groot (1959).

we have, differentiating this equation,

$$\operatorname{cov}\left(t, \sum \frac{\partial \log f}{\partial \theta}\right) = E\left(t \sum \frac{\partial \log f}{\partial \theta}\right) = 1 + b'(\theta). \tag{24.91}$$

Then, by the Cauchy–Schwarz inequality

$$(\operatorname{var} t) E\left(\sum \frac{\partial \log f}{\partial \theta}\right)^2 \geq 1 + b'(\theta)^2,$$

and hence, by (24.90),

$$\operatorname{var} t \geq \frac{\{1 + b'(\theta)\}^2}{E(n) E\{(\frac{\partial \log f}{\partial \theta})^2\}}, \tag{24.92}$$

which is Wolfowitz's form of the lower bound to the variance in sequential estimation. It consists simply of putting $E(n)$ for n in the MVB (17.22). Wolfowitz (1947) also gives an extension of the result to the simultaneous estimation of several parameters. For further discussion of the sequential estimation, see Ghosh (1987).

Example 24.12 (The sequential minimum variance bound for the binomial)

Consider the Bernoulli distribution

$$f(x, \pi) = \pi^x (1 - \pi)^{1-x}, \qquad x = 0, 1.$$

We have

$$\frac{\partial \log f}{\partial \pi} = \frac{x}{\pi} - \frac{1-x}{1-\pi}, \qquad E\left\{\left(\frac{\partial \log f}{\partial \pi}\right)^2\right\} = \frac{1}{\pi(1-\pi)}.$$

If p is any unbiased estimator of π in a sample from this distribution, we then have

$$\operatorname{var} p \geq \frac{\pi(1-\pi)}{E(n)}.$$

From (24.3), $E(n) = k/\pi$, so

$$\operatorname{var} p \geq \pi^2(1-\pi)/k. \tag{24.93}$$

Comparing this with (24.9) for the estimator (24.5), we see that the first term in (24.9) is the bound (24.93), the other terms being of lower order in k. In general, for distributions in the exponential family, given in **24.19**, the leading term in the variance will match that given by (24.92), as seen in the example.

24.29 If n is large the theory of sequential estimation is simplified considerably, due to a general result of Anscombe (1949b; 1952; 1953). This result states that for statistics where a central limit effect is present, the formulae for standard errors are the same for sequential samples as for samples of fixed size. We might argue this heuristically from (24.92). n varies about its mean n_0 with standard deviation of order $n_0^{-1/2}$ and thus formulae accurate to order n^{-1} remain accurate to that order if we use n_0 instead of n. More formally, we have the following theorem.

Theorem 24.2.

Let Y_n, $n = 1, 2, \ldots$ be a sequence of random variables. Let there exist a real number θ, a sequence of positive numbers $\{w_n\}$, and a distribution function $F(x)$ such that:

(a) Y_n converges to θ in the scale of w_n, namely

$$P\left\{\frac{Y_n - \theta}{w_n} \leq x\right\} \to F(x) \text{ as } n \to \infty; \tag{24.94}$$

(b) Y_n is uniformly continuous in probability, namely given (small) positive ε and η,

$$P\left\{\left|\frac{Y_{n'} - Y_n}{w_n}\right| < \varepsilon \text{ for all } n, n' \text{ such that } |n' - n| < \varepsilon n\right\} > 1 - \eta. \tag{24.95}$$

Let n_r be an increasing sequence of positive integers tending to infinity and N_r be a sequence of random variables taking positive integral values such that $N_r/n_r \to 1$ in probability as $r \to \infty$. Then

$$P\left\{\frac{Y_{N_r} - \theta}{w_{N_r}} \leq x\right\} \to F(x) \text{ as } r \to \infty \tag{24.96}$$

at all continuity points of $F(x)$.

The complexity of the enunciation and the proof are due to the features we have already noticed: end-effects (represented by the relation between N and n_r) and the variation in n_r.

Proof. Let (24.95) be satisfied with v large enough so that, for any $n_r > v$,

$$P\{|N_r - n_r| < cn_r\} > 1 - \eta. \tag{24.97}$$

Consider the event $E = \{|N_r - n_r| < cn_r \text{ and } |Y_{N_r} - Y_{n_r}| < \varepsilon w_{N_r}\}$, and the events

$$A = \{|Y_{n'} - Y_n| < \varepsilon w_n\}, \qquad \text{all } n' \text{ such that } |n' - n| < \varepsilon n,$$
$$B = \{|N_r - n_r| < cn_r\}.$$

Then

$$\begin{aligned} P(E) &\geq P\{A \text{ and } B\} = P(A) - P\{A \text{ and not-}B\} \\ &\geq P(A) - P(\text{ not-}B) \\ &\geq 1 - 2\eta. \end{aligned} \tag{24.98}$$

Also

$$\begin{aligned} P\{Y_{N_r} - \theta \leq xw_{N_r}\} = {} & P\{Y_{N_r} - \theta \leq xw_{n_r} \text{ and } E\} \\ & + P\{Y_{N_r} - \theta \leq xw_{n_r} \text{ and not-}E\}. \end{aligned}$$

Thus, from the definition of E we find

$$\begin{aligned} P\{Y_{n_r} - \theta \leq (x - \varepsilon)w_{n_r}\} - 2\eta &< P\{Y_{n_r} - \theta \leq xw_{n_r}\} \\ &< P\{Y_{n_r} - \theta \leq (x + \varepsilon)w_{n_r}\} + 2\eta, \end{aligned}$$

and (24.96) follows. It should be noted that the proof does not assume N_r and Y_n to be independent.

24.30 To apply this result to sequential estimation, let $x_1, x_2, \ldots$ be a sequence of observations, Y_n an estimator of a parameter θ and D_n an estimator of the scale w_n of Y_n. The sampling rule is: given some constant k, sample until the first time $D_n \leq k$ and then compute Y_n. We show that Y_N is an estimator of θ with scale asymptotically equal to k if k is small.

Let conditions (a) and (b) of **24.29** be satisfied and k_r be a sequence of positive numbers tending to zero. Let N_r be the sequence of random variables such that N_r is the least integer n for which $D_n \leq k_r$; and let n_r be the sequence such that n_r is the least n for which $w_n \leq k_r$. We require two further conditions:

(c) $\{w_n\}$ converges monotonically to zero and $w_n/w_{n+1} \to 1$ as $n \to \infty$;

(d) N_r is a random variable for all r and $N_r/n_r \to 1$ in probability as $r \to \infty$.

Condition (c) implies that $w_{n_r}/k_r \to 1$ as $n \to \infty$. It then follows from our previous result that

$$P\left\{\frac{Y_{N_r} - \theta}{k_r} \leq x\right\} \to F(x) \text{ as } r \to \infty. \tag{24.99}$$

24.31 It may also be shown that if the xs are independently and identically distributed, conditions (a) and (c) – which are easily verifiable – together imply condition (b) and the distribution of their sum tends to a distribution function. In particular, these conditions are satisfied for ML estimators, for estimators based on means of some functions of the observations, and for quantiles.

Siegmund (1978) discussed setting confidence intervals following sequential tests, with special reference to the case of the normal mean with variance known or unknown. Whitehead (1986) considers the bias of the ML estimator in these circumstances, and gives a method of reducing the bias and approximating the standard error.

Example 24.13 (Sequential estimation for the normal mean)

Consider the estimation of the mean μ of a normal distribution with unknown variance σ^2. We require of the estimator a (small) variance k^2.

The obvious statistic is $Y_n = \bar{x}_n$. For fixed n this has variance σ^2/n estimated as

$$D_n^2 = \frac{1}{n(n-1)}\sum(x_i - \bar{x})^2. \tag{24.100}$$

Conditions (a) and (c) are obviously satisfied and thus, from the result quoted in **24.30**, so is condition (b). To show that (d) holds, apply Helmert's transformation

$$\xi_i = \left(x_{i+1} - \frac{1}{i}\sum_{j=1}^{i} x_j\right)\sqrt{\frac{i}{i+1}}.$$

The sample variance, written in sequentially updated form, becomes

$$D_n^2 = \frac{1}{n(n-1)}\sum_{i=1}^{n-1}\xi_i^2.$$

By the strong law of large numbers, given ε, η, there is a v such that

$$P\left\{\left|\frac{1}{n-1}\sum_{i=1}^{n-1}\xi_i^2 - \sigma^2\right| < \varepsilon \text{ for all } n > v\right\} > 1 - \eta. \tag{24.101}$$

If k is small enough, the probability exceeds $1-\eta$ that $D_n \leq k$ for any n in the range $2 \leq n \leq v$. Thus, given $N > v$, (24.101) implies that

$$\left|\frac{N}{\sigma^2/k^2} - 1\right| < \frac{\varepsilon}{\sigma^2}$$

with probability exceeding $1 - \eta$. Hence, as k tends to zero, condition (d) holds.

The rule is, then, that we select k and proceed until $D_n \leq k$. The mean $\bar{x}$ then has variance approximately equal to k^2.

Following this line of argument, if $\Phi(z) - \Phi(-z) = 1 - \alpha$, a fixed width confidence interval with approximately the correct coverage $(1 - \alpha)$ is given by

$$(Y_n - zk, Y_n + zk);$$

the coverage is guaranteed only as $k \to 0$. The derivation is due to Chow and Robbins (1965).

Example 24.14 (Estimation for the Poisson distribution)

Consider the Poisson distribution with parameter equal to λ. If we sample until the variance of the mean, estimated as $\bar{x}/n$, is less than k^2, we have an estimator $\bar{x}$ of λ with variance k^2. This is equivalent to proceeding until the sum of the xs falls below k^2n^2. But we should not use this result for small n.

On the other hand, suppose we wanted to specify in advance not the variance but the coefficient of variation, say l. The rule becomes: sample until $n\bar{x} \geq l^{-2}$.

For the Poisson mean, see Weiler (1972). Starr and Woodroofe (1972) treat the exponential distribution mean, and Binns (1975) the negative binomial mean. For a discussion of sequential estimation using a minimum risk approach, see Mukhopadhyay (1991).

Stein's double sampling method

24.32 At the end of Example 21.7, we observed that for fixed n no similar test of the mean μ of a normal population with unknown variance σ^2 could have power independent of σ^2. This implied (Example 19.2) that no confidence interval of pre-assigned length can be found for μ. However, if we use a sequential method, these statements are no longer true, as Stein (1945) pointed out.

24.33 We consider a normal population with mean μ and variance σ^2 and seek a $1 - \alpha$ confidence interval of length l for μ. We first choose a sample of fixed size n_0, and then a further sample $n - n_0$, where n now depends on the observations in the first sample.

Consider a Student's t variable with $n_0 - 1$ degrees of freedom, and let the probability that it lies in the range $-t_\alpha$ to t_α be $1 - \alpha$. Define

$$\sqrt{z} = \frac{1}{2t_\alpha}. \tag{24.102}$$

Let s^2 be the estimated variance from the sample of n_0 values:

$$s^2 = \frac{1}{n_0 - 1}\sum_{i=1}^{n_0}(x_i - \bar{x})^2. \tag{24.103}$$

We determine n by

$$n = \max\{n_0, 1 + [s^2/z]\}, \tag{24.104}$$

where $[s^2/z]$ means the greatest integer less than s^2/z.

Consider the n observations altogether, and let them have mean Y_n. Then Y_n is distributed independently of s and consequently $(Y_n - \mu)\sqrt{n}$ is independent of s, so that $(Y_n - \mu)\sqrt{n}/s$ is distributed as t with $n_0 - 1$ d.fr. Hence

$$P\left\{\left|\frac{(Y_n - \mu)\sqrt{n}}{s}\right| < t_\alpha\right\} = 1 - \alpha,$$

or

$$P\left\{Y_n - \frac{st_\alpha}{\sqrt{n}} \leq \mu \leq Y_n + \frac{st_\alpha}{\sqrt{n}}\right\} = 1 - \alpha,$$

or

$$P\left\{Y_n - \frac{1}{2}l \leq \mu \leq Y_n + \frac{1}{2}l\right\} \leq 1 - \alpha. \tag{24.105}$$

The appearance of the inequality in (24.105) is due to the end-effect that s^2/z may not be an integer, which in general is small, so that the limits given by $Y_n \pm \frac{1}{2}l$ are close to the exact limits for confidence coefficient $1 - \alpha$. We can, by a device suggested by Stein, obtain exact limits, though the procedure entails rejecting observations and is probably not worthwhile in practice.

Seelbinder (1953) and Moshman (1958) discuss the optimum choice of first sample size in Stein's method – it is clearly not efficient to make it too small, since only the first sample is used to estimate σ^2. Hall (1981) inserts an intermediate re-estimation of n in a triple sampling scheme. Bhattacharjee (1965) shows that Stein's procedure is more sensitive to non-normality than Student's t test and that, as we should expect, non-normality reintroduces the dependence of the interval length (and corresponding test power) upon σ^2.

Chapman (1950) and Ghosh (1975) extend Stein's method to testing the means of two normal variables; the tests are independent of both variances and depend on the distribution of the difference of two t variables; see Exercise 16.26. Cox (1952c) considered the problem of estimation in double sampling, obtaining a number of asymptotic results, and considered corrections to the single and double sampling results to improve the approximations of asymptotic theory – see Exercises 24.15–17.

A variety of other double sampling procedures are available; for a review of recent developments, see Chatterjee (1991).

Distribution-free tests

24.34 We may use order statistics to reduce many procedures to the binomial case. Consider, for example, testing the hypothesis that the mean of a normal distribution is greater than μ_0 (a one-sided test). Replace the mean by the median and variate values by a score of, say, + if the sample value falls above it and − in the opposite case. Under $H_0 : \mu = \mu_0$ these signs will be distributed binomially with $\pi = \frac{1}{2}$. Under $H_1 : \mu = \mu_0 + k\sigma$ the probability of a positive sign is

$$\pi_1 = \frac{1}{\sqrt{2\pi}} \int_{-k}^{\infty} \exp(-\tfrac{1}{2}x^2)\,\mathrm{d}x. \qquad (24.106)$$

We may then set up an SPR test of π_0 against π_1 in the usual way. This test will have a Type I error α and a Type II error β, and this Type II error will be less than or equal to β when $\mu - \mu_0 > k\sigma$. This is a sequential form of the sign test; see Exercises 22.23 and 22.24.

Tests of this kind are often remarkably efficient, and the sacrifice of efficiency may be well worthwhile for the simplicity of application. Armitage (1947) compared this particular test with Wald's t test and came to the conclusion that, as judged by sample number, the optimum test is not markedly superior to the sign test.

A detailed discussion of non-parametric sequential procedures is beyond the scope of this volume; the reader should consult Sen (1991) for an overview.

Group sequential tests

24.35 In a series of papers, Jennison and Turnbull (1989; 1991; see also papers cited therein) have developed group sequential procedures as a compromise between fixed sample and purely sequential methods. The aim is to gain the bulk of the benefits of possible early stopping, without the need for continuous monitoring. The method is useful in those contexts where interim data inspections (e.g. of medical patients) may be expensive, time-consuming or inconvenient.

In the simplest case, we may consider consecutive groups of m observations and perform tests sequentially on groups of total size $m, 2m, \ldots, Km$. Since up to K dependent tests may be performed, the overall probability of a Type I error may clearly exceed the nominal level for a single test. A number of authors have developed tables for particular tests; many of these are summarized in Jennison and Turnbull (1991).

Similar problems arise if we seek to establish confidence intervals each time the data from a new group have been recorded. Detailed procedures are given in Jennison and Turnbull (1989).

Statistical quality control

24.36 Statistical quality control is a key tool for ensuring high quality standards in manufacturing, although the methods are also seeing increasing use in service operations. In general, we consider a process observed at times $t_1, t_2, \ldots$, although the times are usually equally spaced for operating convenience. A measurement on a quality characteristic (random variable, x_t) is observed at each time point; the measurement is often a sample mean (weight, diameter, etc.) for n items selected at time t.

The process is said to be in a state of *statistical control* (Shewhart, 1931) if successive observations come from a common distribution, $F(x)$. The distribution $F(x)$ describes the actual (historical) performance of the process, not the desired performance; see Montgomery (1997, pp. 129–143). In our exposition, we shall assume that successive observations are independent, although time dependence is common and a considerable literature exists (Alwan and Roberts, 1988; Rowlands and Wetherill, 1991). If $F(x)$ is known and we wish to check for possible departures from the state of statistical control, such as a shift in the mean, we may test $H_0 : \mu = \mu_0$

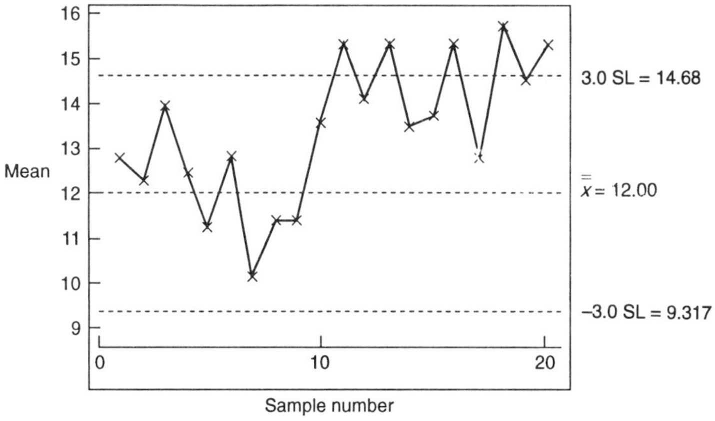

Fig. 24.3 An x-bar chart, based on a mean of 12 and standard deviation of 2, with a shift of 2 units in the mean at time $t = 10$.

against $H_1 : \mu \neq \mu_0$. The classic Shewhart chart assumes that, under H_0, x_t is $N(\mu_0, \sigma^2/n)$, where n is typically a small number such as 5 or 6, and sets upper and lower *control limits* at $\mu_0 \pm d\sigma/\sqrt{n}$. Then the probability of a Type I error is

$$\alpha = P\{|x_t - \mu_0| > d\sigma/\sqrt{n}\};$$

typical values for d are 3.0 ($\alpha = 0.0027$, in the USA) and 3.08 ($\alpha = 0.002$, in the UK).

An example of a Shewhart chart is given in Fig. 24.3. The process was in control up to time $t = 9$, but went out of control at $t = 10$, although the shift was not detected until $t = 11$.

24.37 Since the test is applied repeatedly, the probability that the observation falls outside the control limits *for the first time* at time t is

$$P(t) = (1 - p)^{t-1} p, \qquad t = 1, 2, \ldots, \tag{24.107}$$

where $p = P\{|x_t - \mu_0| > k\sigma/\sqrt{n}|\mu\}$, and $p = \alpha$ when $\mu = \mu_0$. Equation (24.107) defines the *run-length distribution*; it is a geometric distribution (cf. **5.16**) so that the mean time to the first out-of-control signal is

$$E(t) = \frac{1}{p},$$

known as the average run length (ARL). As we would expect, the larger the change in the mean, the shorter the ARL, as shown in Table 24.1. Nevertheless, the ARL may be sizable even for quite large shifts in the mean.

Cusum charts

24.38 If the process mean has shifted to a new level, it is apparent from Fig. 24.3 that the Shewhart chart is not a very powerful procedure since the shift in the mean is only considered on

Table 24.1 Average run length for Shewhart chart ($d = 3$) when $|\mu - \mu_0| = d\sigma/\sqrt{n}$: ARL(2) denotes the two-sided test, ARL(1) denotes the one-tailed test with $d = 2.78$

d	0	0.5	1.0	1.5	2.0	2.5	3.0
ARL(2)	370	155	44	15	6.3	3.2	2.0
ARL(1)	370	88	27	10	4.6	2.6	1.7

a one observation at a time basis. The standard procedure is supplanted in practice by a number of rules such as 'reject H_0 when two successive observations lie outside $\pm 2\sigma/\sqrt{n}$'. For further details, see Montgomery (1997, pp. 146–149).

We may exploit the information contained in successive observations (once a shift has occurred) using cumulative sum (or cusum) charts, first introduced by Page (1954).

The (one-sided) cusum chart is constructed iteratively as

$$y_t = y_{t-1}^+ + z_t, \qquad t = 1, 2, \ldots, \tag{24.108}$$

where $y_t^+ = \max\{y_t, 0\}$, y_0 is some preset value, typically zero and z_t is some function of x_t.

The null hypothesis is rejected when y_t first crosses some threshold, h say. Page's original scheme used

$$z_t = x_t - \mu_0 - k$$

to test $H_0 : \mu = \mu_0$ against $H_1 : \mu = \mu_1 > \mu_0$, where k is known as the *reference* value ($0 < k < \delta = \mu_1 - \mu_0$). Thus, the performance of the test procedure is controlled by the two parameters (h, k), and performance is typically measured by the ARL under H_1 for a given ARL under H_0.

Assume the z_t are independent with common d.f. $G(z)$, and let the probability that the run length is t be

$$p_t(y) = P(t | y_0 = y).$$

Since $y_1, y_2, \ldots$ form a Markov process, we may form the recurrence relation:

$$p_t(y) = \int p_{t-1}(u)\, \mathrm{d}G(y - u), \quad t = 2, 3, \ldots \tag{24.109}$$

where the integral is taken over $-\infty < u < h$ and $u \equiv y_{t-1}^+$; the initial condition is

$$p_1(y) = 1 - G(h - u).$$

24.39 Equations (24.109) may be solved numerically, either by quadrature or using a finite Markov chain approximation; for details, see Rowlands and Wetherill (1991). Once the probabilities are available, the ARL may be computed.

Moustakides (1986) showed that the optimal cusum scheme is always of the form

$$y_t = y_{t-1}^+ + z_t,$$

where z_t is the log probability ratio given in (24.19) for the SPR test. Further, when testing the normal mean, Moustakides (1986) showed that the best choice for k is $(\mu_1 - \mu_0)/2$. Table 24.2

Table 24.2 ARL for one-sided tests, with $|\mu_1 - \mu_0| = d\sigma/\sqrt{n}$

d	0	0.5	1.0	1.5	2.0	2.5
Shewhart	370	88	27	10	4.6	2.6
cusum	370	26	10	6	4.8	4.0
cusum ($h = 10$)	370	27	13	9	6.9	5.8

gives the ARL for the cusum test for the normal mean; recall that the test is one-sided. The one-sided Shewhart values are given for comparison, but it should be remembered that the Shewhart is designed for single-period large shifts rather than a permanent small shift that may take several periods to detect. The cusum values were obtained by simulation.

Extensive tables of the ARL for different (h, k) are given by several authors; see for example, Goel and Wu (1971). Wardell *et al.* (1992) compute the ARL for time-dependent processes.

24.40 Two-sided cusum tests are also available (see Rowlands and Wetherill, 1991); the general theory is more difficult. The review by these authors also covers procedures for mixture distributions and for autocorrelated processes.

Another classical quality control scheme is the double sampling procedure of Dodge and Romig (1944). The underlying ideas were considered in **24.32–33**.

EXERCISES

24.1 In Example 24.1, show by use of Exercise 9.13 that (24.3) implies the biasedness of k/n for π.

24.2 Referring to Example 24.6, sketch the OC curve for a binomial with $\alpha = 0.01$, $\beta = 0.03$, $\pi_1 = 0.1$, $\pi_2 = 0.2$. (The curve is half a bell-shaped curve with a maximum at $\pi = 0$ and zero at $\pi = 1$.) Similarly, sketch the ASN curve for the same binomial.

24.3 Two samples, each of size n, are drawn from populations, P_1 and P_2, with proportions π_1 and π_2 of an attribute. They are paired off in order of occurrence. t_1 is the number of pairs in which there is a success from P_1 and a failure from P_2; t_2 is the number of pairs in which there is a failure from P_1 and a success from P_2. Show that in the (conditional) set of such pairs the probability of a member of t_1 is

$$\pi = (1-\pi_1)\pi_2/\{\pi_1(1-\pi_2) + \pi_2(1-\pi_1)\}.$$

Considering this as an ordinary binomial in the set of $t = t_1 + t_2$ values, show how to test the hypothesis that $\pi_1 \geq \pi_2$ by testing $\pi = \frac{1}{2}$. Hence derive a sequential test for $\pi_1 \geq \pi_2$. If

$$u = \frac{\pi_2(1-\pi_1)}{\pi_1(1-\pi_2)},$$

show that $\pi = u/(1+u)$ and hence derive the following acceptance and rejection numbers:

$$a_t = \frac{\log \frac{\beta}{1-\alpha}}{\log u_1 - \log u_0} + t\frac{\log \frac{1+u_1}{1+u_0}}{\log u_1 - \log u_0},$$

$$r_t = \frac{\log \frac{1-\beta}{\alpha}}{\log u_1 - \log u_0} + t\frac{\log \frac{1+u_1}{1+u_0}}{\log u_1 - \log u_0},$$

where u_i is the value of u corresponding to H_i $(i = 0, 1)$.

(Wald, 1947)

24.4 Referring to the function $h \neq 0$ of **24.14**, show that if z is a random variable such that $E(z)$ exists and is not zero; if there exists a $\delta > 0$ such that $P(e^z < 1-\delta) > 0$ and $P(e^z > 1+\delta) > 0$; and if for any real h, $E(\exp hz) = g(h)$ exists, then

$$\lim_{h\to\infty} g(h) = \infty = \lim_{h\to-\infty} g(h)$$

and that $g''(h) > 0$ for all real values of h. Hence show that $g(h)$ is strictly decreasing over the interval $(-\infty, h^*)$ and strictly increasing over (h^*, ∞), where h^* is the value for which $g(h)$ is a minimum. Hence show that there exists at most one $h \neq 0$ for which $E(\exp hz) = 1$.

(Wald, 1947)

24.5 In **24.27**, deduce the expressions (24.86)–(24.87).

24.6 In Exercise 24.5, show that the third moment of $Z_n - n\mu$ is

$$E(Z_n - n\mu)^3 = \mu_3 E(n) - 3\sigma^2 E\{n(Z_n - n\mu)\},$$

where μ_3 is the third moment of z.

(Wolfowitz, 1947)

24.7 If z is defined as at (24.19), let t be a complex variable such that $E(\exp zt) = \phi(t)$ exists in a certain part of the complex plane. Show that

$$E[\exp(tZ_n)\{\phi(t)\}^{-n}] = 1$$

for any point where $|\phi(t)| \geq 1$.

(Wald, 1947)

24.8 Putting $t = h$ in the previous exercise, show that, if E_b refers to expectation under the restriction that $Z_n \leq -b$ and E_a to the restriction $Z_n \geq a$, then

$$K(h)E_b \exp(hZ_n) + \{1 - K(h)\}E_a \exp(hZ_n) = 1,$$

where K is the OC. Hence, neglecting end-effects, show that

$$K(h) = \begin{cases} \frac{e^{h(a+b)} - e^{hb}}{e^{h(a+b)} - 1}, & h \neq 0, \\ \frac{a}{a+b}, & h = 0. \end{cases}$$

(Girshick, 1946)

24.9 Differentiating the identity of Exercise 24.7 with respect to t and putting $t = 0$, show that

$$E(n) = \frac{a\{1 - K(h)\} - bK(h)}{E(z)}$$

and hence derive equation (24.43).

(Girshick, 1946)

24.10 Assuming, as in the previous exercise, that the identity is differentiable, derive the results of Exercises 24.7 and 24.8.

24.11 In the identity of Exercise 24.7, put

$$-\log \phi(t) = \tau$$

where τ is purely imaginary. Show that if $\phi(t)$ is not singular at $t = 0$ and $t = h$, this equation has two roots $t_1(\tau)$ and $t_2(\tau)$ for sufficiently small values of τ. In the manner of Exercise 24.8, show that the characteristic function of n is given asymptotically by

$$E(e^{n\tau}) = \frac{A^{t_2} - A^{t_1} + B^{t_1} - B^{t_2}}{B^{t_1}A^{t_2} - A^{t_1}B^{t_2}}.$$

(Wald, 1947)

24.12 In the case when z is normal with mean μ and variance σ^2, show that t_1 and t_2 in Exercise 24.11 are

$$t_1 = -\frac{\mu}{\sigma^2} + \frac{1}{\sigma^2}(\mu^2 - 2\sigma^2\tau)^{1/2},$$

$$t_2 = -\frac{\mu}{\sigma^2} - \frac{1}{\sigma^2}(\mu^2 - 2\sigma^2\tau)^{1/2},$$

where the sign of the radical is determined so that the real part of $\mu^2 - 2\sigma^2\tau$ is positive. In the limiting case $B = 0$, A finite (when of necessity $\mu > 0$ if $E(n)$ is to exist), show that the c.f. is A^{-t_1}, and in the case B finite, $A = \infty$ (when $\mu < 0$), show that the c.f. is B^{-t_1}.

(Wald, 1947)

24.13 In the first of the two limiting cases of the previous exercise, show that the distribution of $m = \mu^2 n/2\sigma^2$ has p.d.f.

$$f(m) = \frac{c}{2\Gamma(\frac{1}{2})m^{3/2}} \exp\left(-\frac{c^2}{4m} - m + c\right), \qquad 0 \leq m < \infty,$$

where $c = \mu \log A/\sigma^2 > 0$. This is a special case of the inverse Gaussian distribution of Exercise 11.28 with $c^2/2 = \lambda = 2\mu^2$ there.
For large c show that $2m/c$ is approximately normal with unit mean and variance $1/c$.

(Wald (1947) who also shows that when A, B are finite the distribution of n is the weighted sum of a number of variables of the above type.)

24.14 Values of u are observed from the exponential distribution with p.d.f.

$$f(u) = \lambda e^{-\lambda u}, \quad 0 \leq u < \infty.$$

Show that a sequential test of $\lambda = \lambda_0$ against $\lambda = \lambda_1$ is given by

$$k_1 + (\lambda_1 - \lambda_0)\sum_{j=1}^{n} u_j \leq n \log(\lambda_1/\lambda_0) \leq k_2 + (\lambda_1 - \lambda_0)\sum_{j=1}^{n} u_j,$$

where k_1 and k_2 are constants.
Compare this with the test of Exercise 24.3 in the limiting case when π_1 and π_2 tend to zero so that $\pi_1 t = \lambda_0$ and $\pi_2 t = \lambda_1$ remain finite.

(Anscombe and Page, 1954)

24.15 It is required to estimate a parameter θ with a small variance $a(\theta)/\lambda$ when λ tends to infinity. Given that t_m is an unbiased estimator in samples of fixed size m with variance $v(\theta)/m$; that $\gamma_1(t_m) = O(m^{-1/2})$ and $\gamma_2(t_m) = O(m^{-1})$; and that $a(t_m)$ and $b(t_m)$ can be expanded in series to give asymptotic means and standard errors, consider the double sampling rule:
(a) Take a sample of size $N\lambda$ and let t_1 be the estimate of θ from it.
(b) Take a second sample of size max $\{0, [n_0(t_1) - N\lambda]\}$, where $n_0(t_1) = v(t_1)/a(t_1)$. Let t_2 be the estimate of θ from the second sample.
(c) Let $t = [Nt_1 + \{n_0(t_1) - N\}t_2]/n_0(t_1)$ if $n_0(t_1) \geq N$.
(d) Assume that $N < n_0(\theta)$ and the distribution of $m_0(t_1) = 1/n_0(t_1)$ is such that the event $n_0(t_1) < N$ may be ignored.
Show that under this rule

$$E(t) = \theta + O(\lambda^{-1}),$$
$$\text{var}\, t = a(\theta)\lambda^{-1}\{1 + O(\lambda^{-1})\}.$$

(Cox, 1952c)

24.16 In the previous exercise, take the same procedure except that $n_0(t_1)$ is replaced by

$$n(t_1) = n_0(t_1)\left\{1 + \frac{b(t_1)}{\lambda}\right\}.$$

Show that

$$E(t) = \theta + m_0'(\theta)v(\theta)\lambda^{-1} + O(\lambda^{-2}).$$

Put

$$t' = \begin{cases} t - m_0'(t)v(t)\lambda^{-1} & \text{if } N \leq n(t_1) \\ 0 & \text{otherwise,} \end{cases}$$

and hence show that t' has bias $O(\lambda^{-2})$.
Show further that if we put

$$b(\theta) = n_0(\theta)v(\theta)\{2m_0(\theta)m_0'(\theta)\gamma_1(\theta)v^{-1/2}(\theta) + m_0'^2(\theta) + 2m_0(\theta)m_0''(\theta) + m_0''(\theta)/(2N)\},$$

then

$$\operatorname{var} t' = a(\theta)\lambda^{-1} + O(\lambda^{-3}).$$

(Cox, 1952c)

24.17 Applying Exercise 24.15 to the binomial distribution, with

$$a(\pi) = a\pi^2, \quad v(\pi) = \pi(1-\pi), \quad \gamma_1(\pi) = \frac{(1-2\pi)}{\{\pi(1-\pi)\}^{1/2}},$$

show that the total sample size is

$$n(t_1) = \frac{1-t_1}{at_1} + \frac{3}{t_1(t-t_1)} + \frac{1}{aNt_1}$$

and the estimator $t' = t - (at/(1-t))$.
Thus N should be chosen as large as possible, provided that it does not exceed $(1-\pi)/(a\pi)$.

(Cox, 1952c)

24.18 Referring to Example 24.9, show that

$$K(\sigma) = \left\{\left(\frac{1-\beta}{\alpha}\right)^h - 1\right\} \Big/ \left\{\left(\frac{1-\beta}{\alpha}\right)^h - \left(\frac{\beta}{1-\alpha}\right)^h\right\},$$

where h is given by

$$\sigma\left(\frac{\sigma_1}{\sigma_0}\right)^h = \left\{\frac{h}{\sigma_1^2} - \frac{h}{\sigma_0^2} + \frac{1}{\sigma^2}\right\}^{-1/2},$$

provided that the expression in braces on the right is positive. Hence show how to draw the OC curve.

(Wald, 1947)

24.19 In the previous exercise derive the expression for the ASN,

$$\frac{K(\sigma)\{h_0 - h_1\} + h_1}{\sigma^2 - \gamma},$$

where

$$\gamma = \log(\sigma_1^2/\sigma_0^2) \Big/ \left(\frac{1}{\sigma_0^2} - \frac{1}{\sigma_1^2}\right).$$

(Wald, 1947)

24.20 In Example 24.9, obtain a test based on $\sum(x - \bar{x})^2$ of normal variances when the population mean is unknown.

(Girshick, 1946)

24.21 In **24.10**, $f(x, \theta)$ is replaced by $f(x, \theta, \phi)$, where ϕ is a nuisance parameter, so that H_0 and H_1 are composite. $\hat{\theta}_n$ is the ML estimator of θ after n observations, and $\hat{\phi}_j$ is then the ML estimator of ϕ, given $H_j (j = 0, 1)$. If θ_0 and θ_1 differ from the true value of θ by amounts of order $n^{-1/2}$, show that an SPR test based on

$$t_n = \log L_n(x, \theta_1, \hat{\phi}_1) - \log L_n(x, \theta_0, \hat{\phi}_0)$$

is asymptotically equivalent to using

$$t_n \sim (\theta_1 - \theta_0)[(L_\theta - L_\phi L_{\theta\phi}/L_{\phi\phi}) + \{\hat{\theta}_n - \tfrac{1}{2}(\theta_0 + \theta_1)\}(L_{\theta\theta} - L_{\theta\phi}^2/L_{\phi\phi})],$$

where $L_a = \partial \log L/\partial a$, $L_{ab} = \partial^2 \log L/\partial a \partial b$ $(a, b = \theta, \phi)$, with mean and variance given (writing $E(L_{ab}) = I_{ab}$) by

$$E(t_n) \sim (\theta_1 - \theta_0)\{\theta - \tfrac{1}{2}(\theta_0 + \theta_1)\}(I_{\theta\theta} - I_{\theta\phi}^2/I_{\phi\phi}),$$
$$\text{var}\, t_n \sim (\theta_1 - \theta_0)^2(I_{\theta\theta} - I_{\theta\phi}^2/I_{\phi\phi}).$$

(Bartlett, 1946)

24.22 For a sample from the distribution $f(x|\theta) = g(x)/h(\theta)$, $a \leq x \leq \theta$, show that the SPR test of $H_0 : \theta = \theta_0$ against $H_1 : \theta = \theta_1$ $(0 < \theta_0 < \theta_1)$ has $\alpha = 0$, and has the form 'accept H_1 if $x_n > \theta_0$, $n = 1, 2, \ldots$; accept H_0 if $x_n \leq \theta_0$ and $(\theta_0/\theta_1)^n \leq \beta$; continue sampling otherwise'.
Show directly that when H_0 holds, the sample size is constant at $n_0 = \log \beta / \log(\theta_0/\theta_1)$, and that when H_1 holds the ASN is exactly $(1 - \beta)/(1 - (\theta_0/\theta_1))$ if n_0 is an integer. Verify these formulae for the ASN from (24.43).

24.23 In Example 24.1, consider a random variable t with density

$$f(t) = ct^{k-2}(1-t)^{t-k}, \qquad 0 \leq t \leq 1 - \pi < 1;\ k > 1,$$

where c is therefore the integral in (24.6). Show that

$$E(1-t) < E\{t(1-t)\}E(t^{-1})$$

for any variable on this range, and that if V is the coefficient of variation of p at (24.5) and

$$a = (1 + V^2)/(k-1),$$

then

$$aE(t) = V^2,$$
$$aE(t^2) = \{kV^2 - (1-\pi)\}/2,$$
$$aE(t^{-1}) = \{(k-2)(1-\pi)\}^{-1}.$$

Hence show that

$$\text{var}\, p < \pi^2\{4(k-2)(1-\pi)\}^{-1}[(k-1)\{k-1-2(1-\pi)\} + 4(1-\pi)$$
$$-(k-1)\{\{k-1-2(1-\pi)\}^2 + 8\pi(1-\pi)\}^{1/2}].$$

(Prasad and Sahai, 1982)

CHAPTER 25

TESTS OF FIT

25.1 In our discussion of estimation and test procedures from Chapter 17 onwards, we have concentrated entirely on problems concerning the parameters of distributions of known form. In our classification of hypothesis-testing problems in **20.3** we did indeed define a non-parametric hypothesis, but we have not yet investigated non-parametric hypotheses or estimation problems.

In the present chapter, we confine ourselves to a particular class of procedures that stand slightly apart, but are of sufficient practical importance to justify this special treatment.

Tests of fit

25.2 Let $x_1, x_2, \dots, x_n$ be independent observations on a random variable with distribution function $F(x)$ which is unknown. Suppose that we wish to test the hypothesis

$$H_0 : F(x) = F_0(x) \tag{25.1}$$

where $F_0(x)$ is some particular d.f., which may be continuous or discrete. The problem of testing (25.1) is called a *goodness-of-fit problem*. Any test of (25.1) is called a *test of fit*.

Hypotheses of fit, like parametric hypotheses, divide naturally into simple and composite hypotheses. The hypothesis in (25.1) is a simple one if $F_0(x)$ is completely specified; for example, the hypothesis (a) that the n observations have come from a normal distribution with specified mean and variance is a simple hypothesis. On the other hand, we may wish to test (b) whether the observations have come from a normal distribution whose parameters are unspecified, and this would be a composite hypothesis (in this case it would often be called a 'test of normality'). Similarly, if (c) the normal distribution has its mean, but not its variance, specified, the hypothesis remains composite. This is precisely the distinction we discussed in the parametric case in **20.4**.

25.3 It is clear that (25.1) is no more than a restatement of the general problem of testing hypotheses; we have merely expressed the hypothesis in terms of the distribution function instead of the density function. What is the point of this? Are we not merely retracing our previous steps?

There are several reasons for the new formulation. The parametric hypothesis-testing methods developed earlier were necessarily concerned with hypotheses imposing one or more constraints (cf. **20.4**) in the parameter space; they afford no means whatever of testing a hypothesis like (b) in **25.2**, where no constraint is imposed upon parameters and we are testing the hypothesis that the d.f. is a member of a specified (infinite) family of distributions. In such cases, and even in cases where the hypothesis does impose one or more parametric constraints, as in (a) or (c) of **25.2**, the reformulation of the hypothesis in the form (25.1) leads to new methods. For we would expect the entire sample distribution function to mimic closely that of the true d.f. $F(x)$. It is therefore natural to seek to use the entire sample d.f. directly as a means of testing (25.1), and we shall find that the most important tests of fit do just this. Furthermore, the optimal tests we have devised for parametric hypotheses, H_0, have been recommended by the properties of their power functions against alternative hypotheses that differ from H_0 only in the values of the parameters specified by H_0. It seems at least likely that a test based on the entire sample d.f. will have reasonable power properties against a wider class of alternatives, even though it may not be optimal against any one of them.

The LR and Pearson's X^2 tests of fit for simple H_0

25.4 Two well-known methods of testing goodness of fit depend on a very simple device, which essentially reduces the problem to a parametric one. We consider it first in the case when $F_0(x)$ is completely specified, so that (25.1) is a simple hypothesis.

Suppose that the range of the variate x is arbitrarily divided into k mutually exclusive classes; these need not be, though in practice they usually are taken as, successive intervals in the range of x.[1] Then, since $F_0(x)$ is specified, we may calculate the probability of an observation falling in each class. If these are denoted by p_{0i}, $i = 1, 2, \ldots, k$, and the observed frequencies in the k classes by n_i $(\Sigma_{i=1}^k n_i = n)$, the n_i are multinomially distributed (cf. **7.7**), and from (7.15) we see that the LF is

$$L(n_1, n_2, \ldots, n_k | p_{01}, p_{02}, \ldots, p_{0k}) \propto \prod_{i=1}^{k} p_{0i}^{n_i}. \tag{25.2}$$

On the other hand, if the true distribution function is $F_1(x)$, where F_1 may be any d.f., we may denote the probabilities in the k classes by p_{1i}, $i = 1, 2, \ldots, k$, and the likelihood is

$$L(n_1, n_2, \ldots, n_k | p_{11}, p_{12}, \ldots, p_{1k}) \propto \prod_{i=1}^{k} p_{1i}^{n_i}. \tag{25.3}$$

We may now use **22.1** to find the LR test of the hypothesis (25.1), the composite alternative hypothesis being

$$H_1 : F(x) = F_1(x).$$

The likelihood (25.3) is maximized when we substitute the ML estimators for p_{1i}:

$$\widehat{p}_{1i} = n_i/n.$$

The LR statistic for testing H_0 against H_1 is therefore

$$\begin{aligned} l &= \frac{L(n_1, n_2, \ldots, n_k | p_{01}, p_{02}, \ldots p_{0k})}{L(n_1, n_2, \ldots, n_k | \widehat{p}_{11} \widehat{p}_{12}, \ldots, \widehat{p}_{1k})} \\ &= n^n \prod_{i=1}^{k} (p_{0i}/n_i)^{n_i}. \end{aligned} \tag{25.4}$$

H_0 is rejected when l is small enough, as in (22.6).

The exact distribution of (25.4) is unknown. However, we know from **22.7** that as $n \to \infty$ when H_0 holds, $-2 \log l$ is asymptotically distributed as χ^2 with $k - 1$ degrees of freedom (there being $r = k - 1$ independent constraints on the p_{1i} since $\Sigma_{i=1}^k p_{1i} = 1$).

[1] We discuss the choice of k and of the classes in **25.20–23** and **25.28–30**. For the present, we allow them to be arbitrary.

25.5 However, (25.4) is not the classical *chi-square statistic* put forward by Pearson (1900) for this situation. This procedure, which has been derived already in Example 15.3, uses the asymptotic k-variate normality of the multinomial distribution of the n_i, and the knowledge that, given H_0, the quadratic form in the exponent of this distribution is distributed as χ^2 form with degrees of freedom equal to its rank, $k-1$. In our present notation, this quadratic form was found in Example 15.3 to be[2]

$$X^2 = \sum_{i=1}^{k} \frac{(n_i - np_{0i})^2}{np_{0i}}. \tag{25.5}$$

From (25.4) we have

$$-2\log l = 2\sum_{i=1}^{k} n_i \log(n_i/np_{0i}). \tag{25.6}$$

The two distinct statistics (25.5) and (25.6) thus have the same distribution asymptotically, given H_0. More than this, however, they are asymptotically equivalent statistics when H_0 holds, for if we write $\Delta_i = (n_i - np_{0i})/np_{0i}$, we have

$$\begin{aligned} -2\log l &= 2\sum_i n_i \log(1+\Delta_i) \\ &= 2\sum_i \{(n_i - np_{0i}) + np_{0i}\}\{\Delta_i - \tfrac{1}{2}\Delta_i^2 + O_p(n^{-3/2})\} \\ &= 2\sum_i \left\{(n_i - np_{0i})\Delta_i + np_{0i}\Delta_i - \frac{np_{0i}}{2}\Delta_i^2 + O_p(n^{-1/2})\right\}, \end{aligned}$$

and since $\sum p_{0i}\Delta_i = 0$, we have

$$-2\log l = \sum_i \{np_{0i}\Delta_i^2 + O_p(n^{-1/2})\} = X^2\{1 + O_p(n^{-1/2})\}. \tag{25.7}$$

For small n, the test statistics differ. Pearson's form (25.5) may alternatively be expressed as

$$X^2 = \frac{1}{n}\sum_i \frac{n_i^2}{p_{0i}} - n, \tag{25.8}$$

which is easier to compute; but (25.5) has the advantage over (25.8) of being a direct function of the differences between the observed frequencies n_i and their hypothetical expectations np_{0i}, differences that are themselves of obvious interest.

Wise (1963; 1964) examines the approximations to the multinomial involved in using (25.5) as a χ^2_{k-1} variable, and shows that the error is particularly small when the np_{0i} are equal or nearly so – the latter need not then be large (cf. **25.22** and **25.30**, which also cover composite H_0). Some small-sample tabulations of the distributions of (25.5) and (25.6) in the case when each $p_{0i} = 1/k$ are given by Good *et al.* (1970); Zahn and Roberts (1971) tabulate the case $p_{0i} = 1/n$. See also Chapman (1976).

[2]Following standard practice, we write X^2 for the test statistic and reserve the symbol χ^2 for the distributional form we have so frequently discussed. Earlier writers confusingly wrote χ^2 for the statistic as well as the distribution.

Larntz (1978) shows that for moderate np_{0i} the χ^2_{k-1} approximation is better for X^2 than for the LR statistic (25.6), which rejects H_0 more often than the approximation indicates. Koehler and Larntz (1980) consider the asymptotic distributions of (25.5) and (25.6) as n and k tend to infinity together – they tend to different normal distributions. When the p_{0i} are equal, X^2 has better power, but on the whole, (25.6) performs better when they are unequal. Smith *et al.* (1981) give the approximate mean and variance of (25.6) and (as in **22.9**) adjust (25.6) multiplicatively to have mean $k-1$, which gives a better χ^2_{k-1} approximation when all $p_{0i} = k^{-1}$ and $k \geq 4$.

The statistics (25.5) and (25.6) are special cases of the general statistic

$$t_\lambda = \frac{2}{\lambda(\lambda+1)} \sum_{i=1}^{k} n_i \left\{ \left(\frac{n_i}{np_{0i}} \right)^\lambda - 1 \right\},$$

where $\lambda = 1$ gives X^2 at (25.8) while at $\lambda = 0$ l'Hôpital's rule shows that $t_0 = -2\log l$ as in (25.6). Cressie and Read (1984) study the family t_λ in detail, and show that under H_0 they are all asymptotically equivalent, as they also are for alternative hypotheses of the type in **25.27**. Convergence to the asymptotic moments is fastest for $0.3 \leq \lambda < 2.7$. When all $p_{0i} = k^{-1}$, the χ^2 approximation to the distribution is adequate for $k \leq 6$, $n \geq 10$ and $\frac{1}{3} \leq \lambda \leq \frac{3}{2}$ – for other λ it underestimates test size. The two values of λ that perform best overall are $\lambda = 1$ (i.e. X^2) and, more surprisingly, $\lambda = \frac{2}{3}$, although in accordance with **22.39**, $\lambda = 0$ (i.e. $-2\log l$) is asymptotically best using Bahadur's measure of test efficiency. See also Read (1984). Drost *et al.* (1989) give improved approximations to the distributions.

Choice of critical region for X^2

25.6 Since H_0 is rejected for small values of l, (25.7) implies that when using (25.5) as test statistic, H_0 is to be rejected when X^2 is large. There has been some uncertainty in the literature about this, the older practice being to reject H_0 for small as well as large values of X^2, i.e. to use a two-tailed rather than an upper-tail test. For example, Cochran (1952) supported this practice on the grounds that extremely small X^2 values are likely to have resulted from numerical errors in computation, while on other occasions such values have apparently been due to the frequencies n_i having been biased, perhaps inadvertently, to bring them closer to the hypothetical expectations np_{0i}.

The computational issue may now be laid aside. Cochran's second consideration is more cogent, but it is plain that we are now considering a different and rarer hypothesis (that there has been voluntary or involuntary irregularity in collecting the observations) which must be precisely formulated before we can determine the best critical region to use (cf. Stuart, 1954a). Leaving such irregularities aside, we use the upper tail of the distribution of X^2 as the critical region. This will be justified from the point of view of its asymptotic power in **25.27**.

25.7 The essence of the LR and Pearson tests of fit is the reduction of the problem to one concerning the multinomial distribution. The need to group data into classes clearly involves the sacrifice of a certain amount of information, especially when the underlying variable is continuous. However, this defect is also a virtue: we do not need to know the values of the individual observations, so long as we have k classes for which the hypothetical p_{0i} can be computed. In fact, there need be no underlying variable at all – we may use either of these tests of fit even if the original data refer to a simple categorization. The point is illustrated by Example 25.1.

Table 25.1 Observed and expected frequencies for Mendel's data

Seeds	Observed frequency n_i	Theoretical probability p_{0i}
Round and yellow	315	9/16
Wrinkled and yellow	101	3/16
Round and green	108	3/16
Wrinkled and green	32	1/16
	$n = 556$	1

Example 25.1 (Use of X^2 to test goodness-of-fit)

In some classical experiments on pea-breeding, Mendel observed the frequencies of different kinds of seeds in crosses from plants with round yellow seeds and plants with wrinkled green seeds. These are given in Table 25.1, together with the theoretical probabilities on the Mendelian theory of inheritance.

From (25.8) we obtain

$$X^2 = \tfrac{1}{556} \times 16\{\tfrac{315^2}{9} + \tfrac{101^2}{3} + \tfrac{108^2}{3} + \tfrac{32^2}{1}\} - 556$$
$$= \tfrac{16}{556} \times 19\,337.3 - 556 = 0.47.$$

For $k - 1 = 3$ degrees of freedom, tables of χ^2 show that the probability of a value exceeding 0.47 lies between 0.90 and 0.95, so that the fit of the observations to the theory is very good indeed: a test of any size $\alpha \leq 0.90$ would not reject the null hypothesis.

For the LR statistic, (25.6) gives $-2\log l = 0.48$, very close to the value for X^2.

Composite H_0

25.8 Confining attention to (25.5), we consider the situation when the hypothesis tested is composite – the LR test remains asymptotically equivalent when H_0 holds (cf. Exercise 25.11). Suppose that $F_0(x)$ is specified as to its form, but that some (or perhaps all) of the parameters are left unspecified, as in (b) or (c) of **25.2**. The specification of the form of $F_0(x)$ is essential, since this cannot be estimated from the sample without disruptive effects on the distribution of X^2, for we should be imposing an unknown number of non-linear constraints upon the agreement between n_i and the np_{0i}. In the multinomial formulation of **25.4**, the new feature is that the hypothetical probabilities p_{0i} are not now immediately calculable, since they are functions of the s (assumed less than $k - 1$) unspecified parameters $\theta_1, \theta_2, \ldots, \theta_s$, which we may denote collectively by $\boldsymbol{\theta}$. Thus we must write $p_{0i}(\boldsymbol{\theta})$. In order to make progress, we must estimate $\boldsymbol{\theta}$ by some vector of estimators $\mathbf{t}$, and use (25.5) in the form

$$X^2 = \sum_{i=1}^{k} \frac{\{n_i - np_{0i}(\mathbf{t})\}^2}{np_{0i}(\mathbf{t})}.$$

This clearly changes our distribution problem, for now the $p_{0i}(\mathbf{t})$ are themselves random variables, and it is not obvious that the asymptotic distribution of X^2 will be of the same form as in the case of a simple H_0. In fact, the term $n_i - np_{0i}(\mathbf{t})$ does not necessarily have a zero expectation. We may write X^2 identically as

$$X^2 = \sum_{i=1}^{k} \frac{1}{np_{0i}(\mathbf{t})}[\{n_i - np_{0i}(\boldsymbol{\theta})\}^2 + n^2\{p_{0i}(\mathbf{t}) - p_{0i}(\boldsymbol{\theta})\}^2 \\ -2n\{n_i - np_{0i}(\boldsymbol{\theta})\}\{p_{0i}(\mathbf{t}) - p_{0i}(\boldsymbol{\theta})\}]. \tag{25.9}$$

Now we know from the theory of the multinomial distribution in **7.7** that asymptotically

$$n_i - n_{0i}(\boldsymbol{\theta}) \sim cn^{1/2},$$

so that the first term in the square brackets in (25.9) is of order n. If we also have

$$p_{0i}(\mathbf{t}) - p_{0i}(\boldsymbol{\theta}) = o_p(n^{-1/2}), \tag{25.10}$$

the second and third terms will be of order less than n, and relatively negligible, so that (25.9) asymptotically behaves like its first term. Even this function has the random variable $np_{0i}(\mathbf{t})$ in its denominator, but to the same order of approximation we may replace this by $np_{0i}(\boldsymbol{\theta})$. We thus see that if (25.10) holds, (25.8) behaves asymptotically as (25.5) – it is distributed as χ^2 with $k-1$ degrees of freedom. However, if the $p_{0i}(\mathbf{t})$ are 'well-behaved' functions of $\mathbf{t}$, they will differ from the $p_{0i}(\boldsymbol{\theta})$ by the same order of magnitude as $\mathbf{t}$ does from $\boldsymbol{\theta}$. Then for all practical purposes (25.10) requires that

$$\mathbf{t} - \boldsymbol{\theta} = o_p(n^{-1/2}). \tag{25.11}$$

Equation (25.11) is not customarily satisfied, since we usually have estimators with variances and covariances of order n^{-1} and then

$$\mathbf{t} - \boldsymbol{\theta} = O_p(n^{-1/2}). \tag{25.12}$$

In this 'regular' case, therefore, our argument above does not hold. But it does hold in cases where estimators have variances of order n^{-2}, as we have found to be the case for estimators of parameters that locate the end-points of the range of a variable (cf. **18.19** and Exercises 29.25–26). In such cases, therefore, we may use (25.8) with no new theory required. In the more common case where (25.12) holds, we must investigate further.

25.9 It will simplify our discussion if we first give Fisher's (1922a) alternative and revealing proof of the asymptotic distribution of (25.5) for the simple hypothesis case.

Suppose we have k independent Poisson variates, the ith having parameter np_{0i}, where $n = \sum_{i=1}^{k} n_i$ and $\sum_i p_{0i} = 1$. The probability that the first takes the value n_1, the second n_2 and so on, is

$$P(n_1, n_2, \ldots, n_k, n) = \prod_{i=1}^{k} \mathrm{e}^{-np_{0i}}(np_{0i})^{n_i}/n_i! = \mathrm{e}^{-n}n^n \prod_{i=1}^{k} p_{0i}^{n_i}/n_i!; \tag{25.13}$$

n appears explicitly as a variate in (25.13), although of course the resulting $(k+1)$-variate distribution is singular since $n = \sum_i n_i$. Its marginal distribution $g(n)$ is easily found, since the sum of the k independent Poisson variables is itself (cf. Example 11.17) a Poisson variable with parameter equal to $\sum_{i=1}^{k} np_{0i} = n$. Thus

$$g(n) = e^{-n}n^n/n! \tag{25.14}$$

and the conditional distribution

$$\begin{aligned} h(n_1, n_2, \ldots, n_k|n) &= \frac{P(n_1, n_2, \ldots, n_k, n)}{g(n)} \\ &= \frac{n!}{n_1!n_2!\ldots n_k!} p_{01}^{n_1} p_{02}^{n_2} \cdots p_{0k}^{n_k}. \end{aligned} \tag{25.15}$$

We see at once that (25.15) is precisely the multinomial distribution of the n_i on which our test procedure is based. Thus, as an alternative to the proof of the asymptotic distribution of X^2 given in Example 15.3 (cf. **25.5**), we may obtain it by regarding the n_i as the values of k independent Poisson variables with parameters np_{0i}, conditional upon n being fixed. By Example 4.9, the standardized variable

$$x_i = \frac{n_i - np_{0i}}{(np_{0i})^{1/2}} \tag{25.16}$$

is asymptotically normal as $n \to \infty$. Hence, as $n \to \infty$,

$$X^2 = \sum_{i=1}^{k} x_i^2$$

is the sum of squares of k independent standardized normal variates, subject to the single condition $\sum_i n_i = n$, which is equivalent to $\sum_i (np_{0i})^{1/2} x_i = 0$. By Example 11.6, X^2 therefore has a χ^2 distribution asymptotically, with $k - 1$ degrees of freedom.

25.10 The utility of this alternative proof is that, in conjunction with Example 11.6 to which it refers, it shows clearly that if s further homogeneous linear conditions are imposed on the n_i, the only effect on the asymptotic distribution of X^2 will be to reduce the degrees of freedom from $k - 1$ to $k - s - 1$.

We now return to the composite hypothesis of **25.8** in the case when (25.2) holds. Suppose that we choose as our set of estimators $\mathbf{t}$ of $\boldsymbol{\theta}$ the maximum likelihood (or other asymptotically equivalent efficient) estimators, so that $\mathbf{t} = \widehat{\boldsymbol{\theta}}$. Now the likelihood function L in this case is simply the multinomial (25.15) regarded as a function of the θ_j, on which the p_{0i} depend. Thus, generalizing **18.20** (where $s = 1$),

$$\frac{\partial \log L}{\partial \theta_j} = \sum_{i=1}^{k} n_i \frac{\partial p_{0i}}{\partial \theta_j} \frac{1}{p_{0i}}, \qquad j = 1, 2, \ldots, s, \tag{25.17}$$

and the ML estimators in this regular case are the roots of the s equations obtained by equating (25.17) to zero for each j. Clearly, each such equation is a homogeneous linear relationship among the n_i. We thus see that, in this regular case, we have s additional constraints imposed by the process of efficient estimation of $\boldsymbol{\theta}$ from the multinomial distribution, so that the statistic (25.5) is asymptotically distributed as χ^2 with $k - s - 1$ degrees of freedom – see **25.11**. A more rigorous and detailed proof is given by Cramér (1946) – see also Birch (1964). We shall call $\widehat{\boldsymbol{\theta}}$ the *multinomial* ML estimators.

The effect of estimation on the distribution of X^2

25.11 We may now, following Watson (1959), consider the general problem of the effect of estimating the unknown parameters on the asymptotic distribution of X^2 statistic. We confine ourselves to the regular case, when (25.12) holds, and we write for any estimator $\mathbf{t}$ of $\boldsymbol{\theta}$,

$$\mathbf{t} - \boldsymbol{\theta} = n^{-1/2}\mathbf{A}\mathbf{x} + o_p(n^{-1/2}), \tag{25.18}$$

where $\mathbf{A}$ is an arbitrary $s \times k$ matrix and $\mathbf{x}$ is the $k \times 1$ vector whose ith element is

$$x_i = \frac{n_i - np_{0i}(\boldsymbol{\theta})}{\{np_{0i}(\boldsymbol{\theta})\}^{1/2}}, \tag{25.19}$$

defined just as at (25.16) for the simple hypothesis case; we assume $\mathbf{A}$ to have been chosen so that

$$\mathbf{E}(\mathbf{A}\mathbf{x}) = \mathbf{0}. \tag{25.20}$$

It follows at once from (25.18) and (25.20) that the covariance matrix of $\mathbf{t}$, $\mathbf{V}(\mathbf{t})$, is of order n^{-1}. By a Taylor expansion applied to $\{p_{0i}(\mathbf{t}) - p_{0i}(\boldsymbol{\theta})\}$ in (25.9), we may write

$$X^2 = \sum_{i=1}^{k} y_i^2,$$

where, as $n \to \infty$,

$$y_i = x_i - n^{1/2}\sum_{j=1}^{s}(t_j - \theta_j)\frac{\partial p_{0i}(\boldsymbol{\theta})}{\partial \theta_j}\frac{1}{\{p_{0i}(\boldsymbol{\theta})\}^{1/2}} + o_p(1),$$

or in matrix form,

$$\mathbf{y} = \mathbf{x} - n^{1/2}\mathbf{B}(\mathbf{t} - \boldsymbol{\theta}) + o_p(1), \tag{25.21}$$

where $\mathbf{B}$ is the $k \times s$ matrix whose (i, j)th element is

$$b_{ij} = \frac{\partial p_{0i}(\boldsymbol{\theta})}{\partial \theta_j}\frac{1}{\{p_{0i}(\boldsymbol{\theta})\}^{1/2}}. \tag{25.22}$$

Substituting (25.18) into (25.21), we find

$$\mathbf{y} = (\mathbf{I} - \mathbf{B}\mathbf{A})\mathbf{x} + o_p(1). \tag{25.23}$$

25.12 Now from equations (25.19) the x_i have zero means. As $n \to \infty$, they tend to multinormality, by the multivariate central limit theorem, and their covariance matrix is, temporarily writing p_i for $p_{0i}(\boldsymbol{\theta})$,

$$\mathbf{V}(\mathbf{x}) = \begin{pmatrix} 1-p_1 & -(p_1p_2)^{1/2} & -(p_1p_3)^{1/2} & \cdots & -(p_1p_k)^{1/2} \\ -(p_2p_1)^{1/2} & 1-p_2 & -(p_2p_3)^{1/2} & \cdots & -(p_2p_k)^{1/2} \\ \vdots & \vdots & \vdots & & \vdots \\ -(p_kp_1)^{1/2} & -(p_kp_2)^{1/2} & -(p_kp_3)^{1/2} & \cdots & 1-p_k \end{pmatrix}$$
$$= \mathbf{I} - (\mathbf{p}^{1/2})(\mathbf{p}^{1/2})^T \qquad (25.24)$$

where $\mathbf{p}^{1/2}$ is the $k \times 1$ vector with ith element $\{p_{0i}(\boldsymbol{\theta})\}^{1/2}$. It follows at once from (25.23) and (25.24) that the y_i also are asymptotically normal with zero means and covariance matrix

$$\mathbf{V}(\mathbf{y}) = (\mathbf{I} - \mathbf{BA})\{\mathbf{I} - \mathbf{p}^{1/2}(\mathbf{p}^{1/2})^T\}(\mathbf{I} - \mathbf{BA})^T. \qquad (25.25)$$

Thus $X^2 = \mathbf{y}^T\mathbf{y}$ is asymptotically distributed as the sum of squares of k normal variates with zero means and covariance matrix (25.25). Exercise 15.18 shows that the distribution of X^2 is χ^2 with r degrees of freedom if and only if $\mathbf{V}^3 = \mathbf{V}^2$ and $\text{rank}(\mathbf{V}^2) = \text{tr}(\mathbf{V}) = r$. This implies that all eigenvalues are 0 or 1, and since $\mathbf{V}$ is symmetric this is equivalent to the idempotency of $\mathbf{V}$ by the argument near the end of **15.11**.

25.13 We now consider particular cases of (25.25). First, the case of a simple hypothesis, where no estimation is necessary, is formally obtainable by putting $\mathbf{A} = \mathbf{0}$ in (25.18). The covariance matrix (25.25) then becomes simply

$$\mathbf{V}(\mathbf{y}) = \mathbf{V}(\mathbf{x}) = \mathbf{I} - \mathbf{p}^{1/2}(\mathbf{p}^{1/2})^T. \qquad (25.26)$$

Since $(\mathbf{p}^{1/2})^T\mathbf{p}^{1/2} = \sum_{i=1}^k p_{0i}(\boldsymbol{\theta}) = 1$, (25.26) is seen (on squaring it) to be idempotent, and its trace is $k-1$. Thus X^2 is a χ^2_{k-1} variable in this case, as we already know from two different proofs.

25.14 The composite hypothesis case is not so straightforward. First, suppose as in **25.10** that the multinomial ML estimators $\widehat{\boldsymbol{\theta}}$ are used. We seek the form of the matrix $\mathbf{A}$ in (25.18) when $\mathbf{t} = \widehat{\boldsymbol{\theta}}$. Now we know from **18.26** that the elements of the inverse of the covariance matrix of $\widehat{\boldsymbol{\theta}}$ are asymptotically given by

$$\{\mathbf{V}(\widehat{\boldsymbol{\theta}})\}^{-1}_{jl} = -E\left\{\frac{\partial^2 \log L}{\partial\theta_j\,\partial\theta_l}\right\}, \qquad j, l = 1, 2, \ldots, s. \qquad (25.27)$$

From (25.17), the multinomial ML equations give

$$\frac{\partial^2 \log L}{\partial\theta_j\,\partial\theta_l} = \sum_{i=1}^k \frac{n_i}{p_{0i}}\left(\frac{\partial^2 p_{0i}}{\partial\theta_j\,\partial\theta_l} - \frac{1}{p_{0i}}\frac{\partial p_{0i}}{\partial\theta_j}\frac{\partial p_{0i}}{\partial\theta_l}\right). \qquad (25.28)$$

On taking expectations in (25.28), we find

$$-E\left\{\frac{\partial^2 \log L}{\partial\theta_j\,\partial\theta_l}\right\} = n\left\{\sum_{i=1}^k \frac{1}{p_{0i}}\frac{\partial p_{0i}}{\partial\theta_j}\frac{\partial p_{0i}}{\partial\theta_l} - \sum_{i=1}^k \frac{\partial^2 p_{0i}}{\partial\theta_j\,\partial\theta_l}\right\}. \qquad (25.29)$$

The second term on the right of (25.29) is zero, since it is $(\partial^2/\partial\theta_j\,\partial\theta_l)\sum_{i=1}^k p_{0i}$. Thus, using (25.22),

$$-E\left\{\frac{\partial^2 \log L}{\partial\theta_j\,\partial\theta_l}\right\} = n\sum_{i=1}^{k} b_{ij}b_{il},$$

so that, from (25.27),

$$\{\mathbf{V}(\widehat{\boldsymbol{\theta}})\}^{-1} = n\mathbf{B}^T\mathbf{B}$$

or

$$\mathbf{C} = n\mathbf{V}(\widehat{\boldsymbol{\theta}}) = (\mathbf{B}^T\mathbf{B})^{-1}. \tag{25.30}$$

But from (25.18) and (25.24) we have

$$\mathbf{D} = n\mathbf{V}(\mathbf{t}) = \mathbf{A}\mathbf{V}(\mathbf{x})\mathbf{A}^T = \mathbf{A}\{\mathbf{I} - \mathbf{p}^{1/2}(\mathbf{p}^{1/2})^T\}\mathbf{A}^T. \tag{25.31}$$

Here (25.30) and (25.31) are alternative expressions for the same matrix.

We choose $\mathbf{A}$ to satisfy (25.30) by noting that

$$\mathbf{B}^T(\mathbf{p}^{1/2}) = \mathbf{0} \tag{25.32}$$

(since the jth element of this $s \times 1$ vector is $(\partial/\partial\theta_j)\sum_{i=1}^k p_{0i} = 0$), and hence that if $\mathbf{A} = \mathbf{G}\mathbf{B}^T$ where $\mathbf{G}$ is symmetric and non-singular, (25.30) gives $\mathbf{G}\mathbf{B}^T\mathbf{B}\mathbf{G}^T$. If this is to be equal to (25.30), we obviously have $\mathbf{G} = (\mathbf{B}^T\mathbf{B})^{-1}$, so finally

$$\mathbf{A} = (\mathbf{B}^T\mathbf{B})^{-1}\mathbf{B}^T \tag{25.33}$$

in the case of multinomial ML estimation. The covariance matrix (25.25) then becomes, using (25.32),

$$\begin{aligned}\mathbf{V}(\mathbf{y}) &= \{\mathbf{I} - \mathbf{B}(\mathbf{B}^T\mathbf{B})^{-1}\mathbf{B}^T\}^2 - \mathbf{p}^{1/2}(\mathbf{p}^{1/2})^T\\ &= \mathbf{I} - \mathbf{B}(\mathbf{B}^T\mathbf{B})^{-1}\mathbf{B}^T - \mathbf{p}^{1/2}(\mathbf{p}^{1/2})^T.\end{aligned} \tag{25.34}$$

By squaring, this matrix is shown to be idempotent. Its rank is equal to its trace, which is given by

$$\operatorname{tr}\mathbf{V}(\mathbf{y}) = \operatorname{tr}\{\mathbf{I} - \mathbf{p}^{1/2}(\mathbf{p}^{1/2})^T\} - \operatorname{tr}\{\mathbf{B}^T\cdot\mathbf{B}(\mathbf{B}^T\mathbf{B})^{-1}\}.$$

and, using **25.13**, this is

$$\operatorname{tr}\mathbf{V}(\mathbf{y}) = (k-1) - s.$$

Thus the distribution of X^2 is χ^2_{k-s-1} asymptotically, as we saw in **25.10**.

25.15 Our present approach enables us to enquire further: what happens to the asymptotic distribution of X^2 if some estimators other than $\widehat{\boldsymbol{\theta}}$ are used? This question was first considered in the simplest case by Fisher (1928b). Chernoff and Lehmann (1954) considered a case of particular interest, when the estimators used are the ML estimators based on the n individual observations and not the multinomial ML estimators $\widehat{\boldsymbol{\theta}}$, based on k frequencies n_i, that we have so far discussed. If we have the values of the n observations, it is clearly an efficient procedure to utilize this knowledge in $\boldsymbol{\theta}$, even though we are going to use the k class frequencies alone in carrying out the test of fit. We shall find, however, that the X^2 statistic obtained in this way no longer has an asymptotic χ^2 distribution.

25.16 Let us return to the general expression (25.25) for the covariance matrix. Multiplying it out, we rewrite it

$$\begin{aligned}\mathbf{V}(\mathbf{y}) = \{\mathbf{I}-\mathbf{p}^{1/2}(\mathbf{p}^{1/2})^T\} - \mathbf{BA}\{\mathbf{I}-\mathbf{p}^{1/2}(\mathbf{p}^{1/2})^T\}\\ -\{\mathbf{I}-\mathbf{p}^{1/2}(\mathbf{p}^{1/2})^T\}\mathbf{A}^T\mathbf{B}^T + \mathbf{BA}\{\mathbf{I}-\mathbf{p}^{1/2}(\mathbf{p}^{1/2})^T\}\mathbf{A}^T\mathbf{B}^T.\end{aligned} \tag{25.35}$$

Rather than find the eigenvalues λ_i of $\mathbf{V}(\mathbf{y})$, we consider those of $\mathbf{I}-\mathbf{V}(\mathbf{y})$, which are $1-\lambda_i$. We use (25.31) to write this matrix form

$$\begin{aligned}\mathbf{I}-\mathbf{V}(\mathbf{y}) &= \mathbf{p}^{1/2}(\mathbf{p}^{1/2})^T + \mathbf{B}[\mathbf{A}\{\mathbf{I}-\mathbf{p}^{1/2}(\mathbf{p}^{1/2})^T\} - \tfrac{1}{2}\mathbf{A}\{\mathbf{I}-\mathbf{p}^{1/2}(\mathbf{p}^{1/2})^T\}\mathbf{A}^T\mathbf{B}^T]\\ &\quad +[\{\mathbf{I}-\mathbf{p}^{1/2}(\mathbf{p}^{1/2})^T\}\mathbf{A}^T - \tfrac{1}{2}\mathbf{BA}\{\mathbf{I}-\mathbf{p}^{1/2}(\mathbf{p}^{1/2})^T\}\mathbf{A}^T]\mathbf{B}^T\\ &= \mathbf{p}^{1/2}(\mathbf{p}^{1/2})^T + \mathbf{B}[\mathbf{A}\{\mathbf{I}-\mathbf{p}^{1/2}(\mathbf{p}^{1/2})^T\} - \tfrac{1}{2}\mathbf{DB}^T]\\ &\quad +[\{\mathbf{I}-\mathbf{p}^{1/2}(\mathbf{p}^{1/2})^T\}\mathbf{A}^T - \tfrac{1}{2}\mathbf{BD}]\mathbf{B}^T.\end{aligned} \tag{25.36}$$

Expression (25.36) may be written as the product of two partitioned matrices,

$$\mathbf{I}-\mathbf{V}(\mathbf{y}) = \begin{pmatrix} \mathbf{p}^{1/2} \\ \mathbf{B} \\ \{\mathbf{I}-\mathbf{p}^{1/2}(\mathbf{p}^{1/2})^T\}\mathbf{A}^T - \tfrac{1}{2}\mathbf{BD} \end{pmatrix} \begin{pmatrix} (\mathbf{p}^{1/2})^T \\ \mathbf{A}\{\mathbf{I}-\mathbf{p}^{1/2}(\mathbf{p}^{1/2})^T\} - \tfrac{1}{2}\mathbf{DB}^T \\ \mathbf{B}^T \end{pmatrix}^T.$$

The matrices on the right may be transposed without affecting the non-zero eigenvalues. This step converts their product from a $k \times k$ to a $(2s+1)\times(2s+1)$ matrix, which is reduced, on using (25.30) and (25.32), to

$$\left(\begin{array}{c|cc} 1 & 0 & 0 \\ \hline 0 & & \\ & \mathbf{M} & \\ 0 & & \end{array}\right) = \left(\begin{array}{c|cc} 1 & \mathbf{0}_{1\times s} & \mathbf{0}_{1\times s} \\ \hline \mathbf{0}_{s\times 1} & \mathbf{AB} - \tfrac{1}{2}\mathbf{DC}^{-1} & \mathbf{D} - \tfrac{1}{2}\mathbf{ABD} - \tfrac{1}{2}\mathbf{DB}^T\mathbf{A}^T + \tfrac{1}{4}\mathbf{DC}^{-1}\mathbf{D} \\ \mathbf{0}_{s\times i} & \mathbf{C}^{-1} & \mathbf{B}^T\mathbf{A}^T - \tfrac{1}{2}\mathbf{C}^{-1}\mathbf{D} \end{array}\right). \tag{25.37}$$

Equation (25.37) has one eigenvalue of unity and $2s$ others which are those of the matrix $\mathbf{M}$ partitioned off in its south-east corner. If $k \geq 2s+1$, which is almost invariably the case in applications, this implies that (25.36) has $k-2s-1$ zero eigenvalues, one of unity, and $2s$ others which are roots of $\mathbf{M}$. Thus for $\mathbf{V}(\mathbf{y})$ itself, we have $k-2s-1$ eigenvalues of unity, and one of zero and $2s$ which are the complements to unity of the eigenvalues of $\mathbf{M}$.

25.17 We now consider the problem introduced in **25.15**. Suppose that, to estimate $\boldsymbol{\theta}$, we use the ML estimators based on the n individual observations, which we shall call the 'ordinary ML estimators' and denote by $\widehat{\boldsymbol{\theta}}^*$. We know from **18.26** that if f is the density function of the observations, we have asymptotically

$$\mathbf{D} = n\mathbf{V}(\widehat{\boldsymbol{\theta}}^*) = -\left\{E\left(\frac{\partial^2 \log f}{\partial\theta_j\,\partial\theta_l}\right)\right\}^{-1} \tag{25.38}$$

and that the elements of $\widehat{\boldsymbol{\theta}}^*$ are the roots of

$$\frac{\partial \log L}{\partial \theta_i} = 0, \qquad i = 1, 2, \dots, s,$$

where L is now the ordinary (not the multinomial) likelihood function. Thus, if $\boldsymbol{\theta}_0$ is the true value, we have the Taylor expansion

$$0 = \left[\frac{\partial \log L}{\partial \theta_i}\right]_{\theta_j=\theta_j^*} = \left[\frac{\partial \log L}{\partial \theta_i}\right]_{\theta_j=\theta_{0j}} + (\widehat{\theta}_j^* - \theta_{0j})\left[\frac{\partial^2 \log L}{\partial \theta_i\, \partial \theta_j}\right]_{\theta_j=\theta_{0j}+\varepsilon} \qquad i, j = 1, 2, \dots, s,$$

and as in **18.26** this gives asymptotically, using (25.38),

$$n^{1/2}(\widehat{\boldsymbol{\theta}}^* - \boldsymbol{\theta}) = n^{-1/2}\mathbf{D}\left(\frac{\partial \log \mathbf{L}}{\partial \boldsymbol{\theta}}\right). \tag{25.39}$$

In this case, we evaluate $\mathbf{V}(\mathbf{y})$ directly from (25.21) with $\mathbf{t} = \widehat{\boldsymbol{\theta}}^*$, where we see that asymptotically

$$\begin{aligned} \mathbf{V}(\mathbf{y}) &= E\{\mathbf{x} - \mathbf{n}^{1/2}\mathbf{B}(\widehat{\boldsymbol{\theta}}^* - \boldsymbol{\theta})\}\{\mathbf{x} - \mathbf{n}^{1/2}\mathbf{B}(\widehat{\boldsymbol{\theta}}^* - \boldsymbol{\theta})\}^T \\ &= \mathbf{V}(\mathbf{x}) + n\mathbf{B}\mathbf{V}(\widehat{\boldsymbol{\theta}}^*)\mathbf{B}^T - (\mathbf{T}\mathbf{B}^T + \mathbf{B}\mathbf{T}^T) \end{aligned} \tag{25.40}$$

where

$$\mathbf{T} = E\{\mathbf{x} \cdot \mathbf{n}^{1/2}(\widehat{\boldsymbol{\theta}}^* - \boldsymbol{\theta})^T\} = E\left\{\mathbf{x} \cdot \mathbf{n}^{-1/2}\left(\frac{\partial \log \mathbf{L}}{\partial \boldsymbol{\theta}}\right)^T\right\}\mathbf{D}$$

asymptotically by (25.39). We write this $\mathbf{T} = \mathbf{E}\mathbf{D}$.

If we write $\delta_{ir} = 1$ if the rth observation falls into the ith class, C_i, and $\delta_{ir} = 0$ otherwise, we can write (25.19) as

$$x_i = \sum_{r=1}^{n} (\delta_{ir} - p_{0i})/(np_{0i})^{1/2},$$

so that the characteristic term of $\mathbf{E}$ is

$$\begin{aligned} n^{-1}p_{0i}^{-1/2}E\left\{\sum_{r=1}^{n}(\delta_{ir} - p_{0i}) \cdot \frac{\partial \log L}{\partial \theta_j}\right\} &= p_{0i}^{-1/2}E\left\{(\delta_{ir} - p_{0i})\frac{\partial \log L}{\partial \theta_j}\right\} \\ &= p_{0i}^{-1/2}E\left\{\delta_{ir}\frac{\partial \log f(x_r|\theta)}{\partial \theta_j}\right\}, \end{aligned}$$

where we have used (17.18) and the independence of the observations. Thus we have

$$p_{0i}^{-1/2}\int_{C_i}\frac{\partial \log f}{\partial \theta_j} f\,\mathrm{d}x = p_{0i}^{-1/2}\int_{C_i}\frac{\partial f}{\partial \theta_j}\mathrm{d}x = p_{0i}^{-1/2}\frac{\partial}{\partial \theta_j}\int_{C_i} f\,\mathrm{d}x = p_{0i}^{-1/2}\frac{\partial}{\partial \theta_j}p_{0i}.$$

This is precisely b_{ij} of (25.22). Thus

$$\mathbf{E} = \mathbf{B}, \quad \mathbf{T} = \mathbf{B}\mathbf{D}. \tag{25.41}$$

Using (25.24), (25.38) and (25.41) in (25.40), we obtain

$$\mathbf{V}(\mathbf{y}) = \mathbf{I} - \mathbf{p}^{1/2}(\mathbf{p}^{1/2})^T - \mathbf{BDB}^T. \tag{25.42}$$

25.18 Matrix (25.42) is not idempotent, as may be seen by squaring it. Moreover, (25.41) shows that the non-negative covariance matrix

$$\mathbf{V}\left\{\mathbf{B}^T\mathbf{x} - n^{-1/2}\left(\frac{\partial \log L}{\partial \boldsymbol{\theta}}\right)\right\} = \mathbf{B}^T\mathbf{B} + \mathbf{D}^{-1} - (\mathbf{B}^T\mathbf{E} + \mathbf{E}^T\mathbf{B}) = \mathbf{D}^{-1} - \mathbf{B}^T\mathbf{B} = \mathbf{D}^{-1} - \mathbf{C}^{-1},$$

using (25.30). Thus $\mathbf{C} - \mathbf{D}$ is non-negative and (25.42) may be written in the form

$$\mathbf{V}(\mathbf{y}) = \mathbf{I} - \mathbf{p}^{1/2}(\mathbf{p}^{1/2})^T - \mathbf{BCB}^T + \mathbf{B}(\mathbf{C} - \mathbf{D})\mathbf{B}^T. \tag{25.43}$$

The first two terms on the right of (25.43) are what we obtained in (25.26) for the case when no estimation takes place, when $\mathbf{V}(\mathbf{y})$ has $k - 1$ eigenvalues of unity, and one of zero. The first three terms are (25.34), when, with multinomial ML estimation, $\mathbf{V}(\mathbf{y})$ has $k - s - 1$ eigenvalues of unity and $s + 1$ of zero. Because of the non-negative definiteness of all its terms, reduction of (25.43) to canonical form shows that its eigenvalues are bounded by the corresponding eigenvalues of (25.26) and (25.34). Thus (25.43) has $k - s - 1$ eigenvalues of unity, one of zero, and s between zero and unity, as established by Chernoff and Lehmann (1954).

In general, the values of the s eigenvalues depend upon $\boldsymbol{\theta}$, but if $\boldsymbol{\theta}$ contains only a location and a scale parameter, this is not so – cf. Exercise 25.13. In any case, it follows from the fact that the two sets of ML estimators $\widehat{\boldsymbol{\theta}}$ and $\widehat{\boldsymbol{\theta}}^*$ draw closer together as k increases, so that $\mathbf{D} \to \mathbf{C}$, that the last s eigenvalues tend to zero as $k \to \infty$.

Molinari (1977) shows that if parameters are estimated by the method of moments, X^2 is no longer necessarily bounded above by a χ^2_{k-1} variable.

Chase (1972) shows that if parameters are estimated from an independent sample, there is over-recovery of d.fr., the distribution of X^2 then being bounded below by a χ^2_{k-1} variable whether the ordinary or the multinomial ML estimators are used.

25.19 What we have found, therefore, is that X^2 does not have an asymptotic χ^2 distribution when fully efficient (ordinary ML) estimators are used in estimating parameters – there is partial recovery of the s degrees of freedom lost by the multinomial ML estimators. However, the distribution of X^2 is bounded between a χ^2_{k-1} and a χ^2_{k-s-1} variable, and as k becomes large these are so close together that the difference can be ignored – this is another way of expressing the final deduction in **25.18**. But for k small, the effect of using the χ^2_{k-s-1} distribution for test purposes may lead to serious error; for the probability of exceeding any given value will be greater than we suppose. s is rarely more than 1 or 2, but it is as well to be sure, when ordinary ML estimation is being used, that the critical values of χ^2_{k-s-1} and χ^2_{k-1} are both exceeded by X^2. Tables of χ^2 show that, for a test of size $\alpha = 0.05$, the critical value for $k - 1$ degrees of freedom exceeds that for $k - s - 1$ degrees of freedom, if s is small, by Cs, approximately, where C declines from about 1.5 at $k - s - 1 = 5$ to about 1.2 when $k - s - 1 = 30$. For $\alpha = 0.01$, the corresponding values of C are about 1.7 and 1.3.

The choice of classes for the X^2 test

25.20 The whole of the asymptotic theory of the X^2 test that we have discussed so far is valid however we determine the k classes into which the observations are grouped, so long as they are determined *without reference to the observations*. The italicized condition is essential, for we have made no provision for the class boundaries themselves being random variables. However, it is common practice to determine the class boundaries, and sometimes even to fix k itself, after reference to the general picture presented by the observations. We must therefore discuss the formation of classes, and then consider how far it affects the theory we have developed.

We first consider the determination of class boundaries, leaving the choice of k until later. If there is a natural discreteness imposed by the problem (as in Example 25.1 where there are four natural groups) or if we have a sample of observations from a discrete distribution, the class-boundary problem arises only in the sense that we may decide (in order to reduce k, or in order to improve the accuracy of the asymptotic distribution of X^2 as we shall see in **25.30**) to combine some of the expected frequencies. Indeed, if a discrete distribution has infinite range, like the Poisson, we are forced to combine terms if k is not to be infinite with most of the expected frequencies very small indeed.

But the class-boundary problem arises in its most acute form when we are sampling from a continuous distribution. There are now no natural boundaries at all. If we suppose k to be determined in advance in some way, how are these boundaries to be determined?

In practice, arithmetic convenience is usually allowed to dictate the solution: the classes are taken to cover equal ranges of the variate, except at an extreme where the range of the variate is infinite. The range of a class is roughly determined by the dispersion of the distribution, while the location of the distribution helps to determine where the central class should fall. Thus, if we wished to form $k = 10$ classes for a sample to be tested for normality, we might roughly estimate (perhaps by eye) the mean $\bar{x}$ and the standard deviation s of the sample and take the class boundaries as $\bar{x} \pm \frac{1}{2}js$, $j = 0, 1, 2, 3, 4$. The classes would then be

$$(-\infty, \bar{x}-2s), \quad (\bar{x}-2s, \bar{x}-1.5s), \quad (\bar{x}-1.5s, \bar{x}-s), \quad (\bar{x}-s, \bar{x}-0.5s),$$
$$(\bar{x}-0.5s, \bar{x}), \quad (\bar{x}, \bar{x}+0.5s), \quad (\bar{x}+0.5s, \bar{x}+s), \quad (\bar{x}+s, \bar{x}+1.5s),$$
$$(\bar{x}+1.5s, \bar{x}+2s), \quad (\bar{x}+2s, \infty).$$

25.21 Although this procedure is not very precise, it clearly makes the class boundaries random variables, and it is not obvious that the asymptotic distribution of X^2, calculated for classes formed in this way, is the same as when the classes are fixed. However, intuition suggests that since the asymptotic theory holds for *any* set of k fixed classes, it should hold also when the class boundaries are determined from the sample. That this is so when the class boundaries are determined by consistent estimation of parameters in the regular case was shown for normal distribution by Watson (1957b) and for continuous distributions in general by Roy (1956), Watson (1958; 1959) and Chibisov (1971) – cf. also Dahiya and Gurland (1972; 1973). Moore (1971) and Moore and Spruill (1975) give a unified general treatment of the large-sample theory.

We may thus neglect random variations in the class boundaries so far as the asymptotic distribution of X^2, when H_0 holds, is concerned. Small-sample distributions, of course, will be affected. (We discuss small-sample distributions of X^2 in the fixed boundaries case in **25.30**.)

The equal-probabilities method of constructing classes

25.22 We may now directly face the questions of how class boundaries should be determined, in light of the assurance of the last paragraph of **25.21**. If we now seek an optimum method of boundary determination, it ought to be in terms of the power of the test; we should choose that set of boundaries which maximizes power for a test of given size. Failing this, we seek some means of avoiding the unpleasant fact that there is a multiplicity of possible sets of classes, any one of which will in general give a different result for the same data; we require a rule which is plausible and practical.

One such rule has been suggested by Mann and Wald (1942) and by Gumbel (1943): given k, choose the classes so that the hypothetical probabilities p_{0i} are all equal to $1/k$. This procedure is perfectly definite and unique, but requires that the data should be available ungrouped. The procedure is illustrated in Example 25.2.

Example 25.2 (Goodness-of-fit test using equal probability classes)

Quenouille (1959) gives, apart from a change in location, 1000 random values from the exponential distribution

$$f(x) = \exp(-x), \qquad 0 \leq x < \infty.$$

The first 50 of these, arranged in ascending order, are:

0.01, 0.01, 0.04, 0.17, 0.18, 0.22, 0.22, 0.25, 0.25, 0.29, 0.42, 0.46, 0.47, 0.47, 0.56, 0.59, 0.67, 0.68, 0.70, 0.72, 0.76, 0.78, 0.83, 0.85, 0.87, 0.93, 1.00, 1.01, 1.01, 1.02, 1.03, 1.05, 1.32, 1.34, 1.37, 1.47, 1.50, 1.52, 1.54, 1.59, 1.71, 1.90, 2.10, 2.35, 2.46, 2.46, 2.50, 3.73, 4.07, 6.03.

Suppose that we wished to form four classes for an X^2 test that the underlying distribution is indeed as stated. A natural grouping with equal-width intervals would be as shown in Table 25.2.

We find $X^2 = 3.1$ with 3 d.fr., a value which would not reject the hypothetical distribution for any test of size less than $\alpha = 0.37$; the agreement between observation and hypothesis is therefore satisfactory.

Table 25.2 Observed and expected frequencies for exponential data

Variate values	Observed frequency	Expected frequency
0–0.50	14	19.7
0.51–1.00	13	11.9
1.01–1.50	10	7.2
1.51 and over	13	11.2
	50	50.0

Table 25.3

Variate values	Observed frequency	Expected frequency
0–0.28	9	12.5
0.29–0.69	9	12.5
0.70–1.38	17	12.5
1.39 and over	15	12.5
	50	50.0

Let us now consider how the same data would be treated using equal probability classes. We form Table 25.3 by splitting the data into four equal probability classes at the quartiles 0.288, 0.693 and 1.386.

X^2 is now easier to calculate, since (25.8) reduces to

$$X^2 = \frac{k}{n}\sum_{i=1}^{k} n_i^2 - n \tag{25.44}$$

since all the hypothetical probabilities $p_{0i} = 1/k$. We find here that $X^2 = 4.08$, which would not lead to rejection unless the test size exceeded 0.25.

It will be seen that there is little extra arithmetical work involved in the equal-probabilities method of carrying out the X^2 test. Instead of a constant class width, we have irregular class widths computed so that the expected frequencies are equal.

We have had no parameters to estimate in this example. If parameters must be estimated, we can only equalize the *estimated* expected frequencies.

25.23 Although the equal-probabilities method resolves the class-boundary issue, it may not necessarily increase the power of the test. A goodness-of-fit hypothesis is likely to be most vulnerable at the extremes of the range of the variable, and the equal-probabilities method may well result in a loss of sensitivity at the extremes unless k is rather large. This brings us the question of how k should be chosen, and in order to discuss this question we must consider the power of the X^2 statistic.

The moments of the X^2 test statistic

25.24 We suppose, as before, that we have hypothetical probabilities p_{0i} when H_0 holds, so that our test statistic is, as in (25.8),

$$X^2 = \frac{1}{n}\sum_{i=1}^{k} \frac{n_i^2}{p_{0i}} - n.$$

We confine ourselves to simple hypotheses. Suppose that the true probabilities are p_{1i}, $i = 1, 2, \ldots, k$. The expected value of the test statistic is then

$$E(X^2) = \frac{1}{n}\sum_{i=1}^{k}\frac{1}{p_{0i}}E(n_i^2) - n.$$

From the moments of the multinomial distribution given in (7.18),

$$E(n_i^2) = np_{1i}(1 - p_{1i}) + n^2 p_{1i}^2, \tag{25.45}$$

whence

$$E(X^2) = \sum_{i=1}^{k}\frac{p_{1i}(1 - p_{1i})}{p_{0i}} + n\left\{\sum_{i=1}^{k}\frac{p_{1i}^2}{p_{0i}} - 1\right\}. \tag{25.46}$$

When H_0 holds ($p_{1i} = p_{0i}$), this reduces to

$$E(X^2 | H_0) = k - 1. \tag{25.47}$$

This exact result is known to hold asymptotically, since X^2 is then a χ^2_{k-1} variate. If we differentiate (25.46) with respect to the p_{1i}, subject to $\sum_i p_{1i} = 1$, we find that, as $n \to \infty$, (25.46) has its minimum value when $p_{1i} = p_{0i}$. For any hypothesis H_1 specifying a set of probabilities $p_{1i} \neq p_{0i}$, we therefore have asymptotically

$$E(X^2 | H_1) > k - 1. \tag{25.48}$$

Inequality (25.48), like the asymptotic argument based on the LR statistic in **25.6**, indicates that the critical region for the X^2 test consists of the upper tail, although this indication is not conclusive since the asymptotic distribution of X^2 is not of the χ^2 form when H_1 holds. This alternative distribution is, in fact, a non-central χ^2 under the condition given in **25.27**.

Even the variance of X^2 is a relatively complicated function of the p_{0i} and p_{1i} (cf. Exercise 25.5). However, when H_0 holds it simplifies to

$$\begin{aligned} \text{var}(X^2 | H_0) &= n^{-1}\left\{2(n-1)(k-1) - k^2 + \sum_i p_{0i}^{-1}\right\} \\ &\underset{n\to\infty}{\longrightarrow} 2(k-1), \end{aligned} \tag{25.49}$$

this limit being the asymptotic χ^2_{k-1} variance. When all $p_{0i} = k^{-1}$ expression (25.49) is exactly equal to $2(k-1)(n-1)/n$, but may be much larger than its limiting value $2(k-1)$ if some p_{0i} are very small.

From (25.46) and Exercise 25.3, we also have in the equal-probabilities case ($p_{0i} = k^{-1}$)

$$\begin{aligned} E(X^2 | H_1) &= (k-1) + (n-1)\left(k\sum_i p_{1i}^2 - 1\right) > k - 1, \\ \text{var}(X^2 | H_1) &\sim 4(n-1)k^2\left\{\sum_i p_{1i}^3 - \left(\sum_i p_{1i}^2\right)^2\right\}. \end{aligned} \tag{25.50}$$

Consistency and unbiasedness of the X^2 test

25.25 Equations (25.50) are sufficient to demonstrate the consistency of the equal-probabilities X^2 test. The test consists of comparing the value of X^2 with a fixed critical value, say c_α, in the upper tail of its distribution when H_0 holds. When H_1 holds, the mean value and variance of X^2 are each of order n. By Chebyshev's inequality (3.95),

$$P\{|X^2 - E(X^2)| \geq \lambda[\text{var}(X^2)]^{1/2}\} \leq \frac{1}{\lambda^2}. \tag{25.51}$$

Since c_α is fixed, it differs from $E(X^2)$ by a quantity of order n, so that if we require the probability that X^2 differs from its mean sufficiently to fall below c_α, the multiplier λ on the left of (25.51) must be of order $n^{1/2}$, and the right-hand side is therefore of order n^{-1}. Thus

$$\lim_{n\to\infty} P\{X^2 < c_\alpha\} = 0,$$

and the test is consistent for any H_1 specifying unequal class probabilities.

The general X^2 test, with unequal p_{0i}, is also consistent against alternatives specifying at least one $p_{1i} \neq p_{0i}$, as is intuitively reasonable. A proof is given by Neyman (1949).

25.26 Although the X^2 test is consistent (and therefore asymptotically unbiased), it is not in general for fixed n, as may be seen for the simplest case $k = 2$ in Exercise 25.12. For, from (25.46), we may have $E(X^2|H_1) < k - 1$ in finite samples, and in view of the upper-tail critical region, this indicates the possibility that the test is biased against H_1 for some values of α. Haberman (1988) discusses this point in detail and recommends care when k is large and many p_{0i} are small and variable.

Cohen and Sackrowitz (1975) show that the X^2 test, and also the LR test, are strictly unbiased if we use equal probabilities, and that X^2 minimizes $|\partial^2 P/\partial\theta_i\partial\theta_j|(i, j = 1, 2, \ldots, k-1)$ at H_0. Thompson (1979) shows that every unbiased test against one-sided alternatives has a monotone power function.

The limiting power function of the X^2 test

25.27 Suppose that, as in our discussion of ARE in Chapter 22, we allow H_1 to approach H_0 as n increases, at a rate sufficient to keep the power bounded away from 1. In fact, let $p_{1i} - p_{0i} = c_i n^{-1/2}$, where the c_i are constants not dependent on n. Then the distribution of X^2 is asymptotically a non-central χ^2 with degrees of freedom $k - s - 1$ (where s parameters are estimated by the multinomial ML estimators) and non-central parameter

$$\lambda \sum_{i=1}^{k} \frac{c_i^2}{p_{0i}} = n \sum_{i=1}^{k} \frac{(p_{1i} - p_{0i})^2}{p_{0i}}. \tag{25.52}$$

This result, first obtained by Eisenhart (1938), follows at once from the representation of **25.9–10**; its proof is left to the reader as Exercise 25.4. We now see that the second term on the right of (25.46) is simply λ of (25.52), in accordance with κ_1 in Exercise 22.1. It now follows by using the Neyman–Pearson lemma (20.6) on (22.18) that the best critical region for testing $\lambda = 0$ consists of the upper tail of the distribution of X^2.

The tables described in **22.8** may be used to evaluate the asymptotic power of the X^2 test, given by the integral (22.29). The approximation (22.30) may also be used.

For *fixed* p_{1i}, the limiting distribution as $n \to \infty$ of $\{X^2 - E(X^2)\}/\{\text{var}(X^2)\}^{1/2}$ is standardized normal; when power is large, this gives a better power approximation for simple H_0 than does (25.52) – see Broffitt and Randles (1977).

Example 25.3 (Effect of number of classes on power)

We may illustrate the use of the limiting power function by returning to the problem of Example 25.2 and examining the effect on the power of the equal-probabilities procedure of doubling k. For sense of presentation, we take four classes with slightly unequal probabilities (see Table 25.4).

In the table, the p_{0i} are obtained from the gamma distribution with parameter 1, as before, and the p_{1i} from the gamma distribution with parameter 1.5. For these four classes, and $n = 50$ as in Example 25.2, we evaluate the non-central parameter of (25.52) as $\lambda = 0.2353 \times 50 = 11.8$. With 3 d.fr. for X^2, this gives a power when $\alpha = 0.05$ of 0.83, from Patnaik's table.

Suppose now that we form eight classes by splitting each of the above classes into two, with new p_{0i} as equal as is convenient for the use of the tables. We obtain Table 25.5.

Table 25.4 Goodness-of-fit test for exponential with 4 cells

Values	p_{0i}	p_{1i}	$(p_{1i} - p_{0i})^2$	$(p_{1i} - p_{0i})^2/p_{0i}$
0–0.3	0.259	0.104	0.0240	0.0927
0.3–0.7	0.244	0.190	0.0029	0.0119
0.7–1.4	0.250	0.282	0.0010	0.0040
1.4 and over	0.247	0.424	0.0313	0.1267
				$0.2353 = \lambda/n$

Table 25.5 Goodness-of-fit test for exponential with 8 cells

Values	p_{0i}	p_{1i}	$(p_{1i} - p_{0i})^2$	$(p_{1i} - p_{0i})^2/p_{0i}$
0–0.15	0.139	0.040	0.0098	0.0705
0.15–0.3	0.120	0.064	0.0031	0.0258
0.3–0.45	0.103	0.071	0.0010	0.0097
0.45–0.7	0.141	0.119	0.0005	0.0035
0.7–1.0	0.129	0.134	0.0000	0.0002
1.0–1.4	0.121	0.148	0.0007	0.0058
1.4–2.1	0.125	0.183	0.0034	0.0272
2.1 and over	0.122	0.241	0.0142	0.1163
				$0.2590 = \lambda/n$

For $n = 50$, we now have $\lambda = 13.0$ with 7 d.fr. The approximate power for $\alpha = 0.05$ is now about 0.75 from Patnaik's table. The doubling of k has increased λ, but only slightly. The power is actually reduced, because for given λ the central and non-central χ^2 distributions draw closer together as degrees of freedom increase (cf. Exercise 22.3) and here this effect is stronger than the increase in λ. However, n is too small here for us to place any reliance on the exact values of the power obtained from the limiting power function, and we should perhaps conclude that doubling k has had only a slight effect on the power.

The choice of *k* with equal probabilities

25.28 With the aid of the asymptotic power function of **25.27**, we can get a heuristic indication of how to choose k in the equal-probabilities case. The non-central parameter (25.52) is then, putting $\theta_i = p_{1i} - k^{-1}$,

$$\lambda = nk \sum_{i=1}^{k} \theta_i^2. \tag{25.53}$$

We now assume that $|\theta_i| \leq 1/k$, for all i, and consider what happens when k becomes large. $\theta = \sum_{i=1}^{k} \theta_i^2$, as a function of k, will then be of the same order of magnitude as a sum of squares in the interval $(-1/k, 1/k)$, i.e.

$$\theta \sim a \int_{-1/k}^{1/k} u^2\,\mathrm{d}u = 2a \int_{0}^{1/k} u^2\,\mathrm{d}u. \tag{25.54}$$

The asymptotic power of the test is a function $P\{k, \lambda\}$ which therefore is $P\{k, \lambda(k)\}$; it is a monotone increasing function of λ, and has its stationary values when $\mathrm{d}\lambda(k)/\mathrm{d}k = 0$. We thus put, using (25.53) and (25.54),

$$0 = \frac{1}{n}\frac{\mathrm{d}\lambda}{\mathrm{d}k} \sim \theta + k \cdot 2a\left(\frac{1}{k}\right)^2\left(-\frac{1}{k^2}\right) = \theta - 2ak^{-3}$$

giving

$$k^{-3} \sim \theta/(2a). \tag{25.55}$$

We cannot let $k \to \infty$ without restriction since all θ_i then tend to 0, but we assume k large enough so that both the H_0 and H_1 distributions of X^2 are near normality, and the approximate power function of the test is (cf. (22.137)) therefore

$$P = G\left\{\frac{[\frac{\partial}{\partial\theta}E(X^2|H_1)]_{\theta=0}}{[\mathrm{var}(X^2|H_0)]^{1/2}} \cdot \theta - \lambda_\alpha\right\}, \tag{25.56}$$

where

$$G(-\lambda_\alpha) = \alpha \tag{25.57}$$

determines the size of the test. From (25.49) and (25.50)

$$\left[\frac{\mathrm{d}}{\mathrm{d}\theta}E(X^2|H_1)\right]_{\theta=0} = (n-1)k, \tag{25.58}$$

$$\text{var}(X^2|H_0) = 2(k-1)(n-1)/n, \tag{25.59}$$

and if we insert these values and also (25.55) into (25.56), we obtain approximately

$$P = G\{2^{1/2}a(n-1)k^{-5/2} - \lambda_\alpha\}. \tag{25.60}$$

This is the approximate power function at the point where power is maximized for the choice of k. If we choose a value P_0 at which we wish the maximization to occur, we have, on inverting (25.60),

$$G^{-1}\{P_0\} = 2^{1/2}a(n-1)k^{-5/2} - \lambda_\alpha, \tag{25.61}$$

or

$$k = b\left\{\frac{2^{1/2}(n-1)}{\lambda_\alpha + G^{-1}\{P_0\}}\right\}^{2/5}, \tag{25.62}$$

where $b = a^{2/5}$.

25.29 In the special case $P_0 = \frac{1}{2}$, $G^{-1}(0.5) = 0$ and (25.62) simplifies. In this case, Mann and Wald (1942) obtained (25.62) by a much more sophisticated and rigorous argument – they found $b = 4$ in the case of the simple hypothesis. Our own derivation suggests that the same essential argument applies for the composite hypothesis, but b may be different in this case.

We could conclude that k should be increased in the equal-probabilities case in proportion to $n^{2/5}$, and that k should be smaller if we are interested in the region of high power (when $G^{-1}\{P_0\}$ is large) than if we are interested in the 'neighbouring' region of low power (when $G^{-1}\{P_0\}$ approaches $-\lambda_\alpha$ from above since any test is unbiased).

With $b = 4$ and $P_0 = \frac{1}{2}$, (25.62) leads to much larger values of k than are commonly used. k will be doubled when n increases by a factor of $4\sqrt{2}$. When $n = 200$, $k = 31$ for $\alpha = 0.05$ and $k = 27$ for $\alpha = 0.01$ – these are about the lowest values of k for which the approximate normality assumed in our argument (and also in Mann and Wald's) is at all accurate. In this case, Mann and Wald recommend that the use of (25.62) when $n \geq 450$ for $\alpha = 0.05$ and $n \geq 300$ for $\alpha = 0.01$. It will be seen that n/k, the hypothetical expectation in each class, increases as $n^{3/5}$, and is equal to about 6 and 8 respectively when $n = 200$, $\alpha = 0.05$ and 0.01.

Williams (1950) reported that k can be halved from the Mann–Wald optimum without serious loss of power at the 0.50 point. But it should be remembered that n and k must be substantial before (25.62) produces good results. Example 25.4 illustrates the point, which is also borne out by calculations made by Hamdan (1963) for tests of a normal mean. Hamdan (1968) verifies in the bivariate normal case that the use of (25.62) can result in loss of power compared with the use of equal-length classes. For a restricted class of alternatives, Quine and Robinson (1985) show that asymptotic efficiency is best if not too many classes are used – see also Kallenberg *et al.* (1985) and Kallenberg (1985). The rate of increase of k as $n \to \infty$ is crucial, essentially for the reason discussed in Example 25.3.

Koehler and Gan (1990) also consider the effects of cell selection on power and confirm the need for relatively large values of k. They compare the power of χ^2 and l in (25.4) and find their maximum powers to be similar, although l needs a higher value of k to achieve that maximum.

Koehler and Gan (1990) also show that unequal probability partitions may increase power for specific alternatives.

Example 25.4 (Choice of k with equal probabilities)

Consider again the problem of Example 25.3. We found that we were around the 0.8 value for power. From the normal distribution $G^{-1}(0.8) = 0.84$. With $b = 4$, $\alpha = 0.05$, $\lambda_\alpha = 1.64$, (25.62) gives the optimum k around this point,

$$k = 4\left\{\frac{2^{1/2}(n-1)}{2.48}\right\}^{2/5} = 3.2(n-1)^{2/5}.$$

For $n = 50$, this gives $k = 15$ approximately.

Suppose now that we construct a 15-class grouping with probabilities p_{0i} as nearly equal as is convenient. We obtain Table 25.6.

Table 25.6 Goodness-of-fit test for exponential with 15 classes

Values	p_{0i}	p_{1i}	$(p_{1i} - p_{0i})^2/p_{0i}$
0–0.05	0.049	0.008	0.034
0.05–0.15	0.090	0.032	0.037
0.15–0.20	0.042	0.020	0.012
0.20–0.30	0.078	0.044	0.015
0.30–0.40	0.071	0.047	0.008
0.40–0.50	0.063	0.048	0.004
0.50–0.65	0.085	0.072	0.002
0.65–0.75	0.050	0.047	0.000
0.75–0.90	0.065	0.067	0.000
0.90–1.1	0.074	0.083	0.000
1.1–1.3	0.060	0.075	0.004
1.3–1.6	0.071	0.095	0.008
1.6–2.0	0.067	0.101	0.018
2.0–2.7	0.068	0.116	0.034
2.7 and over	0.067	0.145	0.098
			$0.274 = \lambda/n$

Here $\lambda = 13.7$ and Patnaik's table gives a power of 0.64 for 14 degrees of freedom. λ has again been increased, but power reduced because of the increase in k. We are not at the optimum here. With large k (and hence large n), the effect of increasing degrees of freedom would not offset the increase of λ in this way.

25.30 We must not make k too large, since the multinomial approximation to the multinormal distribution cannot be expected to be satisfactory if the np_{0i} are very small. If the H_0 distribution is unimodal, and equal-length classes are used in a conventional manner, small expected frequencies will occur only at the tails. Yarnold (1970), generalizing earlier recommendations by Cochran

(1952; 1954), concludes from a detailed theoretical investigation of the simple H_0 case that for $\alpha = 0.05$ or 0.01 the *minimum* expected frequency may be as small as $5f$, where f is the proportion of the $k \geq 3$ classes with expectations less than 5 – Lawal and Upton (1980) suggest a lognormal approximation for ν d.fr. when this minimum exceeds $fk\nu^{-3/2}$, which is usually considerably smaller than $5f$. Roscoe and Byars (1971) found in extensive sampling experiments that if $k > 2$, the approximation is good for an *average* expected frequency of at least 2 for a test size $\alpha = 0.05$ and at least 4 for $\alpha = 0.01$.

In the equal-probabilities case, all the expected frequencies will be equal. Slakter (1966; 1968) shows that even for fractional equal expected frequencies, the approximation remains good, but the power is approximately 80 per cent of its nominal value – see also Roscoe and Byars (1971), Kempthorne (1967) and **25.5** for simple H_0.

The Mann–Wald procedure of **25.29** leads to expected frequencies always greater than 5 for $n \geq 200$. It is interesting to note that in Examples 25.3–4, the application of the 5 limit would have ruled out the 15-class procedure, and that the more powerful 8-class procedure, with expected frequencies ranging from 5 to 7, would have been acceptable.

Finally, we remark that the large-sample nature of the distribution theory of X^2 is not a disadvantage in practice, for we do not usually wish to test goodness of fit except in relatively large samples.

Recommendations for the X^2 test of fit

25.31 We summarize the above discussion with a few practical recommendations:

1. Use classes with equal, or nearly equal, probabilities.
2. Determine the number of classes when n exceeds 200 approximately by (25.62) with b between 2 and 4.
3. If parameters are to be estimated, use the ordinary ML estimators in the interest of efficiency, but recall that there is partial recovery of degrees of freedom (**25.19**) so that critical values should be adjusted upwards; if the multinomial ML estimators are used, no such adjustment is necessary.

None of the theory above will hold if the *form* (instead of the parameters alone) of $F_0(x)$ is estimated from the data used to test goodness of fit. Finally, we note that when lack of fit in the tails is of particular interest, recommendation 1 may lead to a considerable loss of power. In such cases, the Kolmogorov D test (**25.37**) will be preferable; see **25.41**. For further discussion of the X^2 test and its properties, see Greenwood and Nikulin (1996).

X^2 tests of independence

25.32 The general theory of the X^2 test may be used to test the fit of hypotheses of independence for categorized variables, where we are given only the frequencies falling into various categories. For the independence implies a set of expected frequencies, and we test the fit of the observed frequencies to these in the usual way.

Example 25.5 (Large-sample test of independence in a 2 × 2 table)

Suppose the members of a population (assumed infinite) are categorized by the presence or absence of attributes A and B. The four possible outcomes will have probabilities

	B	not-B	Totals
A	p_{11}	p_{12}	$p_{1\cdot}$
not-A	p_{21}	p_{22}	$p_{2\cdot}$
Totals	$p_{\cdot 1}$	$p_{\cdot 2}$	1

(25.63)

where a dot suffix denotes summation over that subscript. This is a sample multinomial as in (7.15) with four classes. A random sample of size n from the population has corresponding frequencies arranged in a 2 × 2 table:

n_{11}	n_{12}	$n_{1\cdot}$
n_{21}	n_{22}	$n_{2\cdot}$
$n_{\cdot 1}$	$n_{\cdot 2}$	n

(25.64)

The ML estimates of the p_{ij} are obtained by maximizing the log likelihood, which is $\log L =$ constant $+ \sum_{i=1}^{2}\sum_{j=1}^{2} n_{ij}\log p_{ij}$, subject to $\sum_i \sum_j p_{ij} = 1$. We obtain

$$\widehat{p}_{ij} = n_{ij}/n, \tag{25.65}$$

the generalization of the binomial result in Example 17.9 and **18.5**. Hence we have

$$\left.\begin{aligned}\widehat{p}_{i\cdot} &= n_{i\cdot}/n,\\ \widehat{p}_{\cdot j} &= n_{\cdot j}/n.\end{aligned}\right\} \tag{25.66}$$

Now consider the hypothesis that the two attributes are independently distributed over the population, i.e. that there is the same probability of an A among the Bs as among the not-Bs. The null hypothesis is

$$H_0 : p_{11}/p_{\cdot 1} = p_{12}/p_{\cdot 2} \tag{25.67}$$

which implies

$$p_{11}/p_{\cdot 1} = p_{1\cdot}/1,$$

or

$$p_{11} = p_{1\cdot}p_{\cdot 1},$$

and similarly any

$$p_{ij} = p_{i\cdot}p_{\cdot j}. \tag{25.68}$$

We require the ML estimators of the p_{ij} under the constraint imposed by H_0. We can save ourselves the routine algebra by observing that the marginal probabilities $p_{i\cdot}$ and $p_{\cdot j}$ are quite unaffected by the hypothesis of independence, which concerns only the relations between the probabilities p_{ij} in the body of the table. Thus the ML estimators under H_0 of $p_{i\cdot}$ and $p_{\cdot j}$ will remain as at (25.66). In consequence, the estimators of the individual p_{ij} under H_0 will, by (25.68), be

$$\widehat{\widehat{p}}_{ij} = \frac{n_{i\cdot}}{n}\frac{n_{\cdot j}}{n}. \tag{25.69}$$

The expected frequencies corresponding to the n_{ij} will therefore be

$$e_{ij} = n\widehat{\widehat{p}}_{ij} = n_{i\cdot}n_{\cdot j}/n \tag{25.70}$$

and we test the fit by

$$X^2 = \sum_{i=1}^{2}\sum_{j=1}^{2}\frac{(n_{ij}-e_{ij})^2}{e_{ij}}. \tag{25.71}$$

The asymptotic equivalence of X^2 to the LR test $-2\log l$ and the fact that by **22.7** the latter is asymptotically a χ^2 variable with degrees of freedom equal to the number of constraints imposed by H_0 show that X^2 is asymptotically a χ^2 variable with 1 d.fr., since the single constraint (25.67) implies the set of four expected frequencies.

Expression (25.71) may be simplified in this example, for clearly each $(n_{ij} - e_{ij})^2$ takes the same value, say D^2. Thus

$$\begin{aligned} X^2 = D^2\sum_i\sum_j\frac{1}{e_{ij}} &= n\left(n_{11}-\frac{n_{1\cdot}n_{\cdot 1}}{n}\right)^2\left(\frac{1}{n_{1\cdot}n_{\cdot 1}}+\frac{1}{n_{1\cdot}n_{\cdot 2}}+\frac{1}{n_{2\cdot}n_{\cdot 1}}+\frac{1}{n_{2\cdot}n_{\cdot 2}}\right) \\ &= \frac{n(n_{11}n_{12}-n_{12}n_{21})^2}{n_{1\cdot}n_{\cdot 1}n_{2\cdot}n_{\cdot 2}}, \end{aligned} \tag{25.72}$$

a convenient form for computation.

Example 25.6 (Independence in an $r \times c$ table)

The generalization of Example 25.5 to the case where there are r (≥ 2) categories for the attribute A and s (≥ 2) categories for the attribute B is quite straightforward. The probabilities are now

p_{11}	p_{12}	$\cdots$	p_{1c}	$p_{1\cdot}$
p_{21}	p_{22}	$\cdots$	p_{2c}	$p_{2\cdot}$
$\vdots$	$\vdots$		$\vdots$	$\vdots$
p_{r1}	p_{r2}	$\cdots$	p_{rc}	$p_{r\cdot}$
$p_{\cdot 1}$	$p_{\cdot 2}$		$p_{\cdot c}$	1

instead of (25.63), while (25.64) is replaced by the $r \times c$ table

n_{11}	n_{12}	$\cdots$	n_{1c}	$n_{1\cdot}$
n_{21}	n_{22}	$\cdots$	n_{2c}	$n_{2\cdot}$
$\vdots$	$\vdots$		$\vdots$	$\vdots$
n_{r1}	n_{r2}	$\cdots$	n_{rc}	$n_{r\cdot}$
$n_{\cdot 1}$	$n_{\cdot 2}$	$\cdots$	$n_{\cdot c}$	n

As before, the unconstrained ML estimators are given by (25.65)–(25.66). The hypothesis of independence of A and B now states that there is the same set of probabilities in each of the c columns, i.e.

$$H_0: \frac{p_{i1}}{p_{\cdot 1}} = \frac{p_{i2}}{p_{\cdot 2}} = \cdots = \frac{p_{ic}}{p_{\cdot c}}, \qquad i = 1, 2, \ldots, r. \tag{25.73}$$

For each i, H_0 imposes $c-1$ constraints, and (25.73) for any $r-1$ values of i implies (25.73) for the other value of i. Thus there are $(r-1)(c-1)$ constraints imposed by H_0.

As before, the ML estimators of the marginal probabilities are unaffected by the hypothesis of independence. Since (25.73) implies (25.68) as before, we have (25.69) once again for the constrained ML estimators of the individual probabilities. X^2 will now have $(r-1)(c-1)$ d.fr. in its limiting χ^2 distribution, and may be written, from the extension of (25.71) to r rows, c columns, as

$$X^2 = n\left\{\sum_{i=1}^{r}\sum_{j=1}^{c}\frac{n_{ij}^2}{n_{i.}n_{.j}} - 1\right\}. \tag{25.74}$$

Continuity corrections

25.33 Yates (1934) proposed a correction to improve the continuous asymptotic approximation to the discrete distribution of X^2. This is

$$X_c^2 = \frac{n(|n_{11}n_{22} - n_{12}n_{21}| - \frac{1}{2}n)^2}{n_{1.}n_{.1}n_{2.}n_{.2}}, \tag{25.75}$$

equivalent to decreasing (increasing) n_{11} and n_{22} and increasing (decreasing) n_{12} and n_{21} by $\frac{1}{2}$ each when $n_{11}n_{22}$ is greater (less) than $n_{12}n_{21}$.

Whether or not to use (25.75) in practice continues to be a matter for debate; see, for example, Yates (1984), Haviland (1990) and the ensuing discussions on both papers. Several issues have been raised and we will try to deal with these in turn. First of all, however, we note that the X^2 statistic is simply the square of

$$X = [n_{11} - (n_{1.}n_{.1}/n)]/[n_{1.}n_{.1}n_{2.}n_{.2}/n^3]^{1/2}. \tag{25.76}$$

Under H_0, $E(X) = 0$ $\text{var}(X) = 1$ and treating X as a $N(0, 1)$ variate is akin to treating X^2 as $\chi^2(1)$. Further, under H_0, when we condition on *both* sets of marginal totals, n_{11} has a hypergeometric distribution, as in **5.13**. The calculation of probabilities using the hypergeometric yields *Fisher's exact test*, and clearly such calculations are now rather trivial for 2×2 tables, although they remain sufficiently convoluted for high-order tables that simulation is often preferred as the way of evaluating the distribution under H_0.

We now turn to the various issues that have been raised:

(1) *Should we condition on both sets of marginal totals*? Although Pearson originally did not apply a conditional argument, Fisher's arguments have convinced most statisticians to condition on both margins. Nevertheless, there may occasionally be circumstances where repeated samples are taken with one set of margins fixed and the other random. In such cases, the less restrictive conditioning may be appropriate (cf. Haviland, 1990).

(2) *Should we use Fisher's exact test*? At first sight, this question seems trivial, since an exact calculation is surely to be preferred. However, many authors now favour using the mid-p value defined for a single tail as (if $X_{obs} > 0$):

$$\text{mid-}p = P(X > X_{obs}|H_0) + \tfrac{1}{2}P(X = X_{obs}|H_0), \tag{25.77}$$

where X is given by (25.76), with a comparable expression for $X_{obs} < 0$. Clearly (25.77) could be expressed directly in terms of n_{11}. The argument in favour of the mid-p is that the exact test is unduly conservative; see, for example, Yates (1984) and Routledge (1994). When the mid-p value is given by (25.77), the two-tailed value is obtained simply by doubling the single-tail value.

(3) *Should we use Yates' continuity correction*? By design, (25.75) gives a χ^2 approximation that is closer to Fisher's exact test than the original X^2 statistic. However, if we elect to use the mid-p value, the original statistic leads to a better approximation than the corrected version.

At this point, we tentatively conclude that using X^2 without the correction is to be preferred, but since the controversy has continued for almost 100 years, we hesitate to claim a final settlement.

Lack of power and the use of signs of deviations

25.34 Apart from the difficulties we have already discussed in connection with X^2 tests, which are not very serious, they have been criticized on two counts. In each case, the criticism is of the power of the test. Firstly, the fact that the essential underlying device is the reduction of the problem to a multinomial distribution problem itself implies the necessity for grouping the observations into classes. In a broad general sense, we must lose information by grouping in this way, and we suspect that the loss will be greatest when we are testing the fit of a continuous distribution. Secondly, the fact that the X^2 statistic is based on the *squares* of the deviations of observed from expected frequencies implies that the X^2 test will be insensitive to the pattern of signs of these deviations, which is clearly informative. The first of these criticisms is the more radical, since it must clearly lead to the search for other test statistics to replace X^2, but we postpone discussion of such tests until after we have discussed the second criticism.

Let us consider how we should expect the pattern of deviations (of observed from expected frequencies) to behave in some simple cases. Suppose that a simple hypothesis specifies a continuous unimodal distribution with location and scale parameters, say equal to the mean and standard deviation; and suppose that the hypothetical mean is too high. For any set of k classes, the p_{0i} will be too small for low values of the variate, and too large thereafter. Since in large samples the observed proportions will converge stochastically to the true probabilities, the pattern of signs of observed deviations will be a series of positive deviations followed by a series of negative deviations. If the hypothetical mean is too low this pattern is reversed.

Suppose now that the hypothetical value of the scale parameter is too low. The pattern of deviations in large samples is now a series of positives, followed by a series of negatives, followed by positives again. If the hypothetical scale parameter is too high, all these signs are reversed.

Now of course we do not knowingly use the X^2 test for changes in location or scale alone, since we can then find more powerful test statistics. However, when there is error in both location and scale parameters, the situation is essentially unchanged; we shall still have three (or in more complicated cases, somewhat more) 'runs' of signs of deviations. More generally, whenever the parameters have true values differing from their hypothetical values, or when the true distributional form is one differing 'smoothly' from the hypothetical form, we expect the signs of deviations to cluster in this way instead of being distributed normally, as they should be if the expected frequencies were the true ones.

This observation suggests that we may supplement the X^2 test with a test of the number of runs of signs among the deviations, small numbers forming the critical region. The elementary theory of runs necessary for this purpose is given as Exercise 25.8. Before we can use it in any precise way, however, we must investigate the relationship between the 'runs' test and the X^2 test. David (1947), Seal (1948) and Fraser (1950) showed that when H_0 holds the tests are asymptotically independent (cf. Exercise 25.7) and that for testing the simple hypothesis all patterns of signs are equiprobable, so that the distribution theory of Exercise 25.8 can be combined with the X^2 test as indicated in Exercise 25.9.

The supplementation by the 'runs' test is likely to be valuable in increasing sensitivity when testing a simple hypothesis, as in the illustrative discussion above. For composite hypotheses, of particular interest where tests of fit are concerned, when all parameters are to be estimated from the sample, it is of no practical value, since the patterns of signs of deviations, although independent of X^2, are not equiprobable as in the simple hypothesis case, and the distribution theory of Exercise 25.8 is therefore of no use (cf. Fraser, 1950).

Other tests of fit

25.35 We now turn to the discussion of alternative tests of fit. Since these have striven to avoid the loss of information due to grouping suffered by the X^2 test, they cannot avail themselves of multinomial simplicities, and we must expect their theory to be more difficult. Before we discuss the most important test, we remark on a feature they have in common.

It will have been noticed that, when using X^2 to test a simple hypothesis, its distribution is asymptotically χ^2_{k-1} *whatever the composite hypothesis may be*, although its exact distribution does depend on the hypothetical distribution specified. It is clear that this result is achieved because of the intervention of the multinomial distribution and its tendency to joint normality. Moreover, the same is true of the composite hypothesis situation if multinomial ML estimators are used – in this case $X^2 \to \chi^2_{k-s-1}$ *whatever the composite hypothesis may be*, though its exact distribution is even more clearly seen to be dependent on the composite hypothesis concerned. When other estimators are used (even when fully efficient ordinary ML estimators are used) these pleasant asymptotic properties do not hold in general: even the asymptotic distribution of X^2 now depends on eigenvalues of the matrix (25.37), which are in general functions both of the hypothetical distribution and of the values of the parameters θ.

We express these results by saying that, in the first two instances above, the distribution of X^2 is asymptotically *distribution-free* (i.e. free of the influence of the hypothetical distribution's form and parameters), whereas in the third instance it is not asymptotically distribution-free or even *parameter-free* (i.e. free of the influence of the parameters of F_0 without being distribution-free).

25.36 The most important alternative tests of fit all make use, directly or indirectly, of the *probability-integral transform*, which we have encountered on various occasions (e.g. **1.27**) as a way of mapping any known continuous distribution onto the interval (0, 1). In our present context, if we have a simple hypothesis of fit specifying a d.f. $F_0(x)$, to which a density of $f_0(x)$ corresponds, then the variable $y = \int_{-\infty}^{x} f_0(u)\,du = F_0(x)$ is uniformly distributed on (0, 1). Thus if we have a set of n observations x_i and transform them to a new set y_i by the probability-integral transform for a known $F_0(x)$, and use a function of the y_i to test the departure of the y_i from uniformity, the distribution of the test statistic will be distribution-free, not merely asymptotically but for any n.

When the hypothetical distribution is composite, say $F_0(x|\theta_1, \theta_2, \ldots, \theta_s)$ with the s parameters θ_j to be estimated, we must select s functions $t_1, \ldots, t_s$ of the x_i for this purpose. the transformed variables are now

$$y_i = \int_{-\infty}^{x_i} f_0(u|t_1, t_2, \ldots, t_s)\, du,$$

but they are neither independent nor uniformly distributed, and their distribution will depend in general both on the hypothetical distribution F_0 and on the true values of its parameters, as David and Johnson (1948) showed in detail. However (cf. Exercise 25.10), if F has only parameters of location and scale, suitably invariantly estimated, the distribution of the y_i will depend on the form of F but not on its parameters. It follows that for finite n, no test statistic based on y_i can be distribution-free for a composite hypothesis of fit (although it may be parameter-free if only location and scale parameters are involved). Of course, such a test statistic may still be asymptotically distribution-free.

Test of fit based on the sample distribution function: Kolmogorov's D_n

25.37 The tests of fit we now consider are all functions of the cumulative distribution of the sample, or *sample distribution function*, defined by

$$S_n(x) = \begin{cases} 0, & x < x_{(1)}, \\ \frac{r}{n}, & x_{(r)} \le x < x_{(r+1)}, \\ 1, & x_{(n)} \le x. \end{cases} \tag{25.78}$$

The $x_{(r)}$ are the order statistics, discussed in Chapter 14. That is, the observations are arranged so that

$$x_{(1)} \le x_{(2)} \le \cdots \le x_{(n)}.$$

$S_n(x)$ is simply the proportion of the observations not exceeding x. If $F_0(x)$ is the true d.f., fully specified, from which the observations come, we have, for each value of x, from the strong law of large numbers (**8.46**) that

$$\lim_{n\to\infty} P\{S_n(x) = F_0(x)\} = 1, \tag{25.79}$$

and in fact stronger results are available concerning the convergence of the sample d.f. to the true d.f.

In a sense, (25.79) is the fundamental relationship on which all statistical theory is based. If something like it did not hold, there would be no point in random sampling. In our present context, it is clear that a test of fit can be based on any measure of divergence of $S_n(x)$ from $F_0(x)$.

Durbin (1972) surveys the field of tests based on the sample distribution function. The most important is based on deviations of the sample d.f. $S_n(x)$ from the completely specified continuous hypothetical d.f. $F_0(x)$. The measure of deviation used, however, is simple, being the maximum absolute difference between $S_n(x)$ and $F_0(x)$. Thus we define

$$D_n = \sup_x |S_n(x) - F_0(x)|. \tag{25.80}$$

The appearance of the absolute value in the definition (25.80) leads us to expect difficulties in the investigation of the distribution of D_n, but remarkably enough, the asymptotic distribution was obtained by Kolmogorov (1933) when he first proposed the statistic; he showed that:

$$\lim_{n\to\infty} P\{D_n > zn^{-1/2}\} = 2\sum_{r=1}^{\infty}(-1)^{r-1}\exp\{-2r^2z^2\}. \tag{25.81}$$

Smirnov (1948) tabulates (25.81) (actually its complement) for $z = 0.28\ (0.01)\ 2.50\ (0.05)\ 3.00$ to 6 d.p. or more. This is the whole effective range of the limiting distribution.

As well as deriving the limiting result (25.81), Kolmogorov (1933) gave recurrence relations for finite n, which have since been used to tabulate the distribution of D_n. Birnbaum (1952) gives tables of $P\{D_n < c/n\}$ to 5 d.p., for $n = 1\ (1)\ 100$ and $c = 1\ (1)\ 15$, and inverse tables of the values of D_n for which this probability is 0.95 for $n = 2\ (1)\ 5\ (5)\ 30\ (10)\ 100$ and for which the probability is 0.99 for $n = 2\ (1)\ 5\ (5)\ 30\ (10)\ 80$. Miller (1956) gives inverse tables for $n = 1\ (1)\ (100)$ and probabilities 0.90, 0.95, 0.98, 0.99. Massey (1950; 1951) had previously given $P\{D_n < c/n\}$ for $n = 5\ (5)\ 80$ and selected values of $c \leq 9$, and also inverse tables for $n = 1\ (1)\ 20\ (5)\ 35$ and probabilities 0.80, 0.85, 0.90, 0.95, 0.99.

It emerges that the critical values of the asymptotic distribution are $1.3581n^{-1/2}$ for $\alpha = 0.05$ and $1.6276n^{-1/2}$ for $\alpha = 0.01$ and that these are always greater than the exact values for finite n. The approximation for these values of α is satisfactory at $n = 80$.

Confidence limits for distribution functions

25.38 Because the distribution of D_n is distribution-free and adequately known for all n, and because it uses as its measure of divergence the maximum absolute deviation between $S_n(x)$ and $F_0(x)$, we may reverse the procedure of testing for fit and use D_n to set confidence limits for a (continuous) distribution function *as a whole*. For, *whatever the true* $F(x)$, we have, if d_α is the critical value of D_n for test size α,

$$P\{D_n = \sup_x |S_n(x) - F(x)| > d_\alpha\} = \alpha.$$

We may invert this into the confidence statement

$$P\{S_n(x) - d_\alpha \leq F(x) \leq S_n(x) + d_\alpha, \text{ for all } x\} = 1 - \alpha. \tag{25.82}$$

Thus we simply set up a band of width $\pm d_\alpha$ around the sample d.f. $S_n(x)$, and there is a probability $1-\alpha$ that the true $F(x)$ lies *entirely* within this band. This is a remarkably simple and direct method of estimating a distribution function. No other test of fit permits this inversion into confidence limits since none uses so direct and simply interpretable a measure of divergence as D_n.

One can draw useful conclusions from this confidence limits technique as to the sample size necessary to approximate a d.f. closely. For example, from the critical values given at the end of **25.37**, it follows that a sample of 100 observations would have probability 0.95 of having its sample d.f. everywhere within 0.1358 of the true d.f. To be within 0.05 of the true d.f. everywhere, with probability 0.99, would require a sample size of $(1.6276/0.05)^2$, i.e. more than 1000.

Noether (1963) shows that the left-hand side of (25.82) holds with probability at least $1 - \alpha$ for discrete distributions. Thus the D_n t0est is then also conservative. Conover (1972) and Gleser (1985) give exact results for the discrete case and Wood and Altavela (1978) some asymptotic results.

Exercise 25.15 shows how to set somewhat analogous conservative confidence intervals, using X^2, for all the theoretical probabilities p_{0i} simultaneously.

Using the fact that symmetry of a d.f. about a known point θ implies

$$F(x) = 1 - F(2\theta - x),$$

Schuster (1973) gives a Kolmogorov-type test for this case that exactly halves the width of the confidence band (25.82). Rao *et al.* (1975) give an estimator of the centre of symmetry θ based on D_n, and Schuster (1975) an estimator of the d.f. See also Doksum *et al.* (1977), Koziol (1980), Randles *et al.* (1980) and Breth (1982).

25.39 Because it is a modular quantity, D_n does not permit us to set one-sided confidence limits for $F(x)$, but we may consider positive deviations only and define

$$D_n^+ = \sup_x\{S_n(x) - F_0(x)\},$$

as was done by Wald and Wolfowitz (1939) and Smirnov (1939a).

The asymptotic distribution of the variable $y = 2n(D_n^+)^2$ is exponential:

$$f(y) = \exp(-y) \qquad 0 \leq y < \infty.$$

Alternatively, we may express this by saying that $2y = 4n(D_n^+)^2$ is asymptotically a χ^2 variate with 2 d.fr. Evidently, exactly the same theory will hold if we consider only negative deviations and a statistic D_n^-.

Birnbaum and Tingey (1951) give an expression for the exact distribution of D_n^+, and tabulate the values it exceeds with probabilities 0.10, 0.05, 0.01, 0.001, for $n = 5, 8, 10, 20, 40, 50$. As for D_n, the asymptotic values exceed the exact values, and the differences are small for $n = 50$.

We may evidently use D_n^+ to obtain one-sided confidence regions which take the form $P\{S_n(x) - d_\alpha^+ \leq F(x)\} = 1 - \alpha$, where d_α^+ is a critical value of D_n^+.

Stephens (1970; 1974) gives simple modifications of the statistics D_n, D_n^+ which permit very compact tables of critical values (reproduced in the *Biometrika Tables*, Vol. II, Table 54) and makes some power comparisons.

Probability plots

25.40 When H_0 specifies that the distribution function is $F_0(x)$, we may plot the expected quantiles $F_0^{-1}(p_r)$ against the observed $x_{(r)}$, where $p_r = (r + a)/(n + b)$, with a and b chosen to avoid problems at the terminals of the range. Common choices are $(a = -\frac{1}{2}, b = 0)$ or $(a = 0, b = 1)$; the differences are very slight unless n is small.

The advantage of such plots, known as Q–Q plots (Q = quantile) is that the observations should lie close to a straight line when H_0 is true. The confidence limits computed in **25.38** are readily superimposed so that systematic departures from H_0 can be detected.

Example 25.7 (A Q–Q plot for testing normality)

Figure 25.1 shows a Q–Q plot for a sample of $n = 100$ observations taken from a $\chi^2(6)$ distribution. The null hypothesis is that the distribution is normal, and 95 per cent confidence intervals are plotted.

The systematic departures from H_0 are readily evident.

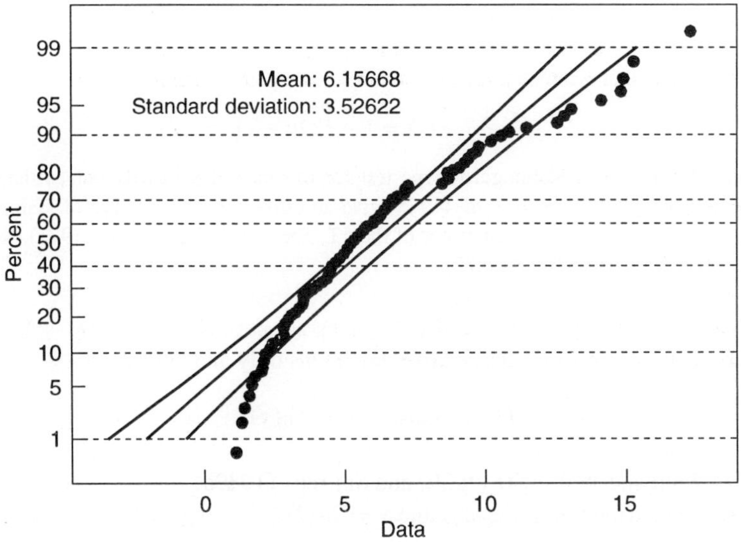

Fig. 25.1 Probability plot for n = 100 observations from chi-square (6)

Gan and Koehler (1990) have examined the use of plots of p_r against $F_0(z_{(r)})$, where $z_{(r)} = (x_{(r)} - \hat{\mu})/\hat{\sigma}$ to make the plots location and scale invariant. Compared to Q–Q plots, tests based upon these P–P plots are more efficient for light to moderate tail-weight alternatives, but less efficient for heavy tails.

Comparison of D_n and X^2 tests

25.41 Using calculations made by Williams (1950), Massey (1951) compared the values of Δ for which the large-sample powers of the X^2 and the D_n tests are at least 0.5. For test size $\alpha = 0.05$, the D_n test can detect with power 0.5 a Δ about half the magnitude of that which the X^2 test can detect with its power; even with $n = 200$, the ratio of Δs is 0.6, and it declines steadily in favour of D_n as n increases. For $\alpha = 0.01$ the relative performances are very similar. Since this comparison is based on a poor lower bound to the power of D_n, we must conclude that D_n is a very much more sensitive test for the fit of a continuous distribution.

Kac *et al.* (1955) point out that if the Mann–Wald equal-probabilities procedure of **25.28–29** is used, the X^2 test requires Δ to be of order $n^{-2/5}$ to attain power $\frac{1}{2}$, whereas D_n requires Δ to be of order $n^{-1/2}$. Thus D_n asymptotically requires sample size to be of order $n^{4/5}$ compared to n for the X^2 test, and is asymptotically very much more efficient – in fact the relative efficiency of X^2 will tend to zero as n increases.

Kallenberg and Kourouklis (1992) show that the D_n test for a simple H_0 is asymptotically efficient in terms of a criterion based upon the Kullback–Liebler information measure. Koehler and Gan (1990) show numerically that the performance of X^2 may approach that of D_n for an optimal (unequal) choice of cell partitions, but is inferior otherwise.

Computation of D_n

25.42 If we are setting confidence limits for the unknown $F(x)$, no computations are required beyond the simple calculation of $S_n(x)$ and the setting of bounds distant $\pm d_\alpha$ from it. In using D_n for testing, however, we have to face the possibility of calculating $F_0(x)$ for every observed value of x.

Although now mostly of historical interest, a considerable saving of labour may be effected in manual calculations, as in the following example due to Birnbaum (1952).

Example 25.8 (Computation of D_n statistic)

Consider a sample of 40 observations arranged in ascending order: 0.0475, 0.2153, 0.2287, 0.2824, 0.3743, 0.3868, 0.4421, 0.5033, 0.5945, 0.6004, 0.6255, 0.6331, 0.6478, 0.7867, 0.8878, 0.8930, 0.9335, 0.9602, 1.0448, 1.0556, 1.0894, 1.0999, 1.1765, 1.2036, 1.2344, 1.2543, 1.2712, 1.3507, 1.3515, 1.3528, 1.3774, 1.4209, 1.4304, 1.5137, 1.5288, 1.5291, 1.5677, 1.7238, 1.7919, 1.8794. We wish to test, with $\alpha = 0.05$, whether the true $F_0(x)$ is normal with mean 1 and variance 1/6. From Birnbaum's (1952) tables we find for $n = 40, \alpha = 0.05$ that $d_\alpha = 0.2101$. Consider the smallest observation, $x_{(1)}$. To be acceptable, $F_0(x_{(1)})$ should lie between 0 and d_α, i.e. in the interval (0, 0.2101). The observed value of $x_{(1)}$ is 0.0475, and from the normal d.f. we find $F_0(x_{(1)}) = 0.0098$, within the above interval, so the hypothesis is not rejected by this observation.

Furthermore it cannot possibly be rejected by the next higher observations until we reach an $x_{(i)}$ for which *either* (a) $i/40 - 0.2101 > 0.0098$, i.e. $i > 8.796$, *or* (b) $F_0(x_{(i)}) > 0.2101 + 1/40$, i.e. $x_{(i)} > 0.7052$ (from the tables again). The 1/40 is added on the right of (b) because we know that $S_n(x_{(i)}) \geq 1/40$ for $i > 1$. Now from the data, $x_{(i)} > 0.7052$ for $i \geq 14$. We next need, therefore, to examine $i = 9$ (from the inequality (a)). We find there the acceptance interval $F_0(x_{(9)})$,

$$(S_9(x) - d_\alpha, S_8(x) + d_\alpha) = (9/40 - 0.2101, 8/40 + 0.2101) = (0.0149, 0.4101).$$

Now $F_0(x_{(9)}) = F_0(0.5945) - 0.1603$, which is acceptable. To reject H_0, we now require either

$$i/40 - 0.2101 > 0.1603, \qquad \text{i.e. } i > 14.82$$

or

$$F_0(x_{(i)}) > 0.4101 + 1/40, \quad \text{i.e. } x_{(i)} > 0.9052, \quad \text{i.e. } i \geq 17.$$

We therefore proceed to $i = 15$, and so on. The reader should verify that only the six values $i = 1, 9, 15, 21, 27, 34$ require computations in this case. The hypothesis is accepted because in every one of these six cases the value of F_0 lies in the confidence interval; it would have been rejected, and computations ceased, if any one value had lain outside the interval.

25.43 For a composite hypothesis of fit with unspecified parameters, Kolmogorov-type tests were investigated by Durbin (1975). By **25.36**, the tests are not now distribution-free, but will be parameter-free if the parameters are those of location and scale.

The most important special cases are the normal distribution, discussed in **25.46**, and the exponential distribution with unknown scale parameter. For the latter, Durbin (1975) tabulated the percentage points of D_n^+, D_n^- and D_n up to $n = 100$ – Iman (1982) gives them in graphical form for $\alpha = 0.01, 0.05, 0.10$ and $n = 5, 10, 20, 30, 50, 100$; Stephens' (1970) approximations for D_n in the composite case are adequate for $n \geq 5$, being $1.09n^{-1/2}$ and $1.31n^{-1/2}$ asymptotically for $\alpha = 0.05, 0.01$, or about four-fifths of the simple H_0 values in **25.37**. See also Margolin and Maurer (1976). Stephens (1976) obtains asymptotic percentage points.

Stephens (1977) and Chandra *et al.* (1981) gives tests for the extreme-value distribution of Exercise 18.6 and the Weibull distributions of **5.33**, and Stephens (1979a) for the logistic distribution.

Other tests using the sample distribution function

25.44 Several other test statistics have been developed that depend upon $S_n(x)$, the sample distribution function. In particular, we may define the Cramér–von Mises family of statistics:

$$W_n = \int_{-\infty}^{\infty} [S_n(x) - F(x)]^2 \psi(x)\, \mathrm{d}F(x), \tag{25.83}$$

where $\psi(x)$ is a non-negative weighting function. The cases that have received the most attention in the literature correspond to $\psi(x) = 1$ (Cramér, 1928; von Mises, 1931) and $\psi(x) = [F(x)\{1 - F(x)\}]^{-1}$ (Anderson and Darling, 1952).

For computational purposes, if we put

$$z_i = F(x_{(i)}), \qquad i = 1, \ldots, n,$$

the Cramér–von Mises statistic is given by

$$W_n(\text{CVM}) = (12n)^{-1} + \sum_{i=1}^{n} [z_i - (2i-1)/n]^2 \tag{25.84}$$

whereas the Anderson–Darling statistic is

$$W_n(\text{AD}) = -n - n^{-1}\left\{\sum_{i=1}^{n} (2i-1)[\ln z_i + \ln(1 - z_{n+1-i})\right\}. \tag{25.85}$$

Volume II of the *Biometrika Tables* (Table 54) provides approximate percentage points for these statistics. The asymptotic values are (multiply table entry by n^{-1}):

	5%	1%
CVM, simple H_0	0.461	0.743
composite H_0 (normal)	0.126	0.178
AD, simple H_0	2.492	3.857
composite H_0 (normal)	0.787	1.092

By composite H_0 (normal), we mean that the test is carried out for the normal distribution using the ML estimators for the unknown parameters. When greater accuracy is required, Czorgo and Faraway (1996) present a first-order correction to the asymptotic distribution of the Cramér–von Mises statistic that is remarkably accurate for values of n as low as 7. Baglivo *et al.* (1992) develop a general procedure for carrying out exact tests using functions of the sample distribution function. Their general procedure is to invert the characteristic function of the distribution of the test statistic using the fast Fourier transform.

See also Durbin (1972) and Stephens (1976) for earlier work and the volume edited by D'Agostino and Stephens (1986).

As may be expected from the form of the weighting function, the Anderson–Darling statistic is more sensitive to departures in the tail areas of the distribution.

Smooth goodness-of-fit tests

25.45 Rayner and Best (1989) develop a class of smooth or score tests of fit, building on the original ideas of Neyman (1937) and Rao (1948); see **22.42**. Given that the p.d.f. under H_0 is $f(x)$, an *order-k alternative* p.d.f. may be written as

$$g_k(x) = C(\theta)\exp\left\{\sum_{i=1}^{k}\theta_i h_i(x)\right\}f(x), \tag{25.86}$$

where $C(\theta)$ is the normalizing constant, the θ_i are parameters (all $\theta_i = 0$ under H_0) and the $h_i(x)$ are orthonormal functions defined with respect to $f(x)$, such as the Chebyshev–Hermite polynomials for the normal, given in **6.14**. Rayner and Best (1989) show that the score statistic for testing (25.85) is

$$S_k = \sum_{i=1}^{k} u_i^2, \tag{25.87}$$

where

$$u_i = \sum_{j=1}^{n} h_i(x_j)/n^{1/2}. \tag{25.88}$$

Under H_0, S_k is asymptotically distributed as χ_k^2. An attraction of the score tests is that they can be designed to be responsive to particular alternatives; Rayner and Best (1989) report encouraging power comparisons for the normal and exponential cases.

Langholz and Kronmal (1991) develop smoothed chi-square tests for composite hypotheses using a Fourier series approach.

Tests of normality

25.46 We now consider the specific question of testing whether the underlying distribution is normal, given the central importance of the normality assumption in statistical inference. Any of the tests discussed earlier may be used, although a number of specific alternatives have been developed. In particular, it is common to test the observed moment ratios $\sqrt{b_1}$ and b_2 against their distributions given the hypothesis of normality (cf. **12.18** and Exercises 12.9–10 and the percentage points given in the *Biometrika Tables*, Vol. II, Tables 16–18) and these are sometimes called 'tests of normality', although they are better called tests of skewness and kurtosis, respectively.

D'Agostino and Pearson (1973) present a combined test of the form

$$K^2 = Z^2(\sqrt{b_1}) + Z^2(b_2) \tag{25.89}$$

where $Z(\cdot)$ represents the *normal equivalent deviate*; that is, the standardized normal deviate corresponding to the percentage point of the observed statistic with its own distribution. Despite the construction, under the null hypothesis of normality K^2 is only asymptotically distributed as

$\chi^2(2)$. Bowman and Shenton (1975) provide graphs that enable the percentage points of K^2 to be evaluated for $\alpha = 0.10$, 0.05 and 0.01. The statistic

$$B = (\sqrt{b_1})^2/\sigma_1^2 + (b_2 - 3)^2/\sigma_2^2, \tag{25.90}$$

where $\sigma_1^2 = 6/n$ and $\sigma_2^2 = 24/n$ are the large-sample variances, is asymptotically $\chi^2(2)$. The statistic B has become quite popular in recent years and is often attributed to Jarque and Bera (1980), although it first appeared in Bowman and Shenton (1975). Bowman and Shenton argued that the convergence to the asymptotic distribution was too slow and recommended normalizing transforms based upon the Johnson curves (**6.27–36**) before forming the sum of squares.

Referring back to the tests considered in **25.45**, we may show that, when $k = 2$, S_2 in (25.87) corresponds to (25.89); note that h_1 and h_2 refer to the third and fourth order polynomials, respectively.

25.47 We now briefly summarize research on the use of standard goodness-of-fit tests for examining the normal hypothesis.

Moore (1971) gives values of X^2 for testing normality with $\alpha = 0.001$, 0.005, 0.01, 0.05, 0.10, 0.25 and $k = 5, 7, 9, 11, 15, 21$ when fully efficient ML estimators are used to determine boundaries with estimated equal probabilities in classes.

Moore (1982) shows that if the observations are normal but positively dependent, the X^2 test tends to reject the hypothesis of normality. This holds also for non-normal distributions – see Gleser and Moore (1983).

Kac *et al.* (1955) discuss the distributions of D_n in testing normality when the two parameters (μ, σ^2) are estimated from the sample by $(\bar{x}, s^2)$. The limiting distributions are parameter-free (because μ and σ are location and scale parameters – cf. **25.36**) but are not obtained explicitly. Lilliefors (1967) gave critical values of D_n in testing normality, recomputed to much greater accuracy by Dallal and Wilkinson (1986) for $\alpha = 0.001$, 0.01, 0.05, (0.05) 0.20 and $n = 4$ (1) 20 (5) 30, 40, 100, 400, 900, with an analytic approximation – Iman (1982) gives graphs for $\alpha = 0.01, 0.05, 0.10$, and $n = 5$ (5) 20, 30, 50, 100. These critical values are roughly two-thirds of the simple H_0 values in **25.37** – e.g. for $\alpha = 0.05$, the value is $0.886n^{-1/2}$ and for $\alpha = 0.01$ it is $1.031n^{-1/2}$.

Shapiro and Wilk (1965) give a criterion for testing normality based on the regression of the order statistics upon their expected values, using the theory of **29.16–17** and extensive sampling experiments to establish its distribution. It is defined by $W = (\sum_{i=1}^n a_i x_{(i)})^2 / \sum_{i=1}^n (x_i - \bar{x})^2$, where the a_i are tabulated coefficients. Small values of W are critical. Leslie *et al.* (1986) study its asymptotic distribution and establish its consistency. See also Verrill and Johnson (1987; 1988), who give tables and large-sample theory for censored samples. Tables of W are given in the *Biometrika Tables*, Vol. II, Table 54. Shapiro and Francia (1972) give a simplified approximate version, W', for large samples which is proved consistent by Sarkadi (1975) – see also Weisberg (1974).

Davis and Stephens (1989) provide computer algorithms for the Kolmogorov, Cramér–von Mises and Anderson–Darling tests for the normal and exponential distributions when the parameters are unknown. Royston (1982; 1995) provides an algorithm for the Shapiro–Wilk test that extends to include censored data.

Power comparisons

25.48 Shapiro *et al.* (1968) and Stephens (1974) make power comparisons from extensive sampling experiments and show that W is usually somewhat superior to the other tests given in this chapter for testing normality, although $\sqrt{b_1}$ and b_2 together are sensitive to non-normality. In sampling experiments using many different populations, Pearson *et al.* (1977) compare the powers of W, W' and tests based on $\sqrt{b_1}$, b_2 and K^2. W does well for symmetric populations with $\beta_2 < 3$, but one-tailed b_2 tests are better; for $\beta_2 > 3$, W performs poorly. Similarly, one-tailed $\sqrt{b_1}$ is recommended for skew populations, though W does well compared with other omnibus tests. The K^2 test appears to do well against a range of alternatives. Saniga and Miles (1979) also recommend b_2 and $\sqrt{b_1}$ against asymmetric stable alternatives.

A simple test of normality against asymmetric alternatives, based on the characteristic independence of the sample mean and variance (cf. Exercise 12.21), is proposed by Lin and Mudholkar (1980); its power compares well with those of the other tests above.

Doksum *et al.* (1977) discuss plots and tests for assessing symmetry, some of which compared favourably in power with W above.

Mudholkar *et al.* (1991) provide a graphical procedure for comparing goodness-of-fit tests which enables the user to determine the areas of relative strength of one test over another for different alternative hypotheses.

Tests for multivariate normality

25.49 Throughout this chapter we have concentrated upon univariate goodness-of-fit tests, but there is increasing interest in multivariate tests, particularly for the multinormal distribution. Looney (1995) reviews a number of existing tests and concludes that the multivariate skewness and kurtosis tests developed by Small (1980) are among the most powerful.

Mudholkar *et al.* (1992) provide a test that exploits the characteristic independence between the mean vector and the covariance matrix and that appears to have reasonable power against heavy-tailed alternatives. Mardia and Kent (1991) develop a score test (cf. **25.45**) for multivariate normality.

EXERCISES

25.1 Show that if, in testing a composite hypothesis, an inconsistent set of estimators $\mathbf{t}$ is used, then the statistic $X^2 \to \infty$ as $n \to \infty$.

(Cf. Fisher, 1924c)

25.2 Using (25.33), show that the matrix $\mathbf{M}$ defined in (25.37) reduces, when the vector of multinomial ML estimators $\widehat{\boldsymbol{\theta}}$ is used, to

$$\mathbf{M} = \begin{pmatrix} \frac{1}{2}\mathbf{I} & \frac{1}{4}\mathbf{C} \\ \mathbf{C}^{-1} & \frac{1}{2}\mathbf{I} \end{pmatrix},$$

and that $\mathbf{M}$ is idempotent with $\operatorname{tr}\mathbf{M} = s$. Hence confirm the result of **25.10** and **25.14** that X^2 is asymptotically distributed as χ^2_{k-s-1} when $\widehat{\boldsymbol{\theta}}$ is used.

(Watson, 1959)

25.3 Show from the limiting joint normality of the n_i that as $n \to \infty$, the variance of the simple hypothesis X^2 statistic in the equal-probabilities case ($p_{0i} = 1/k$) is

$$\operatorname{var}(X^2) = \lim_{n\to\infty}\left[2k^2\left\{\sum_i p_{1i}^2 - \left(\sum_i p_{1i}^2\right)^2\right\} + 4(n-1)k^2\left\{\sum_i p_{1i}^3 - \left(\sum_i p_{1i}^2\right)^2\right\}\right],$$

where $p_{1i}, i = 1, 2, \ldots, k$ are the true class probabilities. Verify that this reduces to the correct value $2(k-1)$ when

$$p_{1i} = p_{0i} = 1/k.$$

(Mann and Wald, 1942)

25.4 Establish the non-central χ^2 result of **25.27** for the alternative hypothesis distribution of the X^2 test statistic.

(Cf. Cochran, 1952)

25.5 Show from the moments of the multinomial distribution (cf. (7.18)) that the exact variance of the simple-hypothesis X^2 statistic is given by

$$n\operatorname{var}(X^2) = 2(n-1)\left\{2(n-2)\sum_i \frac{p_{1i}^3}{p_{0i}^2} - (2n-3)\left(\sum_i \frac{p_{1i}^2}{p_{0i}}\right)^2 \right.$$
$$\left. -2\left(\sum_i \frac{p_{1i}^2}{p_{0i}}\right)\left(\sum_i \frac{p_{1i}}{p_{0i}}\right) + 3\sum_i \frac{p_{1i}^2}{p_{0i}^2}\right\} - \left(\sum_i \frac{p_{1i}}{p_{0i}}\right)^2 + \sum \frac{p_{1i}}{p_{0i}^2}.$$

Show that the limiting results of Exercise 25.3 follow when all $p_{0i} = k^{-1}$. When H_0 holds ($p_{1i} = p_{0i}$), show that

$$n\operatorname{var} X^2 = 2(n-1)(k-1) - k^2 + \sum_i p_{0i}^{-1}$$

exactly, reducing to $n \operatorname{var} X^2 = 2(n-1)(k-1)$ if all $p_{0i} = k^{-1}$.

(Cf. Patnaik, 1949)

25.6 For the same alternative hypothesis as in Example 25.3, namely the gamma distribution with parameter 1.5, compute the p_{1i} for the unequal-probabilities four-class grouping in Example 25.2. Calculate the non-central parameter (25.52) for this case, and show by comparison with Example 25.3 that the unequal-probabilities grouping would require about a 25 per cent larger sample than the equal-probabilities grouping in order to attain the same power against this alternative.

25.7 k independent standardized normal variables x_j are subject to c homogeneous linear constraints. Show that $S = \sum_{j=1}^{k} x_j^2$ is distributed independently of the signs of the x_j. Given that $c = 1$, and that the constraint is $\sum_{j=1}^{k} x_j = 0$, show that all the sequences of signs are equiprobable (except all signs positive, or all signs negative, which cannot occur), but that this is not so generally for $c > 1$. Hence show that any test based on the sequence of signs of the deviations of observed from expected frequencies $n_i - np_{0i}$ is asymptotically independent of the X^2 test when H_0 holds.

(David, 1947; Seal, 1948; Fraser, 1950)

25.8 M elements of one kind and N of another are arranged in a sequence at random ($M, N > 0$). A *run* is defined as a subsequence of elements of one kind immediately preceded and succeeded by elements of the other kind. Let R be the number of runs in the whole sequence ($2 \leq R \leq M + N$). Show that

$$P\{R = 2s\} = 2\binom{M-1}{s-1}\binom{N-1}{s-1}\Big/\binom{M+N}{M},$$

$$P\{R = 2s-1\} = \left\{\binom{M-1}{s-2}\binom{N-1}{s-1} + \binom{M-1}{s-1}\binom{N-1}{s-2}\right\}\Big/\binom{M+N}{M},$$

and that

$$E(R) = 1 + \frac{2MN}{M+N}$$

$$\operatorname{var} R = \frac{2MN(2MN - M - N)}{(M+N)^2(M+N-1)}.$$

(Stevens (1939); Wald and Wolfowitz (1940). Swed and Eisenhart (1943) tabulate the distribution of R for $M \leq N \leq 20$.)

25.9 From Exercises 25.7 and 25.8, show that if there are M positive and N negative deviations ($n_i - np_{0i}$), we may use the runs test to supplement the X^2 test for the simple hypothesis. From Exercise 16.4, show that if P_1 is the probability of a value of X^2 not less than that observed and P_2 is the probability of a value of R not greater than that observed, then $U = -2(\log P_1 + \log P_2)$ is asymptotically distributed as χ^2 with 4 d.fr., large values of U forming the critical region for the combined test.

(David, 1947)

25.10 $x_1, x_2, \ldots, x_n$ are independent random variables with the same distribution $f(x|\theta_1, \theta_2)$. θ_1 and θ_2 are estimated by statistics $t_1(x_1, x_2, \ldots, x_n)$, $t_2(x_1, x_2, \ldots, x_n)$. Show that the random variables

$$y_i = \int_{-\infty}^{x_i} f(u|t_1, t_2)\,du$$

are not independent and that they have a distribution depending in general on f, θ_1 and θ_2; but that if f is of the form $\theta_2^{-1} f\{(x - \theta_1)/\theta_2\}, \theta_2 > 0$, and t_1, t_2 respectively satisfy $t_1(\mathbf{x} + \alpha) = t_1(\mathbf{x}) + \alpha$, $t_2(\beta\mathbf{x}) = \beta t_2(\mathbf{x})$, then $\beta > 0$, then the distribution of y_i is not dependent on θ_1 and θ_2, but on the form of f alone.

(David and Johnson, 1948)

25.11 Show that for testing a composite hypothesis the X^2 test statistic using multinomial ML estimators is asymptotically equivalent to the LR test statistic when H_0 holds.

25.12 Show that when there are $k = 2$ classes and $n = 3$ observations, of which n_1 fall into the first class which has hypothetical probability $p_{01} = \frac{1}{4}$, X^2 has its critical region made up cumulatively of the points $n_1 = 3, n_1 = 2, n_1 = 0, n_1 = 1$ in that order, according to the value of the test size α. Show that if $\alpha < 1$, the critical region is biased.

(Cf. Cohen and Sackrowitz, 1975)

25.13 Show that if the distribution tested is of the form

$$\frac{1}{\theta_2} f\left(\frac{x - \theta_1}{\theta_2}\right), \qquad \theta_2 > 0,$$

the matrix $\mathbf{BDB}^T$ in (25.42) does not depend upon θ_1, θ_2, so that its eigenvalues also do not.

(Cf. Watson, 1958)

25.14 Show that in the equal-probabilities case, X^2 defined in (25.44) varies by multiples of $2k/n$ and hence that we may expect the χ^2 approximation to the distribution of X^2 to be improved by a continuity correction of $-k/n$. Show also that the minimum attainable value of X^2 is

$$\frac{n - k[\frac{n}{k}]}{n}\left\{k\left(\left[\frac{n}{k}\right] + 1\right) - n\right\},$$

where $[z]$ is the integral part of z, and hence is zero only when n is an integral multiple of k.

25.15 If X^2 for the simple or composite H_0 is asymptotically distributed as χ^2 with f degrees of freedom, its $100(1-\alpha)$ percentile being $\chi^2_{f,\alpha}$, show that if $\max|(n_i - np_{0i})/n| = \Delta \leq 0.5$, the minimum possible value of X^2 for a fixed Δ, whatever the n_i and the p_{0i}, is $4n\Delta^2$. Putting $n \geq n_0 = \chi^2_{f,\alpha}/(4\Delta^2)$, show that

$$P\{X^2 \leq 4n\Delta^2\} \geq 1 - \alpha$$

for any set of p_{0i} whatever, and that for sufficiently large n, a conservative set of $100(1-\alpha)$ per cent confidence intervals for all the true p_{0i} simultaneously is given by

$$\left\{\frac{n_i}{n} \pm 0.5(\chi^2_{f,\alpha}/n)^{1/2}\right\}, \qquad i = 1, 2, \ldots, k.$$

(Naddeo, 1968)

25.16 Let $\boldsymbol{\theta}$ be the unspecified parameters in testing a composite hypothesis of fit for n observations $\mathbf{x}$. Suppose that $\mathbf{t}$, with fewer than n components, is minimal sufficient for $\boldsymbol{\theta}$, and that we can make a one-to-one transformation from $\mathbf{x}$ to $(\mathbf{t}, \mathbf{u})$, where $\mathbf{u}$ is distributed independently of $\mathbf{t}$. Show that if the value of $\mathbf{t}$ is discarded, and is replaced by a random observation $\mathbf{t}'$ from its distribution with a known value of $\boldsymbol{\theta}$, then the set of observations $\mathbf{x}'$ obtained by the inverse transformation from $(\mathbf{t}', \mathbf{u})$ is distributed independently of $\boldsymbol{\theta}$, so that the hypothesis of fit becomes simple.

(Durbin, 1961)

25.17 A random sample of n observations u_r is taken from the uniform distribution on the interval (0, 1), dividing that interval into $n + 1$ lengths c_j, where $c_j \geq 0$ and $\sum_{j=1}^{n+1} c_j = 1$. The c_j are ordered so that $c_{(1)} \leq c_{(2)} \leq \cdots \leq c_{(n+1)}$. Show that the non-negative variables

$$\begin{cases} g_1 = (n+1)c_{(1)}, \\ g_j = (n+2-j)(c_{(j)} - c_{(j-1)}), \end{cases} \qquad j = 2, 3, \ldots, n+1; \quad \sum_{j=1}^{n+1} g_j = 1,$$

have the joint density function

$$f(g_1, \ldots, g_n) = n!$$

and that the unordered c_j also have the same distribution, so that

$$f(c_1, \ldots, c_n) = n!$$

(the $(n+1)$th variable being omitted in each case to remove the singularity of the distribution). Hence show that the variables

$$w_r = \sum_{j=1}^{r} g_j, \qquad r = 1, 2, \ldots, n,$$

are distributed exactly as the order statistics of the original sample, $u_{(r)}$, $r = 1, 2, \ldots, n$. Thus any test of fit based on the probability-integral transformation may be applied to the w_r, as well as to the $u_{(r)}$ obtained from the transformation.

(Durbin (1961), who finds from sampling experiments that a one-sided Kolmogorov test (D_n^-) applied to the w_r has better power properties than the ordinary two-sided D_n test for detecting changes in distributional form.)

CHAPTER 26

COMPARATIVE STATISTICAL INFERENCE

26.1 Statistical inference is an inductive process from sample to population. In thinking of a hypothesis (H) and observational data or evidence (E), there is no problem in making probabilistic statements of the form $P(E|H)$; indeed, these are justified by deductive logic once the axioms of probability are specified, and such statements have been used repeatedly throughout these volumes. However, the very existence of inductive statements of the form $P(H|E)$ has been questioned, and many philosophers, notably Sir Karl Popper (1968; 1969), have concluded that such probabilities do not exist. Such probabilities are, of course, the posterior probabilities of the Bayesian approach, so the debate is of vital interest to statisticians. Indeed, it shows no signs of abating, and the interested reader should consult Popper and Miller (1983), Good (1988), Gemes (1989) and Miller (1990) for recent developments.

Any inferential procedure must be based on some more or less rational set of rules, but the rationality of any such given system and the apparent value of the conclusions it allows remain open to debate.

In these volumes, we have adopted the *frequentist* paradigm, sometimes known as the *classical* approach, which has been the dominant school of statistical thought for most of this century. However, the *Bayesian* viewpoint has grown in popularity since the 1950s and several other approaches to inference, some more complete than others, have been developed in recent years.

The original essay by Thomas Bayes (1764) was reproduced in Biometrika in 1958; see the refernces for further details.

26.2 In this chapter, we attempt to outline both the areas of agreement and the principal differences between the major schools; it is not our intent to develop each approach in detail. For example, the companion volume by O'Hagan (1994) provides a comprehensive development of inference from the Bayesian viewpoint. General discussions of comparative inference are given by Barnett (1982), Dawid (1983), and the volume edited by Godambe and Sprott (1971). A more philosophical discussion, supporting the subjective Bayesian approach, appears in Howson and Urbach (1989) and Urbach (1992).

Since our discussion is a rather short appraisal of a large and complex literature, we shall emphasize only the principal points at issue. Thus, we examine 'standard' positions within each school and do not emphasize debates within a school (e.g. choice of axioms for subjective probability). We hope that these broad brush strokes are seen to produce portraits and not caricatures!

A framework for inference

26.3 In general terms, the inferential process contains the following ingredients:

(a) *A measurable (vector) random variable*, X taking on values in the sample space, $\mathcal{X}$.

(b) The *unknown parameter(s)*, θ, which may be partitioned into parameters of direct interest and nuisance parameters, then denoted by θ and ϕ, respectively. The set of possible θ values is defined by the parameter space Ω.

(c) The *population of interest*, which we assume is representable in terms of one of a family of probability distributions $\{F(x, \theta)\}$, indexed by θ. We use

$$F \equiv F(\theta) \equiv F(x, \theta) = P(X \leq x|\theta)$$

interchangeably where no ambiguity arises. The functional form of F may be completely specified or a member of some class of distributions, $\mathcal{F}$.

(d) A *statistical experiment* that produces a set of observations, described by the random vector $X = (X_1, X_2, \ldots, X_n)$, with a particular realization, the *sample data* denoted by $x = \{x_1, x_2, \ldots, x_n\}$. The experimental procedure specifies the mode of sampling and the form of the stopping rule, S, whether or not such information is required.

Our notation will not distinguish vectors and scalars, unless the discussion explicitly requires that the distinction be made.

In addition, there may be historical (or prior) information regarding θ, of either a personal or objective nature, which we summarize by some function $p(\theta)$. Since the specification, use and even existence of such information is a matter for considerable debate, we defer further discussion of this topic until **26.35**. The general form of the inference problem is to use the available information

$$I = \{\mathcal{X}, \Omega, \mathcal{F}, x, S, p\} \tag{26.1}$$

to make inductive statements about θ. We now examine the various approaches to this problem, beginning with an overview of the frequentist approach that we have adopted hitherto. This discussion leads us to consider such issues as ancillarity, conditionality and sufficiency which, in turn, lead to an appraisal of the likelihood approach. Then, after an examination of the fiducial approach, we turn our attention to Bayesian inference and so to decision theory. The chapter concludes with an evaluation of the different approaches and a discussion of attempts at reconciliation between these schools of thought.

The frequentist approach

26.4 The frequency theory of probability, introduced in **8.11**, assumes that it is possible to consider an infinite sequence of independent replications of the same statistical experiment.

We now confine attention primarily to point estimation, but include comments on interval estimation and testing hypotheses as appropriate; we compared Bayesian and fiducial intervals with confidence intervals in **19.48–49**. We may consider a statistic or estimator, $T(X)$, as a summary of the information about θ; for simplicity, we will often restrict attention to a single parameter. In Chapter 17, we identified certain desirable properties for T, such as *consistency* (**17.7**) and *unbiasedness* (**17.9**). Since there is often a multiplicity of estimators satisfying these requirements, we sought measures of *efficiency* (**17.28–29**) and identified desirable estimators as minimum variance unbiased (**17.13–27**). The broader criterion of minimum mean square error (**17.30**) is sometimes felt to be more appropriate and applicable.

Although these criteria may be deemed desirable in themselves, they lack a method for constructing suitable statistics, T. Within the exponential family, the set of sufficient statistics may be identified (**17.31–41**) which leads to the MV unbiased estimator of θ if it exists (**17.35**). More generally, in Chapter 18, we established that the maximum likelihood (ML) estimator obtained as

$$\hat{\theta} = \max_{\theta \in \Omega} L(x|\theta), \tag{26.2}$$

where

$$L(x|\theta) = \prod_{i=1}^{n} f(x_i|\theta), \tag{26.3}$$

is consistent and asymptotically unbiased under mild regularity conditions when the observations are independent and from the same distribution. Furthermore, the ML estimator is a function of the sufficient statistics and is asymptotically MV unbiased.

26.5 Even at this stage, we find some parting of the ways, which the large-sample properties of the ML estimator tend to obscure. If T is an unbiased estimator for θ and g is a non-linear function, then $g(T)$ is not unbiased for $\phi = g(\theta)$, whereas the ML estimator is functionally invariant so that

$$\hat{\theta} = T \Leftrightarrow \hat{\phi} = g(T). \tag{26.4}$$

Example 26.1 (Differences between MV unbiased and ML estimators)

Given a random sample of n observations from a normal population with mean θ and variance 1, consider $\phi = \theta^2$. The ML estimators are

$$\hat{\theta} = \bar{X} \quad \text{and} \quad \hat{\phi} = (\bar{X})^2.$$

However, the MV unbiased estimator for ϕ is

$$T = (\bar{X})^2 - 1/n. \tag{26.5}$$

Although $E(T) = \phi$ and $\phi \geq 0$, it may happen that the observed value of T is negative. Common sense suggests replacing negative values of T by zero, even though this violates the property of unbiasedness. In general, such adjustments produce estimators with smaller m.s.e., so that different criteria may lead to different estimators.

Ad hoc estimators obtained by solving $T = g(\hat{\theta})$, where $E(T) = g(\theta)$, are widely used and justified by appeal to the unbiasedness for $g(\theta)$, even though these estimators are biased for θ unless $g(\theta)$ is a linear function. See, for example, the discussion on the generalized method of moments (GMM) in **18.62–64** and the discussion on unbiased estimating equations in **26.9**.

Stopping rules

26.6 Another component of the information set that affects the unbiasedness property is the stopping rule. Suppose that the random variable describes the presence or absence of an attribute, with $\pi = P(\text{present})$. We know from **9.32** that, if X denotes the number of successes in a sample of fixed size n, the estimator $p = X/n$ has $E(p) = \pi$ and is MV unbiased. However, if we sample until the kth success, observing X failures along the way, the MV unbiased estimator becomes $(k-1)/(k+X-1)$; see Example 9.15. More generally, as we saw in **24.26–31**, it is difficult to obtain exact sampling distributions for sequential procedures.

Censoring mechanisms

26.7 Unbiasedness is clearly a property of the sample space since it depends upon taking the expectation over X. Suppose, for example, that $X \equiv [0, \infty)$, but the exact value of x is recorded only in the range $[0, u]$, otherwise the observation is censored. Any unbiasedness property possessed by an estimator would be destroyed by the censoring.

Example 26.2 (Effect of censoring on unbiasedness)

Suppose that X is exponentially distributed with mean θ. Given a sample of size n, clearly $E(X) = \theta$, but if we now censor the observation at u, it follows that

$$E(X) = g(\theta) = \theta\{1 - \exp(-u/\theta)\}, \tag{26.6}$$

so that the solution of $\bar{X} = g(\hat{\theta})$ now yields an *ad hoc* estimator for θ which is consistent but only asymptotically unbiased.

This led to a famous discourse in Pratt (1962) which we summarize briefly. Suppose that the researcher carries out the experiment as described above with observations censored at u since the measuring instrument will only record in the range $[0, u]$. Then the statistician recommends use of the estimator based on (26.6) even if *no observations were actually censored.* The researcher then reports that, had censoring occurred, another machine was available, but would allow recording up to $u' > u$; the statistician duly modifies the estimator derived from (26.6), with u' in place of u. Finally, the researcher reports that the second machine was broken and the statistician reverts to his first estimator. As Pratt notes, none of these machinations affect what was actually observed and the dependence of the estimator on what was not observed appears unreasonable. This criticism of frequentist methods has been raised by many others over the years (e.g. Jeffreys, 1961).

26.8 In part, the response to this issue relates to the nature of the estimator used. Suppose that n observations are taken with censoring at u and that j observations are actually censored. The likelihood function for $(\mathbf{x}, j)$ may be written as

$$\begin{aligned} L(\mathbf{x}, j|\theta) &= L(\mathbf{x}|j, \theta)L(j|\theta) \\ &= \prod_{i=1}^{n-j}\left\{\frac{f(x_i|\theta)}{1-Q}\right\}\binom{n}{j}Q^j(1-Q)^{n-j}, \end{aligned} \tag{26.7}$$

where

$$Q = Q(\theta) = P(X > u). \tag{26.8}$$

From (26.7), we can see that the statistician has landed in the predicament described by Pratt through use of an *inefficient* estimator; the information contained in $L(j|\theta)$ has been ignored since the estimator based on (26.6) does not depend on j. When the full likelihood in (26.7) is used and $j = 0$, we obtain the original uncensored LF, as can be seen by inspection. Alternatively, one might argue that censoring should only be taken into account when it actually occurs, an application of the conditionality principle discussed in **26.15**.

26.9 The critic of the frequentist approach might justifiably claim that we have saved the 'statistician' only by rejecting the frequency approach. Our response is that the estimator based on (26.6) is not wrong, but *inefficient.* A user of statistical methods must decide upon the properties considered desirable in an estimator and, for example, an overly rigid insistence upon unbiasedness may lead to difficulties.

Nevertheless, the notion of unbiasedness has considerable intuitive appeal and many would be reluctant to abandon it. Therefore, let us take a step backwards and consider the problem in a more general setting. It is evident that many estimation procedures involve finding the global maximum (or minimum) of some function $G(\theta|x)$. Under suitable regularity conditions, this value is given by setting the derivative to zero so that $\hat{\theta}$ is given by

$$G' \equiv G'(\hat{\theta}|x) = 0; \tag{26.9}$$

this is true for both ML and LS estimators, for example. Since $E(G') = 0$ directly from (26.9), we term (26.9) an *unbiased estimating equation* (Godambe, 1960; 1976); see **18.60–61**. This weaker concept of unbiasedness often produces the usual estimators based on fixed size sampling rules; however, *the estimators are unaffected by the stopping rule.*

Example 26.3 (Unbiased estimating equations and uniformly MV unbiased estimators)

The exponential family equation in **17.19** has p.d.f.

$$f(x|\theta) = \exp\{A(\theta)B(x) + C(x) + D(\theta)\}. \tag{26.10}$$

The ML estimators are given by

$$G' = A(\theta)\sum B(x_i) + nD'(\theta) = 0. \tag{26.11}$$

From **17.17**, it is evident that the estimating equation gives the MV unbiased estimator when it exists.

Example 26.4 (Unbiased estimating equations in regression)

The LS estimating equation for the regression model is, from **29.3** and (26.9),

$$\mathbf{X}^T\mathbf{X}\boldsymbol{\theta} - \mathbf{X}^T\mathbf{y} = \mathbf{0}. \tag{26.12}$$

When $\mathbf{X}$ is of full rank and fixed, the standard LS estimator results.

Example 26.5 (Unbiased estimating equation for a Bernoulli process)

Given x successes in n trials for a Bernoulli process with probability of success θ, the likelihood function is

$$L(x|\theta) = \text{const.} - x\log\theta - (n-x)\log(1-\theta),$$

so that $G' = 0$ yields $\hat{\theta} = x/n$. By contrast, the unbiased estimators for fixed and sequential sampling are, respectively, x/n and $(x-1)/(n-1)$; see **9.32** and **9.37**.

26.10 The estimators generated by (26.9) will be optimal when ML is used if there are no nuisance parameters or if it is possible to condition upon statistics that are complete and sufficient for the nuisance parameters (Godambe, 1976). When this approach is not feasible, it is still possible to consider the conditional likelihood (cf. **18.33**). Lindsay (1982) develops a weaker optimality criterion that is met by such estimators.

It might be argued that this revised concept of unbiasedness gives away too much, but since users of unbiased estimators often use them to estimate non-linear functions of the parameters (when unbiasedness is lost), we feel that the estimating equation approach has considerable merit.

26.11 Our discussion has touched on a variety of issues, including the use of conditional arguments and heavy reliance upon the likelihood function. Conditional estimators are most effective when the statistics on which we condition contain no information about θ. This leads to the notion of *ancillary* statistics and paves the way for discussions of other approaches to inference.

Ancillary statistics

26.12 Consider a set of $r+s$ $(r \geq 1, s \geq 0)$ statistics, written (T_r, T_s), that are minimal sufficient for $k+l$ $(k \geq 1, l \geq 0)$ parameters, which we shall write (θ_k, θ_l). Suppose now that the subset T_s has a distribution free of θ_k . (This is only possible if the distribution of (T_r, T_s) is not boundedly complete – cf. Exercise 21.7.) We then have the factorization of the LF into

$$\begin{aligned} L(\mathbf{x}|\theta_k, \theta_l) &= g(T_r, T_s|\theta_k, \theta_l)h(\mathbf{x}) \\ &= g_1(T_r|T_s, \theta_k, \theta_l)g_2(T_s|\theta_l)h(\mathbf{x}). \end{aligned} \tag{26.13}$$

Fisher (e.g. 1956) calls T_s an *ancillary statistic*, while Bartlett (e.g. 1939) calls the conditional statistic $(T_r|T_s)$ a *quasi-sufficient statistic* for θ_k, the term arising from the resemblance of (26.13) when θ_l is known to the factorization of (17.84) that characterizes a sufficient statistic.

It should be noted that some authors do not require (T_r, T_s) to be minimally sufficient. In general, this increases the problems associated with the non-uniqueness of the ancillary statistics, and we shall restrict attention to the minimally sufficient case.

The interest in ancillary statistics derived from the idea that, since the distribution of T_s does not depend on θ_k, we should base inferences about θ_k on the conditional distribution of $T_r|T_s$.

If T_s is sufficient for θ_l when θ_k is known, it follows from (26.13) that

$$L(\mathbf{x}|\theta_k, \theta_l) = g_1(T_r|T_s, \theta_k)g_2(T_s|\theta_l)h(\mathbf{x}); \tag{26.14}$$

thus, we may confine ourselves to functions of $(T_r|T_s)$ in making inferences about θ_k.

However, the real question is whether we should confine ourselves to the conditional statistic when T_s is *not* sufficient for θ_l. The difficulty is essentially that only the *marginal* distribution of T_s is free of θ_k; T_s remains a component of the set of minimal sufficient statistics for all the parameters, whose *joint* distribution depends on θ_k.

26.13 If we denote the information matrix (17.88) for θ_k by $I(\mathbf{x})$ and that for inferences based on $T_r|T_s$ by $I(T_r|T_s)$, we know that when (26.14) holds

$$E_{T_s}\{I(T_r|T_s)\} = I(\mathbf{x}), \tag{26.15}$$

so that use of T_s allows complete recovery of the information. In any particular case, the conditional information may be more or less than the expected information.

Example 26.6 (Conditional versus unconditional test procedures (Cox and Hinkley, 1974))

An experiment involves drawing a sample from an $N(\theta, \sigma^2)$ population where σ^2 is known. The sample size, A, is either n or kn (k an integer, $k > 1$) and is selected by spinning a fair coin. The conditional information is n/σ^2, given $A = n$, and kn/σ^2, given $A = kn$; and $\hat{\theta} = \bar{X}$ in both cases. The UMP test, given A, is of the usual form, but power considerations based upon the unconditional experiment lead to a different and less intuitively appealing test procedure.

Example 26.7 (Ancillary statistics in regression analysis)

When (X, Y) are binormal with joint p.d.f.

$$f(x, y) = f_1(x) f_2(y|x),$$

where $E(X) = \mu$, $\mathrm{var}(X) = \sigma^2$, $E(Y|x) = \beta_0 + \beta_1 x$, and $\mathrm{var}(Y|x) = \omega^2$, we may partition the parameters into

$$\theta = (\beta_0, \beta_1, \omega^2), \qquad \phi = (\mu, \sigma^2)$$

and the minimal sufficient statistics into

$$T = \{\bar{Y}, \hat{\beta}_1, \sum(Y - \hat{Y})^2\}, \qquad A = \{\bar{X}, \sum(X - \bar{X})^2\}.$$

It is easily verified that (26.14) holds. As we shall see in Chapter 29, inferences for θ are usually made conditionally and the conditional information matrix is

$$\omega^{-2}\begin{bmatrix} n & \sum x \\ \sum x & \sum x^2 \end{bmatrix}.$$

These examples suggest that conditioning is desirable, but the argument is not one-sided. For example, Welch (1939) gave an example (Exercise 22.23) which showed that the conditional test based on $(T_r|T_s)$ may be uniformly less powerful than an alternative (unconditional) test. Furthermore, the ancillary statistics may not be uniquely determined; see Exercise 26.1 and also Basu (1964), Cox (1971) and Dawid (1975).

26.14 Cox (1971) suggested that the usefulness of an ancillary statistic should be measured by the variation in the conditional information, $I(T|A)$; he recommends choosing A to maximize some measure of the variability in $I(T|A)$ such as the variance; see also Becker and Gordon (1983). Barnard and Sprott (1971) argue that ancillary statistics should be chosen to preserve the invariance properties of the likelihood function.

Brown (1990) provides an interesting ancillarity paradox in the context of linear regression that relates to the James–Stein estimator (described in **26.56**). For a recent general discussion of ancillarity, see Lehmann and Scholz (1992).

The conditionality principle

26.15 The preceding discussion and Example 26.2 in particular suggest that when some random mechanism independent of θ is used to select a statistical experiment from a set of possible procedures, only the experiment actually performed is relevant to subsequent inferences. That is, the selection mechanism is ancillary, so that we should condition upon the outcome of that initial selection procedure.

This principle is often stated as a slightly weaker form, due to Basu (1975) but implicit in Cox (1958a). Denote the jth experiment by

$$E_j = (X_j, \theta, \{f_j\}), \qquad j = 1, 2,$$

where $f_j \equiv f_j(x_j|\theta)$, and the mixed experiment by

$$E^* = (X^*, \theta, f^*),$$

wherein $J = 1$ or 2 is observed with $P(J = 1) = P(J = 2) = \frac{1}{2}$; further, let $X^* = (J, X_J)$ and $f^* \equiv f^*(\{j, x_j\}) = \frac{1}{2}f(x_j|\theta)$. Only θ need be common to E_1 and E_2. Finally, the *evidence* obtained from the experiment (the term is used to indicate potentially greater generality than *information* as previously defined) is denoted by $Ev(E, x)$ for experiment E.

The *weak conditionality principle* (WCP) is as follows (cf. Birnbaum, 1962, p. 172). Suppose there are two experiments, E_1 and E_2, with common θ, and a mixed experiment E^* containing E_1 and E_2 as its two constituents. Then, for each outcome (E_j, x_j) of E^*, we have

$$Ev(E^*, x^*) = Ev(E_j, x_j); \tag{26.16}$$

that is, the evidence from E^* is exactly that from E_j when E_j is actually performed.

The WCP is sometimes paraphrased as stating 'the irrelevance of unobserved outcomes'.

The WCP seems undemanding, yet its acceptance runs counter to the purer frequentist approaches as we would have to rule out some of the inferential procedures based upon randomization arguments; see Godambe (1982) and Genest and Schervish (1985).

Helland (1995) gives several examples that he claims contradict the conditionality principle; a number of authors respond defending the priciple in the *American Statistician*, **50**, pp. 382–386.

26.16 Since conditional arguments often arise in hypothesis testing, we now examine this aspect in greater detail. First, we consider a familiar example from this viewpoint.

Example 26.8 (Weak conditionality and a test of the normal mean)

We have seen (Example 17.17) that in normal samples the pair $(\bar{x}, s^2)$ is jointly sufficient for (μ, σ^2), and we know that the distribution of s^2 does not depend on μ. Thus we have

$$L(\mathbf{x}|\mu, \sigma^2) = g_1(\bar{x}, |s^2, \mu, \sigma^2)g_2(s^2|\sigma^2)h(\mathbf{x}),$$

a case of (26.13) with $k = l = r = s = 1$. The conditionality principle states that the statistic $\bar{x}|s^2$ is to be used in drawing inferences about μ. (It happens that $\bar{x}$ is actually independent of s^2 in this case, but this is merely a simplification irrelevant to the general argument.) But s^2 is not a sufficient statistic for the nuisance parameter σ^2, so that the distribution of $\bar{x}|s^2$ is not free of σ^2. If we have no prior distribution given for σ^2 we can only make progress by integrating out σ^2 in some more or less arbitrary way. If we are prepared to use its fiducial distribution and integrate over that, we arrive back at the discussion of Example 19.10, where we found that this gives the same result as that obtained from the standpoint of maximizing power in Examples 21.7 and 21.14, namely that Student's t distribution should be used.

Another conditional test principle

26.17 Another principle may be invoked (cf. Cox, 1958a) to suggest the use of $(T_r|T_s)$ whenever $l \geq 1$ and T_s is sufficient for θ_l when θ_k is known, irrespective of whether its distribution depends on θ_k, for then we have

$$L(\mathbf{x}|\theta_k, \theta_l) = g_1(T_r, |T_s, \theta_k)g_2(T_s|\theta_k, \theta_l)h(\mathbf{x}), \qquad (26.17)$$

so that the conditional statistic is distributed independently of the nuisance parameter θ_l. Here again, we have no obvious reason to suppose that the procedure is optimal in any sense.

The justification of conditional tests

26.18 The results of **21.30–32** enable us to see that if the distribution of the sufficient statistics (T_r, T_s) is of the exponential form (21.73), the use of the conditional distribution of T_r for given T_s will give UMPU tests. This results follows directly for, in our previous notation, the statistic T_r is $s(\mathbf{x})$ and T_s is $T(\mathbf{x})$, and we have seen that the UMPU tests are always based on the distribution of T_r for given T_s. If the sufficient statistics are not distributed in the form (21.73) (e.g. in the case of a distribution with range depending on the parameters) this justification is no longer valid. However, following Lindley (1971), we may derive a further justification of the conditional statistic $T_r|T_s$ in **26.17** provided that $g_2(T_s|\theta_k, \theta_l)$ in (26.17) is boundedly complete when θ_k is known. For then, by **21.19**, every size-α critical region similar with respect to θ_l will consist of a fraction α of all surfaces of constant T_s. Thus any similar test on θ_k will be a conditional test based on $T_r|T_s$, and any optimum conditional test will be an optimum similar test.

Welch's (1939) counter-example, which is given in Exercise 21.23, falls within the scope of neither of our justifications of the use of conditional test statistics. There, the range of $f(x)$ depends on the only parameter θ, and $l = 0$. The two-component minimal sufficient statistic (M, R), a one-to-one function of $(x_{(1)}, x_{(n)})$, is not boundedly complete, since

$$E\left(R - \frac{n-1}{n+1}\right) = 0,$$

so that Exercise 21.7 does not preclude R from being an ancillary statistic.

Basu (1977) gives a general review of methods of eliminating nuisance parameters. Kiefer (1977) discusses and extends the literature on conditional confidence interval and test procedures, where it is decided in advance to make certain types of inference given that certain random events occur, with confidence coefficients that are conditional upon those events.

Given a vector of p sufficient statistics for a scalar parameter, θ, McCullagh (1984) shows that this may be partitioned into a locally sufficient statistic and $p-1$ locally ancillary statistics, but that these are not unique. For a recent discussion of conditioning and frequentist inference, see Berger (1985a).

The sufficiency principle

26.19 A second principle that is widely accepted is based upon the property of *sufficiency* that has been widely used throughout these volumes. A *weak* version of the *sufficiency principle* (WSP) may be stated as follows (Dawid, 1977). Consider an experiment E and suppose that the statistic $T(X)$ is sufficient for θ. Then, if $T(x_1) = T(x_2)$,

$$Ev(E, x_1) = Ev(E, x_2). \tag{26.18}$$

Example 26.9 (The sufficiency principle and the normal mean)

Two sets of five observations from $N(\theta, 1)$ populations result in the observations $(-1.5, -0.8, 0, 0.8, 1.5)$ and $(-2, -2, -2, -2, 8)$. In each case, the WSP leads to $\hat{\theta} = 0$. Although the second sample appears to have come from a possibly skewed population with a much larger variance, model criticism is precluded.

26.20 The *strong sufficiency principle* requires that (26.18) should hold even when two different experiments (e.g. with different stopping rules) are performed.

Example 26.10 (The strong sufficiency principle and stopping rules)

If a series of attribute trials produces x successes and $n-x$ failures, the likelihood function is proportional to

$$\theta^x(1-\theta)^{n-x}$$

and the stopping rule does not affect the inferences; cf. Example 26.5. Note that the use of unbiased estimating equations (26.12) is compatible with the strong sufficiency principle.

It is clear that discussions of the conditionality and sufficiency principles rely heavily on the likelihood function; we now consider its role in inference in more detail.

Barnard and Godambe (1982) present criticisms of the sufficiency principle, reporting that it ignores relevant features of an experiment.

The likelihood principle

26.21 In **18.50**, we considered Fisher's recommendation that the LF should be used as an information summary. However, it is possible to take this line of reasoning further and to argue that all inferential procedures should be based solely upon the LF. This view may be stated formally as the *likelihood principle* (LP), which also comes in weak and strong forms. The weak principle (WLP) states that all the information about θ obtained from statistical experiment, E, is contained in the LF, $L(x|\theta)$. If two replications, yielding observations x_1 and x_2, lead to proportional likelihoods:

$$L(x_1|\theta) = c(x_1, x_2)L(x_2|\theta),$$

where the function c is independent of θ, x_1 and x_2 provide the same information about θ, or

$$Ev(E, x_1) = Ev(E, x_2). \tag{26.19}$$

The strong form (SLP) extends the principle to include two different experiments, E_1 and E_2, so that

$$Ev(E_1, x_1) = Ev(E_2, x_2).$$

In particular, the SLP implies the irrelevance of the stopping rule, as in Example 26.10. The likelihood approach to inference has been explored in detail by Berger and Wolpert (1988) and Edwards (1992); Edwards (1974) traces the history of the LP.

Example 26.11 (Tests using the likelihood principle)

Consider a random sample of size n from an $N(\theta, 1)$ population. The standard test of H_0 : $\theta = 0$ against the two-sided alternative yields the critical region $\{|\bar{X}| > zn^{-1/2}\}$. The LP leads to the rejection of H_0 when $|\bar{X}| > (2k)n^{-1/2}$ if we employ the rule

$$\text{reject } H_0 \text{ when } \lambda = \log\{L(\theta = 0|\mathbf{x})/(L(\hat{\theta}|\mathbf{x})\} < k;$$

however, there is no basis for linking the value of k to that of z. In general, k will be determined by the ordinate of the distribution rather than the tail area. When σ^2 is unknown, one possibility is to eliminate the nuisance parameter by using the LR statistic as in Example 22.1. Again, the ordinate of the t distribution would be used in the decision rule.

An alternative approach is the use of a pivotal method, wherein a probability statement can be made for the random variable $Z = g(t, \theta)$ that does not involve the nuisance parameters. This statement may then be used as the basis for tests of hypotheses or to generate interval estimates, as in Example 19.2 where we developed a confidence interval for the mean using Student's t distribution. By contrast, the Bayesian approach is to integrate out the nuisance parameter; cf. Examples 19.13 and 19.14.

One of the conclusions to be drawn from this example is that different approaches will often produce similar numerical results, but the basis upon which the inferences are drawn is quite different.

26.22 The unrestrained use of the LP lacks appeal for many statisticians who find it intuitively unacceptable to ignore the sample space when making inferences, but a result of Birnbaum (1962) makes serious consideration of the LP necessary. Although it has undergone several subsequent modifications, the essence of Birnbaum's result is that

$$\text{WCP} + \text{WSP} \Leftrightarrow \text{WLP}. \tag{26.20}$$

Our demonstration of the result assumes that the sample space is discrete; the proof follows Berger and Wolpert (1988, pp. 27–28).

Proof of $\Rightarrow$. The WCP implies that

$$Ev(E^*, \{j, x_j\}) = Ev(E_j, x_j). \tag{26.21}$$

Experiment E^* is defined as for the WCP in **26.15** and has random outcome $X_J^* = (J, X_J)$. Consider the statistic

$$T(X_J^*) = \begin{cases} (1, x_{10}) & \text{if } J = 2,\ X_2 = x_{20}, \\ (J, X_J) & \text{otherwise,} \end{cases}$$

where x_{10} and x_{20} are particular outcomes for E_1 and E_2. Clearly, $(1, x_{10})$ and $(2, x_{20})$ result in the same value of T. It follows that T is sufficient for θ since

$$P(X^* = (j, x_j) | T = t \neq (1, x_{10})) = \begin{cases} 1, & \text{if } (j, x_j) = t \\ 0, & \text{otherwise} \end{cases}$$

and

$$\begin{aligned} P(X^* = (1, x_{10}) | T = (1, x_{10})) &= 1 - P(X^* = (2, x_{20}) | T = (1, x_{10})) \\ &= \frac{\frac{1}{2} f_1(x_{10}|\theta)}{\frac{1}{2} f_1(x_{10}|\theta) + \frac{1}{2} f_2(x_{20}|\theta)}, \end{aligned}$$

which is independent of θ from (26.19). Thus, the WSP implies that

$$Ev(E^*, (1, x_1^*)) = Ev(E^*, (2, x_2^*)). \tag{26.22}$$

Expressions (26.21) and (26.22) combine to imply that

$$Ev(E_1, x_1) = Ev(E_2, x_2),$$

so the WLP holds.

Proof of $\Leftarrow$. To establish that WLP $\Rightarrow$ WCP, we note that, for E^*,

$$L(\theta | (j, x_j)) = \tfrac{1}{2} f_j(x_j|\theta)$$

which is clearly proportional to $L(\theta | x_j)$, the likelihood for E_j when x_j is observed. Hence,

$$Ev(E^*, (j, x_j)) = Ev(E_j, x_j).$$

Finally, if the sufficient statistic T yields $T(x_1) = T(x_2)$ for realizations x_1 and x_2, it follows that the likelihoods must be proportional so that WLP $\Rightarrow$ WSP. If the stopping rule is ignored so that the strong conditionality and sufficiency principles may be assumed, (26.20) may be restated as an equivalence with the SLP. Birnbaum (1972) produced another proof wherein the SP was replaced by a logically weaker axiom of *mathematical equivalence*.

26.23 When we turn to continuous sample spaces, the principles must be restated to allow possible exceptions of measure zero; a relative likelihood principle may then be established; see Berger and Wolpert (1988, pp. 28–36) for details. Bjørnstad (1996) extends the LP to a more general framework that includes predictive likelihood, and shows that the more general version of Birnbaum's theorem continues to hold.

26.24 Birnbaum's result has led to a considerable discussion. Durbin (1970) pointed out that the theorem does not hold if we first reduce the problem to consideration of the *minimal* sufficient statistics. However, Birnbaum (1970), Savage (1970), and Berger and Wolpert (1988, pp. 45–46) claim that this restriction is inappropriate in general. Kalbfleisch (1975) pointed out that the proof fails if sufficiency is held to be inapplicable to simple mixture experiments; see also the discussion following his paper.

Akaike (1982) demonstrated that, in some cases, the notion of mathematical equivalence is equivalent to the LP. In the discussion on Birnbaum (1962), Kempthorne asks whether it is not the case that the conditionality principle implies the sufficiency principle so that, in fact, the conditionality and likelihood principles are equivalent; Birnbaum concurred. This has been demonstrated rigorously by Evans *et al.* (1985; 1986); these authors go on to argue forcibly against the LP providing examples that suggest information may be suppressed when SP and CP are applied uncritically. Joshi (1989) provides a counter-example for the case of a finite sample space.

Dawid (1991) re-examines Fisher's framework and demonstrates that inferential models that respect the LP may be shown to have an asymptotic sampling theory justification. The key to this result is the selection of an appropriate frame of reference; that is, an appropriate set of conditioning arguments, in general accordance with a conditionality principle.

26.25 It is interesting to note that in his later writings, Birnbaum concluded that the LP was inadequate as a basis for inference since it conflicted with the *confidence principle*, which he formulated (Birnbaum, 1977, p. 24) as follows:

> a concept of statistical evidence is not plausible unless it finds 'strong evidence' for H_2 as against H_1 with small probability (α) when H_1 is true and with much larger probability $(1 - \beta)$ when H_2 is true.

This principle is, of course, an excellent foundation for frequentist inference. For further discussion, see the special volume of *Synthèse*, **36**(1), 1977, devoted to the foundations of statistical inference.

Fiducial inference

26.26 The construction of fiducial intervals for one- and two-sample problems was examined at some length in **19.44–47**. Therefore, our present discussion concentrates on the underlying assumptions rather than detailing further examples.

As noted in **19.45**, if t is the minimal sufficient statistic for the single parameter θ, with d.f. $F(t|\theta)$, the fiducial distribution of θ given t has probability density function

$$g(\theta|t) = \frac{\partial G(\theta|t)}{\partial\theta} = \frac{\partial F(t|\theta)}{\partial\theta}, \tag{26.23}$$

provided F is monotone decreasing in θ. Some of the difficulties with this approach include how to proceed in the absence of a sufficient statistic, lack of uniqueness (in multiparameter cases) and the lack of a frequency interpretation.

26.27 Let us consider the last concern first. Fisher's writings in this area were evidently influenced by Keynes (1921), Carnap (1962) and others who sought to develop an epistemic view of probability that would measure the 'degree of rational credibility' of a hypothesis H relative to data or evidence E. Thus, although the initial development of fiducial probability was confused, the ultimate aim was clear: to make probabilistic statements of the form $P(H|E)$ or, in our present context, to develop a d.f. $G(\theta|t)$. By construction, and intent, G is designed to provide statements about θ *for a single trial*, so the absence of a frequency interpretation is hardly surprising.

It is clear that the fiducial approach is seeking to deliver a quite different inductive statement than is available from the frequency viewpoint. Although it is possible, as Barnard (1950) has shown, to justify the Fisher–Behrens solution of the two-means problem from a different frequency standpoint, as he himself goes on to argue, the idea of a fixed 'reference set', in terms of which frequencies are to be interpreted, is really foreign to the fiducial approach. Thus, the statistician must choose between confidence intervals, which make precise frequency-interpretable statements and which may on exceptional occasions be trivial, and other methods, which forgo frequency interpretations in the interests of what are, perhaps intuitively, felt to be more relevant inferences.

For the present, we now accept the declared aims of the fiducial approach and examine the methods in greater detail.

Paradoxes and restrictions in fiducial theory

26.28 The principal difficulties with the fiducial approach may be illustrated in the following way. Assume that (t_1, t_2) are jointly sufficient for (θ_1, θ_2), and write the alternative factorizations

$$\begin{aligned} L(x|\theta_1,\theta_2) \propto g(t_1,t_2|\theta_1,\theta_2) &= g_1(t_1|t_2,\theta_1,\theta_2)g_2(t_2|\theta_1,\theta_2) \\ &= g_3(t_2|t_1,\theta_1,\theta_2)g_4(t_1|\theta_1,\theta_2). \end{aligned}$$

We may distinguish two special structures of the sufficient statistics:

(a) One of the statistics depends on only one parameter. The factorization becomes

$$\begin{array}{ll} \text{either} & L \propto g_1(t_1|t_2,\theta_1,\theta_2)g_2(t_2|\theta_2) \\ \text{or} & L \propto g_3(t_2|t_1,\theta_1,\theta_2)g_4(t_1|\theta_1). \end{array} \tag{26.24}$$

(b) One of the conditional distributions depends on only one parameter, giving

$$\begin{array}{ll} \text{either} & L \propto g_1(t_1|t_2,\theta_1)g_2(t_2|\theta_1,\theta_2) \\ \text{or} & L \propto g_3(t_2|t_1,\theta_2)g_4(t_1|\theta_1,\theta_2). \end{array} \tag{26.25}$$

If the first line of (26.25) holds, t_2 is singly sufficient for θ_2 when θ_1 is known; if the second line holds, t_1 is singly sufficient for θ_1 when θ_2 is known; note that (26.24) and (26.25) correspond to (26.13) and (26.17) respectively, in a slightly different notation.

Either line of (26.24), or of (26.25), permits a joint fiducial distribution to be constructed by first obtaining the fiducial distribution of one parameter from the factor in which it appears alone, and then obtaining the conditional fiducial distribution of the other parameter (the value of the first parameter being fixed) from the factor in which both parameters appear. The product of these distributions is taken as the joint fiducial distribution, on the analogy of the multiplication theorem for probabilities. Expressions (26.24) and (26.25) were used in this way by Fisher (1956) and Quenouille (1958), respectively.

Referring back to the discussion for the means of one and two samples, in Examples 19.9–10, it will be seen that both (26.24) and (26.25) hold, because the sample mean (or difference of means) t_1 is distributed independently of the sample variance(s) t_2. In general, however, even these special sufficiency structures are not enough to guarantee the uniqueness of the joint fiducial distribution, as Tukey (1957b) and Brillinger (1962) showed by counter-examples. The non-uniqueness arises precisely because *both* lines of (26.24), or of (26.25), can hold simultaneously, and the joint fiducial distribution may depend on which line we use to construct it. See also Mauldon (1955) and Dempster (1963).

26.29 One critical difficulty of fiducial theory is exemplified by the derivation of Student's distribution in fiducial form given in Example 19.9. It appears to us that this particular matter has been widely misunderstood, except by Jeffreys (1961). Since Student's distribution gives the same result for fiducial theory as for confidence theory, whereas the two methods differ on the problem of two means, both sides seem to have sought their basic differences in the second, not in the first. But in our view *c'est le premier test qui coûte*. If the logic of this is agreed, the more general Fisher–Behrens result follows by a very simple extension. This is also evident from the Bayes–Jeffreys approach, in which Example 19.13 is an obvious extension of Example 19.12 for two independent samples.

The question, as noted in Example 19.9, is whether, given the joint distribution of $\bar{x}$ and s (which are independent in the ordinary sense), we can replace $d\bar{x}\, ds$ by $d\mu\, d\sigma/\sigma$. It appears to us that this is not obvious and, indeed, requires a new postulate, just as (19.164) is a new postulate. On this point, the paper by Yates (1939a) is explicit.

As noted in **19.46**, the key to success in fiducial inference is the identification of suitable *pivotal* elements. Sprott (1990) develops a general framework of approximately linear pivotals, using transformations where necessary to induce an approximately Gaussian shape for the likelihood function.

Seidenfeld (1992) and Zabell (1992) give a comprehensive discussion of the current state of fiducial theory. Barnard (1995) describes the history of the fiducial approach and further develops pivotal methods.

Structural inference

26.30 Fraser (1968) gives a general theory of *structural inference*, essentially fiducial in character, which is more simply illustrated and discussed in Fraser (1976) and the comments following it.

The approach involves the specification of a structural equation that relates the measurement, X, to the physical quantity of interest, θ. For example, we may consider

$$X = \theta + \epsilon,$$

where ϵ denotes the random error, which is $N(0, \sigma^2)$. Then, given n observations $(x_1, x_2, \ldots, x_n)$, we can transform to

$$u_1 = \theta + \epsilon_1, \quad u_j = \epsilon_j - \epsilon_{j-1} = x_j - x_{j-1}, \quad j \geq 2.$$

The variates $u_j (j \geq 2)$ have distributions independent of θ and form ancillary statistics. We are thus led to the *reduced* distribution for $\bar{\epsilon} = \bar{X} - \theta$; $\bar{\epsilon}$ is then used as a pivotal function to provide inferences about θ. In this case, the structural argument is very similar to the fiducial, but it may be extended to include general transformations of the form $X = T(\theta)\epsilon$.

The LR as a credibility measure

26.31 The difficulties with the fiducial approach are seen to stem from a lack of uniqueness once we move from the special case of a single parameter and a single sufficient statistic. Since the aim of the fiducial approach is to condition upon the ancillary statistics and then to produce a measure that may be interpreted as a degree of credibility, a somewhat similar end could be reached by considering the ML estimator(s) which are at least 'asymptotically sufficient' in the sense noted in **18.16**. If the parameter set (θ, ϕ) contains nuisance parameters ϕ, these may be replaced by their ML estimators as functions of θ, $\hat{\phi}(\theta)$ say. That is, we might consider the LR in the form

$$g(\theta|x) = L\{x|\theta, \hat{\phi}(\theta)\}/L(x|\hat{\theta}, \hat{\phi}). \tag{26.26}$$

Birnbaum (1962) refers to

$$\alpha(\theta) = 1/\{1 + g(\theta|x)\} \tag{26.27}$$

as the *intrinsic significance level* associated with θ, implying that it could be used for testing hypotheses in this way. Since α is a monotone function of g, the suggestion is comparable to the use of the LR test, although the critical regions will differ since (26.27) bases the test on the ordinate of g rather than a tail area.

26.32 If we are willing to define $g(\theta|x)$ for all $\theta \in \Omega$, the integral over θ, suitably scaled, could be used as a measure of credibility; of course, the denominator in (26.26) is then redundant. Compared to the fiducial argument, this approach has the benefit that there is no problem with reference sets since the use of g is in accordance with the conditionality principle.

This approach is developed more formally in Barndorff-Nielsen (1980, 1983), Cox and Reid (1987) and Fraser and Reid (1989).

Example 26.12 (LF credibility measure for a test of the normal mean)

Consider again the case of a sample of size n from a normal population with mean μ and variance σ^2 unknown. The likelihood reduces to

$$L(x|\mu, \sigma^2) \propto \sigma^{-n} \exp\{-\tfrac{1}{2}\sum(x_i - \mu)^2/\sigma^2\};$$

since

$$n\hat{\sigma}^2(\mu) = \sum(x_i - \bar{x})^2 + n(\bar{x} - \mu)^2 = ns^2 + n(\bar{x} - \mu)^2,$$

(26.27) yields

$$g(\mu|\bar{x}) \propto \left\{1 + \frac{(\bar{x} - \mu)^2}{s^2}\right\}^{-n/2}. \tag{26.28}$$

Expression (26.28) differs from the Student's t form by having n in place of $n - 1$. The exact Student's t form could be obtained by using the marginal likelihood for

$$t = (\bar{x} - \mu)/s.$$

In general, the likelihood-based argument produces a measure of credibility that is equal (at least asymptotically) to the fiducial d.f. and avoids the construction difficulties noted earlier. In Exercise 26.2, the reader is asked to confirm that (26.28) recovers the Behrens–Fisher solution to the two-sample problem in Example 19.11 with n_i in place of $n_i - 1$.

26.33 In order to make this approach more operational, we may use Taylor's theorem to observe that, when $l(\theta) = \log L(\theta)$,

$$l(\theta) \doteq l(\hat{\theta}) + (\theta - \hat{\theta})\left(\frac{\partial l}{\partial \theta}\right)_{\hat{\theta}} + \frac{1}{2}(\theta - \hat{\theta})^2\left(\frac{\partial^2 l}{\partial \theta^2}\right)_{\hat{\theta}}. \tag{26.29}$$

Since $(\partial l/\partial \theta)_{\hat{\theta}} = 0$, by construction, it follows that

$$L(\theta)/L(\hat{\theta}) \doteq \exp\left\{\frac{1}{2}(\theta - \hat{\theta})^2\left(\frac{\partial^2 l}{\partial \theta^2}\right)_{\hat{\theta}}\right\}. \tag{26.30}$$

Since $I(\hat{\theta}) = (\partial^2 l/\partial \theta^2)_{\hat{\theta}}$ is independent of θ, it follows directly from (26.30) that, for large samples, the LR plotted in the parameter space is shaped like the normal p.d.f., where $z = (\theta - \hat{\theta})\{I(\hat{\theta})\}^{1/2}$ is $N(0, 1)$. The quadratic approximation in (26.29) may be improved using transformations, cf. **32.37–44**.

26.34 This argument extends readily to cover several parameters and the elimination of nuisance parameters when (26.30) becomes the usual LR criterion and the resulting distribution of $-2\log l$ (LR) is chi-square; cf. **22.7**. Using the d.f. thus constructed, we have a LR-based significance test that would reject the hypothesis $H_0 : \theta = \theta_0$ if θ_0 did not fall in the interval $[\theta_1, \theta_2]$, where

$$P^*[\theta_1 \leq \theta \leq \theta_2] = 1 - \alpha; \tag{26.31}$$

P^* denotes the measure of credibility derived from (26.26). Interestingly, this has elements of a pure significance test introduced by Fisher before the Neyman–Pearson theory was developed, yet anchored by the ML estimator.

The difficulties with such an approach are, as before, the lack of a frequency interpretation for P^* or, indeed, any direct interpretation of the function. Here, as elsewhere, the statistician must decide whether he or she is willing to make the logical leap in order to justify inferential statements that relate to single experiments.

Bayesian inference

26.35 The Bayesian approach to the problem of induction is to suppose that a *prior* distribution may be specified for parameter θ, $p(\theta)$ say, defined on the parameter space $\theta \in \Omega$. Given the likelihood function, $L(x|\theta)$, it follows from an application of Bayes theorem (cf. **8.7**) that the *posterior* distribution for θ is

$$P(\theta|x) \propto p(\theta)L(x|\theta). \tag{26.32}$$

Several examples are given in Chapter 8 and the accompanying exercises. Once the notion of specifying a prior distribution for θ is accepted, the framework of Bayesian inference may be developed deductively from one of several systems of axioms (e.g. Ramsey, 1931; Good, 1950; Savage, 1954; DeGroot, 1970); for a detailed evaluation, see Fishburn (1986). We leave a detailed exposition of this subject to the companion volume by O'Hagan (1994).

26.36 Thus, the key question is how to specify the prior distribution. Three possible approaches may be considered:

(i) as a frequency distribution, based on past experience;
(ii) as an objective representation of rational initial beliefs regarding the parameter;
(iii) as a subjective statement about what *you* (a specific individual) believe before the data are collected.

26.37 We shall consider alternative (i) only briefly. In keeping with the frequency approach, we would need to assume an underlying process that generates the parameter values that is stable, or at least predictable. Examples include industrial production runs where a prior distribution for the proportion of defectives, say, may be assessed from past records. More generally, state-space models in time series assume that the parameter (state) develops over time according to a state equation such as

$$\theta_t = \theta_{t-1} + \delta_t, \tag{26.33}$$

where δ_t represents a random disturbance at time t. See Kendall and Ord (1990, Chapter 9) for further discussion.

In some ways, this may seem to be mixing oil and water and the counter-claim could be made that prior information is not allowable in the frequentist scheme. Such a claim is indeed made by critics of the frequency approach, but seems to represent an overly literal interpretation of that viewpoint. Indeed, it should be noted that, even though the prior is specified in frequentist terms, (26.33) still requires that we are willing to consider the posterior distribution for θ.

Objective probability

26.38 Objective, or logical, probability was developed, notably by Jeffreys (1961) and others, to provide a substantive measure of the weight of evidence favouring a given hypothesis in light of the data. That is, an agreed prior distribution was sought that would allow posterior probability statements to be made on the basis of a particular trial.

Much of Jeffreys' work focused on the specification of a prior distribution in situations where nothing is known about the parameter before the statistical experiment takes place. Interestingly, most subjective Bayesians, such as Lindley (1971), would now argue that there is always *some* information available and that the specification of prior ignorance is a non-issue. When the number of values of θ in Ω is finite, it is feasible to make use of *Bayes' postulate* (also known as the *principle of insufficient reason* or the *principle of indifference*) and to assign equal prior probabilities to each possible value. This requires that a satisfactory base of possible parameter values can be established, not always a trivial task.

Example 26.13 (Use of the principle of insufficient reason)

An urn contains an unknown number of balls of equal size and weight that are made of the same material. What is the prior probability that a white ball will be selected on the first drawing when you are told that the urn contains balls that are

(a) white or not-white
(b) white, red or blue?

The principle of insufficient reason leads us to conclude $p = 1/2$ in case (a), but $p = 1/3$ in case (b).

Despite this example, the principle may often serve as a reasonable starting point. One implication of the principle is *Laplace's law of succession* (Example 8.8) which shows that if we start from

$$\Omega = \{0, 1/N, 2/N, \ldots, (N-1)/N, 1\} \tag{26.34}$$

and assign equal prior probabilities $1/(N+1)$ to each state, then

$$P\{(m+1)\text{th trial is a success}|\text{first } m \text{ trials are successes}\} = \frac{m+1}{m+2} \tag{26.35}$$

for any m and $N \geq 1$.

Example 26.14 (Prior beliefs and the principle of insufficient reason)

If a coin is tossed m times and comes down heads each time, would we accept that the probability of the next spin yielding a head was given by (26.35)?

The answer is probably not, because we are drawing on a lot of past experience that tells us coins have a head and a tail and that either side is 'equally likely' to fall face uppermost. However, this does not violate the principle, rather it tells us that assigning equal probabilities to the values in (26.34) was not an accurate statement of prior beliefs. Conversely, if there are three coins: one with two heads, one standard, and one with two tails, specifying equal probabilities on (26.34) with $N = 2$ would be very plausible. Note that we do not require a coin to be selected at random, but only that we are ignorant of the selection process.

26.39 Now suppose that Ω is a continuum; even if the prior for θ is uniform over some finite interval, that for any non-linear transform of $g(\theta)$ will not be. This led Jeffreys to propose the use of the prior

$$p(\theta) \propto \{I(\theta)\}^{1/2}, \tag{26.36}$$

where $I(\theta) = -E(\partial^2 \log L/\partial\theta^2)$. He arrived at (26.36) by selecting that form of $g(\theta)$ for which $p\{g(\theta)\}$ is uniform, even if improper in some cases; the function of $g(\theta)$ then corresponds to a location parameter for the distribution, at least locally. Jeffreys (1961) termed priors given by (26.36) *invariant.* The resulting functions $g(\theta)$ are precisely those we obtain when considering variance stabilizing transforms (see **32.38–40**):

Distribution	$g(\theta)$	$p(\theta)$
normal, $N(\theta, 1)$	θ	1
normal, $N(0, \theta)$	$\log\theta$	θ^{-1}
Poisson, $P(\theta)$	$\theta^{1/2}$	$\theta^{-1/2}$
binomial, $B(n, \theta)$	$\sin^{-1}(\theta^{1/2})$	$\{\theta(1-\theta)\}^{-1/2}$
negative binomial, $NB(k, \theta)$	$\sinh^{-1}(\theta^{1/2})$	$\theta^{-1}(1-\theta)^{-1/2}$.

The two binomials have different priors for the same parameter, a property that would appear to be in violation of the SLP. Nevertheless, several authors have recommended vague priors that depend on the stopping rule (e.g. Jaynes, 1968; Box and Tiao, 1973; Zellner, 1977; Bernardo, 1979) and claim that such a step is indeed desirable.

When the number of parameters increases, so does the difficulty of specifying a vague prior and we must introduce new postulates such as the independence of prior beliefs about different parameters; see Examples 19.12–13.

26.40 Although the concept that Jeffreys was trying to make operational is an attractive one, it does not seem possible to develop it in a consistent fashion; see Barnett (1982, Chapter 6) and Howson and Urbach (1989, Chapter 9) for recent critiques. It is interesting to speculate whether Jeffreys would have adopted (26.36) if its results had not matched existing ones.

Exercise 26.5 shows that a paradox may arise if an improper prior is used; the problem disappears once a proper prior is used.

Subjective probabilities

26.41 We now leave the objectivist position behind and accept that prior probabilities are necessarily personal and based upon our own experience. In order to make such a scheme operational, it is necessary that:

(a) you have beliefs about the parameter of interest, which can be expressed in the form of probabilities;

(b) your probabilities may be compared one with another (though they need not be comparable with anyone else's);

(c) your probabilities can be assessed by some scheme of hypothetical bets.

If your betting behaviour is internally consistent, it follows that your probabilities satisfy the standard rules of probability and you are said to be *coherent*; otherwise, you are *incoherent* and a Bayesian could make bets with you in such a way that you would be bound to lose money. This is the *principle of coherence*, which states that your system of bets should be internally consistent. Presumably, coherence was used to avoid confusion with Fisher's use of consistency in estimation and testing. Clearly, non-Bayesians do not have a monopoly of virtuous keywords!

26.42 The key requirement is now the assessment of the prior distribution. Most subjectivists (e.g. Ramsey, 1931; Savage, 1954) use some method of assessing fair bets either directly for the phenomenon under study or by comparison with some standardized experiment (e.g. an urn scheme). It is assumed that such assessments can be made directly for the probabilities, uncontaminated by the relative utilities of different outcomes.

26.43 Once you have established your prior distributions, subjective Bayesian analysis proceeds straightforwardly, although it will often be desirable to use conjugate priors (see **8.31**) to simplify the algebra. If the set of parameters is (θ, ϕ), where ϕ denotes nuisance parameter(s), the standard approach is to examine the *marginal posterior*

$$\begin{aligned} P(\theta|x) &= \int P(\theta, \phi|x)\,\mathrm{d}\phi \\ &= \int L(x|\theta, \phi)p(\theta, \phi)\,\mathrm{d}\phi. \end{aligned} \tag{26.37}$$

Explicit evaluation of (26.37) can prove very difficult for higher-dimensional problems. However, innovative numerical integration procedures known as Markov chain Monte Carlo methods have greatly contributed to the feasibility of this approach; see Gelfand and Smith (1990) or Cowles and Carlin (1996) for further details.

For more general updating rules, see Diaconis and Zabell (1982).

Bayesian estimation

26.44 Point estimation is usually based upon either the mode or the mean of the posterior distribution. The *posterior mode* is given by $\tilde{\theta}$, where

$$P(\tilde{\theta}|x) = \max_{\theta} P(\theta|x); \tag{26.38}$$

when the prior distribution is uniform, $\tilde{\theta}$ will be equivalent to the ML estimator $(\tilde{\theta})$.

The *posterior mean*, given by

$$\bar{\theta} = E(\theta|x), \tag{26.39}$$

will be equal to the ML estimator, if at all, only for distribution-specific choices of the prior.

Example 26.15 (Bayesian inference for Bernoulli trials)

Let π denote the probability of success in a Bernoulli trial with prior density

$$p(\pi) \propto \pi^{a-1}(1-\pi)^{b-1}.$$

Given n trials with x successes, the posterior is

$$P(\pi|x) \propto \pi^{a+x-1}(1-\pi)^{b+n-x-1},$$

whence

$$\tilde{\theta} = \frac{a+x-1}{n+a+b-2} \quad \text{and} \quad \bar{\theta} = \frac{a+x}{n+a+b},$$

compared to $\hat{\theta} = x/n$. Upon inspection, $\tilde{\theta} = \hat{\theta}$ for the uniform prior ($a = b = 1$), whereas $\bar{\theta} = \hat{\theta}$ when $a = b = 0$, a degenerate choice that is not feasible.

26.45 Interval estimates may be obtained from the posterior distribution directly; Bayesian inference allows the statement 'with probability $1-\alpha$, θ lies between the values θ_1 and θ_2' or

$$P(\theta_1 \leq \theta \leq \theta_2) = P(t_2|x) - P(t_1|x) = 1 - \alpha. \tag{26.40}$$

The interval $[\theta_1, \theta_2]$ is known as a $100(1-\alpha)$ per cent *credible* region. Parallel to the notion of the physically shortest interval (**19.13**), we may choose the set of θ values, Ω_1, such that (26.40) is satisfied and

$$\left\{\theta \in \Omega_1 : \frac{\partial P(\theta)}{\partial \theta} \geq c\right\}. \tag{26.41}$$

Such an interval (or region) is known as the *highest posterior density* credible region.

Example 26.16 (Highest posterior density credible region for the normal mean)

For a random sample of size n from a normal population with known variance, $N(\mu, \sigma^2)$ say, consider the prior distribution $N(\phi, \tau^2)$. From **8.30**, the posterior distribution for μ is $N(\mu_P, \sigma_P^2)$, where

$$\mu_P = \frac{\phi\sigma^2 + n\bar{x}\tau^2}{\sigma^2 + n\tau^2} \quad \text{and} \quad \sigma_P^2 = \frac{\sigma^2\tau^2}{\sigma^2 + n\tau^2}.$$

The highest posterior density credible region for μ is

$$\mu_P \pm z_{1-\alpha/2}\sigma_P,$$

where z represents the percentage points of $N(0, 1)$. In this example, $\tilde{\theta} = \bar{\theta}$ and these will be equal to $\hat{\theta}$ for the uniform, improper prior given by letting $\tau \to \infty$; the credible and confidence intervals will then be identical (numerically speaking!).

Bayesian tests

26.46 The two one-sided hypotheses

$$H_0 : \theta \leq \theta_0 \quad \text{and} \quad H_1 : \theta > \theta_0$$

are readily compared by computing their posterior probabilities

$$P(H_0) = P(\theta_0|x), \quad P(H_1) = 1 - P(\theta_0|x). \tag{26.42}$$

However, the comparison of

$$H_0 : \theta = \theta_0 \quad \text{and} \quad H_1 : \theta \neq \theta_0$$

raises some difficulties. Jeffreys (1961, Chapter 5) argued that the value of θ_0 is distinguished from all other θ values and so a prior probability may be assigned to that point:

$$p_0 = p(\theta_0) > 0.$$

The posterior odds in favour of H_0 are then

$$P(\theta_0|x) \Big/ \int_{\Omega-\theta_0} \mathrm{d}P(\theta_0|x). \tag{26.43}$$

Such an assumption is clearly plausible in some cases, such as testing whether a regression coefficient is zero, but depends heavily on the value of p_0 selected. The frequentist view would be that the null hypothesis often is deserving of special attention, but that there is no reasonable way of arriving at an appropriate value of p_0.

A similar device is to expand H_0 to

$$H_0 : \theta_0 - a \leq \theta \leq \theta_0 + b, \qquad a, b > 0,$$

and then consider the posterior odds as before; unfortunately, this depends heavily on the selection of a and b.

Using (26.37), we may formulate (26.43), in the absence of nuisance parameters, as

$$p(\theta_0)L(x|\theta_0)/m(x)$$

where

$$m(x) = \int_{\Omega-\theta_0} L(x|\theta)p(\theta)\,\mathrm{d}\theta.$$

The term

$$B_p(x) = L(x|\theta_0)/m(x) \tag{26.44}$$

is known as the *Bayes Factor.* The function $m(x)$ is maximized if the prior distribution is degenerate at $\hat{\theta}$, so that the LR forms a lower bound, or:

$$B_p(x) \geq L(x|\theta_0)/L(x|\hat{\theta}).$$

Berger and Mortera (1991) argue that the Bayes factor, even if used only through the lower bound, provides a better indication of the odds for H_0 than use of a p-value. Berger and Mortera (1991) develop more stringent bounds for various classes of prior distributions. Their approach is close to the use of the LR as a credibility measure, previously outlined in **26.31**.

26.47 Bernardo (1980) examined the structure of Bayesian tests and concluded that there are no problems when H_0 and H_1 have the same dimensionality. In other cases, it appears that the conclusions to be drawn from such tests are clearly interpretable only when $p(\theta_0)$ depends on the overall prior $p(\theta)$, $\theta \neq \theta_0$.

In general, tests of hypotheses now tend to receive less attention from Bayesians, who would tend to favour the use of decision theory; see **26.52–57**.

The relationship between Bayesian and fiducial approaches

26.48 Lindley (1958) obtained a simple yet far-reaching result that not only illuminates the relationship between fiducial and Bayesian arguments, but also limits the claims of fiducial theory to provide a general method of inference, consistent with and combinable with Bayesian methods. In fact, Lindley shows that the fiducial argument is consistent with Bayesian methods if and only if it is applied to a random variable x and a parameter θ which may be (separately) transformed to u and τ respectively so that τ is a location parameter for u; and in this case, it is equivalent to a Bayesian argument with a uniform prior distribution for τ. The criticism applies equally to 'confidence distributions' defined at the end of **19.6**, in so far as they coincide with fiducial distributions.

26.49 Using (26.23), we write for the fiducial distribution of θ

$$g_x(\theta) = -\frac{\partial}{\partial\theta}F(x|\theta), \tag{26.45}$$

while the posterior distribution for θ given a prior distribution $p(\theta)$ is, by Bayes' theorem,

$$P(\theta|x) = p(\theta)f(x|\theta)\Big/\int p(\theta)f(x|\theta)\,d\theta, \tag{26.46}$$

where $f(x|\theta) = \partial F(x|\theta)/\partial x$, the density function. Writing $r(x)$ for the denominator on the right of (26.46), we thus have

$$P(\theta|x) = \frac{p(\theta)}{r(x)}\frac{\partial F(x|\theta)}{\partial x}. \tag{26.47}$$

If there is some prior distribution $p(\theta)$ for which the fiducial distribution is equivalent to a Bayes posterior distribution, (26.45) and (26.47) will be equal, or

$$\frac{-\frac{\partial}{\partial\theta}F(x|\theta)}{\frac{\partial}{\partial x}F(x|\theta)} = \frac{p(\theta)}{r(x)}. \tag{26.48}$$

Expression (26.48) shows that the ratio on the left-hand side must be a product of a function of θ and a function of x. We rewrite it

$$\frac{1}{r(x)}\frac{\partial F}{\partial x} + \frac{1}{p(\theta)}\frac{\partial F}{\partial\theta} = 0. \tag{26.49}$$

For given $p(\theta)$ and $r(x)$, we solve (26.49) for F. The only non-constant solution is

$$F = G\{R(x) - P(\theta)\}, \tag{26.50}$$

where G is an arbitrary function and R, P are respectively the integrals of r, p with respect to their arguments. If we write $u = R(x)$, $\tau = P(\theta)$, (26.50) becomes

$$F = G\{u - \tau\}, \tag{26.51}$$

so that τ is a location parameter for u. Conversely, if (26.51) holds, (26.48) is satisfied with u and τ for x and θ and $p(\tau)$ a uniform distribution. Thus (26.51) is a necessary and sufficient condition for (26.48) to hold, i.e. for the fiducial distribution to be equivalent to some Bayes posterior distribution.

26.50 Now consider the situation where we have two independent samples, summarized by sufficient statistics x, y, from which to make an inference about θ. We can do this in two ways:

(a) We may consider the combined evidence of the two samples simultaneously, and derive the fiducial distribution $g_{x,y}(\theta)$.

(b) We may derive the fiducial distribution $g_x(\theta)$ from the first sample above, and use this as the prior distribution in a Bayesian argument on the second sample, to produce a posterior distribution $P(\theta|x, y)$.

If the fiducial argument is consistent with Bayesian arguments, (a) and (b) are logically equivalent and we should have $g_{x,y}(\theta) = P(\theta|x, y)$.

Take the simplest case, where x and y have the same distribution. Since it admits a single sufficient statistic for θ, the density function is of the form (17.83), from which we may assume (cf. Exercise 17.14) that the distribution of x itself is of form

$$f(x|\theta) = f(x)h(\theta)\exp(x\theta), \tag{26.52}$$

and similarly for y in the other sample. Moreover, in the combined samples, $x + y$ is evidently sufficient for θ, and thus the combined fiducial distribution $g_{x,y}(\theta)$ is a function of $x + y$ and θ only. We now ask for the conditions under which $P(\theta|x, y)$ is also a function of $x + y$ and θ only. Since by Bayes' theorem

$$P(\theta|x, y) = \frac{g_x(\theta)f(y|\theta)}{\int g_x(\theta)f(y|\theta)\,d\theta},$$

if $P(\theta|x, y)$ is a function of $x + y$ and θ only, so also will be the ratio for two different values of θ,

$$\frac{P(\theta|x, y)}{P(\theta'|x, y)} = \frac{g_x(\theta)f(y|\theta)}{g_x(\theta')f(y|\theta')}. \tag{26.53}$$

Thus (26.53) must be invariant under interchange of x and y. Using (26.52) in (26.53), we therefore have

$$\frac{g_x(\theta)h(\theta)}{g_x(\theta')h(\theta')}\exp\{y(\theta - \theta')\} = \frac{g_y(\theta)h(\theta)}{g_y(\theta')h(\theta')}\exp\{x(\theta - \theta')\},$$

so that

$$\frac{g_x(\theta)}{g_x(\theta')} \cdot \frac{g_y(\theta')}{g_y(\theta)} = \exp\{(x-y)(\theta-\theta')\}$$

or

$$g_x(\theta) = \frac{g_x(\theta')\mathrm{e}^{-x\theta'} \cdot g_y(\theta)\mathrm{e}^{-y\theta}}{g_y(\theta')\mathrm{e}^{-y\theta'}}\mathrm{e}^{x\theta}, \tag{26.54}$$

and if we regard θ' and y as constants, we may write (26.54) as

$$g_x(\theta) = A(x) \cdot B(\theta)\mathrm{e}^{x\theta}, \tag{26.55}$$

where A and B are arbitrary functions. Using (26.45), (26.52) and (26.55), we have

$$\frac{-\frac{\partial}{\partial\theta}F(x|\theta)}{\frac{\partial}{\partial x}F(x|\theta)} = \frac{g_x(\theta)}{f(x|\theta)} = \frac{A(x)B(\theta)}{f(x)h(\theta)}. \tag{26.56}$$

But (26.56) is precisely the condition (26.48), for which we saw (26.51) to be necessary and sufficient. Thus we have $g_{x,y}(\theta) = P(\theta|x, y)$ if and only if x and θ are transformable to (26.51) with τ a location parameter for u, and $p(\tau)$ a uniform distribution. Thus the fiducial argument is consistent with Bayes' theorem if and only if the problem is transformable into a location parameter problem, the prior distribution of the parameter then being uniform. An example where this is not so is given as Exercise 26.3.

Lindley goes on to show that in the exponential family of distributions (17.83), the normal and the gamma distributions are the only ones obeying the condition of transformability to (26.51): this explains the identity of the results obtained by fiducial and Bayesian methods in these cases (cf. Example 19.11). Sprott (1960; 1961) shows that these remain the only such distributions if x and y are differently distributed.

Welch and Peers (1963), Welch (1965), and Peers (1965) examine the problem of correspondence of Bayesian and confidence intervals with special reference to asymptotic solutions. Thatcher (1964) examines this correspondence for binomial predictions. Geisser and Cornfield (1963) and Fraser (1964) display further difficulties with fiducial distributions in the multivariate case. See also the ISI Symposium (*Bulletin of the International Statistical Institute*, 1964, **40**(2), pp. 833–939).

Fraser (1962) proposes a modification of the fiducial method which extends the range of its consistency with Bayesian methods.

Empirical Bayes methods

26.51 An interesting variation on the Bayesian approach is the empirical Bayes scheme developed by Robbins (1956; 1964); see Maritz and Lwin (1989) for a detailed exposition. Suppose that a sample of n observations is available with p.d.f. $f(x_i|\theta_i)$, where θ_i represents a random drawing from prior $p(\theta|\phi)$ and ϕ represents the parameters for the prior distribution. We may then consider the marginal distribution (or mixture, cf. **5.20–24**):

$$f(x|\phi) = \int f(x|\theta)p(\theta|\phi)\,\mathrm{d}\phi \tag{26.57}$$

and use ML methods to estimate ϕ. The posterior distribution for θ_i is then approximated by

$$P(\theta_i|x_i) \propto f(x_i|\theta_i)p(\theta_i|\hat{\phi}). \tag{26.58}$$

In particular cases (e.g. with conjugate priors), explicit determination of (26.57) may be possible; otherwise, numerical procedures must be used.

This approach is something of an amalgam of Bayesian and frequentist ideas and has had a mixed reception. For example, Neyman (1962) hailed it as a breakthrough, whereas Lindley (1971) regards it as involving no new point of principle.

Decision theory

26.52 Abraham Wald's work on sequential analysis (cf. Chapter 24) led also to the development of a general theory of decision-making. Consider a situation where, given the data, it is necessary to make a decision; further, assume that the consequences of these decisions are known and that they can be evaluated numerically. These are not trivial assumptions; for example, Neyman and Pearson concluded that such information was *unlikely* to be available in their development of hypothesis tests (see Chapters 20–22). Given the necessary background, the problem is to decide on optimum decision rules with reference to some performance measure. We now proceed to outline the basics of such a theory; for more detailed expositions, see Wald (1950), Blackwell and Girshick (1954), Ferguson (1967) and DeGroot (1970), among others.

26.53 Suppose we can specify a set of possible actions $A = \{a\}$ and a *decision rule* $d(x)$ that specifies the action to be taken when x is observed. The consequence of taking that action is to incur a *loss* $L[d(x), \theta]$ when the parameter value is θ. Some authors use a utility function rather than a loss function; for most purposes, loss can be viewed as negative utility, although utility may be deemed to be bounded whereas loss functions are often allowed to be unbounded.

The expected loss is known as the *risk function*:

$$R(d, \theta) = \int L[d(x), \theta] f(x|\theta)\,dx. \tag{26.59}$$

A decision rule, d, is *admissible* if there is no rule d' such that

$$R(d', \theta) \leq R(d, \theta) \qquad \text{for all } \theta, \tag{26.60}$$

with strict inequality for at least some θ. Quite generally, nothing is lost by restricting attention to the class of admissible decision rules, although this class may be large.

26.54 In order to select a particular decision rule, we may use a criterion such as *minimax*; that is, we choose the rule $d(x)$ that minimizes risk taken over all θ:

$$\min_d \max_\theta R(d, \theta). \tag{26.61}$$

Example 26.17 (Admissibility of estimators for the normal mean)

Suppose we use the squared error loss function

$$L[d(x), \theta] = \{d(x) - \theta\}^2$$

and consider five estimators for the location parameter of a normal distribution given a sample of size n:

(i) sample mean, $d_1(x) = \bar{x}$;
(ii) sample median, $d_2(x) = m$;
(iii) the first observation, $d_3(x) = x_1$;
(iv) the number 631, $d_4(x) = 631$;
(v) the weighted mean, $d_5(x) = \sum w_i x_i, \sum w_i = 1$.

It follows that the risk functions are, with $R_i = R(d_i, \theta)$,

$$R_1 = \sigma^2/n, \qquad R_2 \doteq 1.57\sigma^2/n, \qquad R_3 = \sigma^2$$
$$R_4 = (631 - \theta)^2, \qquad R_5 = \sigma^2\left\{\frac{1}{n} + \sum(w_i - \bar{w})^2\right\}.$$

By inspection, d_2, d_3 and d_5 are inadmissible; d_4 would be ruled inferior to all others by the minimax criterion even though it is admissible since R_4 is smallest when θ is near 631.

26.55 The partial ordering induced by admissibility or selection by a criterion such as minimax are as much as can be achieved in the absence of a prior distribution for θ. Thus, most modern research in decision theory follows the Bayesian path and assumes the existence of a prior.

Given $p(\theta)$, we may compute the expected risk:

$$\begin{aligned} E(R) &= \int R(d, \theta) p(\theta)\, \mathrm{d}\theta \\ &= \int_\theta \int_x L[d(x), \theta] f(x|\theta) p(\theta)\, \mathrm{d}x\, \mathrm{d}\theta; \end{aligned} \tag{26.62}$$

(26.62) may be re-expressed as

$$E(R) = \int_x \left\{ \int_\theta L[d(x), \theta] P(\theta|x)\, \mathrm{d}\theta \right\} f(x)\, \mathrm{d}x \tag{26.63}$$

provided that the order of integration can be reversed. From (26.63), it is apparent that we may minimize $E(R)$ by choosing $d(x)$ to minimize the inner integral, for each x. The resulting decision rule is known as the *Bayes rule* and its expected risk is termed the *Bayes risk*. Any admissible rule is a Bayes rule for *some* prior distribution (a result due to Wald).

Example 26.18 (Bayes rule for squared error loss)

When squared error loss is used, it follows that, provided the inner integral in (26.63) is bounded, it is minimized by selecting the posterior mean, whatever the distribution.

The James–Stein estimator

26.56 Suppose now that we have a sample of size n from each of K normal populations, $x_i \sim N(\mu_i, \sigma^2)$; we take σ^2 to be known and assume that the μ_i have common prior $N(\phi, \tau^2)$.

Let the decision rule $d(t)$, $t = (t_1, \dots, t_K)$, assign estimate t_i to parameter μ_i, $i = 1, \dots, K$, and let the loss function be

$$L[d(t), \mu] = (t_1 - \mu_1)^2 + \cdots + (t_K - \mu_K)^2. \tag{26.64}$$

The obvious (and ML) estimator is $t_{ML} = (\bar{x}_1, \dots, \bar{x}_K)$ for which

$$E(R) = K\sigma^2/n. \tag{26.65}$$

However, from Example 26.16, the Bayes rule is given by the posterior means:

$$t_i^* = \frac{\phi\sigma^2 + n\bar{x}_i\tau^2}{\sigma^2 + n\tau^2} \tag{26.66}$$

for which

$$E(R) = \frac{K\tau^2\sigma^2}{\sigma^2 + n\tau^2},$$

which is clearly less than the expected risk in (26.65).

The reduced risk associated with (26.66) is now considered unremarkable, since such *shrinkage* estimators have become very popular and encompass a variety of other procedures, including ridge regression (cf. **32.48**). What distinguishes the result is that in his original derivation, Stein (1956) did not use any kind of Bayesian formulation. Nevertheless, he was able to demonstrate that t_{ML} is inadmissible for $K \geq 3$. James and Stein (1961) considered the improved class of estimators (for ϕ known) of the form $t_{JS} = (\tilde{t}_1, \dots, \tilde{t}_K)$ where

$$\tilde{t}_i = 1 - \frac{c\sigma^2(\bar{x}_i - \phi)}{n\sum(\bar{x}_j - \phi)^2};$$

such that $0 < c < K - 2$. If we write

$$V = \sum(\bar{x}_j - \phi)^2,$$

Baranchik (1970) showed that the estimators

$$\tilde{t}_i = \left\{1 - \frac{r(V)}{V}\right\}(\bar{x}_i - \phi)$$

are admissible, provided that $r(V)$ is non-decreasing in V and $0 < r(\cdot) < K - 2$.

For an excellent discussion of these shrinkage estimators and simple proofs of the admissibility results, see Stigler (1990).

26.57 These results have been used by some to cast doubt upon the value of ML estimators. Alternatively, they might be used to sound a strong note of caution about using decision theory unless one is very sure of the loss function and the prior distribution.

For a review of the historical development of decision theory and an extensive bibliography, see Fishburn (1989).

Discussion

26.58 There has been so much controversy about the various methods of estimation we have described that, at this point, we shall have to leave our customary objective standpoint. The remainder of this chapter is an expression of personal views. We think that it is the correct viewpoint; and it represents the result of many years' reflection on the issues involved, a serious attempt to understand what the protagonists say, and an even more serious attempt to divine what they mean.

26.59 We have, then, to examine six major approaches, although some are more closely related than others: frequency, likelihood, fiducial, objective and subjective Bayesian, and decision theory. We must not be misled by, though we may derive some comfort from, the similarity of the results to which they lead in certain simple cases. We shall, however, develop the thesis that, where they differ, the basic reason is not that one or more are wrong, but that they are consciously or unconsciously either answering different questions or resting on different postulates.

26.60 In setting out the differences, it is useful to adopt Lakatos' (1974) concept of competing research programmes, and to establish the *hard core* of assumptions underlying each theory. Supporting each theory is a *protective layer* of auxiliary assumptions, so the conclusions that may be drawn then follow deductively from these foundations. We are not concerned to debate at length which assumptions are major and which auxiliary, but rather to use this as a framework for our discussions.

26.61 The hard core underlying the frequency theory may be summarized as follows:

(a) The Kolmogorov axioms.
(b) Well-defined random sampling procedures, that include specification of the sample space and stopping rule.
(c) The frequency interpretation of probability.
(d) A version of the repeated sampling principle (Cox and Hinkley, 1974, p. 45) which states that statistical procedures are to be assessed by their behaviour in hypothetical repetitions under the same conditions. This is Cox and Hinkley's strong version; the weak version requires only that we should not follow procedures that would be misleading for some parameter combinations (most of the time, in hypothetical repetitions). This principle is essentially the same as Birnbaum's *confidence principle*, introduced in **26.25**. As noted earlier, it is the conflict between this principle and the likelihood principle that is at the root of the debate between frequentists and Bayesians.

The protective belt then includes such concepts as consistency, unbiasedness, efficiency, sufficiency, power, etc. discussed in earlier chapters of this volume. In his discussion of maximum likelihood and decision theory, Efron (1982, p. 343) refers to these concepts as 'ingenious evasions' used by Fisher to avoid a decision-theoretic approach. However, it should be noted that Fisher, like Neyman and Pearson, was at pains to avoid strong assumptions regarding the existence and form of loss functions and prior distributions; the notions are certainly ingenious but form part of an alternative paradigm, not an evasion.

26.62 The frequency approach leads then to point and interval estimates and tests of hypotheses that are keyed to an interpretation of performance *in the long run*. The other approaches we have described constitute several attempts to develop an additional, or alternative, notion of probability that enables the investigator to make inferential statements conditionally upon the data recorded in a *particular statistical experiment*.

26.63 The protective belt of any theory evolves over time as, for example, when the Neyman–Pearson approach to testing hypotheses supplanted Fisher's pure tests of significance; note that we are not claiming such changes are either instantaneous or free from controversy! Another possible modification would be the use of unbiased estimating equations in place of unbiasedness (cf. **26.9**).

Two of the difficulties facing the frequency approach in practice are the specification of the sample space and need to ensure random sampling. Johnstone (1989) argues that it is not necessary that we know that the sample was drawn at random; '[A]ll that is necessary logically is that we *not have knowledge to the contrary*' (his emphasis). Following Fisher, Johnstone terms this a *postulate of ignorance* which is distinct from Bayes' postulate in that it applies to the sample space rather than the parameter space; hence, it is also distinct from our earlier argument in **9.9**, which presumed random sampling.

Johnstone's ideas are clearly open to abuse but, used carefully, have considerable merit. For example, the distribution of the error term in a regression equation applied to some macroeconomic aggregate has a much more plausible interpretation when Johnstone's interpretation is used.

26.64 The frequency approach is quite general in that it may be applied to any sampling situation once the sampling process is fully specified. However, there may be difficulties in execution. For example, when no single sufficient statistic exists, the confidence intervals may be imaginary or otherwise nugatory (cf. **19.12** and Example 19.4). Thus, the approach is not entirely free of the need for sufficiency, although it is not required. Perhaps it would be better to say that problems of interpretation may exist when nested and simply connected intervals cannot be obtained.

The principal argument in favour of the frequency theory of probability is that it does without any assumptions concerning prior distributions such as are essential to the Bayesian approach. This, in our opinion, is undeniable. But it is fair to ask whether it achieves this economy of basic assumption without losing something which the Bayesian theory possesses. Our view is that it does lose something on occasion, and that this something may be important for the purposes of estimation.

Prior information

26.65 Consider the case where we are estimating the mean μ of a normal population with known variance, and suppose that we *know* that μ lies between 0 and 1. According to Bayes' postulate, we should have

$$P(\mu|\bar{x}) = \frac{\exp\{-\frac{n}{2}(\mu-\bar{x})^2\}}{\int_0^1 \exp\{-\frac{n}{2}(\mu-\bar{x})^2\}\,d\mu}, \tag{26.67}$$

and the problem of setting limits to μ, though not free from mathematical complexity, is determinate. What has confidence-interval theory to say on this point? It can do no more than reiterate statements like

$$P\{\bar{x} - 1.96/\sqrt{n} \leq \mu \leq \bar{x} + 1.96/\sqrt{n}\} = 0.95.$$

These are still true in the required proportion of cases, but the statements take no account of our prior knowledge about the range of μ and may occasionally be idle. It may be true, but is absurd, to assert $-1 \leq \mu \leq 2$ if we know already that $0 \leq \mu \leq 1$. Of course, we may truncate our interval to accord with the prior information. In our example, we could assert only that $0 \leq \mu \leq 1$: the observations would have added nothing to our knowledge.

Thus, it appears that the frequency theory has the defect of its principal virtue: it attains its generality at the price of being unable to incorporate prior knowledge into its statements. When we make our final judgement about μ, we have to synthesize the information obtained from the observations with our prior knowledge. Bayes' theorem attempts this synthesis at the outset. Frequency theory leaves it until the end (and, we feel bound to remark, in most current expositions ignores the point completely).

26.66 Fiducial theory, as we have remarked, has been confined by Fisher to the case where sufficient statistics are used, or quite generally, to cases where all the information in the likelihood function can be utilized. No systematic exposition has been given of the procedure to be followed when prior information is available, but there seems no reason why a similar method to that exemplified by equation (26.67) should not be used. That is, if we derive the fiducial distribution $f(\mu)$ over a general range but have the supplementary information that the parameter must lie in the range μ_0 to μ_1 (within that general range), we modify the fiducial distribution by truncation to .

$$f(\mu) \Big/ \int_{\mu_0}^{\mu_1} f(\mu)\, \mathrm{d}\mu.$$

Falsificationism

26.67 One final observation is relevant concerning the frequentist approach. Its development was paralleled by the development of *falsificationism* in the philosophy of science, spearheaded by Sir Karl Popper (cf. Popper, 1968). The basis of Popper's theory is that evidence may or may not refute a theory but does not sustain it; that is, science progresses by performing experiments that challenge theories. This view of science does not allow the results of an experiment to provide explicit corroboration for a theory; it is symbolized by the stricture that we speak of 'not rejecting H_0' rather than 'accepting H_0'. More fundamentally, the frequentist approach does not seek to provide measures of corroboration, and it is the search for such measures that has, in part, fuelled the development of alternative paradigms for statistical inference. Indeed, all the other approaches described in this chapter allow corroborative statements to be made on the basis of the experiment just performed and conditionally upon the observations.

Likelihood-based inference

26.68 The likelihood function is recognized by all schools of thought as being a comprehensive summary of the data. Indeed, as we saw in **18.50**, Fisher (1956) suggested plotting the LF against θ; others (e.g. Efron, 1982) also strongly support the use of the LF as an effective *summary*. Edwards (1992), arguing that the LF described the relative *support* for different values of θ, went further and suggested that inferences be made on the basis of these support values. Clearly, such an approach is consistent with the SLP, although it is incomplete unless supplemented with some procedure for handling nuisance parameters, such as the use of conditional likelihood (cf. **18.33**) or the approaches discussed in **26.31–34**. Such methods have the advantage that prior information may be incorporated through a prior likelihood function.

Bayesian procedures are always consistent with the SLP, although empirical Bayes methods need not be. Fiducial inference may violate the SLP, although such violations tend to be uncommon.

Probability as a degree of belief

26.69 In the Bayesian and fiducial arguments, we must first assume the existence of a different concept of probability that measures the degree of belief or credibility in a hypothesis or theory. Carnap (1962) termed this probability$_1$, as distinct from the frequency concept, probability$_2$. Viewed in this light, the failure (?) of the frequency approach to deliver statements on the credibility of a hypothesis is almost axiomatic, since frequentists are unwilling to accept any probability$_1$ concepts that do not have a frequency interpretation.

The fiducialist argument rests on the assumption that probability$_2$ can be converted into probability$_1$ by means of a pivoting operation. Following the discussion in **26.26–29** we know that the process is possible; the key question is whether the resulting probability measure is meaningful.

26.70 The hard core for Bayesian inference is an axiomatic development that provides the framework for specifying prior probabilities and updating such probabilities by Bayes' theorem. For the objectivist, this means that there should be an agreed process by which a prior may be generated that is acceptable to all. Such a rule is necessarily mechanistic, since subjective interpretation is not admissible; yet, if the performance of the rule cannot be judged by either frequency or subjective criteria, its meaning remains rather obscure. Indeed, we are asked to accept that the prior be flat on $(-\infty, \infty)$ but inversely proportional to θ on $(0, \infty)$. Sophisticated arguments concerning the distinction somehow fail to impress us as touching the root of the problem. Further, it is found that working with uninformative priors can lead to some theoretical difficulties; see Stone (1976) and Exercise 26.5.

26.71 The subjectivists' hard core requires the individual to be willing to bet on anything, but in a logical fashion, as noted in **26.41**. Given this framework, you can certainly begin with your prior and derive your posterior probability statement regarding the plausibility of a hypothesis. Let us begin by considering the process of specifying the prior.

If the prior distribution is specified in conjugate form, such as the normal mean being $N(\phi, \tau^2)$, we are faced with a potential infinite regress in specifying the prior for (ϕ, τ^2) and so on. This is resolved only by claiming knowledge of the (hyper)parameters at some stage (cf. Lindley and Smith, 1972). If the prior distribution is determined within a framework of bets, you must be able to specify your utility function. Once this is available, an axiomatic development such as that of

Savage (1954) shows that coherent behaviour leads to degrees of belief that satisfy the axioms of probability.

Once the prior is available, you may proceed to make inferences in a manner that is consistent with the SLP (and, therefore, possibly inconsistent with the confidence principle). Whether this is a source of strength or weakness lies in the eye of the beholder, but the fact remains that all the inferences made are subjective, your own assessments.

Whether such individual statements are acceptable is problematic. When making a decision in a context that lacks opportunities for replication, the use of your probabilities seems reasonable when you are responsible for the decision. However, we believe that many, if not most, statistical analyses cannot be reasonably fitted into a decision-theoretic framework. Furthermore, an expression of personal beliefs has not proved acceptable as a way of reporting the results of an investigation.

Reconciliation?

26.72 As might be expected, there have been a number of attempts to reconcile the different approaches to statistical inference; we shall review some of these briefly. We begin by noting that, in large samples, all methods are consistent with the strong likelihood principle.

26.73 We saw in **8.25** that the use of Bayes' theorem with a uniform prior distribution gives a posterior mode that is equal to the ML estimator. Even if a non-uniform prior distribution is used, the methods are *asymptotically* equivalent. Equation (8.17) may be written in our present notation as

$$P(\theta|x) \propto p(\theta)L(x|\theta). \tag{26.68}$$

To maximize this with respect to θ is equivalent to maximizing its logarithm,

$$\log p(\theta) + \log L(x|\theta) = \sum_{i=1}^{n}\{\log f(x_i|\theta) + (1/n)\log p(\theta)\}. \tag{26.69}$$

As $n \to \infty$, the second term in braces on the right is negligible provided $p(\theta) > 0$, and we are effectively maximizing $\log L(x|\theta)$ to obtain the ML estimator. We may express this by saying that, given enough observations, the prior distribution becomes irrelevant; this is known as the principle of *stable estimation*. For small n, however, there may be wide differences between the ML and Bayesian estimates – cf. Exercise 26.4.

Diaconis and Freedman (1986a; 1986b) show that when the parameter space is high-dimensional (or infinite-dimensional as in some non-parametric problems), the prior may swamp the data no matter how many observations are available. In this sense, Bayesian estimators may lack consistency; the discussion following the 1986a paper should also be consulted.

Several papers in recent years have worked 'backwards' to develop a prior distribution that provides a match between frequentist and posterior probabilities, at least to $O_p(n^{-1})$; see, for example, Pierce and Peters (1994), and Datta and Ghosh (1995). Although such results are illuminating and may be a source of comfort to some, the interpretations remain distinct, so such research does not really 'bridge the gap'.

Samaniego and Reneau (1994) explore the idea of separating the space of operational priors into 'good' and 'bad' priors, according to whether the assumed prior information benefits the inferential process. They conclude that, as would be expected, the choice between procedures depends crucially on the quality of available information. Bayesian and frequentist procedures each emerge as preferable under certain well-defined and complementary conditions.

Goutis and Casella (1995) characterize frequentist methods as largely 'pre-data'; that is, their validity depends upon inferential statements formulated before the data are collected. By contrast, Bayesian methods are 'post-data' in that they make statements about the parameters given the specific set of observations. Goutis and Casella go on to show how the use of various conditioning arguments has enabled frequentists to construct a reasonable framework for 'post-data' inference.

26.74 One aspect of the Bayesian approach is, as we have suggested on occasion, that it demands too much. For example, we need to be able to specify the functional form of the LF and to list all the variables of interest. Yet, much of the appeal of procedures such as cross-validation and bootstrapping derives from their application in circumstances where it may not be possible to specify the LF precisely. Likewise, randomization in experimental design safeguards against factors that may not have been recognized.

Following this theme, Durbin (1988) points out that the overall complexity of many models makes specification of the LF, and therefore application of the likelihood principle, impractical. Nevertheless, simple diagnostic tests often guide the model builder well, and Durbin suggests that the practical effects of philosophical differences are often small compared to the need for effective statistical modelling.

26.75 Box (1980) identifies two components in statistical modelling: *criticism* and *estimation*. Starting from (26.32), Box would use the posterior distribution for θ for estimation, but the predictive distribution

$$f(x) = \int p(\theta)L(x|\theta)\,\mathrm{d}\theta \tag{26.70}$$

for model criticism. Although $f(x)$ is derived on the assumption that the prior $p(\theta)$ is available, Box recommends frequency procedures for the criticism part of the modelling process. This is similar in spirit to Durbin's comments in **26.74** and also to our earlier discussion in **26.31–34**, save that there θ was removed by maximization rather than integration.

26.76 Giere (1977) distinguishes between *testing* and *information* in statistical inference, suggesting that the information criterion allows a direct measure of evidence for a hypothesis so that the Bayesian approach may be invoked. In the testing framework, no such measure exists, as noted by many frequentist writers from Neyman and Pearson onwards. Giere goes on to argue for probability as a measure of propensity that would allow statements to be made for single experiments.

26.77 Good (cf. 1992 and earlier papers cited therein) calls for a Bayesian–non-Bayesian compromise from a different viewpoint. For Good, frequentist methods often represent a collection of *ad hoc* procedures, and he would accept frequentist procedures whenever they match up sufficiently well with the Bayesian solution. While such an approach may serve to reduce contention, 'compromise' is perhaps an inappropriate descriptor.

26.78 Putting these several considerations together, we see that good statistical practice can often emerge from different paradigms and that, indeed, different notions of probability may be appropriate in different circumstances. Nevertheless, the frequentist approach remains firmly rooted in the Popperian tradition of falsificationism, and any attempt to go beyond that requires recognition of some other concept of probability.

It may be tempting to think in terms of Kuhn's (1962) notion of a scientific revolution whereby the current (frequentist) paradigm is challenged by the newcomer (Bayesian), from which a new orthodoxy will emerge. However, this view is somewhat inappropriate; rather we should recognize that the Bayesian approach seeks to deliver more but, in order to do so, requires stronger assumptions.

In conclusion, it is fitting to quote some words written long ago (Kendall, 1949):

> The frequentist seeks for objectivity in defining his probabilities by reference to frequencies; but he has to use a primitive idea of randomness or equiprobability in order to calculate the probability in any given practical case. The non-frequentist begins by taking probabilities as a primitive idea but he has to assume that the values which his calculations give to a probability reflect, in some way, the behaviour of events ... *Neither party can avoid using the ideas of the other in order to set up and justify a comprehensive theory.*

EXERCISES

26.1 In the multinomial distribution of Example 18.10, show that the four-component minimal sufficient statistic for θ can be written either as $(a+b, c+d, b, d)$ or as $(a+c, b+d, c, d)$, and that the first two components in each case form an ancillary pair of statistics.

26.2 Consider the likelihood function for the two-means problem when the variates are $N(\mu_i, \sigma_i^2), i = 1, 2$. Use (26.26) to show that the likelihood ratio credibility measure is algebraically the same as the Fisher–Behrens distribution, with n_i in place of $n_i - 1$.

26.3 Show that if the distribution of a sufficient statistic x is

$$f(x|\theta) = \frac{\theta^2}{\theta+1}(x+1)\mathrm{e}^{-x\theta}, \qquad x > 0,\ \theta \geq 0,$$

the fiducial distribution of θ combined samples with sufficient statistics x, y, is

$$g_{x,y}(\theta) = \frac{\mathrm{e}^{-z\theta}}{(\theta+1)^3}[\theta^3(2z^2 + \tfrac{4}{3}z^3 + \tfrac{1}{6}z^4) + \theta^4(z^2 + z^3 + \tfrac{1}{6}z^4)],$$

where $z = x + y$, while that for a single sample is

$$g_x(\theta) = \frac{\theta x \mathrm{e}^{-x\theta}}{(\theta+1)^2}[1 + (1+\theta)(1+x)].$$

(Note that the minus sign in (26.23) is unnecessary here, since $F(x|\theta)$ is an increasing function of θ.) Hence show that the Bayes posterior distribution from the second sample, using $g_x(\theta)$ as prior distribution, is

$$P(\theta|y;x) \propto \mathrm{e}^{-z\theta}\left(\frac{\theta}{\theta+1}\right)^3 x(1+y)[1+(1+\theta)(1+x)],$$

so that

$$P(\theta|y;x) \neq g_{x,y}(\theta).$$

Note that $P(\theta|y;x) \neq P(\theta|x;y)$ also.

(Lindley, 1958)

26.4 If the prior distribution of $\nu\theta^{-1}$ is χ^2 with ν d.fr., and an observation x is distributed normally with mean zero and variance θ, show (cf. Exercise 16.25) that the posterior distribution of θ is

$$P(\theta|x) \propto \exp\{-(\nu + x^2)/(2\theta)\}\theta^{-(\nu+3)/2}.$$

Hence show that the Bayes estimate of θ must *always* exceed the pre-assigned constant $k_\nu = \nu/(\nu+3)$, however small $|x|$ is, although the prior probability that $\theta < k_\nu$ tends to 0.5 as $\nu \to \infty$, exceeding 0.2 at $\nu = 10$. Show that the ML estimator of θ is $\hat{\theta} = x^2$.

26.5 An amount x is deposited into a box, B_1, and $2x$ into a second box B_2. The boxes appear indistinguishable from the outside to a contestant, C, who is allowed to choose one box. Having made a choice, C is then offered the opportunity to switch boxes. C reckons that if the selected box contains y, the other box is equally likely to contain $y/2$ or $2y$ so that the switch is justified. Having made the switch, C repeats the argument and switches again, and so on.

Let Y_S, Y_N denote the expected amount from opening the selected and non-selected boxes, respectively. Show that

$$E(Y_N|Y_S = y) = y\frac{\frac{1}{2}p(y/2) + 2p(y)}{p(y/2) + p(y)}, \qquad (*)$$

where $p(x)$ is the prior density for C for the amount placed in B_1. Also show that $E(Y_N) = E(Y_S) = 3E(x)/2$ when $E(x)$ exists, whatever the form of $p(x)$. Note that if the improper prior

$$p(x) = \text{constant}, \qquad 0 < x < \infty$$

is used in $(*)$, C appears justified in switching.

26.6 Consider the linear model $\mathbf{y} = \mathbf{X}\beta + \epsilon$, where ϵ has a multinormal distribution with mean $\mathbf{0}$ and covariance matrix $\sigma^2\mathbf{I}$, and $\mathbf{X}$ is of rank $k < p$.
Given that $\boldsymbol{\beta}$ has a p-dimensional multinormal prior distribution with mean $\boldsymbol{\gamma}$ and $V(\boldsymbol{\beta}) = \mathbf{C}^{-1}$, show that $\boldsymbol{\beta}$ is estimable provided $(\mathbf{X}^T\mathbf{X} + \sigma^2\mathbf{C})$ is of rank p. Setting $\mathbf{C} = \sigma^{-2}\mathbf{B}^T\mathbf{B}$ and $\mathbf{a} = \mathbf{B}\boldsymbol{\gamma}$, compare this estimator with that given in **29.9**.

CHAPTER 27

STATISTICAL RELATIONSHIP: LINEAR REGRESSION AND CORRELATION

27.1 In the next six chapters, we consider relationships between two or more variables. We have already discussed bivariate and multivariate distributions and their properties; in particular, the normal case was examined in Chapters 15 and 16. However, a systematic discussion of the relationships between variables was deferred until the theory of estimation and testing hypotheses had been explored. Even so, the area that we are about to study is very large and we restrict attention to correlation and linear models, leaving a complete treatment of multivariate analysis to Krzanowski and Marriott (1994; 1995). We begin with a general overview.

27.2 Most of our work stems from an interest in the joint distribution of two or more random variables: we may describe this as the problem of *statistical relationship*. There is a quite distinct field of interest concerning relationships of a strictly functional kind between variables, such as those of classical physics; this subject is of statistical interest because the functionally related variables are subject to observational or instrumental errors. This topic is known as the problem of *functional relationship*, and is discussed by Van Ness and Cheng (1999). In these chapters, we are concerned with the problem of statistical relationship alone, where the variables are not (except in degenerate cases) functionally related, although they may also be subject to observational or instrumental errors.

Within the field of statistical relationship we may be interested either in the *interdependence* between a number of variables or in the *dependence* of one or more variables upon others. For example, we may be interested in whether there is a relationship between length of arm and length of leg; put this way, it is a problem of interdependence. But if we are interested in using leg length measurements to convey information about arm length, we are considering the dependence of the latter upon the former. This is a case in which either interdependence or dependence may be of interest. On the other hand, there are situations when only dependence is of interest. The relationship of crop yields and rainfall is an example where non-statistical considerations make it clear that there is an essential asymmetry in the situation: we say, loosely, that rainfall ‘causes’ crop yield to vary, and we are quite certain that crop yields do not affect the rainfall, so we measure the dependence of yield upon rainfall. We refer to rainfall as an *input*, or *regressor*, variable. The term *independent* variable is also widely used, but we feel that the use of the word ‘independent’ in this context is a source of confusion rather than illumination and shall endeavour to avoid such usage.

The statistical terminology is not always clear-cut. For example, in Chapter 28, the interdependence of two variables, with the effects of other variables eliminated, is known as *partial correlation*, whereas the dependence of a single variable upon a group of others is termed *multiple correlation*. Nevertheless, it is true in the main that the study of *interdependence* leads to the theory of correlation, whereas the study of *dependence* leads to the theory of regression. We shall examine the theory for two random variables in this chapter and that for $k \geq 2$ variables in

Chapter 28. Chapter 29 considers the general linear model. Chapters 30 and 31 cover the analysis of variance and Chapter 32 considers model diagnostics.

Causality in regression

27.3 Before considering the theory of correlation (largely developed around the beginning of the twentieth century by Karl Pearson and by Yule), we make one final general point. A statistical relationship, however strong and however suggestive, can never *establish* a causal connection: our ideas on causation must come from outside statistics, ultimately from some theory or other. Even in the simple example of crop yield and rainfall discussed in **27.2**, we had no *statistical* reason for dismissing the idea of dependence of rainfall upon crop yield: the dismissal is based on quite different considerations. Even if rainfall and crop yields were in perfect functional correspondence, we should not dream of reversing the 'obvious' causal connection. Statistical relationships, of whatever kind, cannot logically imply causation without additional assumptions.

George Bernard Shaw made this point brilliantly in his Preface to the *The Doctor's Dilemma* (1906):

> Even trained statisticians often fail to appreciate the extent to which statistics are vitiated by the unrecorded assumptions of their interpreters.... It is easy to prove that the wearing of tall hats and the carrying of umbrellas enlarges the chest, prolongs life, and confers comparative immunity from disease.... A university degree, a daily bath, the owning of thirty pairs of trousers, a knowledge of Wagner's music, a pew in church, anything, in short, that implies more means and better nurture ...can be statistically palmed off as a magic-spell conferring all sort of privileges.... The mathematician whose correlations would fill a Newton with admiration, may, in collecting and accepting data and drawing conclusions from them, fall into quite crude errors by just such popular oversights as I have been describing.

Although Shaw was on this occasion supporting a characteristically doubtful cause, his logic was valid. In the first flush of enthusiasm for correlation techniques, it was easy for early followers of Karl Pearson and Yule to be incautious. It was not until 20 years after Shaw wrote that Yule (1926) frightened statisticians by adducing cases of very high correlations which were obviously not causal: for example, the annual suicide rate was highly correlated with the membership of the Church of England. Most of these 'nonsense' correlations operate through concomitant variation in time, but they had the salutary effect of bringing home to the statistician that causation cannot be deduced from such observational studies.

An interesting form of 'spurious' correlation arises when two variables are scaled by a common third variable; see Exercise 27.18.

27.4 The issue of causality cannot, however, be overlooked. It behoves statisticians to indicate when, if at all, the results of a statistical investigation may be admissible as evidence in support of a causal relationship.

One step we can take is to distinguish between controlled and observational investigations. In controlled, or experimental, studies, the input (or regressor) variables can be manipulated by the investigator and other sources of variation effectively controlled. It is then reasonable to argue that changes in the *dependent* or (*response*) variables are caused by the changes in the input variables. Ultimately, this depends upon the investigator's success in achieving the stated experimental conditions. *Observational* studies, where designed experiments are infeasible, will rarely provide evidence of causation. We now consider this argument in greater detail. Our discussion is based upon Holland (1986).

27.5 In order to explore the issue of causality further, we simplify matters by considering a finite population of N individuals $P = \{I_1, I_2, \ldots, I_N\}$ who may be placed into one of two groups: t = treatment or c = control. By focusing upon a population, we are able to sidestep any sampling issues. In a statistical experiment, each individual is assigned at random to a group, by a process that does not depend upon the characteristics under study.

Before the assignment to groups takes place, we may conceive of two possible measurements for individual $I_j : y_j(t)$ and $y_j(c)$, representing the possible values under t and c, respectively. The assignment of individuals into treatment group $T = \{I_{(1)}, I_{(2)}, \ldots, I_{(M)}\}$ and control group $C = \{I_{(M+1)}, \ldots, I_{(N)}\}$ creates two non-overlapping groups such that $T \cup C = P$. We will, of course, observe $y_j(t)$ only if $I_j \in T$, and $y_j(c)$ only if $I_j \in C$.

What we would like to do is to compare

$$\bar{y}(t) = \sum_{j=1}^{N} y_j(t)/N \quad \text{and} \quad \bar{y}(c) = \sum_{j=1}^{N} y_j(c)/N,$$

but these quantities are not observable. However, if both measurements are possible and the logically antecedent process of partitioning the individuals does not depend upon values $y_j(\cdot)$, then we may estimate the two means unbiasedly by

$$m(t) = \sum_{j \in T} y_j(t)/M \quad \text{and} \quad m(c) = \sum_{j \in C} y_j(c)/(N - M).$$

Then, if the difference between $m(t)$ and $m(c)$ is sufficiently large, according to some appropriate decision rule, we may conclude that the treatment has *caused* that difference.

The key ingredients of this argument are:

(1) we measure the *effects of causes* (i.e. a treatment) in a controlled study;
(2) each member of the population may be assigned to either treatment group;
(3) the randomization procedure is independent of the treatments (and of any factors that may be correlated with the treatments).

This brief description outlines the basis of the argument presented in Holland (1986). In particular, Holland stresses that attempts to discern the *causes of effects* (the basis of the most observational studies) will generally founder because of the multiplicity of possible cases.

This framework for inducing causality may seem overly narrow since observational studies, or even studies using non-assignable factors such as the sex of the individual, are ruled out by guidelines (1)–(3); indeed, some of the discussion following Holland's paper seeks to broaden the framework. In practice, researchers always have and will no doubt continue to make causal inferences when some of (1)–(3) are not strictly satisfied. However, any investigation attempting to induce causality should strive to satisfy these conditions at the design stage as far as possible, and make clear any shortcomings in this respect.

Rosenbaum and Rubin (1983) present a theory of propensity scores for dealing with covariates in observational studies. Rubin (1990; and earlier papers cited therein), provides a systematic framework for causal inference. Granger (1969) describes a form of causality based on a clear time ordering of the variables, such as a sales increase following a price cut; this alternative is also examined in Holland (1986) and the ensuing discussion.

Conditional expectation and covariance

27.6 In Chapter 1 (Tables 1.15, 1.23 and 1.24) we gave some examples of bivariate distributions arising in practice. Table 27.1 provides a further example that will be used for illustrative purposes.

For a moment, we treat these data as populations, leaving aside the question of sampling until later in the chapter.

For univariate distributions, we constructed summary measures such as the mean and variance. We now seek measures of interdependence.

Denote the two variables by x and y. For any given value of x, x_0 say, we may consider the conditional density of y, $f(y|x_0)$. The conditional expectation of y, given $x = x_0$, is

$$E(y|x_0) = \sum yf(y|x_0) \tag{27.1}$$

or, for continuous random variables,

$$\begin{aligned} E(y|x_0) &= \int yf(y|x_0)\,\mathrm{d}y \\ &= \frac{\int yf(x_0, y)\,\mathrm{d}y}{f_x(x_0)}, \end{aligned} \tag{27.2}$$

where $f_x(x_0) = \int f(x_0, y)\,\mathrm{d}y$ and the x subscript is used to indicate the marginal density of x. All the integrals are taken over $(-\infty, \infty)$ and (27.2) exists provided $f_x(x_0) \neq 0$ and the integral in the numerator is finite. When the conditional expectation is defined for all x values, the function $E(y|x)$ is known as the *regression function* or, simply, the *regression* of y on x. The regression of x on y is similarly defined as $E(x|y)$.

The existence of two regression lines sometimes serves as a source of confusion. However, the definitions in (27.1)–(27.2) show that we are dealing with quite distinct subpopulations. For example, in Table 27.1, $E(y|x = 64)$ describes the average weight of women 64 inches tall and is computed from the 1454 entries in that column. From Table 27.1, $E(y|x = 64) = 134.6$; similarly, $E(x|y = 134.5) = 63.84$ computed from the 521 entries in the appropriate row. Although the (x, y) points in Fig 27.1 almost coincide, this is purely fortuitous as only 175 subjects (in the class $x = 64$, $y = 134.5$) are included in both calculations. It is clear from Fig. 27.1 that the pairs of conditional expectations may be quite different.

When both regression functions are linear, the angle between them may be found using (27.17).

27.7 A natural measure of dependence is the product moment μ_{11} known as the *covariance* of x and y, defined (cf. **3.27**) as

$$\begin{aligned} \mu_{11} &= \int_{-\infty}^{\infty}\int_{-\infty}^{\infty} (x - \mu_x)(y - \mu_y)\,\mathrm{d}F(x, y) \\ &= E\{(x - \mu_x)(y - \mu_y)\} \equiv E(xy) - E(x)E(y). \end{aligned} \tag{27.3}$$

Table 27.1 Distribution of weight and stature for 4995 women in Great Britain, 1951. Reproduced by permission, from *Women's Measurements and Sizes*, London, HMSO, 1957

Weight (y) class midpoints, in pounds	Stature (x): class midpoints, in inches											Total
	54	56	58	60	62	64	66	68	70	72	74	
278.5						1						1
272.5												0
266.5						1						1
260.5							1					1
254.5												0
248.5					1	1						2
242.5							1					1
236.5							1					1
230.5					2				1			3
224.5					1	2	1					4
218.5			1		2	1		1				5
212.5				2	1	6		1	1			11
206.5				2	2	3	2		1			10
200.5			4	2	6	2						14
194.5				1	3	7	7	4	1			23
188.5			1	5	14	8	12	3	1	2		46
182.5			1	7	12	26	9	5		1	2	63
176.5			5	8	18	21	15	11	7		2	87
170.5			2	11	17	44	21	13	3	1		112
164.5		1	3	12	35	48	30	15	5	3		152
158.5			8	17	52	42	36	21	9			185
152.5		1	7	30	81	71	58	21	2	2		273
146.5		2	13	36	76	91	82	36	8	1		345
140.5		1	6	55	101	138	89	50	8			448
134.5			15	64	95	175	122	45	5			521
128.5		1	19	73	155	207	101	25	3			584
122.5		3	34	91	168	200	81	12	1	1		591
116.5		3	24	108	184	184	50	8				561
110.5		5	33	119	165	124	22	4				472
104.5	1	3	33	87	95	35	6					260
98.5	1	5	29	59	45	16	3					159
92.5		6	10	21	9							46
86.5		1	5	3								9
80.5	2	1	1									4
Total	5	33	254	813	1340	1454	750	275	56	11	4	4995

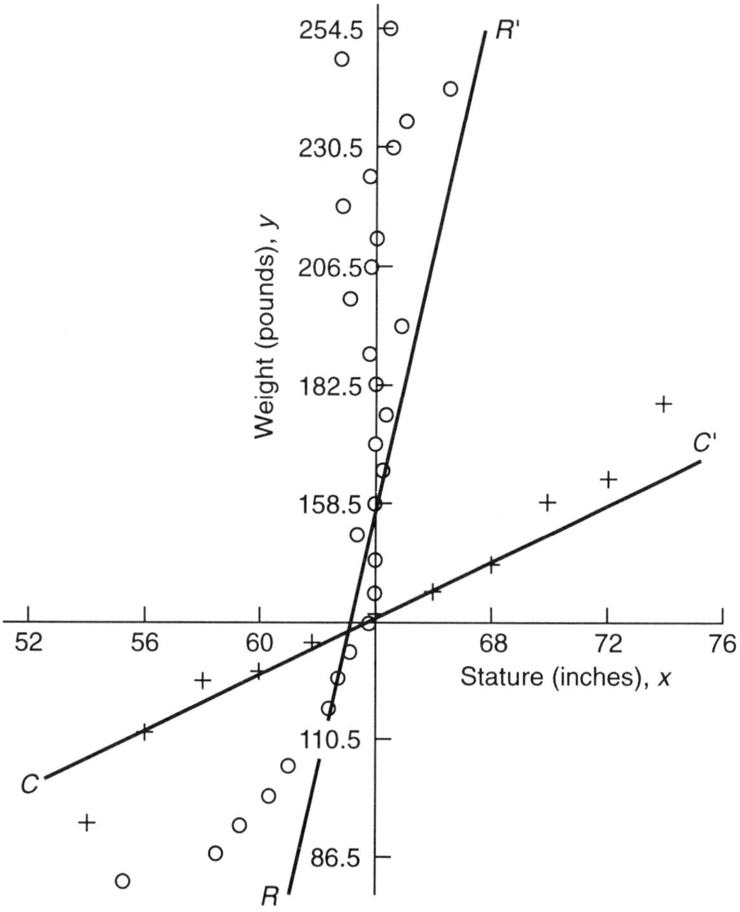

Fig. 27.1 Regressions for data of Table 27.1

In turn, (27.3) may be expressed as

$$\mu_{11} = E_x[x\{E(y|x) - E(y)\}]$$

so that the covariance represents a weighted average of the conditional means, measured from the unconditional mean. If the variates x, y are independent, $E(xy) = E(x)E(y)$, so that

$$\mu_{11} = \kappa_{11} = 0, \qquad (27.4)$$

The converse is not generally true: (27.4) does not generally imply independence, which requires

$$\kappa_{rs} = 0 \qquad \text{for all } r \times s \neq 0. \qquad (27.5)$$

For a *binormal* distribution, however, we know that $\kappa_{rs} = 0$ for all $r+s > 2$, so that κ_{11} is the only non-zero product cumulant. Thus (27.4) implies (27.5) and independence for normal variables. It may also do so for other specified distributions, but it does not do so in general. Example 27.1 gives a non-normal distribution for which $\kappa_{11} = 0$ implies independence; Example 27.2 gives one where it does not.

Example 27.1 (Independence implied by zero correlation)

If x and y are binormal standardized variables, the joint characteristic function of x^2 and y^2 is, using (16.45),

$$\phi(t,u) = \int_{-\infty}^{\infty} (2\pi)^{-1/2} \exp(-\tfrac{1}{2}x^2 + x^2 it) \\ \times \left[\int_{-\infty}^{\infty} \{2\pi(1-\rho^2)\}^{-1/2} \exp\{-\tfrac{1}{2}[(y-\rho x)^2/(1-\rho^2)] + y^2 iu\}\mathrm{d}y\right] \mathrm{d}x.$$

The inner integral is the c.f. of $(1-\rho^2)$ times a $\chi'^2(1, \rho^2 x^2/(1-\rho^2))$ variate, which by Exercise 22.1 is

$$\{1 - 2iu(1-\rho^2)\}^{-1/2} \exp\left\{\frac{\rho^2 x^2 iu}{1 - 2iu(1-\rho^2)}\right\}.$$

Hence

$$\phi(t,u) = \{1 - 2iu(1-\rho^2)\}^{-1/2} \\ \times \int_{-\infty}^{\infty} (2\pi)^{-1/2} \exp\{-\tfrac{1}{2}x^2 + x^2[it + \rho^2 iu/\{1 - 2iu(1-\rho^2)\}]\} \,\mathrm{d}x.$$

The integral is the c.f. of a central χ_1^2 variate with it replaced by the expression in square brackets. Thus

$$\phi(t,u) = \{1 - 2iu(1-\rho^2)\}^{-1/2}\{1 - 2[it + \rho^2 iu/\{1 - 2iu(1-\rho^2)\}]\}^{-1/2} \\ = [(1-2it)(1-2iu) + 4\rho^2 tu]^{-1/2}.$$

We see that

$$\phi(t,0) = (1-2it)^{-1/2}, \\ \phi(0,u) = (1-2iu)^{-1/2},$$

so that the marginal distributions are chi-squared with one degree of freedom, as we know. By differentiating the logarithm of $\phi(t,u)$ we find

$$\mu_{11} = \kappa_{11} = \left[\frac{\partial^2 \log \phi(t,u)}{\partial(it)\,\partial(iu)}\right]_{t=u=0} = 2\rho^2.$$

Now when $\rho = 0$, we see that

$$\phi(t,u) = \phi(t,0)\phi(0,u),$$

a necessary and sufficient condition for independence of x^2 and y^2 from **4.16–17**. Thus $\mu_{11} = 0$ implies independence in this case.

Example 27.2 (Independence not implied by zero correlation)

Consider a bivariate distribution with uniform probability over a unit disc centred at the means of x and y. We have

$$f(x, y) = \frac{1}{\pi} \qquad 0 \le x^2 + y^2 \le 1,$$

so that

$$\begin{aligned}\mu_{11} &= \int\!\!\int xy\,\mathrm{d}F = \frac{1}{\pi}\int\!\!\int xy\,\mathrm{d}x\,\mathrm{d}y \\ &= \frac{1}{\pi}\int x\left[\int y\,\mathrm{d}y\right]\mathrm{d}x \\ &= \frac{1}{\pi}\int x\left[\tfrac{1}{2}y^2\right]_{-(1-x^2)^{1/2}}^{+(1-x^2)^{1/2}}\mathrm{d}x = 0,\end{aligned}$$

as is otherwise obvious. But clearly x and y are not independent, since the range of variation of each depends on the value of the other.

The correlation coefficient

27.8 A meaningful measure of interdependence should be invariant to changes in location and scale, as well as being symmetric in the random variables. These conditions are met by the *correlation coefficient*

$$\operatorname{corr}(x, y) = \rho = \mu_{11}/\sigma_1\sigma_2, \tag{27.6}$$

where σ_1^2 and σ_2^2 denote the variances of x and y, respectively.

By the Cauchy–Schwarz inequality (**2.31**)

$$\begin{aligned}\mu_{11}^2 &= \left\{\int\!\!\int (x-\mu_x)(y-\mu_y)\,\mathrm{d}F\right\}^2 \\ &\le \left\{\int\!\!\int (x-\mu_x)^2\,\mathrm{d}F\right\}\left\{\int\!\!\int (y-\mu_y)^2\,\mathrm{d}F\right\} = \sigma_1^2\sigma_2^2\end{aligned}$$

so that

$$0 \le \rho^2 \le 1, \tag{27.7}$$

the upper inequality in (27.7) holding if and only if x and y are in strict linear functional relationship. For certain inequalities among sets of correlations, see Exercises 27.7–8.

Example 27.3 (Correlation for a bivariate chi-squared distribution)

Since x^2 and y^2 in Example 27.1 are each chi-squared with one degree of freedom, it follows that their variances are both 2 and

$$\operatorname{corr}(x^2, y^2) = 2\rho^2/(2\cdot 2)^{1/2} = \rho^2.$$

A perfect linear relationship requires $x^2 = c^2y^2$ or $x = \pm cy$.

From (27.7) and Example 27.2 we see that although independence of x and y implies $\mu_{11} = \rho = 0$, the converse does not generally apply. It does apply for jointly normal variables, and sometimes for others (Example 27.1). Indeed, ρ is a coefficient of *linear* interdependence, and it does not capture more complex forms of interdependence. In general, joint variation is too complex to be summarized by a single coefficient.

To *express* a quality, moreover, is not the same as to *measure* it. If $\rho = 0$ implies independence, we know that as $|\rho|$ increases, the interdependence also increases until when $|\rho| = 1$ we have the limiting case of linear functional relationship. Even so, it remains an open question which function of ρ should be used as a *measure* of interdependence: we see from (27.18) that ρ^2 is more directly interpretable than ρ itself, being the ratio of the variance of the fitted line to the overall variance. Leaving this point aside, ρ gives us a measure in such cases, though there may be better measures.

On the other hand, if $\rho = 0$ does not imply independence, it is difficult to interpret ρ as a measure of interdependence, and perhaps wiser to use it as an indicator rather than as a precise measure. In applications, we recommend the use of ρ as a *measure* of interdependence only in cases of normal or near-normal variation.

Linear regression

27.9 If the regression of y on x is exactly linear, we have the expression

$$E(y|x) = \alpha_2 + \beta_2 x, \tag{27.8}$$

and we now wish to find α_2 and β_2. Note that

$$\begin{aligned} E(y) &= \int\int y \, \mathrm{d}F(x, y) \\ &= \int\int \{y \, \mathrm{d}F(y|x)\} \, \mathrm{d}F(x) \\ &= E_x\{E_{y|x}(y|x)\} \end{aligned} \tag{27.9}$$

provided both sides exist, as we shall now assume. For a counter-example, see Exercise 27.24. Thus, taking expectations on both sides of (27.8) with respect to x, we see that

$$\mu_y = \alpha_2 + \beta_2 \mu_x. \tag{27.10}$$

If we subtract (27.10) from (27.8), multiply both sides by $(x - \mu_x)$ and take expectations with respect to x, we find

$$E\{(x - \mu_x)(y - \mu_y)\} = \beta_2 E\{(x - \mu_x)^2\},$$

or

$$\beta_2 = \mu_{11}/\sigma_1^2. \tag{27.11}$$

Similarly, we obtain

$$\beta_1 = \mu_{11}/\sigma_2^2 \tag{27.12}$$

for the coefficient in an exactly linear regression of x on y. Expressions (27.11) and (27.12) define the (linear) *regression coefficients*[1] of y on $x(\beta_2)$ and of x on $y(\beta_1)$. Using (27.8), (27.10) and (27.11), we have the *linear regression equations*

$$E(y|x) - \mu_y = \beta_2(x - \mu_x). \qquad (27.13)$$

and

$$E(x|y) - \mu_x = \beta_1(y - \mu_y). \qquad (27.14)$$

We have already encountered a case of exact linear regressions in our discussion of the bivariate normal distribution in **16.23**. It follows from (27.6), (27.11) and (27.12) that, when both regressions are linear,

$$\beta_1\beta_2 = \rho^2. \qquad (27.15)$$

Example 27.4 (Linear regression for a bivariate chi-squared distribution)

The regressions of x^2 and y^2 on each other in Example 27.1 are strictly linear. For, from **16.23**, putting $\sigma_1 = \sigma_2 = 1$ in (16.46), we have

$$\begin{aligned} E(y|x) &= \rho x, \\ \text{var}(y|x) &= 1 - \rho^2. \end{aligned} \qquad (27.16)$$

Thus

$$\begin{aligned} E(y^2|x) &\equiv \text{var}(y|x) + \{E(y|x)\}^2 \\ &= 1 - \rho^2 + \rho^2 x^2. \end{aligned}$$

To each value of x^2 there correspond values $+x$ and $-x$ which occur with equal probability. Thus, since $E(y^2|x)$ is a function of x^2 only,

$$E(y^2|x^2) = \tfrac{1}{2}\{E(y^2|x) + E(y^2|-x)\} = E(y^2|x) = 1 - \rho^2 + \rho^2 x^2,$$

which we may rewrite, in the form of (27.13),

$$E(y^2|x^2) - 1 = \rho^2(x^2 - 1),$$

and the regression of y^2 on x^2 is strictly linear. Similarly for x^2 on y^2. Since we saw in Example 27.1 that $\mu_{11} = 2\rho^2$, and we know that the variances are equal to 2, we may confirm from (27.11) and (27.12) that ρ^2 is the regression coefficient in each of the linear regression equations.

Exercise 27.28 extends this result to sums of squares.

---

[1]The notation β_1, β_2 is unconnected with the symbolism for skewness and kurtosis in **3.31–32**; they are unlikely to be confused, since they arise in different contexts.

Example 27.5 (Regressions with no linear component)

(a) In Example 27.2 it is easily seen that $E(x|y) = E(y|x) = 0$, so that we have linear regressions here, too, the coefficients being zero.

(b) Consider the variables x and y^2 in Example 27.4. We saw there that the regression of y^2 is linear on x^2, with coefficient ρ^2, and it is therefore not linear on x when $\rho \neq 0$. However, since $E(y|x) = \rho x$ we have

$$E(y|x^2) = \tfrac{1}{2}\{E(y|x) + E(y|-x)\} = 0,$$

so that the regression of y on x^2 is linear with regression coefficient zero.

Exercise 27.2 describes a bivariate F distribution with linear regression functions.

27.10 The reader is asked in Exercise 27.13 to show that the angle between the two regression lines (27.13) and (27.14) is

$$\theta = \arctan\left\{\frac{\sigma_1\sigma_2}{\sigma_1^2+\sigma_2^2}\left(\frac{1}{\rho} - \rho\right)\right\}, \tag{27.17}$$

so that as ρ varies over its range from -1 to $+1$, θ increases steadily from 0 to its maximum of $\frac{1}{2}\pi$ when $\rho = 0$, and then decreases steadily to 0 again. Thus, if and only if x and y are in strict linear functional relationship, the two regression lines coincide ($\rho^2 = 1$). If and only if $\rho = 0$, when x and y are said to be (linearly) *uncorrelated*, the regression lines are at right angles to each other.

Further, it may be shown that

$$\rho^2 = \text{var}(\alpha_2 + \beta_2 x)/\sigma_2^2 = \text{var}(\alpha_1 + \beta_1 y)/\sigma_1^2; \tag{27.18}$$

see Exercise 27.23. The linear function with maximum variance is known as the first principal component; see Exercise 27.1.

The terms ‘regression’, ‘lines of regression’ and ‘correlation’ were first used by Galton in 1886–88 (his 1877 synonym for ‘regression’ was ‘reversion’); ‘coefficient of correlation’ was first used by Edgeworth in 1892. The term ‘correlation’ arose naturally from ‘co-relation’, but ‘regression’ requires some explanation. Galton found that the average stature of adult offspring increased with parents’ stature, but not by as much, and he called this a ‘regression to mediocrity’. The term has stuck firmly, but the apparently dramatic result is theoretically trivial, for, as we have seen, if $\sigma_1 = \sigma_2$ (as we may reasonably suppose here), then $\beta_1 = \beta_2 = \rho \leq 1$. Thus the same phenomenon would be observed if we interchanged the roles of parents and offspring, although, as noted in **27.2**, there is an essential asymmetry in the variables for the geneticist.

Approximate linear regression: least squares

27.11 Examples 27.4 and 27.5 give instances where one or both regression functions are exactly linear. Suppose now that we are dealing with a joint distribution whose true regression structure is unknown. Although the regression is unlikely to be exactly linear, a linear approximation may be deemed appropriate.

When there are no sampling considerations involved, the choice of a method of fitting is essentially arbitrary. As we are considering the dependence of y on x, we choose the ‘best-fitting’ regression line of y on x,

$$y = \alpha_2 + \beta_2 x.$$

That is, we select α_2 and β_2 to minimize the expected value of the squared deviations from the fitted regression line:

$$S = E\{(y - \alpha_2 - \beta_2 x)^2\}. \tag{27.19}$$

Thus, our approach uses the method of least squares, introduced in **18.52**. Differentiating (27.19) with respect to α_2 and β_2, we obtain the pair of equations

$$-\frac{1}{2}\frac{\partial S}{\partial \alpha_2} = E(y) - \alpha_2 - \beta_2 E(x) = 0$$

and

$$-\frac{1}{2}\frac{\partial S}{\partial \beta_2} = E(xy) - \alpha_2 E(x) - \beta_2 E(x^2) = 0.$$

These equations yield

$$\beta_2 = \mu_{11}/\sigma_1^2 \quad \text{and} \quad \alpha_2 = \mu_y - \beta_2 \mu_x$$

so that the fitted least squares regression line is

$$y - \mu_y = \beta_2(x - \mu_x)$$

as in (27.13). Thus, we may conclude that the evaluation of an approximate regression line by the method of least squares gives results that are the same as those that arise when the regression function is exactly linear. It should be noted that the expectation in (27.19) is taken over both x and y; this is sometimes known as the *unconditional* case.

Sample coefficients

27.12 We now turn to the consideration of sampling problems for correlation and regression coefficients. The sample coefficients are:

$$\begin{aligned}
b_1 &= m_{11}/s_2^2 = \frac{1}{n}\sum(x-\bar{x})(y-\bar{y})\Big/\frac{1}{n}\sum(y-\bar{y})^2,\\
b_2 &= m_{11}/s_1^2 = \frac{1}{n}\sum(x-\bar{x})(y-\bar{y})\Big/\frac{1}{n}\sum(x-\bar{x})^2,\\
r &= m_{11}/s_1 s_2 = \frac{1}{n}\sum(x-\bar{x})(y-\bar{y})\Big/\left\{\frac{1}{n}\sum(x-\bar{x})^2\frac{1}{n}\sum(y-\bar{y})^2\right\}^{1/2}.
\end{aligned} \tag{27.20}$$

For computational purposes these expressions simplify to

$$b_1 = \frac{\sum xy - (\sum x)(\sum y)/n}{\sum y^2 - (\sum y)^2/n},$$

$$b_2 = \frac{\sum xy - (\sum x)(\sum y)/n}{\sum x^2 - (\sum x)^2/n}, \tag{27.21}$$

$$r = \frac{\sum xy - (\sum x)(\sum y)/n}{[\{\sum x^2 - (\sum x)^2/n\}\{\sum y^2 - (\sum y)^2/n\}]^{1/2}}.$$

As before, $-1 \leq r \leq +1$.

Example 27.6 (Evaluation of sample regression lines)

For the data in Table 27.1, we find

$$\bar{x} = 63.06, \bar{y} = 132.82,$$
$$s_1^2 = 7.25, s_2^2 = 507.46 \qquad \text{and } m_{11} = 19.52,$$

so that $r = 0.322$ and the estimated linear regression equations are

$$x \text{ on } y : x - 63.06 = 0.0385(y - 132.82) \text{ or}$$
$$x = 0.0385y + 57.95,$$
$$y \text{ on } x : y - 132.82 = 2.692(x - 63.06) \text{ or}$$
$$y = 2.692x - 36.96.$$

These lines appear in Fig. 27.1 as RR' and CC', respectively.

As is evident from Fig. 27.1, the same regression lines would result if each single observation were replaced by the group mean of all observations with the same values of the regressor variable. Any method of grouping the observations that is based only upon the values of the regressor variables will continue to provide unbiased estimators for the regression coefficients. However, the correlation coefficient lacks such invariance and increases in absolute value as the level of aggregation increases; see Exercise 27.6. The misuse of correlations derived from aggregated data to represent the correlation for individuals is sometimes known as the 'fallacy of ecological correlation'.

Example 27.7 (Sample results showing effects of aggregation)

Table 27.2 shows the yields of wheat and of potatoes in 48 counties of England in 1936. We find that

$$\sum x = 758.0, \qquad \sum y = 291.1,$$
$$\sum x^2 = 12\,170.48, \qquad \sum y^2 = 1791.03, \qquad \sum xy = 4612.64.$$

Hence $r = 0.219$ and the estimated linear regression equations are

$$x \text{ on } y : x = 0.612y + 12.08$$
$$y \text{ on } x : y = 0.078x + 4.83.$$

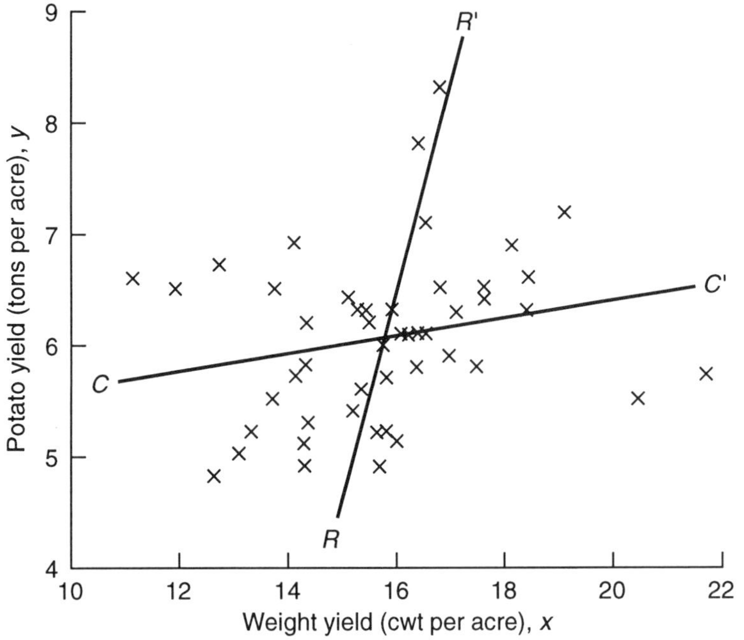

Fig. 27.2 Fitted regression lines for the data of Table 27.2

The data and the regression lines are shown diagrammatically in Fig. 27.2, known as a *scatter plot* or *scatter diagram*; its use is strongly recommended, since it conveys quickly and simply an idea of the adequacy of the fitted regression lines (not very good in our example). Indeed, a scatter diagram, plotted in advance of the analysis, will often make it clear whether developing a regression model is worthwhile.

Exercise 27.29 describes a graphical estimator for ρ^2 using the scatter diagram.

The correlation coefficient for the data given in Table 27.2 is indeed subject to the fallacy of ecological correlation, since the counties are merely convenient reporting units that could, and have been modified over time. To illustrate, we ordered the observations by the potato yield (y) and then combined successive values into 24 sets of 2, 16 sets of 3, 8 sets of 6 and 4 sets of 12. The analysis is not exact since the values of the regression variable differ slightly, which shows up as changes in the regression coefficient. The results are given in Table 27.3 and show a steady increase in r as the level of aggregation is increased. This feature is discussed further in Exercise 27.6.

In general, we also note that an application of Sheppard's corrections (see **3.18**) to the moments always increases the correlation (Exercise 27.16), whereas data errors depress the correlation (Exercise 27.17).

Table 27.2 Yields of wheat and potatoes in 48 counties in England in 1936

County	Wheat (cwt per acre) x	Potatoes (tons per acre) y	County	Wheat (cws per acre) x	Potatoes (tons per acre) y
Bedfordshire	16.0	5.3	Northamptonshire	14.3	4.9
Huntingdonshire	16.0	6.6	Peterborough	14.4	5.6
Cambridgeshire	16.4	6.1	Buckinghamshire	15.2	6.4
Ely	20.5	5.5	Oxfordshire	14.1	6.9
Suffolk, West	18.2	6.9	Warwickshire	15.4	5.6
Suffolk, East	16.3	6.1	Shropshire	16.5	6.1
Essex	17.7	6.4	Worcestershire	14.2	5.7
Hertfordshire	15.3	6.3	Gloucestershire	13.2	5.0
Middlesex	16.5	7.8	Wiltshire	13.8	6.5
Norfolk	16.9	8.3	Herefordshire	14.4	6.2
Lincs (Holland)	21.8	5.7	Somersetshire	13.4	5.2
Lincs (Kesteven)	15.5	6.2	Dorsetshire	11.2	6.6
Lincs (Lindsey)	15.8	6.0	Devonshire	14.4	5.8
Yorkshire (East Riding)	16.1	6.1	Cornwall	15.4	6.3
Kent	18.5	6.6	Northumberland	18.5	6.3
Surrey	12.7	4.8	Durham	16.4	5.8
Sussex, East	15.7	4.9	Yorkshire (North Riding)	17.0	5.9
Sussex, West	14.3	5.1	Yorkshire (West Riding)	16.9	6.5
Berkshire	13.8	5.5	Cumberland	17.5	5.8
Hampshire	12.8	6.7	Westmorland	15.8	5.7
Isle of Wight	12.0	6.5	Lancashire	19.2	7.2
Nottinghamshire	15.6	5.2	Cheshire	17.7	6.5
Leicestershire	15.8	5.2	Derbyshire	15.2	5.4
Rutland	16.6	7.1	Staffordshire	17.1	6.3

Table 27.3 Correlation and regression slope for wheat and potatoes, when observations are grouped

Group size	Number of observations	Correlation	Slope
1	48	0.219	0.612
2	24	0.285	0.604
3	16	0.368	0.652
6	8	0.661	0.690
12	4	0.775	0.513

Standard errors

27.13 The standard errors of the coefficients (27.20) are easily obtained. In fact, we have already obtained the large-sample variance of r in Example 10.6, where we saw that it is, in general, an expression involving all the second-order and fourth-order moments of the population sampled. In the binormal case, however, we found that it is simplified to

$$\text{var}\, r \doteq (1-\rho^2)^2/n, \qquad (27.22)$$

though (27.22) is of little value in practice since the distribution of r tends to normality so slowly (cf. **16.29**): it is unwise to use it for $n < 500$. The difficulty is of no practical importance, since, as we saw in **16.33**, the simple transformation of r,

$$z = \frac{1}{2}\log\left(\frac{1+r}{1-r}\right) = \text{arctanh}\, r, \qquad (27.23)$$

is for normal samples much more closely normally distributed with approximate mean

$$E(z) \doteq \frac{1}{2}\log\left(\frac{1+\rho}{1-\rho}\right) \qquad (27.24)$$

and variance approximately

$$\text{var}\, z \doteq \frac{1}{n-3}, \qquad (27.25)$$

independent of ρ. For $n > 50$, the use of this standard error for z is adequate; closer approximations are given in **16.33**.

Subrahmaniam *et al.* (1981) give series expansions for the distribution of z and other transformations of r.

For the sample regression coefficient of y on x,

$$b_2 = m_{11}/s_1^2,$$

the use of (10.17) gives, just as in Example 10.16 for r, the large-sample approximation

$$\text{var}\, b_2 \doteq \left(\frac{\mu_{11}}{\sigma_1^2}\right)^2 \left\{\frac{\text{var}\, m_{11}}{\mu_{11}^2} + \frac{\text{var}(s_1^2)}{\sigma_1^4} - \frac{2\,\text{cov}(m_{11}, s_1^2)}{\mu_{11}\sigma_1^2}\right\}.$$

Substituting for the variances and covariances from (10.23) and (10.24), yields

$$\text{var}\, b_2 \doteq \frac{1}{n}\left(\frac{\mu_{11}}{\sigma_1^2}\right)^2 \left\{\frac{\mu_{22}}{\mu_{11}^2} + \frac{\mu_{40}}{\sigma_1^4} - \frac{2\mu_{31}}{\mu_{11}\sigma_1^2}\right\}. \qquad (27.26)$$

For a binormal population, we substitute the moments relations given in (3.71) and obtain

$$\begin{aligned}\text{var}\, b_2 &\doteq \frac{1}{n}\left(\frac{\mu_{11}}{\sigma_1^2}\right)^2 \left\{\frac{(1+2\rho^2)\sigma_1^2\sigma_2^2}{\mu_{11}^2} + \frac{3\sigma_1^4}{\sigma_1^4} - \frac{6\rho\sigma_1^3\sigma_2}{\mu_{11}\sigma_1^2}\right\} \\ &= \frac{1}{n}\left(\frac{\mu_{11}}{\sigma_1^2}\right)^2 \left\{\frac{\sigma_1^2\sigma_2^2}{\mu_{11}^2} - 1\right\}\end{aligned}$$

$$= \frac{1}{n}\frac{\sigma_2^2}{\sigma_1^2}(1-\rho^2). \tag{27.27}$$

Similarly, for the regression coefficient of x on y,

$$\operatorname{var} b_1 \doteq \frac{1}{n}\frac{\sigma_1^2}{\sigma_2^2}(1-\rho^2). \tag{27.28}$$

Expressions (27.27) and (27.28) are rather more useful for standard error purposes (when, of course, we substitute s_1^2, s_2^2 and r for σ_1^2, σ_2^2 and ρ in them) than (27.22), since, as we saw in **16.35**, the exact distribution of b_2 is symmetrical about β_2: it is left to the reader as Exercise 27.9 to show from (16.86) that (27.27) is exact when multiplied by a factor $n/(n-3)$, and that the distribution of b_2 tends to normality rapidly, its measure of kurtosis being of order $1/n$.

As noted in **16.37**, this result may be also derived by a conditional argument.

The estimation of ρ in normal samples

27.14 The sampling theory of the binormal distribution was developed in **16.23–36**: we may now discuss, in particular, the problem of estimating ρ from a sample, in the light of our results in estimation theory (Chapters 17–18).

In Example 18.14 we saw that the likelihood function contains the observations only in the form of the five statistics $\bar{x}$, $\bar{y}$, s_1^2, s_2^2, r. These are therefore a set of sufficient statistics for the five parameters μ_1, μ_2, σ_1^2, σ_2^2, ρ, and their distribution is complete by **21.10**. When the other four parameters are known we have the pair of sufficient statistics for ρ,

$$\left\{\sum(x-\mu_1)(y-\mu_2),\ \frac{\sum(x-\mu_1)^2}{\sigma_1^2}+\frac{\sum(y-\mu_2)^2}{\sigma_2^2}\right\}.$$

In Example 18.14 we saw that the maximum likelihood estimator of ρ takes a different form according to which other parameters, if any, are being simultaneously estimated: the ML estimator is always a function of the set of five sufficient statistics, but it is a *different* function in different situations. When ρ alone is being estimated, the ML estimator is the root of a cubic equation (Example 18.3); when all five parameters are being estimated, the ML estimator is the sample correlation coefficient r (Example 18.14). In practice, the latter is by far the most common case, and therefore we now consider the estimation of functions of ρ by functions of r.

27.15 The exact distribution of r, which depends only upon ρ, is given by (16.61) or, more conveniently, by (16.66). Its mean value is given by (16.73). Expanding the gamma functions in (16.73) by Stirling's series (3.64) and taking the two leading terms of the hypergeometric function, we find, as stated there,

$$E(r) = \rho\left\{1 - \frac{(1-\rho^2)}{2n} + O(n^{-2})\right\}. \tag{27.29}$$

Thus r is a biased estimator of ρ when $0 \neq \rho^2 \neq 1$ and it is interesting to enquire whether this bias can be removed.

27.16 We may approach the problem in two ways. First, we may ask: is there a function $g(r)$ such that

$$E\{g(r)\} = g(\rho) \tag{27.30}$$

holds identically in ρ? Under certain regularity conditions Hotelling (1953) showed that if g is not dependent on n, then $g(r)$ could only be a linear function of $\arcsin r$, and Harley (1956; 1957) showed that in fact

$$E(\arcsin r) = \arcsin \rho, \tag{27.31}$$

a simple proof of Harley's result being given by Daniels and Kendall (1958). If we relax (27.30) to

$$E\{a(n)g(r) + b(n)\} = g(\rho),$$

it is satisfied by $g(r) = r^2/(1 - r^2)$; the result is given in Exercise 16.15 and in (28.99).

27.17 The second, more direct, approach, is to seek a function of r unbiased for ρ itself. Since r is a function of a set of complete sufficient statistics, the unbiased function of r must be unique (cf. **21.9**). Olkin and Pratt (1958) found the unbiased estimator of ρ, say r_u, to be the hypergeometric function

$$r_u = rF(\tfrac{1}{2}, \tfrac{1}{2}, \tfrac{1}{2}(n-2), (1 - r^2)) \tag{27.32}$$

which, expanded as a series, gives

$$r_u = r\left\{1 + \frac{1 - r^2}{2(n-2)} + \frac{9(1 - r^2)^2}{8n(n-2)} + O(n^{-3})\right\}. \tag{27.33}$$

No term in the series is negative, so that

$$|r_u| \geq |r|,$$

the equality holding only if $r^2 = 0$ or 1. Since $F(\frac{1}{2}, \frac{1}{2}, \frac{1}{2}(n-2), 0) = 1$ and r_u is an increasing function of r, we have $0 \leq r^2 \leq r_u^2 \leq 1$.

Evidently, the first correction term in (27.33) is counteracting the downward bias of the term in $1/n$ in (27.29). Olkin and Pratt recommend the use of the two-term expansion

$$r_u^* = r\left\{1 + \frac{1 - r^2}{2(n-4)}\right\}. \tag{27.34}$$

The term in braces in (27.34) gives r_u/r within 0.01 for $n \geq 8$ and within 0.001 for $n \geq 18$, uniformly in r.

Olkin and Pratt give exact tables of r_u for $n = 2\ (2)\ 30$ and $|r| = 0\ (0.1)\ 1$ which show that for $n \geq 14$, $|r_u|$ never exceeds $|r|$ by more than 5 per cent.

Finally, we note that as n appears only in the denominators of the hypergeometric series, $r_u \to r$ as $n \to \infty$, so that it has the same limiting distribution as r, namely a normal distribution with mean ρ and variance $(1-\rho^2)^2/n$. A similar analysis of the estimation of ρ^2 by r^2 is to be found in **28.34**.

Confidence intervals and tests for ρ

27.18 For testing that $\rho = 0$, the tests based on r are UMPU; this is not so when we are testing a non-zero value of ρ. However, if we confine ourselves to test statistics which are invariant under changes in location and scale, one-sided tests based on r are UMP invariant, as Lehmann (1986) shows.

Alternatively, we may make an approximate test using Fisher's z-transformation, using the results given in **27.13**. To test $H_0 : \rho = \rho_0$, we compute z and assume that under H_0, z is approximately normally distributed with mean (27.24) and variance (27.25). A one- or two-tailed test is appropriate according to whether the alternative is one- or two-sided. For interval estimation purposes, we may use David's (1938) charts or Odeh's tables, described in **19.34**, which Subrahmaniam and Subrahmaniam (1983) extend up to $n = 50$. By the duality between confidence intervals and tests noted in **20.9**, these charts and tables may also be used to read off the values of ρ to be rejected by a size-α test, i.e. all values of ρ not covered by the confidence interval for that α. David (1937) has shown that this test is slightly biased (and the confidence intervals correspondingly so). This slight bias in the equal-tails r test is a reflection of the approximate nature of the z-transformation.

Exercise 27.15 shows that the LR test is biased on r, but is not equal-tails except when testing $\rho = 0$.

27.19 We may use the z-transformation to test the composite hypothesis that the correlation parameters of two independently sampled bivariate normal populations are the same. For if so, the two transformed statistics z_1, z_2 will each be approximately normally distributed so that $z_1 - z_2$ will have zero mean and variance $1/(n_1-3) + 1/(n_2-3)$, where n_1 and n_2 are the sample sizes. Exercises 27.19–21 show that $z_1 - z_2$ is exactly the likelihood ratio statistic when $n_1 = n_2$, and approximately so when $n_1 \neq n_2$. In either case, however, the test is approximate, being a standard-error test. However, when $n_1 \neq n_2$, a weighted average of z_1 and z_2 will *not* provide the ML estimator; see Exercise 27.22.

The more general composite hypothesis, that the two correlation parameters ρ_1, ρ_2 differ by an amount Δ, cannot be tested in this way. For then

$$E(z_1 - z_2) = \frac{1}{2}\log\left(\frac{1+\rho_1}{1-\rho_1}\right) - \frac{1}{2}\log\left(\frac{1+\rho_2}{1-\rho_2}\right) = \frac{1}{2}\log\left\{\left(\frac{1+\rho_1}{1-\rho_1}\right)\left(\frac{1-\rho_2}{1+\rho_2}\right)\right\}$$

is not a function of $|\rho_1 - \rho_2|$ alone. The z-transformation could be used to test

$$H_0 : \frac{1+\rho_1}{1-\rho_1} = a\left(\frac{1+\rho_2}{1-\rho_2}\right)$$

for any constant a, but this is not a hypothesis of interest. Except in the large-sample case, when we may use standard errors, there seems to be no way of testing $H_0 : \rho_1 - \rho_2 = \Delta$, for the exact distribution of the difference $r_1 - r_2$ has not been investigated.

Dunn and Clark (1969; 1971) examine methods of testing the equality of correlation parameters when the sample coefficients are not independent. See also Elston (1975).

Test of independence and regression tests

27.20 In the particular case when we wish to test $\rho = 0$, i.e. the *independence* of the normal variables, we may use the exact result of **16.28**, that

$$t = r\{(n-2)/(1-r^2)\}^{1/2} \tag{27.35}$$

follows Student's t distribution with $n-2$ degrees of freedom. t^2 is essentially the LR test statistic for $\rho = 0$ – cf. Exercise 27.15 – and this is equivalent to an equal-tails test on t.

Essentially, we are testing here that the population covariance is zero, and clearly this implies that the population regression coefficients β_1, β_2 are zero. In **16.36**, we showed that

$$t = (b_2 - \beta_2)\left\{\frac{s_1^2(n-2)}{s_2^2(1-r^2)}\right\}^{1/2} \tag{27.36}$$

has Student's t distribution with $n-2$ degrees of freedom. When $\beta_2 = 0$, (27.36) is seen to be identical with (27.35). Thus the test of independence may be regarded as a test that a regression coefficient is zero, a special case of the general test (27.36) for hypotheses concerning β_2. It will be noted that the exact test for any hypothetical value of β_2 is of a much simpler form than that for ρ.

Tests of independence can be made without assuming that the random variables are normally distributed, as we shall see in **27.22**.

Other measures of correlation

27.21 Hitherto, we have restricted attention to the linear correlation coefficient r defined in (27.20). However, with respect to tests for independence, a wide variety of possibilities is available. Daniels (1944) defined a class of correlation coefficients by the expression

$$r_D = \frac{\sum_{(2)} a_{ij} b_{ij}}{\{\sum_{(2)} a_{ij}^2 \sum_{(2)} b_{ij}^2\}^{1/2}}, \tag{27.37}$$

where $\sum_{(2)}$ denotes $\sum_{i=1}^{n}\sum_{j=1}^{n}$ and a_{ij}, b_{ij} depend on (x_i, x_j) and (y_i, y_j), respectively. The product moment correlation r in (27.20) is a special case of (27.37) when $a_{ij} = x_i - x_j$, $b_{ij} = y_i - y_j$. Rather than base the (a_{ij}, b_{ij}) values on the actual observations, we could use measures based on the *ranks*. That is, if the sample values are ordered so that (ignoring ties for the moment)

$$x_{(1)} < x_{(2)} < \cdots < x_{(n)}$$

with the y values being ordered as

$$y_{(1)} < y_{(2)} < \cdots < y_{(n)},$$

we define the ranks of the pair (x_i, y_i) by their relative positions within these two sets of order statistics

$$\operatorname{rank}(x_{(i)}) = i, \qquad \operatorname{rank}(y_{(j)}) = j$$

or

$$\operatorname{rank}(x_i) = A_i, \qquad \operatorname{rank}(y_i) = B_i, \tag{27.38}$$

where $A_i = j$ if $x_i = x_{(j)}$ and B_i is similarly defined. Clearly, (A_i, B_i) are not restricted to the ranks as any set of coefficients might be employed in (27.38) corresponding to the pattern of dependence it is desired to detect. We leave the detailed discussion of non-parametric statistics to the monograph by Hettmansperger and McKean (1998), providing only a few examples of principal alternatives to r in (27.22).

For more general discussion of dependence measures, see Kent (1983).

Permutation distributions

27.22 We now restrict attention to the special case of (27.37)

$$r = \sum A_i B_i / \{\sum A_i^2 \sum B_i^2\}^{1/2}, \tag{27.39}$$

where A_i and B_i are adjusted so that $\sum A_i = \sum B_i = 0$, and all summations are taken over $i = 1, \ldots, n$. Under the hypothesis that the two random variables are independent, we may regard the $n!$ distinct assignments of $\{B_{(i_1)}, B_{(i_2)}, \ldots, B_{(i_n)}\}$ to $\{A_1, \ldots, A_n\}$ as equally likely; note that there is no loss of generality in fixing the order of one set of coefficients and then permuting the other. It follows directly that the denominator of (27.39) is invariant under these permutations so that the expected value of the numerator, taken with respect to the set of $n!$ possible assignments (denoted by E_P), is

$$\begin{aligned} E_P\left(\sum A_i B_i\right) &= \sum A_i \cdot \frac{1}{n}\{B_1 + \cdots + B_n\} \\ &= n^{-1}\left(\sum A_i\right)\left(\sum B_i\right) = 0. \end{aligned}$$

It follows immediately that

$$E_P(r) = 0. \tag{27.40}$$

Similarly, keeping the order of the As fixed, we have that the variance of $\sum A_i B_i$ over the set of random permutations is

$$\operatorname{var}_P\left(\sum A_i B_i\right) = \sum A_i^2 \operatorname{var}_P(\tilde{B}_i) + \sum_{i \neq j}\sum A_i A_j \operatorname{cov}_P(\tilde{B}_i, \tilde{B}_j), \tag{27.41}$$

where the random variable $\tilde{B}_i$ describes the coefficient assigned to the ith pair so that

$$P(\tilde{B}_i = B_k) = \frac{1}{n}, \qquad k = 1, \ldots, n$$
$$P(\tilde{B}_i = B_k, \tilde{B}_j = B_l) = 1/\{n(n-1)\}, \qquad k, l = 1, \ldots, n;\ k \neq l$$

and so on. It follows that

$$\mathrm{var}_P\left(\sum A_i B_i\right) = S_{AA}S_{BB}/(n-1),$$

where $S_{AA} = \sum A_i^2$ and $S_{BB} = \sum B_i^2$. Thus, (27.41) leads to

$$\mathrm{var}_P(r) = 1/(n-1), \tag{27.42}$$

regardless of the actual values of the coefficients (A_i, B_i).

27.23 We have seen at (27.40) and (27.42) that the first two moments of r are independent of the actual values of (x, y) observed. By similar methods, it may be shown that

$$E_P(r^3) = \frac{n-2}{n(n-1)^2}\left(\frac{k_3}{k_2^{3/2}}\right)\left(\frac{k_3'}{(k_2')^{3/2}}\right)$$
$$E_P(r^4) = \frac{3}{n^2-1}\left\{1 + \frac{(n-2)(n-3)}{3n(n-1)^2}\left(\frac{k_4}{k_2^2}\right)\left(\frac{k_4'}{(k_2')^2}\right)\right\}, \tag{27.43}$$

where the ks are the k-statistics of the observed xs and the k's the k-statistics of the ys. Neglecting the differences between k-statistics and sample cumulants, we may rewrite (27.43) as

$$E(r^3) \doteq \frac{(n-2)}{n(n-1)^2} g_1 g_1'$$
$$E(r^4) \doteq \frac{3}{n^2-1}\left\{1 + \frac{(n-2)(n-3)}{3n(n-1)^2} g_2 g_2'\right\}, \tag{27.44}$$

where g_1, g_2 are the measures of skewness and kurtosis of the xs, and g_1', g_2' those of the ys. If these are fixed, the skewness and kurtosis measures for r are approximately

$$\gamma_1(r) \doteq g_1 g_1'/n^{1/2} \tag{27.45}$$

and

$$\gamma_2(r) \doteq (-6 + g_2 g_2')/n.$$

Thus, as $n \to \infty$, we have approximately

$$E(r^3) = 0$$
$$E(r^4) = \frac{3}{n^2-1}. \tag{27.46}$$

The moments (27.40), (27.42), and (27.46) are precisely those of (16.62), the exact distribution of r in samples from a bivariate normal distribution with $\rho = 0$, as may be verified by evaluating the expected values of r^2 and r^4 for the p.d.f. given in (16.62). Thus, to a close approximation, the permutation distribution of r is also

$$f(r) = \frac{1}{B\{\frac{1}{2}(n-2), \frac{1}{2}\}}(1-r^2)^{(n-4)/2}, \qquad -1 \le r \le 1, \tag{27.47}$$

and we may therefore use (27.47), or equivalently the fact that $t = \{(n-2)r^2/(1-r^2)\}^{1/2}$ has Student's t distribution with $n-2$ degrees of freedom, to carry out our tests. The approximation is accurate even for small n, as we might guess from the agreement of its first two moments with those of the permutation distribution.

The convergence of the permutation and normal-theory distributions to a common limiting normal distribution has been rigorously proved by Hoeffding (1952).

Rank correlation coefficients

27.24 Returning to the original discussion on the use of ranks in **27.21**, we may assign $A_i = i - \frac{1}{2}(n+1)$ by ordering the x values so that (27.39) becomes

$$r_s = \sum_{i=1}^{n} R_{yi}\{i - \tfrac{1}{2}(n+1)\}/\{\tfrac{1}{12}n(n^2-1)\}, \tag{27.48}$$

where R_{yi} denotes the rank of the y value in the ith pair and

$$\sum i = \sum R_{yi} = n(n+1)/2$$
$$S_{AA} = S_{BB} = n(n^2-1)/12$$

when ranks are used. Further, since

$$\sum_{i=1}^{n} iR_{yi} = \tfrac{1}{6}n(n+1)(2n+1) - \tfrac{1}{2}\sum_{i=1}^{n}(i - R_{yi})^2,$$

(27.48) becomes

$$r_s = 1 - \frac{6}{n(n^2-1)}\sum_{i=1}^{n}(i - R_{yi})^2. \tag{27.49}$$

This statistic is usually known as Spearman's rank correlation coefficient, after the eminent psychologist who first introduced it in 1906. It follows from (27.40), (27.42) and (27.46) that, under independence,

$$E(r_s) = E(r_s^3) = 0$$
$$\text{var}(r_s) = (n-1)^{-1}$$

and

$$E(r_s^4) = \frac{3}{n^2-1}\left\{1 + \frac{12(n-2)(n-3)}{25n(n-1)^2}\right\}. \tag{27.50}$$

Kendall and Gibbons (1990) give tables of the frequency function of $\sum(i - R_{yi})^2$ for $n = 4\ (1)\ 10$; the tail values are reproduced in the *Biometrika Tables*, Vol. I, Table 45. Owen (1962) gives tables for $n = 11$ and Franklin (1988) for $n = 12\ (1)\ 18$. Zar (1972) gives tables based on an intermediate (Pearson Type II) approximation for $n = 12\ (1)\ 50\ (2)\ 100$. The approximation afforded by (27.47) is adequate for $n \geq 30$ and test size α down to 0.005.

27.25 The coefficient r_s is but one of many possible rank coefficients, selected primarily for its simplicity. Daniels (1948) showed that all such coefficients, defined by (27.37), are essentially coefficients of disarray, in the sense that if a pair of values of y are interchanged to bring them into the same order as the corresponding values of x, the value of any coefficient of this class will increase. We now consider the question of measuring disarray among the ranks of x and y directly.

First, consider the expression:

$$\tau = P\{(x_i - x_j)(y_i - y_j) > 0\} - P\{(x_i - x_j)(y_i - y_j) < 0\}. \tag{27.51}$$

The statistic

$$t = \frac{2}{n(n-1)} \sum\sum_{i<j} \text{sign}[(x_i - x_j)(y_i - y_j)] \tag{27.52}$$

is an unbiased estimator for τ for any continuous distribution. If we let $q_{ij} = 1$ when $(x_i - x_j)(y_i - y_j) < 0$, that is, when the order of the ys is opposite to that of the xs, and put $q_{ij} = 0$ otherwise, we may write

$$t = 1 - \frac{4Q}{n(n-1)}, \tag{27.53}$$

where $Q = \sum\sum_{i<j} q_{ij}$.

The coefficient (27.53) was discussed by several early writers (Fechner, Lipps) around the year 1900 and subsequently by several other writers, notably Lindeberg, in the 1920s – historical details are given by Kruskal (1958), but first became widely used after a series of papers by Kendall starting in 1938 and consolidated in a monograph (Kendall, 1962; latest edition, Kendall and Gibbons, 1990). It is usually known as Kendall's rank correlation coefficient.

27.26 The statistic t may be represented in the form of (27.37) if we set

$$a_{ij} = \begin{cases} 1 & \text{if } x_j > x_i, \\ 0 & \text{otherwise;} \end{cases} \qquad b_{ij} = \begin{cases} 1 & \text{if } y_j > y_i, \\ 0 & \text{otherwise.} \end{cases}$$

It follows from (27.51) and (27.52) that $E(t) = \tau$. Under independence,

$$E(t) = 0, \qquad \text{var}(t) = \frac{2(2n+5)}{9n(n-1)} \tag{27.54}$$

and the distribution of t tends to normality rapidly with mean and variance given by (27.54). Kendall and Gibbons (1990) give the exact distribution for $n = 4\ (1)\ 10$. Beyond this point, the asymptotic normal distribution may be used with little loss of accuracy. Under the weaker null hypothesis $H_0 : \tau = 0$ we may use t as a test of association. The distribution of t is no longer distribution-free, but Samara and Randles (1988) provide a robust test procedure based on t that remains distribution-free under the stronger null hypothesis of independence.

r_s and t have the same asymptotic relative efficiency in all cases; compared to r, Konijn (1956) showed the AREs to be

distribution	normal	uniform	double exponential
ARE(r_s or t; r)	0.912	1.0	1.266

Intraclass correlation

27.27 On occasion, we may wish to measure the similarity between individuals that share common membership of some class, such as individuals within a family or persons in the same neighbourhood. It is natural to consider all pairs within each class, but there is no natural ordering in the pair. Consider, for example, the heights of brothers within a family; any attempt to order the individuals by age, size, or any other criterion would lead to a potentially different concept of association than the symmetric within-class measure we are seeking to define. To resolve the problem, we define *intraclass* correlation directly: suppose there are p classes with $k_1, \ldots, k_p$ members and the random variable y_{ij} refers to the jth member of the ith class. We assume that

$$E(y_{ij}) = \mu, \qquad \text{var}(y_{ij}) = \sigma^2 \tag{27.55}$$

and

$$\text{cov}(y_{ij}, y_{i'j'}) = \begin{cases} \rho_I \sigma^2, & i = i', j \neq j', \\ 0, & i \neq i', \end{cases} \tag{27.56}$$

where ρ_I is the intraclass correlation coefficient.

27.28 When $k_i = k$, for all i, and the variates are normally distributed, the ML estimators for the parameters in (27.56) are

$$\widehat{\mu} = \sum_i \sum_j y_{ij}/kp = \sum_i \bar{y}_{i\cdot}/p = \bar{y}_{\cdot\cdot},$$

$$S_W^2 = \sum_i \sum_j (y_{ij} - \bar{y}_{i\cdot})^2/p(k-1) \tag{27.57}$$

$$S_B^2 = k \sum_i (\bar{y}_{i\cdot} - \bar{y}_{\cdot\cdot})^2/(p-1), \tag{27.58}$$

where

$$E(S_W^2) = \sigma^2(1 - \rho_I) \tag{27.59}$$

and

$$E(S_B^2) = \sigma^2\{1 + (k-1)\rho_I\}. \tag{27.60}$$

The ratio $F = S_B^2/S_W^2$ has Fisher's F distribution (see **16.15–22**) with $(p-1, p(k-1))$ d.fr. when $\rho_I = 0$. Indeed, the F statistic is precisely that used to test

$$H_0 : \rho_I = 0 \qquad \text{vs } H_1 : \rho_I > 0$$

in the one-way random effects model

$$y_{ij} = \mu + \theta_i + \epsilon_{ij}$$

where $\theta_i \sim N(0, \sigma^2 \rho_I)$, $\epsilon_{ij} \sim N(0, \sigma^2(1-\rho_I))$ with the usual assumptions of independence; see **31.12**. It is remarkable that intraclass correlation was introduced long before the random effects model.

Equations (27.57)–(27.60) yield the estimator

$$r_I = (F-1)/(k-1+F). \tag{27.61}$$

It follows immediately from (27.61) that $-(k-1)^{-1} \leq r_I \leq 1$. r_I is a biased estimator for ρ_I; an unbiased estimator is given by Olkin and Pratt (1958), but has seen limited application.

27.29 A confidence interval for ρ_I is readily constructed using the F distribution. Starting from the statement

$$P\{F_L \leq F^* \leq F_U\} = 1-\alpha,$$

where $F^* = F(1-\rho_I)/\{1+(k-1)\rho_I\}$, we obtain the $100(1-\alpha)$ per cent confidence interval

$$\frac{F-F_U}{F+(k-1)F_U} \leq \rho_I \leq \frac{F-F_L}{F+(k-1)F_L}. \tag{27.62}$$

These results were first obtained by Fisher (1921c); see Exercise 27.14.

27.30 When the class sizes vary, the form of the estimators becomes more involved; see Shoukri and Ward (1984). However, an exact test of $H_0 : \rho_I = \rho_0$ against $H_1 : \rho_I > \rho_0$ is still available using the F statistic with

$$S_W^2 = \sum_i \sum_j (y_{ij} - \bar{y}_{i\cdot})^2 / \sum_i (k_i - 1), \tag{27.63}$$

$$S_B^2 = \sum_i g_i (\bar{y}_{i\cdot} - \bar{y}_0)^2/(p-1), \tag{27.64}$$

where

$$\bar{y}_0 = \sum g_i \bar{y}_{i\cdot} / \sum g_i$$

and

$$g_i = \frac{n_i(1-\rho_0)}{1-\rho_0+n_i\rho_0}; \tag{27.65}$$

cf. Donner *et al.* (1989). Exact confidence intervals are not available for unequal k; various approximations are considered by Donner and Wells (1986). Paul (1990) and Gleser (1992) discuss estimation procedures for the general case.

Tetrachoric correlation

27.31 We now discuss the estimation of ρ in a binormal population when the data are grouped. First, we consider an extreme case exemplified by Table 27.4, based on the data for cows according to age and milk yield given in Table 1.24, Exercise 1.4. Suppose that, instead of being given that table, we had only Table 27.4.

Table 27.4 Cows by age and milk yield

	Age 6 and over	Age 3–5 year	Total
Yield 8–18 gal	1078	1407	2485
Yield 19 gal and over	1546	881	2427
Total	2624	2288	4912

If we assume that the underlying distribution is binormal, how can we estimate ρ from this table? In general, consider a 2×2 table with frequencies

$$\begin{array}{ccc} a & b & a+b \\ c & d & c+d \\ \hline a+c & b+d & a+b+c+d=n \end{array} \tag{27.66}$$

In (27.66) we shall always take d to be such that neither of its marginal totals contains the median value.

If this table is derived by a double dichotomy of the binormal distribution we can find h' such that

$$\int_{-\infty}^{h'} \int_{-\infty}^{\infty} f(x, y)\, dx\, dy = \frac{a+c}{n}. \tag{27.67}$$

Putting $h = (h' - \mu_1)/\sigma_1$, (27.67) becomes

$$F(h) = (2\pi)^{-1/2} \int_{-\infty}^{h} \exp(-\tfrac{1}{2}x^2)\, dx = \frac{a+c}{n}. \tag{27.68}$$

Likewise k is given by

$$F(k) = (2\pi)^{-1/2} \int_{-\infty}^{k} \exp(-\tfrac{1}{2}y^2)\, dy = \frac{a+b}{n}. \tag{27.69}$$

The arrangement of the table in (27.66) ensures that h and k are never negative.

Having fitted univariate normal distributions to the marginal frequencies of the table in this way, we now estimate ρ using

$$\frac{d}{n} = \int_{h}^{\infty} \int_{k}^{\infty} \frac{1}{2\pi(1-\rho^2)^{1/2}} \exp\left\{\frac{-1}{2(1-\rho^2)}(x^2 - 2\rho xy + y^2)\right\} dy\, dx. \tag{27.70}$$

After some algebra, we find that (27.70) may be expressed in terms of the tetrachoric functions which were defined in (6.44),

$$\frac{d}{n} = \sum_{j=0}^{\infty} \rho^j \tau_j(h)\tau_j(k), \tag{27.71}$$

where $\tau_0(x) = 1 - F(x)$. The reader is asked to derive this result as Exercise 27.27.

27.32 Formally, (27.71) provides a soluble equation for ρ, although series (27.71) may converge very slowly. The reader is asked to demonstrate convergence in Exercise 27.26.

The estimate of ρ derived from a sample of n in this way is known as *tetrachoric r* and is due to Pearson. We shall denote it by r_t.

Example 27.8 (Evaluation of the tetrachoric correlation)

For the data of Table 27.4 we find the solution of $F(h) = 2624/4912 = 0.5342$ is $h = 0.086$, and similarly $F(k) = 2485/4912 = 0.5059$ yields $k = 0.015$. We have also for d/n the value $881/4912 = 0.1794$.

From the tables, we find for varying values of h, k and ρ the following values of d:

		$h = 0$	$h = 0.1$			$h = 0$	$h = 0.1$
$\rho = -0.30$	$k = 0$	0.2015	0.1818	$\rho = -0.35$	$k = 0$	0.1931	0.1735
	$k = 0.1$	0.1818	0.1639		$k = 0.1$	0.1735	0.1555

Linear interpolation gives for $h = 0.086$, $k = 0.015$, the result $\rho = -0.32$ approximately. In Table 27.4, we have inverted the order of columns, and taking account of this change yields $r_t = +0.32$. (The product moment coefficient for Table 1.24 is $r = 0.22$.)

27.33 Tetrachoric r_t has been used mainly by psychologists, whose material is often of the 2×2 type. Karl Pearson (1913) gave an asymptotic expression for its standard error.

Since the estimation procedure equates observed frequencies with the corresponding probabilities (say, $\boldsymbol{\theta}$) in the bivariate normal distribution, it is using the ML estimator $\hat{\boldsymbol{\theta}}$. Now the ML estimator of ρ, which is a function $\rho(\boldsymbol{\theta})$, is $\rho(\hat{\boldsymbol{\theta}})$. Thus r_t is the ML estimator of ρ from a 2×2 table. Its approximate large-sample variance, obtained from **18.18** by Hamdan (1970), is

$$\left\{ n^2 \left(\frac{1}{a} + \frac{1}{b} + \frac{1}{c} + \frac{1}{d} \right) f^2(h, k|r_t) \right\}^{-1},$$

where $f(x, y|\rho)$ is the standardized binormal density function.

Brown (1977) gives a computer algorithm for r_t and its standard error. Digby (1983) suggests various simplifying approximations.

Biserial correlation

27.34 We now consider a binormal distribution for (x, y) where the y variable is coded as $Y = 1$ or 2 according as y is greater or less than k_0, an unknown point of dichotomy for y. The regression of x on y is linear:

$$E(x|y) - \mu_x = \rho\sigma_x(y - \mu_y)/\sigma_y. \tag{27.72}$$

Given two values y_1, y_2, we have that

$$\rho = \left\{\frac{E(x|y_1) - E(x|y_2)}{\sigma_x}\right\} \Big/ \left(\frac{y_1 - y_2}{\sigma_y}\right). \tag{27.73}$$

If $Y = 2$ for n_2 out of n observations, we may estimate k_0 by k in

$$1 - F(k) = (2\pi)^{-1/2} \int_k^\infty \exp(-\tfrac{1}{2}u^2)\, du = \frac{n_2}{n_1 + n_2} = p. \tag{27.74}$$

Further, $y_2 = E(y|Y = 2)$ may be estimated from

$$\begin{aligned} \frac{y_2 - \mu_y}{\sigma_y} &= (2\pi)^{-1/2} \int_k^\infty u \exp\left(-\frac{1}{2}u^2\right) du \Big/ (2\pi)^{-1/2} \int_k^\infty \exp\left(-\frac{1}{2}u^2\right) du \\ &= (2\pi)^{-1/2} \exp(-\tfrac{1}{2}k^2) \Big/ \left(\frac{n_1}{n_1 + n_2}\right) = f_k/p, \end{aligned} \tag{27.75}$$

where f_k denotes the standard normal density function at k; likewise, the value corresponding to $Y = 1$ is $-f_k/q$, $q = (1 - p)$. If we estimate $E(x|y_i)$ by $\overline{x_i}$, $i = 1, 2$, we arrive at the biserial estimator for ρ, from (27.73), as

$$r_b = \frac{\overline{x}_2 - \overline{x}_1}{s_x} \frac{pq}{f_k}. \tag{27.76}$$

Example 27.9 (Evaluation of the biserial correlation (Pearson, 1909))

Table 27.5 shows the returns for 6156 candidates for the London University Matriculation Examination for 1908–9. The average ages for the two highest age groups have been estimated.

Table 27.5 Examination results by age

Age of candidate	Passed $Y = 2$	Failed $Y = 1$	Totals
16	583	563	1146
17	666	980	1646
18	525	868	1393
19–21	383	814	1197
22–30 (mean 25)	214	439	653
over 30 (mean 33)	40	81	121
Totals	2411	3745	6156

We have

$$\bar{x}_2 = 18.4280, \qquad \bar{x}_1 = 18.9877, \qquad s_x^2 = (3.2850)^2, \qquad p = 2411/6156 = 0.3917.$$

From (27.74), $1 - F(k) = 0.3917$, so that $k = 0.275$ and $f_k = 0.384$. Hence, from (27.76)

$$r_b = -\frac{0.5597}{3.2850} \frac{3.3917 \times 0.6083}{0.384} = -0.11.$$

The estimated correlation between age and success is small.

27.35 As for r_t, the assumption of underlying normality is crucial to r_b. The distribution of biserial r_b is not known, but Soper (1914) derived the expression for its variance in normal samples

$$\text{var}\, r_b \doteq \frac{1}{n}\left[\rho^4 + \rho^2\left\{\frac{pqk^2}{f_k^2} + (2p-1)\frac{k}{f_k} - \frac{5}{2}\right\} + \frac{pq}{f_k^2}\right], \tag{27.77}$$

and showed that (27.77) is generally well approximated by

$$\text{var}\ r_b \doteq \frac{1}{n}\left[r_b^2 - \frac{(pq)^{1/2}}{f_k}\right]^2.$$

Maritz (1953) and Tate (1955) showed that in normal samples r_b is asymptotically normally distributed with mean ρ and variance (27.77); they also considered the ML estimation of ρ in biserial data. It appears, as might be expected, that the variance of r_b is least, for fixed ρ, when the dichotomy is at the middle of the dichotomized variate's range ($y = \mu_y$). When $\rho = 0$, r_b is an efficient estimator of ρ, but when $\rho^2 \to 1$ the efficiency of r_b tends to zero. Tate also tabulates Soper's formula (27.77) for var r_b. Cf. Exercises 27.10–12.

Terrell (1983) gives tables of percentage points for r_b. Kraemer (1981) considers a modified form that does not assume normality; its asymptotic standard error is given by Koopman (1983). Relationships between ρ and the correlation within part of the dichotomy are given in Exercises 27.3–4.

27.36 An alternative form of biserial correlation is due to Lord (1963). We first rank the n observations as $x_{(1)} \leq x_{(2)} \leq \cdots \leq x_{(n)}$ and then define

$$\bar{x}_M = [x_{(n_1+1)} + \cdots + x_{(n)}]/n_2.$$

By construction, $(\bar{x}_M - \bar{x})/s_x$ is a consistent estimator for

$$\int_k^\infty u f(u)\,du \Big/ \int_k^\infty f(u)\,du \tag{27.78}$$

for *any* distribution with finite mean and variance. But (27.78) is the general version of the right-hand side of (27.75). Thus, an estimator for (27.73) is

$$\begin{aligned} r_b^* &= q(\bar{x}_2 - \bar{x}_1)/(\bar{x}_M - \bar{x}) \\ &= (\bar{x}_2 - \bar{x})/(\bar{x}_M - \bar{x}). \end{aligned} \tag{27.79}$$

From the second form, it is clear that $|r_b^*| \leq 1$.

Bedrick (1990; 1992) shows that r_b^* is asymptotically normally distributed given that ρ is positive or negative (but not zero) for any distribution. Bedrick's numerical studies suggest that r_b^* is reasonably efficient and preferable to other modifications of the biserial coefficient.

Point-biserial correlation

27.37 By way of contrast, we now consider another coefficient, the *point-biserial correlation*, which we denote by ρ_{pb}, and by r_{pb} for a sample. Suppose that y is truly a binary variable (values 0 or 1, say) rather than a partition of a normal distribution. For example, in Table 27.5 it is not implausible to suppose that success in the examination is a dichotomy of a normal distribution of ability to pass it. But if y represented a binary variable, such as a candidate's sex, a different approach is necessary.

We must now consider correlation between a binary variable y and the variable x. If π is the true proportion of values of y with $y = 1$, we have from the binomial distribution

$$E(y) = \pi, \qquad \sigma_y^2 = \pi(1-\pi)$$

so that

$$\rho_{pb} = \frac{\mu_{11}}{\sigma_x\sigma_y} = \frac{E(xy) - \pi E(x)}{\sigma_x\{\pi(1-\pi)\}^{1/2}}. \qquad (27.80)$$

We estimate $E(xy)$ by $m_{11} = \sum_{i=1}^{n_1} x_i/(n_1+n_2)$, $E(x)$ by $\bar{x}$, σ_x by s_x, and π by $p = n_2/(n_1+n_2)$, obtaining

$$r_{pb} = \frac{p\bar{x}_2 - p(p\bar{x}_2 + q\bar{x}_1)}{s_x(pq)^{1/2}}$$
$$= \frac{(\bar{x}_2 - \bar{x}_1)(pq)^{1/2}}{s_x}. \qquad (27.81)$$

27.38 r_{pb} in (27.81) may be compared with the biserial r_b defined in (27.76). We have

$$\frac{r_{pb}}{r_b} = \frac{f_k}{(pq)^{1/2}}. \qquad (27.82)$$

it has been shown by Tate (1953) by a consideration of Mills' ratio (cf. **5.38**) that the expression on the right of (27.82) is less than or equal to $(2/\pi)^{1/2}$ and the values of the coefficients will thus, in general, be appreciably different. In Exercise 27.5, the reader is asked to show that $|r_{pb}| \leq 1$.

Tate (1954) shows that r_{pb} is asymptotically normally distributed with mean ρ_{pb} and variance

$$\operatorname{var} r_{pb} \sim \frac{(1-\rho_{pb}^2)^2}{n}\left(1 - \frac{3}{2}\rho_{pb}^2 + \frac{\rho_{pb}^2}{4pq}\right), \qquad (27.83)$$

which is a minimum when $p = q = \frac{1}{2}$.

Apart from the measurement of correlation, it is clear from (26.81) that, in effect, for a point-biserial situation, we are simply comparing the means of two samples of a variate x, the y classification being no more than a labelling of the samples. In fact

$$\frac{r_{pb}^2}{1-r_{pb}^2} = \frac{(\bar{x}_1 - \bar{x}_2)^2\{(n_1n_2)/(n_1+n_2)\}}{\sum(x_{1i}-\bar{x}_1)^2 + \sum(x_{2i}-\bar{x}_2)^2} = \frac{t^2}{n_1+n_2-2}, \qquad (27.84)$$

where t is the usual Student's t test used for comparing the means of two normal populations with equal variance (cf. Example 21.18). Thus if the distribution of x is normal for $y = 0, 1$, the point-biserial coefficient is a simple transformation of the t^2 statistic, which may be used to test it.

27.39 When $\rho_{pb} \neq 0$, the statistic

$$t = \frac{(n_1 + n_2 - 2)^{1/2}(r_{pb} - \rho_{pb})}{\sqrt{(1 - r_{pb}^2)(1 - \rho_{pb}^2)}} \tag{27.85}$$

is asymptotically normally distributed with mean zero and variance, from (27.83),

$$1 + \rho_{pb}^2[(1 - 6pq)/4pq].$$

Finally, we note that the point-biserial correlation coefficient does not require an assumption of normality for its validity, unlike the biserial coefficient.

Circular correlation

27.40 Measures of correlation for random variables defined on the circle or on the sphere are considered by Stephens (1979b) and by Fisher and Lee (1986). Measures of correlation between circular and linear random variables are considered by Mardia and Sutton (1978) and by Liddell and Ord (1978).

Criteria for linearity of regression

27.41 Throughout much of this chapter we have focused upon linear regression, with dependence assessed by a single (correlation) parameter. It behoves us to ask under what distributional assumptions linear regression is assured. We first provide a necessary and sufficient condition for linearity and then find conditions under which linear regression leads to binormality.

Let $\psi(t_1, t_2) = \log \phi(t_1, t_2)$ be the joint cumulant generating function of x and y. We first prove the following result.

Theorem 27.1.

If the regression of y upon x is linear, so that

$$\mu'_{1x} = E(y|x) = \beta_0 + \beta_1 x, \tag{27.86}$$

then

$$\left[\frac{\partial \psi(t_1, t_2)}{\partial t_2}\right]_{t_2=0} = \mathrm{i}\beta_0 + \beta_1 \frac{\partial \psi(t_1, 0)}{\partial t_1}; \tag{27.87}$$

and conversely, if a completeness condition is satisfied, then (27.87) is sufficient as well as necessary for (27.86).

Proof. Write the joint c.f. as

$$\phi(t_1, t_2) = \int_{-\infty}^{\infty}\int_{-\infty}^{\infty} \exp(\mathrm{i}t_1 x + \mathrm{i}t_2 y) g(x) h_x(y)\, \mathrm{d}y\, \mathrm{d}x. \tag{27.88}$$

If we first differentiate with respect to t_2 and then set $t_2 = 0$, we arrive at

$$\left[\frac{\partial\phi(t_1,\ t_2)}{\partial t_2}\right]_{t_2=0} = \mathrm{i}\int_{-\infty}^{\infty} \exp(\mathrm{i}t_1x)g(x)\mu'_{1x}\,\mathrm{d}x. \tag{27.89}$$

If (27.86) holds, this reduces to

$$\begin{aligned}\left[\frac{\partial\phi(t_1,\ t_2)}{\partial t_2}\right]_{t_2=0} &= \mathrm{i}\int_{-\infty}^{\infty} \exp(\mathrm{i}t_1x)g(x)(\beta_0+\beta_1x)\,\mathrm{d}x \\ &= \mathrm{i}\beta_0\phi(t_1,\ 0) + \beta_1\frac{\partial}{\partial t_1}\phi(t_1, 0).\end{aligned} \tag{27.90}$$

Putting $\psi = \log\phi$ in (27.90), and dividing through by $\phi(t_1, 0)$, we obtain (27.87).
Conversely, if (27 87) holds, we rewrite it, using (27.89), in the form

$$\mathrm{i}\int_{-\infty}^{\infty} \exp(\mathrm{i}t_1x)(\beta_0+\beta_1x-\mu'_{1x})g(x)\,\mathrm{d}x = 0. \tag{27.91}$$

We now see that (27.91) implies

$$\beta_0+\beta_1x-\mu'_{1x} = 0 \tag{27.92}$$

identically in x if $\exp(\mathrm{i}t_1x)g(x)$ is complete, and hence (27.86) follows.

27.42 If all cumulants exist, we have the definition of bivariate cumulants in (3.74),

$$\psi(t_1, t_2) = \sum_{r,s=0}^{\infty} \kappa_{rs}\frac{(\mathrm{i}t_1)^r}{r!}\frac{(\mathrm{i}t_2)^s}{s!},$$

where κ_{00} is defined to be equal to zero. Hence

$$\begin{aligned}\left[\frac{\partial\psi(t_1, t_2)}{\partial t_2}\right]_{t_2=0} &= \left[\mathrm{i}\sum_{r=0}^{\infty}\sum_{s=1}^{\infty}\kappa_{rs}\frac{(\mathrm{i}t_1)^r}{r!}\frac{(\mathrm{i}t_2)^{s-1}}{(s-1)!}\right]_{t_2=0} \\ &= \mathrm{i}\sum_{r=0}^{\infty}\kappa_{r1}\frac{(\mathrm{i}t_1)^r}{r!}.\end{aligned} \tag{27.93}$$

Using (27.93), (27.87) gives

$$\sum_{r=0}^{\infty}\kappa_{r1}\frac{(\mathrm{i}t_1)^r}{r!} = \beta_0 + \beta_1\sum_{r=0}^{\infty}\kappa_{r0}\frac{(\mathrm{i}t_1)^{r-1}}{(r-1)!}. \tag{27.94}$$

Identifying coefficients of t^r in (27.94) gives

$$(r=0) \qquad \kappa_{01} = \beta_0 + \beta_1\kappa_{10}, \tag{27.95}$$

as is obvious from (27.86), and

$$(r \geq 1) \qquad \kappa_{r1} = \beta_1 \kappa_{r+1,0}. \tag{27.96}$$

Condition (27.96) for linearity of regression is also due to Wicksell (1934). Expressions (27.95) and (27.96) together are sufficient, as well as necessary, for (27.87) and (given the completeness of $\exp(it_1 x)g(x)$, as before) for linearity condition (27.86).

If we express (27.87) in terms of the c.f. ϕ, instead of its logarithm ψ, as in (27.90), and carry through the process leading to (27.96), we find for the central moments,

$$\mu_{r1} = \beta_1 \mu_{r+1,0}.$$

If the regression of x on y is also linear, of the form

$$x = \beta_0' + \beta_1' y, \tag{27.97}$$

we also have

$$\kappa_{1r} = \beta_1' \kappa_{0,r+1}, \qquad r \geq 1. \tag{27.98}$$

When $r = 1$, (27.96) and (27.98) give

$$\kappa_{11} = \beta_1 \kappa_{20} = \beta_1' \kappa_{02},$$

leading to

$$\beta_1 \beta_1' = \kappa_{11}^2 / \kappa_{20}\kappa_{02} = \rho^2, \tag{27.99}$$

which is (27.15) again.

27.43 We now impose a further restriction on our variables: we suppose that the conditional distribution of y about its mean value (which, as before, is a function of the fixed value of x) is the same for any x, i.e. that only the mean of y changes with x. That is, we may write

$$y = \mu_{1x}' + \epsilon, \tag{27.100}$$

where the $\{\epsilon\}$ are independent of each other, and identically distributed. In turn, this requirement implies that ϵ is independent of x.

In particular, if the regression is linear, (27.100) is

$$y = \beta_0 + \beta_1 x + \epsilon. \tag{27.101}$$

If the $\{\epsilon\}$ are independently and identically distributed, then the joint distribution of x and y becomes

$$f(x, y) = g(x)h(\epsilon) \tag{27.102}$$

where h is now the conditional distribution of ϵ.

The c.f. for x and y is

$$\begin{aligned}\phi(t_1, t_2) &= E\{\exp(\mathrm{i}t_1 x + \mathrm{i}t_2 y)\}\\ &= E\{\exp(\mathrm{i}t_1 x + \mathrm{i}t_2\beta_0 + \mathrm{i}t_2\beta_1 x + \mathrm{i}t_2\epsilon)\}\\ &= \phi_g(t_1 + t_2\beta_1)\phi_h(t_2)\exp(\mathrm{i}t_2\beta_0).\end{aligned} \tag{27.103}$$

Note that if $\beta_1 = 0$, then (27.103) shows that x and y are *independent*: linearity of regression, identical errors and a zero regression coefficient imply independence, as is intuitively obvious.

Exercises 27.30–31 provide examples of non-normal linear regression functions.

A characterization of the bivariate normal distribution

27.44 We may now prove a remarkable result.

Theorem 27.2.

If the regression of y on x and of x on y are both linear with independently and identically distributed errors, then x and y are binormally distributed unless (a) they are independent of each other, or (b) they are functionally related.

Proof. Given the assumptions, we have, taking logarithms in (27.103),

$$\psi(t_1, t_2) = \psi_g(t_1 + t_2\beta_1) + \psi_h(t_2) + \mathrm{i}t_2\beta_0, \tag{27.104}$$

and similarly, from the regression of x on y,

$$\psi(t_1, t_2) = \psi_{g'}(t_2 + t_1\beta_1') + \psi_{h'}(t_1) + \mathrm{i}t_1\beta_0', \tag{27.105}$$

where primes are used as in (27.97). Equating (27.104) and (27.105), and considering successive powers of t_1 and t_2, we find, denoting the r th cumulant of g by κ_{r0}, that of g' by κ_{0r}, that of h by λ_{r0} and that of h' by λ_{0r}.

First order:

$$\kappa_{10}\mathrm{i}(t_1 + t_2\beta_1) + \lambda_{10}\mathrm{i}t_2 + \mathrm{i}t_2\beta_0 = \kappa_{01}\mathrm{i}(t_2 + t_1\beta_1') + \lambda_{01}\mathrm{i}t_1 + \mathrm{i}t_1\beta_0',$$

or, equating coefficients of t_1 and of t_2,

$$\kappa_{10} = \kappa_{01}\beta_1' + \lambda_{01} + \beta_0', \tag{27.106}$$

$$\kappa_{10}\beta_1 + \lambda_{10} + \beta_0 = \kappa_{01}. \tag{27.107}$$

Without loss of generality, we assume that the errors have zero means. Equations (27.106) and (27.107) correspond to the expected value expression (27.95) for x and y, respectively.

Second order:

$$\kappa_{20}(t_1 + t_2\beta_1)^2 + \lambda_{20}t_2^2 = \kappa_{02}(t_2 + t_1\beta_1')^2 + \lambda_{02}t_1^2,$$

which, on equating coefficients of t_1^2, t_1t_2 and t_2^2, gives

$$\kappa_{20} = \kappa_{02}(\beta_1')^2 + \lambda_{02}, \quad \kappa_{20}\beta_1 = \kappa_{02}\beta_1', \quad \kappa_{20}\beta_1^2 + \lambda_{20} = \kappa_{02}.$$

These expressions give relations between g, h, g' and h'; in particular, β_1/β_1' is equal to κ_{02}/κ_{20}, the ratio of the variances.

Third order:

$$\kappa_{30}\{\mathrm{i}(t_1 + t_2\beta_1)\}^3 + \lambda_{30}(\mathrm{i}t_2)^3 = \kappa_{03}\{\mathrm{i}(t_2 + t_1\beta_1')\}^3 + \lambda_{03}(\mathrm{i}t_1)^3.$$

The terms in $t_1^2 t_2$ and $t_1 t_2^2$ give us

$$\kappa_{30}\beta_1 = \kappa_{03}(\beta_1')^2, \qquad \kappa_{30}\beta_1^2 = \kappa_{03}\beta_1'.$$

Unless $\beta_1, \beta_1' = 0$ or $\beta_1\beta_1' = 1$, we see that these conditions imply $\kappa_{30} = \kappa_{03} = 0$. Similarly, if we take the fourth and higher powers, we find that all the higher cumulants κ_{r0}, κ_{0r} must vanish. Then it follows from equations such as those obtained from the terms in t_1^3, t_2^3 in the third-order equation, namely

$$\kappa_{30} = \kappa_{03}(\beta_1')^3 + \lambda_{03}, \qquad \kappa_{30}\beta_1^3 + \lambda_{30} = \kappa_{03},$$

that the cumulants after the second of h, h' also must vanish. Thus all the distributions g, h, g', h' are normal and it follows that $\psi(t_1, t_2)$ is quadratic in t_1, t_2, and hence that x, y are binormal.

In the exceptional cases we have neglected, this is no longer true. If $\beta_1\beta_1' = 1$, the correlation between x and y is ± 1 by (26.99) and x is a strict linear function of y. On the other hand, if β_1 or $\beta_1' = 0$, the variables x, y are independent.

The first result of this kind appears to be due to Bernstein (1928). For a proof under general conditions see Féron and Fourgeaud (1952). For more recent discussions, see Kagan *et al.* (1976). These results extend to the case of several regressor variables considered in Chapter 28. Exercises 27.32–36 provide several other characterizations of linear regression and binormality.

Testing the linearity of regression

27.45 In order to assess the linearity of a regression function, we may consider the *correlation ratio* for y on x:

$$\eta^2 = \operatorname{var}\{E(y|x)\}/\operatorname{var}(y). \tag{27.108}$$

When the regression function is linear,

$$\operatorname{var}\{E(y|x)\} = E_x\{(E(y|x) - E(y))^2\} = \beta_2^2\sigma_1^2 \tag{27.109}$$

by (27.13), so that

$$\eta^2 = \beta_2^2\sigma_1^2/\sigma_2^2 = \mu_{11}^2/\sigma_1^2\sigma_2^2 = \rho^2. \tag{27.110}$$

However, by the Cauchy–Schwarz inequality

$$\begin{aligned}\mu_{11}^2 &= (E_{x,y}[\{x - E(x)\}\{y - E(y)\}])^2 \\ &= (E_x[\{x - E(x)\}\{E(y|x) - E(y)\}])^2 \\ &\le \operatorname{var}(x)\operatorname{var}\{E(y|x)\}\end{aligned} \tag{27.111}$$

so that, in general,

$$0 \leq \rho^2 \leq \eta^2 \leq 1, \tag{27.112}$$

with (27.110) holding if and only if the regression is linear. It should be noted that (27.108) takes no account of the x-ordering of the conditional means so that it does not measure any particular form of non-linearity.

27.46 We now suppose that $n = n_1 + \cdots + n_k$ observations are available from a bivariate distribution with n_i replicates for x_i denoted by y_{ij}, $j = 1, \ldots, n_i$. If

$$n_i \bar{y}_i = \sum_j y_{ij} \quad \text{and} \quad n\bar{y} = \sum_i n_i \bar{y}_i$$

we can write

$$SS_Y = \sum_i \sum_j (y_{ij} - \bar{y})^2 = \sum_i \sum_j (y_{ij} - \bar{y}_i)^2 + \sum_i n_i (\bar{y}_i - \bar{y})^2 \tag{27.113}$$

and

$$\sum_i n_i (y_i - \bar{y})^2 = \sum_i n_i \{\bar{y}_i - \bar{y} - b(x_i - \bar{x})\}^2 + b^2 \sum_i n_i (x_i - \bar{x})^2, \tag{27.114}$$

where b is the sample regression coefficient for y on x. We now define the sample correlation ratio to be

$$e^2 = \sum n_i (\bar{y}_i - \bar{y})^2 / SS_Y \tag{27.115}$$

and it then follows that (26.114) may be written as

$$e^2 SS_Y = (e^2 - r^2) SS_Y + r^2 SS_Y. \tag{27.116}$$

Under the hypothesis H_0 that (x, y) is binormal (and, therefore that the regression is linear), it follows from Cochran's theorem (**15.16**) that the two sums of squares in (27.114) are independently distributed as χ^2 with $k - 2$ and $n - k$ degrees of freedom, respectively. Thus, under H_0, the test statistic

$$F = \frac{(e^2 - r^2)/(k - 2)}{(1 - e^2)/(n - k)} \tag{27.117}$$

has an F distribution with $(k - 2, n - k)$ d.fr.

When replicates are unavailable, but the sample size is reasonably large, this F test may be applied after grouping the data.

27.47 Tests for the general linear model are explored more systematically in Chapters 29 and 32.

EXERCISES

27.1 If (x, y) have zero means, variances σ_1^2 and σ_2^2 and correlation ρ, show that the linear function

$$l = ax + by$$

subject to $a^2 + b^2 = 1$ has maximum variance at $a = \cos\theta$, $b = \sin\theta$, where

$$\tan 2\theta = 2\rho\sigma_1\sigma_2/[\sigma_1^2 - \sigma_2^2],$$

reducing to $a = b\,\mathrm{sign}(\rho)$ when $\sigma_1 = \sigma_2$. Note that when $\rho = 0$, $a = 1$ (or 0) for $\sigma_1^2 >$ (or $<$) σ_2^2. At the maximum, l is known as the first principal component

(Krzanowski and Marriott, 1994, pp. 75–78)

27.2 If x_i/σ^2 $(i = 1, 2, 3)$ are independent χ^2 variates with ν_i degrees of freedom, $y_1 = x_1/x_3$ and $y_2 = x_2/x_3$, show that the joint distribution of y_1 and y_2 is

$$g(y_1, y_2) = \frac{\Gamma(\frac{1}{2}\nu)}{\prod_{i=1}^3 \Gamma(\frac{1}{2}\nu_i)} \frac{y_1^{(\nu_1-1)/2} y_2^{(\nu_2-1)/2}}{(1 + y_1 + y_2)^{\nu/2}}, \qquad 0 < y_1, y_2 < \infty,$$

where $\nu = \sum_{i=1}^3 \nu_i$. Show that the regression of y_1 on y_2, and of y_2 on y_1, is linear. Show that if $\nu_3 > 4$, their correlation coefficient is

$$\rho = [\nu_1\nu_2/\{(\nu_1 + \nu_3 - 2)(\nu_2 + \nu_3 - 2)\}]^{1/2}.$$

Assuming that x_1/σ^2, x_2/σ^2 are non-central $\chi'^2(\lambda_i, \nu_i)$ show, using Exercise 29.17, that this generalizes to

$$\rho^2 = \prod_{i=1}^{2} \left\{ \frac{(\nu_i + \lambda_i)^2}{(\nu_i + \lambda_i)^2 + (\nu_3 - 2)(\nu_i + 2\lambda_i)} \right\}.$$

27.3 A binormal distribution is dichotomized at some value of y, say y_0. The variance of x for the whole distribution is known to be σ_x^2. Given $y > y_0$, denote the conditional variance of x by σ_1^2 and the conditional correlation for x and y by γ. Show that the correlation of x and y in the whole distribution is given by θ, where

$$\rho^2 = 1 - \frac{\sigma_1^2}{\sigma_x^2}(1 - \gamma^2).$$

27.4 In the previous exercise, given that σ_y^2 is the variance of y in the whole distribution and σ_2^2 is the conditional variance when $y > y_0$, show that ρ is given by

$$\rho^2 = \frac{\gamma^2\sigma_y^2}{\sigma_2^2 + \gamma^2(\sigma_y^2 - \sigma_2^2)}.$$

27.5 Show that tetrachoric r_t and point-biserial r_{pb} can never exceed unity in absolute value, but that biserial r_b may do so.

27.6 Suppose that $y_{ij} = \alpha + \beta x_i + \epsilon_{ij}$, where the ϵ_{ij} are independent and identically distributed with zero means and variances σ^2, and $i = 1, \ldots, m$, $j = 1, \ldots, K$. Let $z_i = \sum_j y_{ij}/K$. Show that the least squares estimators of α and β remain unbiased, but that

$$E\left\{\sum(z_i - \bar{z})^2\right\} = \beta^2 \sum(x_i - \bar{x})^2 + \frac{\sigma^2(m-2)}{K}.$$

Hence, show that, at least in large samples, the correlation of x and z, $|r| \to 1$ as K increases.

27.7 A set of variables $x_1, x_2, \ldots, x_n$ are distributed so that the product moment correlation of x_i and x_j is ρ_{ij}. They all have the same variance. Show that the average value of ρ_{ij} defined by

$$\bar{\rho} = \frac{1}{n(n-1)} \sum_{i=1}^{n} \sum_{j=1}^{n} \rho_{ij}, \quad i \neq j,$$

must be not less than $-1/(n-1)$.

27.8 In the previous exercise show that $|\rho|$, the determinant of the correlation matrix, is nonnegative. Hence show that

$$\rho_{12}^2 + \rho_{13}^2 + \rho_{23}^2 \leq 1 + 2\rho_{12}\rho_{13}\rho_{23}.$$

27.9 Show from (16.86) that in samples from a binormal population the sampling distribution of b_2, the regression coefficient of y on x, has exact variance

$$\operatorname{var} b_2 = \frac{1}{n-3} \frac{\sigma_2^2}{\sigma_1^2}(1-\rho^2), \qquad n \geq 4,$$

and that its skewness and kurtosis coefficients are

$$\gamma_1 = 0, \qquad n \geq 5,$$
$$\gamma_2 = \frac{6}{n-5}, \qquad n \geq 6.$$

Show that when $\rho = 0$, $\operatorname{var} b_2$ is the expectation of the variance given for fixed xs in Example 29.1, which is in our present notation $\operatorname{var}(b_2|x) = \sigma_2^2/ns_1^2$.

27.10 Let $\psi\{(x-\mu)/\sigma, y\}$ denote the binormal density with means of x and y equal to μ and 0 respectively, variances equal to σ^2 and 1 respectively, and correlation ρ. Define

$$\xi(x, \omega) = \int_{\omega}^{\infty} \psi \, dy, \qquad \eta(x, \omega) = \int_{-\infty}^{\omega} \psi \, dy.$$

Given that z_i is a random variable taking the values 0, 1 according as $y < \omega$ or $y \geq \omega$, show that in a biserial table the likelihood function may be written

$$L(x, y|\omega, \rho, \mu, \sigma) = \prod_{i=1}^{n} \left\{ z_i \xi\left(\frac{x_i - \mu}{\sigma}, \omega\right) + (1 - z_i)\eta\left(\frac{x_i - \mu}{\sigma}, \omega\right) \right\}.$$

If ∂^2 represents a partial differential of second order with respect to any pair of parameters, show that

$$E(\partial^2 \log L) = n\{1 - p(x)\}E_0(\partial^2 \log \eta) + np(x)E_1(\partial^2 \log \xi),$$

where

$$p(x) = \int_{x}^{\infty} (2\pi)^{-1/2} \exp(\tfrac{1}{2}t^2) \, dt,$$

and E_0, E_1 are conditional expectations with respect to x for $y < \omega$, $y \geq \omega$, respectively. Hence derive the inverse of the covariance matrix for the maximum likelihood estimators of the four parameters (the order of rows and columns being the same as the order of the parameters in the LF):

$$\mathbf{V}^{-1} = \frac{n}{(1-\rho^2)}\begin{pmatrix} a_0 & \frac{\rho\omega a_0 - a_1}{1-\rho^2} & \frac{\rho a_0}{\sigma} & \frac{\rho a_1}{\sigma} \\ & \frac{a_2 - 2\rho\omega a_1 + \rho^2\omega^2 a_0}{(1-\rho^2)^2} & \frac{\rho^2\omega a_0 - \rho a_1}{\sigma(1-\rho^2)} & \frac{\rho^2\omega a_1 - \rho a_2}{\sigma(1-\rho^2)} \\ & & \frac{1-\rho^2+\rho^2 a_0}{\sigma^2} & \frac{\rho^2 a_1}{\sigma^2} \\ & & & \frac{2(1-\rho^2)+\rho^2 a_2}{\sigma^2} \end{pmatrix},$$

where

$$a_s = \int_{-\infty}^{\infty} x^s g(x, \omega, \rho)\, dx,$$

$$g(x, \omega, \rho) = (2\pi)^{-1/2}\exp(-\tfrac{1}{2}x^2)\phi\left(\frac{\omega - \rho x}{(1-\rho^2)^{1/2}}\right)\phi\left(\frac{\rho x - \omega}{(1-\rho^2)^{1/2}}\right),$$

and

$$\phi(x) = (2\pi)^{-1/2}\exp(-\tfrac{1}{2}x^2)/\{1 - p(x)\}.$$

By inverting this matrix, derive the asymptotic variance of the maximum likelihood estimator $\hat{\rho}_b$ in the form

$$\operatorname{var}\hat{\rho}_b = \frac{(1-\rho^2)^3}{n}\left\{\frac{\int_{-\infty}^{\infty} g\, dx}{\int_{-\infty}^{\infty} g\, dx \int_{-\infty}^{\infty} x^2 g\, dx - (\int_{-\infty}^{\infty} xg\, dx)^2}\right\} + \frac{\rho^2(1-\rho^2)}{n}.$$

(Tate, 1955)

27.11 In Exercise 27.10, show that when $\rho = 0$,

$$\operatorname{var}\hat{\rho}_b = \frac{2\pi p(\omega)\{1 - p(\omega)\}}{n\exp(-k^2)}.$$

By comparing this with the large-sample formula (27.77), show that when $\rho = 0$, r_b is a fully efficient estimator.

(Tate, 1955)

27.12 In Exercise 27.10, show that $n \operatorname{var}\hat{\rho}_b$ tends to zero as $|\rho|$ tends to unity, and from (27.77) that $n \operatorname{var}\rho_b$ does not, and hence that r_b is of zero efficiency near $|\rho| = 1$.

(Tate (1955); the results of Exercise 27.10–12 are extended to the multinormal distribution by Hannan and Tate (1965))

27.13 Establish equation (27.17).

27.14 Writing l for the sample intraclass correlation coefficient (27.61) and λ for the population value, show that the exact distribution of l is given by

$$f(l) \propto \frac{(1-l)^{p(k-1)/2-1}\{1 + (k-1)l\}^{(p-3)/2}}{\{1 - \lambda + \lambda(k-1)(1-l)\}^{(kp-1)/2}},$$

reducing in the case $k = 2$ to

$$f(l) = \frac{\Gamma(p - \frac{1}{2})}{\Gamma(p-1)(2\pi)^{1/2}}\operatorname{sech}^{p-1/2}(z - \xi)\exp\{-\tfrac{1}{2}(z - \xi)\}$$

where $l = \tanh z$, $\lambda = \tanh \xi$. Hence show that, for $k = 2$, $z - \xi$ is nearly normal with mean zero and variance $1/(n - 3/2)$.

(Fisher, 1921c)

27.15 Show that for testing $\rho = \rho_0$ in a bivariate normal population, the likelihood ratio statistic is given by

$$l^{1/n} = \frac{(1 - r^2)^{1/2}(1 - \rho_0^2)^{1/2}}{(1 - r\rho_0)},$$

so that $l^{1/n} = (1 - r^2)^{1/2}$ when $\rho_0 = 0$, and when $\rho_0 \neq 0$ we have

$$(1 - \rho_0^2)^{-1/2} l^{1/n} = 1 + r\rho_0 + r^2(\rho_0^2 - \tfrac{1}{2}) + \cdots.$$

27.16 Show that the effect of applying Sheppard's corrections to the moments is always to increase the absolute value of the correlation coefficient (cf. **3.18**).

27.17 Show that if x and y are respectively subject to errors of observation u, v, where u and v are uncorrelated with x, y and each other, the correlation coefficient is attenuated by a factor

$$\left\{\left(1 + \frac{\sigma_u^2}{\sigma_x^2}\right)\left(1 + \frac{\sigma_v^2}{\sigma_y^2}\right)\right\}^{1/2}.$$

27.18 If x_i $(i = 1, 2, 3)$ are mutually independent variates with means μ_i, variances σ_i^2 and coefficients of variation $V_i = \sigma_i/\mu_i$, show that the correlation between x_1/x_3 and x_2/x_3 is

$$\rho = \frac{\mu_1\mu_2}{\left\{\sigma_1^2\left(1 + \frac{1}{V_4^2}\right) + \mu_1^2\right\}^{1/2}\left\{\sigma_2^2\left(1 + \frac{1}{V_4^2}\right) + \mu_2^2\right\}^{1/2}}$$

exactly, where V_4 is the coefficient of variation of $1/x_3$. Thus $\rho = 0$ if either μ_1 or $\mu_2 = 0$; if neither is, ρ takes the sign of their product, and

$$|\rho| = \left\{1 + V_1^2\left(1 + \frac{1}{V_4^2}\right)\right\}^{-1/2}\left\{1 + V_2^2\left(1 + \frac{1}{V_4^2}\right)\right\}^{-1/2}.$$

Show that if V_4^2 is small, we obtain approximately

$$|\rho| \doteq V_3^2/\{(V_1^2 + V_3^2)(V_2^2 + V_3^2)\}^{1/2},$$

since $V_3 \doteq V_4$ by the argument of **10.7**.

(The approximation was differently obtained by Pearson (1897) who dealt with the case where μ_1, μ_2 and therefore ρ are positive; he also treated the case where x_1, x_2 and x_3 are correlated. He called ρ a 'spurious' correlation because the original x_i are uncorrelated, but the term is inapt if one is fundamentally interested in the ratios.)

27.19 Given that two binormal populations have $\rho_1 = \rho_2 = \rho$, the other parameters being unspecified, show that the maximum likelihood estimator of ρ is

$$\hat{\rho} = \frac{n(1 + r_1 r_2) - \{n^2(1 - r_1 r_2)^2 - 4n_1 n_2(r_1 - r_2)^2\}^{1/2}}{2(n_1 r_2 + n_2 r_1)},$$

where n_i, r_i are the sample sizes and correlation coefficients ($i = 1, 2$) and $n = n_1 + n_2$. If $n_1 = n_2 = \frac{1}{2}n$, show that if z_1, z_2 are defined by (27.23), and

$$\hat{\zeta} = \frac{1}{2}\log\left(\frac{1+\hat{\rho}}{1-\hat{\rho}}\right),$$

then

$$\hat{\zeta} = \frac{1}{2}(z_1 + z_2)$$

exactly.

27.20 Using the result of the previous exercise, show that the likelihood ratio test of $\rho_1 = \rho_2$ when $n_1 = n_2$ uses the statistic

$$l^{1/n} = \operatorname{sech}\{\tfrac{1}{2}(z_1 - z_2)\},$$

so that it is a one-to-one function of $z_1 - z_2$, the statistic suggested in **27.19**.

(Brandner, 1933)

27.21 In Exercise 27.19, show that if $n_1 \neq n_2$, we have *approximately* for the ML estimator of ζ

$$\hat{\zeta} = \frac{1}{n}(n_1 z_1 + n_2 z_2),$$

and hence that the LR test of $\rho_1 = \rho_2$ uses the statistic

$$l = \left[\operatorname{sech}\left\{\frac{n_1}{n}(z_1 - z_2)\right\}\right]^{n_2}\left[\operatorname{sech}\left\{\frac{n_2}{n}(z_1 - z_2)\right\}\right]^{n_1}$$

approximately, again a one-to-one function of $(z_1 - z_2)$.

(Brandner, 1933)

27.22 To estimate a common value of ρ for two binormal populations, show that

$$\hat{\zeta}^* = \frac{(n_1 - 3)z_1 + (n_2 - 3)z_2}{n_1 + n_2 - 6}$$

is the linear combination of z_1 and z_2 with minimum variance as an estimator of ζ, but that when $n_1 \neq n_2$ this does not give the maximum likelihood estimator of ρ given in Exercise 27.19.

27.23 Show that the correlation coefficient between x and y, ρ_{xy}, satisfies

$$\rho_{xy}^2 = \frac{\operatorname{var}(\alpha_2 + \beta_2 x)}{\sigma_2^2} \equiv 1 - \frac{E\{[y - (\alpha_2 + \beta_2 x)]^2\}}{\sigma_2^2}$$

and hence establish (27.7)

27.24 Let $f(x, y) = \alpha(\alpha + 1)(x + y)^{-(\alpha+2)}$, $\alpha > 0$, $1 \leq x, y < \infty$. Show that $E(y|x) = (\alpha + 1 + x)/\alpha$, $\alpha > 0$, whereas $E(y) = (\alpha + 1)/(\alpha - 1)$, $\alpha > 1$. (Note that $E(y|x)$ exists for all x, but $E(y)$ does not when $0 < \alpha \leq 1$.)

27.25 Show from (27.3) that the covariance

$$\mu_{11} = \int_{-\infty}^{\infty}\int_{-\infty}^{\infty}\{F(x, y) - F(x, \infty)F(\infty, y)\}\,dx\,dy.$$

27.26 Prove that the tetrachoric series (27.71) always converges for $|\rho| < 1$.

27.27 By using the bivariate form of the inversion theorem for the characteristic function given in **4.17** show that (27.70) reduces to (27.71).

27.28 In a standardized binormal distribution, show that on any (elliptical) contour of constant probability

$$y = \rho x \pm \{(k - x^2)(1 - \rho^2)\}^{1/2}, \qquad k > 0,$$

and that y attains its maximum $k^{1/2}$ at $x = \rho k^{1/2}$. Hence, show that the ratio of the squared vertical length of the ellipse at $x = 0$ (h^2) to its maximum squared overall vertical length (H^2) is exactly $1 - \rho^2$.

Show that if all, or nearly all, of the points in a sample scatter plot are surrounded by an ellipse, and the regression of y on x is linear and homoscedastic, h^2/H^2 is approximately the ratio of the residual variance about the regression line to the total variance of y; that is, it is approximately equal to $1 - r^2$ irrespective of bivariate normality. Apply this method to the data of Fig. 27.2 and compare the resulting approximation to r^2 with the true value given in Example 27.7.

(Cf. Châtillon, 1984)

27.29 Consider the statistics $u = \sum_{i=1}^{n} x_i^2$ and $v = \sum_{i=1}^{n} y_i^2$ based on a sample of size n from a standard binormal distribution with correlations ρ. Using Example 27.2 and the results in **27.41–42**, derive the regression functions

$$E(v|u) = \rho^2 u + n(1 - \rho^2)$$
$$E\{\text{var}(v|u)\} = 2(1 - \rho^2)\{2\rho^2 u + n(1 - \rho^2)^2\}.$$

27.30 The bivariate distribution of x and y is uniform over the parallelogram bounded by the lines $x = 3(y-1)$, $x = 3(y+1)$, $x = y+1$, $x = y-1$. Show that the regression of y on x is linear, but that the regression of x on y consists of sections of three straight lines joined together.

27.31 From (27.89), show that if the marginal distribution of a bivariate distribution is of the Gram–Charlier form

$$f = \alpha(x)\{1 + a_3 H_3 + a_4 H_4 + \cdots\},$$

then the regression of y on x is

$$\mu'_{1x} = \frac{\sum_{r=0}^{\infty}\sum_{s=0}^{\infty} \frac{\kappa_{r1}}{r!} a_s H_{r+s}(x)}{1 + \sum_{r=3}^{\infty} a_r H_r(x)}.$$

(Wicksell, 1917)

27.32 If, for each fixed y, the conditional distribution of x is normal, show that their bivariate distribution must be of the form

$$f(x, y) = \exp\{-(a_1 x^2 + a_2 x + a_3)\},$$

where the a_i are funtions of y. Show that if, in addition, the equiprobable contours of $f(x, y)$ are similar concentric ellipses, f must be binormal.

(Bhattacharyya, 1943)

27.33 Show that if the regression of x on y is linear, if the conditional distribution of x for each fixed y is normal and homoscedastic and if the marginal distribution of y is normal, then $f(x, y)$ must be binormal.

(Bhattacharyya, 1943)

27.34 Show that if the conditional distributions of x for each fixed y, and of y for each fixed x, are normal, and one of these conditional distributions is homoscedastic, then $f(x, y)$ is binormal.

(Bhattacharyya, 1943)

27.35 Show that if every non-degenerate linear function of x and y is normal, the $f(x, y)$ is binormal.

(Bhattacharyya, 1943)

27.36 Show that if the regressions of x on y and of y on x are both linear, and the conditional distribution of each for every fixed value of the other is normal, the $f(x, y)$ is either binormal or may (with suitable choice of origin and scale) be written in the form

$$f = \exp\{-(x^2 + a^2)(y^2 + b^2)\}.$$

(Bhattacharyya, 1943)

CHAPTER 28

PARTIAL AND MULTIPLE CORRELATION

28.1 As we saw in **27.8**, the correlation parameter ρ may be used as a measure of interdependence when the joint distribution is, at least approximately, binormal. When we come to interpret interdependence in practice, however, we often meet difficulties of the kind discussed in **27.3**: if a variable is correlated with a second variable, this may be merely incidental to the fact that both are correlated with another variable or set of variables. This consideration leads us to examine the correlations between variables when other variables are held constant, i.e. conditionally upon those other variables taking certain fixed values. These functions are known as the *partial correlations*.

If we find that holding another variable fixed reduces the correlation between two variables, we infer that their interdependence arises in part through that other variable; and, if the partial correlation is zero, we infer that their interdependence is entirely attributable to that variable. Conversely, if the partial correlation is larger than the original correlation between the variables we infer that the other variable was obscuring the stronger connection or masking the correlation. But it must be remembered that we cannot assume a causal connection. As we noted in **27.4** for ordinary product moment correlations, claims of causality must, ultimately, rely upon extra-statistical criteria.

28.2 In this branch of the subject, it is difficult at times to arrive at a notation which is unambiguous and flexible without being impossibly cumbrous. We shall rely upon matrix notation for demonstrations of the basic results, but revert to Yule's (1907) notation when discussing individual coefficients.

We develop the theory assuming that the variables are jointly normally distributed. Such an approach is not necessary, but simplifies the development and provides a basis for discussing sampling distributions later in the chapter.

Partial correlation

28.3 Let $\mathbf{x}$ denote a vector of p random variables that are multinormally distributed. We exclude the singular case (cf. **15.2**) and, without loss of generality, standardize the variables so that the covariance matrix becomes the correlation matrix $\mathbf{C}$. Thus, from (15.19), the joint density function is

$$f(\mathbf{x}) = (2\pi)^{-p/2}|\mathbf{C}|^{-1/2}\exp\{-\tfrac{1}{2}\mathbf{x}^T\mathbf{C}^{-1}\mathbf{x}\}, \tag{28.1}$$

where $\mathbf{C}$ has elements

$$\begin{aligned} c_{ii} &= 1 \\ c_{ij} &= \rho_{ij} = \operatorname{corr}(x_i, x_j), \qquad i \neq j. \end{aligned}$$

From (15.20), the c.f. is

$$\phi(\mathbf{t}) = \exp(-\tfrac{1}{2}\mathbf{t}^T\mathbf{C}\mathbf{t}). \tag{28.2}$$

If $\mathbf{x}$ is now partitioned into $\mathbf{x}_1$ $(k \times 1)$ and $\mathbf{x}_2$ $((p-k) \times 1)$, we may partition $\mathbf{C}$ conformably into

$$\mathbf{C} = \begin{pmatrix} \mathbf{C}_{11} & \mathbf{C}_{12} \\ \mathbf{C}_{21} & \mathbf{C}_{22} \end{pmatrix}, \tag{28.3}$$

where $\mathbf{C}_{12} = \mathbf{C}_{21}^T$. It follows directly from (28.2) that $\mathbf{x}_2$ is multinormal with mean zero and covariance matrix $\mathbf{C}_{22}$, or $\mathbf{x}_2 \sim N(\mathbf{0}, \mathbf{C}_{22})$.

28.4 The conditional distribution of $\mathbf{x}_1$ given $\mathbf{x}_2$ has density function

$$f(\mathbf{x}_1|\mathbf{x}_2) = f(\mathbf{x})/f_2(\mathbf{x}_2), \tag{28.4}$$

following the usual notation. In order to evaluate (28.4) explicitly, we first note the matrix identity

$$\mathbf{G}^{-1} = \begin{pmatrix} \mathbf{A} & \mathbf{B}^T \\ \mathbf{B} & \mathbf{D} \end{pmatrix}^{-1} = \begin{pmatrix} \mathbf{H}^{-1} & -\mathbf{H}^{-1}\mathbf{B}^T\mathbf{D}^{-1} \\ -\mathbf{D}^{-1}\mathbf{B}\mathbf{H}^{-1} & \mathbf{D}^{-1}+\mathbf{D}^{-1}\mathbf{B}\mathbf{H}^{-1}\mathbf{B}^T\mathbf{D}^{-1} \end{pmatrix}, \tag{28.5}$$

where $\mathbf{A}$ and $\mathbf{D}$ are symmetric and

$$\mathbf{H} = \mathbf{A} - \mathbf{B}^T\mathbf{D}^{-1}\mathbf{B}. \tag{28.6}$$

Expression (28.5) may be demonstrated by checking that $\mathbf{G}\mathbf{G}^{-1} = \mathbf{I}$.

28.5 It follows from (28.1) that (28.4) may be written as

$$f(\mathbf{x}_1|\mathbf{x}_2) = \text{constant} \times \exp\{-\tfrac{1}{2}(\mathbf{x}^T\mathbf{C}^{-1}\mathbf{x} - \mathbf{x}_2^T\mathbf{C}_{22}^{-1}\mathbf{x}_2)\}. \tag{28.7}$$

We now restrict attention to minus twice the exponent in (28.7), which is

$$\begin{pmatrix} \mathbf{x}_1^T & \mathbf{x}_2^T \end{pmatrix} \begin{pmatrix} \mathbf{C}_{11} & \mathbf{C}_{12} \\ \mathbf{C}_{21} & \mathbf{C}_{22} \end{pmatrix}^{-1} \begin{pmatrix} \mathbf{x}_1 \\ \mathbf{x}_2 \end{pmatrix} - \mathbf{x}_2^T\mathbf{C}_{22}^{-1}\mathbf{x}_2.$$

Using (28.5), this expression becomes

$$\mathbf{x}_2^T(\mathbf{C}_{22}^{-1} + \mathbf{C}_{22}^{-1}\mathbf{C}_{21}\mathbf{H}^{-1}\mathbf{C}_{12}\mathbf{C}_{22}^{-1})\mathbf{x}_2 - 2\mathbf{x}_1^T\mathbf{H}^{-1}\mathbf{C}_{12}\mathbf{C}_{22}^{-1}\mathbf{x}_2 + \mathbf{x}_1^T\mathbf{H}^{-1}\mathbf{x}_1 - \mathbf{x}_2^T\mathbf{C}_{22}^{-1}\mathbf{x}_2$$

which reduces to

$$(\mathbf{x}_1 - \mathbf{C}_{12}\mathbf{C}_{22}^{-1}\mathbf{x}_2)^T\mathbf{H}^{-1}(\mathbf{x}_1 - \mathbf{C}_{12}\mathbf{C}_{22}^{-1}\mathbf{x}_2). \tag{28.8}$$

By inspection of (28.7) and (28.8), it follows that the conditional distribution of $\mathbf{x}_1$ given $\mathbf{x}_2$ is multinormal, or

$$\mathbf{x}_1|\mathbf{x}_2 \sim N(\mathbf{C}_{12}\mathbf{C}_{22}^{-1}\mathbf{x}_2, \mathbf{H}), \tag{28.9}$$

where $\mathbf{H} = \mathbf{C}_{11} - \mathbf{C}_{12}\mathbf{C}_{22}^{-1}\mathbf{C}_{21}$. The constant of integration in (28.7) must be of the form

$$(2\pi)^{-k/2}|\mathbf{H}|^{-1/2};$$

so that, from (28.1) and (28.4),

$$|\mathbf{H}| = |\mathbf{C}|/|\mathbf{C}_{22}|. \tag{28.10}$$

We see from (28.9) that **H** describes the partial correlation structure for elements of $\mathbf{x}_1$, given $\mathbf{x}_2$. Further, the regression function is

$$E(\mathbf{x}_1|\mathbf{x}_2) = \mathbf{C}_{12}\mathbf{C}_{22}^{-1}\mathbf{x}_2. \tag{28.11}$$

Marsaglia (1964) shows that (28.11) and, indeed, (28.9) hold for the singular multinormal distribution provided that $\mathbf{C}_{22}^{-1}$ is replaced by the generalized inverse

$$\mathbf{C}_{22}^{+} = \mathbf{T}^T(\mathbf{T}\mathbf{T}^T)^{-2}\mathbf{T},$$

where $\mathbf{C}_{22} = \mathbf{T}^T\mathbf{T}$.

28.6 We now set $k = 2$ so that $\mathbf{x}_1 = (x_1, x_2)^T$ and examine the resulting expressions for different values of $q = p-k$. When $q = 0$, **H** yields $\text{corr}(x_1, x_2) = \rho_{12}$, as before. When $q = 1$, $\mathbf{C}_{22} \equiv 1$ and (28.6) yields

$$\begin{aligned}\mathbf{H} &= \begin{pmatrix} 1 & \rho_{12} \\ \rho_{12} & 1 \end{pmatrix} - \begin{pmatrix} \rho_{13} \\ \rho_{23} \end{pmatrix} \begin{pmatrix} \rho_{13} & \rho_{23} \end{pmatrix} \\ &= \begin{pmatrix} 1-\rho_{13}^2 & \rho_{12}-\rho_{13}\rho_{23} \\ \rho_{12}-\rho_{13}\rho_{23} & 1-\rho_{23}^2 \end{pmatrix}\end{aligned} \tag{28.12}$$

so that

$$\text{corr}(x_1, x_2|x_3) = \rho_{12.3} = \frac{\rho_{12}-\rho_{13}\rho_{23}}{\{(1-\rho_{13}^2)(1-\rho_{23}^2)\}^{1/2}}. \tag{28.13}$$

If we revert to the use of the covariance matrix, we obtain

$$\mathbf{H} = \begin{pmatrix} \sigma_1^2 & \sigma_{12} \\ \sigma_{12} & \sigma_2^2 \end{pmatrix} - \frac{1}{\sigma_3^2}\begin{pmatrix} \sigma_{13} \\ \sigma_{23} \end{pmatrix} \begin{pmatrix} \sigma_{13} & \sigma_{23} \end{pmatrix}$$

so that

$$\text{var}(x_1|x_3) = \sigma_{1.3}^2 = \sigma_1^2 - \sigma_{13}^2/\sigma_3^2 \tag{28.14}$$

and

$$\text{cov}(x_1, x_2|x_3) = \sigma_{12.3} = \sigma_{12} - \sigma_{13}\sigma_{32}/\sigma_3^2. \tag{28.15}$$

28.7 If we first derive the conditional distribution of (x_1, x_2, x_3) given $\mathbf{x}_m$, where $\mathbf{x}_m$ denotes any subset of $(x_4, \ldots, x_p)$, it follows that the conditional distribution will be of the form (28.9) with the elements of **C**, after restandardizing the variables, given by

$$\begin{aligned} C_{ij} &= 1, \qquad i = j \\ C_{ij} &= \rho_{ij.m} \equiv \text{corr}(x_i, x_j|x_m). \end{aligned}$$

Henceforth, the subscript m will be used to denote any subset of conditioning variables. We may then repeat steps (28.12)–(28.13) to arrive at

$$\begin{aligned}\text{corr}(x_1, x_2|x_3, \mathbf{x}_m) &= \rho_{12\cdot 3m} \\ &= \frac{\rho_{12\cdot m} - \rho_{13\cdot m}\rho_{23\cdot m}}{\{(1-\rho_{13\cdot m}^2)(1-\rho_{23\cdot m}^2)\}^{1/2}}.\end{aligned} \tag{28.16}$$

If we redefine **C** as the covariance matrix, the extension of (28.15) is

$$\begin{aligned}\text{cov}(x_1, x_2|x_3, \mathbf{x}_m) &= \sigma_{12\cdot 3m} \\ &= \sigma_{12\cdot m} - \sigma_{13\cdot m}\sigma_{23\cdot m}/\sigma^2_{3\cdot m}.\end{aligned} \tag{28.17}$$

The subscripts $(3, 4, \ldots, p)$ may, of course, be permuted into any order.

28.8 Although the results in **28.4–7** have been derived assuming multinormality, we shall define these to be the partial correlations whatever the underlying distribution.

When the random variables are multinormal and $\rho_{ij.m} = 0$, we say that x_i and x_j are *conditionally independent*, given $\mathbf{x}_m$. When the joint distributions are not normal, we may say only that x_i and x_j are *conditionally uncorrelated.*

Linear regression

28.9 When $k = 1$, (28.11) yields

$$E(x_1|\mathbf{x}_2) = E(x_1|x_2, \ldots, x_p) = \sum_{j=2}^{p} \beta_j x_j, \tag{28.18}$$

where

$$\boldsymbol{\beta} = (\beta_2, \ldots, \beta_p)^T = \mathbf{C}_{22}^{-1}\mathbf{C}_{21}. \tag{28.19}$$

The coefficients β_j, also denoted by $\beta_{1j\cdot m(j)}$, where $m(j) = \{2, 3, \ldots, j-1, j+1, \ldots, p\}$, are known as the *(partial) regression coefficients*. From (28.6), the conditional variance is

$$\begin{aligned}\text{var}(x_1|x_2) &= \sigma^2_{1\cdot 2\cdots p} \\ &= c_{11} - \mathbf{C}_{12}\mathbf{C}_{22}^{-1}\mathbf{C}_{21} \\ &= c_{11} - \mathbf{C}_{12}\boldsymbol{\beta}_1 \\ &= c_{11} - \boldsymbol{\beta}_1^T\mathbf{C}_{22}\boldsymbol{\beta}_1.\end{aligned} \tag{28.20}$$

Now taking **C** as the covariance matrix with elements $\{\sigma_{ij}\}$, with $\sigma_{ii} = \sigma_i^2$, (28.20) yields

$$\sigma^2_{1\cdot 2\cdots p} = \sigma_1^2 - \sum_{j=2}^{p} \beta_j \sigma_{1j}. \tag{28.21}$$

The error variance (28.21) is independent of $(x_2, \ldots, x_p)$ if the β_j are independent of these values. The conditional distribution of x_1 is then said to be *homoscedastic*; otherwise, it is *heteroscedastic*. All conditional distributions derived from the multinormal are homoscedastic. In other cases, we must make due allowance for heteroscedasticity; the partial regression coefficients are then, perhaps, best regarded as *averages* taken over all possible values of $\mathbf{x}_2$. This interpretation is justified in **28.12** below.

28.10 When $k = 1$ and $q = 2$, it follows from (28.19) that

$$\boldsymbol{\beta} = \begin{pmatrix} \sigma_{22} & \sigma_{23} \\ \sigma_{23} & \sigma_{33} \end{pmatrix}^{-1} \begin{pmatrix} \sigma_{12} \\ \sigma_{13} \end{pmatrix},$$

so that

$$\beta_{12\cdot 3} = (\sigma_{33}\sigma_{12} - \sigma_{23}\sigma_{13})/(\sigma_{22}\sigma_{33} - \sigma_{23}^2) \tag{28.22}$$

$$= \frac{\beta_{12} - \beta_{13}\beta_{32}}{1 - \beta_{23}\beta_{32}}, \tag{28.23}$$

where $\beta_{12} = \sigma_{12}/\sigma_{22}$ and so on. By a similar argument to that in **28.7**, this result extends to

$$\beta_{12\cdot 3m} = \frac{\beta_{12\cdot m} - \beta_{13\cdot m}\beta_{32\cdot m}}{1 - \beta_{23\cdot m}\beta_{32\cdot m}}, \tag{28.24}$$

so that the regression coefficients of a given order may be expressed in terms of the next lowest order. This recurrence relation avoids the need for matrix inversion and is particularly useful in the context of stepwise regression; see **32.26**. Finally, we note from direct use of (28.9) that

$$\begin{aligned} \beta_{12\cdot m} &= \sigma_{12\cdot m}/\sigma_{22\cdot m} \\ \beta_{21\cdot m} &= \sigma_{12\cdot m}/\sigma_{11\cdot m} \end{aligned} \tag{28.25}$$

so that $\rho_{12\cdot m}^2 = \beta_{12\cdot m}\beta_{21\cdot m}$ for any subset m.

Other recurrence relations are given in Exercise 28.1. The total number of partial correlations of order s is given in Exercise 28.2. When all the correlations of zero order are equal to ρ, all the partials of order s are equal (obvious by symmetry); the general form is given in Exercise 28.3.

28.11 Since

$$f(x_1|x_2, \mathbf{x}_m) = \frac{f(x_1, x_2|\mathbf{x}_m)}{f(x_2|\mathbf{x}_m)}, \tag{28.26}$$

it follows from (28.17) that

$$\begin{aligned} \sigma_{1\cdot 2m}^2 &= \sigma_{1\cdot m}^2 - (\sigma_{12\cdot m}^2/\sigma_{2\cdot m}^2) \\ &= \sigma_{1\cdot m}^2(1 - \rho_{12\cdot m}^2). \end{aligned} \tag{28.27}$$

Applying (28.27) to successively smaller subsets of the random variables and substituting back into the original expression, we obtain eventually

$$\sigma_{1\cdot 2\cdots p}^2 = \sigma_1^2(1 - \rho_{12\cdot 3\cdots p}^2)(1 - \rho_{13\cdot 4\cdots p}^2)\cdots(1 - \rho_{1,(p-1)\cdot p}^2)(1 - \rho_{1p}^2); \tag{28.28}$$

changing the order of the subscripts yields

$$\sigma_{1\cdot 2\cdots p}^2 = \sigma_1^2(1 - \rho_{12}^2)(1 - \rho_{13\cdot 2}^2)\cdots(1 - \rho_{1p\cdot 23\cdots(p-1)}^2). \tag{28.29}$$

From (28.29), it is evident that the conditional variance will decrease when an additional variable is included, unless x_1 and x_p are conditionally uncorrelated ($\rho_{1p\cdot 23\cdots(p-1)} = 0$) when it is unchanged.

Approximate linear regression

28.12 In **28.9–11**, we have assumed the regression relationship to be exactly linear, as in (28.18). We now consider the evaluation of approximate linear regression functions when the joint distribution is non-normal. We consider the linear function

$$x_1 = \boldsymbol{\beta}^T \mathbf{x} = \sum_{j=2}^{p} \beta_j x_j \tag{28.30}$$

and following the principle of least squares (see **18.52**), we choose $\boldsymbol{\beta}$ to minimize the expected sum of squares

$$S = E\{(x_1 - \boldsymbol{\beta}^T \mathbf{x})^2\}. \tag{28.31}$$

The minimum value for S is achieved when

$$E(\mathbf{x}\mathbf{x}^T)\boldsymbol{\beta} = E(\mathbf{x}^T x_1)$$

or

$$\boldsymbol{\beta} = \mathbf{C}_{22}^{-1}\mathbf{C}_{21}, \tag{28.32}$$

where $\mathbf{C}_{22} = \text{var}(\mathbf{x})$ and $\mathbf{C}_{21} = \text{cov}(\mathbf{x}, x_1)$. Expectations are taken over *all* $(x_1, \ldots, x_p)$, so that (28.32) represents the *unconditional* solution. If we consider the conditional distribution for $x_{i1}|(x_{i2}, \ldots, x_{ip})$, $i = 1, 2, \ldots, n$, then we are led to minimize

$$\begin{aligned} S &= E\left\{\sum_{i=1}^{n}(x_{i1} - \boldsymbol{\beta}^T \mathbf{x}_i)^2\right\} \\ &= E\{(\mathbf{x}_1 - \mathbf{X}\boldsymbol{\beta})^T(\mathbf{x}_1 - \mathbf{X}\boldsymbol{\beta})\}, \end{aligned} \tag{28.33}$$

where

$$\mathbf{x}_1^T = (x_{11}, x_{21}, \ldots, x_{n1})$$

and

$$\mathbf{X}^T = (\mathbf{x}_2, \mathbf{x}_3, \ldots \mathbf{x}_n) = \{x_{ij}\}^T.$$

The minimum value of (28.33) is given by

$$\boldsymbol{\beta} = (\mathbf{X}^T\mathbf{X})^{-1}\mathbf{X}^T E(\mathbf{x}_1); \tag{28.34}$$

the LS estimator for $\boldsymbol{\beta}$ is given by putting $\mathbf{x}_1$ in place of $E(\mathbf{x}_1)$ in (28.34).

28.13 Although there is a difference in interpretation between the linear regressions in **28.9** and **28.12**, we see that the functional form of $\boldsymbol{\beta}$ is the same and that the LS estimators derived from (28.32) or (28.34) are the same as the ML estimators we obtain for the multinormal case. It follows that the results of this chapter apply whenever the least squares argument can be invoked. In particular, the approximate linear regression derived in **28.12** is often considered to be *the* regression function, without further qualification; this may be a questionable assumption.

This equivalence is also used to justify the use of partial correlations, whatever the underlying distribution. However, it should be noted that whereas zero partial correlations imply conditional independence for the multinormal, this is *not* true in general. Korn (1984) gives limits for the partial correlations under conditional independence.

Sample coefficients

28.14 If we are using a sample of n observations and fit regressions by least squares, all the relationships we have discussed will hold between the sample coefficients. Following our usual convention, we shall use r instead of ρ, b instead of β, and s^2 instead of σ^2 to distinguish the sample coefficients from their population equivalents. The bs are determined by minimizing

$$ns^2_{1\cdot23\cdots p} = (\mathbf{x}_1 - \mathbf{X}\boldsymbol{\beta})^T(\mathbf{x}_1 - \mathbf{X}\boldsymbol{\beta}) \tag{28.35}$$

and the LS estimator is, from (28.34),

$$\mathbf{b} = (\mathbf{X}^T\mathbf{X})^{-1}\mathbf{X}^T\mathbf{x}_1. \tag{28.36}$$

$\mathbf{b}$ is the MV unbiased estimator for $\boldsymbol{\beta}$. Further, the sample analogue of (28.25) is

$$r^2_{ij\cdot m} = b_{ij\cdot m}b_{ji\cdot m},$$

while the analogues of (28.16), (28.24) and (28.29) also hold.

From (28.36), we may define the sample *residuals*

$$\mathbf{x}_{1\cdot m} = \mathbf{x}_1 - \mathbf{X}\mathbf{b} = \mathbf{x}_1 - \mathbf{X}(\mathbf{X}^T\mathbf{X})^{-1}\mathbf{X}^T\mathbf{x}_1 = \mathbf{M}\mathbf{x}_1, \tag{28.37}$$

where the matrix $\mathbf{M} = \mathbf{I} - \mathbf{X}(\mathbf{X}^T\mathbf{X})^{-1}\mathbf{X}^T$ is symmetric and *idempotent*, i.e. $\mathbf{M}^T = \mathbf{M} = \mathbf{M}^2$. Thus,

$$\mathbf{x}^T_{1\cdot m}\mathbf{x}_{1\cdot m} = \mathbf{x}^T_1\mathbf{M}^T\mathbf{M}\mathbf{x}_1 = \mathbf{x}^T_1\mathbf{M}\mathbf{x}_1 = \mathbf{x}^T_{1\cdot m}\mathbf{x}_1. \tag{28.38}$$

Further, if we let $\mathbf{x}_m$ comprise only $(x_{k+1}, \ldots, x_p)^T$ and then obtain the regression functions for x_1 on $\mathbf{x}_m$ and for x_2 on $\mathbf{x}_m$, we have, from (28.37),

$$\mathbf{x}_{i\cdot m} = \mathbf{M}\mathbf{x}_i$$

and

$$\mathbf{x}^T_{1\cdot m}\mathbf{x}_{2\cdot m} = \mathbf{x}^T_1\mathbf{M}\mathbf{x}_2 = \mathbf{x}^T_1\mathbf{x}_{2\cdot m}. \tag{28.39}$$

Finally, since the estimators $\mathbf{b}$ are obtained by differentiating (28.35) with respect to $\boldsymbol{\beta}$, we see that the j th estimating equation becomes

$$\mathbf{x}^T_j(\mathbf{x}_1 - \mathbf{X}\mathbf{b}) = 0 \tag{28.40}$$

which yields, from (28.37)

$$\mathbf{x}^T_j\mathbf{M}\mathbf{x}_1 = 0, \qquad j = 2, \ldots p. \tag{28.41}$$

Relations like (28.38) and (28.39) hold for the population errors as well as the sample residuals, but we shall find them of use mainly in sampling problems, which is why we have expressed them in terms of residuals. Exercise 28.5 gives the most general rule for the omission of common secondary subscripts in summing products of residuals or of errors.

Multinormal distributions for $x_{ij\cdot m}$ are considered in Exercises 28.4 and 28.6.

Geometrical interpretation of partial correlation

28.15 From our results, it is clear that the whole complex of partial regressions, correlations and variances or covariances of errors or residuals is completely determined by the variances and correlations, or by the variances and regressions, of zero order. It is interesting to consider this result from the geometrical point of view.

Suppose in fact that we have n observations on p $(< n)$ variates

$$x_{11}, \ldots, x_{1p}; x_{21}, \ldots, x_{2p}; \ldots; x_{n1}, \ldots, x_{np}.$$

Consider a (Euclidean) sample space of n dimensions. The observations $\mathbf{x}_k^T = (x_{1k}, \ldots, x_{nk})$ on the kth variate determine one point in this space, and there are p such points, one for each variate. Call these points $Q_1, Q_2, \ldots, Q_p$. We will assume that the xs are measured about their means, and take the origin to be P.

The quantity ns_1^2 may then be interpreted as the square of the length of the vector joining the point Q_1 (with coordinates $\mathbf{x}_l$) to P. Similarly, r_{lm} may be interpreted as the cosine of the angle $Q_l P Q_m$, for

$$r_{lm} = \frac{\mathbf{x}_l^T \mathbf{x}_m}{(\mathbf{x}_l^T \mathbf{x}_l \mathbf{x}_m^T \mathbf{x}_m)^{1/2}} \tag{28.42}$$

which is the cosine of the angle between PQ_1 and PQ_m.

Our result may then be expressed by saying that all the relations connecting the p points in the n-space are expressible in terms of the lengths of the vectors PQ_i and of the angles between them; and the theory of partial correlation and regression is thus exhibited as formally identical with the trigonometry of an n-dimensional constellation of points.

28.16 The reader who prefers the geometrical way of looking at this branch of the subject will have no difficulty in translating the foregoing equations into trigonometrical terminology. We will indicate only the more important results required for later sampling investigations.

Note in the first place that the p points Q_i and the point P determine (except perhaps in degenerate cases) a subspace of p dimensions in the n-space. Consider the point $Q_{1\cdot 2\cdots p}$ whose coordinates are the n residuals $x_{1\cdot 2\cdots p}$. From (28.40), the vector $PQ_{1\cdot 2\cdots p}$ is orthogonal to each of the vectors $PQ_{2,\ldots}$, PQ_p and hence to the space of $p-1$ dimensions spanned by P, $Q_2, \ldots, Q_p$.

Consider now the residual vectors $Q_{1\cdot m}$, $Q_{2\cdot m}$, where m represents the secondary subscripts $3, 4, \ldots, p-1$. The cosine of the angle between them, say θ, is $r_{12\cdot m}$ and each is orthogonal to the space spanned by P, $Q_3, \ldots, Q_{p-1}$. In Fig. 28.1, let M be the foot of the perpendicular from $Q_{1\cdot m}$ on to PQ_p and $Q'_{2\cdot m}$ a point on $PQ_{2\cdot m}$ such that $Q'_{2\cdot m}M$ is also perpendicular to PQ_p. Then $MQ_{1\cdot m}$ and $MQ'_{2\cdot m}$ are orthogonal to the space spanned by P, $Q_3, \ldots, Q_p$, and the cosine of the angle between them, say ϕ, is $r_{12\cdot mp}$. Thus, to express $r_{12\cdot mp}$ in terms of $r_{12\cdot m}$ we have to express ϕ in terms of θ, or the angle between the vectors $PQ_{1\cdot m}$ and $PQ'_{2\cdot m}$ in terms of that between their projections on the hyperplane perpendicular to PQ_p. We now drop the prime in $Q'_{2\cdot m}$ for convenience. By Pythagoras' theorem,

$$\begin{aligned}(Q_{1\cdot m}Q_{2\cdot m})^2 &= PQ_{1\cdot m}^2 + PQ_{2\cdot m}^2 - 2PQ_{1\cdot m} \cdot PQ_{2\cdot m}\cos\theta \\ &= MQ_{1\cdot m}^2 + MQ_{2\cdot m}^2 - 2MQ_{1\cdot m} \cdot MQ_{2\cdot m}\cos\phi.\end{aligned}$$

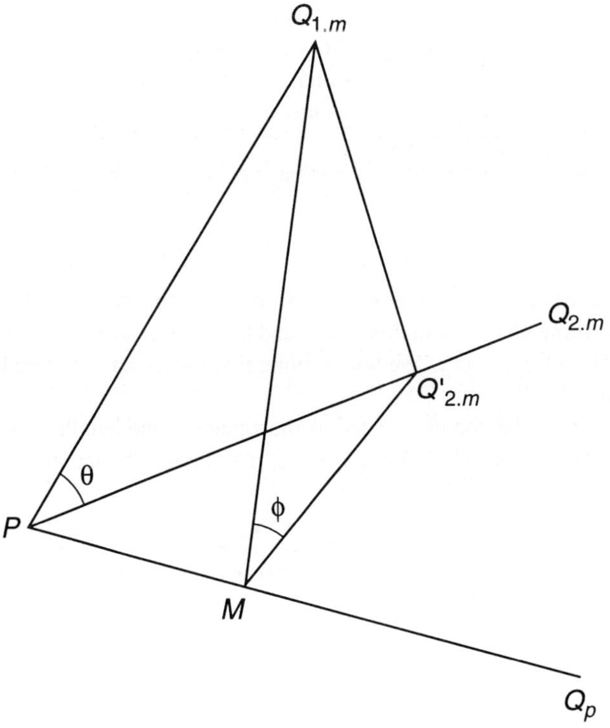

Fig. 28.1 The geometry of partial correlation

Further,

$$PQ_{1\cdot m}^2 = PM^2 + MQ_{1\cdot m}^2$$

and

$$PQ_{2\cdot m}^2 = PM^2 + MQ_{2\cdot m}^2,$$

and hence we find

$$MQ_{1\cdot m}MQ_{2\cdot m}\cos\phi = PQ_{1\cdot m}PQ_{2\cdot m}\cos\theta - PM^2$$

or

$$\frac{MQ_{1\cdot m}}{PQ_{1\cdot m}} \cdot \frac{MQ_{2\cdot m}}{PQ_{2\cdot m}} \cos\phi = \cos\theta - \frac{PM}{PQ_{1\cdot m}} \cdot \frac{PM}{PQ_{2\cdot m}}. \qquad (28.43)$$

Now $MQ_{1\cdot m}/PQ_{1\cdot m}$ and $PM/PQ_{1\cdot m}$ are the sine and cosine of the angle between PQ_p and $PQ_{1\cdot m}$. Since $PQ_{1\cdot m}$ is orthogonal to the space spanned by $P, Q_3, \ldots, Q_{p-1}$, its angle with PQ_p is unchanged if the latter is projected orthogonally to that space, i.e. if we replace PQ_p by $PQ_{p\cdot m}$. The cosine of the angle between $PQ_{1\cdot m}$ and $PQ_{p\cdot m}$ is $r_{1p\cdot m}$, and hence $PM/PQ_{1\cdot m} = r_{1p\cdot m}$, $MQ_{1\cdot m}/PQ_{1\cdot m} = (1 - r_{1p\cdot m}^2)^{1/2}$. The same result holds with the subscript 2 replacing 1. Thus, substituting in (28.43)

$$r_{12\cdot mp} = \frac{r_{12\cdot m} - r_{1p\cdot m}r_{2p\cdot m}}{\{(1 - r^2_{1p\cdot m})(1 - r^2_{2p\cdot m})\}^{1/2}}, \tag{28.44}$$

which is (28.16) again. We thus see that the expression of a partial correlation in terms of that of next lower order may be represented as the projection of an angle in the sample space on to a subspace orthogonal to the variable held fixed in the higher-order coefficient alone.

28.17 We now give an example to illustrate how partial correlations may be interpreted.

Example 28.1 (Interpretation of partial autocorrelations)

In some investigations into the variation of crime in 16 large cities in the USA, Ogburn (1935) found a correlation of -0.14 between crime rate (x_1) as measured by the number of known offences per thousand of the population and church membership (x_5) as measured by the number of church members aged 13 years or over per 100 of total population aged 13 years or over. The obvious inference is that religious belief acts as a deterrent to crime. Let us consider this more closely.

If x_2 is the percentage of males, x_3 the percentage of total population who are male immigrants, and x_4 the number of children under 5 years old per 1000 married women between 15 and 44 years old, Ogburn finds the values

$$\begin{aligned} r_{12} &= +0.44, & r_{24} &= -0.19, \\ r_{13} &= -0.34, & r_{25} &= -0.35, \\ r_{14} &= -0.31, & r_{34} &= +0.44, \\ r_{15} &= -0.14, & r_{35} &= +0.33, \\ r_{23} &= +0.25, & r_{45} &= +0.85. \end{aligned}$$

From these data it may be shown that the regression of x_1 on the other four variates, measured in standard form, is

$$\begin{aligned} \hat{x}_1 = 0.742&x_2 - 0.512x_3 - 0.681x_4 + 0.868x_5 \\ (3.27)& \quad (-2.20) \quad (-1.75) \quad (2.20) \end{aligned} \tag{28.45}$$

where the numbers in parentheses are the t statistics calculated from (28.50) below.

The partial correlations include

$$\begin{aligned} r_{15\cdot 3} &= -0.03, \\ r_{15\cdot 4} &= +0.25, \\ r_{15\cdot 34} &= +0.23. \end{aligned}$$

From (28.45), we see that when the other factors are constant x_1 and x_5 are positively related, i.e. church membership appears to be positively associated with crime. How does this effect come to be masked so as to give a negative correlation in the coefficient of zero order r_{15}?

First of all, the partial correlation between crime and church membership when the effect of x_3, the percentage of male immigrants, is eliminated, is near zero. Also, the partial correlations given x_4, the number of young children, and given both x_3 and x_4, are positive. It appears from the regression equation that a high percentage of immigrants and a high proportion of children are negatively associated with the crime rate. Both these factors are positively correlated with church membership, which masks the positive association with crime of church membership among other members of the population. Since $n = 16$ in this example, the results should be viewed as illustrative only.

Path analysis

28.18 It is apparent from Example 28.1 that disentangling the effects of multiple possible 'causes' is no easy matter. Sewall Wright, the geneticist, developed the method of path analysis (Wright, 1923; 1934) for 'working out the logical consequences of a hypothesis as to the causal relations in a system of correlated variables' (Wright, 1923, p. 254). In genetics, such well-ordered systems often exist, although the method has seen wide application in less structured areas.

To begin, we note that a path diagram for the variables $(x_1, x_2, \ldots, x_p)$ is constructed so that the pair $(x_i, x_j, i < j)$ or simply (i, j) is: unconnected, if there is no direct dependency; connected by a one-directional arrow (from j to i) if x_i depends directly on x_j; connected by a two-directional arrow if x_i and x_j are directly related but no direction has been established.

We shall suppose that the variates are decomposed into two groups $\mathbf{x}_1^T = (x_1, \ldots, x_k)$ and $\mathbf{x}_2^T = (x_{k+1}, \ldots, x_p)$ such that x_i represents the *endogenous* variables of direct interest, whereas $\mathbf{x}_2$ denotes the exogenous or explanatory variables. We assume that no hypothesis has been advanced about the structure of $\mathbf{x}_2$ so that the relations between its components may be connected by two-directional arrows. Finally, we assume that the variables in $\mathbf{x}_1$ may be ordered so that x_i depends only on x_j $(j > i)$. These assumptions give rise to a *recursive* system of equations such as

$$\begin{aligned} x_1 - \beta_{12}x_2 - \beta_{13}x_3 - \cdots - \beta_{1k}x_k - \beta_{1,k+1}x_{k+1} - \cdots - \beta_{1p}x_p &= u_1 \\ x_2 - \beta_{23}x_3 - \cdots - \beta_{2k}x_k - \beta_{2,k+1}x_{k+1} - \cdots - \beta_{2p}x_p &= u_2 \\ \vdots \qquad\qquad \vdots \qquad\qquad \vdots \qquad\qquad & \vdots \\ x_k - \beta_{k,k+1}x_{k+1} - \cdots - \beta_{kp}x_p &= u_p \end{aligned} \qquad (28.46)$$

or

$$(\mathbf{I} - \mathbf{A})\mathbf{x}_1 = \mathbf{B}\mathbf{x}_2 + \mathbf{u}, \qquad (28.47)$$

where

$$\mathbf{A} = \begin{pmatrix} 0 & \beta_{12} & \cdots & \beta_{1k} \\ & & \ddots & \vdots \\ & & & \beta_{k-1,k} \\ 0 & & & 0 \end{pmatrix} \text{ and } \mathbf{B} = \begin{pmatrix} \beta_{1,k+1} & \cdots & \beta_{1p} \\ \vdots & & \vdots \\ \beta_{k,k+1} & & \beta_{kp} \end{pmatrix}.$$

The $\{\beta_{ij}\}$ are known as the *path coefficients*, and $\mathbf{u}$ denotes a vector of uncorrelated random errors. The system of equations in (28.47) is known as *recursive* or sometimes as a *causal chain* (Wold, 1960); note that the implication of causality comes from the prior specification of

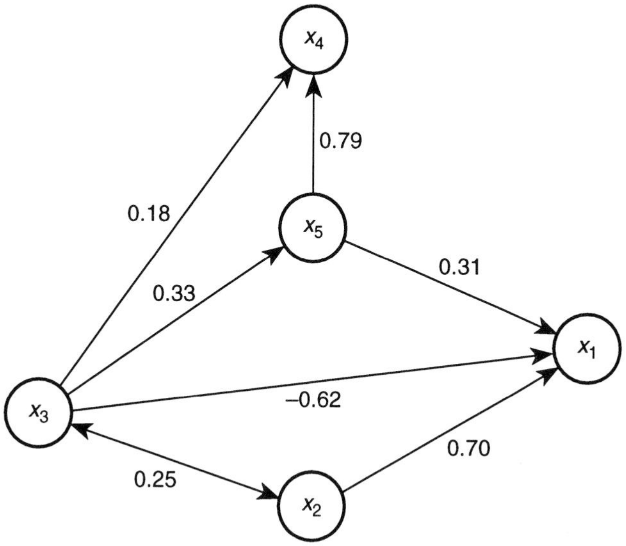

Fig. 28.2 Path diagram for data of Example 28.1, with estimated path coefficients shown

the system, and the role of statistical analysis is to examine whether or not the data support the hypothesis. In general, the hypothesis will include additional statements regarding the relationship between variables, leading to certain pairs being unconnected ($\beta_{ij} = 0$).

Example 28.2 (Path analysis for crime data)

Returning to the data of Ogburn (1935), presented in Example 28.1, we treat (x_1, x_4, x_5) as endogenous and (x_2, x_3) as exogenous. A possible structure for the variables is suggested by the following equations, which are equivalent to the path diagram shown in Fig. 28.2:

$$\begin{aligned} x_1 - \beta_{15}x_5 - \beta_{12}x_2 - \beta_{13}x_3 &= u_1 \\ x_4 - \beta_{45}x_5 - \beta_{43}x_3 &= u_2 \\ x_5 - \beta_{53}x_3 &= u_3 \end{aligned}$$

That is, we consider church membership to be affected by the levels of immigration and numbers of children to be affected by both variables. Finally, crime is assumed to be directly affected by the total number of males, immigration, and church membership.

28.19 Wold (1960) showed that recursive systems like (28.47) may be estimated by the method of least squares. Further, if the u_i are taken to be independent and normally distributed, these estimators are precisely those given by ML. The estimation is usually carried out after standardizing the variables to have zero mean and unit variance so that the path coefficients may be computed from the correlation matrix and are dimensionless, making comparisons easier. The significance of individual coefficients may be tested using the t statistic given in (28.50) below, as was done for the original regression equation (28.45).

Example 28.3 (A recursive systems model for crime data)

Following Example 28.2, the estimated equations are given by

$$\begin{aligned}\hat{x}_5 &= \underset{(1.37)}{0.33x_3}\\ \hat{x}_4 &= \underset{(1.22)}{0.179x_3} + \underset{(5.40)}{0.791x_5}\end{aligned}$$

and

$$\hat{x}_1 = \underset{(2.87)}{0.703x_2} - \underset{(-2.54)}{0.618x_3} + \underset{(1.23)}{0.310x_5}.$$

The path coefficients are also included in Fig. 28.2. The equation for x_1 should be compared with that given in (28.45).

28.20 When **A** in (28.47) has all upper triangular elements present, the system is said to be *complete.* When some coefficients are preset to zero, the system is termed *incomplete.* Since the motive for studying such a system is to interpret the *correlation* structure, it is encouraging to know that when the system is complete, the covariance matrix is unrestricted and so the sample covariance matrix is the ML estimator. However, when the system is incomplete, the ML estimator for the covariance matrix does not necessarily coincide with the observed covariance matrix whenever (i, j) are connected. The heuristic estimator proposed by Wright (1934) for the incomplete case cannot, therefore, be justified in general. Conditions under which Wright's rule does correspond to ML estimation are given by Wermuth (1980).

Although the direct interpretation of the partial correlations may be lost, path diagrams are more readily interpreted by means of models like (28.47). When the system is non-recursive, more complex estimators are required (cf. Joreskög, 1981).

Sampling distributions of partial correlation and regression coefficients in the normal case

28.21 We now consider the sampling distributions of the partial correlation and regression coefficients in the normal case. For large samples, the standard errors appropriate to zero-order coefficients (cf. **27.13**) may be used with obvious adjustments. Writing m for a set of secondary subscripts, we have, from (27.22),

$$\operatorname{var} r_{12\cdot m} \doteq \frac{1}{n}(1-\rho_{12\cdot m}^2)^2, \tag{28.48}$$

and from (27.28)

$$\operatorname{var} b_{12\cdot m} \doteq \frac{1}{n}\frac{\sigma_{1\cdot m}^2}{\sigma_{2\cdot m}^2}(1-\rho_{12\cdot m}^2) = \frac{1}{n}\frac{\sigma_{1\cdot 2m}^2}{\sigma_{2\cdot m}^2}, \tag{28.49}$$

by (28.27). The proof of (28.48) and (28.49) by the direct methods of Chapter 10 is very tedious. They follow directly, however, from noting that the joint distribution of any two errors $x_{1\cdot m}$ and $x_{2\cdot m}$ is bivariate normal with correlation coefficient $\rho_{12\cdot m}$. It follows, as Yule (1907) pointed out, that the sample correlation and regressions between the corresponding residuals have at least the large-sample distribution of a zero-order coefficient. We now show that in a sample of size n, the exact distribution of $r_{12\cdot m}$ is that of a zero-order correlation based on $n-d$ observations, where d is the number of secondary subscripts in m.

28.22 Consider now the geometrical representation of **28.15–16**. Suppose that we have three vectors PQ_1, PQ_2, PQ_3, representing n observations on x_1, x_2, x_3. As we saw in **28.16**, the partial correlation $r_{12\cdot 3}$ is the cosine of the angle between PQ_1 and PQ_2 projected on to the subspace orthogonal to PQ_3, which is of dimension $n-1$. If we make an orthogonal transformation (i.e. a rotation of the coordinate axes), the correlations, being cosines of angles, are unaffected; moreover, if the n original observations on the three variables are independent of each other, the n observations on the orthogonally transformed variables will also be. (This is a generalization of the result of Examples 11.2 and 11.3 and of **15.27** for independent x_1, x_2, x_3, and its proof is left for the reader as Exercise 28.7; it is geometrically obvious from the radial symmetry of the standardized multinormal distribution.) If PQ_3 is taken as one of the new co-ordinate axes in the orthogonal transformation, the distribution of $r_{12\cdot 3}$ is at once seen to be the same as that of a zero-order coefficient based on $n-1$ independent observations. By repeated application of this argument, it follows that the distribution of a correlation coefficient of order d based on n observations is that of a zero-order coefficient based on $n-d$ observations: each secondary subscript involves a projection in the sample space orthogonal to that variable and a loss of one degree of freedom. The result is due to Fisher (1924a).

The results of the previous chapter are thus immediately applicable to partial correlations, with this adjustment. If d is small compared with n, the distribution of partial correlations as n increases is effectively the same as that of zero-order coefficients, confirming the approximation (28.48) to the standard error.

It also follows for partial regression coefficients that the zero-order coefficient distribution (16.86) persists when the set m of secondary subscripts is added throughout, with n replaced by $(n-d)$. In particular, Student's t statistic for (27.36) becomes, for the regression of x_1 on x_2,

$$t = (b_{12\cdot m} - \beta_{12\cdot m})\left\{\frac{s^2_{2\cdot m}(n-d-2)}{s^2_{1\cdot m}(1-r^2_{12\cdot m})}\right\}^{1/2} \qquad (28.50)$$

with $n-d-2$ degrees of freedom. If the set m consists of all $p-2$ other variates, there are $n-p$ degrees of freedom. Since the regression coefficients are functions of distances (variances) as well as angles in the sample space, the distribution of b_{12} itself, unlike that of r, is not directly preserved under projection with only degrees of freedom being reduced; the statistics $s^2_{1\cdot m}$, $s^2_{2\cdot m}$ in (28.50) make the necessary 'distance' adjustments for the projections. When all correlations of order zero are equal to ρ, a special form of t statistic is available; see Exercises 28.8–10.

The multiple correlation coefficient

28.23 The variance in the population of x_1 about its regression on the other variates (28.18) is

$\sigma^2_{1\cdot 2\cdots p}$, defined in **28.9**. We now define the *multiple correlation coefficient*[1] $\mathbf{R}_{1(2\cdots p)}$ between x_1 and $x_2, \ldots, x_p$ by

$$1 - \mathbf{R}^2_{1(2\cdots p)} = \sigma^2_{1\cdot 2\cdots p}/\sigma^2_1. \tag{28.51}$$

From (28.51) and (28.29),

$$0 \leq \mathbf{R}^2 \leq 1.$$

We shall define $\mathbf{R}$ as the positive square root of $\mathbf{R}^2$: it is always non-negative. $\mathbf{R}$ is evidently not symmetric in its subscripts, and it is, indeed, a measure of the *dependence* of x_1 upon $x_2, \ldots, x_p$.

To justify its name, we have to show that it is in fact a correlation coefficient.. We have, from **28.9** and **28.14,**

$$\sigma^2_{1\cdot 2\cdots p} = E(x^2_{1\cdot 2\cdots p}), \tag{28.52}$$

and by the population analogue of (28.38),

$$E(x^2_{1\cdot 2\cdots p}) = E(x_1 x_{1\cdot 2\cdots p}). \tag{28.53}$$

Equations (28.52) and (28.53) give, since $E(x_{1\cdot 2\cdots p}) = 0$,

$$\sigma^2_{1\cdot 2\cdots p} = \text{var}(x_{1\cdot 2\cdots p}) = \text{cov}(x_1, x_{1\cdot 2\cdots p}). \tag{28.54}$$

If we now consider the correlation between x_1 and its conditional expectation

$$E(x_1|x_2, \ldots, x_p) = x_1 - x_{1\cdot 2\cdots p},$$

we find that this is

$$\frac{\text{cov}(x_1, x_1 - x_{1\cdot 2\cdots p})}{\{\text{var}\, x_1\, \text{var}(x_1 - x_{1\cdot 2\cdots p})\}^{1/2}} = \frac{\text{var}\, x_1 - \text{cov}(x_1, x_{1\cdot 2\cdots p})}{[\text{var}\, x_1\{\text{var}\, x_1 + \text{var}\, x_{1\cdot 2\cdots p} - 2\,\text{cov}(x_1, x_{1\cdot 2\cdots p})\}]^{1/2}},$$

and, using (28.54), this is

$$\frac{\sigma^2_1 - \sigma^2_{1\cdot 2\cdots p}}{\{\sigma^2_1(\sigma^2_1 - \sigma^2_{1\cdot 2\cdots p})\}^{1/2}} = \left\{\frac{\sigma^2_1 - \sigma^2_{1\cdot 2\cdots p}}{\sigma^2_1}\right\}^{1/2} = \mathbf{R}_{1(2\cdots p)} \tag{28.55}$$

by (28.51). Thus $\mathbf{R}_{1(2\cdots p)}$ is the ordinary product moment correlation coefficient between x_1 and the conditional expectation $E(x_1|x_2, \ldots, x_p)$. Since the sum of squared errors (and therefore their mean, $\sigma^2_{1\cdot 2\cdots p}$) is minimized in finding the LS regression, which is identical with $E(x_1|x_2, \ldots, x_p)$ (cf. **28.12**), it follows from (28.55) that $\mathbf{R}_{1(2\cdots p)}$ is the correlation between x_1 and the 'best-fitting' linear combination of $x_2, \ldots, x_p$. No other linear function of $x_2, \ldots, x_p$ will have greater correlation with x_1.

[1]We use a bold-face $\mathbf{R}$ for the population coefficient, and will later use an italic R for the corresponding sample coefficient: we are reluctant to use the Greek capital for the population coefficient, in accordance with our usual convention, because it resembles a capital P, which might be confusing.

28.24 From (28.51) and (28.29), we have

$$1 - \mathbf{R}^2_{1(2\cdots p)} = (1-\rho^2_{12})(1-\rho^2_{13\cdot 2})\cdots(1-\rho^2_{1p\cdot 23\cdots(p-1)}), \tag{28.56}$$

expressing the multiple correlation in terms of the partial correlations. Since permutation of the subscripts other than 1 is allowed in (28.29), it follows at once from (28.56) that, since each factor on the right is in the interval [0, 1],

$$1 - \mathbf{R}_{1(2\cdots p)} \leq 1 - \rho^2_{1j\cdot s}$$

where $\rho_{1j\cdot s}$ is any partial or zero-order coefficient having 1 as a primary subscript. Thus

$$\mathbf{R}_{1(2\cdots p)} \geq |\rho_{1j\cdot s}|;$$

the multiple correlation coefficient is no less in value than the absolute value of any correlation coefficient with a common primary subscript. It follows that if $\mathbf{R}^2_{1(2\cdots p)} = 0$, all the corresponding $\rho_{1j\cdot s} = 0$ also, so that x_1 is completely uncorrelated with all the other variables. On the other hand, if $\mathbf{R}^2_{1(2\cdots p)} = 1$, at least one $\rho_{1j\cdot s}$ must be 1 also to make the right-hand side of (28.56) equal to zero (Exercise 28.18 shows that all zero-order ρ_{1j} may nevertheless be arbitrarily small). In this case, (28.51) shows that $\sigma^2_{1\cdot 2\cdots p} = 0$, so that all points in the distribution of x_1 lie on the regression line, and x_1 is a strict linear function of $x_2, \ldots, x_p$.

Thus $\mathbf{R}_{1(2\cdots p)}$ is a measure of the linear dependence of x_1 upon $x_2, \ldots, x_p$. When $x_2, \ldots, x_p$ are all uncorrelated, (28.56) simplifies as shown in Exercise 28.17.

28.25 So far, we have considered the multiple correlation coefficient between x_1 and all the other variates, but we may evidently also consider the multiple correlation of x_1 and any subset. Thus we define

$$\mathbf{R}^2_{1(s)} = 1 - \frac{\sigma^2_{1\cdot s}}{\sigma^2_1} \tag{28.57}$$

for any set of subscripts s. It now follows immediately from (28.29) that

$$\sigma^2_{1\cdot s} \leq \sigma^2_{1\cdot r}, \tag{28.58}$$

where r is any subset of s: the error variance cannot be increased by the addition of a further variate. We thus have, from (28.57) and (28.58), relations of the type

$$\mathbf{R}^2_{1(2)} \leq \mathbf{R}^2_{1(23)} \leq \mathbf{R}^2_{1(234)} \leq \cdots \leq \mathbf{R}^2_{1(2\cdots p)}, \tag{28.59}$$

expressing the fact that the multiple correlation coefficient can never be reduced by adding to the set of variables upon which the dependence of x_1 is to be measured.

In the particular case $p = 2$, we have from (28.56)

$$\mathbf{R}^2_{1(2)} = \rho^2_{12}, \tag{28.60}$$

so that $\mathbf{R}_{1(2)}$ is the absolute value of the ordinary correlation coefficient between x_1 and x_2.

Geometrical interpretation of multiple correlation

28.26 We may interpret $\mathbf{R}_{1(2\ldots p)}$ in the geometrical terms of **28.15–16**. Consider first the interpretation of the LS regression (28.18): by **28.23**, it is that linear function of the variables $x_2, \ldots, x_p$ which minimizes the sum of squares (28.31). Thus we choose the vector PV in the $(p-1)$-dimensional subspace spanned by $P, Q_2, \ldots, Q_p$, which minimizes the distance Q_1V, i.e. which minimizes the angle between PQ_1 and PV. By (28.55), $\mathbf{R}_{1(2\cdots p)}$ is the cosine of this minimized angle. But this means that $\mathbf{R}_{1(2\cdots p)}$ is the cosine of the angle between PQ_1 and the $(p-1)$-dimensional subspace itself, for otherwise the angle would not be minimized.

If $\mathbf{R}_{1(2\cdots p)} = 0$, PQ_1 is orthogonal to the $(p-1)$-subspace so that x_1 is uncorrelated with $x_2, \ldots, x_p$ and with any linear function of them. If, on the other hand, $\mathbf{R}_{1(2\cdots p)} = 1$, PQ_1 lies in the $(p-1)$-subspace, so that x_1 is a strict linear function of $x_2, \ldots, x_p$. These are the results we obtained in **28.24**.

We shall find this geometrical interpretation helpful in deriving the distribution of the sample coefficient $\mathbf{R}_{1(2\cdots p)}$ in the normal case. It is a direct generalization of the representation used in **16.24** to obtain the distribution of the ordinary product moment correlation coefficient r which, as we observed at (28.60), is essentially the signed value of $\mathbf{R}_{1(2)}$.

Canonical correlation

28.27 We saw in **28.23** that $\mathbf{R}$ is the maximal correlation between x_1 and $(x_2, \ldots, x_p)$. This argument may be extended as follows. As in **28.2**, partition $\mathbf{x}$ into $\mathbf{x}_1$ and $\mathbf{x}_2$ with corresponding covariance matrix (28.3). Let

$$\mathbf{z}_1 = \mathbf{L}^T\mathbf{x}_1, \qquad \mathbf{z}_2 = \mathbf{M}^T\mathbf{x}_2 \tag{28.61}$$

where $\mathbf{L}$ and $\mathbf{M}$ are $k \times t$ and $(p-k) \times t$ matrices with $t \leq k$, and we may take $k \leq p-k$ without loss of generality, as our final function of interest may be derived for $\mathbf{z}_1|\mathbf{z}_2$ or $\mathbf{z}_2|\mathbf{z}_1$.

From **16.49**, the generalized variance for a vector $\mathbf{x}$ with covariance matrix $\mathbf{C}$ is $|\mathbf{C}|$; thus, we may define the generalized variance ratio (GVR) for the conditional distribution of $\mathbf{z}_1|\mathbf{z}_2$, relative to the unconditional distribution for $\mathbf{z}_1$, as

$$\text{GVR} = |\mathbf{V}(\mathbf{z}_1) - \mathbf{V}(\mathbf{z}_1|\mathbf{z}_2)|/|\mathbf{V}(\mathbf{z}_1)|;. \tag{28.62}$$

GVR corresponds to $\mathbf{R}^2$ when $k = 1$; by construction $0 \leq \text{GVR} \leq 1$, with its maximum occurring when $\mathbf{V}(\mathbf{z}_1|\mathbf{z}_2) \equiv \mathbf{0}$.

From **28.5**,

$$\begin{aligned}\mathbf{V}(\mathbf{z}_1) &= \mathbf{L}^T\mathbf{C}_{11}\mathbf{L} \\ &= \mathbf{A},\end{aligned} \tag{28.63}$$

say, and

$$\begin{aligned}\mathbf{V}(\mathbf{z}_1|\mathbf{z}_2) &= \mathbf{L}^T\mathbf{C}_{11}\mathbf{L} - \mathbf{L}^T\mathbf{C}_{12}\mathbf{M}(\mathbf{M}^T\mathbf{C}_{22}\mathbf{M})^{-1}\mathbf{M}^T\mathbf{C}_{21}\mathbf{L} \\ &= \mathbf{A} - \mathbf{H},\end{aligned} \tag{28.64}$$

say, so that

$$\text{GVR} = |\mathbf{H}|/|\mathbf{A}|. \tag{28.65}$$

It follows that (the proof is left to the reader as Exercise 28.22)

$$\max_{L,M} \text{GVR} = |\mathbf{D}| \tag{28.66}$$

where

$$\mathbf{GL} = \mathbf{LD} \tag{28.67}$$
$$\mathbf{G} = \mathbf{C}_{11}^{-1}\mathbf{C}_{12}\mathbf{C}_{22}^{-1}\mathbf{C}_{21} \tag{28.68}$$
$$\text{and } \mathbf{M} = \mathbf{C}_{22}^{-1}\mathbf{C}_{21}\mathbf{L}. \tag{28.69}$$

That is, **D** denotes the diagonal matrix of the t largest eigenvalues of **G** and **L** is the corresponding matrix of eigenvectors. Writing

$$|\mathbf{D}| = d_1 \cdot d_2 \cdot \ldots \cdot d_t$$

where $d_1 \geq d_2 \geq \cdots \geq d_t$, the $\{d_j\}$ are the canonical correlations for $\mathbf{x}_1$ and $\mathbf{x}_2$. The process of construction ensures that we select the jth pair $z_{1(j)}$ and $z_{2(j)}$ to maximize the correlation between them (d_j), subject to each of $z_{1(j)}$ and $z_{2(j)}$ being uncorrelated with all $z_{1(i)}$ and $z_{2(i)}$, $i < j$. For a detailed discussion of canonical correlation analysis, see Krzanowski and Marriott (1994, Chapter 4).

The sample multiple correlation coefficient and its conditional distribution

28.28 We now define the sample analogue of $\mathbf{R}^2_{1(2\cdots p)}$ by

$$1 - R^2_{1(2\cdots p)} = \frac{s^2_{1\cdot 2\cdots p}}{s^2_1}, \tag{28.70}$$

and all the relations of **28.23–25** hold with the appropriate substitutions of r for ρ, and s for σ. We proceed to discuss the sampling distribution of R^2 in detail. Since, by **28.23**, it is a correlation coefficient, whose value is independent of location and scale, its distribution will be free of location and scale parameters.

First, consider the conditional distribution of R^2 when the values of $x_2, \ldots, x_p$ are fixed. From **28.14** we may write

$$\begin{aligned}\mathbf{x}_1^T\mathbf{x}_1 &\equiv \mathbf{x}_1^T(\mathbf{I}-\mathbf{M})\mathbf{x}_1 + \mathbf{x}_1^T\mathbf{M}\mathbf{x}_1 \\ &= \mathbf{x}_1^T(\mathbf{I}-\mathbf{M})\mathbf{x}_1 + \mathbf{x}_{1\cdot m}^T\mathbf{x}_{1\cdot m}.\end{aligned} \tag{28.71}$$

Since $ns_1^2 = \mathbf{x}_1^T\mathbf{x}_1$ and $ns^2_{1\cdot 2\cdots p} = \mathbf{x}_1^T\mathbf{x}_{1\cdot m}$, it follows from (28.70) and (28.71) that

$$\begin{aligned}ns_1^2 &\equiv n(s_1^2 - s^2_{1\cdot 2\cdots p}) + ns^2_{1\cdot 2\cdots p}, \\ &\equiv ns_1^2R^2_{1(2\cdots p)} + ns_1^2(1 - R^2_{1(2\cdots p)}).\end{aligned} \tag{28.72}$$

If the observations on $\mathbf{x}_1$ are independent normal variates, so that $R^2_{1(2\cdots p)} = 0$, and we standardize them, the left-hand side of (28.72) has a chi-squared distribution with $n-1$ degrees of freedom, and the quadratic forms in $\mathbf{x}_1$ on the right of (28.71) may be shown to have ranks $p-1$ and $n-p$ respectively, since $\mathbf{M}$ is idempotent with rank $p-1$. It follows by Cochran's theorem (15.16) that they are independently distributed as χ^2 with these degrees of freedom and that the ratio

$$F = \frac{R^2_{1(2\cdots p)}/(p-1)}{(1-R^2_{1(2\cdots p)})/(n-p)} \tag{28.73}$$

has the F distribution with $(p-1, n-p)$ degrees of freedom, a result first given by Fisher (1924b).

When $\mathbf{R}^2 \neq 0$, the F statistic given in (28.73) has a non-central F distribution with non-centrality parameter

$$\begin{aligned}\lambda &= \sum_i \boldsymbol{\beta}_i^T \mathbf{x}_i \mathbf{x}_i^T \boldsymbol{\beta}/\{\sigma_1^2(1-\mathbf{R}^2)\} \\ &= \boldsymbol{\beta}^{\mathbf{T}}\mathbf{V}\boldsymbol{\beta}/\{\sigma_1^2(1-\mathbf{R}^2)\},\end{aligned} \tag{28.74}$$

where $\boldsymbol{\beta}$ is the $(p-1)\times 1$ vector of regression coefficients and $\mathbf{V}$ is the $(p-1)\times(p-1)$ observed covariance matrix of $x_2, \ldots, x_p$. We will discuss this distribution in greater detail in **29.26**.

The multinormal (unconditional) case

28.29 If we now allow the values of $x_2, \ldots, x_p$ to vary also, and suppose that we are sampling from a multinormal population, we find that the distribution of R^2 is unchanged if $\mathbf{R}^2 = 0$, but quite different otherwise from that of R^2 with $x_2, \ldots, x_p$ fixed. Thus the power function of the test of $\mathbf{R}^2 = 0$ is different in the two cases, although the same test is valid in each case. As $n \to \infty$, however, the limiting results are identical in the two cases.

We derive the multinormal result for $\mathbf{R}^2 = 0$ geometrically, and proceed to generalize it in **28.30**.

Consider the geometrical representation of **28.26**. R is the cosine of the angle, say θ, between PQ_1 (the x_1 vector) and the vector PV, in the $(p-1)$-dimensional space S_{p-1} of the other variables, which makes the minimum angle with PQ_1. If $\mathbf{R} = 0$, x_1 is, since the population is multinormal, independent of $x_2, \ldots, x_p$, and the vector PQ_1 will then, because of the radial symmetry of the normal distribution, be randomly directed with respect to S_{p-1}, which we may therefore regard as fixed in the subsequent argument. (We therefore see how it is that the conditional and unconditional results coincide when $\mathbf{R}^2 = 0$.)

We have to consider the relative probabilities with which different values of θ may arise. For fixed variance s_1^2, the probability density of the sample of n observations is constant on the $(n-2)$-dimensional surface of an $(n-1)$-dimensional hypersphere. If θ and PV are fixed, PQ_1 is constrained to lie on a hypersphere of $(n-2)-(p-1) = n-p-1$ dimensions, whose content is proportional to $\sin^{n-p-1}\theta$ (cf. **16.24**). Now consider what happens when PV varies. PV is free to vary within S_{p-1}, where by radial symmetry it will be equiprobable on the $(p-2)$-dimensional surface of a $(p-1)$-sphere. This surface has content proportional to $\cos^{p-2}\theta$. For fixed θ, therefore, we have the probability element $\sin^{n-p-1}\theta\cos^{p-2}\theta\,\mathrm{d}\theta$. Putting $R = \cos\theta$, and $\mathrm{d}\theta \propto \mathrm{d}(R^2)/\{R(1-R^2)^{1/2}\}$, we find for the distribution of R^2 the beta distribution

$$\mathrm{d}F \propto (R^2)^{(p-3)/2}(1-R^2)^{(n-p-2)/2}\,\mathrm{d}(R^2), \qquad 0 \le R^2 \le 1. \tag{28.75}$$

The constant of integration is easily seen to be $1/B\{\frac{1}{2}(p-1), \frac{1}{2}(n-p)\}$. The transformation (28.73) applied to (28.75) then gives us exactly the same F distribution as that derived for $x_2, \ldots, x_p$ fixed in **28.28**. When $p = 2$, (28.73) reduces to (16.62), which is expressed in terms of $\mathrm{d}R$ rather than $\mathrm{d}(R^2)$.

28.30 We now turn to the case when $\mathbf{R} \neq 0$. The distribution of R in this case was first given by Fisher (1928a) by a considerable development of the geometrical argument of **28.29**. We give a much simpler derivation due to Moran (1950).

We may write (28.56) for the sample coefficient as

$$1 - R^2_{1(2\cdots p)} = (1 - r^2_{12})(1 - T^2), \tag{28.76}$$

say, where T is the multiple correlation coefficient between $x_{1\cdot 2}$ and $x_{3\cdot 2}, x_{4\cdot 2}, \ldots, x_{p\cdot 2}$. Now $\mathbf{R}_{1(2\cdots p)}$ and the distribution of $R_{1(2\cdots p)}$ are unaffected if we make an orthogonal transformation of $x_2, \ldots, x_p$ so that x_2 itself is the linear function of $x_2, \ldots, x_p$ which has maximum correlation with x_1 in the population, i.e. $\rho_{12} = \mathbf{R}_{1(2\cdots p)}$. It then follows from (28.56) that

$$\rho^2_{13\cdot 2} = \rho^2_{14\cdot 23} = \cdots = \rho^2_{1p\cdot 23\cdots(p-1)} = 0,$$

and since subscripts other than 1 may be permuted in (28.56), it follows that *all* partial coefficients of the form $\rho_{1j\cdot 2s}$ are zero. Thus $x_{1\cdot 2}$ is uncorrelated with (and since the variation is normal, independent of) $x_{3\cdot 2}, x_{4\cdot 2}, \ldots, x_{p\cdot 2}$, and T in (28.76) is distributed as a multiple correlation coefficient, based on $n-1$ observations (since we lose one dimension by projection for the residuals), between one variate and $p-2$ others, with $\mathbf{R} = 0$. Moreover, T is distributed independently of r_{12}, for all the variates $x_{j\cdot 2}$ are orthogonal to x_2 by (28.41). Thus the two factors on the right of (28.76) are independently distributed. The distribution of r_{12}, say $f_1(r)$, is (16.60) with $\rho = \mathbf{R}_{1(2\cdots p)}$ integrated for β over its range, while that of T^2, say $f_2(R^2)$, is (28.75) with n and p each reduced by 1. We therefore have from (28.76) the distribution of R^2,

$$\mathrm{d}F = \int_{-R}^{R} \left\{ f_2\left(\frac{R^2 - r^2}{1 - r^2}\right) \right\} \mathrm{d}F_1(r), \tag{28.77}$$

which, dropping all subscripts for convenience, is

$$\begin{aligned}
&= \frac{n-2}{\pi}(1-\mathbf{R}^2)^{(n-1)/2} \int_{-R}^{R} (1-r^2)^{(n-4)/2} \int_0^\infty \frac{\mathrm{d}\beta}{(\cosh\beta - \mathbf{R}r)^{n-1}} \\
&\quad \times \left\{ \frac{1}{B\{\frac{1}{2}(p-2), \frac{1}{2}(n-p)\}} \left(\frac{R^2-r^2}{1-r^2}\right)^{(p-4)/2} \left(\frac{1-R^2}{1-r^2}\right)^{(n-p-2)/2} \frac{\mathrm{d}R^2}{(1-r^2)} \right\} \mathrm{d}r \\
&= \frac{n-2}{\pi} \frac{(1-\mathbf{R}^2)^{(n-1)/2}(1-R^2)^{(n-p-2)/2}\,\mathrm{d}(R^2)}{B\{\frac{1}{2}(p-2), \frac{1}{2}(n-p)\}} \\
&\quad \times \int_{-R}^{R} (R^2-r^2)^{(p-4)/2} \left[\int_0^\infty \frac{\mathrm{d}\beta}{(\cosh\beta - \mathbf{R}r)^{n-1}}\right] \mathrm{d}r.
\end{aligned} \tag{28.78}$$

We can substitute for the inner integral as in (16.64)–(16.65). If in (28.78) we put $r = R\cos\psi$, and write the integral with respect to β from $-\infty$ to ∞, dividing by 2 to compensate for this, we obtain Fisher's form of the distribution,

$$dF = \frac{\Gamma(\frac{1}{2}n)(1-\mathbf{R}^2)^{(n-1)/2}}{\pi\Gamma\{\frac{1}{2}(p-2)\}\Gamma\{\frac{1}{2}(n-p)\}}(R^2)^{(p-3)/2}(1-R^2)^{(n-p-2)/2}\,d(R^2)$$
$$\times\int_0^{\pi}\sin^{p-3}\psi\left\{\int_{-\infty}^{\infty}\frac{d\beta}{(\cosh\beta-\mathbf{R}R\cos\psi)^{n-1}}\right\}d\psi. \qquad (28.79)$$

28.31 The distribution (28.79) may be expressed as a hypergeometric function. Expanding the integrand in a uniformly convergent series of powers of $\cos\psi$, it becomes, since odd powers of $\cos\psi$ will vanish on integration from 0 to π,

$$\sum_{j=0}^{\infty}\binom{n+2j-2}{2j}\frac{\sin^{p-3}\psi\cos^{2j}\psi}{\cosh^{n-1+2j}\beta}(\mathbf{R}R)^{2j}$$

and since

$$\int_0^{\pi}\cos^{2j}\psi\sin^{p-3}\psi\,d\psi = B\{\tfrac{1}{2}(p-2),\tfrac{1}{2}(2j+1)\}$$

and

$$\int_{-\infty}^{\infty}\frac{d\beta}{\cosh^{n-1+2j}\beta} = B\{\tfrac{1}{2},\tfrac{1}{2}(n+2j-1)\},$$

the integral in (28.79) becomes

$$\sum_{j=0}^{\infty}\binom{n+2j-2}{2j}B\{\tfrac{1}{2}(p-2),\tfrac{1}{2}(2j+1)\}B\{\tfrac{1}{2},\tfrac{1}{2}(n+2j-1)\}(\mathbf{R}R)^{2j};$$

writing this out in terms of gamma functions and simplifying, it becomes

$$= \frac{\pi\Gamma\{\frac{1}{2}(p-2)\}\Gamma\{\frac{1}{2}(n-1)\}}{\Gamma\{\frac{1}{2}n\}\Gamma\{\frac{1}{2}(p-1)\}}F\{\tfrac{1}{2}(n-1),\tfrac{1}{2}(n-1),\tfrac{1}{2}(p-1),\mathbf{R}^2R^2\}. \qquad (28.80)$$

Substituting (28.80) for the integral in (28.79), we obtain

$$dF = \frac{(R^2)^{(p-3)/2}(1-R^2)^{(n-p-2)/2}d(R^2)}{B\{\frac{1}{2}(p-1),\frac{1}{2}(n-p)\}}\cdot(1-\mathbf{R}^2)^{(n-1)/2}$$
$$\times F\{\tfrac{1}{2}(n-1),\tfrac{1}{2}(n-1),\tfrac{1}{2}(p-1),\mathbf{R}^2R^2\}. \qquad (28.81)$$

This unconditional distribution should be compared with the conditional distribution of R^2 given in Exercise 28.13. Exercise 28.14 shows that as $n\to\infty$, both yield the same non-central χ^2 distribution for nR^2. An alternate form of (28.81) for $n-p$ even is given in Exercise 28.16.

The first factor on the right of (28.81) is the distribution (28.75) when $\mathbf{R}=0$, the second factor then being unity. Expression (28.81) generally converges slowly, for the first two arguments in the hypergeometric function are $\frac{1}{2}(n-1)$. Lee (1971) gives recurrence relations for the p.d.f. and d.f. of R^2 (cf. Exercise 16.14 for r) which make the computation of its distribution straightforward when p is even.

Ding and Bargmann (1991a; 1991b) provide computer algorithms for the evaluation of the p.d.f. and of its quantities. Exercises 28.19–21 derive results for the distribution of $R^2/(1-R^2)$ in the multinormal case from the conditional distribution in **28.28**.

The moments and limiting distributions of R^2

28.32 From (28.81), it may be shown (cf. Wishart, 1931) that the mean value of R^2 in the multinormal case is

$$E(R^2) = 1 - \frac{n-p}{n-1}(1-\mathbf{R}^2)F\{1, 1, \tfrac{1}{2}(n+1), \mathbf{R}^2\}, \tag{28.82}$$

$$= \mathbf{R}^2 + \frac{p-1}{n-1}(1-\mathbf{R}^2) - \frac{2(n-p)}{n^2-1}\mathbf{R}^2(1-\mathbf{R}^2) + O(n^{-2}). \tag{28.83}$$

In particular, when $\mathbf{R}^2 = 0$, (28.83) reduces to

$$E(R^2|\mathbf{R}^2 = 0) = \frac{p-1}{n-1}, \tag{28.84}$$

an exact result also obtainable directly from (28.75).

Similarly, the variance may be shown to be

$$\text{var}(R^2) = \frac{(n-p)(n-p+2)}{n^2-1}(1-\mathbf{R}^2)^2 F(2, 2, \tfrac{1}{2}(n+3), \mathbf{R}^2) - \{E(R^2)-1\}^2 \tag{28.85}$$

$$= \frac{n-p}{(n^2-1)(n-1)}(1-\mathbf{R}^2)^2 \times\left[2(p-1) + \frac{4\mathbf{R}^2\{(n-p)(n-1)+4(p-1)\}}{n+3} + O(\mathbf{R}^4/n^2)\right], \tag{28.86}$$

which may be written

$$\text{var}(R^2) = \frac{4\mathbf{R}^2(1-\mathbf{R}^2)^2(n-p)^2}{(n^2-1)(n+3)} + O(n^{-2}), \tag{28.87}$$

so that if $\mathbf{R}^2 \neq 0$

$$\text{var}(R^2) \doteq 4\mathbf{R}^2(1-\mathbf{R}^2)^2/n. \tag{28.88}$$

But if $\mathbf{R}^2 = 0$, (28.87) is of no use, and we return to (28.86), finding

$$\text{var}(R^2) = \frac{2(n-p)(p-1)}{(n^2-1)(n-1)} \doteq 2(p-1)/n^2, \tag{28.89}$$

the exact result in (28.89) being obtainable from (28.75). The reader is asked to derive these results in Exercise 28.11.

28.33 The different orders of magnitude of the asymptotic variances (28.88) and (28.89) when $\mathbf{R} \neq 0$ and $\mathbf{R} = 0$ reflect the fundamentally different behaviour of the distribution of R^2 in the two circumstances. Although (28.84) shows that R^2 is a biased estimator of $\mathbf{R}^2$, it is clearly consistent; for large n, $E(R^2) \to \mathbf{R}^2$ and var $(R^2) \to 0$. When $\mathbf{R} \neq 0$, the distribution of R^2 is asymptotically normal with mean $\mathbf{R}^2$ and variance given by (28.88) (cf. Exercise 28.15). When $\mathbf{R} = 0$, however, R, which is confined to the interval [0, 1], is converging to the value 0 at the lower extreme of its range, and this alone is enough to show that its distribution is not normal in this case (cf. Exercises 28.14–15). It is no surprise in these circumstances that its variance is of order n^{-2}: the situation is analogous to the estimation of a terminal of a distribution, where we saw in Exercise 17.36 that variances of order n^{-2} occur.

The distribution of R behaves similarly in respect of its limiting normality to that of R^2, though we shall see that its variance is always of order $1/n$.

One direct consequence of the singularity in the distribution of R at $\mathbf{R}^2 = 0$ should be mentioned. It follows from (28.88) that

$$\text{var}\, R \sim (1 - \mathbf{R}^2)^2/n, \tag{28.90}$$

which is the same as the asymptotic expression for the variance of the product moment correlation coefficient (cf. (27.22))

$$\text{var}\, r \sim (1 - \rho^2)^2/n.$$

It is natural to apply the variance-stabilising z-transformation of **16.33** (cf. also Exercise 16.18) to R also, obtaining a transformed variable $z = \text{arctanh}\, R$ with variance close to $1/n$, independent of the value of $\mathbf{R}$. But this will not do near $\mathbf{R} = 0$, as Hotelling (1953) pointed out, since (28.90) breaks down there; its asymptotic variance then will be given by (28.84) as

$$\text{var}\, R = E(R^2) - \{E(R)\}^2 \sim (p - 1)/n, \tag{28.91}$$

agreeing with the value $1/n$ obtained from (28.90) only for $p = 2$, when $R = |r|$. Lee (1971) investigates the approximation numerically, finds it inadequate, and proposes better ones.

Unbiased estimation of $\mathbf{R}^2$ in the multinormal case

28.34 Since, by (28.83), R^2 is a biased estimator of $\mathbf{R}^2$, we may wish to adjust it for the bias. Olkin and Pratt (1958) show that an unbiased estimator of $\mathbf{R}^2_{1(2\cdots p)}$, is

$$t = 1 - \frac{n-3}{n-p}(1 - R^2_{1(2\cdots p)})F(1, 1, \tfrac{1}{2}(n - p + 2), 1 - R^2_{1(2\cdots p)}), \tag{28.92}$$

where $n > p \geq 3$. The reader is asked to verify that $E(t) = \mathbf{R}^2$ in Exercise 28.12. t is the unique unbiased function of R^2 since it is a function of the complete sufficient statistics. Expression (28.92) may be expanded as the series

$$t = R^2 - \frac{p-3}{n-p}(1 - R^2) - \left\{\frac{2(n-3)}{(n-p)(n-p+2)}(1 - R^2)^2 + O(n^{-2})\right\}, \tag{28.93}$$

whence it follows that $t \leq R^2$. If $R^2 = 1, t = 1$ also. When R^2 is zero or small, on the other hand, t is negative, as we might expect. We cannot (cf. **17.9**) find an unbiased estimator of $\mathbf{R}^2$ (i.e. an estimator whose expectation is $\mathbf{R}^2$ whatever the true value of $\mathbf{R}^2$) which takes only non-negative values, even though we know that $\mathbf{R}^2$ is non-negative. We may remove the absurdity of negative estimates by using as our estimator

$$t' = \max\{t, 0\}, \tag{28.94}$$

but (28.94) is no longer unbiased.

28.35 Lehmann (1986) shows that for testing $\mathbf{R}^2 = 0$ in the multinormal case, tests rejecting large values of R^2 are UMP among test statistics which are invariant under location and scale changes.

Ezekiel and Fox (1959) and Kramer (1963) give charts and tables for constructing confidence intervals for $\mathbf{R}^2$ from the value of R^2. Lee (1972) gives 4 d.p. tables of the upper 5 per cent and 1 per cent points of the d.f. of R for $\mathbf{R} = 0\ (0.1)\ 0.9$; $p - 1 = 1\ (1)\ 10, 12, 15, 20, 24, 30, 40$; and $(n-p)^{1/2} = 60/\nu$ where $\nu = 1\ (1)\ 6\ (2)\ 20$. The *Biometrika Tables*, Vol. II, Table 52, give a 3 d.p. table by Kramer of the lower and upper 5 and 1 per cent points, for the same values of $\mathbf{R}$, and $p - 1 = 2\ (2)\ 12\ (4)\ 24, 30, 34, 40$; $n - p = 10\ (10)\ 50$.

Estimation of $\mathbf{R}^2/(1 - \mathbf{R}^2)$

28.36 The relative complexity of the estimator (28.92) does not persist if instead we seek to estimate $\theta = \mathbf{R}^2/(1 - \mathbf{R}^2)$. From **28.28**, the conditional distribution of (28.73) given $\mathbf{x} = (x_2, \ldots, x_p)$ is a non-central $F'(p-1, n-p, \lambda)$, and thus by Exercise 29.17 its conditional expectation is

$$E(F'|\mathbf{x}) = \frac{(p-1+\lambda)(n-p)}{(p-1)(n-p-2)}, \qquad n - p \geq 3;$$

hence that of $u = R^2/(1 - R^2)$ is

$$E(u|\mathbf{x}) = \left(\frac{p-1}{n-p}\right) E(F'|\mathbf{x}) = \frac{p-1+\lambda}{n-p-2}.$$

The unconditional expectation of u is therefore

$$E(u) = (n-p-2)^{-1}\{p - 1 + E(\lambda)\}. \tag{28.95}$$

To find $E(\lambda)$, we see from **28.28** that if $\mathbf{x}$ is allowed to vary multinormally, λ/θ is the sum of squares of n independent normal variables subject to a single constraint, and is thus distributed as χ^2 with $n - 1$ d.fr. Hence, $E(\lambda/\theta) = n - 1$ and (28.95) becomes

$$E(u) = (n-p-2)^{-1}\{p - 1 + (n-1)\theta\} \tag{28.96}$$

so that an unbiased estimator for θ is given by

$$V = (n-1)^{-1}\{(n-p-2)u - (p-1)\}, \tag{28.97}$$

a linear function of the corresponding statistic u. Although it is the unique MV unbiased estimator of θ, it may be negative and is subject to the remarks concerning t in **28.34**. Muirhead (1985) shows that its m.s.e. may be substantially reduced by 'shrinking' it as in Example 17.14 to aV, where

$$a = \frac{(n-1)(n-p-4)}{(n+1)(n-p-2)}, \qquad n - p \geq 5 \tag{28.98}$$

and truncating it at zero as at (28.94). For the bivariate case $p = 2$, (28.97) gives

$$E[(n-1)^{-1}\{(n-4)r^2/(1-r^2) - 1\}] = \rho^2/(1-\rho^2), \tag{28.99}$$

which was otherwise obtained in Exercise 16.15.

EXERCISES

28.1 Show that

$$\beta_{12\cdot34\cdots(p-1)} = \frac{\beta_{12\cdot34\cdots p} + \beta_{1p\cdot23\cdots(p-1)}\beta_{p2\cdot13\cdots(p-1)}}{1 - \beta_{1p\cdot23\cdots(p-1)}\beta_{p1\cdot23\cdots(p-1)}},$$

and that

$$\rho_{12\cdot34\cdots(p-1)} = \frac{\rho_{12\cdot34\cdots p} + \rho_{1p\cdot23\cdots(p-1)}\rho_{2p\cdot13\cdots(p-1)}}{\{(1-\rho^2_{1p\cdot23\cdots(p-1)})(1-\rho^2_{2p\cdot13\cdots(p-1)})\}^{1/2}}.$$

(Yule, 1907)

28.2 Show that for p variates there are $\binom{p}{2}$ correlation coefficients of order zero and $\binom{p-2}{s}\binom{p}{2}$ of order s. Show further that there are $\binom{p}{2}2^{p-2}$ correlation coefficients altogether and $\binom{p}{2}2^{p-1}$ regression coefficients.

28.3 If the correlations of zero order among a set of variables are all equal to ρ, show that every partial correlation of sth order is equal to $\rho/(1+s\rho)$.

28.4 From (28.30), define the error term

$$x_{1\cdot q(1)} = x_1 - \boldsymbol{\beta}^T\mathbf{x},$$

where $\boldsymbol{\beta}$ is given by (28.32) and $q(1)$ denotes all subscripts other than 1. Let $\mathbf{C}$ be the correlation matrix and denote the elements of $\mathbf{C}^{-1}$ by $\{c^{ij}\}$ Show that

$$g_{ij} = \text{cov}(x_{i\cdot q(i)}, x_{j\cdot q(j)}) = \sigma_i\sigma_j/c^{ij}$$

and hence that the coefficient of $x_i x_j$ in the exponent of the multinormal distribution of $(x_1, x_2, \ldots, x_p)$ is $1/g_{ij}$.

28.5 Show from (28.39) that in summing the product of two residuals, any or all of the secondary subscripts may be omitted from a residual *all* of whose secondary subscripts are included among those of the other residual, i.e. that

$$\sum x_{1\cdot stu}x_{2\cdot st} = \sum x_{1\cdot stu}x_{2\cdot s} = \sum x_{1\cdot stu}x_2,$$

but that

$$\sum x_{1\cdot stu}x_{2\cdot st} \neq \sum x_{1\cdot su}x_{2\cdot st},$$

where s, t, u are sets of subscripts. (The same result holds for products of errors.)

(Chandler, 1950)

28.6 Using the notation in Exercise 28.5, apply the transformation

$$y_1 = x_1,$$
$$y_2 = x_{2\cdot1},$$
$$y_3 = x_{3\cdot21},$$

etc., to show that the multinormal distribution has density function

$$\frac{1}{(2\pi)^{p/2}\sigma_1\sigma_{2\cdot1}\sigma_{3\cdot12}\cdots}\exp\left\{-\frac{1}{2}\left(\frac{x_1^2}{\sigma_1^2}+\frac{x_2^2}{\sigma_{2\cdot1}^2}+\frac{x_{3\cdot12}^2}{\sigma_{3\cdot12}^2}+\cdots\right)\right\}$$

so that the residuals $x_1, x_{2\cdot1}, \ldots$ are independent of each other. Hence show that any two residuals $x_{j\cdot r}$, and $x_{k\cdot r}$ (where r is a set of common subscripts) are distributed binormally with correlation $\rho_{jk\cdot r}$.

28.7 Show that if an orthogonal transformation is applied to a set of n independent observations on p multinormal variates, the transformed set of n observations will also be independent.

28.8 Show that in a sample $(x_1, \ldots, x_n)$ of one observation from an n-variate multinormal population with all means μ, all variances σ^2 and all correlations equal to ρ, the statistic

$$t^2 = \frac{(\bar{x} - \mu)^2}{\sum_{i=1}^{n}(x_i - \bar{x})^2/\{n(n-1)\}} \left\{ \frac{1-\rho}{1+(n-1)\rho} \right\}$$

has Student's t^2 distribution with $n-1$ degrees of freedom. When $\rho = 0$, this reduces to the ordinary test of a mean of n independent normal variates.

(Walsh, 1947)

28.9 If $x_0, x_1, \ldots, x_n$ are normal variates with common variance α^2, $x_1, \ldots, x_n$ being independent of each other and x_0 having zero mean and correlation λ with each of the others, show that the n variates

$$y_i = x_i - ax_0, \qquad i = 1, 2, \ldots, n,$$

are multinormally distributed with all correlations equal to

$$\rho = (a^2 - 2a\lambda)/(1 + a^2 - 2a\lambda)$$

and all variances equal to

$$\sigma^2 = \alpha^2/(1-\rho).$$

(Stuart, 1958)

28.10 Use the result of Exercise 28.9 to establish that of Exercise 28.8.

(Stuart, 1958)

28.11 Establish (28.83) and (28.85), the expressions for $E(R^2)$ and $\text{var}(R^2)$.

(Wishart, 1931)

28.12 Verify that (28.92) is an unbiased estimator of $\mathbf{R}^2$.

28.13 Show from the non-central F distribution of F in (28.73) when $\mathbf{R}^2 \neq 0$, that the distribution of R^2 in this case, when $x_2, \ldots, x_p$ are fixed, is

$$dF = \frac{1}{B\{\frac{1}{2}(p-1), \frac{1}{2}(n-p)\}}(R^2)^{(p-3)/2}(1-R^2)^{(n-p-2)/2}\, dR^2 \cdot \exp\{-\tfrac{1}{2}(n-p)\mathbf{R}^2\}$$
$$\times \sum_{j=0}^{\infty} \frac{\Gamma\{\frac{1}{2}(n-1+2j)\}\Gamma\{\frac{1}{2}(p-1)\}}{\Gamma\{\frac{1}{2}(n-1)\}\Gamma\{\frac{1}{2}(p-1+2j)\}} \frac{\{\frac{1}{2}(n-p)\mathbf{R}^2 R^2\}^j}{j!}.$$

(Fisher, 1928a)

28.14 Show from (28.71) that for $n \to \infty$, p fixed, the distribution of $nR^2 = B^2$ is

$$dF = \frac{(B^2)^{(p-3)/2}}{2^{(p-1)/2}\Gamma\{\frac{1}{2}(p-1)\}} \exp(-\tfrac{1}{2}\beta^2 - \tfrac{1}{2}B^2)$$
$$\times \left\{1 + \frac{\beta^2 B^2}{(p-1)\cdot 2} + \frac{(\beta^2 B^2)^2}{(p-1)(p+1)\cdot 2 \cdot 4} + \cdots \right\} d(B^2),$$

where $\beta^2 = n\mathbf{R}^2$, and hence that nR^2 is a non-central χ^2 variate of form (22.18) with $\nu = p-1$, $\lambda = n\mathbf{R}^2$. Show that the same result holds for the conditional distribution of nR^2, from Exercise 28.13.

(Fisher, 1928a)

28.15 In Exercise 28.14, use the c.f. of a non-central χ^2 variate given in Exercise 22.1 to show that as $n \to \infty$ for fixed p, R^2 is asymptotically normally distributed when $\mathbf{R} \neq 0$, but not when $\mathbf{R} = 0$. Extend the result to R.

28.16 Show that the distribution function of R^2 in multinormal samples may be written, if $n-p$ is even, in the form

$$(1-\mathbf{R}^2)^{(n-1)/2} R^{p-1} \sum_{j=0}^{(n-p-2)/2} \frac{\Gamma\{\frac{1}{2}(p-1+2j)\}}{\Gamma\{\frac{1}{2}(p-1)\}} \frac{(1-R^2)^j}{(1-\mathbf{R}^2R^2)^{(n-1+2j)/2}}$$
$$\times F\{-j, -\tfrac{1}{2}(n-p), \tfrac{1}{2}(p-1), \mathbf{R}^2R^2\}.$$

(Fisher, 1928a)

28.17 Show that if each pair from $x_2, \ldots, x_p$ is uncorrelated,

$$\mathbf{R}^2_{1(2\cdots p)} = \sum_{s=2}^{p} \rho^2_{1s} = \sum_{s=2}^{p} (\beta_{1s\cdot q}\sigma_s)^2/\sigma_1^2.$$

28.18 Consider variables x_1, x_2, x_3 for which

$$\rho_{13} = 0, \quad \rho_{12} = \cos\theta, \quad \rho_{23} = \sin\theta, \quad 0 < \theta < \pi/2.$$

Show that $\rho_{12\cdot 3} = 1$ and hence that $R^2_{1(23)} = 1$. By letting $\theta \to \pi/2$, show that $R^2_{1(23)} = 1$ is consistent with $\rho_{13} = 0$, $\rho_{12} = \varepsilon$, for any $\varepsilon > 0$. Interpret the result geometrically.

28.19 In **28.28**, show that when $x_2, \ldots, x_p$ vary multinormally, $\lambda(1-\mathbf{R}^2)/\mathbf{R}^2$ is distributed as χ^2 with $n-1$ d.fr., hence, using the c.f. of χ^2 in Exercise 22.1, that the distribution of $R^2/(1-R^2)$ in the multinormal case is that of the ratio of independent variables y, z, with c.f.s

$$\phi_y(t) = (1-2it)^{(n-p)/2}\{1-2it(1+\theta)\}^{-(n-1)/2}$$
$$\phi_z(t) = (1-2it)^{-(n-p)/2}$$

where $\theta = \mathbf{R}^2/(1-\mathbf{R}^2)$, and hence that y may be represented as

$$y = \chi^2_{p-2} + (\chi_1 + \theta^{1/2}\chi_{n-1})^2,$$

all the variables being independent. (Cf. Exercise 16.6 when $p = 2$.)

(Gurland, 1968; Lee, 1971)

28.20 In Exercise 28.19, show that if $n - p = 2k$ is even,

$$\phi_y(t) = \{1-2(1+\theta)it\}^{-(p-1)/2}(1+\theta)^{-k} \sum_{j=0}^{k} \binom{k}{j} \left(\frac{\theta}{1-2(1+\theta)it}\right)^j,$$

and hence that $R^2/(1-R^2)$ has the d.f.

$$G(x) = \sum_{j=0}^{k} \binom{k}{j} \left(\frac{\theta}{1+\theta}\right)^j (1+\theta)^{j-k} H_{p-1+2j,2k}\left(\frac{x}{1+\theta}\right),$$

where $H_{a,b}(x)$ is the d.f. of a ratio of independent χ^2 variates with a and b degrees of freedom.

(Gurland (1968), who also gives two infinite series for $G(x)$, one for odd $(n-p)$, $\mathbf{R}^2 < \frac{1}{2}$, and one for all n, p and $\mathbf{R}$. A general class of series is given by Gurland and Milton (1970).)

28.21 In Exercise 28.19, approximate the distribution of $\mathbf{R}^2/(1-\mathbf{R}^2)$ by assuming that it is a multiple g of the ratio of independent central χ^2 variates with ν and $n-p$ degrees of freedom, and equate the first two moments with the exact ones, as in **19.34**. Show that $g = B/A$ and $\nu = A^2/B$, where

$$A = (n-1)\theta + p - 1 \text{ and } B = (n-1)\theta(\theta+2) + p - 1.$$

(Gurland (1968); the result is exact when $R^2 = \theta = 0$, and the approximation seems generally good. See also Gurland and Milton (1970).)

28.22 Show that, without loss of generality, we can take $\mathbf{L}$ to be orthogonal in (28.61) so that $\mathbf{A}$ becomes diagonal. Thus, maximizing the GVR is equivalent to minimizing $\mathbf{N}^T\mathbf{C}\mathbf{N}$ where $\mathbf{N}^T = (\mathbf{L}^T, \mathbf{M}^T)$, subject to (28.63).
Hence show that the maximum GVR is given by the solution (28.67)–(28.69).

CHAPTER 29

THE GENERAL LINEAR MODEL

29.1 Our discussion in Chapters 27 and 28 assumed that the joint distribution of all the random variables was known. While such an assumption leads to interesting theoretical conclusions, it may well be an unreasonable requirement in practice. Furthermore, in experimental statistics, it is often the case that the regressor variables are set to pre-assigned levels, so that even the idea of random variation among such variables is unreasonable. In such circumstances, it is preferable to carry out the analysis *conditionally*, given the values of the regressor variables. That is, we consider the dependent variable, y, to be a function of p input or regressor variables, $\mathbf{x} = (x_1, \ldots, x_p)^T$ and a random error, ϵ; that is,

$$y = f(\mathbf{x}, \boldsymbol{\beta}, \epsilon) \tag{29.1}$$

where $\boldsymbol{\beta}$ denotes a vector of unknown parameters. The analysis will proceed *conditionally* upon the $\mathbf{x}$-values.

We shall assume throughout most of this chapter that expression (29.1) denotes a *linear* model with an additive error term so that, if we have n observations, the expression for the ith observation is

$$y_i = \beta_1 x_{1i} + \beta_2 x_{2i} + \cdots + \beta_p x_{pi} + \epsilon_i, \qquad i = 1, \ldots, n, \tag{29.2}$$

where $x_{1i} \equiv 1$ in most cases, to allow for a constant (or intercept) term.

It should be noted that the adjective *linear* will always be understood to refer to the parameter structure and not to the regressor variables: thus,

$$y = \beta_1 + \beta_2 x + \beta_3 x^2 + \epsilon \tag{29.3}$$

is a linear model, whereas

$$y = \beta_1 + \beta_1^2 x + \epsilon \tag{29.4}$$

is not.

29.2 If we write $\mathbf{y} = (y_1, \ldots, y_n)^T$, the set of equations in (29.2) may be rewritten in matrix form as

$$\mathbf{y} = \mathbf{X}\boldsymbol{\beta} + \boldsymbol{\epsilon}, \tag{29.5}$$

where $\boldsymbol{\beta}$ is a $p \times 1$ vector of parameters, or regression parameters, $\mathbf{X}$ is an $n \times p$ matrix of known values, and $\boldsymbol{\epsilon}$ is an $n \times 1$ vector of random errors. Usually all the elements in the first column of $\mathbf{X}$ are set equal to one, as noted above.

We analyse (29.5) using the principle of least squares (LS) introduced in **18.52**. That is, we seek to estimate $\boldsymbol{\beta}$ by minimizing

$$S = \boldsymbol{\epsilon}^T \boldsymbol{\epsilon} = \sum_{i=1}^{n} \epsilon_i^2. \tag{29.6}$$

The basic assumptions underlying our analysis are as follows:

(i) the model is linear (in the parameters);

(ii) the error structure is additive;

(iii) the random errors have zero means, equal variances, and are mutually uncorrelated, written as

$$E(\boldsymbol{\epsilon}) = \mathbf{0}, \quad \mathbf{V}(\boldsymbol{\epsilon}) = \sigma^2 \mathbf{I}, \tag{29.7}$$

where $\mathbf{I} \equiv \mathbf{I}_n$ denotes the $n \times n$ identity matrix;

(iv) the matrix $\mathbf{X}$ is of full rank p, so that $\mathbf{X}^T\mathbf{X}$ is positive definite.

Conditions (i) and (ii) are implied by (29.5), whereas the assumptions in (29.7) are necessary if the estimators are to have certain optimal properties, as we shall see in **29.5**. Condition (iv) is needed to ensure that the LS solution is unique; it also implies that $n \geq p$, which we shall assume henceforth. Violations of the assumptions and the subsequent effects upon the analysis are discussed in **32.46–64**.

29.3 We are now in a position to derive the LS estimators for $\boldsymbol{\beta}$ and to demonstrate some of the basic properties.

Theorem 29.1.

Given model (29.5) and assumptions (i)–(iv):

(i) The LS estimators for $\boldsymbol{\beta}$ *are*

$$\hat{\boldsymbol{\beta}} = (\mathbf{X}^T\mathbf{X})^{-1}\mathbf{X}^T\mathbf{y}. \tag{29.8}$$

Further,

(ii)

$$E(\hat{\boldsymbol{\beta}}) = \boldsymbol{\beta},$$

so the estimators are unbiased;

(iii)

$$\mathbf{V}(\hat{\boldsymbol{\beta}}) = \sigma^2(\mathbf{X}^T\mathbf{X})^{-1}.$$

Proof. (i) Criterion (29.6) becomes

$$S = \boldsymbol{\epsilon}^T \boldsymbol{\epsilon} = (\mathbf{y} - \mathbf{X}\boldsymbol{\beta})^T(\mathbf{y} - \mathbf{X}\boldsymbol{\beta}) \tag{29.9}$$

so that

$$\frac{\partial S}{\partial \boldsymbol{\beta}} = 2(\mathbf{X}^T\mathbf{X}\boldsymbol{\beta} - \mathbf{X}^T\mathbf{y}) \tag{29.10}$$

and

$$\frac{\partial^2 S}{\partial \boldsymbol{\beta}\, \partial \boldsymbol{\beta}^T} = 2\mathbf{X}^T\mathbf{X}. \tag{29.11}$$

Using assumption (iv), setting $\partial S/\partial \boldsymbol{\beta} = \mathbf{0}$ yields

$$\hat{\boldsymbol{\beta}} = (\mathbf{X}^T\mathbf{X})^{-1}\mathbf{X}^T\mathbf{y} \tag{29.12}$$

as required. Since $\mathbf{X}^T\mathbf{X}$ is positive definite, it follows from (29.11) that solution (29.12) is a minimum and clearly it is unique.

(ii) From (29.5) and (29.12) we have that

$$\begin{aligned} E(\hat{\boldsymbol{\beta}}) &= E[(\mathbf{X}^T\mathbf{X})^{-1}\mathbf{X}^T\mathbf{y}] \\ &= E[(\mathbf{X}^T\mathbf{X})^{-1}\mathbf{X}^T(\mathbf{X}\boldsymbol{\beta} + \boldsymbol{\epsilon})] \\ &= \boldsymbol{\beta} + (\mathbf{X}^T\mathbf{X})^{-1}\mathbf{X}^T E(\boldsymbol{\epsilon}) \\ &= \boldsymbol{\beta} \end{aligned} \tag{29.13}$$

since $\mathbf{X}$ is fixed, $\boldsymbol{\beta}$ is a vector of fixed parameters and $E(\boldsymbol{\epsilon}) = \mathbf{0}$ by assumption (iii).

(iii) From (29.5), (29.12), and (29.13)

$$\begin{aligned} \hat{\boldsymbol{\beta}} - \boldsymbol{\beta} &= (\mathbf{X}^T\mathbf{X})^{-1}\mathbf{X}^T\mathbf{y} - \boldsymbol{\beta} \\ &= (\mathbf{X}^T\mathbf{X})^{-1}\mathbf{X}^T\boldsymbol{\epsilon}, \end{aligned} \tag{29.14}$$

so that

$$\begin{aligned} \mathbf{V}(\hat{\boldsymbol{\beta}}) &= E[(\hat{\boldsymbol{\beta}} - \boldsymbol{\beta})(\hat{\boldsymbol{\beta}} - \boldsymbol{\beta})^T] \\ &= E[(\mathbf{X}^T\mathbf{X})^{-1}\mathbf{X}^T\boldsymbol{\epsilon}\boldsymbol{\epsilon}^T\mathbf{X}(\mathbf{X}^T\mathbf{X})^{-1}] \\ &= (\mathbf{X}^T\mathbf{X})^{-1}\mathbf{X}^T E(\boldsymbol{\epsilon}\boldsymbol{\epsilon}^T)\mathbf{X}(\mathbf{X}^T\mathbf{X})^{-1} \\ &= (\mathbf{X}^T\mathbf{X})^{-1}\mathbf{X}^T(\sigma^2\mathbf{I})\mathbf{X}(\mathbf{X}^T\mathbf{X})^{-1} \\ &= \sigma^2(\mathbf{X}^T\mathbf{X})^{-1} \end{aligned} \tag{29.15}$$

using assumption (iii).

Example 29.1 (Form of least squares estimators)

Consider the model with $p = 2$:

$$\mathbf{y} = (\,\mathbf{1} \quad \mathbf{x}_2\,)\begin{pmatrix}\beta_1 \\ \beta_2\end{pmatrix} + \boldsymbol{\epsilon}, \tag{29.16}$$

where **1** is the $n \times 1$ vector of ones. From (29.12),

$$\hat{\boldsymbol{\beta}} = \begin{pmatrix} n & \sum x_2 \\ \sum x_2 & \sum x_2^2 \end{pmatrix}^{-1} \begin{pmatrix} \sum y \\ \sum x_2 y \end{pmatrix}$$

$$= \frac{1}{\{n \sum x_2^2 - (\sum x_2)^2\}} \cdot \begin{pmatrix} \sum x_2^2 \sum y - \sum x_2 \sum x_2 y \\ -\sum x_2 \sum y + n \sum x_2 y \end{pmatrix}, \qquad (29.17)$$

so that

$$\hat{\beta}_2 = \frac{n \sum x_2 y - \sum x_2 \sum y}{n \sum x_2^2 - (\sum x_2)^2} = \frac{\sum (x_2 - \bar{x}_2)(y - \bar{y})}{\sum (x_2 - \bar{x}_2)^2}, \qquad (29.18)$$

and

$$\hat{\beta}_1 = \frac{\sum x_2^2 \sum y - \sum x_2 \sum x_2 y}{n \sum x_2^2 - (\sum x_2)^2} = \bar{y} - \hat{\beta}_2 \bar{x}_2. \qquad (29.19)$$

Further, since

$$(\mathbf{X}^T\mathbf{X})^{-1} = \frac{1}{\sum (x_2 - \bar{x}_2)^2} \begin{pmatrix} (1/n) \sum x_2^2 & -\bar{x}_2 \\ -\bar{x}_2 & 1 \end{pmatrix},$$

we have from (29.15) that

$$\text{var}(\hat{\beta}_1) = \sigma^2 \sum x_2^2 / [n \sum (x_2 - \bar{x}_2)^2], \qquad \text{var}(\hat{\beta}_2) = \sigma^2 / \sum (x_2 - \bar{x}_2)^2,$$
$$\text{cov}(\hat{\beta}_1, \hat{\beta}_2) = -\sigma^2 \bar{x}_2 / \sum (x_2 - \bar{x}_2)^2.$$

$\hat{\beta}_1$ and $\hat{\beta}_2$ are uncorrelated if and only if $\bar{x}_2 = 0$. If x_2 is measured about its mean, not only are the two estimators uncorrelated, but also $\hat{\beta}_1 = \bar{y}$. More general results for the introduction of a constant term into the model are given in Exercises 29.1 and 29.2.

29.4 It is evident from Theorem 29.1 that the key computation in the analysis is the inversion of $\mathbf{X}^T\mathbf{X}$. When p is large, or when the matrix is *ill-conditioned*, rounding errors may cause sizeable computational errors. We do not consider numerical problems in detail; the interested reader should consult Goodall (1993). However, we do note one particular approach that has had widespread application. The *Householder decomposition* of $\mathbf{X}$ defines an orthogonal matrix $\mathbf{P}$ such that

$$\mathbf{P}^T\mathbf{X} = \begin{pmatrix} \mathbf{U} \\ \mathbf{0} \end{pmatrix},$$

where $\mathbf{U}$ is a non-singular $p \times p$ upper triangular matrix, so that (29.12) becomes

$$\hat{\boldsymbol{\beta}} = (\mathbf{U}^T\mathbf{U})^{-1} \begin{pmatrix} \mathbf{U} \\ \mathbf{0} \end{pmatrix}^T \mathbf{P}^T\mathbf{y} = \mathbf{U}^{-1} \begin{pmatrix} \mathbf{I} \\ \mathbf{0} \end{pmatrix} \mathbf{P}^T\mathbf{y}, \qquad (29.20)$$

and $\mathbf{U}$ may be inverted more accurately than $\mathbf{U}^T\mathbf{U} = \mathbf{X}^T\mathbf{X}$.

Generally, when one or more of the eigenvalues of $\mathbf{X}^T\mathbf{X}$ is close to zero, some elements on the main diagonal of its inverse will be large, indicating that the corresponding estimators are imprecise (have large variances). This situation, known as (near) *multicollearity*, is considered in detail in **32.46–50**.

Optimum properties of least squares

29.5 We now show that the minimum variance (MV) unbiased linear estimators of any set of linear functions of the parameters β_i are given by the LS method. Plackett (1949), in a discussion of the origins of LS theory, makes it clear that the fundamental results are due to Gauss, as does a more detailed historical review by Seal (1967) – see also Plackett (1972). The result is usually known as the *Gauss–Markov theorem.*

Theorem 29.2.

Let

$$\mathbf{t} = \mathbf{T}\mathbf{y} \tag{29.21}$$

be any linear function of the observations that provides an unbiased estimator for $\mathbf{C}\boldsymbol{\beta}$*; that is* $E(\mathbf{t}) = \mathbf{C}\boldsymbol{\beta}$*, where* $\mathbf{C}$ *is an arbitrary matrix of constants.*

The form of (29.21) with minimum variance is given by

$$\mathbf{T} = \mathbf{C}(\mathbf{X}^T\mathbf{X})^{-1}\mathbf{X}^T; \tag{29.22}$$

that is, the MV unbiased estimator is

$$\hat{\mathbf{t}} = \mathbf{C}\hat{\boldsymbol{\beta}}. \tag{29.23}$$

Proof. If $\mathbf{t}$ is to be unbiased for $\mathbf{C}\boldsymbol{\beta}$ where $\mathbf{C}$ is a known matrix of coefficients, we must have $E(\mathbf{t}) = E(\mathbf{T}\mathbf{y}) = \mathbf{C}\boldsymbol{\beta}$, which from (29.5) gives

$$E\{\mathbf{T}(\mathbf{X}\boldsymbol{\beta} + \boldsymbol{\epsilon})\} = \mathbf{C}\boldsymbol{\beta}. \tag{29.24}$$

Since (29.24) must hold identically in $\boldsymbol{\beta}$ and $E(\boldsymbol{\epsilon}) = \mathbf{0}$, we have that

$$\mathbf{T}\mathbf{X} = \mathbf{C}. \tag{29.25}$$

The identity (29.25) is a necessary and sufficient condition for $\mathbf{C}\boldsymbol{\beta}$ to be unbiasedly estimable by $\mathbf{T}\mathbf{y}$.

The covariance matrix of $\mathbf{t}$ is

$$\mathbf{V}(\mathbf{t}) = E\{(\mathbf{t} - \mathbf{C}\boldsymbol{\beta})(\mathbf{t} - \mathbf{C}\boldsymbol{\beta})^T\} \tag{29.26}$$

and since, from (29.14), (29.5) and (29.25), $\mathbf{t} - \mathbf{C}\boldsymbol{\beta} = \mathbf{T}\boldsymbol{\epsilon}$, (29.26) becomes

$$\mathbf{V}(\mathbf{t}) = E(\mathbf{T}\boldsymbol{\epsilon}\boldsymbol{\epsilon}^T\mathbf{T}^T) = \sigma^2\mathbf{T}\mathbf{T}^T. \tag{29.27}$$

We wish to minimize the diagonal elements of (29.27), which are the variances of our set of estimators.

We now write $\mathbf{t} = (\mathbf{t} - \hat{\mathbf{t}}) + \hat{\mathbf{t}}$, where

$$\hat{\mathbf{t}} = \mathbf{C}\hat{\boldsymbol{\beta}} = \mathbf{C}(\mathbf{X}^T\mathbf{X})^{-1}\mathbf{X}^T\mathbf{y}. \tag{29.28}$$

From (29.28), we see that $\hat{\mathbf{t}}$ has $\mathbf{T} = \mathbf{C}(\mathbf{X}^T\mathbf{X})^{-1}\mathbf{X}^T$, so that (29.25) is always satisfied and *any* linear function $\mathbf{C}\hat{\boldsymbol{\beta}}$ is unbiasedly estimable by $\hat{\mathbf{t}}$. We have

$$\mathbf{V}(\mathbf{t}) = \mathbf{V}(\mathbf{t} - \hat{\mathbf{t}}) + \mathbf{V}(\hat{\mathbf{t}}) + 2\mathbf{D}(\mathbf{t} - \hat{\mathbf{t}}, \hat{\mathbf{t}}), \tag{29.29}$$

where $\mathbf{D}(\mathbf{u}, \mathbf{w})$ is the matrix of covariances of the elements of $\mathbf{u}$ and the elements of $\mathbf{w}$. Thus

$$\begin{aligned}\mathbf{D}(\mathbf{t}-\hat{\mathbf{t}}, \hat{\mathbf{t}}) &= E\{(\mathbf{t}-\hat{\mathbf{t}})(\hat{\mathbf{t}}-\mathbf{C}\boldsymbol{\beta})^T\} \\ &= E\{[\mathbf{T}-\mathbf{C}(\mathbf{X}^T\mathbf{X})^{-1}\mathbf{X}^T]\boldsymbol{\epsilon}\boldsymbol{\epsilon}^T[\mathbf{C}(\mathbf{X}^T\mathbf{X})^{-1}\mathbf{X}^T]^T\} \\ &= \sigma^2[\mathbf{T}-\mathbf{C}(\mathbf{X}^T\mathbf{X})^{-1}\mathbf{X}^T]\mathbf{X}(\mathbf{X}^T\mathbf{X})^{-1}\mathbf{C}^T = \mathbf{0},\end{aligned} \tag{29.30}$$

using (29.25). From (29.29) and (29.30),

$$\mathbf{V}(\mathbf{t}) = \mathbf{V}(\mathbf{t}-\hat{\mathbf{t}}) + \mathbf{V}(\hat{\mathbf{t}}), \tag{29.31}$$

so that each diagonal element of $\mathbf{V}(\mathbf{t})$ cannot be less than the corresponding element of $\mathbf{V}(\hat{\mathbf{t}})$. Thus (29.23) has MV among unbiased linear estimators of $\mathbf{C}\boldsymbol{\beta}$. Its covariance matrix is, from (29.27),

$$\mathbf{V}(\hat{\mathbf{t}}) = \sigma^2\mathbf{C}(\mathbf{X}^T\mathbf{X})^{-1}\mathbf{C}^T. \tag{29.32}$$

The reader should observe the resemblance of (29.30)–(29.31) to the result at the end of **17.27** and in Exercise 17.11, which is not confined to linear estimators.

Although the LS estimators are MV unbiased in the linear model among linear functions of $\mathbf{y}$, they are not so in general if non-linear functions of $\mathbf{y}$ are admitted as estimators – this depends on the distribution of the ϵ_i. If the latter are normal (and hence independent), the stronger property follows from the fact that the LS estimators (which are then ML, by **18.52**) are functions of the $p+1$ minimal sufficient statistics $(\hat{\boldsymbol{\beta}}, s^2)$ for the parameters $(\boldsymbol{\beta}, \sigma^2)$ – see **17.39**. Anderson (1962b) shows that if all possible distributions of the ϵ_i are considered, LS estimators are very rarely MV unbiased among all estimators.

Cox and Hinkley (1968) give the efficiency of the LS estimators for some special cases.

If biased estimators are permitted, it does not follow (cf. **17.30**) that LS estimators will have minimum m.s.e. properties – cf. **29.9** and Exercise 29.21.

29.6 A direct consequence of the result in **29.5** is that the LS estimator $\hat{\boldsymbol{\beta}}$ minimizes the value of the generalized variance for linear estimators of $\hat{\boldsymbol{\beta}}$. This result, which is due to Aitken (1948), is exact, unlike the equivalent asymptotic result proved for ML estimators in **18.28**.

The result of Theorem 29.2, specialized to the estimation of a single linear function $\mathbf{c}^T\hat{\boldsymbol{\beta}}$, where $\mathbf{c}^T$ is a $1 \times p$ vector, is that

$$\text{var}(\mathbf{c}^T\hat{\boldsymbol{\beta}}) \leq \text{var}(\mathbf{c}^T\mathbf{u}), \tag{29.33}$$

where $\hat{\boldsymbol{\beta}}$ is the LS estimator, and $\mathbf{u}$ any other linear estimator, of $\boldsymbol{\beta}$. Thus by the argument leading from (18.66) to (18.67), we have

$$|\mathbf{V}| \leq |\mathbf{D}|, \tag{29.34}$$

the required result.

A geometrical interpretation

29.7 It is instructive to display the MV property geometrically, following Durbin and Kendall

(1951). We shall here discuss only their simplest case, where we are estimating a single parameter β from a sample of n observations, all with mean β and variance σ^2. Thus $y_j = \beta + \epsilon_j$, $j = 1, 2, \ldots, n$, which is (29.5) with $p = 1$ and $\mathbf{X}$ an $n \times 1$ vector of ones. We consider linear estimators

$$t = \sum_j c_j y_j, \tag{29.35}$$

the simplest case of (29.21). The unbiasedness condition (29.25) becomes

$$\sum_j c_j = 1. \tag{29.36}$$

Consider an n-dimensional Euclidean space with one coordinate for each c_j. We call this the estimator space. Expression (29.36) is a hyperplane in this space, and any point P in the hyperplane corresponds uniquely to one unbiased estimator. Now since the y_j are uncorrelated, we have, from (29.35),

$$\operatorname{var} t = \sigma^2 \sum_j c_j^2 \tag{29.37}$$

so that the variance of t is $\sigma^2 OP^2$, where O is the origin in the estimator space. It follows at once that t has MV when P is the foot of the perpendicular from O to the hyperplane. By symmetry, we must then have every $c_j = 1/n$ and $t = \bar{y}$, the sample mean.

Now consider the usual n-dimensional sample space, with one coordinate for each y_j. The bilinear form (29.35) establishes a duality between this and the estimator space. For any fixed t, a point in one space corresponds to a hyperplane in the other, while for varying t a point in one space corresponds to a family of parallel hyperplanes in the other. To the hyperplane (29.36) in the estimator space there corresponds the point $(t, t, \ldots, t)$ lying on the equiangular vector in the sample space. If a vector through the origin is orthogonal to a hyperplane in one space, the corresponding hyperplane and vector are orthogonal in the other space.

It now follows that the MV unbiased estimator will be given in the sample space by the hyperplane orthogonal to the equiangular vector at the point $L = (\bar{y}, \bar{y}, \ldots, \bar{y})$. If Q is the sample point, we drop a perpendicular from Q on to the equiangular vector to find L, i.e. we minimize $QL^2 = \sum_j (y_j - t)^2$. Thus we minimize a sum of squares in the sample space and consequently minimize the variance (another sum of squares) in the estimator space as a result of the duality established between them

Kruskal (1961) gives a completely geometrical approach to LS theory.

Estimation of the variance

29.8 To complete the estimation phase of the analysis, we must estimate the variance of the errors, σ^2. We first define the set of regression residuals

$$\mathbf{e} = \mathbf{y} - \mathbf{X}\hat{\boldsymbol{\beta}} = \{\mathbf{I} - \mathbf{X}(\mathbf{X}^T\mathbf{X})^{-1}\mathbf{X}^T\}\mathbf{y}, \tag{29.38}$$

which follows directly from (29.12). We now assume that $n > p$, since $\mathbf{e} \equiv \mathbf{0}$ when $n = p$. Using (29.5) and (29.38) reduces to

$$\begin{aligned} \mathbf{e} &= \{\mathbf{I} - \mathbf{X}(\mathbf{X}^T\mathbf{X})^{-1}\mathbf{X}^T\}\boldsymbol{\epsilon} \\ &= \mathbf{M}\boldsymbol{\epsilon}, \end{aligned} \tag{29.39}$$

say, where $\mathbf{M}$ denotes the term in the braces. We note that $\mathbf{M}$ is idempotent and symmetric ($\mathbf{M}^2 = \mathbf{M} = \mathbf{M}^T$) so that the *residual sum of squares*, from (29.6) is

$$\begin{aligned} S &= (\mathbf{y} - \mathbf{X}\hat{\boldsymbol{\beta}})^T(\mathbf{y} - \mathbf{X}\hat{\boldsymbol{\beta}}) \\ &= \mathbf{e}^T\mathbf{e} = \boldsymbol{\epsilon}^T\mathbf{M}^T\mathbf{M}\boldsymbol{\epsilon} = \boldsymbol{\epsilon}^T\mathbf{M}\boldsymbol{\epsilon}. \end{aligned} \tag{29.40}$$

The expected value of S is

$$\begin{aligned} E(S) &= E(\boldsymbol{\epsilon}^T\mathbf{M}\boldsymbol{\epsilon}) = E[\mathrm{tr}(\mathbf{M}\boldsymbol{\epsilon}\boldsymbol{\epsilon}^T)] \\ &= \mathrm{tr}[\mathbf{M}E(\boldsymbol{\epsilon}\boldsymbol{\epsilon}^T)] = \sigma^2\,\mathrm{tr}(\mathbf{M}), \end{aligned} \tag{29.41}$$

using the properties of the trace operator (tr) and assumptions (29.7). Now

$$\begin{aligned} \mathrm{tr}(\mathbf{M}) &= \mathrm{tr}[\mathbf{I}_n - \mathbf{X}(\mathbf{X}^T\mathbf{X})^{-1}\mathbf{X}^T] \\ &= \mathrm{tr}(\mathbf{I}_n) - \mathrm{tr}[\mathbf{X}(\mathbf{X}^T\mathbf{X})^{-1}\mathbf{X}^T] \\ &= \mathrm{tr}(\mathbf{I}_n) - \mathrm{tr}[(\mathbf{X}^T\mathbf{X})^{-1}(\mathbf{X}^T\mathbf{X})] \\ &= \mathrm{tr}(\mathbf{I}_n) - \mathrm{tr}(\mathbf{I}_p) \\ &= n - p. \end{aligned} \tag{29.42}$$

Thus, an unbiased estimator for σ^2 is

$$\begin{aligned} \hat{\sigma}^2 \text{ (or } s^2) &= \mathbf{e}^T\mathbf{e}/(n-p) \\ &= (\mathbf{y} - \mathbf{X}\hat{\boldsymbol{\beta}})^T(\mathbf{y} - \mathbf{X}\hat{\boldsymbol{\beta}})/(n-p); \end{aligned} \tag{29.43}$$

that is, the residual sum of the squares divided by the degrees of freedom. From (29.43), we immediately obtain the unbiased estimator for $\mathbf{V}(\hat{\boldsymbol{\beta}})$ in (29.15) by using s^2 in place of σ^2.

The sufficiency argument given near the end of **29.5** ensures that s^2 is the MV unbiased estimator of σ^2 when the ϵ_i are normally distributed; the multiple of s^2 with smallest m.s.e. is $(n-p)s^2/(n-p+2)$ by exactly the argument of Exercise 17.16. The MV unbiased estimator for σ^2 when $\mathbf{X}^T\mathbf{X}$ is singular is considered in Exercises 29.7 and 29.8. An unbiased estimator for quadratic form $\boldsymbol{\beta}^T\mathbf{A}\boldsymbol{\beta}$ is considered in Exercise 29.9.

Finally, note that the residuals generated by (29.38) are necessarily correlated since their covariance matrix $\sigma^2\mathbf{M}$ is of rank $n-p$. One procedure for generating a reduced set of uncorrelated residuals is considered in Exercise 29.10. A second approach is to generate estimators recursively from the first observations ($\hat{\boldsymbol{\beta}}_n$) and then to form the recursive residuals

$$e_{n+1} = y_{n+1} - \mathbf{x}_{n+1}^T\hat{\boldsymbol{\beta}}_n$$

for $n = p+1, p+2, \ldots$. The recursive estimators are considered in Exercise 29.11; the recursive residuals are used as a diagnostic in **32.64**.

Ridge regression

29.9 The LS estimators are MV unbiased among linear estimators, but they are not in general minimum m.s.e. in that class, and hence the squared Euclidean distance between $\hat{\boldsymbol{\beta}}$ and $\boldsymbol{\beta}$, say

$\mathbf{D}^2(\hat{\boldsymbol{\beta}})$, which is the sum of the m.s.e.s of the components of $\hat{\boldsymbol{\beta}}$, is not minimized. Hoerl and Kennard (1970a,b) generalize $\hat{\boldsymbol{\beta}}$ to the class of *ridge regression* estimators

$$\boldsymbol{\beta}_c^* = (\mathbf{X}^T\mathbf{X} + c\mathbf{I})^{-1}\mathbf{X}^T\mathbf{y}, \tag{29.44}$$

where $\boldsymbol{\beta}_0^* = \hat{\boldsymbol{\beta}}$) and choose the positive scalar c to minimize $\mathbf{D}^2(\boldsymbol{\beta}_c^*)$. Since

$$E(\boldsymbol{\beta}_c^*) = \{\mathbf{I} - c(\mathbf{X}^T\mathbf{X} + c\mathbf{I})^{-1}\}\boldsymbol{\beta} \tag{29.45}$$

and

$$\mathbf{V}(\boldsymbol{\beta}_c^*) = \sigma^2(\mathbf{X}^T\mathbf{X} + c\mathbf{I})^{-1}\mathbf{X}^T\mathbf{X}(\mathbf{X}^T\mathbf{X} + c\mathbf{I})^{-1}, \tag{29.46}$$

the m.s.e.s of the elements of $\boldsymbol{\beta}_c^*$ are in the main diagonal of

$$\begin{aligned}\mathbf{B}(\boldsymbol{\beta}_c^*) &= \mathbf{V}(\boldsymbol{\beta}_c^*) + (E(\boldsymbol{\beta}_c^*) - \boldsymbol{\beta})(E(\boldsymbol{\beta}_c^*) - \boldsymbol{\beta})^T \\ &= (\mathbf{X}^T\mathbf{X} + c\mathbf{I})^{-1}(\sigma^2\mathbf{X}^T\mathbf{X} + c^2\boldsymbol{\beta}\boldsymbol{\beta}^T)(\mathbf{X}^T\mathbf{X} + c\mathbf{I})^{-1},\end{aligned} \tag{29.47}$$

and

$$\begin{aligned}\mathbf{D}^2(\boldsymbol{\beta}_c^*) = \operatorname{tr}\mathbf{B}(\boldsymbol{\beta}_c^*) &= \operatorname{tr}\{(\mathbf{X}^T\mathbf{X} + c\mathbf{I})^{-2}(\sigma^2\mathbf{X}^T\mathbf{X} + c^2\boldsymbol{\beta}\boldsymbol{\beta}^T)\} \\ &= \sum_{j=1}^{p} \frac{\sigma^2\lambda_j + c^2 P_j}{(\lambda_j + c)^2},\end{aligned} \tag{29.48}$$

where the λ_j are the eigenvalues of $\mathbf{X}^T\mathbf{X}$ and $P_j \geq 0$. This quantity could be made less than $\mathbf{D}^2(\boldsymbol{\beta}_0^*) = \sigma^2 \sum \lambda_j^{-1}$ by choosing $c < \sigma^2/\max P_j$ if the latter were known. If instead $\boldsymbol{\beta}_c^*$ is computed for a range of values from 0 upwards, we can *estimate* the minimizing c, subject of course to sampling error, which may lead in some cases to a value worse than the zero of the LS theory. $\mathbf{D}^2(\boldsymbol{\beta}_0^*)$ is large when there are some small eigenvalues λ_j, for example, when the columns of $\mathbf{X}$ are highly correlated, the near-multicollinear situation of **27.4**.

The stability of the components of $\boldsymbol{\beta}_c^*$ can be examined by plotting them against c in the so-called *ridge trace* given by (29.48).

Dempster *et al.* (1977b) and Gunst and Mason (1977) give the results of extensive sampling experiments comparing this and other methods, including *shrunken* estimators (cf. Example 17.14), to LS.

Smith and Campbell (1980) criticize ridge regression methods on several grounds, including the use of a simple sum of m.s.e.s as a criterion. See also their discussion following their paper, and Exercise 29.12 below, which demonstrates the importance of the parametrization adopted. In particular, it is recommended that the regressor variables be standardized (to have zero means and unit variance) before the method is applied.

A different approach to using biased estimators is to restrict attention to a k-dimensional subspace of $\mathbf{X}$. Exercise 29.14 provides an example of this approach using principal components.

The singular case

29.10 In (29.2) we assumed $\mathbf{X}^T\mathbf{X}$ to be non-singular, so that (29.12) was valid, and $n > p$, so that (29.43) could be valid. If $n = p$, (29.12) still holds if $(\mathbf{X}^T\mathbf{X})^{-1}$ exists, but (29.43) is useless since the sum of squared residuals is identically zero. If $n < p$, the rank of $\mathbf{X}$ (and that of $\mathbf{X}^T\mathbf{X}$, which is the same) is less than p, so $\mathbf{X}^T\mathbf{X}$ has no inverse.

We now let $\mathbf{X}$ (and $\mathbf{X}^T\mathbf{X}$) have rank $r < p$ and suppose that $n \geq r$. The LS estimation problem must be discussed afresh, since $\mathbf{X}^T\mathbf{X}$ has no inverse and (29.12) is invalid. The present treatment follows Plackett (1950); a more general approach using projections is given in Chapter 30, see especially **30.2–3**.

Condition (29.25) is still necessary and sufficient for $\mathbf{C}\boldsymbol{\beta}$ to be unbiasedly estimable by $\mathbf{Ty}$. In the singular case, it cannot be satisfied if we wish to estimate $\boldsymbol{\beta}$ itself when it becomes

$$\mathbf{TX} = \mathbf{I}. \tag{29.49}$$

Since $\mathbf{X}$ is of rank r, we partition it into

$$\mathbf{X} = \left(\begin{array}{c|c} \mathbf{X}_{r,r} & \mathbf{X}_{r,p-r} \\ \hline \mathbf{X}_{n-r,r} & \mathbf{X}_{n-r,p-r} \end{array}\right) \tag{29.50}$$

The subscripts of the matrix elements of (29.50) indicate the numbers of rows and columns. We assume, without loss of generality, that $\mathbf{X}_{r,r}$ is non-singular, and therefore has inverse $\mathbf{X}_{r,r}^{-1}$. The last $n-r$ rows of $\mathbf{X}$ are linearly dependent upon the first r rows, so that $\mathbf{X}_{n-r,r} = \mathbf{G}\mathbf{X}_{r,r}$ and $\mathbf{X}_{n-r,p-r} = \mathbf{G}\mathbf{X}_{r,p-r}$ for some $(n-r) \times r$ matrix $\mathbf{G}$. Define a new matrix of order $p \times (p-r)$,

$$\mathbf{D} = \begin{pmatrix} \mathbf{X}_{r,r}^{-1} \cdot \mathbf{X}_{r,p-r} \\ -\mathbf{I}_{p-r} \end{pmatrix}, \tag{29.51}$$

where $\mathbf{I}_{p-r}$ is the identity matrix of that order. Evidently, $\mathbf{D}$ is of rank $p-r$. If we form the product $\mathbf{XD}$, we see at once that

$$\mathbf{XD} = \mathbf{0}. \tag{29.52}$$

If we postmultiply (29.49) by $\mathbf{D}$, we obtain, using (29.52),

$$\mathbf{D} = \mathbf{TXD} = \mathbf{0}. \tag{29.53}$$

This contradicts the fact that $\mathbf{D}$ has rank $p-r$. Hence (29.49) cannot hold.

Of course, some linear functions $\mathbf{C}\boldsymbol{\beta}$ may still be estimated unbiasedly, for (29.53) is then generalized to $\mathbf{CD} = \mathbf{0}$, which does not contradict the rank $p-r$ of $\mathbf{D}$ if $\mathbf{C} \neq \mathbf{I}$.

Alaouf and Stylan (1979) give various characterizations of estimability. Kourouklis and Paige (1981) give numerically stable methods of deriving estimators, their variances and their covariances.

29.11 We may proceed as in the non-singular case if we first introduce a set of $p-r$ linear constraints upon the parameters

$$\mathbf{a} = \mathbf{B}\boldsymbol{\beta}, \tag{29.54}$$

where $\mathbf{a}$ is a $(p-r) \times 1$ vector of constants and $\mathbf{B}$ is a known $(p-r) \times p$ matrix, of rank $p-r$. We now seek and estimator of the form $\mathbf{t} = \mathbf{Ly} + \mathbf{Na}$. Condition (29.49) becomes

$$\mathbf{I} = \mathbf{LX} + \mathbf{NB}. \tag{29.55}$$

Provided that the determinant

$$|\mathbf{BD}| \neq 0, \tag{29.56}$$

the matrix $\mathbf{B}$, of rank $p - r$, makes up the deficiency in rank of $\mathbf{X}$. In fact, we may treat $\mathbf{a}$ as a vector of dummy observations and consider the augmented model

$$\begin{pmatrix} \mathbf{y} \\ \mathbf{a} \end{pmatrix} = \begin{pmatrix} \mathbf{X} \\ \mathbf{B} \end{pmatrix} \boldsymbol{\beta} + \begin{pmatrix} \boldsymbol{\epsilon} \\ \mathbf{0} \end{pmatrix}. \tag{29.57}$$

The matrix

$$\begin{pmatrix} \mathbf{X} \\ \mathbf{B} \end{pmatrix}^T \begin{pmatrix} \mathbf{X} \\ \mathbf{B} \end{pmatrix} = \mathbf{X}^T\mathbf{X} + \mathbf{B}^T\mathbf{B}$$

is positive definite, for a non-null vector $\mathbf{d}$ makes $\mathbf{d}^T\mathbf{X}^T\mathbf{X}\mathbf{d} = (\mathbf{X}\mathbf{d})^T\mathbf{X}\mathbf{d}$ equal to zero only if $\mathbf{X}\mathbf{d} = \mathbf{0}$, whence $\mathbf{d}$ must be a column of $\mathbf{D}$. But (29.56) ensures that $\mathbf{B}\mathbf{d} \neq \mathbf{0}$, so that $\mathbf{d}^T\mathbf{B}^T\mathbf{B}\mathbf{d} > 0$. Thus $\mathbf{X}^T\mathbf{X} + \mathbf{B}^T\mathbf{B}$ is strictly positive definite and may be inverted. Expression (29.57) therefore yields the estimators

$$\mathbf{C}\hat{\boldsymbol{\beta}} = \mathbf{C}(\mathbf{X}^T\mathbf{X} + \mathbf{B}^T\mathbf{B})^{-1}(\mathbf{X}^T\mathbf{y} + \mathbf{B}^T\mathbf{a}), \tag{29.58}$$

which, as before, is the MV linear unbiased estimator of $\boldsymbol{\beta}$. Using (29.27), its covariance matrix, is, since $\mathbf{a}$ is constant,

$$\mathbf{V}(\mathbf{C}\hat{\boldsymbol{\beta}}) = \sigma^2\mathbf{C}(\mathbf{X}^T\mathbf{X} + \mathbf{B}^T\mathbf{B})^{-1}\mathbf{X}^T\mathbf{X}(\mathbf{X}^T\mathbf{X} + \mathbf{B}^T\mathbf{B})^{-1}\mathbf{C}^T. \tag{29.59}$$

The matrix $\mathbf{B}$ in (29.54) is arbitrary, subject to (29.56). In fact, if for $\mathbf{B}$ we substitute $\mathbf{UB}$, where $\mathbf{U}$ is any non-singular $(p-r) \times (p-r)$ matrix, (29.58) and (29.59) are unaltered in value. Thus we may choose $\mathbf{B}$ for convenience in computation in any particular case. An alternate form for the variance is given in Exercise 29.7.

Exercise 29.8 shows that σ^2 is estimated unbiasedly by the sum of squared residuals divided by $n - r$ if $n > r$.

Chipman (1964) gives a detailed discussion of LS theory in the singular case; see also Rao (1974).

Example 29.2 (LS estimation in the singular case)

As a simple example of a singular situation, suppose that we have

$$\boldsymbol{\beta} = \begin{pmatrix} \beta_1 \\ \beta_2 \\ \beta_3 \end{pmatrix} \qquad \mathbf{X} = \begin{pmatrix} 1 & 1 & 0 \\ 1 & 0 & 1 \\ 1 & 1 & 0 \\ 1 & 0 & 1 \end{pmatrix}.$$

Here $n = 4$, $p = 3$ and $\mathbf{X}$ has rank $2 < p$ because of the linear relation between its column vectors

$$\mathbf{x}_1 - \mathbf{x}_2 - \mathbf{x}_3 = \mathbf{0}$$

We first verify that $\boldsymbol{\beta}$ cannot be unbiasedly estimated, as we saw in **29.10**.

The matrix $\mathbf{D}$ at (29.51) is of order 3×1, being

$$\mathbf{D} = \begin{pmatrix} \begin{pmatrix} 1 & 1 \\ 1 & 0 \end{pmatrix}^{-1} \cdot \begin{pmatrix} 0 \\ 1 \end{pmatrix} \\ -1 \end{pmatrix} = \begin{pmatrix} 1 \\ -1 \\ -1 \end{pmatrix},$$

expressing the linear relation. We now introduce the matrix of order 1×3

$$\mathbf{B} = (1 \quad 0 \quad 0),$$

which satisfies (29.56) since $\mathbf{BD} = 1$, a scalar in this case of a single linear relation. Hence (29.54) is

$$c = (1 \quad 0 \quad 0)\begin{pmatrix} \beta_1 \\ \beta_2 \\ \beta_3 \end{pmatrix} = \beta_1;$$

that is, we assume that β_1 is known to be equal to c. From (29.58), the LS estimator is

$$\begin{pmatrix} \hat{\beta}_1 \\ \hat{\beta}_2 \\ \hat{\beta}_3 \end{pmatrix} = \left[\begin{pmatrix} 1 & 1 & 1 & 1 \\ 1 & 0 & 1 & 0 \\ 0 & 1 & 0 & 1 \end{pmatrix}\begin{pmatrix} 1 & 1 & 0 \\ 1 & 0 & 1 \\ 1 & 1 & 0 \\ 1 & 0 & 1 \end{pmatrix} + \begin{pmatrix} 1 \\ 0 \\ 0 \end{pmatrix}(1\,0\,0)\right]^{-1}$$

$$\times \left[\begin{pmatrix} 1 & 1 & 1 & 1 \\ 1 & 0 & 1 & 0 \\ 0 & 1 & 0 & 1 \end{pmatrix}\begin{pmatrix} y_1 \\ y_2 \\ y_3 \\ y_4 \end{pmatrix} + \begin{pmatrix} 1 \\ 0 \\ 0 \end{pmatrix} c\right]$$

$$= \begin{pmatrix} 5 & 2 & 2 \\ 2 & 2 & 0 \\ 2 & 0 & 2 \end{pmatrix}^{-1}\begin{pmatrix} y_1 + y_2 + y_3 + y_4 + c \\ y_1 + y_3 \\ y_2 + y_4 \end{pmatrix}$$

$$= \begin{pmatrix} 1 & -1 & -1 \\ -1 & \frac{3}{2} & 1 \\ -1 & 1 & \frac{3}{2} \end{pmatrix}\begin{pmatrix} \sum y + c \\ y_1 + y_3 \\ y_2 + y_4 \end{pmatrix} = \begin{pmatrix} c \\ \frac{1}{2}(y_1 + y_3) - c \\ \frac{1}{2}(y_2 + y_4) - c \end{pmatrix}.$$

Since we chose $\mathbf{B}$ so that $c = \beta_1$, we can obviously make no progress in estimating $\boldsymbol{\beta}$ itself. However, by (29.25), any set of linear functions $\mathbf{C}\boldsymbol{\beta}$ is unbiasedly estimated by $\mathbf{Ty}$ if $\mathbf{TX} = \mathbf{C}$. Thus $\beta_1 + \beta_2$ and $\beta_1 + \beta_3$ are estimable, since

$$\mathbf{C} = \begin{pmatrix} 1 & 1 & 0 \\ 1 & 0 & 1 \end{pmatrix}$$

satisfies (29.25) with

$$\mathbf{T} = \begin{pmatrix} \frac{1}{2} & 0 & \frac{1}{2} & 0 \\ 0 & \frac{1}{2} & 0 & \frac{1}{2} \end{pmatrix}.$$

The estimator of $\beta_1 + \beta_2$ is therefore $\frac{1}{2}(y_1 + y_3)$ and that of $\beta_1 + \beta_3$ is $\frac{1}{2}(y_2 + y_4)$. It may be verified that $\mathbf{CD} = \mathbf{0}$ as stated in **29.10**.

From (29.59),

$$\mathbf{V}(\hat{\boldsymbol{\beta}}) = \sigma^2 \begin{pmatrix} 1 & -1 & -1 \\ -1 & \frac{3}{2} & 1 \\ -1 & 1 & \frac{3}{2} \end{pmatrix} \begin{pmatrix} 4 & 2 & 2 \\ 2 & 2 & 0 \\ 2 & 0 & 2 \end{pmatrix} \begin{pmatrix} 1 & -1 & -1 \\ -1 & \frac{3}{2} & 1 \\ -1 & 1 & \frac{3}{2} \end{pmatrix}$$

$$= \sigma^2 \begin{pmatrix} 0 & 0 & 0 \\ 0 & \frac{1}{2} & 0 \\ 0 & 0 & \frac{1}{2} \end{pmatrix}$$

so that

$$\text{var}(\hat{\beta}_1 + \hat{\beta}_2) = \text{var}\,\hat{\beta}_2 = \text{var}(\hat{\beta}_1 + \hat{\beta}_3) = \text{var}\,\hat{\beta}_3 = \sigma^2/2,$$

as is evident from the fact that each estimator is a mean of two observations with variance σ^2. Also

$$\text{cov}(\hat{\beta}_1 + \hat{\beta}_2, \hat{\beta}_1 + \hat{\beta}_3) = 0,$$

a useful property which is due to the orthogonality of the second and third columns of **X**. When we discuss the analysis of variance in Chapter 30, we use projections to develop estimators for singular models subject to constraints.

LS with known linear contraints

29.12 Suppose that, when $\mathbf{X}^T\mathbf{X}$ is singular, (29.54) represents a set of q (instead of $p - r$) linear relations among the p parameters $\boldsymbol{\beta}$ that are known *a priori* to hold. Provided that the augmented model (29.57) is of full rank p, (29.58)–(29.59) follow as before, while $\boldsymbol{\beta}$ itself may now be unbiasedly estimable. The analysis of variance considered in Chapter 30 often involves situations of this kind, which also arise naturally in a Bayesian framework – see Exercise 26.6. In particular, we see that the ridge regression estimator is given by putting $\mathbf{a} = \mathbf{0}$ and $\mathbf{B}^T\mathbf{B} = c\mathbf{I}$ in (29.58), although its motivation in **29.9** was different.

Extension of a linear model to include further parameters

29.13 Suppose that a linear model (singular or non-singular)

$$\mathbf{y} = \mathbf{X}\boldsymbol{\beta} + \boldsymbol{\epsilon} \tag{29.60}$$

has been fitted to the observations, and that the LS estimator of $\boldsymbol{\beta}$ is $\hat{\hat{\boldsymbol{\beta}}} = \mathbf{T}\mathbf{y}$, where $\mathbf{T} = (\mathbf{X}^T\mathbf{X})^{-1}\mathbf{X}^T$ in the non-singular case. The sum of squared residuals is

$$\begin{aligned}(\mathbf{Y} - \mathbf{X}\hat{\hat{\boldsymbol{\beta}}})^T(\mathbf{y} - \mathbf{X}\hat{\hat{\boldsymbol{\beta}}}) &= \{(\mathbf{I} - \mathbf{XT})\mathbf{y}\}^T\{(\mathbf{I} - \mathbf{XT})\mathbf{y}\} \\ &= \mathbf{y}^T(\mathbf{I} - \mathbf{XT})\mathbf{y},\end{aligned} \tag{29.61}$$

since the matrix $\mathbf{I} - \mathbf{XT}$ is idempotent.

Now consider the extended model

$$\mathbf{y} = \mathbf{X}\boldsymbol{\beta} + \mathbf{Z}\boldsymbol{\gamma} + \boldsymbol{\epsilon}. \tag{29.62}$$

The LS estimators, $\hat{\boldsymbol{\beta}}$ and $\hat{\boldsymbol{\gamma}}$, are given by the solutions of the two estimating equations

$$\begin{aligned} \mathbf{X}^T\mathbf{X}\boldsymbol{\beta} + \mathbf{X}^T\mathbf{Z}\boldsymbol{\gamma} - \mathbf{X}^T\mathbf{y} = \mathbf{0}, \\ \mathbf{Z}^T\mathbf{X}\boldsymbol{\beta} + \mathbf{Z}^T\mathbf{Z}\boldsymbol{\gamma} - \mathbf{Z}^T\mathbf{y} = \mathbf{0}, \end{aligned} \tag{29.63}$$

whence

$$\hat{\boldsymbol{\beta}} = \mathbf{T}(\mathbf{y} - \mathbf{Z}\hat{\boldsymbol{\gamma}}),$$

so that, provided the model is non-singular,

$$\begin{aligned} \hat{\boldsymbol{\gamma}} &= \{\mathbf{Z}^T(\mathbf{I} - \mathbf{XT})\mathbf{Z}\}^{-1}\mathbf{Z}^T(\mathbf{I} - \mathbf{XT})\mathbf{y}, \\ \mathbf{V}(\hat{\boldsymbol{\gamma}}) &= \sigma^2\{\mathbf{Z}^T(\mathbf{I} - \mathbf{XT})\mathbf{Z}\}^{-1}. \end{aligned} \tag{29.64}$$

29.14 The residuals from (29.62) are

$$\mathbf{e} = \mathbf{y} - \mathbf{X}\hat{\boldsymbol{\beta}} - \mathbf{Z}\hat{\boldsymbol{\gamma}} = (\mathbf{I} - \mathbf{XT})\mathbf{y} - \mathbf{A}(\mathbf{A}^T\mathbf{A})^{-1}\mathbf{A}^T\mathbf{y}, \tag{29.65}$$

where $\mathbf{A} = (\mathbf{I} - \mathbf{XT})\mathbf{Z}$. It follows that the residual sum of squares is

$$\mathbf{e}^T\mathbf{e} = \mathbf{y}^T(\mathbf{I} - \mathbf{XT})\mathbf{y} - \mathbf{y}^T\mathbf{A}(\mathbf{A}^T\mathbf{A})^{-1}\mathbf{A}^T\mathbf{y}. \tag{29.66}$$

Compared with (29.61), the reduction in the sum of squares is

$$\mathbf{y}^T\mathbf{A}(\mathbf{A}^T\mathbf{A})^{-1}\mathbf{A}^T\mathbf{y} = \hat{\boldsymbol{\gamma}}^T\mathbf{Z}^T(\mathbf{I} - \mathbf{XT})\mathbf{y}. \tag{29.67}$$

The difference between (29.61) and (29.67) is

$$(\mathbf{y} - \mathbf{Z}\hat{\boldsymbol{\gamma}})^T(\mathbf{I} - \mathbf{XT})\mathbf{y}, \tag{29.68}$$

the sum of squared residuals when the extended model (29.62) has been fitted.

It is easy to see that a reduction of exactly the same form as (29.67) applies to the minimized SS under any constraints upon the elements of $\boldsymbol{\beta}$: the only change is that $\mathbf{I} - \mathbf{XT}$ is replaced by the matrix $\mathbf{Q}$ of the quadratic form of the minimized SS. The analogues of (29.67)–(29.68) are then $\hat{\boldsymbol{\gamma}}^T\mathbf{Z}^T\mathbf{Q}\mathbf{y}$ and $(\mathbf{y} - \mathbf{Z}\hat{\boldsymbol{\gamma}})^T\mathbf{Q}\mathbf{y}$, respectively.

A more general linear model

29.15 We have assumed throughout that (29.7) holds, i.e. that the errors are uncorrelated and have constant variance. There is no difficulty in generalizing the linear model to the situation where the covariance matrix of errors is $\sigma^2\mathbf{V}$, $\mathbf{V}$ being positive definite (it is always non-negative definite by **15.3**), and we find (see **31.33–4**) that (29.28) generalizes to

$$\hat{\mathbf{t}} = \mathbf{C}(\mathbf{X}^T\mathbf{V}^{-1}\mathbf{X})^{-1}\mathbf{X}^T\mathbf{V}^{-1}\mathbf{y}, \tag{29.69}$$

which is the MV unbiased estimator of $\mathbf{C}\boldsymbol{\beta}$. The Gauss–Markov theorem remains true and (29.32) becomes

$$\mathbf{V}(\hat{\mathbf{t}}) = \sigma^2\mathbf{C}(\mathbf{X}^T\mathbf{V}^{-1}\mathbf{X})^{-1}\mathbf{C}^T. \tag{29.70}$$

In Exercise 29.6 we consider the special case when $\mathbf{V}$ is diagonal but not equal to $\mathbf{I}$, so that the ϵ_i are uncorrelated but with unequal variances. Exercise 29.4 considers the mean and variance of a quadratic form for general $\mathbf{V}$.

Expression (29.69) coincides with the simple form (29.28) whenever $\mathbf{y} = \mathbf{Xz}$, i.e. y is in the space spanned by the columns of $\mathbf{X}$. Thus a necessary and sufficient condition for (29.69) and (29.28) to be identical is that they coincide for $\mathbf{y}$ orthogonal to $\mathbf{X}$, i.e. that $\mathbf{X}^T\mathbf{y} = \mathbf{0}$ implies $\mathbf{X}^T\mathbf{V}^{-1}\mathbf{y} = \mathbf{0}$. McElroy (1967) shows that if $\mathbf{X}$ contains a unit vector (so that one parameter is a constant term, as in Exercise 29.1), (29.69) coincides with (29.28) if the errors have equal variances and are all equally non-negatively correlated.

It should be observed that (29.28) will remain unbiased in the present more general model, as may be seen from Theorem 29.1, part (ii). However, the variance estimator is generally biased; Swindel (1968) gives attainable bounds for the bias. Kariya (1980) shows that the variance estimator, as well as $\hat{\mathbf{t}}$, remains unchanged if and only if

$$\mathbf{V} = \mathbf{XPX}^T + \mathbf{I} - \mathbf{X}(\mathbf{X}^T\mathbf{X})^{-1}\mathbf{X}^T$$

for some positive definite matrix $\mathbf{P}$. Bloomfield and Watson (1975) and Knott (1975) obtain the minimum efficiency of (29.28) compared with (29.69).

Baksalary and Kala (1983) discuss the optimum combination of two vectors of estimators.

To use (29.69)–(29.70), of course, we need to know $\mathbf{V}$. In practical cases this is usually unknown; if an estimate of $\mathbf{V}$ is available from past experience, or from replicated observations on the model, it can be substituted for $\mathbf{V}$, but the optimum properties of (29.69) no longer necessarily hold; see Rao (1967). Bement and Williams (1969) discuss the effect of such estimation on the variances in (29.70) in the case when $\mathbf{V}$ is diagonal – see also Fuller and Rao (1978) and Rao (1980). Rao (1970) and Chew (1970) discuss the estimation of the elements of $\mathbf{V}$ in the diagonal and other cases. Strand (1974) gives a bound for the m.s.e. incurred by using an incorrect $\mathbf{V}$ – see also Horn *et al.* (1975).

A naive estimator for $\mathbf{V}$ would be the diagonal of the singular matrix $\mathbf{ee}^T$; clearly such an estimator is not consistent for $\mathbf{V}$ and only $n - p$ degrees of freedom. Nevertheless, Gourieroux *et al.* (1984) show that the resulting estimators for $\boldsymbol{\beta}$ and the variance *are* consistent. However, the efficiency appears to be low.

Arnold (1979) shows that if the errors are multinormal with equal variances and equal correlations ρ, neither σ^2 nor ρ is separately estimable, although $\sigma^2(1-\rho)$ is, and that the general mean in the linear model is also not estimable. Procedures concerning $\boldsymbol{\beta}$ are otherwise unaffected, and retain optimum properties in this more general model. Jobson and Fuller (1980) discuss the case when the error variances are functions of $\boldsymbol{\beta}$. Pfefferman (1984) reviews results in the case when $\boldsymbol{\beta}$ is a random vector.

A conceptually different model, but one that fits into the same general framework, arises when we consider the two models

$$\mathbf{y}_1 = \mathbf{X}_1\boldsymbol{\beta}_1 + \boldsymbol{\epsilon}_1 \quad \text{and} \quad \mathbf{y}_2 = \mathbf{X}_2\boldsymbol{\beta}_2 + \boldsymbol{\epsilon}_2$$

that are connected only through dependence of the error terms (typically $\text{cov}(\epsilon_{j1}, \epsilon_{k2}) \neq 0$ if $j = k$). This structure is known as the problem of *seemingly unrelated regressions*, introduced by Zellner (1962). Further details appear in Exercise 29.23.

Ordered LS estimation of location and scale parameters

29.16 A particular situation in which (29.69) and (29.70) are of value is in the estimation of location and scale parameters from the order statistics, i.e. the sample observations ordered according to magnitude. The results are due to Lloyd (1952) and Downton (1953).

We denote the order statistics, as previously (see **14.1**), by $y_{(1)}, y_{(2)}, \ldots, y_{(n)}$. As usual, we write μ and σ for the location and scale parameters to be estimated and

$$z_{(r)} = (y_{(r)} - \mu)/\sigma, \qquad r = 1, 2, \ldots, n. \tag{29.71}$$

Let

$$\begin{aligned} E(\mathbf{z}) &= \boldsymbol{\alpha}, \\ \mathbf{V}(\mathbf{z}) &= \mathbf{V}, \end{aligned} \tag{29.72}$$

where $\mathbf{z}$ is the $n \times 1$ vector of the $z_{(r)}$. Since $\mathbf{z}$ has been defined by (29.71), $\boldsymbol{\alpha}$ and $\mathbf{V}$ are independent of location and scale.

Now, from (29.71) and (29.72),

$$E(\mathbf{y}) = \mu\mathbf{1} + \sigma\boldsymbol{\alpha}, \tag{29.73}$$

where $\mathbf{y}$ is the vector of $y_{(r)}$ and $\mathbf{1}$ is the unit vector, while

$$\mathbf{V}(\mathbf{y}) = \sigma^2\mathbf{V}. \tag{29.74}$$

We may now apply (29.69) and (29.70) to find the LS estimators of μ and σ. We have

$$\begin{pmatrix} \hat{\mu} \\ \hat{\sigma} \end{pmatrix} = \{(\mathbf{1}\ \boldsymbol{\alpha})^T\mathbf{V}^{-1}(\mathbf{1}\ \boldsymbol{\alpha})\}^{-1}(\mathbf{1}\ \boldsymbol{\alpha})^T\mathbf{V}^{-1}\mathbf{y} \tag{29.75}$$

and

$$\mathbf{V}\begin{pmatrix} \hat{\mu} \\ \hat{\sigma} \end{pmatrix} = \sigma^2\{(\mathbf{1}\ \boldsymbol{\alpha})^T\mathbf{V}^{-1}(\mathbf{1}\ \boldsymbol{\alpha})\}^{-1}. \tag{29.76}$$

Now

$$\begin{aligned} \{(\mathbf{1}\ \boldsymbol{\alpha})^T\mathbf{V}^{-1}(\mathbf{1}\ \boldsymbol{\alpha})\}^{-1} &= \begin{pmatrix} \mathbf{1}^T\mathbf{V}^{-1}\mathbf{1} & \mathbf{1}^T\mathbf{V}^{-1}\boldsymbol{\alpha} \\ \mathbf{1}^T\mathbf{V}^{-1}\boldsymbol{\alpha} & \boldsymbol{\alpha}^T\mathbf{V}^{-1}\boldsymbol{\alpha} \end{pmatrix}^{-1} \\ &= \frac{1}{\Delta}\begin{pmatrix} \boldsymbol{\alpha}^T\mathbf{V}^{-1}\boldsymbol{\alpha} & -\mathbf{1}^T\mathbf{V}^{-1}\boldsymbol{\alpha} \\ -\mathbf{1}^T\mathbf{V}^{-1}\boldsymbol{\alpha} & \mathbf{1}^T\mathbf{V}^{-1}\mathbf{1} \end{pmatrix}, \end{aligned} \tag{29.77}$$

where

$$\Delta = \{(\mathbf{1}^T\mathbf{V}^{-1}\mathbf{1})(\boldsymbol{\alpha}^T\mathbf{V}^{-1}\boldsymbol{\alpha}) - (\mathbf{1}^T\mathbf{V}^{-1}\boldsymbol{\alpha})^2\}. \tag{29.78}$$

From (29.75) and (29.77),

$$\begin{aligned} \hat{\mu} &= -\boldsymbol{\alpha}^T\mathbf{V}^{-1}(\mathbf{1}\boldsymbol{\alpha}^T - \boldsymbol{\alpha}\mathbf{1}^T)\mathbf{V}^{-1}\mathbf{y}/\Delta, \\ \hat{\sigma} &= \mathbf{1}^T\mathbf{V}^{-1}(\mathbf{1}\boldsymbol{\alpha}^T - \boldsymbol{\alpha}\mathbf{1}^T)\mathbf{V}^{-1}\mathbf{y}/\Delta. \end{aligned} \tag{29.79}$$

From (29.76) and (29.77),

$$\begin{aligned} \operatorname{var}\hat{\mu} &= \sigma^2\boldsymbol{\alpha}^T\mathbf{V}^{-1}\boldsymbol{\alpha}/\boldsymbol{\Delta}, \\ \operatorname{var}\hat{\sigma} &= \sigma^2\mathbf{1}^T\mathbf{V}^{-1}\mathbf{1}/\boldsymbol{\Delta}, \\ \operatorname{cov}(\hat{\mu}, \hat{\sigma}) &= -\sigma^2\mathbf{1}^T\mathbf{V}^{-1}\boldsymbol{\alpha}/\boldsymbol{\Delta}. \end{aligned} \tag{29.80}$$

29.17 Now since $\mathbf{V}$ and $\mathbf{V}^{-1}$ are positive definite, we may write

$$\begin{aligned}\mathbf{V} &= \mathbf{T}\mathbf{T}^T,\\ \mathbf{V}^{-1} &= (\mathbf{T}^{-1})^T\mathbf{T}^{-1},\end{aligned} \tag{29.81}$$

so that for an arbitrary vector $\mathbf{b}$

$$\mathbf{b}^T\mathbf{V}\mathbf{b} = \mathbf{b}^T\mathbf{T}\mathbf{T}^T\mathbf{b} = (\mathbf{T}^T\mathbf{b})^T(\mathbf{T}^T\mathbf{b}) = \sum_{i=1}^{n} h_i^2,$$

where h_i is the ith row element of the vector $\mathbf{T}^T\mathbf{b}$.

Similarly, for a vector $\mathbf{c}$,

$$\mathbf{c}^T\mathbf{V}^{-1}\mathbf{c} = (\mathbf{T}^{-1}\mathbf{c})^T(\mathbf{T}^{-1}\mathbf{c}) = \sum_{i=1}^{n} k_i^2,$$

k_i being the element of $\mathbf{T}^{-1}\mathbf{c}$. Now by the Cauchy–Schwarz inequality,

$$\sum h_i^2 \sum k_i^2 = \mathbf{b}^T\mathbf{V}\mathbf{b}\cdot\mathbf{c}^T\mathbf{V}^{-1}\mathbf{c} \geq \left(\sum h_i k_i\right)^2 = \{(\mathbf{T}^T\mathbf{b})^T(\mathbf{T}^{-1}\mathbf{c})\}^2 = (\mathbf{b}^T\mathbf{c})^2. \tag{29.82}$$

In (29.82), put

$$\begin{aligned}\mathbf{b} &= (\mathbf{V}^{\ 1} - \mathbf{I})\mathbf{1},\\ \mathbf{c} &= \boldsymbol{\alpha}.\end{aligned} \tag{29.83}$$

We obtain

$$\mathbf{1}^T(\mathbf{V}^{-1} - \mathbf{I})\mathbf{V}(\mathbf{V}^{-1} - \mathbf{I})\mathbf{1}\cdot\boldsymbol{\alpha}^T\mathbf{V}^{-1}\boldsymbol{\alpha} \geq \{\mathbf{1}^T(\mathbf{V}^{-1} - \mathbf{I})\boldsymbol{\alpha}\}^2. \tag{29.84}$$

If μ and σ^2 are the mean and variance, we find that

$$\begin{aligned}E(\mathbf{1}^T\mathbf{z}) &= \mathbf{1}^T\boldsymbol{\alpha} = 0,\\ \mathbf{V}(\mathbf{1}^T\mathbf{z}) &= \mathbf{1}^T\mathbf{V}\mathbf{1} = n = \mathbf{1}^T\mathbf{1}.\end{aligned} \tag{29.85}$$

Using (29.85) in (29.84), it becomes

$$(\mathbf{1}^T\mathbf{V}^{-1}\mathbf{1} - n)\boldsymbol{\alpha}^T\mathbf{V}^{-1}\boldsymbol{\alpha} \geq (\mathbf{1}^T\mathbf{V}^{-1}\boldsymbol{\alpha})^2,$$

which we may rewrite, using (29.80) and (29.78),

$$\text{var}\,\hat{\mu} \leq \sigma^2/n = \text{var}\,\bar{y}. \tag{29.86}$$

Expression (29.86) is obvious enough, since $\bar{y}$, the sample mean, is a linear estimator and therefore cannot have variance less than the MV estimator $\hat{\mu}$. But the point of the above argument is that it enables us to determine when (29.86) becomes a strict equality. This happens when inequality (29.82) becomes an equality, i.e. when $h_i = \lambda k_i$ for some constant λ, or

$$\mathbf{T}^T\mathbf{b} = \lambda\mathbf{T}^{-1}\mathbf{c}.$$

From (29.83) this is

$$\mathbf{T}^T(\mathbf{V}^{-1} - \mathbf{I})\mathbf{1} = \lambda\mathbf{T}^{-1}\boldsymbol{\alpha},$$

or

$$\mathbf{T}\mathbf{T}^T(\mathbf{V}^{-1} - \mathbf{I})\mathbf{1} = \lambda\boldsymbol{\alpha}. \tag{29.87}$$

Using (29.81), (29.87) finally becomes

$$(\mathbf{I} - \mathbf{V})\mathbf{1} = \lambda\boldsymbol{\alpha}, \tag{29.88}$$

the condition that var $\hat{\mu}$ = var $\bar{y} = \sigma^2/n$. If (29.88) holds, we must also have, by the uniqueness of the LS solution,

$$\hat{\mu} = \bar{y}, \tag{29.89}$$

and this may be verified by using (29.88) on $\hat{\mu}$ in (29.79).

If the distribution is symmetrical, the situation simplifies. For then the vector of expectations

$$\boldsymbol{\alpha} = \begin{pmatrix} E(z_{(1)}) \\ \vdots \\ E(z_{(n)}) \end{pmatrix} = \begin{pmatrix} \alpha_1 \\ \vdots \\ \alpha_n \end{pmatrix}$$

has

$$\alpha_i = -\alpha_{n+1-i} \qquad \text{for all } i, \tag{29.90}$$

as follows immediately from (14.2). Hence

$$\boldsymbol{\alpha}^T\mathbf{V}^{-1}\mathbf{1} = \mathbf{1}^T\mathbf{V}^{-1}\boldsymbol{\alpha} = 0 \tag{29.91}$$

and thus (29.79) becomes

$$\begin{aligned} \hat{\mu} &= \mathbf{1}\mathbf{V}^{-1}\mathbf{y}/\mathbf{1}^T\mathbf{V}^{-1}\mathbf{1}, \\ \hat{\sigma} &= \boldsymbol{\alpha}^T\mathbf{V}^{-1}\mathbf{y}/\boldsymbol{\alpha}^T\mathbf{V}^{-1}\boldsymbol{\alpha}, \end{aligned} \tag{29.92}$$

while (29.80) simplifies to

$$\begin{aligned} \text{var}\,\hat{\mu} &= \sigma^2/\mathbf{1}^T\mathbf{V}^{-1}\mathbf{1}, \\ \text{var}\,\hat{\sigma} &= \sigma^2/\boldsymbol{\alpha}^T\mathbf{V}^{-1}\boldsymbol{\alpha}, \\ \text{cov}(\hat{\mu}, \hat{\sigma}) &= 0. \end{aligned} \tag{29.93}$$

Thus the ordered LS estimators $\hat{\mu}$ and $\hat{\sigma}$ are uncorrelated if the distribution is symmetrical, an analogous result to that for ML estimators obtained in **18.37**.

Example 29.3 (LS estimation for the uniform, using order statistics)

We wish to estimate the mid-range (or mean) μ and range σ of the uniform distribution

$$f(y) = \sigma^{-1}, \qquad \mu - \tfrac{1}{2}\sigma \leq y \leq \mu + \tfrac{1}{2}\sigma.$$

Using (29.17), it is easy to show that, standardizing as in (29.71),

$$\alpha_r = E(z_{(r)}) = \{r/(n+1)\} - \tfrac{1}{2}, \tag{29.94}$$

and from (14.2) that the elements of the covariance matrix $\mathbf{V}$ of the $z_{(r)}$ are (cf. Example 11.4 for the variances)

$$V_{rs} = r(n-s+1)/\{(n+1)^2(n+2)\}, \qquad r \leq s. \tag{29.95}$$

The inverse of $\mathbf{V}$ is

$$\mathbf{V}^{-1} = (n+1)(n+2)\begin{pmatrix} 2 & -1 & & & \\ -1 & 2 & & & \mathbf{0} \\ & & \ddots & & \\ & \mathbf{0} & & & -1 \\ & & & -1 & 2 \end{pmatrix}. \tag{29.96}$$

From (29.96),

$$\mathbf{1}^T\mathbf{V}^{-1} = (n+1)(n+2)\begin{pmatrix} 1 \\ 0 \\ 0 \\ \vdots \\ 0 \\ 0 \\ 1 \end{pmatrix}; \tag{29.97}$$

and, from (29.94) and (29.96),

$$\boldsymbol{\alpha}^T\mathbf{V}^{-1} = \tfrac{1}{2}(n+1)(n+2)\begin{pmatrix} -1 \\ 0 \\ 0 \\ \vdots \\ 0 \\ 0 \\ 1 \end{pmatrix}. \tag{29.98}$$

Using (29.97) and (29.98), (29.92) and (29.93) give

$$\begin{aligned} \hat{\mu} &= \tfrac{1}{2}(y_{(1)} + y_{(n)}), \\ \hat{\sigma} &= (n+1)(y_{(n)} - y_{(1)})/(n-1), \\ \operatorname{var}\hat{\mu} &= \sigma^2/\{2(n+1)(n+2)\}, \\ \operatorname{var}\hat{\sigma} &= 2\sigma^2/\{(n-1)(n+2)\}, \\ \operatorname{cov}(\hat{\mu}, \hat{\sigma}) &= 0. \end{aligned} \tag{29.99}$$

Apart from the bias correction to σ^2, these are essentially the results we obtain by the ML method. The agreement is to be expected, since $y_{(1)}$, and $y_{(n)}$ are a pair of jointly sufficient statistics for μ and σ, as we saw in effect in Example 17.21.

As will have been made clear by Example 29.3, in order to use these results we must determine the covariance matrix $\mathbf{V}$ of the standardized order statistics, and this is a function of the form of the underlying distribution. This is in direct contrast with the general LS theory using unordered observations, discussed previously.

29.18 Before moving on from estimation to the testing of hypotheses, we examine a version of maximum likelihood estimation that is particularly important for the general linear model when the covariance matrix must also be estimated. These estimators are known as restricted maximum likelihood (REML) estimators and were first proposed by Thompson (1962), although our development uses a different justification due to Smyth and Verbyla (1996). A particular attraction of the REML approach is that it builds in an adjustment for the degrees of freedom when multiple paramenters are estimated, something that the original ML procedure did not do. This property is demonstrated in Example 29.4.

Restricted maximum likelihood

29.19 Suppose the parameters partition into $(\boldsymbol{\beta}, \boldsymbol{\theta})$, where we are directly interested only in $\boldsymbol{\theta}$. One approach would be to determine the ML estimators $\hat{\boldsymbol{\beta}}(\boldsymbol{\theta})$ and then to consider the resulting reduced likelihood function condition on these values. That is, we define the *reduced likelihood* as

$$L(\boldsymbol{\theta}|\hat{\boldsymbol{\beta}}, \boldsymbol{\beta}) \propto f(\mathbf{y}|\boldsymbol{\beta}, \boldsymbol{\theta})/f(\hat{\boldsymbol{\beta}}|\boldsymbol{\beta}, \boldsymbol{\theta}). \quad (29.100)$$

Since this approach is primarily used for mixed effects ANOVA models with normal errors (see Chapter 31), we consider (29.100) only in the context of the general linear model. Assume that y is mulitvariate normal, with

$$E(\mathbf{y}) = \mathbf{X}\boldsymbol{\beta}, \qquad \mathbf{V}(\mathbf{y}) = \mathbf{V}(\boldsymbol{\theta}),$$

so that

$$\log f(\mathbf{y}|\boldsymbol{\beta}, \boldsymbol{\theta}) = \text{const.} - \tfrac{1}{2}\log|\mathbf{V}(\boldsymbol{\theta})| - \tfrac{1}{2}(\mathbf{y} - \mathbf{X}\boldsymbol{\beta})^T[\mathbf{V}(\boldsymbol{\theta})]^{-1}(\mathbf{y} - \mathbf{X}\boldsymbol{\beta}). \quad (29.101)$$

The estimators follow in the usual way (cf. (29.15)) as

$$\hat{\boldsymbol{\beta}} = (\mathbf{X}^T\mathbf{V}^{-1}\mathbf{X})^{-1}\mathbf{X}^T\mathbf{V}^{-1}\mathbf{y}, \quad (29.102)$$

where we write $\mathbf{V} = \mathbf{V}(\boldsymbol{\theta})$ when no ambiguity arises. We know that, given $\boldsymbol{\theta}$, the distribution of $\hat{\boldsymbol{\beta}}$ is $N[\boldsymbol{\beta}, (\mathbf{X}^T\mathbf{V}^{-1}\mathbf{X})^{-1}]$, so the reduced likelihood in (29.100) becomes

$$\begin{aligned}\log L(\boldsymbol{\theta}|\hat{\boldsymbol{\beta}}, \boldsymbol{\beta}) = \text{const.} &- \tfrac{1}{2}\log|\mathbf{V}| - \tfrac{1}{2}\log|\mathbf{X}^T\mathbf{V}^{-1}\mathbf{X}| - \tfrac{1}{2}(\mathbf{y} - \mathbf{X}\boldsymbol{\beta})^T\mathbf{V}^{-1}(\mathbf{y} - \mathbf{X}\boldsymbol{\beta}) \\ &+ \tfrac{1}{2}(\hat{\boldsymbol{\beta}} - \boldsymbol{\beta})^T(\mathbf{X}^T\mathbf{V}^{-1}\mathbf{X})(\hat{\boldsymbol{\beta}} - \boldsymbol{\beta}). \quad (29.103)\end{aligned}$$

In turn, this expression reduces to

$$\log L(\boldsymbol{\theta}|\hat{\boldsymbol{\beta}}, \boldsymbol{\beta}) = \text{const.} - \tfrac{1}{2}\log|\mathbf{V}| - \tfrac{1}{2}\log|\mathbf{X}^T\mathbf{V}^{-1}\mathbf{X}| - \tfrac{1}{2}\mathbf{y}^T\mathbf{V}^{-1}(\mathbf{I} - \mathbf{Q})\mathbf{y}, \quad (29.104)$$

where

$$\mathbf{Q} = \mathbf{X}(\mathbf{X}^T\mathbf{V}^{-1}\mathbf{X})^{-1}\mathbf{X}^T\mathbf{V}^{-1}.$$

From (29.104), we observe that the reduces likelihood does not depend upon $\boldsymbol{\beta}$. Such a property is peculiar to the normal distribution, but is particularly valuable nonetheless.

The estimators for $\boldsymbol{\theta}$ determined from (29.104) are known as the REML estimators. In addition to the present development, taken from Smyth and Verbyla (1996), a number of other derivations have appeared, including a Bayesian argument by Harville (1974), who integrates out $\boldsymbol{\beta}$ to obtain (29.104) as a marginal likelihood (cf. **18.32**).

Example 29.4 (REML estimation for the general linear model)

Consider the simple case of the linear model where $\mathbf{V}(\theta) = \theta\mathbf{I}$. It follows from (29.104) that the reduced likelihood is

$$\log L = \text{const.} - \frac{n}{2}\log\theta + [(p/2)\log\theta - \tfrac{1}{2}\log|\mathbf{X}^T\mathbf{X}|] - \frac{1}{2\theta}\mathbf{y}^T(\mathbf{I}-\mathbf{Q})\mathbf{y},$$

where $\mathbf{Q} = \mathbf{X}(\mathbf{X}^T\mathbf{X})^{-1}\mathbf{X}^T$. Thus the REML estimator is $\hat{\theta} = (n-p)^{-1} \times$ (sum of squared errors) and provides a corresction for the degrees of freedom, making $\hat{\theta}$ unbiased in this case.

The canonical form of the general linear model

29.20 We now consider testing hypotheses in the linear model. In order to use the results of Chapter 22 more readily, we reduce the linear model to a *canonical form* so that the null hypothesis specifies the values of a number of means from independent distributions. The main result is expressed in terms of the following theorem. An alternative derivation using projections is given in **30.35**.

Theorem 29.3.

Consider the linear model

$$\mathbf{y} = \mathbf{X}\boldsymbol{\beta} + \boldsymbol{\epsilon}, \tag{29.105}$$

where $E(\boldsymbol{\epsilon}) = \mathbf{0}$, $\mathbf{V}(\boldsymbol{\epsilon}) = \sigma^2\mathbf{I}_n$ *and* $\mathbf{X}$ *is of full rank p, for which we wish to test the null hypothesis*

$$H_0 : \mathbf{A}\boldsymbol{\beta} = \mathbf{c}_0. \tag{29.106}$$

The model may be expressed in the canonical form

$$\mathbf{z} = \boldsymbol{\mu} + \boldsymbol{\delta}$$

or

$$\begin{pmatrix} \mathbf{z}_1 \\ \mathbf{z}_2 \\ \mathbf{z}_3 \end{pmatrix} = \begin{pmatrix} \boldsymbol{\mu}_1 \\ \boldsymbol{\mu}_2 \\ \mathit{0} \end{pmatrix} + \begin{pmatrix} \boldsymbol{\delta}_1 \\ \boldsymbol{\delta}_2 \\ \boldsymbol{\delta}_3 \end{pmatrix}, \tag{29.107}$$

where $E(\boldsymbol{\delta}) = \mathbf{0}$, $\mathbf{V}(\boldsymbol{\delta}) = \sigma^2\mathbf{I}_n$ *and the hypothesis reduces to*

$$H_0 : \boldsymbol{\mu}_1 = \mathbf{c}_1,$$

to be tested against $H_1 : \boldsymbol{\mu}_1 \neq \mathbf{c}_1$.

Note: since the result is proved by construction, the elements $\mathbf{z}$, $\boldsymbol{\mu}$, $\boldsymbol{\delta}$, and $\mathbf{c}_1$ are defined in the course of the proof.

Proof. Since $\mathbf{A}$ is of rank r, we may reparametrize the model using

$$\begin{aligned}\boldsymbol{\gamma}_1 &= \mathbf{A}\boldsymbol{\beta}\\ \boldsymbol{\gamma}_2 &= \boldsymbol{\beta}_2\end{aligned}$$

or

$$\boldsymbol{\gamma} = \mathbf{H}\boldsymbol{\beta}, \tag{29.108}$$

where $\boldsymbol{\gamma} = \binom{\boldsymbol{\gamma}_1}{\boldsymbol{\gamma}_2}$ and $\boldsymbol{\beta} = \binom{\boldsymbol{\beta}_1}{\boldsymbol{\beta}_2}$ are conformable partitions,

$$\mathbf{H} = \begin{pmatrix} \mathbf{A}_1 & \mathbf{A}_2 \\ \mathbf{0} & \mathbf{I}_{k-r} \end{pmatrix},$$

and $\mathbf{A} = (\mathbf{A}_1, \mathbf{A}_2)$. $\mathbf{A}_1$ is an $r \times r$ matrix of full rank, which always exists, although some reordering of the columns of $\mathbf{A}$ may be necessary.

The model becomes

$$\mathbf{y} = \mathbf{W}\boldsymbol{\gamma} + \boldsymbol{\epsilon}, \tag{29.109}$$

where $\mathbf{W} = \mathbf{XH}$ and the null hypothesis reduces to

$$H_0 : \boldsymbol{\gamma}_1 = \mathbf{c}_0. \tag{29.110}$$

Now, for some $\mathbf{C}$ (square and of full rank p) and some $\mathbf{F}$ ($(n-p) \times n$ and of rank $n-p$), let

$$\begin{aligned}\mathbf{z}_0 &= \mathbf{C}(\mathbf{W}^T\mathbf{W})^{-1}\mathbf{W}^T\mathbf{y}\\ &= \mathbf{CNy},\end{aligned} \tag{29.111}$$

say, and

$$\mathbf{z}_3 = \mathbf{F}(\mathbf{I}_n - \mathbf{WN})\mathbf{y}. \tag{29.112}$$

Substituting (29.109) into (29.111) and (29.112) yields $E(\mathbf{z}_0) = \mathbf{C}\boldsymbol{\gamma}$ and $E(\mathbf{z}_3) = \mathbf{0}$ for any $\mathbf{F}$. Further,

$$\mathbf{V}(\mathbf{z}_0) = \sigma^2\mathbf{C}(\mathbf{W}^T\mathbf{W})^{-1}\mathbf{C}^T, \tag{29.113}$$

$$\mathbf{V}(\mathbf{z}_3) = \sigma^2\mathbf{F}(\mathbf{I} - \mathbf{WN})\mathbf{F}^T \tag{29.114}$$

and

$$\mathbf{C}(\mathbf{z}_0, \mathbf{z}_3) = \mathbf{0}. \tag{29.115}$$

The derivations are straightforward and are left to the reader as Exercise 29.15. Since $\mathbf{C}$ is square, $\mathbf{V}(\mathbf{z}_0) = \sigma^2\mathbf{I}_p$ if and only if

$$\mathbf{C}(\mathbf{W}^T\mathbf{W})^{-1}\mathbf{C}^T = \mathbf{I}_p$$

or

$$\mathbf{W}^T\mathbf{W} = \mathbf{C}^T\mathbf{C}. \tag{29.116}$$

Further, there is always at least one choice of $\mathbf{F}$ that will make $\mathbf{V}(\mathbf{z}_3) = \sigma^2\mathbf{I}_{n-p}$; the proof is left as Exercise 29.15. From (29.116), we may choose $\mathbf{C}$ to be lower triangular,

$$\mathbf{C} = \begin{pmatrix} \mathbf{C}_{11} & \mathbf{0} \\ \mathbf{C}_{21} & \mathbf{C}_{22} \end{pmatrix},$$

so that (29.111) may be written in the form:

$$\mathbf{z}_0 = \begin{pmatrix} \mathbf{z}_1 \\ \mathbf{z}_2 \end{pmatrix} = \begin{pmatrix} \mathbf{C}_{11}\boldsymbol{\gamma}_1 \\ \mathbf{C}_{21}\boldsymbol{\gamma}_1 + \mathbf{C}_{22}\boldsymbol{\gamma}_2 \end{pmatrix} + \begin{pmatrix} \boldsymbol{\delta}_1 \\ \boldsymbol{\delta}_2 \end{pmatrix}, \tag{29.117}$$

where $\boldsymbol{\delta}$ has been constructed by implication in (29.111)–(29.115) so that $E(\boldsymbol{\delta}) = \mathbf{0}$ and $\mathbf{V}(\boldsymbol{\delta}) = \sigma^2\mathbf{I}_p$. Finally, we may write (29.117) as

$$\begin{pmatrix} \mathbf{z}_1 \\ \mathbf{z}_2 \end{pmatrix} = \begin{pmatrix} \boldsymbol{\mu}_1 \\ \boldsymbol{\mu}_2 \end{pmatrix} + \begin{pmatrix} \boldsymbol{\delta}_1 \\ \boldsymbol{\delta}_2 \end{pmatrix}, \tag{29.118}$$

and the hypothesis under test becomes

$$H_0 : \boldsymbol{\mu}_1 = \mathbf{C}_{11}\boldsymbol{\gamma}_1 = \mathbf{c}_1,$$

say, since $\boldsymbol{\gamma}_1 = \mathbf{c}_0$ is equivalent to $\boldsymbol{\mu}_1 = \mathbf{C}_{11}\mathbf{c}_0$. Combining (29.118) with the results for $\mathbf{z}_3$ in (29.112) and (29.114) we have the required form

$$\mathbf{z} = \begin{pmatrix} \mathbf{z}_1 \\ \mathbf{z}_2 \\ \mathbf{z}_3 \end{pmatrix} = \begin{pmatrix} \boldsymbol{\mu}_1 \\ \boldsymbol{\mu}_2 \\ \mathbf{0} \end{pmatrix} + \begin{pmatrix} \boldsymbol{\delta}_1 \\ \boldsymbol{\delta}_2 \\ \boldsymbol{\delta}_3 \end{pmatrix}. \tag{29.119}$$

29.21 We note that the theorem is capable of considerable extension. If $\mathbf{V}(\boldsymbol{\epsilon}) = \sigma^2(\mathbf{T}^T\mathbf{T})^{-1}$, we may define $\mathbf{y}_1 = \mathbf{T}\mathbf{y}$ and reformulate (29.105) in terms of $\mathbf{y}_1$. If $\mathbf{X}$ is not of full rank, the canonical form may be developed using a generalized inverse or projections (cf. **30.2–3**).

Tests of hypotheses

29.22 In order to develop tests of hypotheses, we need to make assumptions about the distribution of the errors in the linear model (29.119); specifically, we take each ϵ_i to be normal and hence, since they are uncorrelated, independent. It follows that the z_i are independently normally distributed. Their joint distribution therefore gives the LF

$$\begin{aligned} L(\boldsymbol{\mu}, \sigma^2|\mathbf{z}) &= (2\pi\sigma^2)^{-n/2}\exp\left\{-\frac{1}{2\sigma^2}(\mathbf{z}-\boldsymbol{\mu})^T(\mathbf{z}-\boldsymbol{\mu})\right\} \\ &= (2\pi\sigma^2)^{-n/2}\exp\left[-\frac{1}{2\sigma^2}\{(\mathbf{z}_1-\boldsymbol{\mu}_1)^T(\mathbf{z}_1-\boldsymbol{\mu}_1)+(\mathbf{z}_2-\boldsymbol{\mu}_2)^T(\mathbf{z}_2-\boldsymbol{\mu}_2)+\mathbf{z}_3^T\mathbf{z}_3\}\right]. \end{aligned} \tag{29.120}$$

We saw in Example 21.14 that if we have only one constraint ($r = 1$) there is a UMPU test of H_0 against H_1, since we are then testing the mean of a single normal population with unknown variance: the UMPU test is, as we saw in Example 21.14, the ordinary equal-tailed Student's t test for this hypothesis.

Kolodziejczyk (1935), to whom the first general results concerning the linear model are due, demonstrated the impossibility of a UMP test with more than one constraint, and showed that there is a pair of one-sided UMP similar tests when $r = 1$: these are the one-sided Student's t tests (cf. Example 21.7). We have just seen that there is a two-sided Student's t test which is

UMPU for $r = 1$, but the critical region of this test is different according to which of the μ_i is being tested: thus there is no common UMPU critical region for $r > 1$.

Since there is no 'optimum' test in any sense that we have so far discussed, we use the LR method to give an intuitively reasonable test.

29.23 The derivation of the LR statistic is straightforward. Under H_1 the ML estimators are

$$\hat{\mu}_i = z_i, \qquad i = 1, 2, \ldots, p,$$
$$\sigma^2 = \frac{1}{n}\sum_{i=p+1}^{n} z_i^2,$$

whence

$$(\mathbf{z} - \hat{\boldsymbol{\mu}})^T(\mathbf{z} - \hat{\boldsymbol{\mu}}) = n\hat{\sigma}^2.$$

Thus the unconditional maximum of the LF is

$$L(\hat{\boldsymbol{\mu}}, \hat{\sigma}^2|\mathbf{z}) = (2\pi\hat{\sigma}^2 e)^{-n/2} = \left(\frac{2\pi e}{n}\sum_{i=p+1}^{n} z_i^2\right)^{-n/2}. \tag{29.121}$$

When H_0 holds, the ML estimators are

$$\hat{\hat{\mu}}_i = z_i, \qquad i = r+1, r+2, \ldots, p,$$
$$\hat{\hat{\sigma}}^2 = \frac{1}{n}\left\{\sum_{i=p+1}^{n} z_i^2 + \sum_{i=1}^{r}(z_i - c_{0i})^2\right\},$$

whence

$$(\mathbf{z} - \hat{\hat{\boldsymbol{\mu}}})^T(\mathbf{z} - \hat{\hat{\boldsymbol{\mu}}}) = n\hat{\hat{\sigma}}^2,$$

so that the conditional maximum of the LF is

$$L(\mathbf{c}_0, \hat{\hat{\boldsymbol{\mu}}}, \hat{\hat{\sigma}}^2|\mathbf{z}) = (2\pi\hat{\hat{\sigma}}^2 e)^{-n/2} = \left[\frac{2\pi e}{n}\left\{\sum_{i=p+1}^{n} z_i^2 + \sum_{i=1}^{r}(z_i - c_{0i})^2\right\}\right]^{-n/2}. \tag{29.122}$$

From (29.121) and (29.122) the LR statistic l is given by

$$l^{2/n} = \frac{\hat{\sigma}^2}{\hat{\hat{\sigma}}^2} = \frac{1}{1+W}, \tag{29.123}$$

where

$$\begin{aligned} W &= (\mathbf{z}_1 - \mathbf{c}_0)^T(\mathbf{z}_1 - \mathbf{c}_0)/\mathbf{z}_3^T\mathbf{z}_3 \\ &= \frac{\sum_{i=1}^{r}(z_i - c_{0i})^2}{\sum_{i=p+1}^{n} z_i^2} \\ &= \frac{\hat{\hat{\sigma}}^2 - \hat{\sigma}^2}{\hat{\sigma}^2}. \end{aligned} \tag{29.124}$$

Reverting to the original notation for the linear model, as given in (29.105), we find that $(\mathbf{z}-\boldsymbol{\mu})^T(\mathbf{z}-\boldsymbol{\mu})$ corresponds to the partition

$$\begin{aligned} S = \boldsymbol{\epsilon}^T\boldsymbol{\epsilon} &= (\mathbf{y}-\mathbf{X}\boldsymbol{\beta})^T(\mathbf{y}-\mathbf{X}\boldsymbol{\beta}) \\ &= (\mathbf{y}-\mathbf{X}\hat{\boldsymbol{\beta}})^T(\mathbf{y}-\mathbf{X}\hat{\boldsymbol{\beta}}) + R, \end{aligned} \tag{29.125}$$

where $R = \{\mathbf{X}(\hat{\boldsymbol{\beta}}-\boldsymbol{\beta})\}^T\{\mathbf{X}(\hat{\boldsymbol{\beta}}-\boldsymbol{\beta})\}$.

From (29.125), $S = n\hat{\sigma}^2$ arises when we minimize S by setting $\boldsymbol{\beta} = \hat{\boldsymbol{\beta}}$; under H_0, we obtain $S = n\hat{\hat{\sigma}}^2$ with $\boldsymbol{\beta}$ estimated subject to the original condition $\mathbf{A}\boldsymbol{\beta} = \mathbf{c}_0$.

29.24 Since l is a monotone decreasing function of W, the LR test is equivalent to rejecting H_0 when W is large. If we divide the numerator and denominator of W by σ^2, we see that when H_0 holds, W is the ratio of the sum of squares of $n-p$ such variates, i.e. is the ratio of two independent χ^2 variates with r and $n-p$ degrees of freedom. Thus when H_0 holds, $F = [(n-p)/r]W$ has an F distribution (cf. **16.15**) with $(r, n-p)$ degrees of freedom, large values of F forming the critical region for the test.

Example 29.5 (F test for the linear model)

As a special case of particular importance, consider the null hypothesis

$$H_0 : \boldsymbol{\beta}_1 = \mathbf{0},$$

where $\boldsymbol{\beta}_1$ is an $r \times 1$ subvector of $\boldsymbol{\beta}$ in (29.105). We may therefore rewrite (29.105) as

$$\mathbf{y} = (\mathbf{X}_1\mathbf{X}_2)\begin{pmatrix}\boldsymbol{\beta}_1\\ \boldsymbol{\beta}_2\end{pmatrix} + \boldsymbol{\epsilon}$$

where $\mathbf{X}_1$ is of order $n \times r$ and $\mathbf{X}_2$ is of order $n \times (p-r)$. Then H_0 becomes equivalent to specifying

$$\mathbf{y} = \mathbf{X}_2\boldsymbol{\beta}_2 + \boldsymbol{\epsilon}.$$

From (29.125) the minimum under H_0 is

$$n\hat{\hat{\sigma}}^2 = \mathbf{y}^T\{\mathbf{I} - \mathbf{X}_2(\mathbf{X}_2^T\mathbf{X}_2)^{-1}\mathbf{X}_2^T\}\mathbf{y},$$

whereas under H_1 it is

$$n\hat{\sigma}^2 = \mathbf{y}^T\{\mathbf{I} - \mathbf{X}(\mathbf{X}^T\mathbf{X})^{-1}\mathbf{X}^T\}\mathbf{y},$$

where $\mathbf{X} = (\mathbf{X}_1\ \mathbf{X}_2)$. The F statistic is

$$F = \frac{n-p}{r}\left(\frac{\hat{\hat{\sigma}}^2 - \hat{\sigma}^2}{\hat{\sigma}^2}\right),$$

having an F distribution with $(r, n-p)$ degrees of freedom.

An F test for the comparison of estimators from two different samples is considered in Exercise 29.22.

29.25 The LR test is based on the statistic (29.124), which may be rewritten as

$$W = \frac{\sum_{i=1}^{r}(z_i - c_{0i})^2/\sigma^2}{\sum_{i=p+1}^{n} z_i^2/\sigma^2}.$$

Whether or not H_0 holds, the denominator of W is distributed as χ^2 with $n-p$ degrees of freedom. When H_0 holds, as we saw, the numerator is also a χ^2 variate with r degrees of freedom, but when H_0 does not hold this is no longer so: in general, the numerator follows a non-central χ^2 distribution (cf. **22.4**) with r degrees of freedom and non-centrality parameter

$$\lambda = \sum_{i=1}^{r}(c_{0i} - \mu_i)^2/\sigma^2 = (\boldsymbol{\mu}_1 - \mathbf{c}_0)^T(\boldsymbol{\mu}_1 - \mathbf{c}_0)/\sigma^2, \tag{29.126}$$

where μ_i is the true mean of z_i. Only when H_0 holds is λ equal to zero, giving a central χ^2 distribution. Since we wish to investigate the distribution of W (or equivalently of F) when H_0 is not true, so that we can evaluate the power of the LR test, we are led to the study of the ratio of a non-central to a central χ^2 variate.

The non-central *F* distribution

29.26 Consider first the ratio of two variates z_1, z_2, independently distributed as non-central χ^2 (22.18) with degrees of freedom ν_1, ν_2 and non-centrality parameters λ_1, λ_2, respectively. Using (11.74), the distribution of $u = z_1/z_2$ is given by

$$h(u) = \int_0^\infty \frac{e^{-(uv+\lambda_1)/2}(uv)^{(\nu_1/2)-1}}{2^{\nu_1/2}} \sum_{r=0}^{\infty} \frac{\lambda_1^r (uv)^r}{2^{2r} r!\Gamma(\frac{1}{2}\nu_1 + r)} \times \frac{e^{-(v+\lambda_2/2)} v^{(\nu_2/2)-1}}{2^{\nu_2/2}} \sum_{s=0}^{\infty} \frac{\lambda_2^s v^s}{2^{2s} s!\Gamma(\frac{1}{2}\nu_2 + s)} v\, dv. \tag{29.127}$$

If we write $\lambda = \lambda_1 + \lambda_2$, $\nu = \nu_1 + \nu_2$, and simplify, this becomes

$$h(u) = \frac{e^{-\lambda/2}}{2^{\nu/2}} \sum_{r=0}^{\infty}\sum_{s=0}^{\infty} \frac{\lambda_1^r \lambda_2^s}{2^{2r+2s} r!s!\Gamma(\frac{1}{2}\nu_1 + r)\Gamma(\frac{1}{2}\nu_2 + s)} \times \left\{\int_0^\infty e^{-v(1+u)/2} v^{(\nu/2)+r+s-1}\, dv\right\} u^{(\nu_1/2)+r-1}. \tag{29.128}$$

The integral in (29.128) is equal to $\Gamma(\frac{1}{2}\nu + r + s)/((1+u)/2)^{(\nu/2)+r+s}$. Thus

$$h(u) = e^{-\lambda/2} \sum_{r=0}^{\infty}\sum_{s=0}^{\infty} \frac{(\frac{1}{2}\lambda_1)^r}{r!} \frac{(\frac{1}{2}\lambda_2)^s}{s!} u^{(\nu_1/2)+r-1} \left(\frac{1}{1+u}\right)^{(\nu/2)+r+s} \frac{1}{B(\frac{1}{2}\nu_1 + r, \frac{1}{2}\nu_2 + s)}, \tag{29.129}$$

a result first obtained by Tang (1938). If we put

$$F'' = \frac{z_1/\nu_1}{z_2/\nu_2} = \frac{\nu_2}{\nu_1} u$$

in (29.129), we obtain the doubly non-central F distribution, the computation of whose d.f. is considered by Bulgren (1971). Harter and Owen (1974) give tables of the d.f. due to M.L. Tiku.

29.27 If now we put $\lambda_2 = 0$ and $\lambda_1 = \lambda$, F'' is only singly non-central, and we write it F'. From (29.129), its distribution is

$$g(F') = e^{-\lambda/2} \sum_{r=0}^{\infty} \frac{\frac{(\frac{1}{2}\lambda)^r}{r!} \left(\frac{\nu_1}{\nu_2}\right)^{(\nu_1/2)+r}}{B(\frac{1}{2}\nu_1 + r, \frac{1}{2}\nu_2)} \frac{(F')^{(\nu_1/2)+r-1}}{\left(1 + \frac{\nu_1}{\nu_2} F'\right)^{r+(\nu_1+\nu_2)/2}}. \tag{29.130}$$

Expression (29.130) is called the non-central F distribution with degrees of freedom ν_1, ν_2 and non-centrality parameter λ. We sometimes write it $F'(\nu_1, \nu_2, \lambda)$. Like (22.18), it was first discussed by Fisher (1928a); formulae for its moments are in Exercise 29.17. For further details, see Johnson *et al.* (1995, Chapter 30).

With $\lambda_2 = 0$, (29.129) is a Poisson mixture of beta distributions of the second kind **(6.16)** – cf. Exercise 22.19 for the parallel result for the non-central case. Thus the c.f. of F' is the same Poisson mixture of central F c.f.s – see **16.7** and Phillips (1982).

The 50, 75, 90 and 95 per cent points of F' are tabulated by Wallace and Toro-Vizcarrondo (1969) for ν_1 and $\nu_2 = 1$ (1) 30, 40, 60, 120, 200, 400, 1000 and $\lambda = 1$ only (they use $\frac{1}{2}\lambda$ as parameter).

As $\nu_2 \to \infty$, $\nu_1 F'(\nu_1, \nu_2, \lambda) \to \chi^2(\nu_1, \lambda)$ defined at (22.18) – this result when $\lambda = 0$ has already been noted at **16.22**(6). Cf. Exercise 29.16.

The power function of the LR test

29.28 It follows at once from (29.23)–(29.25) and (29.27) that the power function of the LR test is

$$P = \int_{F'_\alpha(\nu_1,\nu_2,0)}^{\infty} g\{F'(\nu_1, \nu_2, \lambda)\} \, dF' \tag{29.131}$$

where F'_α is the $100(1-\alpha)$ per cent point of the distribution, $\nu_1 = r$, $\nu_2 = n - p$, and λ is defined in (29.126). We show in Exercise 30.18 that the power function is increasing in λ and, therefore, that the test is unbiased.

Numerous tables and charts have been constructed:

(1) Tang (1938) gives $1 - P$ (i.e. the Type II error β) to 3 d.p. for test sizes $\alpha = 0.01, 0.05$; ν_1 (his f_1) = 1 (1) 8; ν_2 (his f_2) = 2 (2) 6 (1) 30, 60, ∞; and $\phi = \{\lambda/(\nu_1 + 1)\}^{1/2} = 1$ (0.5) 3 (1) 8.

(2) Tiku (1967; 1972) extends Tang's tables (1) to 4 d.p. for $\alpha = 0.005, 0.001, 0.025, 0.05, 0.10$; $\nu_1 = 1$ (1) 10, 12; $\nu_2 = 2$ (2) 30, 40 60, 120, ∞; and $\phi = 0.5$, 1.0 (0.2) 2.2 (0.4) 3.0.

(3) Lehmer (1944) gives inverse tables of ϕ for $\alpha = 0.01, 0.05$; $\nu_1 = 1$ (1) 10, 12, 15, 20, 24, 30, 40, 60, 80, 120, ∞; $\nu_2 = 2$ (2) 20, 24, 30, 40, 60, 80, 120, 240, ∞; and P (her β) = 0.7, 0.8.

(4) Dasgupta (1968) gives tables of $\frac{1}{2}\lambda$ (his δ) for ν_1 (his M) = 1 (1) 10; ν_2 (his N) = 10 (5) 50 (10) 100, ∞; $\alpha = 0.01, 0.05$ and P (his β) = 0.1 (0.1) 0.9.

(5) Kastenbaum *et al.* (1970a) give tables of $\tau = \{2\lambda(\nu_1 + 1)/(\nu_1 + \nu_2 + 1)\}^{1/2}$ to 3 d.p. for $\alpha = 0.01, 0.05, 0.1, 0.2$; $\beta = 1 - P = 0.005, 0.01, 0.05, 0.1, 0.2, 0.3$; ν_1 (their $k - 1$) =1 (1) 5; and $(\nu_1 + \nu_2 + 1)/(\nu_1 + 1)$ (their N) = 2(1) 8(2) 30, 40 (20) 100, 200, 500, 1000. See also their related smaller tables (1970b) and Bowman (1972).

(6) Pearson and Hartley (1951) give eight charts of the power function, one for each value of ν_1 from 1 to 8. Each chart shows the power for $\nu_2 = 6\ (1)\ 10, 12, 15, 20, 30, 60, \infty$; $\alpha = 0.01, 0.05$; $\nu_1 = 1\ (1)\ 8$. The charts are reproduced in the *Biometrika Tables*, Vol. II, Table 30, with two more for $\nu_1 = 12, 24$.

(7) Fox (1956) gives inverse charts, one for each of the combinations of $\alpha = 0.01, 0.05$, with power P (his β) $= 0.5, 0.7, 0.8, 0.9$. Each chart shows, for $\nu_1 = 3\ (1)\ 10\ (2)\ 20\ (20)\ 100, 200, \infty$; $\nu_2 = 4\ (1)\ 10\ (2)\ 20\ (20)\ 100, 200, \infty$, the contours of constant ϕ. He also gives a nomogram for each α to facilitate interpolation in β.

(8) Duncan (1957) gives two charts, one for $\alpha = 0.01$ and one for $\alpha = 0.05$. Each shows, for $\nu_2 = 6\ (1)\ 10, 12, 15, 20\ (10)\ 40, 60, \infty$, the values of ν_1 (ranging from 1 to 8) and ϕ required to attain power $P = 1 - \beta = 0.50$ and 0.90.

Approximation to the power function of the LR test

29.29 We may obtain a simple approximation to the power function using the results of our approximation to the non-central χ^2 distribution in **22.5**. If z_1 is a non-central χ^2 variate with ν_1 degrees of freedom and non-central parameter λ, we have from (22.21) that $z_1/[(\nu_1+2\lambda)/(\nu_1+\lambda)]$ is approximately a central χ^2 variate with degrees of freedom $\nu^* = (\nu_1+\lambda)^2/(\nu_1+2\lambda)$. That is,

$$z_1\left\{\nu^*\left(\frac{\nu_1+2\lambda}{\nu_1+\lambda}\right)\right\} = z_1/(\nu_1+\lambda)$$

is approximately a central χ^2 variate divided by its degrees of freedom ν^*. Hence z_1/ν_1 is approximately a multiple $(\nu_1+\lambda)/\nu_1$ of such a variate. If we now define the non-central F' variate

$$F' = \frac{z_1/\nu_1}{z_2/\nu_2},$$

where z_2 is a central χ^2 variate with ν_2 degrees of freedom, it follows at once that approximately

$$F' = \frac{\nu_1+\lambda}{\nu_1}F, \tag{29.132}$$

where F is a central F variate with degrees of freedom $\nu^* = (\nu_1+\lambda)^2/(\nu_1+2\lambda)$ and ν_2. The first two moments of (29.132) are exact, because those of the χ'^2 approximation in **22.5** are, through the exact moment relationship given in Exercise 29.17.

The simple approximation (29.132) is surprisingly effective. By making comparisons with Tang's (1938) exact tables, Patnaik (1949) shows that the power function calculated by the use of (29.132) is generally accurate to two significant figures; it will therefore suffice for all practical purposes.

To calculate the power of the LR test for the linear hypothesis, we therefore replace (29.132) by the approximate central F integral

$$P = \int_{(\nu_1/(\nu_1+\lambda))F_\alpha(\nu_1,\nu_2)}^{\infty} g\left\{F\left(\frac{(\nu_1+\lambda)^2}{(\nu_1+2\lambda)}, \nu_2\right)\right\}\mathrm{d}F, \tag{29.133}$$

the size of the test being determined by putting $\lambda = 0$.

A central F approximation due to Tiku (1965; 1966) obtained by equating three moments is even more accurate – see also Pearson and Tiku (1970).

The non-central *t* distribution

29.30 When $\nu_1 = 1$, the non-central F distribution (29.130) reduces to the non-central t^2 distribution, just as for the central distributions (cf. **16.15**), and as before we write t' instead of t to indicate non-centrality. Evidently, from the derivation of the non-central χ^2 as the sum of squared normal variates with non-zero means, we may write

$$t' = (z + \delta)/w^{1/2}, \tag{29.134}$$

where z is a standardized normal variate and w is independently distributed as χ^2/f with f degrees of freedom (we write f instead of ν_2 in (29.130), and $\delta^2 = \lambda$, in this case and sometimes write the variates as $t'(f, \delta)$). Our discussion of the F' distribution covers the t'^2 distribution, but the t' distribution has received special attention because of its importance in applications. It is derived in Exercise 29.18, its d.f. is given in Exercise 29.19, and its moments are given in Exercise 29.20. For further details see Johnson *et al.* (1995, Chapter 31).

Johnson and Welch (1939) studied the distribution and gave tables for finding $100(1-\alpha)$ per cent points of the distribution of t' for α or $1-\alpha = 0.005, 0.01, 0.025, 0.05, 0.1\ (0.1)\ 0.5$, $f = 4\ (1)\ 9$, $16, 36, 144, \infty$, and any δ; and conversely for finding δ for given values of t'. Resnikoff (1962) gives additional tables.

Resnikoff and Lieberman (1957) have given tables of the density and the d.f. of t' to 4 d.p., at intervals of 0.05 for $t/f^{1/2}$, for $f = 2\ (1)\ 24\ (5)\ 49$, and for the values defined by

$$\int_{\delta/(f+1)^{1/2}}^{\infty} (2\pi)^{-1/2} \exp(-\tfrac{1}{2}x^2)\, dx = \alpha,$$

$\alpha = 0.001, 0.0025, 0.004, 0.01, 0.025, 0.04, 0.065, 0.10, 0.15, 0.25$. They, and also Scheuer and Spurgeon (1963), give some percentage points of the distributions. Locks *et al.* (1963) give similar tables at intervals of 0.2 for t', with $f = 1\ (1)\ 20\ (5)\ 40$ and δ defined by $\delta(f+1)^{-1/2}$ or $\delta(f+2)^{-1/2} = 0\ (0.25)\ 3$. Owen (1963) gives very extensive tables of percentage points from which the *Biometrika Tables*, Vol. II, Table 26, provides tables (a) for seven values of α, to compute critical values of t' given δ and f; and (b) for four values of α, inverse tables to compute δ in term of f and the critical value t'. Hogben *et al.* (1961) give a method of obtaining moments, with tables (reproduced in the *Biometrika Tables*, Vol. II, Table 28) for the first four – simple formulae are given in Exercise 29.20. Amos (1964) studies series approximations of the distribution.

Krishnan (1967; 1968) and Bulgren and Amos (1968) study and tabulate the double non-central t distribution analogous to F'' of **29.26**. Extensive tables due to Bulgren are given in Harter and Owen (1974).

29.31 A particularly important application of the t' distribution is in evaluating the power of a Student's t test for which the critical region is in one tail only (the equal-tailed case, of course, corresponds to the t'^2 distribution). The test is that $\delta = 0$ in (29.134), the critical region being determined from the central t distribution. Its power is evidently just the integral of the non-central t distribution over the critical region. It has been tabulated by Neyman *et al.* (1935), who give, for $\alpha = 0.05, 0.01$, f (their n) $= 1\ (1)\ 30, \infty$ and δ (their p) $= 1\ (1)\ 10$, tables and charts for the complement $1-P$ of the power of the test, together with the values of δ for which $P = 1-\alpha$. Owen (1965) gives 5 d.p. tables of δ for $\alpha = 0.05, 0.025, 0.01$ and 0.005; $f = 1\ (1)\ 30\ (5)\ 100\ (10)\ 200, \infty$ and $1-P = 0.01, 0.05, 0.10\ (0.10)\ 0.90$. The Pearson–Hartley chart of **29.24** (6) for $\nu_1 = 1$ applies to equal-tailed tests on t' and is reproduced in each volume of the *Biometrika Tables*, Vol I, Table 10 and Vol. II, Table 30. Hodges and Lehmann (1967) give an asymptotic series for the power and use it (Hodges and Lehmann, 1968) to construct an extremely compact table.

Optimum properties of the LR test

29.32 We saw in **29.22** that, apart from the case $r = 1$, there is no UMPU test of the general linear hypothesis. Nevertheless, the LR test of that hypothesis has certain optimum properties which we now proceed to develop, making use of simplified proofs due to Wolfowitz (1949) and Lehmann (1950).

In **29.23** we derived the ML estimators of the $p-r+1$ unspecified parameters when H_0 holds. They are the components

$$\mathbf{t} = (\hat{\hat{\mu}}, \hat{\hat{\sigma}}^2),$$

which are defined above (29.122). When H_0 holds, the components of $\mathbf{t}$ are a set of $p-r+1$ sufficient statistics for the unspecified parameters. By **21.10**, their distribution is complete. Thus, by **21.19**, every similar size-α critical region w for H_0 will consist of a fraction α of every surface $\mathbf{t}$ = constant. Here every component of $\mathbf{t}$ is to be constant, and in particular the component $\hat{\hat{\sigma}}^2$. Let

$$n\hat{\hat{\sigma}}^2 = \sum_{i=p+1}^{n} z_i^2 + \sum_{i=1}^{r}(z_1 - c_{0i})^2 = a^2, \tag{29.135}$$

where a is a constant.

Now consider a fixed value of λ, defined at (29.126), say $\lambda = d^2 > 0$. The power of any similar region on this surface will consist of the aggregate of its power on (29.135) for all a. For a fixed a, the power on the surface $\lambda = d^2$ is

$$P(w|\lambda, a) = \int_{\lambda=d^2} L(\mathbf{z}|\boldsymbol{\mu}, \sigma^2)\,\mathbf{dz} \tag{29.136}$$

where L is the LF defined in (29.119). We may write this out fully as

$$\begin{aligned} P(w|\lambda, a) = (2\pi\sigma^2)^{-n/2} \int_{\lambda=d^2} \exp\Big\{-\frac{1}{2\sigma^2}[\{(\mathbf{z}_1 - \mathbf{c}_0) - (\boldsymbol{\mu}_1 - \mathbf{c}_0)\}^T\{(\mathbf{z}_1 - \mathbf{c}_0) \\ -(\boldsymbol{\mu}_1 - \mathbf{c}_0)\} + (\mathbf{z}_2 - \boldsymbol{\mu}_2)^T(\mathbf{z}_2 - \boldsymbol{\mu}_2) + \mathbf{z}_3^T\mathbf{z}_3]\Big\}\,\mathbf{dz}. \end{aligned} \tag{29.137}$$

Using (29.135) and (29.126), (29.137) becomes

$$P(w|\lambda, a) = (2\pi\sigma^2)^{-(n-p+r)/2} \exp\left\{-\frac{1}{2}\left(d^2 + \frac{a^2}{\sigma^2}\right)\right\} \int_{\lambda=d^2} \exp\{(\mathbf{z}_1 - \mathbf{c}_0)^T(\boldsymbol{\mu}_1 - \mathbf{c}_0)\}\,\mathbf{dz}_1 \tag{29.138}$$

the vector $\mathbf{z}_3$ having been integrated out over its whole range since its distribution is free of λ and independent of a. The only non-constant factor in (29.138) is the integral, which is to be maximized to obtain the critical region w with maximum P. The integral is over the surface $\lambda = d^2$ or $(\boldsymbol{\mu}_1 - \mathbf{c}_0)^T(\boldsymbol{\mu}_1 - \mathbf{c}_0)$ = constant; it is clearly a monotone increasing function of $|\mathbf{z}_1 - \mathbf{c}_0|$, i.e. of

$$(\mathbf{z}_1 - \mathbf{c}_0)^T(\mathbf{z}_1 - \mathbf{c}_0) = \sum_{i=1}^{r}(z_i - c_{0i})^2.$$

Now if $\sum_{i=1}^{r}(z_i - c_{0i})^2$ is maximized for fixed a in (29.135), W defined in (29.124) is also maximized. Thus for any fixed λ and a, the maximum value of $P(w|\lambda, \mathbf{a})$ is attained when w consists of larger values of W. Since this holds for each a, it holds when the restriction of fixed a is removed. We have therefore established that on any surface $\lambda = d^2 > 0$, the LR test, which consists of rejecting large values of W, has maximum power, a result due to Wald (1942).

An immediate consequence is Hsu's (1941) result, that the LR test is UMP among all test whose power is a function of λ only.

Further properties of the LR tests for the general linear model are considered in the context of invariance in **23.29–31**.

Generalized linear models

29.33 For the linear models in **29.2–3**, we assumed that the expectation of $\mathbf{y}$ is linear in $\boldsymbol{\beta}$ with known coefficients, and for testing purposes that the distribution of $\mathbf{y}$ is normal, with a specified form of covariance matrix. We now consider generalizing these assumptions.

Clearly, the linearity in $\boldsymbol{\beta}$ is an essential element in the relatively simple analysis of the model, but there is no obvious reason to choose the expectation of $\mathbf{y}$ as the linear function – more generally, we may allow a function of it to be linear in $\boldsymbol{\beta}$. Further, as we saw in Example 5.8, the normal distribution is a member of the natural exponential family, which may therefore be used as a generalization of normality. Thus we arrive at the *generalized linear model* (so called by Nelder and Wedderburn, 1972) in which

(a) writing $E(\mathbf{y}) = \boldsymbol{\mu}$, there is a function g such that $g(\boldsymbol{\mu}) = \boldsymbol{\eta}$ and $\boldsymbol{\eta} = \mathbf{X}\boldsymbol{\beta}$;

(b) each observation $y/a(\phi)$ is independently distributed in the same natural exponential form, where $a(\phi)$ is a scaling constant, possibly depending on a further parameter ϕ.

The function g is called the *link function*, because it connects the expectation of $\mathbf{y}$ with the function $\boldsymbol{\eta}$, which is linear in $\boldsymbol{\beta}$. It is assumed to be monotonic and differentiable.

29.34 We write the distribution of each y in the form

$$\exp\left\{\frac{y\theta - b(\theta)}{a(\phi)} + c(y, \phi)\right\}; \tag{29.139}$$

$y/a(\phi)$ in (29.139) plays the role of x in (5.117), and $-b(\theta)/a(\phi)$ that of $D(\theta)$ there. Thus from (5.121) we have

$$\kappa_r\{y/a(\phi)\} = \frac{\partial^r}{\partial\theta^r} b(\theta)/a(\phi),$$

so that for y we have

$$\kappa_r = b^{(r)}(\theta)\{a(\phi)\}^{r-1}, \tag{29.140}$$

whence

$$\begin{aligned} \mu = E(y) &= b'(\theta) \\ \operatorname{var} y &= b''(\theta)a(\phi). \end{aligned} \tag{29.141}$$

Thus the mean of y does not depend on the scaling constant, and the variance does so only as a multiplicative factor of the variance function

$$V = b''(\theta).$$

The substitutions

$$b(\theta) = \tfrac{1}{2}\theta^2, \qquad a(\phi) = \sigma^2$$

reduce (29.139) to the $N(\theta, \sigma^2)$ distribution, as the cumulants (29.141) make clear.

29.35 Ignoring constants,

$$\log L(y|\theta) = \left\{\sum y_i\theta_i - \sum b(\theta_i)\right\}/a(\phi) \tag{29.142}$$

with

$$\boldsymbol{\eta} = g(\boldsymbol{\mu}) = g\{b'(\theta_1), b'(\theta_2), \ldots, b'(\theta_n)\} = \mathbf{X}\boldsymbol{\beta}$$

using (29.141). In general, we cannot simplify $\log L$ further. However, if the link function g has the property that $\boldsymbol{\eta} = (\theta_1, \theta_2, \ldots, \theta_n)^T$ (in which case the link is called *canonical*), the first term in (29.142) is equal to $\mathbf{y}^T\mathbf{X}\boldsymbol{\beta}$, whence it is clear that the k statistics $\mathbf{y}^T\mathbf{x}_j$ (where x_j is the jth column of $\mathbf{X}$) are (minimal) sufficient for the k parameters in $\boldsymbol{\beta}$. If $b(\theta) = \frac{1}{2}\theta^2$ (the normal case), then (29.142) may be written

$$\log L = \{\mathbf{y}^T\mathbf{X}\boldsymbol{\beta} - \tfrac{1}{2}(\mathbf{X}\boldsymbol{\beta})^T\mathbf{X}\boldsymbol{\beta}\}/a(\phi)$$

so that putting $\partial \log L/\partial\boldsymbol{\beta} = \mathbf{0}$ leads back to the usual LS solution (29.8), as it must since we are now satisfying the conditions that we used there.

29.36 Because of its greater generality, we cannot expect the generalized linear model to give closed-form solutions as the ordinary linear model does. However, iterative ML estimation may be carried out using the methods given in **18.21** and **18.27**. The iterative procedure here becomes a weighted form of LS analysis; general details are given by McCullagh and Nelder (1983). Smyth and Verbyla (1996) provide an REML approach.

Non-linear least squares

29.37 In general, we may apply the method of least squares to non-linear functions in the parameters of the form:

$$y = g(x, \boldsymbol{\beta}) + \varepsilon,$$

as implied in **18.52**. Estimation typically requires numerical methods similar to those employed in **18.21**. As would be expected from the discussion on ML estimators, the non-linear LS estimators are consistent under fairly mild conditions but they are not unbiased. For further discussion, see **32.84–87**.

EXERCISES

29.1 In the linear model (29.5), suppose that a further parameter β_0 is introduced, so that we have the new model

$$\mathbf{y} = \mathbf{X}\boldsymbol{\beta} + \mathbf{1}\beta_0 + \boldsymbol{\epsilon},$$

where $\mathbf{1}$ is an $n \times 1$ vector of units. Show that the LS estimator in the new model of $\boldsymbol{\beta}$, the original vector of p parameters, remains of exactly the same form (29.8) as in the original model, with y_i replaced by $y_i - \bar{y}$ and x_{ij} by $x_{ij} - \bar{x}_j$ for $i = 1, 2, \ldots, n$ and $j = 1, 2, \ldots, p$.

29.2 Generalizing Exercise 29.1, show that if we extend (29.5) to

$$\mathbf{y} = \mathbf{X}\boldsymbol{\beta} + \mathbf{Z}\boldsymbol{\beta}_0 + \boldsymbol{\epsilon},$$

where $\boldsymbol{\beta}_0$ is a vector of additional parameters, the new LS estimator of the original parameter $\boldsymbol{\beta}$ is the solution of

$$(\mathbf{I} - \mathbf{M})\mathbf{y} = (\mathbf{I} - \mathbf{M})\mathbf{X}\boldsymbol{\beta},$$

where $\mathbf{M} = \mathbf{Z}(\mathbf{Z}^T\mathbf{Z})^{-1}\mathbf{Z}^T$. Show that if $\boldsymbol{\beta}_0$ is a scalar and $\mathbf{Z}$ is a vector whose first n_1 elements are unity and the remaining $n_2 = n - n_1$ elements zero, $\hat{\boldsymbol{\beta}}$ is as in the original model, except that for $i = 1, 2, \ldots, n_1$, y_i is replaced by

$$\left(y_i - \frac{1}{n_1}\sum_{i=1}^{n_1} y_i\right)$$

and x_{ij} by

$$\left(x_{ij} - \frac{1}{n_1}\sum_{i=1}^{n_1} x_{ij}\right).$$

In Example 29.1, show that $\hat{\beta}_2$ is thus changed from

$$\frac{\operatorname{cov}(x_2, y)}{\operatorname{var} x_2}$$

to

$$\hat{\beta}_2 = \frac{n_1 \operatorname{cov}_1(x_2, y) + n_2 \operatorname{cov}_2(x_2, y)}{n_1 \operatorname{var}_1 x_2 + n_2 \operatorname{var}_2 x_2},$$

where cov_1, var_1 refer to the first n_1 observations and cov_2, var_2 to the other n_2.

29.3 If, in the linear model (29.5), we replace the simple covariance matrix (29.7) by a non-singular covariance matrix $\sigma^2\mathbf{V}$ that allows correlations and unequal variances among the ϵ_i, show by putting $\mathbf{w} = \mathbf{T}^T\mathbf{y}$, where $\mathbf{T}\mathbf{T}^T = \mathbf{V}^{-1}$, that the LS and MV unbiased estimator of $\mathbf{C}\boldsymbol{\beta}$ is

$$\mathbf{C}\hat{\boldsymbol{\beta}} = \mathbf{C}(\mathbf{X}^T\mathbf{V}^{-1}\mathbf{X})^{-1}\mathbf{X}^T\mathbf{V}^{-1}\mathbf{y}.$$

(Cf. Aitken, 1935; Plackett, 1949)

29.4 For any matrix $\mathbf{B}$, show that if $E(\boldsymbol{\epsilon}\boldsymbol{\epsilon}^T) = \sigma^2\mathbf{V}$, $E(\boldsymbol{\epsilon}^T\mathbf{B}\boldsymbol{\epsilon}) = \sigma^2 \operatorname{tr}(\mathbf{BV})$. Show further that $\operatorname{var}(\boldsymbol{\epsilon}^T\mathbf{B}\boldsymbol{\epsilon}) = 2\sigma^4 \operatorname{tr}(\mathbf{BVBV})$ if $\boldsymbol{\epsilon}$ is multinormal.

29.5 In Exercise 29.3, show that, generalizing (29.32),

$$\mathbf{V}(\mathbf{C}\hat{\boldsymbol{\beta}}) = \sigma^2\mathbf{C}(\mathbf{X}^T\mathbf{V}^{-1}\mathbf{X})^{-1}\mathbf{C}^T,$$

and that the generalization of (29.41) is

$$E\{(\mathbf{y} - \mathbf{X}\hat{\boldsymbol{\beta}})^T\mathbf{V}^{-1}(\mathbf{y} - \mathbf{X}\hat{\boldsymbol{\beta}})\} = E\{\boldsymbol{\epsilon}^T[\mathbf{V}^{-1} - \mathbf{V}^{-1}\mathbf{X}(\mathbf{X}^T\mathbf{V}^{-1}\mathbf{X})^{-1}\mathbf{X}^T\mathbf{V}^{-1}]\boldsymbol{\epsilon}\} = (n-p)\sigma^2.$$

29.6 In Exercises 29.3 and 29.5, show that (29.31) remains true.
When $p = 1, \mathbf{X} = \mathbf{1}, \mathbf{C} = 1$ and the covariance matrix $\mathbf{V}$ is diagonal with elements V_j, show that the LS estimator of β, the common mean of the uncorrelated observations y_j, is

$$\hat{\mathbf{t}} = \sum V_j^{-1} y_j / \sum V_j^{-1} = \sum w_j y_j,$$

say, with variance $(\sum V_j^{-1})^{-1} = \sum w_j V_j$.
Let $t = \mathbf{u}^T \mathbf{y}$ be any other linear estimator with $\mathbf{u}$ a random variable distributed completely independently of $\mathbf{y}$ and satisfying (29.31), $\mathbf{u}^T \mathbf{u} = 1$. Using (29.31), show that

$$\text{var}\, t = \text{var}\, \hat{t} \left\{ 1 + \sum_j w_j^{-1} E(u_j - w_j)^2 \right\},$$

depending on the distribution of $\mathbf{u}$ only through the individual m.s.e. of the u_j, even though the latter may be dependent. Show that if the u_j are consistent estimators of the w_j, $\text{var}\, t \sim \text{var}\, \hat{t}$.
Given that all V_j are equal, show that

$$\text{var}\, t = \text{var}\, \hat{t}(1 + U^2),$$

where the U is the coefficient of variation of the u_j.

(Cf. Rubin and Weisberg, 1975)

29.7 In **29.11**, show using (29.52) that $(\mathbf{X}^T \mathbf{X} + \mathbf{B}^T \mathbf{B})^{-1} \mathbf{B}^T \mathbf{B} = \mathbf{D}(\mathbf{B}\mathbf{D})^{-1} \mathbf{B}$ and hence that (29.59) gives

$$\mathbf{V}(\hat{\boldsymbol{\beta}})(\mathbf{X}^T \mathbf{X}) = \sigma^2 \{\mathbf{I}_k - \mathbf{D}(\mathbf{B}\mathbf{D})^{-1} \mathbf{B}\}.$$

(Plackett, 1950)

29.8 Using the first result of Exercise 29.7, modify the argument of **29.8** to show that the unbiased estimator of σ^2 in the singular case is $[1/(n - r)](\mathbf{y} - \mathbf{X}\hat{\boldsymbol{\beta}})^T (\mathbf{y} - \mathbf{X}\hat{\boldsymbol{\beta}})$.

(Plackett, 1950)

29.9 For the linear model in (29.5), show using **29.8** that the quadratic form $\boldsymbol{\beta}^T \mathbf{A} \boldsymbol{\beta}$ is unbiasedly estimated by $\hat{\boldsymbol{\beta}}^T \mathbf{A} \hat{\boldsymbol{\beta}} - s^2 \,\text{tr}\{(\mathbf{X}^T \mathbf{X})^{-1} \mathbf{A}\}$.

29.10 In Section **29.8**, show that the LS residuals $\mathbf{y} - \mathbf{X}\hat{\boldsymbol{\beta}}$ have covariance matrix $\{\mathbf{I}_n - \mathbf{X}(\mathbf{X}^T \mathbf{X})^{-1} \mathbf{X}^T\}\sigma^2$, so that they are correlated. Show that if the linear model (29.5) is fitted to the first m of N observations only, the m residuals thus obtained are then uncorrelated with the $N - m$ residuals obtained from the other observations when (29.5) is fitted to all N observations, and that this result holds also if the model is singular. Hence, if we successively put $m = 1, 2, \ldots, n - 1$ and $N = m + 1$, we generate a set of n uncorrelated residuals (independent if the ϵ_i are normal).

(Cf. Hedayat and Robson, 1970)

29.11 Let $\mathbf{A} = \mathbf{X}^T \mathbf{X}$ in (29.8), and let the row vector $\mathbf{x}^T$ contain one further observation on each variable. Show that the inverse of $\mathbf{B} = \mathbf{A} + \mathbf{x}\mathbf{x}^T$ is

$$\mathbf{B}^{-1} = \mathbf{A}^{-1} - (\mathbf{A}^{-1} \mathbf{x}\mathbf{x}^T \mathbf{A}^{-1})/(1 + \mathbf{x}^T \mathbf{A}^{-1} \mathbf{x}).$$

Hence show that a quadratic form $\mathbf{x}^T \mathbf{A}^{-1} \mathbf{x}$ may be evaluated by

$$1 + \mathbf{x}^T \mathbf{A}^{-1} \mathbf{x} = |\mathbf{A} + \mathbf{x}\mathbf{x}^T| / |\mathbf{A}|$$

and that the estimator $\hat{\boldsymbol{\beta}}$ at (29.8) is adjusted by

$$\hat{\boldsymbol{\beta}}_{n+1} = \hat{\boldsymbol{\beta}}_n + \mathbf{A}^{-1}\mathbf{x}(y_{n+1} - \mathbf{x}^T\hat{\boldsymbol{\beta}}_n)/(1 + \mathbf{x}^T\mathbf{A}^{-1}\mathbf{x}).$$

(Cf. Bartlett, 1951)

29.12 In **29.9** show that the ridge regression estimate of $\mathbf{A}\boldsymbol{\beta} = \boldsymbol{\phi}$ obtained from $\mathbf{y} = \mathbf{X}\mathbf{A}^{-1}\boldsymbol{\phi} + \boldsymbol{\epsilon}$ is

$$[(\mathbf{X}\mathbf{A}^{-1})^T\mathbf{X}\mathbf{A}^{-1} + c\mathbf{I}]^{-1}(\mathbf{X}\mathbf{A}^{-1})^T\mathbf{y}$$

and hence that $\boldsymbol{\beta}$ is estimated by

$$\boldsymbol{\beta}_c^* = (\mathbf{X}^T\mathbf{X} + c\mathbf{A}^T\mathbf{A})^{-1}\mathbf{X}^T\mathbf{y},$$

agreeing with $\boldsymbol{\beta}_c^*$ in **29.9** if and only if $\mathbf{A}^T\mathbf{A} = \mathbf{I}$, i.e. $\mathbf{A}$ is orthogonal. Show that $\boldsymbol{\beta}_c^*$ or any choice of $\mathbf{A}$ satisfies

$$(\boldsymbol{\beta}_c^* - \tfrac{1}{2}\boldsymbol{\beta}_o^*)^T\mathbf{X}^T\mathbf{X}(\boldsymbol{\beta}_c^* - \tfrac{1}{2}\boldsymbol{\beta}_0^*) \leq \tfrac{1}{4}(\mathbf{X}\boldsymbol{\beta}_0^*)^T(\mathbf{X}\boldsymbol{\beta}_0^*),$$

where $\boldsymbol{\beta}_0^*$ is the LS estimator.

(Leamer (1981) shows that within this bound, choice of parametrization can seriously affect inference.)

29.13 Show by using a vector of Lagrangian multipliers that the minimization of (29.6) subject to p linear constraints (29.54) yields the LS estimator

$$\hat{\boldsymbol{\beta}} = (\mathbf{X}^T\mathbf{X})^{-1}[\mathbf{X}^T\mathbf{y} - \mathbf{B}^T\{\mathbf{B}(\mathbf{X}^T\mathbf{X})^{-1}\mathbf{B}^T\}^{-1}\{\mathbf{B}(\mathbf{X}^T\mathbf{X})^{-1}\mathbf{X}^T\mathbf{y} - \mathbf{a}\}].$$

(Kreuger and Neudecker (1977) treat the singular case.)

29.14 Show that the linear function $\mathbf{c}_1^T\hat{\boldsymbol{\beta}}$ $(\mathbf{c}_1^T\mathbf{c}_1 = 1)$ has the smallest minimized variance (29.32) when $\mathbf{c}_1$ is the eigenvector corresponding to the largest eigenvalue of $\mathbf{X}^T\mathbf{X}$, and similarly that each of the set of linear functions $\{\mathbf{c}_j^T\hat{\boldsymbol{\beta}}\}$, $j = 1, 2, \ldots, k$, has the smallest minimized variance subject to orthogonality to others when the $\mathbf{c}_j$ are eigenvectors corresponding to the p largest eigenvalues of $\mathbf{X}^T\mathbf{X}$.
Show that $\mathbf{c}_j^T\hat{\boldsymbol{\beta}}$ is the LS estimator of λ_j in the model

$$\mathbf{Y} = \mathbf{X}\mathbf{C}^T\boldsymbol{\lambda} + \boldsymbol{\eta},$$

where $\boldsymbol{\lambda}$ is $k \times 1$ and $\mathbf{C}$ is the $k \times p$ matrix whose rows are the $\mathbf{c}_j^T$.

(Greenberg (1975); the $\mathbf{c}_j^T\mathbf{X}$ are the first k principal components of $\mathbf{X}$.)

29.15 Show that the canonical variables defined (29.111)–(29.112) have variances and covariance as given by (29.113)–(29.114). Show that the matrix $\mathbf{I} - \mathbf{W}\mathbf{N}$ in (29.112) is idempotent and has $n - p$ eigenvalues equal to one, p equal to zero. Hence construct a form for $\mathbf{F}$, in terms of the matrix of eigenvalues, to ensure that $\mathbf{V}(\mathbf{z}_2) = \sigma^2\mathbf{I}_{n-p}$.

29.16 For the LR test of the general linear hypothesis based on (29.123), show that the asymptotic non-central χ^2 distribution of $-2\log l$, given in **22.7**, agrees with the asymptotic result for $\nu_1 F'(\nu_1, \nu_2, \lambda)$ at the end of **29.27**.

29.17 As in **29.27**, write the non-central F variate in the form

$$F'(\nu_1, \nu_2, \lambda) = \frac{z_1/\nu_1}{z_2/\nu_2},$$

where z_1 is a non-central $\chi'^2(\nu_1, \lambda)$ variate and z_2 is an independent central $\chi^2(\nu_2)$ variate. Let z_3 be a central $\chi^2(1)$ variate, independent of z_1 and z_2. Show that

$$E\{(F')^r\} = E\left\{\left(\frac{z_3}{z_2/\nu_2}\right)^r\right\}E(z_1^r)/\{\nu_1^r E(z_3^r)\},$$

whence symbolically

$$\mu_r'\{F'(\nu_1, \nu_2, \lambda)\} = \frac{\mu_r'\{F(1, \nu_2)\}\mu_r'\{\chi'^2(\nu_1, \lambda)\}}{\nu_1^r \mu_r'\{\chi^2(1)\}}$$

if $2r < \nu_2$, so that the central F moments exist by Exercise 16.1. Hence, using (16.4)–(16.5), (16.28) and (22.19), show that $F'(\nu_1, \nu_2, \lambda)$ has mean and variance given by

$$\mu_1' = (\nu_1 + \lambda)\nu_2/\{\nu_1(\nu_2 - 2)\}, \qquad \nu_2 > 2,$$

$$\mu_2 = \frac{2\nu_2^2}{\nu_1^3(\nu_2 - 2)^2(\nu_2 - 4)}\{(\nu_1 + 2\lambda)(\nu_1 + \nu_2 - 2) + \lambda^2\}, \qquad \nu_2 > 4.$$

(Cf. Bain (1969); Pearson and Tiku (1970) give μ_3 and μ_4.)

29.18 In (29.134), write the distribution of the numerator $u = z + \delta$ as

$$p_\delta(u) = \exp(-1/2\delta^2 + \delta u)p_0(u),$$

and hence show that the non-central t' distribution is

$$h(t') = \{\Gamma(\tfrac{1}{2})\Gamma(\tfrac{1}{2}f)f^{1/2}\exp(\tfrac{1}{2}\delta^2)\}^{-1}\sum_{s=0}^{\infty}\frac{\{(2/f)^{1/2}\delta t'\}^s}{s!}\frac{\Gamma\{\frac{1}{2}(f+s+1)\}}{(1+t'^2/f)^{(f+s+1)/2}},$$

reducing to the central Student's t distribution (16.15) when $\delta = 0$. Show that the distribution of t'^2 obtained from $h(t')$ agrees with (29.130) with $\nu_1 = 1$, $\nu_2 = f$, $\lambda = \delta^2$ and $F = t'^2$.

29.19 Using the transformation $\xi = (1 + t'^2/f)^{-1}$ as in **16.11** for the central case, show that the d.f. of the non-central t' distribution in Exercise 29.18 is given by

$$F(t') = 1 - \tfrac{1}{2}\exp(-\tfrac{1}{2}\delta^2)\sum_{s=0}^{\infty}\frac{(\delta/2^{1/2})^s}{\Gamma(\frac{1}{2}s + 1)}I_\xi(\tfrac{1}{2}f, \tfrac{1}{2}(s + 1)),$$

reducing to (16.17) when $\delta = 0$.
Show from (29.134) that $t' > 0$ if and only if $z + \delta > 0$ so that $1 - F(0)$ is $G(\delta)$, where G is the standard normal d.f.

(Hawkins, 1975)

29.20 Apply the method of Exercise 29.17 to the non-central t' variate defined at (29.134) by multiplying and dividing $E\{(t')^r\}$ by $E\{z_3^{r/2}\}$, and show that

$$\mu_r'\{t'(f, \delta)\} = \frac{\mu_r'\{N(\delta, 1)\}\mu_{r/2}'\{F(1, f)\}}{\mu_{r/2}'\{x^2(1)\}}, \qquad r < f,$$

where $N(\delta, 1)$ is a normal variate with mean δ and variance 1. If r is even, say $2s$, show that this may be written

$$\mu_{2s}'\{t'(f, \delta)\} = \mu_{2s}'\{N(\delta, 1)\}\mu_{2s}\{t(f)\}/\mu_{2s}\{N(0, 1)\},$$

where $t(f)$ is the central t distribution with f degrees of freedom.

(Cf. Bain, 1969)

29.21 In the linear model $\mathbf{y} = \mathbf{X}_1\boldsymbol{\beta}_1 + \mathbf{X}_2\boldsymbol{\beta}_2 + \boldsymbol{\epsilon}$, show that the LS estimators may be written as

$$\hat{\boldsymbol{\beta}}_1 = (\mathbf{X}_1^T\mathbf{X}_1)^{-1}\mathbf{X}_1^T(\mathbf{y} - \mathbf{X}_2\hat{\boldsymbol{\beta}}_2), \qquad \hat{\boldsymbol{\beta}}_2 = (\mathbf{X}_2^T\mathbf{D}\mathbf{X}_2)^{-1}\mathbf{X}_2^T\mathbf{D}\mathbf{y},$$

where

$$\mathbf{D} = \mathbf{I} - \mathbf{X}_1(\mathbf{X}_1^T\mathbf{X}_1)^{-1}\mathbf{X}_1^T.$$

If $\boldsymbol{\beta}_1$ is first estimated from

$$\mathbf{y} = \mathbf{X}_1\boldsymbol{\beta}_1 + \boldsymbol{\epsilon}^* \tag{A}$$

and $\boldsymbol{\beta}_2$ is then estimated, using the residuals $\mathbf{y}_r$ in (A) as though they were uncorrelated, from

$$\mathbf{y}_r = \mathbf{X}_2\boldsymbol{\beta}_2 + \boldsymbol{\eta},$$

show that the estimators obtained are

$$\boldsymbol{\beta}_1^* = (\mathbf{X}_1^T\mathbf{X}_1)^{-1}\mathbf{X}_1^T\mathbf{y}, \qquad \boldsymbol{\beta}_2^* = (\mathbf{X}_2^T\mathbf{X}_2)^{-1}\mathbf{X}_2^T\mathbf{D}\mathbf{y},$$

and that $\boldsymbol{\beta}_1^*$ and $\boldsymbol{\beta}_2^*$ are biased unless $\mathbf{X}_1^T\mathbf{X}_2 = \mathbf{0}$ or $\boldsymbol{\beta}_2 = \mathbf{0}$. When β_2 is a scalar, show that

$$\beta_2^* = (1 - R^2)\hat{\beta}_2,$$

where R is the multiple correlation coefficient of the single variable x_2 upon all the variables in $\mathbf{X}_1$. In the case $y_j = \beta_1 x_{1j} + \beta_2 x_{2j} + \epsilon_j$, show that the mean square errors of the biased two-step estimators β_1^*, β_2^* are less than the variances of the unbiased LS estimators $\hat{\beta}_1, \hat{\beta}_2$ if $\beta_2^2/V(\hat{\beta}_2) < 1$.

(Cf. Freund *et al.*, 1961; Goldberger and Jochems, 1961; Goldberger, 1961; Zyskind, 1963 and Wallace, 1964)

29.22 Suppose that we have available a set of n_1 observations on $(y, x_2, \ldots, x_p)$ satisfying the model

$$\mathbf{y}_1 = \mathbf{X}_1\boldsymbol{\beta}_1 + \boldsymbol{\epsilon}_1, \qquad \boldsymbol{\epsilon}_1 \sim N(\mathbf{0}, \sigma^2\mathbf{I}).$$

A further set of n_2 observations becomes available for which

$$\mathbf{y}_2 = \mathbf{X}_2\boldsymbol{\beta}_2 + \boldsymbol{\epsilon}_2, \qquad \boldsymbol{\epsilon}_2 \sim N(\mathbf{0}, \sigma^2\mathbf{I}).$$

To test $H_0 : \boldsymbol{\beta}_1 = \boldsymbol{\beta}_2$ against $H_1 : \boldsymbol{\beta}_1 \neq \boldsymbol{\beta}_2$, show that the LR procedure is based on the statistic

$$F = (\hat{\boldsymbol{\beta}}_1 - \hat{\boldsymbol{\beta}}_2)^T[(\mathbf{X}_1^T\mathbf{X}_1)^{-1} + (\mathbf{X}_2^T\mathbf{X}_2)^{-1}]^{-1}(\hat{\boldsymbol{\beta}}_1 - \hat{\boldsymbol{\beta}}_2)/ks^2,$$

where $(n_1 + n_2 - 2p)s^2 = (n_1 - p)s_1^2 + (n_2 - p)s_2^2$. When H_0 is true, show that the statistic has a central F distribution with $(p, n_1 + n_2 - 2p)$ d.fr.

(Chow (1960). A similar test is available when $\mathbf{V}(\epsilon_1) \neq \mathbf{V}(\epsilon_2)$, see Zellner (1962); for improved approximations to the distribution in that case, see Ali and Silver (1985).)

29.23 (*Seemingly unrelated regressions*). Suppose that

$$\begin{pmatrix} \mathbf{y}_1 \\ \mathbf{y}_2 \end{pmatrix} = \begin{pmatrix} \mathbf{X}_1 & \mathbf{0} \\ \mathbf{0} & \mathbf{X}_2 \end{pmatrix} \begin{pmatrix} \boldsymbol{\beta}_1 \\ \boldsymbol{\beta}_2 \end{pmatrix} + \begin{pmatrix} \boldsymbol{\epsilon}_1 \\ \boldsymbol{\epsilon}_2 \end{pmatrix},$$

where

$$E(\boldsymbol{\epsilon}) = \mathbf{0} \qquad \text{and} \qquad \mathbf{V}(\boldsymbol{\epsilon}) = \mathbf{V} = \begin{pmatrix} \sigma_1^2\mathbf{I} & \rho\sigma_1\sigma_2\mathbf{I} \\ \rho\sigma_1\sigma_2\mathbf{I} & \sigma_2^2\mathbf{I} \end{pmatrix}.$$

Show that

$$(\mathbf{X}^T\mathbf{V}^{-1}\mathbf{X})^{-1} = \begin{pmatrix} \mathbf{A}_{11} & \mathbf{A}_{12} \\ \mathbf{A}_{21} & \mathbf{A}_{22} \end{pmatrix}$$

has submatrices such as

$$\mathbf{A}_{11} = \sigma_1^2(1-\rho^2)[\mathbf{X}_1^T\mathbf{X}_1 - \rho^2\mathbf{X}_1^T\mathbf{X}_2(\mathbf{X}_2^T\mathbf{X}_2)^{-1}\mathbf{X}_2^T\mathbf{X}_1]^{-1}.$$

In particular, when $\mathbf{X}_1^T\mathbf{X}_2 = \mathbf{0}$, show that the relative efficiency of these estimators compared to LS is $(1-\rho^2)^{-1}$.

(Zellner, 1962; Binkley and Nelson, 1988)

29.24 Show that in the case of a symmetrical distribution, the condition that the ordered LS estimator $\hat{\mu}$ in (29.126) is equal to the sample mean $\bar{y} = \mathbf{1}^T\mathbf{y}/\mathbf{1}^T\mathbf{1}$ is that

$$\mathbf{V1} = \mathbf{1},$$

i.e. that the sum of each row of the covariance matrix is unity. Show (cf. Exercise 14.15) that this property holds for the univariate normal distribution.

(Lloyd (1952); Stephens (1975) shows that asymptotically $\mathbf{V}\boldsymbol{\alpha} = \frac{1}{2}\boldsymbol{\alpha}$ in the normal case.)

29.25 For the exponential distribution

$$f(y) = \exp\left\{-\left(\frac{y-\mu}{\sigma}\right)\right\}\Big/\sigma, \qquad \sigma > 0;\ \mu \le y < \infty,$$

show that in (29.72) the elements of $\boldsymbol{\alpha}$ are

$$\alpha_r = \sum_{i=1}^{r}(n-i+1)^{-1}$$

and that those of $\mathbf{V}$ are

$$V_{rs} = \sum_{i=1}^{m}(n-i+1)^{-2},$$

where $m = \min(r, s)$. Hence verify that $\mathbf{V}^{-1}$ has elements all zero except

$$V_{r,r+1}^{-1} = V_{r+1,r}^{-1} = -(n-r)^2,$$
$$V_{r,r}^{-1} = (n-r)^2 + (n-r+1)^2.$$

29.26 In Exercise 29.25, show from (29.79) that the MV unbiased estimators are

$$\hat{\mu} = y_{(1)} - (\bar{y} - y_{(1)})/(n-1) \text{ and } \hat{\sigma} = n(\bar{y} - y_{(1)})/(n-1).$$

Compare these with the ML estimators of the same parameters.

(Sarhan, 1954)

CHAPTER 30

FIXED EFFECTS ANALYSIS OF VARIANCE

30.1 In this chapter we derive the basic theory for fixed effects analysis of variance (ANOVA). We also apply these results to regression models. In the next chapter we further apply these results to analysis of covariance models. In a fixed effects ANOVA model we have individuals in different categories on each of whom we have measured some continuous response. We assume that the responses are independently normally distributed with constant variance so that the only difference between the distributions of observations in different classes occurs in the means.

For a simple example of an ANOVA model, suppose we conduct an experiment to find the effect of exercise on weight loss. We recruit four overweight men for each of five exercise programmes (for a total of 20 men). Each man's weight loss is recorded after 15 weeks on the programme. Let Y_{ij} be the weight loss of the jth man on the ith exercise programme, $i = 1, \dots, 5$; $j = 1, \dots, 4$. We assume that the Y_{ij} are independent, $Y_{ij} \sim N(\mu_i, \sigma^2)$, where μ_i is the mean weight loss on the i the exercise programme. We call this model a *one-way* ANOVA model because we have categorized each man in one way, by his exercise programme. (Note that the one-way ANOVA model is the natural generalization of the model used for the two-sample t test.) This model is called *balanced* because we have the same number of observations (4) in each category. We first test whether there is any significant difference between the mean weight losses for the different programmes (i.e. whether there is any significant difference between the μ_i). If we reject this hypothesis we want to determine which particular exercise programmes had significantly different mean weight losses. The problem of determining which μ_i are significantly different is called the *multiple comparisons* problem.

As a more complicated experiment, we might choose five exercise programmes and three diets. For each of the 15 possible pairs of exercise programme and diet we recruit four men (for a total of $5 \times 3 \times 4 = 60$ men). After 20 weeks on the assigned diet and exercise programme we measure each man's weight loss. Let Y_{ijk} be the weight loss of the kth man getting the ith exercise programme and the jth diet, $i = 1, \dots, 5$; $j = 1, \dots, 3$; $k = 1, \dots, 4$. We assume that the Y_{ijk} are independent, $Y_{ijk} \sim N(\mu_{ij}, \sigma^2)$, where μ_{ij} is the mean weight loss for a man on the ith exercise programme and the jth diet. We call this model a *two-way crossed* ANOVA model because we have categorized the men in two ways (by exercise programme and diet) and we have observations for every possible exercise–diet pair. The model is *balanced* because each pair has the same number of men. If we assume that $\mu_{ij} = \gamma_i + \tau_j$ for some constants γ_i and τ_j we say that this model is an *additive model.* (In words, a model is additive if the effect due to a particular exercise programme and a particular diet is the sum of an effect due to the exercise programme and an effect due to the diet.) In an additive two-way model we are interested in two hypotheses:

(a) that μ_{ij} does not depend on i (i.e. that exercise programme has no significant effect on weight loss);

(b) that μ_{ij} does not depend on j (i.e. that diet had no significant effect on weight loss).

Often in a two-way model we do not want to assume additivity. For the possibly non-additive model there are three hypotheses which are often tested:

(a) that the model is additive;

(b) that there is no average effect on weight loss due to exercise;

(c) that there is no average effect on weight loss due to diet.

There are some interpretational difficulties with the last two hypothesis-testing problems, as we shall discuss in **30.20**.

In some two-way models, it is natural to assume that $\mu_{ij} = \mu_{ji}$. This assumption may be sensible if the i and j effects are genetic effects from the mother and father of a person. If we make this assumption we say that we have a *symmetric* two-way model. In an additive symmetric two-way model, we would want to test that μ_{ij} does not depend on either i or j.

Consider now a different experiment in which we have three types of exercise programmes (aerobic, weight lifting, etc.) and for each type of programme we have four particular programmes (jogging, bicycling, etc.), for a total of 12 programmes. Suppose that for each of these 12 programmes we recruit 15 men and measure their weight losses after 20 weeks on the programme. Let Y_{ijk} be the weight loss for the kth man on the jth programme of the ith type $i = 1, \ldots, 3$; $j = 1, \ldots, 4$; $k = 1, \ldots, 15$. We again assume that the Y_{ijk} are independent, $Y_{ijk} \sim N(\mu_{ij}, \sigma^2)$, where μ_{ij} represents the mean weight loss for a man on the jth programme of the ith type. This model is called a *twofold nested* model because each exercise programme occurs in only one type of exercise programme. For example, jogging is an aerobic exercise and not a weight lifting exercise. We are interested in two hypotheses for this model:

(a) that there is no significant difference in weight loss between exercise programmes of the same type;

(b) that there is no significant difference between the average weight losses of different types of exercise.

There are again some interpretational difficulties with this last hypothesis, as we discuss in **30.26**.

It is clear that we could make many more complicated experiments. For example, we could have three-way crossed or threefold nested ANOVA models. We could even have models with both crossed and nested factors. In fact the class of possible fixed effects ANOVA models is quite large.

One way to attack these models is to consider each model separately and find optimal procedures. Our approach in this chapter, however, is to state a general version of the model, derive optimal procedures for that version and then interpret these results for many different ANOVA models. There are several ways to derive procedures simultaneously for many ANOVA models. One approach is to treat each of these models as regression models and use the theory developed in Chapter 29. However for most ANOVA models, there is no natural full rank $\mathbf{X}$ matrix to use. A second approach is to generalize the theory of regression to allow an $\mathbf{X}$ matrix which does not have full rank, using the theory of generalized inverses (see Searle, 1971; Graybill, 1976). A third approach is to derive a formula for the sum of squares for error (SSE) for a model and then find other sums of squares as differences between the SSEs under various hypotheses (see Scheffé, 1959).

In this chapter we use a fourth approach in which the statistics are written as projections on appropriate subspaces. This fourth approach is often convenient because it allows the computation of the statistics by least squares and involves less complicated matrix arguments. In particular, it is not necessary to determine an $\mathbf{X}$ matrix at all with this approach. It also leads to formulae for test statistics in terms of projections which often have a more intuitive interpretation than many of the complicated formulae from regression analysis. Furthermore, in **30.4–6**, we show that the formulae for regression are really just a particular case of those from the projection approach. In fact we shall see that it is often easier to derive optimality results for regression using projections. In **30.5** and **30.7** we also show that the generalized inverse approach and the approach of taking differences of SSEs are also special cases of the projection approach. In fact all three of these other approaches are just ways to compute the projections on the appropriate subspaces. One final nice aspect of the projection approach is that it leads to a very simple method for determining procedures for many orthogonal ANOVA models as described in **30.15–17**.

Although projections have been used for many years to motivate least squares procedures, apparently the first paper to use projections for derivations for linear models was Kruskal (1961). This projection approach is often called the *coordinate-free* approach, because the appropriate statistics are written as lengths of projections and hence it is not necessary to choose a coordinate system in order to compute them.

In **30.2–3**, we present some elementary results about subspaces and projections. In **30.4–7**, we define the general linear model, derive the basic results for it (including least squares estimators, F tests and Scheffé simultaneous confidence intervals) and apply these results to the multiple regression model and also to the generalized inverse approach. In **30.8** we apply these results to the one-way model. In **30.9–12**, we discuss Tukey, Dunnett, Hsu's MCB and Bonferroni simultaneous confidence intervals. In **30.13–14** we look at multiple comparisons procedures. In **30.15–17**, we derive some additional results for models which are orthogonal designs. These results often make it possible to write down immediately the appropriate formulae for F tests and simultaneous confidence intervals for many ANOVA models. In **30.18–28**, we use the results about orthogonal designs to derive procedures for various orthogonal ANOVA models. In **30.29–34**, we look at some ANOVA models which are not orthogonal designs. In **30.35** we derive the so-called canonical form for the general linear model. In **30.36–37**, we consider further testing problems related to the general linear model. In **30.38**, we consider the sensitivity of the procedures to the assumptions used in their derivation.

The F tests in this chapter were developed primarily by Fisher. See, for example, Fisher (1925b; 1935a). A general approach to these models was developed in Scheffé (1959).

Subspaces and projections

30.2 In this section, we present some basic facts about subspaces and projections. We do not prove these results here. Many of the proofs are sketched in the exercises. For additional proofs, see Arnold (1981 pp. 32–39).

A set $V \subset R^n$ is a *subspace* if it is closed under the operation of taking finite linear combinations, i.e. if

$$\mathbf{v}_i \in V, a_i \in R \Rightarrow \sum a_i \mathbf{v}_i \in V.$$

Let $\mathbf{v}_i \in V$. The $\mathbf{v}_i$ span V if every vector in V is a finite linear combination of the $\mathbf{v}_i$, and the $\mathbf{v}_i$ are *linearly independent* if $\sum a_i \mathbf{v}_i = 0 \Rightarrow a_i = 0$. The $\mathbf{v}_i$ are a *basis* for V if they are linearly independent and they span V. The $\mathbf{v}_i$ are an *orthonormal basis* for V if they are a basis and in addition $\mathbf{v}_i^T \mathbf{v}_i = 1$ and $\mathbf{v}_i^T \mathbf{v}_j = 0$ when $i \neq j$. (Note that orthonormal vectors must be linearly independent. Therefore to establish that a set is an orthonormal basis, it is only necessary to establish their orthonormality and that they span V.) It is well known that every subspace has an orthonormal basis and that any two bases for a subspace have the same number of vectors. This number is called the *dimension* of V. A *basis matrix* $\mathbf{X}$ for a p-dimensional subspace V is an $n \times p$ matrix whose columns form a basis for V. If the columns of $\mathbf{X}$ are an orthonormal basis then we say that $\mathbf{X}$ is an *orthonormal basis matrix* for V. If $\mathbf{X}$ is a basis matrix, then its columns are linearly independent and hence $\mathbf{X}^T\mathbf{X}$ is invertible. If $\mathbf{X}$ is an orthonormal basis matrix, then $\mathbf{X}^T\mathbf{X} = \mathbf{I}$.

Let $\mathbf{u}, \mathbf{v} \in R^n$ and let U and V be subspaces of R^n. Then $\mathbf{u}$ and $\mathbf{v}$ are *orthogonal* (written $\mathbf{u} \perp \mathbf{v}$) if $\mathbf{u}^T\mathbf{v} = 0$, $\mathbf{u}$ is orthogonal to V (written $\mathbf{u} \perp V$) if $\mathbf{u}^T\mathbf{v} = 0$ for all $\mathbf{v} \in V$ and U and V are *orthogonal* (written $U \perp V$) if $\mathbf{u}^T\mathbf{v} = 0$ for all $\mathbf{u} \in U, \mathbf{v} \in V$.

Let $W \subset V \subset R^n$ be subspaces. The *orthogonal complement* of V (written $V^\perp$) is the set of all vectors in R^n which are orthogonal to V. The *relative orthogonal complement* of W in V (written $V|W$) is the set of all vectors in V which are orthogonal to W. That is

$$V|W = V \cap W^\perp.$$

It can be shown that $V^\perp$ and $V|W$ are subspaces and that

$$\dim(V^\perp) = n - \dim(V), \quad \dim(V|W) = \dim(V) - \dim(W),$$
$$(V^\perp)^\perp = V, \quad V|(V|W) = W.$$

Note that

$$V^\perp = R^n|V$$

so that results stated here and below for $V^\perp$ are special cases of those for $V|W$.

Example 30.1 (Subspace in R^3)

In R^3, the space consisting only of the $\mathbf{0}$ vector is a subspace, whose dimension is defined to be 0. Lines through the origin are subspaces with dimension 1 and planes through the origin are subspaces of dimension two. If V is a two-dimensional subspace of R^3, then $V^\perp$ is the line through the origin which is orthogonal to the plane V. Note that $\dim(V^\perp) = 3 - \dim(V) = 1$. If W is a one-dimensional subspace of V, then $V|W$ is the line in the plane V which is orthogonal to the line W in V. Note that $\dim(V|W) = \dim(V) - \dim(W) = 2 - 1 = 1$.

The following lemma is often useful for finding basis matrices for $V^\perp$ and $V|W$.

Lemma 30.1.

(a) Let $\mathbf{C}$ be an $n \times q$ matrix of rank q and let V be the subspace all vectors $\mathbf{t} \in R^n$ such that $\mathbf{C}^T\mathbf{t} = \mathbf{0}$. Then $\mathbf{C}$ is a basis matrix for $V^\perp$ and $\dim(V^\perp) = q$.

(b) Let V be a subspace of R^n and let $\mathbf{C}$ be an $n \times q$ matrix whose columns are linearly independent vectors in V. Let W be the subspace of all vectors $\mathbf{v} \in V$ such that $\mathbf{C}^T\mathbf{v} = \mathbf{0}$. Then $\mathbf{C}$ is a basis matrix for $V|W$ and $\dim(V|W) = q$.

Example 30.2 (Some subspaces and basis matrices)

Let V be the subspace of R^3 consisting of vectors $\mathbf{v} = (v_1, v_2, v_3)^T$ such that

$$(1, 2, 3)\mathbf{v} = v_1 + 2v_2 + 3v_3 = 0.$$

Note that $\dim(V) = 3 - 1 = 2$. Let

$$\mathbf{v}_1 = (1, 1, -1)^T, \quad \mathbf{v}_2 = (-1, 2, -1)^T.$$

Then $\mathbf{v}_1$ and $\mathbf{v}_2$ are linearly independent vectors in V and are hence a basis for V. Therefore a basis matrix for V is

$$(\mathbf{v}_1, \mathbf{v}_2) = \begin{pmatrix} 1 & -1 \\ 1 & 2 \\ -1 & -1 \end{pmatrix} = \mathbf{X}.$$

Note also that by the lemma above, $\mathbf{v}_3 = (1, 2, 3)^T$ is a basis matrix for $V^\perp$. Now let W be the subspace of V in which in addition

$$(4, 1, -2)\mathbf{v} = 4v_1 + v_2 - 2v_3 = 0.$$

Note that $\dim(W) = 1$ and $\mathbf{v}_2 \in W$, so that $\mathbf{v}_2$ is a basis matrix for W. Finally, note that $\mathbf{v}_4 = (4, 1, -2)^T \in V$. Hence, by the lemma above, $\mathbf{v}_4$ is a basis matrix for $V|W$.

30.3 For any subspace $V \subset R^n$, there is a unique $n \times n$ matrix $\mathbf{P}_V$ such that for all $\mathbf{y} \in R^n$

$$\mathbf{P}_V\mathbf{y} \in V, \quad \mathbf{y} - \mathbf{P}_V\mathbf{y} \perp V.$$

We call $\mathbf{P}_V\mathbf{y}$ the (orthogonal) *projection* of $\mathbf{y}$ onto V. $\mathbf{P}_V$ is the (orthogonal) *projection matrix* of V. The projection is illustrated by Fig. 30.1.

We now state some elementary properties of projections.

1. (Least squares.) $\|\mathbf{y} - \mathbf{v}\|^2 \geq \|\mathbf{y} - \mathbf{P}_V\mathbf{y}\|^2$, for all $\mathbf{v} \in V$, with equality only if $\mathbf{v} = \mathbf{P}_V\mathbf{y}$, so that $\mathbf{P}_V\mathbf{y}$ is the unique vector in V which is closest to $\mathbf{y}$.
2. Let $\mathbf{X}$ be a basis matrix and $\mathbf{Q}$ be an orthonormal basis matrix for V. Then

$$\mathbf{P}_V = \mathbf{X}(\mathbf{X}^T\mathbf{X})^{-1}\mathbf{X}^T = \mathbf{Q}\mathbf{Q}^T.$$

   so that $\mathbf{P}_V$ is a symmetric idempotent matrix and $\mathrm{tr}(\mathbf{P}_V) = \mathrm{rank}(\mathbf{P}_V) = \dim(V)$.
3. If $\mathbf{y} \in V$, then $\mathbf{P}_V\mathbf{y} = \mathbf{y}$. If $\mathbf{y} \perp V$, then $\mathbf{P}_V\mathbf{y} = \mathbf{0}$.
4. If $W \subset V$, then $\mathbf{P}_W\mathbf{P}_V = \mathbf{P}_V\mathbf{P}_W = \mathbf{P}_W$.
5. If $W \perp V$, then $\mathbf{P}_W\mathbf{P}_V = \mathbf{P}_V\mathbf{P}_W = \mathbf{0}$.
6. $\mathbf{P}_{V^\perp} = \mathbf{I} - \mathbf{P}_V$, $\|\mathbf{P}_{V^\perp}\mathbf{Y}\|^2 = \|\mathbf{Y} - \mathbf{P}_V\mathbf{Y}\|^2 = \|\mathbf{Y}\|^2 - \|\mathbf{P}_V\mathbf{Y}\|^2$.
7. If $W \subset V$, then $\mathbf{P}_{V|W} = \mathbf{P}_V - \mathbf{P}_W$, $\|\mathbf{P}_{V|W}\mathbf{Y}\|^2 = \|\mathbf{P}_V\mathbf{Y} - \mathbf{P}_W\mathbf{Y}\|^2 = \|\mathbf{P}_V\mathbf{Y}\|^2 - \|\mathbf{P}_W\mathbf{Y}\|^2$.

Properties 1 and 2 are often used to compute projections. Property 3 implies that the projection matrix is singular unless $V = R^n$. Properties 6 and 7 explain why there are so many different formulae for the same statistic in ANOVA models. Note that $\mathbf{P}_V$ is often called the hat matrix. See **32.66**.

Example 30.3 (Some projection matrices)

We continue with Example 30.2 above in which we found that a basis matrix for V is

$$\mathbf{X} = \begin{pmatrix} 1 & -1 \\ 1 & 2 \\ -1 & -1 \end{pmatrix}$$

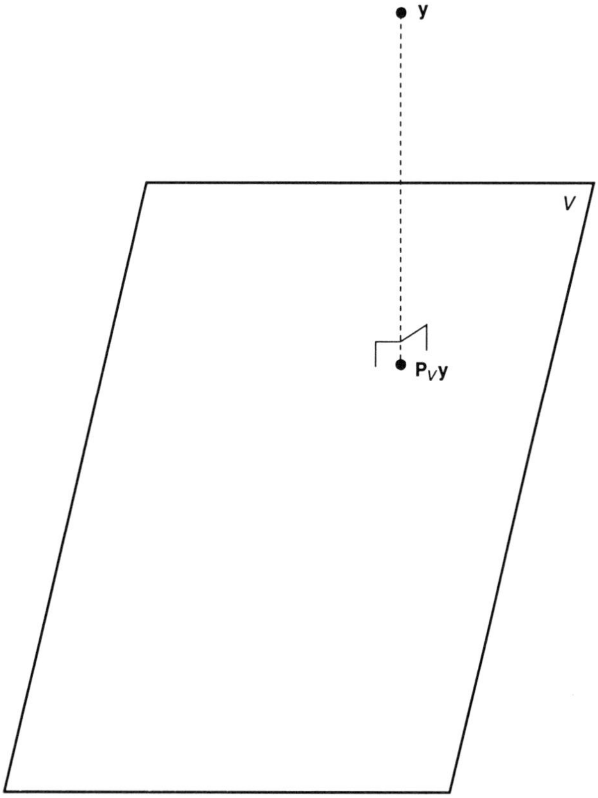

Fig. 30.1 The (orthogonal) projection $\mathbf{P}_V\mathbf{y}$ of $\mathbf{y}$ onto V

and hence

$$\mathbf{P}_V = \mathbf{X}(\mathbf{X}^T\mathbf{X})^{-1}\mathbf{X}^T = \frac{1}{14}\begin{pmatrix} 13 & -2 & -3 \\ -2 & 10 & -6 \\ -3 & -6 & 5 \end{pmatrix}.$$

It is easily verified that $\mathbf{P}_V$ is a symmetric idempotent matrix and that $\text{tr}(\mathbf{P}_V) = 2 = \dim(V)$. Note also that $\mathbf{v}_3 = (1, 2, 3)^T$ is a basis for $V^\perp$ and hence

$$\mathbf{P}_{V^\perp} = \mathbf{v}_3(\mathbf{v}_3^T\mathbf{v}_3)^{-1}\mathbf{v}_3^T = \frac{1}{14}\begin{pmatrix} 1 & 2 & 3 \\ 2 & 4 & 6 \\ 3 & 6 & 9 \end{pmatrix}.$$

It is again easily verified that $\mathbf{P}_{V^\perp}$ is a symmetric idempotent matrix, that $\mathbf{P}_{V^\perp} = \mathbf{I} - \mathbf{P}_V$ and that $\text{tr}(\mathbf{P}_{V^\perp}) = 1 = \dim(V^\perp)$. Now, let W be the subspace of V in which, in addition,

$$(4, 1, -2)\mathbf{v} = 4v_1 + v_2 - 2v_3 = 0.$$

We have seen that $\mathbf{v}_2 = (-1, 2, -1)^T$ is a basis for W. Therefore

$$\mathbf{P}_W = \mathbf{v}_2(\mathbf{v}_2^T\mathbf{v}_2)^{-1}\mathbf{v}_2^T = \frac{1}{6}\begin{pmatrix} 1 & -2 & 1 \\ -2 & 4 & -2 \\ 1 & -2 & 1 \end{pmatrix}$$

Let $\mathbf{v}_4 = (4, 1, -2)^T$. Then we have seen that $\mathbf{v}_4$ is a basis for $V \mid W$ and hence

$$\mathbf{P}_{V|W} = \mathbf{v}_4(\mathbf{v}_4^T\mathbf{v}_4)^{-1}\mathbf{v}_4^T = \frac{1}{21}\begin{pmatrix} 16 & 4 & -8 \\ 4 & 1 & -2 \\ -8 & -2 & 4 \end{pmatrix}$$

Note that $\mathbf{P}_{V|W} = \mathbf{P}_V - \mathbf{P}_W$.

We finish this section with some basic distribution theory for lengths of projections. We write $\mathbf{Y} \sim N_q(\mathbf{a}, \mathbf{B})$ to mean that $\mathbf{Y}$ has a q-dimensional multivariate normal distribution with mean vector $\mathbf{a}$ and covariance matrix $\mathbf{B} > \mathbf{0}$. Let $\mathbf{Y} \sim N_q(\mathbf{a}, \mathbf{I})$, $U = \|\mathbf{Y}\|^2$, $c = \|\mathbf{a}\|^2$. Then U has a χ^2 distribution with q degrees of freedom and non-centrality parameter c, written $U \sim \chi_q^2(c)$. If $c = 0$, then we say U has a central χ^2 distribution with q degrees of freedom and write $U \sim \chi_q^2$. One elementary property of the χ^2 distribution used frequently is that

$$U \sim \chi_q^2(c) \Rightarrow EU = q + c$$

(See **22.4** for a discussion of the χ^2 distribution.)

Theorem 30.1.
Let $\mathbf{Y} \sim N_n(\boldsymbol{\mu}, \sigma^2\mathbf{I})$.
(a) Let V be a p-dimensional subspace of R^n. Then

$$\frac{\|\mathbf{P}_V\mathbf{Y}\|^2}{\sigma^2} \sim \chi_p^2(\delta), \quad \delta = \frac{\|\mathbf{P}_V\boldsymbol{\mu}\|^2}{\sigma^2}.$$

(b) Let W and V be orthogonal subspaces of R^n. Then $\|\mathbf{P}_V\mathbf{Y}\|^2$ and $\|\mathbf{P}_W\mathbf{Y}\|^2$ are independent.

Proof. (a) Let $\mathbf{X}$ be an orthonormal basis matrix for V and let

$$\mathbf{Z} = \sigma^{-1}\mathbf{X}^T\mathbf{Y}, \qquad \boldsymbol{\nu} = \sigma^{-1}\mathbf{X}^T\boldsymbol{\mu}.$$

Note that $\mathbf{Z}$ is $p \times 1$ and $\mathbf{X}^T\mathbf{X} = \mathbf{I}$. By the usual result on linear combinations of normal random variables we see that

$$\mathbf{Z} \sim N_p(\boldsymbol{\nu}, \mathbf{I}) \Rightarrow U = \|\mathbf{Z}\|^2 \sim \chi_p^2(\delta), \quad \delta = \|\boldsymbol{\nu}\|^2.$$

However,

$$U = \mathbf{Z}^T\mathbf{Z} = \frac{\mathbf{Y}^T\mathbf{X}\mathbf{X}^T\mathbf{Y}}{\sigma^2} = \frac{\|\mathbf{P}_V\mathbf{Y}\|^2}{\sigma^2}, \quad \delta = \frac{\|\mathbf{P}_V\boldsymbol{\mu}\|^2}{\sigma^2}.$$

(b) If $W \perp V$ then $\mathbf{P}_V\mathbf{P}_W = \mathbf{0}$. Hence $\mathbf{P}_V\mathbf{Y}$ and $\mathbf{P}_W\mathbf{Y}$ are independent and so therefore are their lengths.

We have said that projection matrices are idempotent. Therefore part (a) of Theorem 30.1 is just a restatement of the well known result on idempotent matrices. However the proof for projections is more elementary, using only the existence of an orthonormal basis matrix rather than properties of eigenvalues. Also the interpretation of the degrees of freedom as the dimension of the subspace is a nice feature of this approach. Furthermore, in Exercise 30.4 it is shown that any idempotent matrix is a projection matrix. This implies that the only quadratic forms which have χ^2 distributions are lengths of projections.

Part (b) of this theorem gives an easily interpretable condition for when χ^2 random variables are independent. In Exercise 30.8 it is shown that $\mathbf{P}_V\mathbf{P}_W = \mathbf{0}$ only if $V \perp W$. This implies that the only χ^2 quadratic forms which are independent are lengths of projections on orthogonal subspaces.

The general linear model – least squares estimators, F tests and Scheffé simultaneous confidence intervals

30.4 In the *general linear model*, we observe

$$\mathbf{Y} \sim N_N(\boldsymbol{\mu}, \sigma^2\mathbf{I}), \qquad \boldsymbol{\mu} \in V,\ \sigma^2 > 0,$$

where $\boldsymbol{\mu}$ and σ^2 are unknown parameters and V is a known p-dimensional subspace of R^N. This model implies that the components of $\mathbf{Y}$ are independently normally distributed with equal variances σ^2 but with possibly different means. As we shall see, multiple regression, fixed effects ANOVA and analysis of covariance are special cases of this model for particular choices of the subspace V.

Define

$$\widehat{\boldsymbol{\mu}} = \mathbf{P}_V\mathbf{Y}, \widehat{\sigma}^2 = \frac{\|\mathbf{P}_{V^\perp}\mathbf{Y}\|^2}{N-p}.$$

Since $\widehat{\boldsymbol{\mu}}$ is the vector in V which minimizes $Q(\boldsymbol{\mu}) = \|\mathbf{Y} - \boldsymbol{\mu}\|^2$, we call $\widehat{\boldsymbol{\mu}}$ the *least squares estimator* of $\boldsymbol{\mu}$. In the following theorem, $\mathbf{A}$ is an arbitrary matrix with n columns.

Theorem 30.2.

(a) $(\widehat{\boldsymbol{\mu}}, \widehat{\sigma}^2)$ is a complete sufficient statistic.

(b) $\mathbf{A}\widehat{\boldsymbol{\mu}}$ and $\widehat{\sigma}^2$ are the minimum variance unbiased estimators of $\mathbf{A}\boldsymbol{\mu}$ and σ^2.

(c) $\mathbf{A}\widehat{\boldsymbol{\mu}}$ and $(N-p)\widehat{\sigma}^2/N$ are the maximum likelihood estimators of $\mathbf{A}\boldsymbol{\mu}$ and σ^2.

Proof. (a) The likelihood function for this model is

$$L_{\mathbf{Y}}(\boldsymbol{\mu}, \sigma^2) \approx \sigma^{-N} \exp\left(\frac{-\|\mathbf{Y} - \boldsymbol{\mu}\|^2}{2\sigma^2}\right) = \sigma^{-N} \exp\left(\frac{(N-p)\widehat{\sigma}^2 + \|\widehat{\boldsymbol{\mu}} - \boldsymbol{\mu}\|^2}{2\sigma^2}\right).$$

By the factorization criterion $(\widehat{\boldsymbol{\mu}}, \widehat{\sigma}^2)$ is sufficient. To show completeness, it is necessary to coordinatize the subspace. The details are outlined in Exercise 30.12.

(b) Note that $\boldsymbol{\mu} \in V$. Therefore

$$E\mathbf{A}\widehat{\boldsymbol{\mu}} = \mathbf{A}\mathbf{P}_V E\mathbf{Y} = \mathbf{A}\mathbf{P}_V\boldsymbol{\mu} = \mathbf{A}\boldsymbol{\mu}.$$

Hence by the Lehmann–Scheffé theorem, $\mathbf{A}\widehat{\boldsymbol{\mu}}$ is the MV unbiased estimator of $\mathbf{A}\boldsymbol{\mu}$. Furthermore,

$$E\widehat{\sigma}^2 = \frac{E\mathbf{Y}^T\mathbf{P}_{V^\perp}\mathbf{Y}}{N-p} = \frac{\sigma^2\operatorname{tr}(\mathbf{P}_{V^\perp}) + \boldsymbol{\mu}^T\mathbf{P}_{V^\perp}\boldsymbol{\mu}}{N-p}.$$

However,

$$\boldsymbol{\mu} \in V \perp V^\perp \Rightarrow \mathbf{P}_{V^\perp}\boldsymbol{\mu} = 0,$$
$$\operatorname{tr}(\mathbf{P}_{V^\perp}) = \dim(V^\perp) = N - \dim(V) = N - p.$$

Therefore, $\widehat{\sigma}^2$ is an unbiased estimator of σ^2 and by the Lehmann–Scheffé theorem is the MV unbiased estimator of σ^2.

(c) Since $\widehat{\boldsymbol{\mu}} = \mathbf{P}_V\mathbf{Y}$ is the point in V closest to $\mathbf{Y}$, we see that

$$\|\mathbf{Y} - \widehat{\boldsymbol{\mu}}\|^2 \leq \|\mathbf{Y} - \boldsymbol{\mu}\|^2$$

for all $\boldsymbol{\mu} \in V$ (with equality only if $\boldsymbol{\mu} = \widehat{\boldsymbol{\mu}}$). Hence $\widehat{\boldsymbol{\mu}}$ is the ML estimator for $\boldsymbol{\mu}$. The ML estimator for σ^2 follows directly from differentiation of the log likelihood. The ML estimator for $\mathbf{A}\boldsymbol{\mu}$ follows from the invariance principle for ML estimators.

Note that the proof of unbiasedness given in part (b) does not use the normal assumption. These estimators are unbiased as long as $\boldsymbol{\mu} \in V$ and $\operatorname{cov}(\mathbf{Y}) = \sigma^2\mathbf{I}$. (In fact, the Gauss–Markov theorem says that $\mathbf{A}\widehat{\boldsymbol{\mu}}$ is the best linear unbiased estimator of $\mathbf{A}\boldsymbol{\mu}$ even for non-normal distributions as long as $E\mathbf{Y} = \boldsymbol{\mu}$ and $\operatorname{cov}(\mathbf{Y}) = \sigma^2\mathbf{I}$. See Exercise 30.15 for a proof.)

In Exercise 30.14, it is shown that the unbiased estimator $\widehat{\sigma}^2$ is a consistent estimator of σ^2, but that the ML estimator for σ^2 is not consistent unless $p/N \to 0$. (Note that many scientists have rules in which they have some fixed number of observations, perhaps 20, per parameter. In this case, $p/N \to 1/20 \neq 0$.)

Example 30.4 (Multiple regression: sufficiency and estimation)

In the multiple regression model, we observe

$$\mathbf{Y} \sim N_N(\mathbf{X}\boldsymbol{\beta}, \sigma^2\mathbf{I}),$$

where $\mathbf{X}$ is a known $N \times p$ matrix of rank p and $\boldsymbol{\beta} \in R^p$, $\sigma^2 > 0$ are unknown parameters. We have seen in **29.5** that a sufficient statistic for this model is $(\widehat{\boldsymbol{\beta}}, \widehat{\sigma}_R^2)$, where

$$\widehat{\boldsymbol{\beta}} = (\mathbf{X}^T\mathbf{X})^{-1}\mathbf{X}^T\mathbf{Y}, \qquad \widehat{\sigma}_R^2 = \frac{\|\mathbf{Y} - \mathbf{X}\widehat{\boldsymbol{\beta}}\|^2}{N-p}.$$

Now, let $\boldsymbol{\mu} = \mathbf{X}\boldsymbol{\beta}$ and let V be the p-dimensional subspace spanned by the columns of $\mathbf{X}$ (so that $\mathbf{X}$ is a basis matrix for V). An equivalent version of this model is the linear model

$$\mathbf{Y} \sim N_n(\boldsymbol{\mu}, \sigma^2\mathbf{I}), \qquad \boldsymbol{\mu} \in V,\ \sigma^2 > 0.$$

Note that

$$\boldsymbol{\beta} = (\mathbf{X}^T\mathbf{X})^{-1}\mathbf{X}^T\mathbf{X}\boldsymbol{\beta} = (\mathbf{X}^T\mathbf{X})^{-1}\mathbf{X}^T\boldsymbol{\mu},$$

so that $\boldsymbol{\mu}$ is an invertible function of $\boldsymbol{\beta}$. Further,

$$\widehat{\boldsymbol{\mu}} = \mathbf{P}_V\mathbf{Y} = \mathbf{X}(\mathbf{X}^T\mathbf{X})^{-1}\mathbf{X}^T\mathbf{Y} = \mathbf{X}\widehat{\boldsymbol{\beta}}, \quad \widehat{\boldsymbol{\beta}} = (\mathbf{X}^T\mathbf{X})^{-1}\mathbf{X}^T\mathbf{X}\widehat{\boldsymbol{\beta}} = (\mathbf{X}^T\mathbf{X})^{-1}\mathbf{X}^T\widehat{\boldsymbol{\mu}},$$

so that $\widehat{\boldsymbol{\mu}}$ is an invertible function of $\widehat{\boldsymbol{\beta}}$. Furthermore, the theorem above implies that $\widehat{\boldsymbol{\beta}}$ is the MV unbiased and ML estimator of $\boldsymbol{\beta}$. (The Gauss–Markov theorem implies that $\widehat{\boldsymbol{\beta}} = (\mathbf{X}^T\mathbf{X})^{-1}\mathbf{X}^T\widehat{\boldsymbol{\mu}}$ is the best linear unbiased estimator of $\boldsymbol{\beta} = (\mathbf{X}^T\mathbf{X})^{-1}\mathbf{X}\boldsymbol{\mu}$ even for non-normal observations; see also **29.5–6**.) In addition

$$(N-p)\widehat{\sigma}_R^2 = \|\mathbf{Y} - \mathbf{X}\widehat{\boldsymbol{\beta}}\|^2 = \|\mathbf{Y} - \mathbf{P}_V\mathbf{Y}\|^2 = \|\mathbf{P}_{V^\perp}\mathbf{Y}\|^2 = (n-p)\widehat{\sigma}^2,$$

so that $\widehat{\sigma}_R^2 = \widehat{\sigma}^2$ is the same for both versions of the model.

30.5 The problem of *testing the general linear hypothesis* is the problem of testing the null hypothesis that $\boldsymbol{\mu} \in W$, $\sigma^2 > 0$ against the alternative hypothesis that $\boldsymbol{\mu} \in V$, $\sigma^2 > 0$, where $W \subset V$ are subspaces of R^N such that

$$k = \dim(W) < p = \dim(V) < N.$$

Let

$$\begin{aligned} SSH &= \|\mathbf{P}_{V|W}\mathbf{Y}\|^2, \quad dfh = \dim(V|W) = p - k, \\ SSE &= \|\mathbf{P}_{V^\perp}\mathbf{Y}\|^2, \quad dfe = \dim(V^\perp) = N - p, \\ MSH &= \frac{SSH}{dfh}, \quad MSE = \frac{SSE}{dfe}, \quad F = \frac{MSH}{MSE}. \end{aligned}$$

We call *SSH* and *SSE* the *sum of squares for the hypothesis* and the *sum of squares for error.* We call *dfh* and *dfe* the *degrees of freedom for the hypothesis* and the *degrees of freedom for error.* We call *MSH* and *MSE* the *mean square for the hypothesis* and the *mean square for error.* We call F the F *statistic* for this testing problem. Note that

$$MSE = \widehat{\sigma}^2$$

and these two notations will be used interchangeably in future sections.

We call the null model in which $\boldsymbol{\mu} \in W$ the *reduced* model and call the alternative model in which $\boldsymbol{\mu} \in V$ the *full* model. Note that both these models are linear models as described above. Note that the sum of squares for error for the reduced model is just $\|\mathbf{P}_{W^\perp}\mathbf{Y}\|^2$, and that

$$SSH = \|P_{V|W}\mathbf{Y}\|^2 = \|P_{W^\perp|V^\perp}\mathbf{Y}\|^2 = \|\mathbf{P}_{W^\perp}\mathbf{Y}\|^2 - \|\mathbf{P}_{V^\perp}\mathbf{Y}\|$$

is *SSE* for the reduced model minus *SSE* for the full model. This formula is often used for deriving *SSH* (see Scheffé, 1959).

We now describe some other useful formulae for *SSH*. Let $\widehat{\boldsymbol{\mu}} = \mathbf{P}_V\mathbf{Y}$ and $\widehat{\widehat{\boldsymbol{\mu}}} = \mathbf{P}_W\mathbf{Y}$ be the LS estimators of $\boldsymbol{\mu}$ under the full and reduced models. Then

$$\begin{aligned} SSH &= \|\mathbf{P}_{V|W}\mathbf{Y}\|^2 = \|\mathbf{P}_V\mathbf{Y} - \mathbf{P}_W\mathbf{Y}\|^2 = \|\widehat{\boldsymbol{\mu}} - \widehat{\widehat{\boldsymbol{\mu}}}\|^2 \\ &= \|\mathbf{P}_V\mathbf{Y}\|^2 - \|\mathbf{P}_W\mathbf{Y}\|^2 = \|\widehat{\boldsymbol{\mu}}\|^2 - \|\widehat{\widehat{\boldsymbol{\mu}}}\|^2. \end{aligned}$$

Before stating the basic theorem on testing the general linear hypothesis, we recall the following definition. Let U and V be independent random variables, $U \sim \chi_k^2(\delta)$, $V \sim \chi_m^2$ and let

$$F = \frac{U/k}{V/m}.$$

Then we say that F has an F distribution with k and m degrees of freedom and non-centrality parameter δ and write $F \sim F_{k,m}(\delta)$. If $\delta = 0$, we say that F has a central F distribution and write $F \sim F_{k,m}$. We write $F_{k,n}^{\alpha}$ for the upper α critical point of the central F distribution with k and m degrees of freedom. When the degrees of freedom should be clear we often abbreviate this expression to F^{α}. See Section **29.26–29** for a discussion of this distribution.

Theorem 30.3.
(a) Let $\xi = \|\mathbf{P}_{V|W}\boldsymbol{\mu}\|^2$. *Then*

$$F \sim F_{dfh,dfe}(\xi/\sigma^2).$$

and hence, under the null hypothesis $F \sim F_{dfh,dfe}$.
(b) Also

$$EMSH = \sigma^2 + \frac{\xi}{dfh}, \qquad EMSE = \sigma^2.$$

(c) The size-α likelihood ratio test of the general linear hypothesis rejects if

$$F > F_{dfh,dfe}^{\alpha}.$$

Proof. (a) Recall that $SSH = \|\mathbf{P}_{V|W}\mathbf{Y}\|^2, \quad SSE = \|\mathbf{P}_{V^\perp}\mathbf{Y}\|^2$. Now

$$V|W \subset V \perp V^{\perp}.$$

Therefore, by Theorem 30.1, SSH and SSE are independent. Note also that

$$\|\mathbf{P}_{V^\perp}\boldsymbol{\mu}\|^2 = 0, \quad \dim(V^{\perp}) = dfe.$$

By Theorem 30.1 again, $SSE/\sigma^2 \sim \chi_{dfe}^2(0)$. Now $\dim(V|W) = dfh$. Therefore, by Theorem 30.1 yet again, $SSH/\sigma^2 \sim \chi_{dfh}^2(\xi/\sigma^2)$ and hence $F \sim F_{dfh,dfe}(\xi/\sigma^2)$.

(b) Note also that

$$EMSH = \frac{\sigma^2}{dfh}\left(\frac{E(SSH)}{\sigma^2}\right) = \frac{\sigma^2}{dfh}\left(dfh + \frac{\xi}{\sigma^2}\right) = \sigma^2 + \frac{\xi}{dfh},$$
$$EMSE = \frac{\sigma^2}{dfe}\left(\frac{E(SSE)}{\sigma^2}\right) = \frac{\sigma^2}{dfe}(dfe) = \sigma^2.$$

(c) The ML estimators for $\boldsymbol{\mu}$ and σ^2 under the alternative hypothesis $\boldsymbol{\mu} \in V$ are

$$\mathbf{P}_V\mathbf{Y} \text{ and } \frac{(N-p)\widehat{\sigma}^2}{N} = \frac{SSE}{N}.$$

Under the null hypothesis that $\boldsymbol{\mu} \in W$, the ML estimators for $\boldsymbol{\mu}$ and σ^2 are

$$\mathbf{P}_W\mathbf{Y} \text{ and } \frac{\|\mathbf{P}_{W^\perp}\mathbf{Y}\|^2}{N} = \frac{SSH + SSE}{N}.$$

Therefore, the likelihood ratio test statistic is

$$\Lambda = \frac{L_{\mathbf{Y}}\left(\mathbf{P}_W\mathbf{Y}, \frac{SSE+SSH}{N}\right)}{L_{\mathbf{Y}}\left(\mathbf{P}_V\mathbf{Y}, \frac{SSE}{N}\right)} = \left(\frac{SSE + SSH}{SSE}\right)^{-\frac{N}{2}} = \left(1 + \frac{dfh\, F}{dfe}\right)^{-\frac{N}{2}}.$$

Hence

$$\Lambda < k \Leftrightarrow F > k^*$$

for some k^* so that the LR test rejects if F is too large. The size-α critical value follows from part (a).

For the remainder of this chapter, when we talk about the size-α F test for testing a linear hypothesis we mean the test which rejects if

$$F > F^{\alpha}_{dfh,dfe}.$$

In Exercise 30.18, it is shown that power function of the F test is an increasing function of the non-centrality parameter and hence that the F test is unbiased. In **23.31**, the canonical form for testing the general linear hypothesis is used to show that the F test for testing the general linear hypothesis is the UMP invariant size-α test for this problem. The test is also known to be admissible (see Lehmann and Stein, 1953; Ghosh, 1964) and to possess certain minimaxity properties (see Lehmann, 1986, pp. 519–527). Sensitivity of this test to its assumptions is discussed in **30.38**. Its power function is discussed in **29.28–29**.

Example 30.5 (Multiple regression: testing)

We return to the example of multiple regression. Consider testing the null hypothesis that $\mathbf{A}\boldsymbol{\beta} = \mathbf{0}$ against the unrestricted alternative $\boldsymbol{\beta} \in R^p$, where $\mathbf{A}$ is a $q \times p$ matrix of rank q. Note that

$$\mathbf{A}\boldsymbol{\beta} = \mathbf{C}^T\boldsymbol{\mu}, \mathbf{C} = \mathbf{X}(\mathbf{X}^T\mathbf{X})^{-1}\mathbf{A}^T.$$

Therefore, let

$$W = \{\boldsymbol{\mu} \in V : \mathbf{C}^T\boldsymbol{\mu} = \mathbf{0}\}.$$

We are testing the null hypothesis that $\boldsymbol{\mu} \in W$ against the alternative hypothesis that $\boldsymbol{\mu} \in V$, so that this is a special case of testing the general linear hypothesis. In Exercise 30.17, it is shown that $\mathbf{C}$ is a basis matrix for $V|W$ (and hence $\dim(V|W) = q$). Therefore,

$$\begin{aligned} SSH &= \|\mathbf{P}_{V|W}\widehat{\boldsymbol{\mu}}\|^2 = \widehat{\boldsymbol{\mu}}^T\mathbf{C}(\mathbf{C}^T\mathbf{C})^{-1}\mathbf{C}^T\widehat{\boldsymbol{\mu}} \\ &= (\mathbf{A}\widehat{\boldsymbol{\beta}})^T(\mathbf{A}(\mathbf{X}^T\mathbf{X})^{-1}\mathbf{A}^T)^{-1}(\mathbf{A}\widehat{\boldsymbol{\beta}}), \qquad dfh = q. \end{aligned}$$

As always, $MSE = \widehat{\sigma}^2$, $dfe = N - p$. Hence

$$F = \frac{(\mathbf{A}\widehat{\boldsymbol{\beta}})^T(\mathbf{A}(\mathbf{X}^T\mathbf{X})^{-1}\mathbf{A})^{-1}(\mathbf{A}\widehat{\boldsymbol{\beta}})}{q\widehat{\sigma}^2} \sim F_{q,N-p}(\xi/\sigma^2),$$

where

$$\xi = \|\mathbf{P}_{V|W}\boldsymbol{\mu}\|^2 = (\mathbf{A}\boldsymbol{\beta})^T(\mathbf{A}(\mathbf{X}^T\mathbf{X})^{-1}\mathbf{A}^T)^{-1}(\mathbf{A}\boldsymbol{\beta}).$$

Of course, F is the same F statistic derived for this hypothesis in Chapter 29.

30.6 Now, let

$$\boldsymbol{\delta} = \mathbf{P}_{V|W}\boldsymbol{\mu}, \qquad \widehat{\boldsymbol{\delta}} = \mathbf{P}_{V|W}\widehat{\boldsymbol{\mu}} = \mathbf{P}_{V|W}\mathbf{Y}.$$

(Note that $V|W \subset V$ and hence $\mathbf{P}_{V|W}\mathbf{P}_V = \mathbf{P}_{V|W}$.) Then

$$\boldsymbol{\mu} \in W \Leftrightarrow \boldsymbol{\delta} = \mathbf{0}, \text{ and } F = \frac{\|\widehat{\boldsymbol{\delta}}\|^2}{dfh\,\widehat{\sigma}^2},$$

so the F test given above is a sensible test.

We now find a pivotal quantity and a confidence region for $\boldsymbol{\delta}$. Define

$$F(\boldsymbol{\delta}) = \frac{\|\widehat{\boldsymbol{\delta}} - \boldsymbol{\delta}\|^2}{dfh\,\widehat{\sigma}^2}.$$

(Note that the F statistic defined in the last section satisfies $F = F(\mathbf{0})$.)

Theorem 30.4.

(a) $F(\boldsymbol{\delta})$ is a pivotal quantity for $\boldsymbol{\delta}$,

$$F(\boldsymbol{\delta}) \sim F_{dfh,dfe}.$$

(b) Also

$$F(\boldsymbol{\delta}) \le F^{\alpha} \Leftrightarrow \mathbf{d}^T\boldsymbol{\delta} \in \mathbf{d}^T\widehat{\boldsymbol{\delta}} \pm (dfh\,MSE\|\mathbf{d}\|^2F^{\alpha})^{1/2}\ \forall \mathbf{d} \in V|W.$$
$$F(0) \le F^{\alpha} \Leftrightarrow 0 \in \mathbf{d}^T\widehat{\boldsymbol{\delta}} \pm (dfh\,MSE\|\mathbf{d}\|^2F^{\alpha})^{1/2}\ \forall \mathbf{d} \in V|W.$$

Proof. (a) This proof is the same as in the previous theorem. Note that

$$\|\widehat{\boldsymbol{\delta}} - \boldsymbol{\delta}\|^2 = \|\mathbf{P}_{V|W}(\mathbf{Y} - \boldsymbol{\mu})\|^2, \quad SSE = \|\mathbf{P}_{V^\perp}(\mathbf{Y} - \boldsymbol{\mu})\|^2$$

and $\mathbf{Y} - \boldsymbol{\mu} \sim N_n(\mathbf{0}, \sigma^2\mathbf{I})$.

(b) We use the fact that

$$\|\widehat{\boldsymbol{\delta}} - \boldsymbol{\delta}\|^2 = \max_{\mathbf{d}\in V|W, \mathbf{d}\neq\mathbf{0}} \frac{(\mathbf{d}^T(\widehat{\boldsymbol{\delta}} - \boldsymbol{\delta}))^2}{\|\mathbf{d}\|^2}$$

(see Exercise 30.19). Therefore

$$F(\boldsymbol{\delta}) \le F^{\alpha} \Leftrightarrow \|\widehat{\boldsymbol{\delta}} - \boldsymbol{\delta}\|^2 \le dfh\,MSE\,F^{\alpha} \Leftrightarrow$$
$$(\mathbf{d}^T(\widehat{\boldsymbol{\delta}} - \boldsymbol{\delta}))^2 \le dfh\,MSE\,\|\mathbf{d}\|^2F^{\alpha}\ \forall \mathbf{d} \in V|W \Leftrightarrow$$
$$\mathbf{d}^T\boldsymbol{\delta} \in \mathbf{d}^T\widehat{\boldsymbol{\delta}} \pm (dfh\,MSE\|\mathbf{d}\|^2F^{\alpha})^{1/2}\ \forall \mathbf{d} \in V|W.$$

(Note that when $\mathbf{d} = \mathbf{0}$ the last two inequalities are trivially satisfied.) Substituting $\boldsymbol{\delta} = \mathbf{0}$, we see that

$$F(\mathbf{0}) \leq F^\alpha \Leftrightarrow 0 = \mathbf{d}^T\mathbf{0} \in \mathbf{d}^T\widehat{\boldsymbol{\delta}} \pm (dfh\,MSE\|\mathbf{d}\|^2 F^\alpha)^{1/2}, \ \forall \mathbf{d} \in V|W.$$

We define a *contrast* associated with testing the null hypothesis that $\boldsymbol{\mu} \in W$ against the alternative hypothesis that $\boldsymbol{\mu} \in V$ to be a function of the form $\mathbf{d}^T\boldsymbol{\delta}, \mathbf{d} \in V|W$. Its associated *estimated contrast* is $\mathbf{d}^T\widehat{\boldsymbol{\delta}}$. By this theorem, we see that a $1-\alpha$ confidence region for $\boldsymbol{\delta}$ is

$$\{\boldsymbol{\delta} : F(\boldsymbol{\delta}) \leq F^\alpha\} = \{\boldsymbol{\delta} : \mathbf{d}^T\boldsymbol{\delta} \in \mathbf{d}^T\widehat{\boldsymbol{\delta}} \pm (dfh\,MSE\|\mathbf{d}\|^2 F^\alpha)^{1/2}, \ \forall \mathbf{d} \in V|W\}.$$

Therefore, a set of simultaneous $1-\alpha$ confidence intervals for the set of all contrasts is

$$\mathbf{d}^T\boldsymbol{\delta} \in \mathbf{d}^T\widehat{\boldsymbol{\delta}} \pm (dfh\,MSE\|\mathbf{d}\|^2 F^\alpha)^{1/2}, \ \forall \mathbf{d} \in V|W.$$

Note that $\mathbf{d}^T\boldsymbol{\delta} = \mathbf{d}^T\boldsymbol{\mu}, \mathbf{d}^T\widehat{\boldsymbol{\delta}} = \mathbf{d}^T\widehat{\boldsymbol{\mu}}$ (see Exercise 30.20), so that an equivalent version of these intervals is

$$\mathbf{d}^T\boldsymbol{\mu} \in \mathbf{d}^T\widehat{\boldsymbol{\mu}} \pm (dfh\,MSE\|\mathbf{d}\|^2 F^\alpha)^{1/2}, \ \forall \mathbf{d} \in V|W.$$

These intervals were first derived in Scheffé (1953) and are called the *Scheffé simultaneous confidence intervals* for the set of all contrasts. Note that these simultaneous $1-\alpha$ confidence intervals together are just a way of representing the $1-\alpha$ confidence region $\{\boldsymbol{\delta} : F(\boldsymbol{\delta}) \leq F^\alpha\}$. We call this region the confidence region *associated with* the Scheffé simultaneous confidence intervals.

We say that a particular contrast $\mathbf{d}^T\boldsymbol{\delta}$ is *significant (at size α)* if its Scheffé simultaneous confidence interval does not contain 0. The second identity in part (b) of this theorem implies that we reject the null hypothesis with the F test if and only if at least one contrast is significant. After rejecting with the F test, we often follow up by searching for significant contrasts. We think of such contrast as the ones that are causing the hypothesis to be rejected.

In **23.31**, the canonical form for the problem of testing the general linear hypothesis is used to show that the confidence region for $\boldsymbol{\delta}$ associated with the Scheffé simultaneous confidence intervals is the UMA equivariant $1-\alpha$ confidence region for $\boldsymbol{\delta}$ and is hence unbiased. This region is also known to be admissible (see Arnold, 1984).

Example 30.6 (Multiple regression: simultaneous confidence intervals)

We return to multiple regression and find the Scheffé simultaneous confidence intervals associated with testing that $\mathbf{A}\boldsymbol{\beta} = \mathbf{0}$. Let $\boldsymbol{\gamma} = \mathbf{A}\boldsymbol{\beta}, \widehat{\boldsymbol{\gamma}} = \mathbf{A}\widehat{\boldsymbol{\beta}}$. We have seen that $\mathbf{C} = \mathbf{X}(\mathbf{X}^T\mathbf{X})^{-1}\mathbf{A}^T$ is a basis matrix for $V|W$ and that $\mathbf{C}^T\boldsymbol{\mu} = \boldsymbol{\gamma}$ and $\mathbf{C}^T\widehat{\boldsymbol{\mu}} = \widehat{\boldsymbol{\gamma}}$. Therefore, $\mathbf{d} \in V|W$ if and only if $\mathbf{d} = \mathbf{C}\mathbf{s}$ for some $\mathbf{s} \in R^q$. Hence the contrasts and estimated contrasts have the form

$$\mathbf{d}^T\boldsymbol{\mu} = \mathbf{s}^T\mathbf{C}^T\boldsymbol{\mu} = \mathbf{s}^T\boldsymbol{\gamma}, \quad \mathbf{d}^T\widehat{\boldsymbol{\mu}} = \mathbf{s}^T\widehat{\boldsymbol{\gamma}}, \quad \|\mathbf{d}\|^2 = \mathbf{s}^T\mathbf{C}^T\mathbf{C}\mathbf{s} = \mathbf{s}^T\mathbf{A}(\mathbf{X}^T\mathbf{X})^{-1}\mathbf{A}^T\mathbf{s}.$$

and the Scheffé simultaneous confidence intervals are given by

$$\mathbf{s}^T\boldsymbol{\gamma} \in \mathbf{s}^T\widehat{\boldsymbol{\gamma}} \pm (q\mathbf{s}^T\mathbf{A}(\mathbf{X}^T\mathbf{X})^{-1}\mathbf{A}\mathbf{s}\widehat{\sigma}^2 F^\alpha)^{1/2}, \ \forall \mathbf{s} \in R^q.$$

If $\mathbf{A} = \mathbf{I}$, then $q = p, \boldsymbol{\gamma} = \boldsymbol{\beta}, \widehat{\boldsymbol{\gamma}} = \widehat{\boldsymbol{\beta}}$ and the intervals reduce to

$$\mathbf{s}^T\boldsymbol{\beta} \in \mathbf{s}^T\widehat{\boldsymbol{\beta}} \pm (p\mathbf{s}^T(\mathbf{X}^T\mathbf{X})^{-1}\mathbf{s}\widehat{\sigma}^2 F^\alpha)^{1/2}, \ \forall \mathbf{s} \in R^p,$$

the usual confidence band for the response surface.

Generalized inverses

30.7 As mentioned above, multiple regression and the general linear model are two different ways of writing the same model. Another approach to linear models is the generalized inverse approach indicated in the following example (see also **29.10**).

Example 30.7 (Multiple regression with a singular **X** *matrix)*

Now suppose we observe $\mathbf{Y} \sim N_N(\mathbf{X}\boldsymbol{\beta}, \sigma^2\mathbf{I})$, where $\mathbf{X}$ is a known $N \times r$ matrix of rank $p < r$ and $\sigma^2 > 0$ and $\boldsymbol{\beta} \in R^r$ are unknown parameters. (Note that if $p = r$ then this is just the multiple regression model discussed above.) Let $(\mathbf{X}^T\mathbf{X})^-$ be a symmetric generalized inverse of $\mathbf{X}^T\mathbf{X}$, i.e. a symmetric matrix which satisfies

$$\mathbf{X}^T\mathbf{X}(\mathbf{X}^T\mathbf{X})^-\mathbf{X}^T\mathbf{X} = \mathbf{X}^T\mathbf{X}.$$

Such a matrix always exists but is not unique when $p < r$ (see Exercise 30.21). (The well known Moore–Penrose generalized inverse is a particular symmetric generalized inverse, but does not seem to have any advantage over other symmetric generalized inverses for this problem. See Graybill (1976, pp. 23–34) or Searle (1971, pp. 1–31), for more properties of generalized inverses.) We say that an estimator $\widehat{\boldsymbol{\beta}}$ is a *least squares* estimator of $\boldsymbol{\beta}$ if

$$\|\mathbf{Y} - \mathbf{X}\widehat{\boldsymbol{\beta}}\|^2 \le \|\mathbf{Y} - \mathbf{X}\mathbf{b}\|^2, \qquad \forall \mathbf{b} \in R^r.$$

The LS estimator of $\boldsymbol{\beta}$ is not unique. In Exercise 30.22 it is shown an estimator $\widehat{\boldsymbol{\beta}}$ is an LS estimator if and only if $\widehat{\boldsymbol{\beta}}$ satisfies the normal equations

$$\mathbf{X}^T\mathbf{X}\widehat{\boldsymbol{\beta}} = \mathbf{X}^T\mathbf{Y},$$

A particular LS estimator is

$$\widehat{\boldsymbol{\beta}} = (\mathbf{X}^T\mathbf{X})^-\mathbf{X}^T\mathbf{Y},$$

but not all LS estimators have this form. Let

$$\boldsymbol{\mu} = \mathbf{X}\boldsymbol{\beta}, \quad V = \{\boldsymbol{\mu} = \mathbf{X}\boldsymbol{\beta};\ \boldsymbol{\beta} \in R^r\}.$$

That is, V is the p-dimensional subspace spanned by the columns of $\mathbf{X}$. We call $\mathbf{X}$ a *spanning* matrix for V. (Note that $\mathbf{X}$ is not a basis matrix because $p < r$.) Then

$$\mathbf{Y} \sim N_N(\boldsymbol{\mu}, \sigma^2\mathbf{I}), \qquad \boldsymbol{\mu} \in V,\ \sigma^2 > 0,$$

so that this is a linear model. As usual, let $\widehat{\boldsymbol{\mu}} = \mathbf{P}_V\mathbf{Y}$ and $\widehat{\sigma}^2 = \|\mathbf{Y} - \widehat{\boldsymbol{\mu}}\|^2/(N-p)$. (Note that $\widehat{\boldsymbol{\mu}}$ and $\widehat{\sigma}^2$ are unique.) In Exercise 30.22, it is shown that

$$\widehat{\boldsymbol{\mu}} = \mathbf{X}(\mathbf{X}^T\mathbf{X})^-\mathbf{X}^T\mathbf{Y}$$

(the obvious generalization from the full-rank case) and that $\widehat{\boldsymbol{\beta}}$ is an LS estimator of $\boldsymbol{\beta}$ if and only if

$$\mathbf{X}\widehat{\boldsymbol{\beta}} = \widehat{\boldsymbol{\mu}} \Rightarrow \widehat{\sigma}^2 = \frac{\|\mathbf{Y} - \mathbf{P}_V\mathbf{Y}\|^2}{N-p} = \frac{\|\mathbf{Y} - \mathbf{X}\widehat{\boldsymbol{\beta}}\|^2}{N-p}.$$

We say that a linear function $\mathbf{c}^T\boldsymbol{\beta}$ is *estimable* if there is a linear unbiased estimator $\mathbf{a}^T\mathbf{Y}$ of $\mathbf{c}^T\boldsymbol{\beta}$ (if and only if there exists $\mathbf{a} \in R^N$ such that $\mathbf{c}^T\boldsymbol{\beta} = \mathbf{a}^T\boldsymbol{\mu}$). Now let $\boldsymbol{\gamma} = \mathbf{H}\boldsymbol{\beta}$ be a $(p-k)$-dimensional vector of linearly independent estimable functions let $\widehat{\boldsymbol{\gamma}} = \mathbf{H}\widehat{\boldsymbol{\beta}}$. (Note that $\boldsymbol{\gamma}$ and $\widehat{\boldsymbol{\gamma}}$ are uniquely defined even though $\boldsymbol{\beta}$ and $\widehat{\boldsymbol{\beta}}$ are not). Consider testing that $\boldsymbol{\gamma} = \mathbf{0}$. Let

$$W = \{\boldsymbol{\mu} = \mathbf{X}\boldsymbol{\beta};\ \boldsymbol{\beta} \in R^r,\ \mathbf{H}\boldsymbol{\beta} = \mathbf{0}\}$$

so that we are testing the null hypothesis that $\boldsymbol{\mu} \in W$ against the alternative hypothesis that $\boldsymbol{\mu} \in V$, i.e. testing a case of the general linear hypothesis. In Exercise 30.22 it is shown that

$$\mathbf{H}\boldsymbol{\beta} = \mathbf{C}^T\boldsymbol{\mu},\ \mathbf{C} = \mathbf{X}(\mathbf{X}^T\mathbf{X})^{-}\mathbf{H}$$

and that $\mathbf{C}$ is a basis matrix for $V|W$. Hence $\dim(V|W) = p - k$ and

$$F = \frac{\|\mathbf{P}_{V|W}\mathbf{Y}\|^2}{(p-k)\widehat{\sigma}^2} = \frac{\mathbf{Y}^T\mathbf{C}(\mathbf{C}^T\mathbf{C})^{-1}\mathbf{C}^T\mathbf{Y}}{(p-k)\widehat{\sigma}^2} = \frac{\widehat{\boldsymbol{\gamma}}^T(\mathbf{H}(\mathbf{X}^T\mathbf{X})^{-}\mathbf{H})^{-1}\widehat{\boldsymbol{\gamma}}}{(p-k)\widehat{\sigma}^2} \sim F_{p-k,N-p}(\xi/\sigma^2),$$
$$\xi = \boldsymbol{\gamma}^T(\mathbf{H}(\mathbf{X}^T\mathbf{X})^{-}\mathbf{H})^{-1}\boldsymbol{\gamma}.$$

Finally, $\mathbf{C}$ is a basis matrix for $V|W$, and

$$\mathbf{s}^T\mathbf{C}^T\boldsymbol{\mu} = \mathbf{s}^T\boldsymbol{\gamma},\ \mathbf{s}^T\mathbf{C}^T\widehat{\boldsymbol{\mu}} = \mathbf{s}^T\widehat{\boldsymbol{\gamma}},\ \|\mathbf{C}\mathbf{s}\| = \mathbf{s}^T\mathbf{H}(\mathbf{X}^T\mathbf{X})^{-}\mathbf{H}^T\mathbf{s}.$$

(See Exercise 30.22.) Therefore the Scheffé simultaneous confidence intervals for contrasts associated with this hypothesis are given by

$$\mathbf{s}^T\boldsymbol{\gamma} \in \mathbf{s}^T\widehat{\boldsymbol{\gamma}} \pm ((p-k)\widehat{\sigma}^2 F^\alpha \mathbf{s}^T\mathbf{H}(\mathbf{X}^T\mathbf{X})^{-}\mathbf{H}^T\mathbf{s})^{1/2}.$$

Note that $\boldsymbol{\mu}$ is an identified parameter (different values for $\boldsymbol{\mu} = E\mathbf{Y}$ lead to different distributions for the data $\mathbf{Y}$) although $\boldsymbol{\beta}$ is not when $p < r$. (If $\mathbf{X}\mathbf{b}_1 = \mathbf{X}\mathbf{b}_2$, the distribution is the same for both $\boldsymbol{\beta} = \mathbf{b}_1$ and $\boldsymbol{\beta} = \mathbf{b}_2$). For this reason we can only draw inference about estimable functions of $\boldsymbol{\beta}$. (Note that a linear function $\mathbf{c}^T\boldsymbol{\beta}$ is estimable if and only if $\mathbf{c}^T\boldsymbol{\beta} = \mathbf{a}^T\boldsymbol{\mu}$ for some $\mathbf{a}$, i.e. if and only if the function of $\boldsymbol{\beta}$ is also a function of $\boldsymbol{\mu}$.)

As we have commented earlier, fixed effects ANOVA models are particular examples of regression models, but unfortunately there is typically no natural basis matrix. Hence the formulae for regression often are not helpful for deriving the formulae for these models. However, there is typically a nice spanning matrix for ANOVA models. Several books on linear models use this matrix and the results stated in the previous example for deriving procedures for ANOVA models (see Searle, 1971; Graybill, 1976.) From the projections perspective we see that this approach is just another way to find the projection on the appropriate subspace. In this book we shall not use this generalized inverse approach to finding the projections but shall instead use the LS property of those projections.

The one-way analysis of variance model

30.8 In the one-way ANOVA model we observe Y_{ij}, independent

$$Y_{ij} \sim N(\mu_i, \sigma^2), \qquad i = 1, \ldots, p;\ j = 1, \ldots, n_i.$$

Note that we are assuming that the means are possibly different in different classes but that the variances are the same in all classes. If the n_i are all equal we say that the model is *balanced.*

To put this model in the form of a linear model, let

$$\mathbf{Y} = (Y_{11}, \ldots, Y_{1n_1}, Y_{21}, \ldots, Y_{pn_p})^T, \quad \boldsymbol{\mu} = E\mathbf{Y} = (\mu_1, \ldots, \mu_1, \mu_2, \ldots, \mu_p)^T, \quad N = \sum n_i.$$

and let V be the p-dimensional subspace of R^N in which the first n_1 elements are the same, the next n_2 elements are the same, etc. (i.e. V is the space of possible values for $\boldsymbol{\mu}$.) Then

$$Y \sim N_N(\boldsymbol{\mu}, \sigma^2\mathbf{I}), \qquad \boldsymbol{\mu} \in V, \ \sigma^2 > 0,$$

so that this model is a linear model.

To find the estimators of the parameters we must minimize

$$\|\mathbf{Y} - \boldsymbol{\mu}\|^2 = \sum\sum(Y_{ij} - \mu_i)^2.$$

By direct differentiation we see that this expression is minimized by

$$\begin{aligned}\widehat{\mu}_{ij} = \widehat{\mu}_i = \frac{\sum_j Y_{ij}}{n_i} = \overline{Y}_{i.} \Rightarrow \widehat{\sigma}^2 = \frac{\|\mathbf{Y} - \widehat{\boldsymbol{\mu}}\|^2}{N-p} = \frac{\sum\sum(Y_{ij} - \overline{Y}_i)^2}{N-p} \\ = \frac{\|\mathbf{Y}\|^2 - \|\widehat{\boldsymbol{\mu}}\|^2}{N-p} = \frac{\sum\sum Y_{ij}^2 - \sum n_i \overline{Y}_{i.}^2}{N-p}.\end{aligned}$$

The null hypothesis we typically want to test with this model is that the $\mu_i = \theta$ for some θ (i.e. that the μ_i are all equal). Therefore, let W be the one-dimensional subspace of R^N consisting of vectors all of whose components are equal. Then we are testing the null hypothesis that $\boldsymbol{\mu} \in W$ against the alternative hypothesis that $\boldsymbol{\mu} \in V$. Hence this problem is testing a general linear hypothesis. To find the LS estimator for the reduced model we must minimize

$$\|\mathbf{Y} - \boldsymbol{\mu}\|^2 = \sum\sum(Y_{ij} - \theta)^2,$$

obtaining

$$\begin{aligned}\widehat{\widehat{\mu}}_{ij} = \widehat{\widehat{\theta}} = \frac{\sum\sum Y_{ij}}{N} = \overline{Y}_{..} \Rightarrow SSH = \|\widehat{\boldsymbol{\mu}} - \widehat{\widehat{\boldsymbol{\mu}}}\|^2 = \sum\sum(\widehat{\mu}_{ij} - \widehat{\widehat{\mu}}_{ij})^2 \\ = \sum n_i(\overline{Y}_{i.} - \overline{Y}_{..})^2 = \|\widehat{\boldsymbol{\mu}}\|^2 - \|\widehat{\widehat{\boldsymbol{\mu}}}\|^2 = \sum n_i \overline{Y}_{i.}^2 - N\overline{Y}_{..}^2, \\ dfh = \dim(V) - \dim(W) = p - 1, \ MSH = \frac{\sum n_i(\overline{Y}_{i.} - \overline{Y}_{..})^2}{p-1}.\end{aligned}$$

Also

$$MSE = \widehat{\sigma}^2 = \frac{\sum\sum(Y_{ij} - \overline{Y}_{i.})^2}{N-p}, \qquad F = \frac{MSH}{MSE},$$

and the F test rejects if $F > F^{\alpha}_{p-1,N-p}$. By similar arguments we see that

$$\|\mathbf{P}_{V|W}\boldsymbol{\mu}\|^2 = \|\boldsymbol{\mu} - \mathbf{P}_W\boldsymbol{\mu}\|^2 = \sum\sum(\mu_i - \overline{\mu})^2 = \sum n_i(\mu_i - \overline{\mu})^2 = \xi,$$

where $\overline{\mu} = \sum n_i \mu_i / N$. Therefore

$$EMSH = \sigma^2 + \frac{\xi}{p-1}, \quad EMSE = \sigma^2, \quad F \sim F_{p-1,N-p}\left(\frac{\xi}{\sigma^2}\right).$$

We next find the Scheffé simultaneous confidence intervals associated with this hypothesis. Note that a vector $\mathbf{d} = (d_{11}, \ldots, d_{1n_1}, d_{21}, \ldots, d_{pn_p}) \in V$ if and only if $d_{ij} = d_i$ does not depend on j. Further a vector $\mathbf{w} = (w_{11}, \ldots, w_{1n_1}, w_{21}, \ldots, w_{pn_p}) \in W$ if and only if $w_{ij} = w$ does not depend on either i or j. Therefore, $\mathbf{d} \in V$ and $\mathbf{d} \perp W$ if and only if $d_{ij} = d_i$ and

$$0 = \mathbf{d}^T \mathbf{w} = w \sum n_i d_i, \ \forall \mathbf{w} \in W.$$

That is, $\mathbf{d} \in V|W$ if and only if $d_{ij} = d_i = c_i / n_i$, where $\sum n_i d_i = \sum c_i = 0$. For such a $\mathbf{d}$,

$$\mathbf{d}^T \boldsymbol{\mu} = \sum n_i d_i \mu_i = \sum c_i \mu_i, \quad \mathbf{d}^T \widehat{\boldsymbol{\mu}} = \sum c_i \overline{Y}_{i.}, \quad \|\mathbf{d}\|^2 = \sum n_i d_i^2 = \sum \frac{c_i^2}{n_i}.$$

Therefore a contrast associated with this hypothesis has the form $\sum c_i \mu_i$, where $\sum c_i = 0$ and the Scheffé simultaneous confidence intervals for these contrasts are given by

$$\sum c_i \mu_i \in \sum c_i \overline{Y}_{i.} \pm \left((p-1) F^{\alpha}_{p-1,N-p} \widehat{\sigma}^2 \sum \frac{c_i^2}{n_i}\right)^{1/2}, \quad \forall c_i \ni \sum c_i = 0.$$

We can also treat this model from a regression point of view. Let $\boldsymbol{\beta} = (\mu_1, \ldots, \mu_p)^T$ and let $\mathbf{X}$ be the $n \times p$ matrix given by $\boldsymbol{\mu} = \mathbf{X}\boldsymbol{\beta}$. (Note that $\boldsymbol{\mu}$ is $N \times 1$ and $\boldsymbol{\beta}$ is $p \times 1$.) In Exercise 30.23 it is shown that

$$\widehat{\boldsymbol{\beta}} = (\mathbf{X}^T\mathbf{X})^{-1}\mathbf{X}^T\mathbf{Y} = \begin{pmatrix} \overline{Y}_{1.} \\ \overline{Y}_{2.} \\ \vdots \\ \overline{Y}_{p.} \end{pmatrix}, \qquad \widehat{\sigma}^2 = \frac{\|\mathbf{Y} - \mathbf{X}\widehat{\boldsymbol{\beta}}\|^2}{N-p} = \frac{\sum\sum (Y_{ij} - \overline{Y}_{i.})^2}{N-p}.$$

(Note that $\widehat{\sigma}^2$ derived by regression is the same as $\widehat{\sigma}^2$ derived above for the linear model.) Furthermore, testing the equality of the μ_i is the same as testing that $\mathbf{A}\boldsymbol{\beta} = \mathbf{0}$, where $\mathbf{A}$ is the $p-1 \times p$ matrix given by

$$\mathbf{A} = \begin{pmatrix} 1 & 0 & \cdots & 0 & -1 \\ 0 & 1 & \cdots & 0 & -1 \\ \vdots & \vdots & \ddots & 0 & -1 \\ 0 & 0 & \cdots & 1 & -1 \end{pmatrix}$$

It can be shown (although the derivation is somewhat complicated) that

$$(\mathbf{A}\widehat{\boldsymbol{\beta}})^T (\mathbf{A}(\mathbf{X}^T\mathbf{X})^{-1}\mathbf{A})^{-1}\mathbf{A}\widehat{\boldsymbol{\beta}} = \sum n_i (\overline{Y}_{i.} - \overline{Y}_{..})^2,$$

so that the F test is the same whether derived with matrices for regression or derived by least squares for the linear model. Note that the formulae are easier to derive by least squares and their expression in terms of projections is more easily interpreted than the rather complicated regression formulae.

Tukey, Dunnett, MCB and Bonferroni simultaneous confidence intervals

30.9 In the previous section we have derived Scheffé simultaneous confidence intervals for the set of all contrasts in a (possibly unbalanced) one-way ANOVA model. In the next few sections we present some alternative procedures for simultaneous confidence intervals for smaller sets. We say a set of simultaneous confidence intervals is an *exact* set of $1-\alpha$ confidence intervals if the simultaneous coverage probability is exactly $1-\alpha$ and a *conservative* set if the simultaneous coverage probability is at least $1-\alpha$.

We first look at the Tukey simultaneous confidence intervals which give exact intervals for the set of all comparisons between two means, then at Dunnett simultaneous confidence intervals which give exact simultaneous confidence intervals for the set of all comparisons of treatments to a control. Note that these comparisons are particular cases of contrasts, so that the Scheffé intervals can be used to give conservative simultaneous confidence intervals for these sets. We then look at Hsu's MCB intervals which are a set of conservative $1-\alpha$ confidence intervals for the set of all $\mu_i - \max_{j\neq i}\mu_j$. These intervals can be used to determine a set of treatments such that the probability that they will contain the treatment with the largest μ_i is at least $1-\alpha$. Note that the $\mu_i - \max_{j\neq i}\mu_j$ are not contrasts so that the Scheffé intervals cannot be used to give conservative intervals for this situation. Finally, we look at Bonferroni simultaneous confidence intervals which give conservative $1-\alpha$ simultaneous confidence intervals for a finite set of contrasts. Note that the set of comparisons and the set of comparisons of treatments to a control are both finite sets of contrasts but the set of all contrasts is not a finite set. For the Tukey, Dunnett and MCB intervals, we assume that the design is a balanced one-way model, $n_i = n$. For the Bonferroni intervals, we allow an unbalanced one-way model. For further discussion of these and other intervals for balanced and unbalanced data, see Hsu (1996).

Let $Z_1, \ldots, Z_k, V$ be independent random variables with $Z_i \sim N(0,1)$, $V \sim \chi^2_r$, and let

$$Q = \frac{\max_{i\neq j}|Z_i - Z_j|}{\sqrt{V/r}}.$$

Then we say that Q has a studentized range distribution and write $Q \sim q_{k,r}$. As usual, we let $q^\alpha_{k,r}$ be the upper α percentile of the distribution of Q. For simplicity, we shall often write q^α for $q^\alpha_{k,r}$.

Now, consider a balanced one-way model with p treatments and n replications. In that case, it is easily seen that

$$Q(\boldsymbol{\mu}) = \frac{\max_{i\neq j}|\overline{Y}_{i.} - \overline{Y}_{j.} - (\mu_i - \mu_j)|}{\sqrt{MSE/n}} \sim q_{p,p(n-1)}.$$

Therefore a sensible size-α test for the null hypothesis of equality of the μ_i rejects if

$$Q(\mathbf{0}) = \frac{\max_{i\neq j}|\overline{Y}_{i.} - \overline{Y}_{j.}|}{\sqrt{MSE/n}} > q^\alpha.$$

We call this test the *studentized range test* for this hypothesis. The F test derived above is the likelihood ratio test, is UMP invariant and unbiased. The studentized range test is not so well understood. One difficulty with this test is that its power function is quite complicated. It depends on the actual values of all the μ_i. (Note that the power function of the F test depends only on the non-centrality parameter $n\sum(\mu_i - \overline{\mu})^2/\sigma^2$.)

We can also use $Q(\boldsymbol{\mu})$ to find simultaneous confidence intervals. A *comparison* in the μ_i is a contrast of the form $\mu_i - \mu_j$. It is easily seen that

$$Q(\boldsymbol{\mu}) \le q^\alpha \Leftrightarrow \mu_i - \mu_j \in \overline{Y}_{i.} - \overline{Y}_{j.} \pm q^\alpha \sqrt{\frac{MSE}{n}}, \qquad \forall i \ne j.$$

Therefore, a set of simultaneous $1-\alpha$ confidence intervals for the set of all comparisons is

$$\mu_i - \mu_j \in \overline{Y}_{i.} - \overline{Y}_{j.} \pm q^\alpha \sqrt{\frac{MSE}{n}}, \qquad \forall i \ne j.$$

These intervals were first suggested in Tukey (1953) and are called the Tukey simultaneous confidence intervals for comparisons.

Note that $\mu_i - \mu_j = \sum c_k \mu_k$ with $c_i = 1$, $c_j = -1$. Hence the Scheffé simultaneous confidence intervals for comparisons are

$$\mu_i - \mu_j \in \overline{Y}_{i.} - \overline{Y}_{j.} \pm \left((p-1)MSE\frac{2}{n}\ F^\alpha\right)^{1/2}.$$

Note that the Tukey simultaneous confidence intervals for comparisons are exact, but that the Scheffé simultaneous confidence intervals for contrasts are exact. Hence the Scheffé simultaneous confidence intervals for comparisons (which are a subset of the contrasts) are not exact but are conservative. For each comparison, the centre of the Tukey interval is the same as the centre of the Scheffé interval, all the Tukey intervals have the same length $(2q^\alpha\sqrt{MSE/n})$ and all the Scheffé intervals for comparisons have the same length $(2\sqrt{(p-1)2F^\alpha MSE/n})$. Therefore, either all the Tukey intervals are contained in the Scheffé intervals or all the Scheffé intervals for comparisons are contained in all the Tukey intervals. Since the Tukey intervals for the comparisons are exact but the Scheffé intervals for comparison are conservative, the Tukey intervals must be shorter than the Scheffé intervals.

It can be shown that (see Exercise 30.27)

$$\mu_i - \mu_j \in \overline{Y}_{i.} - \overline{Y}_{j.} \pm q^\alpha \sqrt{\frac{MSE}{n}},\ \forall i \ne j \Leftrightarrow$$
$$\sum c_i \mu_i \in \sum c_i \overline{Y}_{i.} \pm \sqrt{\frac{MSE}{n}} q^\alpha \sum_i \frac{|c_i|}{2},\ \forall c_i \ni \sum c_i = 0,$$

and hence these intervals are a set of simultaneous $1-\alpha$ confidence intervals for the set of all contrasts in the μ_i, called the Tukey simultaneous $1-\alpha$ confidence intervals for contrasts. The Scheffé intervals for contrasts can be written as

$$\sum c_i \mu_i \in \sum c_i \overline{Y}_{i.} \pm \sqrt{\frac{MSE}{n}} \sqrt{(p-1)\sum c_i^2\ F^\alpha},\ \forall c_i \ni \sum c_i = 0.$$

Note that the centres of the Tukey and Scheffé intervals are the same. Further,

$$\frac{\text{length(Tukey)}}{\text{length(Scheffé)}} = \frac{q^\alpha \sum |c_i|}{2\sqrt{(p-1)F^\alpha \sum c_i^2}}.$$

Which intervals are longer depends on the c_i. The Tukey intervals are shorter for relatively simple contrasts (contrast in which most of the c_i are 0) and the Scheffé intervals are shorter for more complicated contrasts. See Scheffé (1959, p. 77), for some numerical comparisons.

There do not appear to be any exact simultaneous confidence intervals for the set of comparisons in the unbalanced case. However, Tukey (1953) and Kramer (1956) suggested the obvious approximate intervals

$$\mu_i - \mu_j \in \overline{Y}_i - \overline{Y}_j \pm q^\alpha \sqrt{\frac{MSE}{2}\left(\frac{1}{n_i} + \frac{1}{n_j}\right)}, \qquad q^\alpha = q^\alpha_{p,N-p}.$$

Hayter (1984) showed that these intervals are in fact conservative intervals for the set of all comparisons. In later sections of this chapter, however, we shall only discuss Tukey confidence intervals for situations in which they are exact.

We finish this section with a result about the studentized range distribution (derived in Exercise 30.28) which is useful for deriving Tukey procedures for the balanced incomplete block model, various mixed models and repeated measures models.

Theorem 30.5.

Let U and $\mathbf{Z} = (Z_1, \ldots, Z_k)$ be independent, $U \sim \chi^2_m$ and the Z_i jointly normally distributed with

$$EZ_i = 0, \quad \text{var}(Z_i) = 1, \quad \text{cov}(Z_i, Z_j) = a.$$

Let

$$q = \frac{\max_{i \neq j} |Z_i - Z_j|}{\sqrt{(1-a)U/m}}.$$

Then

$$q \sim q_{k,m}.$$

Note that when $a = 0$, this result is just the definition of the studentized range distribution.

30.10 Sometimes in a one-way model, one treatment is a control treatment and we want to compare the other treatments to it. Without loss of generality, we may assume that the control is the first treatment and that we want to compare the other μ_i to μ_1. That is, we want simultaneous confidence intervals for the set $\mu_2 - \mu_1, \ldots, \mu_p - \mu_1$. We now discuss a method for finding such confidence intervals. We again assume that the model is balanced, $n_i = n$.

Let $Z_0, Z_1, \ldots, Z_k, V$ be independent, $Z_i \sim N(0, 1)$, $V \sim \chi^2_r$, and let

$$D = \frac{\max_{i \neq 0} |Z_i - Z_0|}{\sqrt{2V/r}}.$$

Then we say that D has a Dunnett distribution with k and r degrees of freedom and write $D \sim D_{k,r}$. As usual, we write $D^\alpha_{k,r}$ for the upper α percentile of this distribution, which we often abbreviate to D^α when the degrees of freedom should be obvious. (In comparing the formulae for the studentized range distribution and the Dunnett distribution, note the extra $\sqrt{2}$ in the denominator of the Dunnett distribution. The inclusion of this term in the Dunnett distribution but not in the studentized range distribution is, of course, arbitrary. The formulae we have used are the ones used in the literature and hence compatible with existing tables.)

It is easily verified that

$$D(\boldsymbol{\mu}) = \frac{\max_{i\neq 1} |\overline{Y}_{i.} - \overline{Y}_{1.} - (\mu_i - \mu_1)|}{\sqrt{2MSE/n}} \sim D_{p-1,p(n-1)}.$$

Therefore a sensible size-α test of the null hypothesis that $\mu_i = \mu_1$ rejects if

$$D(\mathbf{0}) = \frac{\max_{i\neq 1} |\overline{Y}_{i.} - \overline{Y}_{1.}|}{\sqrt{2MSE/n}} > D^{\alpha}.$$

Unfortunately, the power function of this test is again quite complicated.

It is easily seen that

$$D(\boldsymbol{\mu}) \leq D^{\alpha} \Leftrightarrow$$
$$\mu_i - \mu_1 \in \overline{Y}_{i.} - \overline{Y}_{1.} \pm \sqrt{\frac{2MSE}{n}} D^{\alpha}, \qquad \forall i \neq 1.$$

Therefore, these intervals are a set of simultaneous $1 - \alpha$ confidence intervals for the set of all $\mu_i - \mu_1$. These intervals were suggested by Paulson (1952b) and Dunnett (1955) and are called Dunnett's simultaneous (two-sided) confidence intervals for comparing a set of treatments to a control. By an argument similar to that given above comparing Tukey and Scheffé intervals for comparisons, we can show that Dunnett's intervals for the $\mu_i - \mu_1$ are shorter than the Tukey or Scheffé intervals. We would declare the ith treatment significantly different from the control treatment if the interval given above for $\mu_i - \mu_1$ does not contain 0.

Often in an experiment we are only interested in finding which treatments are better (have higher mean) than a control treatment. In such situations, one-sided Dunnett intervals are often useful. As above, let $Z_0, \ldots, Z_k, V$ be independent, $Z_i \sim N(0, 1)$, $V \sim \chi^2_r$. Define

$$OD = \frac{\max_{i\neq 0}(Z_i - Z_0)}{\sqrt{2V/r}}.$$

Then we say that OD has a one-sided Dunnett distribution and write $OD \sim OD_{k,r}$. As usual, we write $OD^{\alpha}_{k,r}$ (or OD^{α}) for the upper α critical point for this distribution. (Note that D defined above is not $|OD|$ and hence $OD^{\alpha} \neq D^{2\alpha}$. Separate tables are needed for OD^{α} and D^{α}.)

For the balanced one-way model, it is easily seen that

$$OD(\boldsymbol{\mu}) = \frac{\max_{i\neq 1}(\overline{Y}_i - \overline{Y}_1 - (\mu_i - \mu_1))}{\sqrt{2MSE/n}} \sim OD_{p-1,p(n-1)}$$

and therefore a set of simultaneous one-sided $1-\alpha$ confidence intervals for the set of all $\mu_i - \mu_1$ is given by

$$\mu_i - \mu_1 \geq \overline{Y}_{i.} - \overline{Y}_{1.} - OD^{\alpha}\sqrt{\frac{2MSE}{n}}.$$

(since $OD^{1-\alpha} = -OD^{\alpha}$) We call these intervals the one-sided Dunnett intervals for comparing treatments to a control. We could then say that the ith treatment is better than the control as long as the one-sided Dunnett interval for $\mu_i - \mu_1$ does not contain 0. Sometimes we want to see which treatments have lower means than the control and the following one-sided simultaneous $1-\alpha$ intervals in the other direction are useful:

$$\mu_i - \mu_1 \le \overline{Y}_i - \overline{Y}_1 + OD^\alpha \sqrt{\frac{2MSE}{n}}.$$

In the unbalanced case Hsu (1996, p. 55), gives an exact (but computationally intensive) method for constructing one-sided Dunnett confidence intervals for comparisons of treatments to a control. Because this procedure is apparently not usable at the present time, in future sections we shall only discuss Dunnett procedures in situations in which they are exact.

In Exercise 30.28, it is shown that the Dunnett and one-sided Dunnett distributions can also be extended to the case in which the Z_i are dependently jointly normally distributed with equal correlation coefficients, similar to the extension of the studentized range distribution in the previous section.

30.11 We now assume that the treatment with the highest μ_i is the best treatment. We want to find which of the treatments could possibly be the best. (Note that the treatment with the highest sample mean ($\overline{Y}_i$) is not necessarily the treatment with the highest true mean (μ_i).) We continue in the balanced case $n_i = n$.

We want a set of simultaneous confidence intervals for the set of

$$\mu_i - \max_{j \neq i} \mu_i.$$

Let k^* be the treatment with the highest mean (if there is a tie, then any of the tied subscripts may be used). Note that we do not know k^* since we do not know the μ_i. In some sense we are trying to estimate k^*. We recall the notation that $a^+ = \max(a, 0)$ and $-a^- = \min(a, 0)$. Let

$$Q = OD^\alpha \sqrt{\frac{2MSE}{n}}$$

(where $OD^\alpha = OD^\alpha_{p-1,p(n-1)}$ is the critical value for the one-sided Dunnett interval in **30.10**). Using the one-sided Dunnett simultaneous confidence intervals with treatment k^* as control, we see that

$$P(\mu_i - \mu_{k^*} \ge \overline{Y}_i - \overline{Y}_{k^*} - Q \ \forall i \neq k^*) = 1 - \alpha.$$

Now,

$$\begin{aligned}\mu_i - \mu_{k^*} \ge \overline{Y}_i - \overline{Y}_{k^*} - Q \ \forall i \neq k^* &\Rightarrow \mu_{k^*} - \max_{j \neq k^*} \mu_j \le \overline{Y}_{k^*} - \max_{j \neq k^*} \overline{Y}_j + Q \\ &\Rightarrow \mu_i - \max_{j \neq i} \mu_j \le (\overline{Y}_i - \max_{j \neq i} \overline{Y}_j + Q)^+ \ \forall i.\end{aligned}$$

Similarly

$$\begin{aligned}\mu_i - \mu_{k^*} \ge \overline{Y}_i - \overline{Y}_{k^*} - Q \ \forall i \neq k^* &\Rightarrow \mu_i - \max_{j \neq i} \mu_j \ge \overline{Y}_i - \max_{j \neq i} \overline{Y}_j - Q \ \forall i \neq k^* \\ &\Rightarrow \mu_i - \max_{j \neq i} \mu_j \ge -(\overline{Y}_i - \max_{j \neq i} \overline{Y}_j - Q)^- \ \forall i.\end{aligned}$$

(See Exercise 30.82 for details.) Therefore,

$$\mu_i - \mu_{k^*} \geq \overline{Y}_i - \overline{Y}_{k^*} - Q \ \forall i \neq k^* \Rightarrow$$
$$-(\overline{Y}_i - \max_{j \neq i} \overline{Y}_j - Q)^- \leq \mu_i - \max_{j \neq i} \mu_j \leq (\overline{Y}_i - \max_{j \neq i} \overline{Y}_j + Q)^+ \ \forall i$$

and hence

$$P(-(\overline{Y}_i - \max_{j \neq i} \overline{Y}_j - Q)^- \leq \mu_i - \max_{j \neq i} \mu_j \leq (\overline{Y}_i - \max_{j \neq i} \overline{Y}_j + Q)^+ \ \forall i)$$
$$\geq P(\mu_i - \mu_{k^*} \geq \overline{Y}_i - \overline{Y}_{k^*} - Q \ \forall i \neq k^*) = 1 - \alpha.$$

Hence a set of conservative $1-\alpha$ simultaneous confidence intervals for the set of $\mu_i - \max_{j \neq i} \mu_j$ is given by

$$-\left(\overline{Y}_i - \max_{j \neq i} \overline{Y}_j - OD^\alpha \sqrt{\frac{2MSE}{n}}\right)^- \leq \mu_i - \max_{j \neq i} \mu_j$$
$$\leq \left(\overline{Y}_i - \max_{j \neq i} \overline{Y}_j + OD^\alpha \sqrt{\frac{2MSE}{n}}\right)^+ \ \forall i.$$

These intervals were derived by Hsu (1984) and are called Hsu's *multiple comparison with the best* (MCB) simultaneous $1 - \alpha$ confidence intervals. Note that there are only three possibilities for each interval. Either its lower limit is 0 or its upper limit is 0 or the upper limit is positive and the lower limit is negative. (That is, it is not possible for both limits to be positive or negative). For this reason, Hsu (1996) calls these intervals *constrained* MCB intervals. See Hsu (1996, p. 82) for a discussion of why such constrained intervals are desirable.

Let $\overline{Y}_k = \max \overline{Y}_i$. (Unfortunately, as commented above, k^* need not equal k.) If for treatment k the lower limit is 0, then all the other intervals have upper limit 0 and we declare treatment k to be the best. If the lower limit for treatment k is not 0, then we say that the best treatment lies in the set of treatments whose upper limit is not 0 (which includes treatment k and at least one other treatment). These two rules can be summarized in one rule by saying that a treatment is in the class of possibly best treatments if and only if its MCB interval contains positive numbers. Note that treatment k is always a potentially best treatment. An equivalent version of this rule is that treatment i is a possibly best treatment as long as

$$\overline{Y}_i > \overline{Y}_k - OD^\alpha \sqrt{\frac{2MSE}{n}}.$$

That is, the class of potentially best treatments contains the sample best treatment (k) and any other treatments whose sample mean is sufficiently close to $\overline{Y}_k$. Note that the probability (before we collect the data) that this class of possibly best treatments contains the treatment with highest μ_i is at least $1 - \alpha$.

In a similar way we can show that a set of conservative $1-\alpha$ simultaneous confidence intervals for the set of all $\mu_i - \min_{j \neq i} \mu_j$ is given by

$$-\left(\overline{Y}_i - \min_{j \neq i} \overline{Y}_j - Q\sqrt{\frac{MSE}{n}}\right)^- \leq \mu_i - \min_{j \neq i} \mu_j \leq \left(\overline{Y}_i - \min_{j \neq i} \overline{Y}_j + Q\sqrt{\frac{MSE}{n}}\right)^+.$$

These intervals can be used to find a subset of the treatments which contains the treatment with the smallest μ_i with probability at least $1 - \alpha$.

30.12 We return to the possibly unbalanced one-way model. Let S be a finite set of contrasts $\sum c_i \mu_i$ and let s be the number of contrasts in S. Then

$$P\left(\sum c_i \mu_i \in \sum c_i \overline{Y}_i \pm t^{\frac{\alpha}{2s}} \left(MSE \sum \frac{c_i^2}{n_i}\right)^{1/2} \forall \{c_i\} \in S\right) \geq 1 - \alpha$$

(see Exercise 30.32). Therefore the set of intervals

$$\sum c_i \mu_i \in \sum c_i \overline{Y}_i \pm t^{\frac{\alpha}{2s}} \left(MSE \sum \frac{c_i^2}{n_i}\right)^{1/2}$$

is a set of conservative $1 - \alpha$ confidence regions for the set of all functions in S. We call these intervals the Bonferroni simultaneous confidence intervals. Note that we can only have Bonferroni confidence intervals for a finite set of contrasts, so that it is not possible to have Bonferroni intervals for the set of all contrasts.

There are a total of $p(p-1)/2$ comparisons. Therefore the set of intervals

$$\mu_i - \mu_j \in \overline{Y}_i - \overline{Y}_j \pm t^{\alpha^*/2}\left(MSE\left(\frac{1}{n_i} + \frac{1}{n_j}\right)\right)^{1/2} \quad \forall i \neq j, \qquad \alpha^* = \frac{2\alpha}{p(p-1)},$$

is a set of conservative simultaneous $1 - \alpha$ confidence intervals for the set of all comparisons. In the balanced case it is easily seen that the Tukey intervals for the set of all comparisons must be shorter than the Bonferroni intervals, but that the Bonferroni intervals can be applied to unbalanced models where no exact Tukey intervals exist.

Similarly, if the first treatment is a control then there are $p - 1$ possible comparisons for a treatment to the control. Therefore a set of conservative simultaneous $1 - \alpha$ confidence intervals for the set of all comparisons of a treatment to a control is

$$\mu_i - \mu_1 \in \overline{Y}_i - \overline{Y}_1 \pm t^{\alpha^{**}/2}\left(MSE\left(\frac{1}{n_i} + \frac{1}{n_1}\right)\right)^{1/2}, \qquad \forall i \neq 1, \ \alpha^{**} = \frac{\alpha}{p-1}.$$

It is again apparent that the Dunnett intervals are shorter for balanced data, but the Bonferroni intervals can be applied to unbalanced data when the Dunnett intervals cannot.

Finally, note that the $\mu_i - \max_{j \neq i} \mu_j$ are not contrasts so that no Bonferroni alternative exists for the MCB procedure.

Multiple comparisons

30.13 We continue with the one-way model discussed in previous sections. A *multiple comparisons* procedure is a procedure which not only decides whether there is a significant difference between the means but also decides which means are significantly different.

Perhaps the most obvious approach to this problem is to perform individual size-α t tests comparing each pair of treatments. However, if we have many treatments, the probability that we find at least one t test significant, even when there are no differences in the means, is much greater than α. It is often said in this situation that we have not controlled the experiment-wide error rate. The procedures we consider here are attempts to control this experiment-wide error rate. (The procedure described in this paragraph is often called *Fisher's unprotected least significant difference* procedure, because we have no protection for the experiment-wide error rate.)

Earlier, we discussed three multiple comparisons procedures. One is to declare two means significantly different if the Scheffé confidence interval for the comparison between the two means does not contain 0. A second procedure replaces the Scheffé simultaneous confidence intervals with the Bonferroni intervals and a third procedure (applicable only when the sample sizes are equal) replaces the Scheffé simultaneous confidence intervals with the Tukey simultaneous confidence intervals. Since the simultaneous coverage probabilities of these sets of intervals are all at least $1 - \alpha$, the overall probability of declaring two identical means to be significantly different with any of these procedure must be at most α.

We say the a multiple comparisons procedure makes a *Type I* error whenever it finds two equal means to be significantly different and it makes a *Type II* error whenever it finds unequal means to be insignificantly different. It is possible to make both a Type I error and a Type II error at the same time. For example if $\mu_1 = \mu_2 \neq \mu_3$ and the procedure determined that μ_1 and μ_2 were significantly different, but μ_2 and μ_3 were not significantly different, then it would be making a Type I error about μ_1 and μ_2, and making a Type II error about μ_2 and μ_3. Our goal in this section is to limit the probability of Type I errors and then, from procedures with low probability of Type I errors, to find procedures which minimize probabilities of Type II errors.

For a multiple comparison procedure **P**, we define

$$\alpha_{\mathbf{P}}(\boldsymbol{\mu}, \sigma^2) = P_{\boldsymbol{\mu},\sigma^2}(\text{at least 1 Type I error with } \mathbf{P}).$$

Note that if the μ_i are all different, then the procedure cannot make a Type I error and hence $\alpha_{\mathbf{P}}(\boldsymbol{\mu}, \sigma^2) = 0$ in this case. We define the *size* $\alpha_{\mathbf{P}}$ and the *generalized size* $\alpha^*_{\mathbf{P}}$ for a procedure **P** by

$$\alpha_{\mathbf{P}} = \sup_{\mu_1 = \cdots = \mu_p} \alpha_{\mathbf{P}}(\boldsymbol{\mu}, \sigma^2), \quad \alpha^*_{\mathbf{P}} = \sup \alpha_{\mathbf{P}}(\boldsymbol{\mu}, \sigma^2).$$

That is, the size of the procedure is the supremum of the probability of Type I error when the μ_i are all equal, and the generalized size is the supremum of the probability of any Type I error. Early work on the multiple comparisons problem focused on the size of a procedure (see Miller, 1966), but more recent work emphasizes the generalized size (see Gabriel, 1969). In this chapter we shall look for procedures whose generalized size is less than a number α. Out of such procedures, we shall look for procedures which make relatively few Type II errors.

It is not obvious how to define the power of a multiple comparison procedure. Several possible definitions have been suggested which have led to different conclusions on relative performance of procedures. (See, for example, Einot and Gabriel (1975) and Ramsey (1978).) For this section, we shall say that one procedure $\mathbf{P}_1$ is *more powerful* than another procedure $\mathbf{P}_2$ if any significant difference for $\mathbf{P}_2$ is also significant for $\mathbf{P}_1$. In this situation $\mathbf{P}_1$ would have higher power than $\mathbf{P}_2$ for any sensible definition of power.

Since the Scheffé, Bonferroni and Tukey simultaneous confidence intervals for comparisons all have simultaneous false coverage probability less than or equal to α, these multiple comparisons procedures have generalized size at most α. Since the Tukey intervals are contained in the Scheffé intervals and the Bonferroni intervals, any significant difference with Scheffé or Bonferroni would also be significant with Tukey, and hence the Tukey procedure is more powerful than the Scheffé and Bonferroni procedures. (However, the Tukey procedure is only applicable for balanced models.) Note that the multiple comparisons procedure based on the Tukey confidence intervals is often called the *honest significant difference* to distinguish it from the least significant difference procedure described above.

Another early multiple comparisons procedure is due to Fisher (1935a) and is called Fisher's *protected least significant difference*. In this procedure, to control the experiment-wide error rate, we first perform a size-α F test of the null hypothesis of the equality of the μ_i. If we accept this hypothesis, we declare there to be no significant differences. However, if we reject the F test, we then do individual size-α t tests (using MSE for the denominator for all of them) to determine which pairs are significantly different. Since we find a significant difference only if we reject the F test and the F test has size α, the size of this procedure is at most α. However, this procedure has generalized size considerably greater than α. To see why, suppose that $\mu_1 = \mu_2 = \mu_3 = 0$, $\mu_4 = 10^{10}$. Because μ_4 is so far from the other means, we are nearly certain to reject the F test. Once we have rejected the F test, we do three size-α t tests to compare μ_1 and μ_2, to compare μ_1 and μ_3 and to compare μ_2 and μ_3. We make a Type I error if we reject any of these three tests. Therefore, the probability of a Type I error is essentially the probability that we reject at least one of the three size-α t tests, which is considerably greater than α. This example illustrates why it is not enough to restrict attention to procedures whose size is α. We can defeat the experiment-wide control of the least significant difference procedure by including data from another experiment to get past the F test and get to multiple t tests.

30.14 We next define three classes of multiple comparisons procedures. Let $\mathbf{S}$ be a subset of $1, \ldots, p$ with $s > 1$ elements and let

$$\overline{Y}^{\mathbf{S}}_{..} = \frac{\sum_{\mathbf{S}} n_i \overline{Y}_{i.}}{\sum_{\mathbf{S}} n_i}, \qquad F^{\mathbf{S}} = \frac{\sum_{\mathbf{S}} n_i (\overline{Y}_{i.} - \overline{Y}^{S}_{..})^2}{(s-1)MSE}$$

Note that if the means in $\mathbf{S}$ are all equal, then

$$F^{\mathbf{S}} \sim F_{s-1,N-p}.$$

We say that the subset $\mathbf{S}$ is α_s F-significant if

$$F^{\mathbf{S}} > F^{\alpha_s}_{s-1,N-p},$$

so that if the means in $\mathbf{S}$ are all equal the probability that $\mathbf{S}$ is α_s F-significant is α_s. A *multiple F test* with sizes $\alpha_2, \ldots, \alpha_p$ declares two means to be significantly different if every subset $\mathbf{S}$ containing those means is α_s F-significant (where s is the number of elements in $\mathbf{S}$). Different multiple F procedures differ only in their choice for α_s. Note that the numerator of $F^{\mathbf{S}}$ is the numerator of the obvious F statistic for testing the equality of the μ_i, $i \in \mathbf{S}$, but the denominator is always MSE.

Now suppose that the $n_i = n$ are all equal. For any set $\mathbf{S}$, define

$$Q^{\mathbf{S}} = \frac{\max_{i \in \mathbf{S}, j \in \mathbf{S}} |\overline{Y}_{i.} - \overline{Y}_{j.}|}{\sqrt{MSE/n}}.$$

Note that if the means in $\mathbf{S}$ are all equal, then

$$Q^{\mathbf{S}} \sim q_{s,N-p}.$$

We say this subset is α_s q-significant with this procedure if

$$Q^{\mathbf{S}} > q_{s,N-p}^{\alpha_s},$$

so that if the means in $\mathbf{S}$ are all the same, the probability that $\mathbf{S}$ is α_s q-significant is α_s. Note again that the numerator is just that for the obvious studentized range test, but the denominator is always $\sqrt{MSE}$. A *multiple range test* with sizes α_s, $s = 2, \ldots, p$, declares two means to be significantly different if every subset $\mathbf{S}$ containing those two means is α_s q-significant. Note that the procedure is completely determined by the α_s, $s = 2, \ldots, p$. Note also that if

$$q_{s,N-p}^{\alpha_s} \leq q_{p,N-p}^{\alpha}, \qquad \forall s,$$

then the multiple range test is more powerful than the Tukey procedure (see Exercise 30.33). Typically, this inequality is satisfied and the multiple range procedure is more powerful than the Tukey procedure. (For all multiple range procedures, $s < p$, but often $\alpha_s < \alpha$, so that it is hard to tell whether this equality is satisfied until we have actually computed the critical values.)

We now give a third class of multiple comparisons procedure. We return to the possibly unbalanced case. For any set $\mathbf{S}$ with s elements, define

$$B_{\mathbf{S}} = \max_{i\in\mathbf{S}, j\in\mathbf{S}} \frac{|\overline{Y}_{i.} - \overline{Y}_{j.}|}{\sqrt{MSE(\frac{1}{n_i} + \frac{1}{n_j})}}.$$

We say that the set $\mathbf{S}$ is α_s Bonferroni significant if

$$B_{\mathbf{S}} \geq t_{N-p}^{\alpha_s/s(s-1)}.$$

If the means in $\mathbf{S}$ are all equal, then the probability that $\mathbf{S}$ is α_s Bonferroni significant is less than or equal to α_s. A *multiple Bonferroni procedure* with sizes $\alpha_2, \ldots, \alpha_p$ declares two means to be significantly different if every subset containing those two means is α_s Bonferroni significant.

The first multiple range, multiple F and multiple Bonferroni tests were the *Student–Newman–Keuls* (SNK) procedures (see Newman, 1938; Keuls, 1952). In these procedures, we take $\alpha_s = \alpha$. With the SNK procedures we find a significant difference only if we find the whole set significant, so that the size of these procedures is no greater than α. However, their generalized size may be greater than α, at least when $p > 3$. To see why, let $\mu_1 = \mu_2 = 0$, $\mu_3 = \mu_4 = 10^{10}$. Since these two clusters are so far apart, we are essentially sure of finding the whole set significant. In addition, any subset with three means must contain at least one from each set, so that we are again essentially sure to find the sets with three means significant. Therefore, the probability of Type I error in this setting is essentially the probability of finding either $(1, 2)$ or $(3, 4)$ significantly different, which would be considerably greater than α, probably closer to 2α. Initially, one advantage of the SNK multiple range test over other multiple range tests is that we only need the critical values for $\alpha = 0.05$ in order to do a 0.05 SNK test. Other procedures need other critical points which for a long time were untabled. Now that these procedures are implemented on computers, this advantage for the SNK procedure is not so important.

Before discussing additional multiple range, multiple F and multiple Bonferroni tests, we state the following lemma which is useful for showing that such procedures have generalized size at most α. A proof is outlined in Exercise 30.36.

Lemma 30.2.

Consider a multiple range, multiple F or multiple Bonferroni procedure with sizes $\alpha_2, \ldots, \alpha_s$.

(a) If

$$\sum s_i \le p \Rightarrow \sum \alpha_{s_i} \le \alpha$$

then the procedure has generalized size at most α.

(b) If

$$\sum s_i \le p \Rightarrow \prod (1 - \alpha_{s_i}) \ge 1 - \alpha$$

then the procedure has generalized size at most α.

The next procedures were suggested in Ryan (1960) and are called the *Ryan* multiple range, multiple F and multiple Bonferroni tests. For these procedures, we take

$$\alpha_s = \frac{s}{p}\alpha.$$

Note that

$$\sum s_i < p \Longrightarrow \sum \alpha_{s_i} = \frac{\sum s_i}{p}\alpha \le \alpha$$

so that, by Lemma 30.7, this procedure has generalized size less than or equal to α.

The next procedures, due to Duncan (1955), are called the *Duncan* multiple range, multiple F and multiple Bonferroni tests. These procedures choose

$$\alpha_s = 1 - (1 - \alpha)^{\frac{s-1}{p-1}}.$$

Note that

$$\sum s_i \le p \Rightarrow \prod (1 - \alpha_{s_i}) = \prod (1 - \alpha)^{\frac{s_i - 1}{p-1}} = (1 - \alpha)^{\frac{\sum (s_i - 1)}{p-1}} \ge 1 - \alpha,$$

so that by Lemma 30.7, these procedures also have generalized size at most α.

Duncan uses a Bayesian argument to suggest using

$$\alpha = 1 - (0.95)^{p-1}$$

and labels it $\alpha = 0.05$ for this procedure. This choice for α can be quite large. For example, for $p = 10$, $\alpha = 0.37$. In this case, Duncan calls the procedure a 0.05 procedure when its size and generalized size are really about 0.37. This inflated size may account for the great popularity of this procedure in practice.

The last multiple range, multiple F or multiple Bonferroni procedures discussed in this section were first suggested by Einot and Gabriel (1975) who called the procedures modified Ryan procedures. They are now often called the *Einot–Gabriel* or *Ryan–Einot–Gabriel* (REG) procedures. These procedures choose

$$\alpha_s = 1 - (1-\alpha)^{\frac{s}{p}}.$$

Note that

$$\sum s_i < p \Rightarrow \prod (1-\alpha_{s_i}) = \prod (1-\alpha)^{\frac{s_i}{p}} = (1-\alpha)^{\frac{\sum s_i}{p}} \geq 1-\alpha,$$

so that these procedures also have generalized size at most α.

Furthermore,

$$1-(1-\alpha)^{\frac{s}{p}} > \frac{s}{p}\alpha, \qquad 1-(1-\alpha)^{\frac{s}{p}} \geq 1-(1-\alpha)^{\frac{s-1}{p-1}}$$

(see Exercise 30.37), so that the REG procedures are more powerful than either the Ryan or the Duncan procedures for any fixed α. Therefore, the REG procedures seem to be the best of these multiple F, multiple range and multiple Bonferroni procedures. They have correct generalized size and are more powerful than the other suggested procedures with the correct generalized size.

Note that there are a total of

$$b_p = 2^p - p - 1$$

subsets of the integers $1, \ldots, p$ that have at least two elements. When we do a multiple range, multiple Bonferroni or multiple F test, we may have to check all these sets. (Note that if the whole set is not significant, then we can stop. We only need to check all the sets if the means are well separated.) There is a nice algorithm for multiple range and multiple Bonferroni tests in which it is only necessary to evaluate the multiple test for

$$a_p = \binom{p}{2} = \frac{p(p-1)}{2}$$

subsets. (See Arnold (1981, pp. 190–191) for a description of this procedure.) For the multiple F test, we may need to compute the F statistics for all b_p subsets. For example, if we have $p = 20$ different treatments,

$$a_{20} = 190, \quad b_{20} = 1\,048\,555.$$

That is, if we have 20 well-separated means we need to look at most 190 subsets for a multiple range or multiple Bonferroni test but may need to look at over a million subsets for a multiple F test. This computational reduction was initially a big advantage of multiple range and multiple Bonferroni tests over multiple F tests. However, now that these procedures are implemented on high-speed computers, this advantage seems less relevant.

Several studies have been done comparing different multiple comparisons procedures. Early studies compared procedures with the same size but different generalized sizes, and typically found that procedures with high generalized size have high power. More recent papers have concentrated on procedures with the same generalized size (for example, see Einot and Gabriel, 1975; Ramsey 1978). These papers draw somewhat different conclusions because of different definitions of power. Loosely, speaking, however, these papers indicate that multiple F procedures have higher power than multiple range procedures. Note also that multiple range procedures are only correct for the balanced case but multiple F procedures are always valid.

Orthogonal designs

30.15 In the one-way model it is fairly straightforward to compute the least squares estimators of the parameters for both the full and reduced models. However, for many more complicated ANOVA models it is difficult to find the estimators for the full or reduced model. Many of the models considered in this chapter are orthogonal designs which are a particular kind of linear model for which derivations can be made more easily. In this section, we study the implications of the basic theorems on linear models for orthogonal designs. The results are first stated in terms of vectors and subspaces. These results are applied to the one-way ANOVA model. The results are then restated in components in a form which is readily applicable to many ANOVA models.

An *orthogonal design* is a model in which we observe

$$\mathbf{Y} \sim N_N(\boldsymbol{\delta}^{1} + \cdots + \boldsymbol{\delta}^{r}, \sigma^2\mathbf{I}), \boldsymbol{\delta}^{q} \in T_q, \qquad q = 1, \ldots, r,\ \sigma^2 > 0,$$

where $\boldsymbol{\delta}^{1}, \ldots, \boldsymbol{\delta}^{r}$ and σ^2 are unknown parameters and the T_q are known subspaces satisfying

$$T_q \perp T_s, \quad q \neq s, \quad \dim(T_q) = d_q.$$

To make this model into a linear model, let

$$\boldsymbol{\mu} = \boldsymbol{\delta}^{1} + \cdots + \boldsymbol{\delta}^{r}, \quad V = \left\{\mathbf{v} = \sum \mathbf{t}_q : \mathbf{t}_q \in T_q\right\}.$$

V is called the *(orthogonal) direct sum* of the T_i (written $V = T_1 \oplus \cdots \oplus T_r = \bigoplus_{q=1}^{r} T_q$). Note that V is a subspace and

$$\dim(V) = \sum \dim(T_q) = \sum d_q, \quad \|\mathbf{P}_V\mathbf{y}\|^2 = \sum \|\mathbf{P}_{T_q}\mathbf{y}\|^2$$

(see Exercise 30.39). Then $\mathbf{Y} \sim N_n(\boldsymbol{\mu}, \sigma^2\mathbf{I})$, $\boldsymbol{\mu} \in V$, so that this is a linear model.

Let

$$Q(\boldsymbol{\delta}^{1}, \ldots, \boldsymbol{\delta}^{r}) = \|\mathbf{Y} - (\boldsymbol{\delta}^{1} + \cdots + \boldsymbol{\delta}^{r})\|^2 = \|\mathbf{Y} - \boldsymbol{\mu}\|^2,$$

and let $\widehat{\boldsymbol{\delta}}^{1}, \ldots, \widehat{\boldsymbol{\delta}}^{r}$ minimize $Q(\boldsymbol{\delta}^{1}, \ldots, \boldsymbol{\delta}^{r})$. Finally, let $\widehat{\boldsymbol{\mu}} = \widehat{\boldsymbol{\delta}}^{1} + \cdots + \widehat{\boldsymbol{\delta}}^{r}$. Then $\widehat{\boldsymbol{\mu}}$ is the point in V closest to $\mathbf{Y}$ and hence $\widehat{\boldsymbol{\mu}} = \mathbf{P}_V\mathbf{Y}$. Note also that

$$\boldsymbol{\delta}^{q} = \mathbf{P}_{T_q}\boldsymbol{\mu}, \widehat{\boldsymbol{\delta}}^{q} = \mathbf{P}_{T_q}\widehat{\boldsymbol{\mu}} = \mathbf{P}_{T_q}\mathbf{Y}$$

(see Exercise 30.39). Hence $\widehat{\boldsymbol{\delta}}_q$ is the ML and MV unbiased estimator of $\boldsymbol{\delta}_q$.

For this model

$$SSE = \|\mathbf{P}_{V^\perp}\mathbf{Y}\|^2 = \|\mathbf{Y} - \widehat{\boldsymbol{\mu}}\|^2, \quad dfe = n - \dim(V) = n - \sum d_i, \quad MSE = \frac{SSE}{dfe}.$$

Note that $(\widehat{\boldsymbol{\delta}}^{1}, \ldots, \widehat{\boldsymbol{\delta}}^{r}, MSE)$ is an invertible function of $(\widehat{\boldsymbol{\mu}}, MSE)$ and hence $(\widehat{\boldsymbol{\delta}}^{1}, \ldots, \widehat{\boldsymbol{\delta}}^{r}, MSE)$ is a complete sufficient statistic.

Example 30.8 (One-way ANOVA as orthogonal design: estimation)

Consider the one-way ANOVA model in which we observe Y_{ij} independent, $Y_{ij} \sim N(\mu_i, \sigma^2)$, $i = 1, \ldots, p; j = 1, \ldots, n_i$. As in **30.8**, let

$$\mathbf{Y} = (Y_{11}, \ldots, Y_{1n_1}, Y_{21}, \ldots, Y_{pn_p})^T, \quad \boldsymbol{\mu} = E\mathbf{Y} = (\mu_1, \ldots, \mu_1, \mu_2, \ldots, \mu_p)^T$$

(where μ_i is repeated n_i times). To make this model into an orthogonal design, let

$$N = \sum n_i, \quad \theta = \frac{\sum n_i \mu_i}{N}, \quad \alpha_i = \mu_i - \theta.$$

Then

$$\mu_i = \theta + \alpha_i, \quad \sum n_i \alpha_i = 0.$$

Let

$$\boldsymbol{\delta}^1 = (\theta, \ldots, \theta, \theta, \ldots, \theta)^T, \quad \boldsymbol{\delta}^2 = (\alpha_1, \ldots, \alpha_1, \alpha_2, \ldots, \alpha_p)^T$$

(where θ is repeated N times and α_i is repeated n_i times). Then

$$\boldsymbol{\mu} = \boldsymbol{\delta}^1 + \boldsymbol{\delta}^2, \quad \boldsymbol{\delta}^{1T} \boldsymbol{\delta}^2 = \sum \sum \theta \alpha_i = \theta \sum n_i \alpha_i = 0$$

so that $\boldsymbol{\delta}^1$ and $\boldsymbol{\delta}^2$ are orthogonal. (Note that the constraint $\sum n_i \alpha_i$ is the only constraint which makes these effects orthogonal.) There are one θ and $p - 1$ linearly independent α_i. Therefore

$$d_1 = 1, \quad d_2 = p - 1, \quad dfe = N - (p - 1) - 1 = N - p.$$

To find the LS estimators of θ and α_i we must minimize

$$Q(\theta, \alpha_1, \ldots, \alpha_p) = \|\mathbf{Y} - \boldsymbol{\delta}^1 - \boldsymbol{\delta}^2\|^2 = \sum \sum (Y_{ij} - \theta - \alpha_i)^2$$

subject to $\sum n_i \alpha_i = 0$. Note that the constraint on the α_i implies that

$$\frac{\partial}{\partial \theta} Q(\theta, \alpha_1, \ldots, \alpha_p) = -2 \sum \sum (Y_{ij} - \theta - \alpha_i) = -2N(\overline{Y}_{..} - \theta).$$

Therefore the LS estimator of θ is $\widehat{\theta} = \overline{Y}_{..}$. Also

$$\frac{\partial}{\partial \alpha_i} Q(\theta, \alpha_1, \ldots, \alpha_p) = -2 \sum_j (Y_{ij} - \theta - \alpha_i) = -2n_i(\overline{Y}_{i.} - \theta - \alpha_i).$$

Therefore the LS estimator of α_i is $\widehat{\alpha}_i = \overline{Y}_{i.} - \widehat{\theta} = \overline{Y}_{i.} - \overline{Y}_{..}$. (Note that $\sum n_i \widehat{\alpha}_i = 0$, so that these estimators satisfy the constraint.) Hence the LS estimator of $\mu_{ij} = \theta + \alpha_i$ is

$$\widehat{\mu}_{ij} = \widehat{\theta} + \widehat{\alpha}_i = \overline{Y}_{i.} \Rightarrow SSE = \|\mathbf{Y} - \widehat{\boldsymbol{\mu}}\|^2 = \sum \sum (Y_{ij} - \overline{Y}_{i.})^2.$$

30.16 Now consider testing the null hypothesis that $\boldsymbol{\delta}^q = \mathbf{0}$ for some q against the unrestricted alternative hypothesis. Let

$$W_q = \{\boldsymbol{\mu} = \sum \boldsymbol{\delta}^s : \boldsymbol{\delta}^s \in T_s, s = 1, \ldots, r, \boldsymbol{\delta}^q = \mathbf{0}\} = \bigoplus_{j \neq q} T_j$$

be the subspace for the null hypothesis. We are testing the null hypothesis that $\boldsymbol{\mu} \in W_q$ against the alternative that $\boldsymbol{\mu} \in V$. In Exercise 30.40 it is shown that

$$V|W_q = T_q.$$

Therefore, the sum of squares (*SSq*), degrees of freedom (*dfq*), mean square (*MSq*) and expected mean square (*EMSq*) for this hypothesis are

$$SSq = \|\mathbf{P}_{V|W_q}\mathbf{Y}\|^2 = \|\mathbf{P}_{T_q}\mathbf{Y}\|^2 = \|\widehat{\boldsymbol{\delta}}^q\|^2,$$

$$dfq = \dim(V|W_q) = \dim(T_q) = d_q, \quad MSq = \frac{SSq}{dfq}, \quad EMSq = \sigma^2 + \frac{\|\boldsymbol{\delta}^q\|^2}{dfq}.$$

The F statistic for this hypothesis is

$$F_q = \frac{MSq}{MSE} \sim F_{dfq,dfe}(\xi_q/\sigma^2), \quad \xi_q = \|\mathbf{P}_{V|W_q}\boldsymbol{\mu}\|^2 = \|\boldsymbol{\delta}^q\|^2.$$

Let the *total sum of squares* $SST = \|\mathbf{Y}\|^2$. Note that

$$SST = \|\mathbf{Y}\|^2 = \|\mathbf{P}_V\mathbf{Y}\|^2 + \|\mathbf{P}_{V^\perp}\mathbf{Y}\|^2 = \sum \|\mathbf{P}_{Tq}\mathbf{Y}\|^2 + \|\mathbf{P}_{V^\perp}\mathbf{Y}\|^2 = \sum SSq + SSE.$$

The individual sums of squares add up to the total sum of squares if and only if the model is an orthogonal design.

Let $\mathbf{d}^q \in V|W_q = T_q$. In Exercise 30.41 it is shown that

$$\mathbf{d}^{qT}\boldsymbol{\mu} = \mathbf{d}^{qT}\boldsymbol{\delta}^q, \quad \mathbf{d}^{qT}\widehat{\boldsymbol{\mu}} = \mathbf{d}^{qT}\widehat{\boldsymbol{\delta}}^q.$$

Therefore, the Scheffé simultaneous confidence intervals for contrasts associated with this hypothesis are

$$\mathbf{d}^{qT}\boldsymbol{\delta}^q \in \mathbf{d}^{qT}\widehat{\boldsymbol{\delta}}^q \pm (d_q\, MSE\|\mathbf{d}^q\|^2 F^\alpha)^{1/2}, \ \forall \mathbf{d}^q \in T_q.$$

Example 30.9 (One-way ANOVA as orthogonal design: inference)

We continue with the one-way ANOVA model discussed above. For this model there are one θ and $p-1$ α_i. Therefore,

$$SS\theta = \|\widehat{\boldsymbol{\delta}}^1\|^2 = \sum\sum \widehat{\theta}^2 = N\overline{Y}_{..}^2, \quad df\theta = 1, \quad MS\theta = SS\theta, \quad EMS\theta = \sigma^2 + N\theta^2$$

$$SS\alpha = \|\widehat{\boldsymbol{\delta}}^2\|^2 = \sum\sum \widehat{\alpha}_i^2 = \sum n_i(\overline{Y}_{i.} - \overline{Y}_{..})^2, \quad df\alpha = p-1,$$

$$MS\alpha = \frac{SS\alpha}{df\alpha}, \quad EMS(\alpha) = \sigma^2 + \frac{\|\boldsymbol{\delta}^2\|^2}{df\alpha} = \sigma^2 + \frac{\sum n_i\alpha_i^2}{p-1}.$$

In particular, to test that the $\alpha_i = 0$, we reject if

$$F = \frac{MS\alpha}{MSE} > F^{\alpha}_{p-1,N-p}.$$

(This test is, of course, the same test as that derived earlier without orthogonal designs.) To find the simultaneous confidence intervals for contrasts in the α_i, we let $\mathbf{d} = (d_{11}, \ldots, d_{1n_1}, d_{21}, \ldots, d_{pn_p})^T$ exist in the same subspace as $\boldsymbol{\delta}^{\,2}$. That is, the d_{ij} satisfy the same conditions as the $\delta^2_{ij} = \alpha_i$, namely, $d_{ij} = a_i$ depends only on i and $\sum n_i a_i = 0$. In that case

$$\mathbf{d}^T \boldsymbol{\delta}^{\,2} = \sum\sum d_{ij}\delta^2_{ij} = \sum n_i a_i \alpha_i, \quad \mathbf{d}^T \widehat{\boldsymbol{\delta}}^{\,2} = \sum n_i a_i \widehat{\alpha}_i, \quad \|\mathbf{d}\|^2 = \sum n_i a_i^2.$$

Hence the Scheffé simultaneous confidence intervals for contrast in the α_i are given by

$$\sum n_i a_i \alpha_i \in \sum n_i a_i \widehat{\alpha}_i \pm \left((p-1)F^{\alpha}_{p-1,N-p} MSE \sum n_i a^2_{ij}\right)^{1/2}, \qquad \forall a_i \ni \sum n_i a_i = 0.$$

To put these intervals into a simpler form let $c_i = n_i a_i$, giving the intervals

$$\sum c_i \alpha_i \in \sum c_i \widehat{\alpha}_i \pm \left((p-1)F^{\alpha}_{p-1,N-p} MSE \sum \frac{c_i^2}{n_i}\right)^{1/2}, \qquad \forall c_i \ni \sum c_i = 0.$$

(Note that $\sum c_i \alpha_i = \sum c_i \mu_i$ and $\sum c_i \widehat{\alpha}_i = \sum c_i \overline{Y}_{i.}$ so that these intervals are the same as those derived earlier without orthogonal designs.)

30.17 Most ANOVA models do not come naturally in vector form. In this section, we reinterpret the results given above in components. (Statements in parentheses are applications of the principles to the one-way model.) ANOVA models often have many subscripts. In what follows, we use $\mathbf{m}$ as a symbol for all the subscripts. Suppose we observe N independent random variables $Y_{\mathbf{m}}$ such that

$$Y_{\mathbf{m}} \sim N(\mu_{\mathbf{m}}, \sigma^2), \qquad \mu_{\mathbf{m}} = \delta^1_{\mathbf{m}} + \cdots + \delta^r_{\mathbf{m}}.$$

Let $\mathbf{Y}$, $\boldsymbol{\mu}$ and $\boldsymbol{\delta}^{\,q}$ be an ordering of the $Y_{\mathbf{m}}$, $\mu_{\mathbf{m}}$ and $\delta^q_{\mathbf{m}}$ into N-dimensional vectors. Then

$$\mathbf{Y} \sim N_N(\boldsymbol{\mu}, \sigma^2\mathbf{I}), \qquad \boldsymbol{\mu} = \boldsymbol{\delta}^{\,1} + \cdots + \boldsymbol{\delta}^{\,r}.$$

1. To check that this model is an orthogonal design we verify that

$$0 = \boldsymbol{\delta}^{qT} \boldsymbol{\delta}^{\,s} = \sum_{\mathbf{m}} \delta^q_{\mathbf{m}} \delta^s_{\mathbf{m}}, \qquad \text{for } q \neq s.$$

   (Verify that $\sum\sum \theta\alpha_i = 0$.)

2. We find the LS estimators $\widehat{\delta}^q_{\mathbf{m}}$ of $\delta^q_{\mathbf{m}}$ by minimizing

$$Q = \|\mathbf{Y} - (\boldsymbol{\delta}^{\,1} + \cdots + \boldsymbol{\delta}^{\,r})\|^2 = \sum_{\mathbf{m}} (Y_{\mathbf{m}} - (\delta^1_{\mathbf{m}} + \cdots + \delta^r_{\mathbf{m}}))^2.$$

   Then $\widehat{\mu}_{\mathbf{m}} = \widehat{\delta}^1_{\mathbf{m}} + \cdots + \widehat{\delta}^r_{\mathbf{m}}$. (Minimize $\sum\sum(Y_{ij} - \theta - \alpha_i)^2$, to obtain $\widehat{\theta} = \overline{Y}_{..}$, $\widehat{\alpha}_i = \overline{Y}_{i.} - \overline{Y}_{..}$, $\widehat{\mu}_{ij} = \overline{Y}_{i.}$.)

3. We then find $d_q = dfq$, the number of linearly independent parameters in $\boldsymbol{\delta}^q$ and

$$SSq = \|\widehat{\boldsymbol{\delta}}^q\|^2 = \sum_{\mathbf{m}} (\widehat{\delta}^q_{\mathbf{m}})^2.$$

($df\theta = 1, df\alpha = p-1, SS\theta = \sum\sum \widehat{\theta}^2 = N\overline{Y}_{..}^2, SS\alpha = \sum\sum \widehat{\alpha}_i^2 = \sum n_i(\overline{Y}_{i.} - \overline{Y}_{..})^2$.)

4. We then find $dfe = N - (d_1 + \cdots + d_q)$ and

$$SSE = \|\mathbf{Y} - \widehat{\boldsymbol{\mu}}\|^2 = \sum_{\mathbf{m}} (Y_{\mathbf{m}} - \widehat{\mu}_m)^2 = SST - \sum_q SSq$$

where $SST = \|\mathbf{Y}\|^2 = \sum_{\mathbf{m}} Y^2_{\mathbf{m}}$. ($dfe = N-(p-1)-1 = N-p, SSE = \sum\sum(Y_{ij} - \overline{Y}_{i.})^2 = \sum\sum Y_{ij}^2 - \sum n_i \overline{Y}_{i.}^2$.)

5. The non-centrality parameter γ and EMS for the F test for $\boldsymbol{\delta}^q$ are

$$\gamma = \frac{\xi_q}{\sigma^2}, \quad EMSq = \sigma^2 + \frac{\xi_q}{dfq}, \quad \xi_q = \|\boldsymbol{\delta}^q\|^2 = \sum (\delta^q_{\mathbf{m}})^2.$$

($\gamma_\theta = N\theta^2/\sigma^2, EMS\theta = \sigma^2 + N\theta^2, \gamma_\alpha = \sum n_i\alpha_i^2/\sigma^2, EMS\alpha = \sigma^2 + \sum n_i \alpha_i^2/(p-1)$.)

6. Let $d^q_{\mathbf{m}}$ have the same subscripts and satisfy the same constraints as the $\delta^q_{\mathbf{m}}$. Then the Scheffé simultaneous confidence intervals for contrasts associated with $\boldsymbol{\delta}^q$ are

$$\sum_{\mathbf{m}} d^q_{\mathbf{m}} \delta^q_{\mathbf{m}} \in \sum_{\mathbf{m}} d^q_{\mathbf{m}} \widehat{\delta}^q_{\mathbf{m}} \pm \left(d_q \, MSE \sum_{\mathbf{m}} (d^q_{\mathbf{m}})^2 \; F^\alpha\right)^{1/2}.$$

($d_{ni} = a_i$ where $\sum n_i a_i = 0, \sum n_i a_i \alpha_i \in \sum n_i a_i \widehat{\alpha}_i \pm \left((p-1) MSE \sum n_i a_i^2 \; F^\alpha\right)^{1/2}$.)

In using the rules given above, it is important to remember two things. The first is that they only work for orthogonal designs. The second is that any sum over **m** is a sum over all the subscripts in the model, not just those subscripts which occur in the summand. (For example, do not use $\sum_i a_i\alpha_i$ for a contrast but use $\sum_i \sum_j a_i \alpha_i = \sum_i n_i a_i \alpha_i$. Also we do not use $\sum_i \widehat{\alpha}_i^2$ for $SS\alpha$, but use instead $\sum_i \sum_j \widehat{\alpha}_i^2 = \sum_i n_i \widehat{\alpha}_i^2$.)

Once we have calculated the various sums of squares and degrees of freedom, we can calculate the other statistics by

$$MSq = \frac{SSq}{d_q}, \quad MSE = \frac{SSE}{dfe}, \quad F_q = \frac{MSq}{MSE}.$$

For numerical examples, these quantities are often arranged into an ANOVA table in the following form

effect	*df*	*SS*	*MS*	*F*
$\boldsymbol{\delta}^1$	d_1	$SS1$	$MS1$	F_1
⋮	⋮	⋮	⋮	⋮
$\boldsymbol{\delta}^r$	d_r	SSr	MSr	F_r
error	*dfe*	*SSE*	*MSE*	
total	N	*SST*		

Note that for each row of this table, the number in the *MS* column is just the number in the *SS* column divided by the number in the *df* column, and the number in the *F* column is just the number in the *MS* column, divided by *MSE*, so that once we have specified the various degrees of freedom and sums of squares, the remainder of the table can be established immediately. Therefore when presenting tables of formulae we replace the last two columns by a column of expected mean squares. For example, the ANOVA table for the one-way model is

One-way fixed effects model

effect	*df*	*SS*	*EMS*
θ	1	$N\overline{Y}_{..}^2$	$\sigma^2 + N\theta^2$
α	$p-1$	$\sum n_i(\overline{Y}_{i.} - \overline{Y}_{..})^2$	$\sigma^2 + \sum h_i\alpha_i^2$
error	$N-p$	$\sum\sum(Y_{ij} - \overline{Y}_{i.})^2$	σ^2
total	N	$\sum\sum Y_{ij}^2$	

where $h_i = n_i/(p-1)$. (Note that many authors do not include the row for the intercept (θ) in the ANOVA table and report *SST* as $\sum\sum Y_{ij}^2 - SS\theta = \sum\sum(Y_{ij} - \overline{Y}_{..})^2$. In this book we include the intercept in ANOVA tables.)

We make the following conventions for the remainder of this chapter. When we replace a subscript by a . (dot), we mean the sum over that subscript. If in addition, we put a bar over the figure, we mean the average over that subscript. For example if a_{ijk} has three subscripts, $i = 1, \dots, r$; $j = 1, \dots, c$; $k = 1, \dots, n_{ij}$, then

$$a_{ij.} = \sum_{k=1}^{n_{ij}} a_{ijk}, \quad \overline{a}_{ij.} = \frac{a_{ij.}}{n_{ij}}, \quad a_{i..} = \sum_{j=1}^{c}\sum_{k=1}^{n_{ij}} a_{ijk} = \sum_{j=1}^{c} a_{ij.},$$

$$n_{i.} = \sum_{j=1}^{c} n_{ij}, \quad \overline{a}_{i..} = \frac{a_{i..}}{n_{i.}} = \frac{\sum_{j=1}^{c} n_{ij}\overline{a}_{ij.}}{n_{i.}}.$$

Note that the average is always an unweighted average of the original variables, not necessarily an unweighted average of the averages.

Balanced two-way crossed models

30.18 In the balanced two-way model we observe Y_{ijk} independent,

$$Y_{ijk} \sim N(\mu_{ij}, \sigma^2), \qquad i = 1, \dots, r; \qquad j = 1, \dots, c; \qquad k = 1, \dots, n.$$

We call i the *row* of the observation and j the *column* of the observation. We also talk of the observations in the (i, j)th *cell*. This model is called a *crossed* model because we have observations for every possible pair (i, j). It is *balanced* because we have the same number n of observations in each cell. In an *additive* model we make the additional assumption that

$$\mu_{ij} = \gamma_i + \tau_j$$

for some constants γ_i and τ_j. We are interested in testing the equality of the γ_i (that the μ_{ij} do not depend on the row i of the observation) and testing the equality of the τ_j (that the μ_{ij} do not depend on the column j of the observation).

To make this additive model into an orthogonal design, let

$$\theta = \overline{\gamma} + \overline{\tau}, \quad \alpha_i = \gamma_i - \overline{\gamma}, \quad \beta_j = \tau_j - \overline{\tau},$$

so that

$$\mu_{ij} = \theta + \alpha_i + \beta_j, \quad \sum \alpha_i = 0, \quad \sum \beta_j = 0.$$

We call the α_i *row effects* and call the β_j *column effects*. Note also that

$$\gamma_i \text{ are equal} \Leftrightarrow \alpha_i = 0, \quad \tau_j \text{ are equal} \Leftrightarrow \beta_j = 0.$$

This is an orthogonal design with $\delta^1_{ijk} = \theta$, $\delta^2_{ijk} = \alpha_i$, $\delta^3_{ijk} = \beta_j$. To see the orthogonality of these effects note that

$$\sum\sum\sum \delta^1_{ijk}\delta^2_{ijk} = nc\theta \sum \alpha_i = 0, \quad \sum\sum\sum \delta^1_{ijk}\delta^3_{ijk} = nr\theta \sum \beta_j = 0,$$
$$\sum\sum\sum \delta^2_{ijk}\delta^3_{ijk} = n\left(\sum \alpha_i\right)\left(\sum \beta_j\right) = 0.$$

To find the LS estimators of the parameters, we must minimize

$$\sum\sum (Y_{ijk} - \theta - \alpha_i - \beta_j)^2 \quad \text{subject to} \sum \alpha_i = 0, \quad \sum \beta_j = 0.$$

It is easily verified that the LS estimators are

$$\widehat{\theta} = \overline{Y}_{...}, \quad \widehat{\alpha}_i = \overline{Y}_{i..} - \overline{Y}_{...}, \quad \widehat{\beta}_j = \overline{Y}_{.j.} - \overline{Y}_{...}.$$

Therefore,

$$\widehat{\mu}_{ij} = \widehat{\theta} + \widehat{\alpha}_i + \widehat{\beta}_j = \overline{Y}_{i..} + \overline{Y}_{.j.} - \overline{Y}_{...}.$$

Hence

$$SS\theta = \sum\sum\sum \widehat{\theta}^2 = ncr\overline{Y}^2_{...}, \quad SS\alpha = \sum\sum\sum \widehat{\alpha}_i^2 = nc\sum(\overline{Y}_{i..} - \overline{Y}_{...})^2,$$
$$SS\beta = \sum\sum\sum \widehat{\beta}_j^2 = nr\sum(\overline{Y}_{.j.} - \overline{Y}_{...})^2,$$
$$SSE = \sum\sum\sum (Y_{ijk} - \widehat{\mu}_{ij})^2 = \sum\sum\sum (Y_{ijk} - \overline{Y}_{i..} - \overline{Y}_{.j.} + \overline{Y}_{...})^2.$$

In addition, there are one θ, $r - 1$ linearly independent α_i and $c - 1$ linearly independent β_j (and hence $dfe = nrc - (r-1) - (c-1) - 1$). Therefore (by the rules discussed previously) the ANOVA table for this model is given by

Balanced two-way additive fixed effects model

effect	*df*	*SS*	*EMS*
θ	1	$ncr\overline{Y}^2_{...}$	$\sigma^2 + ncr\theta^2$
α_i	$r - 1$	$nc\sum(\overline{Y}_{i..} - \overline{Y}_{...})^2$	$\sigma^2 + g\sum\alpha_i^2$
β_j	$c - 1$	$nr\sum(\overline{Y}_{.j.} - \overline{Y}_{...})^2$	$\sigma^2 + h\sum\beta_j^2$
error	$rcn - r - c + 1$	$\sum\sum\sum(Y_{ijk} - \overline{Y}_{i..} - \overline{Y}_{.j.} + \overline{Y}_{...})^2$	σ^2
total	rcn	$\sum\sum\sum Y_{ijk}^2$	

where $g = nc/(r-1)$, $h = nr/(c-1)$.

Let $d_{ijk} = a_i$, where $\sum a_i = 0$. (That is, the d_{ijk} have the same subscripts and satisfy the same constraints as the $\delta^2_{ijk} = \alpha_i$). Then

$$\sum\sum\sum d_{ijk}\delta^2_{ijk} = nc\sum a_i\alpha_i, \quad \sum\sum\sum d_{ijk}\widehat{\delta}_{ijk} = nc\sum a_i\widehat{\alpha}_i = nc\sum a_i\overline{Y}_{i.},$$
$$\sum\sum\sum(d_{ijk})^2 = nc\sum a_i^2.$$

Hence the Scheffé simultaneous confidence intervals for contrasts in the α_i are given by

$$nc\sum a_i\alpha_i \in nc\sum a_i\widehat{\alpha}_i \pm \left((r-1)MSEnc\sum a_i^2\ F^\alpha\right)^{1/2}, \qquad \forall a_i \ni \sum a_i = 0.$$

Dividing both sides by nc, we see that an equivalent form for these intervals is

$$\sum a_i\alpha_i \in \sum a_i\widehat{\alpha}_i \pm \left((r-1)MSE\sum \frac{a_i^2}{nc}\ F^\alpha\right)^{1/2}, \qquad \forall a_i \ni \sum a_i = 0.$$

In a similar way, we can show that the simultaneous confidence intervals for contrasts in the β_j are

$$\sum b_j\beta_j \in \sum b_j\widehat{\beta}_j \pm \left((c-1)MSE\sum \frac{b_j^2}{nr}\ F^\alpha\right)^{1/2}, \qquad \forall b_j \ni \sum b_j = 0.$$

Note also that the $\overline{Y}_{i..} \sim N(\theta + \alpha_i, \sigma^2/nc)$ are independent, so that

$$Q(\boldsymbol{\alpha}) = \frac{\max_{i\neq j}|\overline{Y}_{i..} - \overline{Y}_{j..} - (\alpha_i - \alpha_j)|}{\sqrt{MSE/nc}} \sim q_{r,dfe}$$

(where $q_{s,t}$ is the studentized range distribution discussed in **30.5**). Hence we can construct Tukey simultaneous confidence intervals for contrasts in the α_i as discussed in **30.9**. Similarly,

$$D(\boldsymbol{\alpha}) = \frac{\max_{i\neq 1}|\overline{Y}_{i..} - \overline{Y}_{1..} - (\alpha_i - \alpha_1)|}{\sqrt{2MSE/nc}} \sim D_{r-1,dfe},$$
$$OD(\boldsymbol{\alpha}) = \frac{\max_{i\neq 1}(\overline{Y}_{i..} - \overline{Y}_{1..} - (\alpha_i - \alpha_1))}{\sqrt{2MSE/nc}} \sim OD_{r-1,dfe}.$$

(Dunnett's two-sided and one-sided distributions) so that we can also construct the Dunnett simultaneous confidence intervals and the MCB procedure is also available for this model as discussed in **30.10–11**. Multiple range, F and Bonferroni procedures are also available, as discussed in **30.12–13**.

30.19 We continue with the balanced two-way crossed model in which we observe Y_{ijk} independent,

$$Y_{ijk} \sim N(\mu_{ij}, \sigma^2), \qquad i = 1, \ldots, r;\ j = 1, \ldots, c;\ k = 1, \ldots, n.$$

We no longer assume that the model is additive. Let

$$\theta = \overline{\mu}_{..}, \quad \alpha_i = \overline{\mu}_{i.} - \overline{\mu}_{..}, \quad \beta_j = \overline{\mu}_{.j} - \overline{\mu}_{..}, \quad \gamma_{ij} = \mu_{ij} - \overline{\mu}_{i.} - \overline{\mu}_{.j} + \overline{\mu}_{..}.$$

Then it is easily verified that

$$\mu_{ij} = \theta + \alpha_i + \beta_j + \gamma_{ij}, \quad \sum \alpha_i = 0, \sum \beta_j = 0, \quad \sum_i \gamma_{ij} = 0, \quad \sum_j \gamma_{ij} = 0.$$

Let $\delta^1_{ijk} = \theta$, $\delta^2_{ijk} = \alpha_i$, $\delta^3_{ijk} = \beta_j$ and $\delta^4_{ijk} = \gamma_{ij}$. Then it is also easily verified that these parameters are orthogonal and that this model is an orthogonal design. As in **30.18**, we call the α_i and the β_j the *row* and *column effects*. We also call these parameters *main effects*. We call the γ_{ij} *interaction effects*.

Note that

$$\gamma_{ij} = 0 \Leftrightarrow \text{model is additive},$$
$$\alpha_i = 0 \Leftrightarrow \overline{\mu}_{i.} \text{ are equal}, \qquad \beta_j = 0 \Leftrightarrow \overline{\mu}_{.j} \text{ are equal}.$$

Therefore when we test for no interaction, we are testing that the model is additive. When we test for no row effect, we are testing that the average effect for different rows is the same; and when we test for no column effect, we are testing that the average effect in each column is the same.

By straightforward differentiation, we see that the LS estimators of the parameters are

$$\widehat{\theta} = \overline{Y}_{...}, \quad \widehat{\alpha}_i = \overline{Y}_{i..} - \overline{Y}_{...}, \quad \widehat{\beta}_j = \overline{Y}_{.j.} - \overline{Y}_{...}, \quad \widehat{\gamma}_{ij} = \overline{Y}_{ij.} - \overline{Y}_{i..} - \overline{Y}_{.j.} + \overline{Y}_{...},$$

and therefore

$$\widehat{\mu}_{ij} = \widehat{\theta} + \widehat{\alpha}_i + \widehat{\beta}_j + \widehat{\gamma}_{ij} = \overline{Y}_{ij.}.$$

There are one θ, $r-1$ linearly independent α_i, $c-1$ linearly independent β_j, and $(r-1)(c-1)$ linearly independent γ_{ij}. Therefore, the ANOVA table for this model is given by

Balanced two-way fixed effects model with interaction

effect	*df*	*SS*	*EMS*
θ	1	$ncr\overline{Y}^2_{...}$	$\sigma^2 + ncr\theta^2$
α_i	$r-1$	$nc\sum(\overline{Y}_{i..} - \overline{Y}_{...})^2$	$\sigma^2 + g\sum \alpha_i^2$
β_j	$c-1$	$nr\sum(\overline{Y}_{.j.} - \overline{Y}_{...})^2$	$\sigma^2 + h\sum \beta_j^2$
γ_{ij}	$(r-1)(c-1)$	$n\sum\sum(\overline{Y}_{ij.} - \overline{Y}_{i..} - \overline{Y}_{.j.} + \overline{Y}_{...})^2$	$\sigma^2 + k\sum\sum \gamma_{ij}^2$
error	$rc(n-1)$	$\sum\sum\sum(Y_{ijk} - \overline{Y}_{ij.})^2$	σ^2
total	rcn	$\sum\sum\sum Y^2_{ijk}$	

where $g = nc/(r-1)$, $h = nr/(c-1)$ and $k = n/(r-1)(c-1)$.

The simultaneous confidence intervals (including Tukey, Dunnett and MCB) for contrasts in α_i are the same as those discussed in **30.18** for the additive model, except that we use *dfe* and *MSE* derived for this model. Multiple range, multiple F and multiple Bonferroni procedures for testing for main effects can be constructed in the obvious way.

To find the simultaneous confidence intervals for contrasts in the γ_{ij}, let $d_{ijk} = c_{ij}$, where

$\sum_i c_{ij} = 0$, $\sum_j c_{ij} = 0$. Then direct substitution (and dividing through by n) gives the intervals

$$\sum\sum c_{ij}\gamma_{ij} \in \sum_{ij}\sum c_{ij}\widehat{\gamma}_{ij} \pm \left((r-1)(c-1)MSE\sum\sum \frac{c_{ij}^2}{n}F^\alpha\right),$$

$$\forall c_{ij} \ni \sum_i c_{ij} = 0, \qquad \sum_j c_{ij} = 0.$$

Note that a contrast in the γ_{ij} must have at least four terms. There are no Tukey, Dunnett or MCB type simultaneous confidence intervals for contrasts in the interactions nor any multiple comparisons procedures for them.

Interpreting main effects – proportional sampling

30.20 The test for main effects in the presence of possible interactions is often hard to interpret in practice. We discuss the difficulty in this section. We focus on testing for row effects, but we could just as easily look at column effects.

The difficulty is that the test for no row effects is testing the equality of the

$$\overline{\mu}_{i.} = \frac{\sum_j \mu_{ij}}{c}.$$

This average may not represent the 'average' effect of the ith row (averaged across the columns). For example, the columns could represent various ethnic groups. In this case, it may not be reasonable to assign equal weights to all the columns when computing the average effect for a particular treatment.

For another example, we could be testing some drugs for treating a disease. If the columns represented the gender of the respondent and 10 per cent of the men had the disease, but only 1 per cent of the women did, we might not want to treat the columns equally. We might want to put more weight on the men than on the women. If $j = 1$ represented men and $j = 2$ represented women, we might want to evaluate the average performance of the ith drug by

$$\overline{\mu}^*_{i.} = \frac{10\mu_{i1} + \mu_{i2}}{11}.$$

We might prefer to compare the $\overline{\mu}^*_{i.}$ instead of comparing the $\overline{\mu}_{i.}$. For example, suppose we had $r = 2$ drugs with means

$$\mu_{11} = 10, \quad \mu_{12} = 4, \quad \mu_{21} = 6, \quad \mu_{22} = 8.$$

(where high numbers mean that the drug is better). Then

$$\overline{\mu}_{1.} = 7 = \overline{\mu}_{2.}$$

so that if we use $\overline{\mu}_{i.}$ as measure of 'average' performance of the drug, we would say that the two drugs performed equally well. However,

$$\overline{\mu}^*_{1.} = \frac{140}{11} \approx 12.72, \quad \overline{\mu}^*_{2.} = \frac{68}{11} \approx 6.18$$

so that in terms of $\overline{\mu}_{i.}^{*}$ the first drug is much better than the second drug.

Consideration of the example above suggests that the researcher interested in testing that there is no 'average' effect due to the rows should have in mind some weights v_j which measure the relative importance of the columns and then should test the equality of the

$$\overline{\mu}_{i.}^{\mathbf{v}} = \frac{\sum_j v_j \mu_{ij}}{\sum_j v_j}.$$

Without any such weights, the hypothesis of no average effect due to rows is not defined. Note that in **30.19** we tacitly assumed that all columns had equal weight. However, in many experiments, it seems that such an assumption would not be warranted. Note also that in the balanced case, it is necessary to take equal weights in order to get an orthogonal design. However, in **30.21** we discuss an alternative sampling scheme in which an orthogonal design can be constructed for any weights. In **30.33**, we discuss procedures for testing for no row effects for balanced or unbalanced models for any weights.

In practice, of course, one rarely has exact weights at one's fingertips to use to define the hypothesis of no row effects. Several procedures have been suggested to deal with this difficulty.

1. Never test this hypothesis of no average row effects. If we imagine that interactions are possible, then we test the hypothesis that μ_{ij} does not depend on the row i, i.e. that $\alpha_i + \gamma_{ij} = 0$. This hypothesis is really what we mean by 'no row effect'. Searle (1971) advocates this approach, but it seems a bit restrictive in practice.
2. First test the null hypothesis that the interactions are 0, i.e. that the model is additive.

   (a) If we reject this hypothesis, we stop and declare there to be a 'row effect'. We could then follow up with individual one-way analyses on the rows for each column treatment (i.e. compare the medicines separately for men and women).

   (b) If there is no significant interaction, we then assume the model is an additive model and test the hypothesis that the $\alpha_i = 0$ for that model. Note that for the additive model, there is no problem interpreting the test for no average row effects. It is merely testing that μ_{ij} does not depend on i. (Another perspective on this fact is that $\mu_{ij} = \zeta_i + \delta_j$ implies that $\overline{\mu}_{i.}^{\mathbf{v}} = \xi_i + \sum v_j \delta_j / \sum v_j$ which implies that $\overline{\mu}_{i.}^{\mathbf{v}}$ are equal if and only if ξ_i are equal which does not depend on the weights.)

   This method is one of the earlier answers to this problem (see, for example, Snedecor and Cochran, 1967). The trouble with this two-stage procedure is that accepting a null hypothesis is not proof that its is true, but is merely lack of proof that it is false. For example, if the test for no interactions had a p-value of 0.07, we would accept the null hypothesis with a 0.05 test, but we probably would not believe that model was additive. We would just be unable to prove it was not additive. It does not seem sensible to use the additive model to test for row effects in this case. Therefore in using this procedure, we should not choose small α for the preliminary test of additivity.
3. The third approach is to take equal weights all the time. This is the approach used in Graybill (1976). These are the weights which lead to an orthogonal design in the case of balanced sampling. In addition, if we know nothing about the relative importance of the

columns, it is hard to imagine choosing any other weights (These are the weights which correspond to Type III sums of squares in SAS.) One disadvantage of this approach is the following. Suppose we started with three columns, but after looking at the data we decided to collapse the second and third columns into one column, so that we now only have two columns in the experiment. What we have done is increased the weight on the first column from $\frac{1}{3}$ to $\frac{1}{2}$. In fact it seems clear that, in any system of weights, we should add the weights for two columns when they are combined into one column. (Note that this last statement is incompatible with always using equal weights.)

4. The fourth approach is to make the researcher choose the weights for the columns (perhaps with equal weights as a default). This approach would force the researcher at least to think about the relative importance of the columns. If he has no weights, then the hypothesis of no row effects is not really defined and probably should not be tested.

The fourth approach to this problem seems reasonable. However, at present it is not practical, because the software programmes do not allow it. In the absence of predetermined weights for the columns (or the appropriate software for implementing the weights) it seems that the second procedure is the most sensible, although for this procedure the choice of α for the preliminary test is problematical.

30.21 Suppose now we have a two-way model in which we observe Y_{ijk} independent,

$$Y_{ijk} \sim N(\mu_{ij}, \sigma^2), \qquad i = 1, \ldots, r; \; j = 1, \ldots, c; \; k = 1, \ldots, n_{ij}.$$

Suppose we also have relative weights v_j for the columns and relative weights w_i for the rows. Let

$$\overline{\mu}^{\mathbf{v}}_{i.} = \frac{\sum v_j \mu_{ij}}{\sum v_j}, \quad \overline{\mu}^{\mathbf{w}}_{.j} = \frac{\sum w_i \mu_{ij}}{\sum w_i}$$

be the appropriate weighted averages across rows and columns. Suppose we are interested in testing three hypotheses: testing the equality of the weighted row averages $\overline{\mu}^{\mathbf{v}}_{i.}$, testing the equality of the weighted column averages $\overline{\mu}^{\mathbf{w}}_{.j}$ and testing the additivity of the μ_{ij}. In this section, we present a design which makes these hypotheses orthogonal.

We say that we have *proportional sampling* if

$$n_{ij} = k w_i v_j$$

for some k. (Note that if the v_j are all equal and the w_i are all equal, then this is just the balanced model discussed previously. Note also that this sampling scheme has more observations for those columns or rows which have higher weights, which are presumably the more important rows or columns for the experiment.) It is easily shown that we have proportional sampling for some set of weights if and only if

$$n_{ij} = \frac{n_{i.} n_{.j}}{N}.$$

where $N = n_{..}$ is the total number of observations in the experiment (see Exercise 30.46). Note that for proportional sampling,

$$\frac{\sum_j n_{.j}\mu_{ij}}{N} = \frac{\sum_j kw_{.}v_j\mu_{ij}}{kw_{.}v_{.}} = \frac{\sum_j v_j\mu_{ij}}{\sum_j v_j} = \overline{\mu}_{i.}^{\mathbf{v}} \qquad \frac{\sum n_{i.}\mu_{ij}}{N} = \overline{\mu}_{.j}^{\mathbf{w}}.$$

Define

$$\theta = \frac{\sum\sum w_i v_j \mu_{ij}}{\sum w_i \sum v_j} = \frac{\sum\sum n_{ij}\mu_{ij}}{N} = \frac{\sum_i w_i \overline{\mu}_{i.}^{\mathbf{v}}}{\sum_i w_i} = \frac{\sum_j v_j \overline{\mu}_{.j}^{\mathbf{w}}}{\sum_j v_j},$$

$$\alpha_i = \overline{\mu}_{i.}^{\mathbf{v}} - \theta,\ \beta_j = \overline{\mu}_{.j}^{\mathbf{w}} - \theta, \qquad \gamma_{ij} = \mu_{ij} - \alpha_i - \beta_j - \theta.$$

Then it is easily verified that

$$\mu_{ij} = \theta + \alpha_i + \beta_j + \gamma_{ij},\ \sum_i n_{ij}\alpha_i = kv_j \sum_i w_i\alpha_i = 0 \qquad \forall j,$$

$$\sum_j n_{ij}\beta_j = 0\ \forall i, \qquad \sum_i n_{ij}\gamma_{ij} = 0\ \forall j, \qquad \sum_j n_{ij}\gamma_{ij} = 0\ \forall i.$$

Also

$$\overline{\mu}_{i.}^{\mathbf{v}} \text{ are equal } \Leftrightarrow \alpha_i = 0, \quad \overline{\mu}_{.j}^{\mathbf{w}} \text{ are equal } \Leftrightarrow \beta_j = 0,$$
$$\text{the } \mu_{ij} \text{ are additive } \Leftrightarrow \gamma_{ij} = 0.$$

Using the above constraints it is easily verified that this model is an orthogonal design.

Using the constraints, it is also easily verified by direct differentiation that the LS estimators for this model are

$$\widehat{\theta} = \overline{Y}_{...}, \quad \widehat{\alpha}_i = \overline{Y}_{i..} - \overline{Y}_{...}, \quad \widehat{\beta}_j = \overline{Y}_{.j.} - \overline{Y}_{...},$$
$$\widehat{\gamma}_{ij} = \overline{Y}_{ij.} - \overline{Y}_{i..} - \overline{Y}_{.j.} + \overline{Y}_{...}, \quad \widehat{\mu}_{ij} = \overline{Y}_{ij.}.$$

As in the balanced case, there are one θ, $r-1$ linearly independent α_i, $c-1$ linearly independent β_j and $(r-1)(c-1)$ linearly independent γ_{ij}. Hence

$$dfe = N - (1 + r - 1 + c - 1 + (r-1)(c-1)) = N - rc.$$

Therefore the ANOVA table for this model is

Two-way fixed effects model with proportional sampling

effect	*df*	*SS*	*EMS*
θ	1	$N\overline{Y}_{...}^2$	$\sigma^2 + N\theta^2$
α_i	$r-1$	$\sum n_{i.}(\overline{Y}_{i..} - \overline{Y}_{...})^2$	$\sigma^2 + \sum g_i\alpha_i^2$
β_j	$c-1$	$\sum n_{.j}(\overline{Y}_{.j.} - \overline{Y}_{...})^2$	$\sigma^2 + \sum h_j\beta_j^2$
γ_{ij}	$(r-1)(c-1)$	$\sum\sum n_{ij}(\overline{Y}_{ij.} - \overline{Y}_{i..} - \overline{Y}_{.j.} + \overline{Y}_{...})^2$	$\sigma^2 + \sum\sum k_{ij}\gamma_{ij}^2$
error	$N-rc$	$\sum\sum\sum(Y_{ijk} - \overline{Y}_{ij.})^2$	σ^2
total	N	$\sum\sum\sum Y_{ijk}^2$	

where $g_i = n_{i.}/(r-1)$, $h_j = n_{.j}/(c-1)$ and $k_{ij} = n_{ij}/(r-1)(c-1)$.

The Scheffé simultaneous confidence intervals for contrasts in the α_i (computed from the method in **30.16** and then simplified by replacing $n_{i.}a_i$ by c_i) are given by

$$\sum_i c_i\alpha_i \in \sum_i c_i\widehat{\alpha}_i \pm \left((r-1)MSE\sum_i \frac{c_i^2}{n_{i.}}F^\alpha\right)^{1/2} \qquad \forall c_i \ni \sum_i c_i = 0.$$

The simultaneous confidence intervals for the β_j are constructed similarly.

We now find the Scheffé confidence intervals for contrasts in the γ_{ij}. From the basic principles for constructing confidence intervals and replacing $n_{ij}c_{ij}$ by d_{ij}, we obtain the intervals

$$\sum_i\sum_j d_{ij}\gamma_{ij} \in \sum_i\sum_j d_{ij}\widehat{\gamma}_{ij} \pm \left((r-1)(c-1)MSE\sum_i\sum_j \frac{d_{ij}^2}{n_{ij}}F^\alpha\right)^{1/2}$$

$$\forall d_{ij} \ni \sum_i d_{ij} = 0, \quad \sum_j d_{ij} = 0.$$

Note that the imbalance in this model implies that Tukey, Dunnett and MCB procedures cannot be used, nor can multiple range procedures. However, multiple F procedures are available.

The formulae in this section seem to be the natural generalizations of those in the balanced case, and it may be conjectured that for any unbalanced two-way crossed design, there are some constraints on the parameters which would also lead to these formulae. However, it can be shown that if the design is not proportional sampling (i.e. if $n_{ij} \neq n_{i.}n_{.j}/N$), then there are no constraints on the parameters which make the design orthogonal. The formulae for sums of squares, etc., are also much worse (when formulae exist at all) unless we have proportional sampling with the weights

$$v_j = kn_{i.}, \qquad w_i = k^* n_{.j},$$

for some k and k^*. (Recall that balanced sampling with equal weights is a special case of proportional sampling.) See **30.33** for details.

Higher-way crossed models – Latin squares

30.22 The results of the previous sections can be extended in a straightforward way to balanced higher-way crossed models. We indicate this extension with a balanced three-way model. In that model, we observe Y_{ijkm} independent,

$$Y_{ijkm} \sim N(\mu_{ijk}, \sigma^2), \qquad i = 1, \ldots, r;\ j = 1, \ldots, c;\ k = 1, \ldots, d;\ m = 1, \ldots, n.$$

We can always write the μ_{ijk} in the form

$$\mu_{ijk} = \theta + \alpha_i + \beta_j + \gamma_k + \delta_{ij} + \varepsilon_{ik} + \eta_{jk} + \xi_{ijk}, \quad \sum_i \alpha_i = 0, \quad \sum_j \beta_j = 0,$$

$$\sum_k \gamma_k = 0, \quad \sum_i \delta_{ij} = 0, \quad \sum_j \delta_{ij} = 0, \quad \sum_i \varepsilon_{ik} = 0, \quad \sum_k \varepsilon_{ik} = 0,$$

$$\sum_j \eta_{jk} = 0, \quad \sum_k \eta_{jk} = 0, \quad \sum_i \xi_{ijk} = 0, \quad \sum_j \xi_{ijk} = 0, \quad \sum_k \xi_{ijk} = 0.$$

We call the α_i, β_j and γ_k *main effects*. We call the α_i the effects of the first factor, the β_j the effects of the second factor, etc. The δ_{ij}, ε_{ik} and η_{jk} are called second-order interactions and the ξ_{ijk} are called third-order interactions. It is easily verified that this model is an orthogonal design. Furthermore,

$$\alpha_i = 0 \Leftrightarrow \overline{\mu}_{i..} \text{ are equal, } \delta_{ij} = 0 \Leftrightarrow \overline{\mu}_{ij.} \text{ are additive}$$
$$\xi_{ijk} = 0 \Leftrightarrow \mu_{ijk} = \lambda_{ij} + \varpi_{ik} + \rho_{jk}$$

for some constants λ_{ij}, ϖ_{ik} and ρ_{jk}. Similar statements could be made about testing that the other main effects and two-way interactions are 0. Note that all the hypotheses except testing the three-way interactions tacitly assume equal weights and may not be appropriate for many practical problems in which the levels of the factors are not equally important.

It is easily verified that the LS estimators for this model are

$$\widehat{\theta} = \overline{Y}_{....}, \quad \widehat{\alpha}_i = \overline{Y}_{i...} - \overline{Y}_{....}, \quad \widehat{\delta}_{ij} = \overline{Y}_{ij..} - \overline{Y}_{i...} - \overline{Y}_{.j..} + \overline{Y}_{....}$$
$$\widehat{\xi}_{ijk} = \overline{Y}_{ijk.} - \overline{Y}_{ij..} - \overline{Y}_{i.k.} - \overline{Y}_{..jk.} + \overline{Y}_{i...} + \overline{Y}_{.j..} + \overline{Y}_{..k.} - \overline{Y}_{....}.$$

The other estimated main effects and two-way interactions can be computed similarly, leading to

$$\widehat{\mu}_{ijk} = \overline{Y}_{ijk.}.$$

Therefore we see that

$$SS\theta = rcdn\widehat{\theta}^{\,2}, \quad df\theta = 1, \quad SS\alpha = cdn\sum \widehat{\alpha}_i^{\,2}, \quad df(\alpha) = r - 1$$
$$SS\delta = dn\sum\sum \widehat{\delta}_{ij}^2, \quad df\delta = (r-1)(c-1), \quad SS\xi = n\sum\sum\sum \widehat{\xi}_{ijk}^2,$$
$$df\xi = (r-1)(c-1)(d-1), \quad SSE = \sum\sum\sum\sum (Y_{ijkm} - \overline{Y}_{ijk\cdot})^2, \quad dfe = rcd(n-1).$$

Other sums of squares can be computed similarly. Hypotheses about the main effects and interactions can be tested in the obvious ways. The ANOVA table for this model can be derived in a straightforward manner, but it has 10 rows and is not given here. Scheffé simultaneous confidence intervals are also available, as are Tukey, Dunnett and MCB confidence intervals for the main effects as well as multiple range, F and Bonferroni tests for these effects.

Now, suppose we have weights v_i, w_j and u_k for the various factors. Then we can obtain an orthogonal design by letting the number of observations in the (i, j, k)th cell be

$$n_{ijk} = K v_i w_j u_k.$$

Then we can set up a model similar to that given earlier for proportional sampling in the two-way model. Note that for proportional sampling in a three-way model, we need that

$$n_{ijk} = \frac{n_{i..} n_{.j.} n_{..k}}{N^2},$$

where $N = n_{...}$ is the total number of observations in the experiment.

30.23 An $m \times m$ Latin square is an $m \times m$ matrix such that in each row each integer between 1 and m occurs exactly once and in each column each such integer occurs exactly once. An example of a 3×3 Latin square is

$$\begin{pmatrix} 1 & 2 & 3 \\ 2 & 3 & 1 \\ 3 & 1 & 2 \end{pmatrix}.$$

Latin squares can be easily constructed for any m.

Now suppose we have a fixed $m \times m$ Latin square with k_{ij} as the number in the ith row and the jth column. We can test three factors on a two-way model with one observation in each of $m \times m$ cells by letting the observation in the (i, j)th cell get the ith level of the first factor, the jth level of the second factor and the kth level of the third factor where $k = k_{ij}$. Using this design, we need only m^2 observations to analyse a three-way model. Note that if we used a three-way model it would take m^3 total observations (even with only one observation in each cell). However, to use a Latin square model, we need to have the same number of levels (m) of each of the three factors, and we need to assume that the model is additive. Note also that if any observations in a Latin square are lost, it becomes very difficult to analyse the remaining observations.

In the Latin square model, we observe Y_{ij} independent,

$$Y_{ij} \sim N(\mu_{ij}, \sigma^2), \qquad i = 1, \ldots, m; \; j = 1, \ldots, m.$$

We assume that

$$\mu_{ij} = \delta_i + \varepsilon_j + \eta_{k_{ij}},$$

for some constants δ_i, ε_j and η_k. In order to make this model into an orthogonal model, let

$$\theta = \overline{\delta} + \overline{\varepsilon} + \overline{\eta}, \quad \alpha_i = \delta_i - \overline{\delta}, \quad \beta_j = \varepsilon_j - \overline{\varepsilon}, \quad \gamma_k = \eta_k - \overline{\eta}.$$

Then

$$\mu_{ij} = \theta + \alpha_i + \beta_j + \gamma_{k_{ij}}, \quad \sum \alpha_i = 0, \quad \sum \beta_j = 0, \quad \sum \gamma_k = 0.$$

We call the α_i the effects of the first factor, the β_j the effects of the second factor and the γ_k the effects of the third factor. We want to test the null hypothesis that $\alpha_i = 0$ (no effect due to the first factor), that $\beta_j = 0$ (no effect due to the second factor) and that $\gamma_k = 0$ (no effect due to the third factor). The conditions of this experiment imply that

$$\sum_i \gamma_{k_{ij}} = 0, \quad \sum_j \gamma_{k_{ij}} = 0.$$

This model is easily seen to be an orthogonal design. We verify one of the equalities necessary to establish this orthogonality:

$$\sum_i \sum_j \alpha_i \gamma_{k_{ij}} = \sum_i \alpha_i \sum_j \gamma_{k_{ij}} = 0.$$

Furthermore, the LS estimators of the parameters are given by

$$\widehat{\theta} = \overline{Y}_{..}, \quad \widehat{\alpha}_i = \overline{Y}_{i.} - \overline{Y}_{..}, \quad \widehat{\beta}_j = \overline{Y}_{.j} - \overline{Y}_{..}, \quad \widehat{\gamma}_k = \overline{Y}_{..k} - \overline{Y}_{...},$$

where $\overline{Y}_{..k}$ is the average of all the Y_{ij} such that $k_{ij} = k$ (that is, the average of all the observations whose third factor level is k). Therefore

$$\widehat{\mu}_{ij} = \widehat{\theta} + \widehat{\alpha}_i + \widehat{\beta}_j + \widehat{\gamma}_{k_{ij}} = \overline{Y}_{i.} + \overline{Y}_{.j} + \overline{Y}_{..k_{ij}} - 2\overline{Y}_{...}.$$

Also

$$dfe = m^2 - 3(m-1) - 1 = (m-1)(m-2).$$

Hence the ANOVA table for this model is given by

Latin square fixed effects model

effect	*df*	*SS*	*EMS*
θ	1	$m^2\overline{Y}^2_{..}$	$\sigma^2 + m^2\theta^2$
α_i	$m-1$	$m\sum(\overline{Y}_{i.} - \overline{Y}_{..})^2$	$\sigma^2 + h\sum\alpha_i^2$
β_j	$m-1$	$m\sum(\overline{Y}_{.j} - \overline{Y}_{..})^2$	$\sigma^2 + h\sum\beta_j^2$
γ_k	$m-1$	$m\sum(\overline{Y}_{..k} - \overline{Y}_{..})^2$	$\sigma^2 + h\sum\gamma_k^2$
error	$(m-1)(m-2)$	$\sum\sum(Y_{ij} - \overline{Y}_{i.} - \overline{Y}_{.j} - \overline{Y}_{..k_{ij}} + 2\overline{Y}_{..})^2$	σ^2
total	m^2	$\sum\sum Y_{ij}^2$	

where $h = m/(m-1)$.

Scheffé simultaneous confidence intervals for contrasts in the α_i and contrasts in the β_j can be constructed in the obvious way. To find the confidence interval for contrasts in the γ_k, let $d_{ij} = c_{k_{ij}}$, where c_k are constants such that $\sum c_k = 0$ (so that they have the same form as the $\delta_{ij} = \gamma_{k_{ij}}$). Then

$$\sum\sum d_{ij}\gamma_{k_{ij}} = m\sum c_k\gamma_k, \quad \sum\sum d_{ij}\widehat{\gamma}_{k_{ij}} = m\sum c_k\overline{Y}_{..k}, \quad \sum\sum d_{ij}^2 = m\sum c_k^2,$$

so that the Scheffé simultaneous confidence intervals for the γ_k are (after dividing through by m)

$$\sum c_k\gamma_k \in \sum c_k\overline{Y}_{..k} \pm \left((m-1)MSE\sum\frac{c_k^2}{m}F^\alpha\right)^{1/2}.$$

Note also that the $\overline{Y}_{i.}$ are independent, as are the $\overline{Y}_{.j}$ and the $\overline{Y}_{..k}$ and

$$\overline{Y}_{i.} \sim N\left(\theta + \alpha_i, \frac{\sigma^2}{m}\right), \quad \overline{Y}_{.j} \sim N\left(\theta + \beta_j, \frac{\sigma^2}{m}\right), \quad \overline{Y}_{..k} \sim N\left(\theta + \gamma_k, \frac{\sigma^2}{m}\right)$$

so that Tukey, Dunnett and MCB simultaneous confidence intervals for any of these factors may be constructed in the obvious way, as can multiple range, F and Bonferroni tests.

Symmetric two-way crossed models – Diallel cross

30.24 In a balanced *symmetric* two-way crossed model we observe Y_{ijk} independent, $i = 1, \ldots, r$; $j = 1, \ldots, r$; $k = 1, \ldots, n$, where

$$Y_{ijk} \sim N(\mu_{ij}, \sigma^2), \qquad \mu_{ij} = \mu_{ji}.$$

(Note that the symmetry assumption forces the number of rows and the number of columns to be the same.) In this section we assume, in addition that the model is additive, i.e. that

$$\mu_{ij} = \tau_i + \tau_j$$

for some constants $\tau_1, \ldots, \tau_r$. (A symmetric model with symmetric interaction is treated in Exercise 30.53 and a test for symmetry in Exercise 30.54). We are primarily interested in testing that the τ_q are all equal, i.e. that there is no effect due to the (column and row) treatments. (Note that we use the subscript i for the row and j for the column, but the subscript q when it could be either.)

In order to make this model into an orthogonal design, let

$$\theta = 2\overline{\tau}, \quad \alpha_q = \tau_q - \overline{\tau},$$

so that

$$\mu_{ij} = \theta + \alpha_i + \alpha_j, \quad \sum \alpha_q = 0, \quad \tau_q \text{ equal } \Leftrightarrow \alpha_q = 0.$$

This model has two effects, $\delta^1_{ijk} = \theta$ and $\delta^2_{ijk} = \alpha_i + \alpha_j$. To see that this model is an orthogonal design, note that

$$\sum\sum\sum \theta(\alpha_i + \alpha_j) = nr\theta\left(\sum \alpha_i + \sum \alpha_j\right) = 0,$$

so that the θ effect is orthogonal to the α effect.

It is straightforward to show that the LS estimator of θ is $\widehat{\theta} = \overline{Y}_{....}$. Note also that the constraint on the α_i implies that

$$\begin{aligned}
-\frac{1}{2}\frac{\partial}{\partial \alpha_q}\sum\sum\sum(Y_{ijk} - \theta - \alpha_i - \alpha_j)^2 &= -\frac{1}{2}\frac{\partial}{\partial \alpha_q}\sum_{j \neq q}\sum_k (Y_{qjk} - \theta - \alpha_q - \alpha_j)^2 \\
-\frac{1}{2}\frac{\partial}{\partial \alpha_q}\sum_{i \neq q}\sum_k (Y_{iqk} - \theta - \alpha_i - \alpha_q)^2 &- \frac{1}{2}\frac{\partial}{\partial \alpha_q}\sum_k (Y_{qqk} - \theta - 2\alpha_q)^2 \\
&= nr(\overline{Y}_{q..} + \overline{Y}_{.q.} - 2\theta - 2\alpha_q),
\end{aligned}$$

which implies that

$$\widehat{\alpha}_q = \frac{\overline{Y}_{q..} + \overline{Y}_{.q.}}{2} - \overline{Y}_{...} \Rightarrow \widehat{\mu}_{ijk} = \widehat{\theta} + \widehat{\alpha}_i + \widehat{\alpha}_j = \frac{\overline{Y}_{i..} + \overline{Y}_{j..} + \overline{Y}_{.i.} + \overline{Y}_{.j.}}{2} - \overline{Y}_{....}$$

Also

$$\sum\sum\sum(\widehat{\alpha}_i + \widehat{\alpha}_j)^2 = 2nr\sum \widehat{\alpha}_q^2.$$

Note that there are one θ and $r-1$ α_i. Therefore an ANOVA table for this model is

Two-way symmetric fixed effects model

effect	*df*	*SS*	*EMS*
θ	1	$nr^2\overline{Y}^2_{\cdots}$	$\sigma^2 + nr^2\theta^2$
α	$r-1$	$2nr\sum\left(\frac{\overline{Y}_{q..}+\overline{Y}_{.q.}}{2} - \overline{Y}_{..}\right)^2$	$\sigma^2 + h\sum\alpha_q^2$
error	$r(nr-1)$	$\sum\sum\sum(Y_{ijk} - \widehat{\mu}_{ijk})^2$	σ^2
total	r^2n	$\sum\sum\sum Y_{ijk}^2$	

where $h = 2nr/(r-1)$.

Now, let $d_{ijk} = a_i + a_j$, where $\sum a_i = 0$ (i.e. the a_i have the same structure as the α_i). Then

$$\sum\sum\sum(a_i + a_j)(\alpha_i + \alpha_j) = 2nr\sum a_q\alpha_q,$$
$$\sum\sum\sum(a_i + a_j)^2 = 2nr\sum a_q^2.$$

Therefore the Scheffé simultaneous confidence intervals for contrasts in the α_q (after dividing by $2nr$) are given by

$$\sum a_q\alpha_{qi} \in \sum a_q\widehat{\alpha}_q \pm \left((r-1)F^\alpha MSE\sum\frac{a_q^2}{2nr}\right)^{1/2}.$$

Note that the symmetry prevents the existence of Tukey, Dunnett or MCB procedures for this model.

30.25 The *Diallel cross* model is a variation on the balanced additive symmetric two-way model in which there are no observations in the diagonal cells. In the Diallel cross model we observe Y_{ijk} independent, $i = 1, \ldots, r$; $j = 1, \ldots, r$; $j \neq i$; $k = 1, \ldots, n$,

$$Y_{ijk} \sim N(\mu_{ij}, \sigma^2), \qquad \mu_{ij} = \theta + \alpha_i + \alpha_j, \sum\alpha_q = 0.$$

(A Diallel cross model with interaction is considered in Exercise 30.57.) As above, we want to test that the α_q are 0. To see that this model is an orthogonal design, note that

$$\sum_i\sum_{j\neq i}\sum_k\theta(\alpha_i + \alpha_j) = n(r-1)\theta\left(\sum_i\alpha_i + \sum_j\alpha_j\right) = 0.$$

A useful identity for this model is

$$\sum a_q = 0 = \sum b_q \Rightarrow \sum_{j\neq i} a_j = -a_i, \quad \sum_{j\neq i} b_j = -b_i$$

and therefore

$$\sum_i\sum_{j\neq i}\sum_k(a_i + a_j)(b_i + b_j) = 2n(r-2)\sum_q a_qb_q,$$

$$\sum_i \sum_{j \neq i} \sum_k (a_i + a_j)^2 = 2n(r-2) \sum_q a_q^2.$$

It is easily verified by direct differentiation that $\widehat{\theta} = \overline{Y}_{\dots}$. To find the estimator of α_q note that

$$\begin{aligned} -\frac{1}{2}\frac{\partial}{\partial \alpha_q} \sum_i \sum_{j \neq i} \sum_k (Y_{ijk} - \theta - \alpha_i - \alpha_j)^2 \\ = 2n(r-1)\left(\frac{\overline{Y}_{q..} + \overline{Y}_{.q.}}{2} - \theta\right) - 2n(r-2)\alpha_q \end{aligned}$$

(see Exercise 30.56) and hence

$$\widehat{\alpha}_q = \frac{r-1}{r-2}\left(\frac{\overline{Y}_{q..} + \overline{Y}_{.q.}}{2} - \overline{Y}_{\dots}\right), \qquad \widehat{\mu}_{ij} = \widehat{\theta} + \widehat{\alpha}_i + \widehat{\alpha}_j.$$

Also

$$\sum_i \sum_{j \neq i} \sum_k \widehat{\theta}^{\,2} = nr(r-1)\overline{Y}^2_{\dots}, \qquad \sum_i \sum_{j \neq i} \sum_k (\widehat{\alpha}_i + \widehat{\alpha}_j)^2 = 2n(r-2) \sum_q \widehat{\alpha}_q^2.$$

The expected mean squares can be computed similarly. Therefore the ANOVA table for this model is

Two-way Diallel cross fixed effects model

effect	*df*	*SS*	*EMS*
θ	1	$nr(r-1)\overline{Y}^2_{\dots}$	$\sigma^2 + nr(r-1)\theta^2$
α	$r-1$	$2n(r-2)\sum \widehat{\alpha}_q^2$	$\sigma^2 + h\sum \alpha_q^2$
error	$nr(r-1)-r$	$\sum\sum\sum (Y_{ijk} - \widehat{\mu}_{ij})^2$	σ^2
total	$nr(r-1)$	$\sum\sum\sum Y_{ijk}^2$	

where $h = 2n(r-2)/(r-1)$. The hypothesis that $\alpha_i = 0$ may be tested in the obvious way.

Now, let $d_{ijk} = a_i + a_j$, where $\sum a_q = 0$. Then

$$\begin{aligned} \sum_i \sum_{j \neq i} \sum_k (a_i + a_j)(\alpha_i + \alpha_j) &= 2n(r-2) \sum a_q \alpha_q, \\ \sum_i \sum_{j \neq i} \sum_k (a_i + a_j)^2 &= 2n(r-2) \sum a_q^2. \end{aligned}$$

Therefore the Scheffé simultaneous confidence intervals for contrasts in the α_i are given by (after dividing by $2n(r-2)$)

$$\sum a_q \alpha_q \in \sum a_q \widehat{\alpha}_q \pm \left((r-1)F^{\alpha} MSE \sum \frac{a_q^2}{2n(r-2)}\right)^{1/2}.$$

Orthogonal nested models

30.26 All the previous models have been crossed models. However, not all experiments are crossed. We now describe a two fold nested model. In such a model we have k classes, and within the ith class we have c_i sub-classes. We call such a situation a *nested* model because the sub-classes are nested inside the classes. We number the classes $i = 1, \dots, r$, and number the sub-classes of the ith class $j = 1, \dots, c_i$. The observations in the jth sub-class of the ith class are numbered $1, \dots, n_{ij}$. Note that we assume that there is no relationship between jth sub-classes of different classes in nested models.

In fact, the sub-classes in the ith class are often numbered $j = C_{i-1} + 1, \dots, C_i$, where $C_i = c_1 + \cdots + c_i$. That is, the sub-classes of class 1 are labelled $j = 1, \dots, c_1$, those of class 2 are numbered $j = c_1 + 1, \dots, c_1 + c_2$, etc. This sequential numbering of sub-classes emphasizes that the sub-classes are distinct and avoids confusion with the crossed models. For example, if we have three car manufacturers, the first of which makes three different cars, the second two types of car and the third one type of car, then there are a total of six types of car which we label

$$(1,1), \quad (1,2), \quad (1,3), \quad (2,1), \quad (2,2), \quad (3,1) \quad \text{or}$$
$$(1,1), \quad (1,2), \quad (1,3), \quad (2,4), \quad (2,5), \quad (3,6).$$

Although the second labelling more accurately represents the nested structure of the data, the first labelling is more common and is the one which we shall adopt. It is important to remember that in a nested model, unlike a crossed model, we assume that there is no relationship between the first type of car for each of the manufacturers, i.e. between the (1, 1), (2, 1) and (3, 1) cells.

In the nested model, we assume that we observe Y_{ijk} independent,

$$Y_{ijk} \sim N(\mu_{ij}, \sigma^2), \qquad i = 1, \dots, r; \; j = 1, \dots, c_i; \; k = 1, \dots, n_{ij}.$$

Let

$$\overline{\mu}_{i.} = \frac{\sum_j n_{ij}\mu_{ij}}{n_{i.}}, \quad \overline{\mu}_{..} = \frac{\sum\sum n_{ij}\mu_{ij}}{N} = \frac{\sum n_{i.}\overline{\mu}_{i.}}{N}.$$

(Note that this definition does not follow the convention described earlier and is only used in this section.) We want to test the hypotheses that there is no effect due to sub-classes (μ_{ij} does not depend on j), and to test that there is no average effect due to classes ($\overline{\mu}_{i.}$ does not depend on i).

To put this model into an orthogonal design, let

$$\theta = \overline{\mu}_{..}, \quad \alpha_i = \overline{\mu}_{i.} - \overline{\mu}_{..}, \quad \delta_{ij} = \mu_{ij} - \overline{\mu}_{i}.$$

Then

$$\mu_{ij} = \theta + \alpha_i + \delta_{ij}, \quad \sum_i n_{i.}\alpha_i = 0, \quad \sum_j n_{ij}\delta_{ij} = 0$$
$$\mu_{ij} \text{ does not depend on } j \Leftrightarrow \delta_{ij} = 0, \quad \overline{\mu}_{i.} \text{ does not depend on } i \Leftrightarrow \alpha_i = 0.$$

It is again easily verified that this model is an orthogonal design. (Note that these constraints on the parameters are the only ones which make the design orthogonal.) The LS estimators of the parameters are easily seen to be

$$\widehat{\theta} = \overline{Y}_{...}, \quad \widehat{\alpha}_i = \overline{Y}_{i..} - \overline{Y}_{...}, \quad \widehat{\delta}_{ij} = \overline{Y}_{ij.} - \overline{Y}_{i..}, \quad \widehat{\mu}_{ij} = \widehat{\theta} + \widehat{\alpha}_i + \widehat{\delta}_{ij} = \overline{Y}_{ij.}.$$

There are one θ, $r-1$ linearly independent α_i and $C-r$ linearly independent δ_{ij} (where $C = c.$ is the total number of sub-classes for this experiment). By the usual calculations for orthogonal designs, we see that the ANOVA table for this model is

Twofold nested fixed effects model

effect	*df*	*SS*	*EMS*
θ	1	$N\overline{Y}^2_{...}$	$\sigma^2 + N\theta^2$
α_i	$r-1$	$\sum n_{i.}(\overline{Y}_{i..} - \overline{Y}_{...})^2$	$\sigma^2 + \sum g_i\alpha_i^2$
δ_{ij}	$C-r$	$\sum\sum n_{ij}(\overline{Y}_{ij.} - \overline{Y}_{i..})^2$	$\sigma^2 + \sum\sum h_{ij}\delta_{ij}^2$
error	$N-C$	$\sum\sum\sum(Y_{ijk} - \overline{Y}_{ij.})^2$	σ^2
total	N	$\sum\sum\sum Y_{ijk}^2$	

(where $g_i = n_{i.}/(r-1)$ and $h_{ij} = n_{ij}/(C-r)$).

To find the simultaneous confidence intervals for the contrasts in the α_i, let $d_{ijk} = a_i$, where $\sum n_{i.}a_i = 0$, and then replace $n_{i.}a_i$ with b_i to obtain

$$\sum b_i\alpha_i \in \sum b_i\widehat{\alpha}_i \pm \left((r-1)MSE\sum \frac{b_i^2}{n_{i.}} F^\alpha\right)^{1/2}, \qquad \forall b_i \ni \sum b_i = 0.$$

Similarly, to find the simultaneous confidence intervals for contrast in the δ_{ij}, use the basic formula with $d_{ijk} = d_{ij}$, $\sum_j n_{ij}d_{ij} = 0$ and replace $n_{ij}d_{ij}$ with c_{ij} to obtain

$$\sum\sum c_{ij}\delta_{ij} \in \sum\sum c_{ij}\widehat{\delta}_{ij} \pm \left((C-r)MSE\sum\sum \frac{c_{ij}^2}{n_{ij}} F^\alpha\right)^{1/2}, \ \forall c_{ij} \ni \sum_j c_{ij} = 0.$$

Note that $\delta_{11} - \delta_{21}$ is not a contrast in the δ_{ij}, but that $\delta_{11} - \delta_{12}$ is a contrast. The most significant contrast in the δ_{ij} is given by $\sum\sum n_{ij}\widehat{\delta}_{ij}\delta_{ij}$.

If the $n_{i.} = n$ are all equal then

$$\overline{Y}_{i..} \sim N\left(\theta + \alpha_i, \frac{\sigma^2}{n}\right)$$

so that Tukey, Dunnett and MCB procedures can be used, as can multiple range tests for inference about the class effect. However, if the $n_{i.}$ are unequal, then the sample means have different variances and these procedures are unavailable. Multiple F and Bonferroni tests are, of course, always available for this inference. For inference about the sub-class effects only Scheffé and Bonferroni simultaneous confidence intervals exist.

The hypothesis of no effect due to class is equality of the $\overline{\mu}_{i.}$. Note that these averages are weighted averages of the sub-class means, weighted by the sample sizes. Therefore, the hypothesis tested when we test for no 'class' effect actually depends on the sample sizes n_{ij}

for the sub-classes. Two people studying the same phenomenon with different sample sizes are actually testing different hypotheses when they test for no class effects with different sample sizes. In fact, it seems that a sensible person would choose some appropriate weights w_{ij} for the sub-classes and then choose the n_{ij} proportional to the w_{ij} . Of course, in practice it is often not easy to choose those weights. However, if we have no weights then the we really have no hypothesis of 'no class effect' and probably should not test that hypothesis. If we have weights, but use a sample scheme for which those weights do not lead to an orthogonal design, we can still test the hypothesis by methods described later in the chapter.

Suppose, for example, we are comparing fuel consumption for cars; the classes might be companies (General Motors, Ford, etc.) and the sub-classes might be the particular car that the company makes (Lincoln, Escort, Taurus, etc.) If we always use unweighted averages, the companies would subdivide the smaller cars into many sub-classes and get a small 'average' fuel consumption. From the other side, it would be unfair to count Lincolns the same as Escorts, since there are so many more Escorts than Lincolns, and the Lincolns would presumably have the worst consumption. As in the two-way crossed model, we see that in order to specify the 'class' effect carefully, we need weights w_{ij} which establish the relative importance of each sub-class within a class. We would then define the 'average' performance of each class by

$$\overline{\mu}_{i.}^{\mathbf{w}} = \frac{\sum_j w_{ij}\mu_{ij}}{\sum_j w_{ij}}.$$

We would then want to compare these weighted averages to see if there is any difference between the classes.

Note that for any nested model, there is always a set of constraints which makes it an orthogonal design. However, for a crossed model, there is often no set of constraints to make the model an orthogonal design. Note also that the appropriate constraints to make a nested model orthogonal may not be relevant to the particular experiment. Nested models with other constraints are not orthogonal designs. See **30.34** for an analysis of such models.

30.27 It is possible to have nested designs with more than two levels. They can often be analysed in an obvious way, as we now indicate with a threefold nested design. In such a design we have r classes, with c_i sub-classes of the ith class and d_{ij} sub-sub-classes of the jth sub-class of the ith class. We have n_{ijk} observations in the kth sub-sub-class of the jth sub-class of the ith class. The model we assume is that Y_{ijkm} are independent,

$$Y_{ijkm} \sim N(\mu_{ijk}, \sigma^2), \qquad i = 1, \ldots, r;\ j = 1, \ldots, c_i;\ k = 1, \ldots, d_{ij};\ m = 1, \ldots, n_{ij}.$$

Let $N = n_{...}$ be the total number of observations, $D = d_{..}$ be the total number of sub-sub-classes and $C = c_{.}$ be the total number of sub-classes. As in **30.26**, we define

$$\overline{\mu}_{ij.} = \frac{\sum_k n_{ijk}\mu_{ijk}}{n_{ij.}}, \quad \overline{\mu}_{i..} = \frac{\sum_j \sum_k n_{ijk}\mu_{ijk}}{n_{i..}} = \frac{\sum_j n_{ij.}\overline{\mu}_{ij.}}{n_{i..}}$$

$$\overline{\mu}_{...} = \frac{\sum\sum\sum n_{ijk}\mu_{ijk}}{n_{...}} = \frac{\sum\sum n_{ij.}\overline{\mu}_{ij.}}{n_{...}} = \frac{\sum n_{i..}\overline{\mu}_{i..}}{n_{...}}.$$

We are interested in testing three hypotheses: that the μ_{ijk} do not depend on k (no sub-sub-class effect); that the $\overline{\mu}_{ij.}$ do not depend on j (no sub-class effect); and that the $\overline{\mu}_{i..}$ do not depend on i (no class effect). To put this model into an orthogonal design, let

$$\theta = \overline{\mu}_{...}, \quad \alpha_i = \overline{\mu}_{i..} - \overline{\mu}_{...}, \quad \delta_{ij} = \overline{\mu}_{ij.} - \overline{\mu}_{i..}, \quad \xi_{ijk} = \mu_{ijk} - \overline{\mu}_{ij.}.$$

Then

$$\mu_{ijk} = \theta + \alpha_i + \delta_{ij} + \xi_{ijk}, \quad \sum_i n_{i..}\alpha_i = 0, \sum_j n_{ij.}\delta_{ij} = 0, \quad \sum_k n_{ijk}\xi_{ijk} = 0.$$

This model is an orthogonal design. In fact these constraints are the only constraints which make the model an orthogonal model. Furthermore,

$$\alpha_i = 0 \Leftrightarrow \overline{\mu}_{i..} \text{ does not depend on i;} \qquad \delta_{ij} = 0 \Leftrightarrow \overline{\mu}_{ij.} \text{ does not depend on j;}$$
$$\xi_{ijk} = 0 \Leftrightarrow \mu_{ijk} \text{ does not depend on } k.$$

The LS estimators of the parameters are given by

$$\widehat{\theta} = \overline{Y}_{....}, \quad \widehat{\alpha}_i = \overline{Y}_{i...} - \overline{Y}_{....}, \quad \widehat{\delta}_{ij} = \overline{Y}_{ij..} - \overline{Y}_{i...},$$
$$\widehat{\xi}_{ijk} = \overline{Y}_{ijk.} - \overline{Y}_{ij..}, \quad \widehat{\mu}_{ijk} = \overline{Y}_{ijk.}.$$

Using these estimators, the sums of squares and degrees of freedom, the ANOVA table can be computed in the obvious way:

Threefold nested fixed effects model

effect	*df*	*SS*	*EMS*
θ	1	$N\overline{Y}^2_{...}$	$\sigma^2 + N\theta^2$
α	$r-1$	$\sum n_{i..}(\overline{Y}_{i...} - \overline{Y}_{....})^2$	$\sigma^2 + \sum g_i\alpha_i^2$
δ	$C-r$	$\sum\sum n_{ij.}(\overline{Y}_{ij..} - \overline{Y}_{i...})^2$	$\sigma^2 + \sum\sum h_{ij}\delta_{ij}^2$
ξ	$D-C$	$\sum\sum\sum n_{ijk}(\overline{Y}_{ijk.} - \overline{Y}_{ij..})^2$	$\sigma^2 + \sum\sum\sum q_{ijk}\xi_{ijk}^2$
error	$N-D$	$\sum\sum\sum\sum(Y_{ijkm} - \overline{Y}_{ijk\cdot})^2$	σ^2
Total	N	$\sum\sum\sum\sum Y_{ijkm}^2$	

where $g_i = n_{i..}/(r-1)$, $h_{ij} = n_{ij.}/(C-r)$ and $q_{ijk} = n_{ijk}/(D-C)$. Simultaneous confidence intervals can also be computed in a straightforward way.

30.28 Sometimes we have models with both crossed and nested effects. Suppose we have a balanced model in which we observe Y_{ijkm} independent, where the i and j effects are crossed and the k effect is nested in the (i, j)th cell. (The m represents replication.) We might then assume that

$$Y_{ijkm} \sim N(\mu_{ijk}, \sigma^2), \qquad i = 1, \ldots, r; \; j = 1, \ldots, c; \; k = 1, \ldots, d; \; m = 1, \ldots, n,$$
$$\mu_{ijk} = \theta + \alpha_i + \beta_j + \gamma_{ij} + \delta_{ijk},$$
$$\sum_i \alpha_i = 0, \quad \sum_j \beta_j = 0, \quad \sum_i \gamma_{ij} = 0, \quad \sum_j \gamma_{ij} = 0, \quad \sum_k \delta_{ijk} = 0.$$

Note that the α and β effects are crossed main effects and the γ effect is their interaction. The δ effect is nested in the α and β effects. This model is again an orthogonal design and may be analysed in the fashion for previous models. Note that for this model it is routine to show

$$\widehat{\mu}_{ijk} = \overline{Y}_{ijk\cdot}, \quad dfe = cdr(n-1).$$

A different model in the above situation would have the i and j effects crossed, as above, but would have the k effect nested in the j effect. In that case we would get a different model in which

$$\mu_{ijk} = \theta + \alpha_i + \beta_j + \gamma_{ij} + \delta_{jk}$$
$$\sum_i \alpha_i = 0, \quad \sum_j \beta_j = 0, \quad \sum_i \gamma_{ij} = 0, \quad \sum_j \gamma_{ij} = 0, \quad \sum_k \delta_{jk} = 0.$$

Note that this model is the same as the previous model except that in this model δ does not depend on i. Note that for this model

$$\widehat{\mu}_{ijk} = \overline{Y}_{ij..} + \overline{Y}_{.jk.} - \overline{Y}_{.j..}, \quad dfe = ncdr - (rc + c(d-1)),$$

so that SSE and dfe are different from the model in the last paragraph. Obviously $SS\delta$ and $df\delta$ would also be different. In practice it is very important to keep track of which effects are nested within which other effects.

Non-orthogonal designs

30.29 We now discuss linear models which are not orthogonal designs. Typically balanced models are orthogonal designs, as we have seen. We have also seen that the unbalanced one-way model can be analysed as an orthogonal model. The two-way model can only be analysed as an orthogonal design for balanced or proportional sampling. Even in these settings for the non-additive two-way model, the orthogonal analysis only makes sense if the numbers of observations in each cell are appropriately chosen for the hypotheses of interest. (For the additive model, the weights do not affect the hypothesis, so that the orthogonal analysis is appropriate for any additive two-way model with balanced or proportional sampling.) Also, unbalanced nested models can be analysed as orthogonal models. However, the hypotheses tested for the orthogonal design may be inappropriate because they involve weighted averages of the means, where the weights depend on the cell sizes. In **30.31–33**, we look at non-orthogonal two-way models including the balanced incomplete blocks model. In **30.34**, we look at the twofold nested models. The results on two-way crossed model and the twofold nested model can be extended to more complicated models in a straightforward way.

In the linear model, we observe $\mathbf{Y} \sim N_n(\boldsymbol{\mu}, \sigma^2\mathbf{I})$ and are testing the null hypothesis that $\boldsymbol{\mu} \in W$ against the alternative hypothesis that $\boldsymbol{\mu} \in V$, where $W \subset V$ are subspaces of R^n,

$$k = \dim(W) < p = \dim(V).$$

To find the appropriate test statistic for this hypothesis we minimize

$$\|\mathbf{Y} - \boldsymbol{\mu}\|^2 = \sum_{\mathbf{m}} (Y_{\mathbf{m}} - \mu_{\mathbf{m}})^2$$

under the alternative and null hypotheses to find

$$\widehat{\boldsymbol{\mu}} = \mathbf{P}_V\mathbf{Y} \text{ and } \widehat{\widehat{\boldsymbol{\mu}}} = \mathbf{P}_W\mathbf{Y}.$$

Then

$$SSH = \|\mathbf{P}_{V|W}\mathbf{Y}\|^2 = \|\widehat{\boldsymbol{\mu}} - \widehat{\widehat{\boldsymbol{\mu}}}\|^2 = \sum(\widehat{\mu}_{\mathbf{m}} - \widehat{\widehat{\mu}}_{\mathbf{m}})^2 = \|\widehat{\boldsymbol{\mu}}\|^2 - \|\widehat{\widehat{\boldsymbol{\mu}}}\|^2 = \sum \widehat{\mu}_{\mathbf{m}}^2 - \sum \widehat{\widehat{\mu}}_{\mathbf{m}}^2,$$
$$SSE = \|\mathbf{P}_{V^\perp}\mathbf{Y}\|^2 = \|\mathbf{Y} - \widehat{\boldsymbol{\mu}}\|^2 = \sum(y_{\mathbf{m}} - \widehat{\mu}_{\mathbf{m}})^2 = \|\mathbf{Y}\|^2 - \|\widehat{\boldsymbol{\mu}}\|^2 = \sum y_{\mathbf{m}}^2 - \sum \widehat{\mu}_{\mathbf{m}}^2,$$
$$dfH = p - k, \quad dfe = n - p, \quad MSH = \frac{SSH}{dfH}, \quad MSE = \frac{SSE}{dfe}, \quad F = \frac{MSH}{MSE}.$$

Another formula for *SSH* which is often useful is

$$SSH = \|\mathbf{P}_{V|W}\mathbf{Y}\|^2 = \max_{\mathbf{d}\in V|W, \mathbf{d}\neq \mathbf{0}} \frac{(\mathbf{d}^T\widehat{\boldsymbol{\mu}})^2}{\|\mathbf{d}\|^2}$$

(see Exercise 30.19(d)). This formula is the basic result used earlier to derive the Scheffé simultaneous confidence intervals.

To find the non-centrality parameter and expected mean squares, we must first compute $\boldsymbol{\mu}^* = \mathbf{P}_W\boldsymbol{\mu}$. Then the non-centrality parameter γ and expected mean squares are

$$\gamma = \frac{\xi}{\sigma^2}, \quad EMS = \sigma^2 + \frac{\xi}{dfh}, \quad \xi = \|\mathbf{P}_{V|W}\boldsymbol{\mu}\|^2 = \|\boldsymbol{\mu} - \boldsymbol{\mu}^*\|^2 = \sum(\mu_{\mathbf{m}} - \mu_{\mathbf{m}}^*)^2.$$

The Scheffé simultaneous confidence intervals for contrasts associated with this hypotheses are given by

$$\mathbf{d}^T\boldsymbol{\mu} \in \mathbf{d}^T\widehat{\boldsymbol{\mu}} \pm ((dfh)\, MSE\|\mathbf{d}\|^2 F^\alpha)^{1/2} \; \forall \mathbf{d} \in V|W.$$

Note that the set $V|W$ is often difficult to determine in non-orthogonal models.

30.30 As we shall see in future sections, the formulae for procedures in non-orthogonal designs are often quite complicated, when they exist at all. However, the necessary equations are often routinely solved by a computer, so that the lack of simple formulae is not today a serious drawback to the use of non-orthogonal designs. In fact, even when formulae exist, it may be easier to compute the appropriate statistics by minimizing $\sum(Y_{\mathbf{m}} - \mu_{\mathbf{m}})^2$ under null and alternative hypotheses. Note that this is just a quadratic function of the parameters to be estimated and that it is only necessary to find the estimators for the $Y_{\mathbf{m}}$ actually observed.

A more serious drawback is that when a design is non-orthogonal the effects tested may be confounded. To take an extreme case, consider a two-way model with two rows and two columns in which there are 100 observations in the (1, 1)th cell and 100 observations in the (2, 2)th cell but no observations in the remaining two cells. Any difference between the means for the (1, 1)th cell and the (2, 2)th cell could be caused by the fact that the cells had different row treatments or that they had different column treatments. In this situation, the row and column treatments are totally confounded. In most designs, the various effects are not confounded as long as there is at least one observation in each cell. (This condition, while typically sufficient, is not necessary for non-confoundedness. For a model which has empty cells with no confounding, see the balanced incomplete block model discussed below.)

However, even if there is at least one observation in each cell, there may be difficulties. To see this consider a two-way model with two rows and two columns and 100 observations in each of the (1, 1)th and (2, 2)th cells but only one observation in each of the (1, 2)th and (2, 1)th cells. In this situation, small changes in the single observations in the (1, 2)th and (2, 1)th cells can dramatically affect the conclusion. The situation is similar to multicollinearity in regression. There, if the observations are completely collinear, there is no way to analyse the data. However, if they are 'nearly' collinear, then the data can be analysed, but the conclusions are very unstable. In non-orthogonal models, data which are 'nearly' totally confounded can still be analysed but the conclusions are very unstable. For this reason, it is often sensible to use designs which are orthogonal or nearly so. In orthogonal designs the various effects are orthogonal to each other and hence cannot be confounded.

Unbalanced two-way models – balanced incomplete blocks

30.31 In the unbalanced two-way model we observe Y_{ijk} independent,

$$Y_{ijk} \sim N(\mu_{ij}, \sigma^2), \qquad i = 1, \dots, r;\ j = 1, \dots, c;\ k = 1, \dots, n_{ij}.$$

We allow the possibility that $n_{ij} = 0$ for some i and j, i.e. that there are no observations in the (i, j)th cell. If $n_{ij} = 0$, we say that the (i, j)th cell is *empty*. Let $N = n_{..}$ be the total number of observations.

We first look at the additive model, in which we assume that $\mu_{ij} = \xi_i + \eta_j$ for some constants ξ_i and η_j. In that setting, we can always write

$$\mu_{ij} = \theta + \alpha_i + \beta_j, \quad \sum_i n_{i.}\alpha_i = 0, \quad \sum_j n_{.j}\beta_j = 0$$

(see Exercise 30.62). Note also that

$$\alpha_i = 0 \Leftrightarrow \mu_{ij} \text{ does not depend on } i,$$
$$\beta_j = 0 \Leftrightarrow \mu_{ij} \text{ does not depend on } j.$$

(We have seen that we can choose any constraints we like for the parameters without affecting the hypotheses of interest. We choose these because they make the θ effect orthogonal to the other effects. Unless we have proportional sampling, no constraint makes the α and β effects orthogonal.)

We assume that the α_i and the β_j can be estimated. This implies certain conditions on the empty cells. Sampling schemes satisfying these conditions are called 'connected'. A sufficient (but not necessary) condition for the sampling to be connected is that there be no empty cells. A necessary (but not sufficient) condition is that the $n_{i.} > 0$ and the $n_{.j} > 0$, so that $\overline{Y}_{i..}$, $\overline{Y}_{.j.}$ and $\overline{Y}_{...}$ are all defined. See Graybill (1976, pp. 549–550) for more details on this issue.

To find the estimators under the alternative hypothesis, we must minimize

$$\sum\sum\sum(Y_{ijk} - \theta - \alpha_i - \beta_j)^2.$$

Routine differentiation (with respect to θ) and the constraints on the α_i and β_j imply that $\widehat{\theta} = \overline{Y}_{...}$.

If we differentiate with respect to α_i, we see that

$$0 = \sum_j \sum_k (Y_{ijk} - \theta - \alpha_i - \beta_j) \Rightarrow \widehat{\alpha}_i = \overline{Y}_{i..} - \overline{Y}_{...} - \frac{\sum_j n_{ij}\widehat{\beta}_j}{n_{i.}}.$$

Similarly,

$$\widehat{\beta}_j = \overline{Y}_{.j.} - \overline{Y}_{...} - \frac{\sum_i n_{ij}\widehat{\alpha}_i}{n_{.j}}.$$

These $r + c$ equations must be solved subject also to $\sum n_{i.}\alpha_i = 0$ and $\sum n_{.j}\beta_j = 0$ to find the LS estimators. (Note that two of the equations are redundant so that there are really $r + c$ equations in $r + c$ unknowns to solve.) Although the solution to these equations does not have a general formula, these are linear equations which can be easily solved numerically on a computer. As long as the sampling is connected, the solution to these equations is unique. (An exact formula for the solution to these equations is given in the next section for a particular case called a *balanced incomplete block design*.) Then the estimator under the alternative hypothesis is

$$\widehat{\mu}_{ij}^A = \widehat{\theta} + \widehat{\alpha}_i + \widehat{\beta}_j.$$

We call this estimator $\widehat{\mu}_{ij}^A$ (for the estimated mean under the additive model). As in the balanced case, there are $r + c - 1$ parameters in this model. Therefore

$$SSE = \sum\sum\sum (Y_{ijk} - \widehat{\mu}_{ij}^A)^2, \quad dfe = N - r - c + 1.$$

Under the null hypothesis that $\alpha_i = 0$, it is easily seen that the LS estimators are

$$\widehat{\widehat{\theta}} = \overline{Y}_{...},\ \widehat{\widehat{\alpha}}_i = 0,\ \widehat{\widehat{\beta}}_j = \overline{Y}_{.j.} - \overline{Y}_{...},\ \widehat{\widehat{\mu}}_{ij} = \widehat{\widehat{\theta}} + \widehat{\widehat{\alpha}}_i + \widehat{\widehat{\beta}}_j = \overline{Y}_{.j.}.$$

There are $1 + (c - 1) = c$ parameters under the null hypothesis. Therefore, for testing that $\alpha_i = 0$ in the additive two way model

$$SSH = \sum\sum\sum (\widehat{\mu}_{ij} - \widehat{\widehat{\mu}}_{ij})^2 = \sum\sum n_{ij}(\widehat{\mu}_{ij}^A - \overline{Y}_{.j.})^2,\ dfh = r + c - 1 - c = r - 1.$$

To compute the non-centrality parameter and EMS for this hypothesis, we must first find the projection of the mean vector in which $\mu_{ij} = \theta + \alpha_i + \beta_j$ onto the subspace in which μ_{ij} does not depend on i. That is, we must find μ_j^* which minimizes

$$\sum\sum\sum (\theta + \alpha_i + \beta_j - \mu_j^*)^2 = \sum\sum n_{ij}(\theta + \alpha_i + \beta_j - \mu_j^*).$$

By direct differentiation, we see that

$$\mu_j^* = \theta + \overline{\alpha}_j + \beta_j,\ \overline{\alpha}_j = \frac{\sum_i n_{ij}\alpha_i}{n_{.j}}.$$

Therefore the non-centrality parameter γ and expected mean squares are given by

$$\gamma = \frac{\xi}{\sigma^2}, \qquad EMSH = \sigma^2 + \frac{\xi}{r - 1}, \qquad EMSE = \sigma^2,$$
$$\xi = \sum\sum\sum (\theta + \alpha_i + \beta_j - (\theta + \overline{\alpha}_j + \beta_j))^2 = \sum\sum n_{ij}(\alpha_i - \overline{\alpha}_j).$$

Similar derivations could be made for testing that the $\beta_j = 0$. Therefore we see that a partial ANOVA table for this model is

Unbalanced additive two-way fixed effects model

effect	*df*	*SS*	*EMS*
α_i	$r-1$	$\sum\sum n_{ij}(\widehat{\mu}_{ij}^A - \overline{Y}_{.j.})^2$	$\sigma^2 + \sum\sum g_{ij}(\alpha_i - \overline{\alpha}_j)^2$
β_j	$c-1$	$\sum\sum n_{ij}(\widehat{\mu}_{ij}^A - \overline{Y}_{i..})^2$	$\sigma^2 + \sum\sum h_{ij}(\beta_j - \overline{\beta}_i)^2$
error	$N-r-c+1$	$\sum\sum\sum(Y_{ijk} - \widehat{\mu}_{ij}^A)^2$	σ^2

where $g_{ij} = n_{ij}/(r-1)$, $h_{ij} = n_{ij}/(c-1)$. We have not included the effect due to θ, the overall mean, because we are rarely interested in testing hypotheses about it and we have not included the total sum of squares because the various sums of squares do not add up to the total sum of squares in a non-orthogonal model.

The Scheffé simultaneous confidence intervals for contrasts associated with the hypotheses $\alpha_i = 0$ for this model are derived in Exercise 30.63. They are too complicated to be of much use. See Graybill (1976, p. 553) for some simpler conservative intervals.

30.32 We now look at a particular class of unbalanced additive two-way models which have enough balance so that we can derive formulae for the procedures.

To motivate the balanced incomplete block model, consider an experiment in which we have four teaching methods which we want to compare. We have eight high schools we can use for the experiment, but each high school can only use three different teaching methods. We might use the following design

school \ *method*	1	2	3	4
1	×	×	×	
2	×	×		×
3	×		×	×
4		×	×	×
5	×	×	×	
6	×	×		×
7	×		×	×
8		×	×	×

where an × in the (i, j)th cell means that school i used teaching method j. Note that this design is balanced in the following senses. It has the same number of methods (3) for each school, the same number of schools (6) for each method. Also each pair of methods occurs together the same number of times (4). (Note that methods 1 and 3 occur together in schools 1, 3, 5 and 7.)

With this example in mind, we define a *balanced incomplete block* design as a design with a total of r rows and c columns in which each cell has either 1 or 0 observations. The occurrence of these observations satisfies the following balancing conditions:

1. Each row occurs in $t < c$ columns.
2. Each column occurs in $b < r$ rows.
3. Each pair of columns occurs together in λ rows.

In a balanced incomplete block design we are primarily interested in inferences about the column effect. In fact, in these designs, the columns are often called treatments and the rows are often called blocks. In this design each treatment occurs in b blocks and each block gets t treatments. There are a total of c treatments and r blocks. Each pair of treatments occurs together in λ blocks. Note that the total number of observations

$$N = bc = \text{tr}.$$

Note also that λ is a function of r, c, b and t. In fact

$$\lambda = \frac{b(t-1)}{c-1}.$$

To see this, consider a particular column (treatment). In the rows (blocks) in which this treatment occurs there are a total of tb observations, so that there are $b(t-1)$ observations in these rows which do not contain the particular treatment. Also, each other column appears λ times with the particular treatment, so that the number of observations in these rows, not including the particular one, must also be $\lambda(c-1)$ and the result follows.

Balanced incomplete blocks do not exist for arbitrary choices of r, c, b and t. A necessary condition is that λ given above is an integer, but it is not sufficient. Even when a balanced incomplete block exists, it may not be easy to find. A fairly extensive list of them for various situations is given in Cochran and Cox (1957).

For reasons discussed below, the *efficiency* of a balanced incomplete block design is defined to be

$$H = 1 - \frac{b-\lambda}{bt} = \frac{c\lambda}{bt} = \frac{c(t-1)}{t(c-1)} < 1$$

(since $t < c$).

We now give some useful identities for balanced incomplete block designs. Let n_{ij} be the number of observations in the (i, j)th cell (so that n_{ij} is 0 or 1 depending on whether there is an observation in the (i, j)th cell or not). Then the definition of balanced incomplete blocks implies that

$$\sum_i n_{ij}^2 = b, \qquad \sum_i n_{ij}n_{ij'} = \lambda \quad \text{if } j \neq j'.$$

Also, let $\sum c_j = 0$, $\sum d_j = 0$, and define $\bar{c}_{i.} = \sum_j n_{ij}c_j/t$. Then using the above identities we can show

$$t\sum n_{ij}\bar{c}_{i.} = (b-\lambda)c_j, \qquad \sum\sum n_{ij}(c_j - \bar{c}_{i.})^2 = bH\sum c_j^2,$$
$$\sum\sum n_{ij}d_j(c_j - \bar{c}_{i.}) = bH\sum d_jc_j$$

(see Exercise 30.65).

We assume an additive model for this situation. That is, we assume that if Y_{ij} is an observation in the (i, j)th cell, then

$$Y_{ij} \sim N(\mu_{ij}, \sigma^2), \qquad \mu_{ij} = \theta + \alpha_i + \beta_j, \ \sum \alpha_i = 0, \ \sum \beta_j = 0.$$

This model is a special case of the unbalanced additive model discussed in **30.31**. In that section, we saw that the LS estimators satisfied

$$\widehat{\theta} = \overline{Y}_{..}, \quad \widehat{\beta}_j = \overline{Y}_{.j} - \overline{Y}_{..} - \sum_i \frac{n_{ij}\widehat{\alpha}_i}{b}, \qquad \widehat{\alpha}_i = \overline{Y}_{i.} - \overline{Y}_{..} - \overline{\widehat{\beta}}_i, \quad \overline{\widehat{\beta}}_i = \frac{\sum_j n_{ij}\widehat{\beta}_j}{t}$$

Let

$$B_j = \frac{\sum_i n_{ij}\overline{Y}_{i.}}{b}$$

be the average of all the observations in the rows that receive the jth treatment. Substituting the equation for $\widehat{\alpha}_i$ into the equation for $\widehat{\beta}_j$, we see that

$$\widehat{\beta}_j = \overline{Y}_{.j} - \overline{Y}_{..} - \sum_i \frac{n_{ij}(\overline{Y}_{i.} - \overline{Y}_{..})}{b} + \sum_i \frac{n_{ij}\overline{\widehat{\beta}}_i}{b} = \overline{Y}_{.j} - B_j + \frac{(b-\lambda)}{bt}\widehat{\beta}_j,$$

$$\Rightarrow \widehat{\beta}_j = \frac{\overline{Y}_{.j} - B_j}{H}.$$

Note that for this model the equation for β_j does not involve any other parameters. Then

$$\widehat{\mu}_{ij}^A = \widehat{\theta} + \widehat{\alpha}_i + \widehat{\beta}_j = \overline{Y}_{i.} - \overline{\widehat{\beta}}_i + \widehat{\beta}_j.$$

Under the null hypothesis that $\beta_j = 0$, $\widehat{\widehat{\mu}}_{ij} = \overline{Y}_{i.}$. Therefore, for this model

$$SSH = \sum_i \sum_j n_{ij}(\widehat{\mu}_{ij}^A - \overline{Y}_{i.})^2 = \sum_i \sum_j n_{ij}(-\overline{\widehat{\beta}}_i + \widehat{\beta}_j)^2 = bH \sum_j \widehat{\beta}_j^2.$$

Also

$$SSE = \|\mathbf{P}_{V^\perp}\mathbf{Y}\|^2 = \|\mathbf{P}_{W^\perp}\mathbf{Y}\|^2 - \|\mathbf{P}_{V|W}\mathbf{Y}\|^2 = \sum_i \sum_j n_{ij}(Y_{ij} - \overline{Y}_{i.})^2 - SSH,$$

$$dfh = c - 1, \quad dfe = N - r - c + 1.$$

The F statistic can be computed directly.

We now find the non-centrality parameter for this model. We first must find the projection of the mean vector onto the subspace in which the mean only depends on the row. That is, we must find μ_i^* which minimizes

$$\sum\sum n_{ij}(\theta + \alpha_i + \beta_j - \mu_i^*)^2.$$

By a similar argument to that for the estimated mean, we see that this projection is

$$\mu_i^* = \frac{\sum_j n_{ij}\mu_{ij}}{t} = \theta + \alpha_i + \overline{\beta}_i, \qquad \overline{\beta}_i = \frac{\sum_j n_{ij}\beta_j}{t}.$$

Therefore, the non-centrality parameter γ is given by

$$\gamma = \frac{\xi}{\sigma^2}, \qquad \xi = \sum_i \sum_j n_{ij}(\mu_{ij} - \mu_{ij}^*)^2 = \sum_i \sum_j n_{ij}(\beta_j - \overline{\beta}_i)^2 = bH \sum_j \beta_j^2.$$

Note that if we had a complete two-way model with one observation in each cell, then the non-centrality parameter would be $b\sum \beta_j^2/\sigma^2$. Therefore, the non-centrality parameter for the balanced incomplete block model is H times the non-centrality parameter for the complete model, which is why H is called the efficiency of the model. Note also that for this model

$$EMSH = \sigma^2 + \frac{\xi}{c-1}, \quad EMSE = \sigma^2.$$

Therefore a partial ANOVA table for this model is the following:

Balanced incomplete block fixed effects model

effect	*df*	*SS*	*EMS*
β_j	$c-1$	$b\sum(\overline{Y}_{.j} - B_j)^2/H$	$\sigma^2 + g\sum\beta_j^2$
error	$N-r-c+1$	$\sum\sum(Y_{ij} - \overline{Y}_{i.})^2 - b\sum(\overline{Y}_{.j} - B_j)^2/H$	σ^2

where $g = bH/(c-1)$.

To find the Scheffé simultaneous confidence intervals for contrasts, we must first characterize the contrasts. Note that a vector $\mathbf{d}$ with components d_{ij} is in the alternative space V if and only if $d_{ij} = a_i + b_j$ for some constants a_i and b_j such that $\sum b_j = 0$, and that this vector is orthogonal to the null space if for all constants μ_i,

$$0 = \sum\sum n_{ij}(a_i + d_j)\mu_i = \sum_i \mu_i\Big(a_i t + \sum_j n_{ij} d_j\Big)$$

which happens if and only if

$$a_i = -\overline{d}_i = -\frac{\sum_j n_{ij} d_j}{t}.$$

In this case (see Exercise 30.66),

$$\sum\sum n_{ij}(a_i + d_j)(\theta + \alpha_i + \beta_j) = \sum\sum n_{ij}(-\overline{d}_i + d_j)\beta_j = bH\sum d_j\beta_j,$$
$$\sum\sum n_{ij}(a_i + d_j)(\widehat{\theta} + \widehat{\alpha}_i + \widehat{\beta}_j) = bH\sum d_j\widehat{\beta}_j, \quad \sum\sum n_{ij}(a_i + d_j)^2 = bH\sum d_j^2.$$

Therefore the Scheffé simultaneous confidence intervals for contrasts in this model are given by

$$bH\sum d_j\beta_j \in bH\sum d_j\widehat{\beta}_j \pm \big((c-1)MSE\, bH\sum d_j^2\ F^\alpha\big)^{1/2}.\ \forall d_j \ni \sum d_j = 0.$$

Dividing through by bH, we get the intervals

$$\sum d_j\beta_j \in \sum d_j\widehat{\beta}_j \pm \Big((c-1)\, MSE \sum \frac{d_j^2}{bH} F^\alpha\Big)^{1/2}, \ \forall d_j \ni \sum d_j = 0.$$

Multiple F tests may be constructed using this fact. In Exercise 30.67, Tukey and Dunnett simultaneous confidence intervals are constructed for this model . Multiple range tests may be constructed similarly. (Note that this model is unbalanced since some cells have one observation and some have none. However, this unbalance is 'balanced' enough so that exact formulae and studentized range procedures exist, which are not available for most other unbalanced additive models.)

30.33 Now consider the possibly non-additive unbalanced two-way model in which we make no assumption about the μ_{ij}. In order to avoid confounding of hypotheses, we assume that there are no empty cells so that $\overline{Y}_{ij.}$, $\overline{Y}_{i..}$, $\overline{Y}_{.j.}$ and $\overline{Y}_{...}$ are defined. There does not seem to be any advantage in writing the μ_{ij} in terms of θ, α_i, β_j and γ_{ij} in the unbalanced non-additive case so we shall phrase all hypotheses, non-centrality parameters, etc., in terms of μ_{ij}. (Recall that the row effects, interactions, etc., were artificial parameters introduced to establish orthogonality of certain hypotheses. No such orthogonality exists for this model.)

To find the LS estimators under the alternative hypothesis, we must minimize

$$\sum\sum\sum(Y_{ijk} - \mu_{ij})^2.$$

By direct differentiation, we see that the LS estimators under the alternative hypothesis are given by

$$\widehat{\mu}_{ij} = \overline{Y}_{ij.}.$$

There are a total of rc parameters in this model. Therefore

$$SSE = \sum\sum\sum(Y_{ijk} - \overline{Y}_{ij.})^2, \quad dfe = N - rc.$$

First consider testing the null hypothesis that the model is additive. The LS estimator under the null hypothesis is $\widehat{\mu}^A_{ij}$ defined in **30.31**. There are a total of $p = rc$ parameters under the alternative hypothesis and $k = r + c - 1$ parameters under the null hypothesis. Therefore, for this testing problem, we see that

$$SSH = \sum\sum n_{ij}(\overline{Y}_{ij.} - \widehat{\widehat{\mu}}^A_{ij})^2, \quad dfH = rc - (r + c - 1) = (r - 1)(c - 1).$$

The F statistic can be computed immediately. The non-centrality parameter γ and EMS are given by

$$\gamma = \frac{\xi}{\sigma^2}, \quad EMS = \sigma^2 + \frac{\xi}{(r-1)(c-1)}, \quad \xi = \sum\sum n_{ij}(\mu_{ij} - \mu^A_{ij})^2,$$

where the $\boldsymbol{\mu}^A$ is the projection of $\boldsymbol{\mu}$ on the additive subspace.

Now, consider testing the hypothesis of no row effect. We recall that this hypothesis is not defined unless we have a set of weights w_j for the columns. (For unbalanced models, the derivations are no easier for the case of equal weights so we make no assumption about the weights in this section.). The hypothesis we are testing is the equality of the

$$\overline{\mu}^{\mathbf{w}}_{i.} = \frac{\sum_j w_j \mu_{ij}}{\sum_j w_j}.$$

To find the estimators under the null hypothesis we must minimize

$$\sum\sum\sum(Y_{ijk} - \mu_{ij})^2 \text{ subject to } \overline{\mu}^{\mathbf{w}}_{i.} \text{ all equal.}$$

This minimization can be accomplished on a computer to get the estimators under the null hypothesis. Once we have these estimators, we can compute the SSH in the usual way. Note that for this hypothesis there are rc parameters satisfying $r - 1$ constraints. Therefore under the null hypothesis there are $rc - (r - 1)$ Hence

$$dfh = rc - (rc - (r - 1)) = r - 1.$$

There is a closed form for SSH for this problem which we now describe. (This formula is derived below). Let

$$A_i = \frac{\sum_j w_j \overline{Y}_{ij}}{\sum_j w_j}.$$

Then the A_i are the optimal (ordinary LS, ML and MV unbiased) estimators of the $\overline{\mu}_i^{\mathbf{w}}$. Furthermore, the A_i are independent,

$$A_i \sim N\left(\overline{\mu}_{i.}^{\mathbf{w}}, \frac{\sigma^2}{Q_i}\right), \qquad \frac{1}{Q_i} = \frac{\sum_j w_j^2/n_{ij}}{\left(\sum_j w_j\right)^2}.$$

Let

$$\overline{A} = \frac{\sum Q_i A_i}{\sum Q_i}.$$

(Note that $\overline{A}$ is the weighted average of the A_i, with the weights inversely proportional to the variances.) Then we show below that for this problem

$$SSH = \sum Q_i (A_i - \overline{A})^2.$$

(Note that this formula reduces to the formula derived earlier when we have proportional sampling and the weights w_j are proportional to the $n_{.j}$.)

By similar calculations, we can show the non-centrality parameter γ and expected mean squares for this problem to be

$$\gamma = \frac{\xi}{\sigma^2}, \quad EMSH = \sigma^2 + \frac{\xi}{r - 1}, \quad \xi = \sum Q_i (\overline{\mu}_{i.}^{\mathbf{W}} - \overline{\mu}^{\mathbf{W}})^2, \quad \overline{\mu}^{\mathbf{W}} = \frac{\sum Q_i \overline{\mu}_i^{\mathbf{W}}}{\sum Q_i}.$$

Similarly, let v_i be relative weights for the rows. Testing for no column effect is testing the equality of

$$\overline{\mu}_{.j}^{\mathbf{V}} = \frac{\sum v_i \mu_{ij}}{\sum v_i}.$$

As above, let

$$B_j = \frac{\sum_i v_i \overline{Y}_{ij}}{\sum_i v_i}, \quad \frac{1}{R_j} = \frac{\sum_i v_i^2/n_{ij}}{\left(\sum_i v_i\right)^2}, \quad \overline{B} = \frac{\sum_j R_j B_j}{\sum_j R_j},$$

so that

$$B_j \sim N(\mu_{.j}^{\mathbf{V}}, \sigma^2/R_j).$$

As above, for testing for no column effect

$$SSH = \sum R_j (B_j - \overline{B})^2, \quad dfh = c - 1, \quad EMS = \frac{\sum R_j (\overline{\mu}_{.j}^{\mathbf{V}} - \overline{\mu}^{\mathbf{V}})^2}{c - 1}, \quad \overline{\mu}^{\mathbf{V}} = \frac{\sum R_j \overline{\mu}_{.j}^{V}}{\sum R_j}.$$

Therefore a partial ANOVA table for this model is

General two-way fixed effects model with interaction

effect	*df*	*SS*	*EMS*
rows	$r-1$	$\sum Q_i(A_i - \overline{A})^2$	$\sigma^2 + \sum g_i(\overline{\mu}_{i.}^{\mathbf{W}} - \overline{\mu}^{\mathbf{W}})^2$
columns	$c-1$	$\sum R_j(B_j - \overline{B})^2$	$\sigma^2 + \sum h_j(\overline{\mu}_{.j}^{\mathbf{V}} - \overline{\mu}^{\mathbf{V}})^2$
interaction	$(r-1)(c-1)$	$\sum\sum n_{ij}(\overline{Y}_{ij.} - \widehat{\mu}_{ij}^{A})^2$	$\sigma^2 + \sum\sum k_{ij}(\mu_{ij} - \mu_{ij}^{A})^2$
error	$N - rc$	$\sum\sum\sum(Y_{ijk} - \overline{Y}_{ij.})^2$	σ^2

where $g_i = Q_i/(r-1)$, $h_j = R_j/(c-1)$ and $k_{ij} = n_{ij}/(r-1)(c-1)$.

The simultaneous confidence intervals associated with the hypothesis of no row effect are now derived. A vector $\mathbf{d}$ is in the alternative space V if and only if $d_{ijk} = d_{ij}$ does not depend on k. It is perpendicular to the null space W if and only if

$$\sum\sum\sum d_{ij}\mu_{ij} = 0\ \forall \mu_{ij} \ni \overline{\mu}_{i.}^{\mathbf{W}} \text{ are equal,}$$

which happens if and only if

$$d_{ij} = \frac{b_i w_j}{n_{ij}\sum w_j} \quad \text{for some } b_i \ni \sum b_i = 0$$

(see Exercise 30.68). In this case

$$\mathbf{d}^T\boldsymbol{\mu} = \sum\sum\sum d_{ij}\mu_{ij} = \sum_i b_i\overline{\mu}_T^{\mathbf{W}}, \quad \mathbf{d}^T\widehat{\boldsymbol{\mu}} = \sum\sum\sum d_{ij}\widehat{\mu}_{ij} = \sum_i b_i A_i,$$

$$\mathbf{d}^T\mathbf{d} = \sum\sum\sum d_{ij}^2 = \sum_i \frac{b_i^2}{Q_i}.$$

Therefore the Scheffé simultaneous confidence intervals for contrasts in the $\overline{\mu}_i^{\mathbf{W}}$ are given by

$$\sum_i b_i\overline{\mu}_{i.}^{\mathbf{W}} \in \sum_i b_i A_i \pm \left((r-1)\,MSE\sum_i \frac{b_i^2}{Q_i}F^{\alpha}\right)^{1/2} \quad \forall b_i \ni \sum_i b_i = 0.$$

We can use the Scheffé simultaneous confidence intervals to construct the numerator of the F test for testing for no row effects. We recall that

$$SSH = \|P_{V|W}\widehat{\boldsymbol{\mu}}\|^2 = \max_{\mathbf{d}\in V|W,\mathbf{d}\neq\mathbf{0}} \frac{(\mathbf{d}^T\widehat{\boldsymbol{\mu}})^2}{\|\mathbf{d}\|^2} = \max \frac{\left(\sum b_i A_i\right)^2}{\sum \frac{b_i^2}{Q_i}}, \text{ subject to } \sum b_i = 0.$$

Since the expression being maximized is unchanged if all the b_i are multiplied by the same constant, we can also assume that $\sum \frac{b_i^2}{Q_i} = 1$. Note also that since $\sum b_i = 0$, $\sum b_i A_i = \sum b_i(A_i - \overline{A})$. We can therefore compute SSH as

$$SSH = \max_{b_i}\left(\sum b_i(A_i - \overline{A})\right)^2, \text{ subject to } \sum b_i = 0 \text{ and } \sum \frac{b_i^2}{Q_i} = 1.$$

Using Lagrange multipliers in the obvious way, we see that the maximum is attained when

$$b_i = \frac{\pm Q_i(A_i - \overline{A})}{\sqrt{\sum Q_i(A_i - \overline{A})^2}}$$

(see Exercise 30.70). Therefore

$$SSH = \sum Q_i(A_i - \overline{A})^2.$$

Note that the A_i are independently distributed as $N(\overline{\mu}_i^{\mathbf{W}}, \sigma^2/Q_i)$. Therefore, if the Q_i are equal, (even if the design is unbalanced) we can construct Tukey and Dunnett simultaneous confidence intervals and multiple range tests in the obvious way. These procedures are not available when the Q_i are different, but multiple F tests can still be computed.

The Scheffé confidence intervals associated with testing for no column effects can be derived in a similar way. The Scheffé simultaneous confidence intervals associated with the hypothesis of no interaction are derived in Exercise 30.71.

Note that the results in this section are also used for balanced models or proportional sampling when we are testing for main effects with weights that do not lead to an orthogonal design.

The results of this section can be extended to unbalanced higher-way crossed models. Note if all the interactions are contained in the model then *SSE* is just the sum of squares of deviations from the mean in each cell. Exact formulae for *SSH* only exist for main effects. The other sums of squares must be computed numerically.

The general twofold nested model

30.34 In a twofold nested model we observe Y_{ijk} independent random variables such that

$$Y_{ijk} \sim N(\mu_{ij}, \sigma^2), \qquad i = 1, \dots, r;\ j = 1, \dots, c_i;\ k = 1, \dots, n_{ij}.$$

Recall that we think of i as representing the class of the observation, j as representing the sub-class and k as representing the replication. Note that even the unbalanced nested model is an orthogonal design for a particular set of weights. However, those weights may not be appropriate for a particular experiment. In this section we derive procedures for such models for arbitrary weights.

Under the alternative hypothesis, we make no assumptions about the μ_{ij}. Therefore $\widehat{\mu}_{ij} = \overline{Y}_{ij}.$ and

$$SSE = \sum\sum\sum(Y_{ijk} - \overline{Y}_{ij.})^2, \quad dfe = N - C$$

(where as usual $N = n_{..}$ is the total number of observations, $C = c_.$ is the total number of sub-classes).

We first consider testing the null hypothesis that there is no sub-class effect, i.e. that μ_{ij} does not depend on j. For this null hypothesis,

$$\widehat{\widehat{\mu}}_{ij} = \overline{Y}_{i..}, \quad SSH = \sum\sum\sum(\overline{Y}_{ij.} - \overline{Y}_{i..})^2 = \sum\sum n_{ij}(\overline{Y}_{ij.} - \overline{Y}_{i..})^2, \quad dfh = C - r,$$

$$EMSH = \sigma^2 + \frac{\sum\sum n_{ij}(\mu_{ij} - \overline{\mu}_{i.})^2}{C - r}.$$

The simultaneous confidence intervals associated with this hypothesis are

$$\sum\sum d_{ij}\mu_{ij} \in \sum\sum d_{ij}\overline{Y}_{ij.} \pm \left((C-r)MSE\sum\sum\frac{d_{ij}^2}{n_{ij}}F^{\alpha}\right)^{1/2} \quad \forall d_{ij} \ni \sum_j d_{ij} = 0$$

(see Exercise 30.73). Note that the formulae given for testing for no sub-class effect and the associated simultaneous confidence intervals are the same as those derived using orthogonal designs, because the hypothesis of no sub-class effect is unaffected by the weights used to define the parameters in that model.

Now consider the problem of testing that there is no average effect due to the class of the observation. Recall that this hypothesis is not defined without some weights w_{ij} for the sub-classes. If we have such weights, let

$$\overline{\mu}_{i.}^{\mathbf{W}} = \frac{\sum_j w_{ij}\mu_{ij}}{\sum_j w_{ij}}.$$

We think of $\overline{\mu}_{i.}^{\mathbf{W}}$ as representing the average mean in the ith class and want to test their equality. To find the estimators under the null hypothesis, we must minimize

$$\sum\sum\sum(Y_{ijk}-\mu_{ij})^2 \text{ subject to equality of } \overline{\mu}_i^{\mathbf{W}}.$$

Once we have such estimators, we can compute *SSH* in the obvious way. Note that *dfh* for this hypothesis is $r-1$.

An exact formula for *SSH* is now described. The estimator of $\overline{\mu}_i^{\mathbf{W}}$ is

$$A_i = \frac{\sum_j w_{ij}\overline{Y}_{ij}}{\sum_j w_{ij}} \sim N\left(\overline{u}_i^{\mathbf{W}}.\frac{\sigma^2}{Q_i}\right), \qquad \frac{1}{Q_i} = \frac{\sum_j w_{ij}^2/n_{ij}}{\left(\sum_j w_{ij}\right)^2},$$

and these are independent. In Exercise 30.74, it is shown that for testing this hypothesis

$$SSH = \sum Q_i(A_i-\overline{A})^2, \quad \overline{A} = \frac{\sum Q_iA_i}{\sum Q_i}, \quad dfh = r-1,$$

$$EMS = \sigma^2 + \frac{\sum Q_i(\overline{\mu}_{i.}^{\mathbf{W}}-\overline{\mu}^{\mathbf{W}})^2}{r-1}, \quad \overline{\mu}^{W} = \frac{\sum Q_i\overline{\mu}_{i.}^{\mathbf{W}}}{\sum Q_i}.$$

(Note the similarity of these formulae to those for the unbalanced non-additive two-way model.) Therefore a partial ANOVA table for this model is

General twofold nested fixed effects model

effect	*df*	*SS*	*EMS*
class	$r-1$	$\sum Q_i(A_i-\overline{A})^2$	$\sigma^2+\sum g_i(\overline{\mu}_{i.}^{\mathbf{W}}-\overline{\mu}^{\mathbf{W}})^2$
subclass	$C-r$	$\sum\sum n_{ij}(\overline{Y}_{ij.}-\overline{Y}_{i.})^2$	$\sigma^2+\sum\sum h_{ij}(\mu_{ij}-\overline{\mu}_{i.})^2$
error	$N-C$	$\sum\sum\sum(Y_{ijk}-\overline{Y}_{ij.})^2$	σ^2

where $g_i = Q_i/(r-1)$, $h_{ij} = n_{ij}/(C-r)$.

To find the simultaneous confidence intervals for contrasts for the hypothesis of no class effect, we must first characterize the contrasts. Note that a vector $\mathbf{d}$ with components d_{ijk} is in V if and only if d_{ijk} does not depend on k. In Exercise 30.74, it is shown that $\mathbf{d} \perp W$ if and only if

$$d_{ij} = \frac{w_{ij} b_i}{n_{ij} \sum_j w_{ij}}, \qquad \text{where } \sum b_i = 0.$$

In this case

$$\mathbf{d}^T \boldsymbol{\mu} = \sum b_i \overline{\mu}_i^{\mathbf{W}}, \quad \mathbf{d}^T \widehat{\boldsymbol{\mu}} = \sum b_i A_i, \quad \|\mathbf{d}\|^2 = \sum \frac{b_i^2}{Q_i},$$

and therefore the Scheffé simultaneous confidence intervals for contrasts are

$$\sum b_i \overline{\mu}_i^{\mathbf{W}} \in \sum b_i A_i \pm \left((r-1) MSE \sum \frac{b_i^2}{Q_i} F^{\alpha} \right)^{1/2} \quad \forall b_i \ni \sum b_i = 0.$$

Note that in the orthogonal analysis we used $w_{ij} = n_{ij}$. In that case $Q_i = n_{i.}$ and the formulae in this section reduce to those in the section on the orthogonal model. Otherwise, the formulae for testing that there is no class effect in this section are different from those for the orthogonal model.

Note also that as long as the Q_i are equal then Tukey and Dunnett simultaneous confidence intervals and multiple range test are available. If the Q_i are unequal multiple F tests are available.

The results in this section can be extended in a straightforward way to more complicated nested models. Notice that the test for the smallest sub-class effect is always the same as for the orthogonal model, but that the other tests depend on the constraints or weights used to define the hypothesis.

The canonical form for testing the general linear hypothesis

30.35 In the problem of testing the general linear hypothesis we observe $\mathbf{Y} \sim N_N(\boldsymbol{\mu}, \sigma^2 \mathbf{I})$, and we want to test the null hypothesis that $\boldsymbol{\mu} \in W$ against the alternative hypothesis that $\boldsymbol{\mu} \in V$ where $W \subset V$ are k-dimensional and p-dimensional subspaces of R^n. Note that the F test for this hypothesis rejects if

$$F = \frac{\|\mathbf{P}_{V|W} \mathbf{Y}\|^2 / (p-k)}{\|\mathbf{P}_{V^\perp} \mathbf{Y}\|^2 / (n-p)} > F^{\alpha}_{p-k,n-p}.$$

The Scheffé simultaneous confidence intervals associated with this testing problem are given by

$$\mathbf{d}^T \boldsymbol{\mu} \in \mathbf{d}^T \widehat{\boldsymbol{\mu}} \pm \left((p-k) \frac{\|\mathbf{P}_{V^\perp} \mathbf{Y}\|^2}{n-p} \|\mathbf{d}\|^2 F^{\alpha}_{p-k,n-p} \right)^{\frac{1}{2}}, \qquad \forall \mathbf{d} \in V|W.$$

In **30.5**, we showed that the F test is the likelihood ratio test for this problem, and in the exercises we show that the power function of the F test is an increasing function of the non-centrality parameter which implies that the F test is unbiased. For most other properties of the F test and simultaneous confidence intervals it is convenient to transform the model into canonical form, which we now describe. (For example, in **23.31** we use this canonical form to show that the F test is the UMP invariant size-α test and the confidence region associated with the Scheffé simultaneous confidence intervals is the UMA equivariant $1-\alpha$ confidence region.)

We use orthonormal basis matrices extensively in this transformation, so we first give a brief review of their elementary properties. Let Q be an arbitrary q-dimensional subspace of R^N. As discussed earlier, an orthonormal basis matrix for Q is an $N \times q$ matrix $\mathbf{X}$ whose columns form an orthonormal basis for Q. (Recall that every subspace has an orthonormal basis matrix.) Some basic properties of orthonormal basis matrices are the following:

$$\mathbf{X}^T\mathbf{X} = \mathbf{I}, \mathbf{X}\mathbf{X}^T = \mathbf{P}_Q, \quad \|\mathbf{P}_Q\mathbf{Y}\|^2 = \|\mathbf{X}^T\mathbf{Y}\|^2.$$

We now return to the problem of testing the general linear hypothesis discussed above. Let $\mathbf{X}_1$, $\mathbf{X}_2$ and $\mathbf{X}_3$ be orthonormal basis matrices for W, $V|W$ and $V^\perp$, and let

$$\mathbf{X} = (\mathbf{X}_1\,\mathbf{X}_2\,\mathbf{X}_3), \quad \mathbf{Z} = \begin{pmatrix}\mathbf{Z}_1\\ \mathbf{Z}_2\\ \mathbf{Z}_3\end{pmatrix} = \mathbf{X}^T\mathbf{Y}, \quad \boldsymbol{\nu} = \begin{pmatrix}\boldsymbol{\nu}_1\\ \boldsymbol{\nu}_2\\ \boldsymbol{\nu}_3\end{pmatrix} = \mathbf{X}^T\boldsymbol{\mu}$$

where $\mathbf{Z}_i$ and $\boldsymbol{\nu}_i$ are $d_i \times 1$ vectors,

$$d_1 = k, \quad d_2 = p-k, \quad d_3 = N-p.$$

Note that $\mathbf{X}$ is an $N \times (k + (p-k) + (N-p)) = N \times N$ matrix and that the orthogonality of the subspaces and orthonormality of the basis matrices implies that $\mathbf{X}^T\mathbf{X} = \mathbf{I}$. Therefore, $\mathbf{X}$ is an invertible matrix and hence $\mathbf{Z}$ is an invertible function of $\mathbf{Y}$.

Theorem 30.6.

(a) $\mathbf{Z}_1$, $\mathbf{Z}_2$ *and* $\mathbf{Z}_3$ *are independent,*

$$\mathbf{Z}_i \sim N_{d_i}(\boldsymbol{\nu}_i, \sigma^2\mathbf{I}).$$

(b) The null and alternative hypotheses can be written as

$$\boldsymbol{\mu} \in V \Leftrightarrow \boldsymbol{\nu}_3 = \mathbf{0}; \quad \boldsymbol{\mu} \in W \Leftrightarrow \boldsymbol{\nu}_2 = \mathbf{0}, \boldsymbol{\nu}_3 = \mathbf{0}.$$

(c) The F *statistic can be written as*

$$F = \frac{\|\mathbf{Z}_2\|^2/d_2}{\|\mathbf{Z}_3\|^2/d_3} \sim F_{d_1,d_2}\left(\frac{\|\boldsymbol{\nu}_2\|^2}{\sigma^2}\right).$$

(d) The Scheffé simultaneous confidence intervals for contrasts can be written as

$$\mathbf{h}^T\boldsymbol{\nu}_2 \in \mathbf{h}^T\mathbf{Z}_2 \pm \left(d_2 F^\alpha \|\mathbf{h}\|^2 \frac{\|\mathbf{Z}_3\|}{d_3}\right)^{1/2}, \qquad \forall \mathbf{h} \in R^{d_2}$$

Proof. (a) Since $\mathbf{X}^T\mathbf{X} = \mathbf{I}$,

$$\mathbf{Z} \sim N_N(\boldsymbol{\nu}, \sigma^2\mathbf{I})$$

and this result follows.

(b) Note that

$$\|\boldsymbol{\nu}_3\|^2 = \|\mathbf{X}_3^T \boldsymbol{\mu}\|^2 = \|\mathbf{P}_{V^\perp}\boldsymbol{\mu}\|^2, \quad \|\boldsymbol{\nu}_2\|^2 = \|\mathbf{P}_{V|W}\boldsymbol{\mu}\|^2, \quad \|\boldsymbol{\nu}_1\|^2 = \|\mathbf{P}_W\boldsymbol{\mu}\|^2.$$

Therefore

$$\begin{aligned}
&\boldsymbol{\mu} \in V \Leftrightarrow 0 = \|\mathbf{P}_{V^\perp}\boldsymbol{\mu}\|^2 = \|\boldsymbol{\nu}_3\|^2 \Leftrightarrow \boldsymbol{\nu}_3 = 0.\\
&\boldsymbol{\mu} \in W \Leftrightarrow 0 = \|\mathbf{P}_{W^\perp}\boldsymbol{\mu}\|^2 = \|\mathbf{P}_{V^\perp}\boldsymbol{\mu}\|^2 + \|\mathbf{P}_{V|W}\boldsymbol{\mu}\|^2 = \|\boldsymbol{\nu}_2\|^2 + \|\boldsymbol{\nu}_3\|^2\\
&\qquad \Leftrightarrow \boldsymbol{\nu}_2 = 0, \boldsymbol{\nu}_3 = 0,
\end{aligned}$$

and the result follows.

(c) As in part (b),

$$\|\mathbf{Z}_3\|^2 = \|\mathbf{P}_{V^\perp}\mathbf{Y}\|^2, \quad \|\mathbf{Z}_2\|^2 = \|\mathbf{P}_{V|W}\mathbf{Y}\|^2, \quad \|\mathbf{Z}_1\|^2 = \|\mathbf{P}_W\mathbf{Y}\|^2.$$

The result follows directly. (To see the distribution of F, note that $\boldsymbol{\nu}_3 = \mathbf{0}$ under both null and alternative hypotheses.)

(d) Recall that by the definition of a basis matrix

$$\mathbf{d} \in V|W \Leftrightarrow \mathbf{d} = \mathbf{X}_2\mathbf{h}$$

for some $\mathbf{h} \in R^{d_2}$. In this case

$$\mathbf{d}^T\boldsymbol{\mu} = \mathbf{h}^T\boldsymbol{\nu}_2, \quad \mathbf{d}^T\widehat{\boldsymbol{\mu}} = \mathbf{h}^T\mathbf{Z}_2, \quad \|\mathbf{d}\|^2 = \|\mathbf{h}\|^2, \quad \|\mathbf{P}_{V^\perp}\mathbf{Y}\|^2 = \|\mathbf{Z}_3\|^2$$

(see Exercise 30.75). The result follows from direct substitution of these results into the formula for the Scheffé simultaneous confidence intervals.

Therefore we have transformed the problem of testing the general linear hypothesis into one in which we observe $\mathbf{Z}_1$, $\mathbf{Z}_2$ and $\mathbf{Z}_3$ independent, $\mathbf{Z}_i \sim N_{d_i}(\boldsymbol{\nu}_i, \sigma^2\mathbf{I})$, $i = 1, 2, 3$, and we are testing

$$\begin{aligned}
H_0 &: \boldsymbol{\nu}_1 \in R^{d_1}, \quad \boldsymbol{\nu}_2 = 0, \quad \boldsymbol{\nu}_3 = 0, \quad \sigma^2 > 0,\\
H_1 &: \boldsymbol{\nu}_1 \in R^{d_1}, \quad \boldsymbol{\nu}_2 \in R^{d_2}, \quad \boldsymbol{\nu}_3 = 0, \quad \sigma^2 > 0.
\end{aligned}$$

That is, in this version of the model, $\sigma^2 > 0$ and $\boldsymbol{\nu}_1$ are unrestricted under both the null and alternative hypotheses and $\boldsymbol{\nu}_3$ is assumed equal to $\mathbf{0}$ under both hypotheses. We are testing the null hypothesis that $\boldsymbol{\nu}_2 = \mathbf{0}$ against the alternative that $\boldsymbol{\nu}_2$ is unrestricted. We have also written the F statistic and the Scheffé simultaneous confidence intervals in terms of the transformed version of the model.

The canonical form for the general linear hypothesis has been used to prove various optimality results for the F test. In particular, the test has been shown to be admissible (see Lehmann and Stein, 1953; Ghosh, 1964). It has also been shown to possess certain minimaxity properties (see Lehmann, 1986, pp. 519–527).

One-sided tests

30.36 Consider the general linear hypothesis-testing problem of testing the null hypothesis that $\boldsymbol{\mu} \in W$ against the alternative hypothesis that $\boldsymbol{\mu} \in V$, where W and V are subspaces of R^n. Let $\mathbf{X}$ be a basis matrix for $V|W$ and let $\boldsymbol{\eta} = \mathbf{X}^T \boldsymbol{\mu}$ and $\widehat{\boldsymbol{\eta}} = \mathbf{X}^T \widehat{\boldsymbol{\mu}}$. By arguments similar to those in **30.35**, we could reinterpret this problem as testing that $\boldsymbol{\eta} = \mathbf{0}$ against $\boldsymbol{\eta} \neq \mathbf{0}$. If $dfh = 1$ then η is a scalar and we could also consider the one-sided problem of testing that $\eta = 0$ against $\eta > 0$. In this case, it is easily seen that

$$\widehat{\eta} \sim N(\eta, \sigma^2 \|\mathbf{X}\|^2) \Rightarrow t = \frac{\widehat{\eta}}{\widehat{\sigma}\|\mathbf{X}\|} \sim t_{n-p}(\theta), \quad \theta = \frac{\eta}{\sigma\|\mathbf{X}\|}.$$

where $t_a(\theta)$ is the non-central t distribution with a degrees of freedom and non-centrality parameter θ (see **29.30–31**).

Therefore a sensible test for this problem is one which rejects if

$$t > t_{n-p}^{\alpha}.$$

In Exercise 30.76, this test is shown to be the LR test for this problem. It can also be shown to be UMP invariant and UMPU. Now consider the two-sided problem of testing $\eta = 0$ against $\eta \neq 0$ (still in the case $dfh = 1$). As mentioned above, this is just a general linear hypothesis problem. Let F be the F statistic for testing this hypothesis. In Exercise 30.76, it is also shown that

$$F = t^2$$

so that the optimal two-sided test rejects if

$$F > F_{1,n-p}^{\alpha} \Leftrightarrow t > t_{n-p}^{\alpha/2} \quad \text{or} \quad t < -t_{n-p}^{\alpha/2},$$

the obvious two-sided test based on t. Note that we have shown that the two-sided test is the LR test, is unbiased and is the UMP invariant size-α test. It can also be shown to be the UMPU size-α test when $dfh = 1$. When $dfh > 1$, there is no UMPU size-α test.

Consider first a one-way model with only two classes in which we observe Y_{ij} independent

$$Y_{ij} \sim N(\mu_i, \sigma^2), \qquad i = 1, 2; \; j = 1, \ldots, n_i.$$

Consider testing that $\mu_1 = \mu_2$. Let $\mathbf{Y}$ be an ordering of the Y_{ij} into an N-dimensional vector ($N = n_1 + n_2$). Let $\boldsymbol{\mu} = E\mathbf{Y}$, let $\widehat{\boldsymbol{\mu}}$ be the LS estimator of $\boldsymbol{\mu}$ (so that $\widehat{\boldsymbol{\mu}}$ has $\overline{Y}_{i.}$ in the (i, j)th place) and let $\widehat{\sigma}^2 = \sum\sum(Y_{ij} - \overline{Y}_{i.})^2/(N-2)$ be the estimator for σ^2. Finally, let $\mathbf{X}$ be a vector whose element in the (i, j)th place is

$$x_{ij} = \frac{(-1)^i}{n_i}.$$

(Note that $\mathbf{X} \in V$, the two-dimensional subspace of vectors such that the component in the (i, j)th place does not depend on j, and that $\mu_1 = \mu_2$ if and only if $\mathbf{X}^T \boldsymbol{\mu} = \mathbf{0}$ so that $\mathbf{X}$ is a basis matrix for $V|W$.) Then

$$\mathbf{X}^T \boldsymbol{\mu} = \mu_1 - \mu_2, \quad \mathbf{X}^T \widehat{\boldsymbol{\mu}} = \overline{Y}_{1.} - \overline{Y}_{2.}, \quad \|\mathbf{X}\|^2 = \frac{1}{n_1} + \frac{1}{n_2}.$$

Therefore, the t statistic for this problem is

$$t = \frac{\overline{Y}_{1.} - \overline{Y}_{2.}}{\widehat{\sigma}\sqrt{\frac{1}{n_1} + \frac{1}{n_2}}},$$

the usual two-sample t statistic. Therefore, the one-sided test for this null hypothesis is just the usual one-sided, two-sample t test for the problem. By the comments above, we see that the usual two-sided, two-sample t test is the same as the F test for the one-way ANOVA with two classes.

Now consider an additive two-way model with only $n = 1$ observation in each cell and only two rows. For the hypothesis of no row effect, $dfh = 1$, so that there is an associated one-sided t test. In Exercise 30.77, it is shown that this t test is just the usual paired test based on differences between the two observations in each column. We therefore see that the two-sided paired t test is the same test as the F test of no row effect in the additive two-way ANOVA model with two rows and one observation in each cell.

The case of known variance

30.37 We now consider the general linear model with known variance in which we observe

$$\mathbf{Y} \sim N_N(\boldsymbol{\mu}, a^2\mathbf{I}), \qquad \boldsymbol{\mu} \in V,$$

where V is a known p-dimensional subspace of R^N and $a^2 > 0$ is a known constant. The derivations for this model are essentially the same as (but a bit easier than) the derivations in **30.4–6** for the case of unknown variance. Therefore, these derivations are left to the reader (see Exercise 30.78) and the results merely stated here.

Let

$$\widehat{\boldsymbol{\mu}} = \mathbf{P}_V\mathbf{Y}$$

be the usual LS estimator of $\boldsymbol{\mu}$. For the model with known variance, $\widehat{\boldsymbol{\mu}}$ is a complete sufficient statistic and is the ML and MV estimator of $\boldsymbol{\mu}$.

Consider the problem of testing the null hypothesis that $\boldsymbol{\mu} \in W$ against the alternative hypothesis that $\boldsymbol{\mu} \in V$, where $W \subset V$ are k-dimensional and p-dimensional subspaces of R^n. As usual, let

$$SSH = \|\mathbf{P}_{V|W}\mathbf{Y}\|^2, \quad dfh = p - k.$$

Then the LR test for this problem rejects if

$$\frac{SSH}{a^2} > \chi^{2\alpha}_{dfh}.$$

The associated simultaneous confidence intervals are given by

$$\mathbf{d}^T\boldsymbol{\mu} \in \mathbf{d}^T\widehat{\boldsymbol{\mu}} \pm (a^2\|\mathbf{d}\|^2\chi^{2\alpha}_{dfh})^{1/2}, \qquad \forall \mathbf{d} \in V|W.$$

Note that in these last two formulae, we have merely taken the formulae for the model with unknown variance and replaced the critical value F^α by $\chi^{2\alpha}$ and replaced $dfh\,MSE$ by the known variance a^2. (Note that the dropping of dfh occurs because in going from the χ^2 distribution to the F distribution we have to divide by the numerator degrees of freedom.)

Sensitivity to assumptions

30.38 In the linear model we observe $\mathbf{Y} \sim N_N(\boldsymbol{\mu}, \sigma^2\mathbf{I})$, $\boldsymbol{\mu} \in V$, $\sigma^2 > 0$, where V is a known p-dimensional subspace of R^N. Let $\mathbf{e} = \mathbf{Y} - \boldsymbol{\mu} = (e_1, \ldots, e_N)^T$. There are four different assumptions about $\mathbf{e}$ for this model.

1. $Ee_i = 0$.
2. $\text{var}(e_i) = \sigma^2 < \infty$ for all i.
3. The e_i are independent.
4. The e_i are normally distributed.

We now briefly discuss the sensitivity of procedures in this chapter to each of these assumptions and remedies for situations when they are not met (see also **32.29–30**).

1. $Ee_i = 0$. The procedures in this chapter are typically quite sensitive to this assumption. If it is not satisfied then often both *SSH* and *SSE* have non-central χ^2 distributions even under the null hypothesis. Even the expected mean squares are no longer correct. This assumption is true if we have included all the appropriate variables in the model. If it is not satisfied then in principle we should just add more variables. In practice this is often difficult. Probably the best way to deal with this assumption is to design the experiment in such a way that the assumption seems reasonable.

2. $\text{var}(e_i) = \sigma^2 < \infty$. The procedures are also often sensitive to this equal-variance assumption. As an illustration consider a one-way model in which we observe Y_{ij} independent

$$EY_{ij} = \mu_i, \quad \text{var}(Y_{ij}) = \sigma_i^2, \quad i = 1, \ldots, p; \; j = 1, \ldots, n_i.$$

The usual F test for the equality of the μ_i is based on

$$MSH = \frac{\sum n_i(\overline{Y}_{i.} - \overline{Y}_{..})^2}{p-1} = \frac{\sum n_i \overline{Y}_{i.}^2 - N\overline{Y}_{..}^2}{p-1}, \quad MSE = \frac{\sum\sum(Y_{ij} - \overline{Y}_{i.})^2}{N-p}.$$

In Exercise 30.79 it is shown that for unequal variances

$$EMSH = \frac{\sum(N - n_i)\sigma_i^2}{N(p-1)}, \quad EMSE = \frac{\sum(n_i - 1)\sigma_i^2}{N-p}$$

under the null hypothesis that the μ_i are equal. Note that these two *EMS*s are the same if either the variances are equal or the sample sizes are equal. This result indicates that the F test may be less sensitive to the assumption of equal variances if the sample sizes are all equal. However, it should be emphasized that the distribution will not be correct or asymptotically correct, even when the sample sizes are equal, unless the variances are equal (or $p = 2$).

One way to handle data with possibly unequal variances is to do a variance stabilizing transform before applying the ANOVA procedure. Such transformations are typically appropriate for models in which the variance in a cell is related to the mean in that cell.

Another way to handle data with possibly unequal variances is to use a more general linear model to find the optimal procedure for known possibly unequal variances. Then substitute the sample variances for the assumed known variances in these formulae (see **31.33** for some examples of this approach). If the number of observations in each cell goes to ∞ this procedure would be asymptotically correct so that it should work well when the number of observations in each cell is large. Such an approach would work in any ANOVA model in which there are many observations in each cell.

The procedures in this chapter are also sensitive to the finite-variance assumption. However since most practical data will be bounded, this assumption is usually satisfied.

3. The e_i are independent. The procedures in this chapter are also quite sensitive to this assumption. To illustrate this sensitivity suppose we observe Y_{ij}, $i = 1, \ldots, p$; $j = 1, \ldots, n$, such that

$$EY_{ij} = \mu_i, \quad \text{var}(Y_{ij}) = \sigma^2, \quad \text{cov}(Y_{ij}, Y_{ik}) = \rho\sigma^2, \quad \text{cov}(Y_{ij}, Y_{i^T j^T}) = 0, \quad i \neq i^T.$$

Let *MSH* and *MSE* be the usual mean squares for the one-way model. In Exercise 30.80, it is shown that, under the null hypothesis of equal means,

$$EMSH = \sigma^2(1 + (n-1)\rho), \quad EMSE = \sigma^2(1-\rho),$$

so that these procedures would be completely inappropriate for this dependency structure. In many data sets, we take several measurements on each of several individuals. In such situations, we can often use repeated measures models (see **31.20–32**). If no exact procedures are available for a particular data set we can often first use the results for a more general linear model to find the optimal procedure for a known covariance matrix and substitute an estimator for the covariance matrix into the formulae derived (see **31.33**). This procedure often works as long as the number of observations is large compared to the number of unknown parameters in the covariance matrix.

4. The e_i are normally distributed. The F test and Scheffé simultaneous confidence are fairly robust against this assumption. Arnold (1981) shows, under a fairly general condition due to Huber (1973), that the size and power of the F test and the confidence coefficient of the associated simultaneous confidence intervals are asymptotically correct as long as the e_i are independently and identically distributed with finite variance. A sufficient condition for Huber's condition to be satisfied in ANOVA models is that the number of observations in each cell must go to ∞. Therefore if there are fairly many observations in each cell the procedures should not be too sensitive to the normal assumption. Another indication of the robustness of the F test to the normal assumption is given in Exercise 30.81, where it is shown that the size of the F test and the confidence coefficient for the associated confidence intervals are exactly correct as long as the distribution of $\mathbf{e}$ is spherical. If the e_i are clearly non-normally distributed, there are also non-parametric alternatives to many of the procedures in this chapter (see Hettmansperger and McKean, 1998 and Akritas, Arnold and Brunner, 1997).

In Arnold (1981, pp. 194–195), it is also shown that the size of the studentitzed range test and the confidence coefficient of associated Tukey simultaneous confidence intervals are also asymptotically unaffected by non-normality. (Note that these procedures are only available for balanced models so that as the total number of observations goes to ∞, the number in each cell must also go to ∞. Hence Huber's condition is automatically satisfied when these procedures are used.) In a similar way we could show that procedures based on Dunnett's distribution are also asymptotically insensitive to the normal assumption used in their derivation.

EXERCISES

30.1 (a) Prove part (a) of Lemma 30.1. (Let Q be the subspace spanned by columns of $\mathbf{C}$. Then $V = Q^{\perp}$ and hence $V^{\perp} = (Q^{\perp})^{\perp} = Q$.)
(b) Prove part (b) of Lemma 30.1. (Let Q be subspace spanned by columns of $\mathbf{C}$. Then $Q \subset V$ and $W = V|Q$ so that $V|W = Q$.)

30.2 Let $\mathbf{v} \in V$ a subspace of R^n and let $\mathbf{y} \in R^n$. Show that $\|\mathbf{y} - \mathbf{v}\|^2 = \|\mathbf{y} - \mathbf{P}_V\mathbf{y}\|^2 + \|\mathbf{P}_V\mathbf{y} - \mathbf{v}\|^2 \geq \|\mathbf{y} - \mathbf{P}_V\mathbf{y}\|$, with equality only if $\mathbf{v} = \mathbf{P}_V\mathbf{y}$.

30.3 Let $\mathbf{X}$ be basis matrix for a subspace $V \subset R^n$ and let $\mathbf{y} \in R^n$.
(a) Show that $\mathbf{v} \in V$ if and only if $\mathbf{v} = \mathbf{Xb}$ for some vector $\mathbf{b}$.
(b) Let $\mathbf{u} = \mathbf{X}(\mathbf{X}^T\mathbf{X})^{-1}\mathbf{X}^T\mathbf{y}$. Show that $\mathbf{u} \in V$ and that $\mathbf{y} - \mathbf{u} \perp V$ and hence that $\mathbf{u} = \mathbf{P}_V\mathbf{y}$.

30.4 (a) Show that $(\mathbf{P}_V)^2 = \mathbf{P}_V$ and $(\mathbf{P}_V)^T = \mathbf{P}_V$.
(b) Suppose that $\mathbf{A}$ is an $n \times n$ matrix such that $\mathbf{A}^2 = \mathbf{A}$ and $\mathbf{A}^T = \mathbf{A}$. Let V be the set of all $\mathbf{Ab}$, $\mathbf{b} \in R^n$. Let $\mathbf{y} \in R^n$ and let $\mathbf{u} = \mathbf{Ay}$. Show that $\mathbf{u} \in V$, $\mathbf{y} - \mathbf{u} \perp V$ so that $\mathbf{A} = \mathbf{P}_V$.

30.5 (a) Suppose that $\mathbf{P}_V\mathbf{y} = \mathbf{y}$. Show that $\mathbf{y} \in V$.
(b) Suppose that $\mathbf{y} \in V$. Use the definition of projection to show that $\mathbf{P}_V\mathbf{y} = \mathbf{y}$.

30.6 (a) Suppose that $\mathbf{P}_V\mathbf{y} = \mathbf{0}$. Show that $\mathbf{y} \perp V$.
(b) Suppose that $\mathbf{y} \perp V$. Use the definition of projection to show that $\mathbf{P}_V\mathbf{y} = \mathbf{0}$.

30.7 Let $W \subset V \subset R^n$ be subspaces and let $\mathbf{y} \in R^n$.
a. Show that $\mathbf{P}_V(\mathbf{P}_W\mathbf{y}) = \mathbf{P}_W\mathbf{y}$ and hence $\mathbf{P}_V\mathbf{P}_W = \mathbf{P}_W$. (Note that $\mathbf{P}_W\mathbf{y} \in W \subset V$.)
(b) Show that $\mathbf{P}_W\mathbf{P}_V = \mathbf{P}_W$. ($\mathbf{P}_W\mathbf{P}_V = (\mathbf{P}_V\mathbf{P}_W)^T$. Why?)

30.8 (a) Suppose that $W \perp V$ are subspaces of R^n and let $\mathbf{y} \in R^n$. Show that $\mathbf{P}_W(\mathbf{P}_V\mathbf{y}) = \mathbf{0}$ and hence $\mathbf{P}_W\mathbf{P}_V = \mathbf{0}$. (Note that $\mathbf{P}_V\mathbf{y} \in V \perp W$.)
(b) Let W and V be subspaces such that $\mathbf{P}_W\mathbf{P}_V = \mathbf{0}$. Show that $V \perp W$. (Let $\mathbf{w} \in W$. Then $\mathbf{P}_V\mathbf{w} = \mathbf{P}_V\mathbf{P}_W\mathbf{w} = \mathbf{0}$, and hence $\mathbf{w} \perp V$.)

30.9 Let $V \subset R^n$ be a p-dimensional subspace and let $\mathbf{y} \in R^n$.
(a) Let $\mathbf{u} = \mathbf{y} - \mathbf{P}_V\mathbf{y}$. Show that $\mathbf{u} \in V^{\perp}$ and that $\mathbf{y} - \mathbf{u} \perp V^{\perp}$, and hence that $\mathbf{u} = \mathbf{P}_{V^{\perp}}\mathbf{y}$ and $\mathbf{P}_{V^{\perp}} = \mathbf{I} - \mathbf{P}_V$.
(b) Show that $\|\mathbf{Y}\|^2 = \|\mathbf{P}_{V^{\perp}}\mathbf{y}\|^2 + \|\mathbf{P}_V\mathbf{y}\|^2$.
(c) Let $\mathbf{x}_1, \ldots, \mathbf{x}_p$ be an orthonormal basis for V and let $\mathbf{x}_{p+1}, \ldots, \mathbf{x}_{p+r}$ be an orthonormal basis for $V^{\perp}$. Show that $\mathbf{x}_1, \ldots, \mathbf{x}_{p+r}$ is an orthonormal basis for R^n and hence $n = \dim(V^{\perp}) + \dim(V)$. (By part (a), $\mathbf{y} = \mathbf{P}_V\mathbf{y} + \mathbf{P}_{V^{\perp}}\mathbf{y}$.)

30.10 Let $W \subset V \subset R^n$ be k-dimensional and p-dimensional subspaces respectively and let $\mathbf{y} \in R^n$.
(a) Let $\mathbf{u} = \mathbf{P}_V\mathbf{y} - \mathbf{P}_W\mathbf{y}$. Show that $\mathbf{u} \in V|W$ and $\mathbf{y} - \mathbf{u} \perp V|W$ so that $\mathbf{u} = \mathbf{P}_{V|W}\mathbf{y}$ and $\mathbf{P}_{V|W} = \mathbf{P}_V - \mathbf{P}_W$.
(b) Show that $\|\mathbf{P}_V\mathbf{y}\|^2 = \|\mathbf{P}_{V|W}\mathbf{y}\|^2 + \|\mathbf{P}_W\mathbf{y}\|^2$.
(c) Let $\mathbf{x}_1, \ldots, \mathbf{x}_k$ and $\mathbf{x}_{k+1}, \ldots, \mathbf{x}_{k+r}$ be orthonormal bases for W and $V|W$. Show that $\mathbf{x}_1, \ldots, \mathbf{x}_{k+r}$ is an orthonormal basis for V and hence that $\dim(V) = \dim(W) + \dim(V|W)$. (By part (a), $\mathbf{v} \in V$ implies that $\mathbf{v} = \mathbf{P}_V\mathbf{v} = \mathbf{P}_W\mathbf{y} + \mathbf{P}_{V|W}\mathbf{y}$.)

30.11 (a) Let V and W be subspaces of R^n such that $\mathbf{P}_V = \mathbf{P}_W$. Show that $V = W$. (Let $\mathbf{y} \in W$. Then $\mathbf{y} = \mathbf{P}_W\mathbf{y} = \mathbf{P}_V\mathbf{y}$, which implies that $\mathbf{y} \in V$.)
(b) Show that $(V^{\perp})^{\perp} = V$ and $V|(V|W) = W$.

30.12 Let $\mathbf{X}$ be an orthonormal basis matrix for the p-dimensional subspace V for the linear model defined in **30.4** and let $\boldsymbol{\mu} = \mathbf{X}\boldsymbol{\beta}$ for some $\boldsymbol{\beta} \in R^p$. Let $\mathbf{T} = \mathbf{X}^T\mathbf{Y}$ and let $U = \mathbf{Y}^T\mathbf{Y}$.
(a) Show that the likelihood for this model can be written as $L_Y(\boldsymbol{\beta}, \sigma^2) \approx \sigma^{-\frac{n}{2}} \exp\left(-\frac{\boldsymbol{\beta}^T\boldsymbol{\beta}}{2\sigma^2}\right) \times \exp\left(-\frac{U - 2\mathbf{T}^T\boldsymbol{\beta}}{2\sigma^2}\right)$.

(b) Show that $(\mathbf{T}, U)$ is a complete sufficient statistic for this version of the linear model.
(c) Show that $(\widehat{\boldsymbol{\mu}}, \widehat{\sigma}^2)$ is a complete sufficient statistic for the linear model.

30.13 Verify that $(N - p)\widehat{\sigma}^2/N$ is the maximum likelihood estimator of σ^2 for the linear model.

30.14 (a) Verify that $\widehat{\sigma}^2$ is a consistent estimator of σ^2 for the linear model.
(b) Verify that if $p/N \to d > 0$, then the ML estimator of σ^2 is a consistent estimator of $\sigma^2(1-d)$ and hence is an inconsistent estimator of σ^2.

30.15 (Gauss–Markov.) Suppose only that $E\mathbf{Y} = \boldsymbol{\mu} \in V$, $\text{cov}(\mathbf{Y}) = \sigma^2\mathbf{I}$. We say that $T(\mathbf{Y})$ is a linear unbiased estimator of $\mathbf{a}^T\boldsymbol{\mu}$ if $\mathbf{T}(\mathbf{Y}) = \mathbf{b}^T\mathbf{Y}$ for some vector $\mathbf{b}$ and $E\mathbf{T}(\mathbf{Y}) = \mathbf{a}^T\boldsymbol{\mu}$.
(a) Show that $\mathbf{b}^T\mathbf{Y}$ is an unbiased estimator of $\mathbf{a}^T\boldsymbol{\mu}$ if and only if $(\mathbf{b}-\mathbf{a})^T\boldsymbol{\mu} = \mathbf{0}$ for all $\boldsymbol{\mu} \in V$ if and only if $\mathbf{b} - \mathbf{a} \in V^{\perp}$.
(b) Show that if $\mathbf{b}^T\mathbf{Y}$ is an unbiased estimator of $\mathbf{a}^T\boldsymbol{\mu}$, then $\text{var}(\mathbf{b}^T\mathbf{Y}) = \sigma^2\|\mathbf{b}-\mathbf{P}_V\mathbf{a}\|^2 + \sigma^2\|\mathbf{P}_V\mathbf{a}\|^2$. (By part (a), $\mathbf{P}_V\mathbf{b} - \mathbf{P}_V\mathbf{a} = \mathbf{0}$.)
(c) Show that if $\mathbf{b}^T\mathbf{Y}$ is an unbiased estimator of $\mathbf{a}^T\boldsymbol{\mu}$, then $\text{var}(\mathbf{b}^T\mathbf{Y}) \geq \text{var}(\mathbf{a}^T\widehat{\boldsymbol{\mu}})$, with equality only if $\mathbf{b}^T\mathbf{Y} = \mathbf{a}^T\widehat{\boldsymbol{\mu}}$.
(d) Now let $\mathbf{BY}$ be a linear unbiased estimator of $\mathbf{A}\boldsymbol{\mu}$. Show that $\text{cov}(\mathbf{BY}) - \text{cov}(\mathbf{A}\widehat{\boldsymbol{\mu}}) \geq 0$. (Note that $\mathbf{s}^T(\text{cov}(\mathbf{BY}) - \text{cov}(\mathbf{A}\widehat{\boldsymbol{\mu}}))\mathbf{s} = \text{var}(\mathbf{s}^T\mathbf{BY}) - \text{var}(\mathbf{s}^T\mathbf{A}\widehat{\boldsymbol{\mu}})$. Use part (c).

30.16 Verify that if $W \subset V$, then $V^{\perp} \subset W^{\perp}$ and $V|W = W^{\perp}|V^{\perp}$.

30.17 Let $\mathbf{C}$ and W be defined as in Example 30.5.
(a) Verify that $\mathbf{C}$ is a basis matrix for $V|W$. (Use Lemma 30.1.)
(b) Verify that $\widehat{\boldsymbol{\mu}}^T\mathbf{C}(\mathbf{C}^T\mathbf{C})^{-1}\mathbf{C}^T\widehat{\boldsymbol{\mu}} = (\mathbf{A}\widehat{\boldsymbol{\beta}})^T(\mathbf{A}(\mathbf{X}^T\mathbf{X})^{-1}\mathbf{A}^T)^{-1}(\mathbf{A}\widehat{\boldsymbol{\beta}})$.

30.18 (a) Let $Z \sim N(\mu, 1)$, $\mu > 0$. Show that for any $a > 0$, $P(|Z|^2 > a)$ is an increasing function of μ.
(b) Let $U \sim \chi^2_{s-1}$ independently of Z and let $S = U + Z^2 \sim \chi^2_s(\delta)$, $\delta = \mu^2$. Show that $P(S > b)$ is an increasing function of δ for all $b > 0$. (*Hint*: $P(S > b) = E(P(Z^2 > b - U|U))$.
(c) Let $T \sim \chi^2_t$ independently of S and let $F = \frac{S/s}{T/t} \sim F_{s,t}(\delta)$. Show that $P(F > c)$ is an increasing function of δ for all $c > 0$. (*Hint*: $P(F > c) = E(P(S > csT/t|T))$.

30.19 In the notation of Theorem 30.4, let $h(\mathbf{d}) = (\mathbf{d}^T(\widehat{\boldsymbol{\delta}} - \boldsymbol{\delta}))^2/\|\mathbf{d}\|^2$.
(a) Show that if $\mathbf{d} \in V|W$, $\mathbf{d} \neq \mathbf{0}$, then $h(\mathbf{d}) \leq \|\widehat{\boldsymbol{\delta}} - \boldsymbol{\delta}\|^2$.
(b) Show that if $\mathbf{d} = \widehat{\boldsymbol{\delta}} - \boldsymbol{\delta}$, then $h(\mathbf{d}) = \|\widehat{\boldsymbol{\delta}} - \boldsymbol{\delta}\|^2$.
(c) Show that $\|\widehat{\boldsymbol{\delta}} - \boldsymbol{\delta}\|^2 = \max_{\mathbf{d} \in V|W, \mathbf{d} \neq \mathbf{0}} h(\mathbf{d})$.
(d) Show that $SSH = \max_{\mathbf{d} \in V|W, \mathbf{d} \neq \mathbf{0}} (\mathbf{d}^T\widehat{\boldsymbol{\mu}})^2/\|\mathbf{d}\|^2$. (Recall that $SSH = \|\widehat{\boldsymbol{\delta}} - \mathbf{0}\|^2$. Use Exercise 30.20).

30.20 Show that if $\mathbf{d} \in V|W$, than $\mathbf{d}^T\boldsymbol{\mu} = \mathbf{d}^T\boldsymbol{\delta}$ and $\mathbf{d}^T\widehat{\boldsymbol{\mu}} = \mathbf{d}^T\widehat{\boldsymbol{\delta}}$. (*Hint*: $\mathbf{d} = \mathbf{P}_{V|W}(\mathbf{d})$).

30.21 Let $\mathbf{X}$ be an $n \times r$ matrix of rank p.
(a) Show that $\mathbf{X}^T\mathbf{X} \geq \mathbf{0}$ and hence that there exists an $r \times p$ matrix $\mathbf{A}$ of rank p such that $\mathbf{X}^T\mathbf{X} = \mathbf{A}\mathbf{A}^T$.
(b) Let $\mathbf{B} = \mathbf{A}(\mathbf{A}^T\mathbf{A})^{-1}\mathbf{A}$. Show that $\mathbf{B}$ is a symmetric generalized inverse of $\mathbf{X}^T\mathbf{X}$.
(c) Show that $\mathbf{B}$ is the Moore–Penrose inverse of $\mathbf{X}^T\mathbf{X}$ (i.e. show that $\mathbf{B}\mathbf{X}^T\mathbf{X}\mathbf{B} = \mathbf{B}$ and $\mathbf{B}\mathbf{X}^T\mathbf{X}$ and $\mathbf{X}^T\mathbf{X}\mathbf{B}$ are both symmetric).
(d) Let $\mathbf{C} \neq \mathbf{0}$ be an $r \times q$ matrix such that $\mathbf{A}^T\mathbf{C} = \mathbf{0}$. (Why must such a matrix exist?) Let $\mathbf{F} = \mathbf{B} + \mathbf{C}\mathbf{C}^T$. Show that $\mathbf{F}$ is also a symmetric generalized inverse of $\mathbf{X}^T\mathbf{X}$, and hence that the symmetric generalized inverse is not unique.

30.22 We follow the notation of Example 30.7.
(a) Show that $\mathbf{X}^T\mathbf{X}\mathbf{C} = \mathbf{0}$ implies that $\mathbf{X}\mathbf{C} = \mathbf{0}$. (Hint: there exist $N \times p$ and $p \times r$ matrices $\mathbf{A}$ and $\mathbf{B}$ of rank p such that $\mathbf{X} = \mathbf{A}\mathbf{B}$. Note that $\mathbf{A}^T\mathbf{A}$ and $\mathbf{B}\mathbf{B}^T$ are invertible.)
(b) Show that $\mathbf{X}(\mathbf{X}^T\mathbf{X})^-\mathbf{X}^T\mathbf{X} = \mathbf{X}$. (Use part (a) with $\mathbf{C} = (\mathbf{X}^T\mathbf{X})^-\mathbf{X}^T\mathbf{X} - \mathbf{I}$.)
(c) Let V be the subspace spanned by the columns of $\mathbf{X}$. Show that $\mathbf{v} \in V$ if and only if $\mathbf{v} = \mathbf{X}\mathbf{b}$ for some vector $\mathbf{b}$.

(d) Let $\mathbf{y} \in R^N$. Let $\mathbf{u} = \mathbf{X}(\mathbf{X}^T\mathbf{X})^-\mathbf{X}^T\mathbf{y}$. Show that $\mathbf{u} \in \mathbf{V}$ and $\mathbf{y} - \mathbf{u} \perp V$ so that $\mathbf{P}_V = \mathbf{X}(\mathbf{X}^T\mathbf{X})^-\mathbf{X}^T$ and $\widehat{\boldsymbol{\mu}} = \mathbf{X}(\mathbf{X}^T\mathbf{X})^-\mathbf{X}^T\mathbf{y}$.
(e) Show that $\widehat{\boldsymbol{\beta}}$ is a least squares estimator of $\boldsymbol{\beta}$ if and only if $\mathbf{X}\widehat{\boldsymbol{\beta}} = \widehat{\boldsymbol{\mu}}$.
(f) Show that $(\mathbf{X}^T\mathbf{X})^-\mathbf{X}^T\mathbf{Y} + (\mathbf{I} - (\mathbf{X}^T\mathbf{X})^-\mathbf{X}^T\mathbf{X})h(\mathbf{Y})$ is an LS estimator of $\boldsymbol{\beta}$ for any function $h(\mathbf{Y})$.
(g) Show that $\mathbf{c}^T\boldsymbol{\beta}$ is an estimable function if and only if $\mathbf{c} = \mathbf{X}^T\mathbf{a}$ for some $\mathbf{a}$.
(h) Suppose that $\mathbf{H}\boldsymbol{\beta}$ is a vector of estimable functions. Show that $\mathbf{H}(\mathbf{X}^T\mathbf{X})^-\mathbf{X}^T\mathbf{X} = \mathbf{H}$. (Note that by part (g), $\mathbf{H} = \mathbf{BX}$ for some $\mathbf{B}$. Use part (b) above.)
(i) Show that $\mathbf{C}^T\boldsymbol{\mu} = \mathbf{H}\boldsymbol{\beta}$ and that $\mathbf{C}$ is a basis matrix for $V|W$. (To see that the columns of $\mathbf{C}$ are linearly independent note that $\mathbf{Cb} = \mathbf{0}$ implies that $\mathbf{0} = \mathbf{XCb} = \mathbf{H}^T\mathbf{b}$ and the rows of $\mathbf{H}$ are assumed linearly independent.)
(j) Show that $\mathbf{Y}^T\mathbf{C}(\mathbf{C}^T\mathbf{C})^{-1}\mathbf{C}^T\mathbf{Y} = \widehat{\boldsymbol{\gamma}}^T(\mathbf{H}(\mathbf{X}^T\mathbf{X})^-\mathbf{H})^{-1}\widehat{\boldsymbol{\gamma}}$.
(k) Show that $\|\mathbf{Cs}\|^2 = \mathbf{s}^T\mathbf{H}(\mathbf{X}^T\mathbf{X})^-\mathbf{H}^T\mathbf{s}$.

30.23 (a) Let $\mathbf{Y}$ and $\boldsymbol{\mu}$ be defined as in the one-way model of **30.8**. Let $\boldsymbol{\beta} = (\mu_1, \ldots, \mu_p)^T$ and let $\boldsymbol{\mu} = \mathbf{X}\boldsymbol{\beta}$. What is $\mathbf{X}$?
(b) Show that $\mathbf{X}^T\mathbf{X} = \begin{pmatrix} n_1 & 0 & \cdots & 0 \\ 0 & n_2 & \cdots & 0 \\ \vdots & \vdots & \ddots & \vdots \\ 0 & 0 & \cdots & n_p \end{pmatrix}$ and that $\mathbf{X}^T\mathbf{Y} = \begin{pmatrix} n_1\overline{Y}_1 \\ n_2\overline{Y}_2 \\ \vdots \\ n_p\overline{Y}_p \end{pmatrix}$.
(c) Show that $\widehat{\boldsymbol{\beta}} = (\overline{Y}_1 \;\; \overline{Y}_2 \;\; \cdots \;\; \overline{Y}_p)^T$, $\|\mathbf{Y} - \mathbf{X}\widehat{\boldsymbol{\beta}}\|^2 = \sum\sum(Y_{ij} - \overline{Y}_i)^2$.

30.24 In the one-way ANOVA model, consider testing the null hypothesis that the μ_i are all 0.
a. Set this testing problem up as problem of testing a linear hypothesis.
(b) Show that for this problem $SSH = \sum n_i\overline{Y}_{i.}^2$, $dfh = p$. What are MSE and dfe for this problem?
(c) Show that the Scheffé simultaneous confidence intervals for contrasts associated with this hypothesis are $\sum c_i\mu_i \in \sum c_i\overline{Y}_{i.} \pm \left(pF^\alpha MSE \sum c_i^2/n_i\right)^{1/2}$ for all possible c_i.

30.25 (a) In the notation of **30.9**, verify that $Q(\boldsymbol{\mu}) \sim q_{p,p(n-1)}$.
(b) Verify that $Q(\boldsymbol{\mu}) \leq q^\alpha$ if and only if $\mu_i - \mu_j \in \overline{Y}_{i.} - \overline{Y}_{j.} \pm q^\alpha\sqrt{MSE/n}$.

30.26 Let $a_1, \ldots, a_k$ be a set of numbers such that $|a_i - a_j| \leq b$ for all i and j. Let $c_1, \ldots, c_k$ be a set of numbers such that $\sum c_i = 0$ and $\sum |c_i| > 0$. Let P be set of subscripts such that $c_i > 0$ and let N be the set of subscripts such that $c_i < 0$. Let $g = \sum_P c_i = -\sum_N c_i = \sum |c_i|/2$.
(a) Show that $g\sum c_ia_i = \sum_{i\in P}\sum_{j\in N}(-c_ic_j)(a_i - a_j)$.
(b) Show that $|\sum c_ia_i| \leq b\sum |c_i|/2$.

30.27 (a) Let a_i be a set of numbers. Show that $|a_i - a_j| \leq b$ for all i and j if and only if $|\sum c_ia_i| \leq b\sum |c_i|/2$ for all c_i such that $\sum c_i = 0$. (Use Exercise 30.26.)
(b) Verify the Tukey simultaneous confidence intervals for contrasts in the one-way model.

30.28 Let $U \sim \chi_m^2$ independently of $\mathbf{Z} = (Z_1, \ldots, Z_k)$, where the Z_i are jointly normally distributed with $EZ_i = 0$, $\text{var}(Z_i) = 1$ and $\text{cov}(Z_i, Z_j) = a$. Let

$$Q = \frac{\max|Z_i - Z_j|}{\sqrt{(1-a)U/m}}, \qquad D = \frac{\max|Z_i - Z_j|}{\sqrt{2(1-a)U/m}}.$$

(a) When $a < 0$, show that $Q \sim q_{k,m}$ and $D \sim D_{k-1,m}$. (*Hint*: Let T be independent of U and $\mathbf{Z}$, $T \sim N(0, -a)$ and let $Z_i^* = Z_i + T$. Then $Z_i - Z_j = Z_i^* - Z_j^*$ and Z_i^* are independent, $Z_i^* \sim N(0, 1-a)$. Why?)
(b) When $a > 0$, show that $Q \sim q_{k,m}$ and that $D \sim D_{k-1,m}$. (*Hint*: Let W_i and T be independent (and also independent of U), $W_i \sim N(0, 1-a)$ and $T \sim N(0, a)$ and let $Z_i = W_i + T$. Then Z_i have the indicated distribution and $Z_i - Z_j = W_i - W_j$. Why?)

30.29 Let $Z_1, \ldots, Z_k$, U be independent, $Z_i \sim N(0, 1)$, $U \sim \chi_m$. Let $M = \max |Z_i|/\sqrt{U/m}$. We say that M has a studentized maximum modulus distribution and write $M \sim M_{k,m}$.
(a) Use this distribution to find a set of simultaneous confidence intervals for the set of all μ_i in a balanced one-way ANOVA model.
(b) Let a_i be a set of numbers. Show that $\max |a_i| \leq d$ if and only if $|\sum c_i a_i| \leq d \sum |c_i|$ for all c_i.
(c) Find a set of simultaneous confidence intervals for the set of all $\sum c_i \mu_i$.

30.30 (a) In the notation of **30.10**, verify that $D(\boldsymbol{\mu}) \sim D_{p-1,p(n-1)}$.
(b) Verify that $D(\boldsymbol{\mu}) \leq D^\alpha$ if and only if $\mu_i - \mu_1 \in \overline{Y}_{i.} - \overline{Y}_{1.} \pm D^\alpha \sqrt{MSE/n}$.

30.31 (a) (Boole's inequality.) Let B_i be subsets of a sample space. Show that $P(\cup B_i) \leq \sum P(B_i)$.
(b) (Bonferroni's inequality.) Let A_i be subsets of a sample space. Show that $P(\cap A_i) \geq 1 - \sum P(A_i^c)$ (Use Boole's inequality with $B_i = A_i^c$.

30.32 Verify that the Bonferroni intervals in **30.11** are a set of conservative $1 - \alpha$ simultaneous confidence intervals for the finite set of contrasts.

30.33 Show that if $q_{s,N-p}^{\alpha_s} \leq q_{p,N-p}^{\alpha}$, the multiple range test with sizes α_s is more powerful than the Tukey procedure.

30.34 Show that if a set S contains means which are all the same, then the probability that the set is α_s Bonferroni significant is less than or equal to α_s. (Use Bonferroni's inequality.)

30.35 (Generalized Bonferroni inequality.) Let $g_1(x), \ldots, g_k(x)$ be non-negative increasing functions from R to R and let X be a random variable such that $\mu_i = Eg_i(X)$ is finite.
(a) Show that $E(g_1(X) - \mu_1)(g_2(X) - \mu_2) \geq 0$. (Let a be such that $g_1(X) \geq \mu_1$ if $X > a$ and $g_1(X) \leq \mu_1$ if $X < a$. Then $(g_1(X) - \mu_1)(g_2(X) - g_2(a)) \geq 0$ and $E(g_1(X) - \mu_1)(g_2(X) - \mu_2) = E(g_1(X) - \mu_1)(g_2(X) - a)$.)
(b) Show that $E(\prod g_i(X)) \geq \prod Eg_i(X)$. (Use part (a) and induction. Note that the product of non-negative increasing functions is a non-negative increasing function.)

30.36 Consider a multiple comparisons problem. Divide the set of means in the following way. Let S_0 be the set of subscripts of means which are unequal to any other means and let $S_1, S_2, \ldots, S_q$ be sets of subscripts with equal means.
(a) Show that a multiple range, multiple F or multiple Bonferroni procedure makes a Type I error only if it finds any of the S_i significant so that the generalized size is less than or equal to $P(\cup S_i \text{ significant})$.
(b) Use Boole's inequality (Exercise 30.31(a)) to prove part (a) of Lemma 30.2.
(c) Show that for multiple F, multiple range and multiple Bonferroni procedures, $P(\cap S_i \text{ insignificant}) \geq \prod P(S_i \text{ insignificant})$. (Note that $P(\cap S_i \text{ insignificant}) = E(P(\cap S_i \text{ insignificant}|MSE)) = E\left(\prod P(S_i \text{ insignificant}|MSE)\right)$ because the S_i are disjoint and hence conditionally on MSE the events S_i significant are independent. Finally, note that $P(S_i \text{ insignificant}|MSE)$ is an increasing function of MSE. Use Exercise 30.35.)
(d) Prove part (b) of Lemma 30.2.

30.37 Prove that if $s \leq p$ then $1 - (1 - \alpha)^{\frac{s}{p}} > 1 - (1 - \alpha)^{\frac{s-1}{p-1}}$ and that $1 - (1 - \alpha)^{\frac{s}{p}} > \frac{s}{p}\alpha$.

30.38 Show that a set with p elements has $2^p - p - 1$ subsets with at least two elements.

30.39 Let T_i be orthogonal subspaces of R^N and let $V = \oplus T_i$.
(a) Verify that V is a subspace and that $T_i \subset V$.
(b) Let $\mathbf{y} \in R^N$ and let $\mathbf{u} = \sum P_{T_i}\mathbf{y}$. Show that $\mathbf{u} \in V$ and $\mathbf{y} - \mathbf{u} \perp V$ so that $\mathbf{u} = \mathbf{P}_V\mathbf{y}$.
(c) Show $\|\mathbf{P}_V\mathbf{y}\|^2 = \sum \|\mathbf{P}_{T_i}\mathbf{y}\|^2$.
(d) Suppose that $\mathbf{P}_V\mathbf{y} = \sum \mathbf{t}_i$, $\mathbf{t}_i \in T_i$. Show that $\mathbf{t}_i = \mathbf{P}_{T_i}\mathbf{P}_V\mathbf{y} = \mathbf{P}_{T_i}\mathbf{y}$. (Note that $\mathbf{P}_{T_i}\mathbf{t}_i = \mathbf{t}_i$ and that $\mathbf{P}_{T_i}\mathbf{t}_j = \mathbf{0}$ if $j \neq i$.)
(e) Show that $\dim(V) = \sum \dim(T_i)$. (Let $\mathbf{X}_i$ be an orthonormal basis matrix for T_i. Then $\mathbf{X} = (\mathbf{X}_1 \ \cdots \ \mathbf{X}_q)$ is an orthonormal basis matrix for V. Why?)

30.40 Let T_i be orthogonal subspaces of R^n, let $V = \oplus_i T_i$ and let $W_j = \oplus_{i \neq j} T_i$. Show that $\mathbf{P}_{V|W_j} = \mathbf{P}_{T_j}$ and hence that $V|W_j = T_j$. ($\mathbf{P}_{V|W_j} = \mathbf{P}_V - \mathbf{P}_{W_j}$.)

30.41 Let $\boldsymbol{\delta} = \mathbf{P}_{V|W}\boldsymbol{\mu}$ and let $\mathbf{d} \in V|W$. Show that $\mathbf{d}^T\boldsymbol{\delta} = \mathbf{d}^T\boldsymbol{\mu}$.

30.42 In the one-way model, show that if $\sum c_i = 0$ then $\sum c_i\mu_i = \sum c_i\alpha_i$ and $\sum c_i\overline{Y}_{i.} = \sum c_i\widehat{\alpha}_i$.

30.43 Verify the least squares estimators for the parameters and the degrees of freedom in the balanced two-way additive model of **30.18**.

30.44 (a) In the two-way additive model, verify that $Q(\boldsymbol{\alpha}) \sim q_{r,dfe}$ and use this result to construct Tukey simultaneous $1-\alpha$ confidence intervals for comparison in the α_i.
(b) Verify that $D(\boldsymbol{\alpha}) \sim D_{r-1,dfe}$ and use this result to derive Dunnet simultaneous $1-\alpha$ confidence intervals for comparing the treatments to the control (μ_1).

30.45 (a) Verify that the balanced two-way model of **30.19** is an orthogonal design.
(b) Verify the least squares estimators of the parameters for this model.

30.46 (a) Show that if $n_{ij} = kw_iv_j$, then $n_{ij} = n_{i.}n_{.j}/N$.
(b) Show that if $n_{ij} = n_{i.}n_{.j}/N$, then $n_{ij} = kw_iv_j$ for some weights w_i and v_j.

30.47 (a) Verify that the proportional sampling model of **30.21** is an orthogonal design.
(b) Verify the least squares estimators of the parameters for this model.
(c) Verify the ANOVA table for this model.
(d) Verify the Scheffé simultaneous confidence intervals for contrasts in the α_i and contrasts in the γ_{ij}.

30.48 Consider an additive three-way model in which we observe $Y_{ijkm} \sim N(\mu_{ijk}, \sigma^2)$, independent, $i = 1, \ldots, r$; $j = 1, \ldots, c$; $k = 1, \ldots, d$; $m = 1, \ldots, n$ where $\mu_{ijk} = \theta + \alpha_i + \beta_j + \gamma_k$, and $\sum \alpha_i = 0$, $\sum \beta_j = 0$ and $\sum \gamma_k = 0$.
(a) Verify that this model is an orthogonal design.
(b) Find the least squares estimators of the parameters and the degrees of freedom. Write down the ANOVA table for this model.
(c) Find the Scheffé simultaneous confidence intervals for contrasts in the α_i.
(d) Find the Tukey simultaneous confidence intervals for comparisons in the α_i.

30.49 Consider the three-way ANOVA model with interactions described in **30.22**.
(a) Verify that the δ_{ij} are orthogonal to the α_i and the and ξ_{ijk}.
(b) Verify the least squares estimators for θ, α_i, δ_{ij} and ξ_{ijk}.
(c) Verify the sums of squares and degrees of freedom for θ, α_i, δ_{ij}, ξ_{ijk} and error.

30.50 Verify that $n_{ijk} = kv_iw_ju_k$ if and only if $n_{ijk} = n_{i..}n_{.j.}n_{..k}/N^2$.

30.51 Consider the Latin square model in **30.23**.
(a) Verify the formulae for the least squares estimators and the degrees of freedom.
(b) Verify the ANOVA table.
(c) What are the Tukey simultaneous confidence intervals for comparisons in the γ_i?

30.52 Consider the two-way symmetric model discussed in **30.24**.
(a) Verify that $\sum\sum\sum(\widehat{\alpha}_i + \widehat{\alpha}_j)^2 = 2nr\sum\widehat{\alpha}_i^2$.
(b) Verify that $\sum\sum\sum(a_i + a_j)(\alpha_i + \alpha_j) = 2nr\sum a_i\alpha_j$.

30.53 Consider the two-way symmetric model with interaction in which we observe $Y_{ijk} \sim N(\mu_{ij}, \sigma^2)$, independent, $i = 1, \ldots, r$; $j = 1, \ldots, r$; $k = 1, \ldots, n$, where $\mu_{ij} = \theta + \alpha_i + \alpha_j + \gamma_{ij}$, $\gamma_{ij} = \gamma_{ji}$, $\sum \alpha_q = 0$ and $\sum_i \gamma_{ij} = 0$.
(a) Show that any μ_{ij} such that $\mu_{ij} = \mu_{ji}$ can be written in the form above.
(b) Verify that this model is an orthogonal design.
(c) Find the least squares estimators of the parameters. $(\widehat{\gamma}_{ij} = (\overline{Y}_{ij.} + \overline{Y}_{ji.})/2 - \widehat{\theta} - \widehat{\alpha}_i - \widehat{\alpha}_j.)$
(d) Find the ANOVA table for this model $(df\gamma = r(r-1)/2.)$.

30.54 Consider the general two-way model in which we observed $Y_{ijk} \sim N(\mu_{ij}, \sigma^2)$, independent, $i = 1, \ldots, r$; $j = 1, \ldots, r$; $k = 1, \ldots, n$. We want to test that $\mu_{ij} = \mu_{ji}$.
(a) Show that we can write $\mu_{ij} = \eta_{ij} + \delta_{ij}$, where $\eta_{ij} = \eta_{ji}$ and $\delta_{ij} = -\delta_{ji}$.
(b) Show that the η effect and the δ effect are orthogonal.
(c) Find the ANOVA table for this model. How would you test that $\mu_{ij} = \mu_{ji}$?

30.55 For the Diallel cross model of **30.25**, verify that if $\sum a_i = 0$ and $\sum b_j = 0$ then $\sum_i \sum_{j \neq i} \sum_k (a_i + a_j)(b_i + b_j) = 2n(r-2) \sum_q a_q b_q$.

30.56 Consider the Diallel cross model of **30.25**.
(a) Verify the least squares estimator for $\widehat{\alpha}_q$.
(b) Verify the ANOVA table.
(c) Verify the Scheffé simultaneous confidence intervals for contrasts in the α_i.

30.57 Consider the two-way Diallel cross model with interaction in which we observe $Y_{ijk} \sim N(\mu_{ij}, \sigma^2)$, independent, $i = 1, \ldots, r$; $j = 1, \ldots, r$; $j \neq i$; $k = 1, \ldots, n$, where $\mu_{ij} = \theta + \alpha_i + \alpha_j + \gamma_{ij}$ and $\gamma_{ij} = \gamma_{ji}$, $\sum \alpha_i = 0$, $\sum_{j \neq i} \gamma_{ij} = 0$.
(a) Verify that this model is an orthogonal design.
(b) Find the least squares estimators of the parameters.
(c) Find the ANOVA table for this model.

30.58 Consider the twofold nested model of **30.26**.
(a) Verify the least squares estimators for the parameters.
(b) Verify the ANOVA table.

30.59 Consider the threefold nested model of **30.27**.
(a) Verify the least squares estimators of the parameters.
(b) Verify the ANOVA table.

30.60 Find the ANOVA table for the first model in **30.30**.

30.61 Find the ANOVA table for the second model in **30.30**.

30.62 Suppose that $\mu_{ij} = \xi_i + \eta_j$. Let $\bar{\xi} = \sum n_{i.} \xi_i / N$, $\bar{\eta} = \sum n_{.j} \eta_j / n$, $\theta = \bar{\xi} + \bar{\eta}$, $\alpha_i = \xi_i - \bar{\xi}$, $\beta_j = \eta_j - \bar{\eta}$. Show that $\mu_{ij} = \theta + \alpha_i + \beta_j$ and that $\sum n_{i.} \alpha_i = 0$ and $\sum n_{.j} \beta_j = 0$.

30.63 Consider the problem of testing that the $\alpha_i = 0$ in the unbalanced additive two-way model.
(a) Show that the set of contrasts associated with this problem is the set of $\mathbf{d}^T \boldsymbol{\mu}$, where $d_{ijk} = a_i - \bar{a}_j$, with $\bar{a}_j = \sum_i n_{ij} a_i / n_{.j}$ and $\sum n_{i.} a_i = 0$. (Show that the set of such contrasts works and has the correct dimension.)
(b) What are the Scheffé simultaneous confidence intervals associated with this hypothesis? (Do not try to simplify these intervals.)

30.64 Verify that in a balanced incomplete block model $1 - (b - \lambda)/bt = c\lambda/bt = c(t-1)/t(c-1)$.

30.65 Consider the balanced incomplete block model of **30.32**. Suppose that $\sum c_j = 0$ and $\sum d_j = 0$. Let $\bar{c}_i = \sum_j n_{ij} c_j / t$.
(a) Show that $t \sum_i n_{ij} \bar{c}_i = (b - \lambda) c_j$. ($t \sum_i n_{ij} \bar{c}_i = \sum_{j'} c_{j'} \sum_i n_{ij} n_{ij'}$.)
(b) Show that $\sum \sum n_{ij} (c_j - \bar{c}_i)^2 = bH \sum c_j^2$.
(c) Show that $\sum \sum n_{ij} d_j (c_j - \bar{c}_i) = bH \sum d_j c_j$.

30.66 In the context of the balanced incomplete block model.
(a) Verify that $\sum \sum n_{ij} (\beta_j - \bar{\beta})^2 = bH \sum \beta_j^2$.
(b) Verify that if $\sum b_j = 0$ and $a_i = -\sum_j n_{ij} b_j / t$, then $\sum \sum n_{ij} (a_i + b_j) \beta_j = bH \sum b_j \beta_j$ and $\sum \sum n_{ij} (a_i + b_j)^2 = bH \sum b_j^2$.

30.67 In the balanced incomplete block design, symmetry in the treatments implies that $\text{var}(\widehat{\beta}_j)$ is the same for all j and that $\text{cov}(\widehat{\beta}_j, \widehat{\beta}_{j'})$ is the same for all j and j'.

(a) Verify that if $\sum b_j = 0$, then $\mathrm{var}(\sum b_j \widehat{\beta}_j) = \sum b_j^2 \sigma^2/bH$. (*Hint*: by the previous exercise, $\sum b_j \widehat{\beta}_j = \sum \sum n_{ij}(a_i + b_j)\widehat{\mu}_{ij}/bH$. Recall that if $\mathbf{d} \in V$, then $\mathrm{var}(\mathbf{d}^T \widehat{\mu}) = \|\mathbf{d}\|^2\sigma^2$.)
(b) Verify that $\mathrm{cov}(\widehat{\beta}_j, \widehat{\beta}_{j'}) = -\mathrm{var}(\widehat{\beta}_j)/(c-1)$. (Note that $\mathrm{var}(\sum_j \widehat{\beta}_j) = 0$. Why?)
(c) Verify that $\mathrm{var}(\widehat{\beta}_j) = \sigma^2(c-1)/cbH$. (Apply parts (a) and (b) to $\widehat{\beta}_j - \widehat{\beta}_{j'}$.)
(d) Construct Tukey simultaneous confidence intervals for the set of all comparisons in the β_j.

30.68 In the context of the unbalanced two-way model:
(a) Verify that $\sum\sum\sum d_{ij}\mu_{ij} = 0$ for all μ_{ij} such that $\overline{\mu}_i^{\mathbf{W}} = 0$ if and only if $d_{ij} = b_i w_j/n_{ij}w.$ for some b_i such that $\sum b_i = 0$. (*Hint*: show that the subspace of such d_{ij} satisfies the constraints and has the same dimension as $V \mid W$.)
(b) Show that for such d_{ij}, $\sum\sum\sum d_{ij}\mu_{ij} = \sum b_i \overline{\mu}_i^{\mathbf{W}}$ and $\sum\sum\sum d_{ij}^2 = \sum b_i^2/Q_i$.

30.69 In the unbalanced two-way model:
(a) Verify that A_i are the minimum variance unbiased estimators of the $\overline{\mu}_i^{\mathbf{W}}$.
(b) Verify that $A_i \sim N(\overline{\mu}_i^{\mathbf{W}}, \sigma^2/Q_i)$.

30.70 In the context of the unbalanced two-way model:
(a) Verify that the maximum of $\sum_i b_i(A_i - \overline{A})^2$ subject to $\sum b_i = 0$ and $\sum b_i^2/Q_i = 1$ is given by $b_i = \pm Q_i(A_i - \overline{A})/\sqrt{\sum Q_i(A_i - \overline{A})^2}$. (Use Lagrange multipliers.)
(b) Verify that the maximum of this function is $\sum Q_i(A_i - \overline{A})^2$.

30.71 In the context of the unbalanced two-way model, consider the Scheffé simultaneous confidence intervals associated with testing that the model is additive.
(a) Show that the set of contrasts associated with this hypothesis have the form $\mathbf{d}^T\boldsymbol{\mu}$, where $d_{ijk} = c_{ij}/n_{ij}$, with $\sum_i c_{ij} = 0$ and $\sum_j c_{ij} = 0$.
(b) Find the Scheffé simultaneous confidence intervals for this hypothesis.

30.72 Suppose that the Q_i are all equal. Find Tukey simultaneous $1-\alpha$ intervals for comparisons in the $\overline{\mu}_i^{\mathbf{W}}$ in the unbalanced two-way model.

30.73 For the unbalanced twofold nested model:
(a) Verify that the degrees of freedom for testing that the μ_{ij} do not depend on j are $C - r$.
(b) Show that the contrasts associated with this testing problem have the form $\sum\sum c_{ij}\mu_{ij}$, where $\sum_j c_{ij} = 0$.
(c) Verify the Scheffé simultaneous confidence intervals associated with this hypothesis.

30.74 Consider the problem of testing the null hypothesis of the equality of the $\overline{\mu}_i^{\mathbf{W}}$ in the unbalanced twofold nested model.
(a) Verify that the degrees of freedom associated with this hypothesis are $R - 1$.
(b) Show that the contrasts associated with this hypothesis have $d_{ijk} = w_{ij}b_{i\cdot}/n_{ij}w_{i.}$ where $\sum b_i = 0$.
(c) Find the Scheffé simultaneous confidence intervals for this hypothesis.
(d) Use Lagrange multipliers to show that for this hypothesis, $SSH = \sum Q_i(A_i - \overline{A})^2$ and $EMS = \sum Q_i(\overline{\mu}_i^{\mathbf{W}} - \overline{\mu}^{\mathbf{W}})^2$.

30.75 Let $\mathbf{X}$ be an orthonormal basis matrix for $V \mid W$, let $v_2 = \mathbf{X}^T\boldsymbol{\mu}$, let $\mathbf{Z}_2 = \mathbf{X}_2^T\mathbf{Y}$ and let $\mathbf{d} = \mathbf{X}\mathbf{h} \in V|W$. Show that $\mathbf{d}^T\boldsymbol{\mu} = \mathbf{h}^T v_2$, $\mathbf{d}^T\widehat{\boldsymbol{\mu}} = \mathbf{d}^T\mathbf{Y}$, and $\|\mathbf{d}\|^2 = \|\mathbf{h}\|^2$. (*Hint*: because $V|W \subset V$, $\mathbf{P}_V\mathbf{X}_2 = \mathbf{X}_2$. Why?)

30.76 We follow the notation of **30.36**.
(a) Under the null hypothesis that $\eta = 0$ (i.e. that $\boldsymbol{\mu} \in W$) show that the ML estimator of $(\boldsymbol{\mu}, \sigma^2)$ is $(\mathbf{P}_W\mathbf{Y}, \|\mathbf{P}_{W\perp}\mathbf{Y}\|^2/N)$.
(b) Under the alternative hypothesis that $\eta > 0$, show that the ML estimator is $(\widehat{\boldsymbol{\mu}}, \|\mathbf{P}_{V\perp}\mathbf{Y}\|^2/N)$ if $\mathbf{X}^T\widehat{\boldsymbol{\mu}} \geq 0$ and $(\mathbf{P}_W\mathbf{Y}, \|\mathbf{P}_{W\perp}\mathbf{Y}\|^2/n)$ if $\mathbf{X}^T\widehat{\boldsymbol{\mu}} < 0$.

(c) Show that the LR test statistic is $\Lambda = (1 + t^2/(n-p))^{-n/2}$ if $t \geq 0$ and $\Lambda = 1$ if $t < 0$.
(d) Show that $\Lambda < k < 1$ if and only if $t > k^*$ for some k^*.
(e) Show that $t^2 = \|\mathbf{P}_{V|W}\widehat{\boldsymbol{\mu}}\|^2/\widehat{\sigma}^2$. (Recall that $\mathbf{X}$ is a basis matrix for $V|W$.)

30.77 Consider an additive two-way model with $n = 1$ and $r = 2$. We consider testing that there is no row effect. Let $\mathbf{Y}$ be a $2c$-dimensional vector of the observations and let $\boldsymbol{\mu} = E\mathbf{Y}$, let $\widehat{\boldsymbol{\mu}}$ be the least squares estimator of $\boldsymbol{\mu}$ and let V be the $(c+1)$-dimensional subspace of possible $\boldsymbol{\mu}$. Finally, let $\mathbf{X}$ be a vector with $(-1)^i/c$ in the (i, j)th place.
(a) Show that $\mathbf{X}^T\boldsymbol{\mu} = \alpha_1 - \alpha_2$, $\mathbf{X}^T\widehat{\boldsymbol{\mu}} = \overline{Y}_{1.} - \overline{Y}_{2.}$ and that $\|\mathbf{X}\|^2 = 2/c$.
(b) Show that $\mathbf{X}$ is a basis matrix for the hypothesis of no row effect (i.e., show that $\mathbf{X} \in V$ and there is no row effect if and only if $\mathbf{X}^T\boldsymbol{\mu} = 0$.
(c) Show that the t statistic for testing the null hypothesis $\alpha_1 = \alpha_2$ against the alternative hypothesis $\alpha_1 > \alpha_2$ is given by $t = (\overline{Y}_{1.} - \overline{Y}_{2.})/\sqrt{2MSE/c}$.
(d) Let $Z_j = Y_{1j} - Y_{2j}$, and let $\overline{Z}$ and S^2 be the sample mean and the sample variance of the Z_j. Show that $t = \sqrt{c}\overline{Z}/S$.

30.78 For the linear model with known variance:
(a) Show that $\widehat{\boldsymbol{\mu}}$ is a complete sufficient statistic. (Recall that for the model with unknown σ^2, we had to coordinatize the model to establish completeness.)
(b) Show that $\widehat{\boldsymbol{\mu}}$ is the ML and the MV unbiased estimator for $\boldsymbol{\mu}$.
(c) Show that the LR test statistic for testing that the null hypothesis that $\boldsymbol{\mu} \in W$ against the alternative hypothesis that $\boldsymbol{\mu} \in V$ is given by $\Lambda = \exp -(SSH/2a^2)$ and show that the size-α LR test rejects if $SSH/a^2 > \chi^{2\,\alpha}$. (Note that the LR test statistic for this problem is not a power of the ratio of variance estimators as it is for many other problems.)
(d) Show that $P(\mathbf{d}^T\boldsymbol{\mu} \in \mathbf{d}^T\widehat{\boldsymbol{\mu}} \pm (a^2\|\mathbf{d}\|^2\chi^{2\alpha})^{1/2}\ \forall \mathbf{d} \in V|W) = 1 - \alpha$ and that 0 is in all these intervals if and only if $SSH/a^2 \leq \chi^{2\,\alpha}$.

30.79 Suppose we observe Y_{ij} independent, $i = 1, \ldots, p$; $j = 1, \ldots, n_i$; $N = \sum n_i$, such that $EY_{ij} = \mu_i$, $\text{var}(Y_{ij}) = \sigma_i^2$. Let *MSH* and *MSE* be the mean square for the hypothesis and the mean square for error for the usual one-way model. Show that under the null hypothesis that the μ_i are equal, $EMSH = \sum(N - n_i)\sigma_i^2/N(p-1)$ and $EMSE = \sum(n_i - 1)\sigma_i^2/(N-p)$. (*Hint*: $E\overline{Y}_{i.}^2 = \mu^2 + \sigma_i^2/n_i$, $E\overline{Y}_{..}^2 = \mu^2 + \sum n_i\sigma_i^2/N^2$.)

30.80 Suppose we observe Y_{ij}, $i = 1, \ldots, p$; $j = 1, \ldots, n$, such that $EY_{ij} = \mu_i$, $\text{var}(Y_{ij}) = \sigma^2$, $\text{cov}(Y_{ij}, Y_{ij^T}) = \rho\sigma^2$, $j \neq j^T$ and $\text{cov}(Y_{ij}, Y_{i^Tj^T}) = 0$, $i \neq i^T$. Let *MSH* and *MSE* be the mean squares for the hypothesis and error for the usual balanced one-way model. Show that under the null hypothesis of equality of the μ_i, $EMSH = \sigma^2(1 + (n-1)\rho)$ and $EMSE = \sigma^2(1 - \rho)$. (*Hint*: $E\overline{Y}_{i.}^2 = \mu^2 + \sigma^2(1 + (n-1)\rho)/n$, $E\overline{Y}_{..}^2 = \mu^2 + \sigma^2(1 + (n-1)\rho)/np$.)

30.81 Consider a linear model in which we observe $\mathbf{Y} = \boldsymbol{\mu} + \mathbf{e}$ and we want to test the null hypothesis that $\boldsymbol{\mu} \in W$ against the alternative hypothesis that $\boldsymbol{\mu} \in V$, where, as usual, $W \subset V \subset R^n$ are subspaces with $k = \dim(W) < p = \dim(V) < n$. Suppose that the distribution of $\mathbf{e}$ is spherical about 0, i.e. that $\boldsymbol{\Gamma}\mathbf{e} \sim \mathbf{e}$ for all orthogonal matrices $\boldsymbol{\Gamma}$. Let $\mathbf{T} = \|\mathbf{e}\|^{-1}\mathbf{e}$.
(a) Show that $\mathbf{T}$ is uniformly distributed on the surface of an n-dimensional unit sphere.
(b) Let

$$F = \frac{\|\mathbf{P}_{V|W}\mathbf{e}\|^2/(p-k)}{\|\mathbf{P}_{V\perp}\mathbf{e}\|^2/(n-p)}.$$

Show that F depends on $\mathbf{e}$ only thorough $\mathbf{T}$ and hence that the null distribution of F is the same for any spherical distribution (and hence has an F distribution for any spherical distribution for the errors).

(c) Show that the null distribution of the F statistic and the confidence coefficient of the Scheffé simultaneous confidence intervals are unaffected by non-normality for any spherical distribution.

30.82 Consider the setting for Hsu's MCB procedure of **30.11** and let $\mu_{k^*} = \max(\mu_i)$ and $\overline{Y}_k = \max(\overline{Y}_i)$. (Note that unfortunately k^* and k need not be the same.)
(a) Show that if $i \neq k^*$, then $\max_{j \neq i} \mu_j = \mu_{k^*}$ and $\max_{j \neq i} \overline{Y}_j \geq \overline{Y}_{k^*}$.
(b) Show that

$$\mu_i - \mu_{k^*} \geq \overline{Y}_i - \overline{Y}_{k^*} - Q \ \forall i \neq k^* \Rightarrow \mu_{k^*} - \max_{j \neq k^*} \mu_j \leq \overline{Y}_{k^*} - \max_{j \neq k^*} \overline{Y}_j + Q$$
$$\Rightarrow \mu_i - \max_{j \neq i} \mu_j \leq (\overline{Y}_i - \max_{j \neq i} \overline{Y}_j + Q)^+ \ \forall i.$$

(c) Show that

$$\mu_i - \mu_{k^*} \geq \overline{Y}_i - \overline{Y}_{k^*} - Q \ \forall i \neq k^* \Rightarrow \mu_i - \max_{j \neq i} \mu_j \geq \overline{Y}_i - \max_{j \neq i} \overline{Y}_j - Q \ \forall i \neq k^*$$
$$\Rightarrow \mu_i - \max_{j \neq i} \mu_j \geq -(\overline{Y}_i - \max_{j \neq i} \overline{Y}_j - Q)^- \ \forall i.$$

(d) Show that

$$\overline{Y}_k - \max_{j \neq k} \overline{Y}_j - Q \geq 0 \Rightarrow \overline{Y}_i - \max_{j \neq i} \overline{Y}_j + Q \leq 0 \ \forall i \neq k$$

(so that if one interval has lower end-point 0, then all other intervals have upper end-point 0).
(e) Show that

$$\overline{Y}_k - \max_{j \neq k} \overline{Y}_j + Q \geq 0$$

(so that at least one interval has a positive upper end-point).

CHAPTER 31

OTHER ANALYSIS OF VARIANCE MODELS

31.1 In this chapter we consider some alternative models for analysis of variance (ANOVA) settings, namely analysis of covariance (ANCOVA), random effects and mixed models, univariate and multivariate repeated measures models and a more general linear model. In **31.2–9**, we present the theory of ANCOVA together with some particular ANCOVA models. In **31.10–20**, we present results for various crossed and nested random effects and mixed models. In **31.21–33** we present results for various crossed and nested univariate and multivariate repeated measures models. In **31.34–35** we consider a more general linear model. For most of the models in this chapter, we show how to take results from an 'associated' fixed effects ANOVA model to obtain results for the model considered in this chapter. Throughout the chapter when we talk about an 'associated ANOVA model', we are always referring to a fixed effects ANOVA model. The results in this chapter are based on those in the previous chapter and the same notation is used in both chapters.

The ANCOVA model is a special case of the general linear model discussed in the previous chapter and we merely indicate how to interpret results for the general linear model to ANCOVA. The optimality of those procedures (ML estimator, LR test, UMP invariance, etc.) follows from the optimality derived for the general linear model. The other models considered in this chapter are not linear models as described in the previous chapter. For the (univariate) repeated measures model and a more general linear model, we develop a general theory first and apply this theory to various examples. For the random effects and mixed models, there does not appear to be a general theory, so that we derive procedures separately for each particular model.

Analysis of covariance – theory

31.2 Consider an experiment to compare five exercise programmes. Suppose we observed weight losses for four men in each of the five programmes. Suppose for each man we also know x_{ij}, the number of cigarettes he smokes in a day. We might assume now that the Y_{ij} are independent, $Y_{ij} \sim N(\mu_i + x_{ij}\gamma, \sigma^2)$. In this case we say that x_{ij} is a *covariate* and call this model a one-way *analysis of covariance* (ANCOVA) model. (Note that the slope is the same for each exercise programme but that the intercepts are different.) For this model there are two different hypotheses we would want to test: that there is no effect due to the exercise programme; and that there is no effect due to the covariate. These hypotheses can be confounded unless the covariate is nearly balanced across the classes (i.e. unless the average number of cigarettes is approximately the same in each exercise programme). See **31.9** for discussion of this problem.

In this chapter we show how one or more covariates can be added to an arbitrary fixed effects ANOVA model. In **31.3–5**, we derive procedures for general ANCOVA models and apply them to the one-way model with a single covariate discussed above. In **31.6**, we extend this model to a two-way model with a single covariate. In **31.7** we consider a one-way model with two covariates, and in **31.8** we consider a one-way model with different slopes. From these examples it should be clear how to allow for covariates in any ANOVA model. In **31.9** we discuss confounding between the covariate and the ANOVA side of the model.

Fisher (1932) may have been the first person to use covariates.

31.3 In an ANCOVA model, we have observations which are taken in different categories as in an ANOVA model and we also have some other variables which are measured continuously. In the general form of this model, we observe

$$\mathbf{Y} \sim N_N(\boldsymbol{\delta} + \mathbf{X}\boldsymbol{\gamma}, \sigma^2\mathbf{I}), \qquad \boldsymbol{\delta} \in Q,\ \boldsymbol{\gamma} \in R^s, \sigma^2 > 0,$$

where $\mathbf{X}$ is a known $N \times s$ matrix and Q is a known q-dimensional subspace. We assume that $\boldsymbol{\delta}$, $\boldsymbol{\gamma}$ and σ^2 are unknown parameters. Typically Q is a subspace from analysis of variance. We call the analysis of variance model with subspace Q the *associated* ANOVA model and call $\mathbf{X}$ the matrix of *covariates*. The results of this section show how to modify the procedures for the associated ANOVA model (which we presume are known) to get the correct procedures for the ANCOVA model.

Let

$$\mathbf{X}^* = \mathbf{P}_{Q^\perp}\mathbf{X}, \qquad \mathbf{Y}^* = \mathbf{P}_{Q^\perp}\mathbf{Y}.$$

In order to guarantee that the covariates are not confounded with the ANOVA part of the model we need to assume that $\mathbf{X}^*$ has full column rank, that is, that

$$\text{rank}(\mathbf{X}^*) = s.$$

(Note that it is not enough that $\mathbf{X}$ has full column rank in order to have a possible covariance analysis.)

Let U be the s-dimensional subspace spanned by the columns of $\mathbf{X}^*$ and let V be the subspace of possible values for $\boldsymbol{\mu}$. That is

$$V = \{\boldsymbol{\mu} = \boldsymbol{\delta} + \mathbf{X}\boldsymbol{\gamma};\ \boldsymbol{\delta} \in Q,\ \boldsymbol{\gamma} \in R^s\}$$

Then

$$\mathbf{Y} \sim N_N(\boldsymbol{\mu}, \sigma^2\mathbf{I}), \qquad \boldsymbol{\mu} \in V,\ \sigma^2 > 0,$$

so that this model is a linear model as described in **30.4**. We now interpret the theory developed in that section for this model. Note that V is the orthogonal direct sum of Q and U,

$$V = Q \oplus U$$

(see Exercise 31.1). Therefore

$$\hat{\boldsymbol{\mu}} = \mathbf{P}_V\mathbf{Y} = \mathbf{P}_Q\mathbf{Y} + \mathbf{P}_U\mathbf{Y} = \hat{\boldsymbol{\delta}}_u + \mathbf{X}^*\hat{\boldsymbol{\gamma}}, \qquad \hat{\boldsymbol{\delta}}_u = \mathbf{P}_Q\mathbf{Y},$$
$$\hat{\boldsymbol{\gamma}} = (\mathbf{X}^{*T}\mathbf{X}^*)^{-1}\mathbf{X}^{*T}\mathbf{Y} = (\mathbf{X}^{*T}\mathbf{X}^*)^{-1}\mathbf{X}^*\mathbf{Y}^*.$$

($\mathbf{X}^{*T}\mathbf{Y}^* = \mathbf{X}^{*T}\mathbf{Y}$. See Exercise 31.1.) Note that $\hat{\boldsymbol{\delta}}_u$ is just the estimator of $\boldsymbol{\delta}$ based on the associated ANOVA model and $\hat{\boldsymbol{\gamma}}$ is just the least squares estimator of $\boldsymbol{\gamma}$ found by regressing $\mathbf{Y}$ (or $\mathbf{Y}^*$) on $\mathbf{X}^*$. Further results discussed earlier for orthogonal direct sums imply that for this model

$$\|\hat{\mu}\|^2 = \|\hat{\delta}_u\|^2 + \|\mathbf{X}^*\hat{\gamma}\|^2,$$
$$p = \dim(V) = \dim(Q) + \dim(S) = q + s.$$

Now $\hat{\gamma} = (\mathbf{X}^{*T}\mathbf{X}^*)^{-1}\mathbf{X}^{*T}\hat{\mu}$ is the optimal (best unbiased and ML) estimator of $\gamma = (\mathbf{X}^{*T}\mathbf{X}^*)^{-1}\mathbf{X}^*\mu$, but

$$E\hat{\delta}_u = \delta + \mathbf{P}_Q\mathbf{X}\gamma = \delta + (\mathbf{X} - \mathbf{X}^*)\gamma.$$

Therefore, $\hat{\delta}_u$ is the optimal estimator of $\delta + (\mathbf{X} - \mathbf{X}^*)\gamma$. Let

$$\hat{\delta}_a = \hat{\delta}_u - (\mathbf{X} - \mathbf{X}^*)\hat{\gamma}.$$

Then $\hat{\delta}_a$ is the optimal estimator of δ for the ANCOVA model (see Exercise 31.2). We call $\hat{\delta}_u$ and $\hat{\delta}_a$ the *unadjusted* and *adjusted* estimators of δ because $\hat{\delta}_a$ has been adjusted to account for the covariate. (Throughout this the discussion of ANCOVA, we use the subscript 'u' to indicate the appropriate expression for the associated ANOVA model (unadjusted for the covariate) and use the subscript 'a' to indicate the appropriate expression for the ANCOVA model (adjusted for the covariate).)

Therefore we see that the sum of squares and degrees of freedom for error, SSE_{a} and dfe_{a}, for the ANCOVA model are given by

$$SSE_{\mathrm{a}} = \|\mathbf{Y} - \hat{\mu}\|^2 = \|\mathbf{Y}\|^2 - (\|\hat{\delta}\|^2 + \|\mathbf{X}^*\hat{\gamma}\|^2) = SSE_{\mathrm{u}} - \hat{\gamma}^T\mathbf{X}^{*T}\mathbf{X}^*\hat{\gamma},$$
$$dfe_a = N - \dim(V) = N - (q + s) = dfe_u - s,$$

where SSE_{u} and dfe_u are the sum of squares and degrees of freedom for error for the associated ANOVA model. We can compute $MSE_a = SSE_{\mathrm{a}}/dfe_a$ in the obvious way.

Before looking at an example, we make an elementary comment. In order to apply the results of this section, it is necessary to find $\mathbf{X}^*$. Let $\hat{\delta}_u = \mathbf{P}_Q\mathbf{Y}$ be the estimator of δ under the associated ANOVA model. Then the residual vector for the full ANOVA model is given by

$$\mathbf{Y} - \hat{\delta}_u = \mathbf{P}_{Q^\perp}\mathbf{Y} = \mathbf{Y}^*.$$

To find $\mathbf{X}^* = \mathbf{P}_{Q^\perp}\mathbf{X}$, we apply the same operation to the xs that we apply to the ys when we find the residuals $\mathbf{Y}^*$ for the associated ANOVA model.

Example 31.1 (One-way ANCOVA: estimation)

We look at a one-way model with a single covariate. In this model, we observe Y_{ij} independent,

$$Y_{ij} \sim N(\mu_{ij}, \sigma^2), \qquad \mu_{ij} = \delta_i + \gamma x_{ij},\ i = 1, \ldots, p;\ j = 1, \ldots, n_i,$$

where the x_{ij} are known constants, and δ_i, γ and $\sigma^2 > 0$ are unknown constants. As in the one-way ANOVA model, let

$$N = \sum n_i, \qquad \theta = \frac{\sum n_i\delta_i}{N}, \qquad \alpha_i = \delta_i - \theta$$

so that

$$\mu_{ij} = \theta + \alpha_i + \gamma x_{ij}, \qquad \sum n_i \alpha_i = 0.$$

The associated ANOVA model is the one-way ANOVA model which occurs when $\gamma = 0$ in this model. For this associated model, the estimator of $\delta_i = \theta + \alpha_i$ is $\hat{\delta}_{iu} = \overline{Y}_{i.}$ and $Y_{ij} - \hat{\delta}_{iu} = Y_{ij} - \overline{Y}_{i.}$. Therefore, as indicated above,

$$Y_{ij}^* = Y_{ij} - \overline{Y}_{i.}, \quad x_{ij}^* = x_{ij} - \overline{x}_{i.}.$$

In order to keep from confounding the covariate with the class effect we need to assume that the x_{ij}^* are not all 0, that is, that the covariate x_{ij} does not depend only on i. To find $\hat{\gamma}$, we regress the Y_{ij} (or Y_{ij}^*) on x_{ij}^*, obtaining

$$\hat{\gamma} = \frac{\sum\sum x_{ij}^* Y_{ij}}{\sum\sum x_{ij}^{*2}} = \frac{\sum\sum x_{ij}^* Y_{ij}^*}{\sum\sum x_{ij}^{*2}}.$$

Therefore,

$$SSE_{\text{a}} = SSE_{\text{u}} - \hat{\gamma}^T \mathbf{X}^{*T} \mathbf{X}^* \hat{\gamma} = \sum\sum (Y_{ij} - \overline{Y}_{i.})^2 - \hat{\gamma}^2 \sum\sum x_{ij}^{*2},$$
$$dfe_a = dfe_u - 1 = N - p - 1, \quad MSE_a = \frac{SSE_{\text{a}}}{dfe_a}.$$

31.4 Now consider testing the null hypothesis that $\boldsymbol{\delta} \in T$, a known t-dimensional subspace of Q. Let

$$\mathbf{X}^{**} = \mathbf{P}_{T\perp}\mathbf{X}, \qquad \mathbf{Y}^{**} = \mathbf{P}_{T\perp}\mathbf{Y}.$$

(Note that if $\mathbf{X}^*$ has full rank then, so does $\mathbf{X}^{**}$. See Exercise 31.3.) As above, $\mathbf{Y}^{**}$ is the vector of residuals for the reduced form of the associated ANOVA model and $\mathbf{X}^{**}$ is the same function of the columns of $\mathbf{X}$ as $\mathbf{Y}^{**}$ is of $\mathbf{Y}$. Let W be the subspace of possible values for $\boldsymbol{\mu}$ under the null hypothesis. By the same argument as above, we see that

$$\hat{\hat{\boldsymbol{\mu}}} = \mathbf{P}_W \mathbf{Y} = \hat{\hat{\boldsymbol{\delta}}}_u + \mathbf{X}^{**}\hat{\hat{\boldsymbol{\gamma}}}, \qquad \hat{\hat{\boldsymbol{\delta}}}_u = \mathbf{P}_T \mathbf{Y},$$
$$\hat{\hat{\boldsymbol{\gamma}}} = (\mathbf{X}^{**T}\mathbf{X}^{**})^{-1}\mathbf{X}^{**T}\mathbf{Y} = (\mathbf{X}^{**T}\mathbf{X}^{**})^{-1}\mathbf{X}^{**T}\mathbf{Y}^{**},$$
$$\|\hat{\hat{\boldsymbol{\mu}}}\|^2 = \|\hat{\hat{\boldsymbol{\delta}}}_u\|^2 + \|\mathbf{X}^{**}\hat{\hat{\boldsymbol{\gamma}}}\|^2,$$
$$k = \dim(W) = t + s.$$

Therefore, for testing $\boldsymbol{\delta} \in T$ for the ANCOVA model, the sum of squares and degrees of freedom for the hypothesis, SSH_a and dfh_a are given by

$$SSH_a = \|\hat{\boldsymbol{\mu}}\|^2 - \|\hat{\hat{\boldsymbol{\mu}}}\|^2 = \|\hat{\boldsymbol{\delta}}_u\|^2 - \|\hat{\hat{\boldsymbol{\delta}}}_u\|^2 + \|\mathbf{X}^*\hat{\boldsymbol{\gamma}}\|^2 - \|\mathbf{X}^{**}\hat{\hat{\boldsymbol{\gamma}}}\|^2$$
$$= SSH_u + \|\mathbf{X}^*\hat{\boldsymbol{\gamma}}\|^2 - \|\mathbf{X}^{**}\hat{\hat{\boldsymbol{\gamma}}}\|^2 = SSH_u + \|\mathbf{X}^*\hat{\boldsymbol{\gamma}} - \mathbf{X}^{**}\hat{\hat{\boldsymbol{\gamma}}}\|^2,$$
$$dfh_a = q + s - (t + s) = q - t = dfh_u,$$

where SSH_u and dfh_u are the sum of squares and degrees of freedom for the hypothesis in the associated ANOVA model. As usual, the mean square $MSH_a = SSH_a/dfh_a$. Note that the degrees of freedom for the hypothesis in the ANCOVA model are the same as the degrees of freedom for the hypothesis in the associated ANOVA model and that if $\mathbf{X}^{**} = \mathbf{X}^*$, then the sum of squares and mean square for the hypothesis in the ANCOVA are the same as the sum of squares and mean square for the hypothesis in the associated ANOVA model. As usual, the optimal size-α test for this hypothesis rejects if

$$F = \frac{MSH_a}{MSE_a} > F^{\alpha}_{dfh_a, dfe_a}.$$

Example 31.2 (One-way ANCOVA: testing for the treatment)

We return to the one-way model with a single covariate. Consider testing that the $\alpha_i = 0$. For the reduced version of the associated ANOVA model the LS estimator of δ is $\overline{Y}_{..}$ and therefore

$$Y_{ij}^{**} = Y_{ij} - \overline{Y}_{..}, \qquad x_{ij}^{**} = x_{ij} - \overline{x}_{..}, \qquad \hat{\hat{\gamma}} = \frac{\sum\sum x_{ij}^{**} Y_{ij}}{\sum\sum x_{ij}^{**2}} = \frac{\sum\sum x_{ij}^{**} Y_{ij}^{**}}{\sum\sum x_{ij}^{**2}}.$$

Therefore

$$\begin{aligned} SSH_a &= SSH_u + \|\mathbf{X}^* \hat{\gamma}\|^2 - \|\mathbf{X}^{**} \hat{\hat{\gamma}}\|^2 \\ &= \sum n_i (\overline{Y}_{i.} - \overline{Y}_{..})^2 + \hat{\gamma}^2 \sum\sum x_{ij}^{*2} - \hat{\hat{\gamma}}^2 \sum\sum x_{ij}^{**2}, \\ dfh_a &= dfh_u = p - 1, \quad MSH_a = \frac{SSH_a}{dfh_a}. \end{aligned}$$

(Note that $SSH_a = SSH_u$ as long as $x_{ij}^{**} = x_{ij}^*$, $\Leftrightarrow \overline{x}_{i.} = \overline{x}_{..}$, that is, as long as the average value of the covariate is the same in each class.) The optimal test rejects if

$$F = \frac{MSH_a}{MSE_a} > F^{\alpha}_{p-1, N-p-1}.$$

We now return to the general ANCOVA model and find the Scheffé simultaneous confidence intervals associated with testing that $\boldsymbol{\delta} \in T$. In Exercise 31.4, it is shown that

$$\mathbf{g} \in V|W \Leftrightarrow \mathbf{g} = \mathbf{d} + \mathbf{X}^*\mathbf{c}, \qquad \mathbf{d} \in Q|T, \qquad \mathbf{c} = -(\mathbf{X}^{*T}\mathbf{X}^*)^{-1}\mathbf{X}^T\mathbf{d}.$$

In this case,

$$\mathbf{g}^T \boldsymbol{\mu} = \mathbf{d}^T \boldsymbol{\delta}, \qquad \mathbf{g}^T \hat{\boldsymbol{\mu}} = \mathbf{d}^T \hat{\boldsymbol{\delta}}_a = \mathbf{d}^T \hat{\boldsymbol{\delta}}_u - \mathbf{d}^T \mathbf{X} \hat{\boldsymbol{\gamma}}, \qquad \|\mathbf{g}\|^2 = \|\mathbf{d}\|^2 + \mathbf{d}^T \mathbf{X}(\mathbf{X}^{*T}\mathbf{X}^*)^{-1}\mathbf{X}^T\mathbf{d}.$$

Therefore, the Scheffé simultaneous confidence intervals for contrasts associated with this problem are given by

$$\mathbf{d}^T \boldsymbol{\delta} \in \mathbf{d}^T \hat{\boldsymbol{\delta}}_u - \mathbf{d}^T \mathbf{X} \hat{\boldsymbol{\gamma}} \pm ((q-t)MSE(\|\mathbf{d}\|^2 + \mathbf{d}^T \mathbf{X}(\mathbf{X}^{*T}\mathbf{X}^*)^{-1}\mathbf{X}^T\mathbf{d})F^{\alpha}_{q-t, N-q+s})^{\frac{1}{2}},$$

for all $\mathbf{d} \in Q|T$.

Comparing this formula to the formula for the associated ANOVA model, we note the following:

1. The contrasts are the same for the ANCOVA model as for the associated ANOVA model.
2. To find the estimated contrasts for the ANCOVA we merely replace the estimated contrast $\mathbf{d}^T \hat{\boldsymbol{\delta}}_u$ for the associated ANOVA model with the adjusted estimated contrast $\mathbf{d}^T \hat{\boldsymbol{\delta}}_a$ or with $\mathbf{d}^T \hat{\boldsymbol{\delta}}_u - \mathbf{d}^T \mathbf{X} \hat{\boldsymbol{\gamma}}$.

3. We replace MSE_u and dfe_u for the associated ANOVA model with MSE_a and dfe_a for the ANCOVA model.
4. Finally, we replace $\|\mathbf{d}\|^2$ for the associated ANOVA model with $\|\mathbf{d}\|^2 + \mathbf{d}^T\mathbf{X}(\mathbf{X}^{*T}\mathbf{X}^*)^{-1}\mathbf{X}\mathbf{d}$.

In Exercise 31.3 it is shown that if $\mathbf{X}^* = \mathbf{X}^{**}$ then the only difference between the Scheffé simultaneous confidence intervals for the ANOVA and ANCOVA model is the replacement in *MSE* and *dfe* in 3 above.

Example 31.3 (One-way ANCOVA: simultaneous confidence intervals for the treatment)

We return to the one-way model with a single covariate. Note that for the associated (one-way) ANOVA model, the contrast $\mathbf{d}^T\boldsymbol{\delta}$ has $d_{ij} = c_i/n_i$ in the position associated with the jth observation in the ith class, where $\sum c_i = 0$. Note also that for the associated ANOVA model

$$\mathbf{d}^T\boldsymbol{\delta} = \sum c_i\alpha_i, \qquad \mathbf{d}^T\hat{\boldsymbol{\delta}}_u = \sum c_i\overline{Y}_{i.}, \qquad \mathbf{d}^T\mathbf{d} = \sum \frac{c_i^2}{n_i},$$

and also that for the ANCOVA model

$$\mathbf{d}^T\mathbf{X} = \sum\sum \frac{c_i}{n_i}x_{ij} = \sum c_i\overline{x}_{i.}, \qquad \mathbf{X}^{*T}\mathbf{X}^* = \sum\sum x_{ij}^{*2}.$$

Therefore the simultaneous confidence intervals for this hypothesis are

$$\sum c_i\alpha_i \in \sum c_i(\overline{Y}_{i.} - \overline{x}_{i.}\hat{\gamma}) \pm \left((p-1)MSE_a F^{\alpha}_{p-1N-p-1}\left(\sum \frac{c_i^2}{n_i} + \frac{(\sum c_i\overline{x}_{i.})^2}{\sum\sum x_{ij}^{*2}}\right)\right)^{1/2},$$

for all c_i such that $\sum c_i = 0$. (Note that if the $\overline{x}_{i.}$ are equal, then $\sum c_i\overline{x}_{i.} = 0$. In this case the only difference between these intervals and the intervals for the associated ANOVA model is the replacement of MSE_a and dfe_a for MSE_u and dfe_u.) In Exercise 31.5 Tukey, Dunnett and MCB simultaneous confidence intervals are derived when the $\overline{x}_{i.}$ are all the same and the model is balanced. In this case, multiple range procedures are also available. If these conditions are not met, then such procedures are not available, but multiple F-tests are always available.

31.5 Let

$$\boldsymbol{\theta} = \mathbf{A}\boldsymbol{\gamma}, \qquad \hat{\boldsymbol{\theta}} = \mathbf{A}\hat{\boldsymbol{\gamma}}$$

for a known $r \times s$ matrix of rank r. Inference about $\boldsymbol{\theta}$ can be based on the pivotal quantity

$$\frac{(\hat{\boldsymbol{\theta}} - \boldsymbol{\theta})^T(\mathbf{A}(\mathbf{X}^{*T}\mathbf{X}^*)^{-1}\mathbf{A}^T)^{-1}(\hat{\boldsymbol{\theta}} - \boldsymbol{\theta})}{rMSE_a} \sim F_{r,dfe_a}.$$

If $\mathbf{A} = \mathbf{I}$, this pivotal quantity reduces to

$$\frac{(\hat{\boldsymbol{\gamma}} - \boldsymbol{\gamma})\mathbf{X}^{*T}\mathbf{X}^*(\hat{\boldsymbol{\gamma}} - \boldsymbol{\gamma})}{sMSE_a} \sim F_{s,dfe_a}.$$

Tests based on these pivotal quantities are just particular examples of the F test for testing the general linear hypothesis (see Exercise 31.6) and are therefore optimal. The pivotal quantities can be remembered by

$$\hat{\boldsymbol{\gamma}} \sim N_s(\boldsymbol{\gamma}, \sigma^2(\mathbf{X}^{*T}\mathbf{X}^*)^{-1}) \Rightarrow \hat{\boldsymbol{\theta}} \sim N_r(\boldsymbol{\theta}, \sigma^2\mathbf{A}(\mathbf{X}^{*T}\mathbf{X}^*)^{-1}\mathbf{A}).$$

Example 31.4 (One-way ANCOVA: inference about the covariate)

We return once again to the one-way model with a single covariate. Inference for γ for this model can be based on the pivotal quantity

$$F(\gamma) = \frac{(\hat{\gamma} - \gamma)^2 \sum\sum x_{ij}^{*2}}{MSE_a} \sim F_{1,N-p-1}.$$

Since γ is univariate, the one-sided pivotal quantity

$$t(\gamma) = \frac{(\hat{\gamma} - \gamma)\sqrt{\sum\sum x_{ij}^{*2}}}{\sqrt{MSE_a}} \sim t_{N-p-1}$$

may also be used. For example, to test the null hypothesis that $\gamma = a$ against the alternative that $\gamma > a$ we reject if $t(a) > t^{\alpha}$. A $1 - \alpha$ confidence interval for γ is given by

$$\gamma \in \hat{\gamma} \pm t^{\frac{\alpha}{2}} \frac{\sqrt{MSE_a}}{\sqrt{\sum\sum x_{ij}^{*2}}}.$$

Examples

31.6 We now consider a balanced two-way model with interaction and one covariate. In that model we observe Y_{ijk} independent,

$$Y_{ijk} \sim N(\mu_{ijk}, \sigma^2), \qquad \mu_{ijk} = \delta_{ij} + x_{ijk}\gamma,\ i = 1, \ldots, r;\ j = 1, \ldots, c;\ k = 1, \ldots, n.$$

Since in a non-additive two-way ANOVA model, we make no assumptions about the δ_{ij}, we see that for this model

$$Y_{ijk}^* = Y_{ijk} - \hat{\delta}_{ij} = Y_{ijk} - \overline{Y}_{ij.} \Rightarrow x_{ijk}^* = x_{ijk} - \overline{x}_{ij.}.$$

(Note that to avoid confounding, we need to assume that the x_{ijk} do not just depend on (i, j).) Therefore

$$SSE_{\text{a}} = SSE_{\text{u}} + \hat{\gamma}^T\mathbf{X}^{*T}\mathbf{X}^*\hat{\gamma} = \sum\sum\sum(Y_{ijk} - \overline{Y}_{ij.}) - \hat{\gamma}^2\sum\sum\sum x_{ijk}^{*2},$$

$$\hat{\gamma} = \frac{\sum\sum\sum x_{ijk}^* Y_{ijk}}{\sum\sum\sum x_{ijk}^{*2}} = \frac{\sum\sum\sum x_{ijk}^* Y_{ijk}^*}{\sum\sum\sum x_{ijk}^{*2}},$$

$$dfe_a = dfe_u - 1 = rc(n-1) - 1, \quad MSE_a = \frac{SSE_{\text{a}}}{dfe_a}.$$

As in the associated ANOVA model, let

$$\delta_{ij} = \theta + \alpha_i + \beta_j + \eta_{ij}, \qquad \sum_i \alpha_i = 0,\ \sum_j \beta_j = 0,\ \sum_i \eta_{ij} = 0,\ \sum_j \eta_{ij} = 0.$$

As discussed previously, the δ_{ij} can always be written in this fashion.

Consider first testing that there is no interaction, that is, that the $\eta_{ij} = 0$. For the reduced ANOVA model,

$$Y_{ijk}^{**} = Y_{ijk} - \hat{\hat{\delta}}_{ij} = Y_{ijk} - \overline{Y}_{i..} - \overline{Y}_{.j.} + \overline{Y}_{...} \Rightarrow x_{ijk}^{**} = x_{ijk} - \overline{x}_{i..} - \overline{x}_{.j.} + \overline{x}_{...},$$
$$\hat{\hat{\gamma}} = \frac{\sum\sum\sum x_{ijk}^{**} Y_{ijk}}{\sum\sum\sum x_{ijk}^{**2}} = \frac{\sum\sum\sum x_{ijk}^{**} Y_{ijk}^{**}}{\sum\sum\sum x_{ijk}^{**2}}.$$

Hence

$$SSH_a = SSH_u + \|\mathbf{X}^*\hat{\gamma}\| - \|\mathbf{X}^{**}\hat{\hat{\gamma}}\| = n\sum\sum(\overline{Y}_{ij.} - \overline{Y}_{i..} - \overline{Y}_{.j.} + \overline{Y}_{...})^2$$
$$+\hat{\gamma}^2\sum\sum\sum x_{ijk}^{*2} - \hat{\hat{\gamma}}^2\sum\sum\sum x_{ijk}^{**2},$$
$$dfh_a = dfh_u = (r-1)(c-1), \qquad MSH = \frac{SSH}{dfh}.$$

The F statistic is computed in the obvious way.

To find the Scheffé simultaneous confidence intervals associated with this test, note that the contrasts for the associated ANOVA model have the form $\mathbf{d}^T\boldsymbol{\delta}$, where $\mathbf{d}$ has element $d_{ijk} = c_{ij}$ in the place associated with the kth observation in the ith row and j th column, where $\sum_i c_{ij} = 0$, $\sum_j c_{ij} = 0$, and that for the associated ANOVA model

$$\mathbf{d}^T\boldsymbol{\delta} = n\sum\sum c_{ij}\eta_{ij}, \qquad \mathbf{d}^T\hat{\boldsymbol{\delta}}_u = n\sum\sum c_{ij}\overline{Y}_{ij.}, \qquad \mathbf{d}^T\mathbf{d} = n\sum\sum c_{ij}^2.$$

Note also that

$$\mathbf{d}^T\mathbf{X} = n\sum\sum c_{ij}\overline{x}_{ij.}, \quad \mathbf{X}^{*T}\mathbf{X}^* = \sum\sum\sum x_{ijk}^{*2}.$$

Therefore the simultaneous confidence intervals for this hypothesis are (after dividing by n)

$$\sum\sum c_{ij}\gamma_{ij} \in \sum\sum c_{ij}(\overline{Y}_{ij.} - \overline{x}_{ij.}\hat{\gamma})$$
$$\pm\left((r-1)(c-1)MSE_a F^\alpha\left(\frac{\sum\sum c_{ij}^2}{n} + \frac{(\sum\sum c_{ij}\overline{x}_{ij})^2}{\sum\sum\sum x_{ijk}^{*2}}\right)\right)^{1/2},$$

for all c_{ij} such that $\sum_i c_{ij} = \sum_j c_{ij} = 0$.

Now consider testing the null hypothesis that the $\alpha_i = 0$. For the associated ANOVA model we see that

$$Y_{ijk}^{**} = Y_{ijk} - \hat{\hat{\delta}}_{ij} = Y_{ijk} - \overline{Y}_{ij.} + \overline{Y}_{.j.} \Rightarrow x_{ijk}^{**} = x_{ijk} - \overline{x}_{ij.} + \overline{x}_{.j.},$$
$$\hat{\hat{\gamma}} = \frac{\sum\sum\sum x_{ijk}^{**} Y_{ijk}}{\sum\sum\sum x_{ijk}^{**2}} = \frac{\sum\sum\sum x_{ijk}^{**} Y_{ijk}^{**}}{\sum\sum\sum x_{ijk}^{**2}}.$$

(Note that x_{ijk}^{**}, Y_{ijk}^{**} and $\hat{\hat{\gamma}}$ are different for this hypothesis than in the previous paragraph for testing for 0 interactions.) As usual,

$$SSH_a = nc\sum(\overline{Y}_{i..} - \overline{Y}_{...})^2 + \hat{\gamma}^2\sum\sum\sum x_{ijk}^{*2} - \hat{\hat{\gamma}}\sum\sum\sum x_{ijk}^{**2}, \qquad dfh_a = dfh_u = r-1.$$

The F statistic for this hypothesis may be computed directly. The simultaneous confidence intervals associated with this hypothesis are (after dividing by nc)

$$\sum a_i\alpha_i \in \sum a_i(\overline{Y}_{i..} - \overline{x}_{i..}\hat{\gamma}) \pm \left((r-1)F^{\alpha}MSE_a\left(\sum \frac{a_i^2}{nc} + \frac{(\sum a_i\overline{x}_{i..})^2}{\sum\sum\sum x_{ijk}^{*2}}\right)\right)^{1/2}$$

The hypothesis $\beta_j = 0$ could be tested similarly. To draw inference about γ use the pivotal quantity

$$\frac{(\hat{\gamma} - \gamma)\sqrt{\sum\sum\sum x_{ijk}^{*2}}}{\sqrt{MSE_a}} \sim t_{dfe}.$$

From these two examples it should be fairly clear that we can routinely find procedures for any ANCOVA model with a single covariate for which we have procedures for the associated ANOVA model.

31.7 We next consider the one-way ANCOVA model with two covariates in which we observe Y_{ij} independent

$$Y_{ij} \sim N(\mu_{ij}, \sigma^2), \qquad \mu_{ij} = \delta_i + x_{1ij}\gamma_1 + x_{2ij}\gamma_2, \; i = 1, \ldots, p; \; j = 1, \ldots, n_i.$$

As before, let $N = \sum n_i$. We have seen above that for this model

$$Y_{ij}^* = Y_{ij} - \overline{Y}_{i.} \Rightarrow x_{kij}^* = x_{kij} - \overline{x}_{ki.}, \qquad k = 1, 2$$

(so that we again must assume that the X_{kij} do not depend only on (k, i)). We find $\hat{\boldsymbol{\gamma}} = (\hat{\gamma}_1, \hat{\gamma}_2)^T$ by regressing the Y_{ij} (or Y_{ij}^*) on the x_{1ij}^* and the x_{21j}^*. Therefore for this model

$$SSE_{\mathrm{a}} = SSE_{\mathrm{u}} - \|\mathbf{X}^*\hat{\boldsymbol{\gamma}}\|^2 = \sum\sum(Y_{ij} - \overline{Y}_{i.})^2 - \sum\sum(x_{1ij}^*\hat{\gamma}_1 + x_{2ij}\hat{\gamma}_2)^2$$
$$dfe_a = dfe_u - 2 = N - p - 2, \quad MSE_a = \frac{SSE_{\mathrm{a}}}{dfe_a}.$$

Now let $\delta_i = \theta + \alpha_i$, where $\sum n_i\alpha_i = 0$, and consider testing that the $\alpha_i = 0$. As before, we see that

$$Y_{ij}^{**} = Y_{ij} - \overline{Y}_{..} \Rightarrow x_{kij}^{**} = x_{kij} - \overline{x}_{k...}.$$

We find $\hat{\hat{\boldsymbol{\gamma}}} = (\hat{\hat{\gamma}}_1, \hat{\hat{\gamma}}_2)^T$, by regressing the Y_{ij} (or Y_{ij}^{**}) on the x_{1jk}^{**} and the x_{2jk}^{**}. Then

$$\begin{aligned} SSH_a &= SSH_u + \|\mathbf{X}^*\hat{\boldsymbol{\gamma}}\|^2 - \|\mathbf{X}^{**}\hat{\hat{\boldsymbol{\gamma}}}\|^2 \\ &= \sum n_i(\overline{Y}_{i.} - \overline{Y}_{..})^2 + \sum\sum(x_{1ij}^*\hat{\gamma}_1 + x_{2ij}^*\hat{\gamma}_2)^2 - \sum\sum(x_{1ij}^{**}\hat{\hat{\gamma}}_1 + x_{2ij}\hat{\hat{\gamma}}_2)^2, \\ dfh_a &= dfh_u = p - 1, \qquad MSH = \frac{SSH_a}{dfh_a}. \end{aligned}$$

The F statistic for this hypothesis may be computed immediately. The Scheffé simultaneous confidence intervals associated with this hypothesis are also straightforward.

We can also test many hypotheses about $\boldsymbol{\gamma}$ (e.g. $\boldsymbol{\gamma} = \mathbf{0}$, $\gamma_1 = 0$, $\gamma_1 = \gamma_2$) which can all be put in the framework of testing that $\mathbf{A}\boldsymbol{\gamma} = \mathbf{0}$ for some matrix (or vector) $\mathbf{A}$.

From this example it should be apparent that it is straightforward to find procedures for models with several covariates for any ANCOVA model for which we have procedures for the associated ANOVA model.

31.8 We next consider a one-way model with different slopes in each class. In this model we observe Y_{ij} independent,

$$Y_{ij} \sim N(\mu_{ij}, \sigma^2), \qquad \mu_{ij} = \delta_i + x_{ij}\gamma_i, \; i = 1, \ldots, p; \; j = 1, \ldots, n_i,$$

where the x_{ij} are known constants, and δ_i, γ_i and σ^2 are unknown parameters. Note that this is a model with p covariates and that $\boldsymbol{\gamma} = (\gamma_1, \ldots, \gamma_p)^T$. Hence the $\mathbf{X}$ matrix for this model is $N \times p$. The entry in the kth column of the row associated with the jth individual in the ith class is x_{ij} when $k = i$ and is 0 otherwise.

Let N, θ and α_i be defined as in the one-way model with a single covariate discussed above, so that

$$\mu_{ij} = \theta + \alpha_i + x_{ij}\gamma_i, \qquad \sum n_i\alpha_i = 0.$$

Note that the associated ANOVA model is the one-way model discussed above. Therefore the x^*_{ij} and Y^*_{ij} are the same, that is, $Y^*_{ij} = Y_{ij} - \overline{Y}_{i.}$, $\quad x^*_{ij} = x_{ij} - \overline{x}_{i.}$. To find the $\hat{\gamma}_i$ we must minimize

$$\sum\sum(Y_{ij} - x^*_{ij}\gamma_i)^2,$$

so that the estimators of the γ_i are given by

$$\hat{\gamma}_i = \frac{\sum_j x^*_{ij}Y_{ij}}{\sum_j x^{*2}_{ij}} = \frac{\sum_j x^*_{ij}Y^*_{ij}}{\sum_j x^{*2}_{ij}}.$$

Therefore

$$SSE_{\mathrm{a}} = SSE_{\mathrm{u}} - \hat{\boldsymbol{\gamma}}^T\mathbf{X}^{*T}\mathbf{X}^*\hat{\boldsymbol{\gamma}} = \sum\sum(Y_{ij} - \overline{Y}_{i.}) - \sum_i\Big(\hat{\gamma}_i^2\sum_j x^{*2}_{ij}\Big)$$

$$dfe_a = dfe_u - p = N - 2p, \quad MSE_a = \frac{SSE_{\mathrm{a}}}{dfe_a}.$$

Now consider testing the null hypothesis that the γ_i are all equal. This model can be put in the form of testing that $\mathbf{A}\boldsymbol{\gamma} = \mathbf{0}$ for a $(p-1) \times p$ contrast matrix $\mathbf{A}$. However, it is somewhat easier to derive the appropriate test from first principles. Under the null hypothesis, this model is the same as the one-way model with a single covariate. Therefore, we see that

$$\hat{\mu}_{ij} = \overline{Y}_i + \hat{\gamma}_i x^*_{ij}, \qquad \hat{\hat{\mu}}_{ij} = \overline{Y}_{i.} + \hat{\hat{\gamma}} x^*_{ij}, \qquad \hat{\hat{\gamma}} = \frac{\sum\sum x^*_{ij}Y_{ij}}{\sum\sum x^{*2}_{ij}}.$$

Therefore the sum of squares, mean square and degrees of freedom for this hypothesis are given by

$$SSH = \sum\sum(\hat{\mu}_{ij} - \hat{\hat{\mu}}_{ij}) = \sum_i((\hat{\gamma}_i - \hat{\hat{\gamma}})^2 \sum_j x_{ij}^{*2})$$

$$dfh = 2p - (p+1) = p - 1, \qquad MSH = \frac{SSH}{dfh},$$

and the optimal test rejects if

$$F = \frac{MSH}{MSE} > F^{\alpha}_{p-1,N-2p}.$$

It is also possible to test that the α_i are all 0 (i.e. that the intercepts are the same) for this model. Note, however, that if the slopes are different then the lines cross each other. In this situation we are rarely interested in the intercepts. (Note that if the slopes are assumed the same (as in the one-way model with a single covariate discussed above) then testing that the intercepts are equal is really testing that the lines are the same.) For this reason we do not give the appropriate test here. In Exercise 31.9 a test is derived for testing that the lines are the same for this model (i.e., that the $\alpha_i = 0$ and that the γ_i are the same).

Further comments

31.9 Unfortunately, as we have indicated above, it is possible for the covariates to be confounded with the ANOVA part of the model. This occurs whenever the $\mathbf{X}^*$ matrix does not have full column rank. For example, consider a one-way model with two cells and a single covariate in which the covariate takes the value 1 for all the observations in the first cell, and 0 for all the observations in the second cell. Suppose that the means for the two cells are different. Then there is no way to tell from the data whether the differences are due to the different cells or to the difference in the covariate. In this case, it turns out that $\mathbf{X}^* = \mathbf{0}$, and the theory above cannot be applied. In a less extreme case, we might have the values for the covariate all near 100 for the first cell and near 0 for the second cell. If the values in each cell were not equal, then the $\mathbf{X}^*$ matrix has the appropriate rank, and the procedure can be applied. However, the ANOVA part of the model would be 'nearly' confounded with the covariate, and hence the conclusions could be very sensitive to small changes in the observations.

We say that the covariate is *balanced* for a particular hypothesis if for that hypothesis $\mathbf{X}^{**} = \mathbf{X}^*$. In this case the covariate and the hypothesis are orthogonal and hence no confounding can occur. For most hypotheses, the covariates are balanced if the average value for the covariate is the same in every cell. This indicates that in practice, when we do a covariance analysis, we should try to have the average value of the covariate the same (or nearly the same) in each cell.

The ANCOVA model is a particular linear model so that the discussion in **30.38** on sensitivity to assumptions applies also to ANCOVA models. For a one-way ANCOVA model with a single covariate, Huber's condition (the necessary and sufficient condition for asymptotic insensitivity to normal assumptions) is satisfied as long as the number of observations in each cell go to ∞ and

$$\frac{\max(x_{ij}^{*2})}{\sum\sum x_{ij}^{*2}} \to 0.$$

Similar conditions guarantee asymptotic insensitivity to the normal assumption for more complicated ANCOVA models.

Random effects and mixed models

31.10 In Chapter 30 we have considered many ANOVA models. For all these models we have assumed that the various factor levels were chosen (fixed) in advance so that the effect of the factor level could be represented by an unknown constant shift factor. In this situation we say that the model is a *fixed effects* model. However, often the factor levels are a sample from some larger set of factor levels and we are interested in inference for the population from which the factor levels were selected, not in the particular factor levels selected. In this case we often model the effects due to the factors as random variables and call the model a *random effects* model. If a model has both fixed and random factors, we call the model a *mixed* model. Random effects are assumed independent, which implies that the random effects are sampled from an infinitely large population. In practice, this means that to use random effects and mixed models the random effects must be sampled from a population which is much larger than the sample. Also the factor levels for a particular random effect are assumed to have identical distributions, which implies that populations sampled from must be fairly homogeneous.

For an example of a one-way random effects model, suppose that we want to compare some sort of scholastic achievement test (SAT) scores from different high schools in a city with many similar high schools. We first find a sample of 10 high schools in the city and from each high school select 20 students. Let Y_{ij} represent the SAT score for the jth student at the ith school, $i = 1, \dots, 10$; $j = 1, \dots, 20$. Let M_i be the expected value of the SAT score for a randomly chosen student from the ith school selected. (Note that M_i is an unobserved random variable for this setting.) We assume that conditionally on the M_i, $Y_{i1}, \dots, Y_{i20}$ are independently and identically distributed (i.i.d.), $Y_{ij}|M_i \sim N(M_i, \sigma_e^2)$, where σ_e^2 is a measure of the randomness between students in the ith school. (As in the fixed effects model, we assume that the variance does not depend on the school selected and hence is not a random variable but an unknown parameter.) Since the schools are selected randomly we assume that the M_i are themselves i.i.d. random variables, $M_i \sim N(\theta, \sigma_a^2)$, where θ is an unknown constant (parameter) representing the average SAT score averaged across the all the schools in the city and σ_a^2 represents the variation between schools in the city. Note that $\text{var}(Y_{ij}) = \sigma_a^2 + \sigma_e^2$ so that σ_a^2 and σ_e^2 are called *components of variance* for this model. The parameter σ_a^2 is the parameter of primary interest for this model. There is unfortunately no non-negative unbiased estimator or pivotal quantity for σ_a^2. We shall, however, derive the ML estimator and the restricted maximum likelihood (REML) estimator for σ_a^2. (One nice property of an ML estimator or REML estimator is that it must lie in the parameter space and hence must be non-negative if the parameter being estimated is non-negative.) We shall also develop a test that $\sigma_a^2 = 0$ (that there is no difference between schools) and a pivotal quantity for σ_a^2/σ_e^2. In the exercises an approximate pivotal quantity for σ_a^2 is derived.

Now suppose we want to do a similar analysis at a national level, so we randomly select five cities and in each selected city randomly select 10 schools. In each selected school the SAT scores for each of 20 randomly selected students are recorded. Let Y_{ijk} be the SAT score for the kth student chosen in the jth school of the ith city, $i = 1, \dots, 5$; $j = 1, \dots, 10$; $k = 1, \dots, 20$. Let M_{ij} be the expected value of the SAT scores from the jth school selected in the ith city and let P_i be the expected SAT score for the ith city (averaged across all the schools in the city, not just those in the sample). Given the M_{ij}, we assume that $Y_{ij1}, \ldots, Y_{ij20}$ are i.i.d., $Y_{ijk}|M_{ij} \sim N(M_{ij}, \sigma_e^2)$, where σ_e^2 is a measure of variation between students in the same school. Given P_i, we assume that $M_{i1}, \ldots, M_{i10}$ are i.i.d., $M_{ij}|P_i \sim N(P_i, \sigma_d^2)$, where σ_d^2 is a measure of variation between school districts in the same city. Finally, we assume that the P_i are i.i.d., $P_i \sim N(\theta, \sigma_a^2)$, where θ represents the expected SAT score for the whole country and σ_a^2 represents the variation between cites in the state. Note that

$$\text{var}(Y_{ijk}) = \sigma_a^2 + \sigma_d^2 + \sigma_e^2$$

so that these parameters are again called components of variance. We shall develop tests that $\sigma_d^2 = 0$ (no difference between schools in the same city) and that $\sigma_a^2 = 0$ (no difference between cities). We shall also discuss ML and REML estimators for this model and find pivotal quantities for σ_d^2/σ_e^2 and $\sigma_a^2/(n\sigma_d^2 + \sigma_e^2)$. There are no non-negative unbiased estimators or pivotal quantities for σ_a^2 or σ_d^2, which are typically the parameters of primary interest. We call this a *twofold nested random effects* model because the schools are nested in the cities and both the cities and schools are sampled from larger classes of schools and cities.

In the example above, we treated cities as a random effect, which means that there must have been a large number of cities in the country. Suppose, however, that in the above example the country only had five large cities so that the cities could not be considered as sampled from a larger class. In this case, let Y_{ijk} and M_{ij} be defined as above but let μ_i be the mean SAT score for the ith city (an unknown constant (parameter) now because the cities are not selected at random). As before we assume that conditionally on the M_{ij}, the Y_{ijk} are i.i.d., $Y_{ijk}|M_{ij} \sim N(M_{ij}, \sigma_e^2)$, and that the M_{ij} are independent, $M_{ij} \sim N(\mu_i, \sigma_d^2)$. We consider testing that $\sigma_d^2 = 0$ (no difference between schools in the same city) and that the μ_i are equal (no difference between cities). We find a pivotal quantity for σ_d^2/σ_e^2 and simultaneous confidence intervals for contrasts in the μ_i. We call this model a *twofold nested mixed* model because the districts are nested in the cities, the districts are chosen at random but the cites are fixed before the experiment.

Note that for the nested random effects model the parameters are $(\theta, \sigma_a^2, \sigma_d^2, \sigma_e^2)$, for the mixed model the parameters are $(\mu_1, \ldots, \mu_5, \sigma_d^2, \sigma_e^2)$ and for the fixed effects model they are $(\mu_{11}, \ldots, \mu_{5,10}, \sigma^2)$. Note also that we are not interested in confidence intervals for contrasts in random effects because we are not interested in drawing inference for the particular effects chosen for the experiment but rather we are interested in inference for the population from which those effects were sampled. If we are interested in the particular effects sampled, we should treat those effects as fixed effects.

Our next example is a *two-way crossed random effects* model. Suppose a city wants to see if there is any effect on SAT scores due to teachers and school districts. One possible experiment is to let each of five randomly chosen teachers teach an SAT training class in each of 10 randomly chosen schools, obtaining scores for each of 20 students in each class. This model is a two-way crossed model because we obtain observations for each teacher in each school. It is a random effects model because both the teachers and the schools have been randomly chosen from larger populations. The hypotheses we want to test are that there is no average effect due to teacher, no average effect due to schools and that there is no interaction effect between the school and the teacher. The assumptions we use in this model are more complicated than those for the previous models and will not be given here; see **31.16** for details.

Suppose now in the previous example that the city only had five teachers capable of teaching the appropriate class. Then the teachers would not have been sampled from any population but set before the experiment. In this case we would treat the schools as a random effect but treat the teachers as a fixed effect, giving a *two-way crossed mixed* model. Actually two variations on this model are used, one called the *restricted* model because the interactions are restricted to sum to 0 across the fixed effect (teachers) and one called the *unrestricted* model because the interactions are all assumed independent. As we shall see, derivations are easier for the unrestricted model but the restrictive model leads to more natural interpretations for the procedures.

In this chapter we only consider balanced models. There do not appear to be any optimal or even exact procedures for most unbalanced models. For some sensible approximate procedures for unbalanced models, see Searle (1988).

Typically for random effects and mixed models the observations are not independent, but have a very complicated multivariate normal distribution for which it is very difficult to determine the likelihood directly. In the next section we give a result which allows us to find the likelihood for many balanced random effects and mixed models. In later sections we look at the one-way random effects model, the twofold nested random effects and mixed models and two-way crossed random effects and mixed models. The extension to higher-way crossed and nested random effects and mixed models is immediate but will not be discussed here.

For each model, after reinterpreting the model in terms of unobserved independent random variables, we write down the sums of squares for the associated fixed effects model. The joint distribution of these sums of squares for the random effects or mixed model is derived, leading to new ANOVA tables. These tables are used to derive sensible F tests for parameters of interest. These tables can also be used to find pivotal quantities for many functions of the parameters. Typically, however, there is no pivotal quantity for the parameters of most interest. In the models we consider, there is always a sensible test of no difference due to a particular effect. However, in the three-way crossed random effects model, there is no test for differences between main effects (see Exercise 31.24). (A method for constructing approximate pivotal quantities and test statistics for these situations, due to Satterthwaite (1946), is discussed in Exercise 31.16.) The likelihood is then determined and complete sufficient statistics are derived. Typically, we shall find that there are no non-negative unbiased estimators for the most interesting parameters. Also the parameter space for these models is often quite complicated, leading to complicated formulae for ML estimators. Therefore formulae for the ML estimators are given only in the one-way case. For other cases the estimators can be easily determined numerically for the particular data set observed. Harville (1977) suggested for random effects and mixed models computing estimators and tests from only the joint likelihood of the sums of squares for the effects (rather than from the joint likelihood of the original observations). These estimators and tests are called *restricted ML (REML) estimators* and *restricted LR tests*. These estimators and tests are often simpler and more stable than the ML estimators and LR tests. Using REML estimators and restricted LR tests instead of ML estimators and LR tests for random effects and mixed models is essentially the same as using Wishart likelihood estimators and Wishart likelihood ratio tests instead of ML estimators and LR tests for multivariate problems.

Random effects and fixed effects models were treated as one model for many years. For example, Fisher often viewed ANOVA models from a random effects perspective. See, for example, Chapter 7 of Fisher (1925b). Eisenhart (1947) seems to have been the first person to distinuguish between random effects and fixed effects, and even he implied that the analysis would be the same for the two models.

For the random effects and mixed models we employ the following convention. We use Greek letters for unobserved parameters (fixed effects) and Latin letters for unobserved random variables (random effects). With this convention, it is apparent from the model equation which effects are assumed random and which are assumed fixed.

31.11 It is often difficult to write down the likelihood for random effects and mixed models directly because the observations have rather strangely patterned joint covariance matrices which can be hard to invert and whose determinants are not obvious. One method for deriving the likelihood for these models is to transform the data to independent observations whose likelihood can be computed easily. However, this transformation is often quite messy. (For some examples, see Arnold, 1981, pp. 242–275.) The following theorem (a slight generalization of a result due to Graybill and Hultquist, 1961) allows a derivation of the likelihood from some properties we need to derive for other reasons and is therefore somewhat easier to use. (See Graybill, 1976, pp. 643–647, for an alternative statement of this theorem for random effects models.)

Suppose we start with a fixed effects model which is an orthogonal design. Let $\mathbf{Y}$ be the $N \times 1$ vector of observations for this model. Let the sums of squares and degrees of freedom for this model be given by

$$SS_1(\mathbf{Y}), \ldots, SS_q(\mathbf{Y}), \quad df_1, \ldots, df_q.$$

where $SS_q(\mathbf{Y}) = SSE$ and $df_q = dfe$. (Note that $SS_i(\mathbf{Y})$ is the ith sum of squares evaluated at $\mathbf{Y}$ and hence $SS(\mathbf{Y} - \boldsymbol{\mu})$ is the ith sum of squares evaluated at $\mathbf{Y} - \boldsymbol{\mu}$.)

Theorem 31.1.

Suppose that for some other model in which $\mathbf{Y} \sim N_N(\boldsymbol{\mu}, \boldsymbol{\Sigma})$, the following two conditions are satisfied:

(a) The $SS_i(\mathbf{Y})$ are independent.

(b) There exist γ_i and δ_i such that

$$\frac{SS_i(\mathbf{Y})}{\gamma_i} \sim \chi^2_{df_i}(\delta_i).$$

Then the likelihood for this second model is

$$L_{\mathbf{Y}} \approx \Pi \gamma_i^{-df_i/2} \exp\left(-\sum SS_i(\mathbf{Y} - \boldsymbol{\mu})/2\gamma_i\right)$$

Proof. Since $\mathbf{Y} \sim N_n(\boldsymbol{\mu}, \boldsymbol{\Sigma})$, the likelihood for this model has the form

$$L \approx |\boldsymbol{\Sigma}|^{-1/2} \exp(-(\mathbf{Y} - \boldsymbol{\mu})^T \boldsymbol{\Sigma}^{-1} (\mathbf{Y} - \boldsymbol{\mu})/2).$$

Let T_i be the subspace associated with $SS_i(\mathbf{Y})$, so that $SS_i(\mathbf{Y}) = \|\mathbf{P}_{T_i}\mathbf{Y}\|^2$. Let $\mathbf{X}_i$ be an orthonormal basis matrix for T_i and let $\mathbf{X} = (\mathbf{X}_1, \ldots, \mathbf{X}_q)$. Since the subspaces are orthogonal and the $\mathbf{X}_i$ are orthonormal,

$$\mathbf{X}^T\mathbf{X} = \mathbf{I}.$$

Furthermore, $\mathbf{X}_i$ is $N \times df_i$. Since the design is orthogonal, $\sum df_i = N$. Therefore $\mathbf{X}$ is an $N \times N$ orthogonal matrix. Condition (a) above implies that for $i \neq j$,

$$\mathbf{X}_i\mathbf{X}_i^T \boldsymbol{\Sigma}\mathbf{X}_j\mathbf{X}_j^T = \mathbf{P}_{T_i}\boldsymbol{\Sigma}\mathbf{P}_{T_j} = 0 \Rightarrow \mathbf{X}_i^T\boldsymbol{\Sigma}\mathbf{X}_j = 0.$$

(Recall that $\mathbf{X}_i^T\mathbf{X}_i = \mathbf{I}$.) Condition (b) implies that

$$\gamma_i^{-1}\mathbf{X}_i\mathbf{X}_i^T\boldsymbol{\Sigma}\gamma_i^{-1}\mathbf{X}_i\mathbf{X}_i^T = \gamma_i^{-1}\mathbf{X}_i\mathbf{X}_i^T \Rightarrow \mathbf{X}_i^T\boldsymbol{\Sigma}\mathbf{X}_i = \gamma_i\mathbf{I}_{df_i}.$$

Therefore

$$\mathbf{X}^T\boldsymbol{\Sigma}\mathbf{X} = \begin{pmatrix} \gamma_1\mathbf{I}_{df_1} & 0 & \cdots & 0 \\ 0 & \gamma_2\mathbf{I}_{df_2} & \cdots & 0 \\ \vdots & \vdots & \ddots & \vdots \\ 0 & 0 & \cdots & \gamma_q\mathbf{I}_{df_q} \end{pmatrix} = \mathbf{D} \Rightarrow \boldsymbol{\Sigma} = \mathbf{X}\mathbf{D}\mathbf{X}^T.$$

Hence

$$\begin{aligned}(\mathbf{Y}-\boldsymbol{\mu})^T\boldsymbol{\Sigma}^{-1}(\mathbf{Y}-\boldsymbol{\mu}) &= (\mathbf{X}^T(\mathbf{Y}-\boldsymbol{\mu}))^T\mathbf{D}^{-1}(\mathbf{X}^T(\mathbf{Y}-\boldsymbol{\mu})) \\ &= \sum\frac{(\mathbf{Y}-\boldsymbol{\mu})^T\mathbf{X}_i\mathbf{X}_i^T(\mathbf{Y}-\boldsymbol{\mu})}{\gamma_i} = \sum\frac{\|\mathbf{P}_{T_i}(\mathbf{Y}-\boldsymbol{\mu})\|^2}{\gamma_i} \\ &= \sum\frac{SS_i(\mathbf{Y}-\boldsymbol{\mu})}{\gamma_i},\end{aligned}$$

and

$$|\boldsymbol{\Sigma}| = |\mathbf{D}| = \Pi\gamma_i^{df_i}.$$

The result follows directly.

In using this theorem we shall take the sums of squares for the associated fixed effects model, show they are still independent for the random effects or mixed model and then show that have the appropriate rescaled χ^2 distribution. As in the previous chapter, we let

$$MS_i = \frac{SS_i}{df_i} \Rightarrow \quad EMS_i = \gamma_i + \frac{\kappa_i}{df_i}$$

when $SS_i/\gamma_i \sim \chi^2_{df_i}(\kappa_i/\gamma_i)$. In later sections of this chapter we shall often write down ANOVA tables similar to those for fixed effects models, giving the effects, together with their associated degrees of freedom, SSs and EMSs. Note that such a table only makes sense after we have shown that the various SSs are independent and have rescaled (possibly non-central) χ^2 distributions with the indicated degrees of freedom.

The one-way random effects model

31.12 In the balanced one-way random effects model we observe Y_{ij}, the jth observation in the ith class, $i = 1, \ldots, p$; $j = 1, \ldots, n$. We assume that the classes are sampled from a set of larger classes. Let M_i be the expected value of an observation in the ith chosen class (note that M_i is random since the class is randomly chosen) and let $\theta = EM_i$ be the mean of a randomly chosen observation averaged across all the potential classes. We assume the following:

1. Conditionally on the $\{M_i\}$, the Y_{ij} are independent, $Y_{ij}|\{M_i\} \sim N(M_i, \sigma_e^2)$, where σ_e^2 is a measure of variation which is independent of i (and hence an unknown constant).
2. The M_i are independent, $M_i \sim N(\theta, \sigma_a^2)$.

The parameter space for this model is

$$\theta \in R, \qquad \sigma_a^2 \geq 0, \ \sigma_e^2 > 0.$$

(Note that we assume that $\sigma_e^2 > 0$ but allow the possibility that $\sigma_a^2 = 0$. If $\sigma_a^2 = 0$, then there is no effect due to the class.) The parameters of most interest in this model are σ_a^2 and σ_a^2/σ_e^2.

Let $a_i = M_i - \theta$, $e_{ij} = Y_{ij} - M_i$. Then

$$Y_{ij} = \theta + a_i + e_{ij}, \qquad i = 1, \ldots, p; \ j = 1, \ldots, n,$$

where the a_i and e_{ij} are all independent, unobserved random variables such that

$$a_i \sim N(0, \sigma_a^2), \qquad e_{ij} \sim N(0, \sigma_e^2).$$

(Note that conditionally on M_i, the e_{ij} are i.i.d., $e_{ij}|\{M_i\} \sim N(0, \sigma_e^2)$, and therefore the e_{ij} are independent of the M_i and hence of the a_i.)

Note that

$$EY_{ij} = \theta, \qquad \operatorname{var}(Y_{ij}) = \sigma_a^2 + \sigma_e^2,$$

$$\operatorname{cov}(Y_{ij}, Y_{ij'}) = \sigma_a^2, \qquad \rho = \operatorname{corr}(Y_{ij}, Y_{ij'}) = \frac{\sigma_a^2}{\sigma_a^2 + \sigma_e^2}.$$

ρ is called the *intraclass correlation coefficient* for this model.

In models considered in other chapters of this book, there are observed random variables and unobserved parameters. Other models do not have unobserved random variables the way the random effects and mixed models do. To write the model in a more traditional form we could let $\mathbf{Y}$ be the vector of the observed Y_{ij} and let $\boldsymbol{\mu} = E\mathbf{Y}$, $\boldsymbol{\Sigma} = \operatorname{cov}(\mathbf{Y})$. (Note that $\boldsymbol{\mu} = \theta\mathbf{1}$ (where $\mathbf{1}$ is an $N \times 1$ vector of 1s) and $\boldsymbol{\Sigma}$ is the function of σ_a^2 and σ_e^2 determined by the expression above.) For example, if $p = 2$ and $n = 3$, then

$$\boldsymbol{\mu} = \begin{pmatrix} \theta \\ \theta \\ \theta \\ \theta \\ \theta \\ \theta \end{pmatrix}, \qquad \boldsymbol{\Sigma} = (\sigma_a^2 + \sigma_e^2)\begin{pmatrix} 1 & \rho & \rho & 0 & 0 & 0 \\ \rho & 1 & \rho & 0 & 0 & 0 \\ \rho & \rho & 1 & 0 & 0 & 0 \\ 0 & 0 & 0 & 1 & \rho & \rho \\ 0 & 0 & 0 & \rho & 1 & \rho \\ 0 & 0 & 0 & \rho & \rho & 1 \end{pmatrix}.$$

To be compatible with other models in this book we should say we observe $\mathbf{Y} \sim N_N(\boldsymbol{\mu}, \boldsymbol{\Sigma})$, where $\boldsymbol{\mu}$ and $\boldsymbol{\Sigma}$ have the appropriate forms. However, the version given above with the artificial random variables a_i and e_{ij} is easier to write down and also easier to analyse.

31.13 We next write down the appropriate sums of squares for the balanced one-way fixed effects model, which are

$$SS\theta = SS_1(\mathbf{Y}) = np\overline{Y}_{..}^2, \qquad SSa = SS_2(\mathbf{Y}) = n\sum(\overline{Y}_{i.} - \overline{Y}_{..})^2,$$
$$SSe = SS_3(\mathbf{Y}) = \sum\sum(Y_{ij} - \overline{Y}_{i.})^2.$$

Let $H_i = a_i + \overline{e}_{i.}$. Then it is easily seen that

$$SS\theta = np(\overline{H} + \theta)^2, \qquad SSa = n\sum(H_i - \overline{H})^2, \qquad SSe = \sum\sum(e_{ij} - \overline{e}_{i.})^2.$$

By the usual results of independence of sample mean and sample variance, $\overline{e}_{i.}$ is independent of SSe and hence *SSE* is independent of SSa and $SS\theta$. The H_i are i.i.d. By the independence of sample mean and sample variance (applied now to the H_i), SSa and $SS\theta$ are independent so that all three sums of squares are independent. Furthermore, from results for the fixed effects model

$$SSe/\sigma_e^2 \sim \chi^2_{p(n-1)}.$$

Also

$$\sqrt{n}H_i \sim N(0, \gamma), \qquad \sqrt{np}(\overline{H} + \theta) \sim N(\sqrt{np}\theta, \gamma), \qquad \gamma = n\sigma_a^2 + \sigma_e^2.$$

Therefore

$$\frac{SSa}{\gamma} \sim \chi^2_{p-1}, \qquad \frac{SS\theta}{\gamma} \sim \chi_1^2\left(\frac{np\theta^2}{\gamma}\right).$$

We can summarize these results in the following ANOVA table:

Balanced one-way random effects model

effect	*df*	*SS*	*EMS*
θ	1	$np\overline{Y}_{..}$	$\sigma_e^2 + n\sigma_a^2 + np\theta^2$
a_i	$p-1$	$n\sum(\overline{Y}_{i.} - \overline{Y}_{..})^2$	$\sigma_e^2 + n\sigma_a^2$
error	$p(n-1)$	$\sum\sum(Y_{ij} - \overline{Y}_{i.})^2$	σ_e^2
total	np	$\sum\sum Y_{ij}^2$	

Note that the sums of squares and degrees of freedom are the same as in the associated fixed effects model. Also the distribution of SSe is the same as for the fixed effects model (note that σ_e^2 in this model is playing the role of σ^2 in the fixed effects model), but that the distribution of $SS\theta$ and SSa are different for the two models.

Now consider testing the null hypothesis that $\sigma_a^2 = 0$ against the alternative hypothesis that $\sigma_a^2 > 0$. By the derivations above we see that sensible size-α test for this hypothesis rejects if

$$F = \frac{MSa}{MSe} > F^{\alpha}_{p-1,p(n-1)}.$$

In Exercises 31.11–12 it is shown that this test is the LR test, and the UMP invariant size-α test for this problem. It is also the UMP unbiased size-α test (see Lehmann, 1986, pp. 418–422.) Note that this test is the same as the test for no class effect in the one-way fixed effects model.

The alternative distribution for the fixed effects model is a non-central F distribution but for the random effects model, the alternative distribution is

$$\frac{F}{\delta} \sim F_{p-1,p(n-1)}, \qquad \delta = \frac{\gamma}{\sigma_e^2} = n\frac{\sigma_a^2}{\sigma_e^2} + 1,$$

a rescaled central F distribution. (Note that $\sigma_a^2 = 0 \Leftrightarrow \delta = 1$.)

The quantity F/δ is a pivotal quantity for δ which can be used in the obvious way to construct tests and confidence intervals for δ (and hence tests and confidence intervals for σ_a^2/σ_e^2). Unfortunately no exact confidence intervals or tests exist for σ_a^2 (except for testing $\sigma_a^2 = 0$). See Graybill (1976, pp. 618–619) for a conservative $1-\alpha$ confidence interval for σ_a^2. An approximate confidence interval for σ_a^2 is derived in Exercise 31.17. Note that this approximate confidence interval is useless when the unbiased estimator of σ_a^2 is negative (since it will only contain negative values for the positive parameter σ_a^2).

Pivotal quantities for θ and σ_e^2 in the random effects model are

$$\frac{\sqrt{np}(\overline{Y}_{..} - \theta)}{\sqrt{MSa}} \sim t_{p-1}, \qquad \frac{SSe}{\sigma_e^2} \sim \chi^2_{p(n-1)}$$

which can be used to construct tests and confidence intervals in the obvious way. (Note that for the pivotal quantity for θ in the fixed effects model we use MSE and $dfe = p(n-1)$ instead of MSA and $dfa = p - 1$.)

We can now apply Theorem 31.1 to find the likelihood for this model. Note that the associated ANOVA model is an orthogonal design and that

$$SS_1(\mathbf{Y} - \boldsymbol{\mu}) = np(\overline{Y}_{..} - \theta)^2, \qquad SS_2(\mathbf{Y} - \boldsymbol{\mu}) = SS_2(\mathbf{Y}) = SSa,$$
$$SS_3(\mathbf{Y} - \boldsymbol{\mu}) = SS_3(\mathbf{Y}) = SSe.$$

Therefore the likelihood for this model is

$$L_{\mathbf{Y}} \approx \gamma^{\frac{-p}{2}} \sigma_e^{-p(n-1)} \exp\left(-\frac{(np(\overline{Y}_{..} - \theta)^2 + SSa)}{2\gamma} - \frac{SSe}{2\sigma_e^2}\right)$$
$$= \gamma^{\frac{-p}{2}} \sigma_e^{-p(n-1)} \exp\left(\frac{-np\theta^2}{2\gamma}\right) \exp\left(np\overline{Y}_{..}\frac{\theta}{\gamma} - \frac{(np\overline{Y}_{..}^2 + SSa)}{2\gamma} - \frac{SSe}{2\sigma_e^2}\right).$$

By the exponential criterion we see that $\mathbf{T} = (\overline{Y}_{..}, np\overline{Y}_{..}^2 + SSa, SSe)$ is a complete sufficient statistic for this model and hence $(\overline{Y}_{..}, SSa, SSe)$ (an invertible function of $\mathbf{T}$) is also a complete sufficient statistic.

Now

$$E\overline{Y}_{..} = \theta, \quad EMSa = \gamma, \quad EMSe = \sigma_e^2.$$

Therefore by the Lehmann–Scheffé theorem, $\overline{Y}_{..}$, MSa and MSe are the minimum variance unbiased estimators of θ, γ and σ_e^2. Note also that

$$EU = \sigma_a^2, \qquad U = \frac{MSa - MSe}{n},$$

so that by the Lehmann–Scheffé theorem again, U is the MV unbiased estimator of σ_a^2. Note, however, that MSa and MSe are independent and hence

$$P(U < 0) > 0$$

(i.e. U may be negative). Since $\sigma_a^2 \geq 0$, U does not seem to be a sensible estimator of σ_a^2. It has been suggested that we use the estimator

$$V = \max(U, 0),$$

but this estimator is no longer unbiased. In fact, in Exercise 31.13 it is shown that there is no non-negative unbiased estimator of σ_a^2.

To find the ML estimator of $(\theta, \gamma, \sigma_e^2)$, we must maximize the likelihood subject to the constraints

$$\theta \in R, \qquad \gamma \geq \sigma_e^2 > 0.$$

By direct differentiation we see that $\hat{\theta} = \overline{Y}_{..}$, and that if $(n-1)SSa \geq SSe$, then $\hat{\gamma} = ((p-1)/p)MSa$, $\hat{\sigma}_e^2 = MSe$. If $(n-1)SSa < SSe$, then the maximum must occur on the boundary $\gamma = \sigma_e^2$. If we substitute σ_e^2 for γ in the likelihood and differentiate we see that in this case $\hat{\gamma} = \hat{\sigma}_e^2 = \frac{SSa+SSe}{np}$. Therefore we see that the ML estimator is as follows:

$$\begin{array}{llll} \text{if } (n-1)SSa \geq SSe & \text{then } \hat{\theta} = \overline{Y}_{..}, & \hat{\gamma} = \dfrac{p-1}{p}MSa, & \hat{\sigma}_e^2 = MSe \\ \text{if } (n-1)SSa < SSe & \text{then } \hat{\theta} = \overline{Y}_{..}, & \hat{\gamma} = \hat{\sigma}_e^2 = \dfrac{SSa+SSe}{np}. & \end{array}$$

By the invariance principle for ML estimators we see that the ML estimator for σ_a^2 is given by

$$\begin{aligned} \hat{\sigma}_a^2 = \frac{\hat{\gamma} - \hat{\sigma}_e^2}{n} &= \begin{cases} \frac{(p-1)MSa - pMSe}{np} & \text{if } (n-1)SSa \geq SSe \\ 0 & \text{if } (n-1)SSa < SSe \end{cases} \\ &= \max\left(\frac{(p-1)MSa - pMSe}{np}, 0\right). \end{aligned}$$

Many statisticians believe that in random effects and mixed models more appealing estimators for the components of variance can be computed by maximizing the joint likelihood of the sums of squares for the random effects instead of maximizing the joint likelihood of the original observations (or equivalently maximizing the joint likelihood of the sufficient statistic). In this case, such estimators for σ_a^2 and σ_e^2 can be computed by maximizing the joint likelihood of (SSa, SSe) which has the form

$$L^* \approx \gamma^{-(p-1)/2} \sigma_e^{-p(n-1)} \exp\left(-\frac{SSa}{2\gamma} - \frac{SSe}{2\sigma_e^2}\right)$$

(see Exercise 31.15). Since we are only using a part of the likelihood, these estimators are called REML estimators. This restricted likelihood is maximized by

$$\hat{\gamma}_R = MSa \text{ and } \hat{\sigma}_{eR}^2 = MSe, \text{ when } MSa \geq MSe,$$

$$\hat{\gamma}_R = \hat{\sigma}^2_{eR} = \frac{SSa + SSe}{np - 1}, \text{ when } MSa < MSe.$$

(see Exercise 31.15). Therefore the REML estimator of σ_a^2 is given by

$$\hat{\sigma}^2_{aR} = \frac{\hat{\gamma}_R - \hat{\sigma}^2_{eR}}{n} = \max(U, 0) = V,$$

where U is the MV unbiased estimator of σ_a^2 given above. Similarly, one can compute restricted LR test statistics from the restricted likelihood L^*. For the one-way model, the LR test and the restricted LR test are the same.

One appealing aspect of ML and restricted ML estimators is that they must always be in the parameter space. In particular an ML or REML estimator of a non-negative parameter must itself be non-negative.

Nested random effects and mixed models

31.14 In the twofold nested random effects model, we observe Y_{ijk}, the kth observation in the jth subclass of the ith class, $i = 1, \ldots, p$; $j = 1, \ldots, c$; $k = 1, \ldots, n$. We assume that both the classes and subclasses are randomly chosen from some larger set of classes and subclasses. Let M_{ij} be the mean of an observation in the jth chosen subclass of the ith chosen class, let P_i be the mean of an observation in the ith chosen class averaged over all its subclasses (not just those chosen for the experiment) and let θ be the mean of the randomly chosen observations averaged over all the classes and subclasses. We assume the following:

1. Given the $\{M_{ij}, P_i\}$, the Y_{ijk} are independent, $Y_{ijk}|\{M_{ij}, P_i\} \sim N(M_{ij}, \sigma_e^2)$, where σ_e^2 is an unknown parameter representing the variation in observations in the same subclass.
2. Given the $\{P_i\}$, the M_{ij} are independent, $M_{ij}|\{P_i\} \sim N(P_i, \sigma_d^2)$, where σ_d^2 is an unobserved parameter representing the variation in the subclasses of the same class.
3. The P_i are independent, $P_i \sim N(\theta, \sigma_a^2)$, where σ_a^2 is an unknown parameter representing the variation between classes.

The parameter space for this model is

$$\theta \in R, \qquad \sigma_a^2 \geq 0, \qquad \sigma_d^2 \geq 0, \qquad \sigma_e^2 > 0.$$

If $\sigma_a^2 = 0$, then there is no difference between classes; and if $\sigma_d^2 = 0$, then there is no difference between subclasses of any particular class.

To put this model in simpler form, let

$$a_i = P_i - \theta, \qquad d_{ij} = M_{ij} - P_i, \qquad e_{ijk} = Y_{ijk} - M_{ij}.$$

By arguments similar to those for the one-way model, we see that

$$Y_{ijk} = \theta + a_i + d_{ij} + e_{ijk}, \qquad a_i \sim N(0, \sigma_a^2),\ d_{ij} \sim N(0, \sigma_d^2),\ e_{ijk} \sim N(0, \sigma_e^2)$$

and the a_i, d_{ij} and e_{ijk} are all independent (see Exercise 31.18). Note that

$$EY_{ijk} = \theta, \quad \operatorname{cov}(Y_{ijk}, Y_{i'j'k'}) = 0, \quad i \neq i', \qquad \operatorname{cov}(Y_{ijk}, Y_{ij'k'}) = \sigma_a^2, \quad j \neq j',$$

$$\text{cov}(Y_{ijk}, Y_{ijk'}) = \sigma_a^2 + \sigma_d^2,\ k \neq k', \quad \text{var}(Y_{ijk}) = \sigma_a^2 + \sigma_d^2 + \sigma_e^2.$$

The a_i, d_{ij} and e_{ijk} are again unobserved random variables.

We now write down the sums of squares for the associated fixed effects model

$$SS\theta = ncp\overline{Y}_{...}^2, \qquad SSa = nc\sum(\overline{Y}_{i..} - \overline{Y}_{...})^2,$$
$$SSd = n\sum\sum(\overline{Y}_{ij.} - \overline{Y}_{i..})^2, \qquad SSe = \sum\sum\sum(Y_{ijk} - \overline{Y}_{ij.})^2.$$

By arguments similar to those for the one-way model, we see that these sums of squares are independent,

$$\frac{SS\theta}{\gamma_a} \sim \chi_1^2\left(\frac{ncp\theta^2}{\gamma_a}\right), \qquad \frac{SSa}{\gamma_a} \sim \chi_{p-1}^2, \qquad \gamma_a = nc\sigma_a^2 + n\sigma_d^2 + \sigma_e^2,$$
$$\frac{SSd}{\gamma_d} \sim \chi_{p(c-1)}^2, \qquad \frac{SSe}{\sigma_e^2} \sim \chi_{pc(n-1)}^2, \qquad \gamma_d = n\sigma_d^2 + \sigma_e^2$$

(see Exercise 31.19). These results can be summarized in the following ANOVA table:

Balanced twofold nested random effects model

effect	*df*	*SS*	*EMS*
θ	1	$ncp\overline{Y}_{...}^2$	$\sigma_e^2 + n\sigma_d^2 + nc\sigma_a^2 + ncp\theta^2$
a_i	$p-1$	$nc\sum(\overline{Y}_{i..} - \overline{Y}_{...})^2$	$\sigma_e^2 + n\sigma_d^2 + nc\sigma_a^2$
d_{ij}	$p(c-1)$	$n\sum\sum(\overline{Y}_{ij.} - \overline{Y}_{i..})^2$	$\sigma_e^2 + n\sigma_d^2$
error	$pc(n-1)$	$\sum\sum\sum(Y_{ijk} - \overline{Y}_{ij.})^2$	σ_e^2
total	pcn	$\sum\sum\sum Y_{ijk}^2$	

By these results, we see that to test that $\sigma_a^2 = 0$ against $\sigma_a^2 > 0$ or $\sigma_d^2 = 0$ against $\sigma_d^2 > 0$ we reject if

$$F_a = \frac{MSa}{MSd} > F_{p-1,p(c-1)}^{\alpha} \quad \text{or } F_d = \frac{MSd}{MSe} > F_{p(c-1),pc(n-1)}^{\alpha}.$$

These tests are both UMP unbiased size-α tests (see Lehmann, 1986, pp. 422–427). Note that the test for no subclass effect ($\sigma_d^2 = 0$) is the same as for the associated fixed effects model but the test for no class effect ($\sigma_a^2 = 0$) is different for the two models. Note also that the alternative distributions for these F statistics are rescaled central F distributions for the random effects model and non-central F distributions for the associated fixed effects model.

Let $\delta_a = \sigma_a^2/(n\sigma_d^2 + \sigma_e^2)$, $\delta_d = \sigma_d^2/\sigma_e^2$. Then pivotal quantities for these parameters are

$$\frac{F_a}{p\delta_a + 1} \sim F_{p-1,p(c-1)}, \qquad \frac{F_d}{n\delta_d + 1} \sim F_{p(c-1),pc(n-1)},$$

which can be used in the obvious way to construct confidence intervals and tests for these parameters. (These facts follow directly from the calculations above the ANOVA table above.) There are unfortunately no exact confidence intervals or tests for σ_a^2 or σ_d^2 (except for testing $\sigma_a^2 = 0$ or $\sigma_d^2 = 0$) which are usually the parameters of interest in this model. Approximate confidence intervals are derived in Exercise 31.21.

Using Theorem 31.1 we can now find the likelihood for this model:

$$L \approx \gamma_a^{-\frac{p}{2}} \gamma_d^{-\frac{p(c-1)}{2}} \sigma_e^{-pc(n-1)} \exp\left(-\frac{ncp(\overline{Y}_{\dots} - \theta)^2}{2\gamma_a} - \frac{SSa}{2\gamma_a} - \frac{SSd}{2\gamma_d} - \frac{SSe}{2\sigma_e^2}\right).$$

From this it is straightforward to see that $(\overline{Y}_{\dots}, SSa, SSd, SSe)$ is a complete sufficient statistic for this model.

Since they are unbiased we see that $\overline{Y}_{\dots}$, MSa, MSd and MSe are the MV unbiased estimators of θ, γ_a, γ_d and σ_e^2. MV unbiased estimators of σ_a^2 and σ_d^2 are

$$\frac{MSa - MSd}{nc} \text{ and } \frac{MSd - MSe}{n}.$$

Note that both these estimators may be negative. In fact, using the completeness of the sufficient statistic, we can show that there are no non-negative unbiased estimators for these parameters. (See Exercise 31.20 for derivations of these facts.)

To find the ML estimator of the parameter, we first maximize the likelihood subject to the constraints

$$\theta \in R, \qquad \gamma_a \geq \gamma_d \geq \sigma_e^2 > 0.$$

This maximization leads to consideration of several different cases but is easily solved numerically for the particular Y_{ijk} observed. The ML estimators for σ_a^2 and σ_e^2 can then be found by the invariance principle for these estimators.

The REML estimators for γ_a, γ_d and σ_e^2 are found by maximizing the joint density of SSa, SSd and SSe,

$$L^* \approx \gamma_a^{-\frac{p-1}{2}} \gamma_d^{-\frac{p(c-1)}{2}} \sigma_e^{-pc(n-1)} \exp\left(-\frac{SSa}{2\gamma_a} - \frac{SSd}{2\gamma_d} - \frac{SSe}{2\sigma_e^2}\right)$$

subject to the above constraints. REML estimators of σ_a^2 and σ_d^2 are determined by the invariance principle.

31.15 In the twofold nested mixed model, we observe Y_{ijk}, the kth observation in the jth subclass of the ith class, $i = 1, \dots, p$; $j = 1, \dots, c$; $k = 1, \dots, n$. We assume that the class is fixed in the experiment but that the subclass is chosen randomly from a much larger collection of subclasses. Let M_{ij} be the mean of an observation in the jth chosen subclass of the ith class and let μ_i be the mean of an observation in the ith class (averaged across the subclasses). Note that M_{ij} is a random variable because the subclass is sampled from a larger set but μ_i is an unknown constant (parameter) because the classes are fixed.

We assume the following:

1. Given the $\{M_{ij}\}$, the Y_{ijk} are independent, $Y_{ijk}|\{M_{ij}\} \sim N(M_{ij}, \sigma_e^2)$.
2. The M_{ij} are independent, $M_{ij} \sim N(\mu_i, \sigma_d^2)$.

The parameter space for this model is

$$\boldsymbol{\mu} = (\mu_1, \dots, \mu_p)^T \in R^p, \qquad \sigma_d^2 \geq 0,\ \sigma_e^2 > 0.$$

To put this model in a more familiar form, let

$$e_{ijk} = Y_{ijk} - M_{ij}, \qquad d_{ij} = M_{ij} - \mu_i, \qquad \alpha_i = \mu_i - \overline{\mu}_{.}, \qquad \theta = \overline{\mu}_{.}.$$

Then

$$Y_{ijk} = \theta + \alpha_i + d_{ij} + e_{ijk},$$

where θ and α_i are unknown parameters such that $\sum \alpha_i = 0$, and d_{ij} and e_{ijk} are unobserved independent random variables such that $d_{ij} \sim N(0, \sigma_d^2)$ and $e_{ijk} \sim N(0, \sigma_e^2)$. Note that

$$EY_{ijk} = \theta + \alpha_i, \quad \text{var}(Y_{ijk}) = \sigma_d^2 + \sigma_e^2, \quad \text{cov}(Y_{ijk}, Y_{ijk'}) = \sigma_d^2.$$

(Note that in going from the random effects model to the mixed model the class effect has moved from the covariance matrix to the mean vector of the joint normal distribution of the observations.)

Let $SS\theta$, $SS\alpha$, SSd and SSe be the sums of squares for the associated fixed effects model (given above). Then by arguments similar to those in previous sections we can show that these sums of squares are independent and

$$\frac{SS\theta}{\gamma} \sim \chi_1^2\left(\frac{ncp\theta^2}{\gamma}\right), \quad \frac{SS\alpha}{\gamma} \sim \chi_{p-1}^2\left(\frac{nc\sum \alpha_i^2}{\gamma}\right), \quad \frac{SSd}{\gamma} \sim \chi_{p(c-1)}^2,$$
$$\frac{SSe}{\sigma_e^2} \sim \chi_{pc(n-1)}^2, \quad \gamma = n\sigma_d^2 + \sigma_e^2$$

(see Exercise 31.23). Therefore, we see that the ANOVA table for this model is

Balanced twofold nested mixed model

effect	df	SS	EMS
θ	1	$ncp\overline{Y}_{...}^2$	$\sigma_e^2 + n\sigma_d^2 + ncp\theta^2$
α	$p-1$	$nc\sum(\overline{Y}_{i..} - \overline{Y}_{...})^2$	$\sigma_e^2 + n\sigma_d^2 + h\sum \alpha_i^2$
d	$p(c-1)$	$n\sum\sum(\overline{Y}_{ij.} - \overline{Y}_{i..})^2$	$\sigma_e^2 + n\sigma_d^2$
error	$pc(n-1)$	$\sum\sum\sum(Y_{ijk} - \overline{Y}_{ij.})^2$	σ_e^2
total	ncp	$\sum\sum\sum Y_{ijk}^2$	

where $h = nc/(p-1)$. Therefore we see that to test that $\sigma_d^2 = 0$ or to test that $\alpha_i = 0$, we reject if

$$F_d = \frac{MSd}{MSe} > F^\alpha \text{ or } F_\alpha = \frac{MS\alpha}{MSd} > F^\alpha.$$

Note that the alternative distributions of the F statistics are given by

$$\frac{F_d}{\delta} \sim F_{p(c-1),pc(n-1)}, \quad F_\alpha \sim F_{p-1,p(c-1)}\left(\frac{n\sum \alpha_i^2}{\gamma}\right), \quad \gamma = n\sigma_d^2 + \sigma_e^2, \quad \delta = \frac{\gamma}{\sigma_e^2},$$

so that the alternative distribution of F_d is a rescaled central F distribution while the alternative distribution of F_α is a non-central F distribution. Note that F_d/δ is a pivotal quantity for $\delta = n\sigma_d^2/\sigma_e^2 + 1$ and hence also a pivotal quantity for σ_d^2/σ_e^2.

We now discuss simultaneous confidence intervals for contrasts in the α_i. Note that the $\overline{Y}_{ij.}$ are independent,

$$\overline{Y}_{ij.} \sim N\left(\theta + \alpha_i, \frac{\gamma}{n}\right).$$

This model is a balanced one-way fixed effects model (with p classes and c observations in each class), for which we can construct simultaneous confidence intervals for contrasts in the α_i in the obvious way. Note that the variance estimator for this model is

$$\frac{\sum\sum(\overline{Y}_{ij.} - \overline{Y}_{i..})^2}{p(c-1)} = \frac{MSd}{n}.$$

Therefore the Scheffé simultaneous confidence intervals for contrasts in the α_i are

$$\sum a_i\alpha_i \in \sum a_i\hat{\alpha}_i \pm \left((p-1)MSdF^{\alpha}_{p-1,p(c-1)}\sum\frac{a_i^2}{nc}\right)^{1/2}, \qquad \forall a_i \ni \sum a_i = 0.$$

Note that these intervals are the same as those for the fixed effects nested model except that we have replaced *MSE* and *dfe* for that model with MSd and dfd. Note also that

$$\frac{\max|\overline{Y}_{i..} - \overline{Y}_{i'..} - (\alpha_i - \alpha_{i'})|}{\sqrt{MSd/nc}} \sim q_{p,p(c-1)},$$

so that Tukey type simultaneous confidence intervals can also be constructed (see Exercise 31.25). In addition multiple F and multiple range procedures for multiple comparisons in the class effect can be implemented in the obvious way. Dunnet and MCB simultaneous confidence intervals are similarly available.

We can use Theorem 23.1 to find the likelihood for this model:

$$L \approx \gamma^{-\frac{pc}{2}}\sigma_e^{-pc(n-1)}\exp\left(-\frac{ncp(\overline{Y}_{...} - \theta)^2 + nc\sum(\overline{Y}_{i..} - \overline{Y}_{...} - \alpha_i)^2 + SSd}{2\gamma} - \frac{SSe}{2\sigma_e^2}\right).$$

From this we see that a complete sufficient statistic for this model is $(\overline{Y}_{1..}, \ldots, \overline{Y}_{p..}, SSd, SSe)$ (see Exercise 31.24). We can construct unbiased estimators of $\mu_i, \theta, \alpha_i, \gamma$ and σ_e^2 in the obvious way. There is no non-negative unbiased estimator of σ_d^2. The ML estimator of the parameter is derived in Exercise 31.24.

To find the REML estimator for (σ_d^2, σ_e^2), we work with the joint density of SSd and SSe,

$$L^* \approx \gamma^{-\frac{p(c-1)}{2}}\sigma_e^{-pc(n-1)}\exp\left(-\frac{SSd}{2\gamma} - \frac{SSe}{2\sigma^2}\right).$$

The REML estimator is also derived in Exercise 31.24.

Crossed random effects and mixed models

31.16 In the two-way random effects model we observe Y_{ijk}, the kth observation in the ith row and the jth column, $i = 1, \ldots, r$; $j = 1, \ldots, c$; $k = 1, \ldots, n$. We assume that the rows and columns are independently sampled from some larger sets of column and row treatments.

Let M_{ij} be the mean of an observation in the ith selected row and the jth selected column, let P_i be the mean of an observation in the ith selected row (averaged across all the potential columns) and let Q_j be the mean of an observation in the jth selected column (averaged across all the potential rows). Finally let θ be the mean of all observations (averaged across all the potential rows and columns). Note that M_{ij}, P_i and Q_j are unobserved random variables but that θ is an unobserved parameter.

We make the following assumptions:

1. Given the $\{M_{ij}, P_i, Q_j\}$, we assume that the Y_{ijk} are independent, $Y_{ijk}|\{M_{ij}, P_i, Q_j\} \sim N(M_{ij}, \sigma_e^2)$, where σ_e^2 is an unknown parameter representing the variation of observations in the same cell.
2. Given the $\{P_i, Q_j\}$ we assume that the M_{ij} are independent,
$$M_{ij}|\{P_i, Q_j\} \sim N(P_i + Q_j - \theta, \sigma_d^2),$$
where σ_d^2 is a measure is a measure of variation between cell means.
3. We assume that $P_1, \ldots, P_r$ are independent, $P_i \sim N(\theta, \sigma_a^2)$, and that $Q_1, \ldots, Q_c$ are independent, $Q_j \sim N(\theta, \sigma_b^2)$, where σ_a^2 is a measure of the variation between the row means and σ_b^2 is a measure of the variation between the column means. Since the columns and rows are selected independently we also assume that the P_i are independent of the Q_j.

As shall be shown below, assumption 2 above does not guarantee that the rows and columns are additive. Note that under these assumptions

$$E(M_{ij}|P_i) = E((P_i + Q_j - \theta)|P_i) = P_i + EQ_j - \theta = P_i.$$

Similarly, $E(M_{ij}|Q_j) = Q_j$.

Perhaps a more natural version of assumption 2 (but one which is more difficult to use) is to assume that, conditionally on the $\{P_i\}$, $M_{i1}, \ldots, M_{ic}$ are independent, $M_{ij}|\{P_i\} \sim N(P_i, \tau_1^2)$, and, conditionally on the $\{Q_j\}$, $M_{1j}, \ldots, M_{rj}$ are independent, $M_{ij}|\{Q_j\} \sim N(Q_j, \tau_2^2)$ (where τ_1^2 and τ_2^2 are measures of the variations between cell means in a particular row and between cell means in a particular column), and that M_{ij} and $M_{i'j'}$ are independent when $i \neq i'$, $j \neq j'$. Replacing assumption 2 above with this assumption leads to an equivalent version of the two-way random effects model. (See, for example, Arnold (1981, pp. 264–265) for a derivation of the random effects model from this different set of assumptions.)

The parameter space for this model is

$$\theta \in R, \qquad \sigma_a^2 \geq 0, \qquad \sigma_b^2 \geq 0, \qquad \sigma_d^2 \geq 0, \qquad \sigma_e^2 > 0.$$

If $\sigma_a^2 = 0$, we say that there is no row effect, if $\sigma_b^2 = 0$ we say that there is no column effect, and if $\sigma_d^2 = 0$ we say that there is no interaction effect.

To put this model into a simpler form, let

$$a_i = P_i - \theta, \qquad b_j = Q_j - \theta, \qquad d_{ij} = M_{ij} - P_i - Q_j + \theta, \qquad e_{ijk} = Y_{ijk} - M_{ij}.$$

Lemma 31.1.

(a) $Y_{ijk} = \theta + a_i + b_j + d_{ij} + e_{ijk}$.
(b) a_i, b_j, d_{ij} *and* e_{ijk} *are all independent,*

$$a_i \sim N(0, \sigma_a^2), \quad b_j \sim N(0, \sigma_b^2), \quad d_{ij} \sim N(0, \sigma_d^2), \quad e_{ijk} \sim N(0, \sigma_e^2)$$

Proof.
(a) This is easily verified.
(b) By assumption 1 above, conditionally on the $\{M_{ij}, P_i, Q_j\}$, the e_{ijk} are independent,

$$e_{ijk}|\{M_{ij}, P_i, Q_j\} \sim N(0, \sigma_e^2).$$

Therefore the e_{ijk} are independent of the $\{M_{ij}, P_i, Q_j\}$ and hence of the a_i, b_j and d_{ij} and $e_{ijk} \sim N(0, \sigma_e^2)$. Also, by assumption 3 above, the a_i and b_j are all independent and have the indicated distributions. Conditionally on the P_i and Q_j the d_{ij} are independent,

$$d_{ij}|(P_i, Q_j) \sim N(0, \sigma_d^2).$$

Since this conditional distribution does not involve the P_i or Q_j, it is the same as the marginal joint distribution of the d_{ij}. (That is, the d_{ij} are independent, $d_{ij} \sim N(0, \sigma_d^2)$). Furthermore, this fact implies that the d_{ij} are independent of the P_i and Q_j and hence of the a_i and the b_j. Therefore the a_i, b_j, d_{ij} and e_{ijk} are all independent.

Note that

$$EY_{ijk} = \theta, \quad \text{var}(Y_{ijk}) = \sigma_e^2 + \sigma_d^2 + \sigma_b^2 + \sigma_a^2$$
$$\text{cov}(Y_{ijk}, Y_{ijk'}) = \sigma_d^2 + \sigma_b^2 + \sigma_a^2, \quad k \neq k'; \quad \text{cov}(Y_{ijk}, Y_{ij'k'}) = \sigma_a^2, \quad j \neq j'$$
$$\text{cov}(Y_{ijk}, Y_{i'jk'}) = \sigma_b^2, \quad i \neq i'; \quad \text{cov}(Y_{ijk}, Y_{i'j'k'}) = 0, \quad i \neq i', \quad j \neq j',$$

so that the joint covariance matrix of the Y_{ijk} is quite complicated.

31.17 We next list the sums of squares for the associated fixed effects model:

$$SS\theta = ncr\overline{Y}_{...}^2, \qquad SSa = nc\sum(\overline{Y}_{i..} - \overline{Y}_{...})^2, \qquad SSb = nr\sum(\overline{Y}_{.j.} - \overline{Y}_{...})^2,$$
$$SSd = n\sum\sum(\overline{Y}_{ij.} - \overline{Y}_{i..} - \overline{Y}_{.j.} + \overline{Y}_{...})^2, \qquad SSe = \sum\sum\sum(Y_{ijk} - \overline{Y}_{ij.})^2.$$

Lemma 31.2.
For the random effects model, $SS\theta$, SSa, SSb SSd and SSe are independent.

$$\frac{SS\theta}{\gamma_\theta} \sim \chi_1^2\left(\frac{ncr\theta^2}{\gamma_\theta}\right), \qquad \frac{SSa}{\gamma_a} \sim \chi_{r-1}^2, \qquad \frac{SSb}{\gamma_b} \sim \chi_{c-1}^2,$$
$$\frac{SSd}{\gamma_d} \sim \chi_{(r-1)(c-1)}^2, \qquad \frac{SSe}{\sigma_e^2} \sim \chi_{rc(n-1)}^2,$$

where

$$\gamma_\theta = nc\sigma_a^2 + nr\sigma_b^2 + n\sigma_d^2 + \sigma_e^2, \qquad \gamma_a = nc\sigma_a^2 + n\sigma_d^2 + \sigma_e^2,$$
$$\gamma_b = nr\sigma_b^2 + n\sigma_d^2 + \sigma_e^2, \qquad \gamma_d = n\sigma_d^2 + \sigma_e^2.$$

Proof. It is convenient to introduce some further unobserved random variables. Let

$$G_{ij} = \sqrt{n}(d_{ij} + \bar{e}_{ij.}), \qquad H_i = (nc)^{1/2}(a_i + \bar{d}_{i.} + \bar{e}_{i..}) = (nc)^{1/2}a_i + c^{1/2}\overline{G}_{i.},$$
$$K_j = (nr)^{1/2}(b_j + \bar{d}_{.j} + \bar{e}_{.j.}) = (nr)^{1/2}b_j + r^{1/2}\overline{G}_{.j},$$
$$L = (ncr)^{1/2}(\theta + \bar{a}_. + \bar{b}_. + \bar{d}_{..} + \bar{e}_{...}) = (ncr)^{1/2}(\theta + \bar{a}_.) + c^{1/2}\overline{K}_.$$
$$= (ncr)^{1/2}(\theta + \bar{b}_.) + r^{1/2}\overline{H}_..$$

Then

$$SS\theta = L^2, \quad SSa = \sum(H_i - \overline{H}_.)^2, \quad SSb = \sum(K_j - \overline{K}_.)^2,$$
$$SSd = \sum\sum(G_{ij} - \overline{G}_{i.} - \overline{G}_{.j} + \overline{G}_{..})^2, \quad SSe = \sum\sum\sum(e_{ijk} - \bar{e}_{ij.})^2.$$

Note that by usual results on independence of sample means and sample variances, the $\bar{e}_{ij.}$ are independent of SSe and hence SSe is independent of the other sums of squares. It is straightforward to show that $\overline{G}_{i.}$ is uncorrelated with $G_{ij} - \overline{G}_{i.} - \overline{G}_{.j} + \overline{G}_{..}$. (These variables are linear functions of jointly normally distributed random variables and hence jointly normally distributed so that to establish their independence, it is enough to establish lack of correlation.) Therefore the H_i are independent of SSd and hence SSa and $SS\theta$ are independent of SSd. Similarly SSb is independent of SSd. It is also straightforward to show that L, $H_i - \overline{H}_.$ and $K_j - \overline{K}_.$ are uncorrelated and hence that $SS\theta$, SSa and SSb are independent. Therefore all these sums of squares are independent. The distribution for SSe follows from results for the fixed effects model. Note that the G_{ij} are independent, $G_{ij} \sim N(0, \gamma_d)$. Therefore the distribution of SSd follows directly from results for the additive two-way fixed effects model (with no replication) applied to the G_{ij}. The H_i are also independent, $H_i \sim N(0, \gamma_a)$, so that the distribution of SSa follows from the usual distribution of the sample variance. The distribution of SSb follows similarly. Finally, $L \sim N((ncr)^{1/2}\theta, \gamma_\theta)$. The distribution of $SS\theta$ follows directly.

In the proof of this lemma the unobserved random variables have to be defined carefully. It is necessary to write each SS as a function of random variables which are independent. For example if we had defined $H_i^* = (nc)^{1/2}\overline{Y}_{i..} = (nc)^{1/2}(\theta + a_i + \bar{b}_. + \bar{d}_{i.} + \bar{e}_{i..}) = H_i + (nc)^{1/2}(\bar{b}_. + \theta)$, then $SSa = \sum(H_i^* - \overline{H}^*_.)^2$ as before, but the H_i^* are not independent because they all have $\bar{b}_.$ in common.

The results of this theorem are often summarized in the following ANOVA table:

Balanced two-way crossed random effects model

effect	*df*	*SS*	*EMS*
θ	1	$ncr\overline{Y}_{...}^2$	$\sigma_e^2 + n\sigma_d^2 + nr\sigma_b^2 + nc\sigma_a^2 + ncr\theta^2$
a_i	$r-1$	$nc\sum(\overline{Y}_{i..} - \overline{Y}_{...})^2$	$\sigma_e^2 + n\sigma_d^2 + nc\sigma_a^2$
b_j	$c-1$	$nr\sum(\overline{Y}_{.j.} - \overline{Y}_{...})^2$	$\sigma_e^2 + n\sigma_d^2 + nr\sigma_b^2$
d_{ij}	$(r-1)(c-1)$	$n\sum\sum(\overline{Y}_{ij.} - \overline{Y}_{i..} - \overline{Y}_{.j.} + \overline{Y}_{...})^2$	$\sigma_e^2 + n\sigma_d^2$
error	$rc(n-1)$	$\sum\sum\sum(Y_{ijk} - \overline{Y}_{ij.})^2$	σ_e^2
total	rcn	$\sum\sum\sum Y_{ijk}^2$	

Size-α F tests for several hypotheses can be computed in the obvious way from the ANOVA table above. In particular, to test that $\sigma_a^2 = 0$ or $\sigma_d^2 = 0$, we reject if

$$F_a = \frac{MSa}{MSd} > F^{\alpha}_{r-1,(r-1)(c-1)} \quad \text{or } F_d = \frac{MSd}{MSe} > F^{\alpha}_{(r-1)(c-1),rc(n-1)}.$$

We can find pivotal quantities for many parameters, for example, σ_e^2, σ_d^2/σ_e^2 and $\sigma_a^2/(n\sigma_d^2 + \sigma_e^2)$. There are no pivotal quantities for θ, σ_a^2, σ_b^2 or σ_d^2, which are often the parameters of interest. Approximate pivotal quantities are derived in Exercise 31.27.

Using this lemma, we can find the likelihood in the usual way:

$$L \approx \gamma_\theta^{-\frac{1}{2}} \gamma_a^{-\frac{r-1}{2}} \gamma_b^{-\frac{c-1}{2}} \gamma_d^{-\frac{(r-1)(c-1)}{2}} \sigma_e^{-rc(n-1)}$$
$$\times \exp\left(-\frac{ncr(\overline{Y}_{...} - \theta)^2}{2\gamma_\theta} - \frac{SSa}{2\gamma_a} - \frac{SSb}{2\gamma_b} - \frac{SSd}{2\gamma_d} - \frac{SSe}{2\sigma_e^2} \right)$$

(see Exercise 31.28). Therefore we see that $(\overline{Y}_{...}, SSa, SSb, SSd, SSe)$ is a sufficient statistic. The exponential criterion cannot be used directly to establish completeness. In exponential form there are six terms and there are only five parameters $(\theta, \gamma_a, \gamma_b, \gamma_d, \sigma_e^2)$. Note that

$$\gamma_\theta = \gamma_a + \gamma_b - \gamma_d.$$

However, a more delicate argument can be used to establish completeness in this and other balanced random effects and mixed models. (See Arnold, 1981, pp. 271–272.)

$\overline{Y}_{...}$, MSa, MSb, MSd and MSe are MV unbiased estimators of θ, γ_a, γ_b, γ_d and σ_e^2. MV unbiased estimators of σ_a^2, σ_b^2 and σ_d^2 are given by

$$\frac{MSa - MSd}{nc}, \qquad \frac{MSb - MSd}{nr}, \qquad \frac{MSd - MSe}{n}.$$

There are no non-negative unbiased estimators of these parameters.

To find the ML estimator for this model, we first maximize the likelihood subject to the constraints

$$\gamma_\theta = \gamma_a + \gamma_b - \gamma_d, \qquad \gamma_a \geq \gamma_d, \qquad \gamma_b \geq \gamma_d, \qquad \gamma_d \geq \sigma_e^2 > 0.$$

This maximization is fairly routine on a computer but formulae for the estimators become quite complex due to the many possible boundaries. Also, the ML estimator is somewhat unstable because of the constraint involving γ_θ. ML estimators for σ_a^2, σ_b^2 and σ_e^2 can be determined by the invariance principle.

To find the REML estimator for $(\gamma_a, \gamma_b, \gamma_d, \sigma_e^2)$, we must maximize the joint density of (SSa, SSb, SSd, SSe), given by

$$L^* \approx \gamma_a^{-\frac{r-1}{2}} \gamma_b^{-\frac{c-1}{2}} \gamma_d^{-\frac{(r-1)(c-1)}{2}} \sigma_e^{-rc(n-1)} \exp\left(-\frac{SSa}{2\gamma_a} - \frac{SSb}{2\gamma_b} - \frac{SSd}{2\gamma_d} - \frac{SSe}{2\sigma_e^2}\right)$$

(see Exercise 31.28) subject to the constraints

$$\gamma_a \geq \gamma_d, \qquad \gamma_b \geq \gamma_d, \qquad \gamma_d \geq \sigma_e^2 > 0.$$

REML estimators for σ_a^2, σ_b^2 and σ_d^2 can be computed from these estimators by the invariance principle. The fact that there is no γ_θ in the restricted likelihood L^* implies that the REML estimators are easier to compute than the ML estimators and are also more stable.

31.18 We now consider a balanced two-way mixed model in which we observe Y_{ijk}, the kth observation in the ith row and the jth column, $i = 1, \ldots, r;\ j = 1, \ldots, c;\ k = 1, \ldots, n$. We assume that the columns are sampled from a larger class of columns, but that the rows are fixed. There are two models which are often used in this setting, depending on the structure of the interaction. We first present the 'unrestricted model' whose parameters and random variables are somewhat hard to interpret in practice, but which is easier to study. We then present the 'restricted model' whose parameters and random variables are more easily interpreted but which is more difficult to study directly. Rather than study this model directly, we show that it is essentially a reparametrization of the unrestricted model and use the results derived for the unrestricted model to establish results for the restricted model.

In the unrestricted mixed model, we assume that

$$Y_{ijk} = \theta + \alpha_i + b_j + d_{ij} + e_{ijk},$$

where θ and α_i are unknown parameters and b_j, d_{ij} and e_{ijk} are unobserved independent random variables such that

$$\sum \alpha_i = 0, \qquad b_j \sim N(0, \sigma_b^2), \qquad d_{ij} \sim N(0, \sigma_d^2), \qquad e_{ijk} \sim N(0, \sigma_e^2).$$

Note that the α_i are constants because they depend only on the row chosen, which is assumed fixed (non-random) in this model, but that b_j and d_{ij} depend on the randomly chosen column and are hence random variables. The parameter space for this model is

$$\theta \in R, \qquad \sum \alpha_i = 0, \qquad \sigma_b^2 \geq 0, \qquad \sigma_d^2 \geq 0, \qquad \sigma_e^2 > 0.$$

Testing for no row effect is testing that the $\alpha_i = 0$, testing for no column effect is testing that $\sigma_b^2 = 0$ and testing for no interaction effect is testing that $\sigma_d^2 = 0$.

By similar derivations to those for the two-way random effects model, it can be shown that for this mixed model

$$\frac{SS\theta}{\gamma_b} \sim \chi_1^2\left(\frac{ncr\theta^2}{\gamma_b}\right), \qquad \frac{SSa}{\gamma_d} \sim \chi_{r-1}^2\left(\frac{cn\sum\alpha_i^2}{\gamma_d}\right), \qquad \frac{SSb}{\gamma_b} \sim \chi_{c-1}^2,$$
$$\frac{SSd}{\gamma_d} \sim \chi_{(r-1)(c-1)}^2, \qquad \frac{SSe}{\sigma_e^2} \sim \chi_{rc(n-1)}^2, \qquad \gamma_d = n\sigma_d^2 + \sigma_e^2, \qquad \gamma_a = nr\sigma_b^2 + n\sigma_d^2 + \sigma_e^2$$

and that these sums of squares are independent (see Exercise 31.30). Therefore the ANOVA table for this model is

Balanced two-way crossed mixed model – unrestricted case

effect	*df*	*SS*	*EMS*
θ	1	$ncr\overline{Y}_{...}^2$	$\sigma_e^2 + n\sigma_d^2 + nr\sigma_b^2 + ncr\theta^2$
α_i	$r-1$	$nc\sum(\overline{Y}_{i..} - \overline{Y}_{...})^2$	$\sigma_e^2 + n\sigma_d^2 + h\sum\alpha_i^2$
b_j	$c-1$	$nr\sum(\overline{Y}_{.j.} - \overline{Y}_{...})^2$	$\sigma_e^2 + n\sigma_d^2 + nr\sigma_b^2$
d_{ij}	$(r-1)(c-1)$	$n\sum\sum(\overline{Y}_{ij.} - \overline{Y}_{i..} - \overline{Y}_{.j.} + \overline{Y}_{...})^2$	$\sigma_e^2 + n\sigma_d^2$
error	$rc(n-1)$	$\sum\sum\sum(Y_{ijk} - \overline{Y}_{ij.})^2$	σ_e^2
total	rcn	$\sum\sum\sum Y_{ijk}^2$	

where $h = nc/(r-1)$.

We can use the ANOVA table in the obvious way to test for no row effect, no column effect and no interaction effect. Note that all three of these tests are the same as for the random effects model, but the tests for row and column effects are different from those for the fixed effects model. We can also find pivotal quantities for many parameters (e.g., θ, σ_d^2/σ_e^2 and $\sigma_b^2/(n\sigma_d^2 + \sigma_e^2)$) but not for σ_b^2 or σ_d^2 which are usually the parameters of interest.

We now look at simultaneous confidence intervals for contrasts in the α_i. The Scheffé $1-\alpha$ simultaneous confidence intervals for contrasts in the α_i are given by

$$\sum a_i\alpha_i \in \sum a_i\overline{Y}_{i..} \pm \left((r-1)F_{r-1,(r-1)(c-1)}^{\alpha} MSd \sum\frac{a_i^2}{nc}\right)^{1/2}, \qquad \forall a_i \ni \sum a_i = 0$$

(see Exercise 31.31). Note that these intervals can be computed by substituting MSd and *dfd* for MSe and *dfe* in the formulae in the fixed effects model. In Exercise 31.32, it is shown that a similar result holds for the Tukey simultaneous confidence intervals.

Using Theorem 31.1 above, we see that the likelihood for this model is given by

$$L \approx \gamma_b^{-\frac{c}{2}} \gamma_d^{-\frac{c(r-1)}{2}} \sigma_e^{-rc(n-1)}$$
$$\times \exp\left(-\frac{ncr(\overline{Y} - \theta)^2 + SSb}{2\gamma_b} - \frac{nc\sum(\overline{Y}_{i..} - \overline{Y}_{...} - \alpha_i)^2 + SSd}{2\gamma_d} - \frac{SSe}{2\sigma_e^2}\right)$$

(see Exercise 31.30). It is straightforward to show that a complete sufficient statistic for this model is $(\overline{\mathbf{Y}}, SSb, SSd, SSe)$ where $\overline{\mathbf{Y}} = (\overline{Y}_{1..}, \ldots, \overline{Y}_{r..})$. Therefore $\overline{Y}_{i..}$, $\overline{Y}_{i..} - \overline{Y}_{...}$, MSb, MSd

and MSe are the MV unbiased estimators of $\theta + \alpha_i$, α_i, γ_b, γ_d and σ_e^2. MV unbiased estimators of σ_b^2 and σ_d^2 are given by

$$\frac{MSb - MSd}{nr} \quad \text{and} \quad \frac{MSd - MSe}{n}.$$

There are no non-negative unbiased estimators of these non-negative parameters. To find the ML estimator, we maximize the likelihood subject to the constraints

$$\sum \alpha_i = 0, \qquad \gamma_b \geq \gamma_d \geq \sigma_e^2 > 0.$$

It is easily seen that the ML estimators of θ and α_i are $\overline{Y}_{...}$ and $\overline{Y}_{i..} - \overline{Y}_{...}$. The ML estimators for the other parameters are straightforward to compute for a particular data set but are complicated to write down because of their constraints. To find the REML estimators for γ_a, γ_b and σ_e^2, we maximize the restricted likelihood of (SSb, SSd, SSe) given by

$$L^* \approx \gamma_b^{-\frac{c-1}{2}} \gamma_d^{-\frac{(r-1)(c-1)}{2}} \sigma_e^{-rc(n-1)} \exp\left(-\frac{SSb}{2\gamma_b} - \frac{SSd}{2\gamma_d} - \frac{SSe}{2\sigma_e^2}\right)$$

subject to the same constraints as above (see Exercise 31.30). We can use the invariance principle to find ML and REML estimators for σ_b^2 and σ_d^2.

31.19 We now consider a second form of the balanced two-way mixed model. In that model we assume, as above, that the rows are fixed ahead of the experiment but that the columns are chosen from some larger set of columns. Let M_{ij} be the mean of an observation in the ith row and the jth column chosen. Let μ_i be the mean of an observation in the ith row (averaged across all the potential columns) and let $P_j = \overline{M}_{.j}$ be the mean of an observation in the jth column chosen (averaged across only the rows in the experiment). Note that the μ_i are unobserved parameters but that the P_j are unobserved random variables. Let $\theta^* = \overline{\mu}_. = EP_j$ be the mean of an observation averaged across all potential columns and across all the rows in the experiment.

The assumptions we make for this model are are follows:

1. Given $\{M_{ij}, P_j\}$, the Y_{ijk} are independent, $Y_{ijk}|\{M_{ij}, P_j\} \sim N(M_{ij}, \sigma_e^{*2})$.
2. Given the P_j, the $(M_{1j} - \mu_1, \ldots, M_{rj} - \mu_r)$ are exchangeably jointly normally distributed, $M_{ij} - \mu_i|\{P_j\} \sim N(P_j - \theta^*, \frac{r-1}{r}\sigma_d^{*2})$. Furthermore, if $j \neq j'$, then M_{ij} and $M_{i'j'}$ are independent.
3. $P_1, \ldots, P_c$ are independent, $P_j \sim N(\theta^*, \sigma_b^{*2})$.

Note that we use stars on the parameters in this model to distinguish them from the parameters for the previous mixed model and that we have defined the conditional variance of the M_{ij} to be $(r-1)\sigma_d^{*2}/r$ to make the formulae work out more nicely below.

The parameter space for this model is

$$\boldsymbol{\mu} = (\mu_1, \ldots, \mu_r)^T \in R^r, \qquad \sigma_b^{*2} \geq 0, \qquad \sigma_d^{*2} \geq 0, \qquad \sigma_e^{*2} > 0.$$

Testing for no row effect is testing the equality of the μ_i. Testing for no column effect is testing that $\sigma_b^{*2} = 0$ (i.e., that the P_j are all equal). Testing for no interaction effect is testing that $\sigma_d^{*2} = 0$ (i.e. that $M_{ij} = \mu_i + P_j - \theta$).

Assumption 2 is somewhat less intuitive than the assumptions for other models. However, note that conditionally on P_j, the M_{ij} cannot be independent because $P_j = \overline{M}_{.j}$ and hence

$$\text{var}\Big(\sum_i M_{ij}|P_j\Big) = \text{var}(rP_j|P_j) = 0.$$

Note also that $EM_{ij} = \mu_i$, so that $M_{1j}, \ldots, M_{rj}$ could not be exchangeably distributed conditionally on the P_j. The conditional exchangeability of the $M_{ij} - \mu_i$ seems to be the weakest assumption possible for this model. The lack of intuitively clear assumptions for this model led Scheffé (1959, pp. 261–291) to reject both the restricted and unrestricted version of the mixed model discussed in this chapter and instead to suggest a multivariate alternative formulation.

To put the restricted model in a simpler form, let

$$\alpha_i^* = \mu_i - \theta^*, \qquad b_j^* = P_j - \theta^*, \qquad d_{ij}^* = M_{ij} - \mu_i - P_j + \theta^*, \qquad e_{ijk}^* = Y_{ijk} - M_{ij}.$$

By a proof similar to the one for the two-way random effects model, we can show the following lemma.

Lemma 31.3.

(a) $Y_{ijk} = \theta^* + \alpha_i^* + b_j^* + d_{ij}^* + e_{ijk}^*$.

(b) $\sum_i d_{ij}^* = 0$.

(c) Let $\mathbf{d}_j^* = (d_{1j}^*, \ldots, d_{rj}^*)$. *Then the* b_j^*, *the* $\mathbf{d}_j^*$ *and the* e_{ijk}^* *are all independent,*

$$b_j^* \sim N(0, \sigma_b^{*2}), \qquad e_{ijk}^* \sim N(0, \sigma_e^{*2}).$$

Furthermore, $d_{1j}^*, \ldots, d_{rj}^*$ *are exchangeably distributed*

$$d_{ij}^* \sim N\left(0, \frac{r-1}{r}\sigma_d^{*2}\right).$$

This model is called the restricted version of the mixed model because we have restricted the d_{ij}^* to sum to 0 over i. Note that this assumption guarantees that the d_{ij}^* are not independent. Also the constraint plus the assumed exchangeability guarantee that $\text{cov}(d_{ij}^*, d_{i'j}^*) = -\sigma_d^{*2}/r$ (see Exercise 31.33).

Now consider again the unrestricted version of the mixed model in which

$$Y_{ijk} = \theta + \alpha_i + b_j + d_{ij} + e_{ijk}, \qquad \sum \alpha_i = 0, \qquad \theta \in R,$$

where the b_j, d_{ij} and e_{ijk} are all independent, $b_j \sim N(0, \sigma_b^2)$, $d_{ij} \sim N(0, \sigma_d^2)$ and $e_{ijk} \sim N(0, \sigma_e^2)$. Let

$$\theta^* = \theta, \qquad \alpha_i^* = \alpha_i, \qquad b_j^* = b_j + \overline{d}_{.j}, \qquad d_{ij}^* = d_{ij} - \overline{d}_{.j}, \qquad e_{ijk}^* = e_{ijk}.$$

It is easily verified that

$$Y_{ijk} = \theta^* + \alpha_i^* + b_j^* + d_{ij}^* + e_{ijk}^*, \quad \sum \alpha_i^* = 0, \quad \theta^* \in R, \quad \sum_i d_{ij}^* = 0.$$

Also the b_j, the $\mathbf{d}_j^* = (d_{1j}^*, \ldots, d_{rj}^*)$ and the e_{ijk}^* are all independent, the $d_{1j}^*, \ldots, d_{rj}^*$ are exchangeable and

$$b_j^* \sim N\left(0, \sigma_b^2 + \frac{\sigma_d^2}{r}\right), \qquad d_{ij}^* \sim N\left(0, \frac{r-1}{r}\sigma_d^2\right), \qquad e_{ijk}^* \sim N(0, \sigma_e^2).$$

(see Exercise 31.34). Therefore the restricted and unrestricted versions of the crossed mixed model are just reparametrizations of the same model with

$$\theta^* = \theta, \quad \alpha_i^* = \alpha_i, \quad \sigma_b^{*2} = \sigma_b^2 + \frac{\sigma_d^2}{r}, \quad \sigma_d^{*2} = \sigma_d^2, \quad \sigma_e^{*2} = \sigma_e^2.$$

Therefore the only parameter which is different for these two models is the component of variance for the columns. However, the distributions of the sums of squares are the same as before with

$$\gamma_d = n\sigma_d^2 + \sigma_e^2 = n\sigma_d^{*2} + \sigma_e^{*2}, \qquad \gamma_b = nr\sigma_b^2 + n\sigma_d^2 + \sigma_e^2 = nr\sigma_b^{*2} + \sigma_e^{*2}.$$

Therefore the ANOVA table for this model is

Balanced two-way crossed mixed model-restricted case

effect	*df*	*SS*	*EMS*
θ^*	1	$ncr\overline{Y}_{...}^2$	$\sigma_e^{*2} + nr\sigma_b^{*2} + ncr\theta^{*2}$
α_i^*	$r-1$	$nc\sum(\overline{Y}_{i..} - \overline{Y}_{...})^2$	$\sigma_e^{*2} + n\sigma_d^{*2} + h\sum\alpha_i^{*2}$
b_j^*	$c-1$	$nr\sum(\overline{Y}_{.j.} - \overline{Y}_{...})^2$	$\sigma_e^{*2} + nr\sigma_b^{*2}$
d_{ij}^*	$(r-1)(c-1)$	$n\sum\sum(\overline{Y}_{ij.} - \overline{Y}_{i..} - \overline{Y}_{.j.} + \overline{Y}_{...})^2$	$\sigma_e^{*2} + n\sigma_d^{*2}$
error	$rc(n-1)$	$\sum\sum\sum(Y_{ijk} - \overline{Y}_{ij.})^2$	σ_e^{*2}
total	rcn	$\sum\sum\sum Y_{ijk}^2$	

where $h = nc/(r-1)$. Note that the df and SS columns are the same for the two versions of the two-way mixed model, as are the EMS for rows, interactions and error, but the EMS are different for the intercept and the columns.

Note that $\alpha_i^* = \alpha_i$ so that any inference for the row effects is the same for the two models. In particular, the Scheffé and Tukey simultaneous confidence intervals for contrasts in the row effects are the same for the two models. Similarly, $\sigma_d^{*2} = \sigma_d^2$, so that the hypotheses of no effect due to interaction are tested the same way in the two versions of the mixed model. However, the hypothesis of no effect due to columns is different for the two models and the F tests are different. To test that there is no column effect for the unrestricted model we use the same F statistic as for the random effects model, but for the restricted model we use the same F statistic as for the fixed effects model.

The likelihood and restricted likelihood (in terms of γ_b and γ_d) are also the same for the two models (with θ, α_i and σ_e^2 replaced by θ^*, α_i^* and σ_e^{*2}). Therefore $(\overline{\mathbf{Y}}, SSb, SSd, SSe)$ is a set of complete sufficient statistics for the restricted version also and MV unbiased estimators for most parameters can be determined immediately. In particular, the MV unbiased estimators for σ_b^{*2} and σ_d^{*2} are given by

$$\frac{MSb - MSe}{nr} \quad \text{and} \quad \frac{MSd - MSe}{n}.$$

There are no non-negative unbiased estimators for these non-negative parameters.

The parameter space for the restricted model is slightly larger than that for the unrestricted model, in that we only need

$$\gamma_b \geq \sigma_e^{*2}, \qquad \gamma_d \geq \sigma_e^{*2}$$

(i.e. we do not assume that $\gamma_b \geq \gamma_d$ for the restricted model). Hence the ML estimators and REML estimators may be different for the two models. Again, because of the complicated boundary, formulae are not practical but the estimators may computed easily for a particular data set. ML and REML estimators for σ_b^{*2} and σ_d^{*2} may be computed from the invariance principle.

We have derived the results first for the unrestricted mixed model because the derivations were easier. We then reinterpreted those results for the restricted mixed model. We now discuss why many people prefer the restricted version. Recall that for the fixed effects model, $\theta + \beta_j = \overline{\mu}_{.j}$, the average of the cell means within the jth column averaged over the rows in the experiment, and that for the random effects model, $\theta + b_j$ represented the average of the cells in the j th column averaged over all the potential rows in the experiment. The issue is what $\theta + b_j$ represents in the mixed model. It seems clear that it must represent the average of the cell means in the jth column averaged over the rows in the experiment, that is, $\theta + b_j = \overline{m}_{.j}$. (If, for example, the rows represented different diets there is typically not even a conceptual population from which the rows have been selected.) Using the constraint that $\sum \alpha_i = 0$, we see that $\overline{m}_{.j} = \theta + b_j + \overline{d}_{.j}$. Therefore

$$\theta + b_j = \overline{m}_{.j} \Leftrightarrow \sum_i d_{ij} = 0.$$

Hence the unrestricted version of the mixed model cannot be correct with this interpretation of $\theta + b_j$. In fact, in the unrestricted version of the mixed model it is very difficult to determine what the b_j represent. Therefore the restricted version is preferable to many people. At least for this version it is clear that $\theta + b_j$ represents the obvious mean for the jth column.

Because the only difference between the two versions of the two-way mixed model is in the assumptions about the interaction, it is often convenient to assume that the mixed model is additive, that is that

$$Y_{ijk} = \theta + \alpha_i + b_j + e_{ijk},$$

where θ and α_i are unknown parameters such that $\sum \alpha_i = 0$ and the b_j and e_{ij} are independent random variables such that $b_j \sim N(0, \sigma_b^2)$ and $e_{ijk} \sim N(0, \sigma_e^2)$. For this model, it is straightforward to show that the ANOVA table is given by

Balanced two-way crossed additive mixed model

Effect	*df*	*SS*	*EMS*
θ	1	$ncr\overline{Y}_{...}^2$	$\sigma_e^2 + nr\sigma_b^2 + ncr\theta^2$
α_i	$r - 1$	$nc\sum(\overline{Y}_{i..} - \overline{Y}_{...})^2$	$\sigma_e^2 + h\sum\alpha_i^2$
b_j	$c - 1$	$nr\sum(\overline{Y}_{.j.} - \overline{Y}_{...})^2$	$\sigma_e^2 + nr\sigma_b^2$
error	$nrc - r - c + 1$	$\sum\sum\sum(Y_{ijk} - \overline{Y}_{i..} - \overline{Y}_{.j.} + \overline{Y}_{...})^2$	σ_e^2
total	nrc	$\sum\sum\sum Y_{ijk}^2$	

where $h = nc/(r-1)$. Note that for the additive two-way model, the F test for no row effect is the same for the fixed effects model, the random effects model and the mixed model, as is the F test for no column effect.

Further comments

31.20 Procedures for random effects and mixed models are much less satisfying than for other models in this chapter. We have already seen that there are often no non-negative unbiased estimators or exact confidence intervals for the non-negative parameters of interest. We have also seen that ML estimators are often quite messy. The optimality of the F tests is not obvious. In fact, it appears that optimal procedures may not exist for many of these models.

For most other normal theory models, it has been shown that the procedures are not too dependent on the normal assumption, at least asymptotically. Unfortunately, procedures for random effects and mixed models are typically quite dependent on the normal assumption. To illustrate why this dependence occurs, consider the twofold nested mixed model in which we observe

$$Y_{ijk} = \theta + \alpha_i + d_{ij} + e_{ijk}, \qquad i = 1, \ldots, r;\ j = 1, \ldots, c;\ k = 1, \ldots, n,$$

where now the d_{ij} are a sample from an arbitrary distribution with mean 0 and variance σ_d^2 and the e_{ijk} are a sample from another arbitrary distribution with mean 0 and variance σ_e^2. We consider this model asymptotically as $n \to \infty$, but r and c are fixed. Then it is easily seen that

$$MSe = \frac{\sum\sum\sum(Y_{ijk} - \overline{Y}_{ij.})^2}{rc(n-1)} \xrightarrow{p} \sigma_e^2, \qquad \overline{e}_{ij.} \xrightarrow{p} 0.$$

Therefore, if $\sigma_d^2 > 0$, then

$$\frac{SSd}{n\sigma_d^2 + \sigma_e^2} = \frac{\sum\sum(d_{ij} + \overline{e}_{ij.} - \overline{d}_i - \overline{e}_{i..})^2}{\sigma_d^2 + \sigma_e^2/n} \xrightarrow{d} \frac{\sum\sum(d_{ij} - \overline{d}_i)^2}{\sigma_d^2},$$

which is not $\chi^2_{d(n-1)}$ unless the d_{ij} are normally distributed. (Note that if $\sigma_d^2 = 0$, then $d_{ij} = 0$ (almost surely) and $SSd/\sigma_e^2 \xrightarrow{d} \chi^2_{d(n-1)}$ so that the null distribution of F_d is asymptotically correct even if the alternative distribution is not.) Similarly, if $\sigma_d^2 > 0$, then

$$\frac{SSa}{n\sigma_d^2 + \sigma_e^2} \xrightarrow{d} \frac{\sum c(\alpha_i + \overline{d}_{i.} - \overline{d}_{..})^2}{\sigma_d^2},$$

which has a non-central χ^2 distribution only if the d_{ij} are normally distributed. Note that if the d_{ij} are not normally distributed then neither the numerator nor the denominator of $F_a = MSa/MSd$ has a χ^2 distribution (under either the null hypothesis that $\alpha_i = 0$ or the alternative hypothesis), indicating that the F test for testing that the fixed effect is 0 is asymptotically very sensitive to the distribution of the d_{ij}. (Note, furthermore, that the argument for independence of numerator and denominator sums of squares is only valid if the d_{ij} are normally distributed.)

Actually the other assumptions of a random effects or mixed model may not be appropriate when the observations are not normally distributed. For example, continuing in the twofold mixed model above, we might assume that, as before, $Y_{ij1}, \ldots, Y_{ijn}$ are a sample from a distribution with mean M_{ij} and variance σ_e^2 and that $M_{i1}, \ldots, M_{ic}$ are a sample from a distribution with mean μ_i and variance σ_d^2. If, as above, we let $d_{ij} = M_{ij} - \mu_i$ and $e_{ijk} = Y_{ijk} - M_{ij}$, it is not obvious that d_{ij} and e_{ijk} would be independent, but only that they are uncorrelated. In fact without the normal assumption, it seems that many of the independence assumptions should be replaced by lack of correlation. Similarly, in the derivations of the usual form of the two-way random effects model we established absence of correlation of many of the effects and used the normality to establish independence.

Repeated measures models

31.21 Many ANOVA experiments involve repeated measures on the same individual. We next consider models for such experiments. We assume that we have m individuals (persons, fields, dogs, etc.) on each of which we have r measurements. Let $\mathbf{Y}_k = (Y_{1k}, \ldots, Y_{rk})^T$ be the vector of observations on the kth individual. We assume that the $\mathbf{Y}_k$ are independent, $\mathbf{Y}_k \sim N_r(\boldsymbol{\mu}_k, \boldsymbol{\Sigma})$. We study two different classes of repeated measures models, depending on the assumptions we make about $\boldsymbol{\Sigma}$. If we only assume that $\boldsymbol{\Sigma} > 0$, we call the model a *multivariate repeated measures model*. In order to obtain more powerful procedures, we often make the strong assumption that

$$\text{var}(Y_{ik}) = \sigma^2; \quad \text{cov}(Y_{ik}, Y_{ik'}) = \rho\sigma^2, \quad k \neq k'.$$

That is, we assume that all the observations have the same variance and all pairs of observations on the same individual have the same correlation coefficient. In this case we say that the model is a *univariate repeated measures model*.

In univariate repeated measures models it is important to distinguish between two types of effect. We say that an effect is a *between* effect if it only involves the average effect on the individual, and a *within* effect if the average effect on each individual is 0. In order to obtain a model with exact procedures, we make the further strong assumption that all the effects in the model are either between effects or within effects.

As a simple example of a repeated measures experiment, consider an experiment to compare yields from five different varieties of wheat. We might choose 10 fields, divide each field into five sub-fields and randomly assign each of the varieties of wheat to one of the sub-fields of each field. This is an example of a one-way repeated measures model in which every individual (field) receives every treatment (wheat variety). We could extend this in a elementary way by including three fertilizers in the study, dividing the fields into 15 sub-fields and randomly assigning each variety–fertilizer pair to one sub-field of each field. This is an example of a two-way model in which every individual (field) receives every possible pair of row and column treatments. A similar example can be constructed of a nested design in which each individual (field) receives every subclass of every class. Note that for all the situations described in this paragraph each individual receives the same treatments and hence all the effects of interest are within effects. The only between effect is the overall average of the treatments. In addition, because every individual receives every possible treatment combination, all the treatment combinations must contain the same number of observations (i.e., the experiments must be balanced). We call this design a *single-group* repeated measures design because all the individuals receive the same treatments (and hence the individuals are sampled from a single group).

Now consider a different two-way repeated measures example. Suppose we want to compare the effects of six different exercise programmes on pulse rate. We are also interested in differences between men and women. We could find 20 men and 20 women, and give each person each of the six exercise programmes, and measure the pulse rate after each programme. We could consider the exercise programmes as the rows and the genders as the columns in a two-way model. In this experiment each individual receives every possible row treatment but only one column treatment. Note that in this situation the column effect is a between effect (because it is measuring differences between individuals), but that the row and interaction effects are within effects. We call the design described in this paragraph a *two-group* repeated measures design because the men and women are assumed different (i.e. because there are two independent groups of individuals). This design can be readily extended to a two-way design with r rows and c columns in which each individual receives each row treatment but only one column treatment, to give a *multiple-group* design.

In the previous example we could give different exercise programmes to the men and women, giving us a nested repeated measures model in which each individual receives one class (gender) but receives each subclass of that class. This design is again a two-group design because the men and women are considered different.

We also consider an ANCOVA example indicating that under certain conditions a covariate is a within effect, and under certain other conditions it is a between effect. If neither of these conditions is satisfied then the model cannot be analysed by the methods of this chapter.

Univariate repeated measures designs often arise in agricultural experiments in which an individual is a field which is split into plots which are given different treatments. Such designs are often called *split-plot* designs. For such designs, what we have called between effects are often called *whole-plot* effects and what we have called within effects are often called *sub-plot* effects.

For many years, univariate repeated measures models were analysed as mixed models (for which we have seen there is no general theory). The general approach for univariate repeated measures models given here follows Arnold (1981). In **31.28**, we indicate why the mixed model analysis is essentially the same as the repeated measures models analysis. Note that the slightly enlarged paramter space of the repeated measures model allows for a cleaner analysis for repeated measures models than for mixed models. Note also that the general approach to repeated measures models allows fairly quick derviations for many repeated measures models but each mixed model has to be analysed separately.

In **31.22** we present the basic theory for univariate repeated measures models together with the one-way model. In **31.23–26** we present further examples of univariate repeated measures models and in **31.27** we discuss repeated measures models with a covariate. In **31.28** we show the relationship between univariate repeated measures models and certain mixed models. In **31.29–30** we present some multivariate repeated measures models. In **31.31** we discuss tests for the validity of the univariate repeated measures model and the associated ANOVA model. In **31.32** we make some further comments about repeated measures models.

Univariate repeated measures – theory

31.22 In a repeated measures design, we observe an r-dimensional vector of responses on each of m individuals. Let $\mathbf{Y}_k = (Y_{1k}, \ldots, Y_{rk})^T$ be the vector of responses on the kth individual, $k = 1, \ldots, m$. We assume that the $\mathbf{Y}_k$ are independent, $\mathbf{Y}_k \sim N_r(\boldsymbol{\mu}_k, \boldsymbol{\Sigma})$, where

$\boldsymbol{\mu}_k = (\mu_{1k}, \ldots, \mu_{rk})^T$ is the vector of means for the kth individual and $\boldsymbol{\Sigma}$ is the common covariance matrix for the observations on a particular individual. In this section we further assume that

$$\boldsymbol{\Sigma} = \sigma^2 \begin{pmatrix} 1 & \rho & \cdots & \rho \\ \rho & 1 & \cdots & \rho \\ \vdots & \vdots & \ddots & \vdots \\ \rho & \rho & \cdots & 1 \end{pmatrix} = \sigma^2 \mathbf{A}_r(\rho)$$

for some unknown parameters σ^2 and ρ. That is, we assume that all the observations have the same variance σ^2 and all pairs of observations on the same individual have the same correlation coefficient ρ.

Before defining the remainder of the univariate repeated measures model we give some useful notation. Let $\mathbf{1}_k$ be a k-dimensional vector all of whose elements are 1. For any $p \times q$ matrix $\mathbf{Q}$, we define $\mathbf{Q} \otimes \mathbf{I}_k$ to be the $pk \times qk$ matrix

$$\mathbf{Q} \otimes \mathbf{I}_k = \begin{pmatrix} \mathbf{Q} & \mathbf{0} & \cdots & \mathbf{0} \\ \mathbf{0} & \mathbf{Q} & \cdots & \mathbf{0} \\ \vdots & \vdots & \ddots & \vdots \\ \mathbf{0} & \mathbf{0} & \cdots & \mathbf{Q} \end{pmatrix}.$$

($\mathbf{Q} \otimes \mathbf{I}_k$ is the *Kronecker product* of $\mathbf{Q}$ and $\mathbf{I}_k$.)

Let δ_k be the average of the components of $\boldsymbol{\mu}_k$ and let $\boldsymbol{\gamma}_k = \boldsymbol{\mu}_k - \delta_k \mathbf{1}_r$ be the vector of deviations of the components of $\boldsymbol{\mu}_k$ from their average. We think of the δ_k as representing effects *between* individuals and the $\boldsymbol{\gamma}_k$ as representing effects *within* the individual. Let

$$\mathbf{Y} = \begin{pmatrix} \mathbf{Y}_1 \\ \vdots \\ \mathbf{Y}_m \end{pmatrix}, \quad \boldsymbol{\mu} = E\mathbf{Y} = \begin{pmatrix} \boldsymbol{\mu}_1 \\ \vdots \\ \boldsymbol{\mu}_m \end{pmatrix}, \quad \boldsymbol{\delta} = \begin{pmatrix} \delta_1 \\ \vdots \\ \delta_m \end{pmatrix}, \quad \boldsymbol{\gamma} = \begin{pmatrix} \boldsymbol{\gamma}_1 \\ \vdots \\ \boldsymbol{\gamma}_m \end{pmatrix},$$

so that an equivalent version of the model is that

$$\mathbf{Y} \sim N_{mr}(\boldsymbol{\mu}, \sigma^2 \mathbf{A}_r(\rho) \otimes \mathbf{I}_m).$$

(Note that $\mathbf{Y}$, $\boldsymbol{\mu}$, and $\boldsymbol{\gamma}$ are $mr \times 1$, but that $\boldsymbol{\delta}$ is $m \times 1$.)

The parameter space we assume for this model is

$$\boldsymbol{\delta} \in S, \quad \boldsymbol{\gamma} \in T, \quad \sigma^2 > 0, \quad -\frac{1}{r-1} < \rho < 1.$$

where S is an s-dimensional subspace of R^m and T is a t-dimensional subspace of R^{mr}. That is, we are assuming that all the effects are either effects between the individuals or effects within the individuals. (The restrictions on σ^2 and ρ are equivalent to assuming that $\boldsymbol{\Sigma} = \sigma^2 \mathbf{A}_r(\rho) > 0$. See Exercise 31.35.) Finally, in order to find non-trivial procedures, we assume that

$$s < m, \quad t < r(m-1).$$

We call this model the *univariate* repeated measures model.

Let V be the subspace of possible values for $\boldsymbol{\mu}$. In Exercise 31.36(e), it is shown that

$$\dim(V) = s + t < mr.$$

By the *associated* ANOVA model we mean the model which occurs when we assume that $\rho = 0$ (i.e. that all the observations are independent) in the repeated measures model. That is, the associated ANOVA model is the model in which we observe

$$\mathbf{Y} \sim N_{mr}(\boldsymbol{\mu}, \sigma^2\mathbf{I}), \quad \boldsymbol{\mu} \in V, \quad \sigma^2 > 0.$$

The basic approach of this chapter is to take known results for the associated ANOVA model to generate results for the repeated measures model.

Let

$$SSE = \|\mathbf{P}_{V^\perp}\mathbf{Y}\|^2, \qquad dfe = mr - (s + t)$$

be the usual quantities for the associated ANOVA model. Let $\overline{Y}_k$ be the average of the observations on the kth individual, let $\overline{\mathbf{Y}} = (\overline{Y}_1, \ldots, \overline{Y}_m)^T$ be the vector of individual averages, and let

$$SSB = r\,\|\mathbf{P}_{S^\perp}\overline{\mathbf{Y}}\|^2, \quad SSW = SSE - SSB,$$
$$dfb = m - s, \quad dfw = dfe - dfb.$$

As usual, let $MSB = SSB/dfb$ and $MSW = SSW/dfw$. We call SSB and SSW the *between* and *within sum of squares*, we call dfb and dfw the *between* and *within degrees of freedom* and we call MSB and MSW the *between* and *within mean squares*. Note that we are assuming that

$$dfb = m - s > 0 \quad \text{and} \quad dfw = m(r - 1) - t > 0.$$

Let $\hat{\boldsymbol{\delta}} = (\hat{\delta}_1, \ldots, \hat{\delta}_k)^T$ be the LS estimator of $\boldsymbol{\delta}$ for the associated ANOVA model. In Exercise 31.36(g) it is shown that

$$SSB = r\,\|\overline{\mathbf{Y}} - \hat{\boldsymbol{\delta}}\|^2 = r\sum(\overline{Y}_k - \hat{\delta}_k)^2.$$

Therefore to find these new quantities for the repeated measures model we do the following:

- Compute δ_k, the average of the components in the mean vector for the kth individual, and $\overline{Y}_k$, the average of the observations for the kth individual. If an effect is completely in the δ_k (i.e., not in $\boldsymbol{\gamma}_k = \boldsymbol{\mu}_k - \delta_k\mathbf{1}_r$) the effect is a between effect, and if an effect is not in δ_k it is a within effect.
- Compute SSE, dfe and the LS estimator $\hat{\delta}_k$ for the associated ANOVA model.
- Let r be the number of observations on each individual, let m be the total number of individuals and let s be the number of linearly independent parameters in the δ_k. Then

$$dfb = m - s, \qquad dfw = dfe - dfb.$$

- Compute

$$SSB = r\sum(\overline{Y}_k - \hat{\delta}_k)^2, \qquad SSW = SSE - SSB.$$

In some later examples, there will be more than one subscript for an individual. However, these rules may still be used if we consider k to be a vector representing all the subscripts for the individual.

Example 31.5 (One-way repeated measures model: estimation)

Consider a one-way repeated measures model in which each individual receives every possible treatment. (Note that since each individual receives each treatment level, this model must be balanced.) Let Y_{ik} be the observation for the ith treatment on the kth individual, $i = 1, \ldots, r$; $k = 1, \ldots, m$. Then

$$\mathbf{Y}_k = \begin{pmatrix} Y_{1k} \\ \vdots \\ Y_{rk} \end{pmatrix}, \quad \boldsymbol{\mu}_k = E\mathbf{Y}_k = \begin{pmatrix} \theta + \alpha_1 \\ \vdots \\ \theta + \alpha_r \end{pmatrix},$$

where $\sum \alpha_i = 0$. Therefore δ_k, the average of the components of $\boldsymbol{\mu}_k$, is given by

$$\delta_k = \theta \Rightarrow \boldsymbol{\gamma}_k = \begin{pmatrix} \alpha_1 \\ \vdots \\ \alpha_r \end{pmatrix}.$$

Hence, we see that θ is a between effect and the α_i are a within effect for this model. Furthermore, for the associated one-way ANOVA model we see that

$$SSE = \sum\sum(Y_{ik} - \overline{Y}_{i.})^2, \quad dfe = r(m-1).$$

Now the average of the observations on the kth individual is $\overline{Y}_{.k}$ and the LS estimator of $\delta_k = \theta$ is $\hat{\delta}_k = \hat{\theta} = \overline{Y}_{..}$. Hence $SSE = SSW + SSB$, where

$$SSB = r\sum(\overline{Y}_{.k} - \overline{Y}_{..})^2,$$
$$SSW = \sum\sum(Y_{ik} - \overline{Y}_{i.})^2 - r\sum(\overline{Y}_{.k} - \overline{Y}_{..})^2 = \sum\sum(Y_{ik} - \overline{Y}_{i.} - \overline{Y}_{.k} + \overline{Y}_{..})^2.$$

Furthermore, there is one θ, so that $s = 1$. Therefore

$$dfb = m - 1, \quad dfw = dfe - dfb = r(m-1) - (m-1) = (r-1)(m-1).$$

We now return to the general (univariate) repeated measures model defined above. We define two important parameters for this model. Let

$$\tau_b^2 = \sigma^2(1 + (r-1)\rho), \quad \tau_w^2 = \sigma^2(1-\rho).$$

As we shall see below, these parameters are the variances for between and within effects.

Before stating the basic results for the univariate repeated measures model, we transform it to a simpler model. Let $\mathbf{C}$ be an $(r-1) \times r$ matrix such that

$$\mathbf{B} = \begin{pmatrix} r^{-\frac{1}{2}}\mathbf{1}^T \\ \mathbf{C} \end{pmatrix}$$

is an orthogonal matrix. (The matrix $\mathbf{B}$ is called a *Helmert* matrix.) Let

$$\mathbf{F} = r^{-\frac{1}{2}}\mathbf{1}^T \otimes \mathbf{I}_m, \quad \mathbf{G} = \mathbf{C} \otimes \mathbf{I}_m, \quad \mathbf{\Gamma} = \begin{pmatrix} \mathbf{F} \\ \mathbf{G} \end{pmatrix}.$$

It is easily verified that $\mathbf{\Gamma}$ is an $mr \times mr$ orthogonal matrix. Further, let

$$\mathbf{Z} = \begin{pmatrix} \mathbf{Z}_1 \\ \mathbf{Z}_2 \end{pmatrix} = \mathbf{\Gamma}\mathbf{Y} = \begin{pmatrix} \mathbf{FY} \\ \mathbf{GY} \end{pmatrix}, \quad \boldsymbol{\mu}^* = \mathbf{\Gamma}\boldsymbol{\mu}, \quad \boldsymbol{\delta}^* = r^{1/2}\boldsymbol{\delta}, \quad \boldsymbol{\gamma}^* = \mathbf{G}\boldsymbol{\gamma}$$

(where $\mathbf{Z}_1 = \mathbf{FY}$ and $\mathbf{Z}_2 = \mathbf{GY}$ are $m \times 1$ and $m(r-1) \times 1$) and let T^* and V^* be the sets of possible values for $\boldsymbol{\gamma}^*$ and $\boldsymbol{\mu}^*$. (Note that the space of possible values for $\boldsymbol{\delta}^*$ is still S.)

The following result is the result which makes the univariate repeated measures models work.

Lemma 31.4.

(a) *$\mathbf{Z}_1$ and $\mathbf{Z}_2$ are independent,*

$$\mathbf{Z}_1 \sim N_m(\boldsymbol{\delta}^*, \tau_b^2\mathbf{I}), \quad \mathbf{Z}_2 \sim N_{m(r-1)}(\boldsymbol{\gamma}^*, \tau_w^2\mathbf{I}).$$

(b) *$T^* \subset R^{m(r-1)}$ is a t-dimensional subspace and*

$$(\boldsymbol{\delta} \in S, \boldsymbol{\gamma} \in T) \Leftrightarrow (\boldsymbol{\delta}^* \in S, \boldsymbol{\gamma}^* \in T^*)$$
$$\left(\sigma^2 > 0, -\frac{1}{r-1} < \rho < 1\right) \Leftrightarrow (\tau_b^2 > 0, \tau_w^2 > 0).$$

(c) *In addition,*

$$SSB = \|\mathbf{P}_{S^\perp}\mathbf{Z}_1\|^2, \quad SSW = \|\mathbf{P}_{T^{*\perp}}\mathbf{Z}_2\|^2,$$
$$dfb = \dim(S^\perp), \quad dfw = \dim(T^{*\perp}).$$

Proof. (a) In Exercise 31.35, it is shown that

$$\mathbf{\Gamma}\boldsymbol{\mu} = \begin{pmatrix} \boldsymbol{\delta}^* \\ \boldsymbol{\gamma}^* \end{pmatrix}, \quad \mathbf{\Gamma}(\sigma^2\mathbf{A}(\rho) \otimes \mathbf{I}_m)\mathbf{\Gamma}^T = \begin{pmatrix} \tau_b^2\mathbf{I}_m & \mathbf{0} \\ \mathbf{0} & \tau_w^2\mathbf{I}_{m(r-1)} \end{pmatrix}$$

and the result follows.

(b) In Exercise 31.36b, it is shown that T^* is a t-dimensional subspace. The remainder of this part follows directly from the definitions.

(c) By the definition of $\mathbf{F}$, we see that $\mathbf{Z}_1 = r^{1/2}\overline{\mathbf{Y}}$ and hence

$$SSB = r\|\mathbf{P}_{S^\perp}\overline{\mathbf{Y}}\|^2 = \|\mathbf{P}_{S^\perp}\mathbf{Z}_1\|^2.$$

Note also that

$$\mathbf{P}_V = \mathbf{\Gamma}^T\mathbf{P}_{V^*}\mathbf{\Gamma}, \quad \mathbf{P}_{V^*} = \begin{pmatrix} \mathbf{P}_S & \mathbf{0} \\ \mathbf{0} & \mathbf{P}_{T^*} \end{pmatrix}$$

(see Exercise 31.36). Therefore

$$\|\mathbf{P}_V\mathbf{Y}\|^2 = \|\mathbf{P}_{V^*}\mathbf{Z}\|^2 = \|\mathbf{P}_S\mathbf{Z}_1\|^2 + \|\mathbf{P}_{T^*}\mathbf{Z}_2\|^2.$$

Hence,

$$\begin{aligned} SSE &= \|\mathbf{Y}\|^2 - \|\mathbf{P}_V\mathbf{Y}\|^2 = \|\mathbf{Z}\|^2 - \|\mathbf{P}_{V^*}\mathbf{Z}\|^2 = \|\mathbf{Z}_1\|^2 + \|\mathbf{Z}_2\|^2 - \|\mathbf{P}_S\mathbf{Z}_1\|^2 - \|\mathbf{P}_{T^*}\mathbf{Z}_2\|^2 \\ &= \|\mathbf{P}_{S^\perp}\mathbf{Z}_1\|^2 + \|\mathbf{P}_{T^{*\perp}}\mathbf{Z}_2\|^2 = SSB + \|\mathbf{P}_{T^{*\perp}}\mathbf{Z}_2\|^2, \end{aligned}$$

implying that

$$SSW = SSE - SSB = \|\mathbf{P}_{T^{*\perp}}\mathbf{Z}_2\|^2.$$

Similarly,

$$\begin{aligned} dfb &= m - s = m - \dim(S) = \dim(S^\perp), \quad dfe = \dim(V^\perp) = mr - \dim(V) \\ &= mr - (s + t) = m - s + m(r-1) - t = dfb + \dim(T^{*\perp}). \end{aligned}$$

(Notice that $T^* \subset R^{m(r-1)}$.) Therefore $dfw = dfe - dfb = \dim(T^{*\perp})$.

What we have done with this lemma is transform the univariate repeated measures model into a model which is really two separate linear models, one involving $\mathbf{Z}_1$, $\boldsymbol{\delta}^*$ and τ_b^2 and one involving $\mathbf{Z}_2$, $\boldsymbol{\gamma}^*$ and τ_w^2. Note that part (b) of the lemma says that the parameter space for the transformed model is

$$\boldsymbol{\delta}^* \in S, \quad \tau_b^2 > 0, \quad \boldsymbol{\gamma}^* \in T^*, \quad \tau_w^2 > 0,$$

which implies that the parameters for the two models are unrelated. We call the model involving $\mathbf{Z}_1$ the *between* linear model because it contains the between effects ($\boldsymbol{\delta}^*$) and call the model involving $\mathbf{Z}_2$ the *within* linear model because it contains the within effects ($\boldsymbol{\gamma}^*$). Note that part (c) implies that *SSB* and *SSW* are the error sums of squares for these two models and *dfb* and *dfw* are the degrees of freedom for error for these two models. Hence

$$EMSB = \tau_b^2, \quad EMSW = \tau_w^2.$$

Note also that for the transformed model the likelihood can be written as

$$L_{\mathbf{Z}}(\boldsymbol{\delta}^*, \tau_b^2, \boldsymbol{\gamma}^*, \tau_w^2) = L_{\mathbf{Z}_1}(\boldsymbol{\delta}^*, \tau_b^2) L_{\mathbf{Z}_2}(\boldsymbol{\gamma}^*, \tau_w^2).$$

This representation can be used to derive a complete sufficient statistic, MV unbiased and ML estimators for many parameters (see Exercises 31.37–38).

In most repeated measures settings, interest centres on tests and simultaneous confidence intervals for the between and within effects and the correlation coefficient ρ. The following theorem summarizes those procedures. Let Q be a subspace of S. We consider testing that $\boldsymbol{\delta} \in Q$ against the alternative that $\boldsymbol{\delta} \in S$. Similarly, we consider testing that $\boldsymbol{\gamma} \in P$, a subspace of T, against $\boldsymbol{\gamma} \in T$.

Theorem 31.2.

(*a*) Between effects. *Consider testing the null hypothesis that* $\boldsymbol{\delta} \in Q$ *against the alternative hypothesis that* $\boldsymbol{\delta} \in S$. *Let MSH and dfh be the mean square and degrees of freedom for the hypothesis in the associated ANOVA model. Let*

$$F_B = \frac{MSH}{MSB}.$$

Then, for this hypothesis,

$$F_B \sim F_{dfh,dfb}\left(\frac{\nu}{\tau_b^2}\right), \qquad EMSH = \tau_b^2 + \frac{\nu}{dfh},$$

where ν/σ^2 *is the non-centrality parameter and* $\sigma^2 + \nu/dfh$ *is the EMS for the associated ANOVA model. The size-*α *LR test for this model rejects if*

$$F_B > F^{\alpha}_{dfh,dfb}.$$

The Scheffé simultaneous confidence intervals for this hypothesis for the repeated measures model are computed by replacing MSE and dfe with MSB and dfb in the formulae for the associated ANOVA model.

(*b*) Within effects. *Consider testing the null hypothesis* $\boldsymbol{\gamma} \in P$ *against the alternative that* $\boldsymbol{\gamma} \in T$. *Let MSH and dfh be the mean square and degrees of freedom for the hypothesis in the associated ANOVA model. Let*

$$F_W = \frac{MSH}{MSW}.$$

For this hypothesis

$$F_W \sim F_{dfh,dfw}\left(\frac{\nu}{\tau_w^2}\right), \qquad EMSH = \tau_w^2 + \frac{\nu}{dfh},$$

where ν/σ^2 *is the non-centrality parameter and* $\sigma^2 + \nu/dfh$ *is the EMS for the associated ANOVA model. The size-*α *LR test for this problem rejects if*

$$F_W > F^{\alpha}_{dfh,dfw}.$$

The Scheffé simultaneous confidence intervals for this hypothesis are computed by replacing MSE and dfe with MSW and dfw in the formulae for the associated ANOVA model.

(*c*) *A pivotal quantity for* ρ *is given by*

$$\frac{F}{\psi} \sim F_{dfb,dfw}, \quad F = \frac{MSB}{MSW}, \quad \psi = \frac{1 + (r-1)\rho}{1-\rho}.$$

Proof. (a) We look first at the between linear model in which we observe $\mathbf{Z}_1 \sim N_m(\boldsymbol{\delta}^*, \tau_b^2\mathbf{I})$. Note that we are testing that $\boldsymbol{\delta}^* \in Q$ against $\boldsymbol{\delta}^* \in S$. Let $q = \dim(Q)$,

$$SSHB = \|\mathbf{P}_{S|Q}\mathbf{Z}_1\|^2, \qquad dfhb = s - q, \qquad MSHB = \frac{SSHB}{dfhb}.$$

Then the LR test for the between linear model (involving only $\mathbf{Z}_1$) rejects if

$$F^B = \frac{MSHB}{MSB} > F^{\alpha}_{dfhb,dfb}.$$

(Recall that *MSB* and *dfb* are mean square and degrees of freedom for error for the between linear model.) To see that this test is the LR test for the joint model (involving $\mathbf{Z}_1$ and $\mathbf{Z}_2$), note that by the lemma above, the LR test statistic for the joint problem is

$$\begin{aligned}\Lambda &= \frac{\sup_{\boldsymbol{\delta}^* \in Q, \tau_b^2>0} L_{\mathbf{Z}_1}(\boldsymbol{\delta}^*, \tau_b^2) \sup_{\boldsymbol{\gamma}^* \in T, \tau_w^2>0} L_{\mathbf{Z}_2}(\boldsymbol{\gamma}^*, \tau_w^2)}{\sup_{\boldsymbol{\delta}^* \in S, \tau_b^2>0} L_{\mathbf{Z}_1}(\boldsymbol{\delta}^*, \tau_b^2) \sup_{\boldsymbol{\gamma}^* \in T, \tau_w^2>0} L_{\mathbf{Z}_2}(\boldsymbol{\gamma}^*, \tau_w^2)} \\ &= \frac{\sup_{\boldsymbol{\delta}^* \in Q, \tau_b^2>0} L_{\mathbf{Z}_1}(\boldsymbol{\delta}^*, \tau_b^2)}{\sup_{\boldsymbol{\delta}^* \in S, \tau_b^2>0} L_{\mathbf{Z}_1}(\boldsymbol{\delta}^*, \tau_b^2)},\end{aligned}$$

the same as the LR test statistic for the between problem. By known results for the between linear model,

$$F^B \sim F_{dfhb,dfb}\left(\frac{\|\mathbf{P}_{S|Q}\boldsymbol{\delta}^*\|^2}{\tau_b^2}\right), \qquad EMSHB = \tau_b^2 + \frac{\|\mathbf{P}_{S|Q}\boldsymbol{\delta}^*\|}{dfhb}.$$

Now, let W be the subspace of possible values for $\boldsymbol{\mu}$ when $\boldsymbol{\delta} \in Q$, $\boldsymbol{\gamma} \in T$. For the associated ANOVA model, $SSH = \|\mathbf{P}_{V|W}\mathbf{Y}\|$, $dfh = \dim(V|W)$, the non-centrality parameter and EMS are given by $\|\mathbf{P}_{V|W}\boldsymbol{\mu}\|^2/\sigma^2$ and $\sigma^2 + \|\mathbf{P}_{V|W}\boldsymbol{\mu}\|^2/dfh$. Therefore we need to show that

$$SSHB = SSH, \qquad dfhb = dfh, \qquad \|\mathbf{P}_{S|Q}\boldsymbol{\delta}^*\|^2 = \|\mathbf{P}_{V|W}\boldsymbol{\mu}\|^2.$$

By calculations in the previous proof, we see that

$$\|\mathbf{P}_V\mathbf{Y}\|^2 = \|\mathbf{P}_S\mathbf{Z}_1\|^2 + \|\mathbf{P}_{T^*}\mathbf{Z}_2\|^2.$$

A similar result holds for $\|\mathbf{P}_W\mathbf{Y}\|^2$. Therefore

$$\begin{aligned}SSH &= \|\mathbf{P}_{V|W}\mathbf{Y}\|^2 = \|\mathbf{P}_V\mathbf{Y}\|^2 - \|\mathbf{P}_W\mathbf{Y}\|^2 \\ &= \|\mathbf{P}_S\mathbf{Z}_1\|^2 + \|\mathbf{P}_{T^*}\mathbf{Z}_2\|^2 - (\|\mathbf{P}_Q\mathbf{Z}_1\|^2 + \|\mathbf{P}_{T^*}\mathbf{Z}_2\|^2) \\ &= \|\mathbf{P}_S\mathbf{Z}_1\|^2 - \|\mathbf{P}_Q\mathbf{Z}_1\|^2 = \|\mathbf{P}_{S|Q}\mathbf{Z}_1\|^2 = SSHB.\end{aligned}$$

Similarly,

$$\|\mathbf{P}_{V|W}\boldsymbol{\mu}\|^2 = \|\mathbf{P}_{S|Q}\boldsymbol{\delta}^*\|^2.$$

Finally, $\dim(V) = s + t$. Similarly, $\dim(W) = q + t$. Therefore,

$$dfh = \dim(V) - \dim(W) = s + t - (q + t) = s - q = dfhB.$$

The Scheffé simultaneous confidence intervals for the between model are given by

$$\mathbf{d}^{*T}\boldsymbol{\delta}^* \in \mathbf{d}^{*T}\hat{\boldsymbol{\delta}}^* \pm (dfh F^{\alpha}_{dfh,dfb}\|\mathbf{d}^*\|^2 MSB), \ \forall \mathbf{d}^* \in S|Q.$$

For the associated linear model, the intervals are

$$\mathbf{d}^T\boldsymbol{\mu} \in \mathbf{d}^T\hat{\boldsymbol{\mu}} \pm (dfh F^{\alpha}_{dfh,dfe}\|\mathbf{d}\|^2 MSE), \ \forall \mathbf{d} \in V|W.$$

Now let $\mathbf{d} \in R^{mr}$ and let $\mathbf{c} = \boldsymbol{\Gamma}\mathbf{d}$. In Exercise 31.36(h) it is shown that

$$\mathbf{d} \in V|W \Leftrightarrow \mathbf{c} = \begin{pmatrix} \mathbf{d}^* \\ \mathbf{0} \end{pmatrix}, \qquad \mathbf{d}^* \in S|Q.$$

Furthermore, part (a) of the lemma above implies that

$$\boldsymbol{\mu}^* = \boldsymbol{\Gamma}\boldsymbol{\mu} = E\mathbf{Z} = \begin{pmatrix} \boldsymbol{\delta}^* \\ \boldsymbol{\gamma}^* \end{pmatrix}.$$

Therefore, if $\mathbf{d} \in V|W$,

$$\mathbf{d}^T\boldsymbol{\mu} = \mathbf{c}^T\boldsymbol{\mu}^* = \mathbf{d}^{*T}\boldsymbol{\delta}^*, \quad \mathbf{d}^T\hat{\boldsymbol{\mu}} = \mathbf{d}^{*T}\hat{\boldsymbol{\delta}}^*, \quad \|\mathbf{d}\|^2 = \|\mathbf{c}\|^2 = \|\mathbf{d}^*\|^2,$$

and hence the intervals for the repeated measures model may be computed from those for the associated ANOVA model by replacing *MSE* and *dfe* by *MSB* and *dfb* in the formulae for the associated ANOVA model.

(b) This is essentially the same as part (a) and is derived in Exercise 31.39.

(c) Since $\mathbf{Z}_1$ and $\mathbf{Z}_2$ are independent, *SSB* and *SSW* are independent. Also

$$\frac{SSB}{\tau_b^2} = \frac{SSB}{\sigma^2(1+(r-1)\rho)} \sim \chi^2_{dfb}, \qquad \frac{SSW}{\tau_w^2} = \frac{SSW}{\sigma^2(1-\rho)} \sim \chi^2_{dfw}.$$

This result follows directly.

Parts (a) and (b) of this theorem are quite easy to remember.

- To find the F test or set of simultaneous confidence intervals for contrasts for a between or within effect we merely take the appropriate formula for the associated ANOVA model and replace *MSE* and *dfe* by *MSB* and *dfb* or *MSW* and *dfw* depending on whether the effect is a between or within effect.
- To find the expected mean square or non-centrality parameter, we take the formula from the associated ANOVA model and replace σ^2 with τ_b^2 or τ_w^2, again depending on whether the effect is a between or within effect.

The F tests given in parts (a) and (b) are the UMP invariant size-α tests and the confidence regions associated with the simultaneous confidence intervals are UMA equivariant $1-\alpha$ confidence region. They are also admissible and have certain minimaxity properties (see Arnold, 1981, pp. 220–230 for details).

Example 31.6 (One-way repeated measures model: inference)

We return to the one-way example in which each individual receives all possible treatments. In that case we have seen that the α_i are a within effect. For the associated ANOVA model

$$SS\alpha = m\sum(\overline{Y}_{i.} - \overline{Y}_{..})^2, \quad df\alpha = r - 1.$$

Therefore the F test for testing that the $\alpha_i = 0$ in the univariate repeated measures model rejects if

$$\frac{m\sum(\overline{Y}_{i.} - \overline{Y}_{..})^2}{(r-1)MSW} > F^{\alpha}_{r-1,dfw}.$$

Similarly the Scheffé simultaneous confidence intervals are given by

$$\sum c_i\alpha_i \in \sum c_i\overline{Y}_{i.} \pm \left((r-1)F^{\alpha}MSW\sum\frac{c_i^2}{m}\right)^{1/2}, \qquad \forall c_i \ni \sum c_i = 0.$$

In Exercise 31.41 it is also shown that

$$\frac{\max|\overline{Y}_{i.} - \overline{Y}_{j.} - \alpha_i - \alpha_j|}{\sqrt{MSW/m}} \sim q_{r,dfw},$$

so that we can also compute Tukey simultaneous confidence intervals by substituting MSW and *dfw* for *MSE* and *dfe* in the formulae for the associated model. Note that Dunnett, MCB, multiple range and multiple F tests are also available for this model.

We often make an ANOVA table for repeated measures models. In that table we list first the between effects then the between sum of squares then the within effects and then the within sum of squares and finally a row for the total sum of squares and degrees of freedom. (All the associated ANOVA models considered in this chapter are orthogonal designs so that the sums of squares add to the total sum of squares). For the one-way model above, we obtain

One-way repeated measures model

effect	*df*	*SS*	*EMS*
θ	1	$mr\overline{Y}_{..}^2$	$\tau_b^2 + mr\theta^2$
between	$m-1$	$r\sum(\overline{Y}_{.k} - \overline{Y}_{..})^2$	τ_b^2
α	$r-1$	$m\sum(\overline{Y}_{i.} - \overline{Y}_{..})^2$	$\tau_w^2 + h\sum\alpha_i^2$
within	$(r-1)(m-1)$	$\sum\sum(Y_{ik} - \overline{Y}_{i.} - \overline{Y}_{.k} + \overline{Y}_{..})^2$	τ_w^2
total	rm	$\sum\sum Y_{ik}^2$	

where $h = m/(r-1)$. Note also that

$$\rho = 0 \Leftrightarrow \tau_b^2 = \tau_w^2,$$

and the test for this hypothesis can be read off the ANOVA table in the obvious way.

In this section we have used three different linear models: the associated ANOVA model, the between linear model and the within linear model. However, we use the between and within linear models only to derive the theorem. We shall not use them again in deriving procedures for further repeated measures models. In future sections we shall only need results for the associated ANOVA model. Also much of the notation of this section is used only in the derivations of the results, but is not necessary for their application in later sections. In particular, it is not necessary to find a particular Helmert matrix, to compute $\boldsymbol{\mu}^*$, $\boldsymbol{\delta}^*$ or $\boldsymbol{\gamma}^*$ or to find the subspaces T^* or V^*. In going from the associated ANOVA model to the repeated measures model, we replace the variance σ^2 by either τ_b^2 or τ_w^2, replace *SSE* (*MSE*) by *SSB* (*MSB*) or *SSW* (*MSW*) and replace *dfe* by *dfb* or *dfw* depending on whether the inference is for a between or within effect.

We now present an alternative interpretation for the procedures for between effects. Note that

$$\overline{\mathbf{Y}} \sim N_m(\boldsymbol{\delta}, \varpi^2\mathbf{I}), \qquad \boldsymbol{\delta} \in S,\ \varpi^2 > 0$$

(where $\varpi^2 = \text{var}(\overline{Y}_{.k}) = \tau_1^2/r$). This model is also a linear model. Consider testing that $\boldsymbol{\delta} \in Q$, a q-dimensional subspace of S. For this reduced linear model, we see that the appropriate test statistic is

$$F^* = \frac{\|\mathbf{P}_{S|Q}\overline{\mathbf{Y}}\|^2/(s-q)}{\|\mathbf{P}_{S^\perp}\overline{\mathbf{Y}}\|^2/(m-s)} \sim F_{s-q,m-s}\left(\frac{\|\mathbf{P}_{S|Q}\boldsymbol{\delta}\|^2}{\varpi^2}\right).$$

In Exercise 31.42 it is shown that this statistic is the same as the statistic derived in the theorem above for this hypothesis. In a similar way we could compute the simultaneous confidence intervals for contrasts in the between effects from the model reduced to $\overline{\mathbf{Y}}$ and we would get the same answer as above. In other words, one way to look at tests and simultaneous confidence intervals for between effects is that to gain independence we first replace the vector of observations on each individual by the average value for each individual. These averages are then just a smaller ANOVA model containing only between effects and we test hypotheses and find the confidence intervals for these effects in the obvious way using these averages. These averages are not a sufficient statistic so that it is not obvious that we do not lose information in reducing to the averages. One implication of the theorem above is that for drawing inference about between effects no such information is lost. (However, all the information about the within effects is lost by this reduction.) Unfortunately, there is no similar interpretation for tests about within effects (unless $r = 2$, when we can take the differences to get a linear model in the within effects).

Now suppose that the $\mathbf{Y}_k \sim N_r(\boldsymbol{\mu}_k, \boldsymbol{\Sigma})$ for an arbitrary matrix $\boldsymbol{\Sigma} > 0$. Then $\overline{\mathbf{Y}} \sim N_m(\boldsymbol{\delta}, \varpi^2\mathbf{I})$, where now $\varpi^2 = \text{var}(\overline{Y}_{.k}) = \mathbf{1}_r^T\boldsymbol{\Sigma}\mathbf{1}_r/r^2$. However, this model is still a linear model so that the procedure described in the previous paragraph is still appropriate. Therefore the inference for the between effects is not sensitive to the assumption that $\boldsymbol{\Sigma} = \sigma^2\mathbf{A}(\rho)$, but is valid for any (common) covariance structure on the repeated measures model. In fact, it can be shown that the tests and confidence intervals for between effects are the same for the univariate and multivariate repeated measures model.

Examples

31.23 Now consider a two-way model in which every individual receives every possible row and column treatment. We call this the single-group two-way repeated measures model because every individual receives the same treatments. (Note that since each individual receives each row and column treatment, the associated ANOVA model must be balanced.) Let Y_{ijk} be the observation in the ith row and the jth column for the kth individual, $i = 1, \ldots, d$; $j = 1, \ldots, c$;

$k = 1, \ldots, n$. Let $\mathbf{Y}_k = (Y_{11k}, \ldots, Y_{dck})^T$ be the vector of observations on the kth individual and let $\boldsymbol{\mu}_k = E\mathbf{Y}_k$. We assume that

$$EY_{ijk} = \theta + \alpha_i + \beta_j + \gamma_{ij} \Rightarrow \boldsymbol{\mu}_k = (\theta + \alpha_1 + \beta_1 + \gamma_{11}, \ldots, \theta + \alpha_d + \beta_c + \gamma_{dc})^T,$$
$$\sum_i \alpha_i = 0, \quad \sum_j \beta_j = 0, \quad \sum_i \gamma_{ij} = 0, \quad \sum_j \gamma_{ij} = 0.$$

(Note that $\boldsymbol{\mu}_k$ does not depend on k because the model is a single-group model.) In Exercise 31.43, it is shown that for this model $\delta_k = \theta$ (so that θ is the only between effect) and that

$$r = cd, \quad m = n, \quad s = 1$$
$$SSB = dc \sum (\overline{Y}_{..k} - \overline{Y}_{...})^2,$$
$$SSW = \sum\sum\sum (Y_{ijk} - \overline{Y}_{ij.} - \overline{Y}_{..k} + \overline{Y}_{...})^2,$$
$$dfb = n - 1, \quad dfw = (dc - 1)(n - 1).$$

Therefore the ANOVA table is given by

Single group two-way univariate repeated measures model

effect	*df*	*SS*	*EMS*
θ	1	$ndc\overline{Y}_{...}^2$	$\tau_b^2 + ndc\theta^2$
between	$n-1$	$dc \sum (\overline{Y}_{..k} - \overline{Y}_{...})^2$	τ_b^2
α	$d-1$	$nc \sum (\overline{Y}_{i..} - \overline{Y}_{...})^2$	$\tau_w^2 + h \sum \alpha_i^2$
β	$c-1$	$nd \sum (\overline{Y}_{.j.} - \overline{Y}_{...})^2$	$\tau_w^2 + g \sum \beta_j^2$
γ	$(d-1)(c-1)$	$n \sum\sum (\overline{Y}_{ij.} - \overline{Y}_{i..} - \overline{Y}_{.j.} + \overline{Y}_{...})^2$	$\tau_w^2 + f \sum\sum \gamma_{ij}^2$
within	$(dc-1)(n-1)$	$\sum\sum\sum (Y_{ijk} - \overline{Y}_{ij.} - \overline{Y}_{..k} + \overline{Y}_{...})^2$	τ_w^2
total	ndc	$\sum\sum\sum Y_{ijk}^2$	

where $h = nc/(d-1)$, $g = nd/(c-1)$ and $f = n/(d-1)(c-1)$. We can test for no row effects, no column effects or no interaction effects in the obvious way. We can also test that $\tau_b^2 = \tau_w^2$ (or equivalently that $\rho = 0$). Scheffé simultaneous confidence intervals for contrast in the α_i are given by

$$\sum a_i \alpha_i \in \sum a_i \overline{Y}_{i..} \pm \left((d-1)F^\alpha MSW \sum \frac{a_i^2}{nc}\right)^{1/2}, \qquad \forall a_i \ni \sum a_i = 0.$$

(Note that MSW can be computed from the ANOVA table above.) It can be shown that

$$\frac{\max |\overline{Y}_{i..} - \overline{Y}_{i'..} - (\alpha_i - \alpha_{i'})|}{\sqrt{MSW/nc}} \sim q_{d,dfw},$$

so that Tukey simultaneous confidence intervals for these contrast can be computed in the obvious way from those from the associated ANOVA model (substituting MSW and dfw for MSE and dfe). Dunnett, MCB, multiple range and multiple F procedures are also available. Confidence intervals for contrasts in the β_j are similar. Scheffé simultaneous confidence intervals for contrasts in γ_{ij} are given by

$$\sum\sum d_{ij}\gamma_{ij} \in \sum\sum d_{ij}\overline{Y}_{ij.} \pm \left((d-1)(c-1)F^{\alpha}MSW\sum\sum\frac{d_{ij}^2}{n}\right)^{1/2},$$

for all d_{ij} such that $\sum_i d_{ij} = 0$ and $\sum_j d_{ij} = 0$.

In Chapter 30 we commented that there are interpretational problems with the procedures for main effects in the presence of interactions. In that chapter, we suggested choosing weights and using them before defining main effects. In the model in this section we have chosen equal weights for the rows and columns. In fact, if we choose different weights, then the effects cannot be separated into between and within effects and the whole theory breaks down.

31.24 Now consider a two-way model in which each individual receives each row treatment but only one column treatment. Since individuals receive different column treatments, we call this model the multiple-group two-way repeated measures model. Let Y_{ijk} be the observation on the ith row treatment for the kth person receiving the jth column treatment, $i = 1, \ldots, d$; $j = 1, \ldots, c$; $k = 1, \ldots, n$. Let $\mathbf{Y}_{jk}$ be the vector of observations for the kth person receiving the jth treatment, and let $\boldsymbol{\mu}_{jk} = E\mathbf{Y}_{jk}$. We assume that

$$EY_{ijk} = \theta + \alpha_i + \beta_j + \gamma_{ij} \Rightarrow \boldsymbol{\mu}_{jk} = (\theta + \alpha_1 + \beta_j + \gamma_{1j}, \ldots, \theta + \alpha_d + \beta_j + \gamma_{dj})^T,$$
$$\sum_i \alpha_i = 0, \quad \sum_j \beta_j = 0, \quad \sum_i \gamma_{ij} = 0, \quad \sum_j \gamma_{ij} = 0.$$

(Note that $\boldsymbol{\mu}_{jk}$ is not the same for all the individuals because this is a multiple-group model.) For this model we see that the average of the components of $\boldsymbol{\mu}_{jk}$ is

$$\delta_{jk} = \theta + \beta_j \Rightarrow \boldsymbol{\gamma}_{jk} = (\alpha_1 + \gamma_{1j}, \ldots, \alpha_d + \gamma_{dj})^T.$$

Therefore, we see that θ and β_j are between effects and α_i and γ_{ij} are within effects. (It should be intuitively clear that β_j is a between-individuals effect because each individual only gets one column treatment. It should also be intuitively clear that α_i is a within effect because each individual receives each row treatment.) We also note that $r = d$, the number of observations on each individual, and that $m = nc$, the total number of individuals. Furthermore, $s = c$, the number of linearly independent parameters in δ_{jk}. Therefore,

$$dfb = nc - c = c(n-1),$$
$$dfw = cd(n-1) - c(n-1) = c(d-1)(n-1).$$

In addition, $\overline{Y}_{.jk}$ is the average of the observations on the kth individual getting the jth treatment, and the LS estimator of $\delta_{jk} = \theta + \beta_j$ is $\hat{\delta}_{jk} = \hat{\theta} + \hat{\beta}_j = \overline{Y}_{.j.}$ Hence

$$SSB = d\sum\sum(\overline{Y}_{.jk} - \overline{Y}_{.j.})^2,$$
$$SSW = \sum\sum\sum(Y_{ijk} - \overline{Y}_{ij.})^2 - d\sum\sum(\overline{Y}_{.jk} - \overline{Y}_{.j.})^2$$
$$= \sum\sum\sum(Y_{ijk} - \overline{Y}_{ij.} - \overline{Y}_{.jk} + \overline{Y}_{.j.})^2.$$

Therefore we see that the ANOVA table for this model is

Multiple group two-way univariate repeated measures model

effect	*df*	*SS*	*EMS*
θ	1	$cdn\overline{Y}^2_{...}$	$\tau_b^2 + cdn\theta^2$
β	$c-1$	$nd\sum(\overline{Y}_{.j.} - \overline{Y}_{...})^2$	$\tau_b^2 + g\sum\beta_j^2$
between	$c(n-1)$	$d\sum\sum(\overline{Y}_{.jk} - \overline{Y}_{.j.})^2$	τ_b^2
α	$d-1$	$cn\sum(\overline{Y}_{i..} - \overline{Y}_{...})^2$	$\tau_w^2 + h\sum\alpha_i^2$
γ	$(c-1)(d-1)$	$n\sum\sum(\overline{Y}_{ij.} - \overline{Y}_{i..} - \overline{Y}_{.j.} + \overline{Y}_{...})^2$	$\tau_w^2 + f\sum\sum\gamma_{ij}^2$
within	$c(d-1)(n-1)$	$\sum\sum\sum(Y_{ijk} - \overline{Y}_{ij.} - \overline{Y}_{.jk} + \overline{Y}_{.j.})^2$	τ_w^2
total	cnd	$\sum\sum\sum Y_{ijk}^2$	

(where $h = cn/(d-1)$, $g = nd/(c-1)$ and $f = n/(d-1)(c-1)$.) Note that this table is quite different from the previous one although the associated ANOVA model is the same for both. Appropriate F tests for testing for significant row, column or interaction effects can be written down immediately from this table, as can the F test for testing that $\tau_b^2 = \tau_w^2$, that is, that $\rho = 0$.

The Scheffé simultaneous confidence intervals for contrasts in the β_j are given by

$$\sum b_j\beta_j \in \sum b_j\overline{Y}_{.j.} \pm \left((c-1)F^\alpha MSB\sum\frac{b_j^2}{nd}\right)^{1/2}, \qquad \forall b_j \ni \sum b_j = 0.$$

Scheffé simultaneous confidence intervals for contrast in the α_i are given by

$$\sum a_i\alpha_i \in \sum a_i\overline{Y}_{i..} \pm \left((d-1)F^\alpha MSW\sum\frac{a_i^2}{nc}\right)^{1/2}, \qquad \forall a_i \ni \sum a_i = 0.$$

(Note that *MSB* and *MSW* can also be obtained from the ANOVA table above.) Scheffé simultaneous confidence intervals for contrast in the γ_{ij} can be derived similarly. In Exercise 31.44 it is shown that

$$\frac{\max|\overline{Y}_{i..} - \overline{Y}_{i'..} - (\alpha_i - \alpha_{i'})|}{\sqrt{MSW/nd}} \sim q_{d,dfw}, \qquad \frac{\max|\overline{Y}_{.j.} - \overline{Y}_{.j'.} - (\beta_j - \beta_{j'})|}{\sqrt{MSB/nc}} \sim q_{c,dfb},$$

so that Tukey simultaneous confidence intervals for contrasts in α_i and β_j can be computed in the obvious way from the formulae for the associated two-way ANOVA model (by substituting *MSW* and *dfw* for *MSE* and *dfe* for the α_i and substituting *MSB* and *dfb* for *MSE* and *dfe* for the β_j). Dunnett, MCB, multiple range and multiple F procedures can be similarly computed.

The model in the previous section had to be balanced because every individual received every possible row and column treatment. The model in this section does not have to be balanced. In Exercise 31.45 procedures are derived for a generalization of this model in which n_j individuals receive the jth column treatment and every possible row treatment.

As we have discussed previously, there are problems with interpreting main effects in the presence of interactions. Those problems extend to this repeated measures model.

31.25 Consider a nested repeated measures model in which every individual receives each possible class and subclass of treatment. (Since each individual receives each possible class and subclass, we call this model the single-group twofold nested repeated measures model.) Let Y_{ijk} be the observation on the ith class and the jth subclass for the kth individual, $i = 1, \ldots, d$; $j = 1, \ldots, c$, $k = 1, \ldots, n$. Then $\mathbf{Y}_k = (Y_{11k}, \ldots, Y_{dck})^T$ is the vector of observations on the kth individual. We assume that

$$EY_{ijk} = \theta + \alpha_i + \zeta_{ij} \Rightarrow \boldsymbol{\mu}_k = E\mathbf{Y}_k = (\theta + \alpha_1 + \zeta_{11}, \ldots, \theta + \alpha_d + \zeta_{dc})^T,$$
$$\sum_i \alpha_i = 0, \quad \sum_j \zeta_{ij} = 0.$$

In Exercise 31.46 it is shown that for this model θ is the only between effect and

$$SSB = cd \sum (\overline{Y}_{..k} - \overline{Y}_{...})^2, \quad SSW = \sum\sum\sum (Y_{ijk} - \overline{Y}_{ij.} - \overline{Y}_{..k} + \overline{Y}_{...})^2,$$
$$dfb = n - 1, \quad dfw = (dc - 1)(n - 1).$$

Hence the ANOVA table is given by

Single-group twofold nested univariate repeated measures model

effect	*df*	*SS*	*EMS*
θ	1	$cdn\overline{Y}^2_{...}$	$\tau_b^2 + cdn\theta^2$
between	$n - 1$	$cd \sum (\overline{Y}_{..k} - \overline{Y}_{...})^2$	τ_b^2
α	$d - 1$	$cn \sum (\overline{Y}_{i..} - \overline{Y}_{...})^2$	$\tau_w^2 + h \sum \alpha_i^2$
ζ	$d(c - 1)$	$n \sum\sum (\overline{Y}_{ij.} - \overline{Y}_{i..})^2$	$\tau_w^2 + g \sum\sum \zeta_{ij}^2$
within	$(dc - 1)(n - 1)$	$\sum\sum\sum (Y_{ijk} - \overline{Y}_{ij.} - \overline{Y}_{..k} + \overline{Y}_{...})^2$	τ_w^2
total	dcn	$\sum\sum\sum Y_{ijk}^2$	

From this table F tests can be computed for the various hypotheses of interest. Simultaneous confidence intervals can be constructed in the obvious way.

This model can be extended to allow the possibility that the number of subclasses may be different for different classes. That is, we allow that $j = 1, \ldots, c_i$. (See Exercise 31.47.) Since each individual receives each class and subclass, the number of observations must be the same in each subclass.

As in the associated ANOVA model, there are interpretational problems with the hypothesis of no class effect in the presence of a possible subclass effect.

31.26 Now consider a nested repeated measures model in which each individual receives only one class treatment but gets every subclass treatment for that class. (Since each individual receives only one class effect, this model is called the multiple-group twofold nested repeated measures model.) Let Y_{ijk} represent the observation for the jth subclass for the kth individual receiving the ith class, $i = 1, \ldots, d$; $j = 1, \ldots, c$; $k = 1, \ldots, n$. Then the vector of observations for the kth individual receiving the ith class is $\mathbf{Y}_{ik} = (Y_{i1k}, \ldots, Y_{ick})^T$. We assume that

$$EY_{ijk} = \theta + \alpha_i + \zeta_{ij} \Rightarrow \boldsymbol{\mu}_{ik} = E\mathbf{Y}_{ik} = (\theta + \alpha_i + \zeta_{i1}, \ldots, \theta + \alpha_i + \zeta_{ic})^T,$$

$$\sum_i \alpha_i = 0, \quad \sum_j \zeta_{ij} = 0.$$

For this model we see that the average of the components of $\boldsymbol{\mu}_{ik}$ is $\delta_{ik} = \theta + \alpha_i$, so that θ and α_i are between effects for this model and ζ_{ij} is a within effect. Furthermore, $r = c$, $m = nd$, $s = d$. Therefore,

$$dfb = nd - d = d(n-1), \quad dfw = cd(n-1) - d(n-1) = d(c-1)(n-1).$$

The average of the observations on the kth individual receiving the ith class is $\overline{Y}_{i.k}$ and the LS estimator of $\delta_{ik} = \theta + \alpha_i$ is $\overline{Y}_{i..}$. Therefore

$$SSB = c\sum\sum(\overline{Y}_{i.k} - \overline{Y}_{i..})^2, \quad SSW = \sum\sum\sum(Y_{ijk} - \overline{Y}_{ij.})^2 - c\sum\sum(\overline{Y}_{i.k} - \overline{Y}_{i..})^2$$
$$= \sum\sum\sum(Y_{ijk} - \overline{Y}_{ij.} - \overline{Y}_{i.k} + \overline{Y}_{i..})^2.$$

Hence the ANOVA table for this model is

Multiple-group twofold nested univariate repeated measures model

effect	*df*	*SS*	*EMS*
θ	1	$ncd\overline{Y}_{...}^2$	$\tau_b^2 + ncd\theta^2$
α	$d-1$	$nc\sum(\overline{Y}_{i..} - \overline{Y}_{...})^2$	$\tau_b^2 + h\sum\alpha_i^2$
between	$d(n-1)$	$c\sum\sum(\overline{Y}_{i.k} - \overline{Y}_{i..})^2$	τ_b^2
ζ	$d(c-1)$	$n\sum\sum(\overline{Y}_{ij.} - \overline{Y}_{i..})^2$	$\tau_w^2 + g\sum\zeta_{ij}^2$
within	$d(c-1)(n-1)$	$\sum\sum\sum(Y_{ijk} - \overline{Y}_{ij.} - \overline{Y}_{i.k} + \overline{Y}_{i..})^2$	τ_w^2
total	cdn	$\sum\sum\sum Y_{ijk}^2$	

where $h = nc/(d-1)$ and $g = n/d(c-1)$. F tests for the various hypotheses can be determined immediately from this table. Simultaneous confidence intervals can also be immediately derived.

This model may be easily generalized to allow the number of individuals to be different for different classes. That is, we can let $k = 1, \ldots, n_i$. (See Exercise 31.48.)

As in ANOVA, there are interpretational difficulties with the test for no class effect in the presence of possible subclass effects.

Covariates in repeated measures models

31.27 Consider a one-way repeated measures model in which every individual receives every treatment. Suppose, in addition, there is a covariate in the model. Let Y_{ik} be the observation on the ith treatment for the kth individual and let x_{ik} be the covariate associated with that measurement, $i = 1, \ldots, r$; $k = 1, \ldots, n$. Let $\mathbf{Y}_k = (Y_{1k}, \ldots, Y_{rk})^T$ be the vector of observations on the kth individual. We assume that

$$EY_{ik} = \theta + \alpha_i + x_{ik}\beta \Rightarrow \boldsymbol{\mu}_k = E\mathbf{Y}_k = (\theta + \alpha_1 + x_{ik}\beta, \ldots, \theta + \alpha_r + x_{rk}\beta)^T,$$
$$\sum \alpha_i = 0.$$

Hence for this model

$$\delta_k = \theta + \overline{x}_{.k}\beta, \quad \boldsymbol{\gamma}_k = (\alpha_1 + (x_{1k} - \overline{x}_{.k})\beta, \ldots, \alpha_r + (x_{rk} - \overline{x}_{.k})\beta)^T.$$

Notice that the effect of the covariate is not, in general, a between or within effect, so that this model does not fit in the framework of this chapter. In fact this model is a very difficult model for which sensible exact procedures have not been derived. However, there are two situations that fit into the framework of this chapter:

1. The covariate only depends on the individual ($x_{ik} = x_k \Rightarrow x_{ik} - \overline{x}_{.k} = 0$). In this case the covariate is a between effect.
2. The average for the covariate for each individual is zero ($\overline{x}_{.k} = 0$). In this case the covariate is a within effect.

We look first at the case in which $x_{ik} = x_k$ depends only on k. Note that this condition implies that the average value of the covariate is the same for each treatment, that is, that the treatment and the covariate are orthogonal, which simplifies some of the formulae for the associated ANOVA model. For the associated ANOVA model

$$\hat{\beta} = \frac{\sum(x_k - \overline{x}_.)(\overline{Y}_{.k} - \overline{Y}_{..})}{\sum(x_k - \overline{x}_.)^2}, \quad SSE = \sum\sum(Y_{ik} - \overline{Y}_{i.} - \hat{\beta}(x_k - \overline{x}_.))^2, \quad dfe = r(n-1) - 1.$$

Note also that the LS estimator of $\delta_k = \theta + x_k\beta$ is $\overline{Y}_{..} + (x_k - \overline{x}_.)\hat{\beta}$ and hence

$$\begin{aligned} dfb &= n - 2, \quad dfw = r(n-1) - 1 - n + 2 = (r-1)(n-1), \\ SSB &= \sum(\overline{Y}_{.k} - \overline{Y}_{..} - (x_k - \overline{x}_.)\hat{\beta})^2, \\ SSW = SSE - SSB &= \sum\sum(Y_{ij} - \overline{Y}_{i.} - \overline{Y}_{.k} + \overline{Y}_{..})^2. \end{aligned}$$

Now consider testing the null hypothesis that the $\alpha_i = 0$. Note that for the associated ANOVA model

$$dfh = r - 1, \quad SSH = n\sum(\overline{Y}_{i.} - \overline{Y}_{..})^2.$$

The α_i are a within effect and hence this hypothesis would be tested by

$$F = \frac{MSH}{MSW} \sim F_{dfh,dfw}\left(\frac{n\sum\alpha_i^2}{\tau_w^2}\right).$$

To draw inference for β (a between effect) use the pivotal quantity

$$\frac{(\hat{\beta} - \beta)\sqrt{n\sum(x_k - \overline{x}_.)^2}}{\sqrt{MSB}} \sim t_{dfb}.$$

Now consider the case in which $\overline{x}_{.k} = 0.$ and therefore $\delta_k = \theta$. In this case

$$\begin{aligned} \hat{\beta} &= \frac{\sum\sum(x_{ik} - x_{i.})(Y_{ik} - \overline{Y}_{i.})}{\sum\sum(x_{ik} - \overline{x}_{i.})^2}, \quad dfe = r(n-1) - 1, \\ SSE &= \sum\sum(Y_{ik} - \overline{Y}_{i.} - (x_{ik} - \overline{x}_{i.})\hat{\beta})^2. \end{aligned}$$

For this model the LS estimator of $\delta_k = \theta$ is $\overline{Y}_{..}$ (recalling that $\overline{x}_{..} = 0$ by assumption). Therefore

$$dfb = n - 1, \quad dfw = (r-1)(n-1) - 1, \quad SSB = \sum(\overline{Y}_{.k} - \overline{Y}_{..})^2,$$
$$SSW = SSE - SSB = \sum\sum(Y_{ik} - \overline{Y}_{i.} - \overline{Y}_{.k} + \overline{Y}_{..} - (x_{ik} - x_{i.})\hat{\beta})^2.$$

Now consider testing that the $\alpha_i = 0$. For this hypothesis

$$dfh = r - 1,$$
$$SSH = n\sum(\overline{Y}_{i.} - \overline{Y}_{..})^2 + \hat{\beta}^2\sum\sum(x_{ij} - \overline{x}_{i.})^2 - \hat{\hat{\beta}}\sum\sum(x_{ij} - \overline{x}_{..})^2,$$
$$\hat{\hat{\beta}} = \frac{\sum\sum(x_{ij} - \overline{x}_{..})(Y_{ij} - \overline{Y}_{..})}{\sum\sum(x_{ij} - \overline{x}_{..})^2}.$$

Since the α_i are a within effect, we use the F statistic

$$F = \frac{MSH}{MSW} \sim F_{dfh,dfw}\left(\frac{n\sum\alpha_i^2}{\tau_w^2}\right).$$

Note that for this model, β is also a within effect. For inference about β we use the pivotal quantity

$$\frac{(\hat{\beta} - \beta)\sqrt{\sum\sum(x_{ij} - \overline{x}_{i.})^2}}{\sqrt{MSW}} \sim t_{dfw}.$$

Actually this second model can be generalized slightly to the case in which $\overline{x}_{.k}$ are all the same (but not necessarily 0). See Exercise 31.49 for details. However, if the covariate does not satisfy one of these two conditions ($x_{ik} = x_k$ or $\overline{x}_{.k}$ does not depend on k) then the model cannot be analysed exactly at this time. For an approximate analysis see Exercise 31.63. Note that even though the α_i are a within effect for any covariate even this effect cannot be analysed unless the covariate satisfies one of these two assumptions.

One approach which is sometimes used for data in which the covariate does not satisfy either condition is to assume that

$$EY_{ik} = \theta + \alpha_i + \beta_1\overline{x}_{.k} + \beta_2(x_{ik} - \overline{x}_{.k})$$

(i.e. include two covariates in the model). If the original model is correct we would expect $\beta_1 = \beta_2$, although there is no exact test of this hypothesis. Note that for this model

$$\delta_k = \theta + \beta_1\overline{x}_{.k}, \quad \gamma_k = (\alpha_1 + \beta_2(x_{1k} - \overline{x}_{.k}), \ldots, \alpha_r + \beta_2(x_{rk} - \overline{x}_{.k})),$$

so that the covariate associated with β_1 is a between effect and the covariate associated with β_2 is a within effect.

We could easily extend this analysis to more complicated models (crossed, nested, etc.) with more covariates. (See Exercises 31.50–51 for examples.) The analysis is straightforward as long as each covariate either depends only on the individual or has the same average value for each individual. If the covariate depends on the individual, it is a between effect; and if its average is constant for each individual, it is a within effect. Otherwise it is neither a within nor a between effect and the model cannot be exactly analysed at this time.

Univariate repeated measures models as mixed models

31.28 Consider a one-way repeated measures model in which each individual receives each of r possible treatments. As usual, let Y_{ik} be the observation for the ith treatment on the kth individual. A natural alternative to the univariate repeated measures model is to treat the model as a mixed model in which the individual is a random effect. That is, we might assume that

$$Y_{ik} = \theta + \alpha_i + b_k + e_{ik},$$

where θ and the α_i are unknown parameters such that $\sum \alpha_i = 0$ and the b_k and e_{ik} are unobserved random variables such that

$$b_k \sim N(0, \sigma_b^2), \quad e_{ik} \sim N(0, \sigma_e^2).$$

Now let $\mathbf{Y}_k = (Y_{1k}, \ldots, Y_{rk})^T$ be the vector of observations on the kth individual. Then it is easily seen that

$$\mathbf{\Sigma} = \text{cov}(\mathbf{Y}_k) = \sigma^2 \mathbf{A}_r(\rho), \qquad \sigma^2 = \sigma_b^2 + \sigma_e^2, \qquad \rho = \frac{\sigma_b^2}{\sigma_b^2 + \sigma_e^2}.$$

That is, this mixed model is nearly the same as the univariate repeated measures model. The only difference is that for the mixed model we assume that $0 \leq \rho < 1$, but for the univariate repeated measures model we assume that $-1/(r-1) < \rho < 1$. This enlarged parameter space allows a nicer theory for the repeated measures model. (Note that assuming $\rho \geq 0$ is equivalent to assuming that $\tau_b^2 \geq \tau_w^2$ and that it was restrictions like this on the parameter space that led to the complications for the random effects and mixed models.) However, this difference in the parameter space does not affect the distributions of the various F statistics for testing hypotheses. Note that for this approach,

$$\tau_b^2 = \sigma^2(1 + (r-1)\rho) = r\sigma_b^2 + \sigma_e^2, \quad \tau_w^2 = \sigma^2(1-\rho) = \sigma_e^2.$$

This example illustrates an alternative way to obtain the F tests for a univariate repeated measures model: put in a random effect for the individual and treat the model as a mixed model. This approach works for any of the univariate repeated measures models discussed in this chapter (except the models with covariates for which we have not developed random effects models). Note that we do not include an interaction effect between the random effect and other effects in the model, so that we do not have to worry about whether to use the restricted or unrestricted version of the mixed model.

Multivariate repeated measures models

31.29 We now discuss some multivariate repeated measures models. We first consider single-group repeated measures designs in which each individual receives each possible treatment combination and hence each individual receives the same treatments. As before, we let $\mathbf{Y}_k$ be the vector of observations on the kth individual, $k = 1, \ldots, m$. Since all the individuals receive the same treatments, $E\mathbf{Y}_k$ should be the same for all k. We assume that the $\mathbf{Y}_k$ are independent,

$$\mathbf{Y}_k \sim N_r(\boldsymbol{\mu}, \mathbf{\Sigma}), \qquad \boldsymbol{\mu} \in R^r, \ \mathbf{\Sigma} > 0.$$

If we make no further assumptions about the covariance matrix $\boldsymbol{\Sigma}$ we call the model a *multivariate repeated measures model.* For these multivariate repeated measures models we often test the null hypothesis that $\mathbf{A}\boldsymbol{\mu} = \mathbf{0}$ for some $s \times r$ matrix $\mathbf{A}$ of rank s against the alternative hypothesis that $\boldsymbol{\mu}$ is unrestricted. In this section we indicate how to do this. Since this model is a multivariate one-sample model, discussed more thoroughly in Volumes 1–2 of Kendall's library (Krzanowski and Marriott, 1994; 1995), in this section we merely state the results for the multivariate model, showing how they apply to the repeated measures model. (See also **16.38–51**.)

Let

$$\overline{\mathbf{Y}} = \frac{1}{m}\sum \mathbf{Y}_i, \qquad \mathbf{S} = \frac{1}{m-1}\sum(\mathbf{Y}_i - \overline{\mathbf{Y}})(\mathbf{Y}_i - \overline{\mathbf{Y}})^T.$$

Then $\overline{\mathbf{Y}}$ is an $r \times 1$ vector called the *sample mean vector* and $\mathbf{S}$ is an $r \times r$ matrix called the *sample covariance matrix*. It can be shown that

$$\mathbf{S} > 0 \Leftrightarrow m - 1 \geq r,$$

which we assume throughout the remainder of this section. In this case it can be shown that $(\overline{\mathbf{Y}}, \mathbf{S})$ is a complete sufficient statistic, that $\overline{\mathbf{Y}}$ and $\mathbf{S}$ are the MV unbiased estimators of $\boldsymbol{\mu}$ and $\boldsymbol{\Sigma}$ and that $(\overline{\mathbf{Y}}, ((m-1)/m)\mathbf{S})$ is the ML estimator of $(\boldsymbol{\mu}, \boldsymbol{\Sigma})$ (see Exercise 31.53). Let

$$F = \frac{(m-s)m}{s(m-1)}(\mathbf{A}\overline{\mathbf{Y}})^T(\mathbf{ASA}^T)^{-1}(\mathbf{A}\overline{\mathbf{Y}}).$$

By an argument similar to that in **16.51**, it can be shown that

$$F \sim F_{s,m-s}(\delta), \qquad \delta = m(\mathbf{A}\boldsymbol{\mu})^T(\mathbf{A}\boldsymbol{\Sigma}\mathbf{A}^T)^{-1}(\mathbf{A}\boldsymbol{\mu}).$$

Since $\mathbf{A}\boldsymbol{\mu} = \mathbf{0}$ if and only if $\delta = 0$, a sensible size-α test of the null hypothesis that $\mathbf{A}\boldsymbol{\mu} = \mathbf{0}$ rejects if

$$F > F^{\alpha}_{s,m-s}.$$

This test can be shown to be unbiased, the UMP invariant size-α test and the LR test. Simultaneous confidence intervals for contrasts can also be derived associated with this test. (For example, see Arnold, 1981, pp. 335–342, for derivations of these results.)

Now consider a one-way repeated measures model in which each of m individuals receives each of r treatments so that

$$\boldsymbol{\mu} = (\theta + \alpha_1, \ldots, \theta + \alpha_r)^T.$$

Let $\mathbf{A}$ be the $(r-1) \times r$ matrix given by

$$\mathbf{A} = (\,\mathbf{I}_{r-1} \quad -\mathbf{1}_{r-1}\,).$$

Then

$$\mathbf{A}\boldsymbol{\mu} = (\alpha_1 - \alpha_r, \ldots, \alpha_{r-1} - \alpha_r)^T.$$

Therefore, the $\alpha_i = 0$ if and only if $\mathbf{A}\boldsymbol{\mu} = \mathbf{0}$, so that this hypothesis can be tested with the test described above (with $s = r - 1$).

The test derived here is different from the test derived for the univariate model. Note that for both models the numerator has $r - 1$ degrees of freedom. For the univariate repeated measures model, the denominator has $(r - 1)(m - 1)$ d.fr., but for the multivariate repeated measures model, it has $m - (r - 1)$ d.fr., which is an indication of the loss of power when going from the univariate repeated measures model to the multivariate repeated measures model (unless $r = 2$, in which case the multivariate and univariate tests are the same). In particular, note that if $m < r$, it is not possible to do the multivariate analysis, but the univariate analysis is possible as long as $m > 1$. (Note also that for the associated fixed effects ANOVA model, the numerator also has $r - 1$ d.fr. and the denominator $r(m - 1)$ d.fr. so that there is less loss of power in going from the fixed effects model to the univariate repeated measures model than in going from the univariate repeated measures model to the multivariate repeated measures model.)

From this example it is clear that many single-group repeated measures models can be analysed by the methods described above. (See Exercise 31.54 for a further example.) In particular, the single-group two-way repeated measures model and the single-group twofold nested model fall under the framework of this section. Note that for all these models the hypotheses of interest are within hypotheses. Hence the tests for this multivariate model are different from those for the univariate model and there is considerable loss of power in going from the univariate model to the multivariate model.

31.30 We now consider some two-group multivariate repeated measures models. In particular, we let $\mathbf{Y}_{jk}$ be the observation on the kth individual receiving the jth treatment, $j = 1, 2$; $k = 1, \ldots, n_j$. We assume that the $\mathbf{Y}_{jk}$ are independent

$$\mathbf{Y}_{jk} \sim N_r(\boldsymbol{\mu}_j, \boldsymbol{\Sigma}), \qquad \boldsymbol{\mu}_j \in R^r, \ \boldsymbol{\Sigma} > 0.$$

Let $N = n_1 + n_2$ and let

$$\overline{\mathbf{Y}}_j = \frac{1}{n_j} \sum_k \mathbf{Y}_{jk}, \quad \mathbf{S}_p = \frac{1}{N - 2} \sum_j \sum_k (\mathbf{Y}_{jk} - \overline{\mathbf{Y}}_j)(\mathbf{Y}_{jk} - \overline{\mathbf{Y}}_j)^T.$$

We call $\mathbf{S}_p$ the *pooled sample covariance matrix*. We assume that $N - 2 \geq r$ so that $\mathbf{S}_p > 0$. Then $(\overline{\mathbf{Y}}_1, \overline{\mathbf{Y}}_2, \mathbf{S}_p)$ is a complete sufficient statistic for this model, $\overline{\mathbf{Y}}_j$ and $\mathbf{S}_p$ are the MV unbiased estimators of $\boldsymbol{\mu}_j$ and $\boldsymbol{\Sigma}$ and $(\overline{\mathbf{Y}}_1, \overline{\mathbf{Y}}_2, ((N - 2)/N)\mathbf{S})$ is the ML estimator of $(\boldsymbol{\mu}_1, \boldsymbol{\mu}_2, \boldsymbol{\Sigma})$.

We are interested in testing hypotheses of the form $\mathbf{C}(\boldsymbol{\mu}_1 + \boldsymbol{\mu}_2) = \mathbf{0}$ and hypotheses of the form $\mathbf{B}(\boldsymbol{\mu}_1 - \boldsymbol{\mu}_2) = \mathbf{0}$, where $\mathbf{C}$ and $\mathbf{B}$ are $s \times r$ and $t \times r$ matrices of rank s and t. Let

$$F_+ = \frac{(N - s - 1)}{s(N - 2)(\frac{1}{n_1} + \frac{1}{n_2})} (\mathbf{C}(\overline{\mathbf{Y}}_1 + \overline{\mathbf{Y}}_2))^T (\mathbf{C}\mathbf{S}_p\mathbf{C}^T)^{-1} (\mathbf{C}(\overline{\mathbf{Y}}_1 + \overline{\mathbf{Y}}_2)).$$

By arguments similar to those in **16.51**, it can be shown that

$$F_+ \sim F_{s,N-s-1}(\delta_+), \quad \delta_+ = \frac{1}{\frac{1}{n_1} + \frac{1}{n_2}} (\mathbf{C}(\boldsymbol{\mu}_1 + \boldsymbol{\mu}_2))^T (\mathbf{C}\boldsymbol{\Sigma}\mathbf{A}^T)^{-1} (\mathbf{C}(\boldsymbol{\mu}_1 + \boldsymbol{\mu}_2)).$$

Since $\mathbf{C}(\boldsymbol{\mu}_1 + \boldsymbol{\mu}_2) = \mathbf{0}$ if and only if $\delta_+ = 0$, a sensible test for this hypothesis rejects if

$$F_+ > F^{\alpha}_{s,N-s-1}.$$

Similarly, let

$$F_- = \frac{(N-t-1)}{t(N-2)(\frac{1}{n_1}+\frac{1}{n_2})}(\mathbf{B}(\overline{\mathbf{Y}}_1 - \overline{\mathbf{Y}}_2))^T(\mathbf{B}\mathbf{S}_p\mathbf{B}^T)^{-1}(\mathbf{B}(\overline{\mathbf{Y}}_1 - \overline{\mathbf{Y}}_2)).$$

Then

$$F_- \sim F_{t,N-t-1}(\delta_-), \quad \delta_- = \frac{1}{\frac{1}{n_1}+\frac{1}{n_2}}(\mathbf{B}(\boldsymbol{\mu}_1 - \boldsymbol{\mu}_2))^T(\mathbf{B}\boldsymbol{\Sigma}\mathbf{B}^T)^{-1}(\mathbf{B}(\boldsymbol{\mu}_1 - \boldsymbol{\mu}_2)).$$

Similarly, a sensible test for the null hypothesis that $\mathbf{B}(\boldsymbol{\mu}_1 - \boldsymbol{\mu}_2) = \mathbf{0}$ rejects if

$$F_- > F^{\alpha}_{t,N-t-1}.$$

These two tests are again unbiased, UMP invariant and LR tests and have associated simultaneous confidence intervals. They are also asymptotically insensitive to the normal assumption.

Now consider a two-group two-way repeated measures model in which every individual receives each of r row treatments but only one of two column treatments. (In order to fit in the model of this section, we assume that there are only two possible column treatments (e.g. male and female)). In that case,

$$\boldsymbol{\mu}_j = (\theta + \alpha_1 + \beta_j + \gamma_{ij}, \ldots, \theta + \alpha_r + \beta_j + \gamma_{rj})^T$$

and therefore

$$\boldsymbol{\mu}_1 + \boldsymbol{\mu}_2 = 2(\theta + \alpha_1, \ldots, \theta + \alpha_r)^T,$$
$$\boldsymbol{\mu}_1 - \boldsymbol{\mu}_2 = (\beta_1 - \beta_2 + \gamma_{11} - \gamma_{12}, \ldots, \beta_1 - \beta_2 + \gamma_{r1} - \gamma_{r2})^T.$$

Consider first testing that the $\alpha_i = 0$. This hypothesis is equivalent to testing that $\mathbf{A}(\boldsymbol{\mu}_1 + \boldsymbol{\mu}_2) = \mathbf{0}$, where $\mathbf{A}$ is the contrast matrix defined in the previous section. Therefore for the multivariate repeated measures model we use the test involving F_+ with this $\mathbf{C} = \mathbf{A}$ so that $s = r - 1$. We note again that the numerator degrees of freedom will be the same as for the univariate model, but that the denominator degrees of freedom for this model will be $N - r$ and for the univariate model $(r-1)(N-2)$. (Note that for the ANOVA model the denominator degrees of freedom will be $r(N-2)$, so that there is little loss in power in going from the fixed effects model to the univariate repeated measures model and a large loss of power in going from the univariate repeated measures model to the multivariate repeated measures model.)

Now consider testing that $\beta_j = 0$ or equivalently that $\mathbf{1}_r^T(\boldsymbol{\mu}_1 - \boldsymbol{\mu}_2) = 0$. For this hypothesis, we use F_- with $\mathbf{B} = \mathbf{1}_r^T$. Hence $t = 1$, so that the numerator has 1 d.fr., the same as for the univariate model (recall that there are only two columns). The denominator has $N - 2$ d.fr., also the same as for the univariate repeated measures model. In fact the β_j effect is a between effect for this design and, as we mentioned earlier, for between effects the tests for the univariate and multivariate models are the same. (Note that for the associated ANOVA model the denominator has $r(N-2)$ d.fr., so that there is considerable loss of power between the ANOVA model and the univariate repeated measures model for this between effect but no loss of power between the univariate model and the multivariate model.)

Finally, consider testing that the $\gamma_{ij} = 0$ or equivalently that $\mathbf{A}(\boldsymbol{\mu}_1 - \boldsymbol{\mu}_2) = \mathbf{0}$, where $\mathbf{A}$ is the contrast matrix given above. For testing that $\gamma_{ij} = 0$, we therefore use the test statistic F_{-} with $\mathbf{B}$ replaced by the $(r-1) \times r$ contrast matrix $\mathbf{A}$. We see again that the numerator degrees of freedom $(r - 1)$ are the same for both the univariate and multivariate models but that the denominator degrees of freedom are $N - r$ for the multivariate model and $(r - 1)(N - 2)$ for the univariate model, indicating the loss of power when we use the multivariate model. (For the associated ANOVA model the denominator degrees of freedom are $r(N - 2)$.)

The two-way model described above is limited to two columns. If there are more than two columns, the multivariate repeated measures model can be analysed using procedures for multivariate analysis of variance (MANOVA). See Arnold (1981, pp. 374–378) for details.

Now consider the two-group twofold nested model in which there are two classes each with r subclasses and in which each individual receives only one class, but each subclass of that class. Let $\mathbf{Y}_{ik}$ be the vector of observations on the kth person in the ith class. Then

$$\boldsymbol{\mu}_i = E\mathbf{Y}_{ik} = (\theta + \alpha_i + \delta_{i1}, \ldots, \theta + \alpha_i + \delta_{ir})^T$$

and

$$\mathbf{1}_r^T(\boldsymbol{\mu}_1 - \boldsymbol{\mu}_2) = r(\alpha_1 - \alpha_2).$$

Therefore, to test for no class effect we use F_{-} with $\mathbf{B} = \mathbf{1}_r^T$. The α_i are a between effect so that this test is the same as the test for the univariate repeated measures model. Also

$$\mathbf{A}(\boldsymbol{\mu}_1 \quad \boldsymbol{\mu}_2) = \begin{pmatrix} \delta_{11} - \delta_{1r} & \delta_{21} - \delta_{2r} \\ \vdots & \vdots \\ \delta_{1,r-1} - \delta_{1r} & \delta_{2,r-1} - \delta_{2r} \end{pmatrix}$$

so that testing that the $\delta_{ij} = 0$ is testing that $\mathbf{A}(\boldsymbol{\mu}_1 \quad \boldsymbol{\mu}_2) = \mathbf{0}$, which can be tested by methods of MANOVA.

Testing validity of repeated measures models

31.31 In previous sections we have discussed three different models to apply to a particular repeated measures design: the multivariate repeated measures model; the univariate repeated measures model; and the associated ANOVA model. We have already discussed how to test the null hypothesis that the associated ANOVA model is correct against the alternative that the univariate repeated measures model is correct (i.e. that $\rho = 0$ in the univariate repeated measures model). In this section, we consider the other two testing possibilities: testing the null hypothesis that the ANOVA model is correct against the alternative that the multivariate repeated measures model is correct; and testing the null hypothesis that the univariate repeated measures model is correct against the alternative that the multivariate repeated measures model is correct.

For simplicity, we assume the design is a single-group design in which all the individuals receive the same treatments. (The extension to more complicated designs is immediate.) In that model we observe $\mathbf{Y}_k$ independent, $k = 1, \ldots, m$,

$$\mathbf{Y}_k \sim N_r(\boldsymbol{\mu}, \boldsymbol{\Sigma}), \qquad \boldsymbol{\mu} \in R^r, \ \boldsymbol{\Sigma} > 0.$$

As above, let $\mathbf{S}$ be the sample covariance matrix.

To test the validity of the associated ANOVA model, we test the null hypothesis that $\mathbf{\Sigma} = \sigma^2\mathbf{I}$ for some scalar σ^2 against the alternative that $\mathbf{\Sigma} > 0$. In Exercise 31.56, the LR test statistic for this problem is shown to be

$$\Lambda_1 = \frac{|\mathbf{S}|^{\frac{m}{2}}}{(\text{tr}(\mathbf{S})/r)^{\frac{mr}{2}}}.$$

Using the usual approximation to the distribution of LR test statistics, we see that $-2\log(\Lambda_1)$ is approximately $\chi^2_{r(r+1)/2-1}$ under the null hypothesis. A more refined approximation is given in Anderson (1958, pp. 259–262). (This test is called Mauchly's test for sphericity.)

Now consider testing the null hypothesis that the univariate repeated measures model is correct, that is, that $\mathbf{\Sigma} = \sigma^2\mathbf{A}_r(\rho)$ for some scalars σ^2 and ρ. In Exercise 31.57, it is shown that the LR test statistic for this problem is given by

$$\Lambda_2 = \frac{|\mathbf{S}|^{\frac{m}{2}}}{U^{\frac{m}{2}}((\text{tr}(\mathbf{S}) - U)/(r-1))^{\frac{m(r-1)}{2}}}, \qquad U = \frac{\mathbf{1}_r^T\mathbf{S}\mathbf{1}_r}{r} = \frac{\sum\sum S_{ij}}{r}.$$

By the usual approximation for the LR test, we see that under the null hypothesis $-2\log(\Lambda_2)$ is approximately $\chi^2_{r(r+1)/2-2}$. A more refined approximation is given in Box (1950).

We have seen that the test for between hypotheses is the same for the univariate and multivariate repeated measures models. Furthermore, the test for within hypotheses is correct as long as $\mathbf{C}^T\mathbf{\Sigma}\mathbf{C} = \tau_2^2\mathbf{I}$ for some scalar τ_2^2, where $\mathbf{C}$ is the matrix defined in **31.14**. Therefore, tests for both between and within hypotheses are correct as long as $\mathbf{C\Sigma C}^T = \tau_2^2\mathbf{I}$. (Covariance matrices satisfying this condition are said to satisfy the *Huynh–Feldt condition*. See Huynh and Feldt (1970) for other representations of this condition.) In Exercise 31.57, it is shown that the LR test statistic for testing the null hypothesis that $\mathbf{\Sigma}$ satisfies the Huynh–Feldt condition is given by

$$\Lambda_3 = \frac{|\mathbf{T}|^{\frac{m}{2}}}{(\text{tr}(\mathbf{T})/(r-1))^{\frac{m(r-1)}{2}}}, \qquad \mathbf{T} = \mathbf{CSC}^T.$$

As usual, the null distribution of $-2\log(\Lambda_3)$ is approximately a $\chi^2_{r(r-1)/2-1}$. A more refined approximation to the null distribution is given in Huynh and Feldt (1976).

All three of the tests mentioned in this section are quite sensitive to the normal assumption and should be used with great care. Note that we could reject the univariate repeated measures model in favour of the multivariate repeated measures model because of the non-normality of the distributions. However, the procedures for the univariate model are not too sensitive to the normal assumption so that we would be sacrificing the considerably higher power of the univariate test when it was not necessary to do so.

Another difficulty with using these tests before deciding on what analysis to perform is that accepting a null hypothesis is not statistical proof that the null hypothesis is true but rather lack of statistical proof that it is false. In particular, if we use a 0.05 test and therefore accept the univariate model when the p-value is 0.08, we do not really believe that the univariate model is correct but rather believe it is false but without quite enough evidence to be sure it is false.

Further comments

31.32 There are several strong assumptions for the univariate and multivariate repeated measures model discussed in this chapter. One is that every individual gets the same number (r) of treatments. The theory completely breaks down unless this assumption is satisfied. In practice, however, observations are often missing. In this case some software programmes eliminate the whole vector of observations for any individual missing an observation. This harsh approach leads to exact procedures. Other software programmes use the EM algorithm (see **18.44–46**) to compute the LR test statistics for the tests for various effects. Although the EM algorithm does converge to the true LR test statistic, it is not apparent how to find a sensible critical value for this test (except for very large samples where the usual χ^2 approximation may be used).

A second strong assumption of the univariate model is that all the effects must be either between or within effects. In **31.28**, we discuss an ANCOVA example in which, under certain conditions, the covariate is neither a between effect nor a within effect and the theory breaks down. As another example of this problem, consider a one-way model in which each individual receives only a subset of the treatments (e.g., a balanced incomplete block design). Since different individuals receive different treatments, the treatment effect is no longer a within effect and the theory breaks down again. One approach for dealing with this situation is to obtain an estimator $\hat{\rho}$ for ρ (perhaps by embedding the model in a larger model in which all the effects are either between or within, as we indicated in the covariance example above). We then apply the formulae for the more general linear model (see **31.33**) to this model treating $\hat{\rho}$ as the true value for ρ. As long as $\hat{\rho}$ is consistent the procedures should be asymptotically correct as the number of individuals goes to ∞.

A third strong assumption of the univariate model is the assumption of equal variances and equal covariances. We have shown that the tests for between hypotheses are unaffected by violation of this assumption. In fact, tests of these hypotheses are the same for both the univariate and multivariate repeated measures models. However, tests for the within hypotheses are quite sensitive to this assumption. If this assumption is questionable, the multivariate repeated measures model may be used, but with considerable sacrifice of power (see **31.29**). Several adjustments to the univariate procedures have also been suggested (see Greenhouse and Geisser, 1959; Huynh and Feldt, 1976) for use when the equal variance, equal correlation assumption may be suspect but there are not enough observations to use the multivariate methods.

A fourth assumption is the joint normality of the observations. In Arnold (1980) the univariate procedures for testing between or within effects are shown under fairly general conditions to be fairly robust against the normality assumption as the number of individuals goes to ∞. (Note that the test that $\rho = 0$ is quite sensitive to the normal assumption.) A similar argument can be used to show that the procedures for the multivariate model are robust against the normal assumptions as long as the number of individuals goes to ∞. A non-parametric alternative is given in Akritas and Arnold (1993).

A more general linear model

31.33 In this section, we present the theory for a slight generalization of the (ordinary) linear model discussed in the previous chapter. This model was first studied by Aitken (1935). In fact what we call generalized least squares estimators in this section are often called Aitken estimators.

In this more general linear model, we observe the random vector

$$\mathbf{Y} \sim N_N(\boldsymbol{\mu}, \sigma^2\mathbf{A}), \qquad \boldsymbol{\mu} \in V,\ \sigma^2 > 0,$$

where $\boldsymbol{\mu}$ is an unknown parameter assumed to lie in V, a known p-dimensional subspace of R^N,

σ^2 is an unknown scalar parameter and $\mathbf{A} > 0$ is a known $N \times N$ matrix. Note that if $\mathbf{A} = \mathbf{I}$, then this model is an ordinary linear model as discussed in the previous chapter.

Rather than develop the theory from the beginning for this model, we transform to an ordinary linear model. Let $\mathbf{A}^{-\frac{1}{2}}$ be any $N \times N$ matrix such that

$$\mathbf{A}^{-1/2}\mathbf{A}(\mathbf{A}^{-1/2})^T = \mathbf{I}.$$

Such a matrix always exists because $\mathbf{A} > 0$. (Examples include the inverses of the positive definite and Cholesky square root matrices.) Let

$$\mathbf{Y}^* = \mathbf{A}^{-1/2}\mathbf{Y}, \qquad \boldsymbol{\mu}^* = \mathbf{A}^{-\frac{1}{2}}\boldsymbol{\mu}, \qquad V^* = \{\mathbf{A}^{-\frac{1}{2}}\boldsymbol{\mu},\ \boldsymbol{\mu} \in V\}.$$

(Note that $\mathbf{A}^{-1/2}$ is an invertible matrix which implies that the transformation from $\mathbf{Y}$ to $\mathbf{Y}^*$ is invertible and that V^* is a p-dimensional subspace of R^N.) Using the usual results on linear combinations of normals, we see that we have transformed the generalized linear model to the linear model in which we observe

$$\mathbf{Y}^* \sim N_N(\boldsymbol{\mu}^*, \sigma^2\mathbf{I}), \qquad \boldsymbol{\mu}^* \in V^*,\ \sigma^2 > 0.$$

As in the linear model, let

$$\hat{\boldsymbol{\mu}}_G^* = \mathbf{P}_{V^*}\mathbf{Y}^*, \quad \hat{\sigma}_G^2 = \frac{\|\mathbf{P}_{V^{*\perp}}\mathbf{Y}^*\|^2}{N-p}.$$

Then $\hat{\sigma}_G^2$ is the MV unbiased estimator of σ^2 for this model, but $\hat{\boldsymbol{\mu}}_G^*$ is the MV unbiased estimator of $\boldsymbol{\mu}^* = \mathbf{A}^{-1/2}\boldsymbol{\mu}$. Note that

$$\hat{\boldsymbol{\mu}}_G = \mathbf{A}^{1/2}\hat{\boldsymbol{\mu}}_G^*$$

is the unbiased estimator of $\boldsymbol{\mu} = \mathbf{A}^{1/2}\boldsymbol{\mu}^*$. Now $(\hat{\boldsymbol{\mu}}_G, \hat{\sigma}_G^2)$ is an invertible function of $(\hat{\boldsymbol{\mu}}_G^*, \hat{\sigma}_G^2)$, the complete sufficient statistic for the transformed linear model. Therefore $(\hat{\boldsymbol{\mu}}_G, \hat{\sigma}_G^2)$ is a sufficient statistic for the original generalized linear model.

We now discuss the computation of $\hat{\boldsymbol{\mu}}_G$ and $\hat{\sigma}_G^2$. Note that $\hat{\boldsymbol{\mu}}_G$ minimizes the generalized sum of squares given by

$$q(\boldsymbol{\mu}) = (\mathbf{Y} - \boldsymbol{\mu})^T\mathbf{A}^{-1}(\mathbf{Y} - \boldsymbol{\mu})$$

(see Exercise 31.58) so that $\hat{\boldsymbol{\mu}}_G$ is called the *generalized least squares estimator* of $\boldsymbol{\mu}$. (If $\mathbf{A}$ is a diagonal matrix, the estimator $\hat{\boldsymbol{\mu}}_G$ is often called the *weighted least squares estimator* of $\boldsymbol{\mu}$.) Let $\mathbf{X}$ be a basis matrix for V. Then in Exercise 31.59, it is shown that

$$\hat{\boldsymbol{\mu}}_G = \mathbf{X}(\mathbf{X}^T\mathbf{A}^{-1}\mathbf{X})^{-1}\mathbf{X}^T\mathbf{A}^{-1}\mathbf{Y}.$$

Therefore we have two different methods for computing the generalized LS estimator $\hat{\boldsymbol{\mu}}_G$ (minimize the generalized sum of squares or find a basis matrix) analogous to the two methods used for LS estimators. Note that for either of these methods it is not necessary to find the square root matrix $\mathbf{A}^{-1/2}$. Finally, note that once we have $\hat{\boldsymbol{\mu}}_G$, we can find $\hat{\sigma}_G^2$ by

$$(N-p)\hat{\sigma}_G^2 = (\mathbf{Y} - \hat{\boldsymbol{\mu}}_G)^T\mathbf{A}^{-1}(\mathbf{Y} - \hat{\boldsymbol{\mu}}_G)$$

(see Exercise 28.59).

Example 31.7 (One-way model with unequal variances, known ratios: estimation)

Consider a one-way model in which we observe Y_{ij} independent,

$$Y_{ij} \sim N(\mu_i, \sigma^2/k_i), \qquad i = 1, \dots, p;\ j = 1, \dots, n_i,\ N = \sum n_i,$$

where the k_i are known constants and μ_i and σ^2 are unknown parameters. For this model

$$(\mathbf{Y} - \boldsymbol{\mu})^T \mathbf{A}^{-1} (\mathbf{Y} - \boldsymbol{\mu}) = \sum\sum k_i (Y_{ij} - \mu_i)^2.$$

By direct differentiation we see that this expression is minimized by

$$\hat{\mu}_{iG} = \overline{Y}_{i..},$$

so that the generalized least squares and least squares estimators are the same for this model. Note, however, that for the generalized linear model

$$\hat{\sigma}_G^2 = \frac{\sum\sum k_i (Y_{ij} - \overline{Y}_{i.})^2}{N - p},$$

which is different from the estimator for the one-way linear model discussed earlier.

Now consider testing the null hypothesis that $\boldsymbol{\mu} \in W$, a k-dimensional subspace of V. Let W^* be the subspace of all $\boldsymbol{\mu}^* = \mathbf{A}^{-1/2}\boldsymbol{\mu}$, when $\boldsymbol{\mu} \in W$. Let

$$\hat{\hat{\boldsymbol{\mu}}}_G^* = \mathbf{P}_{W^*}\mathbf{Y}^*, \qquad \hat{\hat{\boldsymbol{\mu}}}_G = \mathbf{A}^{1/2}\hat{\hat{\boldsymbol{\mu}}}^*$$

be the generalized LS estimators of $\boldsymbol{\mu}^*$ and $\boldsymbol{\mu}$ for the reduced model. Let

$$SSH_G = \|\hat{\boldsymbol{\mu}}_G^* - \hat{\hat{\boldsymbol{\mu}}}_G^*\|^2 = (\hat{\boldsymbol{\mu}}_G - \hat{\hat{\boldsymbol{\mu}}}_G)^T \mathbf{A}^{-1} (\hat{\boldsymbol{\mu}}_G - \hat{\hat{\boldsymbol{\mu}}}_G),$$
$$dfh = p - k, \quad MSH = \frac{SSH}{dfh}, \quad dfe = N - p, \quad MSE_G = \hat{\sigma}_G^2.$$

By known results for the transformed linear model, we see that the optimal test for this hypothesis rejects if

$$F_G = \frac{MSH_G}{MSE_G} > F^{\alpha}_{dfh,dfe}.$$

Therefore, to compute the F statistic, we find the generalized LS estimators of $\boldsymbol{\mu}$ for the full and reduced model and substitute into this formula. Note also that for this model

$$EMSH_G = \sigma^2 + (\boldsymbol{\mu} - \boldsymbol{\gamma})^T \mathbf{A}^{-1} (\boldsymbol{\mu} - \boldsymbol{\gamma}), \quad EMSE_G = \sigma^2,$$

where $\boldsymbol{\gamma}$ is found by minimizing $(\boldsymbol{\mu} - \boldsymbol{\gamma})^T \mathbf{A}^{-1} (\boldsymbol{\mu} - \boldsymbol{\gamma})$ over all $\boldsymbol{\gamma} \in W$.

We now discuss the Scheffé simultaneous confidence intervals for contrasts associated with this hypothesis. Let $V \parallel W$ be the set of all vectors $\mathbf{v} \in V$ such that $\mathbf{v}^T \mathbf{A}^{-1} \mathbf{w} = 0$ for all $\mathbf{w} \in W$. It may be shown that $V \parallel W$ is a $(p - k)$-dimensional subspace of V. In Exercise 31.59 it is shown that the set of contrasts associated with this hypothesis is the set of all $\mathbf{d}^T \boldsymbol{\mu}$, $\mathbf{d} \in V \parallel W$, and the Scheffé simultaneous confidence intervals are given by

$$\mathbf{d}^T \mathbf{A}^{-1} \boldsymbol{\mu} \in \mathbf{d}^T \mathbf{A}^{-1} \hat{\boldsymbol{\mu}}_G \pm ((p - k)\ F^{\alpha}\ \mathbf{d}^T \mathbf{A}^{-1} \mathbf{d}\ \hat{\sigma}_G^2)^{1/2}, \qquad \forall \mathbf{d} \in V \parallel W.$$

Example 31.8 (One-way model with unequal variances, known ratios: inference)

We return to the one-way model discussed above and consider testing the equality of all the μ_i. To find the generalized LS estimator for the reduced model we must minimize

$$\sum\sum k_i(Y_{ij}-\mu)^2 \Rightarrow \hat{\mu}_G = \frac{\sum k_i n_i \overline{Y}_{i.}}{\sum k_i n_i} = \overline{Y}_{..G}.$$

Therefore, for this hypothesis

$$SSH = \sum k_i n_i (\overline{Y}_{i.} - \overline{Y}_{..G})^2, dfh = p-1,$$

$$EMSH = \sigma^2 + \sum k_i n_i (\mu_i - \overline{\mu}_G)^2, \quad \overline{\mu}_G = \frac{\sum n_i k_i \mu_i}{\sum n_i k_i}.$$

Therefore a partial ANOVA table for this model is

One-way generalized linear model

effect	*df*	*SS*	*EMS*
classes	$p-1$	$\sum k_i n_i(\overline{Y}_{i.} - \overline{Y}_{..G})^2$	$\sigma^2 + \sum g_i(\mu_i - \overline{\mu}_G)^2$
error	$N-p$	$\sum\sum k_i(Y_{ij} - \overline{Y}_{i.})^2$	σ^2

where $g_i = k_i n_i/(p-1)$.The hypothesis of no difference in the μ_i can be tested in the obvious way. We now find the Scheffé simultaneous confidence intervals associated with this hypothesis. Note that $\mathbf{d} \in V$ if and only if $d_{ij} = d_i$ does not depend on j, and that $\mathbf{w} \in W$ if and only if $w_{ij} = w$ does not depend on i or j. Hence, a vector $\mathbf{d} \in V$ is in $V \parallel W$ if and only if, for all $\mathbf{w} \in W$,

$$0 = \mathbf{d}^T\mathbf{A}^{-1}\mathbf{w} = w\sum\sum d_i k_i = w\sum n_i k_i d_i.$$

Therefore $\mathbf{d} \in V \parallel W$ if and only if $d_{ij} = c_i/n_i k_i$, where $\sum c_i = 0$. Furthermore, in this case,

$$\mathbf{d}^T\mathbf{A}^{-1}\boldsymbol{\mu} = \sum c_i \mu_i, \qquad \mathbf{d}^T\mathbf{A}^{-1}\hat{\boldsymbol{\mu}}_G = \sum c_i \overline{Y}_{i.},$$

$$\mathbf{d}^T\mathbf{A}^{-1}\mathbf{d} = \sum \frac{c_i^2}{n_i k_i}.$$

Therefore the Scheffé simultaneous confidence intervals are given by

$$\sum c_i \mu_i \in \sum c_i \overline{Y}_{i.} \pm \left((p-1)F^\alpha\, \hat{\sigma}_G^2 \sum \frac{c_i^2}{n_i k_i}\right)^{1/2}, \qquad \forall c_i \ni \sum c_i = 0.$$

We close this section with a regression example.

Example 31.9 (The Aitken estimator in regression)

Now consider a model in which we observe $\mathbf{Y} \sim N_n(\mathbf{X}\boldsymbol{\beta}, \sigma^2\mathbf{A})$, where $\boldsymbol{\beta} \in R^p$ and $\sigma^2 > 0$ are unknown parameters, $\mathbf{X}$ is a known $N \times p$ matrix of rank p and $\mathbf{A} > 0$ is a known $N \times N$ matrix. Let $\boldsymbol{\mu} = \mathbf{X}\boldsymbol{\beta}$ and let V be the subspace spanned by the columns of $\mathbf{X}$ (so that $\mathbf{X}$ is a basis matrix for V). Then this model is a generalized linear model as discussed above. Let $\hat{\boldsymbol{\mu}}_G$ be the generalized LS estimator of $\boldsymbol{\mu}$. Then

$$\hat{\boldsymbol{\mu}}_G = \mathbf{X}\hat{\boldsymbol{\beta}}_G, \qquad \hat{\boldsymbol{\beta}}_G = (\mathbf{X}^T\mathbf{A}^{-1}\mathbf{X})^{-1}\mathbf{X}^T\mathbf{A}^{-1}\mathbf{Y}.$$

By the generalized LS property of $\hat{\boldsymbol{\mu}}_G$, we see that

$$(\mathbf{Y} - \mathbf{X}\hat{\boldsymbol{\beta}}_G)^T \mathbf{A}^{-1}(\mathbf{Y} - \mathbf{X}\hat{\boldsymbol{\beta}}_G) \leq (\mathbf{Y} - \mathbf{X}\mathbf{b})^T \mathbf{A}^{-1}(\mathbf{Y} - \mathbf{X}\mathbf{b}).$$

Therefore $\hat{\boldsymbol{\beta}}_G$ is called the *generalized LS estimator* of $\boldsymbol{\beta}$. (It is also called the *Aitken estimator* of $\boldsymbol{\beta}$.) Note that for this generalized linear model

$$(N - p)\hat{\sigma}_G^2 = (\mathbf{Y} - \hat{\boldsymbol{\mu}}_G)^T \mathbf{A}^{-1}(\mathbf{Y} - \hat{\boldsymbol{\mu}}_G) = (\mathbf{Y} - \mathbf{X}\hat{\boldsymbol{\beta}}_G)^T \mathbf{A}^{-1}(\mathbf{Y} - \mathbf{X}\hat{\boldsymbol{\beta}}_G).$$

Consider testing the null hypothesis that $\mathbf{C}\boldsymbol{\beta} = \mathbf{0}$, where $\mathbf{C}$ is a known $s \times p$ matrix of rank s. In Exercise 31.60, it is shown that for this hypothesis

$$SSH_G = (\mathbf{C}\hat{\boldsymbol{\beta}}_G)^T (\mathbf{C}(\mathbf{X}^T \mathbf{A}^{-1}\mathbf{X})^{-1}\mathbf{C}^T)^{-1}(\mathbf{C}\hat{\boldsymbol{\beta}}_G), \qquad dfh = s,$$

so that the optimal test for this hypothesis rejects if

$$\frac{SSH_G}{s\hat{\sigma}_G^2} > F^{\alpha}_{s,N-p}.$$

For an alternative derivation of these results and further discussion, see **29.15**.

31.34 In previous sections of this chapter and in the previous chapter we have derived exact procedures for many models. However, these procedures were derived under fairly strong assumptions (equal variances, independence, etc.). When these strong assumptions are not met, typically exact procedures do not exist. When no exact procedures exist, it is often possible to use the generalized linear model to derive approximate procedure for inference about means, as we now illustrate.

Often in such applications of the generalized linear model we assume that σ^2 is known. In this case, we can assume without loss of generality $\sigma^2 = 1$, that is, that we observe $\mathbf{Y} \sim N_N(\boldsymbol{\mu}, \mathbf{A})$, where $\mathbf{A} > 0$ is a known $N \times N$ matrix and $\boldsymbol{\mu}$ is an unknown parameter assumed to lie in a known p-dimensional subspace V. Let $\hat{\boldsymbol{\mu}}_G$ be the generalized LS estimator of $\boldsymbol{\mu}$. Then $\hat{\boldsymbol{\mu}}_G$ is a sufficient statistic for the generalized linear model with known σ^2. As above, consider testing the null hypothesis that $\boldsymbol{\mu} \in W$, a known k-dimensional subspace of V. Let SSH, dfh be defined as above. Then the optimal test for this hypothesis rejects if

$$SSH > \chi^{2\alpha}_{dfh}.$$

The Scheffé simultaneous confidence intervals associated with this hypothesis are given by

$$\mathbf{d}^T \mathbf{A}^{-1}\boldsymbol{\mu} \in \mathbf{d}^T \mathbf{A}^{-1}\hat{\boldsymbol{\mu}}_G \pm (\chi^{2\alpha}\, \mathbf{d}^T \mathbf{A}^{-1}\mathbf{d})^{1/2}$$

(see Exercise 31.61 for derivations of these facts).

Example 31.10 (One-way model with unknown unequal variances)

Consider a one-way model in which we allow the possibility of unequal variances. That is, we observe Y_{ij} independent, $Y_{ij} \sim N(\mu_i, \sigma_i^2)$, $i = 1, \ldots, p$; $j = 1, \ldots, n_i$; $N = \sum n_i$. We consider testing the null hypothesis that the μ_i are equal. We first assume that the σ_i^2 are known. In this case the model is a general linear model with known σ^2. By the calculations above (with $k_i = 1/\sigma_i^2$), we see that for this problem

$$SSH_G = \sum \frac{n_i(\overline{Y}_{i.} - \overline{Y}_{..G})^2}{\sigma_i^2}, \qquad \overline{Y}_{..G} = \frac{\sum n_i \overline{Y}_{i.}/\sigma_i^2}{\sum n_i/\sigma_i^2}, \qquad dfh = p - 1,$$

and the Scheffé simultaneous confidence intervals are given by

$$\sum c_i\mu_i \in \sum c_i\overline{Y}_{i.} \pm \left(\chi^{2\alpha} \sum \frac{c_i^2\sigma_i^2}{n_i}\right), \qquad \forall c_i \ni \sum c_i = 0.$$

Now suppose that the σ_i^2 are unknown. Let $\hat{\sigma}_i^2 = \sum_j (Y_{ij} - \overline{Y}_{i.})/(n_i - 1)$ be the sample variance of the observations in the ith class. Then an approximate test for this hypothesis for the case of unequal unknown variance rejects if

$$\sum \frac{n_i(\overline{Y}_{i.} - \hat{Y}_{..G})^2}{\hat{\sigma}_i^2} > \chi_{p-1}^{2\alpha}, \qquad \hat{Y}_{..G} = \frac{\sum n_i \overline{Y}_{i.}/\hat{\sigma}_i^2}{\sum n_i/\hat{\sigma}_i^2},$$

and an approximate set of simultaneous confidence intervals for contrasts associated with this hypothesis is

$$\sum c_i\mu \in \sum c_i\overline{Y}_{i.} \pm \left(\chi^{2\alpha} \sum \frac{c_i^2\hat{\sigma}_i^2}{n_i}\right)^{1/2}, \qquad \forall c_i \ni \sum c_i = 0.$$

Note that the size of this approximate test and the confidence coefficient of the confidence intervals are asymptotically correct as long as the $n_i \to \infty$.

As illustrated in the example above, in practical problems for which no exact procedures exist we can often use the generalized linear model (with known σ^2) to find approximate procedures. We first find the optimal procedure assuming that we know all the variances and covariances. We then substitute consistent estimators for the unknown variances and covariances to obtain asymptotically correct procedures for the model with unknown variances and unknown covariances. This approach typically works when the number of unknown parameters in the variance-covariance structure is small compared to the number of observations. (In the example above, there are p unknown variances and N observations). In Exercise 31.62, an optimal procedure is derived for a twofold nested model with known but unequal variances. In practice, we can substitute the sample variance in the (i, j)th cell for the true variance in that cell to get approximate procedures for a twofold nested model with possibly unequal variances.

In Exercise 31.63, the more general linear model is also used to find approximate procedures for a univariate repeated measures model with a covariate which does not satisfy the conditions for an exact procedure to exist. This example indicates that it is typically possible to use the more general linear model to find approximate procedures for many univariate repeated measures models for which exact procedures are not possible.

Now consider a one-way repeated measures model in which every individual receives each possible treatment. Let $\mathbf{Y}_k$ be the vector of observations on the kth individual. Another covariance structure which is often assumed for this situation is that

$$\operatorname{cov}(\mathbf{Y}_k) = \sigma^2\mathbf{B}(\rho),\ \mathbf{B}(\rho) = \begin{pmatrix} 1 & \rho & \rho^2 & \cdots & \rho^{m-1} \\ \rho & 1 & \rho & \cdots & \rho^{m-2} \\ \rho^2 & \rho & 1 & \cdots & \rho^{m-3} \\ \vdots & \vdots & \vdots & \ddots & \vdots \\ \rho^{m-1} & \rho^{m-2} & \rho^{m-3} & \cdots & 1 \end{pmatrix},$$

so that

$$\operatorname{cov}(Y_{ik}, Y_{jk}) = \rho^{|i-j|}\sigma^2.$$

This model is called a *first-order autocorrelation* repeated measures model. It is often used for repeated measures that are being taken over time. Note that the observations closer in time are more correlated than those further away in time. (Note also that the usual univariate assumption of equal correlations between observations on the same individuals may often be unreasonable in such time series situations.) In Exercise 31.64, we ask the reader how to use the more general linear model to give approximate procedures for this repeated measures model for which no exact procedures exist. Similar approximate procedures can be developed for many other more complicated repeated measures assumptions.

The more general linear model is only useful for deriving approximate procedures for inference about mean vectors which have a linear structure. In particular, the more general linear model is rarely helpful for inference about covariance matrices and is therefore not helpful for inference about random effects models.

EXERCISES

31.1 (a) In the notation of **31.3**, show that $V = Q \oplus U$.
(b) Show that $\mathbf{X}^{*T}\mathbf{Y}^* = \mathbf{X}^{*T}\mathbf{Y}$.

31.2 In the notation of **31.3**, show that $\hat{\boldsymbol{\gamma}}$ and $\hat{\boldsymbol{\delta}}_a$ are the best unbiased and maximum likelihood estimators of $\boldsymbol{\gamma}$ and $\boldsymbol{\delta}$.

31.3 In the notation of **31.4**:
(a) Show that $\mathbf{X}^* = \mathbf{P}_{V^\perp}\mathbf{X}^{**}$. (*Hint*: $V^\perp \subset W^\perp$.)
(b) Show that if $\mathbf{X}^*$ has full column rank, then so does $\mathbf{X}^{**}$.

31.4 In the notation of **31.4**:
(a) Show that $\mathbf{g} \in V$ if and only if $\mathbf{g} = \mathbf{d} + \mathbf{X}^*\mathbf{c}$, $\mathbf{d} \in Q$, $\mathbf{c} \in \mathbf{R}^s$.
(b) Show that $\mathbf{g} \in V|W$ if and only if in addition $\mathbf{d} \perp T$ and $\mathbf{c} = -(\mathbf{X}^{*T}\mathbf{X}^*)^{-1}\mathbf{X}^T\mathbf{d}$.
(c) When $\mathbf{g} \in V|W$ show that $\mathbf{g}^T\boldsymbol{\mu} = \mathbf{d}^T\boldsymbol{\delta}$, $\mathbf{g}^T\hat{\boldsymbol{\mu}} = \mathbf{d}^T\hat{\boldsymbol{\delta}}_a = \mathbf{d}^T\hat{\boldsymbol{\delta}}_u - \mathbf{d}^T\mathbf{X}\hat{\boldsymbol{\gamma}}$ and $\|\mathbf{g}\|^2 = \|\mathbf{d}\|^2 + \mathbf{d}^T\mathbf{X}(\mathbf{X}^{*T}\mathbf{X}^*)^{-1}\mathbf{X}^T\mathbf{d}$.
(d) Show that if $\mathbf{X}^* = \mathbf{X}^{**}$ then $\mathbf{X}^T\mathbf{d} = \mathbf{0}$.
(e) Show that if $\mathbf{X} = \mathbf{X}^*$ then $\mathbf{X}^* = \mathbf{X}^{**}$.

31.5 In the balanced one-way ANCOVA model, suppose that the $\overline{x}_{i.}$ are all the same.
(a) Show that $\max_{i \neq j} \sqrt{n}|\overline{Y}_{i.} - \overline{Y}_{j.} - (\alpha_i - \alpha_j)|/\sqrt{MSEa} \sim q_{p,N-p-1}$.
(b) Construct Tukey simultaneous confidence intervals for the set of comparisons in the α_i.
(c) Construct Dunnett simultaneous confidence intervals for comparing treatments to a control.

31.6 (a) In the notation of **31.5**, show that $F(\boldsymbol{\theta}) \sim F_{r,dfe_a}$.
(b) Show that the likelihood ratio test for testing that $\boldsymbol{\theta} = \mathbf{0}$ rejects if $F(\mathbf{0}) > F^\alpha$. (*Hint*: $\mathbf{A}\boldsymbol{\gamma} = \mathbf{C}^T\boldsymbol{\mu}$ and $\mathbf{A}\hat{\boldsymbol{\gamma}} = \mathbf{C}^T\hat{\boldsymbol{\mu}}$ where $\mathbf{C} = \mathbf{X}^*(\mathbf{X}^{*T}\mathbf{X}^*)\mathbf{A}^T$.)

31.7 Consider a twofold nested design with a single covariate.
(a) Write down the model for this design.
(b) What are the least squares estimators of the parameters and *MSE* for this model?
(c) What is the F test for testing for no class effect? What are the associated Scheffé simultaneous confidence intervals?
(d) What is the F test for testing for no subclass effect? What are the associated Scheffé simultaneous confidence intervals?
(e) Under what condition will the only differences between the above procedures and the ones for the associated ANOVA model be in *SSE* and *dfe*?

31.8 Describe appropriate procedures for the twofold nested model with two covariates.

31.9 Consider the one-way model with a single covariate but possibly unequal slopes and intercepts. How would you test the null hypothesis that the slopes are equal and the intercepts are equal? (*Hint*: under the null hypothesis, the model is just a simple regression model.)

31.10 (a) Consider a twofold nested design with a single covariate in which the slope depends on the class. How would you test the null hypothesis that the slopes are equal?
(b) Consider a twofold nested design with a single covariate in which the slope depends on the subclass chosen. How would you test the null hypothesis that the slope depends only on the class chosen?

31.11 In the one-way random effects model, verify that the F test is the LR test for testing the null hypothesis that $\sigma_a^2 = 0$ against the alternative hypothesis that $\sigma_a^2 > 0$.

31.12 (a) Show that the one-way random effects model is invariant under the group of transformations $Y_{ij} \to cY_{ij} + a$, $c > 0$.
(b) Show that the F test is the UMP invariant test for testing the null hypothesis that $\sigma_a^2 = 0$ against the alternative hypothesis that $\sigma_a^2 > 0$.

31.13 In the one-way random effects model let T be an unbiased estimator of σ_a^2.
(a) Show that $E(T|(\overline{Y}_{..}, SSa, SSe)) = MSa$. (Use the fact that $(\overline{Y}_{..}, SSa, SSe)$ is a complete sufficient statistic.)
(b) Show that $P(T > 0) < 1$. (If not, then $P(SSa > 0) = 1$, which is a contradiction.)

31.14 (a) In the one-way random effects model verify that the likelihood is maximized when $\hat{\theta} = \overline{Y}_{..}$, $\hat{\gamma} = (p-1)MSa/p$, $\hat{\sigma}_e^2 = MSE$ provided this value is in the parameter space.
(b) Verify that the likelihood is maximized by $\hat{\theta} = \overline{Y}_{..}$, $\hat{\gamma} = \hat{\sigma}_e^2 = (SSa + SSe)/np$ when we assume that $\gamma = \sigma_e^2$.

31.15 (a) Verify the formula for the restricted likelihood for the one-way random effects model. (*Hint*: $SSa \sim \Gamma((p-1)/2, 2\gamma)$, $SSE \sim \Gamma(p(n-1)/2, 2\sigma_e^2)$ and they are independent.)
(b) Verify the formulae for the REML estimators for γ, σ_e^2 and σ_a^2 in the one-way random effects model.

31.16 (Satterthwaite approximations.) Suppose that we observe $\hat{\gamma}_i$ independent, $d_i\hat{\gamma}_i/\gamma_i \sim \chi^2_{d_i}$ (so that $E\hat{\gamma}_i = \gamma_i$). Let $\tau = \sum a_i\gamma_i$, $\hat{\tau} = \sum a_i\hat{\gamma}_i$. We want to find d such that $W = d\hat{\tau}/\tau$ has approximately the same distribution as $U \sim \chi^2_d$.
(a) Show that $EW = EU$ and that $\text{var}(W) = \text{var}(U) \Leftrightarrow d = \tau^2/\sum(a_i^2\gamma_i^2/d_i)$.
(The Satterthwaite approximation for the distribution of $\hat{\tau}$ is therefore given by $\hat{d}\hat{\tau}/\tau \approx \chi^2_{\hat{d}}$, where $\hat{d} = \hat{\tau}^2/\sum(a_i^2\hat{\gamma}_i^2/d_i)$.)
(b) Use this approximation to find an approximate $1-\alpha$ confidence interval for τ.
(Note that this approximation is purely a method of moments argument. There is really no proof that the distribution should be anywhere near a χ^2 distribution. For example, the Satterthwaite approximations can be silly when some of the $a_i < 0$ but $\tau \geq 0$ (as often happens in random effects and mixed models). In that case, $P(\hat{d}\hat{\tau}/\tau > 0) < 1$. In this case, it is hard to imagine a χ^2 approximation being very useful. For this reason, many people are reluctant to use these approximations, especially when some of the $a_i < 0$.)

31.17 Find a Satterthwaite approximate $1-\alpha$ confidence interval for $\sigma_a^2 = (\gamma_a - \sigma_e^2)/n$ in the one-way random effects model. (Note that this confidence interval will be really silly unless the unbiased estimator of σ_a^2 is positive.)

31.18 (a) In the twofold nested random effects model, verify that $Y_{ijk} = \theta + a_i + d_{ij} + e_{ijk}$.
(b) Verify that the a_i, d_{ij} and e_{ijk} are all independent and that $a_i \sim N(0, \sigma_a^2)$, $d_{ij} \sim N(0, \sigma_d^2)$, $e_{ijk} \sim N(0, \sigma_e^2)$.

31.19 Verify the joint distribution $SS\theta$, SSa, SSd and SSe for the twofold nested random effects model.

31.20 (a) For the twofold nested random effects model, verify the formula for the likelihood and use this formula to verify that $(\overline{Y}_{...}, SSa, SSd, SSe)$ is a complete sufficient statistic.
(b) Use the complete sufficient statistic to verify that there is no non-negative unbiased estimator of σ_a^2 or σ_d^2.
(c) Verify the formula for the restricted likelihood for this model.

31.21 Find Satterthwaite approximate $1-\alpha$ confidence intervals for σ_a^2 and σ_d^2 in the twofold nested random effects model. (Note that these intervals can also be quite silly.)

31.22 (a) In the twofold nested mixed model, verify that $Y_{ijk} = \theta + \alpha_i + d_{ij} + e_{ijk}$.
(b) Verify that $\sum\alpha_i = 0$ and that d_{ij} and e_{ijk} are independent and that $d_{ij} \sim N(0, \sigma_d^2)$, $e_{ijk} \sim N(0, \sigma_e^2)$.

31.23 Verify the joint distribution of $SS\theta$, $SS\alpha$, SSd and SSe for the twofold nested mixed model.

31.24 (a) For the twofold nested mixed model verify the formula for the likelihood and use this formula to verify that $(\overline{Y}_{1..}, \ldots, \overline{Y}_{p..}, SSd, SSe)$ is a sufficient statistic.
(b) Find the ML estimator for this model.

(c) Verify the formula for the restricted likelihood and find the REML estimators of σ_d^2 and σ_e^2.

31.25 (a) For the twofold nested mixed model, verify that

$$\max \sqrt{nc}|\overline{Y}_{i\ldots} - \overline{Y}_{iT\ldots} - (\mu_i - \mu_{iT})|/MSd \sim q_{p,p(c-1)}.$$

(b) Use this fact to construct Tukey simultaneous confidence intervals for comparisons in the μ_i.

31.26 For the two-way crossed random effects model, let G_{ij}, H_i, K_j and L be defined as in Lemma 31.3.
(a) Show that $\overline{G}_{i.}$ is uncorrelated with $G_{ij} - \overline{G}_{i.} - \overline{G}_{.j} + \overline{G}_{..}$.
(b) Show that L, $H_i - \overline{H}$ and $K_j - \overline{K}$ are uncorrelated.
(c) Verify that $SS\theta = L^2$, $SSa = \sum(H_i - \overline{H})^2$, $SSb = \sum(K_j - \overline{K})^2$, $SSd = \sum\sum(G_{ij} - \overline{G}_{i.} - \overline{G}_{.j} + \overline{G}_{..})^2$ and $SSe = \sum\sum\sum(e_{ijk} - \overline{e}_{ij})^2$.

31.27 For the two-way crossed random effects model:
(a) Find a Satterthwaite approximate confidence interval for σ_d^2.
(b) Find a Satterthwaite approximate confidence interval for σ_a^2.

31.28 (a) Verify the formula for the likelihood for the crossed two-way random effects model.
(b) Verify the restricted likelihood for this model.

31.29 Consider a three-way random effects model which can be written as $Y_{ijkm} = \theta + a_i + b_j + c_k + (ab)_{ij} + (ac)_{ik} + (bc)_{jk} + (abc)_{ijk} + e_{ijkm}$, $i = 1, \ldots, r$; $j = 1, \ldots, f$; $k = 1, \ldots, d$; $m = 1, \ldots, n$, where $a_i \sim N(0, \sigma_a^2)$, $(ab)_{ij} \sim N(0, \sigma_{ab}^2)$, $(abc)_{ijk} \sim N(0, \sigma_{abc}^2)$, etc. and random effects are all independent.
(a) Show that $SSa/\gamma_a \sim \chi^2_{r-1}$, $SSab/\gamma_{ab} \sim \chi^2_{(r-1)(f-1)}$, $SSac/\gamma_{ac} \sim \chi^2_{(r-1)(d-1)}$, $SSabc/\gamma_{abc} \sim \chi^2_{(r-1)(f-1)(d-1)}$ where $\gamma_a = nfd\sigma_a^2 + nd\sigma_{ab}^2 + nf\sigma_{ac}^2 + n\sigma_{abc}^2 + \sigma_e^2$, $\gamma_{ab} = nd\sigma_{ab}^2 + n\sigma_{abc}^2 + \sigma_e^2$, $\gamma_{ac} = nf\sigma_{ac}^2 + n\sigma_{abc}^2 + \sigma_e^2$, $\gamma_{abc} = n\sigma_{abc}^2 + \sigma_e^2$. (This implies that there is no natural F test for testing that $\sigma_a^2 = 0$.)
(b) Use a Satterthwaite approximation to obtain an approximate χ^2 to use in the denominator of an approximate F test for testing that $\sigma_a^2 = 0$.
(c) Find another Satterthwaite approximation approximating both the numerator and denominator sums of squares based on the fact that $\sigma_a^2 = 0 \Leftrightarrow \gamma_a + \gamma_{abc} = \gamma_{ab} + \gamma_{ac}$.

31.30 (a) For the unrestricted version of the two-way crossed mixed model, verify the joint distribution of $SS\theta$, $SS\alpha$, SSb, SSd and SSe.
(b) Verify the likelihood for this model and verify that a complete sufficient statistic is $(\overline{Y}_{1\ldots}, \ldots, \overline{Y}_{r\ldots}, SSb, SSd, SSe)$.

31.31 In the unrestricted version of the two-way crossed mixed model, let $M_{ij} = \theta + \alpha_i + d_{ij} + \overline{e}_{ij.}$.
(a) Show that the M_{ij} are independent, $M_{ij} \sim N(\theta + \alpha_i, \tau^2)$, where $\tau^2 = \sigma_d^2 + \sigma_e^2/n$.
(b) Let $Q = \sum\sum(M_{ij} - \overline{M}_{i.} - \overline{M}_{.j} + \overline{M}_{..})^2/(r-1)(c-1)$. Show that

$$P\left(\sum a_i\alpha_i \in \sum a_i\overline{M}_{i.} \pm ((r-1)F^{\alpha}_{r-1,(r-1)(c-1)}Q\sum a_i^2/c)^{1/2}\forall a_i \ni \sum a_i = 0\right) = 1 - \alpha.$$

(Think of the M_{ij} as coming from a two-way crossed model with no replication in which the $\beta_j = 0$.)
(c) Verify the Scheffé simultaneous confidence intervals for contrasts in the α_i. ($\overline{Y}_{ij.} = M_{ij} + \overline{b}_j$, which implies that $\sum a_i\overline{M}_{i.} = \sum a_i\overline{Y}_{i.}$ and $Q = MSd/n$.)

31.32 (a) For the unrestricted version of the two-way mixed model, verify that

$$\max \sqrt{nc}|\overline{Y}_{i..} - \overline{Y}_{i'..} - (\alpha_i - \alpha_{i'})|/MSd \sim q_{r-1,(r-1)(c-1)}.$$

(b) Find Tukey simultaneous confidence intervals for the set of comparisons in the α_i.

31.33 (a) Show that if Q_i are exchangeably distributed and $\sum Q_i = 0$, then $\text{cov}(Q_i, Q_{i'}) = -\text{var}(Q_i)/(r-1)$, provided $\text{var}(Q_i) < \infty$. (*Hint*: $\text{var}(\sum Q_i) = 0$.)
(b) For the restricted version of the two-way crossed mixed model, show that $\text{cov}(d^*_{ij}, d^*_{i'j}) = \sigma_d^{*2}/r$.

31.34 Verify the relationship given between the two versions of the two-way crossed mixed model.

31.35 Following the notation of **31.22**, let $\gamma_k^* = \mathbf{C}^T\gamma$.

(a) Show that $\mathbf{B}\mu_k = \begin{pmatrix} r^{\frac{1}{2}}\delta_k \\ \gamma_k^* \end{pmatrix}$, $\mathbf{B}\sigma^2\mathbf{A}(\rho)\mathbf{B}^T = \begin{pmatrix} \tau_b^2 & \mathbf{0} \\ \mathbf{0} & \tau_w^2\mathbf{I} \end{pmatrix}$.

(*Hint*: $\mathbf{A}(\rho) = (1-\rho)\mathbf{I}_r + \rho\mathbf{1}_r\mathbf{1}_r^T$.)

(b) Show that $\boldsymbol{\Gamma}\mu = \begin{pmatrix} \delta^* \\ \gamma^* \end{pmatrix}$ and that

$$\boldsymbol{\Gamma}(\sigma^2\mathbf{A}(\rho) \otimes \mathbf{I}_m)\boldsymbol{\Gamma}^T = \begin{pmatrix} \tau_b^2\mathbf{I}_m & \mathbf{0} \\ \mathbf{0} & \tau_w^2\mathbf{I}_{m(r-1)} \end{pmatrix}.$$

(c) Show that $\sigma^2\mathbf{A}(\rho) > 0$ if and only if $\tau_b^2 > 0$ and $\tau_w^2 > 0$. (Note that $\mathbf{H} > 0$ if and only if $\mathbf{BHB}^T > 0$. Why?)

31.36 In the notation of **31.22**:
(a) Let $\mathbf{X}$ be an orthonormal basis matrix for V. Show that $\mathbf{X}^* = \boldsymbol{\Gamma}\mathbf{X}$ is an orthonormal basis for V^* and hence that $\dim(V) = \dim(V^*)$ and $\mathbf{P}_{V^*} = \boldsymbol{\Gamma}\mathbf{P}_V\boldsymbol{\Gamma}^T$.
(b) Let $\mathbf{Q}$ be an orthonormal basis matrix for T. Show that $\mathbf{Q}^* = \mathbf{GQ}$ is an orthonormal basis matrix for T^* and hence $\dim(T) = \dim(T^*)$ and $\mathbf{P}_{T^*} = \mathbf{GP}_T\mathbf{G}^T$.
(c) Let $\mathbf{X}_1$ and $\mathbf{X}_2$ be orthonormal basis matrices for S and T^*. Show that

$$\begin{pmatrix} \mathbf{X}_1 & \mathbf{0} \\ \mathbf{0} & \mathbf{X}_2 \end{pmatrix}$$

is an orthonormal basis matrix for V^* and hence $\dim(V^*) = \dim(S) + \dim(T^*)$ and

$$\mathbf{P}_{V^*} = \begin{pmatrix} \mathbf{P}_S & \mathbf{0} \\ \mathbf{0} & \mathbf{P}_{T^*} \end{pmatrix}.$$

(d) Show that $\|\mathbf{P}_{V^*}\mathbf{Z}\|^2 = \|\mathbf{P}_S\mathbf{Z}_1\|^2 + \|\mathbf{P}_{T^*}\mathbf{Z}_2\|^2$.
(e) Show that $\dim(V) = s + t$.
(f) Show that $\|\mathbf{Y} - \mu\|^2 = r\|\overline{\mathbf{Y}} - \delta\| + \|\mathbf{Z}_2 - \gamma^*\|$ and hence that the LS estimator of δ for the associated ANOVA model is $\hat{\delta} = \mathbf{P}_{S\perp}\overline{\mathbf{Y}}$.
(g) Show that $r\|\mathbf{P}_{S\perp}\overline{\mathbf{Y}}\|^2 = r\|\overline{\mathbf{Y}} - \hat{\delta}\|^2$.
(h) Let W and Q be defined as in the proof of part of Theorem 31.5. Show that $\mathbf{d} \in V|W \Leftrightarrow \boldsymbol{\Gamma}\mathbf{d} = \binom{\mathbf{d}^*}{\mathbf{0}}$, $\mathbf{d}^* \in S|Q$.

31.37 In the notation of **31.22**, let $\hat{\delta}^* = \mathbf{P}_S\mathbf{Z}_1$, $\hat{\gamma}^* = \mathbf{P}_{T^*}\mathbf{Z}_2$ and $\hat{\mu} = \mathbf{P}_V\mathbf{Y}$.
(a) Show that a sufficient statistic for the transformed version of the univariate repeated measures model is $\mathbf{T} = (\hat{\delta}^*, \hat{\tau}_b^2, \hat{\gamma}^*, \hat{\tau}_w^2)$. (You may assume that $\mathbf{T}$ is in fact a complete sufficient statistic for this model in part (b).)
(b) Show that $\binom{\hat{\delta}}{\hat{\gamma}} = \boldsymbol{\Gamma}\hat{\mu}$, and hence that a complete sufficient statistic for the univariate repeated measures model is $(\hat{\mu}, \hat{\tau}_b^2, \hat{\tau}_w^2)$. (Note that we have merely replaced $\hat{\sigma}^2$ by $(\hat{\tau}_b^2, \hat{\tau}_w^2)$ in the complete sufficient statistic for the fixed effect ANOVA model.).

31.38 In the transformed version of the univariate repeated measures model
(a) Show that the ML estimator of $(\delta^*, \tau_b^2, \gamma^*, \tau_w^2)$ is $\left(\hat{\delta}^*, \left(\frac{m-s}{m}\right)\hat{\tau}_b^2, \hat{\gamma}^*, \left(\frac{mr-t}{mr}\right)\hat{\tau}_w^2\right)$.

(b) Find the ML estimator of $\boldsymbol{\mu}$, σ^2 and ρ. (Note that $\boldsymbol{\mu} = \boldsymbol{\Gamma}^T\binom{\delta^*}{\gamma^*}$, $\sigma^2 = (\tau_b^2 + (r-1)\tau_w^2)/r$, $\rho = (\tau_b^2 - \tau_w^2)/(\tau_b^2 + (r-1)\tau_w^2)$. Why?)

31.39 Prove part (b) of Theorem 31.5.

31.40 For the univariate one-way repeated measures model, show that $SSE - SSB = \sum\sum(Y_{ij} - \overline{Y}_{i.} - \overline{Y}_{.j} + \overline{Y}_{..})^2$.

31.41 For the univariate one-way repeated measures model

(a) Show that $\max_{i \neq i^T} \sqrt{m}|\overline{Y}_i - \overline{Y}_{i'.} - (\alpha_i - \alpha_{i'})|/\sqrt{MSW} \sim q_{r,dfw}$. (Note that $E\overline{Y}_{i.} = \theta + \alpha_i$, $\text{var}(\overline{Y}_{i.}) = \sigma^2/m$, $\text{cov}(\overline{Y}_{i.}, \overline{Y}_{i'.}) = \rho\sigma^2/m$.)

(b) Use part (a) to construct Tukey simultaneous $1-\alpha$ confidence intervals for the set of all comparisons in the α_i.

31.42 In the notation of Theorem 31.5, verify that $F_b = (m-s)\|\mathbf{P}_{S|Q}\mathbf{Y}\|^2/(s-q)\|\mathbf{P}_{S^\perp}\mathbf{Y}\|^2$.

31.43 (a) For the single group univariate two-way repeated measures model, show that $\delta_k = \theta$, $r = dc$, $m = n$, $s = 1$, $SSB = dc\sum(\overline{Y}_{..k} - \overline{Y}_{...})^2$, $SSW = \sum\sum\sum(Y_{ijk} - \overline{Y}_{ij.} - \overline{Y}_{..k} + \overline{Y}_{...})^2$.

(b) Verify the ANOVA table for this model.

31.44 (a) For the multiple-group univariate two-way repeated measures model, verify that $\sqrt{nc}\max|\overline{Y}_{i..} - \overline{Y}_{i'..} - (\alpha_i - \alpha_{i'})|/\sqrt{MSW} \sim q_{d,dfw}$ and use this fact to construct Tukey simultaneous confidence intervals for the set of all comparisons in the α_i. (Note that $E\overline{Y}_{i.} = \theta + \alpha_i$, $\text{var}(\overline{Y}_{i..}) = \sigma^2/nc$, $\text{cov}(\overline{Y}_{i..}, \overline{Y}_{i'..}) = \rho\sigma^2/nc$.)

(b) Verify that $\sqrt{nd}|\overline{Y}_{.j.} - \overline{Y}_{.j^T.} - (\beta_j - \beta_{j^T})|\sqrt{MSB} \sim q_{c,dfb}$. (Note that $E\overline{Y}_{.j.} = \theta + \beta_j$, $\text{var}(\overline{Y}_{.j.}) = \sigma^2(1 + (d-1)\rho)/dn$, $\text{cov}(\overline{Y}_{.j.}, \overline{Y}_{.j^T.}) = 0$.)

31.45 Now consider a multiple-group univariate two-way repeated measures model in which each individual receives each row treatment but only one column treatment. Suppose that the number of individuals receiving a particular column treatment is n_j, which may be different for different columns. Therefore, let Y_{ijk} be the observation on the ith row treatment for the kth person receiving the jth column treatment, $i = 1, \ldots, d$; $j = 1, \ldots, c$; $k = 1, \ldots, n_j$. Suppose that $EY_{ijk} = \theta + \alpha_i + \beta_j + \gamma_{ij}$, $\sum\alpha_i = 0$, $\sum n_j\beta_j = 0$, $\sum_i \gamma_{ij} = 0$, $\sum_j n_j\gamma_{ij} = 0$. Derive the appropriate ANOVA table for this model. (*Hint*: the associated ANOVA model is a proportional sampling model.)

31.46 For the single-group univariate twofold nested repeated measures model, verify that $\delta_k = \theta$, $SSB = cd\sum(\overline{Y}_{..k} - \overline{Y}_{...})^2$, $SSW = \sum\sum\sum(Y_{ijk} - \overline{Y}_{ij.} - \overline{Y}_{..k} + \overline{Y}_{...})^2$, $dfb = n-1$, $dfw = (dc-1)(n-1)$.

31.47 Consider a single-group univariate twofold nested model in which there are c_i subclasses of class i. (We must have each subclass having the same number n of observations. Why?) Assume that $\sum_j \zeta_{ij} = 0$, $\sum c_i\alpha_i = 0$ (so that the associated ANOVA model is an orthogonal design).

(a) Show that θ is a between effect and that α_i and ζ_{ij} are within effects.

(b) Find the ANOVA table for this model.

(c) Find the Scheffé simultaneous confidence intervals for contrasts in the α_i and contrasts in the ζ_{ij}.

31.48 Consider a multiple-group nested model in which n_i individuals receive the ith class treatment and every subclass treatment of that class. (We must assume that each class has the same number c of subclasses. Why?) Assume the constraints that $\sum_j \zeta_{ij} = 0$ and $\sum n_i\alpha_i = 0$ (so that the associated ANOVA model is an orthogonal design).

(a) Show that θ and α_i are between effects and ζ_{ij} is a within effect.

(b) Find the ANOVA table for this model.

(c) Find the Scheffé simultaneous confidence intervals for contrasts in the α_i and contrasts in the ζ_{ij}.

31.49 Consider a univariate one-way repeated measures model with a covariate. Suppose that the $\overline{x}_{i.} = \overline{x}_{..}$ does not depend on i. Derive the F test for this model. (*Hint*: reparametrize the model with $\theta^* = \theta + \beta\overline{x}_{...}$, $x_{ij}^* = x_{ij} - \overline{x}_{...}$)

31.50 Consider a univariate multiple-group two-way repeated measures model with a covariate that only depends on the individual (i.e. such that $x_{ijk} = x_{jk}$).
(a) Find the F test for testing for no row effect.
(b) Find the F test for testing for no column effect.

31.51 Consider a univariate multiple-group two-way repeated measures model with a covariate whose average over each individual is 0 (i.e. such that $\overline{x}_{.jk} = 0$).
(a) Find the F test for testing for no row effect.
(b) Find the F test for testing for no column effect.

31.52 Suppose we observe a two-way mixed model with no interaction in which $Y_{ik} = \theta + \alpha_i + b_k + e_{ik}$, where θ and α_i are constants and b_k and e_{ik} are independent random variables such that $b_k \sim N(0, \sigma_b^2)$ and $e_{ik} \sim N(0, \sigma_e^2)$. Let $\mathbf{Y}_k = (Y_{1k}, \ldots, Y_{rk})$. Show that $\text{cov}(\mathbf{Y}_k) = \sigma^2 \mathbf{A}_r(\rho), \sigma^2 = \sigma_b^2 + \sigma_e^2$ and $\rho = \sigma_a^2/(\sigma_a^2 + \sigma_e^2)$.

31.53 Consider the single-group multivariate repeated measures model with $m > r$.
(a) Show that the likelihood for this model satisfies

$$L(\boldsymbol{\mu}, \boldsymbol{\Sigma}) \approx |\boldsymbol{\Sigma}|^{-m/2} \exp(-((n-1)\,\text{tr}(\boldsymbol{\Sigma}^{-1}\mathbf{S}) - m(\overline{\mathbf{Y}} - \boldsymbol{\mu})^T \boldsymbol{\Sigma}^{-1} (\overline{\mathbf{Y}} - \boldsymbol{\mu}))/2).$$

(b) Show that $(\overline{\mathbf{Y}}, \mathbf{S})$ is a complete sufficient statistic for this model.
(c) Show that $\overline{\mathbf{Y}}$ and $\mathbf{S}$ are MV unbiased estimators of $\boldsymbol{\mu}$ and $\boldsymbol{\Sigma}$.
(d) Show that the ML estimator for $(\boldsymbol{\mu}, \boldsymbol{\Sigma})$ for this model is $(\overline{\mathbf{Y}}, (\frac{m-1}{m})\mathbf{S})$. (*Hint*: if $\mathbf{A} > 0$ is a known matrix and $\mathbf{Q} > 0$, then $h(\mathbf{Q}) = |\mathbf{Q}|^{-k} \exp(-\,\text{tr}(\mathbf{Q}^{-1}\mathbf{A}))$ is maximized when $\mathbf{Q} = \frac{1}{k}\mathbf{A}$. See, for example, Arnold, 1981, pp. 459–460.)

31.54 Consider a single-group two-way crossed multivariate repeated measures model with two rows and two columns, so that

$$\boldsymbol{\mu} = \begin{pmatrix} \theta + \alpha_1 + \beta_1 + \gamma_{11} \\ \theta + \alpha_1 + \beta_2 + \gamma_{12} \\ \theta + \alpha_2 + \beta_1 + \gamma_{21} \\ \theta + \alpha_2 + \beta_2 + \gamma_{22} \end{pmatrix}.$$

Assume the usual (equal weight) constraints.
(a) Find the vector $\mathbf{a}$ such that $\alpha_i = 0$ if and only if $\mathbf{a}^T \boldsymbol{\mu} = 0$. (Answer $\mathbf{a}^T = (1, 1, -1, -1)$.)
(b) Find the vector $\mathbf{b}$ such that $\beta_j = 0$ if and only if $\mathbf{b}^T \boldsymbol{\mu} = 0$.
(c) Find the vector $\mathbf{c}$ such that $\gamma_{ij} = 0$ if and only if $\mathbf{c}^T \boldsymbol{\mu} = 0$.

31.55 Consider a two-group two-way crossed multivariate repeated measures model with two rows and two columns (in which each individual receives each row treatment but only one column treatment) so that

$$\boldsymbol{\mu}_j = \begin{pmatrix} \theta + \alpha_1 + \beta_j + \gamma_{1j} \\ \theta + \alpha_2 + \beta_j + \gamma_{2j} \end{pmatrix}.$$

Assume the usual constraints.
(a) Find the vector $\mathbf{a}$ such that $\alpha_i = 0$ if and only if $\mathbf{a}^T(\boldsymbol{\mu}_1 + \boldsymbol{\mu}_2) = 0$.
(b) Find the vector $\mathbf{b}$ such that $\beta_j = 0$ if and only if $\mathbf{b}^T(\boldsymbol{\mu}_1 - \boldsymbol{\mu}_2) = 0$.
(c) Find the vector $\mathbf{c}$ such that $\gamma_{ij} = 0$ if and only if $\mathbf{c}^T(\boldsymbol{\mu}_1 - \boldsymbol{\mu}_2) = 0$.

31.56 For the single-group repeated measures model show that the LR test statistic for testing the null hypothesis that $\boldsymbol{\Sigma} = \sigma^2\mathbf{I}$ is given by $r^{mr/2}|\mathbf{S}|^{m/2}/(\text{tr}(\mathbf{S}))^{mr/2}$. (*Hint*: under the null hypothesis, the ML estimator of $(\boldsymbol{\mu}, \sigma^2)$ is $(\overline{\mathbf{Y}}, (m-1)\,\text{tr}(\mathbf{S})/mr)$.)

31.57 In the one-sample multivariate repeated measures model, let $\mathbf{B}$ be a Helmert matrix. Let $\mathbf{Y}_k^* = \mathbf{B}\mathbf{Y}_k$, $\boldsymbol{\mu}^* = \mathbf{B}\boldsymbol{\mu}$ and

$$\boldsymbol{\Sigma}^* = \mathbf{B}\boldsymbol{\Sigma}\mathbf{B}^T = \begin{pmatrix} \boldsymbol{\Sigma}_{11}^* & \boldsymbol{\Sigma}_{12}^* \\ \boldsymbol{\Sigma}_{21}^* & \boldsymbol{\Sigma}_{22}^* \end{pmatrix}, \qquad \mathbf{S}^* = \mathbf{B}\mathbf{S}\mathbf{B}^T = \begin{pmatrix} \mathbf{S}_{11}^* & \mathbf{S}_{12}^* \\ \mathbf{S}_{21}^* & \mathbf{S}_{22}^* \end{pmatrix},$$

where S_{11}^* and Σ_{11}^* are 1×1. Then $\mathbf{Y}_k^* \sim N_r(\boldsymbol{\mu}^*, \boldsymbol{\Sigma}^*)$.
(a) Show that the LR test statistic for the null hypothesis that $\boldsymbol{\Sigma} = \sigma^2\mathbf{A}(\rho)$ is given by $(r-1)^{m(r-1)/2}|\mathbf{S}^*|^{m/2}/(S_{11}^*)^{m/2}(\mathrm{tr}(\mathbf{S}_{22}^*))^{m(r-1)/2}$. (*Hint*:

$$\boldsymbol{\Sigma} = \sigma^2\mathbf{A}(\rho) \Leftrightarrow \boldsymbol{\Sigma}^* = \begin{pmatrix} \tau_b^2 & \mathbf{0} \\ \mathbf{0} & \tau_w^2\mathbf{I} \end{pmatrix}.$$

Under the null hypothesis, the ML estimator of $(\boldsymbol{\mu}, \tau_b^2, \tau_w^2)$ is $\left(\overline{\mathbf{Y}}, \frac{m-1}{m} S_{11}^*, \frac{m-1}{m(r-1)} \mathrm{tr}(\mathbf{S}_{22}^*)\right)$.
(b) Show that $S_{11}^* = \mathbf{1}_r^T\mathbf{S}\mathbf{1}_r/r$ and that $\mathrm{tr}(\mathbf{S}) = \mathrm{tr}(\mathbf{S}^*) = S_{11}^* + \mathrm{tr}(\mathbf{S}_{22}^*)$.
(c) Show that the LR test for testing the Huynh–Feldt condition is

$$(r-1)^{m(r-1)/2}|\mathbf{S}_{22}^*|^{m/2}/(\mathrm{tr}(\mathbf{S}_{22}^*))^{m(r-1)/2}.$$

(*Hint*: the Huynh–Feldt condition is satisfied $\Leftrightarrow \boldsymbol{\Sigma}_{22}^* = \tau_w^2\mathbf{I}$.)

31.58 For the more general linear model show that $\hat{\boldsymbol{\mu}}_G$ minimizes $q(\boldsymbol{\mu}) = (\mathbf{Y} - \boldsymbol{\mu})^T\mathbf{A}^{-1}(\mathbf{Y} - \boldsymbol{\mu})$. (*Hint*: $q(\boldsymbol{\mu}) = \|\mathbf{Y}^* - \boldsymbol{\mu}^*\|^2$.)

31.59 For the more general linear model let $Q = \{\mathbf{A}\mathbf{w} : \mathbf{w} \in W\}$, $W^* = (\mathbf{A}^{-1/2}\mathbf{w} : \mathbf{w} \in W)$.
(a) Show that Q is a subspace and that $\dim(Q) = \dim(W)$.
(b) Show that $V \parallel W$ is a $(p-k)$-dimensional subspace of V. (*Hint*: $V \parallel W = V|Q$.)
(c) For any vector $\mathbf{d}$, let $\mathbf{d}^* = \mathbf{A}^{-1/2}\mathbf{d}$. Show that $\mathbf{d} \in V \parallel W \Longleftrightarrow \mathbf{d}^* \in V^*|W^*$.
(d) Show that $\mathbf{d}^T\mathbf{A}^{-1}\boldsymbol{\mu} = \mathbf{d}^{*T}\boldsymbol{\mu}^*$, $\mathbf{d}^T\mathbf{A}^{-1}\hat{\boldsymbol{\mu}}_G = \mathbf{d}^{*T}\hat{\boldsymbol{\mu}}_G^*$ and $\mathbf{d}^T\mathbf{A}^{-1}\mathbf{d} = \|\mathbf{d}^*\|^2$.
(e) Derive the Scheffé simultaneous confidence intervals for the more general linear model.

31.60 For the regression form of the more general linear model, derive the F test for testing that $\mathbf{C}\boldsymbol{\beta} = \mathbf{0}$.

31.61 Consider the more general linear model in which we observe $\mathbf{Y} \sim N_n(\boldsymbol{\mu}, \mathbf{A})$, where $\mathbf{A} > 0$ is a known matrix. Let $\hat{\boldsymbol{\mu}}_G$, *SSH* and *dfh* be defined as for the model with unknown variance.
(a) Show that $\hat{\boldsymbol{\mu}}_G$ is a sufficient statistic.
(b) Show that the size-α LR test for testing that $\boldsymbol{\mu} \in W$ rejects if $SSH > \chi_{dfh}^{2\alpha}$.
(c) Derive the Scheffé simultaneous confidence intervals associated with this hypothesis.

31.62 (a) Consider a balanced twofold nested ANOVA model with known unequal variances in each cell. Find the χ^2 test for testing for no sub-category effect (i.e. for testing that μ_{ij} does not depend on j).
(b) Describe an approximate χ^2 test for this hypothesis for the case of unknown but possibly unequal variances.

31.63 Consider a one-way model repeated measures model with n individuals but only two rows and a covariate. Assume that the correlation coefficient is known to be Q so that $\mathrm{cov}(\mathbf{Y}) = \sigma^2\mathbf{A}$, where $\mathbf{A} = \mathbf{A}_2(Q) \otimes \mathbf{I}_n$.
(a) Show that $(\mathbf{Y} - \boldsymbol{\mu})^T\mathbf{A}^{-1}(\mathbf{Y} - \boldsymbol{\mu}) =$

$$\frac{\sum_j((Y_{1j} - \mu_1 - \delta x_{1j})^2 + (Y_{2j} - \mu_2 - \delta x_{2j})^2 - 2Q(Y_{1j} - \mu_1 - \delta x_{1j})(Y_{2j} - \mu_2 - \delta x_{2j}))}{1 - Q^2}$$

(b) Show that generalized LS estimator of μ_i is $\hat{\mu}_i = \overline{Y}_{i.} - \hat{\delta}\overline{x}_{i.}$.
(c) Find the generalized LS estimator $\hat{\delta}$ of δ.
(d) Now consider testing that $\mu_1 = \mu_2 = \mu$. Find the generalized LS estimators $\hat{\hat{\mu}}$ and $\hat{\hat{\delta}}$ of μ and δ under the null hypothesis.
(e) Find the χ^2 test for testing the hypothesis that $\mu_1 = \mu_2$.
(f) Suppose now the common correlation coefficient is unknown. Discuss how you would test that $\mu_1 = \mu_2$. (*Hint*: if we assume that $EY_{ij} = \mu_i + \delta_1(x_{ij} - \overline{x}_{.j}) + \delta_2\overline{x}_{.j}$, then the model is a univariate

repeated measures model as described in the text. For such a model it is possible to find the ML estimators for $\tau_1^2 = \sigma^2(1 + (r-1)\rho)$ and $\tau^2 = \sigma^2(1-\rho)$ and hence to find the ML estimator of ρ.)

31.64 Suppose we have a one-way repeated measures design in which each individual receives each possible treatment. As usual, let $\mathbf{Y}_k$ be the vector of observations on the kth individual. Suppose that $\text{cov}(\mathbf{Y}) = \sigma^2\mathbf{B}(\rho)$, where $\mathbf{B}(\rho)$ has the structure given in **31.34**. Describe how you would use results for the more general linear model to test that the treatment means are all equal. (Do not attempt to do the algebra but describe how you would get a computer to make the necessary calculations. Assume that you can get a sensible estimator of ρ.)

CHAPTER 32

ANALYSIS AND DIAGNOSTICS FOR THE LINEAR MODEL

32.1 In Chapters 29 and 30 we developed the inferential framework for the linear model in the contexts of regression analysis and the fixed effects analysis of variance, respectively. The linear model has been used so extensively in many fields that a vast literature exists on how to gain insight from an analysis, how to validate the assumptions and what to do if those assumptions are violated; these topics represent the theme of this chapter. We begin by reformulating the tests of **29.18–20** in an analysis of variance framework and then consider interval estimates. We then consider stepwise regression as a tool for empirical model building before turning to possible violations of the assumptions, and diagnostics to aid in the detection of such problems.

32.2 We shall work in terms of the model introduced in **29.2**, that is,

$$\mathbf{y} = \mathbf{X}\boldsymbol{\beta} + \boldsymbol{\epsilon}, \tag{32.1}$$

where $E(\boldsymbol{\epsilon}) = \mathbf{0}$, $\mathbf{V}(\boldsymbol{\epsilon}) = \sigma^2\mathbf{I}$ and $\mathbf{X}$ is assumed to be of full rank p (after imposing restrictions on $\boldsymbol{\beta}$ if necessary, to include the ANOVA model). Then, as shown in Theorem 29.1, the LS estimators for $\boldsymbol{\beta}$ are

$$\hat{\boldsymbol{\beta}} = (\mathbf{X}^T\mathbf{X})^{-1}\mathbf{X}^T\mathbf{y}, \tag{32.2}$$

where $\hat{\boldsymbol{\beta}}$ is unbiased and its covariance matrix is

$$\mathbf{V}(\hat{\boldsymbol{\beta}}) = \sigma^2(\mathbf{X}^T\mathbf{X})^{-1}. \tag{32.3}$$

Residuals

32.3 The expected values of $\mathbf{y}$ may be estimated by

$$\hat{\mathbf{y}} = \mathbf{X}\hat{\boldsymbol{\beta}}, \tag{32.4}$$

sometimes known as the *fitted* values. We then define the (regression) *residuals*, $\mathbf{e}$, as the differences between observed and fitted values:

$$\mathbf{e} = \mathbf{y} - \hat{\mathbf{y}} = \mathbf{y} - \mathbf{X}\hat{\boldsymbol{\beta}}. \tag{32.5}$$

From (29.39) it follows that

$$E(\mathbf{e}) = E(\mathbf{M}\boldsymbol{\epsilon}) = \mathbf{0}, \tag{32.6}$$

where $\mathbf{M} = \mathbf{I} - \mathbf{X}(\mathbf{X}^T\mathbf{X})^{-1}\mathbf{X}^T$. Further if $x_{1i} = 1$ for all i, the sum of the residuals $\mathbf{1}^T\mathbf{e}$ is zero by construction. The residuals are the basic units that will be used to develop diagnostic procedures later in the chapter.

Finally, from (29.43) we recall that an unbiased estimator for σ^2 is given by s^2, where

$$\begin{aligned}(n-p)s^2 &= \mathbf{e}^T\mathbf{e} = (\mathbf{y}-\mathbf{X}\hat{\boldsymbol{\beta}})^T(\mathbf{y}-\mathbf{X}\hat{\boldsymbol{\beta}}) \\ &= \mathbf{y}^T\mathbf{y} - \hat{\boldsymbol{\beta}}^T\mathbf{X}^T\mathbf{y};\end{aligned} \tag{32.7}$$

$\mathbf{e}^T\mathbf{e}$ is known as the residual sum of squares (*RSS*) and s^2 is the estimated mean squared error (*MSE*).

Tests of hypotheses

32.4 From (29.19)–(29.20), the total sum of squares

$$TSS = \sum_{i=1}^{n}(y_i - \bar{y})^2 = \mathbf{y}^T\mathbf{y} - n\bar{y}^2 \tag{32.8}$$

partitions into two orthogonal components

$$(\mathbf{y}-\hat{\mathbf{y}})^T(\mathbf{y}-\hat{\mathbf{y}}) + (\hat{\mathbf{y}}^T\hat{\mathbf{y}} - n\bar{y}^2), \tag{32.9}$$

where we assume that $\mathbf{X}$ includes a column of ones (this will always be true in the future unless explicitly stated otherwise). From (32.7), we note that the first term in (32.9) is the *RSS*; the second term is termed the explained sum of squares, denoted by *ESS*. These may be expressed as quantities

$$\begin{aligned}RSS &= \boldsymbol{\epsilon}^T\{\mathbf{I} - \mathbf{X}(\mathbf{X}^T\mathbf{X})^{-1}\mathbf{X}^T\}\boldsymbol{\epsilon} \\ &= \boldsymbol{\epsilon}^T\mathbf{M}\boldsymbol{\epsilon}\end{aligned} \tag{32.10}$$

and

$$ESS = \boldsymbol{\epsilon}^T\{\mathbf{X}(\mathbf{X}^T\mathbf{X})^{-1}\mathbf{X}^T\}\boldsymbol{\epsilon} = \boldsymbol{\epsilon}^T(\mathbf{I}-\mathbf{M})\boldsymbol{\epsilon}. \tag{32.11}$$

If, in addition to the assumptions in **32.2**, we assume that the errors are normally distributed, we have from Example 29.3 that a test of the hypothesis

$$H_0 : \boldsymbol{\beta} = \mathbf{0} \text{ versus } H_1 : \boldsymbol{\beta} \neq \mathbf{0} \text{ (not all } \beta_i = 0)$$

is given by the statistic

$$F = \{ESS/(p-1)\}/\{RSS/(n-p)\} \tag{32.12}$$

which has an F distribution with $(p-1, n-p)$ d.fr. under H_0.

32.5 This development was presented from a geometric perspective in **28.16–17** and **28.27–29**. In our present notation, we may write

$$\begin{aligned}R^2 &= ESS/TSS \\ &= (p-1)F/\{(n-p) + (p-1)F\},\end{aligned} \tag{32.13}$$

where R is the sample multiple correlation coefficient. When H_0 is true, it follows from **28.28** that

$$E(R^2|H_0) = (p-1)/(n-1)$$

so that the adjusted coefficient

$$\bar{R}^2 = \frac{(n-1)}{(n-p)}\left[R^2 - \frac{(p-1)}{(n-1)}\right] \tag{32.14}$$

is sometimes preferred, as

$$E(\bar{R}^2|H_0) = 0, \max \bar{R}^2 = 1 \text{ and } \bar{R}^2 \le R^2 \le 1;$$

note that $\bar{R}^2$ may be negative. Whereas R^2 cannot decrease when another variable is added to the model, $\bar{R}^2$ may decrease. In stepwise regression, the criterion max $(\bar{R}^2)$ is sometimes used to choose the final model, but leads to the inclusion of too many terms; see **32.27**.

32.6 The test procedure outlined above is usually presented in the form of an ANOVA table; see Table 32.1.

Table 32.1 Analysis of variance for a regression model

Source	d.fr.	Sum of squares	Mean squares	F
Due to regression (explained)	$p-1$	ESS	$M_E = ESS/(p-1)$	$F = M_E/M_R$
Due to error (residual)	$n-p$	RSS	$M_R = RSS/(n-p)$	
Total	$n-1$	TSS		

If a linear model is fitted using an initial set of p variables with values $\mathbf{X}_1$ and then a further q variables with values $\mathbf{X}_2$ are added to produce the model

$$\mathbf{y} = \mathbf{X}_1\boldsymbol{\beta}_1 + \mathbf{X}_2\boldsymbol{\beta}_2 + \boldsymbol{\epsilon}, \tag{32.15}$$

the hypothesis

$$H_0 : \boldsymbol{\beta}_2 = \mathbf{0}, \text{ given that } \mathbf{X}_1 \text{ is in the model,}$$

may be tested against the general alternative using the statistic

$$F_2 = \{(ESS_2 - ESS_1)/q\}/\{(TSS - ESS_2)/(n-p-q)\}, \tag{32.16}$$

where ESS_2 denotes the explained sum of squares for the complete model. The schematic analysis is laid out in Table 32.2.

32.7 To test the hypothesis $H_0 : \mathbf{c}^T\boldsymbol{\beta} = \mathbf{0}$, we know that $\mathbf{c}^T\hat{\boldsymbol{\beta}}$ is independent of s^2 and that s^2 is χ^2_{n-p} so that, from (16.10) and (29.26),

$$t = (\mathbf{c}^T\hat{\boldsymbol{\beta}} - \mathbf{c}^T\boldsymbol{\beta})/[s^2\mathbf{c}^T(\mathbf{X}^T\mathbf{X})^{-1}\mathbf{c}]^{1/2} \tag{32.17}$$

follows a Student's t distribution with $n-p$ degrees of freedom.

The most common use of (32.17) is to test a hypothesis of the form $H_{0(j)} : \beta_j = 0$, when the statistic reduces to

$$t_j = \hat{\beta}_j/[s^2 a_{jj}]^{1/2}, \tag{32.18}$$

Table 32.2 ANOVA for a regression model with two sets of regressors

Source	d.fr.	Sum of squares	Mean square	F
Due to $\mathbf{X}_1$	$p-1$	ESS_1	$M_1 = ESS_1/(p-1)$	$F_1 = M_1/M_R$
Due to $\mathbf{X}_2$ (after fitting $\mathbf{X}_1$)	q	$ESS_2 - ESS_1$	$M_2 = (ESS_2 - ESS_1)/q$	$F_2 = M_2/M_R$
Due to error	$n-p-q$	RSS	$M_R = RSS/(n-p-q)$	
Total	$n-1$	TSS		

Table 32.3 Data for Example 32.1

Electricity consumption in kW (Y)	Temperatures (°F) Low (x_2)	High (x_3)	Days in school (x_4)	Days in period (x_5)
663	22	56	6	5
1018	14	48	10	7
1407	00	40	0	8
911	02	36	0	6
1758	–06	27	10	8
1165	02	34	10	6
1136	22	44	8	7
1095	10	50	10	7
1809	10	48	16	10
715	02	52	4	4
572	30	62	10	7
704	26	52	9	7
1002	17	40	10	7
848	12	56	10	7
856	12	56	10	7

where a_{jj} is the (j, j)th element of $(\mathbf{X}^T\mathbf{X})^{-1}$.

Example 32.1 (Testing a linear model)

The data in Table 32.3 relate to a household's consumption of electricity in consecutive periods in State College, Pennsylvania, during the winter of 1980/1981; electricity was the sole source of heating. We consider a model in which we regress daily electricity consumption ($y = Y/x_5$) over a particular period on the lowest (x_2) and highest (x_3) temperatures for that period and on school attendance (x_4), taken as the number of days in school (for a household with two children) for that period. The estimated model is

$$\underset{}{y = 219.90 - \underset{(-3.84)}{2.403x_2} - \underset{(-1.86)}{1.276x_3} + \underset{(1.76)}{16.93x_4}}; \quad s = 16.8; \; R^2 = 0.832; \; \bar{R}^2 = 0.787.$$

Table 32.4 ANOVA Table for Example 32.1

Source	d.fr.	SS	MS	F
Regression	3	15 418	5139	18.2
Error	11	3 103	282	–
Total	14	18 521	–	–

The numbers in the brackets under the coefficients are the t ratios calculated from (32.18); from Appendix Table 5, the per cent point is 2.201. The ANOVA is given as Table 32.4.

From Appendix Table 7, the 5 per cent point is 3.59. Overall, the model appears to be effective, but x_3 and x_4 are of marginal value. The signs of the coefficients for x_2 and x_3 are as expected. It might be thought that an increase in x_4 should reduce heating costs, but in fact, x_4 is partly an indicator of time spent away from the house during school holidays. We shall discuss the example further later in the chapter.

Confidence and prediction intervals

32.8 Using the distributional results in **32.7**, the central confidence interval for β_j with coefficient $100(1-\alpha)$ per cent is

$$\hat{\beta}_j \pm t_{1-\alpha/2}\{s^2 a_{jj}\}^{1/2}, \qquad (32.19)$$

where $t_{1-\alpha/2}$ is the $100(1-\frac{1}{2}\alpha)$ percentage point for Student's t distribution with $n-p$ degrees of freedom.

32.9 Suppose that, having fitted the linear regression model, we wish to estimate the expected value of y given $\mathbf{x}_0 = (x_{01}, x_{02}, \ldots, x_{0p})'$, denoted by $E(y|\mathbf{x}_0)$. From (32.4), the MV unbiased estimator is

$$\hat{y} = \mathbf{x}_0^T \hat{\boldsymbol{\beta}}, \qquad (32.20)$$

and it follows that the confidence interval is

$$\hat{y} \pm t_{1-\alpha/2}\{s^2 \mathbf{x}_0^T (\mathbf{X}^T\mathbf{X})^{-1}\mathbf{x}_0\}^{1/2}, \qquad (32.21)$$

with $\nu = n-p$ as before.

32.10 Let y_{n+1} denote a potential new observation, corresponding to $\mathbf{x}_0$. Once again, the estimator is given by (32.4) as

$$\hat{y}_{n+1} = \mathbf{x}_0^T \hat{\boldsymbol{\beta}}, \qquad (32.22)$$

but now we are interested in the single observation rather than its expectation. Since

$$y_{n+1} = \mathbf{x}_0^T \boldsymbol{\beta} + \epsilon_{n+1},$$

we have

$$y_{n+1} - \hat{y}_{n+1} = \mathbf{x}_0^T(\boldsymbol{\beta} - \hat{\boldsymbol{\beta}}) + \epsilon_{n+1}; \qquad (32.23)$$

since the observations are independent and $\hat{\boldsymbol{\beta}}$ does not depend upon y_{n+1}, we have

$$E(y_{n+1} - \hat{y}_{n+1}) = 0,$$

showing the estimator to be unbiased and

$$\begin{aligned} \text{var}(y_{n+1} - \hat{y}_{n+1}) &= \mathbf{x}_0^T \text{var}(\hat{\boldsymbol{\beta}})\mathbf{x}_0 + \sigma^2 \\ &= \sigma^2\{1 + \mathbf{x}_0^T(\mathbf{X}^T\mathbf{X})^{-1}\mathbf{x}_0\} \\ &= \sigma^2 v_{n+1}, \end{aligned} \tag{32.24}$$

say. Thus, the $100(1-\alpha)$ per cent two-sided *prediction* interval for y_{n+1} is

$$\hat{y}_{n+1} \pm t_{1-\alpha/2}\{s^2 v_{n+1}\}^{1/2}. \tag{32.25}$$

If m observations are to be made with $\mathbf{x} = \mathbf{x}_0$, the prediction interval for the mean of these m observations is also given by (32.25), but the first term in v_{n+1} becomes $1/m$ rather than one.

Example 32.2 (Confidence intervals for the model with a single regressor)
In the simple case

$$y_i = \beta_1 + \beta_2 x_i + \epsilon_i, \qquad i = 1, 2, \ldots, n, \tag{32.26}$$

we have seen in Example 29.1 that

$$(\mathbf{X}^T\mathbf{X})^{-1} = \frac{1}{\sum_i (x_i - \bar{x})^2} \begin{pmatrix} \sum x^2/n & -\bar{x} \\ -\bar{x} & 1 \end{pmatrix}. \tag{32.27}$$

If $\mathbf{x}_0 = (1\ x)^T$, the confidence interval for $E(y|\mathbf{x}_0)$ is, from (32.21),

$$\begin{aligned} (\mathbf{x}_0)^T\hat{\boldsymbol{\beta}} &\pm t_{1-\alpha/2}\left\{\frac{s^2}{\sum(x-\bar{x})^2}\begin{pmatrix}1\\x_0\end{pmatrix}^T\begin{pmatrix}\sum x^2/n & -\bar{x}\\ -\bar{x} & 1\end{pmatrix}\begin{pmatrix}1\\x_0\end{pmatrix}\right\}^{1/2} \\ &= (\hat{\beta}_1 + \hat{\beta}_2 x_0) \pm t_{1-\alpha/2}\left\{s^2\left(\frac{1}{n} + \frac{(x_0-\bar{x})^2}{\sum(x-\bar{x})^2}\right)\right\}^{1/2}. \end{aligned} \tag{32.28}$$

If we consider (32.28) as a function of the value x_0 we see that it defines the two branches of a hyperbola of which the fitted regression $(\hat{\beta}_1 + \hat{\beta}_2 x_0)$ is a diameter. The confidence interval obviously has a minimum length when $x_0 = \bar{x}$, the observed mean, and its length increases steadily as $|x_0 - \bar{x}|$ increases, confirming the intuitive notion that we can estimate most accurately near the 'centre' of the observed values of x. Figure 32.1 illustrates the loci of the confidence limits given by (32.28).

32.11 It should be borne in mind that these predictions are conditional upon the assumption that the linear model fitted to the original sample is valid for further observations; that is, we assume that no structural change has taken place.

Finally, we note that the confidence intervals given **32.8–32.9** are often approximately correct because of the central limit theorem. However, we can make no such claim for the prediction

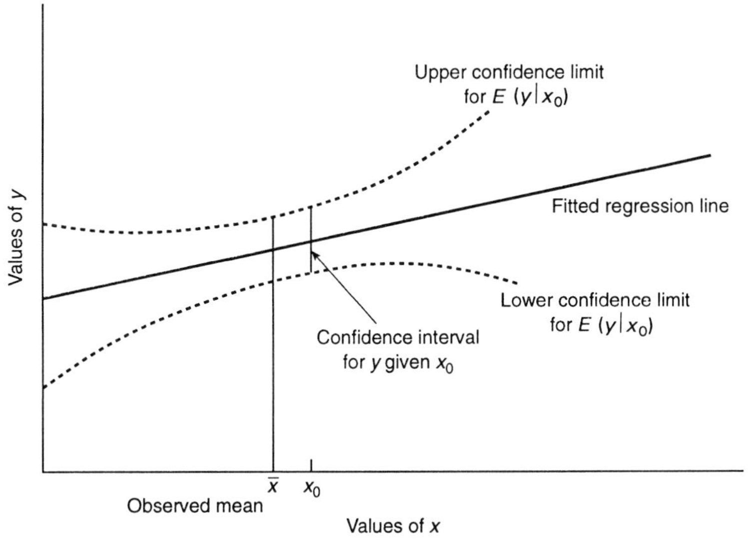

Fig. 32.1 Confidence limits (28.36) for the expected value of y in simple linear regression

interval where a non-normal distribution for the error term may dominate the approximately normal variation in $\hat{\boldsymbol{\beta}}$.

Example 32.3 (Confidence and prediction intervals for Example 32.1)

In order to construct a confidence interval for $E(y|\mathbf{x}_0)$ for the data in Example 32.1, we need only the specific values in $\mathbf{x}_0$. Suppose $\mathbf{x}_0^T = (5, 40, 1.429)$; it follows that $\hat{y} = 181.0$. Using (32.21), the 95 per cent confidence interval is (166.5, 195.5), whereas, from (32.25), the prediction interval is (141.3, 220.7).

Conditional and unconditional inferences

32.12 If $\mathbf{X}$ is the observed value of a set of random variables, the use of the linear model, in which $\mathbf{X}$ is a matrix of known coefficients, is conditional upon the observed $\mathbf{X}$. It is easy to see that conditionally unbiased estimators remain unbiased unconditionally, while conditional confidence intervals and tests remain valid unconditionally, since the value of any fixed conditional probability is unaffected by integrating it over the distribution of $\mathbf{X}$. However, the unconditional efficiency (i.e., selectivity and power) of tests and intervals will generally differ from the conditional (linear model) values – cf. **28.28–31** for the multiple correlation coefficient – because the underlying statistics have different distributions. Similarly, other properties of estimators (e.g. their variances) do not generally persist unconditionally – cf. Exercise 27.9 for the regression coefficient. Sampson (1974) gives a general exposition of the subject.

Design considerations

32.13 From (32.27), it is evident that $\hat{\beta}_1$ and $\hat{\beta}_2$ will be uncorrelated if $\bar{x} = 0$; that is, the x values are measured about their mean. In general, when the matrix $\mathbf{X}^T\mathbf{X}$ is diagonal, the regressor variables are said to be *orthogonal*; this is generally desirable, as it simplifies the interpretation of the effects of each regressor on the response variable, y. Most designed experiments aim for at least some degree of orthogonality; see the designs considered in Chapter 30.

Further, from the example in (32.27), it is apparent that the variances of $\hat{\beta}_1$ and $\hat{\beta}_2$ will be minimized when $\bar{x} = 0$ and $\sum x^2$ is as large as possible. Both orthogonality and minimized sampling variances are therefore achieved if we choose the x_i so that (assuming n to be even)

$$x_1, x_2, \ldots, x_{n/2} = +a,$$
$$x_{n/2+1}, x_{n/2+2}, \ldots, x_n = -a$$

and a is as large as possible. This choice corresponds to the intuitive argument that, if we are certain that the dependence of y upon x is linear with constant variance, we can most efficiently 'fix' the line at its end-points. However, if the dependence were non-linear, we should be unable to detect this if all our observations had been made at two values of x only, and it is therefore usual to distribute the x values more evenly over the range; it is always as well to be able to check the structural assumptions of our model in the course of the analysis.

In the general case, if $\mathbf{X}$ has a column of 1s and we measure each other column about its mean, we make the other regression coefficients orthogonal to the constant term, as here.

32.14 The confidence limits for an expected value of y discussed in Example 32.3, and more generally in **32.9–11**, refer to the value of y corresponding to a particular $\mathbf{x}$; in Fig. 32.1, any particular confidence interval is given by that part of the vertical line through x_0 lying between the branches of the hyperbola. Suppose now that we require a *confidence region for an entire regression line*, i.e., a region R in the (x, y) plane (or, more generally, in the $(\mathbf{x}, y)$ space) such that there is probability $1 - \alpha$ that the true regression line $\mathbf{y} = \mathbf{x}^T\boldsymbol{\beta}$ is contained in R. This is a quite distinct problem from that just discussed; we are now seeking a confidence region, not an interval, and it covers the whole line, not one point in the line. We now consider this problem, first solved in the simplest case by Working and Hotelling (1929) in a remarkable paper; our discussion follows that of Hoel (1951).

Confidence regions for a regression line

32.15 We first treat the simple case of Example 32.2 and assume σ^2 known, restrictions to be relaxed in **32.20–21**. For convenience, we measure the x_j from their mean so that $\bar{x} = 0$; $\operatorname{var}\hat{\beta}_1 = \sigma^2/n$, $\operatorname{var}\hat{\beta}_2 = \sigma^2/\sum x^2$ and $\hat{\beta}_1$ and $\hat{\beta}_2$ are normally and independently distributed, and

$$u = n^{1/2}(\hat{\beta}_1 - \beta_1)/\sigma, \qquad v = \left(\sum x^2\right)^{1/2}(\hat{\beta}_2 - \beta_2)/\sigma, \tag{32.29}$$

are independent standardized normal variates.

Let $g(u^2, v^2)$ be a single-valued even function of u and v, and let

$$g(u^2, v^2) = g_{1-\alpha}, \qquad 0 < \alpha < 1, \tag{32.30}$$

define a family of closed curves in the (u, v) plane such that (a) whenever $g_{1-\alpha}$ decreases, the new curve is contained inside that corresponding to the larger value of $1 - \alpha$; and (b) every interior point of a curve lies on some other curve. To the implicit relation (32.30) between u and v, we assume that there corresponds an explicit relation

$$u^2 = p(v^2)$$

or

$$u = \pm h(v). \tag{32.31}$$

We further assume that $h'(v) = \mathrm{d}h(v)/\mathrm{d}v$ exists for all v and is a monotone decreasing function of v taking all real values.

32.16 We see from (32.29) that for any given set of observations to which a regression has been fitted, there will correspond to the true regression line,

$$y = \beta_1 + \beta_2 x, \tag{32.32}$$

values of u and v such that

$$\beta_1 + \beta_2 x = \left(\hat{\beta}_1 + \frac{\sigma}{n^{1/2}}u\right) + \left(\hat{\beta}_2 + \frac{\sigma}{(\sum x^2)^{1/2}}v\right)x. \tag{32.33}$$

Substituting (32.31) into (32.33), we have two families of regression lines, with v as parameter,

$$\left(\hat{\beta}_1 \pm \frac{\sigma}{n^{1/2}}h(v)\right) + \left(\hat{\beta}_2 + \frac{\sigma}{(\sum x^2)^{1/2}}v\right)x, \tag{32.34}$$

one family corresponding to each sign in (32.31). We now find the envelopes of these families.

Differentiating (32.34) with respect to v and equating the derivative to zero, we obtain

$$x = \mp\left(\frac{\sum x^2}{n}\right)^{1/2} h'(v). \tag{32.35}$$

Substituted into (32.34), (32.35) gives the required envelopes:

$$(\hat{\beta}_1 + \hat{\beta}_2 x) \pm \frac{\sigma}{n^{1/2}}\{h(v) - vh'(v)\}, \tag{32.36}$$

where the functions of v are to be substituted for in terms of x from (32.35). The restrictions placed of $h'(v)$ below (32.31) ensure that the two envelopes in (32.36) exist for all x, are single-valued, and that all members of each family lie on one side only of its envelope. In fact, the curve given taking the upper signs in (32.36) always lies above the curve obtained by taking the lower signs in (32.36), and all members of the two families (32.34) lie between them.

32.17 Any pair of values (u, v) for which

$$g(u^2, v^2) < g_{1-\alpha} \tag{32.37}$$

will correspond to a regression line lying between the pair of envelopes (32.36), because for any fixed v, $u^2 = \{h(v)\}^2$ will be reduced, so that the constant term in (32.34) will be reduced in magnitude as a function of v, while the coefficient of x is unchanged. Thus if u and v satisfy (32.37), the true regression line will lie between the pair of envelopes (32.36). Now choose $g_{1-\alpha}$ so that the continuous random variable $g(u^2, v^2)$ satisfies

$$P\{g(u^2, v^2) < g_{1-\alpha}\} = 1 - \alpha. \tag{32.38}$$

Then we have probability $1 - \alpha$ that (32.37) holds, and the region R between the pair of envelopes (32.36) is a confidence region for the true regression line with confidence coefficient $1 - \alpha$.

32.18 In the original solution to this problem, Working and Hotelling (1929) elected to assume that $g(u^2, v^2)$ describes a circle, so that

$$h(v) = (a^2 - v^2)^{1/2} \tag{32.39}$$

and

$$h'(v) = -v/h(v). \tag{32.40}$$

Their choice was motivated by the observation that, since u and v in (32.29) are independent standardized normal variates, $u^2 + v^2$ is a χ^2 variate with 2 d.fr., and $a^2(= g_{1-\alpha}$ in (32.38)) is simply the $100(1-\alpha)$ per cent point obtained from the tables of that distribution. The boundaries of the confidence region are, putting (32.39) and (32.40) into (32.36),

$$\begin{aligned}(\hat{\beta}_1 + \hat{\beta}_2 x) \pm \frac{\sigma}{n^{1/2}}\left\{h(v) + \frac{v^2}{h(v)}\right\} &= (\hat{\beta}_1 + \hat{\beta}_2 x) \pm \frac{\sigma}{n^{1/2}}[\{h(v)\}^2 + v^2]^{1/2}\left[1 + \frac{v^2}{\{h(v)\}^2}\right]^{1/2} \\ &= (\hat{\beta}_1 + \hat{\beta}_2 x) \pm \frac{\sigma}{n^{1/2}}(g_{1-\alpha})^{1/2}[1 + \{h'(v)\}^2]^{1/2}. \end{aligned} \tag{32.41}$$

Using (32.35), (32.41) becomes

$$(\hat{\beta}_1 + \hat{\beta}_2 x) \pm (g_{1-\alpha})^{1/2}\left\{\frac{\sigma^2}{n} + x^2\frac{\sigma^2}{\sum x^2}\right\}^{1/2}, \tag{32.42}$$

the term in the braces being $\{\text{var}\,\hat{\beta}_1 + x^2\,\text{var}\,\hat{\beta}_2\}$.

If (32.42) is compared with the confidence limits (32.38) for $E(y|\mathbf{x}_0)$ derived in Example 32.2 (where we now put $\bar{x} = 0$, as we have done here), we see that apart from the replacement of s^2 by σ^2, and of $t_{1-\alpha/2}$ by $(g_{1-\alpha})^{1/2}$, the equations are of exactly the same form. Thus the confidence region (32.42) will look exactly like the loci of the confidence limits (32.28) plotted in Fig. 32.1, being a hyperbola with the fitted line as diameter. As might be expected, for given α the branches of the hyperbola (32.42) are farther apart than those of (32.29), for we are now setting a region for the whole line where previously we had loci of limits for a single value on the line. For example, with $\alpha = 0.05$, $t_{1-\alpha/2}$ (with infinite degrees of freedom, corresponding to σ^2 known) $= 1.96$, while $g_{1-\alpha} = 5.99$ for a χ^2 distribution with 2 d.fr., yielding the multiplier 2.45 for (32.42).

32.19 In his extension, Hoel (1951) allowed $h(v)$ to describe a family of ellipses. When $1 - \alpha = 0.95$, the optimal choice is an ellipse with semi-axes 2.62 and 2.32. Hoel found that the length of the interval derived from (32.42) was less than 1 per cent larger than the minimum obtained from the best ellipse.

32.20 If σ^2 is unknown, only slight modifications of the argument are required. Define the variable

$$w^2 = (n-2)s^2/\sigma^2, \qquad (32.43)$$

so that w^2 is the ratio of the sum of squared residuals from the fitted regression to the true error variance, which has a χ^2 distribution with $n-2$ degrees of freedom. From (32.18) and (32.43), we see that each of the statistics

$$u^* = (n-2)^{1/2}u/w = n^{1/2}(\hat{\beta}_1 - \beta_1)/s, \qquad v^* = (n-2)^{1/2}v/w = n^{1/2}(\hat{\beta}_2 - \beta_2)/s$$

has a Student's t distribution with $n-2$ degrees of freedom.

Further, since u^2, v^2 and w^2 are distributed independently of one another as χ^2 with 1, 1 and $n-2$ degrees of freedom respectively, the ratio

$$\left(\frac{u^2+v^2}{2}\right)\Bigg/\left(\frac{w^2}{n-2}\right) = \frac{1}{2}(u^{*2}+v^{*2})$$

has an F distribution with $(2, n-2)$ degrees of freedom. Thus if we replace σ by s in (32.43), and put $g_{1-\alpha}$ equal to twice the $100(1-\alpha)$ per cent point of this F distribution, we obtain the required confidence region. As in **32.18**, we find that the boundaries of the region are always farther apart than the loci of the confidence limits for $E(y|\mathbf{x}_0)$.

32.21 There is no difficulty in extending our results to the case of more than one regressor, a sketch of such a generalization having been given by Hoel (1951). With p regressors we find, generalizing **32.20**, that $(u^{*2}+\sum_{i=1}^{p} v_i^{*2})/(p+1)$ has an F distribution with $(p+1, n-p-1)$ d.fr.

32.22 Wynn and Bloomfield (1971) give tables of the fractional multiplying factors to be applied to the Working–Hotelling region when only part of the line is to be covered for the same α. In general, the confidence region becomes narrower when this restriction is made; that is, the original Working–Hotelling region is conservative. Kanoh (1988) shows how the average width of the region can be further reduced by changing the curvature of the boundaries. Wynn (1984) extends the Working–Hotelling region to the polynomial regression case.

Stepwise regression

32.23 In experimental studies, the regressor variables are well defined, but in observational studies it may happen that a large number of potential regressors is available, with little prior evidence as to the relative merits of these competing variables. One way to reduce the dimensionality of the problem is to reduce the set of variables to a smaller set by use of either factor analysis or principal components analysis; this approach is discussed from a different perspective in **32.49** and is not pursued further here.

A second approach, which is often useful at least in exploratory analyses, is to select best subsets of variables; that is, to choose the best single regressor, then the best pair, then the best triple, and so on. This leads to two questions:

(a) How should the variables be selected to form a subset of size p?

(b) How many variables should be selected (choice of p)?

Since R^2 is used as a measure of overall goodness of fit, a natural answer to (a) is to choose those regressors that maximize R^2 for a given p. Since R^2 is non-decreasing as a function of p, question (b) may be answered by selecting a suitable cut-off value for R^2 as a function of n and p.

32.24 When the total number of regressor variables, P say, is not too large, it is feasible to consider all $\binom{P}{p}$ possible subsets for each p, that is, $2^P - 1$ subsets in all. Current computer programs can handle up to about $P = 20$ and, say, $p \leq 8$ in reasonable time, but larger data sets become rather time-consuming. To some extent, the search process can be improved by using threshold rules such as those developed by Beale *et al.* (1967). Suppose there are s candidate variables from which we wish to choose the best $p - 1$, and denote by j the set of $s - 1$ variables obtained by excluding x_j from the candidates.

Define the *unconditional threshold* of x_j to be

$$T_j = R^2_{1(j)}.$$

This maximizes the multiple correlation obtainable without using x_j; now if, during the search for the best subset of $p - 1$ variables, we achieve for some subset u of size $(p - 1)$ a value of R^2 satisfying

$$R^2_{1(u)} > T_j,$$

we can obviously only obtain a larger R^2 by including x_j in the subset to be considered. Similarly, we may define *conditional thresholds*

$$T_{ji} = R^2_{1(ji)},$$

obtained by excluding the pair (x_j, x_i). If we find an R^2 exceeding T_{ji}, we can only improve it by including one or both of x_j, x_i. By calculating such thresholds, the number of subsets considered may be substantially reduced.

32.25 For larger P, a variety of incomplete search procedures has been developed that check only some of the possible subsets. The simplest such algorithm is the *forward selection* method (cf. Efroymson, 1960) which selects the best single variable (that which maximizes R^2) then *adds* the next best and so on, so that only $\frac{1}{2}P(P + 1) - 1$ different subsets need to be considered for all $1 \leq p \leq P$. A similar method is the *backward elimination* algorithm, which starts all the P variables in the regression and deletes them one at a time in such a way as to minimize the drop in R^2 at each stage; again only $\frac{1}{2}P(P+1)-1$ subsets are considered. As might be expected, these two methods do not produce the same subsets of size p, nor does either procedure necessarily produce the best subset at each stage, except when $p = 1$ (forward) or $p = P - 1$ (backward). Empirical studies suggest that the backward algorithm fares somewhat better for the choices of p of most interest since it is less likely to miss useful combinations of variables. However, the backward procedure may be infeasible because of multicollinearity (most obviously so if $n \leq P$).

32.26 A natural compromise between the two methods is to consider a combined algorithm that operates as follows. Given the initial subset of $s - 1$ regressors:

(1) select the best variable to add to the subset to create a new subset of s variables, $x_{(s)}^F$ say;
(2) select the best variable to delete from the new subset of size, s, $x_{(s)}^B$ say;
(3) if $x_{(s)}^F = x_{(s)}^B$, terminate the search for the best subset of size $s - 1$ and repeat the steps for s; if $x_{(s)}^F \neq x_{(s)}^B$, return to step (1).

The exact number of subsets considered cannot be determined in advance, but it is usually of order P^2 and so fairly quick to compute. Computational speed is improved by noting that if $\mathbf{X}_p$ is the matrix of regressors for p variables

$$\mathbf{X}_{p+1} = (\mathbf{X}_p, \mathbf{x}_{p+1}) \text{ and } \mathbf{M}_p = \mathbf{X}_p^T\mathbf{X}_p,$$

then

$$\mathbf{M}_{p+1} = \mathbf{X}_{p+1}^T\mathbf{X}_{p+1} = \begin{pmatrix} \mathbf{M}_p & \mathbf{m} \\ \mathbf{m}^T & h \end{pmatrix}$$

has inverse

$$\begin{pmatrix} \mathbf{M}_p^{-1} + \alpha\mathbf{d}\mathbf{d}^T & -\alpha\mathbf{d} \\ -\alpha\mathbf{d}^T & \alpha \end{pmatrix},$$

where $\mathbf{d} = \mathbf{M}_p^{-1}\mathbf{m}$ and $\alpha^{-1} = h - \mathbf{m}^T\mathbf{M}_p^{-1}\mathbf{m}$. Thus, variables may be added or deleted without the need to invert $\mathbf{M}_p$ at each stage. For further discussions on computational efficiency, see Broerson (1986) and Ridout (1988).

32.27 Thus far, we have not commented upon the stopping rules to be used in deciding upon p. Many computer packages use percentage points of F based on the simplifying (erroneous) assumption that only a single additional variable is being tested; others allow the user to select such an F value by some private device.

Example 32.4 (Stepwise regression with random data)

To illustrate the potential pitfalls associated with over-optimistic assessments of a stepwise regression analysis, we performed the following experiment. We generated 51 sets of 40 independent identically distributed random variables (i.e., $n = 40$, $P = 50$) and performed stepwise regression using MINITAB, with an algorithm similar to that described in **32.26**. A cut-off value of $F = 4$ was used, which is the default setting. The results for ten replicates are summarized below; in all ten replicates, the results were 'significant' at the 1 per cent level.

	mean	range
Final number of variables selected	3.6	2–9
value of R^2	0.41	0.24–0.77
residual MSE	0.603	0.22–0.93
original MSE	0.926	0.70–1.54

A criterion with more solid credentials may be developed from the prediction variance given in (32.24). Consider n new observations to be recorded at each of $\mathbf{x}_1, \mathbf{x}_2, \ldots, \mathbf{x}_n$. The total prediction variance, based on p-regressors, is

$$\begin{aligned} V_p &= \sigma^2\left\{n + \sum_{i=1}^{n} \mathbf{x}_i^T(\mathbf{X}^T\mathbf{X})^{-1}\mathbf{x}_i\right\} \\ &= \sigma^2\{n + \mathrm{tr}[(\mathbf{X}^T\mathbf{X})^{-1}\sum \mathbf{x}_i\mathbf{x}_i^T]\} \\ &= (n+p)\sigma^2. \end{aligned} \tag{32.44}$$

An unbiased estimator of this quantity, assuming the selected p-regressor model to be correct, is

$$S_p = (n+p)RSS(p)/(n-p). \tag{32.45}$$

We may then select the subset that minimizes S_p over all possible subsets and $1 \leq p \leq P$. Sometimes S_p is known as the final prediction error (FPE) criterion; see Akaike (1969).

A similar procedure, suggested by Mallows (1973), is to examine the quantity

$$C_p = \frac{RSS(p)}{s^2} + 2p - n, \tag{32.46}$$

where s^2 is the unbiased estimator of σ^2 based on all P regressors so that no misspecification error results. If the model with $p < P$ regressors is correctly specified, $E(C_p) \doteq p$. Values of C_p greater than p indicate misspecification, whereas values less than p may be indicative of overkill in the stepwise search.

Additional measures include information criteria of the general form

$$IC = \log\{RSS(p)/n\} + pc(n)/n. \tag{32.47}$$

Akaike (1969) suggested $c(n) = 2$, whereas Schwarz (1978) recommends $c(n) = \log n$ and Hannan and Quinn (1979) consider $c(n) = 2\log(\log n)$.

Shibata (1981; 1984) shows that S_p, C_p and Akaike's information criterion produce equivalent results asymptotically. Breiman and Freedman (1983) demonstrate that criteria such as (32.45) are asymptotically optimal in the sense of minimizing the prediction error. The Schwarz and Hannan–Quinn criteria produce consistent estimators in the sense that the probability of selecting the true model approaches one as $n \to \infty$; the other measures do not achieve this and typically include too many terms.

Wallace and Freeman (1987, 1992) introduce the concept of minimum message length estimation (MMLE). MMLE estimators are based upon Shannon's information measure and provide a natural trade-off between model complexity and goodness of fit.

Example 32.5 (Model selection using information criteria)

Using the data given in Example 32.1, a stepwise increase in R^2 proceeds as in Table 32.5. Many rules would, therefore, terminate the process after step 1 since the next t value does not exceed the 5 per cent level. However, the several criteria we have just discussed all suggest the inclusion of all three variables; see Table 32.6. Stepwise regression is a useful exploratory guide, but not an infallible model-building tool!

Table 32.5 Stepwise regression for data of Example 32.1

Step	Variable entered	R^2	t value of entering variable
1	x_2	0.734	−5.99
2	x_3	0.785	−1.69
3	x_4	0.833	1.76

Table 32.6 Values of selection criteria for different subsets

Subset	S_p	C_p	$IC(c = 2)$	$IC^*(c = 2.71)$
$\{x_2\}$	6439	6.45	6.06	6.16
$\{x_3\}$	9690	15.27	6.47	6.56
$\{x_4\}$	23729	53.32	7.36	7.46
$\{x_2, x_3\}$	5967	5.10	5.98	6.12
$\{x_2, x_4\}$	6116	5.45	6.01	6.15
$\{x_3, x_4\}$	10889	22.19	6.58	6.72
$\{x_2, x_3, x_4\}$	5360	4.00	5.87	6.05

* The Schwarz value; $c \doteq 2$ for the Hannan–Quinn version.

32.28 Several attempts have been made to produce more accurate assessments of the significance levels of the tests involved in stepwise regression (cf. Draper *et al.*, 1971). Wilkinson and Dallal (1981) provide tables for tests based on the forward selection rule, whereas Butler (1984) develops bounds for the p-value of the F test for the next best-fitting regressor variable. Miller (1984) provides a comprehensive review of the field and recommends using the test procedure developed in Spjøtvoll (1972a). Mitchell and Beauchamp (1988) consider the variable selection problem in a Bayesian context.

Narula and Wellington (1979) consider stepwise regression using least absolute deviations rather than least squares. Hoerl *et al.* (1986) show via an extensive simulation study that selection procedures based on ridge regression may be more effective than standard procedures.

Copas (1983) demonstrates that Stein-like 'shrinkage estimators' often give lower prediction mean square errors, especially when stepwise procedures have been used.

Picard and Cook (1984) show that cross-validation procedures may be used to validate tests of models developed by stepwise algorithms. In essence, the data set is divided into two parts; the model is then identified using one subset and its goodness of fit tested using the other subset.

Checking the assumptions

32.29 In **32.2**, we made some very strong assumptions in order to establish the properties of the linear model. Since regression analysis is such a widely used tool in applied statistics, it is clearly necessary to have effective procedures for checking these assumptions, in so far as this is possible. In fact, we need both *diagnostic* procedures to establish whether an assumption has been violated and *extensions* to the basic model to accommodate departures when identified. The rapid advances in statistical computing have opened the way to using a wide variety of computationally intensive diagnostics, mostly based on the residuals, many of which are conceptually simple but were infeasible in times past.

We shall examine the assumptions in **32.2** one at a time, discussing extensions and diagnostics for each potential problem in what seems the most natural sequence. However, we should never lose sight of the fact that all violations may occur simultaneously when working with real data, so that model building is likely to remain very much an art form rather than an exact science.

32.30 Perhaps the most basic assumptions specified in **32.2** are the interwoven threesome: linearity (L), additivity (A) and, later, normality (N). As we observed in **27.44**, these assumptions arise naturally together, since (L+A) implies (N) in some circumstances. When it is not reasonable to assume (L + A + N), either the model must be completely reformulated, as for the non-linear regression models discussed in **32.84–87**, or transformations must be found to restore (L+A+N), at least to a reasonable degree of approximation.

Even when the basic structure (32.1) is justified, a series of other assumptions must be validated, as follows:

Errors have zero means, $E(\boldsymbol{\epsilon}) = \mathbf{0}$. This assumption is essentially untestable in its simplest form. If $E(\epsilon_i) = c$ for all i, this is most likely a measurement error and the data-recording process may be suspect. In more complex cases where the mean may be a function of x_i, the residuals will usually display some pattern of dependence as described below.

Errors have constant variances (are homoscedastic). The errors are said to be *heteroscedastic* if the variances differ. Such departures may be detected using tests based upon the residuals and modified estimators developed.

Errors are uncorrelated. This is the problem of *autocorrelation*, and most commonly arises for time- or space-dependent data. Again, diagnostics are based on the residuals and the patterns of dependence may be built into the extended model.

$\mathbf{X}$ *is of rank p*. At first sight, this is a 'technical condition' which, it might be thought, would not be violated in practice. In a sense this is true as $\mathbf{X}$ is usually of full rank unless the investigator has unwittingly built an exact linear dependence into the set of regressor variables (see Chapter 30). In such circumstances, a suitable constraint will serve to remove the problem, as noted in **29.11**. However, there are many circumstances, particularly in observational studies, where high correlations can exist among the regressors so that the matrix $\mathbf{X}^T\mathbf{X}$ becomes *ill-conditioned*; that is, elements in the inverse become large and numerically unstable. This is the problem of *multicollinearity*; rather different diagnostics and modifications are required when such circumstances are suspected.

Finally, we must not forget that statistical inference requires specification of the error distribution. In general, it is assumed that the errors are normally distributed, after transformation if necessary; an exception to this approach is the *bootstrap* method, described in **32.45**. Tests of normality are considered in **25.46–47**.

Pierce and Kopecky (1979) showed that the usual goodness-of-fit tests may be used for regression residuals, with the same asymptotic distributions. White and MacDonald (1980) and Jarque and Bera (1987) give numerical comparisons.

We shall discuss transformations in **32.31–44** and the bootstrap in **32.45**. The issue of multicolllinearity is addressed in **32.46–56**. We then examine heteroscedasticity in **32.57–60** and autocorrelation in **32.61–64**.

Transformations to the normal linear model

32.31 Following Box and Cox (1964), suppose that we are not prepared uncritically to assume that $\mathbf{y} = \mathbf{X}\boldsymbol{\beta} + \boldsymbol{\epsilon}$; rather, we seek transformations both of $\mathbf{y}$ and of each of the xs so that we have

$$\mathbf{y}_\lambda = \mathbf{X}_\mu \boldsymbol{\beta} + \boldsymbol{\epsilon}, \tag{32.48}$$

where the components of $\boldsymbol{\epsilon}$ are independently normal with zero means and constant variance σ^2. In (32.48), $\lambda = (\lambda_1, \lambda_2, \ldots)$ indexes the transformation of $\mathbf{y}$ within some selected parametric family of transformations, and similarly $\mu = (\mu_1, \mu_2, \ldots, \mu_p)$ indexes the (separate) transformations of the regressors $x_1, x_2, \ldots, x_p$.

32.32 By (32.48), the LF is, in logarithmic form,

$$\log L_{\lambda,\mu}(\mathbf{y}|\boldsymbol{\beta}, \sigma^2) = -\frac{1}{2}n\log(2\pi\sigma^2) - \frac{1}{2\sigma^2}(\mathbf{y}_\lambda - \mathbf{X}_\mu\boldsymbol{\beta})^T(\mathbf{y}_\lambda - \mathbf{X}_\mu\boldsymbol{\beta}) + \log J_\lambda, \tag{32.49}$$

where J_λ is the Jacobian of the inverse transformation from $\mathbf{y}_\lambda$ (the normally distributed variable in (32.48)) to the actually observed $\mathbf{y}$. Now, when the LF (32.49) is maximized for given $\boldsymbol{\lambda}$, $\boldsymbol{\mu}$ with respect to $\boldsymbol{\beta}$ and σ^2, we find that the middle term becomes a constant. If we neglect constants, therefore, we have the conditional maximum for fixed $\boldsymbol{\lambda}$, $\boldsymbol{\mu}$,

$$\log L_{\lambda,\mu}(\mathbf{y}|\hat{\boldsymbol{\beta}}, \hat{\sigma}^2) = -\tfrac{1}{2}n\log\hat{\sigma}^2_{\lambda,\mu} + \log J_\lambda, \tag{32.50}$$

where $n\sigma^2_{\lambda,\mu} = \mathbf{y}_\lambda^T\mathbf{T}\mathbf{y}_\lambda$, say, is the residual SS, again as in **32.3**.

We need now to compute the absolute maximum of the conditional maximum (32.50) over the whole range of $\boldsymbol{\lambda}$, $\boldsymbol{\mu}$. This is a formidable numerical task, except when only one or two transformation indices are involved, for example when

(a) only the dependent variable y is transformed and λ has only one or two components; or
(b) the same transformation is applied to all of, or a subset of, the regressors, so that $\boldsymbol{\mu}$ has only one or two components; or
(c) $\boldsymbol{\lambda}$ has a single component as in (a), and (b) holds with only one component in $\boldsymbol{\mu}$.

In cases (b) and (c), numerical plotting of the contours of (32.49) for all $\boldsymbol{\lambda}$, $\boldsymbol{\mu}$ will generally be necessary. We now confine ourselves to case (a), where only the dependent variable is being transformed. This implies that we can choose proper forms for the regressor variable before considering transformation of the dependent variable. Box and Tidwell (1962) discuss transformations of the general regressors to simpler form (cf. Exercise 32.14).

32.33 In practice, the most useful transformations have been found to be the powers and the logarithm of y, possibly translated by a constant. We therefore consider the family of transformations

$$y_\lambda = \begin{cases} (y+\lambda_2)^{\lambda_1}, & \lambda_1 \neq 0, \\ \log(y+\lambda_2), & \lambda_1 = 0. \end{cases} \tag{32.51}$$

To avoid discontinuity at $\lambda_1 = 0$, we rewrite this equivalently as

$$y_\lambda = \begin{cases} \{(y+\lambda_2)^{\lambda_1} - 1\}/\lambda_1, & \lambda_1 \neq 0, \\ \log(y+\lambda_2), & \lambda_1 = 0. \end{cases} \tag{32.52}$$

Tukey (1957a) studied and charted the structural features of the family (32.51) for $\lambda_1 \leq 1$; Box and Cox (1964) and Dolby (1963) considered properties of the differential equation which it satisfies, namely $(y'_\lambda/y''_\lambda)' = (\lambda_1 - 1)^{-1}$.

32.34 In (32.50), we now have

$$\log J_\lambda = (\lambda_1 - 1)\sum_{i=1}^{n} \log(y_i + \lambda_2), \tag{32.53}$$

and (32.49) can be plotted for selected (λ_1, λ_2) for numerical determination of the absolute maximum. An ANOVA must be carried out for each (λ_1, λ_2) used, to obtain the residual SS in (32.49). In the simplest case when $\lambda_2 = 0$, this can be avoided by equating to zero the first derivative of (32.49) with respect to λ_1. Using (32.52)–(32.53), this gives

$$0 = \frac{\partial \log L_{\lambda_1}(y|\hat{\beta}, \hat{\sigma}^2)}{\partial \lambda_1} = -n\frac{\mathbf{y}_{\lambda_1}^T \mathbf{T}\mathbf{u}_{\lambda_1}}{\mathbf{y}_{\lambda_1}^T \mathbf{T}\mathbf{y}_{\lambda_1}} + n\lambda_1^{-1} + \sum_{i=1}^{n} \log y_i, \tag{32.54}$$

where the elements of $\mathbf{u}$ are $\{\lambda_1^{-1} y_i^{\lambda_1}, \log y_i\}$. In this case, Draper and Cox (1969) examine the precision with which λ_1 is estimated – see also Hinkley (1975) for some corrections.

Carroll (1982) shows that searching for λ over a coarse grid (the values $0, \pm\frac{1}{2}, \pm 1$, are a popular choice) may lead to a considerable increase in the prediction MSE. However, if the true value happens to be very close to a grid value, the prediction MSE may be reduced.

Schlesselman (1971) shows that (32.49) will not in general be invariant under scale changes in y unless the model (32.48) contains a general mean which we now assume.

If the errors are heteroscedastic, the ML estimator for λ given by (32.54) is inconsistent. By contrast, nonlinear LS estimation for the slightly different model

$$y = (1 + \lambda \mathbf{x}^T \boldsymbol{\beta})^{1/\lambda} + \epsilon$$

is always consistent, whether the error is multiplicative or additive (Showalter, 1994). The estimator based upon (32.54) is also sensitive to outliers; Kim *et al.* (1996) provide a more robust estimator.

32.35 In order to test the hypothesis $H_0 : \lambda = \lambda_0$, Andrews (1971) proposed an exact test based upon the following argument. Consider the first-order Taylor series expansion

$$\mathbf{y}_\lambda = \mathbf{y}_0 + (\lambda - \lambda_0)\mathbf{v}, \tag{32.55}$$

where $\mathbf{v} = \mathbf{v}(\mathbf{y}) = [\partial \mathbf{y}_\lambda / \partial \lambda]_{\lambda=\lambda_0}$ and $\mathbf{y}_0$ denotes $\mathbf{y}$ evaluated at λ_0. Using (32.55), (32.48) may be expressed as

$$\mathbf{y}_0 = \mathbf{X}\boldsymbol{\beta} + (\lambda_0 - \lambda)\mathbf{v} + \boldsymbol{\epsilon}. \tag{32.56}$$

Andrews' proposed test reduces to testing the coefficient $\lambda_0 - \lambda$ in (32.56) when $\mathbf{v}$ is replaced by $\hat{\mathbf{v}} = \hat{\mathbf{v}}(\hat{\mathbf{y}})$.

Atkinson (1973) suggested replacing (32.56) by

$$\mathbf{z}_0 = \mathbf{X}\boldsymbol{\beta} + (\lambda_0 - \lambda)\mathbf{w} + \boldsymbol{\epsilon}, \tag{32.57}$$

where $\mathbf{z}_\lambda = \mathbf{y}_\lambda/\{J_\lambda\}^{1/n}$ and $\mathbf{w} = [\partial \mathbf{z}_\lambda/\partial\lambda]_{\lambda=\lambda_0}$; the test procedure is essentially the same. Simulation results (see Atkinson, 1973) suggest that the revised procedure is generally more powerful, although the true probability of Type I error often exceeds the nominal level. Lawrance (1987) provides a modification to the test statistic to correct this problem. It is evident that (32.56) or (32.57) may be used to provide an estimator for λ which, though inefficient, is very easy to compute, a property that will be useful in diagnostic testing; see **32.70**.

LR tests of nested hypotheses

32.36 Box and Cox (1964) present some interesting numerical examples of the application of this method of finding a transformation, and of a parallel Bayesian method of analysis which they develop. In addition, they consider the resolution of the maximized LF into three components corresponding to the normality, the homoscedasticity and the structure of expectation of $\mathbf{y}_\lambda$. Their procedure is of general applicability.

Consider sets of constraints $C_1, C_2, \ldots$, to be applied successively to a mathematical model, and let $\hat{\lambda}_{(s)}$ be the ML estimator of λ when all of $C_1, C_2, \ldots, C_s$ have been applied. $\hat{\lambda}$, without suffix, is the ML estimator when no constraint is imposed. Then, identically, for any s,

$$\begin{aligned} L(y|\hat{\lambda}_{(s)}) &= L(y|\hat{\lambda}) \cdot \frac{L(y|\hat{\lambda}_{(1)})}{L(y|\hat{\lambda})} \cdot \frac{L(y|\hat{\lambda}_{(2)})}{L(y|\hat{\lambda}_{(1)})} \cdots \frac{L(y|\hat{\lambda}_{(s)})}{L(y|\hat{\lambda}_{s-1})} \\ &= L(y|\hat{\lambda}) \cdot l_1 l_2 \cdots l_s, \end{aligned} \tag{32.58}$$

where l_p is the LR test statistic for testing the set of constraints $C_1, C_2, \ldots, C_{p-1}, C_p$ against the set of $C_1, C_2, \ldots, C_{p-1}$ (cf. **22.1**). Each of the l_p lies between 0 and 1, and under regularity conditions, $-2\log l_p$ is asymptotically a non-central χ^2 variable with d.fr. equal to the number of independent constraints upon parameters imposed by C_p (cf. **22.7**). When C_p holds, this becomes a central χ^2 variable, and thus $-2\log l_p$ may be used to test the value of adding C_p to the already imposed $C_1, C_2, \ldots, C_{p-1}$. It should be observed that the l_p are not in general independently distributed, though in particular cases they may be independent under certain hypotheses (cf. Exercises 22.6 and 22.13, and the more general result of Exercise 32.13). The application of the resolution (32.58) to the present problem is left to the reader as Exercise 32.12, since it follows immediately from some results given in Chapter 22.

Spitzer (1978) examined the performance of this method in samples of size 30 and 60. While the estimators were approximately normal with little bias and variance estimators of high efficiency, the t statistics used in tests had long-tailed distributions leading to too frequent rejection of true hypotheses. Models with $\lambda_1 > 0$ generally performed worse both in bias and in variance.

The purposes of transformation

32.37 The virtue of the ML approach discussed in **32.31–35** is that it requires no prior knowledge of the relationship between $\mathbf{y}$ and the regressors, or of the nature of the error distribution of the untransformed $\mathbf{y}$. It starts from the assumption that there exists some transformation in the family considered for which all the conditions of the linear model, including homoscedasticity and normality of the error distribution, are satisfied. In particular cases, of course, this may not be so – cf. Draper and Cox (1969), Hernandez and Johnson (1980) and Bickel and Doksum (1981); but even then, the ML procedure for choice of the transformation must presumably be an improvement on the uncritical use of **y** in its original form. It is a striking fact (evidenced by the numerical examples given by Box and Cox, 1964) that this ML transformation is often very close to what is suggested by non-statistical consideration of the nature of the underlying variables. Such consideration should, of course, be undertaken whenever possible as a supplement and guide to the statistical analysis itself.

Kruskal (1965) gives a method of finding the monotone transformation of the observations which minimizes the residual SS (suitably scaled) from an assumed linear model. No parametric family like (32.52) is required; nor is the normality assumption. He uses his method to reanalyse the Box–Cox examples, with several others. See also Draper and Hunter (1969).

Other approaches to transforming the data to meet the needs of the linear model have been less ambitious. They seek *either* to normalize the errors *or* to stabilize their variance *or* to remove interactions so that the effects are additive; and the hope is that a transformation which affects one of these aims will at least help towards achieving the others. It is remarkable that this indeed often turns out to be the case, and we shall examine some important instances shortly, but it is over-sanguine to expect this to be always so. It is easy to construct examples where the goals of additivity and homoscedasticity conflict; for example, if in a two-way cross-classification the expected value of **y** is additive in row and column effects, but the errors are non-normally distributed with variance a function of $E(\mathbf{y})$, any transformation to remove the heteroscedasticity will destroy exact additivity, whatever may happen to non-normality. Hoyle (1973) reviews the subject and provides a large bibliography. We now examine these different types of transformation in turn.

Variance-stabilizing transformations

32.38 Suppose that a statistic t has mean θ and variance, for fixed sample size n,

$$\operatorname{var} t = D_n^2(\theta). \tag{32.59}$$

To eliminate this dependence of variance on the parameter θ, we seek a function $u(t)$ such that $\operatorname{var} u$ is constant, c. In general, however, we are unlikely to be able to achieve this precisely; we ask only that

$$\operatorname{var}\{u(t)\} = c\{1 + O(R^{-1})\}, \tag{32.60}$$

where R is some known constant which is large enough for R^{-1} to be negligible. In particular, we may have $R = n$, the sample size. We now assume t to be confined to a neighbourhood of its mean θ and that a Taylor expansion may be made as at (10.11), so that we have from (10.14) the approximation

$$\operatorname{var}\{u(t)\} \doteq \left\{\left(\frac{\mathrm{d}u(t)}{\mathrm{d}t}\right)^2\right\}_{t=\theta} \operatorname{var} t. \tag{32.61}$$

If (32.60) and (32.61) are equated, we have the first-order approximation

$$\left\{\left(\frac{\mathrm{d}u(t)}{\mathrm{d}t}\right)^2\right\}_{t=\theta} \doteq \frac{c}{D_n^2(\theta)}. \tag{32.62}$$

Since we are considering only the neighbourhood of θ, we drop the suffix '$t = \theta$', and write θ for t. Thus

$$\frac{du(\theta)}{d\theta} \propto \{D_n^2(\theta)\}^{-1/2}, \tag{32.63}$$

where we drop the constant c without loss, since this is in any case at choice, for multiplication of $u(t)$ by a constant will not affect our purpose of achieving (32.60). We now integrate the equation (32.63), again ignoring the additive constant which results from the indefinite integration without loss, since (32.60) is unaffected. We obtain

$$u(t) \propto \left\{\int \frac{d\theta}{D_n(\theta)}\right\}_{\theta=t}. \tag{32.64}$$

32.39 Although (32.64) was arrived at through approximation, we can check its validity if the theoretical distribution of t is known by computation of the theoretical variance of $u(t)$ to verify its stability as θ varies – it may be found desirable to modify $u(t)$ to improve stability. When, on the other hand, we have only observations upon t and no prior knowledge of its distribution or of the parameter θ of that distribution, we cannot even compute $D_n^2(\theta)$. In such cases, the mean and variance of t in separate groups of observations are calculated, and the latter plotted against the former to give an *estimate* of the relationship (32.59), on which the transformation (32.64) is then based. Here, the approximation is more hazardous, but nevertheless often gives satisfactory results in practice.

Example 32.6 (Variance-stabilizing transform for the Poisson)

If t has the Poisson distribution the mean and variance are equal to θ, so (32.59) becomes

$$D_n^2(\theta) = \text{var}\, t = \theta$$

and (32.64) gives

$$u(t) \propto \left\{\int \theta^{-1/2}\, d\theta\right\}_{\theta=t} \propto t^{1/2}, \tag{32.65}$$

a simple square root transformation. To the first order, by (32.61),

$$\text{var}(t^{1/2}) = \{(\tfrac{1}{2}t^{-1/2})^2\}_{t=\theta}\, \text{var}\, t = \tfrac{1}{4}, \tag{32.66}$$

verifying the variance stabilization to this order.

Bartlett (1936a) pointed out that the variance stabilization could be improved in this case by relocating t before taking the square root. Bartlett suggested that if we define

$$u_c(t) = (t + c)^{1/2},$$

we should use $c = \frac{1}{2}$. Exercise 32.19 shows that $c = \frac{3}{8}$ is a better choice. Table 32.7 gives the variance of $u_c(t)$ as a fraction of its limiting variance as $\theta \to \infty$, for $c = 0$, $\frac{1}{2}$ and $\frac{3}{8}$ – the calculations were done by Bartlett (1936a) and Anscombe (1948).

The inadequacy of the simplest transformation with $c = 0$ is evident for small θ. For $\theta < 3$, the same comparison is made graphically in Fig. 32.2, adapted from Freeman and Tukey (1950),

Table 32.7 Variance of transformed variable for Poisson distribution

θ	Variance of $u_c(t)$ as a fraction of limiting variance		
	$c=0$	$c=\frac{1}{2}$	$c=\frac{3}{8}$
0	0	0	0
0.5	1.240	0.408	
1.0	1.608	0.640	0.717
2.0	1.560	0.856	0.924
3.0	1.360	0.928	0.983
4.0	1.224	0.960	0.999
6.0	1.104	0.980	1.002
9.0	1.052	0.988	
10.0			1.001
12.0	1.036	0.992	
15.0	1.024	0.992	
20.0			1.000

whose own variance-stabilization proposal, $u' = t^{1/2} + (t+1)^{1/2}$, is more stable than $u_{3/8}(t)$ for $\theta \leq 2$, after which either is adequate. u' is within 6 per cent of stability for $\theta \geq 1$, and seems the best choice (cf. Exercise 32.21).

32.40 The variance-stabilization procedure of **32.38** can be repeated if necessary. Suppose that investigation shows the variance of $u(t)$ to be

$$\text{var}\{u(t)\} = c\left\{1 + \frac{p(\theta)}{n}\right\} + o(n^{-1}), \tag{32.67}$$

satisfying (32.60). If we now seek a second transformed variable $v(u)$ such that

$$\text{var}\{v(u)\} = c\{1 + O(n^{-2})\}, \tag{32.68}$$

we have, as at (32.61), the approximation

$$\begin{aligned}\text{var}\{v(u)] &\doteq \left(\frac{dv(u)}{du}\right)^2 \text{var}\, u \\ &= \left(\frac{dv(u)}{du}\right)_\theta^2 c\left\{1 + \frac{p(\theta)}{n}\right\}\end{aligned}$$

by (32.67). Using (32.62), this is

$$\begin{aligned}\text{var}\{v(u)\} &= \left(\frac{dv}{du}\right)_\theta^2 \left(\frac{du}{dt}\right)_\theta^2 D_n^2(\theta)\left\{1 + \frac{p(\theta)}{n}\right\} \\ &= \left(\frac{dv}{dt}\right)_\theta^2 D_n^2(\theta)\left\{1 + \frac{p(\theta)}{n}\right\}.\end{aligned} \tag{32.69}$$

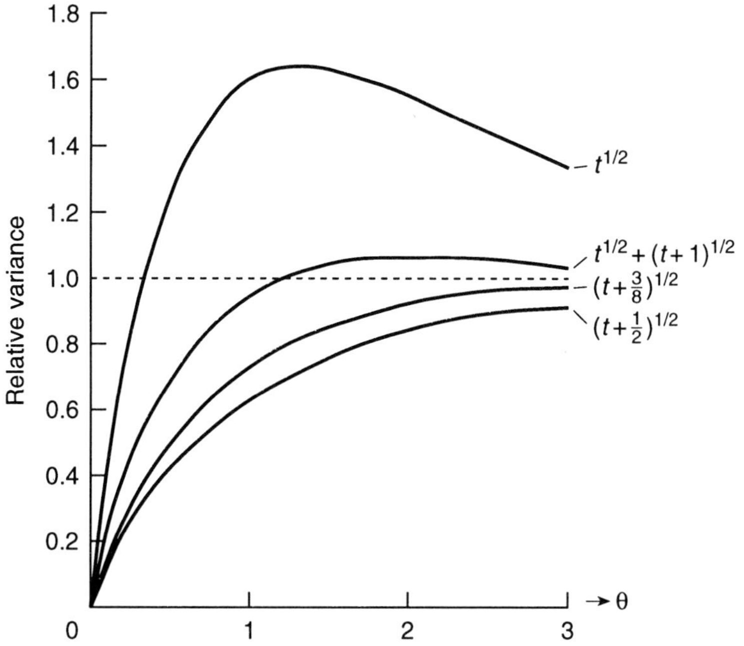

Fig. 32.2 Relative variance of various transformations for the Poisson

Thus, as at (32.64),

$$v(t) \propto \left\{ \int \left[D_n^2(\theta) \left\{ 1 + \frac{p(\theta)}{n} \right\} \right]^{-1/2} \mathrm{d}\theta \right\}_{\theta=t}. \tag{32.70}$$

We have already encountered an instance of this procedure in Hotelling's improved version of Fisher's variance-stabilizing z-transformation – cf. Exercises 16.18–19.

The variance-stabilization procedure could evidently be iterated further if this were necessary.

Exercises 32.15–17 provide the variance-stabilizing transformation for the binomial and negative binomial distributions.

Normalizing transformations

32.41 In **6.25–26** we examined the Cornish–Fisher method of obtaining a normalizing polynomial transformation; and in **6.27–36**, we discussed Johnson's systems of functional transformation to normality.

Curtiss (1943) gives a careful mathematical discussion of the limiting normality of transformations, especially those discussed in our examples and exercises. As noted in **32.37**, a transformation designed to achieve one purpose (here, variance stabilization) often also helps to achieve another (here, normalization). For example, Exercise 32.20 treats the case dealt with in Example 32.6, where this effect occurs. However, the following examples shows that this harmony of purposes is only obtainable by not pressing for optimal achievement in both directions: variance-stabilizing transformations commonly normalize as a by-product, but they do not produce the *optimum* normalization.

Example 32.7 (Transformations for the gamma distribution)

Let z follow a gamma distribution with parameters q and θ, so that

$$f(z) = \theta^{-q} z^{q-1} e^{-z/\theta} / \Gamma(q); \tag{32.71}$$

then

$$E(z) = q\theta \quad \text{and} \quad \text{var}(z) = q\theta^2.$$

If q is fixed, but θ varies between observations, we set

$$D_n^2(\theta) = q\theta^2$$

and (32.64) gives

$$u(z) \propto \left\{ \int \theta^{-1} \, d\theta \right\}_z = \log z, \tag{32.72}$$

the logarithmic transform. However, if θ is fixed and q varies, we obtain $u(z) \propto q^{1/2}$, leading to the square root transform. Using Exercise 32.18, we find that the skewness and kurtosis measures (cf. **3.31–32**) for $\log z$ are

$$\gamma_1 \simeq -q^{-1/2}, \qquad \gamma_2 \simeq 2q^{-1}; \tag{32.73}$$

those for z are

$$\gamma_1 = 2q^{-1/2}, \qquad \gamma_2 = 6q^{-1}, \tag{32.74}$$

so that the variance stabilization, as a by-product, has halved skewness (with a charge of sign) and reduced kurtosis by a factor of 3.

When q varies, the square root transform provides effective variance stabilization, as these values given by Bartlett (1936a) show:

q	0	0.5	1.0	2.0	3.0	4.0	9.0	15.0
$\text{var}(z^{1/2})$	0	0.182	0.215	0.233	0.239	0.242	0.247	0.250

The square root transform is equivalent to Fisher's approximation to the χ^2 distribution, treated in **16.5–6**. It has, by (16.8),

$$\gamma_1 = \tfrac{1}{2} q^{-1/2}, \qquad \gamma_2 = \tfrac{3}{16} q^{-2}, \tag{32.75}$$

distinct improvements over (32.73) for the untransformed variable and better also than (32.73) for its logarithm). Again, the variance stabilization has improved the normalization here. But note that the Wilson–Hilferty cube root transformation in **16.7** has, from (16.13), even better γ_1, of order $q^{-3/2}$, though not such a good γ_2, of order q^{-1}, and (cf. **16.8**) gives the better normal approximation. However, it does not stabilize variance at all, as (16.12) shows. Hernandez and Johnson (1980) show that the cube root is also the asymptotic value of λ_1 obtained by the ML method from (32.54). Thus the best available normalizing transformation sacrifices variance stability, and the square root is a better compromise transformation.

It is interesting that if both γ_1 and γ_2 are to minimized, the appropriate transformation is $z^{5/13}$, but in the Poisson case of Example 32.6, the power of transformation is increased from $\frac{1}{2}$ to $\frac{2}{3}$ if we minimize γ_1 alone, or to $\frac{7}{11}$ if we minimize both γ_1 and γ_2 – cf. Exercise 32.22.

32.42 The reader will see that our discussion of normalization has been couched entirely in terms of skewness and kurtosis. Mathematically, it is taking a good deal for granted to assume that smaller values of γ_1 and γ_2 are equivalent to a closer approach to normality; but we know of no significant example where this assumption misleads us in choosing between normal approximations.

Blom (1954) seeks functional transformations $u(t)$ for which a further polynomial (Cornish–Fisher) transformation as in **6.25–26** has a minimum skewness and is therefore presumably 'nearest' to symmetry and normality. This leads to a differential equation for $u(t)$ which contains all the transformations that we have encountered for the Poisson, gamma, binomial and negative binomial distributions. Haldane (1938) determined λ_1 in (32.51) when $\lambda_2 = 0$ to minimize skewness, and so also in effect determined λ_1 and λ_2 to minimize skewness and kurtosis – cf. Exercise 32.22. See also Hinkley (1975).

Transformations to additivity

32.43 Although in practice it may be important to search for a scale on which effects are additive (i.e., interactions disappear) or nearly so, relatively little work has been done in this area as compared with normalization and variance stabilization. Some general procedures that have been proposed involve minimization, within a class of transformations, of the value of the test statistic used for the hypothesis that interactions between regressor variables are zero. It will be recognized that the test statistic is here being used to carry out a complex estimation procedure, and nothing but intuitive justification has so far been given for this method. Additivity transformations are sometimes suggested by the analysis of residuals.

Generally, if the model appears to be additive in the original variate, but the variance is not stable, the errors should be treated as heteroscedastic; see **32.58–60**. Non-normality may be the result of outliers, which may be detected using the diagnostics described in **32.65–70**; alternatively, robust estimators may be used, as discussed in **32.71–74**.

Removal of transformation bias

32.44 Whatever the purpose of a transformation, it often raises problems of presentation when the analysis is complete. In particular, estimators that are unbiased on the transformed scale will not be so if the inverse transformation is made so that the results may be presented in 'natural' terms (cf. Exercise 32.19) Adjustments of some kind must be made to remove the bias due to transformation; the jackknife method of bias reduction was given in **17.10**. We now discuss an exact method of removing the bias.

Suppose that u is normally distributed with mean μ and variance σ^2, and that the functions of u, $(\hat{\mu}, S^2/\nu)$, are jointly sufficient statistics for these parameters, $\hat{\mu}$ being normally distributed with mean μ and variance $\lambda^2\sigma^2$, and S^2/σ^2, independent of $\hat{\mu}$, a χ^2 variate with ν d.fr. In practice, we usually have $\lambda^2 = 1/n$ and $\nu = n - 1$, where n is sample size. Now consider the function $t(u)$, which in our terms is the *inverse* transformation. Neyman and Scott (1960) (cf. Schmetterer, 1960), using the approach of Exercise 32.23, showed that if $t(u)$ satisfies the

Table 32.8 Bias of transformations

Transformation $u(t)$	Inverse transformation $t(\hat{\mu})$	A	B	Bias $E\{t(\hat{\mu})\} - \theta$	Sign of bias when $\lambda < 1$
$(t+c)^{1/2}$	$\hat{\mu}^2 - c$	2	0	$-1(1-\lambda^2)\sigma^2$	Negative
$\log(t+c)$	$\exp(\hat{\mu}) - c$	c	1	$(\theta+c)[\exp\{-(1-\lambda^2)\sigma^2/2\} - 1]$	$-\operatorname{sgn}(\theta+c)$
$\arcsin(t^{1/2})$	$\sin^2(\hat{\mu})$	2	-4	$(\theta - \frac{1}{2})[\exp\{2(1-\lambda^2)\sigma^2\} - 1]$	$\operatorname{sgn}(\theta - \frac{1}{2})$
$\operatorname{arcsinh}(t^{1/2})$	$\sinh^2(\hat{\mu})$	2	4	$(\theta + \frac{1}{2})[\exp\{-2(1-\lambda^2)\sigma^2\} - 1]$	$-\operatorname{sgn}(\theta + \frac{1}{2})$

second-order differential equation

$$t''(u) = A + Bt(u)$$

for constants A, B, the unique MV unbiased estimator of the mean of the untransformed variable $\theta = E(t)$ is given by

$$\hat{\theta} = \begin{cases} t(\hat{\mu}) + A(1-\lambda^2)S^2/(2\nu), & B = 0, \\ \{t(\hat{\mu}) + A/B\} \sum_{r=0}^{\infty} \frac{\Gamma(\frac{1}{2}\nu)}{r!\Gamma(\frac{1}{2}\nu+r)} \left\{\frac{B(1-\lambda^2)S^2}{4}\right\}^r - \frac{A}{B}, & B \neq 0. \end{cases}$$

This series converges very rapidly, only a few terms usually being required for accuracy.

It follows that the bias of the crude estimator $t(\hat{\mu})$, which is simply the inverse transformation of $\hat{\mu}$, is

$$E\{t(\hat{\mu}) - \theta\} = \begin{cases} -A(1-\lambda^2)\sigma^2/2, & B = 0, \\ \{\theta + (A/B)\}[\exp\{-B(1-\lambda^2)\sigma^2/2\} - 1], & B \neq 0, \end{cases}$$

and its absolute value is always a monotone decreasing function of λ^2. Since usually $\lambda^2 = 1/n$, the bias will increase with sample size.

The most important special cases are shown in Table 32.8. It will be seen that as $\lambda \to 0$ ($n \to \infty$), the bias for the square root transformation tends to $-\sigma^2$. This is the result obtained directly in Exercise 32.19, where $\sigma^2 = \frac{1}{4}$ as at (32.66). The reader may also recall that the results for the logarithmic transformation with $c = 0$, $\lambda^2 = 1/n$, were contained in Exercises 18.7–8. Thöni (1969) gives a table to facilitate computation of $\hat{\theta}$ in this case. The other two transformations in the table are those of Exercises 32.15–16.

Hoyle (1968) evaluates var $\hat{\theta}$ and its MV unbiased estimator for the wider class of transformations treated in Exercise 32.23. One of his results can be obtained directly in Exercise 32.24. Land (1971) considers confidence intervals for linear functions of μ and σ^2 and later (Land, 1974) approximate intervals for general functions. His tables of confidence intervals for $\theta + \frac{1}{2}\sigma^2$, useful in the logarithmic case, are given in Harter and Owen (1975).

Bootstrap

32.45 The discussion thus far has assumed that the errors are normally distributed, after transformation if necessary. An alternative approach is to eschew any distributional assumptions and to work with the *bootstrap*; cf. **10.18**. Let

$$e_{(1)} \leq e_{(2)} \leq \cdots \leq e_{(n)}$$

denote the ordered residuals obtained from the LS fit of model (32.1), leading to the empirical d.f.

$$\tilde{\mathbf{F}}(e) = \frac{j}{n} \qquad \text{if } e_{(j)} \leq e < e_{(j+1)}. \tag{32.76}$$

Then resample from $\tilde{\mathbf{F}}$; that is, draw n observations with replacement, say $\tilde{\mathbf{e}}$, and form

$$\tilde{\mathbf{y}} = \mathbf{X}\hat{\boldsymbol{\beta}} + \tilde{\mathbf{e}}. \tag{32.77}$$

From $(\mathbf{X}, \tilde{\mathbf{y}})$, compute the estimator $\tilde{\boldsymbol{\beta}}$, using LS.

The resampling and estimation process is repeated B times, where B is $O(n \log n)$, to form an empirical sampling distribution for the estimator of any function of $(\boldsymbol{\beta}, \sigma^2)$ that is of interest.

The basic theory for the regression case was developed by Freedman (1981); Freedman and Peters (1984) gave some empirical results showing that the bootstrap estimates may work well and that the asymptotic standard error formulae may substantially understate the required length of the confidence interval. Some authors prefer to use standard residuals, $e_i^* = e_i/\{\text{var}(e_i)\}^{1/2}$ rather than e_i to define $\tilde{\mathbf{F}}$. An extensive theoretical and empirical investigation is given in Wu (1986); see also Hinkley (1988).

Multicollinearity

32.46 If one or more of the eigenvalues of $\mathbf{X}^T\mathbf{X}$ is near zero, this leads to unstable estimators. This is the problem of multicollinearity, which sometimes plagues observational studies; experimental studies will not be so affected if the experiment is well designed. Indeed, $\max(D)$ is often used as a design criterion; see **32.13**.

When $D = 0$, this may be the parametrization used, as in ANOVA, and linear constraints can be imposed to restore a full-rank solution; see **30.18**. We now suppose that such constraints are not available and that D is close to zero. Further, we assume that all variables, including $\mathbf{y}$, are measured about their means so that the constant term may be dropped, and that the regressor variables are expressed in some standard measure. Usually this will mean that $\mathbf{Z} = \frac{1}{n}\mathbf{X}^T\mathbf{X}$ becomes the correlation matrix for $(x_1, \ldots, x_p)$, but other 'standardizations' may be used, provided the resulting variables are measured on a common scale. For ease of notation, we shall assume there are p regressors after deletion of the constant. Henceforth, we assume that $\mathbf{Z}$ denotes the correlation matrix.

Given this specification, it follows from (28.70) that the variance of $\hat{\beta}_j$ is

$$\text{var}(\hat{\beta}_j) = \sigma^2/(1 - \mathbf{R}^2_{(j)}), \tag{32.78}$$

where $\mathbf{R}^2_{(j)}$ is the squared multiple correlation for the regression of x_j on all the regressors other than x_j. To help detect multicollinearity, we may use the *variance inflation factor* (VIF) for each variable

$$\text{VIF}(j) = 1/(1 - \mathbf{R}^2_{(j)}). \tag{32.79}$$

Clearly, VIF $= 1$ if x_j is orthogonal to all other variables and large values indicate potential problems.

Stewart (1987) presents several indices for assessing multicollinearity; the ensuing discussion indicates the lively debate that persists. See also Gunst (1983). Fox and Monette (1992) extend the notion of the VIF to situations where a set of related (e.g. indicator) variables *must* be included in the model.

32.47 We now consider how the effects of multicollinearity might be ameliorated. Let

$$D = \prod_{i=1}^{p} \lambda_i, \tag{32.80}$$

where $\lambda_1 \geq \lambda_2 \geq \cdots \geq \lambda_p \geq 0$ are the eigenvalues of $\mathbf{Z} = (\mathbf{X}^T\mathbf{X})/n$. Since $\text{tr}(\mathbf{Z}) = p = \sum \lambda_i$, it follows immediately that $D \leq 1$. The condition number $k(\mathbf{Z}) = \lambda_1/\lambda_p$ is sometimes used as an index of multicollinearity.

Evidently, D will be increased if we can either remove the smaller roots or increase their values; both options have been considered, as we now describe.

Ridge regression

32.48 If we add a constant c to each λ_j, we have

$$D^* = \prod_{i=1}^{p} (\lambda_i + c). \tag{32.81}$$

In turn, this corresponds to using the biased estimator

$$\hat{\boldsymbol{\beta}}_c = (\mathbf{X}^T\mathbf{X} + c\mathbf{I})^{-1}\mathbf{X}^T\mathbf{y}, \tag{32.82}$$

which is the ridge regression estimator described in **29.9**.

Principal components regression

32.49 It is possible to replace the original regressor variables by a set of p or fewer orthogonal variables. One such set is given by the principal components, defined by the eigenvalues of $\mathbf{X}^T\mathbf{X}$, denoted by $\boldsymbol{\Lambda}$:

$$(\mathbf{X}^T\mathbf{X})\mathbf{U} = \mathbf{U}\boldsymbol{\Lambda}. \tag{32.83}$$

Thus the model becomes

$$\mathbf{y} = \mathbf{Z}\boldsymbol{\alpha} + \boldsymbol{\epsilon}, \tag{32.84}$$

where

$$\boldsymbol{\alpha} = \mathbf{U}^T\boldsymbol{\beta}, \qquad \boldsymbol{\beta} = \mathbf{U}\boldsymbol{\alpha}. \tag{32.85}$$

It follows directly that

$$\begin{aligned}\hat{\boldsymbol{\alpha}} &= (\mathbf{Z}^T\mathbf{Z})^{-1}\mathbf{Z}^T\mathbf{y} \\ &= \boldsymbol{\Lambda}^{-1}\mathbf{Z}^T\mathbf{y}\end{aligned}$$

or

$$\alpha_j = \mathbf{z}_j^T \mathbf{y}/\lambda_j, \qquad j = 1, 2, \ldots, p; \tag{32.86}$$

also

$$\text{var}(\hat{\boldsymbol{\alpha}}) = \sigma^2 \boldsymbol{\Lambda}^{-1}. \tag{32.87}$$

Further, it is evident from (32.87) that the imprecise estimates arise directly from the linear combinations of the regressors for which the corresponding eigenvalues are near zero. If ridge regression is not favoured, we may proceed by taking only the first k ($< p$) principal components, $\mathbf{Z}_k$ say, and employing the estimator

$$\underset{(k \times 1)}{\hat{\boldsymbol{\alpha}}_k} = \underset{(k \times k)}{\boldsymbol{\Lambda}_k^{-1}} \underset{(k \times n)}{\mathbf{Z}_k^T} \underset{(n \times 1)}{\mathbf{y}},$$

where the k-subscript denotes exclusion of components $k + 1, \ldots, p$. Finally from (32.85), we estimate $\boldsymbol{\beta}$ by

$$\underset{(p \times 1)}{\hat{\boldsymbol{\beta}}_k} = \underset{(p \times k)}{\mathbf{U}_k} \underset{(k \times 1)}{\boldsymbol{\alpha}_k}, \tag{32.88}$$

where $\mathbf{U} = (\mathbf{U}_k, \mathbf{U}_k^*)$, effectively imposing the conditions

$$\alpha_{k+1} = \cdots = \alpha_p = 0 \quad \text{or} \quad (\mathbf{U}_k^*)^T \boldsymbol{\beta} = \mathbf{0}. \tag{32.89}$$

As in ridge regression, the reduction in variance must be traded against the bias that is introduced.

Example 32.8 (The effects of multicollinearity on regression estimates (Greenberg, 1975))

Malinvaud (1966) considers a linear regression model for French imports (Y), over eleven years 1949–59. Imports are taken as a function of gross domestic product (x_1), stock formation (x_2) and consumption (x_3). The correlation matrix is

$$\begin{bmatrix} 1.0 & 0.169 & 0.997 \\ & 1.0 & 0.154 \\ & & 1.0 \end{bmatrix}$$

from which the variance inflation factors are

$$\text{VIF}(1) = 193.7, \qquad \text{VIF}(2) = 1.07, \qquad \text{VIF}(3) = 192.8.$$

In this case, the problem is clearly the high correlation between x_1 and x_3, although less obvious patterns will also be detected by the VIF. The eigenvalues and eigenvectors, derived from the correlation matrix, are given in Table 32.9. In Table 32.10, we give estimates of the regression coefficients for the standardized variables, based on one, two and three principal components, respectively. Those for $k = 1$ and $k = 2$ are very similar, while introduction of the third component causes serious problems. A test of the linear conditions $\mathbf{u}_2^T \boldsymbol{\beta} = 0$ and $\mathbf{u}_3^T \boldsymbol{\beta} = 0$ fails to reject either condition at the 5 per cent level. The results using $k = 1$ and $k = 2$ also seem more in accord with economic theory than the estimates given by the full set of variables.

Table 32.9 Eigenvalues and eigenvectors for the data of Example 32.7

λ_j	$1/\lambda_j$	u'_j
2.0472	0.448	(0.692, 0.213, 0.690)
0.9502	1.052	(−0.143, 0.977, −0.158)
0.0026	384.6	(−0.708, 0.010, 0.706)

Table 32.10 Regression derived from components for Example 32.7

	Estimate	Std error		Estimates derived from α		
				$k = 1$	$k = 2$	$k = 3$
$\hat{\alpha}_1$	0.6917	0.0302	$\hat{\beta}_1$	0.478	0.474	−0.648
$\hat{\alpha}_2$	0.0314	0.0430	$\hat{\beta}_2$	0.148	0.178	0.195
$\hat{\alpha}_3$	1.584	0.8516	$\hat{\beta}_3$	0.477	0.472	1.591

32.50 Another natural way to avoid multicolinearity is to use stepwise regression to reduce the set of predictor variables. When the relationship is not well understood and the main purpose of model development is prediction, such an approach can work quite well. However, when the process is well defined and we wish to evaluate the effects of a particular input variable upon the output, such as the effect of price on sales, stepwise regression should be avoided. Belsley (1991) describes various collinearity diagnostics and also provides an extended argument against the use of stepwise regression in the above context.

Polynomial regression: orthogonal polynomials

32.51 The linear model with powers of x as regressors, called the *polynomial regression* model, is

$$y_i = \beta_1 + \beta_2 x_i + \cdots + \beta_p x_i^{p-1} + \epsilon_i, \qquad i = 1, \ldots, n, \tag{32.90}$$

leading to an $\mathbf{X}^T\mathbf{X}$ matrix of the form

$$\mathbf{X}^T\mathbf{X} = n\mathbf{M}_p = n \begin{bmatrix} 1 & 0 & m_2 & \cdots & m_{p-1} \\ 0 & m_2 & m_3 & \cdots & m_p \\ \vdots & \vdots & \vdots & & \vdots \\ m_{p-2} & m_{p-1} & m_p & \cdots & m_{2p-3} \\ m_{p-1} & m_p & m_{p+1} & \cdots & m_{2p-2} \end{bmatrix}, \tag{32.91}$$

where $nm_k = \sum x_i^k$ and we set $m_1 = \bar{x} = 0$ without loss of generality. Clearly, $\mathbf{M}_p$ cannot be diagonal. However, we may choose polynomials of degree j in x, say $\phi_j(x)$, $j = 0, 1, \ldots, p-1$, that are mutually orthogonal so that (32.90) becomes

$$\mathbf{y} = \mathbf{X}\boldsymbol{\beta} + \boldsymbol{\epsilon} = \boldsymbol{\Phi}\boldsymbol{\alpha} + \boldsymbol{\epsilon}, \tag{32.92}$$

where $\Phi = \mathbf{XU}$, $\boldsymbol{\alpha} = \mathbf{U}^{-1}\boldsymbol{\beta}$, and U is lower triangular to ensure that ϕ_j is of degree j. The requirement of orthogonality becomes

$$\boldsymbol{\Phi}^T\boldsymbol{\Phi} = \mathbf{C}, \tag{32.93}$$

where $\mathbf{C}$ is a $p \times p$ diagonal matrix and we may set $C_{00} = 1$ implying $u_{00} = 1$; the diagonal elements of $\mathbf{C}$ are arbitrary and may be set at any convenient value. It follows from (32.92), that the kth-degree polynomial is given by

$$\phi_k(x) = \begin{vmatrix} \mathbf{M}_k & \mathbf{m}_{(k)} \\ \mathbf{x}_{(k)}^T & \end{vmatrix} \Big/ |\mathbf{M}_k|,$$

where

$$\mathbf{x}_{(k)}^T = (1, x, \ldots, x^k) \text{ and}$$

$$\mathbf{m}_{(k)}^T = (m_k, m_{k+1}, \ldots, m_{2k-1}).$$

For example,

$$\phi_1(x) = \begin{vmatrix} 1 & 0 \\ 1 & x \end{vmatrix} \Big/ |1| = x$$

$$\phi_2(x) = \frac{\begin{vmatrix} 1 & 0 & m_2 \\ 0 & m_2 & m_3 \\ 1 & x & x \end{vmatrix}}{\begin{vmatrix} 1 & 0 \\ 0 & m_2 \end{vmatrix}} = x^2 - \frac{m_3}{m_2}x - m_2$$

and so on. A simpler recursive method of obtaining the polynomials is given in Exercise 32.25.

32.52 From (32.92) and (32.93), the LS estimators

$$\hat{\alpha}(\boldsymbol{\Phi}^T\boldsymbol{\Phi})^{-1}\boldsymbol{\Phi}^T y$$

reduce to

$$\hat{\alpha}_j = \sum_{i=1}^{n} y_i\phi_j(x_i) \Big/ \sum_{i=1}^{n}\{\phi_j(x_i)\}^2, \tag{32.94}$$

with the benefit that earlier coefficients are unchanged when higher-degree polynomials are added to the model.

Equally spaced x values

32.53 The most important applications of orthogonal polynomials in regression analysis are to situations where the regressor variable, x, takes values at equal intervals. This is often the case with observations taken at successive times, and with data grouped into classes of equal width. If we have n such equally spaced values of x we measure from their mean and the natural interval as unit, thus obtaining as working values of x : $-\frac{1}{2}(n-1), -\frac{1}{2}(n-3), -\frac{1}{2}(n-5), \ldots, \frac{1}{2}(n-3), \frac{1}{2}(n-1)$. For this simple case, the values of moments in (32.91) can be explicitly calculated:

in fact, apart from the mean which has been taken as origin, these are the moments of the first n natural numbers, obtainable from the cumulants given in Exercise 3.23. The odd moments are zero by symmetry; the even moments are

$$m_2 = (n^2 - 1)/12,$$
$$m_4 = m_2(3n^2 - 7)/20,$$
$$m_6 = m_2(3n^4 - 18n^2 + 31)/112,$$

and so on. Substituting these and higher moments in to (32.91), we obtain for the first six polynomials

$$\begin{aligned}
\phi_0(x) &= 1, \\
\phi_1(x) &= \lambda_{1n} x, \\
\phi_2(x) &= \lambda_{2n}\{x^2 - \tfrac{1}{12}(n^2 - 1)\}, \\
\phi_3(x) &= \lambda_{3n}\{x^3 - \tfrac{1}{20}(3n^2 - 7)x\}, \\
\phi_4(x) &= \lambda_{4n}\{x^4 - \tfrac{1}{14}(3n^2 - 13)x^2 + \tfrac{3}{560}(n^2 - 1)(n^2 - 9)\}, \\
\phi_5(x) &= \lambda_{5n}\{x^5 - \tfrac{5}{18}(n^2 - 7)x^3 + \tfrac{1}{1008}(15n^4 - 230n^2 + 407)x\}, \\
\phi_6(x) &= \lambda_{6n}\{x^6 - \tfrac{5}{44}(3n^2 - 31)x^4 + \tfrac{1}{176}(5n^4 - 110n^2 + 329)x^2 \\
&\quad - \tfrac{5}{14\,874}(n^2 - 1)(n^2 - 9)(n^2 - 25)\}.
\end{aligned} \qquad (32.95)$$

Allan (1930) also gives $\phi_i(x)$ for $i = 7, 8, 9, 10$. Following Fisher (1921b), the arbitrary constant λ_{in} in (32.95) may be determined conveniently so that $\phi_i(x_j)$ is an integer for all $j = 1, 2, \ldots, n$. It will be observed that

$$\phi_{2i}(x) = \phi_{2i}(-x) \text{ and } \phi_{2i-1}(x) = -\phi_{2i-1}(-x);$$

even-degree polynomials are even functions and odd-degree polynomials odd functions.

Tables of orthogonal polynomials

32.54 The *Biometrika Tables*, Vol. I, Table 47, give $\phi_i(x_j)$ for all $j, n = 3\ (1)\ 52$ and $i = 1\ (1)\ \min(6, n-1)$, together with the values of λ_{in} and $\sum_{j=1}^{n} \phi_i^2(x_j)$.

Fisher and Yates (1963, Table XXIII) give $\phi_i(x_j)$ (their ξ_i'), λ_{in} and $\sum_{j=1}^{n} \phi_i^2(x_j)$ for all j, $n = 3\ (1)\ 75$ and $i = 1\ (1)\ \min(5, n-1)$.

The *Biometrika Tables* give references to more extensive tabulations, up to $i = 9, n = 52$, by van der Reyden, and to $i = 5, n = 104$, by Anderson and Houseman.

32.55 There is a large literature on orthogonal polynomials. For theoretical details, the reader should refer to the paper by Fisher (1921b) who first applied them to polynomial regression, to a paper by Allan (1930), and Aitken (1933a; 1933b; 1933c). Rushton (1951) discussed the case of unequally spaced x values and Cox (1958) gave a concise determinantal derivation of general orthogonal polynomials, while Guest (1954; 1956) has considered grouping problems.

Narula (1979) provides efficient numerical procedures for computing orthogonal polynomials when the data are unequally spaced. Studden (1980; 1982) develops optimal designs for polynomial regression where the aim is to provide efficient tests of the hypothesis that an rth-order polynomial is adequate against the alternative that order $s > r$ is required.

The reader should not need to be warned against the dangers of extrapolating from a fitted regression, however close, that has no theoretical basis.

Distributed lags

32.56 A special case of (32.1) of particular interest arises when $x_{jt} = x_{t-j+1}$ so that

$$y_t = \sum_{j=0}^{p} \beta_j x_{t-j+1} + \epsilon_t. \tag{32.96}$$

Often, the series x_t displays high autocorrelation, and the selected value of p is sufficiently large to create numerical difficulties if the LS estimators are used. To overcome this Almon (1965) suggested representing the βs by an rth-order polynomial ($r + 1 < p$)

$$\beta_j = \sum_{j=0}^{r} j^i \delta_i, \qquad j = 0, 1, \ldots, p. \tag{32.97}$$

In matrix terms, the original equation is

$$\mathbf{y} = \mathbf{X}\boldsymbol{\beta} + \boldsymbol{\epsilon}$$

which may now be rewritten as

$$\mathbf{y} = \mathbf{Z}\boldsymbol{\delta} + \boldsymbol{\epsilon},$$

where $\mathbf{Z} = \mathbf{XH}$ and $h_{ij} = j^i$. The reduced number of unknowns will often remove the multicollinearity while allowing the explicit incorporation of the first p lags. Clearly, we need not restrict attention to polynomials, but may use any set of weights in (32.97). Such techniques are known as distributed lag methods and have proved particularly useful in econometrics; see Griliches (1967).

An alternate representation of (32.96) is to use the backshift operator $Bx_t = x_{t-1}$ and write

$$y_t = \phi(B)\, x_t + \epsilon_t,$$

where $\phi(B) = \beta_0 + \beta_1 B + \cdots + \beta_p B^p$. A parsimonious representation is often achieved by letting $\phi(B)$ denote a ratio of polynomials. For example, if

$$\phi(B) = (\beta_0 + \beta_1 B)/(1 - \alpha_1 B)$$

the model becomes

$$(1 - \alpha_1 B)y_t = (\beta_0 + \beta_1 B)x_t + (1 - \alpha_1 B)\epsilon_t$$

or

$$y_t = \alpha_1 y_{t-1} + \beta_0 x_t + \beta_1 x_{t-1} + \epsilon_t^*,$$

where $\epsilon_t^* = \epsilon_t - \alpha_1 \epsilon_{t-1}$. Such models are *transfer* functions, widely used in time series analysis; cf. Kendall and Ord (1990), Box *et al.* (1994).

32.57 If the covariance matrix is not $\mathbf{V}(\boldsymbol{\epsilon}) = \sigma^2 \mathbf{I}$ as previously assumed, there are two principal departures to be considered: *heteroscedasticity*, when $\mathbf{V}(\boldsymbol{\epsilon}) = \sigma^2 \mathbf{V}$, where $\mathbf{V}$ is diagonal and $\mathbf{V} \neq \mathbf{I}$; and *autocorrelation*, when $\mathbf{V}(\boldsymbol{\epsilon}) = \sigma^2 \mathbf{V}$, where at least some off-diagonal elements are non-zero.

We now consider these in turn.

Heteroscedasticity

32.58 If $\mathbf{V}$ is diagonal and known, we may use the weighted LS estimator

$$\hat{\boldsymbol{\beta}} = (\mathbf{X}^T \mathbf{V}^{-1} \mathbf{X})^{-1} \mathbf{X}^T \mathbf{V}^{-1} \mathbf{y} \tag{32.98}$$

whose properties are discussed in **29.15**. When $\mathbf{V}$ must be estimated we must either have replicates for each set of regressor variables or else constrain the elements of $\mathbf{V}$ to satisfy some estimable function. If we write

$$E(y_i) = \mu_i(\boldsymbol{\beta}) \equiv \mu_i = \mathbf{x}_i^T \boldsymbol{\beta} \tag{32.99}$$

and

$$\text{var}(y_i) = \sigma^2 v_i = \sigma^2 \{g(\mu_i, z_i, \theta)\}^2 = \sigma^2 g_i^2, \tag{32.100}$$

the weighted LS criterion would be to minimize

$$\sum_{i=1}^{n} [y_i - \mathbf{x}_i^T \boldsymbol{\beta}]^2 / v_i \tag{32.101}$$

or, equivalently,

$$\sum_{i=1}^{n} [\tilde{y}_i - \tilde{\mathbf{x}}_i^T \boldsymbol{\beta}]^2, \tag{32.102}$$

where $\tilde{y} = y_i/g_i$ and $\tilde{x}_i = x_i/g_i$. Both (32.101) and (32.102) lead directly to (32.98). It is clear from (32.101) that greater weight is assigned to the observations with smaller v_i, thereby making the weighted LS estimator potentially more efficient. Typical forms for g_i in (32.100) are

$$g_i = \mu_i^\theta \quad \text{or} \quad g_i = e^{\theta \mu_i}; \tag{32.103}$$

in some cases $|\mu_i|$ will be more appropriate than μ_i.

32.59 If either version of (32.103) is suspected, a natural diagnostic tool is to plot the LS residuals against μ_i or $\log(\mu_i)$, respectively. Improved plots are discussed in **32.60** when we examine diagnostics more systematically. If the plots provide little evidence of heteroscedasticity, the LS estimators will cause only small losses of efficiency, so we now suppose that sizeable discrepancies have been identified. We may then generate the weighted LS estimators by an interactive process as follows:

1. Using initial consistent estimators of $\boldsymbol{\beta}$ (typically the LS estimators), estimate the weights, $\hat{w}_i = 1/\hat{g}_i$.
2. Generate new estimates of $\boldsymbol{\beta}$ using (32.98).

This cycle may be repeated K times. For any choice of K, the resulting estimators are asymptotically normally distributed,

$$\hat{\boldsymbol{\beta}} \sim N\{\boldsymbol{\beta}, \sigma^2(\mathbf{X}^T\mathbf{V}^{-1}\mathbf{X})^{-1}\}, \tag{32.104}$$

provided the initial estimator is consistent. Although 'well known', this result was not demonstrated formally until the work of Jobson and Fuller (1980) and Carroll and Ruppert (1982).

Full iteration of steps 1 and 2 is a version of *iteratively reweighted least squares* (cf. Green, 1984), although the same end could, in this case, be achieved more efficiently using a Newton–Raphson method. The choice of K has been the subject of some research (see Matloff *et al.*, 1984); the general view is that $K = 2$ or 3 cycles will usually suffice, although increasing K does not lead uniformly to improved estimators.

32.60 The algorithm described in **32.59** does not provide an explicit estimation procedure for the parameters in g_i. Heuristic procedures include plots of $|e_i|$ or $\log|e_i|$ against μ_i or $\log\mu_i$, where the $\{e_i\}$ are the residuals, $\mathbf{e} = \mathbf{y} - \mathbf{X}\hat{\boldsymbol{\beta}}$; Cook and Weisberg (1983) recommend the use of standardized residuals. Such procedures are reasonably satisfactory provided very small residuals are accommodated, for example by using $|e_i^*| = |e_i| + c$, $c > 0$. Yin and Carroll (1990) provide a similar diagnostic, based upon rank statistics, which is less susceptible to outliers. When the errors are normally distributed, ML estimators may be determined, as we now show.

Example 32.9 (ML estimation for heteroscedastic linear models)

Let

$$\mathbf{y} \sim N(\mathbf{X}\boldsymbol{\beta}, \sigma^2\mathbf{V}),$$

where $\mathbf{V} = \text{diag}(v_i)$, $v_i = \mu_i^\theta$ and $\mu_i = \mathbf{x}_i^T\boldsymbol{\beta} > 0$. The log LF is

$$l = \text{const.} - \frac{1}{2}\log|V| - \frac{n}{2}\log\sigma^2 - \frac{1}{2\sigma^2}(\mathbf{y} - \mathbf{X}\boldsymbol{\beta})^T\mathbf{V}^{-1}(\mathbf{y} - \mathbf{X}\boldsymbol{\beta}).$$

The first-order derivatives yield the estimating equations

$$(\mathbf{X}^T\mathbf{V}^{-1}\mathbf{X})\boldsymbol{\beta} - \mathbf{X}^T\mathbf{V}^{-1}\mathbf{y} = \mathbf{0}$$

$$n\sigma^2 - (\mathbf{y} - \mathbf{X}\boldsymbol{\beta})^T\mathbf{V}^{-1}(\mathbf{y} - \mathbf{X}\boldsymbol{\beta}) = 0$$

and

$$-\sum\log\mu_i + \sum e_i^2\mu_i^{-\theta}\log\mu_i = 0,$$

from which $(\hat{\boldsymbol{\beta}}, \hat{\sigma}^2, \hat{\theta})$ may be determined numerically.

For further details of estimation procedures, see Carroll and Ruppert (1988) and Exercise 32.32. Several tests for heteroscedasticity have been proposed; for evaluations, see Kadiyala and Oberhelman (1984) and Lyon and Tsai (1996). Evans (1992) shows that the size of LR-type tests for heteroscedasticity is very susceptible to non-normality of the errors, particularly high kurtosis. Kianiford and Swallow (1996) discuss various tests for heteroscedasticity based upon recursive residuals (cf. **32.64**).

Although the LS estimators for $\boldsymbol{\beta}$ remain consistent in the presence of heteroscedasticity, the estimator for the covariance matrix of $\boldsymbol{\beta}$ is inconsistent. Andrews (1991) and Andrews and Monahan (1992) have developed a class of heteroscedastic and autocorrelation consistent estimators for the covariance matrix using kernel methods.

Autocorrelation

32.61 We now consider the other primary departure from the assumption of independent identical errors, where we allow some of the correlations to be non-zero. Autocorrelation may arise through temporal or spatial patterns in the data or because the same experimental unit is used for several observations (known as *repeated measures*), as the following example illustrates.

Example 32.10 (First-order autocorrelation models)

(a) When data are recorded for successive equal time intervals (monthly, yearly, etc.), the first-order autoregressive model is written as

$$\epsilon_t = \phi\epsilon_{t-1} + u_t, \tag{32.105}$$

where the u_t are taken to be independently distributed with zero mean and variance σ^2. From (32.105), it follows that

$$\text{corr}(\epsilon_t, \epsilon_{t-s}) = \phi^{|s|} \tag{32.106}$$

provided that the series is *stationary* ($|\phi| < 1$ in this case). For further details of autoregressive time series, see Kendall and Ord (1990, Chapter 5).

(b) If data are recorded on a regular grid with cells (i, j), as in remote sensing for example, a possible autoregressive model is

$$\epsilon_{ij} = \phi(\epsilon_{i-1,j} + \epsilon_{i+1,j} + \epsilon_{i,j-1} + \epsilon_{i,j+1}) + u_{ij} \tag{32.107}$$

although the bilateral nature of the spatial dependence may lead to other choices being preferred; see Cliff and Ord (1981, Chapter 6). It is not possible to develop a simple closed form for the autocorrelations.

(c) If k observations are taken on each experimental unit, we may assume that the random errors are independent between units, but not within units. For example, when there is no time ordering, we may assume that, for observations r and s on unit i,

$$E(\epsilon_{ir}\epsilon_{is}) = \phi\sigma^2, \qquad r \neq s \tag{32.108}$$

with $-(k-1)^{-1} < \phi < 1$ and

$$E(\epsilon_{ir}\epsilon_{js}) = 0, \qquad i \neq j; \tag{32.109}$$

that is, we allow for intraclass correlation (cf. **27.27–30**). The effects of this correlation upon the estimation process are illustrated in Exercise 32.28. Repeated measures were considered in detail in **31.21–32**; so, in the remainder of this discussion, we focus primarily upon the time series case.

32.62 The autoregressive structure (32.105) is often a plausible departure from independence and the standard test of $H_0 : \phi = 0$ against $H_1 : \phi \neq 0$ was developed by Durbin and Watson (1950; 1951). Their statistic may be written in terms of the residuals as

$$d = \sum_{t=2}^{n}(e_t - e_{t-1})^2 \Big/ \sum_{t=1}^{n}(e_t - \bar{e})^2. \tag{32.110}$$

When H_0 is true, d has an expected value in the neighbourhood of 2, whereas $d \to 0$ as $\phi \to 1$ and $d \to 4$ as $\phi \to -1$. The first two moments of $Z = \frac{1}{4}d$ are

$$E(Z) = S_1/(n-p) \tag{32.111}$$

$$E(Z^2) = (2S_2 + S_1^2)/\{(n-p)(n-p+2)\}, \tag{32.112}$$

where $S_j = \sum_{i=1}^{n-p} \lambda_i^j$ and $\lambda_1 \geq \lambda_2 \geq \ldots \geq \lambda_{n-p}$ are the non-zero eigenvalues of $\{\mathbf{I} - \mathbf{X}(\mathbf{X}^T\mathbf{X})^{-1}\mathbf{X}^T\}\mathbf{A}$, and $\mathbf{A}$ has elements $a_{ij} = 1$ if $j = i \pm 1$, $a_{ij} = 0$ otherwise.

Evidently, even under H_0, the distribution of d depends upon $\mathbf{X}$. Durbin and Watson devised upper level bounds, d_U and d_L, for the percentage points of d, based upon the eigenvalues of $\mathbf{X}(\mathbf{X}^T\mathbf{X})^{-1}\mathbf{X}^T$; tables from their 1951 paper are reproduced as Appendix Tables 10 and 11. The tables are used in the following way: if $d_U < d < 4 - d_U$, H_0 is not rejected; if $d < d_L$ or $d > 4 - d_L$, reject H_0; otherwise, reserve judgement.

Example 32.11 (Testing residuals for autocorrelation)

For the model given in Example 32.1, we find that $d = 1.64$. With $n = 15$ and $p - 1 = 3$, $d_U = 1.46$ at the 1 per cent level and $d_U = 1.75$ at the 5 per cent level, so the evidence suggests rather weak autocorrelation as $\hat{\rho} = 0.18$. Inclusion of a lagged dependent variable in the model changes the other coefficients only slightly and the coefficient of the lagged term is not significantly different from zero.

Kamat and Satke (1962) show that the statistic

$$d^* = \sum |e_t - e_{t-1}| / \{\sum (e_t - \bar{e})^2\}^{1/2}$$

has an asymptotic relative efficiency of 0.766 with respect to d when the underlying population is normal and may be more efficient for non-normal alternatives. Various alternatives have been proposed to d (cf. Sims, 1975), mainly with a view to finding a statistic with simple distributional properties, but these have usually proved to be somewhat less powerful. In a follow-up paper, Durbin and Watson (1971) suggest using

$$\tilde{d} = 2(1 - \tilde{\rho}),$$

where $\tilde{\rho}$ is the solution with absolute value less than one of the equation

$$(1 + \rho^2)\sum_{1}^{n} e_t^2 - \rho^2(e_1^2 + e_n^2) - 2\rho \sum_{2}^{n} e_t e_{t-1} = 0. \tag{32.113}$$

The distribution of $\tilde{d}$ may be approximated by a beta distribution, with the first two moments given by (32.111) and (32.112). A numerical study by Dent and Cassing (1978) suggests that $\tilde{d}$ has comparable power properties and a somewhat smaller zone of indecision.

Wallis (1972) gives an analogous procedure for fourth-order serial correlation, for use with quarterly (e.g. economic) data. Evans (1992) shows that the Durbin–Watson test is generally quite robust to non-normality in the errors. White (1992) provides a Durbin–Watson test for nonlinear models. Andrews (1991) and Andrews and Monahan (1992) provide kernel-based estimators of the covariance matrix that are consistent in the presence of autocorrelation.

Kohn *et al.* (1993) provide an algorithm for the computation of exact p-values for the Durbin–Watson statistics.

32.63 These arguments may be extended to more general error structures, as follows. Consider the linear regression model

$$y_t = \beta_1 x_{1t} + \beta_2 x_{2t} + \cdots + \beta_k x_{kt} + \epsilon_t \tag{32.114}$$

with an autoregressive error structure of order p

$$\epsilon_t = \phi_1 \epsilon_{t-1} + \cdots + \phi_p \epsilon_{t-p} + u_t \tag{32.115}$$

which we may write as

$$\phi(B)\epsilon_t = u_t, \tag{32.116}$$

where $\phi(B) = 1 - \phi_1 B - \cdots - \phi_p B^p$, using the backward shift operator, B. We use p for the autoregressive scheme in deference to convention and switch temporarily to k for the number of regressors. Combining (32.114) and (32.116), we obtain

$$\phi(B)y_t = \sum \beta_i \phi(B) x_{it} + \phi(B)\epsilon_t$$

or

$$y'_t = \sum \beta_i x'_{it} + u_t, \tag{32.117}$$

where $x'_{it} = \phi(B)x_{it}$ and $y'_t = \phi(B)y_t$. If the coefficients $\{\phi_j\}$ were known, we could estimate the $\{\beta_i\}$ from (32.117) by LS in the usual way. Cochrane and Orcutt (1949) suggested a recursive procedure based on fitting (32.114) by LS, estimating the ϕs from (32.115) and re-estimating the β s from (32.117), then iterating between (32.115) and (32.117).

Durbin (1960) proposed an alternative procedure which yields asymptotically efficient estimators. Writing $\gamma_{ij} = \beta_i \phi_j$, we put (32.117) in the form

$$y_t = \sum_{i=1}^{p} \phi_j y_{t-j} + \sum_{i,j} \gamma_{ij} x_{i,t-j} + u_t. \tag{32.118}$$

If the γs were functionally independent, we could regard this as a regression of y_t on the lagged ys and the xs, and derive the LS estimators of ϕ and γ. If the corresponding estimators of ϕ, β, γ, are f, b, c, it follows that the quantities $f_j - \phi_j$ and $c_{ij} - \phi_j \beta_i$ are asymptotically normal with zero means and ascertainable covariance matrix. We can therefore write down their likelihood and maximize it to obtain the estimators of $\boldsymbol{\phi}$ and $\boldsymbol{\beta}$.

Mizon (1995) cautions against the popular practice of 'autocorrelation correction' whereby the residuals of a linear model such as (32.1) are tested for autocorrelation using (32.110) or a similar statistic and a model such as (32.117) is then fitted if the null hypothesis is rejected. Closer inspection of such a procedure reveals that certain relationships are implied among the parameters when this approach is followed; Mizon (1995) argues that such relationships should be tested in a 'general-to-specific' modelling framework and not just assumed.

Example 32.12 (Potential problems with 'autocorrelation correction' (Mizon, 1995))
Consider the model

$$y_t = \phi y_{t-1} + \epsilon_t, \qquad z_t = \delta_t,$$

where

$$\begin{pmatrix} \epsilon_t \\ \delta_t \end{pmatrix} \sim N\left[\begin{pmatrix} 0 \\ 0 \end{pmatrix}, \begin{pmatrix} \sigma^2 & \rho\sigma\omega \\ \rho\sigma\omega & \omega^2 \end{pmatrix}\right].$$

The reduced model

$$y_t = \beta_1 z_t + u_{1t}$$

has autocorrelated residuals, but the Cochrane–Orcutt procedure produces inconsistent estimates unless $\gamma + \phi\beta = 0$ in the more general model

$$y_t = \phi y_{t-1} + \beta z_t + \gamma z_{t-1} + \epsilon_t.$$

The condition $\gamma + \phi\beta = 0$ should be tested within this more general framework, an approach that fits into the use of (32.118).

In certain cases, the LS estimators are asymptotically efficient – cf. R.L. and T.W. Anderson (1956) and Kramer (1980) – but tests of hypotheses are impaired.

32.64 Modern estimation procedures for models such as (32.114) and (32.115) are fully integrated with time series methods. In particular, the Kalman filter-based approach of Harvey and Phillips (1979) allows efficient estimation of regression models with autoregressive moving average error structures. Their method uses *recursive* residuals, introduced to the statistical literature by Brown *et al.* (1975), although long familiar to engineers (cf. Kailath, 1974); our description restricts attention to the simple case of no autocorrelation.

Given r observations, the LS estimator for $\boldsymbol{\beta}$ is

$$\mathbf{b}_r = (\mathbf{X}_r^T\mathbf{X}_r)^{-1}\mathbf{X}_r^T\mathbf{y}_r; \tag{32.119}$$

we may define the recursive residual for the next observation, y_{r+1}, as

$$w_{r+1} = (y_{r+1} - \mathbf{x}_{r+1}^T\mathbf{b}_r)/\{1 + \mathbf{x}_{r+1}^T(\mathbf{X}_r^T\mathbf{X}_r)^{-1}\mathbf{x}_{r+1}\}^{1/2}, \tag{32.120}$$

which we rewrite as $w_r = z_{r+1}/\{1 + \mathbf{x}_{r+1}^T\mathbf{g}_{r+1}\}$. If follows (cf. Exercise 29.20) that

$$(\mathbf{X}_{r+1}^T\mathbf{X}_{r+1})^{-1} = (\mathbf{X}_r^T\mathbf{X}_r)^{-1} - \mathbf{g}_{r+1}\mathbf{g}_{r+1}^T \tag{32.121}$$

$$\mathbf{b}_{r+1} = \mathbf{b}_r + \mathbf{u}_r\mathbf{z}_{r+1} \tag{32.122}$$

and the residual sum of squares is

$$SS_{r+1} = SS_r + w_{r+1}^2. \tag{32.123}$$

The recursive solutions (32.121)–(32.123) do not require any matrix inversions, so that the calculations can proceed very quickly. Under the hypothesis that the model is correct, successive w_j are independent $(0, \sigma^2)$. This property may be exploited to develop a variety of tests for model specification; see Brown *et al.* (1975). For further discussions of recursive residuals, see Kianiford and Swallow (1996). For a much more detailed treatment of time series models, see Hamilton (1994).

Diagnostics

32.65 Advances in statistical computing have made computationally intensive methods of data analysis easily accessible to the model builder. This particularly true of methods based on the residuals, used to check for departures from the assumptions made in **32.2**. In the following sections, we spell out the principles underlying these techniques, leaving the reader interested in applications to consult the monographs by Belsley *et al.* (1980), Cook and Weisberg (1982) and Atkinson (1985).

Leverage

32.66 From **32.3-4**, the fitted values are

$$\hat{\mathbf{y}} = \mathbf{X}(\mathbf{X}^T\mathbf{X})^{-1}\mathbf{X}^T\mathbf{y} = \mathbf{H}\mathbf{y}, \tag{32.124}$$

where $\mathbf{H} = \mathbf{X}(\mathbf{X}^T\mathbf{X})^{-1}\mathbf{X}^T$ is sometimes called the *hat* matrix. The vector of residuals is

$$\mathbf{r} = \mathbf{y} - \hat{\mathbf{y}} = (\mathbf{I} - \mathbf{H})\mathbf{y}. \tag{32.125}$$

Denote the (i, i)th element of $\mathbf{H}$ by

$$h_i = \mathbf{x}_i^T(\mathbf{X}^T\mathbf{X})^{-1}\mathbf{x}_i; \tag{32.126}$$

h_i provides a measure of the *leverage* of the observation. When h_1 is near one, $\hat{y}_i$ is virtually determined by y_i and the particular observation is relatively isolated in the X space from the remainder. Conversely, if h_i is near zero, y_i has very little impact on $\hat{y}_i$ and the ith observation is close to at least some of the others.

Note that

$$\begin{aligned}\textstyle\sum_{i=1}^n h_i &= \mathrm{tr}\{\textstyle\sum \mathbf{x}_i^T(\mathbf{X}^T\mathbf{X})^{-1}\mathbf{x}_i\}\\ &= \mathrm{tr}\{(\mathbf{X}^T\mathbf{X})^{-1}\textstyle\sum \mathbf{x}_i\mathbf{x}_i^T\} = p\end{aligned} \tag{32.127}$$

so that the average leverage is p/n. Some authors prefer the leverage measure

$$h_i' = h_i/(1 - h_i), \tag{32.128}$$

as this tends to emphasize the highly leveraged observations.

32.67 In order to determine whether a particular observation, the ith say, appears to be in accord with the rest of the sample, a natural approach is to fit the regression model using all the observations except the ith and then to compute the residual based on the difference between y_i and the new fitted value.

Let $\mathbf{X}_{(i)}$ denote the $(n-1) \times p$ matrix after deletion of the ith row. The corresponding fitted value is

$$\hat{y}(i) = \mathbf{x}_i^T \hat{\boldsymbol{\beta}}_{(i)}, \tag{32.129}$$

where

$$\hat{\boldsymbol{\beta}}_{(i)} = (\mathbf{X}_{(i)}^T \mathbf{X}_{(i)})^{-1} \mathbf{X}_{(i)}^T \mathbf{y}_{(i)}. \tag{32.130}$$

Further,

$$\text{var}\{\hat{y}(i)\} = \sigma^2 v_i = \sigma^2 \{1 + \mathbf{x}_i^T (\mathbf{X}_{(i)}^T \mathbf{X}_{(i)})^{-1} \mathbf{x}_i\}. \tag{32.131}$$

We let $s_{(i)}^2$ denote the unbiased estimator for σ^2 after removal of the ith observation. Since y_i and $y_{(i)}$ are independent, we have immediately that

$$r_i^* = \frac{y_i - \hat{y}(i)}{s_{(i)} \sqrt{v_i}} \tag{32.132}$$

follows Student's t distribution with $(n - p - 1)$ degrees of freedom. Since

$$\{\mathbf{X}_{(i)}^T \mathbf{X}_{(i)}\}^{-1} = (\mathbf{X}^T \mathbf{X})^{-1} + (\mathbf{X}^T \mathbf{X})^{-1} \mathbf{x}_i \mathbf{x}_i^T (\mathbf{X}^T \mathbf{X})^{-1} / (1 - h_i) \tag{32.133}$$

(cf. Exercise 29.20), (32.131) reduces to

$$v_i = 1 + \{\mathbf{x}_i^T (\mathbf{X}^T \mathbf{X})^{-1} \mathbf{x}_i\}^2 / (1 - h_i)$$

and

$$\hat{y}(i) = \mathbf{x}_i^T \hat{\boldsymbol{\beta}}_{(i)} = \mathbf{x}_i^T \hat{\boldsymbol{\beta}} - \frac{\mathbf{x}_i^T (\mathbf{X}^T \mathbf{X})^{-1} \mathbf{x}_i r_i}{(1 - h_i)}, \tag{32.134}$$

where $r_i = y_i - \hat{y}_i$ denotes the usual residual. Thus, after some simplification,

$$r_i^* = r_i / \{s_{(i)} (1 - h_i)^{1/2}\}; \tag{32.135}$$

the derivation is essentially that for recursive residuals given in **32.64**. Atkinson (1985) refers to $\{r_i^*\}$ as the *deletion* residuals, which seems a good description, although Cook and Weisberg (1982) prefer the term *externally studentized* in contrast to the residual

$$r_i' = r_i / \{s (1 - h_i)^{1/2}\}, \tag{32.136}$$

which they term *studentized*; r_i' is also known as the *standardized* residual. Either r_i' or r_i^* may be used to determine whether the data set contains outliers, although multiple outliers may cause *masking*; that is, the residuals are distorted by the outliers still left the sample. In principle, this may be overcome by omitting groups of observations, but the computational effort involved is substantial.

An alternative approach, due to Atkinson (1986b), is to use a highly robust but inefficient estimator such as the least median of squares (see **32.74**) and then check for outliers in the usual way.

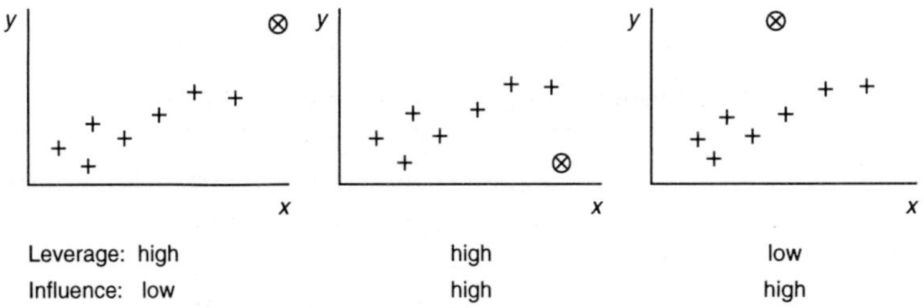

Fig. 32.3 Different patterns of leverage and influence in the single variable model

Miyashita and Newbold (1983) demonstrate that tests for outliers based on the largest studentized residual are very sensitive to heavy tails in the error distribution. Thus, such procedures should probably be used as diagnostic aids rather than formal inferential tools. Sall (1990) provides leverage plots for examining the effect of particular observations upon constraints of the form $\mathbf{C}\boldsymbol{\beta} = \mathbf{0}$.

Influence

32.68 A key question is the effect that a particular observation has upon the location of the regression function. Cook (1977) measured this effect by the quadratic expression

$$D_i = \frac{(\hat{\boldsymbol{\beta}}_{(i)} - \hat{\boldsymbol{\beta}})^T \mathbf{X}^T \mathbf{X} (\hat{\boldsymbol{\beta}}_{(i)} - \hat{\boldsymbol{\beta}})}{ps^2}, \tag{32.137}$$

known as Cook's D. If $\hat{\boldsymbol{\beta}}_{(i)}$ is close to β, omitting the ith observations is said to have little *influence*, reflected in a value of D_i close to zero. Conversely, the larger the value of D_i, the more influential the observation becomes.

It may be shown that

$$\hat{\boldsymbol{\beta}}_{(i)} - \hat{\boldsymbol{\beta}} = -(\mathbf{X}^T\mathbf{X})^{-1}\mathbf{x}_i r_i/(1 - h_i) \tag{32.138}$$

so that

$$\begin{aligned} D_i &= r_i^2 \mathbf{x}_i^T (\mathbf{X}^T\mathbf{X})^{-1}\mathbf{x}_i / \{(1 - h_i)^2 ps^2\} \\ &= r_i^2 h_i / \{(1 - h_i)^2 ps^2\}. \end{aligned} \tag{32.139}$$

From (32.136), this is

$$D_i = (r_i')^2 h_i'/p. \tag{32.140}$$

Thus, the influence measure is a combination of the design effect (leverage) and the deviation of the observation from its expectation; see Fig. 32.3. Belsley *et al.* (1980, 15) prefer a measure known as DFFITS, defined as

$$C_i = (h_i')^{\frac{1}{2}} r_i^*, \tag{32.141}$$

whereas Atkinson (1981) uses

$$C_i' = \{(n - p)/p\}^{1/2} C_i. \tag{32.142}$$

In general, plots of both D_i (or C_i, C_i') and their components should be examined.

Table 32.11 Survey statistics for Example 32.1

Observation	h_i	r'_i	D_i	C_i
1	0.15	1.09	0.05	0.47
2	0.08	−0.22	0.00	−0.07
3	0.46	0.55	0.07	0.51
4	0.47	−1.50	0.45	−1.42
5	0.43	−0.10	0.00	−0.08
6	0.34	−0.41	0.02	−0.29
7	0.26	2.84	0.43	1.67
8	0.14	0.01	0.00	0.00
9	0.16	1.28	0.07	0.55
10	0.38	0.99	0.15	0.68
11	0.30	−0.78	0.07	−0.50
12	0.20	−0.80	0.04	−0.41
13	0.23	−0.60	0.03	−0.33
14	0.14	−0.39	0.01	−0.15
15	0.25	−1.57	0.18	−0.91
Revised values for third observation				
3	0.46	0.55	0.07	0.51
3a	0.53	−0.42	0.05	−0.44
3b	0.63	−1.20	0.58	−1.56
3c	0.78	−2.17	3.21	−4.15
3d	0.46	1.82	0.59	1.68
4d	0.47	−2.82	1.10	−2.67

Table 32.12 Revisions to third observation

	y	x_2	x_3	x_4
original	175.9	0	40	0
3a	175.9	−10	40	0
3b	175.9	−20	40	0
3c	175.9	−40	40	0
3d	205.9	0	40	0

Example 32.13 (Use of regression diagnostics)

Some of the summary statistics for the data of Example 32.1 are presented in Table 32.11. In order to illustrate the performance of these coefficients, we consider the changes to the third observation shown in Table 32.12.

The effects are given at the bottom of Table 32.11. For case $3d$, the effect on the fourth observation (close to observation 3 in the x values) is also listed for comparison.

32.69 The measures described so far rely upon the inclusion or total exclusion of individual observations. Cook (1986) introduced (local) influence graphs for each observation, wherein the ith observation is given variable weight w_i, $0 \leq w_i \leq 1$. Generally, these graphs will be most useful in interactive data analysis where only the plots for 'suspect' observations (e.g., those with high influence) are displayed.

Lawrance (1995) considers measures of influence where two observations are deleted at a time. Peña and Yohai (1995) develop the concept of an *influence matrix* for detecting multiple outliers; see Exercise 32.33.

Influence and transformations

32.70 Upon reflection, it is apparent that an observation may become more or less influential when transformations are used. If only the dependent variable is transformed, the leverage of the observation is unchanged but the (adjusted) residual changes. Thus, a possible outlier may be become acceptable after transformation, although other observations may then become problematical. Atkinson (1983) used the scoring estimator of λ given by (32.57), in conjunction with the influence measure C', to disentangle the effects of influence and transformation; Cook and Wang (1983) developed a similar procedure using D and an approximate ML estimator. Further studies by Atkinson (1986a) indicate that the two approaches typically produce very similar results. Hinkley and Wang (1988) extend the Cook–Wang approach to cover situations where both dependent and regressor variables are transformed. Lawrance (1988) introduces perturbation methods that allow investigation of the local influence of transformations; cf. Cook (1986).

Outliers and robustness

32.71 The identification and treatment (adjustment, elimination, or whatever) of outliers has been a part of statistics since its beginnings, as illustrated in the review by Beckman and Cook (1983). The approach we have taken in **32.65–70** is to identify such observations and to decide whether to modify the model according to the several diagnostic tests applied. In other cases, of course, there may be a recording error and the observation may have to be adjusted or even eliminated.

A different approach to the whole problem is to search for estimation procedures that are *robust*; that is, methods whose performance is not materially affected by the presence of outliers. A detailed discussion of robustness is beyond the scope of this chapter and we content ourselves with a few comments on a certain aspects of that approach in so far as they impinge on our present discussions. For a detailed discussion, see Hettmansperger and McKean (1998).

32.72 The oldest alternative to least squares is the method of least or minimum absolute deviations (LAD or MAD), developed in some detail by Laplace, though originally proposed by Boscovich in 1757; see Stigler (1986, pp. 39–55) for a historical account. To find the LAD estimators, we minimize

$$\sum_{i=1}^{n} |y_i - \mathbf{x}_i^T \boldsymbol{\beta}| \tag{32.143}$$

with respect to $\boldsymbol{\beta}$, a task now easily completed by linear programming methods. The statistical and computational properties of LAD estimators are reviewed by Narula and Wellington (1982). Just as the median is much less susceptible to outliers than the sample mean, (32.143) yields more robust estimators.

32.73 A general form for estimators for the linear model is given by solving the problem

$$\min \rho\{(\mathbf{y} - \mathbf{X}\boldsymbol{\beta})/\sigma\} \tag{32.144}$$

which, under suitable regularity conditions, yields estimators of the form

$$\mathbf{X}^T \psi\{(\mathbf{y} - \mathbf{X}\boldsymbol{\beta})/\sigma\} = \mathbf{0}, \tag{32.145}$$

where $\psi(u) = \rho'(u)$. When $\rho(u) = u^2$ in (32.144), (32.145) yields the LS estimators, whereas $\rho(u) = |u|$ produces the LAD estimators. More general versions of $\psi(u)$ have been proposed by several authors, the best known being Huber's (1964) M-estimators, for which

$$\psi(u) = \begin{cases} |u|, & \text{if } |u| \leq k \\ k \operatorname{sgn}(u), & \text{if } |u| > k, \end{cases}$$

k typically being set at around 1.5σ; clearly, a robust estimator of σ is also needed. However, subsequent research demonstrated that M-estimators were still vulnerable to highly leveraged points, which led to the development of generalized-M estimators (Mallows, 1975) that bound the influence of such observations.

Various other procedures have been developed, notably regression quantile methods (Koenker and Bassett, 1978) and local smoothing methods (cf. Cleveland and Devlin, 1988).

32.74 A rather different approach is to use robust methods to improve the performance of the diagnostic procedures. Carroll and Ruppert (1985; 1987) use the methods described in **32.73** for this purpose, but this can be computationally demanding for larger data sets. A simple approach that produces inefficient but highly robust estimators is the least median of squares (LMS) method developed by Rousseeuw (1984), which selects $\hat{\boldsymbol{\beta}}$ to

$$\text{minimize}_{\hat{\beta}} \ \text{median}_i (r_i^2), \tag{32.146}$$

where r_i denotes the ith residual. For problems with $p > 2$, evaluation of the LMS estimators becomes time-consuming, although Souvaine and Steele (1987) have developed a very effective algorithm. However, for diagnostic purposes, a random search procedure will often suffice. That is, we select p observations without replacement from the original n, to form the set J, say, and then evaluate

$$\hat{\boldsymbol{\beta}}_j = \mathbf{X}_J^{-1} \mathbf{y}_J, \tag{32.147}$$

where subscript j serves to indicate the observations selected in set J. The residuals are then

$$r_{iJ} = Y_i - \mathbf{x}_i^T \hat{\boldsymbol{\beta}}_J \tag{32.148}$$

and will be zero when $i \in J$. The quantity, median (r_{ij}^2) taken over $i \notin J$, may be evaluated for the residuals not contained in set J. By replicating the process a sufficient number of times, a close approximation to (32.146) will yield approximately LMS estimators. This approach has been used by Atkinson (1986b) as the basis for generating robust estimates as inputs to diagnostic procedures (cf. **32.67**).

Heitmann and Ord (1985) show that the usual LS estimator is a weighted average of the estimators (32.147); cf. Exercise 32.31. The weights may be used to develop a weighted LMS estimator which might be expected to improve performance.

Added variable plots

32.75 Thus far, we have assumed that the complete set of regressor variables is available and is included in the model. If some variables are not currently in the model, we may explore their potential using added variable plots, a concept that goes back to Cox (1958) but has been developed and extended by a number of authors; see Cook (1996) for a brief history.

Referring back to **32.6**, we may contrast the models

$$\mathbf{y} = \mathbf{X}\boldsymbol{\theta} + \boldsymbol{\epsilon}_1 \quad \text{and} \quad \mathbf{y} = \mathbf{X}\boldsymbol{\theta} + \mathbf{z}\beta + \boldsymbol{\epsilon}_2.$$

The residuals from the first model are

$$\mathbf{e}_1 = \mathbf{M}\mathbf{y} = [\mathbf{I} - \mathbf{X}(\mathbf{X}^T\mathbf{X})^{-1}\mathbf{X}^T]\mathbf{y}$$

and the regression of $\mathbf{z}$ on $\mathbf{X}$ similarly yields residuals $\mathbf{e}_z = \mathbf{Mz}$. The added variable plot is that of $\mathbf{e}_z$ against $\mathbf{e}_1$.

An alternative to the added variable plot is the component-plus-residual plot of

$$\mathbf{e}_2 + \mathbf{z}\hat{\beta} \quad \text{against} \quad \mathbf{z},$$

where $\{e_2\}$ are the residuals from the second model and $\hat{\beta}$ is the estimate of β for that model. The two plots tell much the same story, although the use of z on the horizontal axis means that the component-plus-residual plot may be more successful than the added variable plot in detecting nonlinear relationships.

Calibration

32.76 In **32.9**, we considered the prediction of y given the values of the regressor variables. We are sometimes faced with the reverse problem: y is observed and we must estimate x. This arises, for example, in the calibration of laboratory instruments where x is the 'true' reading and y denotes the reading obtained from an instrument subject to measurement error. Other examples include preliminary screening tests (e.g., for blood pressure) or sample estimates of the size of population between censuses. The more general question of measurement error in both x and y is briefly discussed in **32.81–83**.

32.77 We now restrict attention to the case where we have n pairs of observations (x_i, y_i), $i = 1, \ldots, n$, and a further y value, y_{n+1}, from which we seek to estimate x_{n+1}. Two cases emerge, which have sometimes been confused in the literature. We refer to these as the *unconditional* and *conditional* models, respectively, the terms being used in a manner that is consistent with our earlier description of regression models in these chapters. Also, we refer to the general question of estimating x_{n+1} as a *calibration* problem, rather than one of *inverse* regression, as the term 'inverse' seems open to misinterpretation.

Unconditional model. Suppose that (x, y) are binormal with parameters $(\mu_1, \mu_2; \sigma_1^2, \sigma_2^2; \rho)$. Given n pairs of observations, the joint log likelihood for the five parameters and x_{n+1} may be partitioned as

$$l = \prod_{i=1}^{n+1} f_y(y_i|\mu_2, \sigma_2^2) f_{x|y}(x_i|y_i; \alpha_1, \beta_1, \omega^2), \tag{32.149}$$

where $\alpha_1 = \mu_1 - \beta_1\mu_2$, $\beta_1 = \rho\sigma_1/\sigma_2$ and $\omega_1^2 = \sigma_1^2(1-\rho^2)$. The ML estimators are

$$\hat{\mu}_2 = \bar{y}_{n+1} = \sum_1^{n+1} y_i/(n+1), \quad \hat{\sigma}_2^2 = \sum_1^{n+1}(y_i - \bar{y}_{n+1})^2/(n+1), \tag{32.150}$$

$$\hat{\alpha}_1 = \bar{x}_n - \hat{\beta}_1\bar{y}_n, \tag{32.151}$$

$$\hat{\beta}_1 = \sum_1^n (x_i - \bar{x}_n)(y_i - \bar{y}_n) / \sum_1^n (y_i - \bar{y}_n)^2 \tag{32.152}$$

and

$$\hat{\omega}_1^2 = \sum_1^n e_i^2/(n+1), \tag{32.153}$$

where $n\bar{x}_n = \sum_1^n x_i$, $n\bar{y}_n = \sum_1^n y_i$ and $e_i = (x_i - \bar{x}_n) - \hat{\beta}_1(y_i - \bar{y}_n)$. Finally, the unconditional estimator for x_{n+1}, denoted by $\hat{x}_u$, is

$$\hat{x}_u = \hat{\alpha}_1 + \hat{\beta}_1 y_{n+1}; \tag{32.154}$$

the conditional variance of the prediction, given $\mathbf{y}$, is

$$\text{var}(\hat{x}_u|\mathbf{y}) = \omega_1^2\left\{1 + 1/n + \frac{(y_{n+1} - \bar{y}_n)^2}{\sum_1^n (y_i - \bar{y}_n)^2}\right\}. \tag{32.155}$$

Further, $E(\hat{x}_u|\mathbf{y}) = x_{n+1}$ and the conditional distribution of x_u given $\mathbf{y}$ is normal.

The unconditional variance is, approximately,

$$\text{var}(\hat{x}_u|y_{n+1}) = \omega_1^2\left\{1 + 1/n + \frac{(y_{n+1} - \mu_2)^2}{n\sigma_2^2}\right\}, \tag{32.156}$$

which would be estimated from (32.153) and (32.155).

Conditional model. If we regard the x_i as fixed, the conditional argument based on the distribution of $\mathbf{y}$ given x, $f_{\mathbf{y}|x}(\mathbf{y}|x, \alpha_2, \beta_2, \omega_2^2)$ leads to the usual estimators

$$\hat{\alpha}_2 = \bar{y}_n - \beta_2\bar{x}_n, \qquad \hat{\beta}_2 = \sum_1^n (x_i - \bar{x}_n)(y_i - \bar{y}_n)^2, \qquad n\hat{\omega}_2^2 = \sum \tilde{e}_i^2,$$

where $\tilde{e}_i = y_i - \bar{y}_n - \hat{\beta}_2(x_i - \bar{x}_n)$, with the usual adjustment to make $\hat{\omega}_2$ unbiased, if desired. It then follows that the conditional estimator, $\hat{x}_c$, is

$$\hat{x}_c = \bar{x}_n + \frac{(y_{n+1} - \hat{\alpha}_2)}{\hat{\beta}_2}. \tag{32.157}$$

Note that $\hat{x}_c$ is the ML estimator, but its mean and variance do not exist, since $\hat{\beta}_2$ is normally distributed and its reciprocal does not have any finite moments. For large samples we have (in the sense of **10.5**)

$$E(\hat{x}_c) \doteq x_{n+1}, \qquad \text{var}(\hat{x}_c) \doteq \sigma_2^2/\beta_2^2. \tag{32.158}$$

32.78 The different properties of $\hat{x}_u$ and $\hat{x}_c$ have led to some debate of their respective merits (cf. Krutchkoff, 1967; Williams, 1969). It seems to us that this debate has largely ignored the distinction between the unconditional and conditional models. Once the model is clearly stated, the choice of estimator follows directly unless, of course, one invokes criteria other than ML to argue for $\hat{x}_u$ when $\hat{x}_c$ is appropriate.

For further discussion of the issues and an extension to the multivariate case, see Brown (1982).

Missing values

32.79 When missing values relate only to the dependent variable, we are back at the prediction problem discussed in **32.9**; the multivariate analogue follows directly from the arguments presented there. When the missing values are among the regressor variables, we have a generalization of the calibration problem. In general, non-iterative solutions may not be available and it is easier to proceed with an iterative solution using the EM algorithm (see **18.44–6**).

32.80 In this case, the M-step corresponds to carrying out the analysis on the 'complete' data, and the E-step requires that we find expected values for the missing regressors. In order to apply this approach, we must assume that the values are *missing at random*; that is, the probability that a value is missing is functionally independent of that missing value. Further, it is usually assumed that $\mathbf{y}$ and $\mathbf{x}_m$, those x with missing values, are jointly normally distributed. For details of the method, see Orchard and Woodbury (1972), Beale and Little (1975), or Little and Rubin (1987, Chapter 8). Simon and Simonoff (1986) provide diagnostic plots to determine the potential effects of the missing observations without the need to assume that they are missing at random. For a comprehensive analysis of missing value, or *imputation* techniques, see Schaffer (1997).

Measurement errors

32.81 A key assumption throughout our discussion of regression analysis has been that the error terms are not related to the regressor variables. Suppose now that the true relationship is, for a single regressor,

$$y = \beta_0 + \beta_1 X + \epsilon \tag{32.159}$$

but that the regressor is measured subject to error as

$$x = X + \delta. \tag{32.160}$$

Thus, the relationship between the observable (x, y) is

$$y = \beta_0 + \beta_1 x + (\epsilon - \beta_1 \delta) \tag{32.161}$$

and the error and the regressor are related through (32.160); we shall assume that δ is independent of X and of ϵ for all observations. This framework leads to the area of *functional and structural relationships*, which is discussed in detail by Van Ness and Cheng (1999). However, a few comments are in order at this stage.

Suppose we have n pairs of observations and we apply the usual LS estimators to (32.161). This leads to an estimating equation for β_1 of the form

$$\sum(x_i - \bar{x})(y_i - \bar{y}) = \hat{\beta}_1 \sum(x_i - \bar{x})^2.$$

Using (32.159) and (32.160), this becomes

$$\hat{\beta}_1 = \frac{\sum\{(X_i - \bar{X}) + (\delta_i - \bar{\delta})\}\{\beta_0 + \beta_1 X_i + \epsilon_i\}}{\sum\{(X_i - \bar{X})^2 + 2(\delta_i - \bar{\delta})(X_i - \bar{X}) + (\delta_i - \bar{\delta})^2\}}. \tag{32.162}$$

From the independence of (ϵ, δ, X), it follows that for large samples, $\hat{\beta}_1$ converges to the limit

$$\beta_1 \frac{S_{XX}}{S_{XX} + \sigma_\delta^2}, \tag{32.163}$$

where $S_{XX} = \lim_{n\to\infty}\{n^{-1}\sum(X_i - \bar{X})^2\}$; that is, the estimator is inconsistent and, if its expectation exists, it is biased towards zero even asymptotically. For further consideration, see Exercise 32.29.

Instrumental variables

32.82 One resolution of the measurement error problem is to find another variable, z say, that is (highly) correlated with X but uncorrelated with δ, so that $n^{-1}\sum z_i\delta_i \to 0$ in probability. Then the estimator

$$\hat{\beta}_1 = \frac{\sum(z_i - \bar{z})(y_i - \bar{y})}{\sum(x_i - \bar{x})(z_i - \bar{z})} \tag{32.164}$$

is consistent. The z variables are known as *instrumental* variables.

Various instrumental variables have been proposed over the years:

(i) an indicator variable dividing the observations into k groups of equal size ($k = 2$, Wald, 1940; $k = 3$, Bartlett, 1949);
(ii) the ranks of the x variables;
(iii) replicate measurements on the regressor where the two measurements are uncorrelated;
(iv) lagged values of the regressors.

In order to make (i) and (ii) operational, it is necessary to assume that the measurement errors do not interfere with the identification of the correct values for the instrumental variables.

These basic ideas clearly extend to multiple regressors; see Exercise 32.30. The set of instrumental variables should be chosen as highly correlated with the regressor variables as possible, subject to being uncorrelated with the measurement errors. Exercise 32.30 also gives the ω variance matrix of the estimators.

32.83 A different form of measurement error arises when the observed values are rounded. Following **3.18–23**, this might suggest applying Sheppard's corrections to the diagonal elements of $\mathbf{X}^T\mathbf{X}$. Dempster and Rubin (1983) use a likelihood argument to show that this step may be justified in some instances, but caution that the appropriate adjustment depends critically upon the distributional assumptions. From a pragmatic viewpoint, Sheppard's corrections reduce the elements on the main diagonal of $\mathbf{X}^T\mathbf{X}$ and may, therefore, worsen any multicollinearity problems.

Nonlinear regression

32.84 Thus far we have assumed that the response, $\mathbf{y}$, is a linear function of the parameters, $\boldsymbol{\beta}$, or at least can be reasonably approximated by such a linear function. Such an assumption may not be plausible, and we now consider departures from linearity. We shall consider the general form:

$$y = \eta(\mathbf{x}, \boldsymbol{\beta}, \epsilon), \tag{32.165}$$

where, as usual, ϵ denotes the random error term. Now examine the following four examples of η in (32.165):

$$\begin{array}{ll} \text{(i)} & \eta = \beta_0 + \beta_1 x + \epsilon \\ \text{(ii)} & \eta = \beta_0 + x e^{\beta_1} + \epsilon \\ \text{(iii)} & \eta = \exp(\beta_0 + \beta_1 x + \epsilon) \\ \text{(iv)} & \eta = \beta_0/(1 + \beta_1 x) + \epsilon. \end{array} \tag{32.166}$$

Version (i) is linear in $\boldsymbol{\beta}$ and (ii) can be made so by use of the transformation $\gamma_i = \exp(\beta_1)$; models such as (ii) are said to display *parameter effects* non-linearity. Model (iii) can be transformed to a linear model for $\log_e y$ and is typical of the schemes discussed in **32.31–44**. However, scheme (iv) is *intrinsically* non-linear since there is no transformation that will induce linearity. The form of the error process is critical here since

$$\eta = \beta_0/(1 + \beta_1 x + \epsilon)$$

clearly can be transformed to a linear model for the reciprocal of $\mathbf{y}$ in (32.165).

32.85 We now assume that (32.165) may be rewritten as

$$y = \eta(\mathbf{x}, \boldsymbol{\beta}) + \epsilon, \tag{32.167}$$

after transformation if necessary, and that the resulting errors satisfy the assumptions made in **32.2**. Then the parameters $\boldsymbol{\beta}$ may be estimated by non-linear least squares; that is by minimizing

$$S = \sum_{i=1}^{n} \{y_i - \eta(\mathbf{x}_i, \boldsymbol{\beta})\}^2 \tag{32.168}$$

with respect to $\boldsymbol{\beta}$. Under suitable regularity conditions, this yields the set of first-order equations:

$$\frac{\delta S}{\delta \beta_j} = 0 = \sum \{y_i - \eta(\mathbf{x}_i, \boldsymbol{\beta})\} v_{ij}, \qquad j = 1, \ldots, p, \tag{32.169}$$

where $v_{ij} = \delta\eta(\mathbf{x_i}, \boldsymbol{\beta})/\delta\beta_j$. A variety of methods is available to determine $\hat{\boldsymbol{\beta}}$, cf. **18.21**.

32.86 Again under fairly general regularity conditions (cf. Gallant, 1987, pp. 253–258), it follows that the non-linear LS estimators are consistent and asymptotically normally distributed with mean vector $\boldsymbol{\beta}$ and covariance matrix given by

$$E\left(\frac{\delta^2 S}{\delta\boldsymbol{\beta}\,\delta\boldsymbol{\beta}^T}\right)^{-1}. \tag{32.170}$$

When errors are normally distributed, these are, of course, the ML results. Since, in this case, the sampling distribution is exactly normal when the model is linear, we may expect 'increasing' non-linearity in the model specification to the source of increased non-normality in the finite sampling distribution. We therefore seek suitable measures of non-linearity as sources of possible correction factors.

32.87 Most asymptotic theory (cf. Chambers, 1973) is based upon the expansion

$$\hat{\eta}_i = \eta(\mathbf{x}_i, \hat{\boldsymbol{\beta}}) = \eta(\mathbf{x}_i, \boldsymbol{\beta}) + \sum_j (\hat{\beta}_j - \beta_j) v_{ij} \tag{32.171}$$

so that a natural way to assess the intrinsic non-linearity of the model is to consider the ratio

$$\hat{\mathbf{N}}(\boldsymbol{\beta}) = \sum_i \left\{\hat{\eta}_i - \eta_i - \sum_j (\hat{\beta}_j - \beta_j) v_{ij}\right\}^2 \Big/ \sum_i (\hat{\eta}_i - \eta_i)^2 \tag{32.172}$$

for a test set of values of $\boldsymbol{\beta}$ in the neighbourhood of $\hat{\boldsymbol{\beta}}$. Beale (1960) introduced $\hat{\mathbf{N}}(\boldsymbol{\beta})$ as an empirical measure of non-linearity, with a second measure $\mathbf{N}(\boldsymbol{\beta})$ defined as the limit of $\hat{\mathbf{N}}(\boldsymbol{\beta})$ as the size of the test set goes to infinity. A third measure introduced by Beale sought the minimum of $\hat{\mathbf{N}}(\boldsymbol{\beta})$ over all transformations $\boldsymbol{\phi} = \boldsymbol{\phi}(\boldsymbol{\beta})$ and is denoted by $\hat{\mathbf{N}}(\boldsymbol{\phi})$, with limiting value $\mathbf{N}(\boldsymbol{\phi})$. It is more convenient to consider $\mathbf{N}(\boldsymbol{\phi})$ together with

$$\mathbf{N}_1(\boldsymbol{\beta}) = \mathbf{N}(\boldsymbol{\beta}) - \mathbf{N}(\boldsymbol{\phi}) \tag{32.173}$$

since this quantity has a minimum value of zero when the most nearly linear transformation has been used.

Bates and Watts (1980, 1981) have shown how Beale's limiting measures may be interpreted geometrically as measures of *curvature*: $4\mathbf{N}(\boldsymbol{\phi})$ is the (mean square) *intrinsic* curvature and $4\mathbf{N}_1(\boldsymbol{\beta})$ is the (mean square) *parameter effects* curvature. For example, in scheme (ii) of (32.166), $\mathbf{N}_1(\boldsymbol{\beta}) > 0$ and $\mathbf{N}(\boldsymbol{\phi}) = 0$, indicating that there is no intrinsic curvature; that is, the non-linearity could be removed by a transformation of the parameters.

The scope of the paper by Bates and Watts (1980) is much more general than we have indicated, as their measures of curvature are not restricted to the quadratic approximations based on (32.171), which may understate the extent of the non-linearity. The reader seeking further details should consult their paper, which includes an algorithm for the evaluation of the curvature measures.

Hougaard (1985; 1988) uses the Bates–Watts curvatures to develop correction factors for the asymptotic distribution of $\hat{\boldsymbol{\beta}}$ and Hamilton and Wiens (1987) provide a corrected F test. Cook *et al.*

(1986) present model diagnostics for nonlinear models and discuss the relationship between bias and curvature.

32.88 Non-linear regression is a vast topic, and a detailed discussion is beyond the scope of this volume. The interested reader is referred to the volumes by Gallant (1987) and Seber and Wild (1989). In the spirit of this chapter, some interesting recent work on diagnostics for non-linear regression models is given in Davison and Tsai (1992) and St Laurent and Cook (1993), among others. Chen and Jennrich (1995a; 1995b) develop improved confidence intervals for the parameters of nonlinear models.

EXERCISES

32.1 Show that if (32.1) and (32.5) hold, but **X** has elements that are non-linear functions of r further parameters $\gamma_1, \ldots, \gamma_r$, making $k + r$ in all, the regression model can be augmented (cf. (32.11)) to

$$\mathbf{y} = (\mathbf{X}, \mathbf{D})\begin{pmatrix} \boldsymbol{\beta} \\ \mathbf{0} \end{pmatrix} + \boldsymbol{\epsilon},$$

where **D** is any $n \times r$ matrix chosen so that (**X**, **D**) is of full rank $k + r$, and **0** is an $r \times 1$ vector of zeros. Hence obtain confidence regions for (a) the complete set of $k + r$ parameters, (b) the r further parameters alone.

(Halperin, 1963; cf. also Hartley, 1964.)

32.2 Show that for interval estimation of β in the linear regression model $y_i = \beta x_i + \epsilon_i$, the interval based on the Student's t variate

$$t = (b - \beta)/\left(s^2/\sum x_i^2\right)^{1/2}$$

is physically shorter for every sample than that based on

$$u = (\bar{y} - \beta\bar{x})/(s^2/n)^{1/2}.$$

32.3 In the regression model

$$y_i = \alpha + \beta x_i + \epsilon_i, \qquad i = 1, 2, \ldots, n.$$

suppose that the observed mean $\bar{x} = 0$ and let x_0 satisfy $\alpha + \beta x_0 = 0$. Use the random variable $\hat{\alpha} + \hat{\beta} x_0$ to set up a confidence statement for a quadratic function, of form

$$P\{Q(x_0) \geq 0\} = 1 - \alpha.$$

Hence derive a confidence statement for x_0 itself, and show that, depending on the coefficients in the quadratic function, this may place x_0:

(i) in a finite interval;
(ii) outside a finite interval;
(iii) in the infinite interval consisting of the whole real line.

(Cf. Lehmann, 1986)

32.4 To determine which of the models

$$y = \beta_0' + \beta_1' x_1 + \epsilon', \qquad y = \beta_0'' + \beta_2' x_2 + \epsilon'',$$

is more effective in predicting **y**, consider the model

$$y_i = \beta_0 + \beta_1 x_{1i} + \beta_2 x_{2i} + \epsilon_i, \qquad i = 1, 2, \ldots, n,$$

with independent normal errors of variance σ^2, estimated by s^2 with $n - 3$ degrees of freedom. Show that the statistics

$$z_s = \sum_i (y_i - \bar{y})(x_{si} - \bar{x}_s) \Big/ \left\{\sum_i (x_{si} - \bar{x}_s)^2\right\}^{1/2}, \qquad s = 1, 2,$$

have

$$\operatorname{var} z_1 = \operatorname{var} z_2 = \sigma^2, \qquad \operatorname{cov}(z_1, z_2) = \sigma^2 r_{12},$$

where r_{12} is the observed correlation between x_1 and x_2. Hence show that $z_1 - z_2$ is exactly normally distributed with mean

$$\beta_1'\left\{\sum_i (x_{1i} - \bar{x}_1)^2\right\}^{1/2} - \beta_2'\left\{\sum_i (x_{2i} - \bar{x}_2)^2\right\}^{1/2}$$

and variance $2\sigma^2(1 - r_{12})$. Using the fact that

$$\sum_i (y_i - \bar{y})^2 - (\beta_s')^2 \sum_i (x_{si} - \bar{x}_s)^2$$

is the sum of squares of deviations from the regression of $\mathbf{y}$ on x_s alone, show that the hypothesis of equality of these two sums of squares may be tested by the statistic $t = (z_1 - z_2)/\{2s^2(1 - r_{12})\}^{1/2}$, distributed as Student's t with $n - 3$ degrees of freedom.

(Hotelling (1940); Healy (1955). See also Williams (1959), Dunn and Clark (1969, 1971) and Choi (1977).)

32.5 Show that if there are two different vectors $\mathbf{y}_1$, $\mathbf{y}_2$ each related to the same set of regressors $\mathbf{x}$ in a linear model, the difference between any pair of corresponding parameters in the models may be tested by applying the method of **32.7** to the differences $(y_{1i} - y_{2i})$.

(Yates (1939b) also considers the case where the regressors are different and the $\mathbf{y}$ vectors correlated.)

32.6 Independent samples of sizes n_i are taken from two regression models

$$y = \alpha_i + \beta_i x + \epsilon, \qquad i = 1, 2,$$

with independently normally distributed errors. The error variance σ^2 is the same in both models. Given that b_1, b_2 are the separate least squares estimators of β_1, β_2, show that $(b_1 - b_2)$ is normally distributed with mean (β_1, β_2) and variance

$$\sigma^2\left\{\left(\sum_{i=1}^{n_1}(x_i - \bar{x}_1)^2\right)^{-1} + \left(\sum_{j=1}^{n_2}(x_j - \bar{x}_2)^2\right)^{-1}\right\},$$

and that

$$t = \{(b_1 - b_2) - (\beta_1 - \beta_2)\}\Big/\left\{s^2\left(\frac{1}{\sum_i (x_i - \bar{x}_1)^2} + \frac{1}{\sum_j (x_j - \bar{x}_2)^2}\right)\right\}^{1/2}$$

has a Student's t-distribution with $n_1 + n_2 - 4$ degrees of freedom, where

$$s^2 = \frac{(n_1 - 2)s_1^2 + (n_2 - 2)s_2^2}{n_1 + n_2 - 4}$$

and s_1^2, s_2^2 are the separate estimators of σ^2 in the two models. Hence show that t may be used to test the hypothesis that $\beta_1 = \beta_2$ against $\beta_1 \neq \beta_2$.

(Cf. Fisher, 1922b)

32.7 For the simple linear model $y = \beta_0 + \beta_1 x + \epsilon$, two independent samples, of sizes m and n, have means $(\bar{y}_m, \bar{x}_m)$ and $(\bar{y}_n, \bar{x}_n)$. Show that $b_1 = (\bar{y}_m - \bar{y}_n)/(\bar{x}_m - \bar{x}_n)$ is an unbiased estimator of β_1, with variance

$$\sigma^2\left(\frac{1}{m} + \frac{1}{n}\right)\Big/(\bar{x}_m - \bar{x}_n)^2.$$

Show that b_1 is not consistent (as $m, n \to \infty$ with m/n fixed) if the two samples were formed by random subdivision of an original sample of $m + n$ observations.

32.8 We are given n observations on the model

$$y = \beta_1 x_1 + \beta_2 x_2 + \epsilon$$

with error variance σ^2, and, in addition, an extraneous unbiased estimator b_1 of β_1 together with an unbiased estimator s_1^2 of its sampling variance σ_1^2. To estimate β_2, consider the regression of $y - b_1 x_1$ on x_2. Show that the estimator

$$b_2 = \sum(y - b_1 x_1)x_2\Big/\sum x_2^2$$

is unbiased, with variance

$$\operatorname{var} b_2 = \left(\sigma^2 + \sigma_1^2 r^2 \sum x_1^2\right)\Big/\sum x_2^2,$$

where r is the observed correlation between x_1 and x_2. Assuming that b_1 is ignored, show that the ordinary least squares estimator of β_2 has variance $\sigma^2/\sum x_2^2(1 - r^2)$ and hence that the use of the extraneous information about β_1 increases efficiency in estimating β_2 if and only if

$$\sigma_1^2 < \frac{\sigma^2}{\sum x_1^2(1 - r^2)},$$

that is, if the variance of b_1 is less than that of the ordinary least squares estimator of β_1. Show that an unbiased estimator of $\operatorname{var} b_2$ is given by

$$\hat{V} = \frac{1}{(n-2)\sum x_2^2}\left[\sum(y - b_1 x_1 - b_2 x_2)^2 + s_1^2 \sum x_1^2\{(n-1)r^2 - 1\}\right],$$

but that if the errors are normally distributed this is not distributed as a multiple of a χ^2 variate.

(Durbin, 1953)

32.9 As a generalization of Exercise 32.8, let $\mathbf{b}_1$ be a vector of unbiased estimators of the h parameters $(\beta_1, \beta_2, \ldots, \beta_h)$, with covariance matrix $\mathbf{V}_1$; and let $\mathbf{b}_2$ be an independently distributed vector of unbiased estimators of the k ($> h$) parameters $(\beta_1, \beta_2, \ldots, \beta_h, \beta_{h+1}, \ldots, \beta_k)$, with covariance matrix $\mathbf{V}_2$. Show that the minimum variance unbiased estimators of $(\beta_1, \ldots, \beta_k)$ which are linear in the elements of $\mathbf{b}_1$ and $\mathbf{b}_2$ are the components of the vector

$$\mathbf{b} = \{(\mathbf{V}_1^{-1})^* + \mathbf{V}_2^{-1}\}^{-1}\{(\mathbf{V}_1^{-1})^*\mathbf{b}_1^* + \mathbf{V}_2^{-1}\mathbf{b}_2\},$$

with covariance matrix

$$\mathbf{V}(\mathbf{b}) = \{(\mathbf{V}_1^{-1})^* + \mathbf{V}_2^{-1}\}^{-1},$$

where an asterisk denotes the conversion of an $h \times 1$ vector into a $k \times 1$ vector or an $h \times h$ matrix into a $k \times k$ matrix by putting it into the leading position and augmenting it with zeros.

Show that $V(b)$ reduces, in the particular case $h = 1$, to

$$V(b) = \sigma^2 \begin{pmatrix} \sum x_1^2 + \sigma^2/\sigma_1^2 & \sum x_1x_2 & \cdots & \sum x_1x_k \\ \sum x_1x_2 & \sum x_2^2 & \cdots & \sum x_2x_k \\ \vdots & \vdots & \ddots & \vdots \\ \sum x_1x_k & \sum x_2x_k & \cdots & \sum x_k^2 \end{pmatrix}^{-1},$$

differing only in its leading term from the usual least squares covariance matrix $\sigma^2(\mathbf{X}^T\mathbf{X})^{-1}$.

(Durbin, 1953)

32.10 A simple graphical procedure may be used to fit an ordinary LS regression of y on x without computations when the x values are equally spaced, say at intervals of s. Let the n observed points on the scatter diagram of (y, x) be $P_1, P_2, \ldots, P_n$ in increasing order of x. Find the point Q_2 on P_1P_2 with x coordinate $\frac{2}{3}s$ above that of P_1; find Q_3 on Q_2P_3 with x coordinate $\frac{2}{3}s$ above that of Q_2; and so on by equal steps, joining each Q point to the next P point and finding the next Q point $\frac{2}{3}s$ above, until finally $Q_{n-1}P_n$ gives the last point, Q_n. Carry out the same procedure backwards, starting from P_nP_{n-1} and determining Q_2', say, $\frac{2}{3}s$ below P_n in x coordinate, and so on until Q_n' on $Q_{n-1}'P_1$ is reached, $\frac{2}{3}s$ below Q_{n-1}'. Then Q_nQ_n' is the LS line. Prove this.

(Askovitz, 1957)

32.11 In the linear model with general parameter given in Exercise 29.1, we take independent samples of sizes n_1, n_2 with possibly different matrices X_1, X_2, and let S_{ml} be the sum of squared residuals in the lth sample when the LS estimators $\hat{\boldsymbol{\beta}}_{\mathbf{m}}$ are based on the mth sample alone ($m, l = 1, 2$). Show that

$$D = S_{12} - S_{22} = (\hat{\boldsymbol{\beta}}_1 - \hat{\boldsymbol{\beta}}_2)^T\mathbf{X}_2^T\mathbf{X}_2(\hat{\boldsymbol{\beta}}_1 - \hat{\boldsymbol{\beta}}_2) \geq 0$$

and that $b = D/\sigma^2$ is a quadratic form in $p + 1$ independent standardized normal variables, whose diagonal matrix contains the eigenvalues of $\mathbf{B} = \mathbf{I} + \mathbf{X}_2^T\mathbf{X}_2(\mathbf{X}_1^T\mathbf{X}_1)^{-1}$ in its leading diagonal. Using Exercise 29.4, show that $E(b) = \sigma^2 \text{tr}\mathbf{B}$, $\text{var}\, b = 2\sigma^4$ tr $(\mathbf{B}^2)$, and that if $r = D/S_{22}$, we have, since b is distributed independently of S_{22},

$$E(r) = \text{tr}(\mathbf{B})/(n_2 - p - 3),$$
$$\text{var}\, r = \frac{2}{(n_2 - p - 3)(n_2 - p - 5)}\left\{\text{tr}(\mathbf{B}^2) + \frac{\{\text{tr}(\mathbf{B})\}^2}{n_2 - k - 3}\right\}.$$

If the sets of x values are the same in each sample, with c times more observations at each in the second sample, show that

$$\mathbf{B} = (1 + c)\mathbf{I}, \quad \text{tr}(\mathbf{B}) = (1 + c)p, \quad \text{tr}(\mathbf{B}^2) = (1 + c)^2 p.$$

(Gardner (1972), who also considered the case where the xs are multinormal.)

32.12 There are G groups of observations, and all observations within a group are normally distributed with common mean and common variance σ_g^2, the model (32.48) holding except for the homoscedasticity condition below it. Consider the sets of constraints

C_1 : all the σ_g^2 are equal ($G - 1$ constraints);

C_2 : r of the k parameters in β are zero.

Working in terms of the variable $z_\lambda = y_\lambda J^{-1/n}$, so that (32.50) reduces to

$$L_\lambda(z|\hat{\boldsymbol{\beta}}, \hat{\sigma}^2) = \{\hat{\sigma}^2_\lambda(z)\}^{-n/2},$$

show that (32.58) gives

$$L(z|\hat{\lambda}_{(2)}) = L(z|\hat{\lambda})l_1(z)l_2(z)$$

where l_1 is the LR test statistic defined at (22.40), and

$$l_2 = \left\{1 + \frac{r}{n-k}F\right\}^{-n/2},$$

where F is the variance ratio test statistic defined generally at (22.90) and for this case in Example 22.3. (This result generalizes Exercise 22.6.)

(Box and Cox, 1964)

32.13 Using Exercise 21.7, show that if in (32.58) l_p is distributed free of certain parameters for which there is a complete sufficient (vector) statistic **t**, and l_q is a function of **t** alone, l_p and l_q are stochastically independent. Apply this result to establish the independence results in Exercises 22.6 and 22.13. Show in Exercise 32.12 that $l_1(z)$ and $l_2(z)$ are independent when C_1 and C_2 both hold.

(Cf. Hogg, 1961)

32.14 Show that if a linear model contains terms $\theta_i x_i^{\mu_i}$, $\mu_i \neq 0$, we have approximately

$$x_i^{\mu_i} = x_i + (\mu_i - 1)x_i \log x_i.$$

Hence show how μ_i can be estimated. The process may be iterated.

(Box and Tidwell, 1962)

32.15 In **32.38**, show that for the binomial distribution of **5.2**, (32.64) gives the variance-stabilizing transformation

$$u\left(\frac{x}{n}\right) = \arcsin\left\{\left(\frac{x}{n}\right)^{1/2}\right\},$$

where x/n is the observed proportion of 'successes'. Show that

$$2\arcsin(y^{1/2}) = \arcsin(2y - 1) + \pi/2,$$

so that $u = \arcsin(2x/n - 1)$ may be used equivalently.

(Anscombe (1948) shows that better variance stabilization is obtained if x/n is replaced by $(x + \frac{3}{8})/(n + \frac{3}{4})$. Freeman and Tukey (1950) suggest

$$\arcsin\left\{\left(\frac{x}{n+1}\right)^{1/2}\right\} + \arcsin\left\{\left(\frac{x+1}{n+1}\right)^{1/2}\right\}$$

(cf. Example 32.5), tabulated by Mosteller and Youtz (1961). See also Laubscher (1961).)

32.16 In **32.38** show that for the negative binomial distribution of **5.16**, (32.64) gives the variance-stabilizing transformation

$$u\left(\frac{x}{n}\right) = \operatorname{arcsinh}\left\{\left(\frac{x}{n}\right)^{1/2}\right\}$$

where x/n is the observed proportion of successes.

(Anscombe (1948) shows that better variance stabilization occurs if x/n is replaced by $(x+\frac{3}{8})/(n-\frac{3}{4})$. See also Laubscher (1961).)

32.17 In Exercise 32.15, show that the alternative transformation

$$u\left(\frac{x}{n}\right) = \operatorname{arcsinh}\left\{\left(\frac{x}{n}\right)^{1/2}\right\}$$

stabilizes the variance near $p = \frac{1}{2}$. Show that this transformation is strictly appropriate when (32.59) is $D_n^2(\theta) = c\theta^2(1-\theta)^2$.

(Cf. Bartlett, 1947)

32.18 If z has the gamma distribution given in (32.69), show that the cumulant generating function for $\log z$ is

$$\psi(w) = \log\Gamma(1+\mathrm{i}w) + \sum_{s=1}^{q}\log(1+\mathrm{i}w/s).$$

Using Exercise 14.4 or otherwise, show that the rth cumulant of $\log z$ is

$$\kappa_r = (-1)^r(r-1)!\sum_{s=q}^{\infty}s^{-r}, \qquad r \geq 2.$$

Finally, show that

$$\gamma_1 = \kappa_3/\kappa_2^{3/2} \simeq -q^{-1/2}$$
$$\gamma_2 = \kappa_4/\kappa_2^2 \simeq 2q^{-1}.$$

(Bartlett and Kendall, 1946)

32.19 In Example 32.6 expand

$$u_c(t) = (\theta+c)^{1/2}\left(1+\frac{t-\theta}{\theta+c}\right)^{1/2}$$

in series and show that

$$E(u_c) = (\theta+c)^{1/2} - \tfrac{1}{8}\theta^{-1/2} + \frac{24c-7}{128}\theta^{-3/2} + o(\theta^{-3/2}),$$

and

$$\operatorname{var} u_c = \frac{1}{4}\left\{1 + \frac{3-8c}{8\theta} + \frac{32c^2-52c+17}{32\theta^2} + o(\theta^{-2})\right\},$$

so that the choice $c = \frac{3}{8}$ removes the term of order θ^{-1} in the variance, reducing it to

$$\operatorname{var} u_{3/8} \sim \frac{1}{4}\left(1+\frac{1}{16\theta^2}\right).$$

Hence show that

$$\{E(u_c)\}^2 - c \sim \theta - \frac{1}{4} - \frac{3-8c}{32\theta},$$

so that if the inverse transformation is used on u_c to obtain an estimator of θ, its downward bias is nearly constant at $\frac{1}{4}$.

(The $c = \frac{3}{8}$ result is due to A.H.L. Johnson; cf. Anscombe (1948).)

32.20 In Exercise 32.19, show that the coefficients of skewness and kurtosis of $u_c(t)$ are

$$\gamma_1 = -\frac{1}{2\theta^2}\left\{1 + \frac{25-48c}{16\theta}\right\} + o(\theta^{-3/2}),$$
$$\gamma_2 = \frac{1}{\theta}\left\{1 + \frac{945-1536c}{256\theta}\right\} + o(\theta^{-2}),$$

compared with

$$\gamma_1 = \theta^{-1/2},$$
$$\gamma_2 = \theta^{-1},$$

for the original Poisson variable t. Thus, whatever c is chosen, γ_1 is approximately halved (with changed sign) and γ_2 unaffected to the first order.

(Anscombe, 1948)

32.21 Using the result for var u_c in Exercise 32.19, show that the transformation

$$u'_\delta = (t + \tfrac{1}{2} + \delta)^{1/2} + (t + \tfrac{1}{2} - \delta)^{1/2}$$

has variance

$$\operatorname{var} u'_\delta = 1 - \frac{1}{8\theta} + \frac{16\delta^2 - 1}{32\theta^2} + o(\theta^{-2}),$$

so that if we choose $\delta = \frac{1}{2}$ to give u' of Example 32.5,

$$\operatorname{var} u'_{1/2} = 1 - \frac{1}{8\theta} + \frac{3}{32\theta^2} + o(\theta^{-2}).$$

32.22 x is a variate with cumulants κ_r all of order n, the sample size. Show that the moments of

$$y = \left(\frac{x}{\kappa_1}\right)^h = \left(1 + \frac{x-\kappa_1}{\kappa_1}\right)^h$$

are

$$\mu'_r(y) = 1 + \frac{(rh)^{(2)}}{2!}\frac{\kappa_2}{\kappa_1^2} + \frac{(rh)^{(3)}}{3!}\frac{\kappa_3}{\kappa_1^3} + \frac{(rh)^{(4)}}{4!}\frac{(\kappa_4 + 3\kappa_2^2)}{\kappa_1^4} + \frac{(rh)^{(5)}}{5!}\frac{10\kappa_3\kappa_2}{\kappa_1^5} + \frac{(rh)^{(6)}}{6!}\frac{15\kappa_2^3}{\kappa_1^6} + O(n^{-4}).$$

Hence show that

$$\mu_3(y) = \frac{h^3}{\kappa_1^4}\{\kappa_3\kappa_1 + 3(h-1)\kappa_2^2\} + \frac{h^3(h-1)}{2\kappa_1^6} \times\{3\kappa_4\kappa_1^2 + 3(7h-10)\kappa_3\kappa_2\kappa_1 + (17h^2 - 55h + 44)\kappa_2^3\} + O(n^{-4}),$$

so that putting $h = 1-\kappa_1\kappa_3/3\kappa_2^2$ makes $\mu_3(y)$ of order n^{-3} and $\gamma_1(y)$ of order $n^{-3/2}$. Show further that by adding a constant to x to make its mean g, and choosing

$$g = \frac{12\kappa_3\kappa_2^2}{20\kappa_3^2 - 9\kappa_4\kappa_2}, \qquad h = \frac{16\kappa_3^2 - 9\kappa_4\kappa_2}{20\kappa_3^2 - 9\kappa_4\kappa_2},$$

the variate $z = (1+(x-\kappa_1)/g)^h$ has γ_1 of order $n^{-3/2}$ and γ_2 of order n^{-2}.
Verify in Example 32.6 that for the χ^2 distribution the first of these results gives the Wilson–Hilferty cube root transformation, while the second gives a power $h = \frac{5}{13}$. Similarly, show in Example 32.6 for the Poisson distribution that we get $h = \frac{2}{3}$ and $h = \frac{7}{11}$ respectively, so that although $h = \frac{1}{2}$ stabilizes variance in each of these examples, normality considerations move h in opposite directions in the two examples.

(Cf. Haldane, 1938)

32.23 In **32.44**, expand $t(u)$ in Taylor series about μ, obtaining

$$\theta = E\{t(u)\} = \sum_{m=0}^{\infty} \frac{t^{(m)}(\mu)}{m!} E(u-\mu)^m = \sum_{m=0}^{\infty} \frac{t^{(2m)}(\mu)}{m!}\left(\frac{\sigma^2}{2}\right)^m;$$

now expand $t^{(2m)}(\mu)$ in Taylor series about zero, so that

$$\theta = \sum_{m=0}^{\infty}\sum_{r=0}^{\infty} \frac{t^{(2m+r)}(0)}{m!r!}\mu^r\left(\frac{\sigma^2}{2}\right)^m,$$

given that these summations converge. Using the result of Exercise 17.10, show that the MV unbiased estimator of θ is

$$\hat{\theta} = \sum_{m=0}^{\infty}\sum_{r=0}^{\infty} \frac{t^{(2m+r)}(0)}{m!r!2^m} \sum_{i=0}^{[r/2]} (-1)^i \frac{r!}{i!(r-2i)!} \frac{\Gamma(\frac{1}{2}v)}{\Gamma\{\frac{1}{2}v+m+i\}} \hat{\mu}^{r-2i}\left(\frac{\lambda^2}{2}\right)^i\left(\frac{s^2}{2}\right)^{m+i}.$$

Hence verify the formulae given in **32.44** for the special case $t''(u) = A + Bt(u)$.

(Neyman and Scott, 1960; Hoyle, 1968)

32.24 For the square root transformation, show that the bias-corrected inverse transformation $\hat{\theta}$ in **32.44** has variance

$$\text{var}(\hat{\theta}) = 4\mu^2\lambda^2\sigma^2 + 2\sigma^4\{(1-\lambda^2)^2/v + \lambda^4\}$$

and that its MV unbiased estimator is

$$\text{est. var}(\hat{\theta}) = \frac{4\hat{\mu}^2\lambda^2 s^2}{v} + \frac{2s^2}{v(v+2)}\left\{\frac{(1-\lambda^2)^2}{v} - \lambda^4\right\}.$$

(Cf. Hoyle, 1968)

32.25 By consideration of the case when $y_j = x_j^k$, $j = 1, 2, \ldots, n$, exactly, show that if the orthogonal polynomials defined by (32.92) are *orthonormal* (i.e., $\mathbf{C} = \mathbf{I}$ in (32.93)) then they satisfy the recurrence relation

$$\phi_k(x_j) = \frac{1}{b_k}\left\{x_j^k - \sum_{i=0}^{k-1} \phi_i(x_j) \sum_{j=1}^{n} x_j^k \phi_i(x_j)\right\},$$

where the normalizing constant b_k is defined by

$$b_k^2 = \sum_{j=1}^{n} \left\{ x_j^k - \sum_{i=0}^{k-1} \phi_i(x_j) \sum_{j=1}^{n} x_j^k \phi_i(x_j) \right\}^2.$$

Hence verify the expressions for $j = 1, 2, 3$, with appropriate adjustments.

(Robson, 1959)

32.26 In fitting orthogonal polynomials of degree k as in **32.51**, the reduction in the total SS associated with the term of degree r is $Q_r = \hat{\alpha}_r^2 \sum_{i=1}^{n} \phi_j^2(x_i)$. Show that the ratios

$$z_r = Q_{k-r+1} \Big/ \sum_{s=k-r+2}^{k+1} Q_s, \qquad r = 1, 2, \ldots, k,$$

where $Q_{k+1} = (n-k)s^2$ is the residual SS, are all independently distributed when the regression coefficients α_r are all zero:

(a) by using the result of Exercise 32.13; and
(b) by using the result of Exercise 21.27.

(This result indicates (cf. Hogg, 1961) that one may independently test the regression coefficients if one starts from the highest order and works downwards, 'pooling' the associated SS of those adjudged zero with the residual SS, until one is adjudged non-zero, when the process stops. All the tests are, of course, $t^2(F)$ tests, and the overall test has size $1 - (1-\alpha)^k \sim k\alpha$ if a test of size α is used at each stage. Anderson (1962a) shows under weak assumptions that this procedure maximizes the probability of correctly locating a non-zero coefficient.)

32.27 Given a set of n observations with linear model

$$\mathbf{y} = \mathbf{X}\boldsymbol{\beta} + \boldsymbol{\epsilon}, \qquad \epsilon_i \sim \text{independent } (0, \sigma^2),$$

suppose that the data are grouped into m groups using the $m \times n$ matrix $\mathbf{G}$ where $g_{ij} = 1$ if the ith observation is in the jth group; $g_{ij} = 0$ otherwise. The regressors may be considered when specifying $\mathbf{G}$, but not the values of $\mathbf{y}$. Show that the resulting estimator for $\boldsymbol{\beta}$ is of weighted LS form (32.98), where $\mathbf{V} = \sigma^2\mathbf{G}\mathbf{G}^T$ is diagonal only when the groups are non-overlapping. Further, if $m > n$, show that a weighted LS estimator based upon the generalized inverse may be used.

(Leech and Cowling, 1982)

32.28 Suppose that the linear model (32.1) has errors with zero means and covariance matrix

$$\mathbf{V} = \sigma^2 \begin{bmatrix} \mathbf{G} & & & & \mathbf{0} \\ & \mathbf{G} & & & \\ & & \mathbf{G} & & \\ & \cdot & & \ddots & \\ \mathbf{0} & & & & \mathbf{G} \end{bmatrix},$$

where $\mathbf{V}$ is $km \times km$ and $\mathbf{G}$ is the $k \times k$ matrix $(1-\phi)\mathbf{I} + \phi\mathbf{1}\mathbf{1}^T$; that is, the errors have an intraclass correlation structure; cf. (32.108)–(32.109). Show that $\mathbf{G}^{-1} = (1-\phi)^{-l}\{\mathbf{I} - c\mathbf{1}\mathbf{1}^T\}$ where $c = \phi/(1-\phi+k\phi)$. Partitioning $\mathbf{X}^T$ into $(\mathbf{X}_1^T, \ldots, \mathbf{X}_m^T)$ and $\mathbf{y}^T$ into $(\mathbf{y}_l^T, \ldots, \mathbf{y}_m^T)$ show that

$$\hat{\boldsymbol{\beta}} = \left(\sum \mathbf{X}_j^T \mathbf{G}^{-1} \mathbf{X}_j\right)^{-1} \left(\sum \mathbf{X}_j^T \mathbf{G}^{-1} \mathbf{y}_j\right)$$

with covariance matrix $\sigma^2(\sum \mathbf{X}_j^T \mathbf{G}^{-1}\mathbf{X}_j)^{-1}$. Further, when $\mathbf{X}_j = \mathbf{1}$ for all j, show that, $\hat{\beta} = \bar{y}$ as with LS, but that its variance is $\sigma^2\{l + (k-1)\phi\}/km$.

32.29 Suppose that there is a linear structural relationship between X and Y, so that

$$Y = \alpha + \beta X.$$

Now assume that X and Y may only be measured, subject to error, as

$$x = X + \delta, \qquad y = Y + \epsilon,$$

where $\delta \sim N(0, \omega^2)$ and $\epsilon \sim N(0, \sigma^2)$ are independent of each other and across all observations. Given that $X \sim N(0, \sigma_x^2)$ and X is independent of δ and ϵ, show that the LS estimator based on sample values $(x_i, y_i), i = 1, 2, \ldots, n$, yields a biased estimator for β since

$$\lim_{n\to\infty} E(\hat{\beta}) = \frac{\beta\sigma_x^2}{\sigma_x^2 + \omega^2}.$$

(*Note*: If the X observations are fixed, the relationship is *functional*; if they are drawn at random from some population, the relationship is *structural*.)

32.30 The instrumental variables estimators for $\boldsymbol{\beta}$ in (32.1) are given by

$$\mathbf{Z}^T(\mathbf{y} - \mathbf{X}\boldsymbol{\beta}) = \mathbf{0},$$

where $\mathbf{Z}$ denotes the $n \times p$ matrix of values of the instrumental variables. Show that the large-sample covariance matrix for $\hat{\boldsymbol{\beta}}$ is $V_I = \sigma^2(\mathbf{Z}^T\mathbf{X})^{-1}(\mathbf{Z}^T\mathbf{Z})(\mathbf{X}^T\mathbf{Z})^{-1}$.

32.31 Consider estimators of the form of (32.147) based on only p observations out of n. For subset J, let

$$w_J = |\mathbf{X}_J|^2/|\mathbf{X}^T\mathbf{X}|.$$

Show that the LS estimator is given by

$$\hat{\boldsymbol{\beta}} = \sum w_J \hat{\boldsymbol{\beta}}_J$$

where the sum is taken over all $\binom{n}{p}$ subsets.

(Heitmann and Ord, 1985)

32.32 Suppose that $\mathbf{y} = (y_1, \ldots, y_n)^T$ satisfies the model

$$y = \mathbf{X}\boldsymbol{\beta} + \boldsymbol{\epsilon},$$

where $E(\boldsymbol{\epsilon}) = \mathbf{0}$ and $\mathbf{V}(\boldsymbol{\epsilon}) = \sigma^2\mathbf{V}$; $\mathbf{V}$ is an unknown diagonal matrix. Let $\mathbf{b}$ denote the ordinary LS estimator for $\boldsymbol{\beta}$ given by (32.2). Show that $\mathbf{b}$ has covariance matrix

$$\mathbf{V}(\mathbf{b}) = \sigma^2(\mathbf{X}^T\mathbf{X})^{-1}\mathbf{X}^T\mathbf{V}\mathbf{X}(\mathbf{X}^T\mathbf{X})^{-1}.$$

If the ordinary LS residuals are $\mathbf{e} = (e_1, \ldots, e_n)^T$ show that $V(\mathbf{b})$ may be consistently estimated by $\mathbf{Q}_n^{-1}\mathbf{S}_n\mathbf{Q}_n^{-1}$, where $n\mathbf{Q}_n = \sum_{j=1}^n \mathbf{x}_j\mathbf{x}_j^T$ and $n\mathbf{S}_n = \sum_{j=1}^n e_j^2\mathbf{x}_j\mathbf{x}_j^T$, under suitable regularity conditions.

(White, 1980)

32.33 Consider the linear model $\mathbf{y} = \mathbf{X}\boldsymbol{\beta} + \boldsymbol{\epsilon}$, where $\boldsymbol{\beta}$ is $p \times 1$ with hat matrix $\mathbf{H} = \mathbf{X}(\mathbf{X}^T\mathbf{X})^{-1}\mathbf{X}^T = (\mathbf{h}_1, \ldots, \mathbf{h}_n)$, and $\text{diag}(\mathbf{H}) = (h_{11}, \ldots, h_n)$.
If $\mathbf{t}_i = \hat{\mathbf{y}} - \hat{\mathbf{y}}_{(i)}$ summarizes the effect of deleting the ith observation, as in **32.67**, show that

$$\mathbf{t}_i = \{e_i/(1 - h_{ii})\}\mathbf{h}_i.$$

Define the matrix $\mathbf{T} = \{\mathbf{t}_1, \ldots, \mathbf{t}_n\}$ and show that

$$\mathbf{T}^T\mathbf{T} = \mathbf{EDHDE} \qquad (= \mathbf{M}_0, \text{ say})$$

where $\mathbf{E} = \text{diag}(e_1, \ldots, e_n)$ and $\mathbf{D} = \text{diag}[(1 - h_{ii})^{-1}]$.
[The matrix $\mathbf{M} = \mathbf{M}/ps^2, s^2 = \mathbf{e}^T\mathbf{e}/(n - p)$, is termed the *influence matrix* and plots of the scores in the eigenvectors corresponding to the largest eigenvalues may be used to detect clusters of influential observations].

(Peña and Yohai, 1995)

APPENDIX TABLES

Appendix Table 1 Density function of the normal distribution $y = \dfrac{1}{\sqrt{2\pi}} e^{-\frac{1}{2}x^2}$ with first and second differences

x	y	$\Delta^1(-)$	Δ^2	x	y	$\Delta^1(-)$	Δ^2
0.0	0.39894	199	−392	2.5	0.01753	395	+79
0.1	0.39695	591	−374	2.6	0.01358	316	+66
0.2	0.39104	965	−347	2.7	0.01042	250	+53
0.3	0.38139	1312	−308	2.8	0.00792	197	+45
0.4	0.36827	1620	−265	2.9	0.00595	152	+36
0.5	0.35207	1885	−212	3.0	0.00443	116	+27
0.6	0.33322	2097	−159	3.1	0.00327	89	+23
0.7	0.31225	2256	−104	3.2	0.00238	66	+17
0.8	0.28969	2360	−52	3.3	0.00172	49	+13
0.9	0.26609	2412	0	3.4	0.00123	36	+10
1.0	0.24197	2412	+46	3.5	0.00087	26	+7
1.1	0.21785	2366	+84	3.6	0.00061	19	+6
1.2	0.19419	2282	+118	3.7	0.00042	13	+4
1.3	0.17137	2164	+143	3.8	0.00029	9	+2
1.4	0.14973	2021	+161	3.9	0.00020	7	+3
1.5	0.12952	1860	+173	4.0	0.00013	4	—
1.6	0.11092	1687	+177	4.1	0.00009	3	—
1.7	0.09405	1510	+177	4.2	0.00006	2	—
1.8	0.07895	1333	+170	4.3	0.00004	2	—
1.9	0.06562	1163	+162	4.4	0.00002	—	—
2.0	0.05399	1001	+150	4.5	0.00002	—	—
2.1	0.04398	851	+137	4.6	0.00001	—	—
2.2	0.03547	714	+120	4.7	0.00001	—	—
2.3	0.02833	594	+108	4.8	0.00000	—	—
2.4	0.02239	486	+91				

Appendix Table 2 Distribution function of the normal distribution

The table shows the area under the curve $y = (2\pi)^{-\frac{1}{2}} e^{-\frac{1}{2}x^2}$ lying to the left of specified deviates x; e.g. the area corresponding to a deviate 1.86 (= 1.5 + 0.36) is 0.9686.

Deviate	0.0+	0.5+	1.0+	1.5+	2.0+	2.5+	3.0+	3.5+
0.00	5000	6915	8413	9332	9772	9^2379	9^2865	9^377
0.01	5040	6950	8438	9345	9778	9^2496	9^2869	9^378
0.02	5080	6985	8461	9357	9783	9^2413	9^2874	9^378
0.03	5120	7019	8485	9370	9788	9^2430	9^2878	9^379
0.04	5160	7054	8508	9382	9793	9^2446	9^2882	9^380
0.05	5199	7088	8531	9394	9798	9^2461	9^2886	9^381
0.06	5239	7123	8554	9406	9803	9^2477	9^2889	9^381
0.07	5279	7157	8577	9418	9808	9^2492	9^2893	9^382
0.08	5319	7190	8599	9429	9812	9^2506	9^2897	9^383
0.09	5359	7224	8621	9441	9817	9^2520	9^2900	9^383
0.10	5398	7257	8643	9452	9821	9^2534	9^303	9^384
0.11	5438	7291	8665	9463	9826	9^2547	9^306	9^385
0.12	5478	7324	8686	9474	9830	9^2560	9^310	9^385
0.13	5517	7357	8708	9484	9834	9^2573	9^313	9^386
0.14	5557	7389	8729	9495	9838	9^2585	9^316	9^386
0.15	5596	7422	8749	9505	9842	9^2598	9^318	9^387
0.16	5636	7454	8770	9515	9846	9^2609	9^321	9^387
0.17	5675	7486	8790	9525	9850	9^2621	9^324	9^388
0.18	5714	7517	8810	9535	9854	9^2632	9^326	9^388
0.19	5753	7549	8830	9545	9857	9^2643	9^329	9^389
0.20	5793	7580	8849	9554	9861	9^2653	9^331	9^389
0.21	5832	7611	8869	9564	9864	9^2664	9^334	9^390
0.22	5871	7642	8888	9573	9868	9^2674	9^336	9^390
0.23	5910	7673	8907	9582	9871	9^2683	9^338	9^404
0.24	5948	7704	8925	9591	9875	9^2693	9^340	9^408
0.25	5987	7738	8944	9599	9878	9^2702	9^342	9^412
0.26	6026	7764	8962	9608	9881	9^2711	9^344	9^415
0.27	6064	7794	8980	9616	9884	9^2720	9^346	9^418
0.28	6103	7823	8997	9625	9887	9^2728	9^348	9^422
0.29	6141	7852	9015	9633	9890	9^2736	9^350	9^425
0.30	6179	7881	9032	9641	9893	9^2744	9^352	9^428
0.31	6217	7910	9049	9649	9896	9^2752	9^353	9^431
0.32	6255	7939	9066	9656	9898	9^2760	9^355	9^433
0.33	6293	7967	9082	9664	9901	9^2767	9^357	9^436
0.34	6331	7995	9099	9671	9904	9^2774	9^358	9^439
0.35	6368	8023	9115	9678	9906	9^2781	9^360	9^441
0.36	6406	8051	9131	9686	9909	9^2788	9^361	9^443
0.37	6443	8078	9147	9693	9911	9^2795	9^362	9^446
0.38	6480	8106	9162	9699	9913	9^2801	9^364	9^448
0.39	6517	8133	9177	9706	9916	9^2807	9^365	9^450
0.40	6554	8159	9192	9713	9918	9^2813	9^366	9^452
0.41	6591	8186	9207	9719	9920	9^2819	9^368	9^454
0.42	6628	8212	9222	9726	9922	9^2825	9^369	9^456
0.43	6664	8238	9236	9732	9925	9^2831	9^370	9^458
0.44	6700	8264	9251	9738	9927	9^2836	9^371	9^459
0.45	6736	8289	9265	9744	9929	9^2841	9^372	9^461
0.46	6772	8315	9279	9750	9931	9^2846	9^373	9^463
0.47	6808	8340	9292	9756	9932	9^2851	9^374	9^464
0.48	6844	8365	9306	9761	9934	9^2856	9^375	9^466
0.49	6879	8389	9319	9767	9936	9^2861	9^376	9^467

Note. Decimal points in the body of the table are omitted. Repeated 9s are indicated by powers, e.g. 9^371 stands for 0.99971.

Appendix Table 3 Quantiles of the d.f. of χ^2

(Reproduced from Table III of Sir Ronald Fisher's *Statistical Methods for Research Workers*, Oliver and Boyd Ltd., Edinburgh, by kind permission of the author and publishers)

$P = 1 - F$	0.99	0.98	0.95	0.90	0.80	0.70	0.50	0.30	0.20	0.10	0.05	0.02	0.01
$\nu =$ 1	0.0^3157	0.0^3628	0.0^2393	0.0158	0.0642	0.148	0.455	1.074	1.642	2.706	3.841	5.412	6.635
2	0.0201	0.0404	0.103	0.211	0.446	0.713	1.386	2.408	3.219	4.605	5.991	7.824	9.210
3	0.115	0.185	0.352	0.584	1.005	1.424	2.366	3.665	4.642	6.251	7.815	9.837	11.345
4	0.297	0.429	0.711	1.064	1.649	2.195	3.357	4.878	5.989	7.779	9.488	11.668	13.277
5	0.554	0.752	1.145	1.160	2.343	3.000	4.351	6.064	7.289	9.236	11.070	13.388	15.086
6	0.872	1.134	1.635	2.204	3.070	3.828	5.348	7.231	8.558	10.645	12.592	15.033	16.812
7	1.239	1.564	2.167	2.833	3.822	4.671	6.346	8.383	9.803	12.017	14.067	16.622	18.475
8	1.646	2.032	2.733	3.490	4.594	5.527	7.344	9.524	11.030	13.362	15.507	18.168	20.090
9	2.088	2.532	3.325	4.168	5.380	6.393	8.343	10.656	12.242	14.684	16.919	19.679	21.666
10	2.358	3.059	3.940	4.865	6.179	7.267	9.342	11.781	13.442	15.987	18.307	21.161	23.209
11	3.053	3.609	4.575	5.578	6.989	8.148	10.341	12.899	14.631	17.275	19.675	22.618	24.725
12	3.571	4.178	5.226	6.304	7.807	9.034	11.340	14.011	15.821	18.549	21.026	24.054	26.217
13	4.107	4.765	5.892	7.042	8.634	9.926	12.340	15.119	16.985	19.812	22.362	25.472	27.688
14	4.660	5.368	6.571	7.790	9.467	10.821	13.339	16.222	18.151	21.064	23.685	26.873	29.141
15	5.229	5.985	7.261	8.547	10.307	11.721	14.339	17.322	19.311	22.307	24.996	28.259	30.578
16	5.812	6.614	7.962	9.312	11.152	12.624	15.338	18.418	20.465	23.542	26.296	29.633	32.000
17	6.408	7.255	8.672	10.085	12.002	13.531	16.338	19.511	21.615	24.769	27.587	30.995	33.409
18	7.015	7.906	9.390	10.865	12.857	14.440	17.338	20.601	22.760	25.989	28.869	32.346	34.805
19	7.633	8.567	10.117	11.651	13.716	15.352	18.338	21.689	23.900	27.204	30.144	33.687	36.191
20	8.260	9.237	10.851	12.443	14.578	16.266	19.337	22.775	25.038	28.412	31.410	35.020	37.566
21	8.897	9.915	11.591	13.240	15.445	17.182	20.337	23.858	26.171	29.615	32.671	36.343	38.932
22	9.542	10.600	12.338	14.041	16.314	18.101	21.337	24.939	27.301	30.813	33.924	37.659	40.289
23	10.196	11.293	13.091	14.848	17.187	19.021	22.337	26.018	28.429	32.007	35.172	38.968	41.638
24	10.856	11.992	13.848	15.659	18.062	19.943	23.337	27.096	29.553	33.196	36.415	40.270	42.980
25	11.524	12.697	14.611	16.473	18.940	20.867	24.337	28.172	30.675	34.382	37.652	41.566	44.314
26	12.198	13.409	15.379	17.292	19.820	21.792	25.336	29.246	31.795	35.563	38.885	42.856	45.642
27	12.879	14.125	16.151	18.114	20.703	22.719	26.336	30.319	32.912	36.741	40.113	44.140	46.963
28	13.565	14.847	16.928	18.939	21.588	23.647	27.336	31.391	34.027	37.916	41.337	45.419	48.278
29	14.256	15.574	17.708	19.768	22.475	24.577	28.336	32.461	35.139	39.087	42.557	46.693	49.588
30	14.953	16.306	18.493	20.599	23.364	25.508	29.336	33.530	36.250	40.256	43.773	47.962	50.892

Note. For values of ν greater than 30 the quantity $\sqrt{2\chi^2}$ may be taken to be distributed normally about mean $\sqrt{2\nu - 1}$ with unit variance.

Appendix Table 4a Distribution of χ^2 for one degree of freedom for values $\chi^2 = 0$ to $\chi^2 = 1$ by steps of 0.01

χ^2	$P = 1 - F$	Δ	χ^2	$P = 1 - F$	Δ
0	1.00000	7966	0.50	0.47950	436
0.01	0.92034	3280	0.51	0.47514	430
0.02	0.88754	2505	0.52	0.47084	423
0.03	0.86249	2101	0.53	0.46661	418
0.04	0.84148	1842	0.54	0.46243	411
0.05	0.82306	1656	0.55	0.45832	406
0.06	0.80650	1516	0.56	0.45426	400
0.07	0.79134	1404	0.57	0.45026	395
0.08	0.77730	1312	0.58	0.44631	389
0.09	0.76418	1235	0.59	0.44242	384
0.10	0.75183	1169	0.60	0.43858	379
0.11	0.74014	1111	0.61	0.43479	374
0.12	0.72903	1060	0.62	0.43105	369
0.13	0.71843	1015	0.63	0.42736	365
0.14	0.70828	974	0.64	0.42371	360
0.15	0.69854	938	0.65	0.42011	355
0.16	0.68916	905	0.66	0.41656	351
0.17	0.68011	874	0.67	0.41305	346
0.18	0.67137	845	0.68	0.40959	343
0.19	0.66292	820	0.69	0.40616	338
0.20	0.65472	795	0.70	0.40278	334
0.21	0.64677	773	0.71	0.39944	330
0.22	0.63904	752	0.72	0.39614	326
0.23	0.63152	731	0.73	0.39288	322
0.24	0.62421	713	0.74	0.38966	318
0.25	0.61708	696	0.75	0.38648	315
0.26	0.61012	679	0.76	0.38333	311
0.27	0.60333	663	0.77	0.38022	308
0.28	0.59670	648	0.78	0.37714	304
0.29	0.59022	634	0.79	0.37410	301
0.30	0.58388	620	0.80	0.37109	297
0.31	0.57768	607	0.81	0.36812	294
0.32	0.57161	595	0.82	0.36518	291
0.33	0.56566	583	0.83	0.36227	287
0.34	0.55983	572	0.84	0.35940	285
0.35	0.55411	560	0.85	0.35655	281
0.36	0.54851	551	0.86	0.35374	278
0.37	0.54300	540	0.87	0.35096	276
0.38	0.53760	530	0.88	0.34820	272
0.39	0.53230	521	0.89	0.34548	270
0.40	0.52709	512	0.90	0.34278	267
0.41	0.52197	503	0.91	0.34011	264
0.42	0.51694	495	0.92	0.33747	261
0.43	0.51199	487	0.93	0.33486	258
0.44	0.50712	479	0.94	0.33228	256
0.45	0.50233	471	0.95	0.32972	253
0.46	0.49762	463	0.96	0.32719	251
0.47	0.49299	457	0.97	0.32468	248
0.48	0.48842	449	0.98	0.32220	246
0.49	0.48393	443	0.99	0.31974	243
0.50	0.47950	436	1.00	0.31731	241

Appendix Table 4b Distribution function of χ^2 for one degree of freedom for values of χ^2 from 1 to 10 by steps of 0.1

χ^2	$P = 1 - F$	Δ	χ^2	$P = 1 - F$	Δ
1.0	0.31731	2304	5.5	0.01902	106
1.1	0.29427	2095	5.6	0.01796	99
1.2	0.27332	1911	5.7	0.01697	94
1.3	0.25421	1749	5.8	0.01603	89
1.4	0.23672	1605	5.9	0.01514	83
1.5	0.22067	1477	6.0	0.01431	79
1.6	0.20590	1361	6.1	0.01352	74
1.7	0.19229	1258	6.2	0.01278	71
1.8	0.17971	1163	6.3	0.01207	66
1.9	0.16808	1078	6.4	0.01141	62
2.0	0.15730	1000	6.5	0.01079	59
2.1	0.14730	929	6.6	0.01020	56
2.2	0.13801	864	6.7	0.00964	52
2.3	0.12937	803	6.8	0.00912	50
2.4	0.12134	749	6.9	0.00862	47
2.5	0.11385	699	7.0	0.00815	44
2.6	0.10686	651	7.1	0.00771	42
2.7	0.10035	609	7.2	0.00729	39
2.8	0.09426	568	7.3	0.00690	38
2.9	0.08858	532	7.4	0.00652	35
3.0	0.08326	497	7.5	0.00617	33
3.1	0.07829	465	7.6	0.00584	32
3.2	0.07364	436	7.7	0.00552	30
3.3	0.06928	408	7.8	0.00522	28
3.4	0.06520	383	7.9	0.00494	26
3.5	0.06137	359	8.0	0.00468	25
3.6	0.05778	337	8.1	0.00443	24
3.7	0.05441	316	8.2	0.00419	23
3.8	0.05125	296	8.3	0.00396	21
3.9	0.04829	279	8.4	0.00375	20
4.0	0.04550	262	8.5	0.00355	19
4.1	0.04288	246	8.6	0.00336	18
4.2	0.04042	231	8.7	0.00318	17
4.3	0.03811	217	8.8	0.00301	16
4.4	0.03594	205	8.9	0.00285	15
4.5	0.03389	192	9.0	0.00270	14
4.6	0.03197	181	9.1	0.00256	14
4.7	0.03016	170	9.2	0.00242	13
4.8	0.02846	160	9.3	0.00229	12
4.9	0.02686	151	9.4	0.00217	12
5.0	0.02535	142	9.5	0.00205	10
5.1	0.02393	134	9.6	0.00195	11
5.2	0.02259	126	9.7	0.00184	10
5.3	0.02133	119	9.8	0.00174	9
5.4	0.02014	112	9.9	0.00165	8
5.5	0.01902	106	10.0	0.00157	8

Appendix Table 5 Quantiles of the d.f. of *t*

(Reproduced from Sir Ronald Fisher and Dr F. Yates: *Statistical Tables for Biological, Medical and Agricultural Research*, Oliver and Boyd Ltd., Edinburgh, by kind permission of the authors and publishers)

$P = 2(1 - F)$	0.9	0.8	0.7	0.6	0.5	0.4	0.3	0.2	0.1	0.05	0.02	0.01	0.001
$\nu = 1$	0.158	0.325	0.510	0.727	1.000	1.376	1.963	3.078	6.314	12.706	31.821	63.657	636.619
2	0.142	0.289	0.445	0.617	0.816	1.061	1.386	1.886	2.920	4.303	6.965	9.925	31.598
3	0.137	0.277	0.424	0.584	0.765	0.978	1.250	1.638	2.353	3.182	4.541	5.841	12.924
4	0.134	0.271	0.414	0.569	0.741	0.941	1.190	1.533	2.132	2.776	3.747	4.604	8.610
5	0.132	0.267	0.408	0.559	0.727	0.920	1.156	1.476	2.015	2.571	3.365	4.032	6.869
6	0.131	0.265	0.404	0.553	0.718	0.906	1.134	1.440	1.943	2.447	3.143	3.707	5.959
7	0.130	0.263	0.402	0.549	0.711	0.896	1.119	1.415	1.895	2.365	2.998	3.499	5.408
8	0.130	0.262	0.399	0.546	0.706	0.889	1.108	1.397	1.860	2.306	2.896	3.355	5.041
9	0.129	0.261	0.398	0.543	0.703	0.883	1.100	1.383	1.833	2.262	2.821	3.250	4.781
10	0.129	0.260	0.397	0.542	0.700	0.879	1.093	1.372	1.812	2.228	2.764	3.169	4.587
11	0.129	0.260	0.396	0.540	0.697	0.876	1.088	1.363	1.796	2.201	2.718	3.106	4.437
12	0.128	0.259	0.395	0.539	0.695	0.873	1.083	1.356	1.782	2.179	2.681	3.055	4.318
13	0.128	0.259	0.394	0.538	0.694	0.870	1.079	1.350	1.771	2.160	2.650	3.012	4.221
14	0.128	0.258	0.393	0.537	0.692	0.868	1.076	1.345	1.761	2.145	2.624	2.977	4.140
15	0.128	0.258	0.393	0.536	0.691	0.866	1.074	1.341	1.753	2.131	2.602	2.947	4.073
16	0.128	0.258	0.392	0.535	0.690	0.865	1.071	1.337	1.746	2.120	2.583	2.921	4.015
17	0.128	0.257	0.392	0.534	0.689	0.863	1.069	1.333	1.740	2.110	2.567	2.898	3.965
18	0.127	0.257	0.392	0.534	0.688	0.862	1.067	1.330	1.734	2.101	2.552	2.878	3.922
19	0.127	0.257	0.391	0.533	0.688	0.861	1.066	1.328	1.729	2.093	2.539	2.861	3.883
20	0.127	0.257	0.391	0.533	0.687	0.860	1.064	1.325	1.725	2.086	2.528	2.845	3.850
21	0.127	0.257	0.391	0.532	0.686	0.859	1.063	1.323	1.721	2.080	2.518	2.831	3.819
22	0.127	0.256	0.390	0.532	0.686	0.858	1.061	1.321	1.717	2.074	2.508	2.819	3.792
23	0.127	0.256	0.390	0.532	0.685	0.858	1.060	1.319	1.714	2.069	2.500	2.807	3.767
24	0.127	0.256	0.390	0.531	0.685	0.857	1.059	1.318	1.711	2.064	2.492	2.797	3.745
25	0.127	0.256	0.390	0.531	0.684	0.856	1.058	1.316	1.708	2.060	2.485	2.787	3.725
26	0.127	0.256	0.390	0.531	0.684	0.856	1.058	1.315	1.706	2.056	2.479	2.779	3.707
27	0.127	0.256	0.389	0.531	0.684	0.855	1.057	1.314	1.703	2.052	2.473	2.771	3.690
28	0.127	0.256	0.389	0.530	0.683	0.855	1.056	1.313	1.701	2.048	2.467	2.763	3.674
29	0.127	0.256	0.389	0.530	0.683	0.854	1.055	1.311	1.699	2.045	2.462	2.756	3.659
30	0.127	0.256	0.389	0.530	0.683	0.854	1.055	1.310	1.697	2.042	2.457	2.750	3.646
40	0.126	0.255	0.388	0.529	0.681	0.851	1.050	1.303	1.684	2.021	2.423	2.704	3.551
60	0.126	0.254	0.387	0.527	0.679	0.848	1.046	1.296	1.671	2.000	2.390	2.660	3.460
120	0.126	0.254	0.386	0.526	0.677	0.845	1.041	1.289	1.658	1.980	2.358	2.617	3.373
∞	0.126	0.253	0.385	0.524	0.674	0.842	1.036	1.282	1.645	1.960	2.326	2.576	3.291

Appendix Table 6 5 per cent points of the distribution of z (values at which the d.f. = 0.95)

(Reproduced from Table VI of Sir Ronald Fisher's *Statistical Methods for Research Workers*, Oliver and Boyd Ltd., Edinburgh, by kind permission of the author and publishers)

Values of ν_2	Values of ν_1: 1	2	3	4	5	6	8	12	24	∞
1	2.5421	2.6479	2.6870	2.7071	2.7194	2.7276	2.7380	2.7484	2.7588	2.7693
2	1.4592	1.4722	1.4765	1.4787	1.4800	1.4808	1.4819	1.4830	1.4840	1.4851
3	1.1577	1.1284	1.1137	1.1051	1.0994	1.0953	1.0899	1.0842	1.0781	1.0716
4	1.0212	0.9690	0.9429	0.9272	0.9168	0.9093	0.8993	0.8885	0.8767	0.8639
5	0.9441	0.8777	0.8441	0.8236	0.8097	0.7997	0.7862	0.7714	0.7550	0.7368
6	0.8948	0.8188	0.7798	0.7558	0.7394	0.7274	0.7112	0.6931	0.6729	0.6409
7	0.8606	0.7777	0.7347	0.7080	0.6896	0.6761	0.6576	0.6369	0.6134	0.5862
8	0.8355	0.7475	0.7014	0.6725	0.6525	0.6378	0.6175	0.5945	0.5682	0.5371
9	0.8163	0.7242	0.6757	0.6450	0.6238	0.6080	0.5862	0.5613	0.5324	0.4979
10	0.8012	0.7058	0.6553	0.6232	0.6009	0.5843	0.5611	0.5346	0.5035	0.4657
11	0.7889	0.6909	0.6387	0.6055	0.5822	0.5648	0.5406	0.5126	0.4795	0.4387
12	0.7788	0.6786	0.6250	0.5907	0.5666	0.5487	0.5234	0.4941	0.4592	0.4156
13	0.7703	0.6682	0.6134	0.5783	0.5535	0.5350	0.5089	0.4785	0.4419	0.3957
14	0.7630	0.6594	0.6036	0.5677	0.5423	0.5233	0.4964	0.4649	0.4269	0.3782
15	0.7568	0.6518	0.5950	0.5585	0.5326	0.5131	0.4855	0.4532	0.4138	0.3628
16	0.7514	0.6451	0.5876	0.5505	0.5241	0.5042	0.4760	0.4428	0.4022	0.3490
17	0.7466	0.6393	0.5811	0.5434	0.5166	0.4964	0.4676	0.4337	0.3919	0.3366
18	0.7424	0.6341	0.5753	0.5371	0.5099	0.4894	0.4602	0.4255	0.3827	0.3253
19	0.7386	0.6295	0.5701	0.5315	0.5040	0.4832	0.4535	0.4182	0.3743	0.3151
20	0.7352	0.6254	0.5654	0.5265	0.4986	0.4776	0.4474	0.4116	0.3668	0.3057
21	0.7322	0.6216	0.5612	0.5219	0.4938	0.4725	0.4420	0.4055	0.3599	0.2971
22	0.7294	0.6182	0.5574	0.5178	0.4894	0.4679	0.4370	0.4001	0.3536	0.2892
23	0.7269	0.6151	0.5540	0.5140	0.4854	0.4636	0.4325	0.3950	0.3478	0.2818
24	0.7246	0.6123	0.5508	0.5106	0.4817	0.4598	0.4283	0.3904	0.3425	0.2749
25	0.7225	0.6097	0.5478	0.5074	0.4783	0.4562	0.4244	0.3862	0.3376	0.2685
26	0.7205	0.6073	0.5451	0.5045	0.4752	0.4529	0.4209	0.3823	0.3330	0.2625
27	0.7187	0.6051	0.5427	0.5017	0.4723	0.4499	0.4176	0.3786	0.3287	0.2569
28	0.7171	0.6030	0.5403	0.4992	0.4696	0.4471	0.4146	0.3752	0.3248	0.2516
29	0.7155	0.6011	0.5382	0.4969	0.4671	0.4444	0.4117	0.3720	0.3211	0.2466
30	0.7141	0.5994	0.5362	0.4947	0.4648	0.4420	0.4090	0.3691	0.3176	0.2419
60	0.6933	0.5738	0.5073	0.4632	0.4311	0.4064	0.3702	0.3255	0.2654	0.1644
∞	0.6729	0.5486	0.4787	0.4319	0.3974	0.3706	0.3309	0.2804	0.2085	0

Appendix Table 7 5 per cent points of the variance ratio F (values at which the d.f. = 0.95)

(Reproduced from Sir Ronald Fisher and Dr F. Yates: *Statistical Tables for Biological, Medical and Agricultural Research*, Oliver and Boyd Ltd., Edinburgh, by kind permission of the authors and publishers)

ν_2 \ ν_1	1	2	3	4	5	6	8	12	24	∞
1	161.40	199.50	215.70	224.60	230.20	234.00	238.90	243.90	249.00	254.30
2	18.51	19.00	19.16	19.25	19.30	19.33	19.37	19.41	19.45	19.50
3	10.13	9.55	9.28	9.12	9.01	8.94	8.84	8.74	8.64	8.53
4	7.71	6.94	6.59	6.39	6.26	6.16	6.04	5.91	5.77	5.63
5	6.61	5.79	5.41	5.19	5.05	4.95	4.82	4.68	4.53	4.36
6	5.99	5.14	4.76	4.53	4.39	4.28	4.15	4.00	3.84	3.67
7	5.59	4.74	4.35	4.12	3.97	3.87	3.73	3.57	3.41	3.23
8	5.32	4.46	4.07	3.84	3.69	3.58	3.44	3.28	3.12	2.93
9	5.12	4.26	3.86	3.63	3.48	3.37	3.23	3.07	2.90	2.71
10	4.96	4.10	3.71	3.48	3.33	3.22	3.07	2.91	2.74	2.54
11	4.84	3.98	3.59	3.36	3.20	3.09	2.95	2.79	2.61	2.40
12	4.75	3.88	3.49	3.26	3.11	3.00	2.85	2.69	2.50	2.30
13	4.67	3.80	3.41	3.18	3.02	2.92	2.77	2.60	2.42	2.21
14	4.60	3.74	3.34	3.11	2.96	2.85	2.70	2.53	2.35	2.13
15	4.54	3.68	3.29	3.06	2.90	2.79	2.64	2.48	2.29	2.07
16	4.49	3.63	3.24	3.01	2.85	2.74	2.59	2.42	2.24	2.01
17	4.45	3.59	3.20	2.96	2.81	2.70	2.55	2.38	2.19	1.96
18	4.41	3.55	3.16	2.93	2.77	2.66	2.51	2.34	2.15	1.92
19	4.38	3.52	3.13	2.90	2.74	2.63	2.48	2.31	2.11	1.88
20	4.35	3.49	3.10	2.87	2.71	2.60	2.45	2.28	2.08	1.84
21	4.32	3.47	3.07	2.84	2.68	2.57	2.42	2.25	2.05	1.81
22	4.30	3.44	3.05	2.82	2.66	2.55	2.40	2.23	2.03	1.78
23	4.28	3.42	3.03	2.80	2.64	2.53	2.38	2.20	2.00	1.76
24	4.26	3.40	3.01	2.78	2.62	2.51	2.36	2.18	1.98	1.73
25	4.24	3.38	2.99	2.76	2.60	2.49	2.34	2.16	1.96	1.71
26	4.22	3.37	2.98	2.74	2.59	2.47	2.32	2.15	1.95	1.69
27	4.21	3.35	2.96	2.73	2.57	2.46	2.30	2.13	1.93	1.67
28	4.20	3.34	2.95	2.71	2.56	2.44	2.29	2.12	1.91	1.65
29	4.18	3.33	2.93	2.70	2.54	2.43	2.28	2.10	1.90	1.64
30	4.17	3.32	2.92	2.69	2.53	2.42	2.27	2.09	1.89	1.62
40	4.08	3.23	2.84	2.61	2.45	2.34	2.18	2.00	1.79	1.51
60	4.00	3.15	2.76	2.52	2.37	2.25	2.10	1.92	1.70	1.39
120	3.92	3.07	2.68	2.45	2.29	2.17	2.02	1.83	1.61	1.25
∞	3.84	2.99	2.60	2.37	2.21	2.09	1.94	1.75	1.52	1.00

Lower 5 per cent points are found by interchange of ν_1 and ν_2, i.e. ν_1 must always correspond to the greater mean square.

Appendix Table 8 1 per cent points of the distribution of z (values at which the d.v. = 0.99)

(Reproduced from Table VI of Sir Ronald Fisher's *Statistical Methods for Research Workers*, Oliver and Boyd Ltd., Edinburgh, by kind permission of the author and publishers)

	Values of ν_1									
Values of ν_2	1	2	3	4	5	6	8	12	24	∞
1	4.1535	4.2585	4.2974	4.3175	4.3297	4.3379	4.3482	4.3585	4.3689	4.3794
2	2.2950	2.2976	2.2984	2.2988	2.2991	2.2992	2.2994	2.2997	2.2999	2.3001
3	1.7649	1.7140	1.6915	1.6786	1.6703	1.6645	1.6569	1.6489	1.6404	1.6314
4	1.5270	1.4452	1.4075	1.3856	1.3711	1.3609	1.3473	1.3327	1.3170	1.3000
5	1.3943	1.2929	1.2449	1.2164	1.1974	1.1838	1.1656	1.1457	1.1239	1.0997
6	1.3103	1.1955	1.1401	1.1068	1.0843	1.0680	1.0460	1.0218	0.9948	0.9643
7	1.2526	1.1281	1.0672	1.0300	1.0048	0.9864	0.9614	0.9335	0.9020	0.8658
8	1.2106	1.0787	1.0135	0.9734	0.9459	0.9259	0.8983	0.8673	0.8319	0.7904
9	1.1786	1.0411	0.9724	0.9299	0.9006	0.8791	0.8494	0.8157	0.7769	0.7305
10	1.1535	1.0114	0.9399	0.8954	0.8646	0.8419	0.8104	0.7744	0.7324	0.6816
11	1.1333	0.9874	0.9136	0.8674	0.8354	0.8116	0.7785	0.7405	0.6958	0.6408
12	1.1166	0.9677	0.8919	0.8443	0.8111	0.7864	0.7520	0.7122	0.6649	0.6061
13	1.1027	0.9511	0.8737	0.8248	0.7907	0.7652	0.7295	0.6882	0.6386	0.5761
14	1.0909	0.9370	0.8581	0.8082	0.7732	0.7471	0.7103	0.6675	0.6159	0.5500
15	1.0807	0.9249	0.8448	0.7939	0.7582	0.7314	0.6937	0.6496	0.5961	0.5269
16	1.0719	0.9144	0.8331	0.7814	0.7450	0.7177	0.6791	0.6339	0.5786	0.5064
17	1.0641	0.9051	0.8229	0.7705	0.7335	0.7057	0.6663	0.6199	0.5630	0.4879
18	1.0572	0.8970	0.8138	0.7607	0.7232	0.6950	0.6549	0.6075	0.5491	0.4712
19	1.0511	0.8897	0.8057	0.7521	0.7140	0.6854	0.6447	0.5964	0.5366	0.4560
20	1.0457	0.8831	0.7985	0.7443	0.7058	0.6768	0.6355	0.5864	0.5253	0.4421
21	1.0408	0.8772	0.7920	0.7372	0.6984	0.6690	0.6272	0.5773	0.5150	0.4294
22	1.0363	0.8719	0.7860	0.7309	0.6916	0.6620	0.6196	0.5691	0.5056	0.4176
23	1.0322	0.8670	0.7806	0.7251	0.6855	0.6555	0.6127	0.5615	0.4969	0.4068
24	1.0285	0.8626	0.7757	0.7197	0.6799	0.6496	0.6064	0.5545	0.4890	0.3967
25	1.0251	0.8585	0.7712	0.7148	0.6747	0.6442	0.6006	0.5481	0.4816	0.3872
26	1.0220	0.8548	0.7670	0.7103	0.6699	0.6392	0.5952	0.5422	0.4748	0.3784
27	1.0191	0.8513	0.7631	0.7062	0.6655	0.6346	0.5902	0.5367	0.4685	0.3701
28	1.0164	0.8481	0.7595	0.7023	0.6614	0.6303	0.5856	0.5316	0.4626	0.3624
29	1.0139	0.8451	0.7562	0.6987	0.6576	0.6263	0.5813	0.5269	0.4570	0.3550
30	1.0116	0.8423	0.7531	0.6954	0.6540	0.6226	0.5773	0.5224	0.4519	0.3481
60	0.9784	0.8025	0.7086	0.6472	0.6028	0.5687	0.5189	0.4574	0.3746	0.2352
∞	0.9462	0.7636	0.6651	0.5999	0.5522	0.5152	0.4604	0.3908	0.2913	0

Appendix Table 9 1 per cent points of the variance ratio *F*
(values at which the d.f. = 0.99)

(Reproduced from Sir Ronald Fisher and Dr F. Yates: *Statistical Tables for Biological, Medical and Agricultural Research*, Oliver and Boyd Ltd., Edinburgh, by kind permission of the authors and publishers)

ν_2 \ ν_1	1	2	3	4	5	6	8	12	24	∞
1	4052	4999	5403	5625	5764	5859	5981	6106	6234	6366
2	98.49	99.00	99.17	99.25	99.30	99.33	99.36	99.42	99.46	99.50
3	34.12	30.81	29.46	28.71	28.24	27.91	27.49	27.05	26.60	26.12
4	21.20	18.00	16.69	15.98	15.52	15.21	14.80	14.37	13.93	13.46
5	16.26	13.27	12.06	11.39	10.97	10.67	10.27	9.89	9.47	9.02
6	13.74	10.92	9.78	9.15	8.75	8.47	8.10	7.72	7.31	6.88
7	12.25	9.55	8.45	7.85	7.46	7.19	6.84	6.47	6.07	5.65
8	11.26	8.65	7.59	7.01	6.63	6.37	6.03	5.67	5.28	4.86
9	10.56	8.02	6.99	6.42	6.06	5.80	5.47	5.11	4.73	4.31
10	10.04	7.56	6.55	5.99	5.64	5.39	5.06	4.71	4.33	3.91
11	9.65	7.20	6.22	5.67	5.32	5.07	4.74	4.40	4.02	3.60
12	9.33	6.93	5.95	5.41	5.06	4.82	4.50	4.16	3.78	3.36
13	9.07	6.70	5.74	5.20	4.86	4.62	4.30	3.96	3.59	3.16
14	8.86	6.51	5.56	5.03	4.69	4.46	4.14	3.80	3.43	3.00
15	8.68	6.36	5.42	4.89	4.56	4.32	4.00	3.67	3.29	2.87
16	8.53	6.23	5.29	4.77	4.44	4.20	3.89	3.55	3.18	2.75
17	8.40	6.11	5.18	4.67	4.34	4.10	3.79	3.45	3.08	2.65
18	8.28	6.01	5.09	4.58	4.25	4.01	3.71	3.37	3.00	2.57
19	8.18	5.93	5.01	4.50	4.17	3.94	3.63	3.30	2.92	2.49
20	8.10	5.85	4.94	4.43	4.10	3.87	3.56	3.23	2.86	2.42
21	8.02	5.78	4.87	4.37	4.04	3.81	3.51	3.17	2.80	2.36
22	7.94	5.72	4.82	4.31	3.99	3.76	3.45	3.12	2.75	2.31
23	7.88	5.66	4.76	4.26	3.94	3.71	3.41	3.07	2.70	2.26
24	7.82	5.61	4.72	4.22	3.90	3.67	3.36	3.03	2.66	2.21
25	7.77	5.57	4.68	4.18	3.86	3.63	3.32	2.99	2.62	2.17
26	7.72	5.53	4.64	4.14	3.82	3.59	3.29	2.96	2.58	2.13
27	7.68	5.49	4.60	4.11	3.78	3.56	3.26	2.93	2.55	2.10
28	7.64	5.45	4.57	4.07	3.75	3.53	3.23	2.90	2.52	2.06
29	7.60	5.42	4.54	4.04	3.73	3.50	3.20	2.87	2.49	2.03
30	7.56	5.39	4.51	4.02	3.70	3.47	3.17	2.84	2.47	2.01
40	7.31	5.18	4.31	3.83	3.51	3.29	2.99	2.66	2.29	1.80
60	7.08	4.98	4.13	3.65	3.34	3.12	2.82	2.50	2.12	1.60
120	6.85	4.79	3.95	3.48	3.17	2.96	2.66	2.34	1.95	1.38
∞	6.64	4.60	3.78	3.32	3.02	2.80	2.51	2.18	1.79	1.00

Lower 1 per cent points are found by interchange of ν_1 and ν_2, i.e. ν_1 must always correspond to the greater mean square.

Appendix Table 10 5 per cent points for the Durbin–Watson *d*-test

	$k' = 1$		$k' = 2$		$k' = 3$		$k' = 4$		$k' = 5$	
n	d_L	d_U	d_L	d_U	d_L	d_U	d_L	d_U	d_L	d_U
15	1.08	1.36	0.95	1.54	0.82	1.75	0.69	1.97	0.56	2.21
16	1.10	1.37	0.98	1.54	0.86	1.73	0.74	1.93	0.62	2.15
17	1.13	1.38	1.02	1.54	0.90	1.71	0.78	1.90	0.67	2.10
18	1.16	1.39	1.05	1.53	0.93	1.69	0.82	1.87	0.71	2.06
19	1.18	1.40	1.08	1.53	0.97	1.68	0.86	1.85	0.75	2.02
20	1.20	1.41	1.10	1.54	1.00	1.68	0.90	1.83	0.79	1.99
21	1.22	1.42	1.13	1.54	1.03	1.67	0.93	1.81	0.83	1.96
22	1.24	1.43	1.15	1.54	1.05	1.66	0.96	1.80	0.86	1.94
23	1.26	1.44	1.17	1.54	1.08	1.66	0.99	1.79	0.90	1.92
24	1.27	1.45	1.19	1.55	1.10	1.66	1.01	1.78	0.93	1.90
25	1.29	1.45	1.21	1.55	1.12	1.66	1.04	1.77	0.95	1.89
26	1.30	1.46	1.22	1.55	1.14	1.65	1.06	1.76	0.98	1.88
27	1.32	1.47	1.24	1.56	1.16	1.65	1.08	1.76	1.01	1.86
28	1.33	1.48	1.26	1.56	1.18	1.65	1.10	1.75	1.03	1.85
29	1.34	1.48	1.27	1.56	1.20	1.65	1.12	1.74	1.05	1.84
30	1.35	1.49	1.28	1.57	1.21	1.65	1.14	1.74	1.07	1.83
31	1.36	1.50	1.30	1.57	1.23	1.65	1.16	1.74	1.09	1.83
32	1.37	1.50	1.31	1.57	1.24	1.65	1.18	1.73	1.11	1.82
33	1.38	1.51	1.32	1.58	1.26	1.65	1.19	1.73	1.13	1.81
34	1.39	1.51	1.33	1.58	1.27	1.65	1.21	1.73	1.15	1.81
35	1.40	1.52	1.34	1.58	1.28	1.65	1.22	1.73	1.16	1.80
36	1.41	1.52	1.35	1.59	1.29	1.65	1.24	1.73	1.18	1.80
37	1.42	1.53	1.36	1.59	1.31	1.66	1.25	1.72	1.19	1.80
38	1.43	1.54	1.37	1.59	1.32	1.66	1.26	1.72	1.21	1.79
39	1.43	1.54	1.38	1.60	1.33	1.66	1.27	1.72	1.22	1.79
40	1.44	1.54	1.39	1.60	1.34	1.66	1.29	1.72	1.23	1.79
45	1.48	1.57	1.43	1.62	1.38	1.67	1.34	1.72	1.29	1.78
50	1.50	1.59	1.46	1.63	1.42	1.67	1.38	1.72	1.34	1.77
55	1.53	1.60	1.49	1.64	1.45	1.68	1.41	1.72	1.38	1.77
60	1.55	1.62	1.51	1.65	1.48	1.69	1.44	1.73	1.41	1.77
65	1.57	1.63	1.54	1.66	1.50	1.70	1.47	1.73	1.44	1.77
70	1.58	1.64	1.55	1.67	1.52	1.70	1.49	1.74	1.46	1.77
75	1.60	1.65	1.57	1.68	1.54	1.71	1.51	1.74	1.49	1.77
80	1.61	1.66	1.59	1.69	1.56	1.72	1.53	1.74	1.51	1.77
85	1.62	1.67	1.60	1.70	1.57	1.72	1.55	1.75	1.52	1.77
90	1.63	1.68	1.61	1.70	1.59	1.73	1.57	1.75	1.54	1.78
95	1.64	1.69	1.62	1.71	1.60	1.73	1.58	1.75	1.56	1.78
100	1.65	1.69	1.63	1.72	1.61	1.74	1.59	1.76	1.57	1.78

k' denotes the number of regressor variables, excluding the constant.

By permission of authors and Biometrika Trust from J. Durbin and G. S. Watson (1951) Testing for Serial Correlation in Least Squares Regression, II. *Biometrika*, **38**, 159.

Appendix Table 11 1 per cent points for the Durbin–Watson *d*-test

	$k'=1$		$k'=2$		$k'=3$		$k'=4$		$k'=5$	
n	d_L	d_U	d_L	d_U	d_L	d_U	d_L	d_U	d_L	d_U
15	0.81	1.07	0.70	1.25	0.59	1.46	0.49	1.70	0.39	1.96
16	0.84	1.09	0.74	1.25	0.63	1.44	0.53	1.66	0.44	1.90
17	0.87	1.10	0.77	1.25	0.67	1.43	0.57	1.63	0.48	1.85
18	0.90	1.12	0.80	1.26	0.71	1.42	0.61	1.60	0.52	1.80
19	0.93	1.13	0.83	1.26	0.74	1.41	0.65	1.58	0.56	1.77
20	0.95	1.15	0.86	1.27	0.77	1.41	0.68	1.57	0.60	1.74
21	0.97	1.16	0.89	1.27	0.80	1.41	0.72	1.55	0.63	1.71
22	1.00	1.17	0.91	1.28	0.83	1.40	0.75	1.54	0.66	1.69
23	1.02	1.19	0.94	1.29	0.86	1.40	0.77	1.53	0.70	1.67
24	1.04	1.20	0.96	1.30	0.88	1.41	0.80	1.53	0.72	1.66
25	1.05	1.21	0.98	1.30	0.90	1.41	0.83	1.52	0.75	1.65
26	1.07	1.22	1.00	1.31	0.93	1.41	0.85	1.52	0.78	1.64
27	1.09	1.23	1.02	1.32	0.95	1.41	0.88	1.51	0.81	1.63
28	1.10	1.24	1.04	1.32	0.97	1.41	0.90	1.51	0.83	1.62
29	1.12	1.25	1.05	1.33	0.99	1.42	0.92	1.51	0.85	1.61
30	1.13	1.26	1.07	1.34	1.01	1.42	0.94	1.51	0.88	1.61
31	1.15	1.27	1.08	1.34	1.02	1.42	0.96	1.51	0.90	1.60
32	1.16	1.28	1.10	1.35	1.04	1.43	0.98	1.51	0.92	1.60
33	1.17	1.29	1.11	1.36	1.05	1.43	1.00	1.51	0.94	1.59
34	1.18	1.30	1.13	1.36	1.07	1.43	1.01	1.51	0.95	1.59
35	1.19	1.31	1.14	1.37	1.08	1.44	1.03	1.51	0.97	1.59
36	1.21	1.32	1.15	1.38	1.10	1.44	1.04	1.51	0.99	1.59
37	1.22	1.32	1.16	1.38	1.11	1.45	1.06	1.51	1.00	1.59
38	1.23	1.33	1.18	1.39	1.12	1.45	1.07	1.52	1.02	1.58
39	1.24	1.34	1.19	1.39	1.14	1.45	1.09	1.52	1.03	1.58
40	1.25	1.34	1.20	1.40	1.15	1.46	1.10	1.52	1.05	1.58
45	1.29	1.38	1.24	1.42	1.20	1.48	1.16	1.53	1.11	1.58
50	1.32	1.40	1.28	1.45	1.24	1.49	1.20	1.54	1.16	1.59
55	1.36	1.43	1.32	1.47	1.28	1.51	1.25	1.55	1.21	1.59
60	1.38	1.45	1.35	1.48	1.32	1.52	1.28	1.56	1.25	1.60
65	1.41	1.47	1.38	1.50	1.35	1.53	1.31	1.57	1.28	1.61
70	1.43	1.49	1.40	1.52	1.37	1.55	1.34	1.58	1.31	1.61
75	1.45	1.50	1.42	1.53	1.39	1.56	1.37	1.59	1.34	1.62
80	1.47	1.52	1.44	1.54	1.42	1.57	1.39	1.60	1.36	1.62
85	1.48	1.53	1.46	1.55	1.43	1.58	1.41	1.60	1.39	1.63
90	1.50	1.54	1.47	1.56	1.45	1.59	1.43	1.61	1.41	1.64
95	1.51	1.55	1.49	1.57	1.47	1.60	1.45	1.62	1.42	1.64
100	1.52	1.56	1.50	1.58	1.48	1.60	1.46	1.63	1.44	1.65

k' denotes the number of regressor variables, excluding the constant.

By permission of authors and Biometrika Trust from J. Durbin and G. S. Watson (1951) Testing for Serial Correlation in Least Squares Regression, II. *Biometrika*, **38**, 159.

REFERENCES

Note. Works by W. G. Cochran, R. A. Fisher, J. Neyman, E. S. Pearson, K. Pearson, "Student", A. Wald, S. S. Wilks and G. U. Yule that are marked with an asterisk are reproduced in the following collections:

W. G. Cochran, *Contributions to Statistics*. Wiley, New York, 1982.

R. A. Fisher, *Contributions to Mathematical Statistics*. Wiley, New York, 1950.

A Selection of Early Statistical Papers of J. Neyman. Cambridge University Press, 1967.

Joint Statistical Papers of J. Neyman and E. S. Pearson. Cambridge University Press, 1967.

The Selected Papers of E. S. Pearson. Cambridge University Press, 1966.

Karl Pearson's Early Statistical Papers. Cambridge University Press, London, 1948.

'Student's' Collected Papers. Biometrika Office, University College London, 1942.

Selected Papers in Statistics and Probability by Abraham Wald. McGraw-Hill, New York, 1955; reprinted by Stanford University Press, 1957.

S. S. Wilks: Collected Papers. Contributions to Mathematical Statistics. Wiley, New York, 1967.

Statistical Papers of George Udny Yule. Griffin, London, 1971.

AITCHISON, J. and SILVEY, S. D. (1958). Maximum-likelihood estimation of parameters subject to restraints. *Ann. Math. Statist.*, **29**, 813.

AITKEN, A. C. (1933a). On the graduation of data by the orthogonal polynomials of least squares. *Proc. Roy. Soc. Edin.*, **A**, **53**, 54.

AITKEN, A. C. (1933b). On fitting polynomials to weighted data by least squares. *Proc. Roy. Soc. Edin.*, **A**, **54**, 1.

AITKEN, A. C. (1933c). On fitting polynomials to data with weighted and correlated errors. *Proc. Roy. Soc. Edin.*, **A**, **54**, 12.

AITKEN, A. C. (1935). On least squares and linear combination of observations. *Proc. Roy. Soc. Edin.*, **A**, **55**, 42.

AITKEN, A. C. (1948). On the estimation of many statistical parameters. *Proc. Roy. Soc. Edin.*, **A**, **62**, 369.

AITKEN, A. C. and SILVERSTONE, H. (1942). On the estimation of statistical parameters. *Proc. Roy. Soc. Edin.*, **A**, **61**, 186.

AKAIKE, H. (1969). Fitting autoregressive models for prediction. *Ann. Inst. Statist. Math.*, **21**, 243.

AKAIKE, H. (1982). On the fallacy of the likelihood principle. *Statist. Prob. Lett.*, **1**, 75.

AKRITAS, M. and ARNOLD, S. (1992). Fully nonparametric hypotheses for factorial designs. I: multivariate repeated measures designs *J. Amer. Statist. Assoc.*, **89**, 336.

AKRITAS, M., ARNOLD, S. and BRUNNER, E. (1997). Nonparametric hypotheses and rank statistics for unbalanced factorial designs. *J. Amer. Statist. Assoc.*, **92**, 258.

ALALOUF, I. S. and STYLAN, G. P. H. (1979). Characterizations of estimability in the general linear model. *Ann. Statist.*, **7**, 194.

ALI, M. M. and SILVER, J. L. (1985). Tests for equality between sets of coefficients in two linear regressions under heteroscedasticity. *J. Amer. Statist. Ass.*, **80**, 730.

ALLAN, F. E. (1930). The general form of the orthogonal polynomials for simple series with proofs of their simple properties. *Proc. Roy. Soc. Edin.*, **A**, **50**, 310.

ALMON, S. (1965). The distributed lag between capital appropriations and expenditures. *Econometrica*, **30**, 178.

ALWAN, L. C. and ROBERTS, H. V. (1988). Time series modelling for statistical process control. *J. Bus. Econ. Statist.*, **6**, 87.

AMARI, S. (1982a). Geometrical theory of asymptotic ancillarity and conditional inference. *Biometrika*, **69**, 1.

AMARI, S. (1982b). Differential geometry of curved exponential families–curvatures and information loss. *Ann. Statist.*, **10**, 357.

AMEMIYA, T. (1980). The n'th order mean squared errors of the maximum likelihood and the minimum logit chi-squared estimators. *Ann. Statist.*, **8**, 488. *Correction*, **12**, 783.

AMOS, D. E. (1964). Representations of the central and non-central t distributions. *Biometrika*, **51**, 451.

ANDERSEN, E. B. (1970a). Sufficiency and exponential families for discrete sample spaces. *J. Amer. Statist. Ass.*, **65**, 1248.

ANDERSEN, E. B. (1970b). Asymptotic properties of conditional maximum likelihood estimators. *J. Roy. Statist. Soc.*, **B**, **32**, 283.

ANDERSON, R. L. and ANDERSON, T. W. (1956). Distribution of the circular serial correlation coefficient for residuals from a fitted Fourier series. *Ann. Math. Statist.*, **21**, 59.

ANDERSON, T. W. (1958). *An Introduction to Multivariate Statistical Analysis*. Wiley, New York.

ANDERSON, T. W. (1960). A modification of the sequential probability ratio test to reduce sample size. *Ann. Math. Statist.*, **31**, 165.

ANDERSON, T. W. (1962a). The choice of the degree of a polynomial regression as a multiple decision problem. *Ann. Math. Statist .*, **33**, 255.

ANDERSON, T. W. (1962b). Least squares and best unbiased estimates. *Ann. Math. Statist.*, **33**, 266.

ANDERSON, T. W. and DARLING, D. A. (1952). Asymptotic theory of certain goodness-of-fit criteria based on stochastic processes. *Ann. Math. Statist.*, **23**, 193.

ANDREWS, D. F. (1971). A note on the selection of data transformations. *Biometrika*, **58**, 249.

ANDREWS, D. W. K. (1991). Heteroscedastic and autocorrelation consistent covariance matrix estimation. *Econometrica*, **59**, 817.

ANDREWS, D. W. K. and MONAHAN, J. C. (1992). An improved heteroscedasticity and autocorrelation consistent covariance matrix estimator. *Econometrica*, **60**, 953.

ANSCOMBE, F. J. (1948). The transformation of Poisson, binomial and negative-binomial data. *Biometrika*, **35**, 246.

ANSCOMBE, F. J. (1949a). Tables of sequential inspection schemes to control fraction defective. *J. Roy. Statist. Soc.*, **A**, **112**, 180.

ANSCOMBE, F. J. (1949b). Large-sample theory of sequential estimation. *Biometrika*, **36**, 455.

ANSCOMBE, F. J. (1950). Sampling theory of the negative binomial and logarithmic series distributions. *Biometrika*, **37**, 358.

ANSCOMBE, F. J. (1952). Large-sample theory of sequential estimation. *Proc. Camb. Phil. Soc.*, **48**, 600.

ANSCOMBE, F. J. (1953). Sequential estimation. *J. Roy. Statist. Soc.*, **B**, **15**, 1.

ANSCOMBE, F. J. and PAGE, E. S. (1954). Sequential tests for binomial and exponential populations. *Biometrika*, **41**, 252.

ANTLE, C., KLIMKO, L. and HARKNESS, W. (1970). Confidence intervals for the parameters of the logistic distribution. *Biometrika*, **57**, 397.

ARMITAGE, P. (1947). Some sequential tests of Student's hypothesis. *J. Roy. Statist. Soc.*, **B**, **9**, 250.

ARNOLD, S. F. (1979). Linear models with exchangeably distributed errors. *J. Amer. Statist. Ass.*, **74**, 194.

ARNOLD, S. F. (1980). The asymptotic validity of invariant procedures for the repeated measures model and multivariate linear model. *J. Multivariate Anal.*, **15**, 325.

ARNOLD, S. F. (1981). *The Theory of Linear Models and Multivariate Analysis.* Wiley, New York.

ARNOLD, S. F. (1984). Pivotal quantities and invariant confidence regions. *Statist. Decisions*, **2**, 257.

ARNOLD, S. F. (1985). Sufficiency and invariance. *Statist. Prob. Lett.*, **3**, 275.

ARNOLD, S. F. (1990). *Mathematical Statistics.* Prentice Hall, Englewood Cliffs, NJ.

ARVESEN, J. N. and LAYARD, M. W. J. (1975). Asymptotically robust tests in unbalanced variance component models. *Ann. Statist.*, **3**, 1122.

ASKOVITZ, S. I. (1957). A short-cut graphic method for fitting the best straight line to a series of points according to the criterion of least squares. *J. Amer. Statist. Ass.*, **52**, 13.

ASPIN, A. A. (1948). An examination and further development of a formula arising in the problem of comparing two mean values. *Biometrika*, **35**, 88.

ASPIN, A. A. (1949). Tables for use in comparisons whose accuracy involves two variances, separately estimated. *Biometrika*, **36**, 290.

ATKINSON, A. C. (1973). Testing transformations to normality. *J. Roy. Statist. Soc.*, **B**, **35**, 473.

ATKINSON, A. C. (1981). Two graphical displays for outlying and influential observations in regression. *Biometrika*, **68**, 13.

ATKINSON, A. C. (1983). Diagnostic regression analysis and shifted power transformations. *Technometrics*, **25**, 23.

ATKINSON, A. C. (1985). *Plots, Transformations and Regression.* Oxford University Press, Oxford.

ATKINSON, A. C. (1986a). Diagnostic tests for transformations. *Technometrics*, **28**, 29.

ATKINSON, A. C. (1986b). Masking unmasked. *Biometrika*, **73**, 533.

BAGLIVO, J., OLIVER, D. and PAGANO, M. (1992). Methods for exact goodness of fit tests. *J. Amer. Statist. Ass.*, **87**, 464

BAHADUR, R. R. (1964). On Fisher's bound for asymptotic variances. *Ann. Math. Statist.*, **35**, 1545.

BAHADUR, R. R. (1967). Rates of convergence of estimates and test statistics. *Ann. Math. Statist.*, **38**, 303.

BAIN, L. J. (1967). Reducing a random sample to a smaller set, with applications. *J. Amer. Statist. Ass.*, **62**, 510.

BAIN, L. J. (1969). The moments of a noncentral t and noncentral F-distribution. *Amer. Statist.*, **23 (4)**, 33.

BAKSALARY, J. K. and KALA, R. (1983). Estimation via linearly combining two given statistics. *Ann. Statist.*, **11**, 691.

BARANCHIK, A. J. (1970). A family of minimax estimators of the mean of a multivariate normal distribution. *Ann. Math. Statist.*, **41**, 642.

BARANKIN, E. W. (1949). Locally best unbiased estimates. *Ann. Math. Statist.*, **20**, 477.

BARANKIN, E. W. and KATZ, M., Jr. (1959). Sufficient statistics of minimal dimension. *Sankhyā*, **21**, 217.

BARANKIN, E. W. and MAITRA, A. P. (1963). Generalization of the Fisher–Darmois–Koopman–Pitman theorem on sufficient statistics. *Sankhyā*, **A**, **25**, 217.

BARLOW, R. E., BARTHOLOMEW, D. J., BREMNER, J. M. and BRUNK, H. D. (1972). *Statistical Inference under Order Restrictions.* Wiley, New York.

BARNARD, G. A. (1946). Sequential tests in industrial statistics. *Suppl. J. Roy. Statist. Soc.*, **8**, 1.

BARNARD, G. A. (1950). On the Fisher–Behrens test. *Biometrika*, **34**, 168.

BARNARD, G. A. (1990). Must clinical trials be large? The interpretation of p-values and the combination of test results. *Statist. Medicine*, **9**, 601.

BARNARD, G. A. (1995). Pivotal models and the fiducial argument. *Int. Statist. Rev.*, **63**, 309.

BARNARD, G. A. and GODAMBE, V. P.. (1982). Allan Birnbaum, A memorial article. *Ann. Statist.*, **10**, 1033.

BARNARD, G. A. and SPROTT, D. A. (1971). A note on Basu's examples of anomalous ancillary statistics (with discussion). In *Foundations of Statistical Inference*, V. P. Godambe and D. A. Sprott (eds). Holt Rinehart and Winston, Toronto, 163.

BARNDORFF-NIELSEN, O. (1980). Conditionality resolutions. *Biometrika*, **67**, 293.

BARNDORFF-NIELSEN, O. (1983). On a formula for the distribution of the maximum likelihood estimator. *Biometrika*, **70**, 343.

BARNDORFF-NIELSEN, O. E. and COX, D. R. (1984). Bartlett adjustment to the likelihood ratio statistic and the distribution of the maximum likelihood estimator. *J. Roy. Statist. Soc.*, **B**, **46**, 483.

BARNDORFF-NIELSEN, O. E. and HALL, P. (1988). On the level-error after Bartlett adjustment of the likelihood ratio statistic. *Biometrika*, **75**, 374.

BARNETT, V. (1982). *Comparative Statistical Inference* (2nd edn). Wiley, Chichester.

BARNETT, V. D. (1966). Evaluation of the maximum-likelihood estimator where the likelihood equation has multiple roots. *Biometrika*, **53**, 151.

BARR, D. R. (1966). On testing the equality of uniform and related distributions. *J. Amer. Statist. Ass.*, **61**, 856.

BARTHOLOMEW, D. J. (1967). Hypothesis testing when the sample size is treated as a random variable. *J. Roy. Statist. Soc.*, **B**, **29**, 53.

BARTLETT, M. S. (1935). The effect of non-normality on the *t*-distribution. *Proc. Camb. Phil. Soc.*, **31**, 223.

BARTLETT, M. S. (1936a). The square root transformation in analysis of variance. *Suppl. J. Roy. Statist. Soc.*, **3**, 68.

BARTLETT, M. S. (1936b). The information available in small samples. *Proc. Camb. Phil. Soc.*, **32**, 560.

BARTLETT, M. S. (1937). Properties of sufficiency and statistical tests. *Proc. Roy. Soc.*, **A**, **160**, 268.

BARTLETT, M. S. (1939). A note on the interpretation of quasi-sufficiency. *Biometrika*, **31**, 391.

BARTLETT, M. S. (1946). The large-sample theory of sequential tests. *Proc. Camb. Phil. Soc.*, **42**, 239.

BARTLETT, M. S. (1947). The use of transformations. *Biometrics*, **3**, 39.

BARTLETT, M. S. (1949). Fitting a straight line when both variables are subject to error. *Biometrics*, **5**, 207.

BARTLETT, M. S. (1951). An inverse matrix adjustment arising in discriminant analysis. *Ann. Math. Statist.*, **22**, 107.

BARTLETT, M. S. (1953a). Approximate confidence intervals. *Biometrika*, **40**, 12.

BARTLETT, M. S. (1953b). Approximate confidence intervals II: More than one unknown parameter. *Biometrika*, **40**, 306.

BARTLETT, M. S. (1955). Approximate confidence intervals III: A bias correction. *Biometrika*, **42**, 201.

BARTLETT, M. S. and KENDALL, D. G. (1946). The statistical analysis of variance-heterogeneity and the logarithmic transformation. *Suppl. J. Roy. Statist. Soc.*, **8**, 128.

BASU, A. P. and GHOSH, J. K. (1980). Asymptotic properties of a solution to the likelihood equation with life-testing applications. *J. Amer. Statist. Ass.*, **75**, 410.

BASU, D. (1955, 1958). On statistics independent of a complete sufficient statistic. *Sankhyā*, **15**, 377. *Correction* **20**, 223.

BASU, D. (1964). Recovery of ancillary information. *Sankhyā*, **A**, **26**, 3.

BASU, D. (1975). Statistical information and likelihood (with discussion). *Sankhyā*, **A**, **37**, 1.

BASU, D. (1977). On the elimination of nuisance parameters. *J. Amer. Statist. Ass.*, **72**, 355.

BATEMAN, G. I. (1949). The characteristic function of a weighted sum of non-central squares of normal variables subject to s linear restraints. *Biometrika*, **36**, 460.

BATES, D. M. and WATTS, D. G. (1980). Relative curvature measures of nonlinearity. *J. Roy. Statist. Soc.*, **B**, **42**, 1.

BATES, D. M. and WATTS, D. G. (1981). Parameter transformations for improved approximate confidence regions in non-linear least squares. *Ann. Statist.*, **9**, 1152.

BAYES, T. (1764). An essay towards solving a problem in the doctrine of chances. *Phil. Trans.*, **53**, 370. (Reprinted in *Biometrika*, **45**, 293 (1958), edited and introduced by G. A. Barnard.)

BEALE, E. M. L. (1960). Confidence regions in non-linear estimation. *J. Roy. Statist. Soc.*, **B**, **22**, 41.

BEALE, E. M. L. and LITTLE, R. J. A. (1975). Missing values in multivariate analysis. *J. Roy. Statist. Soc.*, **B**, **37**, 129.

BEALE, E. M. L., KENDALL, M. G. and MANN, D. W. (1967). The discarding of variables in multivariate analysis. *Biometrika* , **54**, 357.

BECHHOFER, R. (1960). A note on the limiting relative efficiency of the Wald sequential probability ratio test. *J. Amer. Statist. Ass.*, **55**, 660.

BECKER, N. and GORDON, I. (1983). On Cox's criterion for discriminating between alternative ancillary statistics. *Int. Statist. Rev.*, **51**, 89.

BECKMAN, R. J. and COOK, R. D. (1983). Outliers (with discussion). *Technometrics*, **25**, 119.

BEDRICK, E. J. (1990). On the large sample distributions of modified sample biserial correlation coefficients. *Psychometrika*, **55**, 217.

BEDRICK, E. J. (1992). A comparison of generalised and modified sample biserial correlation estimators. *Psychometrika*, **57**, 183.

BEHRENS, W. V. (1929). Ein Beitrag zur Fehlerberechnung bei wenigen Beobachtungen. *Landswirtsch. Jb.*, **68**, 807.

BELSLEY, D. A. (1991). *Conditioning Diagnostics: Collinearity and Weak Data in Regression.* Wiley, New York.

BELSLEY, D. A., KUH, E. and WELSH, R. E. (1980). *Regression Diagnostics.* Wiley, New York.

BEMENT, T. R. and WILLIAMS, J. S. (1969). Variance of weighted regression estimators when sampling errors are independent and heteroscedastic. *J. Amer. Statist. Ass.*, **64**, 1369.

BEMIS, K. G. and BHAPKAR, V. P. (1983). On BAN estimating for chi squared test criteria. *Ann. Statist.*, **11**, 183.

BERGER, J. O. (1985a). The frequentist viewpoint and conditioning. In *Proceedings of the Berkeley Conference in Honor of J. Kiefer and J. Neyman*, L. LeCam and R. Olshen (eds). Wadsworth, Belmont, CA, 15.

BERGER, J. O. (1985b). *Statistical Decision Theory and Bayesian Analysis* (2nd edn). Springer-Verlag, New York.

BERGER, J. O. and MORTERA, J. (1991). Interpreting the stars in precise hypothesis testing. *Int. Statist. Rev.*, **59**, 337.

BERGER, J. O. and WOLPERT, R. L. (1985). *The Likelihood Principle* (2nd edn). Institute of Mathematical Statistics, Hayward, CA.

BERK, R. H. (1975). Comparing sequential and non-sequential tests. *Ann. Statist.*, **3**, 991.

BERK, R. H. (1976). Asymptotic efficiencies of sequential tests. *Ann. Statist.*, **4**, 891.

BERK, R. H. (1978). Asymptotic efficiencies of sequential tests II. *Ann. Statist.*, **6**, 813.

BERK, R. H. and BROWN, L. D. (1978). Sequential Bahadar efficiency. *Ann. Statist.*, **6**, 567.

BERKSON, J. (1938). Some difficulties of interpretation encountered in the application of the chi-square test. *J. Amer. Statist. Ass.*, **33**, 526.

BERKSON, J. (1955). Maximum likelihood and minimum χ^2 estimates of the logistic function. *J. Amer. Statist. Ass.*, **50**, 130.

BERKSON, J. (1956). Estimation of least squares and by maximum likelihood. In *Proceedings of the Third Berkeley Symposium on Mathematical Statistics and Probability*, J. Neyman (ed.). University of California Press, Berkely, **1**, 1.

BERNARDO, J. M. (1979). Reference posterior distributions for Bayesian inference (with discussion). *J. Roy. Statist. Soc.*, **B**, **41**, 113.

BERNARDO, J. M. (1980). A Bayesian analysis of classical hypothesis testing. In *Bayesian Statistics: Proceedings of the First International Meeting*. Valencia University Press, Valencia, Spain.

BERNSTEIN, S. (1928). Fondements géométriques de la théorie des corrélations. *Metron*, **7**(2), 3.

BESAG, J. E. (1975). Statistical analysis of non-lattice data (with discussion). *J. Roy. Statist. Soc.*, **B**, **36**, 192.

BEST, D. J. (1974). The variance of the inverse binomial estimator. *Biometrika*, **61**, 385.

BHATTACHARJEE, G. P. (1965). Effect of non-normality on Stein's two sample test. *Ann. Math. Statist.*, **36**, 651.

BHATTACHARYYA, A. (1943). On some sets of sufficient conditions leading to the normal bivariate distribution. *Sankhyā*, **6**, 399.

BHATTACHARYYA, A. (1946–7–8). On some analogues of the amount of information and their use in statistical estimation. *Sankhyā*, **8**, 1, 201, 315.

BICKEL, P. J. and DOKSUM, K. A. (1981). An analysis of transformations revisited. *J. Amer. Statist. Ass.*, **76**, 296.

BILLARD, L. (1972). Properties of some two-sided sequential tests for the normal distribution. *J. Roy. Statist. Soc.*, **B**, **34**, 417.

BILLINGSLEY, P. (1961). *Statistical Inference for Markov Processes*. University of Chicago Press, Chicago.

BILLINGSLEY, P. (1979). *Probability and measure*. Wiley, New York.

BINKLEY, J. K. and NELSON, C. H. (1988). A note on the efficiency of seemingly unrelated regression. *Amer. Statist.*, **42**, 137.

BINNS, M. (1975). Sequential estimation of the mean of a negative binomial distribution. *Biometrika*, **62**, 433.

BIRCH, M. W. (1964). A new proof of the Pearson–Fisher theorem. *Ann. Math. Statist.*, **35**, 817.

BIRKES, D. (1990). Generalised likelihood ratio tests and uniformly most powerful tests. *Amer. Statist.*, **44**, 163.

BIRNBAUM, A. (1962). On the foundations of statistical inference. *J. Amer. Statist. Ass.*, **57**, 269.

BIRNBAUM, A. (1970). On Durbin's modified principle of conditionality. *J. Amer. Statist. Ass.*, **65**, 402.

BIRNBAUM, A. (1972). More on concepts of statistical evidence. *J. Amer. Statist. Ass.*, **67**, 858.

BIRNBAUM, A. (1977). The Neyman–Pearson theory as decision theory, and as inference theory; with a criticism of the Lindley–Savage argument for Bayesian theory. *Synthèse*, **36**, 19.

BIRNBAUM, Z. W. (1952). Numerical tabulation of the distribution of Kolmogorov's statistic for finite sample size. *J. Amer. Statist. Ass.*, **47**, 425.

BIRNBAUM, Z. W. and TINGEY, F. H. (1951). One-sided confidence contours for probability distribution functions. *Ann. Math. Statist.*, **22**, 592.

BJØRNSTAD, J. F. (1996). On the generalization of the likelihood function and the likelihood principle. *J. Amer. Statist. Ass.*, **91**, 791.

BLACKWELL, D. (1946). On an equation of Wald. *Ann. Math. Statist.*, **17**, 84.

BLACKWELL, D. (1947). Conditional expectation and unbiased sequential estimation. *Ann. Math. Statist.*, **18**, 105.

BLACKWELL, D. and GIRSHICK, M. A. (1954). *Theory of Games and Statistical Decisions*. Wiley, New York.

BLISCHKE, W. R., TRUELOVE, A. J. and MUNDLE, P. B. (1969). On non-regular estimation. I. Variance bounds for estimators of location parameters. *J. Amer. Statist. Ass.*, **64**, 1056.

BLOCH, D. A. (1966). A note on the estimation of the location parameter of the Cauchy distribution. *J. Amer. Statist. Ass.*, **61**, 852.

BLOM, G. (1954). Transformations of the binomial, negative binomial, Poisson and χ^2 distributions. *Biometrika*, **41**, 302.

BLOM, G. (1978). A property of minimum variance estimates. *Biometrika*, **65**, 642.

BLOMQVIST, N. (1950). On a measure of dependence between two random variables. *Ann. Math. Statist.*, **21**, 593.

BLOOMFIELD, P. and WATSON, G. W. (1975). The inefficiency of least squares. *Biometrika*, **62**, 121.

BLYTH, C. R. (1986). Approximate binomial confidence limits. *J. Amer. Statist. Ass.*, **81**, 843.

BLYTH, C. R. and HUTCHINSON, D. W. (1960). Table of Neyman-shortest unbiased confidence intervals for the binomial parameter. *Biometrika*, **47**, 381.

BLYTH, C. R. and HUTCHINSON, D. W. (1961). Table of Neyman-shortest unbiased confidence intervals for the Poisson parameter. *Biometrika*, **48**, 191.

BLYTH, C. R. and STILL, H. A. (1983). Binomial confidence intervals. *J. Amer. Statist. Ass.*, **78**, 108.

BONDAR, J. V. and MILNES, P. (1981). Amenability: a survey of statistical applications of Hunt–Stein and related conditions on groups. *Zeitschr. Wahrsch. Verw. Geb.*, **57**, 103.

BONDESSON, L. (1975). Uniformly minimum variance estimation in location parameter families. *Ann. Statist.*, **3**, 637.

BOOTH, J. G. and HALL, P. (1993). An improvement of the jackknife distribution function estimator. *Ann. Statist.*, **21**, 1476.

BORGAN, O. (1979). Comparison of two sequential tests for two-sided alternatives. *J. Roy. Statist. Soc.*, **B**, **41**, 101.

BOWKER, A. H. (1946). Computation of factors for tolerance limits on a normal distribution when the sample is large. *Ann. Math. Statist.*, **17**, 238.

BOWKER, A. H. (1947). Tolerance limits for normal distributions. In *Selected Techniques of Statistical Analysis*. McGraw-Hill, New York.

BOWMAN, K. O. (1972). Tables of the sample size requirement. *Biometrika*, **59**, 234.

BOWMAN, K. O. and SHENTON, L. R. (1975). Omnibus test contours for departures from normality based on $\sqrt{b_1}$ and b_2. *Biometrika*, **62**, 243.

BOWMAN, K. O. and SHENTON, L. R. (1988). *Properties of Estimators for the Gamma Distribution*. Dekker, New York.

BOX, G. E. P. (1949). A general distribution theory for a class of likelihood criteria. *Biometrika*, **36**, 317.

BOX, G. E. P. (1950). Problems in the anlysis of growth and wear curves. *Biometrics*, **6**, 362-389.

BOX, G. E. P. (1980). Sampling and Bayes inference in scientific modelling and robustness (with discussion). *J. Roy. Statist. Soc.*, **A**, **143**, 383.

BOX, G. E. P. and COX, D. R. (1964). An analysis of transformations. *J. Roy. Statist. Soc.*, **B**, **26**, 211.

BOX, G. E. P. and TIAO, G. C. (1973). *Bayesian Inference in Statistical Analysis*. Addison-Wesley, Reading, MA.

BOX, G. E. P. and TIDWELL, P. W. (1962). Transformation of the independent variables. *Technometrics*, **4**, 531.

BOX, G. E. P., JENKINS, G. M. and REINSEL, G. C. (1994). *Time Series Analysis: Forecasting and Control* (3rd edn). Prentice Hall, Englewood Cliffs, NJ.

BOX, M. J. (1971). Bias in non-linear estimation. *J. Roy. Statist. Soc.*, **B**, **33**, 171.

BOYLES, R. A. (1983). On the convergence of the EM algorithm. *J. Roy. Statist. Soc.*, **B**, **45**, 47.

BRANDNER, F. A. (1933). A test of the significance of the difference of the correlation coefficients in normal bivariate samples. *Biometrika*, **25**, 102.

BREIMAN, L. and FREEDMAN, D. (1983). How many variables should be entered in a regression? *J. Amer. Statist. Ass.*, **78**, 131.

BRETH, M. (1982). Nonparametric estimation for a symmetric distribution. *Biometrika*, **69**, 625.

BRILLINGER, D. R. (1962). Examples bearing on the definition of fiducial probability with a bibliography. *Ann. Math. Statist.*, **33** , 1349.

BRILLINGER, D. R. (1964). The asymptotic behaviour of Tukey's general method of setting approximate confidence limits (the jackknife) when applied to maximum likelihood estimates. *Rev. Int. Statist. Inst.*, **32**, 202.

BROERSON, P. M. T. (1986). Subset regression with stepwise directed search. *Appl. Statist.*, **35**, 168.

BROFITT, J. D. and RANDLES, R. H. (1977). A power approximation for the chi-square goodness-of-fit test: simple hypothesis case. *J. Amer. Statist. Ass.*, **72**, 604.

BROWN, L. D. (1964). Sufficient statistics in the case of independent random variables. *Ann. Math. Statist.*, **35**, 1456.

BROWN, L. D. (1966). On the admissibility of estimators of one or more location parameters, *Ann. Math. Stat.*, **37**, 1087.

BROWN, L. D. (1971). Non-local asymptotic optimality of appropriate likelihood ratio tests. *Ann. Math. Statist.*, **42**, 1206.

BROWN, L. D., (1990). An ancillarity paradox which appears in multiple regression (with discussion). *Ann. Statist.*, **18**, 471.

BROWN, M. B. (1977). Algorithm AS116: The tetrachoric correlation and its asymptotic standard error. *Appl. Statist.*, **26**, 343.

BROWN, P. J. (1982). Multivariate calibration. *J. Roy. Statist. Soc.*, **B**, **44**, 287.

BROWN, R. L., DURBIN, J. and EVANS, J. M. (1975). Techniques for testing the constancy of regression relationships over time. *J. Roy. Statist. Soc.*, **B**, **37**, 149.

BULGREN, W. G. (1971). On representations of the doubly non-central F distribution. *J. Amer. Statist. Ass.*, **66**, 184.

BULGREN, W. G. and AMOS, D. E. (1968). A note on representations of the doubly non-central t distribution. *J. Amer. Statist. Ass.*, **63**, 1013.

BURMAN, J. P. (1946). Sequential sampling formulae for binomial population. *J. Roy. Statist. Soc.*, **B**, **8**, 98.

BUTLER, R. W. (1984). The significance attained by the best-fitting regressor variable. *J. Amer. Statist. Ass.*, **79**, 341.

CARNAP, R. (1962). *Logical Foundations of Probability* (2nd edn). University of Chicago Press, Chicago.

CARROLL, R. J. (1982). Prediction and power transformations when the choice of power is restricted to a finite set. *J. Amer. Statist. Ass.*, **77**, 908.

CARROLL, R. J. and RUPPERT, D. (1982). A comparison between maximum likelihood and generalized least squares in a heteroscedastic linear model. *J. Amer. Statist. Ass.*, **77**, 878.

CARROLL, R. J. and RUPPERT, D. (1985). Transformations: a robust analysis. *Technometrics*, **27**, 1.

CARROLL, R. J. and RUPPERT, D. (1987). Diagnostics and robust estimation when transforming the regression model and the response. *Technometrics*, **29**, 287.

CARROLL, R. J. and RUPPERT, D. (1988). *Transformation and Weighting in Regression*. Chapman & Hall, London and New York.

CHAMBERS, J. R. (1973). Fitting nonlinear models: numerical techniques. *Biometrika*, **60**, 1.

CHANDLER, K. N. (1950). On a theorem concerning the secondary subscripts of deviations in multivariate correlation using Yule's notation. *Biometrika*, **37**, 451.

CHANDRA, M., SINGPURWALLA, N. D. and STEPHENS, M. A. (1981). Kolmogorov statistics for tests of fit for the extreme-value and Weibull distributions. *J. Amer. Statist. Ass.*, **76**, 729.

CHANT, D. (1974). On asymptotic tests of composite hypotheses in non-standard conditions. *Biometrika*, **61**, 291.

CHAPMAN, D. G. (1950). Some two sample tests. *Ann. Math. Statist.*, **21**, 601.

CHAPMAN, D. G. and ROBBINS, H. (1951). Minimum variance estimation without regularity assumptions. *Ann. Math. Statist.*, **22**, 581.

CHAPMAN, J.-A. W. (1976). A comparison of the X^2, $-2 \log R$ and multinomial probability criteria for significance tests when expected frequencies are small. *J. Amer. Statist. Ass.*, **71**, 854.

CHARNES, A., FROME, E. L. and YU, P. L. (1976). The equivalence of generalized least squares and maximum likelihood estimates in the exponential family. *J. Amer. Statist. Ass.*, **71**, 169.

CHASE, G. R. (1972). On the chi-square test when the parameters are estimated independently of the sample. *J. Amer. Statist. Ass.*, **67**, 609.

CHÂTILLON, G. (1984). The balloon rules for a rough estimate of the correlation coefficient. *Amer. Statist.*, **38**, 58.

CHATTAMVELLI, R. and SHANMUGAN, R. (1995). Efficient computation for the noncentral χ^2 distribution. *Commun. Statist.*, **B**, **24**, 675.

CHATTERJEE, S. K. (1991). Two-stage and multi-stage procedures. In *Handbook of Sequential Analysis*, B. K. Ghosh and P. K. Sen (eds). Dekker, New York, 21.

CHEN, J.-S. and JENNRICH, R. I. (1995a). Diagnostics for linearization confidence intervals in nonlinear regression. *J. Amer. Statist. Ass.*, **90**, 1068.

CHEN, J.-S. and JENNRICH, R. I. (1995b). Transformations for improving linearization confidence intervals in nonlinear regression. *J. Amer. Statist. Ass.*, **90**, 1271.

CHENG, R. C. H. and AMIN, N. A. K. (1983). Estimating parameters in continuous univariate distributions with a shifted origin. *J. Roy. Statist. Soc.*, **B**, **45**, 394.

CHERNOFF, H. (1949). Asymptotic studentisation in testing of hypotheses. *Ann. Math. Statist.*, **20**, 268.

CHERNOFF, H. (1951). A property of some Type A regions. *Ann. Math. Statist.*, **22**, 472.

CHERNOFF, H. (1952). A measure of asymptotic efficiency for tests of a hypothesis based on the sum of observations. *Ann. Math. Statist.*, **23**, 493.

CHERNOFF, H. (1954). On the distribution of the likelihood ratio. *Ann. Math. Statist.*, **25**, 573.

CHERNOFF, H. and LEHMANN, E. L. (1954). The use of maximum likelihood estimates in χ^2 tests for goodness of fit. *Ann. Math. Statist.*, **25**, 579.

CHEW, V. (1970). Covariance matrix estimation in linear models. *J. Amer. Statist. Ass.*, **65**, 173.

CHHIKARA, R. S. and FOLKS, J. L. (1976). Optimum test procedures for the mean of first passage time distribution in Brownian motion with positive drift (inverse Gaussian distribution). *Technometrics*, **18**, 189.

CHIBISOV, D. M. (1971). Certain chi-square type tests for continuous distributions. *Theory Prob. Applic.*, **16**, 1.

CHIPMAN, J. S. (1964). On least squares with insufficient observations. *J. Amer. Statist. Ass.*, **59**, 1078.

CHOI, S. C. (1977). Tests of equality of dependent correlation coefficients. *Biometrika*, **64**, 645.

CHOI, S. C. and WETTE, R. (1969). Maximum likelihood estimation of the parameters of the gamma distribution and their bias. *Technometrics*, **11**, 683.

CHOW, G. C. (1960). Test of equality between sets of coefficients in two linear regressions. *Econometrica*, **28**, 591.

CHOW, Y. S. and ROBBINS, H. (1965). On the asymptotic theory of fixed-width sequential confidence intervals for the mean. *Ann. Math. Statist.*, **36**, 457.

CLARK, R. E. (1953). Percentage points of the incomplete beta function. *J. Amer. Statist. Ass.*, **48**, 831.

CLARKE, G. P. Y. (1987a). Approximate confidence limits for a parameter function in non-linear regression. *J. Amer. Statist. Ass.*, **82**, 221.

CLARKE, G. P. Y. (1987b). Marginal curvatures and their usefulness in the analysis of nonlinear regression models. *J. Amer. Statist. Ass.*, **82**, 844.

CLEVELAND, W. S. and DEVLIN, S. J. (1988). Locally weighted regression: An approach to regression analysis by local fitting. *J. Amer. Statist. Ass.*, **83**, 596.

CLIFF, A. D. and ORD, J. K. (1981). *Spatial Processes: Models and Applications*. Pion, London.

CLOPPER, C. J. and PEARSON, E. S. (1934).* The use of confidence or fiducial limits illustrated in the case of the binomial. *Biometrika*, **26**, 404.

COCHRAN, W. G. (1937).* The efficiencies of the binomial series tests of significance of a mean and of a correlation coefficient. *J. Roy. Statist. Soc.*, **100**, 69.

COCHRAN, W. G. (1952).* The χ^2 test of goodness of fit. *Ann. Math. Statist.*, **23**, 315.

COCHRAN, W. G. (1954).* Some methods for strengthening the common χ^2 tests. *Biometrics*, **10**, 417.

COCHRAN, W. G. and COX, D. R. (1957). *Experimental Designs*. Wiley, New York.

COCHRANE, D. and ORCUTT, G. H. (1949). Application of least-squares regression to relationships containing auto-correlated error terms. *J. Amer. Statist. Ass.*, **44**, 32.

COHEN, A. (1972). Improved confidence intervals for the variance of a normal distribution. *J. Amer. Statist. Ass.*, **67**, 382.

COHEN, A. and SACKROWITZ, H. B. (1975). Unbiasedness of the chi-square, likelihood ratio, and other goodness-of-fit tests for the equal cell case. *Ann. Statist.*, **3**, 959.

COHEN, A. and STRAWDERMAN, W. E. (1971). Unbiasedness of tests for homogeneity of variances. *Ann. Math. Statist.*, **42**, 355.

COHEN, J. D. (1988). Noncentral chi-square: some observations on recurrence. *Amer. Statist.*, **42**, 120.

CONOVER, W. J. (1972). A Kolmogorov goodness-of-fit test for discontinuous distributions. *J. Amer. Statist. Ass.*, **67**, 591.

COOK, R. D. (1977). Detection of influential observations in linear regression. *Technometrics*, **19**, 15.

COOK, R. D. (1986). Assessment of local influence. *J. Roy. Statist. Soc.*, **B**, **48**, 133.

COOK, R. D. (1996). Added variable plots and curvature in linear regression. *Technometrics*, **38**, 275.

COOK, R. D. and WANG, P. C. (1983). Transformations and influential cases in regression. *Technometrics*, **25**, 337.

COOK, R. D. and WEISBERG, S. (1982). *Residuals and Influence in Regression*. Chapman & Hall, London and New York.

COOK, R. D. and WEISBERG, S. (1983). Diagnostics for heteroscedasticity in regression. *Biometrika*, **70**, 1.

COOK, R. D. and WITMER, J. A. (1985). A note on parameter-effects curvature. *J. Amer. Statist. Ass.*, **80**, 872.

COOK, R. D., TSAI, C. L. and WEI, B. C. (1986). Bias in non-linear regression. *Biometrika*, **73**, 615.

COPAS, J. B. (1975). On the unimodality of the likelihood for the Cauchy distribution. *Biometrika*, **62**, 701.

COPAS, J. B. (1983). Regression, prediction and shrinkage. *J. Roy. Statist. Soc.*, **B**, **45**, 311.

CORDEIRO, G. M. (1983). Improved likelihood ratio statistics for generalized linear models. *J. Roy. Statist. Soc.*, **B**, **45**, 404.

CORDEIRO, G. M. (1987). On the corrections to the likelihood ratio statistic. *Biometrika*, **74**, 265.

COWLES, M. K. and CARLIN, B. P. (1996). Markov chain Monte Carlo convergence diagnostics: A comparative review. *J. Amer. Statist. Ass.*, **91**, 883.

COX, C. P. (1958). A concise derivation of general orthogonal polynomials. *J. Roy. Statist. Soc.*, **B**, **20**, 406.

COX, D. R. (1952a). Sequential tests for composite hypotheses. *Proc. Camb. Phil. Soc.*, **48**, 290.

COX, D. R. (1952b). A note on the sequential estimation of means. *Proc. Camb. Phil. Soc.*, **48**, 447.

COX, D. R. (1952c). Estimation by double sampling. *Biometrika*, **39**, 217.

COX, D. R. (1956). A note on the theory of quick tests. *Biometrika*, **43**, 478.

COX, D. R. (1958a). Some problems connected with statistical inference. *Ann. Math. Statist.*, **29**, 357.

COX, D. R. (1958b). *Planning of Experiments*. Wiley, London and New York.

COX, D. R. (1961). Tests of separate families of hypotheses. In *Proceedings of the Fourth Berkeley Symposium on Mathematical Statistics and Probability*, J. Neyman (ed.). University of California Press, Berkeley, **1**, 105.

COX, D. R. (1962). Further results on tests of separate families of hypotheses. *J. Roy. Statist. Soc.*, **B**, **24**, 406.

COX, D. R. (1963). Large sample sequential tests for composite hypotheses. *Sankhyā*, **A**, **25**, 5.

COX, D. R. (1967). Fieller's theorem and a generalization. *Biometrika*, **54**, 567.

COX, D. R. (1971). The choice between alternative ancillary statistics. *J. Roy. Statist. Soc.*, **B**, **33**, 251.

COX, D. R. and HINKLEY, D. V. (1968). A note on the efficiency of least-squares estimates. *J. Roy. Statist. Soc.* **B**, **30**, 284.

COX, D. R. and HINKLEY, D. V. (1974). *Theoretical Statistics*. Chapman & Hall, London.

COX, D. R. and REID, N. (1987). Parameter orthogonality and approximate conditional inference. *J. Roy. Statist. Soc.*, **B**, **49**, 1.

COX, D. R. and STUART, A. (1955). Some quick sign tests for trend in location and dispersion. *Biometrika*, **42**, 80.

CRAMÉR, H. (1928). On the composition of elementary errors II: Statistical applications. *Skand. Aktuartidskr.*, **11**, 141.

CRAMÉR, H. (1946). *Mathematical Methods of Statistics*. Princeton University Press, Princeton, NJ.

CRESSIE, N. (1981). Transformations and the jackknife. *J. Roy. Statist. Soc.*, **B**, **43**, 177.

CRESSIE, N. and READ, T. R. C. (1984). Multinomial goodness-of-fit tests. *J. Roy. Statist. Soc.*, B, **46**, 440.

CROW, E. L. (1956). Confidence intervals for a proportion. *Biometrika*, **43**, 423.

CROW, E. L. and GARDNER, R. S. (1959). Confidence intervals for the expectation of a Poisson variable. *Biometrika*, **46**, 441.

CURTIS, J. H. (1943). On transformations used in the analysis of variance. *Ann. Math. Statist.*, **14**, 107.

CZORGO, S. and FARAWAY, J. J. (1996). The exact and asymptotic distributions of Cramer-von Mises statistics. *J. Roy. Statist. Soc.*, **B**, **58**, 221.

D'AGOSTINO, R. B. and PEARSON, E. S. (1973). Tests for departures from normality. Empirical results for the distributions of b_2 and $\sqrt{b_1}$. *Biometrika*, **60**, 613.

D'AGOSTINO, R. B. and STEPHENS, M. A. (1986). *Goodness-of-fit Techniques*. Dekker, New York.

DAHIYA, R. C. (1981). An improved method of estimating an integer-parameter by maximum likelihood. *Amer. Statist.*, **35**, 34.

DAHIYA, R. C. and GURLAND, J. (1972). Pearson chi-squared test of fit with random intervals. *Biometrika*, **59**, 147.

DAHIYA, R. C. and GURLAND, J. (1973). How many classes in the Pearson chi-square test? *J. Amer. Statist. Ass.*, **68**, 707.

DALLAL, G. E. and WILKINSON, L. (1986). An analytical approximation to the distribution of Lilliefors' test statistic for normality. *Amer. Statist.*, **40**, 294.

DANIELS, H. E. (1944). The relation between measures of correlation in the universe of sample permutations. *Biometrika*, **33**, 129.

DANIELS, H. E. (1948). A property of rank correlations. *Biometrika*, **35**, 416.

DANIELS, H. E. (1951–2). The theory of position finding. *J. Roy. Statist. Soc.*, **B**, **13**, 186 and **14**, 246.

DANIELS, H. E. (1961). The asymptotic efficiency of a maximum likelihood estimator. In *Proceedings of the Fourth Berkeley Symposium on Mathematical Statistics and Probability*, J. Neyman (ed.). University of California Press, Berkeley, **1**, 151.

DANIELS, H. E. and KENDALL, M. G. (1958). Short proof of Miss Harley's theorem on the correlation coefficient. *Biometrika*, **45**, 571.

DANTZIG, G. B. (1940). On the non-existence of tests of 'Student's' hypothesis having power functions independent of σ. *Ann. Math. Statist.*, **11**, 186.

DARMOIS, G. (1935). Sur les lois de probabilité à estimation exhaustive. *C. R. Acad. Sci., Paris*, **200**, 1265.

DASGUPTA, P. (1968). Tables of the non-centrality parameter of F-test as a function of power. *Sankhyā*, **B**, **30**, 73.

DASGUPTA, S. and PERLMAN, M. D. (1974). Power of the noncentral F-test: effect of additional variates on Hotelling's T-test. *J. Amer. Statist. Ass.*, **69**, 174.

DASGUPTA, S., ANDERSON, T. and MUDHOLKAR, G. (1964). Monotonicity of the power function of some tests of the multivariate linear hypothesis. *Ann. Math. Statist.*, **35**, 200.

DATTA, C. S. and GHOSH, J. K. (1995). On priors providing frequentist validity for Bayesian inference. *Biometrika*, **82**, 37.

DAVID, F. N. (1937). A note on unbiased limits for the correlation coefficient. *Biometrika*, **29**, 157.

DAVID, F. N. (1938). *Tables of the Correlation Coefficient*. Cambridge University Press, London.

DAVID, F. N. (1947). A χ^2 smooth test for goodness of fit. *Biometrika*, **34**, 299.

DAVID, F. N. (1950). An alternative form of χ^2. *Biometrika*, **37**, 448.

DAVID, F. N. and JOHNSON, N. L. (1948). The probability integral transformation when parameters are estimated from the sample. *Biometrika*, **35**, 182.

DAVIS, A. W. and SCOTT, A. J. (1971). On the k-sample Behrens–Fisher distribution. (Div. of Math. Statist. Tech. Paper No. 33, C.S.I.R.O., Australia.)

DAVIS, C. S. and STEPHENS, M. A. (1989). Algorithm AS248: Empirical distribution function goodness-of-fit tests. *Appl. Statist.*, **38**, 535.

DAVIS, L. (1984). Comments on a paper by T. Amemiya on estimation in a dichotomous logit regression model. *Ann. Statist.*, **12**, 778.

DAVIS, R. C. (1951). On minimum variance in nonregular estimation. *Ann. Math. Statist.*, **22**, 43.

DAVISON, A. C. and TSAI, C.-L. (1992). Regression model diagnostics. *Int. Statist. Rev.*, **60**, 337.

DAVISON, A. C., HINKLEY, D. V. and WORTON, B. J. (1992). Bootstrap likelihoods. *Biometrika*, **79**, 113.

DAWID, A. P. (1975). On the concepts of sufficiency and ancillarity in the presence of nuisance parameters. *J. Roy. Statist. Soc.*, **B**, **37**, 248.

DAWID, A. P. (1977). Conformity of inference patterns. In *Recent Developments in Statistics*, J. R. Varva *et al.* (eds). North-Holland, Amsterdam.

DAWID, A. P. (1984). Statistical theory, the prequential approach (with discussion). *J. Roy. Statist. Soc.*, **A**, **147**, 278.

DAWID, A. P. (1991). Fisherian inference in likelihood and prequential frames of reference (with discussion). *J. Roy. Statist. Soc.*, **B 53**, 79.

DAWID, A. P., STONE, M. and ZIDEK, J. V. (1973). Marginalisation paradoxes in Bayesian and structural inference. *J. Roy. Statist. Soc.*, **B**, **37**, 248.

DE GROOT, M. H. (1959). Unbiased sequential estimation for binomial populations. *Ann. Math. Statist.*, **30**, 80.

DE GROOT, M. H. (1970). *Optimal Statistical Decisions*. McGraw-Hill, New York.

DEMPSTER, A. P. (1963). Further examples of inconsistencies in the fiducial argument. *Ann. Math. Statist.*, **34**, 884.

DEMPSTER, A. P. and RUBIN, D. B. (1983). Rounding error in regression: the appropriateness of Sheppard's corrections. *J. Roy. Statist. Soc.*, **B**, **45**, 51.

DEMPSTER, A. P. and SCHATZOFF, M. (1965). Expected significance level as a sensitivity index for test statistics. *J. Amer. Statist. Ass.*, **60**, 420.

DEMPSTER, A. P., LAIRD, N.M. and RUBIN, D. B. (1977a). Maximum likelihood from incomplete data via the EM algorithm. *J. Roy. Statist. Soc.*, **B**, **39**, 1.

DEMPSTER, A. P., SCHATZOFF, M. and WERMUTH, N. (1977b). A simulation study of alternatives to ordinary least squares (with comments following). *J. Amer. Statist. Ass.*, **72**, 77.

DENNY, J. L. (1967). Sufficient conditions for a family of probabilities to be exponential. *Proc. Nat. Acad. Sci., U.S.A.*, **57**, 1184.

DENNY, J. L. (1972). Sufficient statistics and discrete exponential families. *Ann. Math. Statist.*, **43**, 1320.

DENT, W. T. and CASSING, S. (1978). On modified maximum likelihood estimators of the autocorrelation parameter in linear models. *Biometrika*, **65**, 211.

DIACONIS, P. and FREEDMAN, D. A. (1986a). On the consistency of Bayes estimates (with discussion). *Ann. Statist.*, **14**, 1.

DIACONIS, P. and FREEDMAN, D. A. (1986b). On inconsistent Bayes estimates of location. *Ann. Statist.*, **14**, 68.

DIACONIS, P. and ZABELL, S. L. (1982). Updating subjective probability. *J. Amer. Statist. Ass.*, **77**, 822.

DICICCIO, T. and EFRON, B. (1992). More accurate confidence intervals in exponential families. *Biometrika*, **79**, 231.

DICICCIO, T. J., MARTIN, M. A. and YOUNG, G. A. (1992). Fast and accurate approximate double bootstrap confidence intervals. *Biometrika*, **79**, 285.

DIGBY, P. G. N. (1983). Approximating the tetrachoric correlation coefficient. *Biometrics*, **39**, 753.

DING, C. G. (1992). Computing the non-central χ^2 distribution. *Appl. Statist.*, **41**, 478.

DING, C. G. and BARGMANN, R. E. (1991a). Evaluation of the distribution of the square of the sample multiple correlation coefficient. *Appl. Statist.* **40**, 195.

DING, C. G. and BARGMANN, R. E. (1991b). Quantiles of the distribution of the square of the sample multiple correlation coefficient. *Appl. Statist.*, **40**, 199.

DIXON, W. J. (1953). Power functions of the Sign Test and power efficiency for normal alternatives. *Ann. Math. Statist.*, **24**, 467.

DODGE, H. F. and ROMIG, H. G. (1944). *Sampling Inspection Tables.* Wiley, New York.

DOKSUM, K. A., FENSTAD, G. and AABERGE, R. (1977). Plots and tests for symmetry. *Biometrika*, **64**, 473.

DOLBY, J. L. (1963). A quick method for choosing a transformation. *Technometrics*, **5**, 317.

DONNER, A. (1986). A review of inference procedures for the intraclass correlation coefficient in the one way random effects model. *Int. Statist. Rev.*, **54**, 67.

DONNER, A. and WELLS, G. (1986). A comparison of confidence interval methods for the intraclass correlation coefficient. *Biometrics*, **42**, 401. (Corr. **43**, 1035).

DONNER, A., WELLS, G. A. and ELIASZIW, M. (1989). On two approximations to the F distribution: Application to testing intraclass correlation. *Canad. J. Statist.*, **17**, 209.

DOWNTON, F. (1953). A note on ordered least-squares estimation. *Biometrika*, **40**, 457.

DRAPER, N. R. and COX, D. R. (1969). On distributions and their transformation to normality. *J. Roy. Statist. Soc.*, **B**, **31**, 472.

DRAPER, N. R. and GUTTMAN, I. (1995). Confidence intervals versus regions. *The Statistician*, **44**, 399.

DRAPER, N. R. and HUNTER, W. G. (1969). Transformations: some examples revisited. *Technometrics*, **11**, 23.

DRAPER, N. R., GUTTMAN, I. and KANEMASU, H. (1971). The distribution of certain regression statistics. *Biometrika*, **58**, 295.

DROST, F. C., KALLENBERG, W. C. M., MOORE, D. S. and OOSTERHOFF, J. (1989). Power approximations to multinominal tests of fit. *J. Amer. Statist. Ass.*, **84**, 130.

DUMOUCHEL, W. H. (1983). Estimating the stable index in order to measure tail thickness: Critique. *Ann. Statist.*, **11**, 1019.

DUNCAN, D. B. (1955). Multiple range and multiple F tests, *Biometrics*, **11**, 1.

DUNCAN, A. J. (1957). Charts of the 10% points and 50% points of the operating characteristic curves for fixed effects analysis of variance F-tests, α=0.10 and 0.05. *J. Amer. Statist. Ass.*, **52**, 345.

DUNN, O. J. and CLARK, V. (1969). Correlation coefficients measured on the same individuals. *J. Amer. Statist. Ass.*, **64**, 366.

DUNN, O. J. and CLARK, V. (1971). Comparisons of tests of the equality of dependent correlation coefficients. *J. Amer. Statist. Ass.*, **66**, 904.

DUNNETT, C. W. (1955). A multiple comparisons procedure for comparing several treatments with a control, *J. Amer. Statist. Ass.*, **50**, 1096.

DURBIN, J. (1953). A note on regression when there is extraneous information about one of the coefficients. *J. Amer. Statist. Ass.*, **48**, 799.

DURBIN, J. (1960). Estimation of parameters in time-series regression models. *J. Roy. Statist. Soc.*, **B**, **22**, 139.

DURBIN, J. (1961). Some methods of constructing exact tests. *Biometrika*, **48**, 41.

DURBIN, J. (1970). On Birnbaum's theorem on the relation between sufficiency, conditionality and likelihood. *J. Amer. Statist. Ass.*, **65**, 395.

DURBIN, J. (1972). *Distribution theory for tests based on the sample distribution function.* Society for Industrial and Applied Mathematics.

DURBIN, J. (1975). Kolmogorov–Smirnov tests when parameters are estimated with applications to tests of exponentiality and tests on spacings. *Biometrika*, **62**, 5.

DURBIN, J. (1980). Approximations for densities of sufficient estimators. *Biometrika*, **67**, 311.

DURBIN, J. (1988). Is a philosophical consensus for statistics attainable? *J. Econometrics*, **37**, 51.

DURBIN, J. and KENDALL, M. G. (1951). The geometry of estimation. *Biometrika*, **38**, 150.

DURBIN, J. and WATSON, G. S. (1950). Testing for serial correlation in least squares regression, I. *Biometrika*, **37**, 409.

DURBIN, J. and WATSON, G. S. (1951). Testing for serial correlation in least squares regression, II. *Biometrika*, **38**, 1599.

DURBIN, J. and WATSON, G. S. (1971). Testing for serial correlation in least squares regression, III. *Biometrika*, **58**, 1.

DYER, D. D. and KEATING, J. P. (1980). On the determination of critical values for Bartlett's test. *J. Amer. Statist. Ass.*, **75**, 313.

DYNKIN, E. B. (1951). Necessary and sufficient statistics for a family of probability distributions [in Russian]. *Usp. Mat. Nauk* (N.S.), **6**, No. 1(41), 68.

EATON, M. L. (1983). *Multivariate Statistics*. Wiley, New York.

EDELMAN, D. (1990). A note on UMP two-sided tests. *Amer. Statist.*, **44**, 219.

EDWARDS, A. W. F. (1972). *Likelihood*. Johns Hopkins University Press, Baltimore, MD.

EDWARDS, A. W. F. (1974). The history of likelihood. *Int. Statist. Rev.*, **42**, 9.

EFRON, B. (1967). The power of the likelihood ratio test. *Ann. Math. Statist.*, **38**, 802.

EFRON, B. (1975). Defining the curvature of a statistical problem (with applications to second order efficiency). *Ann. Statist.*, **3**, 1189.

EFRON, B. (1978). The geometry of exponential families. *Ann. Statist.*, **6**, 362.

EFRON, B. (1979). Bootstrap methods: Another look at the jackknife. *Ann. Statist.*, **7**, 1.

EFRON, B. (1981). Nonparametric estimates of standard error: The jackknife, the bootstrap and other methods. *Biometrika*, **68**, 589.

EFRON, B. (1982). Maximum likelihood and decision theory. *Ann. Statist.*, **10**, 341.

EFRON, B. (1985). Bootstrap confidence intervals for a class of parametric problems. *Biometrika*, **72**, 85.

EFRON, B. (1987). Better bootstrap confidence intervals (with discussion). *J. Amer. Statist. Ass.*, **82**, 171.

EFRON, B. (1992). Jackknife after bootstrap standard errors and influence functions (with discussion). *J. Roy. Statist. Soc.*, **B**, **54**, 83.

EFRON, B. (1994). Missing data, imputation and the bootstrap. *J. Amer. Statist. Ass.*, **89**, 463.

EFRON, B. and HINKLEY, D. V. (1978). Assessing the accuracy of the maximum likelihood estimator; observed versus expected Fisher information (with following comments). *Biometrika*, **65**, 457.

EFRON, B. and STEIN, C. (1981). The jackknife estimate of variance. *Ann. Statist.*, **9**, 586.

EFRON, B. and TIBSHIRANI, R. J. (1993). *An Introduction to the Bootstrap*. Chapman & Hall, London.

EFROYMSON, M. A. (1960). Multiple regression analysis. In *Mathematical Models for Digital Computers*, A. Ralston and H. S. Wilf (ed.). Wiley, New York.

EINOT, I. and GABRIEL, K. R. (1975). A study of the power of several methods of multiple comparisons, *J. Amer. Statist. Ass.*, **70**, 574.

EISENBERG, B. and GHOSH, B. K. (1991). The sequential probability ratio test. In *Handbook of Sequential Analysis*, B. K. Ghosh and P. K. Sen (eds). Dekker, New York, 47.

EISENHART, C. (1938). The power function of the χ^2 test. *Bull. Amer. Math. Soc.*, **44**, 32.

EISENHART, C. (1947). The assumptions underlying the analysis of variance. *Biometrics*, **3**, 1.

ELLISON, B. E. (1964). On two-sided tolerance intervals for a normal distribution. *Ann. Math. Statist.*, **35**, 762.

ELSTON, R. C. (1975). On the correlation between correlations. *Biometrika*, **62**, 133.

EPSTEIN, B. and SOBEL, M. (1954). Some theorems relevant to life testing from an exponential distribution. *Ann. Math. Statist.*, **25**, 373.

EVANS, M. (1992). Robustness of size of tests of autocorrelation and heteroscedasticity to non-normality. *J. Econometrics*, **51**, 7.

EVANS, M. J., FRASER, D. A. S. and MONETTE, G. (1985). On the role of principles in statistical inference. In *Statistical Theory and Data Analysis*, K. Matsuita (ed.). North-Holland, Amsterdam.

EVANS, M. J., FRASER, D. A. S. and MONETTE, G. (1986). On principles and arguments to likelihood. *Canad. J. Statist.*, **14**, 181.

EZEKIEL, M. J. B. and FOX, K. A. (1959). *Methods of Correlation and Regression Analysis: Linear and Curvilinear.* Wiley, New York.

FEDER, P. I. (1968). On the distribution of the log likelihood ratio test statistic when the true parameter is 'near' the boundaries of the hypothesis regions. *Ann. Math. Statist.*, **39**, 2044.

FEDER, P. I. (1975). The log likelihood ratio in segmented regression. *Ann. Statist.*, **3**, 84.

FELLER, W. (1938). Note on regions similar to the sample space. *Statist. Res. Mem.*, **2**, 117.

FEND, A. V. (1959). On the attainment of CramÈr–Rao and Bhattacharyya bounds for the variance of an estimate. *Ann. Math. Statist.*, **30**, 381.

FERGUSON, T. S. (1967). *Mathematical Statistics: A Decision Theoretic Approach.* Academic Press, New York.

FERGUSON, T. S. (1978). Maximum likelihood estimation of the parameters of the Cauchy distribution for samples of size 3 and 4. *J. Amer. Statist. Ass.*, **73**, 211.

FERGUSON, T. S. (1982). An inconsistent maximum likelihood estimate. *J. Amer. Statist. Ass.*, **77**, 831.

FÉRON, R. and FOURGEAUD, C. (1952). Quelques propriétés caractéristiques de la loi de Laplace–Gauss. *Publ. Inst. Statist. Paris*, **1**, 44.

FIELLER, E. C. (1940). The biological standardisation of insulin. *Suppl. J. Roy. Statist. Soc.*, **7**, 1.

FIELLER, E. C. (1954). Some problems in interval estimation. *J. Roy. Statist. Soc.*, **B**, **16**, 175.

FINNEY, D. J. (1941). On the distribution of a variate whose logarithm is normally distributed. *Suppl. J. Roy. Statist. Soc.*, **7**, 155.

FIRTH, D. (1993). Bias reduction of maximum likelihood estimates. *Biometrika*, **80**, 27.

FISHBURN, P. C. (1986). The axioms of subjective probability. *Statist. Sci.*, **1**, 335.

FISHBURN, P. C. (1989). Foundations of decision analysis: Along the way. *Management Sci.*, **35**, 387.

FISHER, N. I. and HALL, P. (1990). On bootstrap hypothesis testing. *Austr. J. Statist.*, **32**, 177.

FISHER, N. I. and LEE, A. J. (1986). Correlation coefficients for random variables on a unit sphere or hypersphere. *Biometrika*, **73**, 159.

FISHER, R. A. (1921a).* On the mathematical foundations of theoretical statistics. *Phil. Trans.*, **A**, **222**, 309.

FISHER, R. A. (1921b).* Studies in crop variation. I. An examination of the yield of dressed grain from Broadbalk. *J. Agric. Sci* ., **11**, 107.

FISHER, R. A. (1921c). On the 'probable error' of a coefficient of correlation deduced from a small sample. *Metron*, **1**(4), 3.

FISHER, R. A. (1922a).* On the interpretation of chi-square from contingency tables, and the calculation of *P. J. Roy. Statist. Soc.*, **85**, 87.

FISHER, R. A. (1922b).* The goodness of fit of regression formulae and the distribution of regression coefficients. *J. Roy. Statist. Soc.*, **85**, 597.

FISHER, R. A. (1924a). The distribution of the partial correlation coefficient. *Metron*, **3**, 329.

FISHER, R. A. (1924b). The influence of rainfall on the yield of wheat at Rothamsted. *Phil. Trans.*, **B, 213**, 89.

FISHER, R. A. (1924c).* The conditions under which χ^2 measures the discrepancy between observation and hypothesis. *J. Roy. Statist. Soc.*, **87**, 442.

FISHER, R. A. (1925a).* Theory of statistical estimation. *Proc. Camb. Phil. Soc.*, **22**, 700.

FISHER, R. A. (1925b). *Statistical Methods for Research Workers*. Oliver and Boyd, Edinburgh. (14th edition, 1970.)

FISHER, R. A. (1928a).* The general sampling distribution of the multiple correlation coefficient. *Proc. Roy. Soc.*, **A, 121**, 654.

FISHER, R. A. (1928b).* On a property connecting the χ^2 measure of discrepancy with the method of maximum likelihood. *Atti Congr. Int.. Mat., Bologna*, **6**, 94.

FISHER, R. A. (1932). *Statistical Methods for Research Workers* (4th edn). Oliver and Boyd, Edinburgh.

FISHER, R. A. (1935a). *The Design of Experiments*. Oliver and Boyd, Edinburgh.

FISHER, R. A. (1935b).* The fiducial argument in statistical inference. *Ann. Eugen.*, **6**, 391.

FISHER, R. A. (1939).* The comparison of samples with possibly unequal variances. *Ann. Eugen.*, **9**, 174.

FISHER, R. A. (1941).* The negative binomial distribution. *Ann. Eugen.*, **11**, 182.

FISHER, R. A. (1956). *Statistical Methods and Scientific Inference*. Oliver and Boyd, Edinburgh.

FISHER, R. A. (1966). *The Design of Experiments* (8th edn). Oliver and Boyd, Edinburgh.

FISHER, R. A. and YATES, F. (1963). *Statistical Tables for Biological, Agricultural and Medical Research* (7th edn). Oliver and Boyd, Edinburgh; and Hafner, New York.

FIX, E. (1949). Tables of noncentral χ^2. *University Calif. Publ. Statist.*, **1**, 15.

FOLKS, J. L. and CHHIKARA, R. S. (1978). The inverse Gaussian distribution and its statistical application – a review. *J. Roy. Statist. Soc.*, **B, 40**, 263.

FOUNTAIN, R. L. and RAO, C. R. (1993). Further investigations of Berkson's example. *Commun. Statist.*, **A, 22** , 613.

FOUTZ, R. V. (1977). On the unique consistent root to the likelihood equations. *J. Amer. Statist. Ass.*, **72**, 147.

FOUTZ, R. V. and SRIVASTAVA, R. C. (1977). The performance of the likelihood ratio test when the model is incorrect. *Ann. Statist.*, **5**, 1183.

FOX, J. and MONETTE, G. (1992). Generalised collinearity diagnostics. *J. Amer. Statist. Ass.*, **87**, 178.

FOX, M. (1956). Charts of the power of the F-test. *Ann. Math. Statist.*, **27**, 484.

FRANKLIN, L. A. (1988). The complete exact null distribution of Spearman's rho for $n = 12(1)18$. *J. Statist. Comp. Simul.*, **29**, 255.

FRASER, D. A. S. (1950). Note on the χ^2 smooth test. *Biometrika*, **37**, 447.

FRASER, D. A. S. (1953). Nonparametric tolerance regions. *Ann. Math. Statist.*, **24**, 44.

FRASER, D. A. S. (1962). On the consistency of the fiducial method. *J. Roy. Statist. Soc.*, **B, 22**, 425.

FRASER, D. A. S. (1963). On sufficiency and the exponential family. *J. Roy. Statist. Soc.*, **B, 25**, 115.

FRASER, D. A. S. (1964). Fiducial inference for location and scale parameters. *Biometrika*, **51**, 17.

FRASER, D. A. S. (1968). *The Structure of Inference*. Wiley, New York.

FRASER, D. A. S. (1976). Necessary analysis and adaptive inference. *J. Amer. Statist. Ass.*, **71**, 99.

FRASER, D. A. S. and GUTTMAN, I. (1956). Tolerance regions. *Ann. Math. Statist.*, **27**, 162.

FRASER, D. A. S. and REID, N. (1989). Adjustments to profile likelihood. *Biometrika*, **76**, 477.

FREEDMAN, D. A. (1981). Bootstrapping regression models. *Ann. Statist.*, **9**, 1218.

FREEDMAN, D. A. and PETERS, S. C. (1984). Bootstrapping a regression equation: Some empirical results. *J. Amer. Statist. Ass.*, **79**, 97.

FREEMAN, M. F. and TUKEY, J. W. (1950). Transformations related to the angular and the square root. *Ann. Math. Statist.*, **21**, 607.

FREUND, R. J., VAIL, R. W. and CLUNIES-ROSS, C. W. (1961). Residual analysis. *J. Amer. Statist. Ass.*, **56**, 98 (corrigenda: 1005).

FRYDENBERG, M. and JENSEN, J. L. (1989). Is the improved likelihood ratio statistic really improved in the discrete case? *Biometrika*, **76**, 655.

FU, J. C. (1973). On a theorem of Bahadur on the rate of convergence of point estimators. *Ann. Statist.*, **1**, 745.

FUJINO, Y. (1980). Approximate binomial confidence limits. *Biometrika*, **67**, 677.

FULLER, W. A. (1995). *Introduction to Statistical Times Series* (2nd edn). Wiley, New York.

FULLER, W. A. and RAO, J. N. K. (1978). Estimation for a linear regression model with unknown diagonal covariance matrix. *Ann. Statist.*, **6**, 1149.

GABRIEL, K. R. (1969). Simultaneous test procedures – some theory of multiple comparisons. *Ann. Math. Statist.*, **40**, 224.

GALLANT, A. R. (1987). *Nonlinear Statistical Models*. Wiley, New York.

GAN, F. F. and KOEHLER, K. J. (1990). Goodness of fit tests based on P-P probability plots. *Technometrics*, **32**, 289.

GARDNER, M. J. (1972). On using an estimated regression line in a second sample. *Biometrika*, **59**, 263.

GART, J. J. and PETTIGREW, H. M. (1970). On the conditional moments of the *k*-statistics for the Poisson distribution. *Biometrika*, **57**, 661.

GARWOOD, F. (1936). Fiducial limits for the Poisson distribution. *Biometrika*, **28**, 437.

GEARY, R. C. (1942). The estimation of many parameters. *J. Roy. Statist. Soc.*, **105**, 213.

GEARY, R. C. (1944). Comparison of the concepts of efficiency and closeness for consistent estimates of a parameter. *Biometrika*, **33**, 123.

GEISSER, S. and CORNFIELD, J. (1963). Posterior distributions for multivariate normal parameters. *J. Roy. Statist. Soc.*, **B**, **25**, 368.

GELFAND, A. E. and SMITH, A. F. M. (1990). Sampling-based approaches to calculating marginal densities. *J. Amer. Statist. Ass.*, **85**, 398.

GEMAN, S. and GEMAN, D. (1984). Stochastic relaxation, Gibbs distributions and the Bayesian restoration of images. *IEEE Trans., Pattern Anal. Machine Intell.*, **6**, 721.

GEMES, K. (1989). A refutation of Popperian inductive scepticism. *Brit. J. Phil. Sci.*, **40**, 183.

GENEST, C. and SCHERVISH, M. J. (1985). Resolution of Godambe's paradox. *Canad. J. Statist.*, **4**, 293.

GEYER, C. J. and THOMPSON, E. A. (1992). Constrained Monte Carlo maximum likelihood for dependent data (with discussion). *J. Roy. Statist. Soc.* **B**, **54**, 657.

GHOSH, B. K. (1969). Moments of the distribution of sample size in SPRT. *J. Amer. Statist. Ass.*, **64**, 1560.

GHOSH, B. K. (1975). A two-stage procedure for the Behrens–Fisher problem. *J. Amer. Statist. Ass.*, **70**, 457.

GHOSH, B. K. (1979). A comparison of some approximate confidence intervals for the binomial parameter. *J. Amer. Statist. Ass.*, **74**, 894.

GHOSH, B. K. (1987). On the attainment of the Cramer–Rao bound in the sequential case. *Sequential Anal.*, **6**, 267.

GHOSH, B. K. (1991). A brief history of sequential analysis. In *Handbook of Sequential Analysis*, B. K. Ghosh and P. K. Sen (eds). Dekker, New York, 1.

GHOSH, B. K. and SEN, P. K., (eds). (1991). *Handbook of Sequential Analysis*. Dekker, New York.

GHOSH, J. K. and SINGH, R. (1966). Unbiased estimation of location and scale parameters. *Ann. Math. Statist.*, **37**, 1671.

GHOSH, J. K. and SINHA, B. K. (1981). A necessary and sufficient condition for second order admissibility with applications to Berkson's bioassay problem. *Ann. Statist.*,**9**, 1334.

GHOSH, J. K. and SUBRAMANYAM, K. (1974). Second order efficiency of maximum likelihood estimators. *Sankhyā*, **A**, **36**, 325.

GHOSH, M. N. (1964). On the admissibility of some tests in MANOVA, *Ann. Math. Statist.*, **35**, 789.

GIERE, R. N. (1977). Allan Bimbaum's conception of statistical evidence. *Synthèse*, **36**, 5.

GIRSHICK, M. A. (1946). Contributions to the theory of sequential analysis. *Ann. Math. Statist.*, **17**, 123 and 282.

GJEDDEBAEK, N. F. (1949–61). Contribution to the study of grouped observations. 1–VI. *Skan. Aktuartidskr.*, **32**, 135; **39**, 154; **40**, 20; *Biometrics*, **15**, 433; *Skand. Aktuartidskr.*, **42**, 194;

GLASER, R. E. (1976). Exact critical values for Bartlett's test for homogeneity of variance. *J. Amer. Statist. Ass.*, **71**, 488.

GLASER, R. E. (1980). A characterization of Bartlett's statistic involving incomplete beta functions. *Biometrika*, **67**, 53.

GLESER, L. J. (1985). Exact power of goodness-of-fit tests of Kolmogorov type for discontinuous distributions. *J. Amer. Statist. Ass.*, **80**, 954.

GLESER, L. J. (1992). A note on the analysis of familial data. *Biometrika*, **79**, 412.

GLESER, L. J. and MOORE, D. S. (1983). The effect of dependence on chi-squared and empiric distribution tests of fit. *Ann. Statist.*, **11**, 1100.

GODAMBE, V. P. (1960). An optimum property of regular maximum likelihood estimation. *Ann. Math. Statist.*, **31**, 1208.

GODAMBE, V. P. (1976). Conditional likelihood and unconditional estimating equations. *Biometrika*, **63**, 277.

GODAMBE, V. P. (1982). Ancillarity principle and a statistical paradox. *J. Amer. Statist. Ass.*, **77**, 931.

GODAMBE, V. P. (ed.) (1991). *Estimating Functions*. Clarendon Press, Oxford.

GODAMBE, V. P. and SPROTT, D. A. (eds) (1971). *Foundations of Statistical Inference*. Holt, Rinehart and Winston, Toronto.

GOEL, A. L. and WU, S. M. (1971). Determination of AQL and a contour nomogram for cusum charts to control normal mean. *Technometrics*, **13**, 221.

GOLDBERGER, A. S. (1961). Stepwise least squares: residual analysis and specification error. *J. Amer. Statist. Ass.*, **56**, 998.

GOLDBERGER, A. S. and JOCHEMS, D. B. (1961). Note on stepwise least squares. *J. Amer. Statist. Ass.*, **56**, 105.

GOOD, I. J. (1950). *Probability and the Weighing of Evidence*. Griffin, London.

GOOD, I. J. (1988). The interface between statistics and philosophy of science. *Statist. Sci.*, **3**, 386.

GOOD, I. J. (1992). The Bayes/non-Bayes compromise: A brief review. *J. Amer. Statist. Ass.*, **87**, 597.

GOOD, I. J., GOVER, T. N. and MITCHELL, G. J. (1970). Exact distributions for X^2 and for the likelihood-ratio statistic for the equiprobable multinomial distribution. *J. Amer. Statist. Ass.*, **65**, 267.

GOODALL, C. R. (1993). Computation using QR algorithm. In *Handbook of Statistics Vol. 9: Computational Statistics*, C. R. Rao (ed.). Elsevier, Amsterdam, 467.

GOURIEROUX, C., MONTFORT, A. and TROGNAN, A. (1984). Pseudo maximum likelihood methods: Theory. *Econometrica*, **52**, 681.

GOUTIS, C. and CASELLA, G. (1992). Increasing the confidence in Student's t interval. *Ann. Statist.*, **20**, 1501.

GOUTIS, C. and CASELLA, G. (1995). Frequentist post-data inference. *Int. Statist. Rev.*, **63**, 325.

GRANGER, C. W. J. (1969). Investigating causal relations by econometric models and cross-spectral methods. *Econometrica*, **37**, 424.

GRAYBILL, F. A. (1976). *Theory and Applications of the Linear Model.* Duxbury, North Scituate, MA.

GRAYBILL, F. A. and HULTQUIST, R. A. (1961). Theorems concerning Eisenhart's Model II. *Ann. Math. Stat.*, **32**, 261.

GRAYBILL, F. A. and MARSAGLIA, G. (1957). Idempotent matrices and quadratic forms in the general linear hypothesis. *Ann. Math. Statist.*, **28**, 678.

GREEN, P. J. (1984). Iteratively reweighted least squares for maximum likelihood estimation, and some robust and resistant alternatives (with discussion). *J. Roy. Statist. Soc.*, **B**, **46**, 149.

GREENBERG, E. (1975). Minimum variance properties of principal component regression. *J. Amer. Statist. Ass.*, **70**, 194.

GREENHOUSE, S. W. and GEISSER, S. (1959). On methods in the analysis of profile data. *Psychometrika*, **24**, 95.

GREENWOOD, P. E. and NIKULIN, M. S. (1996). *A Guide to Chi-squared Testing*. Wiley, Chichester.

GRICE, J. V. and BAIN, L. J. (1980). Inference concerning the mean of the gamma distribution. *J. Amer. Statist. Ass.*, **75**, 929.

GRILICHES, Z. (1967). Distributed lags: a survey. *Econometrica*, **35**, 16.

GROENEBOOM, P. and OOSTERHOFF, J. (1981). Bahadur efficiency and small-sample efficiency. *Int. Statist. Rev.*, **49**, 127.

GUENTHER, W. C. (1964). Another derivation of the non-central chi-square distribution. *J. Amer. Statist. Ass.*, **59**, 957.

GUENTHER, W. C. and WHITCOMB, M. G. (1966). Critical regions for tests of interval hypotheses about the variance. *J. Amer. Statist. Ass.*, **61**, 204.

GUEST, P. G. (1954). Grouping methods in the fitting of polynomials to equally spaced observations. *Biometrika*, **41**, 62.

GUEST, P. G. (1956). Grouping methods in the fitting of polynomials to unequally spaced observations. *Biometrika*, **43**, 149.

GUMBEL, E. J. (1943). On the reliability of the classical chi-square test. *Ann. Math. Statist.*, **14**, 253.

GUNST, R. F. (1983). Regression analysis with multicollinear predictor variables: definition, detection and effects. *Commun. Statist.*, **A**, **12**, 2217.

GUNST, R. F. and MASON, R. L. (1977). Biased estimation in regression: An evaluation using mean squared error. *J. Amer. Statist. Ass.*, **72**, 616.

GURLAND, J. (1968). A relatively simple form of the multiple correlation coefficient. *J. Roy. Statist. Soc.*, **B**, **30**, 276.

GURLAND, J. and MILTON, R. (1970). Further consideration of the distribution of the multiple correlation coefficient. *J. Roy. Statist. Soc.*, **B**, **32**, 381.

GUT, A. (1988). *Stopped Random Walks*. Springer-Verlag, New York.

GUTTMAN, I. (1957). On the power of optimum tolerance regions when sampling from normal distributions. *Ann. Math. Statist.*, **28** , 773.

GUTTMAN, I. (1970). *Statistical Tolerance Regions*. Griffin, London

HAAS, G., BAIN, L. and ANTLE, C. (1970). Inferences for the Cauchy distribution based on maximum likelihood estimators. *Biometrika*, **57**, 403.

HABERMAN, S. J. (1988). A warning on the use of chi-squared statistics with frequency tables with small expected cell counts. *J. Amer. Statist. Ass.*, **83**, 555.

HAJNAL, J. (1961). A two-sample sequential t-test. *Biometrika*, **48**, 65.

HALDANE, J. B. S. (1938). The approximate normalization of a class of frequency distributions. *Biometrika*, **29**, 392.

HALDANE, J. B. S. (1955). Substitutes for χ^2. *Biometrika*, **42**, 265.

HALDANE, J. B. S. and SMITH, S. M. (1956). The sampling distribution of a maximum-likelihood estimate. *Biometrika*, **43**, 96.

HALL, P. (1981). Asymptotic theory of triple sampling for sequential estimation of the mean. *Ann. Statist.*, **9**, 1229.

HALL, P. (1982). Improving the normal approximation when constructing one-sided confidence intervals for binomial or Poisson parameters. *Biometrika*, **69**, 647.

HALL, P. (1988). Theoretical comparison of bootstrap confidence intervals (with discussion). *Ann. Statist.*, **16**, 927.

HALL, P. (1992). *The Bootstrap and Edgeworth Expansion.* Springer-Verlag, Berlin and New York.

HALL, W. J., WIJSMAN, R. A. and GHOSH, J. K. (1965). The relationship between sufficiency and invariance with applications in sequential analysis. *Ann. Math. Statist.*, **36**, 575.

HALMOS, P. R. and SAVAGE, L. J. (1949). Application of the Radon–Nikodym theorem to the theory of sufficient statistics. *Ann. Math. Statist.*, **20**, 225.

HALPERIN, M. (1963). Confidence interval estimation in non-linear regression. *J. Roy. Statist. Soc.*, **B**, **25**, 330.

HAMDAN, M. A. (1963). The number and width of classes in the chi-square test. *J. Amer. Statist. Ass.*, **58**, 678.

HAMDAN, M. A. (1968). Optimum choice of classes for contingency tables. *J. Amer. Statist. Ass.*, **63**, 291.

HAMDAN, M. A. (1970). The equivalence of tetrachoric and maximum likelihood estimates of ρ in 2×2 tables. *Biometrika*, **57**, 212.

HAMILTON, D. and WIENS, D. (1987). Correction factors for F-ratios in nonlinear regression. *Biometrika*, **74**, 423.

HAMILTON, D. C., WATTS, D. G. and BATES, D. M. (1982). Accounting for intrinsic nonlinearity in nonlinear regression parameter inference regions. *Ann. Statist.*, **10**, 386.

HAMILTON, J. D. (1994). *Time Series Analysis.* Princeton University Press, Princeton, NJ.

HAMMERSLEY, J. M. (1950). On estimating restricted parameters. *J. Roy. Statist. Soc.*, **B**, **12**, 192.

HAN, C. P. (1975). Some relationships between noncentral chi-squared and normal distributions. *Biometrika*, **62**, 213.

HANNAN, E. J. (1956). The asymptotic power of certain tests based on multiple correlations. *J. Roy. Statist. Soc.*, **B**, **18** , 227.

HANNAN, E. J. and QUINN, B. G. (1979). The determination of the order of an autoregression. *J. Roy. Statist. Soc.*, **B**, **41**, 190.

HANNAN, J. F. and TATE, R. F. (1965). Estimation of the parameters for a multivariate normal distribution when one variable is dichotomized. *Biometrika*, **52**, 664.

HANSEN, L. P. (1982). Large sample properties of generalised method of moments estimators. *Econometrica*, **50**, 1029.

HARLEY, B. I. (1956). Some properties of an angular transformation for the correlation coefficient. *Biometrika*, **43**, 219.

HARLEY, B. I. (1957). Further properties of an angular transformation of the correlation coefficient. *Biometrika*, **44**, 273.

HARRIS, P. (1985). An asymptotic expansion for the null distribution of the efficient score statistic. *Biometrika*, **72**, 653.

HARRIS. P. (1986). A note on Bartlett adjustments to likelihood ratio tests. *Biometrika*, **73**, 735.

HARRIS, P. and PEERS, H. W. (1980). The local power of the efficient score test statistic. *Biometrika*, **67**, 525.

HARSAAE, E. (1969). On the computation and use of a table of percentage points of Bartlett's M. *Biometrika*, **56**, 273.

HARTER, H. L. (1963). Percentage points of the ratio of two ranges and power of the associated test. *Biometrika*, **50**, 187.

HARTER, H. L. (1964). Criteria for best substitute interval estimators, with an application to the normal distribution. *J. Amer. Statist. Ass.*, **59**, 1133.

HARTER, H. L. and OWEN, D. B. (eds) (1970). *Selected Tables in Mathematical Statistics*, Vol I. American Mathematical Society, Providence, RI.

HARTER, H. L. and OWEN, D. B. (eds) (1974). *Selected Tables in Mathematical Statistics*, Vol II. American Mathematical Society, Providence, RI.

HARTER, H. L. and OWEN, D. B. (eds) (1975). *Selected Tables in Mathematical Statistics*, Vol III. American Mathematical Society, Providence, RI.

HARTLEY, H. O. (1964). Exact confidence regions for the parameters in non-linear regression laws. *Biometrika*, **51**, 347.

HARVEY, A. C. and PHILLIPS, G. D. A. (1979). Maximum likelihood estimation of regression models with autoregressive-moving average disturbances. *Biometrika*, **66**, 49.

HARVILLE, D. A. (1974). Bayesian inference for variance components using only error contrasts. *Biometrika*, **61**, 383.

HARVILLE, D. A. (1977). Maximium likelihood approaches to variance component estimation and related problems. *J.Amer. Statist. Ass.*, **72**, 320.

HAVILAND, M. G. (1990). Yates's correction for continuity and the analysis of 2×2 contingency tables (with discussion). *Statist. Medicine*, **9**, 363.

HAWKINS, D. M. (1975). From the noncentral t to the normal integral. *Amer. Statist.*, **29**, 42.

HAYAKAWA, T. (1975). The likelihood ratio criterion for a composite hypothesis under a local alternative. *Biometrika*, **62**, 451.

HAYAKAWA, T. (1977). The likelihood ratio criterion and the asymptotic expansion of its distribution.*Ann. Inst. Statist. Math. Tokyo*, **29**, **A**, 359.

HAYTER, A. J. (1984). A proof of the conjecture that the Tukey–Kramer multiple comparisons procedure is conservative. *Ann. Statist.*, **12**, 61.

HEALY, M. J. R. (1955). A significance test for the difference in efficiency between two predictors. *J. Roy. Statist. Soc.*, **B**, **17**, 266.

HEDAYAT, A. and ROBSON, D. S. (1970). Independent stepwise residuals for testing homoscedasticity. *J. Amer. Statist. Ass.*, **65**, 1573.

HEITMANN, G. J. and ORD, J. K. (1985). An interpretation of the least squares regression surface. *Amer. Statist.*, **39**, 120.

HELLAND, I. S. (1995). Simple counter examples against the conditionality principle. *Amer. Statist.*, **49**, 351.

HERNANDEZ, F. and JOHNSON, R. A. (1980). The large-sample behaviour of transformations to normality. *J. Amer. Statist. Ass.*, **75**, 855.

HERSTEIN, I. N. (1964). *Topics in Algebra*. Blaisdell, New York.

HETTMANSPERGER, T. P. and MCKEAN, J.W. (1998). *Robust Nonparametric Statistical Methods*. Arnold, London.

HILL, B. M. (1963a). The three-parameter lognormal distribution and Bayesian analysis of a point-source epidemic. *J. Amer. Statist. Ass.*, **58**, 72.

HILL, B. M. (1963b). Information for estimating the proportions in mixtures of exponential and normal distributions. *J. Amer. Statist. Ass.*, **58**, 918.

HINKLEY, D. V. (1975). On power transformations to symmetry. *Biometrika*, **62**, 101.

HINKLEY, D. V. (1980). Likelihood as approximate, pivotal distribution. *Biometrika*, **67**, 287.

HINKLEY, D. V. (1988). Bootstrap methods. *J. Roy. Statist. Soc.*, **B**, **50**, 321.

HINKLEY, D. V. and WANG, S. (1988). More about transformations and influential cases in regression. *Technometrics*, **30**, 435.

HOAGLIN, D. C. (1975). The small-sample variance of the Pitman location estimators. *J. Amer. Statist. Ass.*, **70**, 880.

HODGES, J. L., Jr. (1967). Efficiency in normal samples and tolerance of extreme values for some estimates of location. In *Proceedings of the Fifth Berkeley Symposium on Mathematical Statistics and Probability*, L. LeCam and J. Neyman (eds). University of California Press, Berkeley, **1**, 163.

HODGES, J. L., Jr. and LEHMANN, E. L. (1956). The efficiency of some nonparametric competitors of the t-test. *Ann. Math. Statist.*, **27**, 324.

HODGES, J. L., Jr. and LEHMANN, E. L. (1967). Moments of chi and power of t. In *Proceedings of the Fifth Berkeley Symposium on Mathematical Statistics and Probability*, L. LeCam and J. Neyman (eds). University of California Press, Berkeley, **1**, 187.

HODGES, J. L., Jr. and LEHMANN, E. L. (1968). A compact table for power of the t-test. *Ann. Math. Statist.*, **39**, 1629.

HODGES, J. L., Jr. and LEHMANN, E. L. (1970). Deficiency. *Ann. Math. Statist.*, **41**, 783.

HOEFFDING, W. (1952). The large sample power of tests based on permutations of observations. *Ann. Math. Statist.*, **23**, 169.

HOEL, P. G. (1951). Confidence regions for linear regression. In *Proceedings of the Second Berkeley Symposium on Mathematical Statistics and Probability*, J. Neyman (ed.). University of California Press, Berkeley, 75.

HOERL, A. E. and KENNARD, R. W. (1970). Ridge regression: Biased estimation for non-orthogonal problems and applications. *Technometrics*, **12**, 55, 69.

HOERL, R. W., SCHUENMEYER, J. H. and HOERL, A. E. (1986). A simulation of biased estimation and subset selection regression techniques. *Technometrics*, **28**, 369.

HOGBEN, D., PINKHAM, R. S. and WILK, M. B. (1961). The moments of the non-central t-distribution. *Biometrika*, **48**, 465.

HOGG, R. V. (1956). On the distribution of the likelihood ratio. *Ann. Math. Statist.*, **27**, 529.

HOGG, R. V. (1961). On the resolution of statistical hypotheses. *J. Amer. Statist. Ass.*, **56**, 978.

HOGG, R. V. (1962). Iterated tests of the equality of several distributions. *J. Amer. Statist. Ass.*, **57**, 579.

HOGG, R. V. and CRAIG, A. T. (1956). Sufficient statistics in elementary distribution theory. *Sankhyā*, **17**, 209.

HOGG, R. V. and TANIS, E. A. (1963). An iterated procedure for testing the equality of several exponential distributions. *J. Amer. Statist. Ass.*, **58**, 435.

HOLLAND, P. W. (1986). Statistics and causal inference (with discussion). *J. Amer. Statist. Ass.*, **81**, 945.

HORN, S. D., HORN, R. A. and DUNCAN, D. B. (1975). Estimating heteroscedastic variances in linear models. *J. Amer. Statist. Ass.*, **70**, 380.

HOTELLING, H. (1936). Relations between two sets of variates, *Biometrika*, **28**, 321.

HOTELLING, H. (1940). The selection of variates for use in prediction, with some comments on the general problem of nuisance parameters. *Ann. Math. Statist.*, **11**, 271.

HOTELLING, H. (1951). A generalized T-test and measure of multivariate dispersion. *Proceedings of the Second Berkeley Symposium on Mathematical Statistics and Probability*, J. Neyman (ed.). University of California Press, Berkeley, 23.

HOTELLING, H. (1953). New light on the correlation coefficient and its transforms. *J. Roy. Statist. Soc.*, **B**, **15**, 193.

HOUGAARD, P. (1985). The appropriateness of the asymptotic distribution in a nonlinear regression model in relation to curvature. *J. Roy. Statist. Soc.*, **B**, **47**, 103.

HOUGAARD, P. (1988). The asymptotic distribution of nonlinear regression parameter estimates: Improving the approximation. *Int.Statist. Rev.*, **56**, 221.

HOWE, W. G. (1969). Two-sided tolerance limits for normal populations – some improvements. *J. Amer. Statist. Ass.*, **64**, 610.

HOWSON, C. and URBACH, P. (1989). *Scientific Reasoning: The Bayesian Approach.* Open Court, La Salle, IL.

HOYLE, M. H. (1968). The estimation of variances after using a Gaussianating transformation. *Ann. Math. Statist.*, **39**, 1125.

HOYLE, M. H. (1973). Transformations – an introduction and a bibliography. *Int. Statist. Rev.*, **41**, 203.

HSU, J. C. (1984). Constrained two-sided simultaneous confidence intervals for multiple comparisons with the best. *Ann. Statist.*, **12**, 1136.

HSU J. C. (1996). *Multiple Comparisons, Theory and Method,* Chapman & Hall, London.

HSU, P. L. (1941). Analysis of variance from the power function standpoint. *Biometrika*, **32**, 62.

HUBER, P. (1964). Robust estimation of a location parameter. *Ann. Math. Statist.*, **35**, 73.

HUBER, P. J. (1967). The behaviour of maximum likelihood estimates under nonstandard conditions. In *Proceedings of the Fifth Berkeley Symposium on Mathematical Statistics and Probability*, L. LeCam and J. Neyman (eds). University of California Press, Berkeley, **1**, 221.

HUBER, P. J. (1973). Robust regression, asymptotics, conjectures and Monte Carlo. *Ann. Math. Stat.*, **1**, 799.

HUFFMAN, M. D. (1983). An efficient approximate solution to the Kiefer–Weiss problem. *Ann. Statist.*, **11**, 306.

HUNT, G. and STEIN, C. (1946). Most stringent tests of hypotheses. Unpublished.

HUYNH, H. and FELDT, L. S. (1970). Conditions under which means square ratios in repeated measures designs have exact F-distributions. *J.Amer. Statist. Ass.*, **65**, 1582.

HUYNH, H. and FELDT, L. S. (1976). Estimation of the Box correction for degrees of freedom from sample data in randomized block and split block designs. *J. Educ. Statist.*, **1**, 69.

HUZURBAZAR, V. S. (1948). The likelihood equation, consistency and the maxima of the likelihood function. *Ann. Eugen.*, **14**, 185.

HUZURBAZAR, V. S. (1949). On a property of distributions admitting sufficient statistics. *Biometrika*, **36**, 71.

HUZURBAZAR, V. S. (1955). Confidence intervals for the parameter of a distribution admitting a sufficient statistic when the range depends on the parameter. *J. Roy. Statist. Soc.*, **B**, **17**, 86.

IMAN, R. L. (1982). Graphs for use with the Lilliefors test for normal and exponential distributions. *Amer. Statist.*, **36**, 109.

IWASE, K. and SETO, N. (1983). Uniformly minimum variance unbiased estimation for the inverse Gaussian distribution. *J. Amer. Statist. Ass.*, **78**, 660.

JAMES, W. and STEIN, C. (1961). Estimation with quadratic loss. In *Proceedings of the Fourth Berkeley Symposium on Mathematical Statistics and Probability*, J. Neyman (ed.). University of California Press, Berkeley, **1**, 361.

JARQUE, C. M. and BERA, A. K. (1980) Efficient tests for normality, homoscedasticity and serial independence of regression residuals. *Econ. Lett.*, **6**, 255.

JARQUE, C. M. and BERA, A. K. (1987). A test for normality of observations and regression residuals. *Int. Statist. Rev.*, **55**, 163.

JARRETT, R. G. (1984). Bounds and expansions for Fisher information when the moments are known. *Biometrika*, **71**, 101.

JAYNES, E. T. (1968). Prior probabilities. *IEEE Trans. Systems Cybernet.*, **SSC–4**, 227.

JEFFREYS, H. (1961). *Theory of Probability* (3rd edn). Oxford University Press, Oxford.

JENNISON, C. and TURNBULL, B. (1989). Interim analysis: the repeated confidence interval approach. *J. Roy. Statist. Soc.*, **B, 51**, 305.

JENNISON, C. and TURNBULL, B. (1991). Group sequential tests and repeated confidence intervals. In *Handbook of Sequential Analysis*, B. K. Ghosh and P. K. Sen (eds). Dekker, New York, 283.

JENSEN, D. R. and SOLOMON, H. (1972). A Gaussian approximation to the distribution of a definite quadratic form. *J. Amer. Statist. Ass.*, **67**, 898.

JOANES, D. N. (1972). Sequential tests of composite hypotheses. *Biometrika*, **59**, 633.

JOBSON, J. D. and FULLER, W. A. (1980). Least squares estimation when the covariance matrix and parameter vector are functionally related. *J. Amer. Statist. Ass.*, **75**, 176.

JOHN, S. (1975). Tables for comparing two normal variances or two gamma means. *J. Amer. Statist. Ass.*, **70**, 344.

JOHNSON, N. L. (1950). On the comparison of estimators. *Biometrika*, **37**, 281.

JOHNSON, N. L. (1959). On an extension of the connection between Poisson and χ^2 distributions. *Biometrika*, **46**, 352.

JOHNSON, N. L. and PEARSON, E. S. (1969). Tables of percentage points of non-central χ. *Biometrika*, **56**, 255.

JOHNSON, N. L. and WELCH, B. L. (1939). Applications of the non-central t distribution. *Biometrika*, **31**, 362.

JOHNSON, N. L., KOTZ, S. and KEMP, A. W. (1993). *Univariate Discrete Distributions* (2nd edn). Wiley, New York.

JOHNSON, N. L., KOTZ, S. and BALAKRISHNAN, N. (1995). *Continuous Univariate Distributions*. Volume 2 (2nd edn). Wiley, New York.

JOHNSTONE, D. J. (1989). On the necessity for random sampling. *Brit. J. Phil. Sci.*, **40**, 443.

JOINER, B. L. (1969). The median significance level and other small sample measures of test efficacy. *J. Amer. Statist. Ass.*, **64**, 971.

JORESKÖG, K. G. (1981). Analysis of covariance structures (with discussion). *Scand. J. Statist.*, **8**, 65.

JOSHI, V. M. (1976). On the attainment of the Cramér–Rao lower bound. *Ann. Statist.*, **4**, 998.

JOSHI, V. M. (1989). A counter-example against the likelihood principle. *J. Roy. Statist. Soc.*, **B, 51**, 215.

JUOLA, R. C. (1993). More on shortest confidence intervals. *Amer. Statist.*, **47**, 117.

KAC, M., KEIFER, J. and WOLFOWITZ, J. (1955). On tests of normality and other tests of goodness of fit based on distance methods. *Ann. Math. Statist.*, **26**, 189.

KADIYALA, K. R. and OBERHELMAN, H. D. (1984). Alternative tests for heteroscedasticity of disturbances: A comparative study. *Commun. Statist.*, **A**, **13**, 987.

KAGAN, A. M., LINNIK, Yu. V. and RAO, C. R. (1973). *Characterization Problems in Mathematical Statistics*. Wiley, New York.

KAILATH, T. (1974). A view of three decades of linear filtering theory. *IEEE Trans. Info. Theory*, IT-20, 145.

KALBFLEISCH, J. D. (1975). Sufficiency and conditionality (with discussion). *Biometrika*, **62**, 251.

KALBFLEISCH, J. D. and SPROTT, D. A. (1970). Application of likelihood methods to models involving large numbers of parameters (with discussion). *J. Roy. Statist. Soc.*, **B**, **32**, 175.

KALE, B. K. (1961). On the solution of the likelihood equation by iteration processes. *Biometrika*, **48**, 452.

KALE, B. K. (1962). On the solution of likelihood equations by iteration processes. The multiparametric case. *Biometrika*, **49**, 479.

KALLENBERG, W. C. M. (1983). Intermediate efficiency theory and examples. *Ann. Statist.*, **11**, 170.

KALLENBERG, W. C. M. (1985). On moderate and large deviations in multinomial distributions. *Ann. Statist.*, **13**, 1554.

KALLENBERG, W. C. M. and KOUROUKLIS, S. (1992). Hodges–Lehmann optimality of tests. *Statist. Prob. Lett.*, **14**, 31.

KALLENBERG, W. C. M. and LEDWINA, T. (1987). On local and nonlocal measures of efficiency. *Ann. Statist.*, **15**, 1401.

KALLENBERG, W. C. M., OOSTERHOFF, J. and SCHRIEVER, B. F. (1985). The number of classes in chi-squared goodness-of-fit tests. *J. Amer. Statist. Ass.*, **80**, 969.

KAMAT, A. R. and SATKE, Y. S. (1962). Asymptotic power of certain tests criteria (based on first and second differences) for serial correlation between successive differences. *Ann. Math. Statist.*, **33**, 186.

KANOH, S. (1988). The reduction of the width of confidence bands in linear regression. *J. Amer. Statist. Ass.*, **83**, 116.

KARAKOSTAS, K. X. (1985). On minimum variance unbiased estimates. *Amer. Statist.*, **39**, 303.

KARIYA, T. (1980). Note on a condition for equality of sample variances in a linear model. *J. Amer. Statist. Ass.*, **75**, 701.

KASTENBAUM, M. A., HOEL, D. G. and BOWMAN, K. O. (1970a). Sample size requirements: One-way analysis of variance. *Biometrika*, **57**, 421.

KASTENBAUM, M. A., HOEL, D. G. and BOWMAN, K. O. (1970b). Sample size requirements: Randomized block designs. *Biometrika* , **57**, 573.

KATTI, S. K. and GURLAND, J. (1962). Efficiency of certain methods of estimation for the negative binomial and the Neyman type A distributions. *Biometrika*, **49**, 215.

KEATING, J. P., MASON, R. L. and SEN, P. K. (1993). *Pitman's Measure of Closeness*. Society for Industrial and Applied Mathematics, Philadelphia.

KEMPTHORNE, O. (1967). The classical problem of inference – goodness of fit. In *Proceedings of the Fifth Berkeley Symposium on Mathematical Statistics and Probability*, L. LeCam and J. Neyman (eds). University of California Press, Berkeley, **1**, 235.

KENDALL, M. G. (1949). On the reconciliation of theories of probability. *Biometrika*, **36**, 101.

KENDALL, M. G. and GIBBONS, J. D. (1990). *Rank Correlation Methods* (5th edn). Edward Arnold, London; Oxford University Press, New York.

KENDALL, M. G. and ORD, J. K. (1990). *Time Series* (3rd edn). Edward Arnold, London.

KENT, J. T. (1982). Robust properties of likelihood tests. *Biometrika*, **69**, 19.

KENT, J. T. (1983). Information gain and a general measure of correlation. *Biometrika*, **70**, 163.

KENT, J. T. (1986). The underlying structure of non-nested hypothesis tests. *Biometrika*, **73**, 333.

KEULS, M. (1952). The use of the studentized range in connection with analysis of variance. *Euphitica*, **1**, 112.

KEYNES, J. M. (1911). The principal averages and the laws of error which lead to them. *J. Roy. Statist. Soc.*, **74**, 322.

KEYNES, J. M. (1921). *A Treatise on Probability*. Macmillan, London.

KIANIFORD, F. and SWALLOW, W. H. (1996). A review of the development and application of recursive residuals in linear models. *J. Amer. Statist. Ass.*, **91**, 391.

KIEFER, J. (1952). On minimum variance estimators. *Ann. Math. Statist.*, **23**, 627.

KIEFER, J. (1957). Invariance, minimax sequential estimation and continuous time processes. *Ann. Math. Stat.*, **28**, 573.

KIEFER, J. (1977). Conditional confidence statements and confidence estimators. *J. Amer. Statist. Ass.*, **72**, 789.

KIEFER, J. and WEISS, L. (1957). Some properties of generalized sequential probability ratio tests. *Ann. Math. Statist.*, **28**, 57.

KIEFER, J. and WOLFOWITZ, J. (1956). Consistency of the maximum likelihood estimator in the presence of infinitely many incidental parameters. *Ann. Math. Statist.*, **27**, 887.

KIM, C., STORER, B. E. and JEONG, M. (1996). A note on Box–Cox transformation diagnostics. *Technometrics*, **38**, 178.

KIMBALL, B. F. (1946). Sufficient statistical estimation functions for the parameters of the distribution of maximum values. *Ann. Math. Statist.*, **17**, 299.

KNIGHT, W. (1965). A method of sequential estimation applicable to the hypergeometric, binomial, Poisson, and exponential distributions. *Ann. Math. Statist.*, **36**, 1494.

KNOTT, M. (1975). On the minimum efficiency of least squares. *Biometrika*, **62**, 129.

KOEHLER, K. J. and GAN, F. F. (1990). Chi-square goodness of fit tests: Cell selection and power. *Commun. Statist.*, **B**, **19**, 1265.

KOEHLER, K. J. and LARNTZ, K. (1980). An empirical investigation of goodness-of-fit statistics for sparse multinomials. *J. Amer. Statist. Ass.*, **75**, 336.

KOENKER, R. and BASSETT, G. (1978). Robust tests for heteroscedasticity based on regression quantiles. *Econometrica*, **50**, 43.

KOHN, R., SHIVELY, T. S. and ANSLEY, C. F. (1993). Computing p-values for the generalised Durbin–Watson statistics and residual autocorrelations in regression. *Appl. Statist.*, **42**, 249.

KÖLLERSTRÖM, J. and WETHERILL, G. B. (1979). SPRT's for the normal correlation coefficient. *J. Amer. Statist. Ass.*, **74**, 815.

KOLMOGOROV, A. (1933). Sulla determinazione empirica di una legge disitribuzione. *G. Ist. Ital. Attuari*, **4**, 83.

KOLODZIECZYK, S. (1935). On an important class of statistical hypotheses. *Biometrika*, **27**, 161.

KONIJN, H. S. (1956, 1958). On the power of certain tests for independence in bivariate populations. *Ann. Math. Statist.*, **27**, 300 and **29**, 935.

KOOPMAN, B. O. (1936). On distributions admitting a sufficient statistic. *Trans. Amer. Math. Soc.*, **39**, 399.

KOOPMAN, R. F. (1983). On the standard error of the modified biserial correlation. *Psychometrika*, **48**, 639.

KORN, E. L. (1984). The range of limiting values of some partial correlations under conditional independence. *Amer. Statist.*, **38**, 61.

KOSCHAT, M. A. (1987). A characterization of the Fieller solution. *Ann. Statist.*, **15**, 462.

KOUROUKLIS, S. and PAIGE, C. C. (1981). A constrained least squares approach to the general Gauss–Markov linear model. *J. Amer. Statist. Ass.*, **76**, 620.

KOZIOL, J. A. (1978). Exact slopes of certain multivariate tests of hypotheses. *Ann. Statist.*, **6**, 546.

KOZIOL, J. A. (1980). On a Cramer–Von Mises-type statistic for testing symmetry. *J. Amer. Statist. Ass.*, **75**, 161.

KRAEMER, H. C. (1981). Modified biserial correlation coefficients. *Psychometrika*, **46**, 275.

KRAFT, C. H. and LECAM, L. M. (1956). A remark on the roots of the maximum likelihood equation. *Ann. Math. Statist.*, **27**, 1174.

KRAMER, C. Y. (1956). Extension of multiple range tests to group means with unequal number of replications. *Biometrics*, **12**, 309.

KRAMER, K. C. (1963). Tables for constructing confidence limits on the multiple correlation coefficient. *J. Amer. Statist. Ass.*, **58**, 1082.

KRAMER, W. (1980). Finite sample efficiency of ordinary least squares in the linear regression model with autocorrelated errors. *J. Amer. Statist. Ass.*, **75**, 1005.

KREUGER, R. G. and NEUDECKER, H. (1977). Exact linear restrictions on parameters in the general linear model with a singular covariance matrix. *J. Amer. Statist. Ass.*, **72**, 430.

KRISHNAN, M. (1966). Locally unbiased type M test. *J. Roy. Statist. Soc.*, **B**, **28**, 298.

KRISHNAN, M. (1967). The moments of a doubly non-central t distribution. *J. Amer. Statist. Ass.*, **62**, 278.

KRISHNAN, M. (1968). Series representations of the doubly noncentral t-distribution. *J. Amer. Statist. Ass.*, **63**, 1004.

KRUSKAL, J. B. (1965). Analysis of factorial experiments by estimating monotone transformations of the data. *J. Roy. Statist. Soc.*, **B**, **27**, 251.

KRUSKAL, W. (1961). The coordinate-free approach to Gauss–Markov estimation, and its application to missing and extra observations. In *Proceedings of the Fourth Berkeley Symposium on Mathematical Statistics and Probability*, J. Neyman (ed.). University of California Press, Berkeley, **1**, 435.

KRUSKAL, W. H. (1958). Ordinal measures of association. *J. Amer. Statist. Ass.*, **53**, 814.

KRUTCHKOFF, R. G. (1967). Classical and inverse regression methods of calibration. *Technometrics*, **9**, 425.

KRZANOWSKI, W. J. and MARRIOTT, F. H. C. (1994). *Multivariate Analysis, Part 1: Distributions, Ordination and Inference*. Arnold, London.

KRZANOWSKI, W. J. and MARRIOTT, F. H. C. (1995). *Multivariate Analysis, Part 2: Classification, Covariance Structures and Repeated Measures*. Arnold, London.

KUHN, T. S. (1970). *The Structure of Scientific Revolutions* (2nd edn). University of Chicago Press, Chicago.

KULLDORF, G. (1958). Maximum likelihood estimation of the mean/standard deviation of a normal random variable when the sample is grouped. *Skand. Aktuartidskr.*, **41**, 1 and 18.

KULLDORF, G. (1961). *Contributions to the Theory of Estimation from Grouped and Partially Grouped Samples*. Almqvist & Wiksell, Stockholm.

LAI, T. L. (1978). Pitman efficiencies of sequential tests and uniform limit theorems in non-parametric statistics. *Ann. Statist.*, **6**, 1027.

LAI, T.-Z. (1991). Asymptotic optimality of generalized sequential likelihood ratio tests in some classical sequential testing problems. In *Handbook of Sequential Analysis*, B. K. Ghosh, and P. K. Sen (eds). Dekker, New York, 121.

LAIRD, N. (1993). The EM algorithm. In *Handbook of Statistics, Vol. 9: Computational Statistics*, C. R. Rao (ed.). Elsevier, Amsterdam, 509.

LAIRD, N., LANGE, N. and STRAM, D. (1987). Maximum likelihood computations with repeated measures: Applications of the EM algorithm. *J. Amer. Statist. Ass.*, **82**, 97.

LAKATOS, I. (1974). Falsification and the methodology of scientific research programs. In *Criticism and the Growth of Knowledge*, I. Lakatos and A. E. Musgrave (eds). Cambridge University Press, Cambridge, 91.

LAMBERT, D. and HALL, W. J. (1982). Asymptotic lognormality of P-values. *Ann. Statist.*, **10**, 44.

LAND, C. E. (1971). Confidence intervals for linear functions of the normal mean and variance. *Ann. Math. Statist.*, **42**, 1187.

LAND, C. E. (1974). Confidence interval estimation for means after data transformations to normality. *J. Amer. Statist. Ass.*, **69**, 795.

LANGHOLZ, B. and KRONMAL, R. A. (1991). Tests of distributional hypotheses with nuisance parameters using Fourier series methods. *J. Amer. Statist. Ass.*, **86**, 1077.

LARNTZ, K. (1978). Small-sample comparisons of exact levels for chi-squared goodness-of-fit statistics. *J. Amer. Statist. Ass.*, **73**, 253.

LAUBSCHER, N. F. (1961). On stabilizing the binomial and negative binomial variances. *J. Amer. Statist. Ass.*, **56**, 143.

LAWAL, H. B. and UPTON, G. J. G. (1980). An approximation to the distribution of the χ^2 goodness-of-fit statistic for use with small expectations. *Biometrika*, **67**, 447.

LAWLEY, D. (1938). A generalization of Fisher's z-test. *Biometrika*, **30**, 180.

LAWLEY, D. N. (1956). A general method for approximating to the distribution of likelihood ratio criteria. *Biometrika*, **43**, 295.

LAWRANCE, A. J. (1987). The score statistic for regression transformation. *Biometrika*, **74**, 275.

LAWRANCE, A. J. (1988). Regression transformation diagnostics using local influence. *J. Amer. Statist. Ass.*, **83**, 1067.

LAWRANCE, A. J. (1995). Deletion influence and masking in regression. *J. Roy. Statist. Soc.*, **B**, **57**, 181.

LAWTON, W. H. (1965). Some inequalities for central and non-central distributions. *Ann. Math. Statist.*, **36**, 1521.

LEAMER, E. E. (1981). Coordinate-free ridge regression bounds. *J. Amer. Statist. Ass.*, **76**, 842.

LECAM, L. (1953). On some asymptotic properties of maximum likelihood estimates and related Bayes' estimates. *University Calif. Publ. Statist.*, **1**, 277.

LECAM, L. (1970). On the assumptions used to prove asymptotic normality of maximum likelihood estimates. *Ann. Math. Statist.*, **41**, 802.

LEE, A. F. S. and GURLAND, J. (1975). Size and power of tests for equality of means of two normal populations with unequal variances. *J. Amer. Statist. Ass.*, **70**, 933.

LEE, Y. S. (1971). Some results on the sampling distribution of the multiple correlation coefficient. *J. Roy. Statist. Soc.*, **B**, **33**, 117.

LEE, Y. S. (1972). Tables of upper percentage points of the multiple correlation coefficient. *Biometrika*, **59**, 175.

LEECH, D. and COWLING, K. (1982). Generalized regression estimation from grouped observations: A generalization and an application to the relationship between diet and mortality. *J. Roy. Statist. Soc.*, **A**, **145**, 208.

LEHMANN, E. L. (1947). On optimum tests of composite hypotheses with one constraint. *Ann. Math. Statist.*, **18**, 473.

LEHMANN, E. L. (1949). Some comments on large sample tests. In *Proceedings of the First Berkeley Symposium on Mathematical Statistics and Probability*, J. Neyman (ed.). University of California Press, Berkeley, 451.

LEHMANN, E. L. (1950). Some principles of the theory of testing hypotheses. *Ann. Math. Statist.*, **21**, 1.

LEHMANN, E. L. (1983a). Estimation with inadequate information. *J. Amer. Statist. Ass.*, **78**, 624.

LEHMANN, E. L. (1983b). *Theory of Point Estimation*. Wiley, New York.

LEHMANN, E. L. (1986). *Testing Statistical Hypotheses*. (2nd edn). Wiley, New York.

LEHMANN, E. L. (1993). The Fisher and Neyman–Pearson theories of testing hypotheses: One theory or two? *J. Amer. Statist. Ass.*, **88**, 1242.

LEHMANN, E. L. and SCHEFFÉ, H. (1950, 1955). Completeness, similar regions and unbiased estimation. *Sankhyā*, **10**, 305 and **15**, 219.

LEHMANN, E. L. and SCHOLZ, F. W. (1992). Ancillarity. In *Current Issues in Statistical Inference: Essays in Honor of D. Basu*, 32. M. Ghosh and P. K. Pathak (eds), Institute of Mathematical Statistics, Haywood, CA,

LEHMANN, E. L. and STEIN, C. (1948). Most powerful tests of composite hypotheses. I. Normal distributions. *Ann. Math. Statist.*, **19**, 495.

LEHMANN, E. L. and STEIN, C. (1950). Completeness in the sequential case. *Ann. Math. Statist.*, **21**, 376.

LEHMANN, E. L. and STEIN, C. (1953). The admissibility of certain invariant statistical tests involving a translation parameter. *Ann. Math. Stat.*, **24**, 473.

LEHMER, E. (1944). Inverse tables of probabilities of errors of the second kind. *Ann. Math. Statist.*, **15**, 388.

LELE, S. (1991). Jackknifing linear estimating equations: Asymptotic theory and applications in stochastic processes. *J. Roy. Statist. Soc.*, **B**, **53**, 253.

LESLIE, J. R., STEPHENS, M. A. and FOTOPOULOS, S. (1986). Asymptotic distribution of the Shapiro–Wilks *W* for testing normality. *Ann. Statist.*, **14**, 1497.

LEWONTIN, R. C. and PROUT, T. (1956). Estimation of the number of different classes in a population. *Biometrics*, **12**, 211.

LIDDELL, I. G. and ORD, J. K. (1978). Linear-circular correlation coefficients: Some further results. *Biometrika*, **65**, 448.

LILLIEFORS, H. W. (1967). On the Kolmogorov–Smirnov test for normality with mean and variance unknown. *J. Amer. Statist. Ass.*, **62**, 399.

LIN, C. C. and MUDHOLKAR, G. S. (1980). A simple test for normality against asymmetric alternatives. *Biometrika*, **67**, 455.

LINDLEY, D. V. (1950). Grouping corrections and maximum likelihood equations. *Proc. Camb. Phil. Soc.*, **46**, 106.

LINDLEY, D. V. (1953). Statistical inference. *J. Roy. Statist. Soc.*, **B**, **15**, 30.

LINDLEY, D. V. (1958). Fiducial distributions and Bayes' theorem. *J. Roy. Statist. Soc.*, **B**, **20**, 102.

LINDLEY, D. V. (1971). *Bayesian Statistics Review.* Society for Industrial and Applied Mathematics, Philadelphia.

LINDLEY, D. V. and SMITH, A. F. M. (1972). Bayesian estimates for the linear model. *J. Roy. Statist. Soc.*, **B**, **34**, 1.

LINDLEY, D. V., EAST, D. A. and HAMILTON, P. A. (1960). Tables for making inferences about the variance of a normal distribution. *Biometrika*, **47**, 433.

LINDSAY, B. (1982). Conditional score functions: Some optimality results. *Biometrika*, **69**, 503.

LINDSAY, B. G. and ROEDER, K. (1987). A unified treatment of integer parameter models. *J. Amer. Statist. Ass.*, **82**, 758.

LINNIK, Yu, V. (1967). On the elimination of nuisance parameters in statistical problems. In *Proceedings of the Fifth Berkeley Symposium on Mathematical Statistics and Probability*, L. LeCam and J. Neyman (eds). University of California Press, Berkeley, **1**, 267.

LINNIK, Yu. V. (1964). On the Behrens–Fisher problem. *Bull. Int. Statist. Inst.*, **40**(2), 833.

LINSSEN, M. N. (1991). A table for solving the Behrens–Fisher problem. *Statist. Prob. Lett.*, **11**, 359.

LITTLE, R. J. and RUBIN, D. B. (1987). *Statistical Analysis with Missing Data*. Wiley, New York and London.

LLOYD, E. H. (1952). Least-squares estimation of location and scale parameters using order statistics.. *Biometrika*, **39**, 88.

LOCKS, M. O., ALEXANDER, M. J. and BYARS, B. J. (1963). New tables of the noncentral t-distribution. *Aeronaut. Res. Lab. (Ohio)*, no. ARL 63-19.

LOH, W-Y. (1985). A new method for testing separate families of hypotheses. *J. Amer. Statist. Ass.*, **80**, 362.

LOONEY, S. W. (1995). How to use tests for univariate normality to assess multivariate normality. *Amer. Statist.*, **49**, 64.

LORD, F. M. (1963). Biserial estimates of correlation. *Psychometrika*, **28**, 81.

LORDEN, G. (1976). Two SPRT's and the modified Kiefer–Weiss problem of minimizing the expected sample size. *Ann, Statist.*, **4**, 281.

LOUIS, T. A. (1982). Finding the observed information matrix when using the EM algorithm. *J. Roy. Statist. Soc.*, **B**, **44**, 226.

LWIN, T. (1975). Exponential family distribution with a truncation parameter. *Biometrika*, **62**, 218.

LYON, J. D. and TSAI, C.-L. (1996). A comparison of tests for heteroscedasticity. *The Statistician*, **45**, 337.

MACKINNON, W. J. (1964). Tables for both the sign test and the distribution-free confidence intervals of the median for sample sizes to 1000. *J. Amer. Statist. Ass.*, **59**, 935.

MADANSKY, A. (1962). More on length of confidence intervals. *J. Amer. Statist. Ass.*, **57**, 586.

MÄKELÄINEN, T., SCHMIDT, K. and STYAN, G. P. H. (1981). On the existence and uniqueness of the maximum likelihood estimate of a vector-valued parameter in fixed-size samples. *Ann. Statist.*, **9**, 758.

MALINVAUD, E. (1966). *Statistical Methods of Econometrics*. North-Holland, Amsterdam.

MALLOWS, C. L. (1973). Some comments on C_p. *Technometrics*, **15**, 661.

MALLOWS, C. L. (1975). On some topics in robustness. Bell Telephone Labs, Murray Hill, NJ. Unpublished.

MANLY, B. F. J. (1970). The choice of a Wald test on the mean of a normal population. *Biometrika*, **57**, 91.

MANN, H. B. and WALD, A. (1942). On the choice of the number of intervals in the application of the chi-square test. *Ann. Math. Statist.*, **13**, 306.

MARDIA, K. V. and KENT, J. T. (1991). Rao score tests for goodness of fit and independence. *Biometrika*, **78**, 355.

MARDIA, K. V. and SUTTON, T. W. (1978). A model with cylindrical variables with applications. *J. Roy. Statist. Soc.*, **B**, **40**, 229.

MARGOLIN, B. H. and MAURER, W. (1976). Tests of the Kolmogorov–Smirnov type for exponential data with unknown scale, and related problems. *Biometrika*, **63**, 149.

MARITZ, J. S. (1953). Estimation of the correlation coefficient in the case of a bivariate normal population when one of the variables is dichotomised. *Psychometrika*, **18**, 97.

MARITZ, J. S. and LWIN, T. (1989). *Empirical Bayes Methods* (2nd edn). Chapman & Hall, London.

MARSAGLIA, G. (1964). Conditional means and covariances of normal variables with singular covariance matrix. *J. Amer. Statist. Ass.*, **59**, 1203.

MASSEY, F. J., Jr. (1950). A note on the estimation of a distribution function by confidence limits. *Ann. Math. Statist.*, **21**, 116.

MASSEY, F. J., Jr. (1951). The Kolmogorov–Smirnov test of goodness of fit. *J. Amer. Statist. Ass.*, **46**, 68.

MATLOFF, N., ROSE, R. and TAI, R. (1984). A comparison of two methods for estimating optimal weights in regression analysis. *J. Statist. Comp. Simul.*, **19**, 265.

MAULDON, J. G. (1955). Pivotal quantities for Wishart's and related distributions, and a paradox in fiducial theory. *J. Roy. Statist., Soc.*, **B**, **17**, 79.

MCCULLAGH, P. (1984). Local sufficiency. *Biometrika*, **71**, 233.

MCCULLAGH, P. (1992). Conditional inference and Cauchy models. *Biometrika*, **79**, 247.

MCCULLAGH, P. (1993). On the distribution of the Cauchy maximum likelihood estimator. *Proc. Roy. Soc. Lond.*, **A**, **440**, 475.

MCCULLAGH, P. and COX, D. R. (1986). Invariants and likelihood ratio statistics. *Ann. Statist.*, **14**, 1419.

MCCULLAGH, P. and NELDER, J. A. (1983). *Generalized Linear Models*. Chapman & Hall, London and New York.

MCELROY, F. W. (1967). A necessary and sufficient condition that ordinary least-squares estimators can be best linear unbiased. *J. Amer. Statist. Ass.*, **62**, 1302.

MCKENDRICK, A. G. (1926). Applications of mathematics to medical problems. *Proc. Edin. Math. Soc.*, **44**, 98.

MCNOLTY F. (1962). A contour-integral derivation of the non-central chi-square distribution. *Ann. Math. Statist.*, **33**, 796.

MEHTA, J. S. and SRINIVASAN, R. (1970). On the Behrens–Fisher problem. *Biometrika*, **57**, 649.

MENG, X. L. and RUBIN, D. B. (1993). Maximum likelihood estimation via the ECM algorithm: A general framework. *Biometrika*, **80**, 267.

MICKEY, M. R. and BROWN, M. B. (1966). Bounds on the distribution functions of the Behrens–Fisher statistic. *Ann. Math. Statist.*, **37**, 639.

MILLER, A. J. (1984). Selection of subsets of regression variables. *J. Roy. Statist. Soc.*, **A**, **147**, 389.

MILLER, D. (1990). A restoration of Popperian inductive scepticism. *Brit. J. Phil. Sci.*, **41**, 137.

MILLER, L. H. (1956). Table of percentage points of Kolmogorov statistics. *J. Amer. Statist. Ass.*, **51**, 111.

MILLER, R. G., Jr. (1964). A trustworthy jackknife. *Ann. Math. Statist.*, **35**, 1549.

MILLER, R. G. (1966). *Simultaneous Statistical Inference* . McGraw-Hill, New York.

MILLER, R. G. (1974). The jackknife – a review. *Biometrika* , **61**, 1.

MITCHELL, T. J. and BEAUCHAMP, J. J. (1988). Bayesian variable selection in linear regression (with discussion). *J. Amer. Statist. Ass.*, **83**, 1023.

MIYASHITA, H. and NEWBOLD, P. (1983). On the sensitivity to non-normality of a test for outliers in linear models. *Commun. Statist.*, **A**, **12**, 1413.

MIZON, G. E. (1995). A simple message for autocorrelation correctors: Don't. *J. Econometrics*, **69**, 267.

MOLINARI, L. (1977). Distribution of the chi-squared test in nonstandard situations. *Biometrika*, **64**, 115.

MONTGOMERY, D. C. (1997). *Introduction to Statistical Quality Control* (3rd edn). Wiley, New York.

MOORE, D. S. (1971). A chi-square statistic with random cell boundaries. *Ann. Math. Statist.*, **42**, 147.

MOORE, D. S. (1977). Generalized inverses, Wald's method, and the construction of chi-squared tests of fit. *J. Amer. Statist. Ass.*, **72**, 131.

MOORE, D. S. (1982). The effect of dependence on chi-squared tests of fit. *Ann. Statist.*, **10**, 1163.

MOORE, D. S. and SPRUILL, M. D. (1975). Unified large-sample theory of general chi-squared statistics for tests of fit. *Ann. Statist.*, **3**, 599.

MORAN, P. A. P. (1950). The distribution of the multiple correlation coefficient. *Proc. Camb. Phil. Soc.*, **46**, 521.

MORAN, P. A. P. (1970). On asymptotically optimal tests of composite hypotheses. *Biometrika*, **57**, 47.

MOSHMAN, J. (1958). A method for selecting the size of the initial sample in Stein's two sample procedure. *Ann. Math. Statist.*, **29**, 1271.

MOSTELLER, F. and YOUTZ, C. (1961). Tables of the Freeman–Tukey transformations for the binomial and Poisson distributions. *Biometrika*, **48**, 433.

MOUSTAKIDES, G. V. (1986). Optimal stopping times for detecting changes in distributions. *Ann. Statist.*, **14**, 1379.

MUDHOLKAR, G. S., KOLLIA, G. D., LIN, C. T. and PATEL, K. R. (1991). A graphical procedure for computing goodness-of-fit tests. *J. Roy. Statist. Soc.*, B, **53**, 221.

MUDHOLKAR, G. S., MCDERMOTT, M. and SRIVASTAVA, D. K. (1992). A test of p-variate normality. *Biometrika*, **79**, 850.

MUIRHEAD, R. J. (1985). Estimating a particular function of the multiple correlation coefficient. *J. Amer. Statist. Ass.*, **80**, 923.

MUKHOPADHYAY, N. (1991). Parametric sequential point estimation. In *Handbood of Sequential Analysis*, B. K. Ghosh, and P. K. Sen (eds). Dekker, New York, 245.

MULLER, K. E. and BARTON, C. N. (1989). Approximate power for repeated-measures ANOVA lacking sphericity. *J. Amer. Statist. Ass.*, **84**, 549.

MURPHY, R. B. (1948). Non-parametric tolerance limits. *Ann. Math. Statist.*, **19**, 581.

MYERS, M. H., SCHNEIDERMAN, M.A. and ARMITAGE, P. (1966). Boundaries for closed (wedge) sequential t test plans. *Biometrika*, **53**, 431.

NADDEO, A. (1968). Confidence intervals for the frequency function and the cumulative frequency function of a sample drawn from a discrete random variable. *Rev. Int. Statist. Inst.*, **36**, 313.

NAGARSENKER, P. B. (1980). On a test of equality of several exponential distributions. *Biometrika*, **67**, 475.

NAGARSENKER, P. B. (1984). On Bartlett's test for homogeneity of variances. *Biometrika*, **71**, 405.

NARULA, S. C. (1979). Orthogonal polynomial regression. *Int. Statist. Rev.*, **47**, 31.

NARULA, S. C. and WELLINGTON, J. F. (1979). Selection of variables in linear regression using the minimum of weighted absolute errors criterion. *Technometrics*, **21**, 299.

NARULA, S. C. and WELLINGTON, J. F. (1982). The minimum sum of absolute errors regression: A state of the art survey. *Int. Statist. Rev.*, **50**, 317.

NELDER, J. A. and WEDDERBURN, R. W. M. (1972). Generalized linear models. *J. Roy. Statist. Soc.*, **A**, **135**, 370.

NEWEY, W. and WEST, K. (1987). A simple positive definite heteroscedasticity and autocorrelation covariance matrix. *Econometrica* , **55**, 703.

NEWEY, W. and WEST, K. (1994). Automatic lag selection in covariance matrix estimation. *Rev. Econ. Studies*, **61**, 631.

NEWMAN, D. (1938). The distribution of the range in samples from the normal population expressed in terms of an independent estimation of standard deviation. *Biometrika*, **31**, 20.

NEYMAN, J. (1935).* Sur la vérification des hypothèses statistiques composées. *Bull. Soc. Math. France*, **63**, 1.

NEYMAN, J. (1937).* Outline of a theory of statistical estimation based on the classical theory of probability. *Phil. Trans.*, **A**, **236**, 333.

NEYMAN, J. (1938a).* On statistics the distribution of which is independent of the parameters involved in the original probability law of the observed variables. *Statist. Res. Mem.*, **2**, 58.

NEYMAN, J. (1938b). Tests of statistical hypotheses which are unbiased in the limit. *Ann. Math. Statist.*, **9**, 69.

NEYMAN, J. (1949).* Contribution to the theory of the χ^2 test. In *Proceedings of the First Berkeley Symposium on Mathematical Statistics and Probability*, J. Neyman (ed.). University of California Press, Berkeley, 239.

NEYMAN, J. (1959). Optimal asymptotic tests of composite statistical hypotheses. In *Probability and Statistics*, U. Grenander (ed.). Almqvist & Wiksell, Stockholm.

NEYMAN, J. (1962). Two breakthroughs in the theory of statistical decision making. *Rev. Int. Statist. Inst.*, **30**, 11.

NEYMAN, J. and PEARSON, E. S. (1928).* On the use and interpretation of certain test criteria for the purposes of statistical inference. *Biometrika*, **A**, **20**, 175 and 263.

NEYMAN, J. and PEARSON, E. S. (1931).* On the problem of *k* samples. *Bull. Acad. Polon. Sci.*, **3**, 460.

NEYMAN, J. and PEARSON, E. S. (1933a).* On the testing of statistical hypotheses in relation to probabilities a priori. *Proc. Camb. Phil. Soc.*, **29**, 492.

NEYMAN, J. and PEARSON, E. S. (1933b).* On the problem of the most efficient tests of statistical hypotheses. *Phil. Trans.*, **A**, **231**, 289.

NEYMAN, J. and PEARSON, E. S. (1936a).* Sufficient statistics and uniformly most powerful tests of statistical hypotheses. *Statist. Res. Mem.*, **1**, 113.

NEYMAN, J. and PEARSON, E. S. (1936b).* Unbiased critical regions of Type A and Type A_1. *Statist. Res. Mem.*, **1**, 1.

NEYMAN, J. and PEARSON, E. S. (1938).* Certain theorems on unbiased critical regions of Type A, *and* Unbiased tests of simple statistical hypotheses specifying the values of more than one unknown parameter. *Statist. Res. Mem.*, **2**, 25.

NEYMAN, J. and SCOTT, E. L. (1948). Consistent estimates based on partially consistent observations. *Econometrica*, **16**, 1.

NEYMAN, J. and SCOTT, E. L. (1960). Correction for bias introduced by transformation of variables. *Ann. Math. Statist.*, **31**, 643.

NEYMAN, J., IWASEKIEWICZ, K. and KOLODZIEJCZYK, S. (1935).* Statistical problems in agricultural experimentation. *Suppl. J. Roy. Statist. Soc.*, **2**, 107.

NOETHER, G. E. (1955). On a theorem of Pitman. *Ann. Math. Statist.*, 26, 64.

NOETHER, G. E. (1957). Two confidence intervals for the ratio of two probabilities, and some measures of effectiveness. *J. Amer. Statist. Ass.*, **52**, 36.

NOETHER, G. E. (1963). Note on the Kolmogorov statistic in the discrete case. *Metrika*, **7**, 115.

ODEH, R. E. (1982). Critical values of the sample product-moment correlation coefficient in the bivariate normal distribution. *Commun. Statist.*, **B**, **11**, 1.

OGBURN, W. G. (1935). Factors in the variation of crime among cities. *J. Amer. Statist. Ass.*, **30**, 12.

O'HAGAN, A. (1994). *Baysian Inference*. Kendall's Advanced Theory of Statistics, Volume 2B. Arnold, London.

OLIVER, E. H. (1972). A maximum likelihood oddity. *Amer. Statist.* **26**, 43.

OLKIN, I. and PRATT, J. W. (1958). Unbiased estimation of certain correlation coefficients. *Ann. Math. Statist.*, **29**, 201.

ORCHARD, T. and WOODBURY, M. A. (1972). A missing formation principle: Theory and applications. In *Proceedings of the Sixth Berkeley Symposium on Mathematical Statististics and Probability*, L. LeCam, J. Neyman and E. L. Scott (eds). Universityof California Press, Berkeley, **1**, 697.

OSBORNE, C. (1991). Statistical calibration: A review. *Int. Statist. Rev.*, **59**, 309.

OSBORNE, M. R. (1992). Fisher's method of scoring. *Int. Statist. Rev.*, **60**, 99.

OWEN, D. B. (1962). *Handbook of Statistical Tables.* Addison-Wesley, Reading, MA.

OWEN, D. B. (1963). Factors for one-sided tolerance limits and for variables sampling plans. *Sandia Corp. Monogr.* (Off. Techn. Serv., Dept. Commerce, Washington), SCR-607.

OWEN, D. B. (1965). The power of Student's t-test. *J. Amer. Statist. Ass.*, **60**, 320.

OWEN, D. B. and ODEH, R. E. (eds) (1977). *Selected Tables in Mathematical Statistics*, Vol. V. American Mathematical Society, Providence, RI.

PACHARES, J. (1960). Tables of confidence limits for the binomial distribution. *J. Amer. Statist. Ass.*, **55**, 521.

PACHARES, J. (1961). Tables for unbiased tests on the variance of a normal population. *Ann. Math. Statist.*, **32**, 84.

PAGE, E. S. (1954). Continuous inspection schemes. *Biometrika*, **41**, 100.

PARR, W. C. and SCHUCANY, W. R. (1980). The jackknife: A bibliography. *Int. Statist. Rev.*, **48**, 73.

PATIL, G. P. and SHORROCK, R. (1965). On certain properties of the exponential-type families. *J. Roy. Statist. Soc.*, **B**, **27**, 94.

PATNAIK, P. B. (1949). The non-central χ^2- and F-distributions and their applications. *Biometrika*, **36**, 202.

PAUL, S. R. (1990). Maximum likelihood estimation of intraclass correlation in the analysis of familial data. *Biometrika*, **77**, 549.

PAULSON, E. (1941). On certain likelihood-ratio tests associated with the exponential distribution. *Ann. Math. Statist.*, **12**, 301.

PAULSON, E. (1952a). An optimum solution to the k-sample slippage problem for the normal distribution. *Ann. Math. Statist.*, **23**, 610.

PAULSON, E. (1952b). On the comparison of several experimental categories with a control. *Ann. Math. Stat.*, **23**, 239.

PEARSON, E. S. (1959). Note on an approximation to the distribution of non-central χ^2. *Biometrika*, **46**, 364.

PEARSON, E. S. and HARTLEY, H. O. (1951). Charts of the power function for analysis of variance tests derived from the non-central F-distribution. *Biometrika*, **38**, 112.

PEARSON, E. S. and HARTLEY, H. O. (eds) (1966). *Biometrika Tables for Statisticians*, Vol. 1. Cambridge University Press, London.

PEARSON, E. S. and HARTLEY, H. O. (eds) (1972). *Biometrika Tables for Statisticians.* Vol. 2. Cambridge University Press, London.

PEARSON, E. S. and TIKU, M. L. (1970). Some notes on the relationship between the distributions of central and non-central F. *Biometrika*, **57**, 175.

PEARSON, E. S., D'AGOSTINO, R. B. and BOWMAN, K. O. (1977). Tests for departure for normality: Comparison of powers. *Biometrika*, **64**, 231.

PEARSON, K. (1897). On a form of spurious correlation which may arise when indices are used in the measurement of organs. *Proc. Roy. Soc.*, **60**, 489.

PEARSON, K. (1900).* On a criterion that a given system of deviations from the probable in the case of a correlated system of variables is such that it can be reasonably supposed to have arisen in random sampling. *Phil. Mag.*, (**5**), **50**, 157.

PEARSON, K. (1904).* On the theory of contingency and its relation to association and normal correlation. *Drapers' Co. Memoirs, Biometric Series*, No. 1, London.

PEARSON, K. (1909). On a new method for determining correlation between a measured character A and a character B, of which only the percentage of cases wherein B exceeds (or falls short of) a given intensity is recorded for each grade of A. *Biometrika*, **7**, 96.

PEARSON, K. (1913). On the probable error of a correlation coefficient as found from a fourfold table. *Biometrika*, **9**, 22.

PEDDADA, S. D. (1993). Jackknife variance estimation and bias reduction. In *Handbook of Statistics, Volume 9: Computational Statistics*, C. R. Rao (ed.). Elsevier, Amsterdam, 723.

PEERS, H. W. (1965). On confidence points and Bayesian probability points in the case of several parameters. *J. Roy. Statist. Soc.*, **B**, **27**, 9.

PEERS, H. W. (1971). Likelihood ratio and associated test criteria. *Biometrika*, **58**, 577.

PEERS, H. W. (1978). Second-order sufficiency and statistical invariants. *Biometrika*, **65**, 489.

PEÑA, D. and YOHAI, V. J. (1995). The detection of influential subsets in linear regression by using an influence matrix. *J. Roy. Statist. Soc.*, **B 57**, 145.

PEREIRA, B. de B. (1977). Discriminating among separate models: A bibliography. *Int. Statist. Rev.* **45**, 163.

PERLMAN, M.D. (1972). On the strong consistency of approximate maximum likelihood estimates. In *Proceedings of the Sixth Berkeley Symposium on Mathematical Statistics and Probability*, L. LeCam, J. Neyman and E. L. Scott (eds). University of California Press, Berkeley, **1**, 263.

PFANZAGL, J. (1973). Asymptotic expansions related to minimum contrast estimators. *Ann. Statist.*, **1**, 993.

PFANZAGL, J. (1974). On the Behrens–Fisher problem. *Biometrika*, **61**, 39.

PFEFFERMAN, D. (1984). On extensions of the Gauss–Markov theories to the case of stochastic regression coefficients. *J. Roy. Statist. Soc.*, **B**, **46**, 139.

PHILLIPS, P. C. B. (1982). The true characteristic function of the F distribution. *Biometrika*, **69**, 261.

PICARD, R. R. and COOK, R. D. (1984). Cross-validation of regression residuals. *J. Amer. Statist. Ass.*, **79**, 575.

PIERCE, D. A. and KOPECKY, K. J. (1979). Testing goodness of fit for the distribution of errors in regression models. *Biometrika* , **66**, 1.

PIERCE, D. A. and PETERS, D. (1994). Higher-order asymptotics and the likelihood principle: One parameter models. *Biometrika*, **81**, 1.

PIESAKOFF, M. (1950). Transformation of parameters. Unpublished thesis, Princeton University.

PILLAI, K. C. S. (1955). Some new test criteria in multivariate analysis. *Ann. Math. Stat.*, **26**, 117.

PITMAN, E. J. G. (1936). Sufficient statistics and intrinsic accuracy. *Proc. Camb. Phil. Soc.*, **32**, 567.

PITMAN, E. J. G. (1937). The 'closest' estimates of statistical parameters. *Proc. Camb. Phil. Soc.*, **33**, 212.

PITMAN, E. J. G. (1938). The estimation of the location and scale parameters of a continuous population of any given form. *Biometrika*, **30**, 391.

PITMAN, E. J. G. (1939a). A note on normal correlation. *Biometrika*, **31**, 9.

PITMAN, E. J. G. (1939b). Tests of hypotheses concerning location and scale parameters. *Biometrika*, **31**, 200.

PITMAN, E. J. G. (1948). Non-parametric statistical inference. University of North Carolina Institute of Statistics. Mimeographed lecture notes.

PITMAN, E. J. G. (1957). Statistics and science. *J. Amer. Statist. Ass.*, **52**, 322.

PLACKETT, R. L. (1949). A historical note on the method of least squares. *Biometrika*, **36**, 458.

PLACKETT, R. L. (1950). Some theorems in least squares. *Biometrika*, **37**, 149.

PLACKETT, R. L. (1972). The discovery of the method of least squares. *Biometrika*, **59**, 239.

PLANTE, A. (1991). An inclusion-consistent solution to the problem of absurd confidence intervals. I: Consistent exact confidence interval estimation. *Canad. J. Statist.*, **19**, 389.

POPPER, K. R. (1968). *The Logic of Scientific Discovery*. Hutchinson, London.

POPPER, K. R. (1969). *Conjectures and Refutations*. Routledge & Kegan Paul, London.

POPPER, K. R. and MILLER, D. (1983). A proof of the impossibility of inductive probability. *Nature*, **302**, 687.

PORTNOY, S. (1977). Asymptotic efficiency of minimum variance unbiased estimators. *Ann. Statist.*, **5**, 522.

PRASAD, G. and SAHAI, A. (1982). Sharper variance upper bound for unbiased estimation in inverse sampling. *Biometrika*, **69** , 286.

PRATT, J. W. (1961). Length of confidence intervals. *J. Amer. Statist. Ass.*, **56**, 549.

PRATT, J. W. (1962). Discussion on paper by Birnbaum. *J. Amer. Statist. Ass.*, **57**, 314.

PRATT, J. W. (1963). Shorter confidence intervals for the mean of a normal distribution with known variance. *Ann. Math. Statist.*, **34**, 574.

PRATT, J. W. (1976). F. Y. Edgeworth and R. A. Fisher on the efficiency of maximum likelihood estimation. *Ann. Statist.*, **4**, 501.

PRESS, S. J. (1966). A confidence interval comparison of two test procedures proposed for the Behrens–Fisher problem. *J. Amer. Statist. Ass.*, **61**, 454.

PRZYBOROWSKI, J. and WILEŃSKI, M. (1935). Statistical principles of routine work in testing clover seed for fodder. *Biometrika* , **27**, 273.

QUENOUILLE, M. H. (1956). Notes on bias in estimation. *Biometrika*, **43**, 353.

QUENOUILLE, M. H. (1958). *Fundamentals of Statistical Reasoning*. Griffin, London.

QUENOUILLE, M. H. (1959). Tables of random observations from standard distributions. *Biometrika*, **46**, 178.

QUINE, M. P. and ROBINSON, J. (1985). Efficiencies of chi-square and likelihood ratio goodness-of-fit tests. *Ann. Statist.*, **13**, 727.

RAHMAN, M. and SALEH, A. K. M. E. (1974). Explicit form of the distribution of the Behrens–Fisher *d*-statistic. *J. Roy. Statist. Soc.*, **B**, **36**, 54. *Corrections*, 466.

RAMACHANDRAN, K. V. (1958). A test of variances. *J. Amer. Statist. Ass.*, **53**, 741.

RAMSEY, F. P. (1931). Truth and probability. In *The Foundations of Mathematics and Other Essays*. Kegan Paul, Trench, Tubner, London. Reprinted in H. E. Kyburg, Jr. and H. E. Smokler (eds) (1964). *Studies in Subjective Probability*. Wiley, New York, 61.

RAMSEY, P. (1978). Power differences between pairwise multiple comparisons, *J. Amer. Statist. Ass.*, **73**, 479.

RANDLES, R. H., FLIGNER, M. A., POLICELLO, G.E. II and WILFE, D. A. (1980). An asymptotically distribution-free test for symmetry versus asymmetry. *J. Amer. Statist. Ass.*, **75**, 168.

RAO, B. R. (1958). On an analogue of Cramér–Rao's inequality. *Skand. Aktuartidskr.*, **41**, 57.

RAO, C. R. (1945). Information and accuracy attainable in the estimation of statistical parameters. *Bull. Calcutta Math. Soc.*, **37**, 81.

RAO, C. R. (1947). Minimum variance and the estimation of several parameters. *Proc. Camb. Phil. Soc.*, **43**, 280.

RAO, C. R. (1948). Large sample tests of statistical hypotheses concerning several parameters with applications to problems of estimation. *Proc. Camb. Phil. Soc.*, **44**, 50.

RAO, C. R. (1952). *Advanced Statistical Methods in Biometric Research.* Wiley, New York.

RAO, C. R. (1957). Theory of the method of estimation by minimum chi-square. *Bull. Int. Statist. Inst.*, **35**(2), 25.

RAO, C. R. (1961). Asymptotic efficiency and limiting information. *Proceedings of the Fourth Berkeley Symposium on Mathematical Statistics and Probability*, J. Neyman (ed.). University of California Press, Berkeley, **1**, 531.

RAO, C. R. (1962a). Efficient estimates and optimum inference procedures in large samples. *J. Roy. Statist. Soc.*, **B**, **24**, 46.

RAO, C. R. (1962b). Apparent anomalies and irregularities in maximum likelihood estimation. *Sankhyā*, **A**, **24**, 72.

RAO, C. R. (1967). Least squares theory using an estimated dispersion matrix and its application to measurement of signals. In *Proceedings of the Fifth Berkeley Symposium on Mathematical Statistics and Probability*, L. LeCam and J. Neyman (eds). University of California Press, Berkeley, **1**, 355.

RAO, C. R. (1970). Estimation of heteroscadastic variances in linear models. *J. Amer. Statist. Ass.*, **65**, 161.

RAO, C. R. (1974). Projectors, generalized inverses and the BLUE's. *J. Roy. Statist. Soc.*, **B**, **36**, 442.

RAO, C. R. (ed.) (1993). Handbook of Statistics, Vol. 9: *Computational Statistics*. Elsevier, Amsterdam.

RAO, J. N. K. (1980). Estimating the common mean of possibly different normal populations: A simulation study. *J. Amer. Statist. Ass.*, **75**, 447.

RAO, P. V., SCHUSTER, E. F. and LITTELL, R. C. (1975). Estimation of shift and center of symmetry based on Kolmogorov–Smirnov statistics. *Ann. Statist.*, **3**, 862.

RAYNER, J. C. W. and BEST, D. J. (1989). *Smooth Tests of Goodness of Fit.* Oxford University Press, Oxford.

READ, T. R. C. (1984). Small-sample comparisons for the power divergence goodness-of-fit statistics. *J. Amer. Statist. Ass.*, **79** , 929.

REEDS, J. A. (1978). Jackknifing maximum likelihood estimates. *Ann. Statist.*, **6**, 727.

REEDS, J. A. (1985). Asymptotic number of roots of Cauchy location likelihood equations. *Ann. Statist.*, **13**, 775.

REISS, R. D. and RÜSCHENDORF, L. (1976). On Wilks' distribution-free confidence intervals for quantile intervals. *J. Amer. Statist. Ass.*, **71**, 940.

RESNIKOFF, G. J. (1962). Tables to facilitate the computation of percentage points of the non-central t-distribution. *Ann. Math. Statist.*, **33**, 580.

RESNIKOFF, G. J. and LIEBERMAN, G. J. (1957). *Tables of the Non-central t-distribution.* Stanford University Press, Stanford, CA.

RIDOUT, M. S. (1988). An improved branch and bound algorithm for feature subset selection. *Appl. Statist.*, **37**, 139.

RIVEST, L.-P. (1986). Bartlett's, Cochran's and Hartley's tests on variances are liberal when the underlying distribution is long-tailed. *J. Amer. Statist. Ass.*, **81**, 124.

ROBBINS, H. (1944). On distribution-free tolerance limits in random sampling. *Ann. Math. Statist.*, **15**, 214.

ROBBINS, H. (1956). An empirical Bayes approach to statistics. In *Proceedings of the Third Berkeley Symposium on Mathematical Statistics and Probability*, J. Neyman (ed.). University of California Press, Berkeley, **1**, 157.

ROBBINS, H. (1964). The empirical Bayes approach to statistical decision problems. *Ann. Math. Statist.*, **35**, 1.

ROBERTSON, T., WRIGHT, F. T. and DYKSTRA, R. L., (1988). *Order Restricted Inference*. Wiley, New York.

ROBINSON, G. K. (1976). Properties of Student's t and of the Behrens–Fisher solution to the two means problem. *Ann. Statist.*, **4**, 963.

ROBSON, D. S. (1959). A simple method for constructing orthogonal polynomials when the independent variable is unequally spaced. *Biometrics*, **15**, 187.

ROBSON, D. S. and WHITLOCK, J. H. (1964). Estimation of a truncation point. *Biometrika*, **51**, 33.

ROSCOE, J. T. and BYARS, J. A. (1971). An investigation of the restraints with respect to sample size commonly imposed on the use of the chi-square statistic. *J. Amer. Statist. Ass.*, **66**, 755.

ROSENBAUM, P. R. and RUBIN, D. B. (1983). The central role of the propensity score in observational studies for causal effects. *Biometrika*, **70**, 41.

ROSS, W. H. (1987). The expectation of the likelihood ratio criterion. *Int. Statist. Rev.*, **55**, 315.

ROTHE, G. (1981). Some properties of the asymptotic relative Pitman efficiency. *Ann. Statist.*, **9**, 663.

ROTHENBERG, T. J., FISHER, F. M. and TILANUS, C. B. (1964). A note on estimation from a Cauchy sample. *J. Amer. Statist. Ass.*, **59**, 460.

ROUSSEEUW, P. J. (1984). Least median of squares regression. *J. Amer. Statist. Ass.*, **79**, 871.

ROUTLEDGE, R.D. (1994). Practising safe statistics with the mid-p. *Canad. J. Statist.*, **22**, 103.

ROWLANDS, R. J. and WETHERILL, G. B. (1991). Quality control. In *Handbook of Sequential Analysis*, B. K. Ghosh, and P. K. Sen (eds). Dekker, New York, 563.

ROY, A. R. (1956). On χ^2-statistics with variable intervals. Technical Report, Stanford University, Statistics Department.

ROY, K. P. (1957). A note on the asymptotic distribution of likelihood ratio. *Bull. Calcutta Statist. Ass.*, **7**, 73.

ROY, S. N. (1953). On a heuristic method of test construction and its use in multivariate analysis. *Ann. Math. Stat.*, **24**, 513.

ROY, S. N. (1954). Some further results in simultaneous confidence interval estimation. *Ann. Math. Statist.*, **25**, 752.

ROY, S. N. and BOSE, R. C. (1953). Simultaneous confidence interval estimation. *Ann. Math. Statist.*, **24**, 513.

ROYSTON, P. (1982). Algorithm 181: The W test for normality. *Appl. Statist.*, **31**, 176.

ROYSTON, P. (1995). Comments on 'Algorithm 181: The W test for normality.' *Appl. Statist.*, **44**, 547.

RUBIN, D. B. (1990). Formal models of statistical inference for causal effects. *J. Statist. Planning Inf.*, **25**, 279.

RUBIN, D. B. and WEISBERG, S. (1975). The variance of a linear combination of independent estimators using estimated weights. *Biometrika*, **62**, 708.

RUKHIN, A. L. (1986). Improved estimation in lognormal models. *J. Amer. Statist. Ass.*, **81**, 1046.

RUSHTON, S. (1951). On least squares fitting of orthonormal polynomials using the Choleski method. *J. Roy. Statist. Soc.*, **B, 13**, 92.

RYAN, T. A. (1960). Significance tests for multiple comparisons of proportions, variances and other statistics. *Psych. Bull.*, **57**, 318.

SALL, J. (1990). Leverage plots for general linear hypothesis. *Amer. Statist.*, **44**, 308.

SAMANIEGO, F. J. and RENEAU, D. M. (1994). Towards a reconciliation of the Baysian and frequentist approaches to point estimation. *J. Amer. Statist. Ass.*, **89**, 947.

SAMARA, B. and RANDLES, R. H. (1988). A test for correlation based on Kendall's tau. *Commun. Statist.*, **A, 17**, 3191.

SAMPSON, A. R. (1974). A tale of two regressions. *J. Amer. Statist. Ass.*, **69**, 682.

SANIGA, E. M. and MILES, J. A. (1979). Power of some standard goodness-of-fit tests of normality against asymmetric stable alternatives. *J. Amer. Statist. Ass.*, **74**, 861.

SANKARAN, M. (1964). On an analogue of Bhattacharya bound. *Biometrika*, **51**, 268.

SARHAN, A. E. (1954). Estimation of the mean and standard deviation by order statistics. *Ann. Math. Statist.*, **25**, 317.

SARKADI, K. (1975). The consistency of the Shapiro–Francia test. *Biometrika*, **62**, 445.

SATHE, Y. S. and LINGRAS, S. R. (1981). Bounds for the confidence coefficients of outer and inner confidence intervals for quantile intervals. *J. Amer. Statist. Ass.*, **76**, 473.

SATTERTHWAITE, F. E. (1946). An approximate distribution of estimates of variance components. *Biometric Bull.*, **2**, 110.

SAVAGE, L. J. (1954). *The Foundations of Statistics*. Methuen, London.

SAVAGE, L. J. (1970). Comments on a weakened principle of conditionality. *J. Amer. Statist. Ass.*, **65**, 399.

SCHAFER, J. L. (1997). *Analysis of Incomplete Multivariate Data*. Chapman & Hall, London and New York.

SCHEFFÉ, H. (1942a). On the theory of testing composite hypotheses with one constraint. *Ann. Math. Statist.*, **13**, 280.

SCHEFFÉ, H. (1942b). On the ratio of the variances of two normal populations. *Ann. Math. Statist.*, **13**, 371.

SCHEFFÉ, H. (1943). On solutions of the Behrens–Fisher problem based on the t-distribution. *Ann. Math. Statist.*, **14**, 35.

SCHEFFÉ, H. (1944). A note on the Behrens–Fisher problem. *Ann. Math. Statist.*, **15**, 430.

SCHEFFÉ, H. (1953). A method for judging all contrasts in the analysis of variance. *Biometrika*, **40**, 87.

SCHEFFÉ, H. (1959). *The Analysis of Variance*. Wiley, New York.

SCHEFFÉ, H. (1970a). Multiple testing versus multiple estimation. Improper confidence sets. Estimation of directions and ratios. *Ann. Math. Statist.*, **41**, 1.

SCHEFFÉ, H. (1970b). Practical solutions of the Behrens–Fisher problem. *J. Amer. Statist. Ass.*, **65**, 1501.

SCHEFFÉ, H. and TUKEY, J. W. (1945). Non-parametric estimation: I. Validation of order statistics. *Ann. Math. Statist.*, **16**, 187.

SCHEUER, E. M. and SPURGEON, R. A. (1963). Some percentage points of the noncentral t-distribution. *J. Amer. Statist. Ass.*, **58**, 176.

SCHLESSELMAN, J. (1971). Power families: A note on the Box and Cox transformation. *J. Roy. Statist. Soc.*, **B**, **33**, 307.

SCHMETTERER, L. (1960). On a problem of J. Neyman and E. Scott. *Ann. Math. Statist.*, **31**, 656.

SCHUSTER, E. F. (1973). On the goodness-of-fit problem for continuous symmetric distributions. *J. Amer. Statist. Ass.*, **68**, 713. *Corrigenda*, **69**, 288.

SCHUSTER, E. F. (1975). Estimating the distribution function of a symmetric distribution. *Biometrika*, **62**, 631.

SCHWARTZ, R. (1967). Admissible tests in multivariate analysis. *Ann. Math. Statist.*, **38**, 698.

SCHWARZ, G. (1978). Estimating the dimension of a model. *Ann. Statist.*, **6**, 461.

SEAL, H. L. (1948). A note on the χ^2 smooth test. *Biometrika*, **35**, 202.

SEAL, H. L. (1967). The historical development of the Gauss linear model. *Biometrika*, **54**, 1.

SEARLE, S. R. (1971). *Linear Models*, Wiley, New York.

SEARLE, S. R. (1988). *Linear Models for Unbalanced Data,* Wiley, New York.

SEBER, G. A. F. and WILD, C. J. (1989). *Nonlinear Regression*. Wiley, New York.

SEELBINDER, B. M. (1953). On Stein's two-stage sampling scheme. *Ann. Math. Statist.*, **24**, 640.

SEIDENFELD, T. (1992). R. A. Fisher's fiducial argument and Bayes' theorem. *Statist. Science*, **7**, 358.

SELF, S. G. and LIANG, K.-Y. (1987). Asymptotic properties of maximum likelihood estimators and likelihood ratio tests under nonstandard conditions. *J. Amer. Statist. Ass.*, **82**, 605.

SEN, P. K. (1991). Nonparametric methods in sequential analysis. In *Handbook of Sequential Analysis*, B. K. Ghosh and P. K. Sen (eds). Dekker, New York, 331.

SEN, P. K. and GHOSH, B. K. (1976). Comparison of some bounds in estimation theory. *Ann. Statist.*, **4**, 755.

SHAH, B. K. and ODEH, R. E. (1986). *Selected Tables in Mathematical Statistics*, Vol. X. American Mathematical Society, Providence, RI.

SHAO, J. and TU, D. (1995). *The Jackknife and Bootstrap*. Springer-Verlag, New York.

SHAPIRO, S. S. and FRANCIA, R. S. (1972). An approximate analysis of variance test for normality. *J. Amer. Statist. Ass.*, **67**, 215.

SHAPIRO, S. S. and WILK, M. B. (1965). An analysis of variance test for normality (complete samples). *Biometrika*, **52**, 591.

SHAPIRO, S. S., WILK, M. B. and CHEN, H. J. (1968). A comparative study of various tests for normality. *J. Amer. Statist. Ass.*, **63**, 1343.

SHARPE, K. (1970). Robustness of normal tolerance intervals. *Biometrika*, **57**, 71.

SHENTON, L. R. (1949). On the efficiency of the method of moments and Neyman's Type A contagious distribution. *Biometrika*, **36**, 450.

SHENTON, L. R. (1950). Maximum likelihood and the efficiency of the method of moments. *Biometrika*, **37**, 111.

SHENTON, L. R. (1951). Efficiency of the method of moments and the Gram–Charlier Type A distribution. *Biometrika*, **38**, 58.

SHENTON, L. R. and BOWMAN, K. (1963). Higher moments of a maximum-likelihood estimate. *J. Roy. Statist Soc.*, **B**, **25**, 305.

SHENTON, L. R. and BOWMAN, K. O. (1977). *Maximum Likelihood Estimation in Small Samples*. Griffin, London and High Wycombe.

SHEWHART, W. A. (1931). *Economic Control of Manufactured Product*. Van Nostrand Reinhold, New York. (Republished in 1981 by the American Society for Quality Control, Milwaukee, WI).

SHIBATA, R. (1981). An optimal selection of regression variables. *Biometrika*, **68**, 45.

SHIBATA, R. (1984). Approximate efficiency of a selection procedure for the number of regression variables. *Biometrika*, **71**, 43.

SHOUKRI, M. M. and WARD, R. H. (1984). On the estimation of the intraclass correlation coefficient. *Commun. Statist.*, **A**, **13**, 1239.

SHOWALTER, M. (1994) A Monte Carlo investigation of the Box–Cox model and a nonlinear least squares alternative. *Rev. Econ. Statist.*, **76**, 560.

SICHEL, H. S. (1951–2). New methods in the statistical evaluation of mine sampling data. *Trans. Inst. Mining Metallurgy*, **61**, 261.

SIEGEL, A. F. (1979). The noncentral chi-squared distribution with zero degrees of freedom and testing for uniformity. *Biometrika*, **66**, 381.

SIEGMUND, D. (1975). Error probabilities and average sample number of the sequential probability ratio test. *J. Roy. Statist. Soc.*, **B**, **37**, 394.

SIEGMUND, D. (1978). Estimation following sequential tests. *Biometrika*, **65**, 341.

SIEGMUND, D. (1985). *Sequential Analysis: Tests and Confidence Intervals*. Springer-Verlag, New York.

SIEVERS, G. L. (1969). On the probability of large deviations and exact slopes. *Ann. Math. Statist.*, **40**, 1908.

SILVERSTONE, H. (1957). Estimating the logistic curve. *J. Amer. Statist. Ass.*, **52**, 567.

SILVEY, S. D. (1959). The Lagrange multiplier test. *Ann. Math. Statist.*, **30**, 389.

SIMON, G. A. and SIMONOFF, J. S. (1986). Diagnostic plots for missing data in least squares regression. *J. Amer. Statist. Ass.*, **81**, 501.

SIMS, C. A. (1975). A note on exact tests for serial correlation. *J. Amer. Statist. Ass.*, **70**, 162.

SKOVGAARD, I. M. (1985). A second-order investigation of asymptotic ancillarity. *Amer. Statist.*, **13**, 534.

SLAKTER, M. J. (1966). Comparative validity of the chi-square and two modified chi-square goodness-of-fit tests for small but equal expected frequencies. *Biometrika*, **53**, 619.

SLAKTER, M. J. (1968). Accuracy of an approximation to the power of the chi-square goodness of fit test with small but equal expected frequencies. *J. Amer. Statist. Ass.*, **63**, 912.

SMALL, N. J. H. (1980). Marginal skewness and kurtosis in testing multivariate normality. *Appl. Statist.*, **29**, 85.

SMIRNOV, N. V. (1939). On the estimation of the discrepancy between empirical curves of distribution for two independent samples. *Bull. Math. University Moscou, Série Int.*, **2**(2), 3.

SMIRNOV, N. V. (1948). Table for estimating the goodness of fit of empirical distributions. *Ann. Math. Statist.*, **19**, 279.

SMITH, G. and CAMPBELL, F. (1980). A critique of some ridge regression methods. *J. Amer. Statist. Ass.*, **75**, 74.

SMITH, K. (1916). On the 'best' values of the constants in frequency distributions. *Biometrika*, **11**, 262.

SMITH, P. J., RAE, D. S., MANDERSCHEID, R. W. and SILBERGELD, S. (1981). Approximating the moments and distribution of the likelihood ratio statistic for multinomial goodness of fit. *J. Amer. Statist Ass.*, **76**, 737.

SMITH, R. L. (1985). Maximum likelihood estimation in a class of nonregular cases. *Biometrika* **72**, 67.

SMYTH, G. K. and VERBYLA, A. P. (1996). A conditional likelihood approach to residual maximum likelihood estimators in generalised linear models. *J. Roy. Statist. Soc.*, **B**, **58**, 565.

SNEDECOR, G. W. and COCHRAN, W. G. (1967). *Statistical Methods: Applied to Experiments in Agriculture and Biology* (6th edn). Iowa State University Press, Ames.

SOPER, H. E. (1914). On the probable error of the biserial expression for the correlation coefficient. *Biometrika*, **10**, 384.

SOUVAINE, D. L. and STEELE, J. M. (1987). Time and space-efficient algorithms for least median of squares regression. *J. Amer. Statist. Ass.*, **82**, 794.

SPITZER, J. J. (1978). A Monte Carlo investigation of the Box–Cox transformation in small samples. *J. Amer. Statist. Ass.*, **73**, 488.

SPJØTVOLL, E. (1968). Most powerful tests for some non-exponential families. *Ann. Math. Statist.*, **39**, 772.

SPJØTVOLL, E. (1972a). Multiple comparison of regression functions. *Ann. Math. Statist.*, **43**, 1076.

SPJØTVOLL, E. (1972b). Unbiasedness of likelihood ratio confidence sets in cases without nuisance parameters. *J. Roy. Statist. Soc.*, **B**, **34**, 268.

SPROTT, D. A. (1960). Necessary restrictions for distributions *a posteriori*. *J. Roy. Statist. Soc.*, **B**, **22**, 312.

SPROTT, D. A. (1961). An example of an ancillary statistic and the combination of two samples by Bayes' theorem. *Ann. Math. Statist.*, **32**, 616.

SPROTT, D. A. (1990). Inferential estimation, likelihood, and linear pivotals (with discussion). *Canad. J. Statist.*, **18**, 1.

ST LAURENT, R. T. and COOK, R. D. (1993). Leverage, local influence and curvature in nonlinear regression. *Biometrika*, **80**, 99.

STADJE, W. (1985). Estimation problems for samples with measurement errors. *Ann. Statist.*, **13**, 1592.

STARK, A. E. (1975). Some estimators of the integer-valued parameter of a Poisson variate. *J. Amer. Statist. Ass.*, **70**, 685.

STARR, N. and WOODROOFE, M. (1972). Further remarks on sequential estimation: the exponential case. *Ann. Math. Statist.*, **43**, 1147.

STEIN, C. (1945). A two-sample test for a linear hypothesis whose power is independent of the variance. *Ann. Math. Statist.*, **16**, 243.

STEIN, C. (1946). A note on cumulative sums. *Ann. Math. Statist.*, **17**, 498.

STEIN, C. (1956). Inadmissibility of the usual estimator for the mean of a multivariate normal distribution. In *Proceedings of the Third Berkeley Symposium on Mathematical Statistics and Probability*, J. Neyman (eds). University of California Press, Berkeley, **1**, 197.

STEIN, C. (1964). Inadmissibility of the usual estimator for the variance of a normal distribution with unknown mean. *Ann., Inst. Math.*, **16**, 155-160.

STEPHENS, M. A. (1970). Use of the Kolmogorov–Smirnov, Cramér–von Mises and related statistics without extensive tables. *J. Roy. Statist. Soc.*, **B**, **32**, 115.

STEPHENS, M. A. (1974). EDF statistics for goodness of fit and some comparisons. *J. Amer. Statist. Ass.*, **69**, 730.

STEPHENS, M. A. (1975). Asymptotic properties for covariance matrices of order statistics. *Biometrika*, **62**, 23.

STEPHENS, M. A. (1976). Asymptotic results for goodness-of-fit statistics with unknown parameters. *Ann. Statist.*, **4**, 357.

STEPHENS, M. A. (1977). Goodness of fit for the extreme value distribution. *Biometrika*, **64**, 583.

STEPHENS, M. A. (1979a). Tests of fit for the logistic distribution based on the empirical distribution function. *Biometrika*, **66**, 591.

STEPHENS, M. A. (1979b). Vector autocorrelation. *Biometrika*, **66**, 41.

STERNE, T. E. (1954). Some remarks on confidence or fiducial limits. *Biometrika*, **41**, 275.

STEVENS, W. L. (1939). Distribution of groups in a sequence of alternatives. *Ann. Eugen.*, **9**, 10.

STEVENS, W. L. (1950). Fiducial limits of the parameter of a discontinuous distribution. *Biometrika*. **37**, 117.

STEWART, G. W. (1987). Collinearity and least squares regression. *Statist. Sci.*, **2**, 68.

STIGLER, S. M. (1986). *The History of Statistics: The Measurement of Uncertainty before 1900.* Harvard University Press, Cambridge, MA and London.

STIGLER, S. M. (1990). The 1990 Neyman memorial lecture: A Galtonian perspective on shrinkage estimators. *Statist. Sci.*, **5**, 147.

STONE, M. (1976). Strong inconsistency from uniform priors. *J. Amer. Statist. Ass.*, **71**, 114.

STRAND, O. N. (1974). Coefficient errors caused by using the wrong covariance matrix in the general linear model. *Ann. Statist.*, **2**, 935.

STROUD, T. W. F. (1972). Fixed alternatives and Wald's formulation of the noncentral asymptotic behaviour of the likelihood ratio statistic. *Ann. Math. Statist.*, **43**, 447.

STROUD, T. W. F. (1973). Noncentral convergence of Wald's large-sample test statistic in exponential families. *Ann. Statist.*, **1**, 161.

STUART, A. (1954). Too good to be true? *Appl. Statist.*, **3**, 29.

STUART, A. (1955). A paradox in statistical estimation. *Biometrika*, **42**, 527.

STUART, A. (1958). Equally correlated variates and the multinormal integral. *J. Roy. Statist. Soc.*, **B**, **20**, 273.

STUART, A. (1967). The average critical value method and the asymptotic relative efficiency of tests. *Biometrika*, **54**, 308.

STUDDEN, W. J. (1980). D-optimal designs for polynomial regression using continued fractions. *Ann. Statist.*, **8**, 1132.

STUDDEN, W. J. (1982). Some robust-type D-optimal designs in polynomial regression. *J. Amer. Statist. Ass.*, **77**, 916.

'STUDENT' (1908).* The probable error of a mean. *Biometrika*, **6**, 1.

SUBRAHMANIAM, K., GAJJAR, A. V. and SUBRAHMANIAM, K. (1981). Polynomial representations for the distribution of the sample correlation and its transformations. *Sankhyā*, **B**, **43**, 319.

SUBRAHMANIAM, K. and SUBRAHMANIAM, K. (1983). Some extensions to Miss F. N. David's tables of the sample correlation coefficient. *Sankhyā*, **B**, **45**, 75.

SUKHATME, P. V. (1936). On the analysis of *k* samples from exponential populations with special reference to the problem of random intervals. *Statist. Res. Mem.*, **1**, 94.

SUNDRUM, R. M. (1954). On the relation between estimating efficiency and the power of tests. *Biometrika*, **41**, 542.

SWED, F. S. and EISENHART, C. (1943). Tables for testing randomness of grouping in a sequence of alternatives. *Ann. Math. Statist.*, **14**, 66.

SWEETING, T. J. (1992). Parameter-based asymptotics. *Biometrika*, **79**, 219.

SWINDEL, B. F. (1968). On the bias of some least-squares estimators of variance in a general linear model. *Biometrika*, **55** , 313.

TAGUTI, G. (1958). Tables of tolerance coefficients for normal populations. *Rep. Statist. Appl. Res. (JUSE)*, **5**, 73.

TAKEUCHI, K. (1969). A note on the test for the location parameter of an exponential distribution *Ann. Math. Statist.*, **40**, 1838.

TANG, P. C. (1938). The power function of the analysis of variance tests with tables and illustrations of their use. *Statist. Res. Mem.*, **2**, 126.

TATE, R. F. (1953). On a double inequality of the normal distribution. *Ann. Math. Statist.*, **24**, 132.

TATE, R. F. (1954). Correlation between a discrete and a continuous variable. Point-biserial correlation. *Ann. Math. Statist.*, **25**, 603.

TATE, R. F. (1955). The theory of correlation between two continuous variables when one is dichotomised. *Biometrika*, **48**, 205.

TATE, R. F. (1959). Unbiased estimation: Functions of location and scale parameters. *Ann. Math. Statist.*, **30**, 341.

TATE, R. F. and KLETT, G. W. (1959). Optimal confidence intervals for the variance of a normal distribution. *J. Amer. Statist. Ass.*, **54**, 674.

TEICHER, J. (1961). Maximum likelihood characterization of distributions. *Ann. Math. Statist.*, **32**, 1214.

TERRELL, C. D. (1983). Significance tables for the biserial and the point biserial. *Educ. Psych. Meas.*, **42**, 475.

THATCHER, A. R. (1964). Relationships between Bayesian and confidence limits for predictions. *J. Roy. Statist. Soc.*, **B**, **26**, 176.

THOMPSON, J. R. (1968). Some shrinkage techniques for estimating the mean. *J. Amer. Statist. Ass.*, **63**, 113.

THOMPSON, R. (1979). Bias and monotonicity for goodness-of-fit tests. *J. Amer. Statist. Ass.*, **74**, 875.

THOMPSON, W. A. Jr. (1962). The problem of negative estimates of variance components. *Ann. Math. Statist.*, **33**, 273.

THOMPSON, W. R. (1936). On confidence ranges for the median and other expectation distributions for populations of unknown distribution form. *Ann. Math. Statist.*, **7**, 122.

THÖNI, H. (1969). A table for estimating the mean of a lognormal distribution. *J. Amer. Statist. Ass.*, **64**, 632.

THORBURN, D. (1976). Some asymptotic properties of jackknife statistics. *Biometrika*, **63**, 305.

TIKU, M. L. (1965). Laguerre series forms of non-central χ^2 and F distributions. *Biometrika*, **52**, 415.

TIKU, M. L. (1966). A note on approximating the non-central F distribution. *Biometrika*, **53**, 606.

TIKU, M. L. (1967, 1972). Tables of the power of the F-test. *J. Amer. Statist. Ass.*, **62**, 525 and **67**, 709.

TITTERINGTON, D. M. (1984). Recursive parameter estimation using incomplete data. *J. Roy. Statist. Soc.*, **B**, **46**, 257.

TRICKETT, W. H., WELCH, B. L. and JAMES, G. S. (1956). Further critical values for the two-means problem. *Biometrika*, **43**, 203.

TUKEY, J. W. (1947). Non-parametric estimation, II. Statistically equivalent blocks and tolerance regions – the continuous case. *Ann. Math. Statist.*, **18**, 529.

TUKEY, J. W. (1948). Non-parametric estimation, III. Statistically equivalent blocks and multivariate tolerance regions – the discontinuous case. *Ann. Math. Statist.*, **19**, 30.

TUKEY, J. W. (1953). The problem of multiple comparisons. Princeton University. Unpublished.

TUKEY, J. W. (1957a). On the comparative anatomy of transformations. *Ann. Math. Statist.*, **28**, 602.

TUKEY, J. W. (1957b). Some examples with fiducial relevance. *Ann. Math. Statist.*, **28**, 687.

URBACH, P. (1992). Regression analysis: Classical and Bayesian. *Brit. J. Phil. Sci.*, **43**, 311.

U.S. National Bureau of Standards (ed.) (1951). *Tables to Facilitate Sequential t-tests*. Applied Mathematics Series No. 7. U S Government Printing Office, Washington, DC.

VAN DER PARREN, J. L. (1970). Tables for distribution-free confidence limits for the median. *Biometrika*, **57**, 613.

VAN EEDEN, C. (1963). The relation between Pitman's asymptotic relative efficiency of two tests and the correlation coefficient between their test statistics. *Ann. Math. Statist.*, **34**, 1442.

VENABLES, W. (1975). Calculation of confidence intervals for noncentrality parameters. *J. Roy. Statist. Soc.*, **B**, **37**, 406.

VERRILL, S. and JOHNSON, R. A. (1987). The asymptotic equivalence of some modified Shapiro–Wilk statistics – complete and censored sample cases. *Ann. Statist.*, **15**, 413.

VERRILL, S. and JOHNSON, R. A. (1988). Tables and large-sample distribution theory for censored-data correlation statistics for testing normality. *J. Amer. Statist. Ass.*, **83**, 1192.

VON MISES, R. (1931). Neue Grundlagen der Wahrscheinlichkeitsrechnung. *Forsch. Fortschr. Deutsche Wiss.*, **7**, 253.

VOORN, W. J. (1981). A class of variate transformations causing unbounded likelihood. *J. Amer. Statist. Ass.*, **76**, 709.

WALD, A. (1940).* The fitting of straight lines if both variables are subject to error. *Ann. Math. Statist.*, **11**, 284.

WALD, A. (1941).* Asymptotically most powerful tests of statistical hypotheses. *Ann. Math. Statist.*, **12**, 1.

WALD, A. (1942).* On the power function of the analysis of variance test. *Ann. Math. Statist.*, **13**, 434.

WALD, A. (1943a).* Tests of statistical hypotheses concerning several parameters when the number of observations is large. *Trans. Amer. Math. Soc.*, **54**, 426.

WALD, A. (1943b). An extension of Wolks' method for setting tolerance limits. *Ann. Math. Statist.*, **14**, 45.

WALD, A. (1947). *Sequential Analysis*. Wiley, New York.

WALD, A. (1949).* Note on the consistency of the maximum likelihood estimate. *Ann. Math. Statist.*, **20**, 595.

WALD, A. (1950). *Statistical Decision Functions*. Wiley, New York.

WALD, A. and WOLFOWITZ, J. (1939).* Confidence limits for continuous distribution functions. *Ann. Math. Statist.*, **10**, 105.

WALD, A. and WOLFOWITZ, J. (1940).* On a test whether two samples are from the same population. *Ann. Math. Statist.*, **11**, 147.

WALD, A. and WOLFOWITZ, J. (1946).* Tolerance limits for a normal distribution. *Ann. Math. Statist.*, **17**, 208.

WALKER, A. M. (1963). A note on the asymptotic efficiency of an asymptotically normal estimator sequence. *J. Roy. Statist. Soc.*, **B**, **25**, 195.

WALLACE, C. S. and FREEMAN, P. R. (1987). Estimation and inference by compact coding. *J. Roy. Statist. Soc.*, B, **49**, 240.

WALLACE, C. S. and FREEMAN, P. R. (1992). Single factor analysis by minimum message length estimation. *J. Roy. Statist. Soc.*, B, **54**, 195.

WALLACE, D. L. (1958). Asymptotic approximations to distributions. *Ann. Math. Statist.*, **29**, 635.

WALLACE, T. D. (1964). Efficiencies for stepwise regressions. *J. Amer. Statist. Ass.*, **59**, 1179.

WALLACE, T. D. and TORO-VIZCARRONDO, C. E. (1969). Tables for the mean square error test for exact linear restrictions in regression. *J. Amer. Statist. Ass.*, **64**, 1649.

WALLIS, K. F. (1972). Testing for fourth-order autocorrelation in quarterly regression equations. *Econometrica*, **40**, 617.

WALLIS, W. A. (1951). Tolerance intervals for linear regression. In *Proceedings of the Second Berkeley Symposium on Mathematical Statistics and Probability*, J. Neyman (ed.). University of California Press, Berkeley, 43.

WALSH, J. E. (1946). On the power function of the sign test for slippage of means. *Ann. Math. Statist.*, **17**, 358.

WALSH, J. E. (1947). Concerning the effect of intraclass correlation on certain significance tests. *Ann. Math. Statist.*, **18**, 88.

WALTON, G. S. (1970). A note on nonrandomized Neyman-shortest unbiased confidence intervals for the binomial and Poisson parameters. *Biometrika*, **57**, 223.

WARDELL, D. G., MOSKOWITZ, H. and PLANTE, R. D. (1992). Control charts in the presence of data correlation. *Management Sci.*, **38**, 1084.

WATSON, G. S. (1957a). Sufficient statistics, similar regions and distribution-free tests. *J. Roy. Statist. Soc.*, **B**, **19**, 262.

WATSON, G. S. (1957b). The χ^2 goodness-of-fit test for normal distributions. *Biometrika*, **44**, 336.

WATSON, G. S. (1958). On chi-square goodness-of-fit tests for continuous distributions. *J. Roy. Statist. Soc.*, **B**, **20**, 44.

WATSON, G. S. (1959). Some recent results in chi-square goodness-of-fit tests. *Biometrics*, **15**, 440.

WEISBERG, S. (1974). An empirical comparison of the percentage points of W and W'. *Biometrika*, **61**, 644.

WEISS, L. and WOLFOWITZ, J. (1972). An asymptotically efficient sequential equivalent of the t-test. *J. Roy. Statist. Soc.*, **B**, **34**, 456.

WEISS, L. and WOLFOWITZ, J. (1973). Maximum likelihood estimation of a translation parameter of a truncated distribution. *Ann. Statist.*, **1**, 944.

WEISSBERG, A. and BEATTY, G. H. (1960). Tables of tolerance-limit factors for normal distributions. *Technometrics*, **2**, 483.

WELCH, B. L. (1938). The significance of the difference between two means when the population variances are unequal. *Biometrika*, **29**, 350.

WELCH, B. L. (1939). On confidence limits and sufficiency, with particular reference to parameters of location. *Ann. Math. Statist.*, **10**, 58.

WELCH, B. L. (1947). The generalisation of 'Student's' problem when several different population variances are involved. *Biometrika*, **34**, 28.

WELCH, B. L. (1965). On comparisons between confidence point procedures in the case of a single parameter. *J. Roy. Statist. Soc.*, **B**, **27**, 1.

WELCH, B. L. and PEERS, H. W. (1963). On formulae for confidence points based on integrals of weighted likelihood. *J. Roy. Statist. Soc.*, **B**, **25**, 318.

WELLER, H. (1972). Inverse sampling of a Poisson distribution. *Biometrics*, **28**, 959.

WERMUTH, N. (1980). Linear recursive equations, covariance selection and path analysis. *J. Amer. Statist. Ass.*, **75**, 963.

WHITAKER, L. (1914). On Poisson's law of small numbers. *Biometrika*, **10**, 36.

WHITE, H. (1980). A heteroscedastic-consistent covariance matrix estimator and a direct test for heteroscedasticity. *Econometrica*, **48**, 817.

WHITE, H. (1981). Consequences and detection of misspecified nonlinear regression models. *J. Amer. Statist. Ass.*, **76**, 419.

WHITE, H. (1982). Maximum likelihood estimation of misspecified models. *Econmetrica*, **50**, 1.

WHITE, H. and MACDONALD, G. M. (1980). Some large sample tests for non-normality in the linear regression model. *J. Amer. Statist. Ass.*, **75**, 16.

WHITE, K. J. (1992). The Durbin–Watson test for autocorrelation in nonlinear models. *Rev. Econ. Statist.*, **74** , 370.

WHITEHEAD, Y. (1986). On the bias of maximum likelihood estimation following a sequential test. *Biometrika*, **73**, 573.

WHITTAKER, J. (1973). The Bhattacharyya matrix for the mixture of two distributions. *Biometrika*, **60**, 201.

WICKSELL, S. D. (1917). The correlation function of Type A. *Medd. Lunds Astr. Obs.*, Series 2, No. 17.

WICKSELL, S. D. (1934). Analytical theory of regression. *Medd. Lunds Astr. Obs.*, Series 2, No. 69.

WIEAND, H. S. (1976). A condition under which the Pitman and Bahadur approaches to efficiency coincide. *Ann. Statist.*, **4**, 1003.

WIJSMAN, R. (1980). Smallest confidence sets with applications in multivariate analysis. *Multivariate Analysis*, **5**, 483.

WIJSMAN, R. A. (1991). Stopping times: termination, moments, distributions. In Ghosh, B. K. and Sen, P. K. (eds., 1991), 67.

WILKINSON, L. and DALLAL, G. E. (1981). Tests of significance in forward selection regression with an F-to-enter stopping rule. *Technometrics*, **23**, 377.

WILKS, S. S. (1932). Certain generalizations in the analysis of variance. *Biometrika*, **24**, 471.

WILKS, S. S. (1938a).* The large-sample distribution of the likelihood ratio for testing composite hypotheses. *Ann. Math. Statist.*, **9**, 60.

WILKS, S. S. (1938b).* Shortest average confidence intervals from large samples. *Ann. Math. Statist.*, **9**, 166.

WILKS, S. S. (1941).* Determination of sample sizes for setting tolerance limits. *Ann. Math. Statist.*, **12**, 91.

WILKS, S. S. (1942).* Statistical prediction with special reference to the problem of tolerance limits. *Ann. Math. Statist.*, **13**, 400.

WILKS, S. S. and DALY, J. F. (1939).* An optimum property of confidence regions associated with the likelihood function. *Ann. Math. Statist.*, **10**, 225.

WILLIAMS, C. A., Jr. (1950). On the choice of the number and width of classes for the chi-square test of goodness of fit. *J. Amer. Statist. Ass.*, **45**, 77.

WILLIAMS, E. J. (1959). The comparison of regression variables. *J. Roy. Statist. Soc.*, **B**, **21**, 396.

WILLIAMS, E. J. (1969). A note on regression methods in calibration. *Technometrics*, **11**, 189.

WILLIAMSON, J. A. (1984). A note on the proof by H. E. Daniels of the asymptotic efficiency of a maximum likelihood estimator. *Biometrika*, **71**, 651.

WINTERBOTTOM, A. (1979). Cornish–Fisher expansions for confidence limits. *J. Roy. Statist. Soc.*, **B**, **41**, 69.

WISE, M. E. (1963). Multinomial probabilities and the χ^2 and X^2 distributions. *Biometrika*, **50**, 145.

WISE, M. E. (1964). A complete multinomial distribution compared with the X^2 approximation and an improvement to it. *Biometrika*, **51**, 277.

WISHART, J. (1931). The mean and second moment coefficient of the multiple correlation coefficient, in samples from a normal population. *Biometrika*, **22**, 353.

WISHART, J. (1932). A note on the distribution of the correlation ratio. *Biometrika*, **24**, 441.

WOLD, H. O. A. (1960). A generalization of causal chain models. *Econometrica*, **28**, 443.

WOLFOWITZ, J. (1947). The efficiency of sequential estimates and Wald's equation for sequential processes. *Ann. Math. Statist.*, **18**, 215.

WOLFOWITZ, J. (1949). The power of the classical tests associated with the normal distribution. *Ann. Math. Statist.*, **20**, 540.

WOOD, C. L. and ALTAVELA, M. M. (1978). Large-sample results for Kolmogorov–Smirnov statistics for discrete distributions. *Biometrika*, **65**, 235.

WOODCOCK, F. R. and EAMES, A. R. (1970). *Confidence limits for numbers from 0 to 1200 based on the Poisson distribution.* Authority Health and Safety Branch, AHSB(S) R.179. HMSO, London.

WOODROOFE, M. (1978). Large deviations of likelihood ratio statistics with applications to sequential testing. *Ann. Statist.*, **6**, 72.

WOODROOFE, M. (1982). *Nonlinear Renewal Theory in Sequential Analysis.* Society for Industrial and Applied Mathematics, Philadelphia.

WORKING, H. and HOTELLING, H. (1929). The application of the theory of error to the interpretation of trends. *Suppl. J. Amer. Statist. Ass.*, **24**, 73.

WRIGHT, S. (1923). The theory of path coefficients: A reply to Niles' criticism. *Genetics*, **8**, 239.

WRIGHT, S. (1934). The method of path coefficients. *Ann. Math. Statist.*, **5**, 161.

WU, C. F. J. (1983). On the convergence properties of the EM algorithm. *Ann. Statist.*, **11**, 95.

WU, C. F. J. (1986). Jackknife, bootstrap and other resampling plans in regression analysis (with discussion). *Ann. Statist.*, **14** , 1261.

WYNN, H. P. (1984). An exact confidence band for one-dimensional polynomial regression. *Biometrika*, **71**, 375.

WYNN, H. P. and BLOOMFIELD, P. (1971). Simultaneous confidence bands in regression analysis. *J. Roy. Statist. Soc.*, **B**, **33**, 202.

YARNOLD, J. K. (1970). The minimum expectation in X^2 goodness of fit tests and the accuracy of approximations for the null distribution. *J. Amer. Statist. Ass.*, **65**, 864.

YATES, F. (1934). The analysis of multiple classifications with unequal numbers in the different classes. *J. Amer. Statist. Ass.*, **29**, 1.

YATES, F. (1939a). An apparent inconsistency arising from tests of significance based on fiducial distributions of unknown parameters. *Proc. Camb. Phil. Soc.*, **35**, 579.

YATES, F. (1939b). Tests of significance of the differences between regression coefficients derived from two sets of correlated variates. *Proc. Roy. Soc. Edin.*, **A**, **59**, 184.

YATES, F. (1984). Tests of significance for 2×2 contingency tables. *J. Roy. Statist. Soc.*, **A**, **147**, 426.

YIN, Y. and CARROLL, R. J. (1990). A diagnostic for heteroscedasticity based on the Spearman rank correlation coefficient. *Statist. Prob. Letters*, **10**, 69.

YOUNG, G. A. (1994). Bootstrap: More than a stab in the dark? (with discussion). *Statist. Sci.*, **9**, 382.

YULE, G. U. (1907).* On the theory of correlation for any number of variables treated by a new system of notation. *Proc. Roy. Soc.*, **A**, **79**, 182.

YULE, G. U. (1926).* Why do we sometimes get nonsense-correlations between time-series? A study in sampling and the nature of time-series. *J. Roy. Statist. Soc.*, **89**, 1.

ZABELL, S. L. (1992). R. A. Fisher and the fiducial argument. *Statist. Sci.*, **7**, 369.

ZACKS, S. (1970). Uniformly most accurate upper tolerance limits for monotone likelihood ratio families of discrete distributions. *J. Amer. Statist, Ass.*, **65**, 307.

ZAHN, D. A. and ROBERTS, G. C. (1971). Exact χ^2 criterion tables with cell expectations one: An application to Coleman's measure of consensus. *J. Amer. Statist. Ass.*, **66**, 145.

ZAR, J. H. (1972). Significance testing of the Spearman rank correlation coefficient. *J. Amer. Statist. Ass.*, **67**, 578.

ZELLNER, A. (1962). An efficient method of estimating seemingly unrelated regressions and tests for aggregation bias. *J. Amer. Statist. Ass.*, **57**, 348.

ZELLNER, A. (1977). Maximal data information prior distributions. In *New Developments in the Application of Bayesian Methods*, A. Aykae and C. Brumat (eds). North-Holland, Amsterdam.

ZIDEK, J. (1976). A necessary condition for admissibility under convex loss of equivariant estimators. Technical Report no.113, Stanford University, Stanford, CA.

ZYSKIND, G. (1963). A note on residual analysis. *J. Amer. Statist. Ass.*, **58**, 1125.

INDEX OF EXAMPLES IN TEXT

Chapter	17	18	19	20	21	22	23	24	25	26	27	28	29	30	31	32
Examples																
.1	**.7**	**.8**	**.5**	**.7**	**.7**	**.2**	**.7**	**.4**	**.7**	**.5**	**.7**	**.17**	**.3**	**.2**	**.3**	**.7**
.2	**.7**	**.8**	**.6**	**.11**	**.10**	**.2**	**.8**	**.5**	**.22**	**.7**	**.7**	**.18**	**.11**	**.2**	**.4**	**.10**
.3	**.9**	**.9**	**.9**	**.12**	**.14**	**.8**	**.10**	**.9**	**.27**	**.9**	**.8**	**.19**	**.17**	**.3**	**.4**	**.11**
.4	**.10**	**.14**	**.12, .15**	**.12**	**.15**	**.9**	**.11**	**.11**	**.29**	**.9**	**.9**		**.19**	**.4**	**.5**	**.27**
.5	**.12**	**.15**	**.17**	**.13**	**.18**	**.19**	**.11**	**.13**	**.32**	**.9**	**.9**		**.24**	**.5**	**.22**	**.27**
.6	**.17**	**.16**	**.18**	**.14**	**.18**	**.22**	**.12**	**.14**	**.32**	**.13**	**.12**			**.6**	**.22**	**.39**
.7	**.17**	**.16**	**.43**	**.15**	**.20**	**.24**	**.14**	**.15**	**.40**	**.13**	**.12**			**.7**	**.33**	**.41**
.8	**.17**	**.18**	**.46**	**.19**	**.20**	**.27**	**.15**	**.18**	**.42**	**.16**	**.32**			**.15**	**.33**	**.49**
.9	**.17**	**.21**	**.47**	**.19**	**.20**	**.30**	**.15**	**.18**		**.19**	**.34**			**.16**	**.33**	**.60**
.10	**.18**	**.21**	**.47**	**.22**	**.21**	**.33**	**.16**	**.20**		**.20**					**.34**	**.61**
.11	**.22**	**.24**	**.48**	**.23**	**.23**	**.34**	**.17**	**.25**		**.21**						**.62**
.12	**.27**	**.25**	**.48**	**.24**	**.24**	**.42**		**.28**		**.32**						**.63**
.13	**.29**	**.28**	**.48**	**.25**	**.25**			**.31**		**.38**						**.68**
.14	**.30**	**.29**		**.31**	**.33**			**.31**		**.38**						
.15	**.33**	**.30**			**.33**					**.44**						
.16	**.33**	**.31**			**.33**					**.45**						
.17	**.38**	**.33**			**.37**					**.54**						
.18	**.41**	**.35**								**.55**						
.19	**.41**	**.38**														
.20	**.41**	**.39**														
.21	**.41**	**.41**														
.22	**.41**	**.42**														
.23	**.41**	**.45**														
.24		**.48**														
.25		**.58**														
.26		**.63**														

Each entry in the table gives the section number that contains the example numbered in the left margin, in the chapter at the head of the column. Thus the entry **.45**, with coordinates (.23, 18), means that Example 18.23 appears in section **18.45**.

AUTHOR INDEX

SUBJECT INDEX

(References are to chapter sections, displayed at the tops of pages. Examples in the text are indexed by the chapter section in or immediately after which they appear. Exercises appear at the ends of chapters.)